KIRK-OTHMER ENCYCLOPEDIA OF

CHEMICAL TECHNOLOGY

Fifth Edition

VOLUME 2

KIRK-OTHMER ENCYCLOPEDIA OF CHEMICAL TECHNOLOGY, FIFTH EDITION
EDITORIAL STAFF

Vice President, STM Books: **Janet Bailey**

Executive Editor: **Jacqueline I. Kroschwitz**

Editor: **Arza Seidel**

Managing Editor: **Michalina Bickford**

Director, Book Production and Manufacturing: **Camille P. Carter**

Production Manager: **Shirley Thomas**

Senior Production Editor: **Kellsee Chu**

Illustration Manager: **Dean Gonzalez**

Editorial Assistant: **Liam Kuhn**

KIRK-OTHMER ENCYCLOPEDIA OF

CHEMICAL TECHNOLOGY

Fifth Edition

VOLUME 2

Kirk-Othmer Encyclopedia of Chemical Technology
is available Online in full color and with additional content at
http://www3.interscience.wiley.com/cgi-bin/mrwhome/104554789/HOME.

WILEY-INTERSCIENCE

A John Wiley & Sons, Inc., Publication

Library of Congress Cataloging-in-Publication Data:

Kirk-Othmer encyclopedia of chemical technology. – 5th ed.
    p.   cm.
Editor-in-chief, Arza Seidel.
"A Wiley-Interscience publication."
Includes index.
    ISBN 0-471-48494-6 (set) – ISBN 0-471-48521-7 (v. 2)
    1. Chemistry, Technical–Encyclopedias. I. Title: Encyclopedia of chemical technology. II. Kroschwitz, Jacqueline I.
    TP9.K54   2004
    660′.03–dc22   2003021960

Printed in the United States of America

10 9 8 7 6 5 4 3 2 1

CONTENTS

CONTRIBUTORS

Anthony G. Abatjoglou, *Union Carbide Corporation, South Charleston, WV,* Aldehydes

Bijan Amini, *E. I. du Pont de Nemours & Co., Inc., Deepwater, NJ,* Aniline and Its Derivatives

Kazumi Araki, *University of East Asia, Shimonoseki, Japan,* Amino Acids

David A. Atwood, *University of Kentucky, Lexington, KY,* Aluminum Halides and Aluminum Nitrate

Paul T. Barger, *UOP LLC, Des Plaines, IL,* Alkylation

Allen F. Bollmeier, Jr., *ANGUS Chemical Company, Buffalo Grove, IL,* Alkanolamines from Nitro Alcohols

Paul Carmichael, *Imperial College London, London, UK,* Aminophenols

Jeremiah P. Casey, *Air Products and Chemicals, Inc., Allentown, PA,* Amines, Cycloaliphatic

B. Conley, *University of Kentucky, Lexington, KY,* Aluminum Halides and Aluminum Nitrate

Howard B. Cottam, *University of California, San Diego, CA,* Antiaging Agents

David R. Dalton, *Temple University, Philadelphia, PA,* Alkaloids

K. V. Darragh, *Rhone-Poulenc, Inc. Cranbury, NJ,* Aluminum Sulfate and Alums

Maurice Dery, *Akzo Nobel Chemicals, Inc., Dobbs Ferry, NY,* Ammonium Compounds

Ross Dowbenko, *PPG Industries, Allison Park, PA,* Allyl Monomers and Polymers

Alan C. Eachus, *ANGUS Chemical Company, Buffalo Grove, IL,* Alkanolamines from Nitro Alcohols

Martha R. Edens, *The Dow Chemical Company, Freeport, TX,* Alkanolamines from Olefin Oxides and Ammonia

Tim Eggeman, *Neoterics International, Lakewood, CO,* Ammonia

Carolyn A. Ertell, *Rhone-Poulenc, Inc. Cranbury, NJ,* Aluminum Sulfate and Alums

G. W. Grames, *Witco Corporation, Oakland, NJ,* Aluminum Halides and Aluminum Nitrate

Kathryn S. Hayes, *Air Products and Chemicals, Inc., Allentown, PA,* Amines, Lower Aliphatic Amines

William N. Hunter, *Celanes Canada Inc., Edmonton, Alberta, Canada,* Alcohols, Polyhydric

Ingegärd Johansson, *Akzo Chemical Corporation, Stenungsund, Sweden,* Amides, Fatty Acid

James A. Johnson, *UOP LLC, Des Plaines, IL,* Alkylation

Robert J. Keller, *Vinings Industries, Inc. Atlanta, GA,* Aluminates

Corey J. Kenneally, *Proctor & Gamble Company, Cincinnati, OH,* Alcohols, Higher Aliphatic, Survey and Natural Alcohols Manufacture

Joseph A. Kocal, *UOP LLC, Des Plaines, IL,* Alkylation

George R. Lappin, *Ethyl Corporation, Baton Rouge, LA,* Alcohols, Higher Aliphatic, Synthetic Processes

Joseph J. Len, *Vinings Industries, Inc. Atlanta, GA,* Aluminates

John F. Lorenc, *Schenectady Chemicals, Inc. Schenectady, NY,* Alkylphenols

J. Fred Lochary, *The Dow Chemical Company, Freeport, TX,* Alkanolamines from Olefin Oxides and Ammonia

Steven Lowenkron, *The Dow Chemical Corporation, LaPorte, TX,* Aniline and Its Derivatives

Linda M. Macsavage, *Arcadia University,* Alkaloids

George MacZura, *Aluminum Company of America, Pittsburgh, PA,* Alumina, Calcined, Tabular, and Aluminate Cements

Bernard Maisonneuve, *Akzo Chemicals, Inc., Dobbs Ferry, NY,* Amine Oxides

David J. Miller, *Union Carbide Corporation, South Charleston, WV,* Aldehydes

Chanakya Misra, *Aluminum Company of America, Alcoa Center, PA,* Alumnia, Hydrated

Stephen C. Mitchell, *Imperial College London, London, UK,* Aminophenols

Nobuyki Nagato, *Showa Denko K.K., Tokyo, Japan,* Allyl Alcohol and Its Derivatives

Toshitsugu Ozeki, *Kyowa Hakko Kogyo Company, Yokkaichi, Japan,* Amino Acids

Anthony J. Papa, *Union Carbide Corporation, South Charleston, WV,* Amyl Alcohols

John Papcun, *Atotech, Rock Hill, SC,* Ammonium Compounds

Alan Pearson, *Aluminum Company of America, Alcoa Center, PA,* Alumina, Activated

Robert E. Sanders, Jr., *Aluminum Company of America, Alcoa Center, PA,* Aluminum and Aluminum Alloys

William Scheffer, *Schenectady Chemicals, Inc. Schenectady, NY,* Alkylphenols

Steven L. Schilling, *Mobay Corporation, New Martinsville, WV,* Amines by Reduction

Robert J. Schmidt, *UOP LLC, Des Plaines, IL,* Alkylation

T. Shaikh, *University of Kentucky, Lexington, KY,* Aluminum Halides and Aluminum Nitrate

William C. Sleppy, *Aluminum Company of America, Alcoa Center, PA,* Aluminum Compounds, Survey

Michael G. Turcotte, *Air Products and Chemicals, Inc., Allentown, PA,* Amines, Lower Aliphatic Amines

Kenneth Visek, *Akzo Chemicals, Inc., McCook, IL,* Amines, Fatty

Bipin V. Vora, *UOP LLC, Des Plaines, IL,* Alkylation

John D. Wagner, *Ethyl Corporation, Baton Rouge, LA,* Alcohols, Higher Aliphatic, Synthetic Processes

Rosemary Waring, *University of Birmingham, Birmingham, UK,* Aminophenols

Charles W. Weston, *Freeport Research and Engineering Company, Bell Chase, LA,* Ammonium Compounds

Zeno W. Wicks, *Consultant, Louisville, KY,* Alkyd Resins

Laurence L. Williams, *Consultant, Stamford, CT,* Amino Resins and Plastics

Michael Wilson, *Wilson Technologies, Ltd.,* Alkaloids

J. Richard Zietz, *Ethyl Corporation, Baton Rouge, LA,* Alcohols, Higher Aliphatic, Synthetic Processes

CONVERSION FACTORS, ABBREVIATIONS, AND UNIT SYMBOLS

SI Units (Adopted 1960)

The International System of Units (abbreviated SI), is implemented throughout the world. This measurement system is a modernized version of the MKSA (meter, kilogram, second, ampere) system, and its details are published and controlled by an international treaty organization (The International Bureau of Weights and Measures) (1).

SI units are divided into three classes:

BASE UNITS

length	meter[†] (m)
mass	kilogram (kg)
time	second (s)
electric current	ampere (A)
thermodynamic temperature[‡]	kelvin (K)
amount of substance	mole (mol)
luminous intensity	candela (cd)

SUPPLEMENTARY UNITS

plane angle	radian (rad)
solid angle	steradian (sr)

DERIVED UNITS AND OTHER ACCEPTABLE UNITS

These units are formed by combining base units, suplementary units, and other derived units (2–4). Those derived units having special names and symbols are marked with an asterisk in the list below.

[†] The spellings "metre" and "litre" are preferred by ASTM; however, "-er" is used in the *Encyclopedia*.

[‡] Wide use is made of Celsius temperature (t) defined by

$$t = T - T_0$$

where T is the thermodynamic temperature, expressed in kelvin, and $T_0 = 273.15$ K by definition. A temperature interval may be expressed in degrees Celsius as well as in kelvin.

Quantity	Unit	Symbol	Acceptable equivalent
*absorbed dose	gray	Gy	J/Kg
acceleration	meter per second squared	m/s^2	
*activity (of a radionuclide)	becquerel	Bq	1/s
area	square kilometer	km^2	
	square hectometer	hm^2	ha (hectare)
	square meter	m^2	
concentration (of amount of substance)	mole per cubic meter	mol/m^3	
current density	ampere per square meter	A/m^2	
density, mass density	kilogram per cubic meter	kg/m^3	g/L; mg/cm^3
dipole moment (quantity)	coulomb meter	$C \cdot m$	
*dose equivalent	sievert	Sv	J/kg
*electric capacitance	farad	F	C/V
*electric charge, quantity of electricity	coulomb	C	$A \cdot s$
electric charge density	coulomb per cubic meter	C/m^3	
*electric conductance	siemens	S	A/V
electric field strength	volt per meter	V/m	
electric flux density	coulomb per square meter	C/m^2	
*electric potential, potential difference, electromotive force	volt	V	W/A
*electric resistance	ohm	Ω	V/A
*energy, work, quantity of heat	megajoule	MJ	
	kilojoule	kJ	
	joule	J	$N \cdot m$
	electronvolt[†]	eV[†]	
	kilowatt-hour[†]	$kW \cdot h$[†]	
energy density	joule per cubic meter	J/m^3	
*force	kilonewton	kN	
	newton	N	$kg \cdot m/s^2$

[†]This non-SI unit is recognized by the CIPM as having to be retained because of practical importance or use in specialized fields (1).

Quantity	Unit	Symbol	Acceptable equivalent
*frequency	megahertz	MHz	
	hertz	Hz	1/s
heat capacity, entropy	joule per kelvin	J/K	
heat capacity (specific), specific entropy	joule per kilogram kelvin	$J/(kg \cdot K)$	
heat-transfer coefficient	watt per square meter kelvin	$W/(m^2 \cdot K)$	
*illuminance	lux	lx	lm/m^2
*inductance	henry	H	Wb/A
linear density	kilogram per meter	kg/m	
luminance	candela per square meter	cd/m^2	
*luminous flux	lumen	lm	$cd \cdot sr$
magnetic field strength	ampere per meter	A/m	
*magnetic flux	weber	Wb	$V \cdot s$
*magnetic flux density	tesla	T	Wb/m^2
molar energy	joule per mole	J/mol	
molar entropy, molar heat capacity	joule per mole kelvin	$J/(mol \cdot K)$	
moment of force, torque	newton meter	$N \cdot m$	
momentum	kilogram meter per second	$kg \cdot m/s$	
permeability	henry per meter	H/m	
permittivity	farad per meter	F/m	
*power, heat flow rate, radiant flux	kilowatt	kW	
	watt	W	J/s
power density, heat flux density, irradiance	watt per square meter	W/m^2	
*pressure, stress	megapascal	MPa	
	kilopascal	kPa	
	pascal	Pa	N/m^2
sound level	decibel	dB	
specific energy	joule per kilogram	J/kg	
specific volume	cubic meter per kilogram	m^3/kg	
surface tension	newton per meter	N/m	
thermal conductivity	watt per meter kelvin	$W/(m \cdot K)$	
velocity	meter per second	m/s	
	kilometer per hour	km/h	
viscosity, dynamic	pascal second	$Pa \cdot s$	
	millipascal second	$mPa \cdot s$	
viscosity, kinematic	square meter per second	m^2/s	
	square millimeter per second	mm^2/s	

Quantity	Unit	Symbol	Acceptable equivalent
volume	cubic meter	m^3	
	cubic diameter	dm^3	L (liter) (5)
	cubic centimeter	cm^3	mL
wave number	1 per meter	m^{-1}	
	1 per centimeter	cm^{-1}	

In addition, there are 16 prefixes used to indicate order of magnitude, as follows

Multiplication factor	Prefix	symbol	Note
10^{18}	exa	E	
10^{15}	peta	P	
10^{12}	tera	T	
10^9	giga	G	
10^6	mega	M	
10^3	kilo	k	
10^2	hecto	h[a]	[a]Although hecto, deka, deci, and
10	deka	da[a]	centi are SI prefixes, their use
10^{-1}	deci	d[a]	should be avoided except for SI
10^{-2}	centi	c[a]	unit-multiples for area and
10^{-3}	milli	m	volume and nontechnical use of
10^{-6}	micro	μ	centimeter, as for body and
10^{-9}	nano	n	clothing measurement.
10^{-12}	pico	p	
10^{-15}	femto	f	
10^{-18}	atto	a	

For a complete description of SI and its use the reader is referred to ASTM E380 (4) and the article UNITS AND CONVERSION FACTORS which appears in Vol. 24.

A representative list of conversion factors from non-SI to SI units is presented herewith. Factors are given to four significant figures. Exact relationships are followed by a dagger. A more complete list is given in the latest editions of ASTM E380 (4) and ANSI Z210.1 (6).

Conversion Factors to SI Units

To convert from	To	Multiply by
acre	square meter (m^2)	4.047×10^3
angstrom	meter (m)	$1.0 \times 10^{-10\dagger}$
are	square meter (m^2)	$1.0 \times 10^{2\dagger}$
astronomical unit	meter (m)	1.496×10^{11}

†Exact.

To convert from	To	Multiply by
atmosphere, standard	pascal (Pa)	1.013×10^5
bar	pascal (Pa)	$1.0 \times 10^{5\dagger}$
barn	square meter (m²)	$1.0 \times 10^{-28\dagger}$
barrel (42 U.S. liquid gallons)	cubic meter (m³)	0.1590
Bohr magneton (μ_B)	J/T	9.274×10^{-24}
Btu (International Table)	joule (J)	1.055×10^3
Btu (mean)	joule (J)	1.056×10^3
Btu (thermochemical)	joule (J)	1.054×10^3
bushel	cubic meter(m³)	3.524×10^{-2}
calorie (International Table)	joule (J)	4.187
calorie (mean)	joule (J)	4.190
calorie (thermochemical)	joule (J)	$4.184^\dagger$
centipoise	pascal second (Pa·s)	$1.0 \times 10^{-3\dagger}$
centistokes	square millimeter per second (mm²/s)	$1.0^\dagger$
cfm (cubic foot per minute)	cubic meter per second (m³s)	4.72×10^{-4}
cubic inch	cubic meter (m³)	1.639×10^{-5}
cubic foot	cubic meter (m³)	2.832×10^{-2}
cubic yard	cubic meter (m³)	0.7646
curie	becquerel (Bq)	$3.70 \times 10^{10\dagger}$
debye	coulomb meter (C·m)	3.336×10^{-30}
degree (angle)	radian (rad)	1.745×10^{-2}
denier (international)	kilogram per meter (kg/m)	1.111×10^{-7}
	tex‡	0.1111
dram (apothecaries')	kilogram (kg)	3.888×10^{-3}
dram (avoirdupois)	kilogram (kg)	1.772×10^{-3}
dram (U.S. fluid)	cubic meter (m³)	3.697×10^{-6}
dyne	newton (N)	$1.0 \times 10^{-5\dagger}$
dyne/cm	newton per meter (N/m)	$1.0 \times 10^{-3\dagger}$
electronvolt	joule (J)	1.602×10^{-19}
erg	joule (J)	$1.0 \times 10^{-7\dagger}$
fathom	meter (m)	1.829
fluid ounce (U.S.)	cubic meter (m³)	2.957×10^{-5}
foot	meter (m)	$0.3048^\dagger$
footcandle	lux (lx)	10.76
furlong	meter (m)	2.012×10^{-2}
gal	meter per second squared (m/s²)	$1.0 \times 10^{-2\dagger}$
gallon (U.S. dry)	cubic meter (m³)	4.405×10^{-3}
gallon (U.S. liquid)	cubic meter (m³)	3.785×10^{-3}
gallon per minute (gpm)	cubic meter per second (m³/s)	6.309×10^{-5}
	cubic meter per hour (m³/h)	0.2271

†Exact.
‡See footnote on p. xii.

To convert from	To	Multiply by
gauss	tesla (T)	1.0×10^{-4}
gilbert	ampere (A)	0.7958
gill (U.S.)	cubic meter (m^3)	1.183×10^{-4}
grade	radian	1.571×10^{-2}
grain	kilogram (kg)	6.480×10^{-5}
gram force per denier	newton per tex (N/tex)	8.826×10^{-2}
hectare	square meter (m^2)	$1.0 \times 10^{4\dagger}$
horsepower (550 ft · lbf/s)	watt (W)	7.457×10^2
horsepower (boiler)	watt (W)	9.810×10^3
horsepower (electric)	watt (W)	$7.46 \times 10^{2\dagger}$
hundredweight (long)	kilogram (kg)	50.80
hundredweight (short)	kilogram (kg)	45.36
inch	meter (m)	$2.54 \times 10^{-2\dagger}$
inch of mercury (32°F)	pascal (Pa)	3.386×10^3
inch of water (39.2°F)	pascal (Pa)	2.491×10^2
kilogram-force	newton (N)	9.807
kilowatt hour	megajoule (MJ)	$3.6^\dagger$
kip	newton (N)	4.448×10^3
knot (international)	meter per second (m/S)	0.5144
lambert	candela per square meter (cd/m^3)	3.183×10^3
league (British nautical)	meter (m)	5.559×10^3
league (statute)	meter (m)	4.828×10^3
light year	meter (m)	9.461×10^{15}
liter (for fluids only)	cubic meter (m^3)	$1.0 \times 10^{-3\dagger}$
maxwell	weber (Wb)	$1.0 \times 10^{-8\dagger}$
micron	meter (m)	$1.0 \times 10^{-6\dagger}$
mil	meter (m)	$2.54 \times 10^{-5\dagger}$
mile (statue)	meter (m)	1.609×10^3
mile (U.S. nautical)	meter (m)	$1.852 \times 10^{3\dagger}$
mile per hour	meter per second (m/s)	0.4470
millibar	pascal (Pa)	1.0×10^2
millimeter of mercury (0°C)	pascal (Pa)	$1.333 \times 10^{2\dagger}$
minute (angular)	radian	2.909×10^{-4}
myriagram	kilogram (Kg)	10
myriameter	kilometer (Km)	10
oersted	ampere per meter (A/m)	79.58
ounce (avoirdupois)	kilogram (kg)	2.835×10^{-2}
ounce (troy)	kilogram (kg)	3.110×10^{-2}
ounce (U.S. fluid)	cubic meter (m^3)	2.957×10^{-5}
ounce-force	newton (N)	0.2780
peck (U.S.)	cubic meter (m^3)	8.810×10^{-3}
pennyweight	kilogram (kg)	1.555×10^{-3}
pint (U.S. dry)	cubic meter (m^3)	5.506×10^{-4}

†Exact.

To convert from	To	Multiply by
pint (U.S. liquid)	cubic meter (m³)	4.732×10^{-4}
poise (absolute viscosity)	pascal second (Pa·s)	$0.10^{\dagger}$
pound (avoirdupois)	kilogram (kg)	0.4536
pound (troy)	kilogram (kg)	0.3732
poundal	newton (N)	0.1383
pound-force	newton (N)	4.448
pound force per square inch (psi)	pascal (Pa)	6.895×10^{3}
quart (U.S. dry)	cubic meter (m³)	1.101×10^{-3}
quart (U.S. liquid)	cubic meter (m³)	9.464×10^{-4}
quintal	kilogram (kg)	$1.0 \times 10^{-2\dagger}$
rad	gray (Gy)	$1.0 \times 10^{-2\dagger}$
rod	meter (m)	5.029
roentgen	coulomb per kilogram (C/kg)	2.58×10^{-4}
second (angle)	radian (rad)	$4.848 \times 10^{-6\dagger}$
section	square meter (m²)	2.590×10^{6}
slug	kilogram (kg)	14.59
spherical candle power	lumen (lm)	12.57
square inch	square meter (m²)	6.452×10^{-4}
square foot	square meter (m²)	9.290×10^{-2}
square mile	square meter (m²)	2.590×10^{6}
square yard	square meter (m²)	0.8361
stere	cubic meter (m³)	$1.0^{\dagger}$
stokes (kinematic viscosity)	square meter per second (m²/s)	$1.0 \times 10^{-4\dagger}$
tex	kilogram per meter (kg/m)	$1.0 \times 10^{-6\dagger}$
ton (long, 2240 pounds)	kilogram (kg)	1.016×10^{3}
ton (metric) (tonne)	kilogram (kg)	$1.0 \times 10^{3\dagger}$
ton (short, 2000 pounds)	kilogram (kg)	9.072×10^{2}
torr	pascal (Pa)	1.333×10^{2}
unit pole	weber (Wb)	1.257×10^{-7}
yard	meter (m)	$0.9144^{\dagger}$

†Exact.

Abbreviations and Unit Symbols

Following is a list of common abbreviations and unit symbnols used in the Encyclopedia. In general they agree with those listed in *American National Standard Abbreviations for Use on Drawings and in Text* (*ANSI Y1.1*) (6) and *American National Standard Letter Symbols for Units in Science and Technology* (*ANSI Y10*) (6). Also included is a list of acronyms for a number of private and government organizations as well as common industrial solvents, polymers, and other chemicals.

Rules for Writing Unit Symbols (4):

1. Unit symbols are printed in upright letters (roman) regardless of the type style used in the surrounding text.
2. Unit symbols are unaltered in the plural.
3. Unit symbols are not followed by a period except when used at the end of a sentence.
4. Letter unit symbols are generally printed lower-case (for example, cd for candela) unless the unit name has been derived from a proper name, in which case the first letter of the symbol is capitalized (W, Pa). Prefixes and unit symbols retain their prescribed form regardless of the surrounding typography.
5. In the complete expression for a quantity, a space should be left between the numerical value and the unit symbol. For example, write 2.37 lm, *not* 2.37 lm, and 35 mm, *not* 35 mm. When the quantity is used in an adjectival sense, a hyphen is often used, for example, 35-mm film. *Exception:* No space is left between the numerical value and the symbols of degree, minute, and second of plane angle, degree Celsius, and the percent sign.
6. No space is used between the prefix and unit symbol (for example, kg).
7. Symbols, not abbreviations, should be used for units. For example, use "A," not "amp," for ampere.
8. When multiplying unit symbols, use a raised dot:

$$\text{N} \cdot \text{m for newton meter}$$

In the case of W·h, the dot may be omitted, thus:

$$\text{Wh}$$

An exception to this practice is made for computer printouts, automatic typewriter work, etc, where the raised dot is not possible, and a dot on the line may be used.

9. When dividing unit symbols, use one of the following forms:

$$\text{m/s} \quad or \quad \text{m} \cdot \text{s}^{-1} \quad or \quad \frac{\text{m}}{\text{s}}$$

In no case should more than one slash be used in the same expression unless parentheses are inserted to avoid ambiguity. For example, write:

$$\text{J/(mol} \cdot \text{K)} \quad or \quad \text{J} \cdot \text{mol}^{-1} \cdot \text{K}^{-1} \quad or \quad \text{(J/mol)/K}$$

but *not*

$$\text{J/mol/K}$$

10. Do not mix symbols and unit names in the same expression. Write:

$$\text{joules per kilogram} \quad or \quad \text{J/kg} \quad or \quad \text{J} \cdot \text{kg}^{-1}$$

but *not*

$$\text{joules/kilogram} \quad nor \quad \text{Joules/kg} \quad nor \quad \text{Joules} \cdot \text{kg}^{-1}$$

ABBREVIATIONS AND UNITS

A	ampere	AOAC	Association of Official Analytical Chemists
A	anion (eg, HA)		
A	mass number	AOCS	American Oil Chemists' Society
a	atto (prefix for 10^{-18})		
AATCC	American Association of Textile Chemists and Colorists	APHA	American Public Health Association
		API	American Petroleum Institute
ABS	acrylonitrile–butadiene–styrene		
		aq	aqueous
abs	absolute	Ar	aryl
ac	alternating current, *n.*	*ar-*	aromatic
a-c	alternating current, *adj.*	*as-*	Asymmetric(al)
ac-	alicyclic	ASHRAE	American Society of Heating, Refrigerating, and Air Conditioning Engineers
acac	acetylacetonate		
ACGIH	American Conference of Governmental Industrial Hygienists		
		ASM	American Society for Metals
ACS	American Chemical Society	ASME	American Society of Mechanical Engineers
AGA	American Gas Association		
Ah	ampere hour	ASTM	American Society for Testing and Materials
AIChE	American Institute of Chemical Engineers		
		at no.	atomic number
AIME	American Institute of Mining, metallurgical, and Petroleum Engineers	at wt	atomic weight
		av(g)	average
		AWS	American Welding Society
		b	bonding orbital
AIP	American Institute of Physics	bbl	barrel
		bcc	body-centered cubic
AISI	American Iron and Steel Institute	BCT	body-centered tetragonal
		Bé	Baumé
alc	alcohol(ic)	BET	Brunauer-Emmett-Teller (adsorption equation)
Alk	alkyl		
alk	alkaline (not alkali)	bid	twice daily
amt	amount	Boc	*t*-butyloxycarbonyl
amu	atomic mass unit	BOD	biochemical (biological) oxygen demand
ANSI	American National Standards Institute		
		bp	boiling point
AO	atomic orbital	Bq	becquerel

C	coulomb	dil	dilute
°C	degree Celsius	DIN	Deutsche Industrie
C-	denoting attachment to		Normen
	carbon	*dl*-; DL-	racemic
c	centi (prefix for 10^{-2})	DMA	dimethylacetamide
c	critical	DMF	dimethylformamide
ca	circa (Approximately)	DMG	dimethyl glyoxime
cd	candela; current density;	DMSO	dimethyl sulfoxide
	circular dichroism	DOD	Department of Defense
CFR	Code of Federal	DOE	Department of Energy
	Regulations	DOT	Department of
cgs	centimeter-gram-second		Transportation
CI	Color Index	DP	degree of polymerization
cis-	isomer in which	dp	dew point
	substituted groups are	DPH	diamond pyramid
	on some side of double		hardness
	bond between C atoms	dstl(d)	distill(ed)
cl	carload	dta	differential thermal
cm	centimeter		analysis
cmil	circular mil	(*E*)-	entgegen; opposed
cmpd	compound	ϵ	dielectric constant
CNS	central nervous system		(unitless number)
CoA	coenzyme A	*e*	electron
COD	chemical oxygen demand	ECU	electrochemical unit
coml	commerical(ly)	ed.	edited, edition, editor
cp	chemically pure	ED	effective dose
cph	close-packed hexagonal	EDTA	ethylenediaminetetra-
CPSC	Consumer Product Safety		acetic acid
	Commission	emf	electromotive force
cryst	crystalline	emu	electromagnetic unit
cub	cubic	en	ethylene diamine
D	debye	eng	engineering
D-	denoting configurational	EPA	Environmental Protection
	relationship		Agency
d	differential operator	epr	electron paramagnetic
d	day; deci (prefix for 10^{-1})		resonance
d	density	eq.	equation
d-	*dextro*-, dextrorotatory	esca	electron spectroscopy for
da	deka (prefix for 10^{-1})		chemical analysis
dB	decibel	esp	especially
dc	direct current, *n*.	esr	electron-spin resonance
d-c	direct current, *adj*.	est(d)	estimate(d)
dec	decompose	estn	estimation
detd	determined	esu	electrostatic unit
detn	determination	exp	experiment, experimental
Di	didymium, a mixture of all	ext(d)	extract(ed)
	lanthanons	F	farad (capacitance)
dia	diameter	*F*	fraday (96,487 C)

f	femto (prefix for 10^{-15})	hyd	hydrated, hydrous
FAO	Food and Agriculture Organization (United Nations)	hyg	hygroscopic
		Hz	hertz
		^{i}(eg, Pr^i)	iso (eg, isopropyl)
fcc	face-centered cubic	i-	inactive (eg, i-methionine)
FDA	Food and Drug Administration	IACS	international Annealed Copper Standard
FEA	Federal Energy Administration	ibp	initial boiling point
		IC	integrated circuit
FHSA	Federal Hazardous Substances Act	ICC	Interstate Commerce Commission
fob	free on board	ICT	International Critical Table
fp	freezing point		
FPC	Federal Power Commission	ID	inside diameter; infective dose
FRB	Federal Reserve Board		
frz	freezing	ip	intraperitoneal
G	giga (prefix for 10^9)	IPS	iron pipe size
G	gravitational constant $= 6.67 \times 10^{11} N \cdot m^2/kg^2$	ir	infrared
		IRLG	Interagency Regulatory Liaison Group
g	gram		
(g)	gas, only as in $H_2O(g)$	ISO	International Organization Standardization
g	gravitatonal acceleration		
gc	gas chromatography		
gem-	geminal	ITS-90	International Temperature Scale (NIST)
glc	gas–liquid chromatography		
g-mol wt; gmw	gram-molecular weight	IU	International Unit
		IUPAC	International Union of Pure and Applied Chemistry
GNP	gross national product		
gpc	gel-permeation chromatography		
		IV	iodine value
GRAS	Generally Recognized as Safe	iv	intravenous
		J	joule
grd	ground	K	kelvin
Gy	gray	k	kilo (prefix for 10^3)
H	henry	kg	kilogram
h	hour; hecto (prefix for 10^2)	L	denoting configurational relationship
ha	hectare		
HB	Brinell hardness number	L	liter (for fluids only) (5)
Hb	hemoglobin	l-	*levo*-, levorotatory
hcp	hexagonal close-packed	(l)	liquid, only as in $NH_3(l)$
hex	hexagonal	LC_{50}	conc lethal to 50% of the animals tested
HK	Knoop hardness number		
hplc	high performance liquid chromatography	LCAO	linear combnination of atomic orbitals
HRC	Rockwell hardness (C scale)	lc	liquid chromatography
		LCD	liquid crystal display
HV	Vickers hardness number	lcl	less than carload lots

LD_{50}	dose lethal to 50% of the animals tested	N	newton (force)
LED	light-emitting diode	N	normal (concentration); neutron number
liq	liquid	N-	denoting attachment to nitrogen
lm	lumen		
ln	logarithm (natural)	n (as n_D^{20})	index of refraction (for 20°C and sodium light)
LNG	liquefied natural gas		
log	logarithm (common)		
LOI	limiting oxygen index	n (as Bu^n),	normal (straight-chain structure)
LPG	liquefied petroleum gas	n-	
ltl	less than truckload lots	n	neutron
lx	lux	n	nano (prefix for 10^9)
M	mega (prefix for 10^6); metal (as in MA)	na	not available
		NAS	National Academy of Sciences
M	molar; actual mass		
$\overline{M}_w$	weight-average mol wt	NASA	National Aeronautics and Space Administration
$\overline{M}_n$	number-average mol wt		
m	meter; milli (prefix for 10^{-3})	nat	natural
		ndt	nondestructive testing
m	molal	neg	negative
m-	meta	NF	*National Formulary*
max	maximum	NIH	National Institutes of Health
MCA	Chemical Manufacturers' Association (was Manufacturing Chemists Association)	NIOSH	National Institute of Occupational Safety and Health
MEK	methyl ethyl ketone	NIST	National Institute of Standards and Technology (formerly National Bureau of Standards)
meq	milliequivalent		
mfd	manufactured		
mfg	manufacturing		
mfr	manufacturer		
MIBC	Methyl isobutyl carbinol	nmr	nuclear magnetic resonance
MIBK	methyl isobutyl ketone		
MIC	minimum inhibiting concentration	NND	New and Nonofficial Drugs (AMA)
min	minute; minimum	no.	number
mL	milliliter	NOI-(BN)	not otherwise indexed (by name)
MLD	minimum lethal dose		
MO	molecular orbital	NOS	not otherwise specified
mo	month	nqr	nuclear quadruple resonance
mol	mole		
mol wt	molecular weight	NRC	Nuclear Regulatory Commission; National Research Council
mp	melting point		
MR	molar refraction		
ms	mass spectrometry	NRI	New Ring Index
MSDS	material safety data sheet	NSF	National Science Foundation
mxt	mixture		
μ	micro (prefix for 10^{-6})	NTA	nitrilotriacetic acid

NTP	normal temperature and pressure (25°C and 101.3 kPa or 1 atm)	pwd	powder
		py	pyridine
		qv	quod vide (which see)
NTSB	National Transportation Safety Board	R	univalent hydrocarbon radical
O-	denoting attachment to oxygen	(R)-	rectus (clockwise configuration)
o-	ortho	r	precision of data
OD	outside diameter	rad	radian; radius
OPEC	Organization of Petroleum Exporting Countries	RCRA	Resource Conservation and Recovery Act
o-phen	o-phenanthridine	rds	rate-determining step
OSHA	Occupational Safety and Health Administration	ref.	reference
		rf	radio frequency, $n.$
owf	on weight of fiber	r-f	radio frequency, $adj.$
Ω	ohm	rh	relative humidity
P	peta (prefix for 10^{15})	RI	Ring Index
p	pico (prefix for 10^{-12})	rms	root-mean square
p-	para	rpm	rotations per minute
p	proton	rps	revolutions per second
p.	page	RT	room temperature
Pa	Pascal (pressure)	RTECS	Registry of Toxic Effects of Chemical Substances
PEL	personal exposure limit based on an 8-h exposure	^{s}(eg, Bu^{s}); sec-	secondary (eg, secondary butyl)
pd	potential difference	S	siemens
pH	negative logarithm of the effective hydrogen ion concentration	(S)-	sinister (counterclockwise configuration)
		S-	denoting attachment to sulfur
phr	parts per hundred of resin (rubber)	s-	symmetric(al)
p-i-n	positive-intrinsic-negative	S	second
pmr	proton magnetic resonance	(s)	solid, only as in $H_2O(s)$
p-n	positive-negative	SAE	Society of Automotive Engineers
po	per os (oral)		
POP	polyoxypropylene	SAN	styrene-acrylonitrile
pos	positive	sat(d)	saturate(d)
pp.	pages	satn	saturation
ppb	parts per billion (10^9)	SBS	styrene–butadiene–styrene
ppm	parts per milion (10^6)	sc	subcutaneous
ppmv	parts per million by volume	SCF	self-consistent field; standard cubic feet
ppmwt	parts per million by weight		
PPO	poly(phenyl oxide)	Sch	Schultz number
ppt(d)	precipitate(d)	sem	scanning electron microscope(y)
pptn	precipitation		
Pr (no.)	foreign prototype (number)	SFs	Saybolt Furol seconds
pt	point; part	sl sol	slightly soluble
PVC	poly(vinyl chloride)	sol	soluble

soln	solution	*trans-*	isomer in which
soly	solubility		substituted groups are
sp	specific; species		on opposite sides of
sp gr	specific gravity		double bond between
sr	steradian		C atoms
std	standard	TSCA	Toxic Substances Control
STP	standard temperature and		Act
	pressure (0°C and	TWA	time-weighted average
	101.3 kPa)	Twad	Twaddell
sub	sublime(s)	UL	Underwriters' Laboratory
SUs	Saybolt Universal seconds	USDA	United States Department
syn	synthetic		of Agriculture
^{t} (eg, Bu^t),	tertiary (eg, tertiary	USP	*United States*
t-, tert-	butyl)		*Pharmacopeia*
T	tera (prefix for 10^{12}); tesla	uv	ultraviolet
	(magnetic flux density)	V	volt (emf)
t	metric to (tonne)	var	variable
t	temperature	*vic-*	vicinal
TAPPI	Technical Association of	vol	volume (not volatile)
	the Pulp and Paper	vs	versus
	Industry	v sol	very soluble
TCC	Tagliabue closed cup	W	watt
tex	tex (linear density)	Wb	weber
T_g	glass-transition	Wh	watt hour
	temperature	WHO	World Health Organization
tga	thermogravimetric		(United Nations)
	analysis	wk	week
THF	tetrahydrofuran	yr	year
tlc	thin layer chromatography	(Z)-	zusammen; together;
TLV	threshold limit value		atomic number

Non-SI (Unacceptable and Obsolete) Units		Use
Å	angstrom	nm
at	atmosphere, technical	Pa
atm	atmosphere, standard	Pa
b	barn	cm^2
bar†	bar	Pa
bbl	barrel	m^3
bhp	brake horsepower	W
Btu	British thermal unit	J
bu	bushel	m^3; L
cal	calorie	J
cfm	cubic foot per minute	m^3/s
Ci	curie	Bq
cSt	centistokes	mm^2/s
c/s	cycle per second	Hz
cu	cubic	exponential form

†Do not use bar (10^5 Pa) or millibar (10^2 Pa) because they are not SI units, and are accepted internationally only in special fields because of existing usage.

Non-SI (Unacceptable and Obsolete) Units		Use
D	debye	$C \cdot m$
den	denier	tex
dr	dram	kg
dyn	dyne	N
dyn/cm	dyne per centimeter	mN/m
erg	erg	J
eu	entropy unit	J/K
°F	degree Fahrenheit	°C; K
fc	footcandle	lx
fl	footlambert	lx
fl oz	fluid ounce	m^3; L
ft	foot	m
ft · lbf	foot pound-force	J
gf den	gram-force per denier	N/tex
G	gauss	T
Gal	gal	m/s^2
gal	gallon	m^3; L
Gb	gilbert	A
gpm	gallon per minute	(m^3/s); (m^3/h)
gr	grain	kg
hp	horsepower	W
ihp	indicated horsepower	W
in.	inch	m
in. Hg	inch of mercury	Pa
in. H_2O	inch of water	Pa
in.-lbf	inch pound-force	J
kcal	kilo-calorie	J
kgf	kilogram-force	N
kilo	for kilogram	kg
L	lambert	lx
lb	pound	kg
lbf	pound-force	N
mho	mho	S
mi	mile	m
MM	million	M
mm Hg	millimeter of mercury	Pa
$m\mu$	millimicron	nm
mph	miles per hour	km/h
μ	micron	μm
Oe	oersted	A/m
oz	ounce	kg
ozf	ounce-force	N
η	poise	$Pa \cdot s$
P	poise	$Pa \cdot s$
ph	phot	lx
psi	pounds-force per square inch	Pa
psia	pounds-force per square inch absolute	Pa
psig	pounds-force per square inch gage	Pa
qt	quart	m^3; L
°R	degree Rankine	K
rd	rad	Gy
sb	stilb	lx
SCF	standard cubic foot	m^3
sq	square	exponential form
thm	therm	J
yd	yard	m

BIBLIOGRAPHY

1. The International Bureau of Weights and Measures, BIPM (Parc Saint-Cloud, France) is described in Ref. 4. This bureau operates under the exclusive supervision of the International Committee for Weights and Measures (CIPM).
2. *Metric Editorial Guide (ANMC-78-1)*, latest ed., American National Metric Council, 900 Mix Avenue, Suite 1 Hamden CT 06514-5106, 1981.
3. *SI Units and Recommendations for the Use of Their Multiples and of Certain Other Units (ISO 1000-1992)*, American National Standards Institute, 25 W 43rd St., New York, 10036, 1992.
4. Based on IEEE/ASTM-SI-10 *Standard for use of the International System of Units (SI): The Modern Metric System* (Replaces ASTM380 and ANSI/IEEE Std 268-1992), ASTM International, West Conshohocken, PA., 2002. See also www.astm.org
5. *Fed. Reg.*, Dec. 10, 1976 (41 FR 36414).
6. For ANSI address, see Ref. 3. See also www.ansi.org

A

Continued

ALCOHOLS, HIGHER ALIPHATIC, SURVEY

1. Survey and Natural Alcohols Manufacture

Monohydric, aliphatic alcohols with a hydrocarbon chainlength of C6 and above are referred to as higher alcohols. For commercial products, the alcohol group is usually found in the primary position, although secondary alcohols are occasionally seen. The hydrocarbon portion of the molecule is hydrophobic, while the hydroxyl group provides a reactive site for attaching a strong hydrophilic species. The combination of hydrophilic and hydrophobic properties in the same molecule yields an extremely good surfactant, readily biodegradable (see DETERGENCY; SURFACTANTS). Detergent alcohols generally have a hydrocarbon chainlength of C12 and above with at least 35% linear chains. Plasticizer alcohols generally have a chainlength of C6 to C13 (excluding C12 and C13 linear) with a structure that is either linear or highly branched.

Both natural (oleochemical) and synthetic (petrochemical) routes are used to make the higher alcohols. Plasticizer alcohols are made primarily by the synthetic route, with only minor quantities obtained from natural feedstocks. Detergent alcohols are made by both routes, with global capacity currently split 50/50 synthetic/natural. The natural route is based on vegetable and animal fats, principally coconut, palm kernel, and tallow. The products are essentially all even chain length (ie, C12, C14, ...) and 100% linear. The synthetic route is based on ethylene chain growth, and is classified as either Ziegler or oxo. The Ziegler process produces linear, even-chain products, while the oxo process produces a mixture of branched and linear, even- and odd-chain products (see Synthetic Processes section).

1

Global production of plasticizer alcohols was about 3,400,000 tons in 1995, with demand expected to increase 3.1% during the period 1995–2000. Primary end uses are for flexible PVC and for surface coatings applications in construction or automobile manufacturing. Global production of detergent alcohol was about 2,000,000 tons in 1998, with demand expected to increase by 2.5% during the period 1995–2000. Ethoxylates, sulfates, and ethoxysulfate derivatives account for about 84% of U.S. consumption of detergent alcohols in 1995. Various other derivatives account for 11%, with the balance being free alcohol consumption.

1.1. Detergent Range Alcohols.   Detergent range alcohols can be categorized into mid cut (C12–C15) and heavy-cut (C16 and greater) chain lengths, depending on the distillation range where they are separated. Mid cut alcohols are most preferred for detergent use, as these chain lengths provide an optimum tradeoff between surfactancy on the one hand and (lack of) crystallinity on the other. Heavy cut alcohols are used more for health and personal care applications. Mid cut alcohols are produced and distributed as either a pure component or a mixture of two or three components (ie, C12/C13 or C12/C14), while heavy cuts are primarily a single chain length (ie, C16 or C18). Oleyl alcohol (*cis*-9-octadecenol) is the most abundant of the unsaturated alcohols and occurs widely as a wax ester in fish and marine mammal oils. It consitutes 70% of the alcohols in wax esters of sperm body oil.

Long-chain alcohols >C18 are generally of much less commercial importance (1). C20 alcohol [629-96-9] is made commercially through one or more synthetic routes. Applications are as chemical intermediates for rubber, plastics, and textiles. C22 alcohol [661-19-8] is made commercially by reduction of behenic acid or behenic methyl esters. Applications include lubricants and synthetic fibers. Even-chain C24–C32 alcohols are derived from various natural waxes, including carnauba, beeswax, and Chinese insect wax. Applications are similar to that of C22 alcohol. Wool grease from sheep also contains higher alcohols as wax esters, and is a minor commercial source of alcohol.

1.2. Plasticizer Range Alcohols.   C6–C11 and C13 alcohols are available as pure materials or as complex mixtures. In general, the linear alcohols are available as pure materials, and are referred to as "-yl alcohols." There is no single dominant linear alcohol product, as the entire C6–C11 range is about equally important. The branched alcohols are generally available as mixtures, and are referred to as "isoalcohols" or "-anol." 2-Ethyl hexanol is by far the most important branched alcohol.

2. Physical Properties

Table 1 lists some of the basic physical properties of linear and branched higher alcohols (2). Table 2 lists the key thermal, flammability, and critical properties of the linear alcohols (3). In Table 1, it can be seen that specific gravity, boiling point, melt point, and viscosity are all a linear or a polynomial function of chain length. The branched alcohols have properties that are similar to the linears for equal carbon number, with the exception that melt point is

Table 1. Physical Properties of Higher Aliphatic Alcohols

IUPAC name	CAS Registry Number	Molecular formula	Other common names	Specific gravity 20 °C	Bp, °C 101.3 kPa	Mp, °C	Viscosity mPa s	Solubility, in water	% by wt of water	Solubility in other solvents
Primary normal aliphatic										
1-hexanol	[111-27-3]	C$_6$H$_{14}$O	n-hexyl alcohol	0.8212	157	−44	5.9	0.59	7.2	petroleum ether, ethanol
1-heptanol	[111-70-6]	C$_7$H$_{16}$O	n-heptyl alcohol	0.8238	176	−35	7.4	0.1		petroleum ether, ethanol
1-octanol	[111-87-5]	C$_8$H$_{18}$O	n-octyl alcohol	0.8273	195	−15.5	8.4	0.06	4.5	petroleum ether, ethanol
1-nonanol	[143-08-8]	C$_9$H$_{20}$O	n-nonyl alcohol	0.8295	213	−5	11.7			petroleum ether, ethanol
1-decanol	[112-30-1]	C$_{10}$H$_{22}$O	n-decyl alcohol	0.8312	230	7	13.8		2.8	petroleum ether, ethanol; benzene, glacial acetic
1-undecanol	[112-42-5]	C$_{11}$H$_{24}$O	n-undecyl alcohol	0.8339	243	16	17.2	< 0.02	1.3	petroleum ether, ethanol
1-dodecanol	[112-53-8]	C$_{12}$H$_{26}$O	lauryl alcohol	0.881[a]	138[b]	24	18.8	[d]		petroleum ether, ethanol
1-tetradecanol	[112-72-1]	C$_{14}$H$_{30}$O	myristyl alcohol		158[b]	38	53 cAt 60 °C	< 0.02	nil	petroleum ether, ethanol
1-hexadecanol	[36653-82-4]	C$_{16}$H$_{34}$O	cetyl alcohol		177[b]	49	7[c]	0.06	nil	petroleum ether, ethanol
1-octadecanol	[112-92-5]	C$_{18}$H$_{38}$O	stearyl alcohol		203[b]	58		[d]	nil	ethanol, ether
1-eicosanol	[629-96-9]	C$_{20}$H$_{42}$O	arachidyl alcohol		251[b]	66		[d]	nil	benzene, ethanol, ether
1-hexacosanol	[506-52-5]	C$_{26}$H$_{54}$O	ceryl alcohol			80		[d]		
1-triacontanol		C$_{30}$H$_{62}$O	myricyl alcohol	0.777		88				
9-octadecen-1-ol	[143-28-2]	C$_{18}$H$_{36}$O	oleyl alcohol	0.8484	205–210	6				ethanol, diethyl ether
Primary branched aliphatic										
2-methyl-1-pentanol	[105-30-6]	C$_6$H$_{14}$O	2-methylpentyl alcohol	0.8254	148		6.6	0.31	5.4	
2-ethyl-1-butanol	[97-95-0]	C$_6$H$_{14}$O	2-ethylbutyl alcohol	0.8348	147	−114				
2-ethyl-1-hexanol	[104-76-7]	C$_8$H$_{18}$O	2-ethylhexyl alcohol	0.834	184	−70	9.8	0.07	2.6	ethanol, diethyl ether
3,5-dimethyl-1-hexanol	[13501-73-0]	C$_8$H$_{18}$O		0.8297	182.5					ethanol
2,2,4-trimethyl-1-pentanol	[123-44-4]	C$_8$H$_{18}$O		0.839	168	−70				ethanol
3,5,5-trimethyl-1-hexanol		C$_9$H$_{20}$O		0.83	197		15			
iso-octadecanol		C$_{18}$H$_{38}$O	isostearyl alcohol			0	48			

[a] At 24 °C.
[b] At 1.33 kPa pressure.
[c] At 60°C.
[d] insoluble.

Table 2. Thermal, Flammable, and Critical Properties of Higher Aliphatic Alcohols

	n-hexanol	n-octanol	n-decanol	n-dodecanol	n-octadecanol
specific heat, cal/g °C		0.65		0.63	0.6
heat of fusion, kcal/gmol		10.1	10.3	8.3	17.7
heat of vaporization, cal/g	106	98	85	77	67
reference temp, °C	190	224	246	247	242
heat of combusion, kcal/gmol		1262	1577		
surface tension, dyn/cm	24.5	26.1			
thermal conductivity, cal/cm h °C	1.34	1.37	1.41	1.46	1.58
dielectric constant	13.3	10.3			
critical temperature, °C	340	387	425	460	545
critical pressure, MPa	3.5	3.0	2.6	2.3	1.7
flash point, °C	55	85	103	113	127
fire point, °C	68	96	116	138	193
autoignition temperature, °C				244	260

significantly lower for the branched molecules. The plasticizer alcohols are slightly soluble with or in water, but the detergent alcohols are generally insoluble with or in water. Branching does not have a significant impact on water solubility.

In Table 2, it can be seen that heat of fusion, thermal conductivity, critical temperature, and flammability generally increase with chain length. Critical pressure and heat of vaporization are an inverse function of chain length. However, the reference temperature for the heat of vaporization increases over the range of C6–C18 chain lengths. Specific heat is relatively constant over the range of chain lengths listed.

3. Chemical Properties

There are three principal types of reactions involving alcohols (4). These are the reactions involving the O-H bond, reactions involving the C-O bond, and reactions with the alkyl portion of the molecule (ie, the C-H bond of the α carbon). Examples of each type are shown below. Alcohols, like water, can act as both weak acids or weak bases. Reactions are generally done in the liquid phase, using either homogeneous or heterogeneous catalysts. The products of each reaction are useful as valuable chemical intermediates and end products.

3.1. Reactions Involving the O-H bond
Sulfation

$$ROH + SO_3 \longrightarrow ROSO_3H$$
alkyl hydrogen sulfate

$$ROSO_3H + NaOH \longrightarrow ROSO_3Na + H_2O$$
sodium alkyl sulfate

Etherification

$$ROH + H_2C\text{--}CH\text{--}CH_2Cl \longrightarrow ROCH_2CH\text{--}CH_2Cl$$

epichlorohydrin   alkyl chlorohydrin ether

Alkoxidation

$$ROH + M \longrightarrow ROM + \tfrac{1}{2}H_2$$
alkoxide
$$M = Al, Mg, Na, K, etc.$$

3.2. Reactions involving the C-O bond

Esterification

$$ROH + R'COOH \longrightarrow R'COOR + H_2O$$

$$ROH + R'COOR'' \longrightarrow R'COOR + R''OH$$
alkyl ester

Ethoxylation

$$ROH + n\,H_2C\text{--}CH_2 \longrightarrow R(OCH_2CH_2)_nOH$$

polyethoxylated alcohol

Halogenation

$$ROH + HX \longrightarrow RX + H_2O$$
alkyl halide

Dehydration

$$RCH_2CH_2OH \longrightarrow RCH = CH_2 + H_2O$$
olefin

Amination

$$ROH + R'NH_2 \longrightarrow RNHR' + H_2O$$
amine

3.3. Reactions with the α-Carbon

Oxidation

$$RCH_2OH + \tfrac{1}{2}O_2 \longrightarrow RCH{=}O + H_2O$$
aldehyde

$$RCH{=}O + \tfrac{1}{2}O_2 \longrightarrow RCOOH$$
carboxylic acid

Dehydrogenation

$$RCH_2OH \quad \longrightarrow \quad \underset{\text{aldehyde}}{RCH{=}O} + H_2$$

4. Shipment and Storage

Higher alcohols are available in bulk quantities in 208-L (55-gal) drums, 23,000-L (6000-gal) tank trucks, 75,000-L (20,000-gal) tank cars, and marine barges. In addition, some of the plasticizer alcohols are available in bottles and cans, ie, for perfume applications. Higher-melting alcohols (C16 and greater) are also generally available in a flaked form in 22.7-kg (50-lb) polyethylene or paper lined bags. Linear and branched alcohols of 6–9 carbon atoms are classified as combustible for shipment by the U.S. DOT due to their low flash points. Alcohols of C10 and above are classified as nonhazardous. The higher alcohols in anhydrous form do not attack common metals. They may be stored in mild steel, but to avoid iron contamination a liner of zinc silicate, epoxy phenolic, or high baked phenolic may be used (5). For high melting alcohols, and for low melting alcohols under cold climate conditions, insulated tanks and heating coils of stainless steel are required. High temperatures should be avoided during storage to maintain product quality. This can be accomplished by using hot water or low pressure steam on heating coils during meltout to avoid hot spots, and by maintaining temperature at 10°C above the melting point. Carbonyl formation during storage may be minimized with a nitrogen blanket. Moisture in the atmosphere may be excluded by storing the product under a blanket of inert gas or by installing a dehumidifier. In order to minimize any danger of fire in the handling of plasticizer range alcohols, tanks should be grounded, have no interior sources of ignition, be filled from the bottom of the tank to prevent static sparks, and be equipped with flame arrestors.

5. Health and Safety Factors

The higher alcohols are among the less toxic of commonly used industrial or household chemicals. Toxic effects generally decline as chain length increases. Table 3 summarizes the toxicological properties of the higher alcohols (6).

The acute oral toxicity data on the higher alcohols indicate a low order of toxicity by the oral, dermal, or inhalation routes of exposure (6). Higher alcohols are poorly absorbed through the skin. The rate of dermal uptake for the neat material was found to decrease with increasing carbon number. Human repeated skin patch tests with a 1% alcoholic solution of C12–C15 alcohols indicate very slight to mild irritation which is nonfatiguing or sensitizing (6). Primary human skin irritation of C16 and C18 alcohols is nil, as these products have been historically used in cosmetic and personal care products. Inhalation hazard is slight, due to the low vapor pressure of these materials. However, sustained breathing of alcohol vapor or mist is to be avoided, in order to minimize any aspiration hazard.

Table 3. **Toxicological Properties of Higher Alcohols**

	Acute oral rat LD_{50}	Acute dermal rabbit LD_{50}	Skin irritation rabbit	Eye irritation rabbit
n-hexanol	4.7	>5	severe	severe
2-ethylhexanol	3	2.3	slight-moderate	moderate-severe
iso-octanol	2	>2.6	moderate	severe
isodecanol	8	3	moderate	severe
n-undecanol	3	>5	moderate	
n-dodecanol	>11		slight-moderate	
n-tetradecanol	>5	>5		
n-hexadecanol	>8.4	>2.6	slight	slight
n-octadecanol	>8	>3	slight	slight

6. Economic Aspects

Global production of detergent range alcohols in 1998 was about 2,000,000 tons. Production is growing by 3–4% per year in the period 1995 to 2000.

During the 1990s, virtually all new plant capacity added has been based on natural feedstocks, largely in the countries of Malaysia, Indonesia, and Japan. The Philippine producers have been largely overtaken by new Indonesian and Malaysian capacity. This is a reversal of the global trend during the 1960s–1980s, when synthetic alcohol capacity was increasing while natural capacity was flat. The current trend for adding natural alcohol capacity is driven largely by geographical considerations of supply and demand. Demand is increasing 6% in Asia vs. 3–4% in North America and Europe. Asian countries also have an abundant supply of coconut and palm kernel oil, and a desire to harvest and export these commodities. Much of the growth in production between 1995 and 2005 will be for export to developing and Western countries. China will also be an emerging market, for both production and consumption.

Growth in the industry in recent years has been limited by supply (7). By year 2000, new debottlenecked capacity should be on-line in Germany, Italy, and the U.S. New plant construction for synthetics is also planned for Condea and Shell in the U.S. and Sasol in South Africa. Currently, worldwide production capacity is about evenly split between synthetics and naturals. The preponderance of natural producers in Asia is contrasted by a much stronger position by synthetics in North America. Europe is about evenly split. For either approach, cost of production is affected by both feedstock costs and co-product value, and these costs are largely determined in other markets. Figure 1 shows the historical cost for ethylene (8) vs. coconut oil (9), which are the key starting materials for synthetic and natural alcohols, respectively. Due to the cyclic nature of these raw material costs, there are periods of time when natural alcohols have a cost advantage and other periods when synthetic alcohols have the upper hand. Consumer preferences are segmented, with synthetic alcohols preferred for household and industrial applications and natural alcohols preferred for health and personal care. In countries such as Germany and Japan, the naturals are also preferred by consumers for their renewable source of supply. For most applications, linear synthetic alcohols can be used interchangeably with natural

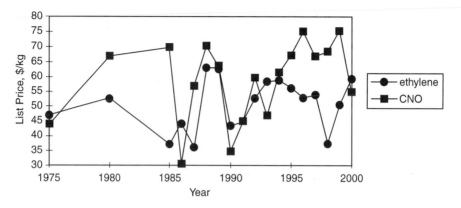

Fig. 1. List price of ethylene vs. conconut oil: 1975–2000.

alcohols except where minor amounts of chain branching or secondary alcohols are absolutely prohibited.

Global producers of detergent range alcohols are numerous. In 1999 there were five in the U.S., ten in Europe, and numerous others in Asia. Table 4 shows relevant information on the five U.S. producers; representative list prices for natural alcohols are given in Table 5 (10). Most major customers have contracts at prices lower than at list. Also, several producers are planning to expand plant capacity in the next few years.

Global production of plasticizer alcohols was 3,400,000 tons in 1995, of which 1,058,000 tons was U.S. About 40% of U.S. consumption is as 2-ethylhexanol. Consumption is equally split among North America, Europe, and Asia. Global demand is projected to increase by 3.1% per year over the period 1997–2002.

Global producers of plasticizer alcohols are numerous. In 1996 there were 11 U.S. producers and at least 15 global producers. Exxon and Eastman

Table 4. Detergent Alcohol Producers in U.S.

Producer	Process	Capacity, tons (1999)	Products	Trade names
P&G	natural methyl ester	80,000	C12–C18	CO, TA
Cognis	natural methyl ester	40,000	C12–C18	Lorol
Shell	synthetic modified oxo	250,000	C12–C15	Neodol
BPAmoco	synthetic modified Ziegler	60,000–95,000	C12–C32	Epal
Condea	synthetic Ziegler	60,000	C12–C28+	Alfol

Table 5. Natural Detergent Alcohol List Pricing (1999)

Alcohol	Price, U.S. $/kg, fob
C12–C14	1.54
C16	1.65
C18	1.65

Table 6. **Plasticizer Alcohol List Pricing (1996)**

Material	Price, US. $/kg, fob	Manufacturer
n-hexanol	2.09	Exxon, Condea, BPAmoco
isohexyl alcohol	1.45	Exxon
n-octanol	2.27	BPAmoco, Condea, P&G, Cognis
2-ethylhexanol	1.23	BASF, Eastman, Union Carbide, Shell
n-decanol	2.02	Exxon, Shell, BPAmoco, P&G, Cognis, Condea
isodecyl alcohol	1.12	Exxon
tridecyl alcohol	1.58	Exxon

Chemicals are the largest U.S. producers. Rather than listing all producers, Table 6 shows typical manufacturers and list prices of some of the common types of plasticizer alcohols (11).

The U.S. market possesses the best economic conditions for continued growth in consumption in the near future. Usage for plasticizers (flexible PVC) and for acrylate/methacrylate esters (surface coating and textiles) are both driving demand. Long term, Asia is the focus of global producers due to emerging strong demand and insufficient local production. Plant capacity has been added in the U.S. during the 1990s by BASF, Condea Vista, and Shell, and globally by Eastman and Exxon.

Particular alcohols are chosen for plasticizer production based on physical properties. In addition to performance, cost is extremely important. The branched chain alcohols are consistently less expensive than the comparable linear alcohols, primarily because propylene feedstock for branched alcohols is consistently less expensive than the ethylene feedstock used for linears. Manufacturers of plasticizer alcohols general sell some of their material to the merchant market and save the rest for internal or "captive" use, largely for production of phthalate plasticizers.

7. Analytical Methods

Higher alcohols analysis is done to determine the chain-length distribution as well as to determine the level of minor components present, which may impact product performance.

Gas chromatography (GC) is used to measure chain-length distribution for mixed products and purity for single-chain products. The composition of both linear and branched alcohols can be characterized by this method. In case of overlapping peaks on the chromatograms, more sophisticated techniques such as GC-mass spec or nuclear magnetic resonance (nmr) can be applied.

Minor components are generally measured by wet titration techniques. This includes % moisture, color, carbonyl value, hydroxyl value, saponification value, acid value, and iodine value (12). Carbonyl value is expressed as ppm of C=O equivalent in the product and is a measure of product oxidation, which is associated with dark color and off-odor. Hydroxyl value is expressed as mg of KOH equivalent to the hydroxyl content of one gram of product and is an indication of the overall alcohol purity. Saponification value is expressed as mg of KOH

Table 7. **Reference Test Methods for Analysis**

Analysis	Reference Test Method No.
acid value	AOCS Te 1a-64
saponification value	AOCS TL 1a-64
moisture	AOCS Tb 2-64
hydroxyl value	AOCS Cd 13-60
iodine value	AOCS Tg 1a-64
peroxide value	AOCS Cd 8-53
color (APHA)	ASTM D 1209
density	ASTM 1298
flash point	ASTM D92

required to saponify esters and acids in one gram of product and is an indication of the sum of the levels of these components in the product. Acidity is commonly expressed as acetic acid equivalents/100 g of product for plasticizer alcohols and as mg KOH/g of product for detergent-grade alcohols. Iodine value is expressed as g I_2/100 g of product and is an indication of the degree of unsaturation (double bonds) present. An iodine value of 0 is completely saturated.

Table 7 shows reference test methods commonly used to analyze the higher alcohols. Methods are from the American Oil Chemists Society (13) and the American Society for Testing and Materials.

8. Specifications and Standards

Commercially available materials are found as both pure components and mixtures. Sales brochures from commercial manufacturers provide further details (14). Even-chain alcohols are produced from natural fats and oils and from the Ziegler process, and are highly linear. Odd-chain alcohols are produced by oxochemistry, and have some branched chains. Linear, detergent-range alcohols are marketed in the U.S. by P&G, Cognis, and Condea. These materials are available as 95–99% pure C12, 14, 16, and 18 alcohols, and also as blends, eg, a 70/30 blend of C12 and C14 alcohols. C16 and C18 alcohols are also available as National Formulary (NF) grade, for cosmetic or pharmaceutical applications. Condea is the only company to produce C20 and C22 alcohols for health and personal care applications. P&G and Cognis exclusively use natural fats and oils, while Condea uses both oleochemical and petrochemical routes. Branched, detergent-range alcohols are marketed in the U.S. by Shell. They are 80% linear, with the balance being random alkyl branched chains. These are typically 50/50 mixtures of even and odd chains, eg, C12 and C13, or C14 and C15. Table 8 displays information

Table 8. **Detergent Alcohol Chain Length, Linerity**

Manufacturer	Product name	Chain length	% Linear
P&G	CO-1270	70/30 blend of C12/C14	100
Cognis	LOROL C1695	95% C16	100
Condea	Alfol 18, Nacol 18	99% C18	>98
Shell	Neodol 45	50/50 blend of C14/C15	80

Table 9. **Properties of Commercial Linear Detergent Range Alcohols**

Descriptive name	Hydroxyl value mg KOH/gr	Saponification value	Carbonyl value, ppm	Acid value	Iodine value	Melting point, °C	Color, APHA	Moisture %	Hydrocarbon %
lauryl (99% C12)	301	0.2		0.02	0.2	22	5	0.03	0.2
lauryl (68% C12)	285	0.2		0.02	0.2	22	3	0.04	0.2
C12–C13	289		40	0.01		20	5	0.02	0.2
cetyl	229	0.4		0.1	0.6	48	5	0.04	0.1
stearyl	206	0.5		0.1	0.6	58	5–10	0.03	0.2
oleyl	206	0.5		0.1	94	4		0.03	

Table 10. **Plasticizer Alcohol Chain Length, Linearity**

Manufacturer	Product name	Chain length	% Linear
Condea	Alfol 6	99.5% C6	>98
PG	CO-898	98% C8	100
Cognis	Lorol C1098	98% C10	100
BPAmoco	Epal 810	50/50 C8/C10	>98
Shell	Linevol 79	97% C7-C9	80
Exxon	Exxal 6	5% ethylbutanol, 60% methyl pentanol, 35% n-hexanol	35
Exxon	Nonanol	80% 3,5,5 trimethyl hexanol	<20
BASF	2-ethylhexanol	99.5% 2-ethyl hexanol	0

on detergent alcohol chain length and linearity, while Table 9 focuses on other physical properties of interest.

Linear, plasticizer alcohols are marketed in the U.S. by P&G, Cognis, Condea, and BPAmoco. These materials are available as 95–99% pure C6, C8, and C10 alcohols, and also as blends, eg, a 45/55 blend of C8 and C10 alcohols. P&G and Cognis utilize natural fats and oils exclusively, while Condea and BPAmoco use petroleum based feedstocks.

A range of branched, plasticizer range alcohols are marketed in the U.S. by Shell and Exxon. Both pure materials and mixtures are available. Shell produces odd-chain C9–C11 alcohols with 80% linearity. Exxon produces highly branched even- and odd-chain C6–C13 alcohols, the major isomers of which are mono-, di-, tri-, or tetramethyl branched molecules. 2-Ethyl hexanol is produced and marketed in the U.S. by BASF, Union Carbide, and Eastman.

Table 10 displays information on chain length and linearity for various plasticizer alcohols, while Table 11 focuses on other physical properties.

9. Manufacture from Fats and Oils

9.1. Feedstocks and Intermediates.
Both methyl esters and fatty acids are suitable intermediates for manufacture of higher natural alcohols.

Table 11. **Properties of Commercial Plasticizer Alcohols**

Descriptive name	Hydroxyl value mg KOH/g	Carbonyl value, ppm	Acidity % as acetic	Color, APHA	Moisture %	Boiling range °C	Flash point °C
hexanol	548		0.001	5	0.05	152–160	63
2-ethyl hexanol	431		0.005	5	0.05	182–186	
isooctyl alcohol	429	100	0.001	5	0.05	186–193	84
isononyl alcohol	380	100	0.001	5	0.05	204–216	91
octanol	431		0.005	5	0.03	184–195	88
decanol	355	50	0.005	5	0.03	226–230	113
tridecanol	283		0.001	5	0.03	254–263	127

Triglycerides from either vegetable or animal sources are the feedstocks used to produce either of these intermediates. Triglycerides consist of three fatty acid chains attached to a glycerine backbone. The fatty acid chain lengths range from C6 to C22. The most useful chain lengths for detergent applications are C12 and C14. Health and personal care applications are more oriented towards C16 and C18. Common triglyceride feedstocks are coconut oil and palm kernel oil for detergent applications, and tallow, palm, or soybean oil for cosmetic applications. Less common is rapeseed oil as a feedstock for behenyl or eurucyl alcohol and castor oil as a feedstock of hydroxy-stearyl alcohol.

As the more widely used intermediate, methyl esters can be made by transesterification of fatty triglycerides or by direct esterification of fatty acids.

Transesterification of triglycerides with methanol is the predominant process for manufacture of methyl esters. The reaction occurs readily at atmospheric pressure at a temperature of 50–70°C using carbon steel equipment and an akaline catalyst. Both triglyceride and methanol are dried to minimize the conversion of alkaline catalyst into soap.

$$C_3H_5(OOCR)_3 + 3\ CH_3OH \longrightarrow 3\ RCOOCH_3 + C_3H_5(OH)_3$$

These mild reaction conditions require that any trace fatty acids present in the fat be neutralized chemically (alkali refining) or removed physically (steam refining) prior to the base catalyzed transesterification. Removal of fatty acids is not required if the reaction is carried out with an acid catalyst, or if it is carried out under pressure (10 MPa) at a temperature of about 240°C with an alkaline catalyst. The latter is advantageous in that lower grades of triglyceride can be used, but this is balanced by the need for a higher excess of methanol and higher equipment pressure ratings than for the conventional process.

The reaction can be done either batchwise or continuous. It is advantageous to have mixing present in the early stages of the reaction to enhance miscibility among the reactants. After reaction, the mixture is settled and the glycerol is recovered in methanol solution in the lower layer. The catalyst and excess methanol are removed from the ester, which can then either be fed directly to the alcohol making process, or distilled to remove impurities or to purify the chain length distribution (15).

Direct esterification of fatty acids to produce methyl esters is done in either a batch or a continuous process.

$$RCOOH + CH_3OH \longrightarrow RCOOCH_3 + H_2O$$

The batch reaction is done at pressures of about 1 MPa and temperatures of 200–250°C. Molar ratios of methanol:fatty acid are about 3–4:1. Water is removed to drive the reaction to a high conversion. The continuous process is done under similar conditions of pressure and temperature in a countercurrent reaction column where methanol is absorbed from the bottom and reacted while water (and excess methanol) are desorbed from the top. The methyl ester is taken off the bottom of the reactor, sent to a flash tank to remove methanol, and then distilled. Advantages of the continuous process are that a lower excess of methanol is required (molar ratio of 1.5:1), and the reaction time is much shorter.

The direct esterification process is advantageous for making esters with a higher chain length purity than the parent triglycerides, eg, separation of the fatty acid into a stearine and an olein fraction.

Fatty acids are less widely used as an intermediate than methyl esters. Most of the equipment has to be made of corrosion-resistant material, the fatty alcohol process has to be designed for a higher pressure and temperature, the catalyst has to be acid resistant, and the finished product tends to be slightly higher in hydrocarbons. On the other hand, fatty acids are more widely available than methyl esters for those who are purchasing, rather than producing, the intermediate material. In addition, the fatty acid process avoids the use of methanol, a flammable solvent subject to local environmental regulations. In commercial processes, the fatty acid material is typically reacted with fatty alcohol to form a wax ester; this intermediate then undergoes hydrogenolysis to the alcohol. Lurgi has licensed the wax ester technology to several companies in Europe and Asia, including Albright and Wilson, PT Aribhawana Utama, Condea Vista, and United Coconut Chemicals (16).

To prepare fatty acid feedstock for fatty alcohol making, triglyceride feedstock must be hydrolyzed with water to fatty acid and glycerine.

$$C_3H_5(OOCR_3 + H_2O \longrightarrow 3\,RCOOH + C_3H_5(OH)_3$$

The hydrolysis can be done at atmospheric pressure with sulfuric or sulfonic acids or at high pressure with or without a catalyst (zinc oxide, magnesium oxide, or lime). High pressure continuous processes are most commonly practiced. The fatty acids are purified by distillation or separated into individual chain lengths by fractional distillation (17).

9.2. High Pressure Hydrogenolysis.   Fatty acids or methyl esters are converted into fatty alcohols in the presence of a heterogeneous catalyst. The reaction for the methyl esters is:

$$RCOOCH_3 + 2\,H_2 \longrightarrow RCH_2OH + CH_3OH$$

There are two major processes in use: slurry based and fixed bed. Procter & Gamble and Kao practice the slurry based process while Cognis and Oleofina practice fixed bed.

Figure 2 shows a process flow diagram for the methyl ester, slurry catalyst process. Dry methyl ester, hydrogen, and catalyst slurry are fed to a series of vertical reactors operated at 250–300°C and 20–30 MPa. The reactors are tubular and contain no packing. Approximately 20 moles of hydrogen per mole of ester are fed to the reactors providing heat, agitation, and chemical reduction. Catalyst is typically copper chromite, which results in the hydrogenation of carbon double bonds as well as reduction of the methyl ester group. Catalyst usage is typically 2 kg/ton of alcohol. The product stream from the last reactor is cooled and separated into liquid and gaseous phases. The gaseous phase, rich in H_2, is recycled back to the reactors. Catalyst is removed from the liquid phase, and most of the catalyst is recycled to the reactors while a small amount is purged and replaced with fresh catalyst. Purged catalyst is landfilled, regenerated, or sold to a reclaimer to recover copper metal. The liquid phase is stripped of

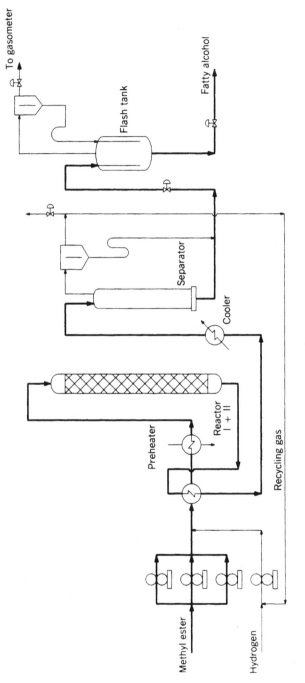

Fig. 2. Methyl ester, slurry catalyst process.

15

methanol, the latter being recycled back to the methyl ester reactors. Stripped crude alcohol is distilled to remove heavy impurities and optionally to fractionate into individual chain lengths. Still bottoms, primarily fatty-fatty ester, are recycled while a small amount are purged. A similar flowsheet is used for the fatty acid based, wax ester process (18).

There are a number of advantages for the fixed bed process compared to the slurry process, including lower catalyst usage (typically 0.8–1.0 kg/tonne of alcohol), improved process reliability, and better alcohol quality, with lower hydrocarbon content and lower saponification values. The low saponification value allows one in many cases to avoid alcohol distillation and rework of the unreacted methyl ester. The key disadvantages of the fixed bed include the need for higher ester feed quality to avoid poisoning the catalyst with sulfur, glycerine, glycerides, or soap. (19). Also, an increase in operating temperature, typically used near the end of the bed lifetime, can compromise product quality and cause elevated levels of hydrocarbon and dialkyl ether. A bed lifetime of 12 mo is typical. Either liquid-phase (trickle-bed) or vapor-phase hydrogenation can be used in the fixed bed. The process conditions for the liquid phase process are 200–225°C and 20.7 MPa, with a molar ratio of hydrogen to methyl ester of 20–100. Supported catalysts such as 20–40% copper chromite on a silica gel carrier are typically used. The catalyst must have high mechanical stability in order to withstand the stresses exerted by the gas–liquid film in the reactor. The vapor-phase process differs in the temperature and pressure requirements (225–250°C and 3.4 MPa); also the molar ratio of hydrogen to methyl ester is 200 or greater. The catalyst is compact pelletized. The high amount of recirculated gas provides fast removal of the heat of reaction, keeping side reactions such as hydrocarbon formation very low.

Figure 3 shows the process flow diagram for the methyl ester, fixed-bed catalyst process. Methyl ester is mixed with fresh and recycled hydrogen and preheated prior to entering the top of the reactor. The effluent is cooled and separated into liquid and gaseous fractions. The gaseous phase, which is rich in hydrogen, is recycled to the reactor, while the liquid phase is expanded into a flash tank to remove methanol. The stripped, crude alcohol generally does not require further distillation, unless fractionation into narrower cuts is desired.

Zinc containing catalysts (ie, zinc chromite) can be used at higher temperatures (275–325°C) and pressures of 25–30 MPa to convert unsaturated esters into unsaturated fatty alcohols using a fixed-bed process. Selectivity is sufficiently high that unsaturated fatty alcohols with iodine values up to 170 are reported (17). Producers of oleyl alcohol are Cognis, Salim Oleochemicals, Witco, and Rhone-Poulenc NA. Unsaturated fatty alcohols can also be made from lauric oils by selective hydrogenation of the unsaturated C18 methyl ester fraction, which has been separated by fractional crystallization (20).

9.3. New Manufacturing Trends. During the 1990s, several new trends emerged in the field of high-pressure hydrogenolysis. This included both new process technology and new catalysts. While many of these developments are by themselves not significant enough to retrofit existing plants, several of them are being incorporated into the construction of new natural alcohol plants.

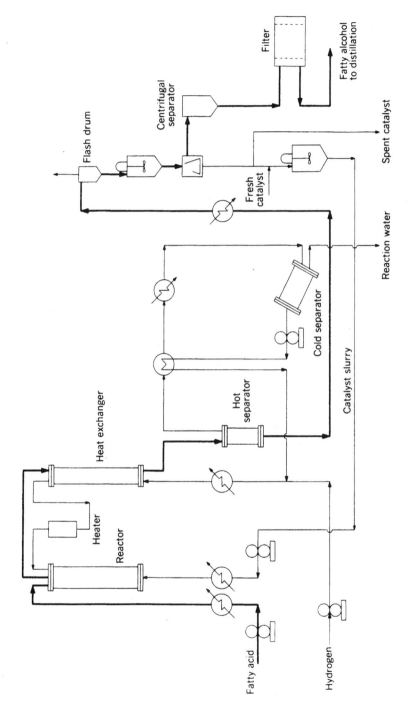

Fig. 3. Methyl ester, fixed-bed catalyst process.

17

9.4. Fatty Acid to Methyl Ester Conversion.   In order to provide flexibility in feedstock usage, Kvaerner (formerly Davy Process Technology) has developed a new continuous countercurrent reactor to convert fatty acids to methyl esters (21). In some parts of the world, fatty acids are more widely available than methyl esters for those producers who do not have equipment for methanolysis of triglycerides. A solid resin acid catalyst developed by Rohm and Haas is used in the Kvaerner process. The resin is reported to be low cost, have long life, be easily separated from the product and have very high ($> 99.5\%$) conversion to the ester. This catalyst avoids some of the issues of previous esterification routes, which used sulfonic acid liquid catalysts that were corrosive to the reactor, were difficult to separate from the product, and thereby led to residual catalyst poisoning of the hydrogenation catalyst.

9.5. Low-Pressure Fixed-Bed Process.   Kvaerner has also developed a totally vapor phase, fixed bed process for methyl ester feedstocks that is lower in pressure (3.4 vs 20.7 MPa) than the conventional fixed bed or slurry process (22). A promoted copper catalyst is used and selectivity is claimed to be 99%, but conversion is not specified. Methyl ester purity requirements are 99.8% as a feedstock for hydrogenation. This purity can be generated in the Kvaerner fatty acid to methyl ester conversion process described above. A very high ratio of hydrogen:methyl ester is used to keep the reaction completely in the vapor phase, and the capital associated with this somewhat offsets the benefits of lower-pressure operation.

Process benefits in the reactor reportedly include better contact between the hydrogen and the methyl ester, better heat contact and removal, and shorter residence time. Product benefits reportedly include lower by-products, such as hydrocarbons, as a result of the lower operating temperature. An integrated fatty acid–methyl ester–fatty alcohol plant located in the Philippines and operated by Primofina Oleochemicals, was successfully started up in 1997, but later shut down due to financial difficulties.

9.6. Direct Hydrogenolysis of Triglycerides.   Several patents (23) were issued in the 1990s by Henkel (now Cognis) concerning direct hydrogenation of triglycerides. On paper, the process should be less costly than either the fatty acid or the methyl ester route, since there are fewer processing steps involved and hence lower capital investment as well as lower labor and utilities. However, propylene glycol, rather than glycerine, is the main co-product generated from the process, and during the period 1981–2000, the average list price for these materials was $1.23/kg and $1.65/kg, respectively (24). Under these conditions, the cost of fatty alcohol is approximately equal for both processes.

9.7. Hydrogenolysis in the Presence of Hydrocarbons under Supercritical Conditions.   Another process technology being developed in the 1990s was hydrogenation under supercritical conditions (25). By adding a mutual solvent, ie, propane or butane, to methyl esters and hydrogen, the whole mixture is brought to a supercritical state. In this way, the gas to liquid mass transfer resistance of hydrogen is minimized. The result is an increase in the rate of hydrogenation by several orders of magnitude. Fatty alcohols have been produced on a pilot scale at about 250°C, 15.2 MPa, and a reaction time of 3 s using a fixed-bed process. Conversion and selectivity were both above 95%. For comparison, conventional reaction times for the fixed bed are on the order of 30 min to 2 h. From a capital standpoint, the advantage of a significant reduction in reactor volume is clearly offset by the requirement to separate and recycle the solvent back to the reactor. The inventors are reportedly looking for a commercial opportunity to scale up the process.

9.8. Biotechnology. Calgene cloned the reductase gene that is responsible for the synthesis of fatty alcohols in plants in 1992. They also reportedly were seeking to clone the ligase gene, which is responsible for the synthesis of long-chain liquid waxes from fatty alcohols in plants. Calgene reportedly plans to genetically engineer the reductase and ligase genes into rapeseed to allow for cost-effective natural production of fatty alcohols and long-chain liquid waxes in plants (26).

9.9. New Catalysts. Most catalysts for fatty alcohol production over the past 50 y have been based on Adkin's copper chromite catalyst (15). During the 1990s, alternatives to copper chromite were being investigated with the objective of replacing chromium as a promoter in the catalyst. This is because chromium is typically not recovered from spent catalyst, while copper is recovered and recycled. Positive results have been found by using zinc as the promoter in silica-supported copper catalysts (27). Another alternative catalyst (Ru-Sn-boride on alumina) was developed for low-pressure (1.5–5 MPa) hydrogenolysis of either saturated or unsaturated methyl esters into corresponding fatty alcohols. Yields of 90% have been demonstrated for saturated fatty alcohols, but only 60–80% yield has been demonstrated for oleyl alcohol (28).

10. Uses of Detergent-Range Alcohols

About 95% of the C12–C18 alcohols are converted to derivatives that are consumed in various end uses, while the remainder are consumed as the free alcohol. Major classes of applications include surfactants, lubricants, cosmetics and personal care, and pharmaceutical and medical products. Surfactants and lubricants tend to use alcohol derivatives, while the other applications tend to use the alcohol as is. Table 12 summarizes the list of applications.

10.1. Surfactants. In 1995, about 63% of the overall consumption of C12–C18 alcohols was as alcohol ethoxylates, half of which is subsequently converted to alcohol ethoxysulfate. Alcohol ethoxylates, ether sulfates, alcohol sulfates, alkyl glyceryl ether sulfonates (AGES), and alkylpolyglucosides (APG) are used as surfactants in household heavy-duty powders and liquids, light duty dishwashing liquids, other household cleaners, personal care products, and a variety of industrial, commercial, and institutional uses. These five surfactants account for about 84% of the 1995 U.S. consumption of detergent alcohols, excluding the C20+ products. Small volumes of fatty nitrogen derivatives are also used in household and industrial surfactants (9).

Consumption of ethoxylates and ether sulfates experienced strong growth during the 1990s due to displacement of LAS from laundry liquids and dishwashing liquids. Alcohol ethoxylates have better compatibility with complex enzymes increasingly used in liquid laundry products. They are also superior for cleaning body oil stains. The ether sulfates are more mild for dish applications. Both surfactants have better tolerance for hard-water ions than any anionic surfactant

Table 12. **Uses of Detergent Range Alcohols**

Industry	Use as alcohol	Use as derivative
detergent	emollient, foam control	surfactant, softener, bleaching agent, degreaser
petroleum and lubrication	drilling mud	emulsifier, lubricant, pour-point depressant
agriculture	evaporation suppressant, sprout inhibitor, debudding agent	pesticide, fungicide, emulsifier, soil conditioner, seed coating
plastics	mold release agent, antifoam, emulsion polymerization agent	stabilizer, uv absorber
textile	lubricant, foam control	emulsifier, softener, lubricant
cosmetics	softener, emollient, skin cleansing, gellant	emulsifier, biocide, hair conditioner, shower gel
pulp and paper	foam control	deinking agent
food	beverage additive	emulsifier, disinfectant
mineral processing	flotation agent	surfactant
pharmaceutical	performance enhancement	antiviral, antiulcer treatment
fuel additive	oxygenating agent	

available. Very low levels of specialty alcohol ethoxylates are used as emulsifiers in cleansing cremes and a few other personal care products. Alcohol ethoxylates are also used in textile processing and metal degreasing. Alcohol sulfates and ether sulfates are used in hard surface, rug and upholstery cleaners, shampoos, bubble baths, toilet soaps, and other personal care products in the U.S. Industrial uses include emulsion polymerization and agricultural emulsifiers. Alkyl glyceryl ether sulfonates and alkylpolyglucosides are small-volume, specialty surfactants. The former is used as a foam-boosting surfactant in light-duty liquid, shampoos, and combination soap-synthetic toilet bars. The latter is a nonionic surfactant with good solubility, foaming, and mildness and can be used in laundry, light-duty liquids, and personal care products. Fatty nitrogen derivatives include fatty amine oxides, ether amines, dialkyldimethylammonium quaternaries, and alkylbenzyldimethylammonium chlorides. Fatty amine oxides are surfactants used in light-duty dishwashing liquids, household cleaners, personal care products, and a few specialized industrial applications.

10.2. Lubricants. Polymethacrylate esters are employed in automotive and aircraft lubricating oils, as well as transmission and hydraulic fluids. They function as viscosity index improvers, pour-point depressants, and polymeric dispersants. Oligoesters of fatty alcohol and a pyrometallic anhydride are useful for waterproofing leather (29). An aqueous belt lubricant composition is based on fatty alcohol polyglycerol ethers (30). A lubricant for cold working of metals is based on monoalkyl ethers of polyethylene glycol and fatty alcohols (31). Lubricating and antifriction characteristics of water-based drilling fluids are improved by additives such as fatty alcohol or esterified or ethoxylated fatty alcohols (32). A corrosion inhibitor for gas pipelines and other steel surfaces is made by adding higher aliphatic alcohols to a mixture containing butanol, urea, various surfactants, and amines (33).

10.3. Cosmetics, Personal Care. C12, C14, and C16 alcohols are used in perfumes and fragrances. C18 alcohol is used in USP ointments (14).

In general, short-chain alcohols, particularly under about C16, tend to be irritants, while longer-chain aliphatic alcohols tend to be non irritating. C1216 alcohol is used as an emollient and bodying agent in hair care formulations. C16, C18, and C22 alcohols are used as a lubricant and bodying agent in skin care and hair care formulations. C20 alcohols are used as a lubricant and bodying agent in skin care formulations and also used in lipsticks, toothpastes, and perfume bases (34). A facial skin cleanser containing C16 and C18 alcohols and their ethoxylates is useful for removing sebum plaque (35). Other personal care applications include a cleansing bar with a moisturizing effect (36), cosmetic cleaning products with superior mildness via use of fatty alcohol polyglycosides (37), and a mild shower gel composition comprising fatty alcohol which imparts improved lathering and thickening properties (38).

Guerbet alcohols (2 alkyl-alcohols) and wax esters (alkyl alkanoates) are also commonly used for personal care applications. C12–C24 Guerbet alcohols are used in skin, hair, and stick applications, while C28–C36 Guerbet alcohols are used in stick and color cosmetic formulations. C24–C44 wax esters are used for cosmetics, pharmaceutical and candy coatings, and candles. The higher chain lengths (C36–C44) may be used as a replacement or extender for montan or carnauba wax.

10.4. Pharmaceutical, Medical. C18 alcohol has been approved as a direct and indirect food ingredient and as an ingredient in over-the-counter drugs. It is generally known that selected alcohols have some physiological activity. More specifically, the longer-chain alcohols, generally greater than C24, have benefits as anti-inflammatories and antiviral agents. C30 alcohol (triacontanol), in a suitable carrier, can be used as a treatment for inflammatory disorders such as herpes simplex, eczema, shingles, psoriasis, etc (39). C27 to C32 aliphatic alcohols are an effective topical anti-inflammatory and may also be used, in suitable carrier compositions, for treatment of virus induced disease and in prevention of infection by disease-causing virus (40). Natural mixtures of higher aliphatic primary alcohols, isolated from beeswax, are useful for antiulcer treatment (41). Perhaps the broadest medical claim associated with the higher alcohols refers to improved physical performance of athletes from ingestion of C24–C30 alcohols, which are active ingredients in wheat germ oil (42).

10.5. Other. Higher alcohol esters of thiodipropionic acid function as antioxidants and are effective in stabilizing polyolefins. Ether amines are often used in mining (ore flotation) applications. Dialkyldimethylammonium quaternaries are used for fabric softeners. Alkylbenzyldimethylammonium chlorides are used as active ingredients in fungicides, biocides, sanitizers, and disinfectants. An aqueous dispersion of C20–C36 alcohols for use in food and drinks offers good acid, salt, and heat resistance (43). C30 alcohol (triacontanol) accelerates growth in plants (44), and accelerates decomposition of sewage and reduces H2S (45). C14–C22 fatty alcohol is useful in dry cement compositions to reduce lime bloom, lime weeping, or crystallization of salts (46). Aqueous fatty alcohol dispersions (C10–C28) are useful as antifoam agents (47). An aqueous-based, solvent-free degreaser composition includes alcohol alkoxylates with a fatty alcohol moiety, alkoxylated fatty alcohol, and a fatty alcohol having an oxyethylate moiety (48).

Table 13. **Uses of Plasticizer Alcohols**

Industry	Use as alcohol	Use as derivative
plastics	emulsion polymerization	plasticizer, coatings
petroleum and lubrication	defoamer	lubricant, diesel additive
agriculture	stabilizer, tobacco sucker control	
mineral processing	extractant, antifoam	extractant, surfactant
paper	plasticizer	deinking agent
metal working	lubricant	lubricant, surfactant
cosmetics	fragrances, perfumes	sun protection products, gel sticks
health and personal care	antiseptics	fragrances

11. Uses of Plasticizer Alcohols

Plasticizer alcohols are either linear or branched C6 to C13 molecules (excluding C12 and C13 linears, which are detergent range). In 1997, about 54% of worldwide production was as 2-ethyl hexanol, 17% as C9 oxo alcohol, 12% as isodecyl alcohol, and 8% as C6–C11 linear alcohols. The remaining 9% is as branched C6 to C8 molecules. The plasticizer alcohols are used primarily in plasticizers, but they also have applications in a wide range of industrial and consumer products, as shown in Table 13 (11). As with the detergent range alcohols, the plasticizer alcohols are mainly consumed as derivatives.

11.1. Plasticizers. Phthalates, adipates, acetates, and trimellitates of C6–C13 alcohols are the lead types of plasticizers in use. Both linear and branched phthalates and adipates are used. Acetates and trimellitates tend to be exclusively branched. When ethylene prices rise in comparison to propylene, consumers tend to substitute branched phthalates for linear phthalates. The C7–C11 linear phthalates thrive better in markets where performance is more critical than price. At similar prices, linear phthalates provide better value than branched phthalates. Linear phthalates are characterized by superior low temperature properties and lower volatility; they are used in coated fabrics and sheet goods and when outdoor weathering resistance is required (swimming pool liners, roofing membranes, tarpaulins, and wire and cable jacketing). Linear phthalates based on alcohols greater than C9 have distinct advantages over branched phthalates in auto upholstery and compartment interiors; while maintaining low temperature flexibility, these linear phthalates reduce window fogging. Branched phthalates are used in applications such as vinyl flooring, carpet backing, and wire and cable jacketing. Adipates are used where low temperature flexibility is important, such as auto accessories, gaskets, hoses, and tubing. Acetates are based on branched alcohols such as isohexyl and isoheptyl and are used in application such as surface coatings. Trimellitates are used for wire and cable compounds, with trioctyl trimellitate and triisononyl trimellitate being the primary molecules (11).

11.2. Other Plastics Applications. 2-Ethyl hexyl esters of trialkyl phosphite serve as thermal stabilizers and antioxidants in plastics. Ba, Cd, and Zn salts of 2-ethyl hexanoic acid are used as PVC stabilizers. 2-Ethyl hexyl acrylate is used in emulsion polymers for pressure-sensitive adhesives. 2-Ethyl hexanol can be used directly or as an acrylate or methacrylate ester for surface coatings. Direct use is as a solvent in the electrodisposition of primer surface coatings in the automotive industry (11).

11.3. Surfactants. A number of surfactants are made from the plasticizer-range alcohols, using processes similar to that of detergent range alcohols, including ethoxylation, sulfation, and amination. Ethoxylated mixtures of C9–C11 linear alchols are used in hard surface cleaners and commercial detergents. Increased solubility and liquidity provide advantages over longer chain alcohols. Ethoxylates are also used for processing textiles, leather, pulp, and paper. C6–C10 and C8–C10 blends are used for ethoxylated/propoxylated/ phosphated surfactants and C6–C10 sulfates and betaines. Automatic dishwasher formulations use propoxylated and ethoxylated linear alcohols. Ethoxylated tridecyl alcohol is used in nonionic surfactants, and the ether sulfate is used in a leading children's shampoo. 2-Ethyl hexanol is used as an additive in dispersing and wetting agents for pigment pastes, and as a sulfosuccinate molecule for wetting and scouring of textiles (11).

11.4. Lube Additives. 2-Ethyl hexyl based esters of C6–C10 dicarboxylic acids, glycols, or polyglycols are useful as greases and lubricants (49). Other branched alcohols are converted to molecules such as zinc diisodithiophosphate (ZDDP) and isodecyl methacrylate for high-temperature stability and cold weather flow in lubricants. 2-Ethyl hexanol is used as an ingredient in a fuel oil composition for diesel engines with decreased levels of particulates (50). 2-Ethyl hexyl nitrate is another type of diesel fuel additive that can be used at low levels to promote ignition (51). A lubricant composition containing 2-ethyl hexanol, dimer fatty acid, and a surfactant is effective in hot steel rolling mill operations (52). C6 and C8 alcohols can also be used for lamp oil and lighting fluids for charcoal grills as a nontoxic, clean-burning fuel with no color or odor (53).

11.5. Agricultural Chemicals. Plasticizer alcohols are used as solvents or as intermediates in the manufacture of insecticides and herbicides, including 2-ethyl hexyl esters of 2,4-dichlorophenoxyacetic acid and 2,4,5-trichlorophenoxyacetic acid. 2-Ethyl hexanol is also used as a carrier in microbiocidal compositions. Blends of C8 and C10 linear alcohols are used a plant growth regulators for tobacco sucker control.

11.6. Mining, Extraction. 2-Ethyl hexanol and various amine derivitives are used as a feedstock in the manufacture of extractants for heavy metals. Di-2-ethylhexyl phosphoric acid (DEHPA) and C8–C10 linear trialkyl amines are used for extraction of uranium ore. C8–C10 linear ether amines and isodecyl alcohol amines are used for flotation of taconite ores (54).

11.7. Solvents. 2-Ethyl hexanol is used as a direct solvent for defoaming in the paper, textile, and oil field industries (55). It is also used as a low-volatility ingredient in solvent blends for dyestuffs and coatings (ie, printing and stamp pad inks, dipping lacquers, etc), as well as an ingredient in solvent compositions to clean soil from dirty articles including metal, ceramic, and glass substrates (56). Quaternary amines used for deinking of paper, surfactants, flotation agents, and biocides also use 2-ethyl hexanol as a diluent (57).

11.8. Other Miscellaneous Applications.   2-Ethyl hexanol is used as a flow and gloss improver in baking finishes (58). Sun protection products are made with esters of 2-ethyl hexanol and *p*-methoxycinnamic acid (59). Benzylammonium chlorides of C8–C10 linear alcohols are used in hard surface cleaners and fungicides in institutional applications. Esters of 2-ethyl hexanol with salicylic acid have a variety of applications in the cosmetic, pharmaceutical, and food industries, as flavor and fragrance chemicals (60).

BIBLIOGRAPHY

"Alcohols, Higher" in *ECT* 1st ed., Vol. 1, pp. 315–321, by H. B. McClure, Carbide and Carbon Chemicals Corporation, Unit of Union Carbide and Carbon Corporation; "Alcohols, Higher, Fatty" in *ECT*, 2nd ed., Vol. 1, pp. 542–559, by K. R. Ericson and H. D. Van Wagenen, The Procter & Gamble Company; "Alcohols, Higher, Synthetic" in *ECT*, 2nd ed., Vol. 1, pp. 560–569, by R. W. Miller, Eastman Chemical Products, Inc. "Alcohols, Higher Aliphatic, Survey and Natural Alcohols Manufacture" in *ECT* 3rd ed., Vol. 1, pp. 716–739, by R. A. Peters, Procter & Gamble Company. "Alcohols, Higher Aliphatic, Survey and Natural Alcohols Manufacture" in *ECT* 4th ed., Vol. 1, pp. 865–893, by R. A. Peters Procter & Gamble Company; "Alcohols, Higher Aliphatic, Survey and Natural Alcohols Manufacture" in *ECT* (online), posting date: December 4, 2000, by R. A. Peters, The Procter & Gample Company.

CITED PUBLICATIONS

1. J. A. Monick, *Alcohols, Their Chemistry, Properties and Manufacture*, Reinhold Book Corp., New York, 1968, pp. 189–193.
2. *CRC Handbook of Chemistry & Physics*, 79th ed., CRC Press, Boca Raton, 1998, pp. 3-1 to 3-330; D. H. Green ed., *Chemical Engineers Handbook*, McGraw-Hill, New York, 1989, pp. 3-25 to 3-44; M. Davies, and B. Kybett, *Faraday Soc. Trans.* **61** (1965); *J. Res. Nat. Bur. of Standards* **13**, 189–192 (Aug. 1934).
3. *Lange's Handbook of Chemistry*, 14 ed., McGraw-Hill, N.Y., 1992, pp. 1–80 to 1–325.
4. J. Roberts and M. Caserio, *Basic Principles of Organic Chemistry*, 2nd ed., Benjamin, New York, 1977, pp. 612–645.
5. Shell Linevol 11 Sales Spec, Shell Chemical Company, 1994.
6. F. Clayton, and G. Clayton, *Patty's Industrial Hygiene and Toxicology*, Vol. 2, Part D, Wiley, N.Y., 1994, pp. 2678–2703; Toxicological Properties of Shell Neodol Alcohols and Derivatives, 1980.
7. *Chemical Marketing Reporter* **253**, 4 (Jan. 26, 1998).
8. K. Al-Husseini, and A Jebens, CEH Marketing Research Report: Petrochemical Industry Overview, SRI International, Menlo Park, Calif., 1997.
9. P. Brown, Personal Correspondence, April 6, 2000.
10. R. Modler, CEH Marketing Research Report: Detergent Alcohols, SRI International, Menlo Park, Calif., 2000.
11. S. Bizzari, CEH Marketing Research Report: Plasticizer Alcohols, SRI International, Menlo Park, Calif., 1998.
12. M. Bockisch, *Fat & Oils Handbook*, AOCS Press, Champaign, 1998, pp. 803–808.
13. *Official Methods and Recommended Practices of the AOCS*, 4th edition, Vols. 1 and 2, 1993.

14. Products from the Chemicals Division, Procter & Gamble Company, Cincinnati, Ohio, 1999; Exxon Chemical Synthetic Alcohol Technical Manual for Surfactants, 1998; Epal Fatty Alcohols, BP Amoco Bulletin OF-7; Condea Vista Cosmetic Ingredient Brochure, August 1999; Typical Physical and Chemical Properties of Shell Neodol Alcohols, Shell Sales Spec on 2–ethylhexanol, BASF Technical Data Sheet, 2–ethylhexanol, March 1998.

15. U. R. Kreutzer, *J. Am. Oil Chem. Soc.* **61**, 343–348 (1984).

16. R. Peters in T. Applewhite, ed., *Proceedings of the World Conference on Oleochemicals–Into the 21st Century*, AOCS Press, Champaign, 1991, p. 183.

17. R. Modler, CEH Marketing Research Report: Fatty Acids, SRI International, Menlo Park, Calif., 1999.

18. T. Voeste and H. Buchold, *J. Am. Oil Chem. Soc.* **61**, 350–352 (1984).

19. R. A. Rieke, D. S. Thakur, B. D. Roberts, and G. T. White, *J. Am. Oil Chem. Soc.* **74**, 333 (1997).

20. European Patent 943596 (Dec. 29, 1999), F. Wieczorek, G. Konetzke, and E. Seifert (to DHW Deut Hydrierwerke GmbH Rodleben).

21. Kvaerner Process Technology (Formerly Davy Process Technology),—"Esterification," 1993.

22. U.S. Patent 5,138,106 (Aug. 11, 1992), M. Wilmott et al., (to Davy McKee, Ltd).

23. U.S. Patent 4,942,266 (July 17, 1990), T. Fleckenstein et al. (to Henkel KgAA); U.S. Patent 4,982,020 (Jan. 1, 1991), F. Carduck, J. Falbe, T. Fleckenstein, and J. Pohl (to Henkel KgAA); U.S. Patent 5,364,986 (November 15, 1994), G. Demmering, S. Heck, and L. Friesenhagen (to Henkel KgAA).

24. R. Martin, CEH Marketing Research Report: Propylene Glycol, SRI International, Menlo Park, Calif., 1997. T Esker, CEH Marketing Research Report: Glycerine, SRI International, Menlo Park, Calif., 1999.

25. S. van den Hark et al., *J. Am. Oil Chem. Soc.* **76**, 1363–1370 (1984).

26. "Calgene Clones Fatty Alcohol Gene," *Appl. Genetics News* **13**, (Oct. 1992).

27. F. Van de Scheur and L. H. Staal, *Appl. Catalysis* **108**, 63–84 (1994).

28. Narasimhan, *Appl. Catalysis* **48**, 11–16 (1984); Narasimhan, *Ind. Eng. Chem. Res.* **28**, 1110–1112 (1989).

29. German Patent 19,644,242 (Apr. 30, 1998), A. Behr, H. Hankwerk, W. Ritter, and R. Zauns-Huber (to Henkel KgaA).

30. U.S. Patent 5,900,392 (May 4, 1999), L. Bernhard (to Loeffler Chemical Corp.).

31. Russian Patent 2,007,439 (Feb. 15, 1994), L. Demina, V. Korovin, and T. Mostovaya (to Mostuvaya TA).

32. Hungarian Patent 47969 (April 28, 1989), T. Balogh et al. (to Koolaj-ES Foldgazbanyas; Koolajkutato Vallalat).

33. Russian Patent 2,023,754 (Nov. 30, 1994), V. Gonchorov, A. Melnik, and I. Zhulov (to Khark Poly; Ukr Natural Gas Res. Inst).

34. Cosmetic Ingredient Brochure, Condea Vista Co., 1999.

35. U.S. Patent 4,495,079 (Jan. 22, 1985), A. Good.

36. British Patent 2,317,396 (Mar. 25, 1998), M. Davey, and P. O'Byrrne (to Cassons Inc.).

37. B. Jackwerth, *Paperboard Packaging*, 36 (Sep. 1995).

38. U.S. Patent 5,866,110 (Feb. 2, 1999), C. Moore, E. Inman, and C. Schell (to Procter & Gamble Co).

39. U.S. Patent 4,670,471 (June 2, 1987), C. Lealand.

40. U.S. Patent 5,070,107 (Dec. 3, 1991), D. Katz (to Lidak Pharmaceuticals).

41. D. Carbajal et al., *J. Pharmacy Pharmacol.* **47**, 731–733 (1995).

42. U.S. Patent 3,031,376 (Apr. 24, 1962), E. Levin.

43. Japanese Patent 11,187,850 (July 13, 1999), K. Atsuko and T. Yoshiharu (to Nippon Oils and Fats Co., Ltd).
44. U.S. Patent 4,150,970 (Apr. 24, 1979), S. Ries and C. Sweeley, (to Michigan State University).
45. U.S. Patent 4,246,100 (Jan. 20, 1981), J. Starr (to Bio-Humus Inc.).
46. World Patent 9,504,008 (Feb. 9, 1995), B. Abdelrazig et al. (to W. R. Grace & Co.).
47. U.S. Patent 5,807,502 (Sep. 15, 1998), H. Wollenweber, R. Hoefer, and H. Schulte (to Henkel KgAA).
48. U.S. Patent 5,880,082 (Mar. 9, 1999), S. Gessner et al. (to BASF Corp.).
49. Russian Patent 2,114,819 (July 10, 1998), V. Kirilovich, T. Leshina, and N. Zakovrya-shina (to Innovation Tech – Econ Res. Commerce).
50. Japanese Patent 9,194,856 (July 29, 1997), K. Keiichi, U. Hiroshi, and I. Tei (to Tonen Corp.).
51. U.S. Patent 5,669,938 (Sep. 23, 1997), S. Schwab (to Ethyl Corp.).
52. U.S. Patent 5,372,736 (Dec. 13, 1994), R. Trivett (to Nalco Chem. Co.).
53. European Patent 874038 (Oct. 28, 1998), J. Falkowski, W. Heck, N. Huebner, and N. Klein (to Henkel KgAA).
54. U.S. Patent 4,319,987 (Mar. 16, 1982), D. Shaw et al. (to Exxon Research and Engineering Co.).
55. U.S. Patent 4,391,722 (July 5, 1983), E. Y. Schwartz, C. Tincher, and J. Maxwell (to BASF Wyandotte Corp.); U.S. Patent 4,612,109 (Sep. 16, 1986), E. Dillon and E. Edmonds (to NL Industries Inc.).
56. U.S. Patent 5,464,557 (Nov. 7, 1995), G. Ferber and G. Smith (to Bush Boake Allen Ltd.).
57. U.S. Patent 5,696,292 (Dec. 9, 1997), C. Cody and N. Martin (to Rheox Inc.).
58. U.S. Patent 3,705,124 (Dec. 5, 1972), R. Selby and D. Williams (to E. I. duPont de Nemours and Co.).
59. U.S. Patent 5,728,865 (Mar. 17, 1998), B. Croitora, A. Ewenson, and A. Shushan (to Bromine Compounds Ltd.).
60. Romanian Patent 102886 (Dec. 30, 1991), C. Barladianu (to Sitiform Intr Medica-mente Coloranti).

COREY J. KENNEALLY
The Procter & Gamble Co.

ALCOHOLS, HIGHER ALIPHATIC, SYNTHETIC PROCESSES

1. Introduction

Higher aliphatic alcohols (C_6–C_{18}) are produced in a number of important industrial processes using petroleum-based raw materials. These processes are summarized in Table 1, as are the principal synthetic products and most important feedstocks (qv). Worldwide capacity for all higher alcohols was approximately 5.3 million metric tons per annum in early 1990, 90% of which was petro-

Table 1. **Synthetic Industrial Processes for Higher Aliphatic Alcohols**

Process	Feedstock(s)	Principal products	Worldwide capacity, millions of tons
Ziegler (organo-aluminum)	ethylene, triethyl-aluminum	primary C_6–C_{18} linear alcohols	0.3
oxo (hydroformylation)	olefins based on ethylene, propylene, butylene, or paraffins	primary alcohols	4.2
aldol	n-butyraldehyde	2-ethylhexanol	[a]
paraffin oxidation	paraffin hydrocarbons	secondary alcohols	0.2
Guerbet	lower primary alcohols	branched primary alcohols	[b]
Total			4.7

[a] Included in oxo process total.
[b] Less than 0.05.

leum-derived. Table 2 lists the major higher aliphatic alcohol producers in the world in early 1990.

By far the largest volume synthetic alcohol is 2-ethylexanol [104-76-7], $C_8H_{18}O$, used mainly in production of the poly(vinyl chloride) plasticizer bis(2-ethylhexyl) phthalate [117-81-7], $C_{24}H_{38}O_4$, commonly called dioctyl phthalate [117-81-7] or DOP (see PLASTICIZERS). A number of other plasticizer primary alcohols in the C_6–C_{11} range are produced, as are large volumes of C_{10}–C_{18} synthetic, mainly primary, alcohols used as intermediates to surfactants (qv)

Table 2. **Major C_6 and Higher Aliphatic Alcohol Producers**[a]

Company and location	Capacity 10^3 t/yr	Alcohol products	Feedstock
		Ziegler process	
Condea Chemie, Brunsbuettel, Germany	70	n-C_6,C_8,C_{10},C_{12},C_{14},C_{16},C_{18},C_{20}	ethylene
Ethyl Corp, Houston, Tex., U.S.	111	n-C_6,C_8,C_{10},C_{12},C_{14},C_{16},C_{18},C_{20}	ethylene
State, Ufa, USSR	48	n-C_6,C_8,C_{10},C_{12},C_{14},C_{16},C_{18},C_{20}	ethylene
Vista Chemical, Lake Charles, La., U.S.	100	n-C_6,C_8,C_{10},C_{12},C_{14},C_{16},C_{18},C_{20}	ethylene
Ziegler subtotal	*329*		
		Guerbet process	
Henkel, Duesseldorf, Germany	2	i-C_{16},C_{18},C_{20},C_{22},C_{24},...,C_{36}	linear alcohols
Guerbet subtotal	*2*		
		Caustic fusion process	
Witco Chemical, Dover, Ohio, U.S.	7	2-octanol	castor oil
Caustic fusion subtotal	*7*		

Table 2 *(Continued)*

Company and location	Capacity 10^3 t/yr	Alcohol products	Feedstock
	Fatty acid hydrogenation processes		
ATOCHEM SA, Lavera, France	7	n-C_7	castor oil
Cocochem, Batangas, Philippines	25	n-$C_8, C_{10}, C_{12}, C_{14}, C_{16}$	coconut oil
Colgate, Barangay, Philippines	4	n-$C_8, C_{10}, C_{12}, C_{14}, C_{16}$	coconut oil
Oleofabrik, Aarhus, Denmark	5	n-C_{16}, C_{18}	palm oil, tallow
State, Kedzierzyn, Poland	10	n-$C_8, C_{10}, C_{12}, C_{14}, C_{16}$	coconut oil
Fatty acid hydrogenation process	*51*		
	Methyl ester hydrogenation process		
ATUL, India	3	n-$C_8, C_{10}, C_{12}, C_{14}, C_{16}$	coconut oil
Aegis, Jalagon, India	5	n-$C_8, C_{10}, C_{12}, C_{14}, C_{16}$	coconut oil
Condea Chemie, Brunsbuettel, Germany	30	n-$C_8, C_{10}, C_{12}, C_{14}, C_{16}$	coconut oil
Henkel, Duesseldorf, Germany	130	n-$C_8, C_{10}, C_{12}, C_{14}, C_{16} C_{18}$	coconut oil, tallow
Henkel, Boussens, France	50	n-$C_8, C_{10}, C_{12}, C_{14}, C_{16}, C_{18}, C_{20}, C_{22}$	coconut oil, other fats
Hüls AG, Marl, Germany	10	n-$C_8, C_{10}, C_{12}, C_{14}, C_{16}$	coconut oil
Kao Corp, Wakayama, Japan	15	n-$C_8, C_{10}, C_{12}, C_{14}, C_{16}$	coconut oil
Marchon (Albright & Wilson), Whitehaven, UK	25	n-$C_8, C_{10}, C_{12}, C_{14}, C_{16}$	coconut oil
New Japan Chemical, Tokushima, Japan	15	n-$C_8, C_{10}, C_{12}, C_{14}, C_{16}$	coconut oil
Philippinas Kao, Jasaan, Philippines	30	n-$C_8, C_{10}, C_{12}, C_{14}, C_{16}$	coconut oil
Procter & Gamble, Kansas City, Kan., U.S.	45	n-$C_8, C_{10}, C_{12}, C_{14}, C_{16}, C_{18}$	coconut oil, tallow
Procter & Gamble, Sacramento, Calif., U.S.	54	n-$C_8, C_{10}, C_{12}, C_{14}, C_{16}, C_{18}$	coconut oil, palm oil
Sherex, Mapleton, Ill., U.S.	7	oleyl alcohol, n-C_{18}	tallow, soybean oil
Sinopec, Shanghai, China	15	n-$C_8, C_{10}, C_{12}, C_{14}, C_{16}$	coconut oil
State, Radleben, Germany	10	n-$C_8, C_{10}, C_{12}, C_{14}, C_{16}$	coconut oil
Synfina-Oleofina, Ertvelde, Belgium	30	n-$C_8, C_{10}, C_{12}, C_{14}, C_{16}$	coconut oil

Table 2   (*Continued*)

Company and location	Capacity 10^3 t/yr	Alcohol products	Feedstock
Methyl ester hydrogenation process subtotal	*474*		
		Oxidation processes	
Japan Catalytic Chemical, Kawasaki, Japan	12	sec-C_{11},C_{12},C_{13},C_{14},C_{15}	n-paraffins
State, Angarsk, USSR	45	i-C_{10},C_{11},C_{12},C_{13},C_{14},C_{15},C_{16},C_{17},C_{18}	n-paraffins
State, Ufa, USSR	90	sec-C_{11},C_{12},C_{13},C_{14},C_{15},C_{16}	n-paraffins
State, Volgodonsk, USSR	45	n-C_{10},C_{11},C_{12},C_{13},C_{14},C_{15},C_{16},C_{17},C_{18}	n-paraffins
Oxidation subtotal	*192*		
		Oxo process	
Enichem, Augusta, Italy	50	n-C_7 to C_{15}	n-paraffins
Exxon Chemical France, Harnes, France	125	i-C_8,C_9,C_{10}; n-C_9,C_{11},C_{13},C_{15}	polygas olefins, alpha olefins
Exxon Chemical Holland, Rozenburg-Europort, Netherlands	200	i-C_8,C_9,C_{10},C_{13},C_{16}	polygas olefins
Exxon Chemical, Baton Rouge, La., U.S.	295	i-C_6 to C_{10},C_{12},C_{13},C_{16}; n-C_7,C_9,C_{11}	polygas olefins, alpha olefins, butene
Hoechst, Oberhausen-Holten, Germany	40	i-C_{10},C_{13}	propylene
ICI, Teeside, United Kingdom	250	i-C_8,C_9,C_{10}; n-C_9 to C_{15}	polygas olefins, alpha olefins
India Nissan Chemical Ind., Baroda, India	13	i-C_7,C_8,C_9,C_{10},C_{11}	polygas olefins
Mitsubishi Kasei, Mizushima, Japan	25	i-C_9	butenes
Mitsubishi Kasei, Mizushima, Japan	30	n-C_7,C_9,C_{11},C_{13},C_{15}	ethylene
Mitsubishi Petrochemical, Yokkaichi, Japan	30	n-C_{12},C_{13},C_{14},C_{15}	n-paraffins
Nippon Oxocol, Ichihara, Japan	85	i-C_7,C_9,C_{10},C_{13}	polygas olefins

Table 2   (*Continued*)

Company and location	Capacity 10^3 t/yr	Alcohol products	Feedstock
Shell Chemical, Stanlow, UK	90	n-$C_{10},C_{11},C_{12},C_{13},C_{14},C_{15}$	ethylene
Shell Chemical, Geismar, La., U.S.	272	n-$C_7,C_8,C_9,C_{10},C_{11},C_{12},C_{13},C_{14},C_{15}$	ethylene
Sterling, Texas City, Tex., U.S.	102	n-C_7,C_9,C_{11},C_{13}	alpha olefins
Unipar, Sao Paulo, Brazil	20	i-C_{10},C_{13}	propylene
Oxo process subtotal	*1627*		
		Oxo/aldol processes	
Aristech, Pasadena, Tex., U.S.	86	2-ethylhexanol	propylene
BASF, Ludwigshafen, Germany[b]	100	i-$C_9;n$-C_9,C_{11},C_{13},C_{15}	butenes, polygas olefins, alpha olefins
BASF, Ludwigshafen, Germany	150	2-ethylhexanol	propylene
BASF, Freeport, Tex., U.S.	30	2-ethylhexanol	propylene
BASF Espanol SA, Tarragona, Spain	30	2-ethylhexanol	propylene
Celanese Mexicana, Celaya, Mexico[c]	70	2-ethylhexanol	acetaldehyde
Chemicke Zavodi, Litwinov, Czechoslovakia	30	2-ethylhexanol	propylene
Chisso, Goi, Japan	50	2-ethylhexanol	propylene
Ciquine, Camacari, Brazil	74	2-ethylhexanol	propylene
Elekieroz do Nordeste, Igarassue, Brazil[c]	15	2-ethylhexanol	acetaldehyde
Hoechst, Oberhausen-Holten, Germany	200	2-ethylhexanol	propylene
Hüs AG, Marl, Germany	200	2-ethylhexanol	propylene
Jilin, Jilin, China	50	2-ethylhexanol	propylene
KII, Koper, Yugoslavia	42	2-ethylhexanol	propylene
Kyowa Yuka, Yokkaichi, Japan	100	2-ethylhexanol	propylene
Lucky, Naju, Korea	120	2-ethylhexanol	propylene
Mitsubishi Kasei, Mizushima, Japan	146	2-ethylhexanol	propylene
National Organic, Bombay, India	8	2-ethylhexanol	propylene
Neste Oxo, Ornskoldsvik, Sweden	10	2-ethylhexanol	n-butyraldehyde

Table 2   (*Continued*)

Company and location	Capacity 10^3 t/yr	Alcohol products	Feedstock
Neste Oxo, Stennungsund, Sweden	126	2-ethylhexanol	propylene
Shell Chemical, Deer Park, Tex., U.S.	27	2-ethylhexanol	propylene
Sinopec, Daqing, China	50	2-ethylhexanol	propylene
Sinopec, Yan Shan, China	20	2-ethylhexanol	propylene
Sinopec, Yueyang-shibequ, China	10	2-ethylhexanol	propylene
Sinopec, Zibo, China	50	2-ethylhexanol	propylene
Societe Oxo-Chemie, Lavera, France	105	2-ethylhexanol	propylene
State, Burgas, Bulgaria	20	2-ethylhexanol	propylene
State, Beijing, China	10	2-ethylhexanol	propylene
State, Leuna, Germany	40	2-ethylhexanol	propylene
State, Schkopau, Germany	40	2-ethylhexanol	propylene
State, Rimnicu Vilcea, Romania	20	2-ethylhexanol	propylene
State, Timisoara, Romania	60	2-ethylhexanol	propylene
State, Angarsk, USSR	45	2-ethylhexanol	propylene
State, Omsk, USSR	45	2-ethylhexanol	propylene
State, Perm, USSR	90	2-ethylhexanol	propylene
State, Saluwat, USSR	45	2-ethylhexanol	propylene
Texas Eastman, Longview, Tex., U.S.	98	2-ethylhexanol	propylene
Tonen, Kawasaki, Japan	50	2-ethylhexanol	propylene
Union Carbide Corp., Texas City, Tex., U.S.	54	2-ethylhexanol	propylene
Zaklady Azotowe, Kedzierzyn, Poland	100	2-ethylhexanol	propylene
Oxo/aldol subtotal	*2616*		
Total world	*5298*		

[a] Data from Refs. 1–6.
[b] Oxo/dimersol process.
[c] Aldol process.

for detergents. Other lower volume synthetic alcohol application areas include solvents and specialty esters.

2. The Ziegler Process

The Ziegler process, based on reactions discovered in the 1950s, produces predominantly linear, primary alcohols having an even number of carbon atoms. The process was commercialized by Continental Oil Company in the United States in 1962, by Condea Petrochemie in West Germany (a joint venture of Continental Oil Company and Deutsche Erdöl, A.G.) in 1964, by Ethyl Corporation in the United States in 1965, and by the USSR in 1983.

Four chemical reactions are used to synthesize alcohols from aluminum alkyls and ethylene (qv).

Triethylaluminum Preparation

$$2\,Al + 3\,H_2 + 6\,C_2H_4 \longrightarrow 2\,(C_2H_5)_3Al$$

Chain Growth

$$(C_2H_5)_3Al + 3x\,C_2H_4 \longrightarrow \left[C_2H_5(C_2H_4)_x\right]_3Al$$

Oxidation

$$2\left[C_2H_5(C_2H_4)_x\right]_3Al + 3\,O_2 \longrightarrow 2\,Al\left[O(C_2H_4)_xC_2H_5\right]_3$$

Hydrolysis

$$2\,Al\left[O(C_2H_4)_xC_2H_5\right]_3 + 3\,H_2O \longrightarrow 6\,C_2H_5(C_2H_4)_xOH + Al_2O_3$$

This process is currently used by Vista Chemical, successor to Continental Oil Company's chemical business, and by Condea. In the Ethyl Corporation process dilute sulfuric acid is used in place of water in the hydrolysis step; producing alum rather than alumina.

2.1. Triethylaluminum Preparation. Triethylaluminum [97-93-8], $C_6H_{15}Al$, can be prepared by a two-step or a one-step process. In the former, aluminum [7429-90-5], Al, powder is added to recycled triethylaluminum and the slurry reacts first with hydrogen [1333-74-0], H_2, to produce diethylaluminum hydride [871-27-2], which in the second step reacts with ethylene [74-85-1], C_2H_4, to produce triethylaluminum. In the one-step process, hydrogen and ethylene are simultaneously fed to the reactor containing the aluminum slurry.

2.2. Chain Growth. Triethylaluminum reacts with ethylene in controlled, highly exothermic, successive addition reactions to produce a spectrum of higher molecular weight alkyls of even carbon number. The distribution of chain lengths in the chain growth mixture corresponds closely to the Poisson equation (7). Side reactions lead to small deviations from the Poisson distribution, greater deviations being observed at higher reaction temperatures. Some

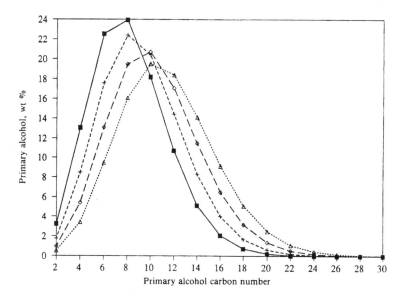

Fig. 1. Ziegler ethylene chain growth. Theoretical (Poisson) distribution of primary alcohols at $(-\blacksquare-)$ 2.5, $(-+-)$ 3.0, $(-\diamond-)$ 3.5, and $(\cdot\cdot\triangle\cdot\cdot)$ 4.0 moles of ethylene per $\frac{1}{3}$ mole aluminum. Courtesy of Ethyl Corporation.

control of the distribution is obtained by adjustment of triethylaluminum–ethylene ratio as shown in Figure 1. In the Ethyl process, steps are taken to produce a longer chain fraction (predominantly C_{12}–C_{18}) that is sent to the oxidation step, and a shorter chain fraction (predominantly C_2–C_{10}) that is recycled for additional chain growth. The final product distribution is about 15–25% C_6–C_{10} and 75–85% C_{12}–C_{18} (8). This approach permits changes in the carbon number distribution of the alcohol product as best fit market demands. A comparison of typical commercial product distributions in the Ethyl and Vista processes is shown in Figure 2.

There are two important side reactions, particularly above 120°C: (1) aluminum alkyls decompose to form dialkylaluminum hydrides and alpha olefins (the dialkylaluminum hydrides rapidly react with ethylene to regenerate a trialkylaluminum);

$$R_2AlCH_2CH_2R' \longrightarrow R_2AlH + CH_2{=\!=}CHR'$$

and (2) alpha olefins can react with trialkylaluminum to produce branched aluminum alkyls and branched olefins.

$$R_2AlCH_2CH_2R' \; + \; CH_2{=}CHR'' \longrightarrow R_2AlCH_2\underset{\underset{R''}{|}}{C}HCH_2CH_2R' \longrightarrow R_2AlH \; + \; CH_2{=}\underset{\underset{R''}{|}}{C}{-}CH_2CH_2R'$$

This second reaction leads to the small amount of branching (usually less than 5%) observed in the alcohol product. The alpha olefins produced by the first reaction represent a loss unless recovered (8). Additionally, ethylene polymerization

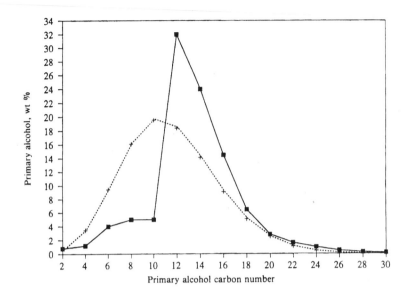

Fig. 2. Estimated primary alcohol distributions for (■) Ethyl Corporation-modified Ziegler and (+) Vista Corporation Ziegler, at 4.0 moles ethylene per $\frac{1}{3}$ mole aluminum. Courtesy of Ethyl Corporation.

during chain growth creates significant fouling problems which must be addressed in the design and operation of commercial production facilities (9).

2.3. Oxidation. Aluminum alkyls are oxidized to the corresponding alkoxides using dry air above atmospheric pressure in a fast, highly exothermic reaction. In general, a solvent is used to help avoid localized overheating and to decrease the viscosity of the solution. By-products include paraffins, aldehydes, ketones, olefins, esters, and alcohols; accidental introduction of moisture increases paraffin formation. To prevent contamination, solvent and by-products must be removed before hydrolysis. Removal can be effected by high temperature vacuum flashing or by stripping.

2.4. Hydrolysis. Aluminum alkoxides are hydrolyzed using either water or sulfuric acid, usually at around 100°C. In addition to the alcohol product, neutral hydrolysis gives high quality alumina (see ALUMINUM COMPOUNDS); the sulfuric acid hydrolysis yields alum. The crude alcohols are washed and then fractionated.

Mild steel is a satisfactory construction material for all equipment in Ziegler chemistry processes except for hydrolysis. If sulfuric acid hydrolysis is employed, materials capable of withstanding sulfuric acid at 100°C are required: lead-lined steel, some alloys, and some plastics. Flow diagrams for the Vista and Ethyl processes are shown in Figures 3 and 4, respectively.

2.5. Environmental Considerations. Environmental problems in Ziegler chemistry alcohol processes are not severe. A small quantity of aluminum alkyl wastes is usually produced and represents the most significant disposal problem. It can be handled by controlled hydrolysis and separate disposal of the aqueous and organic streams. Organic by-products produced in chain growth

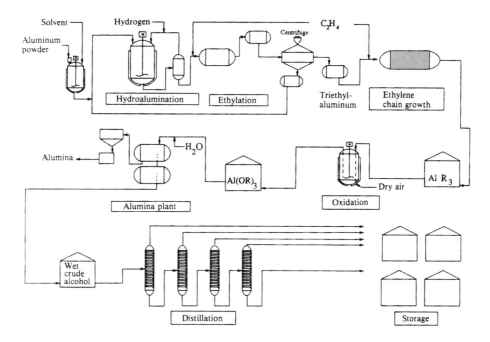

Fig. 3. Flow diagram for the Vista Corporation primary alcohols plant, Lake Charles, Louisiana. Courtesy of Vista Corporation.

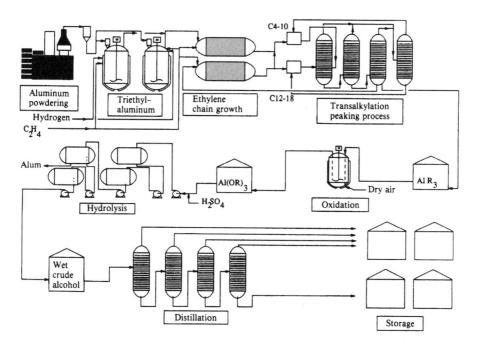

Fig. 4. Flow diagram for the Ethyl Corporation primary alcohols plant, Houston, Texas. Courtesy of Ethyl Corporation.

and hydrolysis can be cleanly burned. Wastewater streams must be monitored for dissolved carbon, such as short-chain alcohols, and treated conventionally when necessary.

3. The Oxo Process

The oxo or hydroformylation reaction was discovered in Germany in 1938 (10) and was first used on a commercial scale by the Enjay Chemical Company (now Exxon) in 1948. By 1990 the total world alcohol capacity based on this general technology was over four million metric tons per year (see OXO PROCESS).

The structures and, hence, the properties of the higher oxo alcohols (C_6–C_{18}) are a function of the oxo process and the olefin employed. All the oxo products are primary alcohols and contain one more carbon atom than the feedstock olefin. They differ in two respects from natural alcohols and from Ziegler products, both of which are linear and of even carbon number. First, depending on the feedstock, they contain either even and odd carbon numbers or all odd carbon numbers. Second, the oxo products all have more branching. Branched olefin gives completely branched products; linear olefin gives some 2-methyl branching, the extent of which is dependent on the process. From a conventional cobalt-catalyzed process, the typical product of a linear olefin is 40–50% branched. Modified catalysts reduce branching to 15–25%.

3.1. Process Technology. In a typical oxo process, primary alcohols are produced from monoolefins in two steps. In the first stage, the olefin, hydrogen, and carbon monoxide [630-08-0], react in the presence of a cobalt or rhodium catalyst to form aldehydes, which are hydrogenated in the second step to the alcohols.

$$RCH{=}CH_2 \ + \ CO \ + \ H_2 \ \xrightarrow{\text{catalyst}} \ RCH_2CH_2CHO \ + \ \underset{\underset{CH_3}{|}}{RCHCHO}$$

$$RCH_2CH_2CHO \ + \ \underset{\underset{CH_3}{|}}{RCHCHO} \ + \ H_2 \ \xrightarrow{\text{catalyst}} \ RCH_2CH_2CH_2OH \ + \ \underset{\underset{CH_3}{|}}{RCHCH_2OH}$$

The oxo catalyst may be modified to function as a hydrogenation catalyst as well and, using a 2:1 ratio of hydrogen to carbon monoxide, alcohols are produced directly.

$$RCH{=}CH_2 \ + \ CO \ + \ 2\,H_2 \ \xrightarrow{\text{catalyst}} \ RCH_2CH_2CH_2OH \ + \ \underset{\underset{CH_3}{|}}{RCHCH_2OH}$$

These reactions are applicable to most monoolefins and are used to obtain a large number of commercial products.

3.2. Cobalt Catalyst, Two-Step, High Pressure Process. The olefin, with recycle and makeup cobalt catalyst at 0.1–1.0% concentration, is preheated and fed continuously to the oxo reactor together with the synthesis gas at a

1–1.2:1 H_2 to CO ratio. The reaction is conducted with agitation at 20,300–30,400 kPa (200–300 atm) and 130–190°C. Liquid hourly space velocity (LHSV) in the reactor is 0.5–1.0. The reaction is highly exothermic, 125 kJ/mol (54,000 Btu/lb-mol), and requires cooling. The intermediate aldehyde is hydrogenated to the alcohol at 5,070–20,300 kPa (50–200 atm) and 150–200°C using a catalyst containing copper, zinc, or nickel. The crude product is then fractionated (Fig. 5). The plant may be operated continuously or on a campaign basis with subsequent blending of the alcohols to give the desired product. The reactor and parts exposed to aldehydes or acids are constructed of alloy steel; the remainder is of carbon steel.

The cobalt catalyst can be introduced into the reactor in any convenient form, such as the hydrocarbon-soluble cobalt naphthenate [61789-51-3], as it is converted in the reaction to dicobalt octacarbonyl [15226-74-1], $Co_2(CO)_8$, the precursor to cobalt hydrocarbonyl [16842-03-8], $HCo(CO)_4$, the active catalyst species. Some of the methods used to recover cobalt values for reuse are (11): conversion to an inorganic salt soluble in water; conversion to an organic salt soluble in water or an organic solvent; treatment with aqueous acid or alkali to recover part or all of the $HCo(CO)_4$ in the aqueous phase; and conversion to metallic cobalt by thermal or chemical means.

3.3. Modified Cobalt Catalyst, One-Step, Low Pressure Process.
The distinguishing feature of this process, as commercialized by Shell, is catalysis by a cobalt–carbonyl–organophosphine complex such as $[Co(CO)_3P(C_4H_9)_3]_2$ (12). The olefin, using recycle and makeup catalyst at about 0.5% concentration, and synthesis gas at a 2–2.5:1 H_2 to CO ratio, react at 6,080–9,120 kPa (60–90 atm) and 170–210°C for detergent range alcohols. Lower pressures (3,040–7,080 kPa) are employed for *n*-butanol [71-36-3] and 2-ethylhexanol production. LHSV in the reactor is 0.1–0.2. The catalyst is highly selective for hydroformylation of 1-olefins at the terminal carbon atom; this results in a product from a linear feedstock which is up to 75–85% linear, having mainly 2-methyl isomers as branched components. The product is alcohol rather than aldehyde, because the modified catalyst promotes hydrogenation; and, because it is such an effective hydrogenation catalyst, approximately 10% of the olefin feed is also converted to paraffins. Because rapid isomerization of intermediates occurs under the reaction conditions, high primary alcohol selectivity can be obtained from internal olefins as well as from alpha olefins. After degassing and vacuum flashing, the crude alcohols are washed with caustic to convert esters to alcohols, water-washed, and distilled. Purified alcohols are then finished by hydrogenation and filtration (13).

Significant differences in this modified process include use of a lower pressure, slightly higher temperature, lower LHSV, formation of alcohol in one processing step, and a higher hydrogenation of the olefins to paraffins. The process is operated commercially by Shell Chemical U.S.A., Shell Chemical UK, and Mitsubishi Petrochemical exclusively for detergent range alcohols. Detergent range alcohols produced by the Shell process are particularly well-suited for downstream production of ethylene oxide adducts, which are major Shell Chemical products. The process schematic is shown in Figure 6.

Rhodium Catalysts. Rhodium carbonyl catalysts for olefin hydroformylation are more active than cobalt carbonyls and can be applied at lower

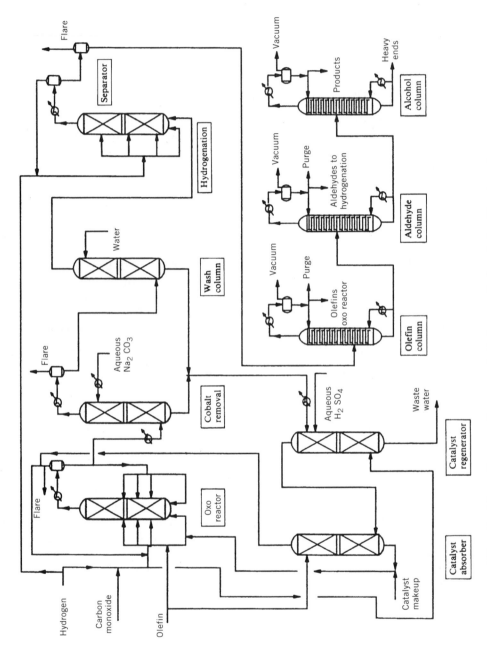

Fig. 5. Flow diagram for oxo alcohol manufactured by the two-stage process. Courtesy of the Ethyl Corporation.

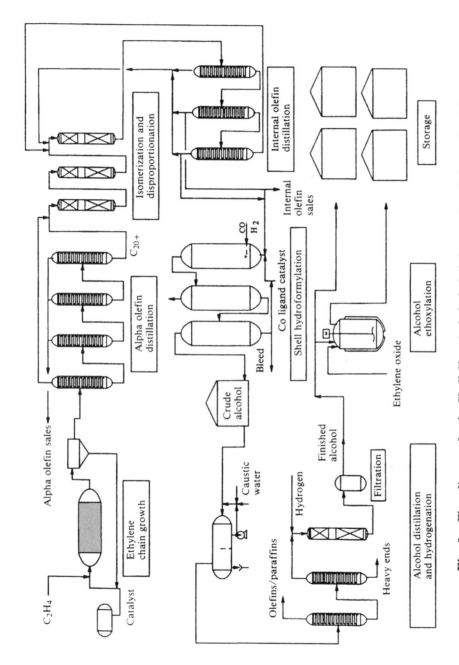

Fig. 6. Flow diagram for the Shell Chemical alcohol-olefin complex, Geismar, Louisiana, and Stanlow, United Kingdom. Courtesy of the Shell Chemical Corporation and the Ethyl Corporation.

temperatures and pressures (14). Rhodium hydrocarbonyl [75506-18-2], $HRh(CO)_4$, results in lower n-butyraldehyde [123-72-8] to isobutyraldehyde [78-84-2] ratios from propylene [115-07-1], C_3H_6, than does cobalt hydrocarbonyl, ie, 50/50 vs 80/20. Ligand-modified rhodium catalysts, $HRh(CO)_2L_2$ or $HRh(CO)L_3$, afford n-/iso-ratios as high as 92/8; the ligand is generally a tertiary phosphine. The rhodium catalyst process was developed jointly by Union Carbide Chemicals, Johnson-Matthey, and Davy Powergas and has been licensed to several companies. It is particularly suited to propylene conversion to n-butyraldehyde for 2-ethylhexanol production in that by-product isobutyraldehyde is minimized.

3.4. Olefin Sources.   The choice of feedstock depends on the alcohol product properties desired, availability of the olefin, and economics. A given producer may either process different olefins for different products or change feedstock for the same application. Feedstocks believed to be currently available are as follows.

Propylene.   2-Ethylhexanol is now produced almost entirely from propylene, with the exception of a minor portion that comes from ethylene-derived acetaldehyde.

Polygas Olefins.   Refinery propylene and butenes are polymerized with a phosphoric acid catalyst at 200°C and 3040–6080 kPa (30–60 atm) to give a mixture of branched olefins up to C_{15}, used primarily in producing plasticizer alcohols (isooctyl, isononyl, and isodecyl alcohol). Since the olefins are branched (75% have two or more CH_3 groups) the alcohols are also branched. Exxon, BASF, Ruhrchemie (now Hoechst), ICI, Nissan, Getty Oil, U.S. Steel Chemicals (now Aristech), and others have all used this olefin source.

Other Dimer Olefins.   Olefins for plasticizer alcohols are also produced by the dimerization of isobutene [115-11-7], C_4H_8, or the codimerization of isobutene and n-butene [25167-67-3]. These highly branched octenes lead to a highly branched isononyl alcohol [68526-84-1] product. BASF, Ruhrchemie, ICI, Nippon Oxocol, and others have used this source.

The Dimersol process (French Petroleum Institute) produces hexenes, heptenes, and octenes from propylene and linear butylene feedstocks. This process is reported to produce olefin with less branching than the corresponding polygas olefins. BASF practices this process in Europe.

Normal Paraffin-Based Olefins.   Detergent range n-paraffins are currently isolated from refinery streams by molecular sieve processes (see ADSORPTION, LIQUID SEPARATION) and converted to olefins by two methods. In the process developed by Universal Oil Products and practiced by Enichem and Mitsubishi Petrochemical, a n-paraffin of the desired chain length is dehydrogenated using the Pacol process in a catalytic fixed-bed reactor in the presence of excess hydrogen at low pressure and moderately high temperature. The product after adsorptive separation is a linear, random, primarily internal olefin. Shell formerly produced n-olefins by chlorination–dehydrochlorination. Typically, C_{11}–C_{14} n-paraffins are chlorinated in a fluidized bed at 300°C with low conversion (10–15%) to limit dichloroalkane and trichloroalkane formation. Unreacted paraffin is recycled after distillation and the predominant monochloroalkane is dehydrochlorinated at 300°C over a catalyst such as nickel acetate [373-02-4]. The product is a linear, random, primarily internal olefin.

Ethylene-Based Olefins

Aluminum Alkyl Chain Growth. Ethyl, Chevron, and Mitsubishi Chemical manufacture higher, linear alpha olefins from ethylene via chain growth on triethylaluminum (15). The linear products are then used as oxo feedstock for both plasticizer and detergent range alcohols; and because the feedstocks are linear, the linearity of the alcohol product, which has an entirely odd number of carbons, is a function of the oxo process employed. Alcohols are manufactured from this type of olefin by Sterling, Exxon, ICI, BASF, Oxochemie, and Mitsubishi Chemical.

Catalytic Oligomerization. Shell Chemical provides C_{11}–C_{14} linear internal olefin feedstock for C_{12}–C_{15} detergent oxo alcohol production from its SHOP (Shell Higher Olefin Process) plant (16,17). C_9–C_{11} alcohols are also produced by this process. Ethylene is first oligomerized to linear, even carbon–number alpha olefins using a nickel complex catalyst. After separation of portions of the α-olefins for sale, others, particularly C_{18} and higher, are catalytically isomerized to internal olefins, which are then disproportionated over a catalyst to a broad mixture of linear internal olefins. The desired C_{11}–C_{14} fraction is separated; the lighter and heavier fractions are recycled to the isomerization/disproportionation section. The SHOP process has been described in detail in the literature (18) and is shown schematically in Figure 6.

4. The Aldol Process

The important solvent and plasticizer intermediate, 2-ethylhexanol, is manufactured from *n*-butyraldehyde by aldol addition in an alkaline medium at 80–130°C and 300–1010 kPa (3–10 atm).

$$2\ CH_3CH_2CH_2CHO \xrightarrow{\text{catalyst}} CH_3CH_2CH_2CH{=}\underset{\underset{CH_2CH_3}{|}}{C}CHO \ + \ H_2O$$

This step is followed by catalytic hydrogenation at 230°C and 5,070–20,300 kPa (50–200 atm).

$$CH_3CH_2CH_2CH{=}\underset{\underset{CH_2CH_3}{|}}{C}CHO \ + \ 2\ H_2 \xrightarrow{\text{catalyst}} CH_3CH_2CH_2CH_2\underset{\underset{CH_2CH_3}{|}}{C}HCH_2OH$$

The *n*-butyraldehyde may be obtained from acetaldehyde [75-07-0] by aldol addition followed by hydrogenation, or from propylene by the oxo process. This latter process is predominantly favored (Fig. 7).

The oxo and aldol reactions may be combined if the cobalt catalyst is modified by the addition of organic–soluble compounds of zinc or other metals. Thus, propylene, hydrogen, and carbon monoxide give a mixture of C_4 aldehydes and 2-ethylhexenaldehyde [123-05-7] which, on hydrogenation, yield the corresponding alcohols.

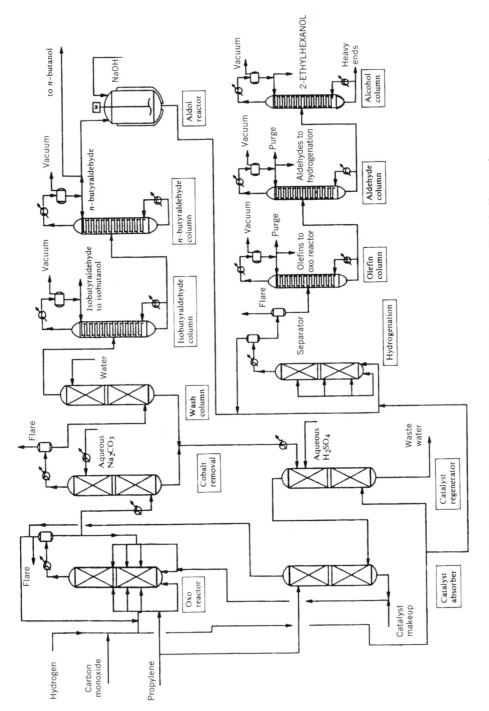

Fig. 7. Flow diagram for the oxo–aldol manufacture of 2-ethylhexanol.

5. The Paraffin Oxidation Process

Secondary alcohols (C_{10}–C_{14}) for surfactant intermediates are produced by hydrolysis of secondary alkyl borate or boroxine esters formed when paraffin hydrocarbons are air-oxidized in the presence of boric acid [10043-35-3] (19,20). Union Carbide Corporation operated a plant in the United States from 1964 until 1977. A plant built by Nippon Shokubai (Japan Catalytic Chemical) in 1972 in Kawasaki, Japan was expanded to 30,000 t/yr capacity in 1980 (20). The process has been operated industrially in the USSR since 1959 (21). Also, predominantly primary alcohols are produced in large volumes in the USSR by reduction of fatty acids, or their methyl esters, from permanganate-catalyzed air oxidation of paraffin hydrocarbons (22). The paraffin oxidation is carried out in the temperature range 150–180°C at a paraffin conversion generally below 20% to a mixture of trialkyl borate, $(RO)_3B$, and trialkyl boroxine, $(ROBO)_3$. Unconverted paraffin is separated from the product mixture by flash distillation. After hydrolysis of residual borate esters, the boric acid is recovered for recycle and the alcohols are purified by washing and distillation (19,20).

The product secondary alcohols from paraffin oxidation are converted to ethylene oxide adducts (alcohol ethoxylates) which are marketed by Japan Catalytic Chemical and BP Chemicals as SOFTANOL secondary alcohol ethoxylates. Union Carbide Chemical markets ethoxylated derivatives of the materials in the United States under the TERGITOL trademark (23).

6. The Guerbet Process

Higher molecular weight branched alcohols are produced by condensation of lower alcohols in the Guerbet reaction.

$$2\,RCH_2CH_2OH \longrightarrow RCH_2CH_2\underset{\underset{R}{|}}{C}HCH_2OH + H_2O$$

In earlier studies (24), the reaction was carried out at temperatures above 200°C under autogenous pressure conditions using alkali metal hydroxide or alkoxide catalysts; significant amounts of carboxylic acid, RCH_2COOH, were formed as were other by-products. More recent reports describe catalysts which minimize by-products: MgO–K_2CO_3–CuC_2O_2 (25), less basic but still requiring high temperatures; Rh, Ir, Pt, or Ru complexes (26); and an alkali metal alkoxide plus Ni or Pd (27), effective at much lower temperatures.

Some 2,000–3,000 t/yr of these specialty alcohols are produced in the United States (Exxon) and in Germany (Henkel) (28). Their high liquidity because of branching permits use of less volatile, higher molecular weight materials, reported to be less irritating than the lower molecular weight linear alcohol materials, in a variety of cosmetic products (29).

BIBLIOGRAPHY

"Alcohols, Higher" in *ECT* 1st ed., Vol. 1, pp. 315–321, by H. B. McClure, Carbide and Carbon Chemicals Corporation; "Alcohols, Higher, Synthetic" in *ECT* 2nd ed., Vol. 1, pp. 560–569, by R. W. Miller, Eastman Chemicals Products, Inc; "Alcohols, Higher Aliphatic, Synthetic in *ECT* 3rd ed., Vol. 1 pp. 740–754 by M. F. Gautreaux, W. T. Davis, and E. D. Travis, Ethyl Corporation.

CITED PUBLICATIONS

1. Asociacion Petroquimica Latinomamericana, Anuario Petroquimico Latino Americano 1985, Buenos Aires, 1985.
2. J-P. Davreux, Synfina–Oleofina, 1988.
3. R. F. Modler, "Detergent Alcohols" in *Chemical Economics Handbook*, SRI International, Menlo Park, Calif., 1987.
4. T. Gibson, "Plasticizer Alcohols" in *Chemical Economics Handbook*, SRI International, Menlo Park, Calif., 1985.
5. T. Gibson, "Oxo Chemicals" in ref. 4.
6. G. R. Lappin, J. D. Wagner, Ethyl Corporation, 1989.
7. H. Weslau, *Justus Liebigs Ann. Chem.* **629**, 198 (1960).
8. U.S. Pat. 3,415,861 (Dec. 10, 1968), W. T. Davis and C. L. Kingrea (to Ethyl Corporation).
9. G. R. Lappin in G. R. Lappin and J. D. Sauer, eds., *Alpha Olefins Applications Handbook*, Marcel Dekker, New York, 1989, p. 36.
10. Ger. Pat. 849,548 (Sept. 15, 1952), O. Roelen (to Chemische Verwertungsgesellschaft Oberhausen GmbH).
11. H. Lemke, *Hydrocarbon Process.* **45**(2), 148 (Feb. 1966).
12. U.S. Pats. 3,239,569; 3,239,570; 3,239,571 (Mar. 8, 1966), L. H. Slaugh and R. D. Mullineaux (to Shell Oil Company); Brit. Pats. 988,941; 988,942; 988,943; 988,944 (Apr. 14, 1965) (to Shell Internationale Research Maatschappij NV).
13. E. D. Heerdt, Shell Development Co., personal communication, 1989.
14. Ger. Pat. 953,605 (Dec. 6, 1956), G. Schiller (to Chemische Verwertungsgesellschaft Oberhausen GmbH).
15. Ref. 3, 51–53.
16. Ref. 3, 54–57.
17. U.S. Pat. 3,647,906 (Mar. 7, 1972), F. F. Farley (to Shell Oil Company); U.S. Pat. 3,726,938 (Apr. 10, 1973), A. J. Berger.
18. E. R. Freitas and C. R. Gum, *Chem. Eng. Prog.*, 73 (Jan. 1979).
19. J. Kurata and K. Koshida, *Hydrocarbon Process.* **57**(1), 145 (Jan. 1978); N. J. Steens and J. R. Livingston, Jr., *Chem. Eng. Prog.* **64**(7), 61 (July 1968).
20. N. Kurata, K. Koshida, H. Yokoyama, and T. Goto, in E. J. Wickson, ed., *Monohydric Alcohols, ACS Symp. Ser. 159*, American Chemical Society Washington, D.C., 1981, 113–157.
21. I. M. Towbin and D. M. Boljanskii, *Maslo. Zhir. Prom.* **32**, 29 (1966).
22. H. Stage, *Seifen, Öle, Fette, Wachse* **99** (6/7), 143; (8), 185; (9), 217; (11), 299 (1973).
23. M. Tsuchino, and co-workers, *Paper no. 54 (I&EC Div.)*, 196th National Meeting of the American Chemical Society, Los Angeles, September 25–30, 1988.
24. M. Guerbet, *J. Pharm. Chim.* **6**, 49 (1913); *Chem. Abstr.* **7**, 1494 (1913).
25. M. N. Dvornikoff and M. W. Farrar, *J. Org. Chem.* **22**, 540 (1957).
26. G. Gregorio and G. F. Pregaglia, *J. Organometall. Chem.* **37**, 385 (1972); P. L. Burk, R. L. Pruett, and K. S. Campo, *J. Mol. Catal.* **33**, 1, 15 (1985).
27. J. Sabadie and G. Descotes, *Bull. Soc. Chim. Fr.* **253** (1983).

28. K. Noweck and H. Ridder, *Ullmann's Encyclopedia of Industrial Chemistry*, 5th ed., VCH Verlagsgesellschaft mbh, Weinheim, Germany, 1987, p. 288.

29. K. Klein, P. E. Bator, and S. Hans, *Cosmetics and Toiletries*, **95**, 70 (1980); A. J. O'Lenick, Jr., and R. E. Bilbo, *Soap, Cosmet. Chem. Spec.* **52** (April, 1987).

GENERAL REFERENCES

J. A. Monick, *Alcohols, Their Chemistry, Properties, and Manufacture*, Reinhold, New York (1968).

Ziegler Chemistry Processes; Triethylaluminum Synthesis

F. Albright, *Chem. Eng.* **74**, 179 (Dec. 4, 1967).
K. Ziegler and co-workers, *Angew. Chem.* **67**, 424 (1955).
K. Ziegler, *Erdöl Kohle* **11**, 766 (1958).
K. Ziegler and co-workers, *Justus Liebigs Ann. Chem.* **629**, 1 (1960).

Ziegler Chemistry Processes: Chain Growth

K. Ziegler, *Angew. Chem.* **64**, 323 (1952).
K. Ziegler, *Brennst. Chem.* **35**, 321 (1954).
K. Ziegler, *Angew. Chem.* **68**, 721 (1956).
K. Ziegler and co-workers, *Justus Liebigs Ann. Chem.* **629**, 121, 172 (1960).
K. Ziegler, *Angew. Chem.* **72**, 829 (1960).
K. Ziegler and H. Hoberg, *Chem. Ber.* **93**, 2938 (1960).

Ziegler Chemistry Processes: Displacement Reactions

K. Ziegler, H. Martin, and F. Krupp, *Justus Liebigs Ann. Chem.* **629**, 14 (1960).
K. Ziegler, W. R. Kroll, W. Larbig, and O. W. Steudel, *Justus Liebigs Ann. Chem.* **629**, 53 (1960).
K. Ziegler, in H. H. Zeiss, ed., *Organometallic Chemistry*, Reinhold, New York, 1960, p. 218 ff.

Ziegler Chemistry Processes: Oxidation

K. Ziegler, F. Krupp, and K. Zosel, *Justus Liebigs Ann. Chem.* **629**, 241 (1960).
K. Ziegler, F. Krupp, and K. Zosel, *Angew. Chem.* **67**, 425 (1955).

Oxo Processes

G. U. Ferguson, *Chem. Ind.* **11**, 451 (1965).
H. Weber and J. Falbe, *Ind. Eng. Chem.* **62**(4), 33 (Apr. 1970).
H. Weber, W. Dimmling, and A. M. Desal, *Hydrocarbon Process.* **55**(4), 127 (1976).

E. J. Wickson and H. P. Dengler, *Hydrocarbon Process.* **51**(11), 69 (1972).

J. Falbe, *Carbon Monoxide in Organic Synthesis*, Springer-Verlag, New York, 1970.

B. Cormels, in J. Falbe, ed., *New Synthesis with Carbon Monoxide*, Springer-Verlag, New York, 1980, p. 1–225.

Plant Locations, Capacities, Feedstocks

Refs. 1 through 6.

JOHN D. WAGNER
GEORGE R. LAPPIN
J. RICHARD ZIETZ
Ethyl Corporation

ALCOHOLS, POLYHYDRIC

1. Introduction

Polyhydric alcohols or polyols contain three or more CH_2OH functional groups. The monomeric compounds have the general formula $R(CH_2OH)_n$, where $n = 3$ and R is an alkyl group or CCH_2OH; the dimers and trimers are also commercially significant. Related species where $n = 2$ are discussed elsewhere (see GLYCOLS).

The most important polyhydric alcohols are shown in Figure 1. Each is a white solid, ranging from thecrystalline pentaerythritols to the waxy trimethylol alkyls. The trihydric alcohols are very soluble in water, as is ditrimethylolpropane. Pentaerythritol is moderately soluble and dipentaerythritol and tripentaerythritol are less soluble. Table 1 lists the physical properties of these alcohols. Pentaerythritol and trimethylolpropane have no known toxic or irritating effects (1,2). Finely powdered pentaerythritol, however, may form explosive dust clouds at concentrations above 30 g/m^3 in air. The minimum ignition temperature is 450°C (3).

2. Reactions

Direct acetylation of pentaerythritol using acetic acid in aqueous solution or in toluene produces a mixture of acetates which can be fairly readily separated by chromatographic methods or distillation (8,9). The final product composition can be varied somewhat by altering the amount of water present. Esters of higher homologues and trimethylolpropane can also be synthesized using this procedure. Acrylate and methacrylate monoesters may be produced (10,11) when protecting groups are placed on three of the pentaerythritol hydroxyls. The protected intermediate then reacts with either acryloyl chloride [814-68-6] or

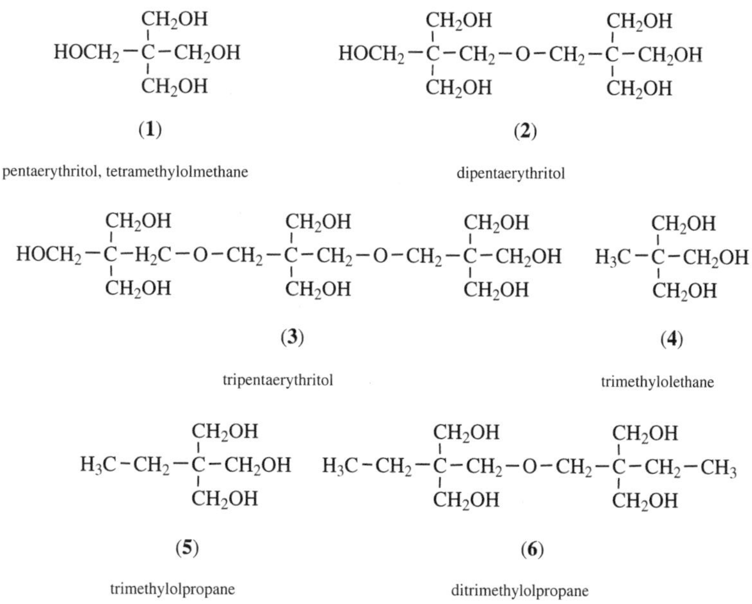

$$
\begin{array}{c}
\text{CH}_2\text{OH} \\
| \\
\text{HOCH}_2-\text{C}-\text{CH}_2\text{OH} \\
| \\
\text{CH}_2\text{OH}
\end{array}
\qquad
\begin{array}{c}
\text{CH}_2\text{OH} \qquad\qquad \text{CH}_2\text{OH} \\
| \qquad\qquad\qquad | \\
\text{HOCH}_2-\text{C}-\text{CH}_2-\text{O}-\text{CH}_2-\text{C}-\text{CH}_2\text{OH} \\
| \qquad\qquad\qquad | \\
\text{CH}_2\text{OH} \qquad\qquad \text{CH}_2\text{OH}
\end{array}
$$

(1)　　　　　　　　　　　　　　　　　(2)

pentaerythritol, tetramethylolmethane　　　　　　　　dipentaerythritol

$$
\begin{array}{c}
\text{CH}_2\text{OH} \qquad\qquad \text{CH}_2\text{OH} \qquad\qquad \text{CH}_2\text{OH} \\
| \qquad\qquad\qquad | \qquad\qquad\qquad | \\
\text{HOCH}_2-\text{C}-\text{H}_2\text{C}-\text{O}-\text{CH}_2-\text{C}-\text{CH}_2-\text{O}-\text{CH}_2-\text{C}-\text{CH}_2\text{OH} \\
| \qquad\qquad\qquad | \qquad\qquad\qquad | \\
\text{CH}_2\text{OH} \qquad\qquad \text{CH}_2\text{OH} \qquad\qquad \text{CH}_2\text{OH}
\end{array}
\qquad
\begin{array}{c}
\text{CH}_2\text{OH} \\
| \\
\text{H}_3\text{C}-\text{C}-\text{CH}_2\text{OH} \\
| \\
\text{CH}_2\text{OH}
\end{array}
$$

(3)　　　　　　　　　　　　　　　　　(4)

tripentaerythritol　　　　　　　　　　　　trimethylolethane

$$
\begin{array}{c}
\text{CH}_2\text{OH} \\
| \\
\text{H}_3\text{C}-\text{CH}_2-\text{C}-\text{CH}_2\text{OH} \\
| \\
\text{CH}_2\text{OH}
\end{array}
\qquad
\begin{array}{c}
\text{CH}_2\text{OH} \qquad\qquad \text{CH}_2\text{OH} \\
| \qquad\qquad\qquad | \\
\text{H}_3\text{C}-\text{CH}_2-\text{C}-\text{CH}_2-\text{O}-\text{CH}_2-\text{C}-\text{CH}_2-\text{CH}_3 \\
| \qquad\qquad\qquad | \\
\text{CH}_2\text{OH} \qquad\qquad \text{CH}_2\text{OH}
\end{array}
$$

(5)　　　　　　　　　　　　　　　　　(6)

trimethylolpropane　　　　　　　　　　ditrimethylolpropane

Fig. 1. Polyhydric alcohols. Systematic names are (**1**) 2,2-bis(hydroxymethyl)-1,3-propanediol; (**2**) 2,2-[oxybis(methylene)]-bis[2-hydroxymethyl]-1,3-propanediol; (**3**) 2,2-bis{[3-hydroxy-2,2-bis(hydroxymethyl)propoxy]methyl}-1,3-propanediol; (**4**) 2-hydroxy-methyl-2-methyl-1-2-methyl-1,3-propanediol; (**5**) 2-ethyl-2-hydroxymethyl-1,3-propanediol; and (**6**) 2,2-[oxybis(methylene)]-bis(2-ethyl)-1,3-propanediol.

methacryloyl chloride [27550-72-7] to give products of the type

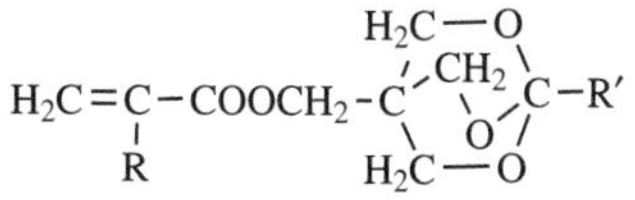

$$
\text{H}_2\text{C}=\text{C}-\text{COOCH}_2-\text{C}
\overset{\text{H}_2\text{C}-\text{O}}{\underset{\text{H}_2\text{C}-\text{O}}{\overset{\diagup\,\text{CH}_2\,\diagdown}{\diagdown\quad\,\text{O}\quad\diagup}}}
\text{C}-\text{R}'
$$
$$
\underset{\text{R}}{|}
$$

An alternative synthesis converts monobromopentaerythritol to the ortho ester, followed by reaction with cuprous acrylate.

Long-chain esters of pentaerythritol have been prepared by a variety of methods. The tetranonanoate is made by treatment of methyl nonanoate [7289-51-2] and pentaerythritol at elevated temperatures using sodium phenoxide alone, or titanium tetrapropoxide in xylene (12). Phenolic esters having good antioxidant activity have been synthesized by reaction of phenols or long-chain aliphatic acids and pentaerythritol or trimethylolpropane (13). Another ester synthesis employs the reaction of a long-chain ketone and pentaerythritol in xylene or chlorobenzene (14). Mixed esters have been produced using mixed isostearic and cyclohexane carboxylic acids in tribromophosphoric acid, followed by reaction with lauric acid (15).

Polyhydric alcohol mercaptoalkanoate esters are prepared by reaction of the appropriate alcohols and thioester using p-toluenesulfonic acid catalyst under nitrogen and subsequent heating (16, 17). Organotin mercapto esters are similarly produced by reaction of the esters with dibutyltin oxide (18).

Table 1. Physical Properties of Polyhydric Alcohols

Property	Penta-erythritol	Dipenta-erythritol	Tripenta-erythritol	Trimethylol-ethane	Trimethylol-propane	Ditrimethylol-propane[a]
CAS Registry Number	[115-77-5]	[126-58-9]	[78-24-0]	[77-85-0]	[77-99-6]	[23235-61-2]
molecular formula	$C_5H_{12}O_4$	$C_{10}H_{22}O_7$	$C_{15}H_{32}O_{10}$	$C_5H_{12}O_3$	$C_6H_{14}O_3$	$C_{12}H_{26}O_5$
melting point, °C	261–262[b]	221–222.5[b]	248–250[b]	202[c]	58.8[d]	112–114
boiling point, °C	276 (4 kPa)			283[c]	289[d]	210 (0.12 kPa)
solubility, g/100 g water						
25°C	7.23[b]	0.28[e]	0.018[e]	soluble	soluble	2.6
50°C	16.1[e]	1.1[e]	0.07[e]			8.3
90°C	51.9[e]	6.1[e]	0.51[e]		>200 (completely soluble)	
flash point, °C (Cleveland open cup)	260[f]				180[d]	>150
density, g/mL	1.396[f]	1.369[f]	1.30[f]		1.09[d]	1.18
refractive index	1.55 (20°C)[g]				1.472 (70°C)[g]	

[a] Data supplied by Perstorp AB.
[b] Ref. 1.
[c] Refs. 4 and 5.
[d] Ref. 2.
[e] Estimated value.
[f] Ref. 6.
[g] Ref. 7.

Pentaerythritol can be oxidized to 2,2-bis(hydroxymethyl)hydracrylic acid [2831-90-5], $C_5H_{10}O_5$,

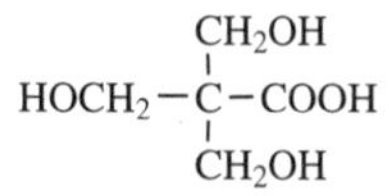

$$\begin{array}{c} CH_2OH \\ | \\ HOCH_2-C-COOH \\ | \\ CH_2OH \end{array}$$

by direct air oxidation in aqueous solution using a palladium–carbon catalyst (19), or by biological oxidation using corynebacterium or arthrobacter cultures (20).

Bromohydrins can be prepared directly from polyhydric alcohols using hydrobromic acid and acetic acid catalyst, followed by distillation of water and acetic acid (21). Reaction conditions must be carefully controlled to avoid production of simple acetate esters (22). The raw product is usually a mixture of the mono-, di- and tribromohydrins.

Borolane products of mixed composition can be synthesized by direct addition of boric acid to pentaerythritol (23).

Reaction between pentaerythritol and phosphorous trichloride [7719-12-2] yields the spirophosphite, 3,9-dichloro-2,4,8,10-tetraoxa-3,9,-diphosphaspiro [5,5]-undecane [3643-70-7], $C_5H_8Cl_2O_4P_2$,

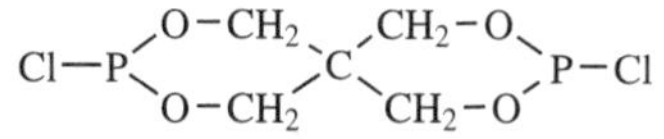

$$Cl-P \underset{O-CH_2}{\overset{O-CH_2}{<}} C \underset{CH_2-O}{\overset{CH_2-O}{>}} P-Cl$$

in the presence of benzene or a methyl acid phosphate catalyst (24,25) followed by removal of hydrogen chloride. Substituents may then replace the remaining chloride on treating the product with an alcohol or phenol (26) in the presence of a hydrogen chloride binding base such as triethylamine. Direct reaction of triethyl phosphite [122-52-1] and pentaerythritol is also possible (27). Pentaerythritol phosphate is similarly prepared by the reaction of pentaerythritol and phosphorus oxychloride [10025-87-3], $POCl_3$, in dioxane (28). Substituted diphosphaspiro compounds are made by reaction of pentaerythritol and either phosphonic anhydrides or trialkyl phosphites and trialkylamine (29,30).

The commercially important explosive pentaerythritol tetranitrate [78-11-5] (PETN), $C_5H_8N_4O_{12}$,

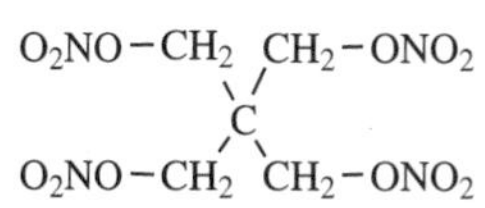

$$\begin{array}{c} O_2NO-CH_2 \quad CH_2-ONO_2 \\ \diagdown \diagup \\ C \\ \diagup \diagdown \\ O_2NO-CH_2 \quad CH_2-ONO_2 \end{array}$$

is produced by direct reaction of pentaerythritol in nitric or nitric–sulfuric acid media (31–33).

Aminoalkoxy pentaerythritols are obtained by reduction of the cyanoethoxy species obtained from the reaction between acrylonitrile, pentaerythritol, and lithium hydroxide in aqueous solution. Hydrogen in toluene over a ruthenium catalyst in the presence of ammonia is used (34). The corresponding aminophenoxyalkyl derivatives of pentaerythritol and trimethylolpropane can also be prepared (35).

Tosylates of pentaerythritol and the higher homologues can be converted to their corresponding tetra-, hexa-, or octaazides by direct reaction of sodium azide (36), and azidobenzoates of trimethylolpropane and dipentaerythritol are prepared by reaction of azidobenzoyl chloride and the alcohols in pyridine medium (37).

Pentaerythritol can be converted to the biscyclic formal, 2,4,8,10-tetra-oxaspiro[5,5]undecane [126-54-5], $C_7H_{12}O_4$,

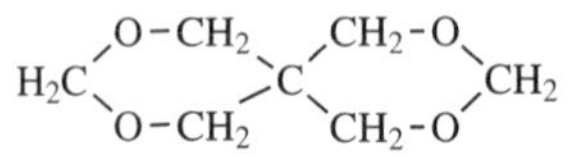

$$H_2C \underset{O-CH_2}{\overset{O-CH_2}{<}} \hspace{-2pt} > \hspace{-2pt} C \underset{CH_2-O}{\overset{CH_2-O}{<}} \hspace{-2pt} CH_2$$

by heating in the presence of formaldehyde or paraformaldehyde and an acid catalyst (38). Alternatively, a cation-exchange resin catalyst may be used and excess water removed by azeotropic distillation (39). Higher aldehydes have also been used to prepare long-chain alkyl and aryl cyclic acetals (40–43).

Simple alkyl and alkenyl ethers of pentaerythritol are produced on direct reaction of the polyol and the required alkyl or alkenyl chloride in the presence of quaternary alkylamine bromide (44). Allyl chloride produces the pentaerythritol tetrallyl ether [1471-18-7],

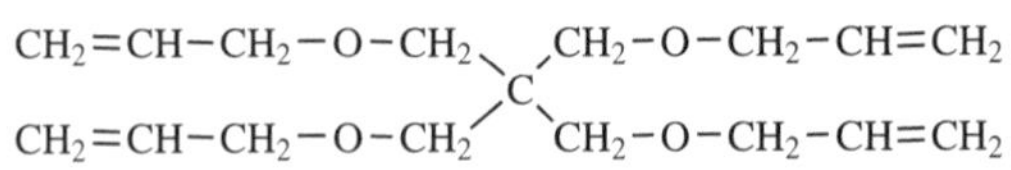

$$\begin{array}{l} CH_2{=}CH{-}CH_2{-}O{-}CH_2 \\ CH_2{=}CH{-}CH_2{-}O{-}CH_2 \end{array} \hspace{-4pt} > \hspace{-2pt} C \hspace{-2pt} < \hspace{-4pt} \begin{array}{l} CH_2{-}O{-}CH_2{-}CH{=}CH_2 \\ CH_2{-}O{-}CH_2{-}CH{=}CH_2 \end{array}$$

in high yield by this method (45,46) or by using sodium hydroxide catalyst (47). Polycyclic crown looped and starburst dendrimer ethers are synthesized utilizing blocking-deblocking and high dilution cyclization techniques in reactions of di-alcohols and ditosylates (48,49).

3. Manufacture

Pentaerythritol is produced by reaction of formaldehyde [50-00-0] and acetaldehyde [75-07-0] in the presence of a basic catalyst, generally an alkali or alkaline-earth hydroxide. Reaction proceeds by aldol addition to the carbon adjacent to the hydroxyl on the acetaldehyde. The pentaerythrose [3818-32-4] so produced is converted to pentaerythritol by a crossed Cannizzaro reaction using formaldehyde. All reaction steps are reversible except the last, which allows completion of the reaction and high yield industrial production.

The main intermediates in the pentaerythritol production reaction have been identified and synthesized (50,51) and the intermediate reaction mechanisms deduced. Without adequate reaction control, by-product formation can easily occur (52,53). Generally mild reaction conditions are favored for optimum results (1,54). However, formation of by-products cannot be entirely eliminated, particularly dipentaerythritol and the linear formal of pentaerythritol, 2,2'-[methylenebis(oxymethylene)]bis(2-hydroxymethyl-1,3-propanediol) [6228-26-8]:

$$
\underset{\substack{\big| \\ \text{CH}_2\text{OH}}}{\overset{\substack{\text{CH}_2\text{OH} \\ \big|}}{\text{HOCH}_2-\text{C}-\text{CH}_2-\text{O}-\text{CH}_2-\text{O}-\text{CH}_2-\text{C}-\text{CH}_2\text{OH}}}
$$

The quantities of formaldehyde and base catalyst required to produce pentaerythritol from 1 mol of acetaldehyde are always in excess of the theoretical amounts of 4 mol and 1 mol, respectively, and mole ratios of formaldehyde to acetaldehyde vary widely. As the mole ratio increases, formation of dipentaerythritol and pentaerythritol linear formal is suppressed. Dipentaerythritol formation may also be reduced by increasing the formaldehyde concentration, although linear formal production increases under those conditions (55,56).

The most common catalysts are sodium hydroxide and calcium hydroxide, generally used at a modest excess over the nominal stoichiometric amount to avoid formaldehyde-only addition reactions. Calcium hydroxide is cheaper than NaOH, but the latter yields a more facile reaction and separation of the product does not require initial precipitation and filtration of the metal formate (57).

A typical flow diagram for pentaerythritol production is shown in Figure 2. The main concern in mixing is to avoid loss of temperature control in this exothermic reaction, which can lead to excessive by-product formation and/or reduced yields of pentaerythritol (55,58,59). The reaction time depends on the reaction temperature and may vary from about 0.5 to 4 h at final temperatures of about 65 and 35°C, respectively. The reactor product, neutralized with acetic or formic acid, is then stripped of excess formaldehyde and water to produce a highly concentrated solution of pentaerythritol reaction products. This is then cooled under carefully controlled crystallization conditions so that the crystals can be readily separated from the liquors by subsequent filtration.

The first stage crystals are rich in pentaerythritol linear formal and may be treated (60, 61) to convert this species to pentaerythritol and formaldehyde, which can then be recovered. The concentrated liquors obtained after redissolving are then recrystallized and filtered prior to drying of the final product.

The exact order of the production steps may vary widely; in addition, some parts of the process may also vary. Metal formate removal may occur immediately after the reaction (62) following formaldehyde and water removal, or by separation from the mother liquor of the first-stage crystallization (63). The metal formate may be recovered to hydroxide and/or formic acid by ion exchange or used as is for deicing or other commercial applications. Similarly, crystallization may include sophisticated techniques such as multistage fractional crystallization, which allows a wider choice of composition of the final product(s) (64,65).

Staged reactions, where only part of the initial reactants are added, either to consecutive reactors or with a time lag to the same reactor, may be used to reduce dipentaerythritol content. This technique increases the effective formaldehyde-to-acetaldehyde mole ratio, maintaining the original stoichiometric one. It also permits easier thermal control of the reaction (66,67). Both batch and continuous reaction systems are used. The former have greater flexibility whereas the product of the latter has improved consistency (55,68).

Dipentaerythritol and tripentaerythritol are obtained as by-products of the pentaerythritol process and may be further purified by fractional crystallization or extraction. Trimethylolethane and trimethylolpropane may be prepared by a similar aldol–cross Cannizzaro reaction scheme using propionaldehyde or butyraldehyde, respectively (58) in place of acetaldehyde. Formaldehyde and catalyst requirements are somewhat reduced because of the lower hydrogen content. Ditrimethylolpropane is obtained as a by-product of the trimethylolpropane synthesis (59).

4. Economic Aspects

Production of pentaerythritol in the United States has been erratic. Demand decreased in 1975 because of an economic recession and grew only moderately to 1980 (69). The range of uses for pentaerythritol has grown rapidly in lubricants (qv), fire-retardant compositions, adhesives, and other formulations where the cross-linking capabilities are of critical importance.

The world's largest producers are Perstorp AB (Sweden, United States, Italy), Hoechst Celanese Corporation (United States, Canada), Degussa (Germany), and Hercules (United States) with estimated 1989 plant capacities of 65,000, 59,000, 30,000, and 22,000 t/yr, respectively. Worldwide capacity for pentaerythritol production was 316,000 t in 1989, about half of which was from the big four companies. Most of the remainder was produced in Asia (Japan, China, India, Korea, and Taiwan), Europe (Italy, Spain), or South America (Brazil, Chile). The estimated rate of production for 1989 was about 253,000 t or about 80% of nameplate capacity.

The world's largest producers of trimethylolpropane are Perstorp AB at 50,000 t/yr, Hoechst Celanese at 23,000 t/yr, and Bayer at 20,000 t/yr. Estimated worldwide capacity is 139,000 t and actual production is on the order of 88,000 t.

Dipentaerythritol is sold by Perstorp AB and by Hercules (United States), ditrimethylolpropane by Perstorp AB both in relatively pure form. Tripentaerythritol is also available; however, the purity is limited. Trimethylolethane is produced commercially by Alcolac (United States) and Mitsubishi Gas Chemicals (Japan).

Pentaerythritol is produced in a variety of grades having differing amounts of dipentaerythritol and small quantities of linear formal. Mono pentaerythritol contains a minimum of 98.0% pentaerythritol with most of the remaining material being dipentaerythritol. Nitration-grade pentaerythritol composition is dependent on the individual customer's demands. The product may be highly pure (>99.5% pentaerythritol) or may contain up to about 1.6% dipentaerythritol. Technical-grade pentaerythritols generally contain at least 8% dipentaerythritol and the normal limit is about 12%, although some specialty products may be even higher. Pure dipentaerythritol may be added to standard technical-grade pentaerythritol. Tripentaerythritol is also present in most technical-grade product.

5. Analytical Methods

Pentaerythritol may be analyzed by nonspecific wet chemical means such as the hydroxyl method, the results of which include all the usual impurities such as dipentaerythritol, tripentaerythritol, and the formals (70), or the benzal method. A number of gas chromatographic methods allowing facile analysis of the more volatile ethers and esters formed by a number of reagents as well as simultaneous determination of other normal constituents are also available. Acetate esters (71), trimethylsilyl ethers (72), and trifluoroacetate esters (73) give highly satisfactory analyses. ASTM methods D2195 (wet chemical) and D2999 (gas chromatographic) are recognized standards for industrial use.

6. Health and Safety Factors

Pentaerythritol and trimethylolpropane are classified as nuisance particulate and dust, respectively. They are both nontoxic to animals by ingestion or inhalation and are essentially nonirritating to the skin or eyes (2,74).

7. Uses

The most important industrial use of pentaerythritol is in a wide variety of paints, coatings, and varnishes, where the cross-linking capability of the four hydroxy groups is critical. Alkyd resins (qv) are produced by reaction of pentaerythritol with organic acids such as phthalic acid or maleic acid and natural oil species.

The resins obtained using pentaerythritol as the only alcohol group supplier are noted for high viscosity and fast drying characteristics. They also exhibit superior water and gasoline resistance, as well as improved film hardness and adhesion. The alkyd resin properties may be modified by replacing all or part of the pentaerythritol by glycols, glycerol, trimethylolethane, or trimethylolpropane, thereby reducing the functionality. Similarly, replacing the organic acid by longer-chain acids such as adipic or altering the quantities of the oil components (linseed, soya, etc) modifies the drying, hardness, and wear characteristics of the final product. The catalyst and the actual cooking procedures also significantly affect overall resin characteristics.

Rosin esters of pentaerythritol prepared from varying amounts of oil yield varnishes of the required oil length for a wide range of outlets such as wood and metal finishing, sealing and jointing formulations, and sand binders for molds. Formulations for some of these complex resins have been described in great detail in the literature, although exact production details are normally guarded jealously by the manufacturers because the order of reaction can also alter the properties of the final product (75–78).

Long-chain esters of pentaerythritol have been used as pour-point depressants for lubricant products, ranging from fuel oils or diesel fuels to the high performance lubricating oils required for demanding outlets such as aviation, power turbines, and automobiles. These materials require superior temperature, viscosity, and aging resistance, and must be compatible with the wide variety of metallic surfaces commonly used in the outlets (79–81).

The explosives and rocket fuels formed by nitration of pentaerythritol to the tetranitrate using concentrated nitric acid (33) are generally used as a filling in detonator fuses. Use of pentaerythritol containing small amounts of dipentaerythritol produces crystallization behavior that results in a high bulk density product having excellent free-flow characteristics, important for fuse burning behavior (82). PETN is also used for medicinal purposes as a vasodilator in the treatment of angina. This product is a dry mixture of pure PETN and an inert carrier, for example, lactose or mannitol, to minimize the usual explosive potential. For the same reason, only small quantities are normally shipped and rigorous packaging is recommended (83).

Pentaerythritol is used in self-extinguishing, nondripping, flame-retardant compositions with a variety of polymers, including olefins, vinyl acetate and alcohols, methyl methacrylate, and urethanes. Phosphorus compounds are added to the formulation of these materials. When exposed to fire, a thick foam is produced, forming a fire-resistant barrier (see FLAME RETARDANTS) (84–86).

Polymer compositions containing pentaerythritol are also used as secondary heat-, light-, and weather-resistant stabilizers with calcium, zinc, or barium salts, usually as the stearate, as the prime stabilizer. The polymers may be in plastic or fiber form (87–89).

Pentaerythritol in rosin ester form is used in hot-melt adhesive formulations, especially ethylene–vinyl acetate (EVA) copolymers, as a tackifier. Polyethers of pentaerythritol or trimethylolethane are also used in EVA and polyurethane adhesives, which exhibit excellent bond strength and water resistance. The adhesives may be available as EVA melts or dispersions (90, 91) or as thixotropic, one-package, curable polyurethanes (92). Pentaerythritol spiro ortho esters have been used in epoxy resin adhesives (93). The EVA adhesives are especially suitable for cellulose (paper, etc) bonding.

Pentaerythritol and trimethylolpropane acrylic esters are useful in solventless lacquer formulations for radiation curing (qv), providing a cross-linking capability for the main film component, which is usually an acrylic ester of urethane, epoxy, or polyester. Some specialty films utilize dipentaerythritol and ditrimethylolpropane (94,95).

Titanium dioxide pigment coated with pentaerythritol, trimethylolpropane, or trimethylolethane exhibits improved dispersion characteristics when used in paint or plastics formulations. The polyol is generally added at levels of 0.1–0.5% (96).

Photocurable materials for photographic films contain pentaerythritol and dialkylamino and/or nitrile compounds, which have good adhesion and peelability of the layers, and produce clear transfer images (97,98).

Electroless plating on metal substrates can be improved by addition of pentaerythritol, either to a photosensitive composition of a noble metal salt (99), or with glycerine to nickel plating solutions (100). Both resolution and covering power of the electrolyte are improved.

Binary mixtures of pentaerythritol with many other polyol species can produce heat or electrical storage media having excellent retention properties based on the solid–solid, crystal–plastic–crystal phase-transition phenomena. The solid solutions of these mixtures can also be tailored, by adjusting the ratios of the individual components, to exhibit phase transformation and energy release under the required conditions. These media are applicable to domestic hot water, solar heat, and industrial process heat storage and recovery (101–103).

BIBLIOGRAPHY

"Pentaerythritol" in *ECT* 1st ed., Vol. 10, pp. 1–6, by H. Weinberger, Fine Organics, Inc., "Other Polyhydric Alcohols" under "Alcohols, Polyhydric," in *ECT* 2nd ed., Vol. 1, pp. 588–598, by E. Berlow, Heyden Newport Chemical Corp.; in *ECT* 3rd ed., Vol. 1, pp. 778–789, by J. Weber, Celanese Canada, Ltd., and J. Daley, Celanese Chemical Company; in *ECT* 4th ed., Vol. 1, pp. 913–925, by William N. Hunter, Celanese Canada, Inc.

CITED PUBLICATIONS

1. E. Berlow, R. H. Barth, and E. J. Snow, *The Pentaerythritols*, Reinhold Publishing Corp., New York, 1958.
2. *Material Safety Data Sheet C41, Trimethylolpropane, Flake*, Celanese Canada Inc., Montreal, 1990.
3. I. Harttmann and J. Nagy, *U.S. Bureau of Mines Report of Investigation No. 3751*, 1944.
4. *Product Bulletin PE-10-55*, Heyden Newport Chemical Corp.
5. *Technical Data Sheets TDS-1 and TB-1*, Commercial Solvents Corp.
6. *Product Bulletin, Pentaerythritol*, Sam Yang Chemical Co., Korea.
7. *Physical Properties Manual*, Celanese Chemical Company, Dallas, Tex., 1981.
8. N. S. Bankar, P. N. Chaudhari, and G. H. Kulkarni, *Indian J. Chem.* **13**, 986–987 (1975).
9. Jpn. Pat. 53/63,306 (June 6, 1978), K. Takeuchi, Y. Matsui, H. Kano, and T. Motohashi (to Ajinomoto Co., Ltd.).
10. A. B. Padias and H. K. Hall, Jr., *Macromolecules* **15**, 217–223 (1982).
11. U.S. Pat. 4,405,798A (Sept. 20, 1983), H. K. Hall and D. R. Wilson (to Celanese Corp.).
12. Brit. Pat. 1,374,263 (Nov. 20, 1974), T. Keating (to Imperial Chemical Industries Ltd.).
13. T. S. Chao, M. Kjonaas, and J. DeJovine, *Preprints Div. Pet. Chem. Am. Chem. Soc.* **24**, 836–846 (1979).
14. U.S. Pat. 3,932,460 (Jan. 13, 1976), D. R. McCoy, T. B. Jordan, and F. K. Ward (to Texaco Inc.).
15. Ger. Pat. 2,758,780 (July 12, 1979), K. H. Hentschel, R. Dhein, H. Rudolph, K. Nuetzel, K. Morche, and W. Krueger (to Bayer A.G.).
16. Jpn. Pat. 57/38,767 A2 (Mar. 3, 1982), (to Asahi Denka Kogyo K.K.).
17. Jpn. Pat. 57/11,959 A2 (Jan. 21, 1982), (to Asahi Denka Kogyo K.K.).
18. Brit. Pat. 1,439,753 (June 16, 1976), H. Coates, J. D. Collins, and I. H. Siddiqui (to Albright and Wilson Ltd.).
19. Jpn. Pat. 52/100,415 (Aug. 23, 1977), T. Kiyoura (to Mitsui Toatsu Chemicals, Inc.).
20. Jpn. Pat. 52/72,882 (June 17, 1977), T. Ooe, T. Nakazato, and R. Yoshikawa (to Mitsubishi Gas Chemical Co., Inc.).
21. Ger. Pat. 2,440,612 (Mar. 6, 1975), Y. Christidis (to Nobel Hoechst Chimie).
22. Ger. Pat. 3,125,338 A1 (Jan. 13, 1983), K. Koenig and M. Schmidt (to Bayer A.G.).
23. A. Kamors, *Tezisy Dokl. Konf. Molodykh Nauchn. Rab. Inst. Neorg. Khim., Akad. Nauk Latv. SSR, 5th*, 17–20.
24. Czech. Pat. 190,732 B (Dec. 15, 1981), J. Holcik, M. Karvas, and J. Masek.
25. Eur. Pat. Appl. EP 113,994 A1 (July 25, 1984), B. E. Johnston and P. R. Napier (to Mobil Oil Corp.).
26. Jpn. Pat. 61,225,191 A2 (Oct. 6, 1986), K. Tajima, M. Takahashi, K. Nishikawa, and T. Takeuchi (to Adeka Argus Chemical Co., Ltd.).

27. Ger. Pat. 2,630,257 (Jan. 12, 1978), H. Habarlein and F. Scheidl (to Hoechst A.G.).
28. U.S. Pat. 4,454,064 A (June 12, 1984), Y. Halpern and R. H. Niswander (to Borg-Warner Corp.).
29. Fr. Pat. 2,489,333 A1 (Mar. 5, 1982), J. Kiefer (to Ciba-Geigy A.G.).
30. U.S. Pat. 4,152,373 (May 1, 1979), M. L. Honig and E. D. Weil (to Stauffer Chemical Co.).
31. L. Desvergnes, *Chim. Ind.* **29**, 1263 (1933).
32. Jpn. Pat. 51/118,708 (Oct. 18, 1976), S. Oinuma and M. Kusakabe (to Agency of Industrial Sciences and Technology).
33. Czech Pat. 221,031 B (Mar. 15, 1986), S. Zeman, M. Dimun, and Z. Cervenka.
34. U.S. Pat. 4,352,919 A (Oct. 5, 1982), E. W. Kluger and C. D. Welch (to Milliken Research Corp.).
35. U.S. Pat. 4,136,044 (Jan. 23, 1979), M. Braid (to Mobil Oil Corp.).
36. W. S. Anderson and H. J. Hyer, *JANNAF Propul. Meet.*, vol. 1, AD-A103 844, CPIA Publ. 340, 1981, 387–398.
37. Jpn. Pat. 56/55,362 (May 15, 1981), (to Teijin Ltd.).
38. U.S. Pat. Appl. 183,707 (Jan. 30, 1981), (to United States National Aeronautics and Space Administration).
39. USSR Pat. 1,035,025 A1 (Aug. 15, 1983), E. B. Smirnova, Z. L. Chilyasova, N. P. Zykova, and L. A. Druganova (to Central Scientific-Research and Design Institute of the Wood Chemical Industry).
40. Ger. Pat. 2,707,875 (Aug. 31, 1978), J. Perner, K. Stork, F. Merger, and K. Oppenlaender (to BASF A.G.).
41. S. Shimizu, Y. Sasaki, and C. Hirai, *Nihon Daigaku Seisankogakubu Hokoku, A*, **15**(1), 47–56 (1982).
42. U.S. Pat. 4,151,211 (Apr. 24, 1979), I. Heckenbleikner and W. P. Enlow (to Borg-Warner Corp.).
43. Ger. Pat. 2,501,285 (July 24, 1975), A Schmidt (to Ciba-Geigy A.G.).
44. Belg. Pat. 885,670 (Feb. 2, 1981), R. Leger, R. Nouguier, J. C. Fayard, and P. Maldonado (to Elf France).
45. Eur. Pat. Appl. EP 46,731 A1 (Mar. 3, 1982), F. Lohse and C. E. Monnier (to Ciba-Geigy A.G.).
46. Jpn. Pat. 63,162,641 A2 (July 6, 1988), G. Watanabe, N. Nakajima, and Y. Ito (to Yokkaichi Chemical Co., Ltd.).
47. Jpn. Pat. 62,223,141 A2 (Oct. 1, 1987), Y. Fujio, Y. Nishi, and T. Nishimoto (to Osaka Soda Co., Ltd.).
48. E. Weber, *J. Org. Chem.* **47**, 3478–3486 (1982).
49. A. B. Padias, H. K. Hall, Jr., D. A. Tomalia, and J. R. McConnell, *J. Org. Chem.* **52**, 5305–5312 (1987).
50. J. E. Vik, *Acta Chem. Scand.* **27**, 239 (1973).
51. T. G. Bonner, E. J. Bourne, and J. Butler, *J. Chem. Soc.*, 301 (1964).
52. J. E. Vik, *Acta Chem. Scand. B* **28**, 325 (1974).
53. P. Werle, E. Busker, and E. Wolf-Heuss, *Liebigs Ann. Chem.* 1082–1087 (1985).
54. D. I. Belkin, *Zh. Prikl. Khim. (Leningrad)* **52**, 237–239 (1979).
55. L. Kovacic-Beck, I. Beck, and F. Anusic, *Nafta (Zagreb)* **34**(3), 131–135 (1983).
56. M. Lichvar, J. Sabados, V. Macho, and L. Komora, *Chem. Prum.* **36**(2), 57–61 (1986).
57. U.S. Pat. 2,612,526 (Sept. 30, 1952), C. W. Gould (to Hercules Powder Co.).
58. Jpn. Pat. 57/139,028 A2 (Aug. 27, 1982), (to Koei Kagaku Kogyo K.K.).
59. U.S. Pat. 3,962,347 (June 8, 1976), K. Herz (to Perstorp AB).
60. Ger. Pat. 2,930,345 (Feb. 19, 1981), P. Werle and G. Pohl (to Degussa).
61. Czech Pat. 183,405 (May 15, 1980), L. Komora and H. Nitoschneiderova.
62. Span. Pat. 467,714 (Oct. 16, 1978), (to Patentes y Novedades S.A.).

63. Eur. Pat. Appl. EP 242,784 A1 (Oct. 28, 1987), H. V. Holmberg and H. E. Larsson (to Perstorp AB).

64. Czech Pat. 220,256 B (Feb. 15, 1986), J. Ziak.

65. Jpn. Pat. 5,808,028 (Jan. 18, 1983), (to Koei Kagaku Kogyo K.K.).

66. Czech Pat. 181,486 (Jan. 15, 1980), M. Lichvar, J. Sabados, J. Vidovenec, and L. Butkovsky.

67. Jpn. Pat. 57/142,929 A2 (Sept. 3, 1982), (to Koei Kagaku Kogyo K.K.).

68. V. V. Pakulin, A. A. Kruglikov, Y. V. Rogachev, and P. E. Gulevich, *Plast. Massy* **3**, 12–13 (1988).

69. *Synthetic Organic Chemicals*, Publication 776, U.S. International Trade Commission.

70. *Technical Bulletin, Pentaerythritol*, Oxyquim S.A.

71. D. S. Wiersma, R. E. Hoyle, and H. Rempis, *Anal. Chem.* **34**, 1533 (1962).

72. R. R. Suchanec, *Anal. Chem.* **37**, 1361 (1965).

73. *Technical Bulletin, Pentaerythrit*, Degussa.

74. *Material Safety Data Sheet C73, Pentaerythritol*, Celanese Canada Inc., 1990.

75. Belg. Pat. 874,241 (June 18, 1979), (to BASF Farben und Fasern A.G.).

76. Jpn. Pat. 55/82,165 (June 20, 1980), (to Nippon Synthetic Chemical Industry Co., Ltd.).

77. U.S. Pat. 4,690,783 A (Sept. 1, 1987), R. W. Johnston, Jr. (to Union Camp Corp.).

78. Czech Pat. 222,510 B (Mar. 15, 1986), K. Hajek and co-workers.

79. Neth. Pat. 77/11852 (May 2, 1979), (to Hercules Inc.).

80. U.S. Pat. 4,229,310 (Oct. 21, 1980), G. Frangatos (to Mobil Oil Corp.).

81. Ger. Pat. 3,328,739 (Feb. 9, 1984), C. J. Dorer and K. Hayashi (to Lubrizol Corp.).

82. Ger. Pat. 1,901,769 (Oct. 9, 1967), H. Thomas and J. Turbet (to Imperial Chemical Industries Ltd.).

83. *U.S. Pharmacopeia: National Formulary XV*, 20th ed., Mack Publishing Co., Easton, Pa., 1980.

84. Eur. Pat. Appl. 17,609 (Oct. 15, 1980), D. Alt, K. D. Hutschgau, A. Schillmoeller, and B. Fischer (to Siemens A.G.).

85. U.S. Pat. 4,762,746 A (Aug. 9, 1988), L. Wesch and E. Weiss (to Odenwald-Chemie GmbH).

86. J. P. Jain, N. K. Saxena, I. Singh, and D. R. Gupta, *Res. Ind.* **30**(1), 20–24 (1985).

87. Ger. Pat. 2,847,628 (May 31, 1979), B. Sallmen, C. A. Sjoegreen, M. O. Maansson, and K. Ogemark (to Perstorp A.B.).

88. U.S. Pat. 4,162,242 (July 24, 1979), R. House (to Chevron Research Co.).

89. Eur. Pat. 219,427 A1 (Apr. 11, 1987), C. D. G. deMezeyrac, R. Fugier, B. Gicquel, and S. Tetard (to Isover Saint-Gobain).

90. Fr. Pat. 2,378,835 (Aug. 25, 1978), (to E. I. du Pont de Nemours & Co., Inc.).

91. Ger. Pat. 3,410,957 A1 (Sept. 26, 1985), K. J. Gardenier and W. Heimbuerger (to Henkel K.-G.a.A.).

92. Belg. Pat. 890,841 A2 (Feb. 15, 1982), S. B. Labelle and J. A. E. Hagquist (to Fuller H.B. Co.).

93. Jpn. Pat. 61,027,987 A2 (Feb. 7, 1986), A. Matsumaya, H. Ozawa, and S. Hirose (to Mitsui Toatsu Chemicals, Inc.).

94. R. Holman, ed., *UV and EB Curing Formulations for Printing Inks, Coatings and Paints*, SITA Technology, London, 1984.

95. C. G. Roffey, *Photopolymerization of Surface Coatings*, John Wiley & Sons, Inc., New York, 1982.

96. Brit. Pat. 896,067 (Aug. 19, 1960), W. R. Whately and G. M. Sheehan (to American Cyanamid Co.).

97. Jpn. Pat. 54/95,688 (July 28, 1979), S. Kondo, A. Matsufuji, and A. Umehara (to Fuji Photo Film Co., Ltd.).

98. Jpn. Pat. 60,237,444 A2 (Nov. 26, 1985), T. Komamura, M. Iwagaki, and T. Masukawa (to Konishiroku Photo Industry Co., Ltd.).

99. Jpn. Pat. 63,115,395 A2 (May 19, 1988), Y. Takeuchi (to Canon K.K.).

100. USSR Pat. 1,310,460 A1 (May 15, 1987), Y. Y. Lukomski, T. V. Mulina, V. V. Vasiliev, and R. V. Kopteva.

101. Jpn. Pat. 61,204,292 A2 (Sept. 10, 1986), S. Anzai, H. Sakaguchi, H. Yamazaki, K. Shiina, and M. Kurodi (to Hitachi, Ltd.).

102. Jpn. Pat. 61,240,627 A2 (Oct. 25, 1986), Y. Kudo, S. Yoshimura, S. Tsuchiya, and T. Kojima (to Matsushita Electric Industrial Co., Ltd.).

103. U.S. Pat. 4,572,864 A (Feb. 25, 1986), D. K. Benson, R. W. Burrows, and Y. D. Shinton (to United States Dept. of Energy).

WILLIAM N. HUNTER
Celanese Canada Inc.

ALDEHYDES

1. Introduction

Aldehydes are carbonyl-containing organic compounds of the general formula RCH=O. The R group represents an aliphatic, aromatic, or heterosubstituted radical, except in formaldehyde, where R represents hydrogen. Aldehydes are inherently reactive compounds. The carbonyl group is susceptible to both oxidation and reduction, yielding acids and alcohols, respectively. Additionally, the carbonyl group is susceptible to nucleophilic addition, providing a means by which to form new chemical bonds. Furthermore, the presence of the carbonyl activates the hydrogens bound to the alpha carbon, and thus provides an additional site of reactivity. Ketones are a related class of compounds having two alkyl groups attached to the carbonyl group $R_1R_2C=O$ (see KETONES).

2. Nomenclature

The common method of naming aldehydes corresponds very closely to that of the related acids (see CARBOXYLIC ACIDS), in the sense that the term *aldehyde* is added to the base name of the acid. For example, formaldehyde (qv) comes from formic acid, acetaldehyde (qv) from acetic acid, and butyraldehyde (qv) from butyric acid. If the compound contains more than two aldehyde groups, or is cyclic, the name is formed using *carbaldehyde* to indicate the functionality. The IUPAC system of aliphatic aldehyde nomenclature is derived by replacing the final *-e* from the name of the parent acyclic hydrocarbon by the suffix *-al*. If two aldehyde functional groups are present, the suffix *-dial* is used. The prefix *formyl* is used with polyfunctional compounds. Examples of nomenclature types are shown in Table 1.

Table 1. **Aldehyde Nomenclature**

Structural formula	Name	CAS Registry Number
H—CHO	formaldehyde or methanal	[50-00-0]
CH$_3$CHO	acetaldehyde or ethanal	[75-07-0]
(CH$_3$)$_2$CHCHO	isobutyraldehyde or 2-methylpropanal or α-methylpropionaldehyde	[78-84-2]
CH$_3$CH=CHCHO	crotonaldehyde or 2-butenal	[4170-30-3]
CH$_3$ | CH$_3$CH$_2$CHCHO	2-methylbutyraldehyde or 2-methylbutanal	[96-17-3]
OHCCH$_2$CH$_2$CH$_2$CHO	glutaraldehyde or pentandial	[111-30-8]
CHO | OHCCH$_2$CHCH$_2$CHO	1,2,3,-propanetricarbaldehyde or formylpentandial	[61703-13-7]
OHCCH$_2$CH$_2$CH$_2$COOH	4-formylbutanoic acid	[5746-02-1]
CH$_2$CH$_2$CH$_2$CH$_2$CHCHO	formylcyclopentane or cyclopentane-carbaldehyde	[872-53-7]

3. Physical Properties

The C$_1$ and C$_2$ carbon aliphatic aldehydes, formaldehyde and acetaldehyde are gases at ambient conditions whereas the C$_3$ (propanal [123-38-6]) through C$_{11}$ (undecanal [112-44-7]) aldehydes are liquids, and higher aldehydes are solids at room temperature. As can be seen from Table 2, the presence of hydrocarbon branching tends to lower the boiling or melting point, as does unsaturation in the carbon skeleton. Generally, an aldehyde has a boiling point between those of the corresponding alkane and alcohol. Aldehydes are usually soluble in common organic solvents and, except for the C$_1$ to C$_5$ aldehydes, are only sparingly soluble in water. The lower, C$_1$ to C$_8$, aldehydes have pungent, penetrating, unpleasant odors, some of which may be attributed to the presence of the corresponding acids that readily form by air oxidation. Above C$_8$, aldehydes have more pleasant odors in their diluted state, and some higher aldehydes are used in the perfume and flavoring industries (see PERFUMES). Interestingly, the C$_9$ aldehyde, nonanal [124-19-6], is reported to possibly be a human sex pheromone (1). Aldehydes must be kept from contact with air (oxygen) to retain purity.

3.1. Spectroscopic Properties. Characteristic spectroscopic data for aldehydes are summarized in Table 3. The infrared (ir) carbonyl stretching frequency between 1720 and 1740 cm^{-1} in saturated aldehydes, is lowered when there is unsaturation in conjugation with the carbon–oxygen double bond. The carbonyl group also exhibits a weak ultraviolet (uv) absorption near 280 nm as a result of the excitation of one of the unshared electrons on the carbonyl oxygen n, to π^* transition).The nuclear magnetic resonance (nmr) absorptions

Table 2. Properties of Aldehydes

Name	CAS Registry Number	Formula	Molecular weight	Melting point, °C	Boiling point, °C	Refractive index at 25°C	Density 25°C, g/mL	Viscosity 25°C, cP	Surface tension at 25°C dyn/cm	Heat capacity at 25°C, J/g°K	Heat of vaporization NBP, kJ/mol	Solubility, at 25°C, g/100g water
Aliphatic aldehydes												
formaldehyde	[50-00-0]	HCHO	30.03	−92	−19.1		0.7328	0.1421	27.3797	2.3425	23.065	miscible
acetaldehyde	[75-07-0]	CH₃CHO	44.05	−123	20.85	1.3283	0.7744	0.2122	20.7644	2.5576	25.731	miscible
propionaldehyde	[123-38-6]	CH₃CH₂CHO	58.08	−103.15	48	1.3593	0.7912	0.3174	21.9551	2.3039	23.497	28.21
butanal (*n*-butyraldehyde)	[123-72-8]	CH₃(CH₂)₂CHO	72.11	−96.4	74.8	1.3766	0.7974	0.4192	24.9484	2.2705	31.056	7.64
2-methylpropanal (isobutyraldehyde)	[78-84-2]	CH₃CH(CH₃)CHO	72.11	−65	64.1	1.3698	0.7835	0.5211	20.3115	2.1608	31.234	6.38
pentanal (*n*-valeraldehyde)	[110-62-3]	CH₃(CH₂)₃CHO	86.13	−91.15	102.9	1.3917	0.8047	0.5054	25.326	2.1932	34.165	1.37
2-methyl butanal	[96-17-3]	CH₃CH₂CH(CH₃)CHO	86.13	<−100	91.3	1.3885	0.8007	0.525			32.35	<1.5
3-methylbutanal (isovaleraldehyde)	[590-86-3]	CH₃CH(CH₃)CH₂CHO	86.13	−105.5	92.9	1.3854	0.7915	0.51	22.6		32.397	<1.5
2,2-dimethylpropanal (pivalaldehyde)	[630-19-3]	CH₃C(CH₃)₂CHO	86.13	3.8	74.3	1.376	0.7749	0.64	21		30.198	1.43

hexanal (caproaldehyde)	[66-25-1]	$CH_3(CH_2)_4CHO$	100.16	−56	128.3	1.4017	0.8097	0.6625	25.9961	2.1560	36.558	0.60
heptanal (heptaldehyde)	[111-71-7]	$CH_3(CH_2)_5CHO$	114.19	−43.35	152.8	1.4094	0.8140	0.8619	26.5381	2.1793	38.827	<0.5
octanal (caprylaldehyde)	[124-13-0]	$CH_3(CH_2)_6CHO$	128.21	−27.15	174	1.4156	0.8181	1.1497	27.5748	2.0971	41.549	<0.5
2-ethylhexanal	[123-05-7]	$CH_3(CH_2)_3CH\text{-}(C_2H_5)CHO$	128.21		164	1.4137	0.8152	1.0527	27.8571	1.9391	38.73	<0.5
nonanal (pelargonaldehyde)	[124-19-6]	$CH_3(CH_2)_7CHO$	142.24	−18	195	1.4208	0.8228	1.3563	28.3954	2.0755	43.67	<0.5
decanal (capraldehyde)	[112-31-2]	$CH_3(CH_2)_8CHO$	156.27	−6	215	1.4251	0.8213	1.5417	28.2958	2.0821	46.468	<0.5
Aromatic Aldehydes												
benzaldehyde	[100-52-7]	C_6H_5CHO	106.12	−57.13	178.7	1.5428	1.0415	1.3761	38.6152	1.6222	41.468	<0.5
phenylacetaldehyde	[122-78-1]	$(C_6H_5)CH_2CHO$	120.15		195		1.0229			1.8240	44.357	<0.5
o-tolualdehyde	[529-20-4]	$CH_3(C_6H_4)CHO$	120.15	−38.15	201	1.546	1.0315	2.2214	40.2099	1.6878	43.522	<0.5
m-tolualdehyde	[620-23-5]	$CH_3(C_6H_4)CHO$	120.15	−22.15	199	1.5389	1.0153	1.6629	37.6454	1.7493	44.782	<0.5
p-tolualdehyde	[104-87-0]	$CH_3(C_6H_4)CHO$	120.15		204		1.0093	1.8282	36.285	1.5101	45.61	<0.5
salicylaldehyde (o-hydroxybenzaldehyde)	[90-02-8]	$HO(C_6H_4)CHO$	122.12	1.6	196.5	1.57017	1.1491	2.4911	42.1437	1.8264	45.565	1.7
p-anisaldehyde (p-methoxybenzaldehyde)	[123-11-5]	$CH_3O(C_6H_4)CHO$	136.15	0	249.1	1.5707	1.1965				51.182	0.2

Table 3. **Spectroscopic Absorptions of Aldehydes**

Compound	ir, cm^{-1a}	uv, nm	^{1}H nmr, ppm^b	^{13}C nmr, ppc^c
acetaldehyde	1730	290	9.80	200
butyraldehyde	1725	283	9.74	202
benzaldehyde	1695	278	10.00	192
2-butenal	1700	301	9.48	193

^{a} Carbonyl stretching frequency.
^{b} Aldehyde proton, relative to TMS.
^{c} Carbonyl carbon, relative to TMS.

of carbonyl proton (^{1}H-nmr) and carbon (^{13}C-nmr) are very characteristic and highly informative. The proton attached to the carbonyl exhibits a strong downfield resonance ($\sim$10 ppm) relative to the standard tetramethylsilane (TMS), as does the carbonyl carbon that appears at $\sim$200 ppm, again relative to TMS. The downfield shift of the nmr resonances and the intensity of the ir stretching frequency are attributable to the polar nature of the carbon–oxygen double bond.

4. Chemical Properties

Aldehydes are very reactive compounds. Reactions generally fall into two classes: those directly involving the carbonyl group and those occurring at the adjacent carbon atom. The polar nature of the carbonyl group($RC\overset{\delta+}{H} = \overset{\delta-}{O}$) lends itself to nucleophilic addition, reduction and oxidation, and also affects the reactivity of its adjacent carbon atom by rendering its hydrogens relatively acidic. Aldehydes having acidic hydrogens (also known as active methylene compounds) must be protected against inadvertent contact with bases, as such contact may result in an exothermic condensation reaction that may become dangerous. There are many available references offering detailed descriptions of aldehyde reactions (2–6). Many of the reactions involving aldehydes have been named for their discoverers and comprehensive reviews of name reactions are also available (7–9).

4.1. Reduction Reactions. Aldehydes can be reduced to the corresponding alcohols by catalytic hydrogenation using heterogeneous, as well as homogeneous catalysts.

$$CH_3CH_2CH_2CHO \xrightarrow[\text{catalyst}]{H_2} CH_3CH_2CH_2CH_2OH$$

Common heterogeneous catalyst compositions contain oxides or salts of platinum, nickel, copper, cobalt, or palladium and are often present as mixtures of more than one metal. Homogeneous catalysts consist of trialkylphosphine complexes of transition metals such as rhodium, cobalt, and ruthenium. Metal hydrides, such as lithium aluminum hydride [16853-85-3] or sodium borohydride [16940-66-2], can also be used to reduce aldehydes. Depending on additional functionalities that may be present in the aldehyde molecule, specialized reducing reagents such as trimethoxyaluminum hydride or alkylboranes (less reactive and more selective) may be used. Other less industrially significant reduction procedures such as the Clemmensen reduction or the modified Wolff–Kishner reduction exist as well.

4.2. Oxidation Reactions. In general, aldehyde are easily oxidized in air to the corresponding carboxylic acid, by a free-radical chain reaction.

$$\text{RCHO} \xrightarrow[\text{catalyst}]{\text{O}_2} \text{RCOOH}$$

Air or oxygen are commonly used, with or without the homogeneous metal catalysts. This is the basis for many industrially significant preparations, eg, acetaldehyde to acetic acid [64-19-7] (see ACETIC ACID AND DERIVATIVES), propionaldehyde to propionic acid [79-09-47], furfural [98-01-1] to furoic acid [26447-28-9], and acrolein toacrylic acid [79-10-7] (see ACRYLIC ACID AND DERIVATIVES). For specialty applications, a variety of oxidizing reagents are available to perform this transformation. Although both chromium and manganese compounds can be used, an aqueous solution of potassium permanganate [7722-64-7] under either acidic or basic conditions is the more commonly employed reagent. Other known reagents for the oxidation of aldehydes include, hydrogen peroxide, nitric acid and a suspension of silver oxide in aqueous alkali. The latter provides a very mild and selective method for this type of oxidation.

Aldehydes can undergo an intermolecular oxidation–reduction (Cannizzaro reaction) by the action of strong base to produce a carboxylic acid salt and an alcohol. This reaction is common to aromatic aldehydes, and aliphatic ones with no α-hydrogen atoms.

$$2\,\text{C}_6\text{H}_5\text{CHO} \xrightarrow{\text{NaOH}} \text{C}_6\text{H}_5\text{CH}_2\text{OH} + \text{C}_6\text{H}_5\text{CO}_2\text{Na}$$

A related oxidation–reduction of aldehydes with α-hydrogens is the Tischenko reaction. This reaction occurs by the action of aluminum ethoxide, and the product is an ester.

$$2\,\text{RCH}_2\text{CHO} \xrightarrow{\text{Al(OCH}_2\text{CH}_3)_3} \text{RCH}_2\text{CH}_2\text{OCCH}_2\text{R} \quad (\overset{\text{O}}{\overset{\|}{})}$$

4.3. Addition Reactions. *Aldol Addition.* The hydrogen atoms on the carbon adjacent to the aldehyde carbonyl are relatively acidic and can be abstracted by strong bases. The resulting carbon anion,

$$\text{CH}_3\text{CH}=\text{O} \underset{\text{catalyst}}{\overset{\text{OH}^-}{\rightleftharpoons}} [^-\text{CH}_2\text{CHO}] + \text{H}_2\text{O} \underset{}{\overset{\text{CH}_3\text{CH}=\text{O}}{\rightleftharpoons}} \underset{\text{O}^-}{\text{CH}_3\text{CHCH}_2\text{CHO}} \underset{-\text{OH}^-}{\overset{\text{H}_2\text{O}}{\rightleftharpoons}} \underset{\text{OH}}{\text{CH}_3\text{CHCH}_2\text{CHO}}$$

also known as enolate, adds to the carbonyl of another aldehyde molecule to form a dimer adduct which is protonated to the alcohol-aldehyde product known as aldol.

Aldehyde dimerization can also occur with acid catalysts. The name aldol was introduced by Wurtz in 1872 to describe the product resulting from the acid-catalyzed reaction of acetaldehyde. The aldol products are not usually isolated; and they readily dehydrate to form α,β-unsaturated aldehydes. This overall transformation is commercially employed to produce 2-ethyl-2-hexenal [26266-68-2] often referred to as ethylpropylacrolein, from butyraldehyde (qv). The ethylpropylacrolein is then hydrogenated to produce 2-ethylhexanol, a commercially significant plasticizer alcohol (see ALCOHOLS, HIGHER ALIPHATIC). The next higher homologue, a C_{10} alcohol mixture containing 2-propylheptanol, can be produced in an analogous fashion from a valeraldehyde product mixture resulting from hydroformylating a mixed butenes stream (see BUTYLENES). This hydroformylation is accomplished using a new generation of highly active phosphite-promoted rhodium catalysts.

Claisen and Perkin Reactions.   These reactions are similar to the aldol addition. The Claisen reaction is different than the better-known Claisen condensation of esters, and is carried out by combining an aromatic aldehyde and an ester in the presence of a strong base, such as sodium hydride, sodium amide, to give a β-hydroxy ester that dehydrates to an α,β-unsaturated ester.

$$C_6H_5CHO + CH_3CO_2C_2H_5 \xrightarrow[0-5°C]{NaH} C_6H_5CH{=}CHCO_2C_2H_5$$

The Perkin reaction, utilizing an aromatic aldehyde, an acid anhydride, and a base such as an acid salt or amine, produces the corresponding α,β-unsaturated acid.

$$C_6H_5CHO + (CH_3CO_2)O \xrightarrow{(CH_3CH_2)_3N} C_6H_5CH{=}CHCOOH + CH_3COOH$$

Analogously, aldehydes react with ammonia [7664-41-7] or primary amines to form Schiff bases. Subsequent reduction produces amine. Other synthetically useful reactions include the addition of hydrogen cyanide [74-90-8], sodium bisulfite [7631-90-5], amines, or thiols to the carbonyl group, and usually requires the use of a catalyst to assist in reaching the desired equilibrium product.

Addition of Alcohols.   This acid catalyzed reaction leads to the formation of a class of compounds known as acetals. The first addition product, a hemiacetal, is unstable, the equilibrium generally favoring the parent aldehyde. Subsequent steps involve protonation of the —OH which leads to a stabilized carbonium ion, followed by reaction of a second alcohol molecule.

Acetals have been used in racing car fuels, gasoline additives, and paint and varnish solvents and

$$RCHO \underset{H^+}{\overset{H^+}{\rightleftharpoons}} RCH_2OH \underset{}{\overset{R'OH}{\rightleftharpoons}} RCH\begin{smallmatrix}OH\\OR'\end{smallmatrix} \underset{H_2O}{\overset{H^+}{\rightleftharpoons}} RCH_2\begin{smallmatrix}+\\OR'\end{smallmatrix} \underset{H^+}{\overset{R'OH}{\rightleftharpoons}} RCH\begin{smallmatrix}OR'\\OR'\end{smallmatrix}$$

hemiacetal   acetal

strippers. Acetals of higher aldehydes have fragrances similar to, but not so pungent as the parent aldehydes. Because they are not as sensitive to alkalies or autoxidation, acetals find use as fragrances for alkaline formulations such as soaps, shampoos and heavy-duty detergents.

The Wittig Reaction.   In this very important reaction initially involving nucleophilic addition to an aldehyde carbonyl, an aldehyde reacts with a phosphorus **ylide** forming a **betaine** intermediate that decomposes to produce an olefin and a tertiary phosphine oxide.

$$
\underset{\text{ylide}}{R-\overset{\overset{\text{O}}{\|}}{C}-H \;+\; R_3\overset{+}{P}-\overset{\overset{-}{}}{\underset{\underset{R'}{|}}{C}}-R''}
\;\longrightarrow\;
\underset{\text{betaine}}{\overset{\overset{R}{|}}{O}-\overset{|}{C}-H}
\;\longrightarrow\;
R-\overset{\overset{H}{|}}{C}=\underset{\underset{R'}{|}}{C}-R'' \;+\; R_3PO
$$

Perhaps the most notable example of this chemistry is in the production of vitamin A [68-26-8], where the β-ionylidenacetaldehyde is condensed with the ester-ylid to obtain the polyene ester. Reduction then yields vitamin A (see VITAMINS).

The Wittig reaction has been extended to include carbanions generated from phosphonates, which is often referred to as the Horner–Wittig or Horner–Emmons reaction.

$$
R-\overset{\overset{\text{O}}{\|}}{C}-H \;+\; (RO)_2\overset{\overset{}{|}}{\underset{\underset{}{\|}}{P}}-\overset{-}{\underset{\underset{R'}{|}}{C}}-R''
\;\longrightarrow\;
R-\overset{\overset{H}{|}}{C}=\underset{\underset{R'}{|}}{C}-R'' \;+\; (RO)_2PO_2^-
$$

The Horner–Emmons reaction has a number of advantages over the conventional Wittig reaction. It occurs with a wider variety of aldehydes and ketones under relatively mild conditions as a result of the higher nucleophilicity of the phosphonate carbanions. The separation of the olefinic product is easier due to the aqueous solubility of the phosphate by-product, and the phosphonates are readily available from the Arbusov reaction. Furthermore, although the reaction itself is not stereospecific, the majority favor the formation of the trans olefin and many produce the trans isomer as the sole product.

A useful synthetic application of aldehydes is as masked 1,3-dithianes. In these aldehyde derivatives, the normal mode of reaction of the carbonyl, which is to react with nucleophiles, is reversed (umpolung) and can react with electrophiles, if a proton in the 1,3-dithiane is first removed with butyllithium. The following sequence of reaction is a practical method for the converting an aldehyde into a ketone.

$$
RCHO \;\xrightarrow{\text{HS(CH}_2)_3\text{SH}}\;
\begin{array}{c}\text{R}\;\;\text{S}\\ \diagdown \diagup \\ \text{C}\\ \diagup \diagdown \\ \text{Li}\;\;\text{S}\end{array}
\;\xrightarrow{\text{BuLi}}\;
\begin{array}{c}\text{R}\;\;\text{S}\\ \diagdown \diagup \\ \text{C}\\ \diagup \diagdown \\ \text{Li}\;\;\text{S}\end{array}
\;\xrightarrow{\text{R'X}}\;
\begin{array}{c}\text{R}\;\;\text{S}\\ \diagdown \diagup \\ \text{C}\\ \diagup \diagdown \\ \text{R'}\;\;\text{S}\end{array}
\;\xrightarrow{\text{H}_2\text{O}}\;
R-\overset{\overset{\text{O}}{\|}}{C}-R'
$$

5. Manufacture

A complete discussion of synthetic procedures that yield aldehydes as products can be found in the literature (10, 11).

Of these methods only a few are used on industrial scale. One important industrial process for manufacture of aldehydes is by hydroformylation of olefins using synthesis gas and transition metal catalysts (oxo synthesis). This reaction was discovered in Germany in 1938 and commercialized in the 1950s for the production propionaldehyde from ethylene and butyraldehydes from propylene using cobalt catalyst (12).

$$RCH{=}CH_2 \ + \ CO \ + \ H_2 \ \xrightarrow{\text{catalyst}} \ \underset{\underset{CH=O}{|}}{RCHCH_3} \ + \ RCH_2CH_2CH{=}O$$

Current commercial processes employ ligand-modified metal catalysts, which operate under milder conditions and provide better control of linear and branched aldehyde selectivities. Recent advances in hydroformylation technology include: catalysts promoted with phosphite ligands, which exhibit high reactivity with less reactive olefins under mild reaction conditions; novel catalyst–product separation technologies, which permit the manufacture of high molecular, weight aldehydes (C_7 to C_{15}) as well as more thermally sensitive aldehydes; fatty aldehydes, generally produced by dehydrogenation of corresponding alcohols in the presence of a suitable catalyst, can now be manufactured using oxo technology.

The direct oxidation of ethylene is used to produce acetaldehyde (qv) in the Wacker-Hoechst process. The catalyst system is an aqueous solution of palladium chloride and cupric chloride. Under appropriate conditions, an olefin can be oxidized to form an unsaturated aldehyde such as the production of acrolein [107-02-8] from propylene (see ACROLEIN AND DERIVATIVES).

Another commercial aldehyde synthesis is the catalytic dehydrogenation of primary alcohols at high temperature in the presence of a copper or a copper-chromite catalyst. Although there are several other synthetic processes employed, these tend to be smaller scale reactions. For example, acyl halides can be reduced to the aldehyde (Rosenmund reaction) using a palladium-on-barium sulfate catalyst. Formylation of aryl compounds, similar to hydroformylation, using HCN and HCl (Gatterman reaction) or carbon monoxide and HCl (Gatterman–Koch reaction) can be used to produce aromatic aldehydes.

$$ArH \ + \ CO \ \xrightarrow[\text{HCl}]{Cu_2Cl_2, \, AlCl_3} \ ArCHO$$

Additionally, Grignard reagent reacts with an alkyl orthoformate to form an acetal that is then hydrolyzed to the corresponding aldehyde using dilute acid.

$$CH_3CH_2MgX \ + \ HC(OR)_3 \ \xrightarrow[-Mg(OR)X]{} \ CH_3CH_2CH(OR)_2 \ \xrightarrow{\text{catalyst}} \ CH_3CH_2CHO \ + \ ROH$$

6. Production and Economic Aspects

From the list of commercially significant aldehydes in Table 4, formaldehyde is produced in the largest volume and has comparable economic value (volume × price) to that of butyraldehyde.

Table 4. **Annual Consumption of Aldehydes by Major Regions**[a]

Aldehyde	Thousands of metric tons per year				2000 prices, $/kg
	U.S. and Canada	Europe	Asia and Japan	*World Total*	
formaldehyde	4630	7036	4504	*17553*	0.28
acetaldehyde	190	651	436		1.00
propionaldehyde	580	53			1.12
butyraldehyde	1566	1892	1901	*5701*	1.14
phenylacetaldehyde					37.9
salicylaldehyde					11
p-anisaldehyde					13.2

[a] *Chemical Economics Handbooks*-SRI international, 1999.

Table 5 contains a listing of representative aldehyde producers. Only the products identified in Table 4 have been included.

7. Characterization

Aldehydes can be characterized qualitatively through the use of Tollens' or Fehling's reagents as well as by spectroscopic means. The use of Tollens' reagent, a solution of silver nitrate in dilute, basic, aqueous ammonia, leads to the deposition of a silver mirror in the presence of an aldehyde. Fehling's reagent, an aqueous solution of copper sulfate, sodium potassium tartrate, and sodium hydroxide, produces a reddish brown precipitate of cuprous oxide in the presence of an aldehyde. Additionally, aldehyde carbonyl groups can be derivatized. Aldehyde oximes, phenylhydrazones, 2,4-dinitrohydrazones, semicarbazones, or sodium bisulfite addition products can be generated, purified, and characterized by their distinctive melting points. Hydrazone derivatives are often useful in isolating the aldehyde as a solid, crystalline material. The carbonyl group may also be oxidized to an acid through the use of hydrogen peroxide or potassium permanganate and the resultant carboxylic acid characterized by its spectroscopic properties and derivatives.

8. Health and Safety Factors

Interest in the toxicity of aldehydes has focused primarily on specific compounds, particularly formaldehyde, acetaldehyde, and acrolein (13). Little evidence exists to suggest that occupational levels of exposure to aldehydes would result in mutations, although some aldehydes are clearly mutagenic in some test systems. There are, however, acute effects of aldehydes.

8.1. Irritation and Sensitization. Low molecular weight aldehydes, the halogenated aliphatic aldehydes, and unsaturated aldehydes are particularly irritating to the eyes, skin, and respiratory tract. The mucous membranes

Table 5. **Aldehyde Producers**[a]

Company	Aldehyde product
United States and Canada	
Aristech Chemical Corp.	butyraldehyde
BASF Corp.	butyraldehyde
Borden Chemical	formaldehyde
Celanese Corporation	butyraldehyde, propionaldehyde, formaldehyde
Eastman	acetaldehyde, propionaldehyde, butyraldehyde
Firmenich Incorporated	phenylacetaldehyde
Georgia-Pacific Resins, Inc.	formaldehyde
Givauden-Roure Corporation	phenylacetaldehyde, *p*-anisaldehyde
Penta Manufacturing	*p*-anisaldehyde
Perstorp Polyols, Inc.	formaldehyde
Reichhold Limited	formaldehyde
Solutia Inc.	formaldehyde
Union Carbide	propionaldehyde, butyraldehyde
Europe	
BASF actiengesellschaft	formaldehyde, propionaldehyde, butyraldehyde, *p*-anisaldehyde, phenylacetaldehyde
Borden Chemical	formaldehyde
Celanese Chemicals Europe	acetaldehyde, butyraldehyde
Degussa- Hüls AG	formaldehyde
Givaudan-Roure SA	phenylacetaldehyde, *p*-anisaldehyde
Harmann & Reimer	phenylacetaldehyde, *p*-anisaldehyde
Hüls	acetaldehyde, butyraldehyde
Lonza	acetaldehyde
Neste Oxo AB	butyraldehyde
NORSOLOR	formaldehyde
Oxochimie SA	butyraldehyde
Perstorp Specialty Chemicals	formaldehyde
East Asia	
Chisso Corp.	butyraldehyde
Daicel Chemical Industries	phenylacetaldehyde
Eastman Chemical Singapore	butyraldehyde
Korea Alcohol Industries Co.	acetaldehyde
Kyowa Yuka Company	acetaldehyde, butyraldehyde
LG Chemical Ltd.	butyraldehyde
Lee Chang Yung Chem. Ind.	acetaldehyde, formaldehyde
Midori Kagaki Co.	*o*-anisaldehyde, *p*-anisaldehyde
Mitsubishi Chemical Corporation	formaldehyde, acetaldehyde, butyraldehyde
Mitsui	formaldehyde
Nippon Shokubai Co.	*p*-anisaldehyde
Petro Oxo Nusantara	butyraldehyde
Showa Denko K. K.	acetaldehyde, phenylacetaldehyde

[a] *Chemical Economics Handbooks*-SRI international, 1999.

of nasal and oral passages and the upper respiratory tract can be affected, producing a burning sensation, an increased ventilation rate, bronchial constriction, choking, and coughing. If exposures are low, the initial discomfort may abate after 5–10 min but will recur if exposure is resumed after an interruption. Furfural, the acetals, and aromatic aldehydes are much less irritating than formaldehyde and acrolein. Reports of sensitization reactions to formaldehyde are numerous.

8.2. Anesthesia. Materials that have unquestionable anesthetic properties arechloral hydrate [302-17-0], paraldehyde, dimethoxymethane [109-87-5], and acetaldehyde diethyl acetal. In industrial exposures, however, any action as an anesthesia is overshadowed by effects as a primary irritant, which prevent voluntary inhalation of any significant quantities. The small quantities that can be tolerated by inhalation are usually metabolized so rapidly that no anesthetic symptoms occur.

8.3. Organ Pathology. The principal pathology experimentally produced in animals exposed to aldehyde vapors is that of damage to the respiratory tract and pulmonary edema. In general, the aldehydes are remarkably free of actions that lead to definite cumulative organic damage to tissues. Thus the aldehydes cannot generally be regarded as potent carcinogens. Noted, however, that in chronic animal studies conducted by inhalation, formaldehyde has induced tumors of the nasal tissue. On the other hand, isobutyraldehyde has not produced a similar carcinogenic effect in similarly designed studies. Moreover, the intolerable irritant properties of the compounds preclude substantial worker exposure under normal conditions.

There is a significant difference in the toxicological effects of saturated and unsaturated aliphatic aldehydes. As can be seen in Table 6, the presence of the double bond considerably enhances toxicity. The precautions for handling reactive unsaturated aldehydes such as acrolein, methacrolein [78-85-3], and crotonaldehyde should be the same as those for handling other highly active eye and pulmonary irritants, as, eg, phosgene.

Material Safety Data Sheets, (MSDS), for individual compounds should be consulted for detailed information. Precautions for the higher aldehydes are essentially those for most other reactive organic compounds and should include: adequate ventilation in areas where high exposures are expected; fire and explosion precautions; and proper instruction of employees in use of respiratory, eye, and skin protection.

Table 6. **Effect of Unsaturation on Toxicity of Aldehydes**

Compound	Formula	LC_{50}, ppm[a]	LD_{50}, mg/kg[b]	TWA[c]
acetaldehyde	CH_3CHO	20,000	1,930	200
propionaldehyde	CH_3CH_2CHO	26,000	1,410	
acrolein	$CH_2{=}CHCHO$	130	25.9	0.1
isobutyraldehyde	$(CH_3)_2CHCHO$	>8,000[d]	2810	
methacrolein	$CH_2{=}C(CH_3)CHO$	250[d]	111	
n-butyraldehyde	$CH_3(CH_2)_2CHO$	60,000	2,490	
crotonaldehyde (2-butenal)	$CH_3CH{=}CHCHO$	1,400	260	2

[a] In rats, an exposure time of 30 min.
[b] In rats, dosage administered orally.
[c] Occupational Safety and Health Administration (OSHA PEL).
[d] Exposure time of 4 h.

9. Uses

Aldehydes find the most widespread use as chemical intermediates. The production of acetaldehyde, propionaldehyde, and butyraldehyde as precursors of the corresponding alcohols and acids are examples. The aldehydes of low molecular weight are also condensed in an aldol reaction to form derivatives that are important intermediates for the plasticizer industry (see PLASTICIZERS). As mentioned earlier, 2-ethylhexanol, produced from butyraldehyde, is used in the manufacture of di(2-ethylhexyl) phthalate [117-87-7]. Aldehydes are also used as intermediates for the manufacture of solvents (alcohols and ethers), resins, and dyes. Isobutyraldehyde is used as an intermediate for production of primary solvents and rubber antioxidants (see ANTIOXIDANTS). Fatty aldehydes C_8 to C_{13} are used in nearly all perfume types and aromas (see PERFUMES). Polymers and copolymers of aldehydes exist and are of commercial significance.

BIBLIOGRAPHY

"Aldehydes" in *ECT* 1st ed., Vol. 1, pp. 334–342 by E. F. Landau, Celanese Corporation of America, E. I. Becker, Polytechnic Institute of Brooklyn, and O. C. Dermer, Oklahoma Agricultural and Mechanical College; in *ECT* 2nd ed., Vol. 1, pp. 639–648 by L. J. Fleckenstein, Eastman Kodak Co.; in *ECT* 3rd ed., Vol. 1, pp. 790–798, by P. D. Sherman, Union Carbide Corporation; in *ECT* 4th ed., Vol. 1, pp. 926–937 by David J. Miller, Union Carbide Chemicals and Plastics Corporation; " Aldehydes" in *ECT* (online), Posting date: December 4, 2000 by David J. Miller, Union Carbide Chemicals and Plastics Corporation.

CITED PUBLICATIONS

1. J. Buckingham, *Dictionary of Organic Compounds*, 5th ed., Chapman and Hall, New York, 1982.
2. S. Patai, *The Chemistry of the Carbonyl Group*, Wiley-Interscience, New York, 1966 and 1970.
3. H. O. House, *Modern Synthetic Reactions*, 2nd ed., W. A. Benjamin, Inc., Menlo Park, Calif., 1972.
4. C. D. Gutche, *The Chemistry of Carbonyl Compounds*, Prentice-Hall, Inc., Englewood Cliffs, N.J., 1967.
5. A. T. Nielson and W. J. Houlihan, *Organic Reactions*, Vol. 16, John Wiley & Sons, Inc., New York, 1968.
6. T. Mukaiyama, *Organic Reactions*, Vol. 28, John Wiley & Sons, Inc., New York, 1982.
7. A. R. Surrey, *Name Reactions in Organic Chemistry*, 2nd ed., Academic Press, Inc., New York, 1961.
8. R. C. Denny, *Named Organic Reactions*, Plenum Press, New York, 1969.
9. H. Krauch and W. Kunz, *Organic Name Reactions*, 2nd ed., translated by J. M. Harkin, John Wiley & Sons, Inc., New York, 1964.
10. G. Hilgetag and A. Martini, eds., *Preparative Organic Chemistry*, 4th ed., Wiley-Interscience, New York, 1972, pp. 301–400.
11. I. T. Harrison, ed., *Compendium of Organic Synthetic Methods*, Vol. 1, 1971, pp. 132–176; Vol. 2, I. T. Harrison, ed., 1974, pp. 53–69; Vol. 3, L. S. Hegedus and L. G. Wade, Jr., eds., 1977, pp. 66–87; Vol. 4, L. G. Wade, Jr., ed., 1980, pp. 73–101; Vol. 5,

L. G. Wade, Jr., ed., 1984, pp. 92–123; Vol. 6, M. B. Smith, ed., 1988, pp. 51–66, Wiley- Interscience, New York.

12. B. Cornils and W. A. Herrmann, eds., *Applied Homogeneous Catalysis with Organometallic Compounds*, Vol. 1. pp. 29–90. VCH Publishers, New York, 1996,

13. F. A. Patty, *Patty's Industrial Hygiene and Toxicology*, John Wiley & Sons, Inc., New York, 1981, pp. 2629–2669.

ANTHONY G. ABATJOGLOU
DAVID J. MILLER
Union Carbide Corporation

ALKALOIDS

1. Introduction

Crude preparations of the naturally occurring materials now known as alkaloids were probably utilized by the early Egyptians and/or Sumarians (1). However, the beginnings of recorded, reproducible isolation from plants of substances with certain composition first took place in the early nineteenth century. Then in close succession, narcotine [128-62-1] (**1**, now called noscopine, $C_{22}H_{23}NO_7$) (2) and morphine [57-27-2] (**2**, R = H) (3) (both from the opium poppy, *Papaver somniferum* L.) were obtained.

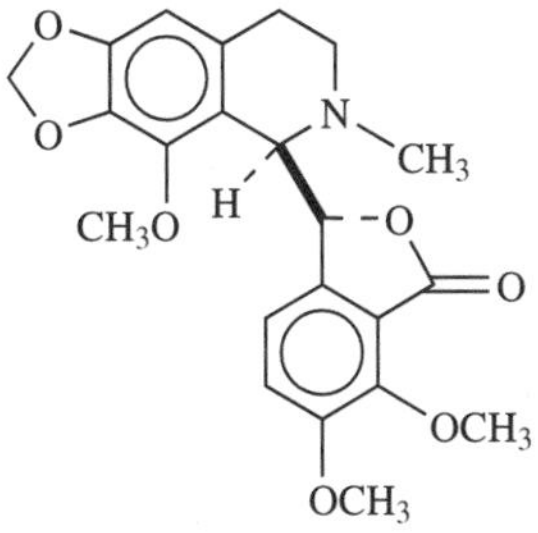

(**1**) 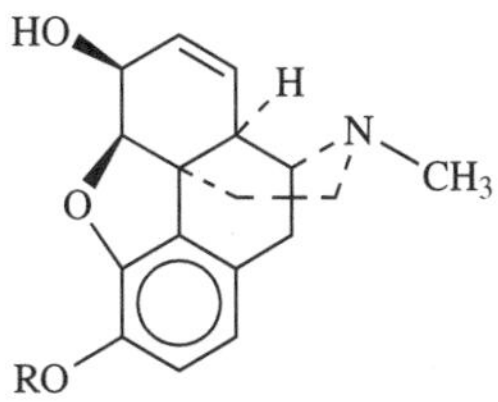   (**2**)

Although their presently accepted structures were unknown, they were characterized with the tools available at the time. Because morphine (**2**, R = H), $C_{17}H_{19}NO_3$, was shown to have properties similar to the basic soluble salts obtained from the ashes of plants (alkali) it was categorized as a vegetable alkali or *alkaloid*, and it is generally accepted that it was for this case that the word was coined.

However, there is currently no simple definition of what is meant by alkaloid. Most practicing chemists working in the field would agree that most alkaloids, in addition to being products of secondary metabolism, are organic nitrogen-containing bases of complex structure, occurring for the most part in seed-bearing plants and having some physiological activity. A 1961 compendium

(4) carefully avoids simple amine bases known to be present in some plants, but does list a variety of compounds such as aristolochic acid I [313-67-7] (3) (from *Aristolochia indica* L., the Dutchman's Pipe) and colchicine [64-86-8] (4) (from *Colchicum autumnala* L., the autumn crocus), neither of which is basic, but both of which are physiologically active. In a later (1975) reference (5), the list of materials called alkaloids had grown and more structures had been elucidated, but the definition was essentially unchanged. Subsequently, a much more sophisticated definition was proposed (6) which, while meritorious, has apparently been found unworkable. The most recent catalog (7), listing nearly 10,000 alkaloids, contains compounds generally fitting within the categories that were used in 1960, but widened still further to include not only nonbasic nitrogen-containing materials from plants, but also substances occurring in animals. Other compounds, the physiological activity of which has not been measured, are also reported (8). Nonetheless, because of their widespread distribution across all forms of life, alkaloids are intimately interwoven into the fabric of existence. Both our understanding of the roles these substances play in their respective sources and the possibility of genomic modification to adjust alkaloid production are being pursued as the twenty-first century dawns.

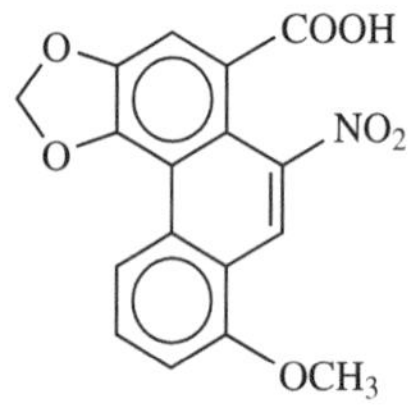

(3) 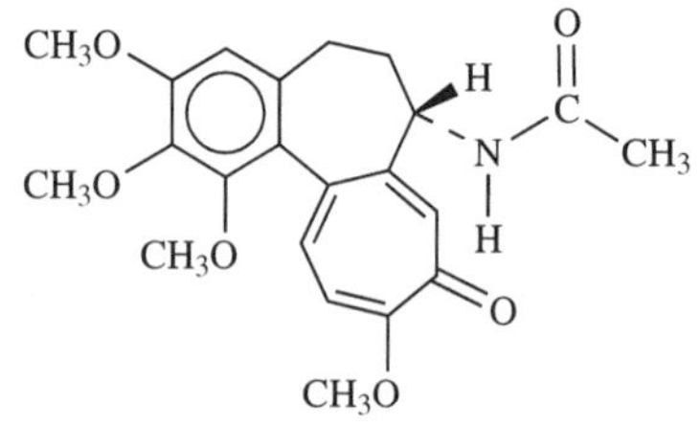   (4)

2. History

From today's perspective, the history of alkaloid chemistry can be divided into four parts. The first part, which doubtlessly developed over aeons prior to the appearance of present-day flora and fauna and about which, with genomic mapping a little is now known, deals with the role alkaloids may really play (as divorced from anthropocentric imaginings) in animal and plant defense, reproduction, etc. Second, in the era prior to ~1800, human use was apparently limited to apothecaries' crude mixtures and folk medicinals that were administered as palliatives, poisons, and potions. Knowledge of this is based on individual or group records or memory. In the third period, ~1800–1950, early analytical and isolation technologies were introduced. Good records were kept and techniques honed, so that the wrenching out of specific materials, in truly minute quantities, from the cellular matrices in which they are held could be reproducibly effected. This time period also saw the beginnings of correlation of the specific structures of those hard-won materials with their properties. Finally, the current era has seen a flowering of structure elucidation as a consequence of the maturation of some analytical techniques, a renaissance in synthetic methods, the introduction of biosynthetic probes, and the application of molecular genetics to biosynthesis (9). The most recent developments build on the newest analytical techniques and the ability to correlate huge quantities of information at high speed.

During the first era some insects developed relationships with the plants on which they fed, which allowed them to incorporate intact alkaloids for storage and subsequent use. This type of relationship apparently continues to exist. Thus in 1892 there was a report (10) that pharmacophagus swallowtail butterflies (*Papilios*) obtain and store poisonous substances from their food plants, and some 75 years later an investigation (11) showed that the warningly colored and potently odoriferous Aristolochia-feeding swallowtail butterfly (*Pachlioptera aristolochiae* Fabr.) is even less acceptable than the unpalatable Danainae to bird predators. Both the plant on which the swallowtail feeds (eg, *Aristolchia indica* L.) and the swallowtail itself contain aristolochic acid I (3), $C_{17}H_{11}NO_7$, and related materials. These materials are presumably ingested as larvae feed on the plant, stored during the pupal stage, and carried into the adult butterfly. With regard to the *Danainae*, the larvae of the butterflies *Danaus plexippus* L. and *Danaus chrysippus* L. feed on *Senecio* spp. which contain, among other compounds, the pyrrolizidine alkaloid senecionine (5) (12). Metabolites of this and other related alkaloids apparently serve in courtship and mating, with the more alkaloid-rich individuals having an advantage (13).

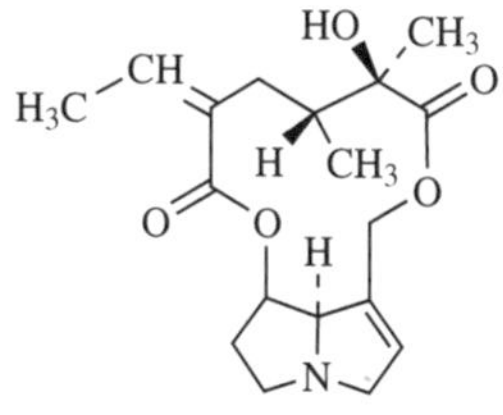

(5) 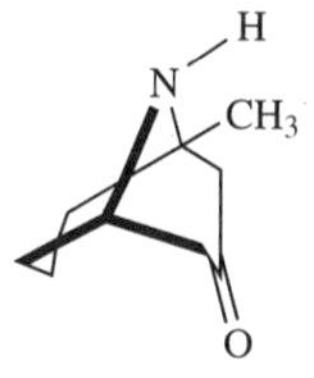   (6) 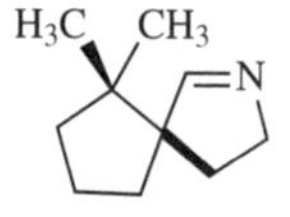   (7) 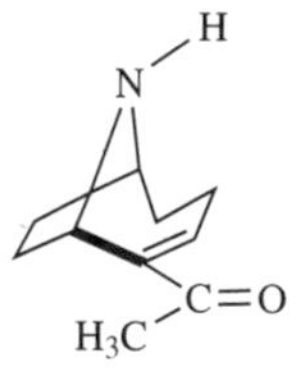   (8)

There are many other examples of insect use of alkaloids, such as the homotropane alkaloid euphococcinine [15486-23-4] (6), $C_9H_{15}NO$, which has been noted as a defensive alkaloid in the blood of the Mexican bean beetle (*Epilachna varivestis*) (14) and the azaspiroalkene polyzonimine [55811-47-7] (7), $C_{10}H_{17}N$, an insect repellent produced by the milliped *Polyzonium rosalbum* (15).

The "very fast death factor" (VFDF), anatoxin-a [64285-06-9] (8), $C_{10}H_{15}NO$, a fish poison, has been isolated from a toxic strain of microalgae *Anabaena flos-aquae* (16). For (6), (7), and (8), little is yet known about the formation (or genesis) of the alkaloid material.

The period prior to ~1800 includes the history of the crude exudate from unripe poppy pods, which, it is now known, contains narcotine (1, noscopine), $C_{22}H_{23}NO_7$, and morphine (2, R = H) along with other closely related materials. Also during this time natives of the Upper Amazon basin were making use of crude alkaloid-containing preparations as arrow poisons. To help their hunting, some tribes developed the red resinous mixture called tubocurare, containing, among others, the alkaloid tubocurarine [57-95-4] (9), $C_{37}H_{41}N_2O_6$, obtained primarily from plants of the *Chondrodendron*; others developed Calabash curare, containing, among others, the alkaloid C-toxiferine [6696-58-8] (10), $C_{40}H_{46}N_4O_2 \cdot 2Cl$, from plants belonging to *Strychnos* spp.

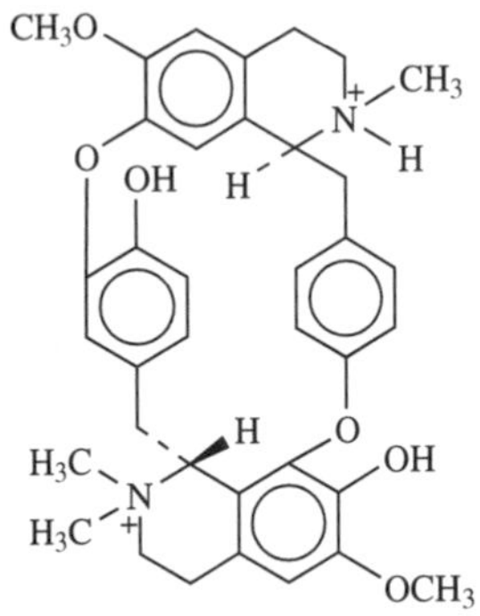

(9) 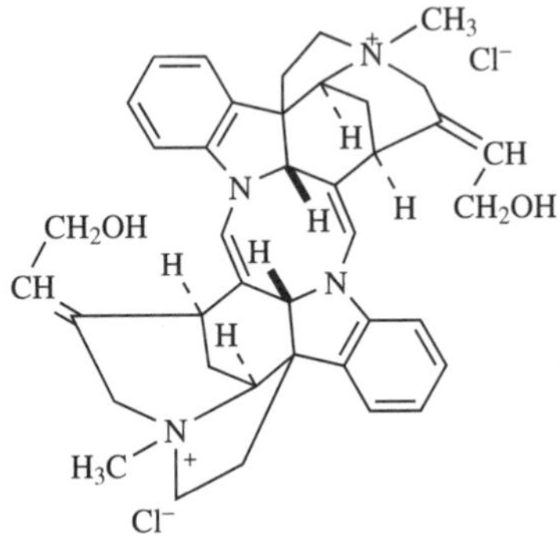   (10)

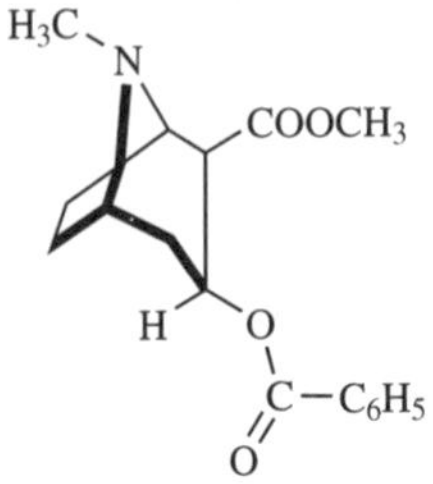

(11) 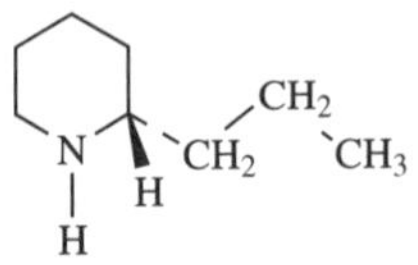   (12)

The natives of Peru were learning to ease their physical pains by chewing the leaves of coca shrub (*Erythroxylon truxillence*, Rusby), which contain, among others, the alkaloid cocaine [50-36-2] (11), and European citizens were recognizing other poisons such as coniine [458-88-8] (12), from the poison hemlock (*Conium maculatum* L.).

With the introduction of improved analytical techniques, starting ~1817, the evaluation of drugs began and, over a span of ~10 years, strychnine [57-24-9] (13, R = H), emetine [283-18-1] (14), brucine [357-57-3] (13, R = OCH₃), piperine [94-62-2] (15), caffeine [58-08-2] (16), quinine [130-95-0] (17, R = OCH₃), colchicine (4), cinchonidine [118-10-5] (17, R = H), and coniine (12) were isolated (17). But, because the science was young and the materials complex, it was not until 1870 that the structure of the relatively simple base coniine (12) was established (18) and not until 1886 that the racemic material was synthesized (19). The correct structure for strychnine (13, R = H) was not confirmed by X-ray crystallography until 1956 (20) and the synthesis was completed in 1963 (21).

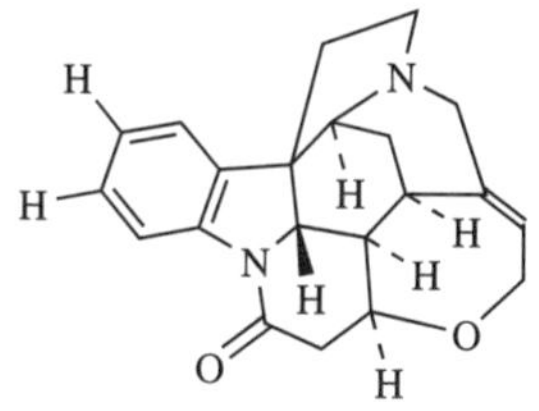

(13) 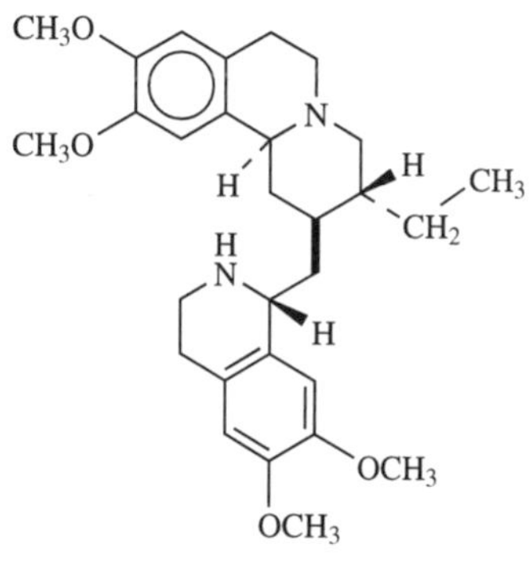   (14)

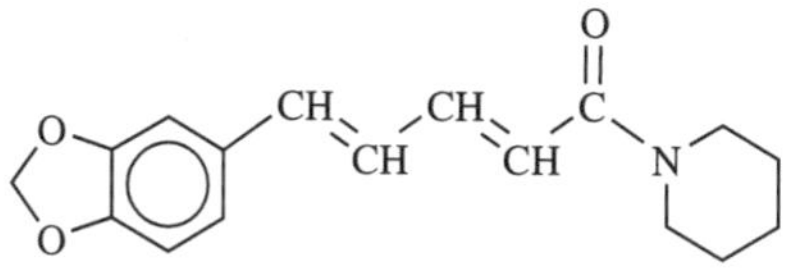

(15) 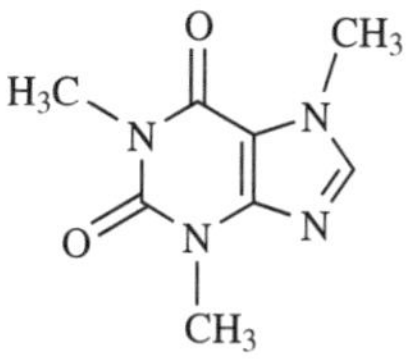   (16) 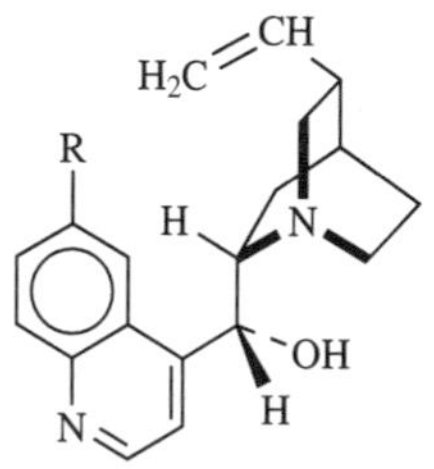   (17)

3. Occurrence, Detection, and Isolation

Given the massive volume of material available, the following discussion is necessarily incomplete and the interested reader is directed to the materials in (7) and (8), in particular, for more detailed information.

The most recent compendium (7) of alkaloids indicates that most alkaloids so far detected occur in flowering plants and it is probably true that the highest concentrations of alkaloids are to be found there. However, as detection methods improve it is almost certain that some concentration of alkaloids will be found almost everywhere. In the higher plant orders, somewhat more than one-half contain alkaloids in easily detected concentrations. Major alkaloid bearing orders are *Campanulales, Centrospermae, Gentianales, Geraniales, Liliflorae, Ranales, Rhoedales, Rosales, Rubiales, Sapindales,* and *Tubiflorae,* and within these orders most alkaloids have been isolated from the families *Amaryllidaceae, Apocynaceae, Euphorbiaceae, Lauraceae, Leguminoseae, Liliaceae, Loganiaceae, Menispermaceae, Papveraceae, Ranuculaceae, Rubiaceae, Rutaceae,* and *Solanaceae.* Alkaloids have also been found in butterflies, beetles, millipedes, and algae and are known to be present in fungi, eg, agroclavine [548-42-5] (18) from the fungus *Claviceps purpurea,* which grows as a parasite on rye and has been implicated, with its congeners, in causing convulsive ergotism (22). They are found in toads (*Bufo vulgaris,* Laur.), eg, bufotenine [487-93-4] (19), an established hallucinogen in humans (23); in frogs (*Epipedobates tricolor*) eg, epibatidine [140111-52-0] (20), and in the musk deer [family *Moschidae* and three species *Moschus moschiferus, M.berezovskii, and M. chrysogaster.*], muscopyridine [501-08-6] (21), $C_{16}H_{25}N$. Even in humans morphine (2, R = H) is a naturally occurring component of cerebrospinal fluid (24).

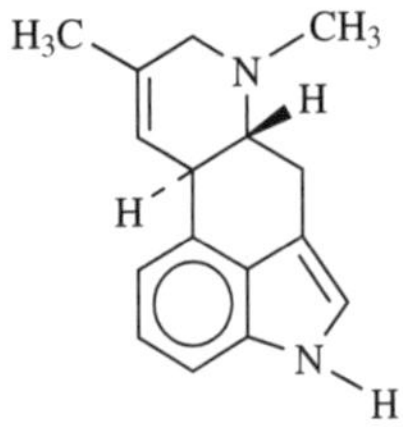

(18) 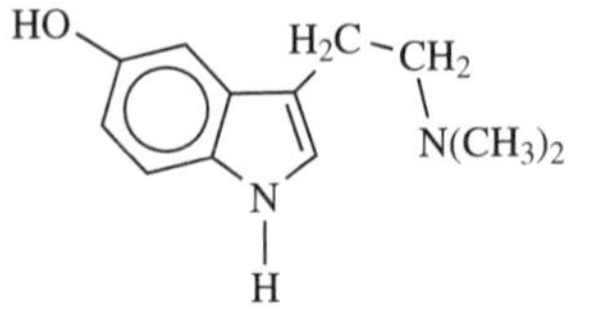   (19) 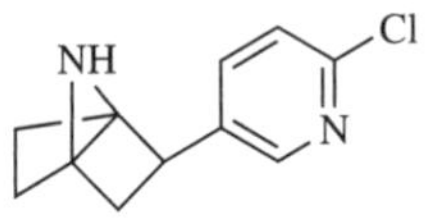   (20)

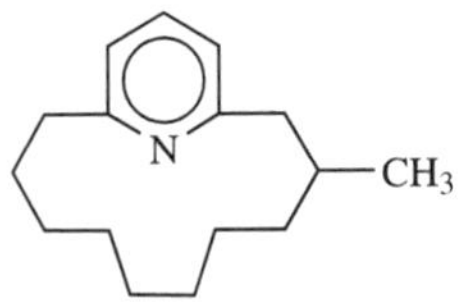

(21) 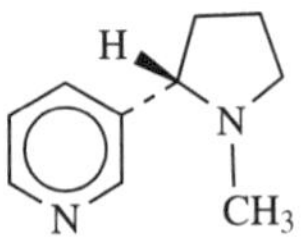   (22)

The concentration of alkaloids, as well as the specific area of occurrence or localization within the plant or animal, can vary enormously. Thus the amount of nicotine [54-11-5] (22), $C_{10}H_{14}N_2$, apparently synthesized in the roots of various species of *Nicotiana* and subsequently translocated to the leaves varies with soil conditions, moisture, extent of cultivation, season of harvest, as well as other factors that may not yet have been evaluated and may be as high as 8% of the dry leaf, whereas the amount of morphine (2, R = H) in cerebrospinal fluid is of the order of 2–339 fmol/mL (24).

Initially, the search for alkaloids in plant material depended largely on reports of specific plant use for definite purposes or observations of the effect specific plants have on indigenous animals among native populations. Historically, tests on plant material have relied on metal-containing reagents such as that of Dragendorff (25), which contains bismuth salts, or Mayer (26), which contains mercury salts. These metal cations readily complex with amines and the halide ions present in their prepared solutions, yielding brightly colored products. Despite false positive and negative responses (27), field testing continues to make use of these solutions. However, it is now clear that newer methods, such as kinetic energy mass spectrometry (MIKE) on whole plant material (28), have the potential to replace these spot tests.

After detection of a presumed alkaloid, large quantities of the specific plant material are collected, dried, and defatted by petroleum ether extraction if seed or leaf is investigated. This process usually leaves polar alkaloidal material behind but removes neutrals. The residue, in aqueous alcohol, is extracted with dilute acid and filtered, and the acidic solution is made basic. Crystallization can occasionally be effected by adjustment of the pH. If such relatively simple purification fails, crude mixtures may be used or, more recently, very sophisticated separation techniques have been employed. Once alkaloidal material has been found, taxonomically related plant material is also examined.

Until separation techniques such as chromatography (29,30) and counter-current extraction had advanced sufficiently to be of widespread use, the principal alkaloids were isolated from plant extracts and the minor constituents were either discarded or remained uninvestigated. With the advent of, first, column, then preparative thin layer, and now high pressure liquid chromatography (hplc), even very low concentrations of materials of physiological significance can be obtained in commercial quantities. The alkaloid leurocristine [57-22-7] (vincristine, **23**, R = CHO), one of the >90 alkaloids found in *Catharanthus roseus* G. Don, from which it is isolated and then used in chemotherapy, occurs in concentrations of ~2 mg/100 kg of plant material.

Most recently, with the advent of enzyme assay and genomic manipulation, the possiblity of utilization of callous or root tissue or even isolated enzymes along with genetic engineering techniques can be employed to enhance or modify production of specific alkaloids (31–36)

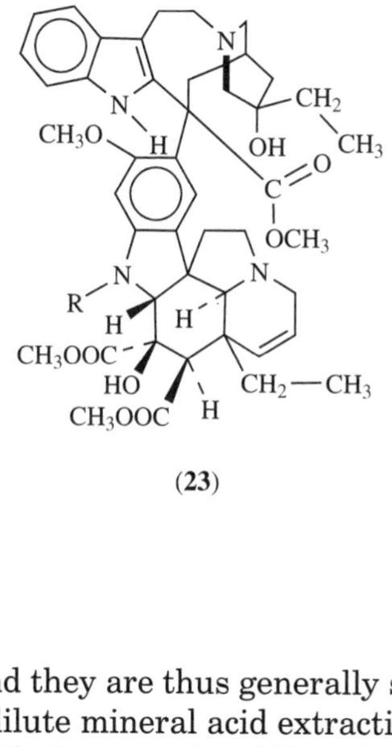

(**23**)

4. Properties

Most alkaloids are basic and they are thus generally separated from accompanying neutrals and acids by dilute mineral acid extraction. The physical properties of most alkaloids, once purified, are similar. Thus they tend to be colorless, crystalline, with definite melting points, and chiral; only one enantiomer is isolated. However, among >10,000 individual compounds, these descriptions are over generalizations and some alkaloids are not basic, some are liquid, some brightly colored, some achiral, and in a few cases both enantiomers have been isolated in equal amounts, ie, the material as derived from the plant is racemic (or racemization has occurred during isolation).

5. Organization

Early investigators grouped alkaloids according to the plant families in which they are found, the structural types based on their carbon framework, or their principal heterocyclic nuclei. However, as it became clear that the alkaloids, as secondary metabolites (37–40), were derived from compounds of primary metabolism (eg, amino acids or carbohydrates), biogenetic hypotheses evolved to link the more elaborate skeletons of alkaloids with their simpler proposed progenitors (41). These hypotheses continue to serve as valuable organizational tools (7,42,43) and in many cases, enzyme catalyzed processes affirming them have been found (36).

The building blocks of primary metabolism, from which biosynthetic studies have shown the large majority of alkaloids to be built, are few and include the common amino acids ornithine (24), lysine (25), phenylalanine (26, R = H), tyrosine (26, R = OH, and tryptophan (27). Others are nicotinic acid (28), anthranilic acid (29), and histidine (30), and the nonnitrogenous acetate-derived fragment mevalonic acid (31). Mevalonic acid (31) is the progenitor of isopentenyl pyrophosphate (32) and its isomer 3,3-dimethylallyl pyrophosphate (33), later referred to as the C_5 fragment. A dimeric C_5 fragment (the C_{10} fragment), ie, geranyl pyrophosphate (34), gives rise to the iridoid loganin [18524-94-2] (35), and the trimer farnesyl pyrophosphate (36). The C_{15} fragment is also considered the precursor to the C_{30} steroid, ie, $2 \times C_{15} = C_{30}$.

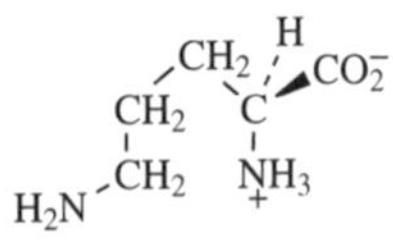

(24) 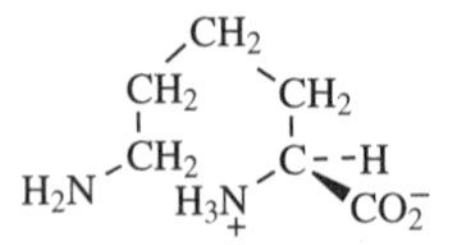   (25) 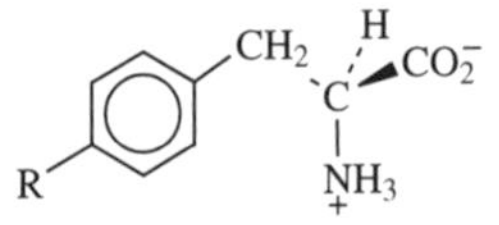   (26)

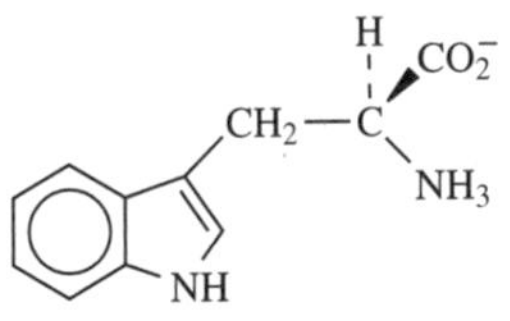

(27) 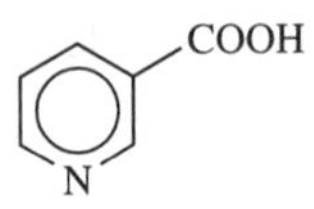   (28) 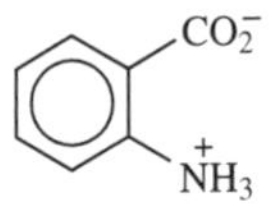   (29)

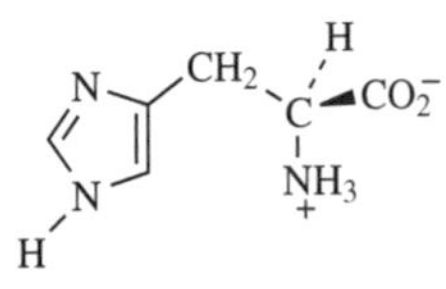

(30) 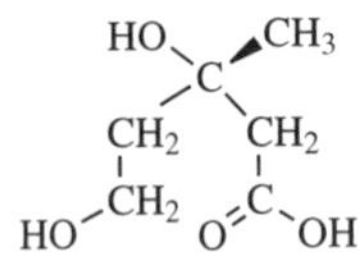   (31) 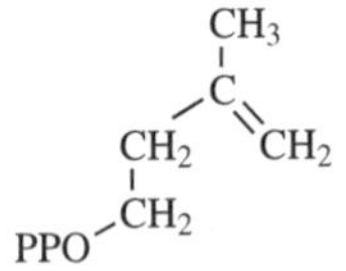   (32) 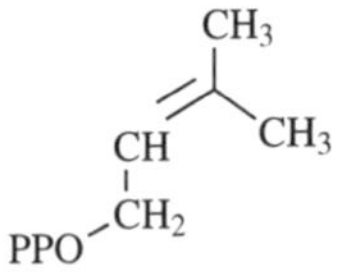   (33)

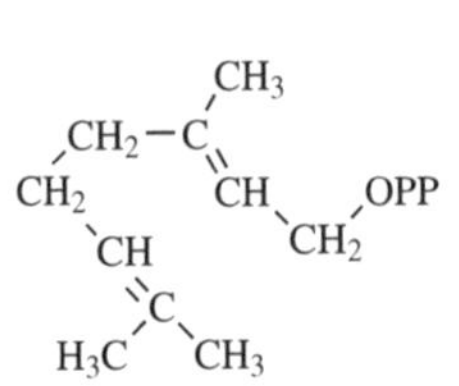

(34) 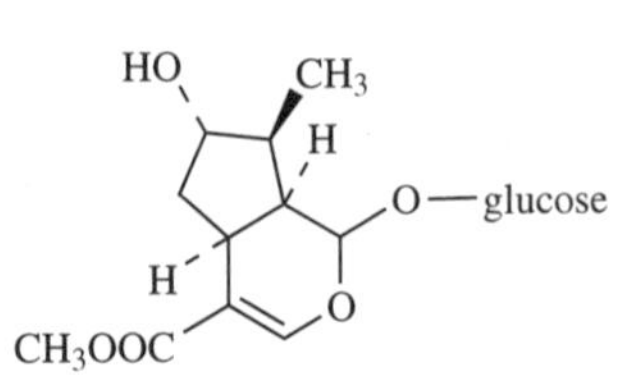   (35)

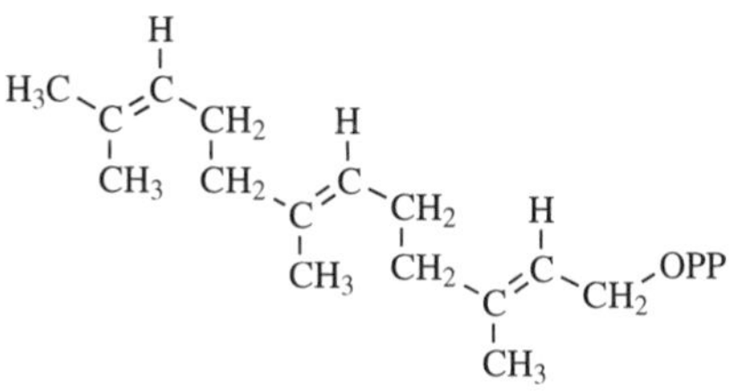

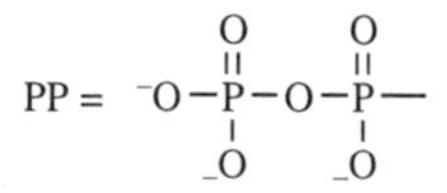

(36)

5.1. Ornithine-Derived Alkaloids (44).

Ornithine (24) undergoes biological reductive decarboxylcation to generate either putrescine [110-60-1] (37), $C_4H_{12}N_2$, or its biological equivalent, and subsequent oxidation and cyclization gives rise to the pyrroline [5724-81-2], (38), C_4H_7N.

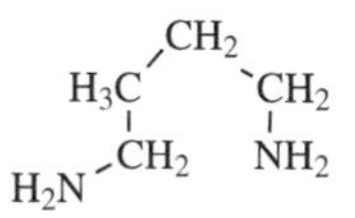

(37) 　　　　　　　(38) 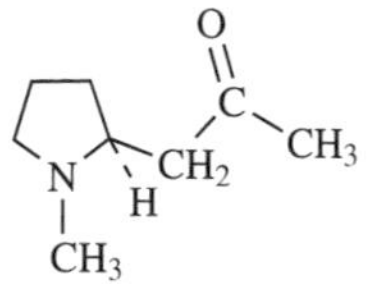　　　　　　　(39)

The details have been confirmed by suitable labeling (^{14}C and ^{15}N) and it is fairly certain that either (38) or something very similar to it is available to react with either acetoacetic acid or its biological equivalent to generate the alkaloid hygrine (39). Hygrine is an oily, distillable base found, along with cocaine (11), in the leaves of the Peruvian coca shrub (Erythroxylon truxillence Rusby).

If, instead of an acetoacetate equivalent, a malonyl derivative such as (40) were involved (45), appropriate condensation reactions would lead to the tropane [280-05-7] skeleton (41), $C_7H_{13}N$.

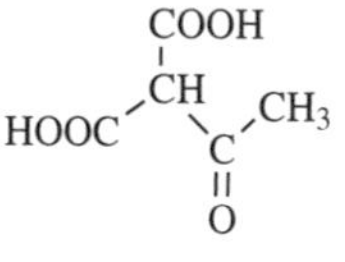

(40) 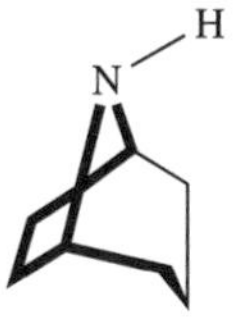　　　　　　(41) 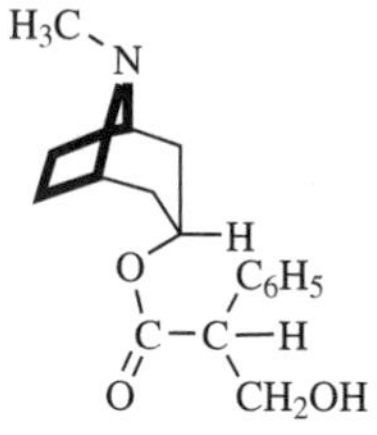　　　　　　(42) 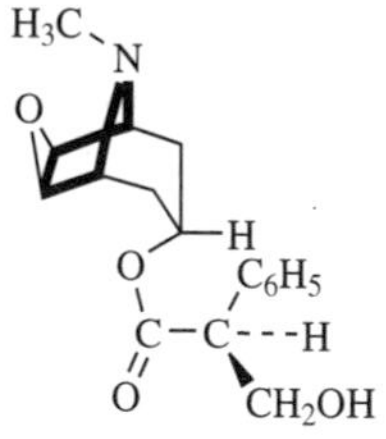　　　　　　(43)

The physiologically and commercially important alkaloids of this group of compounds, occurring widely in the Solanaceae and Convolulaceae as well as the Erythroxylaceae, include not only cocaine (11) but also atropine (42) and scopolamine (43) (33).

Atropine (42), isolated from the deadly nightshade (Atropa belladonna L.) is the racemic form, as isolated, of (−)-hyoscyamine [which is not isolated, of course, from the same plant but is typically found in Solanaceous plants such as henbane (Hyoscyamus niger L.)]. Atropine (42) is used to dilate the pupil of the eye in ocular inflammations and is available both as a parasympatholytic agent for relaxation of the intestinal tract and to suppress secretions of the salivary, gastric, and respiratory tracts. In conjunction with other agents, it is used as part of an antidote mixture for organophosphorus poisons (see CHEMICALS IN WAR).

Scopolamine (**43**), an optically active, viscous liquid, also isolated from *Solanaceae*, eg, *Datura metel* L., decomposes on standing and is thus usually both used and stored as its hydrobromide salt. The salt is employed as a sedative or, less commonly, as a prophylactic for motion sickness. It also has some history of use in conjunction with narcotics as it appears to enhance their analgesic effects. Biogenetically, scopolamine is clearly an oxidation product of atropine, or, more precisely, because it is optically active, of (−)-hyoscyamine.

Cocaine (**11**) had apparently been used by the natives of Peru prior to the European exploration of South America. Stories provided by early explorers suggested that the leaves of, for example, *Erythroxylon coca* Lam. were chewed without apparent addiction by the indigenous peoples and with only mild numbing of the lips and tongue in return for increased endurance. Indeed, the recognition of the anesthetic properties possessed by the leaves and the (unwarranted) assumption that addiction was avoidable led to creation of plantations in Bolivia, Brazil, and Java to ensure a continued supply of this valuable material for medicinal purposes. Although it appears that native populations continue the practice of leaf chewing, the purified base obtained by simple extraction of the leaves has become a substance of abuse in the more civilized world. It is now recognized that the alkaloid (**11**) itself is too toxic to be used as an anesthetic by injection.

Condensation of a pyrroline system (**38**) with a second equivalent of ornithine-derived precursor is presumably an alternative to condensation with acetoacetate- or malonate-derived fragments. Indeed, early feeding experiments with, eg, *Senecio istideus* showed that two equivalents of ornithine (**24**) could be accounted for in the structure of the necine, ie, the 1-azabicyclo[3.3.0]heptane or pyrrolizidine portion of the alkaloids, eg, heliotridine (**44**), containing that ring system (**46**). Generally, the pyrrolizidine alkaloids are found esterified with low molecular weight carboxylic acids [or dicarboxylic acids, as in senecionine, (**5**)] at either or both of the hydroxyl groups of the necine. The acids themselves, called necic acids, are generally not found elsewhere in alkaloids, and, although for some time they were believed to arise from acetate or mevalonate, it is now clear that, at least for the few that have been carefully examined, they are themselves derived from simple amino acids.

In addition to the alkaloids in *Senecio* spp. (including asters and ragworts) commented on earlier, the adaptive use of which by butterflies was noted, members of this widely spread group of compounds are found in different genera (*Heliotropium*, *Trachelanthus*, and *Trichodesma*) within cosmopolitan families (eg, *Boraginaceae* and *Leguminoseae*). Most of these alkaloids are toxic, affecting the liver (an organ lacking in moths, butterflies, etc) and their ingestion is manifested in animals with the onset of symptoms associated with names such as "horse staggers" or "walking disease". The cell biology and metabolic engineering of some alkaloids discussed here have recently been reviewed (47,48).

(**44**)

5.2. Lysine-Derived Alkaloids. Just as putrescine (**37**) derived from ornithine (**24**) is considered the progenitor of the nucleus found in pyrrolidine-containing alkaloids, so cadaverine [462-94-2] (**45**), $C_5H_{14}N_2$, derived from lysine (**25**) is the idealized progenitor of the 1-dehydropiperidine [28299-36-7] nucleus (**46**), C_5H_9N, found in the pomegranate, *Sedum, Lobelia, Lupin,* and *Lycopodium* alkaloids (**49**).

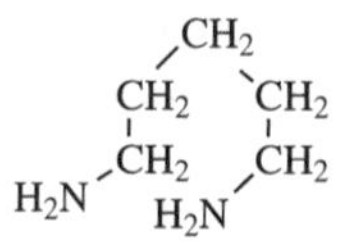

(**45**) 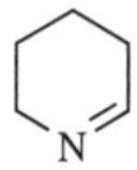   (**46**)

As was the case for the alkaloids derived from ornithine (**34**), if either (**46**), its biological equivalent, or something closely resembling it reacts with acetoacetate or its biological equivalent, the pomegranate alkaloid pelletierine (**47**) can arise; note the resemblence to hygrine (**39**). Simple reduction of the carbonyl, with stereospecificity common to enzyme-mediated reactions, can be accommodated and the Sedum alkaloid sedridine (**48**) results; cyclization and N-methylation produce pseudopelletierine (**49**). There are somewhat more than 600 annual, biennial, or perennial succulents belonging to the *Sedum* genus of the family *Crassulaceae*, many of which are characterized by the ability to grow where little else can.

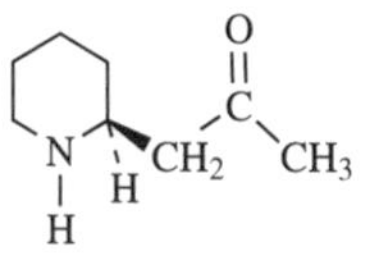

(**47**) 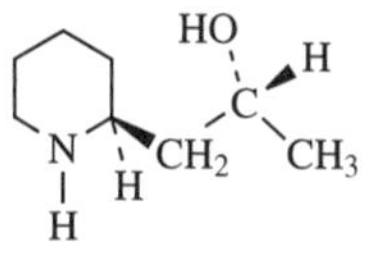   (**48**) 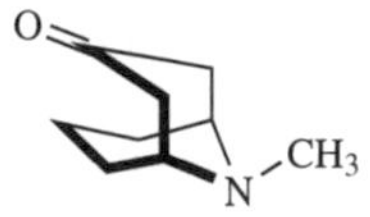   (**49**)

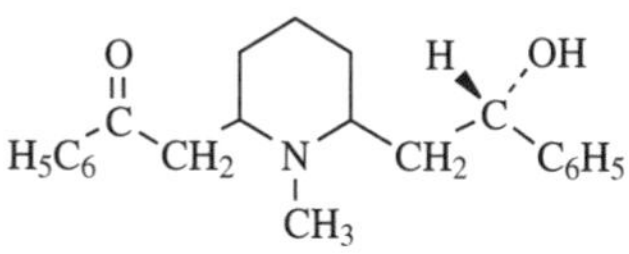

(**50**)    (**51**) 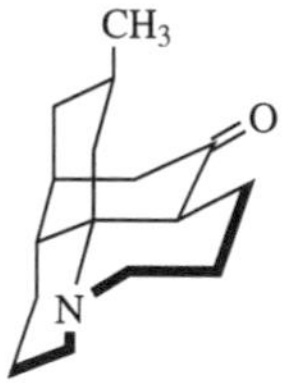   (**52**)

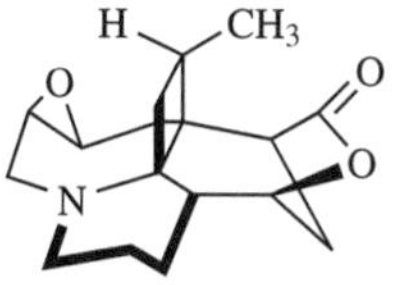

(**53**)

The pomegranate alkaloids, pelletierine (**47**) and pseudopelletierine (**49**) as well as minor accompanying bases, have a long history as salts of tannic acid as an anthelmintic mixture for intestinal pinworms. The alkaloids themselves (as the tannates) are obtained from pomegranate tree (*Punica granatum* L.)

root bark and are among the few bases named after an individual (P. J. Pelletier) rather than a plant.

Isolates from Indian tobacco (*Lobelia inflata* L.), as a crude mixture of bases, have been recognized as expectorants. The same (or similar) fractions were also used both in the treatment of asthma and as emetics. The principal alkaloid in L. inflata is lobeline (**50**), an optically active tertiary amine which, unusual among alkaloids, is reported to readily undergo mutarotation, a process normally associated with sugars. Interestingly, it appears that the aryl-bearing side chains in (**50**) are derived from phenylalanine (**26**, R = H) (**50**).

Feeding experiments utilizing ^{13}C-, ^{15}N-, and ^{2}H-labeled cadaverine (**45**) and lysine (**25**) in *Lupinus augustifolius*, a source of the lupine alkaloids (−)-sparteine (**51**, R = H,H) and (+)-lupanine (**51**, R = O), have been reported that lend dramatic credence to the entire biosynthetic sequence for these and the related compounds discussed above (**51**). That is, the derivation of these bases is in concert with the expected cyclization from the favored all-trans stereo-isomer of the trimer expected on self-condensation of the 1-dehydropiperidine (**46**).

The spores of *Lycopodium clavatum* L. (a club moss), sometimes called vegetable sulfur, have been used medicinally as an absorbent dusting powder; other uses as diverse as additives to gunpowder and suppository coatings have also been recorded. Although for some years the alkaloids common to a number of *Lycopodium* spp., lycopodine (**52**) and annotinine (**53**), were thought to have arisen from suitably folded polyketide chains, it is now accepted that two pelle-tierine (**47**) or pelletierine-like fragments would suffice. The details of feeding experiments with pelletierine and its precursors appear to indicate, however, that only one pelletierine and, separately, a second acetoacetate and a second 1-dehydropiperidine (**46**), which could otherwise be combined to a second pelle-tierine, are used to generate both of these alkaloids.

5.3. Tobacco Alkaloids. The relatively small number of alkaloids derived from nicotinic acid (**28**) (the tobacco alkaloids) are obtained from plants of significant commercial value and have been extensively studied. They are distinguished from the bases derived from ornithine (**24**) and, in particular, lysine (**25**), since the six-membered aromatic substituted pyridine nucleus common to these bases apparently is not derived from (**25**). Current work with isolated enzymes and plant genomic material (32,48,52) confirms and extends earlier work with less pure fractions (53) that led to the early hypotheses.

These alkaloids include the substituted pyridone ricinine [524-40-3] (**54**), $C_8H_8N_2O_2$, which is easily isolated in high yield as the only alkaloid from the castor bean (*Ricinus communis* L.). The castor bean is also the source of castor oil (qv), which is obtained by pressing the castor bean and, rich in fatty acids, has served as a gentle cathartic.

(**54**)   (**55**)   (**56**)   (**57**)

The highly toxic alkaloid S-$(-)$-nicotine (22) and related tobacco bases including such materials as $(-)$anabasine [494-52-0] (55), $C_{10}H_{14}N_2$, are obtained from commercially grown tobacco plants (eg, *Nicotiana tabacum* L.). Various tobaccos have differing amounts of these and other bases, as well as different flavoring constituents, some of which are apparently habituating to some individuals. Currently, the assay of the $(-)$-nicotine (22) content of tobacco, the annual world production of which is in excess of 7 million tons (see below), is desirable and in some countries mandatory, although the toxicity of the unassayed plant bases may be as high as or higher than that of $(-)$-nicotine (22). Interestingly, there appears to be some evidence that cultivation of tobacco increases the alkaloid content, from which it can be argued that increased alkaloid content has insured survival of a particular cultivar.

The pyrrolidine ring of nicotine is derived from ornithine (24), whereas the piperidine ring of anabasine (55) is derived from lysine (25) (54). Also, the carboxylic acid functionality of nicotinic acid (28) is lost (along with the C-6 proton) during the biosynthesis in the roots of the tobacco plants from which the bases are subsequently translocated to the leaves. Curiously, whereas nornicotine [494-97-3] (56), $C_9H_{12}N_2$, frequently accompanies nicotine, the former is apparently derived by demethylation of the latter rather than the latter undergoing methylation to the former. This is in contrast to what usually seems to occur; ie, methylation at nitrogen and oxygen is usually a late-stage process in alkaloid biosynthesis.

Finally, millions of people in the Far East are apparently addicted to chewing ground betel nut, the fruit of the palm tree Areca catechu L., which they mix with lime and wrap in betel leaf (Piper betle L.) for consumption. They are said to experience a feeling of well-being. Among the alkaloids found in betel nut is arecoline [63-75-2] (57), $C_8H_{13}NO_2$, an optically inactive, steam-volatile base that is used commercially as a vermifuge in dogs and is also a potent muscarinic agent. It is reasonable to assume (evidence lacking) that arecoline may be derived from nicotinic acid (28) by a (rare) reductive mechanism.

5.4. Phenylalanine- and Tyrosine-Derived Alkaloids. Carbohydrate metabolism leads via a seven-carbon sugar, ie, a heptulose, derivative to shikimic acid [138-59-0] (58), $C_7H_{10}O_5$, which leads in turn to prephenic acid [126-49-8] (59), $C_{10}H_{10}O_6$ (55).

(58)   (59)

This is the branch-point differentiating phenylalanine (26, R = H) from tyrosine (26, R = OH). Both phenylalanine and tyrosine contain an aryl ring, a three-carbon side chain (a C_6–C_3 fragment), and a nitrogen. Decarboxylation yields a two-carbon side chain (a C_6–C_2 fragment), eg, 2-phenethylamine (60, R = H) from phenylalanine and tyramine (60, R = OH) from tyrosine, although it is not certain that in all cases decarboxylation must precede use in alkaloid construction.

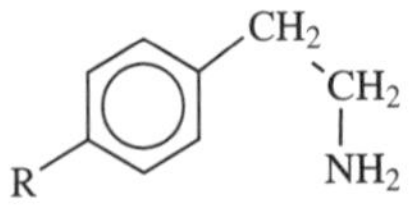

(60) 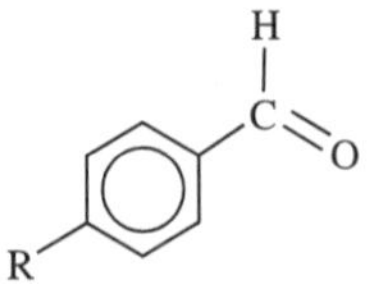   (61)

After the branching point at prephenic acid (**59**), phenylalanine (**26**, R = H) and tyrosine (**26**, R = OH), as well as the amines (**60**), are not interconvertible. Finally, deamination and oxidative cleavage of the presumed (and in some circumstances isolated) resulting alkenes yields the equivalent of benzaldehyde (**61**, R = H), C_7H_6O, and *p*-hydroxybenzaldehyde (**61**, R = OH), ie, aromatics with one aliphatic carbon attached C_6–C_1 fragments).

All of these pieces are used, in conjunction with some of the earlier fragments discussed, as building blocks for alkaloids containing an aromatic ring. In the cases discussed here, a link to either phenylalanine or tyrosine or, in some cases with two aromatic rings to both, has been established by suitable feeding experiments on growing plants.

There is a relatively large number of alkaloids that may be considered as simple phenethylamine [64-04-0] (**60**, R = H), $C_8H_{11}N$, or tyramine [51-67-2] (**60**, R = OH), $C_8H_{11}NO$, derivatives. These include mescaline (**62**) from the small woolly peyotyl cactus *Lophophora williamsii* (Lemaire) Coult., anhalamine (**63**) and lophocerine (**64**) from other *Cactaceae*, and the important antamebic alkaloids (−)-protoemetine (**65**), (−)-ipecoside (**66**), and (−)-emetine (**67**) from the South American straggling bush *Cephaelis ipecacuanha* (Brotero) Rich. All of these bases are derived from tyrosine (**26**, R = OH) and not from phenylalanine (**26**, R = H).

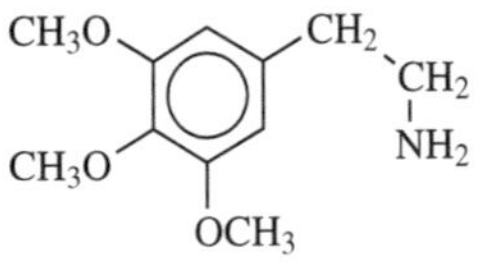

(62) 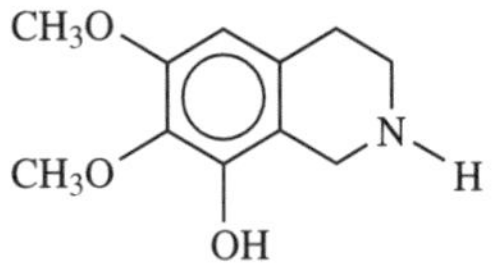   (63) 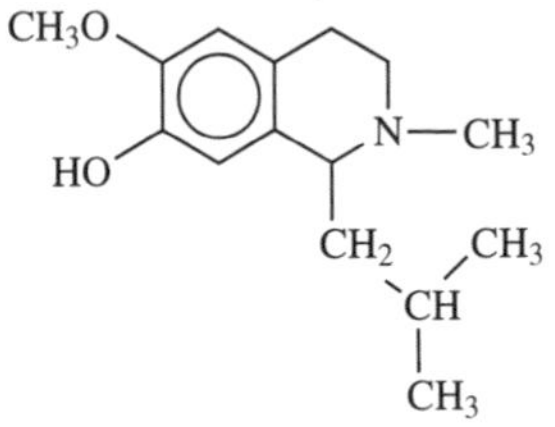   (64)

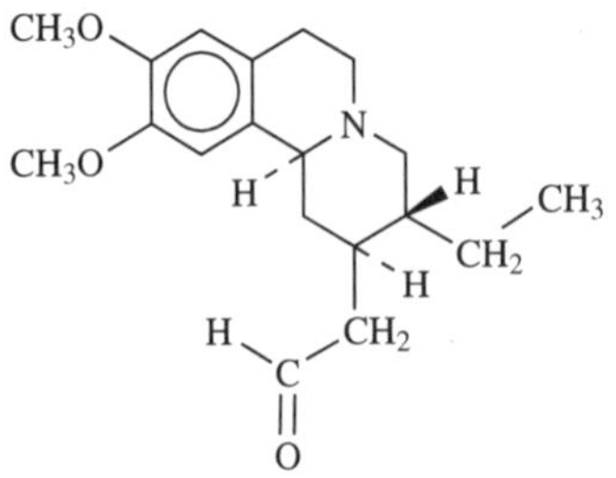

(65) 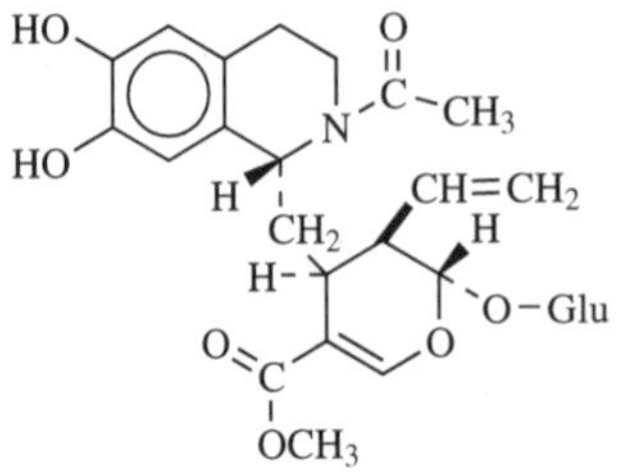   (66)

(67)

Crude preparations of mescaline (62) from peyote were first reported by the Spanish as they learned of its use from the natives of Mexico during the Spanish invasion of that country in the sixteenth century. The colorful history (56) of mescaline has drawn attention to its use as a hallucinogen and even today it is in use among natives of North and South America. Although in connection with drug abuse complaints, mescaline is considered dangerous, it has been reported (57) that it is not a narcotic nor is it habituating. It was also suggested that its sacramental use in the Native American Church of the United States be permitted since it appears to provoke only visual hallucination while the subject retains clear consciousness and awareness.

Both of the alkaloids anhalamine (63) from *Lophophora williamsii* and lophocerine (64) from *Lophocereus schotti* were isolated (after the properties of purified mescaline had been noted) in the search for materials of similar behavior. Interestingly, lophocerine (64), isolated as its methyl ether, after diazomethane treatment of the alkali-soluble fraction of total plant extract, is racemic. It is not known if the alkaloid in the plant is also racemic or if the isolation procedure causes racemization.

The iridoid loganin (35), $C_{17}H_{26}O_{10}$, has been shown to serve as a C_{10} progenitor and, here, C_{10} fragments are apparent in the alkaloids (−)-protoemetine (65), (−)-ipecoside (66), and (−)-emetine (67). It has been shown that loganin is specifically incorporated into each of these bases in *Cephaelis ipecacuanha* (Brot. A. Rich) and that they are apparently formed sequentially, that is, formation of (−)-ipecoside (66) precedes that of (−)-protoemetine (65) and (−)-emetine (67). The crude dried rhizome and roots from *C. ipecacuanha* which is sometimes known as Rio or Brazilian ipecac, contains all three, as well as other related bases, and has a long history based on native Indian reports of use as an emetic. Purification of the crude extract yields the individual bases, and, because it is relatively more stable and is also present in reasonable quantity, led to the use of emetine (67) as its hydrochloride salt in place of the crude plant extract. This use of the pure base rather than crude plant extract has allowed greater certainty in dosage, which is important because, although emetine is quite effective in combating acute amebic hepatitis and is claimed to have some effect against the present scourge of African schistosomiasis, its administration may be accompanied by a rapid drop in blood pressure, irregular heart function, and paralysis of skeletal muscle. The danger of inappropriately large doses, certain to cure the ailment but with the possible death of the patient, is clearly greater with crude extract than with purified alkaloid. Long term, even appropriate dosage may be accompanied by dermatitis, diarrhea, and nausea.

There are only two groups of alkaloids that appear to be derived from tyrosine (**26**, R = OH) utilized as a C_6–C_2 fragment and a C_6–C_1 unit that comes from phenylalanine (**26**, R = H). The first is that small group found only in the *Orchidaceae*, exemplified by cryptostyline I [22324-79-4] (**68**, from *Cryptostylis fulva* Schltr.), $C_{19}H_{21}NO_4$.

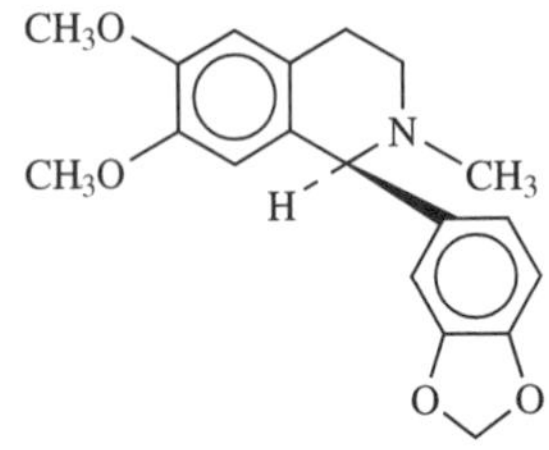

(**68**)

The second, a very large group of compounds, is the alkaloids of the *Amaryllidaceae*. This cosmopolitan family of related compounds includes >100 isolated and characterized members of known structure. In every case examined the C_6–C_2 unit is derived from tyrosine and the C_6–C_1 unit comes from phenylalanine, never from tyrosine. For this large number of compounds it is now believed (48) that a single progenitor derived from the original coupling of the C_6–C_2 unit and a C_6–C_1 unit, ie, norbelladine (**69**, R = H) accounts for all of the compounds isolated. This precursor (**69**, R = H) undergoes a variety of enzyme-catalyzed free-radical intramolecular cyclization reactions, followed by late-stage oxidations, eliminations, rearrangements, and O- and N-alkylations. Working from this generalization as an organizing principle, the majority of known Amaryllidaceae alkaloids can be divided into eight structural classes (59).

Alkaloids typical of the eight classes as shown below. The simple base ismine (**76**), isolated from, eg, *Sprekelia formosissima*, along with numerous other alkaloids, has long been considered a degradation product of other bases and is presumably generated in that way from a suitable member of the pyrrolo-[*de*]phenanthridine or [5,10b]ethanophenanthridine group.

Lycorine (**70**) was recognized as a potent emetic and a moderately toxic base from the time of its initial isolation from *Narcissus pseudonarcissus* L. (in ~1877) (60). Since that time its isolation from many other *Amaryllidaceae*, eg, *Lycoris radiate* Herb., has served to establish it as the most cosmopolitan alkaloid of the family. Typically, as much as 1% of the dry weight of daffodil bulbs may consist of lycorine (**70**), which has been reported to crystallize as colorless prisms directly from aqueous acid extract of crude plant material after basification. A high yield synthesis of the racemic base has been reported (61). Galanthamine (**72**) was originally isolated from the Caucasian snowdrop, *Galanthus woronowii* Vel., and as its hydrobromide salt has been proposed for use in regeneration of sciatic nerve. Galanthamine (**72**) is currently sold as a paliative in the treatment of Alzheimer's disease. In addition to demonstration of powerful cholinergic activity (62), it is reported to have analgesic activity comparable to morphine (2, R = H) (63).

Tazettine (74) has gained some small notoriety since, subsequent to its isolation from *Sprekelia formosissima* or *Narcisus tazetta* and proof that it was generated *in vivo* from haemanthamine [466-75-1] (77), $C_{17}H_{19}NO_4$, in accord with biosynthetic dogma (64), more careful work (65), in which the strongly basic conditions usually employed in alkaloid isolation were avoided, showed that it is an artifact of isolation. Further, it is readily generated from its precursors during the work-up of the plant material. Manthine (75) occurs, along with several homologues, in South African *Haemanthus* species. Manthine is of interest because it appears that, like tazettine (74), it can be easily generated *in vitro* from a derivative of haemanthamine (77) (66).

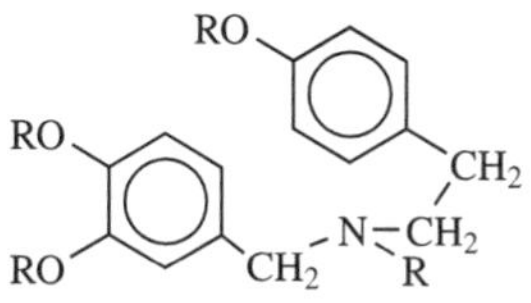

(69) 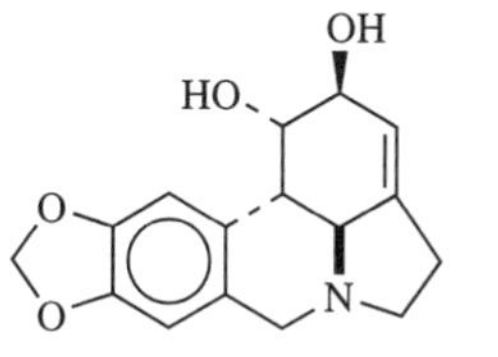 (70) 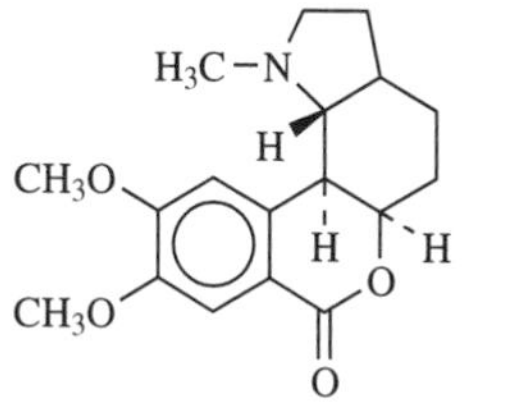 (71)

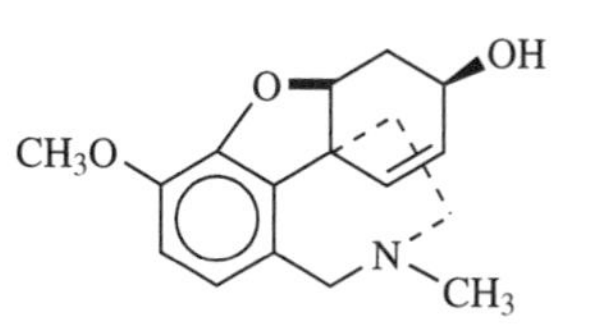

(72) 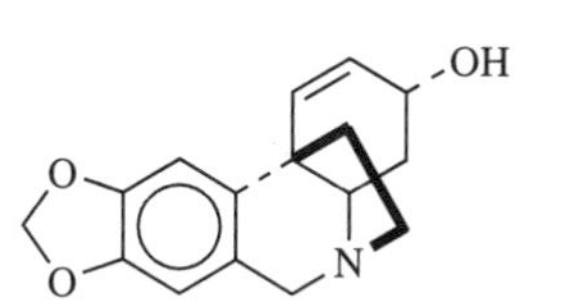 (73) 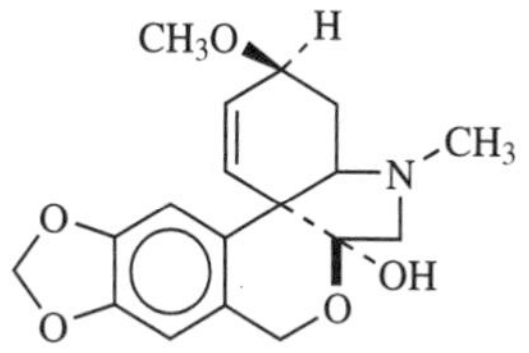 (74)

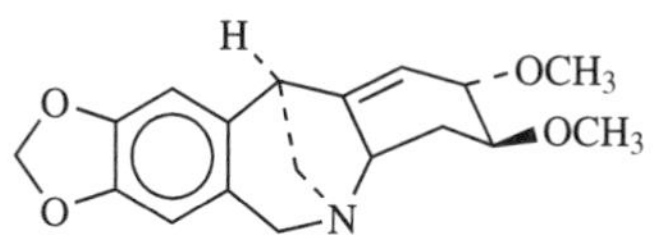

(75) 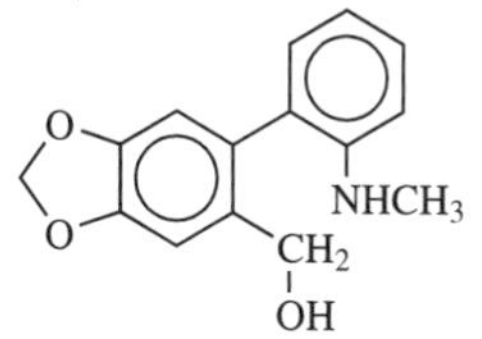 (76) 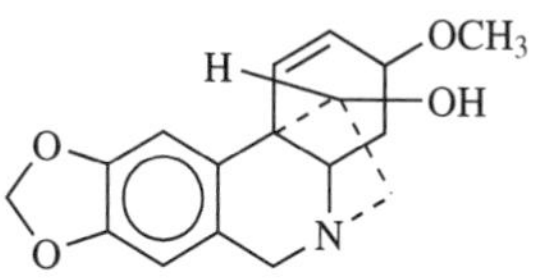 (77)

Just as norbelladine (69, R = H) can be considered as the precursor of C_6–C_2 + C_6–C_1 alkaloids, norlaudanosoline (78, R = H) seems to be the progenitor of the vast number of C_6–C_2 + C_6–C_2 alkaloids. Laudanosine (78, R = CH$_3$), isolated from *Papaver somniferum* L. [along with narcotine (1, noscopine), morphine (2, R = H), and numerous other alkaloids], has been shown to have tyrosine (26, R = OH) as a specific precursor. Labeling experiments with, eg, [2-^{14}C]tyrosine show that two equivalents of this amino acid are incorporated specifically but not to the same extent, implying that some partitioning has occurred prior to alkaloid formation. Papaverine (79), isolated in much greater quantity from *P. somniferum* L. than its tetrahydro-derivative laudanosine (**78**, R = CH$_3$), has a long history of use as an antispasmodic for smooth muscle. It is used as its hydrochloride salt, a more stable material than the free base. It is said to be nonhabit-forming, although it is classified as a narcotic by the U.S. Federal Narcotic Laws. Large doses may produce drowsiness, constipation, and increased excitability; if it is given orally, gastric distress may occur.

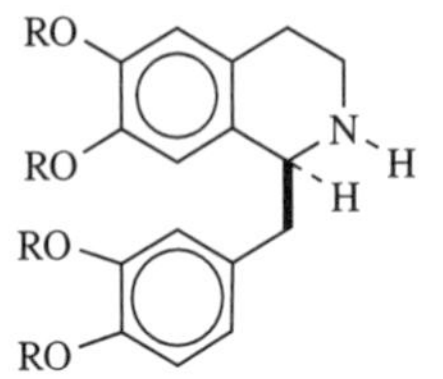

(78) 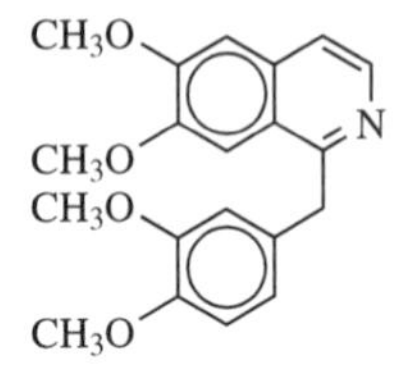   (79)

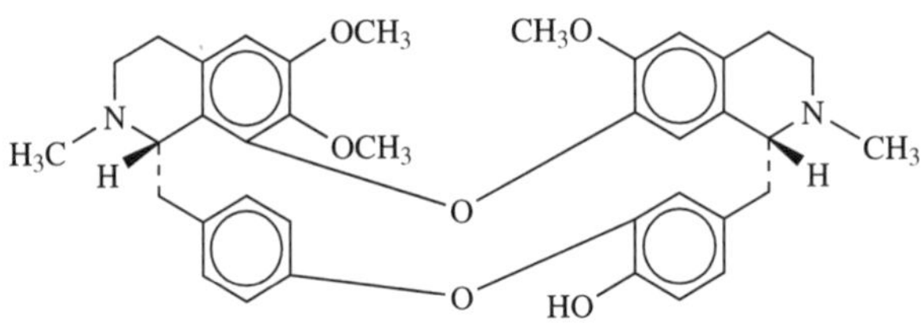

(80)

Phenolic intermolecular coupling (58) of two laudanosoline (**78**, R = H) fragments, which may be preceded or followed by partial O- or N-methylation, gives rise to the dimeric or bis(benzylisoquinoline) alkaloids such as oxyacanthine (**80**), obtained along with related materials from the roots of *Berberis vulgaris* L.

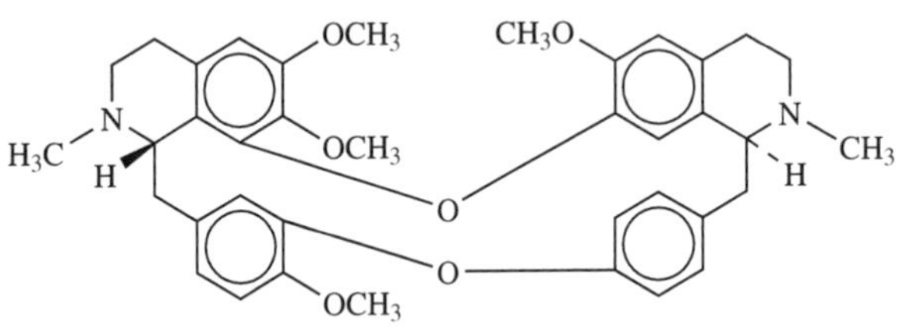

(81) 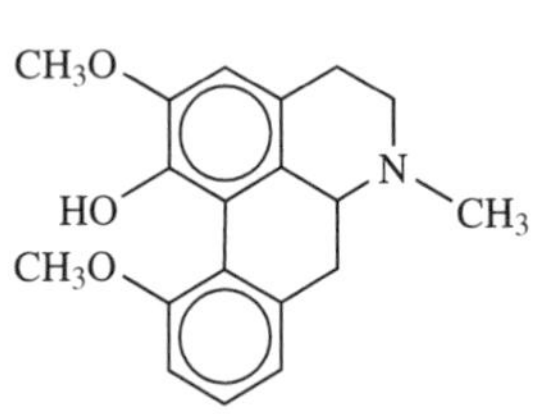   (82)

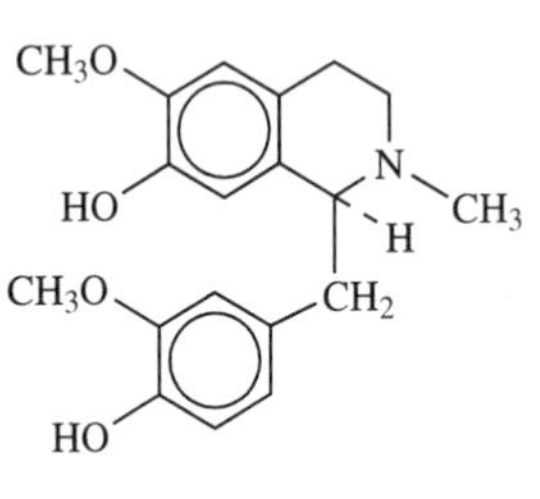

(83) 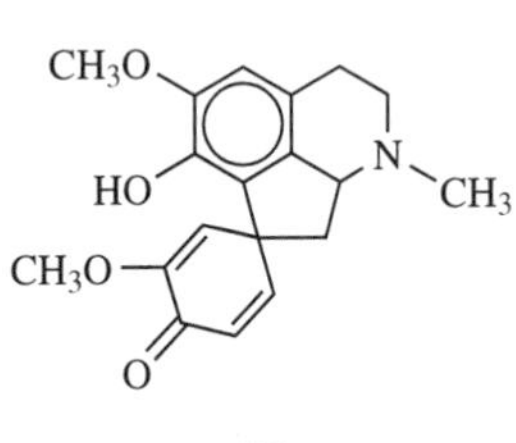   (84) 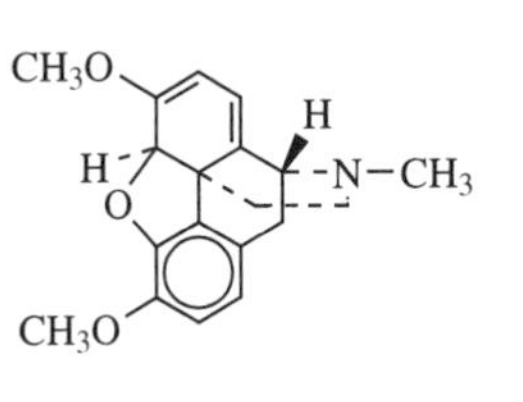   (85)

(86)   (87)

Many other bisbenzylisoquinoline alkaloids, such as tetrandrine (81), from *Cyclea peltata* Hook., are also known. Compound (81), although it causes hypotension and hepatotoxicity in mammals, possessed enough anticancer activity to be considered for preclinical evaluation (66). The arrow poison tubocurare prepared from *Chondrendendron* spp. also contains the bisbenzylisoquinoline alkaloid tubocurarine (9). In this vein it is noteworthy that specific enzymes required for the genesis of some bis(benzylisoquinoline) alkaloids have been isolated (48).

In an early attempt to understand the genesis of alkaloids from amino acids it was postulated (68) that intramolecular phenolic coupling should lead from benzylisoquinoline bases such as laudanosine (78, R = CH₃), before it was completely methylated, to aporphine bases such as isothebaine (82). For example, between a benzylisoquinoline derived from laudanosoline (78, R = H), such as orientaline (83), and an aporphine alkaloid such as isothebaine (82), there should be a proaporphine alkaloid such as orientalinone (84) (68). The isolation of 84 lent credence to the hypothesis. Indeed, the fragile nature of 84 (it readily undergoes the dienone–phenol rearrangement on acid treatment) required unusual skill in obtaining it from total plant extract.

Isothebaine (82), which may be derived from orientalinone (84) in the laboratory, is isolated from the roots of *Papaver orientale* after the period of active growth of the aerial parts and the production of thebaine (85) has ceased. The viscous milky exudate of the unripe seed pods of *P. orientale* as well as the opium poppy *Papaver somniferum* L. is opium. Opium cultivation appears to have spread from Asia Minor to China (via India) and it has been noted that the smoking of opium was common in China and elsewhere in the Far East when trade began in earnest as the eighteenth century closed. Active cultivation was encouraged by the revenues generated from addicts. Today, in the United States, although narcotine (1, noscopine) has some commercial value as a nonaddictive antitussive that occasionally leads to drowsiness, it is morphine (2, R = H), codeine (2, R = CH₃), and thebaine (85), the latter being converted to both of the former, which are of major commercial value.

The importance of morphine (2, R = H) as an analgesic, despite danger of addiction and side effects that include depression of the central nervous system, slowing of respiration, nausea, and constipation, cannot be underestimated, and significant efforts have been expended to improve isolation techniques from crude dried opium extract. Depending on its source, the morphine content of poppy straw or dried exudate may be as high as ~20%. Although the details of current manufacturing processes are closely held secrets, early work (69) has probably not been modified extensively. Usually, the crude opium is extracted with water and filtered, and the aqueous extract concentrated, mixed with ethanol, and made strongly basic with ammonium hydroxide. Morphine usually precipitates, while the other bases remain in solution, and is further purified by crystallization as its sulfate.

Codeine (**2**, R = CH₃) occurs in the opium poppy along with morphine (**2**, R = H) but usually in much lower concentration. Because it is less toxic than morphine (**2**, R = H) and, because its side effects (including depression, etc) are less marked, it has found widespread use in the treatment of minor pain. Much of the morphine (**2**, R = H) found in crude opium is converted to codeine (**2**, R = CH₃). The commercial conversion of morphine to codeine makes use of a variety of methylating agents, among which the most common are trimethylphenyl-ammonium salts. In excess of 200 tons of codeine are consumed annually from production facilities scattered around the world (see below).

The first synthesis of morphine, and therefore also codeine, was completed in 1956 (70). Although an additional 15 or so syntheses have been reported since then, isolation of morphine remains more important than any synthetic process. However, synthetic endeavors continue. These efforts produce new synthetic tools and continue the search for modified analogues that retain the analgesic properties of morphine but are nonaddicting.

Whereas the particular methylated derivative of laudanasoline (**78**, R = H) called (−)-reticuline (**86**) (71) gives rise to thebaine (**85**), codeine (**2**, R = CH₃), and morphine (**2**, R = H), a different derivative of **78** (R = H), ie, (+)-*N*-norproto-sinomenine (**87**), serves as the progenitor of erythraline (**88**), one of the bases found in *Erythrina crista galli* (72). The alkaloids found in all plant parts of *Erythrina* have been intensively studied because many of them produce smooth muscle paralysis, much like tubocurarine (**9**). A significant amount of work on the enzymes involved in the biosynthesis of the opium alkaloids has been summarized (33–36,48,73).

(**88**)

Additional oxidative coupling processes among the various methylated derivatives of laudanosoline yield many other families of bases, including the pavine argemonine (**89**) from *Argemone mexicana* L.; berberine (**90**) from *Hydrastis canadensis* L. which, despite its toxicity, has been used as an antimalarial; protopine (**91**) and chelidonine (**92**) from *Chelidonium majus* L.; rhoeadine (**93**); and the cephalotaxus ester harringtonine (**94**) from Japanese plum yews (*Cephalotaxus* spp.), which is a compound of some significance because it possesses potent antileukemic activity.

(**89**)

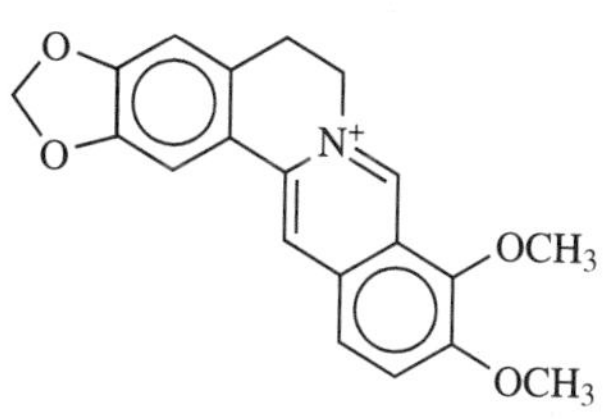

(90) 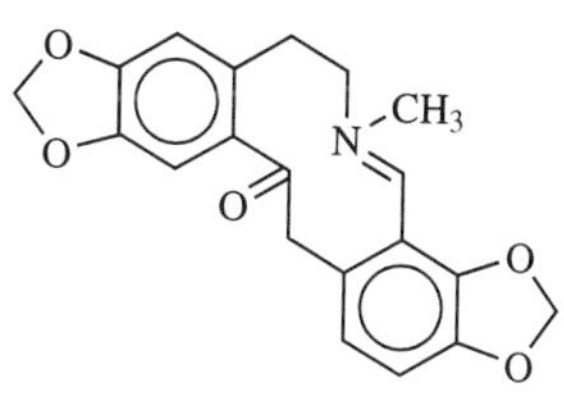   (91) 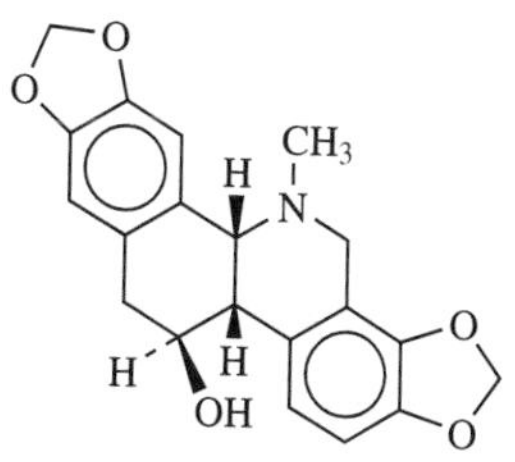   (92)

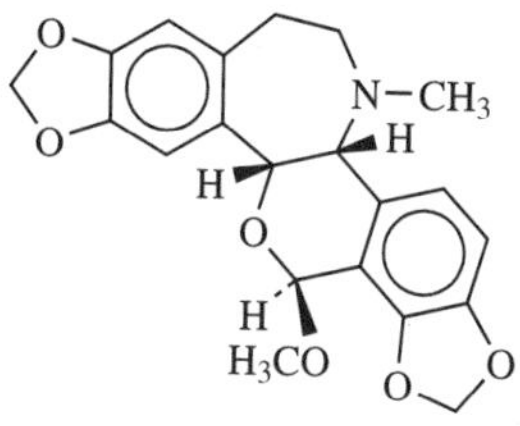

(93) 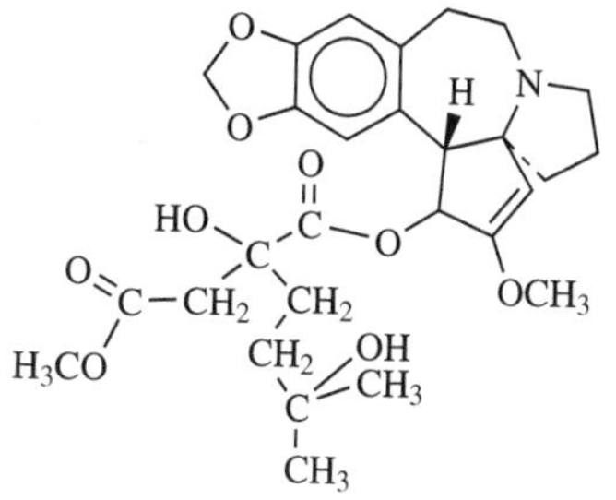   (94)

The last group of compounds from tyrosine and phenylalanine to be discussed here is the group derived from utilization of tyrosine (**26**, R = OH) as the C_6–C_2 fragment and phenylalanine (**26**, R = H) as a C_6–C_3 fragment. They include the 1-phenethyl-tetrahydroisoquinoline autumnaline (95), the homoproaporphine kreysiginone (**96**), which are typical of their kind. Both are isolated from *Colchicum cornigerum* (Schweinf.) and share a genesis similar to the toxic principle of the autumn crocus (*Colchicum autumnale*), colchicine (**4**). In crude form, extracts of *Colchicum* spp. were reportedly known to Dioscorides, a contemporary of Pliny, who served Nero as a physician and was the first to establish systematically the medicinal value of some 600 plants. The use of *Colchicum* spp. extracts in the treatment of gout appears to have begun in the sixteenth century, although it was not until much later that colchicine (**4**) was actually isolated (~1884). Recent interest in colchicine stems from its ability to bring cell division to an abrupt halt at a particular stage.

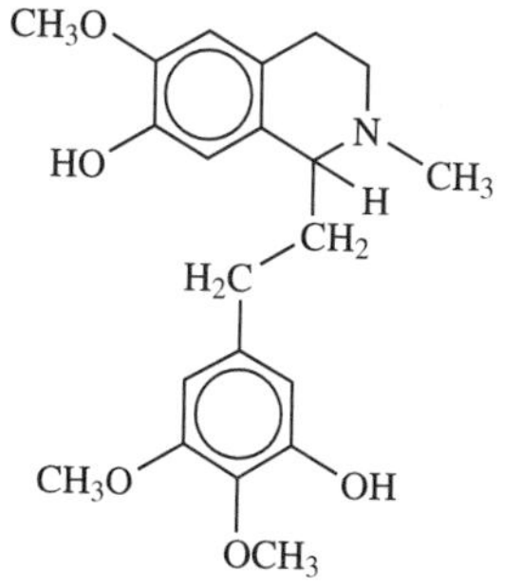

(95) 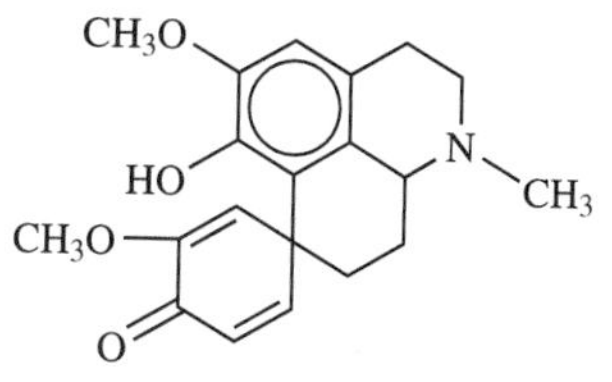   (96)

The structures of the brightly colored (red-violet and yellow) alkaloids found in the order *Centrospermae* (cacti, red beet, etc) remained unknown until the 1960s. In part this was doubtlessly due to the fact that these pigments, called betacyanins or betaxanthins, are relatively unstable and they are water soluble zwitterions. Invariably, in these plants they are found as acetals or ketals of sugars and one of two aglycone fragments called betanidine (**97**, $R_1 = H$, $R_2 = COOH$) and isobetanidine (**97**, $R_2 = H$, $R_1 = COOH$). Furthermore, it would appear that all betacyanins or betaxanthins may simply be imine derivatives (with the appropriate amino acid) of betalamic acid (**98**).

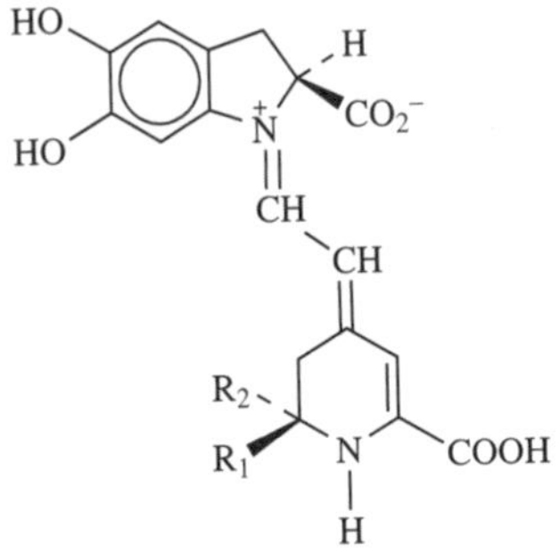

(97) 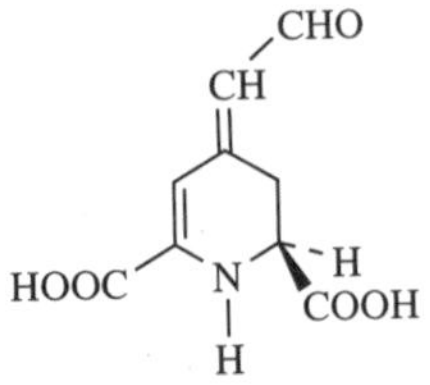 (98)

5.5. Tryptophan-Derived Alkaloids.

The last decade has seen dramatic progress in genetic engineering and related work on many of these materials. There are a few simple indole derivatives that are arguably derived, or have actually been shown to be derived, from tryptophan (**27**) (31,33–36,48,52,73,74). Serotonin (5-hydroxytryptamine [50-67-9], (**99**), R = OH) was first isolated (75) as a vasoconstrictor substance from beef serum and shown to be derived from tryptophan (**27**). It has also been isolated from bananas and the stinging nettle but its genesis in plants has not yet been established. *N,N*-Dimethylserotonin [487-93-4] (bufotenine, **19**) has been found in such widely diverse sources as the shrub *Piptadenia peregrina*, the seeds of which are said to be the source of a ceremonial narcotic snuff; the parotid gland of the toad (*Bufo vulgaris* Laur.); certain fungi (eg, *Amantia mappa* Batsch.); and human urine. The only slightly more complicated base harmine (**100**) is found widely distributed in the *Leguminoseae* and *Rubiaceae*, extracts of which were at one time used therapeutically against tremors in Parkinson's disease. The seeds of the African rue, *Peganum harmala* L., which are rich in harmine and related alkaloids, have also been used as a tapeworm remedy. Harmine has been shown to be derived from tryptamine (99, R = H) by [14]C- and ^{15}N-labeling experiments (76).

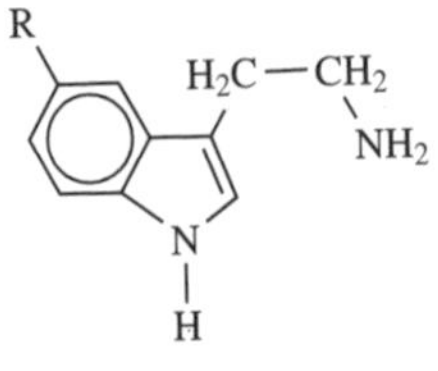

(99) 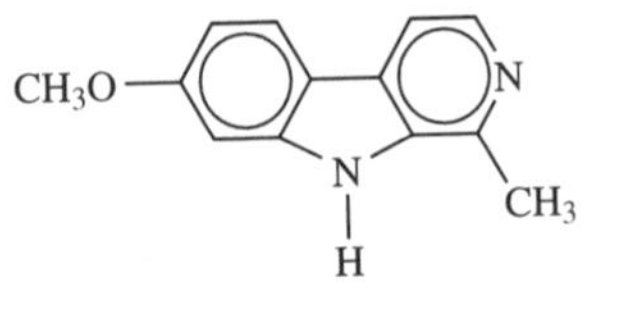 (100)

The C_6 building block, mevalonic acid (**31**), has been shown again and again to lose CO_2 and then serve as the progenitor of C_5, C_{10}, C_{15}, C_{20}, and C_{30} systems via the isomeric pair 3,3-dimethylallyl pyrophosphate (**33**)-isopentenyl pyrophosphate (**32**). This explains the genesis of a variety of bases containing the pattern defined by the five-carbon branched chain common to them and to the derivatives of mevalonic acid (**31**). Included among these are that small group of alkaloids that correspond to a joining of a tryptophan (**27**) and a C_5 unit to produce bases such as the potent uterine stimulant agroclavine (**18**) and its relatives, among which are the peptide amides of lysergic acid (**101**, R = OH).

The pistil of rye and certain other grasses may be infected by the parasitic fungus *Claviceps purpurea* (Fries). Unless the infected grain is sieved, the fungus passes into the flour, and bread made from such contaminated flour apparently retains activity from some of the alkaloids elaborated by and present in the fungus. Thus ingestion of the contaminated flour results in the disease called ergotism (St. Anthony's Fire). Convulsive ergotism causes violent muscle spasms that bend the sufferer into otherwise unattainable positions and frequently leaves physical and mental scars; outbreaks of the disease have been recorded into the twentieth century (22). Agroclavine (**18**) and derivatives of lysergic acid (**101**, R = OH) are considered responsible. Nonetheless, extracts of *C. purpurea* have long been used medicinally since they effect smooth muscle contraction and, even today, compounds related to lysergic acid and agroclavine are used for the same purpose. In the early 1940s, it was discovered that the diethylamide of lysergic acid [**101**, R = N(CH_2CH_3)$_2$] (LSD-25, as the tartrate salt) could be absorbed through the skin with resulting inebriation. In a bold experiment, it was then demonstrated that oral ingestion resulted in symptoms characteristic of schizophrenia which, although temporary, were quite dramatic (77).

There are currently two medicinally valuable alkaloids of commercial import obtained from ergot. Commercial production involves generation parasitically on rye in the field or production in culture because a commercially useful synthesis is unavailable. The common technique today (78) is to grow the fungus in submerged culture. *Claviceps paspali* (Stevens and Hall) is said to be more productive than *C. purpurea* (Fries). In this way, ergotamine (**102**) and ergonovine [**101**, R = NHCH(CH_3)CH_2OH] are produced. Ergotamine (**102**) is obtained from crude extract by formation of an aluminum complex.

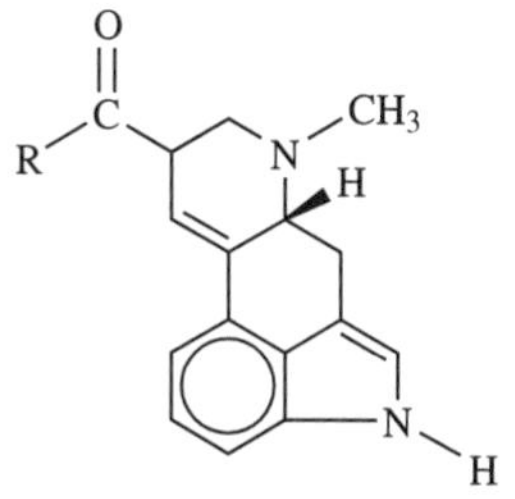

(101) 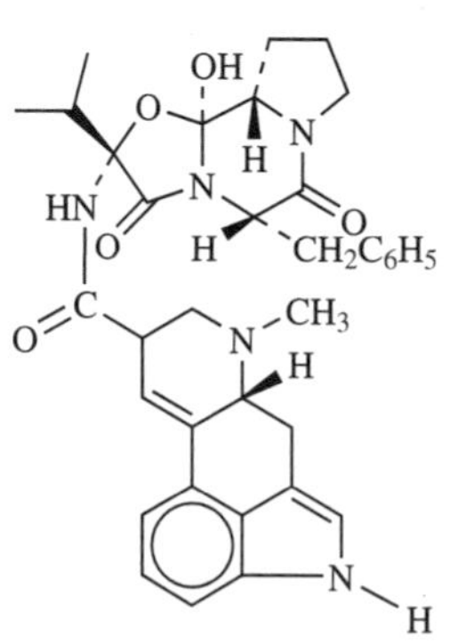   (102)

Destruction of the aluminum complex with ammonia then permits hydrocarbon extraction of the alkaloid. The alkaloid is subsequently both isolated and used as its tartrate salt. This nonnarcotic drug, for which tolerance may develop, is frequently used orally with caffeine (16) for treatment of migraine; it acts to constrict cerebral blood vessels, thus reducing blood flow to the brain.

Ergonovine [101, R = NHCH(CH₃)CH₂OH] was found to yield lysergic acid (101, R = OH) and (+)-2-aminopropanol on alkaline hydrolysis during the early analysis of its structure (79) and these two components can be recombined to regenerate the alkaloid. Salts of ergonovine with, eg, malic acid are apparently the drugs of choice in the control and treatment of postpartum hemorrhage.

Loganin (35), the iridoid derived from the dimer (34) of the isomeric pair of C_5 isoprenoid units, isopentenyl pyrophosphate (32) and 3,3-dimethylallyl pyrophosphate (33), has been recognized for some years (80) as the C_9–C_{10} unit which, along with tryptophan (27), makes up the huge group of bases, nearly 1000 well-characterized compounds, found in the *Corynanthe-Strychnos*, *Cinchona*, *Iboga*, *Aspidosperma*, and *Eburna*. Loganin is known to undergo oxidative cleavage to secologanin [19351-63-4] (103), and this fragment, combined with what appears to be tryptamine [61-54-1] (99, R = H), $C_{10}H_{12}N_2$, or its biological equivalent, leads to compounds whose permuted structures are often novel and quite complicated. Numerous single examples of rearranged, oxidized, and convoluted structures abound, but the subdivision into the families given below is convenient for description of the majority of structural types of bases (81).

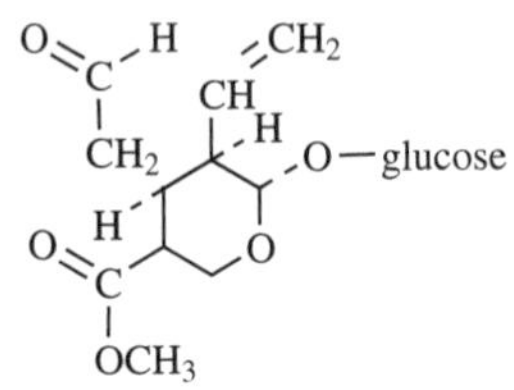

(103)

Thus in the *Corynanthe-Strychnos* are found bases such as ajmalicine (104), yohimbine (105), reserpine (106), ajmaline (107), and strychnine (13); in the *Cinchona*, quinine (17, R = OCH₃) and cinchonidine (17, R = H); in the *Iboga*, catharanthine (108); in the *Aspidosperma*, tabersonine (109); and in the *Eburna*, vincamine (110).

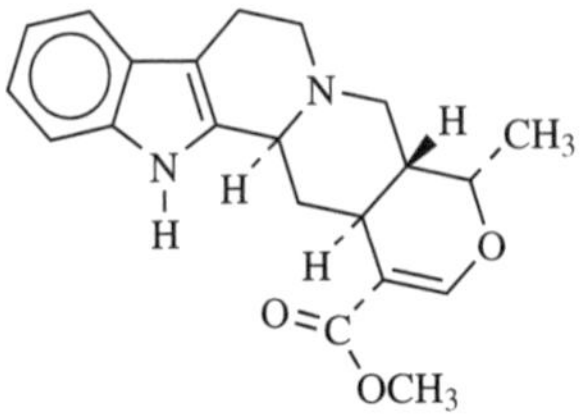

(104) 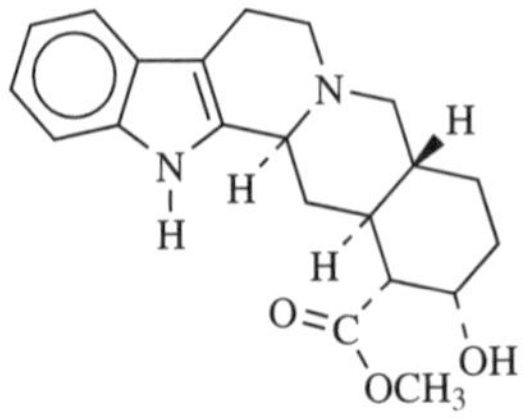   (105)

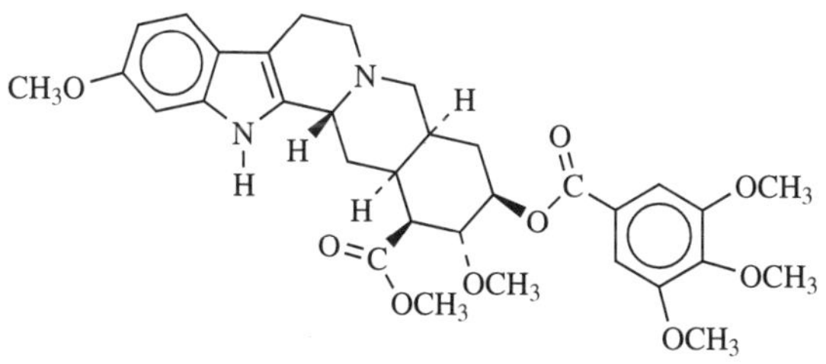

(106)

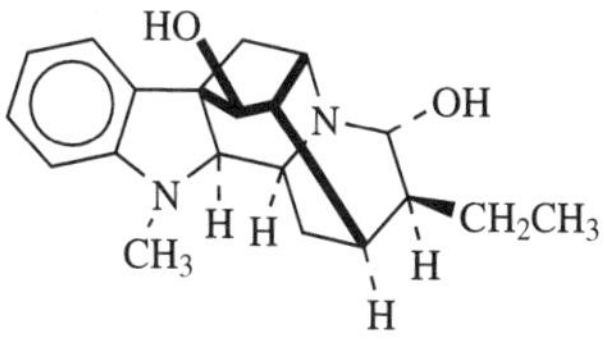

(107)

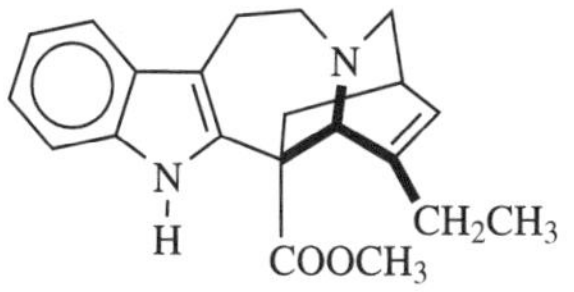

(108)

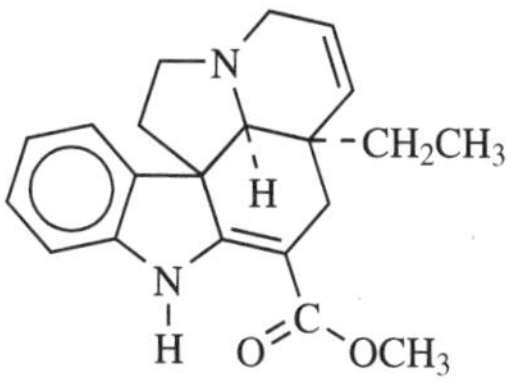

(109)

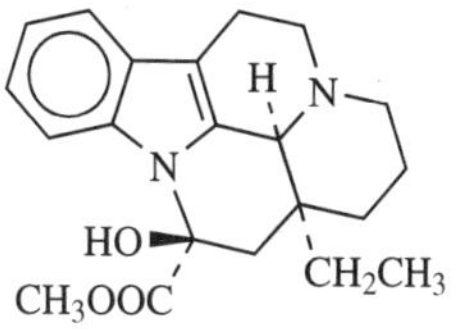

(110)

Ajmalicine (**104**) has been isolated (frequently as the weakest base present) numerous times from a variety of sources, eg, from the bark of *Corynanthe yohimbe* K. Schum. (*Rubiaceae*), from the roots of *Rauwolfia serpentina* (L.) Benth. (*Apocynaceae*), and from many other species of the genus *Rauwolfia*, and it is also found in plants of the genus *Catharanthus* (*Apocynaceae*). It is included here to demonstrate the pattern in secologanin (**103**)–tryptamine (**99**, R = H) coupling, the subsequent elaboration of which will give rise to the other bases to be considered. Thus ajmalicine (**104**) can be visualized, as is actually the case (**52**), as arising from the formation of an imine between the exposed aldehyde in the secologanin (**103**) and the basic nitrogen of tryptamine (**99**, R = H). This is followed by cyclization to the 2-position of the indole nucleus to form the C ring, opening of the glucose-masked acetal, and carbon–carbon bond rotation changing the cis ring junction stereochemistry found in secologanin to the trans stereochemistry of the D/E ring juncture in ajmalicine. The process continues with a second cyclization. That is the freshly exposed aldehyde resulting from opening of the acetal, the latter having swung around so that it is close to the secondary amine of the newly created C ring closes and this is followed by a reduction of the imine so created and a final cyclization of the liberated enol onto the alkene. In short, all 10 carbon atoms of secologanin and the entire tryptamine skeleton, as well as the geometry of the product, have been accounted for. This pathway, broadly painted above, is supported in detail by numerous labeling experiments with isotopes of carbon, hydrogen, and nitrogen, as well as the actual isolation of some of the intermediates described and enzymes involved in their production. All of the work has been summarized (34,48). Ajmalicine (**104**) increases cerebral blood flow and commercial mixtures of ajmalicine with one or more ergot alkaloids have been used in treating vascular disorders and hypertension.

Yohimbine (**105**), also from the bark of *C. yohimbe* K. Schum. and from the roots of *R. serpentina* (L.) Benth., has a folk history (unsubstantiated) of use as an aphrodisiac. Its use has been confirmed experimentally as a local anesthetic, with occasional employment for relief in angina pectoris and arteriosclerosis, but is frequently contraindicated by its undesired renal effects. Yohimbine and some of its derivatives have been reported as hallucinogenic (82). In addition, its pattern of pharmacological activities in a variety of animal models is so broad that its general use is avoided. All 10 carbon atoms of secologanin (103) as well as the entire skeleton of tryptamine (99, R=H) are clearly seen as intact portions of this alkaloid.

Reserpine (106), also from the roots of *R. serpentina* (L.) Benth. and other *Rauwolfia* spp., is currently used as a hypotensive. There are reports in the older popular literature showing its use for a wide variety of ailments in the tropics and subtropics where the plants grow. Apparently, it was originally used to treat both high blood pressure and insanity. The former use has been replaced by substances of greater value. Even its use as a tranquilizer and sedative, which has shown some apparent successes with neuroses, at lower doses (0.05–1.5 mg/day), is no longer in vogue for treatment of psychoses (at 0.5–5.0 mg/day); better materials have been found. Nonetheless, although no analgesic effect has been noted, reserpine does act as a sedative that reduces aggressiveness. At higher doses, reserpine has been reported to cause depression as well as peptic ulceration. There is some evidence that chronic administration in women results in an increased incidence of breast cancer (83). In other experimental systems it has shown antitumor activity (84). Its interesting structure contains, as expected, a tryptamine (99, R=H) unit (this time substituted with a 6-methoxy group), a secologanin (103) C_{10} fragment, and a trimethoxybenzoic acid unit. This is presumably derived by methylation of gallic acid [149-91-7] (111), typically derived oxidatively from shikimic acid (58) and normally associated with tannins in nutgalls, from which it is obtained by hydrolysis. It is not known for *Rauwolfia spp.* if methylation of the gallic acid to produce the trimethoxybenzoic acid unit found in reserpine (106) occurs before the acid is esterified with the remainder of the system or if methylation occurs later. Indeed, the involvement of gallic acid (111), $C_7H_6O_5$, itself is, as noted above, presumptive. The total synthesis of resperpine was a landmark synthesis (85).

(111)

In ajmaline (107), also obtained from the roots of *R. serpentina* (L.) Benth., a more deeply rearranged secologanin (103) fragment is embedded in the molecular framework and there has been a decarboxylation from the masked β-keto carboxylic acid to generate a C_9 unit. Interestingly, the C_9 unit found in ajmaline actually began (34,48,52) as the same C_{10} unit already seen in ajmalicine (104), yohimbine (105), and reserpine (106), but the additional cyclization to the C ring and subsequent bonding to the 3-position on the indole nucleus creating a sixth ring is accompanied by decarboxylation. Ajmaline has aroused some interest because it appears to possess antiarrhythmic activity (**86**), but care is required for this use when there is liver disease.

The synthesis of strychnine was a truly monumental undertaking (**21**). Strychnine (**13**, R = H), although only moderately toxic when compared to other poisons, both naturally occurring and produced synthetically, probably owes its reputation to its literary use. Obtained from the seeds and leaves of *Strychnos nux vomica* L. and other *Strychnos* spp., it has some history of use as a rodenticide. Poisoning is manifested by convulsions, and death apparently results from asphyxia. As little as 30–60 mg has been reported as fatal to humans, although at lower dosage it has received some medical use as an antidote for poisoning by central nervous system depressants, as a circulatory stimulant, and in treatment of delirium tremens. The useful medicinal dosage is normally <4 mg. As was the case for ajmaline (**107**), also derived from tryptamine (**99**, R = H) and secologanin (**103**), the pattern for the formation of strychnine (**13**) has been extended to even more deep seated rearrangement as well as the loss of one carbon atom, the same carboxylate as was lost in ajmaline (**107**). In addition, an acetate (C$_2$) unit has been added (to the indole nitrogen) and another ring created. Novel synthetic techniques continue to be applied to members of this family of alkaloids (**87**).

Quinine (**17**, R = OCH$_3$) and cinchonidine (**17**, R = H), which occur together along with other bases, eg, materials epimeric at the one-carbon bridge joining the aromatic nucleus with the 1-azabicyclo[2.2.2]heptane system, are constituents of the root, bark, and dried stems of various *Cinchona* species, but the main source remains *Cinchona officinalis* L. A crude preparation from this source was introduced into use as a palliative for malaria in the seventeenth century but several hundred years then elapsed before the first mixture of crystalline bases was obtained. Until the second World War quinine (**17**, R = OCH$_3$) and crude *Cinchona* preparations were the only antimalarials available. As supplies became unobtainable, synthetic materials capable of replacing quinine were developed. Recently, however, quinine has again become the treatment of choice for malaria as *Plasmodium falciparum* resistant to other drugs developed. Apparently, resistance to quinine is more difficult for the rapidly changing parasite population to acquire. Nonetheless, because quinine is not a prophylactic drug but rather a material which suppresses the overt manifestations of malaria, work continues on better treatment. The isolation of quinine from *Cinchona* bark generally involves conversion of the salts of the basic alkaloids to the free bases with, eg, calcium hydroxide (**88**), and extraction of the alkaloids into benzene or toluene.

Examination of the structures of the alkaloids (**17**) obtained from *Cinchona* suggests that they probably have a different pattern of formation than those already discussed. However, the differences are more formal than profound. Thus the 1-azabicyclic system is formed from one of the aldehyde equivalent carbons of a secologanin (**103**) bound fragment with the terminal nitrogen of tryptamine (**99**, R = H). Cleavage of that nitrogen away from the indole leaves behind the carbon to which it was bound, formally, at the oxidation level of an aldehyde. Then, oxidative opening of the five-membered ring between the indole nitrogen and the adjacent carbon is followed by recyclization from the aryl amine so liberated to the aldehyde function set free in the previous step. Thus the nine expected carbons, one of the original carbons in the C_{10} fragment having been lost by decarboxylation, remain.

An understanding of the chemistry and structure of catharanthine (**108**), an otherwise minor alkaloid found in *Vinca rosea* Linn. or *Catharanthus roseus* G. Don. which is a potent diuretic in rats, was critical in unraveling the structure of two of the alkaloids cooccurring in *Vinca* which had been shown to be active against leukemia, first in mice and later in humans, ie, leurocristine (vincristine [57-22-7], **23**, R = CHO) and vincaleukoblastine (vinblastine [865-21-4], **23**, R = CH$_3$). Interestingly, the genera *Vinca* and *Catharanthus* (*Apocynaceae*) appear to be used interchangeably, ie, *Vinca rosea* Linn. is frequently called *Catharanthus roseus* G. Don. by some but not by others and vice versa (89) although the latter is now generally preferred (see below). Both catharanthine (**108**) and vincaleukoblastine in concentrated hydrochloric acid, when treated with stannous chloride yielded, among other fragments, the (+)-cleavamine [1674-01-7] (**112**), so called because it is a "broken" or "cleaved" amine.

(**112**)

After the structure and absolute stereochemistry of cleavamine (**112**), $C_{19}H_{24}N_2$, was established, its synthesis was completed and impetus to unravel the structure of the dimeric bases (**23**) was bolstered (90). Again, the C_9 fragment, now only slightly modified from that originally present in secologanin (**103**), is readily seen in catharanthine (**108**).

Tabersonine (**109**), clearly a reduced and simplified version of the "second-half" of the alkaloids **23**, was originally isolated from *Amsonia tabernaemontane* L. and is considered to be a simplified parent of a rather more elaborate subgroup of indole alkaloids.

Among the examples of monoindole bases being discussed, vincamine (**110**) is the principal alkaloid of *Vinca minor* L. and has received some notoriety because it apparently causes some improvement in the abilities of sufferers of cerebral arteriosclerosis (91). It is believed that this is the result of increasing cerebral blood flow with the accompanying increase in oxygenation of tissue as a result of its action as a vasodilator.

Finally, for this group of tryptophan (**27**)-derived bases there are those in which a tryptamine (**99**, R = H) residue is not obvious. Nonetheless, the pyrido-carbazole bases originally isolated from *Ochrosia elliptica* Labill. (*Apocynaceae*) and subsequently from the genus *Aspidosperma*, among others, which include ellipticine (**113**, R$_1$ = CH$_3$, R$_2$ = H) and olivacine (**113**, R$_1$ = H, R$_2$ = CH$_3$), are derived from tryptophan and the normal C_9–C_{10} fragment expected. These alkaloids are known to inhibit proliferation of cells and continue to be of interest in chemotherapeutic treatments. They appear to inhibit nucleic acid synthesis irreversibly by interacting strongly with DNA.

Bisindole Alkaloids from Tryptophan. There are two widely different types of alkaloids derived from two tryptophan (**27**) units. The first is a rather small group of compounds based simply on the dimers of tryptophan that includes compounds such as calycanthine (**114**). isolated from the seeds of the flowering aromatic shrubs Carolina Allspice (*Calycanthus floridus*) and Japanese Allspice (*Chimonanthus fragans*). The second type is that group in which the two halves arise in two distinct ways (92). Both halves may be composed of identical fragments, as in C-toxiferine (**10**), the arrow poison packed in gourds and derived from, eg, *Strychnos froesii* Ducke and *Strychnos toxifera*. The more common and very numerous (nearly 1000 compounds) family is characterized by two halves derived from different fragments, eg, leurocristine (vicristine (**23**), R = CHO) and vincaleukoblastine (vinblastine (**23**), R = CH$_3$), along with nearly 100 other compounds, from *Catharanthus roseus* (L.) G. Don, occasionally referred to in the earlier literature (see above) as *Vinca rosea* L. (93). This second group has in common the genesis of each half from tryptophan (**27**) and at least one fragment derived from mevalonic acid (**31**) itself (ie, a C$_5$ fragment) or derived from a monoterpene such as geraniol (**34**), −OH in place of −OPP), ie, loganin (**35**) or secologanin (**103**), a C$_{10}$ fragment.

(113)   (114)

The search for the bisindole derivatives (**23**) was originally (94) initiated on the basis of folklore. A brew made from Jamaican periwinkle had established itself in local medicine as a treatment for diabetes and it was this material that was investigated and found to contain the cytotoxic compounds (**23**), among others. No materials useful in the treatment of diabetes have been reported from this source.

Although the compounds were isolated in quantities of only a few milligrams per kilogram of crude plant leaves, extensive work on a variety of animal tumor systems led to eventual clinical use of these bases, first alone and later in conjunction with other materials, in the treatment of Hodgkin's disease and acute lymphoblastic leukemia. Their main effect appears to be binding tightly to tubulin, the basic component of microtubules found in eukaryotic cells, thus interfering with its polymerization and hence the formation of microtubules required for tumor proliferation (95).

Initial attempts to synthesize the compounds (**23**) were hampered by the failure to obtain the correct stereochemical configuration about the vindoline–catharanthine linkage, a most difficult problem eventually solved by insight and hard work (96).

5.6. Introduction of Nitrogen into a Terpenoid Skeleton. The acetate-derived fragments (43) mevalonic acid (31), which yields isopentenyl pyrophosphate (32) and its isomer, 3,3-dimethylallyl pyrophosphate (33); a dimeric C_5 fragment, geranyl pyrophosphate (34), which gives rise to the iridoid loganin (35); and the trimer farnesyl pyrophosphate (36), which is also considered the precursor to C_{30} steroids, have already been mentioned. Three of the fragments [(31), the pair (32–33), and (35)] have been invoked as descriptive progenitors of alkaloids such as emetine (67), lysergic acid (101, R = OH), and many other bases already discussed and broadly categorized as monoterpenoid indole alkaloids, eg, ajmalicine (104). The path that links (31) to (36) and hence to the steroids is clear (97) and the details of the relationships with some of the subfragments on that path and the alkaloids resembling them and included here is well defined in some cases (34,49,52) where enzymatic pathways have been detailed. The relatively incomplete examination of alkaloidal material seems to arise only because the techniques are new and the investigators few. Historically, simpler labeling experiments led to less well defined results and were fraught with difficulty. The techniques for feeding suitably labeled precursors such as ^{13}C- and/or ^{14}C-labeled and ^{2}H- and/or ^{3}H-labeled acetate or even larger fragments (those further along on the metabolic pathway to alkaloidal product) such as mevalonic acid (31) and loganin (35) to many actively growing plants have been worked out. This is shown by the incorporation of the labeled material into alkaloids. However, it needs to be appreciated that lack of incorporation does not rule out utilization by the plant of the material fed to it. This is because there is no guarantee that the fed material reached the site of alkaloid synthesis. Thus, a negative result may simply mean that the particular feeding technique, stage of plant growth, feeding cycle, photocycle or some other variable was inappropriate for the specific material fed, rather than implying that the material is not capable of incorporation. In this vein, it is generally true that the larger fragments are more difficult to incorporate. Even though they may enter the plant when they might be actively metabolized, the particular form in which they arrive at the cell wall may be wrong for transport across the wall. Smaller fragments are less likely to have transport problems. Thus mevalonic acid (31) generated endogenously from exogenous labeled acetate is frequently more easily traced than suitably labeled exogenous mevalonic acid itself. However, the value is correspondingly diminished because everything may be thought of as derived from acetate and whatever the precursor, the label will have been incorporated. Thus a balance must be struck between what can be fed as labeled material and what will be incorporated into the plant. Frequently, the largest useful fragment that can be incorporated is mevalonic acid (or its corresponding lactone).

A second experimental problem is that incorporation of a material such as loganin (35), or even an amino acid that seems clearly to be a precursor by some biogenetic hypothesis, does not necessarily prove it is a precursor. The material fed may so completely swamp the normal pathways in the plant that the utilization of what was fed generates an aberrant path that nonetheless produces the same product.

These considerations are particularly important when the description of the alkaloids is based on a presumed biosynthesis from terpene fragments because the experimental work linking the smaller pieces with the larger has yet to succeed. That is, the well worked out paths from acetate, through (31) to the steroids, via geranyl pyrophosphate (34), farnesyl pyrophosphate (36), and the universal steroid precursor squalene [111-02-4] (115), $C_{30}H_{50}$, (97) have not been clearly demonstrated to apply in the higher, alkaloid producing, plants. Furthermore, in almost all of the alkaloids whose presumed biosynthesis derives from an insertion of nitrogen into the mevalonic acid-derived fragment, it is not quite clear at what stage the nitrogen insertion occurs. Introduction of the nitrogen at a very late stage might be an artifact of isolation because basification with ammonia of the acidic extract initially employed to isolate the basic materials is common and reaction of water soluble materials with ammonia, followed by cyclizations, etc, might occur.

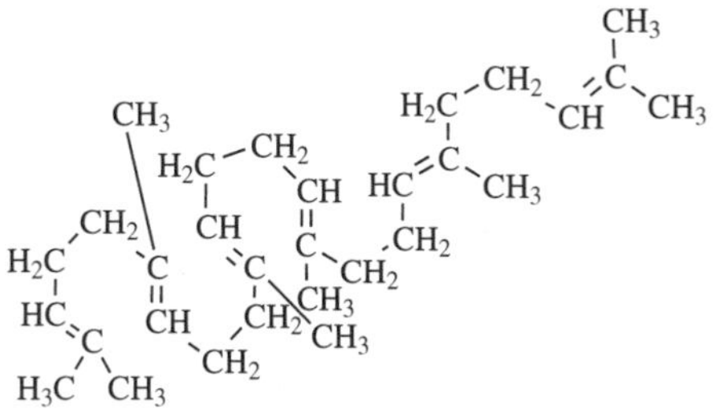

(115)

When racemic mevalonic acid as the corresponding lactone and labeled at C-2 with ^{14}C is fed to the Chilean shrub *Skytanthus acutus* Meyen., labeled β-skytanthine [24282-31-3] (116) is obtained. Skytanthus alkaloids are reputed to be tremorgenic (98).

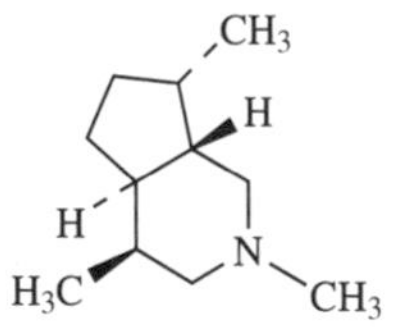

(116) 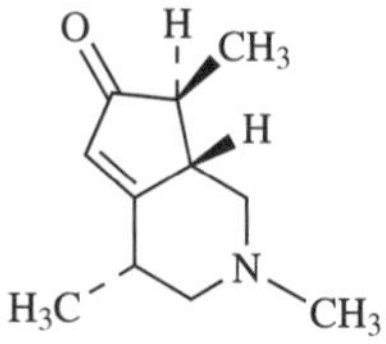   (117) 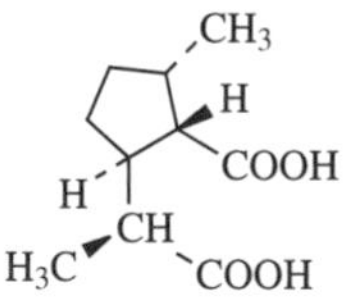   (118) 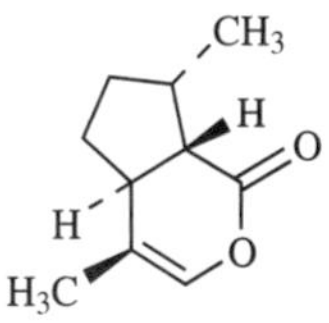   (119)

Tecomanine [6878-83-7] (117), $C_{11}H_{17}NO$, said to be a potent feline attractant (99) and material clearly related to β-skythathine (116), $C_{11}H_{21}N$, as well as to nepetalinic acid [485-06-3] (118), $C_{10}H_{16}O_4$, a degradation product of nepetalactone [490-10-8] (119), $C_{10}H_{14}O_2$, which is a major constituent of volatile oil of catnip (*Nepeta cataria* L.), is obtained from *Tecoma stans* Juss. Extracts of the latter (100) have some history demonstrating antidiabetic properties. Both of these bases (115 and 116) are derived from geranyl pyrophosphate (34) or loganin (35) or suitable similar precursor(s) before cleavage of the precursor to secologanin (103) or its equivalent.

Alternatively, there are those alkaloids, such as gentianine [439-89-4] (120), $C_{10}H_9NO_2$, isolated from *Gentiana tibetica* King, among others, which are presumably derived from secologanin (103) and exhibit antiinflammatory action along with being muscle relaxants.

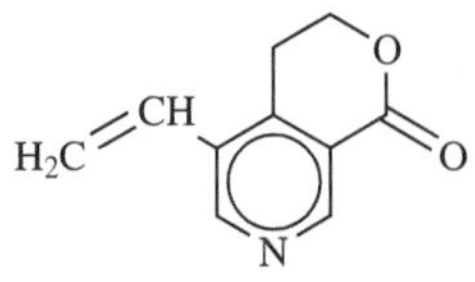

(**120**)

The C_5 trimer farnesyl pyrophosphate (**36**), in addition to serving as a progenitor of steroids via squalene (**115**), is also the progenitor of the C_{15} compounds known as sesquiterpenes. It has been suggested (101) that farnesyl pyrophosphate (**36**) similarly serves as the carbon backbone of alkaloids such as deoxynupharidine (**121**) from *Nuphar japonicum* (*Nymphaceae*) (water lilies) and dendrobine (**122**) from *Dendrobium nobile* Lindl. (*Orchidaceae*). The latter is the source of the Chinese drug Chin-Shih-Hu. Compared to the other families of bases discussed earlier, the numbers of alkaloids supposedly derived from farnesyl pyrophosphate or a close relative is small. However, given the wide variety of plant families containing sesquiterpenes, it is most likely that the numbers of compounds to be found will dramatically increase.

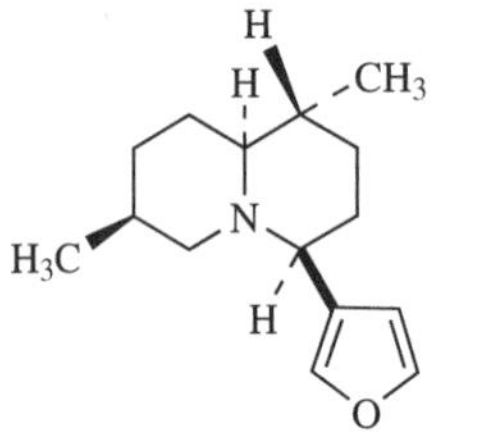

(**121**) 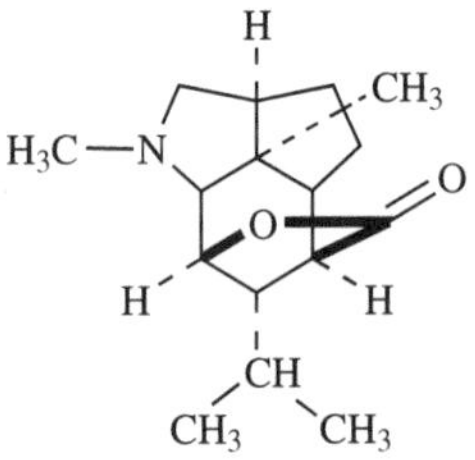   (**122**)

Whereas dimerization of two farnesyl pyrophosphates (**36**) generates squalene (**115**) on the path to steroids (102), the addition of one more C_5 unit, as isopentenyl pyrophosphate (**32**) or its isomer, 3,3-dimethylallyl pyrophosphate (**33**), to the C_{15} compound farnesyl pyrophosphate produces the C_{20} diterpene precursor geranylgeranyl pyrophosphate [6699-20-3] (**123**).

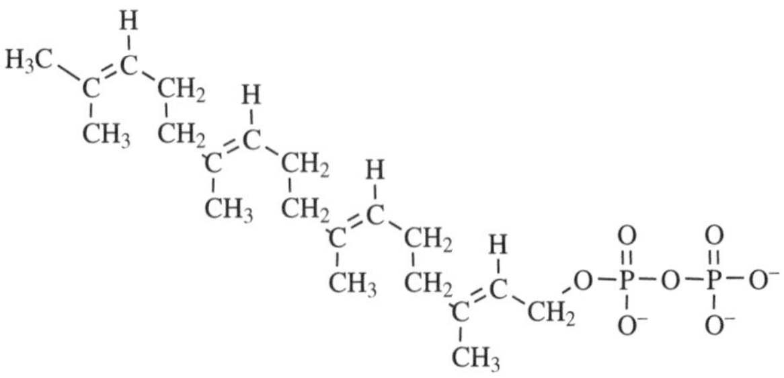

(**123**)

This C_{20} pyrophosphate (**123**), $C_{20}H_{36}O_7P_2$, is thought to provide the carbon framework of the diterpene alkaloids such as veatchine (**124**), atisine (**125**), and aconitine **126**). It is not known at what stage the nitrogen is incorporated into the framework established by the skeleton. The potential for terpene rearrangements and the observation that the alkaloids are frequently found esterified, often by acetic or benzoic acid, as well as free, has led to permutations and combinations producing over 100 such compounds.

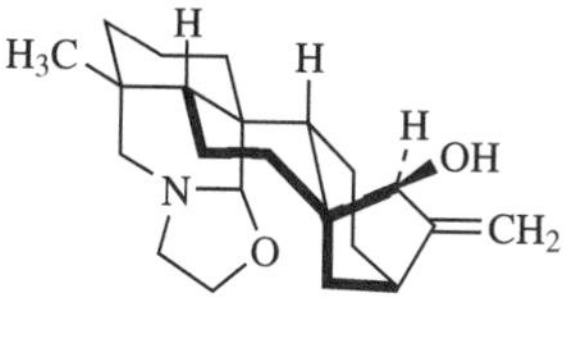

(**124**) 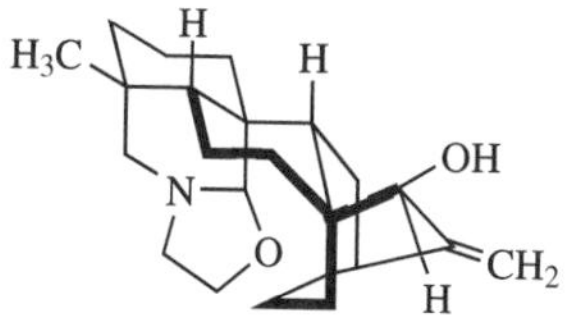   (**125**)

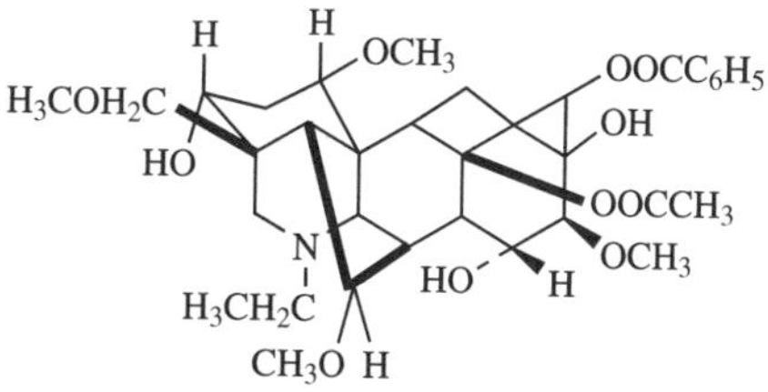

(**126**)

The diterpene alkaloids elaborated by most species of *Aconitum* and *Delphinium* (family *Ranunculaceae*) are apparently not found in other genera (*Ranunculus, Trollius, Anemone,* etc.) in the same family. Similar bases are, however, found in *Garrya* (eg, *Garrya veatchii* Kellog., family *Cornaceae*). Monkshood (occasionally wolf's bane, friar's cowl, or mouse bane) is obtained from the dried tuberous root of *Aconitum napellus* L. agg., and the plant is said to occur wild (103) in England and Wales as well as in the Swiss and Italian Alps. It is considered among the most dangerous of plants, all parts of it being poisonous, although the bases appear to be most concentrated in the roots. As has usually been the case, this alkaloid-bearing plant, along with the others containing diterpene alkaloids, was initially examined based on folklore and in the hope of finding medicinally valuable palliatives. Crude plant material has long been used internally as a febrifuge to lower fever and externally for neuralgia.

The base veatchine (**124**) and related materials are found in the bark of, eg, *G. veatchii* Kellog., and structural elucidation of this complicated and reactive material required massive efforts (104). Its relationship to atisine (**125**) from the roots of the atis plant, *Aconitum heterophyllum* Wall., is clearly seen as that of the well known terpene rearrangement of an *exo*-methylene octa[3.2.1]bicyclic system to that of its [2.2.2] isomer, the remainder of the molecule remaining unchanged. More deep seated rearrangements in the same part of the molecule (ie, a 6-6-5 set of rings with an *exo*-methylene group yielding a 7-5-6 set now incorporating the methylene) generates aconitine (**126**).

The path from squalene (**115**) to the corresponding oxide and thence to lanosterol [79-63-0] (**127**), $C_{30}H_{50}O$, cholesterol [57-88-5] (**128**), and cycloartenol [469-38-5] (**129**) has been demonstrated in nonphotosynthetic organisms. It has not yet been demonstrated that there is an obligatory path paralleling the one known for generation of plant sterols despite the obvious structural relationships of, eg, cycloartenol (**129**), $C_{30}H_{50}O$, to cyclobuxine-D (**130**), $C_{25}H_{42}N_2O$. The latter, obtained from the leaves of *Buxus sempervirens* L., has apparently found use medicinally for many disorders, from skin and venereal diseases to treatment of malaria and tuberculosis. In addition to cyclobuxine-D [2241-90-9] (**130**) from the *Buxaceae*, steroidal alkaloids are also found in the *Solanaceae, Apocynaceae*, and *Liliaceae*.

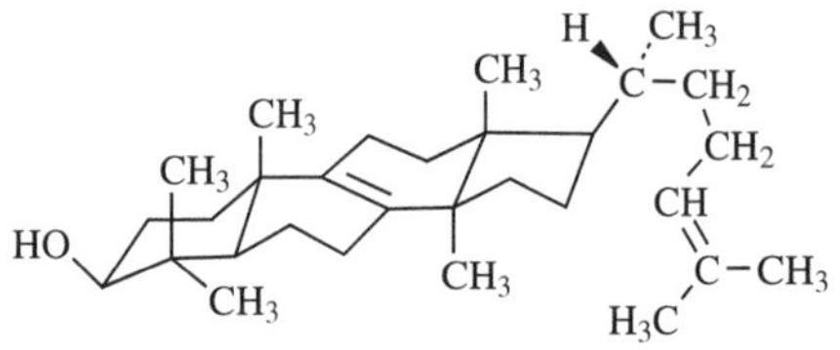

(**127**)

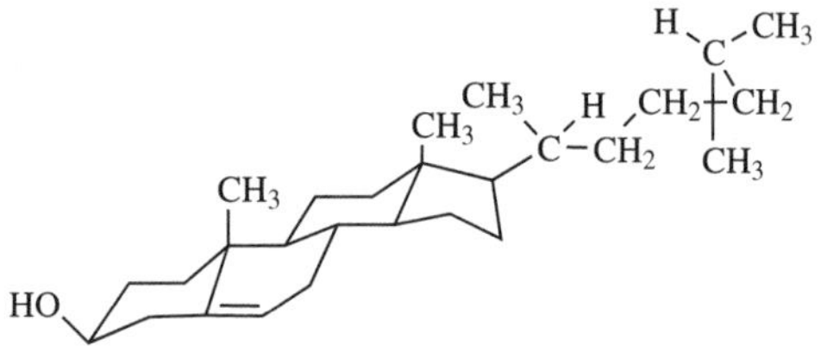

(**128**)

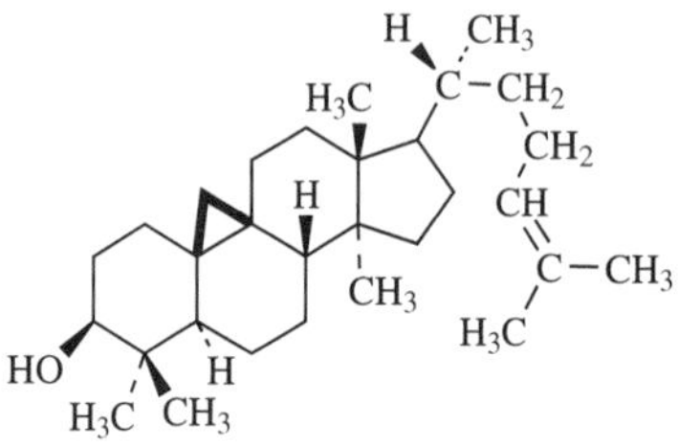

(**129**)

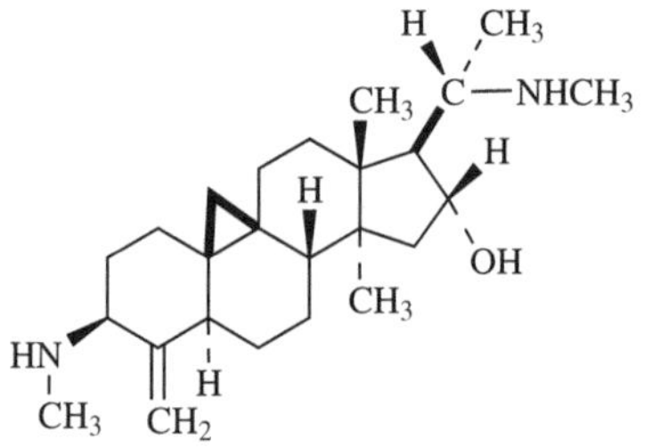

(**130**)

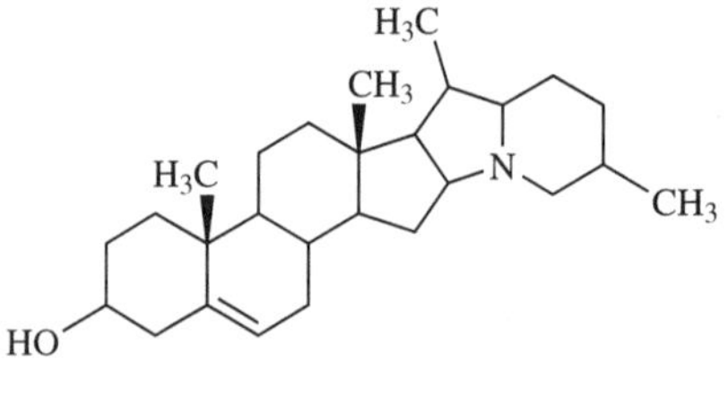

(**131**)

The plants of *Solanum* include, among others, the potato (*Solanum tuberosum* L.) and the tomato (*Solanum lycoperisicum* L.). Frequently, the plant bases occur as the aglycone portion of a glycoalkaloid bonded to one or more six-carbon sugars. Hydrolysis of the sugar portion and, somewhere along the degradative pathway, excision of the nitrogen (usually via a Hoffmann-type elimination) results in a steroid-like fragment, the analysis of which falls back on the large body of accumulated information about steroids and their degradation products. Solanidine [80-78-4] (**131**) is typical of the kind of bases present and has been isolated from a number of Solanum.

Interestingly, feeding experiments in *Solanum chacoense* L. (105) demonstrate that cholesterol (**128**), $C_{27}H_{46}O$, can be incorporated into solanidine (**131**), $C_{27}H_{43}NO$, but the amount of steroid incorporated is very low and there has been more than one suggestion (106) that the route involves initial degradation of the fed cholesterol to acetate, followed by recreation of the entire skeleton.

In addition to the alkaloids such as cyclobuxine-D (**130**) and solanidine (**131**) where the structural similarities to steroids are clear (although it must be remembered that detailed evidence actually linking the compounds is lacking) there are the less obvious (but nonetheless also clearly related) *Veratrum* alkaloids. These compounds, of which protoveratrine A [143-57-7] (**132**), $C_{41}H_{63}NO_{14}$, obtained from the rhizome of *Veratrum album* L. (*Liliaceae*), is a typical example, produce dramatic declines in blood pressure on administration and have been received by the medical community as good antihypertensive agents. Generally, however, the dosage must be individualized (slowly) from ∼2 mg in 200 mL of saline upward. Because the therapeutically valuable dosage is similar to the toxic dose, and even nonlethal large doses may cause cardiac arrhythmias and peripheral vascular collapse, use of these compounds has frequently been limited to extreme cases where close attention can be accorded the patient.

(**132**)

5.7. Purine Alkaloids. The purine skeleton is not derived from histidine (**30**), as might be imagined, nor is it derived from any obvious amino acid progenitor. As has been detailed elsewhere (41,43), the nucleus common to xanthine [69-89-6] (**133**), $C_5H_9N_4O_2$, and found in the bases of caffeine (**16**), theophylline [58-55-9] (**134**, $R_1 = R_2 = CH_3$; $R_3 = H$), and theobromine [83-67-0] (**134**, $R_1 = H$; $R_2 = R_3 = CH_3$), $C_7H_8N_4O_2$, is created from small fragments that are attached to a ribosyl unit during synthesis and can presumably be utilized in a nucleic acid backbone subsequently. All three alkaloids, caffeine, theophylline, and theobromine, occur widely in beverages commonly used worldwide.

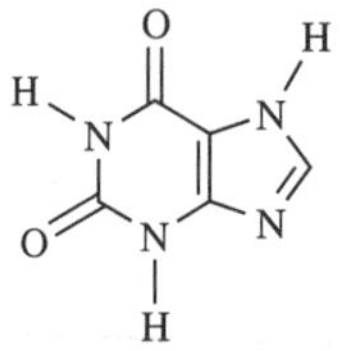

(133) 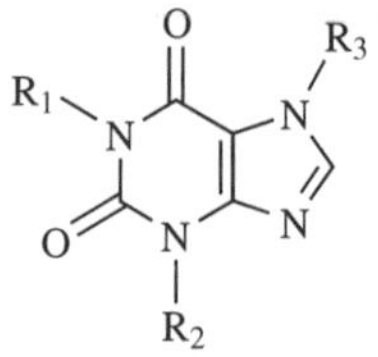   (134)

The leaf and leaf buds of *Cammelia sinensis* (L.) O Kuntze and other related plants and most teas contain, depending on climate, specific variety, time of harvest, etc., somewhat $< 5\%$ caffeine (16) and smaller amounts of theophylline (134, $R_1 = R_2 = CH_3$; $R_3 = H$) and theobromine (134, $R_1 = H$; $R_2 = R_3 = CH_3$). Coffee consists of various members of the genus *Coffea*, although the seeds of *Coffea arabica* L., believed to be indigenous to East Africa, are thought to have been the modern progenitor of the varieties of coffees currently available and generally cultivated in Indonesia and South America. The seeds contain less than ~3% caffeine which, bound to other agents, is set free during the roasting process. The caffeine may be sublimed from the roast or extracted with a variety of agents, such as methylene chloride, ethyl acetate, or dilute acid (eg, an aqueous solution of carbon dioxide) to generate decaffinated material (107). Supercritical extraction with carbon dioxide has also found to be useful.

Two other commonly found sources of caffeine (16) are kola (Cola) from the seeds of, eg, *Cola nitida* (Vent.) Schott and Engl., which contains 1–4% of the alkaloid, but little theophylline or theobromine, and cocoa (from the seeds of *Theobroma cacao* L.), which generally contains ~3% theobromine and significantly less caffeine.

All three of these materials are apparently central nervous system (CNS) stimulants. It is believed that for most individuals caffeine causes greater stimulation than does theophylline. Theobromine apparently causes the least stimulation. There is some evidence that caffeine acts on the cortex and reduces drowsiness and fatigue, although habituation can reduce these effects.

5.8. Miscellaneous Alkaloids.   Shikimic acid (58) is a precursor of anthranilic acid (29) and, in yeasts and *Escherichia coli* (a bacterium), anthranilic acid (*o*-aminobenzoic acid) is known to serve as a precursor of tryptophan (27). A similar but yet unknown path is presumed to operate in higher plants. Nonetheless, anthranilic acid itself is recognized as a precursor to a number of alkaloids. Thus damascenine [483-64-7] (135), $C_{10}H_{13}NO_3$, from the seed coats of *Nigella damascena* has been shown (108) to incorporate labeled anthranilic acid when unripe seeds of the plant are incubated with labeled precursor.

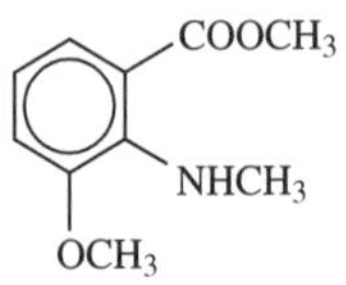

(135)

Similarly, anthranilic acid (**29**) has been suggested as a reasonable precursor and some early labeling studies have been carried out showing that dictamnine [484-29-7] (**136**), $C_{12}H_9NO_2$, from *Dictamnus albus* and skimminanine [83-95-4] (**137**), $C_{14}H_{13}NO_4$, from *Skimmia japonica* incorporate anthranilic acid in a nonrandom fashion (109).

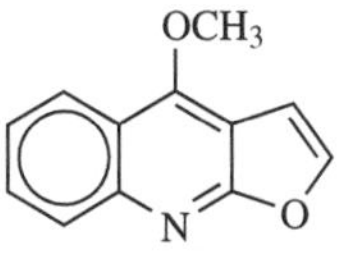

(**136**) 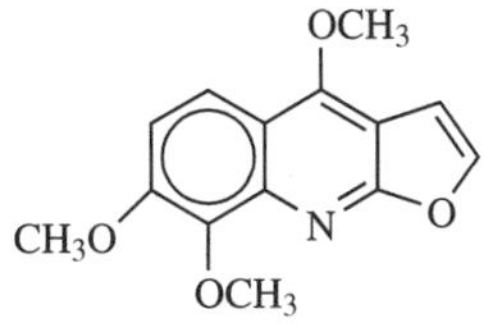   (**137**)

Securinine [3610-40-2] (**138**), $C_{13}H_{15}NO_2$, is the major alkaloid of *Securinega surroticosa* Rehd. and has been shown to arise from two amino acid fragments, lysine (**25**) and tyrosine (**26** R = H) (110).

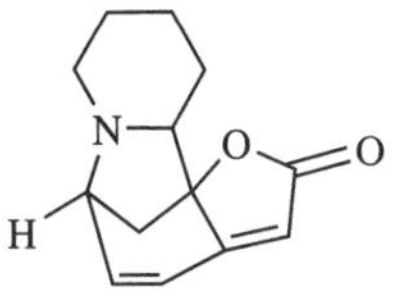

(**138**)

Reactions at the aromatic nucleus that are quite different from the usual mild condensations and rearrangements that apparently generate the typical alkaloids already discussed must be involved. Securinine (**138**) is reported to stimulate respiration and increase cardiac output, as do many other alkaloids, but it also appears generally to be less toxic (111).

Coniine (**12**), implicated by Plato in the death of Socrates, is the major toxic constituent of *Conium maculatum* L. (poison hemlock) and, as pointed out earlier, was apparently the first alkaloid to be synthesized. For years it was thought that coniine was derived from lysine (**25**), as were many of its obvious relatives containing reduced piperidine nuclei and a side chain, eg, pelletierine (**47**). However, it is now known (112) that coniine is derived from a polyketooctanoic acid [7028-40-2] (**139**), $C_8H_{10}O_5$, or some other similar straight-chain analogue.

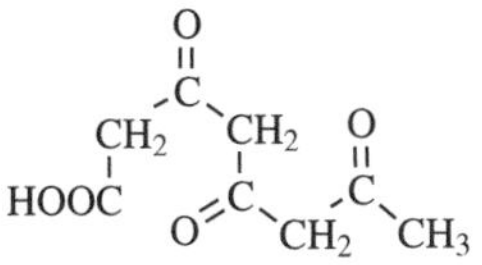

(**139**)

6. Economic Aspects

As the twenty-first century dawns, many alkaloids, such as atropine (**42**) and reserpine (**106**), that have served humanity since early history are being replaced by synthetic materials. Others, such as the *Vinca* bases, eg, vincristine (leurocristine, **23**, R = CHO) remain as powerful medical tools. Replacement of naturally occurring alkaloids is desireable in order to maintain and augment favorable properties while eliminating undesireable properties and effects. Through strides in biochemical research, especially structure–reactivity studies, design of model compounds, combinatorial synthesis, and genomic modification synthetic and semisynthetic materials are under development or have been developed. These new materials, while occasionally related to alkaloids, either because they are derived from alkaloids or alkaloid precursors, or because their structures are similar, are not naturally occurring (unless coming from modified genomic material) and thus are not "alkaloids". However, they are generally much more specific in their action and since they are protected by patents, much more expensive.

There are four broad classes of alkaloids whose general economic aspects are important: (*1*) the opiates such as morphine and codeine (**2**, R = H and R = CH_3, respectively); (*2*) cocaine (**11**) (both licit and illicit); (*3*) caffeine (**16**) and related bases in coffee and tea, and (*4*) the tobacco alkaloids such as nicotine (**22**).

6.1. The Opiates. The International Narcotics Control Board (INCB), Vienna, tracks the licit production of narcotic drugs and annually estimates world requirements for the United Nations.

Their most recent publication (113) contains the estimate for 2001 that the primary sources of opiates are 568,539 kg of poppy straw concentrate and 1,663,576 kg of opium. The 2000 numbers were 603,899 kg and 2,154,304 kg, respectively It is estimated that worldwide opiate requirements for 2001 are ~277,713 kg of morphine (**2**, R = H), 384,539 kg of codeine (**2**, R = CH_3) and 68,952 kg of thebaine (**85**), with lesser amounts of related materials used for medicinal purposes being generated from these.

6.2. Cocaine. Production of cocaine [50-36-2] (**11**) for licit purposes (113) takes place in Bolivia and Peru. In 1988 the former exported 204 t and the latter 47 t of coca (not cocoa) leaves into the United States. The average total licit production of cocaine as a by-product in the extraction of flavoring agents from coca leaves was reported (112) to be only 425 kg in the United States in 1988. INCB estimates that 503 kg of cocaine will be used for licit purposes in 2002. The 2000 use was 923 kg (112) This must be weighed against the estimates by the U.S. Drug Enforcement Administration that the value of illegal annual exports of coca from Bolivia (1987) is $2 billion (114). It has been suggested (114) that this dollar sum is three times as much as the earnings from legal exports of tin, coffee, etc, and that it is a significant support to the Bolivian economy.

6.3. Caffeine. About 3% by weight of the roasted coffee bean is caffeine (**16**). The United Nations (UN) Food and Agriculturial Organization (FAO) reports the 1999 world production of coffee beans as 6,831,537 Mt (115). World coffee consumption is predicted to rise in the foreseeable future at the rate of 1–2% per year and thus the total amount of caffeine and related alkaloids ingested from this source can also be expected to increase. Caffeine and related bases (eg, theophylline) are also found in various teas as well as in cocoa. Again, the UNFAO reports that 3.744,714 Mt of tea and 2,943,169 Mt of cocoa beans were produced in 1999 (115).

6.4. Tobacco. Tobacco is the principal source of the alkaloid nicotine (**22**), which, it is claimed, is at least partially responsible for the addicting properties of tobacco. The FAO data currently available (115) show that over the last decade tobacco production worldwide has remained nearly constant with 7,064,272 Mt produced in 1989 and 6,971,648 Mt in 1999. It is apparently understood that tobacco is an economically valuable crop, the use of which is deleterious to the health of the users and it may be that a steady state has been reached. Its production is advocated on the one hand as it generates revenue for growers, governments, and others, but discouraged on the other hand as health problems for its users commonly result.

BIBLIOGRAPHY

"Alkaloids, Manufacture" in *ECT* 1st ed., Vol. 1, pp. 507–516, by N. Applezweig, Hygrade Laboratories, Inc.; "Alkaloids, History, Preparation, and Use" in *ECT* 2nd ed., Vol. 1, pp. 778–809, by G. H. Svoboda, Eli Lily and Co.; "Alkaloids, Survey" in *ECT* 2nd ed., Vol. 1, pp. 758–778, by W. I. Taylor, Ciba Pharmaceutical Co.; "Alkaloids" in *ECT* 3rd ed., Vol. 1, pp. 883–943, by Geoffrey A. Cordell, College of Pharmacy, University of Illinois."Alkaloids," in *ECT* 4th ed., Vol. 1, pp. 1039–1087, by David R. Dalton, Temple University; "Alkaloids" in *ECT* (online), posting date: December 4, 2000, by David R. Dalton, Temple University.

CITED PUBLICATIONS

1. T. I. Williams, *Drugs from Plants*, Sigma, London, 1947, p. 87; L. S. Goodman and A. Gilman, *The Pharmacological Basis of Therapeutics: A Textbook of Pharmacology*, 3rd ed., Macmillan, New York, 1965, pp. 247–266.
2. C. Derosne, *Ann. Chim. (Paris)* **45**, 257 (1803).
3. F. W. Serturner, *Ann. Chim. Phys.* (2)**5**, 21 (1817).
4. H.-G. Boit, *Ergebnisse der Alkaloid-Chemie Bis 1960*, Akademie-Verlag, Berlin, 1961.
5. J. S. Glasby, *Encyclopedia of the Alkaloids*, Vols. 1–4, Plenum Press, New York, 1975.
6. S. W. Pelletier, *Alkaloids. Chemical and Biological Perspectives*, Vol. 1, John Wiley & Sons, Inc., New York, 1983, pp. 25–27.
7. I. W. Southon and J. Buckingham, eds., *Dictionary of Alkaloids*, Chapman and Hall, New York, 1989.
8. R. H. F. Manske and H. L. Holmes, eds., *The Alkaloids: Chemistry and Physiology*, Vol. 1, Academic Press, Inc., New York, 1950. This series gives a detailed exposition of the chemistry and pharmacology of the alkaloids, by structural class. Vol. 57, G. A. Cordell, ed. was published in 2001.
9. T. M. Kutchan, in G. A. Cordell, ed., *The Alkaloids, Chemistry and Biology*, Vol. 50, Academic Press, New York, 1998, pp. 257–316; D. A. Rathbone, D. L. Lister, and N. C. Bruce in G. A. Cordell, ed., *The Alkaloids, Chemistry and Biology*, Vol. 57, Academic Press, New York, 2001, pp. 1–74.
10. E. Hasse, *Biblotheca Zoologica* **8**, 1 (1892).
11. J. V. Euw, T. Reichstein, and M. Rothschild, *Israel J. Chem.* **6**, 659 (1968).
12. J. A. Edgar, P. A. Cockrum, and J. L. Frahn, *Experientia* **32**, 1535 (1976).

13. J. Meinwald, *Ann. N. Y. Acad. Sci.* **471**, 197 (1986).

14. T. Eisner, M. Goetz, D. Aneshansley, G. Ferstandig-Arnold, and J. Meinwald, *Experientia* **42**, 204 (1986).

15. J. Smolanoff, A. F. Kluge, J. Meinwald, A. McPhail, R. W. Miller, Karen Hicks and T. Eisner, *Science* **188**, 734 (1975).

16. J. J. Tufariello, H. Meckler, and K. P. A. Senaratne, *J. Am. Chem. Soc.* **106**, 7979 (1984).

17. P. J. Pelleiter and J. B. Caventou, *Ann. Chim. Phys.* **8**, 323 (1818); **10**, 142 (1819).

18. J. Geiger, *Berzelius' Jahresber.* **12**, 220 (1870).

19. A. Ladenburg, *Berzelius* **19**, 439 (1886).

20. A. F. Peerdeman, *Acta Crystallogr.* **9**, 824 (1956).

21. R. B. Woodward, M. P. Cava, W. D. Ollis, A. Hunger, H. U. Daeniker, and K. Schenker, *Tetrahedron* **19**, 247 (1963).

22. L. R. Caporael, *Science* **192**, 21 (1976).

23. H. Weiland, F. Konz, and K. Mittasch, *Ann.* **513**, 1 (1934).

24. G. J. Cardinale and co-workers, *Life Sci.* **40**, 301 (1987).

25. O. Dragendorff, *Z. Anal. Chem.* **137** (1866); *The Merck Index*, 5th ed., Merck & Co., Rahway, N. J., 1940, p. 687.

26. H. Mayer, *Am. J. Pharm.* **35**, 20 (1863); *The Merck Index*, 5th ed., Merck & Co., Rahway, N.J., 1940, p. 883.

27. N. R. Farnsworth, *J. Pharm. Sci.* **55**, 225 (1966).

28. R. W. Kondrat, R. G. Cooks, and J. L. McLaughlin, *Science* **199**, 978 (1978).

29. G. Zweig and J. Sherma, eds., *CRC Handbook of Chromatography*, CRC Press, Cleveland, Ohio, 1972.

30. E. Stahl, Dunnschicht-Chromatographie, Springer, Berlin, 1969.

31. V. de Luca in P. M. Dey and J. B. Harborne, eds. *Methods in Plant Biochemistry*, Vol 9, P. J. Lea, ed, *Enzymes of Secondary Metabolism*, Academic Press, New York, 1993, pp. 345–368.

32. T. Hashimoto and Y. Yamada, in P. M. Dey and J. B. Harborne, eds. *Methods in Plant Biochemistry*, Vol 9, P. J. Lea, ed, *Enzymes of Secondary Metabolism*, Academic Press, New York, 1993, pp. 369–379.

33. T. M. Kutchan, in B. E. Ellis, and co-workers, eds., *Genetic Engineering of Plant Secondary Metabolism*, Plenum Press, New York, 1994, pp. 35–59.

34. M. F. Roberts, in M. F. Roberts and M. Wink, eds., *Alkaloids, Biochemistry, Ecology and Medicinal Applications*, Plenum Press, New York, 1998, pp. 109–146.

35. K. Saito and I. Marakoshi, in M. F. Roberts and M. Wink, eds., *Alkaloids, Biochemistry, Ecology and Medicinal Applications*, Plenum Press, New York, 1998, pp. 147–157.

36. J. Berlin and L. F. Fecker, in R. Verpoorte and A. W. Alfermann, eds., *Metabolic Engineering of Plant Secondary Metabolism*, Kluwer Academic, Boston, 2000, pp. 195–216.

37. R. Bentley and I. M. Campbell, in M. Florkin and E. H. Stotz, eds., *Comprehensive Biochemistry*, Vol. 20, Elsevier, New York, 1968, pp. 415ff.

38. E. Winterstein and G. Trier, *Die Alkaloide*, Berntrager, Berlin, 1910.

39. Sir Robert Robinson, *The Structural Relations of Natural Products*, Oxford University Press, Oxford, 1955.

40. A recent redefinition of "secondary metabolism" has been set forth. See, R. Verpoorte, in R. Verpoorte and A. W. Alfermann, eds., *Metabolic Engineering of Plant Secondary Metabolism*, Kluwer Academic, Boston, 2000, p1 ff.

41. I. D. Spenser, in M. Florkin and E. H. Stotz, eds., *Comprehensive Biochemistry*, Vol. 20, Elsevier, New York, 1968, pp. 231ff.

42. G. A. Cordell, *Introduction to Alkaloids: A Biogenetic Approach*, Wiley-Interscience, New York, 1981.

43. D. R. Dalton, *The Alkaloids—A Biogenetic Approach*, Marcel Dekker, New York, 1979.

44. R. W. Herbert, in S. W. Pelletier, ed., *Alkaloids, Chemical and Physiological Perspectives*, Vol. 3, Wiley-Interscience, New York, 1985.

45. E. Leete and S. H. Kim, *J. Am. Chem. Soc.* **110**, 2976 (1988).

46. G. Grue-Sorensen and I. D. Spenser, *J. Am. Chem. Soc.* **105**, 7401 (1983).

47. H. D. Boswell, B. Drager, R. McLauchla, A. Portsteffen, D. J. Robins, R. J. Robins, and N. J. Walton, *Phytochemistry* **52**, 871 (1999).

48. P. J. Facchini, *Ann. Rev. Plant Physiol. Plant Molec. Biol.* **52**, 29. (2001)

49. E. Leistner and I. D. Spenser, *J. Am. Chem. Soc.* **95**, 4715 (1973); T. Hemscheidt and I. D. Spenser, *J. Am. Chem. Soc.* **112**, 6360 (1990).

50. D. G. O'Donovan, D. J. Long, E. Forde, and P. Geary, *J. Chem. Soc. Perkin Trans.* **1**, 415 (1975).

51. W. M. Golebiewski and I. D. Spenser, *J. Am. Chem. Soc.* **106**, 7925 (1984).

52. T. M. Kutchan, *The Plant Cell*, **7**, 1059 (1995).

53. T. Robinson, *Phytochem.*, **4**, 67 (1965).

54. E. Leete and Y.-Y. Liu, *Phytochemistry* **12**, 593 (1973).

55. S. D. Copley and J. R. Knowles, *J. Am. Chem. Soc.* **109**, 5008 (1987); W. J. Guilford, S. D. Copley, and J. R. Knowles, *J. Am. Chem. Soc.* **109**, 5013 (1987).

56. R. E. Schultes and A. Hofmann, *Plants of the Gods*, McGraw-Hill Book Co., Inc., New York, 1979.

57. W. La Barre, D. P. McAllister, J. S. Slotkin, O. C. Stewart, and S. Tax, *Science* **114**, 582 (1952).

58. W. I. Taylor and A. R. Battersby, eds., *Oxidative Coupling of Phenols*, Marcel Dekker, New York, 1967.

59. Ref. 43, pp. 197ff.

60. A. W. Gerrard, *Pharm. J.* **8**, 214 (1877).

61. O. Moller, E. M. Steinberg, and K. Torssell, *Acta Chem. Scand.* **B32**, 98 (1978).

62. J. Bolssier, G. Combes, and J. Pagny, *Ann. Pharm. Fr.* **18**, 888 (1960).

63. T. Kametani and co-workers, *J. Chem. Soc. C*, 1043 (1971).

64. H. M. Fales, J. Mann, and S. H. Mudd, *J. Am. Chem. Soc.* **85**, 2025 (1963).

65. W. C. Wildman and D. T. Bailey, *J. Am. Chem. Soc.* **91**, 150 (1969).

66. W. C. Wildman, in R. H. F. Manske, eds., *The Alkaloids*, Vol. 11, Academic Press, New York, 1968, pp. 308ff.

67. E. H. Herman and D. P. Chadwick, *Pharmacology* **12**, 97 (1974); E. J. Gralla, G. L. Coleman, and A. M. Jonas, *Cancer Chemother. Rep. Part 3* **5**, 79 (1974).

68. D. H. R. Barton and T. Cohen, *Festschrift A. Stoll*, Birkhauser Verlag, Basel, 1957, p. 117.

69. M. A. Barbier, *Ann. Pharm.* **5**, 121 (1947).

70. M. Gates and G. Tschudi, *J. Am. Chem. Soc.* **78**, 1380 (1956).

71. H. I. Parker, G. Blaschke, and H. Rapoport, *J. Am. Chem. Soc.* **94**, 1276 (1972).

72. D. H. R. Barton, C. J. Potter, and D. A. Widdowson, *J. Chem. Soc. Perkin Trans.* **1**, 346 (1974); D. H. R. Barton, R. D. Bracho, C. J. Potter, and D. A. Widdowson, *J. Chem. Soc. Perkin Trans. 1* 2278 (1974).

73. M. H. Zenk, *Special Publication of the Royal Society Chemistry*, 1995, Vol. 148, *Organic Reactivity: Physical and Biological Aspects*, pp. 89–109.

74. J. E. Saxton, ed., *Indoles, Part Four, The Monoterpenoid Indole Alkaloids*, Wiley-Interscience, New York, 1983.

75. M. M. Rapport, A. A. Green, and I. H. Page, *J. Biol. Chem.* **176**, 1243 (1948).

76. K. Stolle and D. Groger, *Arch. Pharm. (Weinheim)* **301**, 561 (1968).

77. A. Hofmann, *Botanical Museum Leaflets*, Vol. 20, Harvard University, Cambridge, Mass., 1963, p. 194.

78. Fr. Add. 91,948 (Aug. 30, 1968), J. Rutschmann and H. Kobel (Sandoz Ltd.).

79. W. A. Jacobs and L. C. Craig, *Science* **82**, 16 (1935).

80. A. R. Battersby, A. R. Burnett, and P. G. Parsons, *Chem. Commun.*, 1280 (1968).

81. I. Kompis, M. Hesse, and H. Schmid *Lloydia* **34**, 269 (1971) (gives a much more elaborate classification scheme).

82. G. Holmberg and S. Gershon, *Psychopharmacologia* **2**, 93 (1961); M. L. Brown, S. Gershon, W. J. Lang, and B. Korol, *Arch. Intern. Pharmacodyn.* **160**, 407 (1966).

83. B. Armstrong, N. Stevens, and R. Doll, *Lancet* **2**, 672 (1974).

84. J. L. Hartwell, *Cancer Treatment Rep.* **60**, 1031 (1976).

85. R. B. Woodward, F. E. Bader, H. Bickel, A. J. Frey, and R. W. Kierstead, *Tetrahedron* **2**, 1 (1958).

86. M. L. Chatterjee and M. S. De, *Bull. Calcutta School Trop. Med.* **5**, 173 (1957); *Chem. Abstr.* **52**, 8356a (1958).

87. S. A. Kozmin, T. Iwama, Y. Huang and V. H. Rawal, *J. Am. Chem. Soc.* **124**, 4628 (2002).

88. J. Schwyzer, *Die Fabrikation Pharmazeutischer and Chemisch, Technischer Produkte*, Springer-Verlag, Berlin, 1931.

89. W. I. Taylor and N. R. Farnsworth, eds., *The Catharanthus Alkaloids, Botany, Chemistry, Pharmacology and Clinical Uses*, Marcel Dekker, New York, 1973; W. I. Taylor and N. R. Farnsworth, eds., *The Vinca Alkaloids, Botany, Chemistry and Pharmacology*, Marcel Dekker, New York, 1973.

90. G. Buchi, P. Kulsa, K. Ogasawara, and R. L. Rosati, *J. Am. Chem. Soc.* **92**, 999 (1970) and references therein.

91. A. Ravina, *Presse Med.* **74**, 525 (1978).

92. G. A. Cordell in Ref. 74, pp. 539ff.

93. W. T. Stearn, *Lloydia* **29**, 196 (1966).

94. W. A. Creasey, in F. Hahn, ed., *Antibiotics*, Vol. 5, Springer-Verlag, Berlin, 1979, p. 414.

95. E. K. Rowinsky, L. A. Cazenave, and R. C. Donehower, *J. Natl. Cancer Inst.* **82**, 1247 (1990) (deals with antimicrotubule agents, albeit with a nonalkaloidal agent).

96. J. P. Kutney and co-workers, *J. Am. Chem. Soc.* **97**, 5013 (1975); M. E. Kuehne and T. C. Zebovitz, *J. Org. Chem.* **52**, 4331 (1987); M. E. Kuehne, T. C. Zebovitz, W. G. Bornmann, and I. Marko, *J. Org. Chem.* **52**, 4340 (1987).

97. G. Popjak and J. W. Cornforth, *Biochem. J.* **101**, 553 (1966); J. W. Cornforth, R. H. Cornforth, A. Pelter, M. G. Horning, and G. Popjak, *Tetrahedron* **5**, 311 (1959); R. B. Clayton in T. W. Goodwing, ed., *Aspects of Terpenoid Chemistry and Biochemistry*, Academic Press, New York, 1971, pp. 1ff.

98. T. Sakan, *Tampakushitsu Kakusan Koso* **12**, 2 (1967); *Chem. Abstr.* **73**, 42351c (1970).

99. G. L. Gatti and M. Marotta, *Ann. Ist. Super. Sanita* **2**, 29 (1966); *Chem. Abstr.* **65**, 14293e (1966).

100. W. C. Wildman, J. LeMen, and K. Wiesner, in W. I. Taylor and A. R. Battersby, eds., *Cyclopentanoid Terpene Derivatives*, Marcel Dekker, New York, 1969, pp. 239ff.

101. O. E. Edwards, in W. I. Taylor and A. R. Battersby, eds., *Cyclopentanoid Terpene Derivatives*, Marcel Dekker, New York, 1969, pp. 357ff.

102. T. T. Tchen and K. Block, *J. Am. Chem. Soc.* **77**, 6085 (1955); R. B. Clayton and K. Block, *J. Biol. Chem.* **218**, 319 (1956); L. J. Goad, *Symp. Biochem. Soc.* **29**, 45 (1970).

103. G. A. Swan, *An Introduction to the Alkaloids*, John Wiley & Sons, Inc., New York, 1967, p. 274.

104. S. W. Pelletier, N. V. Mody, and H. K. Desai, *J. Org. Chem.* **46**, 1840 (1981).

105. H. Ripperger, W. Mortiz, and K. Schreiber, *Phytochemistry* **10**, 2699 (1971).

106. S. J. Jadav, D. K. Salunkhe, R. E. Wyse, and R. R. Dalvi, *J. Food Sci.* **38**, 453 (1973).

107. U. S. Pat. 3,108,876 (Oct. 29, 1963), H. H. Turken and T. P. Daley (to Duncan Coffee Co.); U. S. Pat. 3,361,571 (Jan. 2, 1968), L. Nutting and G. S. Chong (to Hills Bros. Coffee, Inc.).
108. E. J. Miller, S. R. Pinnell, G. R. Martin, and E. Schiffman, *Biochem. Biophys. Res. Commun.* **26**, 132 (1967).
109. E. Monkovic, I. D. Spenser, and A. O. Plunkett, *Can. J. Chem.* **45**, 1935 (1967).
110. M. Matsuo and Y. Kasida, *Chem. Pharm. Bull. (Tokyo)* **14**, 1108 (1966).
111. V. A. Snieckus, *Alkaloids (N.Y.)* **14**, 425 (1973).
112. E. Leete, *Acc. Chem. Res.* **4**, 100 (1971).
113. International Narcotics Control Board—;Vienna, (www.incb.org) *Narcotic Drugs, Estimated World Requirements for 1990, Statistics for 1988*, United Nations Publ. E/F/S.89.XI.3, New York, 1989, pp. 33ff. Current statistics are from (http://www. incb.org/e/index.htm?) Statistics for 2000–2002.
114. L. Mahnke, *Aach. Geog. Arbeit.* **19**, 137 (1987).
115. http://apps. fao. org/page/form?collection=CBD. CropsAndProducts& Domain=CBD &servlet=1&language=EN&hostmane=apps.fao.org&version=default.

David R. Dalton
Temple University
Linda M. Mascavage
Arcadia University
Michael Wilson
Wilson Technologies, Ltd.

ALKANOLAMINES FROM NITRO ALCOHOLS

1. Introduction

The nitro alcohols (qv), obtained by the condensation of nitroparaffins (qv) with formaldehyde [50-00-0], may be reduced to a unique series of alkanolamines (β-amino alcohols):

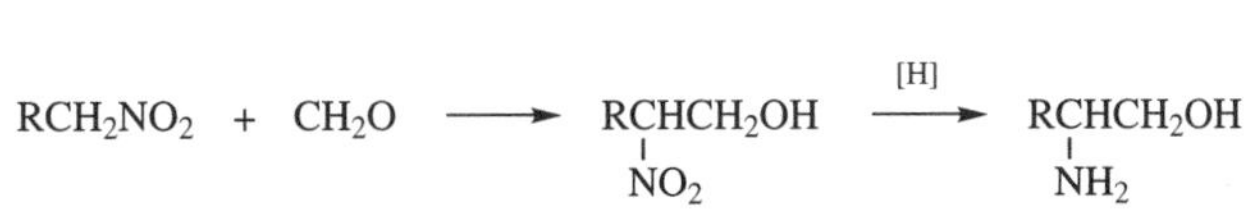

$$RCH_2NO_2 + CH_2O \longrightarrow \underset{NO_2}{RCHCH_2OH} \xrightarrow{[H]} \underset{NH_2}{RCHCH_2OH}$$

The condensation may occur one to three times, depending on the number of available hydrogen atoms on the α-carbon of the nitroparaffin, giving rise to amino alcohols with one to three hydroxyl groups. A comprehensive review of these compounds has been published (1).

Many members of this series are known, based on nitroparaffin condensations with aldehydes of longer chain length than formaldehyde. However, only the five primary amino alcohols discussed in the following are manufactured on a commercially-significant scale. N-Substituted derivatives of these compounds

Table 1. **Commercial Alkanolamines**

Name	Common designation	Molecular formula	Molecular weight	CAS Registry Number	EINECS Number
2-amino-2-methyl-1-propanol	AMP	$C_4H_{11}NO$	89.14	[124-68-5]	204-709-8
2-amino-2-ethyl-1,3-propanediol	AEPD[a]	$C_5H_{13}NO_2$	119.16	[115-70-8]	204-101-2
2-dimethylamino-2-methyl-1-propanol	DMAMP	$C_6H_{15}NO$	117.19	[7005-47-2]	230-279-6
2-amino-2-(hydroxymethyl)-1,3-propanediol[b]	TRIS AMINO[a]	$C_4H_{11}NO_3$	121.14	[77-86-1]	201-064-4
2-amino-2-methyl-1,3-propanediol	AMPD	$C_4H_{11}NO_2$	105.16	[115-69-5]	204-100-7
2-amino-1-butanol	AB[a]	$C_4H_{11}NO$	89.14	[96-20-8]	202-488-2

[a] AB, AEPD, and TRIS AMINO are trademarks of ANGUS Chemical Company.
[b] Common name is tris(hydroxymethyl)aminomethane.

also have been prepared, but only 2-dimethylamino-2-methyl-1-propanol is available in commercial quantities (Table 1).

2. Physical Properties

Physical properties of the six commercial alkanolamines are given in Table 2. Because 2-amino-2-methyl-1-propanol (AMP) and 2-amino-2-ethyl-1,3-propanediol (AEPD) melt near room temperature and usually contain some water, they may be semisolid pastes, rather than crystalline solids. Water-diluted forms of both these alkanolamines are marketed because such solutions remain liquid at lower temperatures than do the pure compounds, eg, AMP-95, which contains 5% water, solidifies at $-2°C$.

These compounds are highly soluble in water. AMP, AB, AEPD, and DMAMP are completely miscible in water at 20°C; the solubility of AMPD is

Table 2. **Properties of Alkanolamines**

Compound	Boiling point, °C	Melting point, °C	Specific gravity	pH of 0.1 M aqueous solution	pK_a at 25°C
AMP	165[a]	30–31	0.934[b]	11.3	9.72
AEPD	152–153[c]	37.5–38.5	1.099[b]	10.8	8.80
DMAMP	160[a]	19	0.90[d]	11.6	10.2
TRIS AMINO	219–220[c]	171–172	N.A.[e]	10.4	8.03
AMPD	151–152[c]	109–111	N.A.[e]	10.8	8.76
AB	178[a]	−2	0.944[b]	11.1	9.52

[a] At 101.3 kPa (1 atm).
[b] At 20/20°C.
[c] At 1.3 kPa (10 mm Hg).
[d] At 25/25°C.
[e] N.A. = Not applicable.

250 g/100 mL H_2O at 20°C. They are generally very soluble in alcohols, slightly soluble in aromatic hydrocarbons, and nearly insoluble in aliphatic hydrocarbons; tris(hydroxymethyl)aminomethane is appreciably soluble only in water (80 g/100 mL at 20°C) and methanol.

Alkanolamines have high boiling points; under normal ambient conditions, their vapor pressures are low. Only DMAMP (see Table 2) forms an azeotrope with water, which boils at 98°C and contains 25% by weight of DMAMP. According to current DOT regulations, AMP, AMP-95, DMAMP, DMAMP-80, AEPD, and AB are all classified as combustible liquids.

DMAMP and AMP are among the most strongly basic commercially-available amines. The dissociation constants of these materials appear in Table 2. All alkanolamines have slight amine odors in the liquid state; the solid products are nearly odorless.

3. Chemical Properties

The alkanolamines discussed here exhibit the chemical reactivity of both amines and alcohols, as is the case with other alkanolamines. Typically, they attack copper, brass, and aluminum, but not steel or iron. Alkanolamines are useful as amination agents; however, the reactivity of both the amino and alcohol group must be considered in attempting any specific synthetic scheme with them.

With mineral acids, the alkanolamines form ammonium salts which hydrolyze readily in the presence of water and dissociate upon heating. Fatty acids, such as oleic acid, give soaps which are highly efficient emulsifying agents with important industrial uses, particularly the soaps of AMP (see EMULSIONS; SURFACTANTS).

On heating, an alkanolamine soap first dehydrates to the amide; this is also obtained from the methyl ester of the fatty acid by heating with the alkanolamine at 60°C in the presence of a catalytic amount of sodium methoxide. Methanol is removed under partial vacuum. At higher temperature, the amide is dehydrated to an oxazoline.

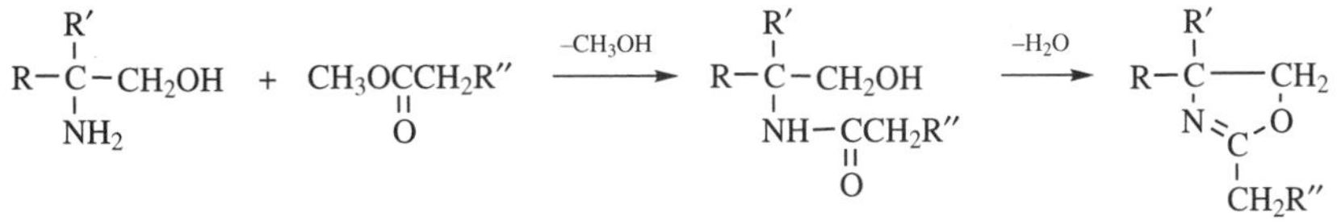

where for AEPD, R = C_2H_5, R′ = CH_2OH and for TRIS AMINO, R = R′ = CH_2OH

These oxazolines have cationic surface-active properties and are emulsifying agents of the water-in-oil type. They are acid acceptors and, in some cases, corrosion inhibitors (see CORROSION). Reactions of AMP with organic acids to create oxazoline functionality are useful as a tool for determination of double-bond location in fatty acids (2), or for use as a carboxylic-acid protective group in synthesis (3,4). The oxazolines from AMPD, AEPD, and TRIS AMINO contain hydroxyl groups that can be esterified easily, giving waxes (qv) with saturated acids and drying oils (qv) with unsaturated acids.

Formaldehyde reacts with the hydrogen on the α-carbon of the fatty acid from which the oxazoline was formed to yield a vinyl monomer that can be polymerized or utilized for synthesis (5). Thus, esters of the oxazoline formed from TRIS AMINO undergo the reaction:

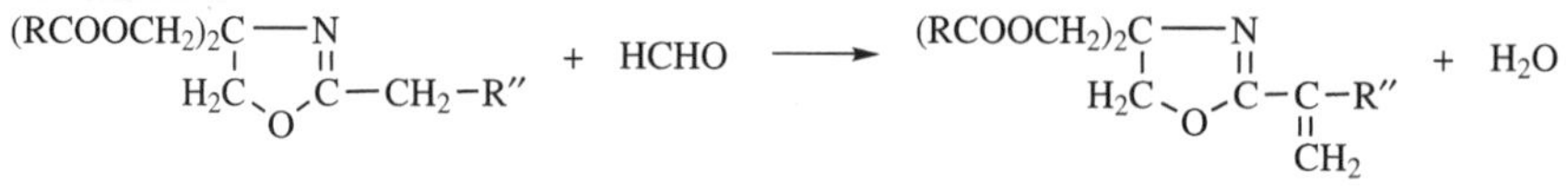

$$(RCOOCH_2)_2C\overset{\displaystyle N}{\underset{\displaystyle H_2C \diagdown O \diagup C-CH_2-R''}{|}} + HCHO \longrightarrow (RCOOCH_2)_2C\overset{\displaystyle N}{\underset{\displaystyle H_2C \diagdown O \diagup \overset{|}{\underset{CH_2}{C}}-\overset{||}{C}-R''}{|}} + H_2O$$

These products are useful for modification of alkyd resins (qv), preparation of paint vehicles, and copolymerization with other monomers.

Substitution on the amino group occurs readily, giving bases stronger than the parent amines.

$$R\text{---}NH_2 + 2\,HCHO \xrightarrow{2\,H_2} R\text{---}N(CH_3)_2 + 2\,H_2O$$

Alkanolamines react with nitro alcohols to form nitrohydroxylamines (6).

$$\underset{NO_2}{R_2C\text{-}CH_2OH} + \underset{CH_2OH}{NH_2\text{-}CR'_2} \longrightarrow \underset{NO_2}{R_2C\text{-}CH_2\text{-}NH\text{-}CR'_2} + H_2O$$

$$(1)$$

Some of these compounds show antibacterial activity. Reduction gives 2-[(2-aminoethyl)amino]ethanols that react with organic acids to form amides that, on further heating, cyclize to imidazolines (7). For example, the diamine obtained by reducing (1) reacts with an organic acid (R''COOH) to give:

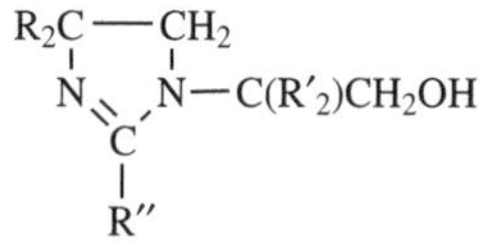

$$\begin{array}{c} R_2C\text{---}CH_2 \\ \underset{\diagdown \overset{\displaystyle C}{\underset{\displaystyle R''}{|}} \diagup}{N} \quad N\text{---}C(R'_2)CH_2OH \end{array}$$

Mercaptothiazolines are obtained from the corresponding sulfate esters and carbon disulfide.

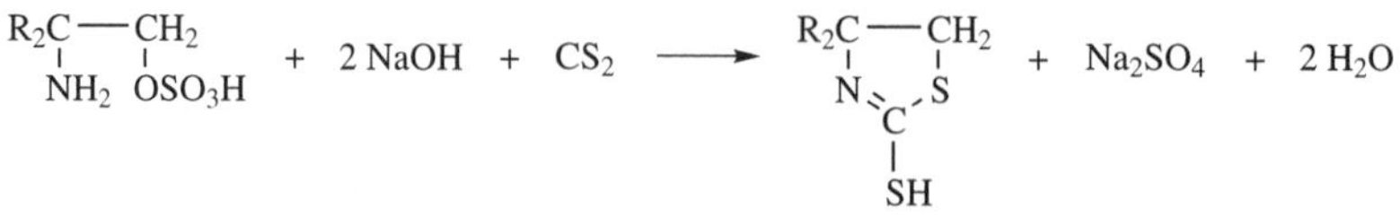

$$\underset{NH_2 \quad OSO_3H}{R_2C\text{---}CH_2} + 2\,NaOH + CS_2 \longrightarrow \underset{\underset{SH}{\overset{|}{C}}}{\overset{R_2C\text{---}CH_2}{N\diagdown \underset{C}{} \diagup S}} + Na_2SO_4 + 2\,H_2O$$

Aldehydes react with monohydric alkanolamines to give monocyclic oxazolidines

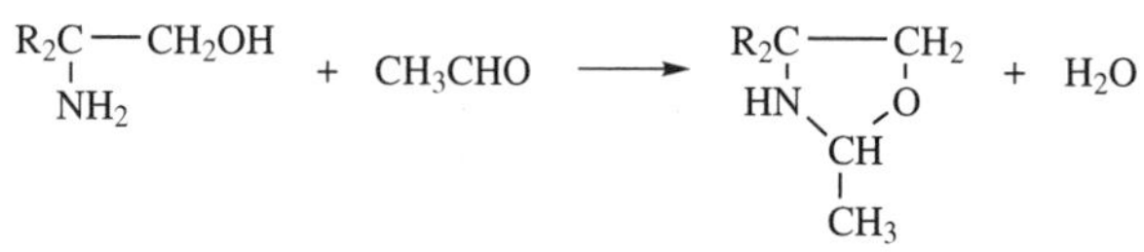

$$\underset{NH_2}{R_2C\text{---}CH_2OH} + CH_3CHO \longrightarrow \underset{HN\diagdown \underset{\underset{CH_3}{\overset{|}{CH}}}{} \diagup O}{R_2C\text{---}CH_2} + H_2O$$

or with polyhydric alkanolamines to give bicyclic oxazolidines (8,9).

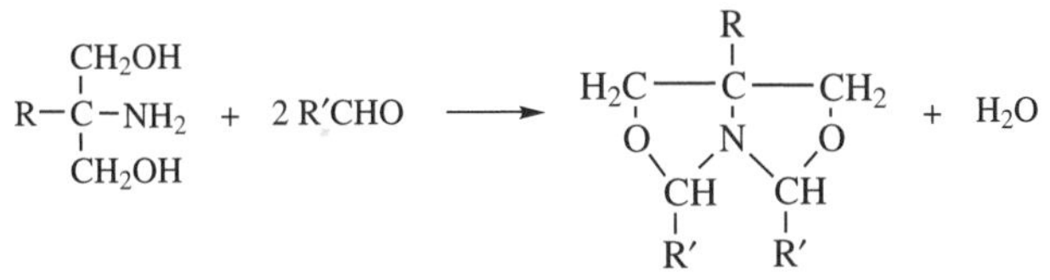

$$
\begin{array}{c}
CH_2OH \\
| \\
R-C-NH_2 \\
| \\
CH_2OH
\end{array}
+ \ 2\,R'CHO \ \longrightarrow \
\begin{array}{c}
R \\
| \\
H_2C-C-CH_2 \\
\diagdown \ \ | \ \ \diagup \\
O \ \ N \ \ O \\
| \ \ \ \ | \\
CH \ \ CH \\
| \ \ \ \ | \\
R' \ \ \ R'
\end{array}
+ \ H_2O
$$

(2)

These can be hydrogenated to N,N-substituted alkanolamines. Thus (2) yields:

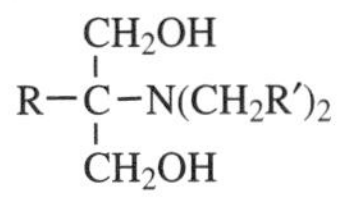

$$
\begin{array}{c}
CH_2OH \\
| \\
R-C-N(CH_2R')_2 \\
| \\
CH_2OH
\end{array}
$$

Oxidation of the hydroxyl group, after protection of the amine group by benzoylation, gives amino acids (8), eg, oxidation of 2-amino-2-methyl-1-propanol to 2-methylalanine [62-57-7], $(CH_3)_2CNH_2COOH$.

4. Manufacture

The reduction of nitro alcohols to alkanolamines is readily accomplished by hydrogenation in the presence of Raney nickel catalyst (1,10,11).

AMP, AEPD, and AB are purified by distillation. TRIS AMINO and AMPD are purified by crystallization. TRIS AMINO concentrate in water (40% assay) is also available.

2-Dimethylamino-2-methyl-1-propanol is manufactured from AMP by hydrogenation in the presence of formaldehyde and purified by distillation. It is marketed primarily as DMAMP-80 (water added), however.

5. Economic Aspects

Production statistics on alkanolamines are not available, but they are sold in 1000-ton quantities and are available for bulk shipment except for DMAMP, AB, TRIS AMINO, and AMPD (the latter two being crystalline solids). TRIS AMINO concentrate is available in bulk. AMPD is manufactured only in low volumes to meet limited demand in certain specialized uses.

ANGUS Chemical Company is the basic manufacturer of technical-grade TRIS AMINO. However, ANGUS and numerous processors offer recrystallized, higher purity grades of this alkanolamine for specialized applications (tris buffer; tromethamine USP).

Table 3 gives the 2002 prices for bulk quantities and the net weights packaged in drums or bags; Table 4 provides specification values.

Table 3. **2002 U.S. Prices of Alkanolamines, $/kg**

Product	Tank car	Carload	Net weight, kg[a]
AMP	6.97	7.19	191.6 (D)
AMP-95	3.81	4.07	191.6 (D)
AEPD		5.08	204.5 (D)
AEPD-85	3.52	3.43	204.5 (D)
TRIS AMINO crystals		17.67	22.7 (B)
TRIS AMINO 40% aq.	5.98	6.09	226.9 (D)
TRIS AMINO ultrapure		22.10	50 (F)
DMAMP-80		13.44	186 (D)
AMPD		98.78	11.3,22.6 (F)
AB		9.90	193.2 (D)

[a] D = 208-L (55-gal) steel drums; B = bags; F = fiber drums.

6. Health and Safety Factors

Alkanolamines are only slightly toxic by ingestion (acute oral LD_{50} in rodents = 1.0–5.5 g/kg).

Undiluted DMAMP, AMP, and AB cause eye burns and permanent damage, if not washed out of the eye immediately. They are also severely irritating to the skin, causing burns upon prolonged or repeated contact. Of these three alkanolamines, only AMP has been studied in subchronic and chronic oral studies. The principal effect noted was the action of AMP on the stomach as a result of its alkalinity. The no-observed-effect level (NOEL) in a 1-year feeding study in dogs was 110 ppm in the diet. In general, the low volatility, and the applications for which these products are used, preclude the likelihood of exposure by inhalation.

Table 4. **Specifications for Alkanolamines**

Product	Neutral equivalent	Color, APHA, max	Water, wt %, max	Melting point, °C, min	Amine assay, %
AMP	88.5–91.0	20	0.8		
AMP-95	93.0–97.0	20	4.8–5.8		
AEPD	124.0, max	2[a,b]	3.8		
AEPD-85	none established	2[a,b]	13.0–15.0		
DMAMP-80		100	18.0–22.0		78.0–82.0[d]
AMPD[c]	103.0–107.0	50[b]	0.5	100.0	
AB		100	0.5		98.0
TRIS AMINO					
40% aqueous		5[a]			38.0–42.0[d]
crystals	121.0–122.0	40[b]	0.5	160.0	
ultrapure[e,f]		20[b]	0.2	170.0–172.0	99.9–100.1

[a] Gardner color.
[b] 20% aq solution.
[c] Residue on ignition = 0.01 wt%, max; insoluble matter = 0.01 wt%, max.
[d] By titration.
[e] Additional specifications for heavy metal content; meets USP and ACS specifications.
[f] Residue on ignition = 0.05 wt%, max; insoluble matter = 0.005 wt%, max.

AEPD is severely irritating to the eyes and should be washed out immediately on contact; it is only mildly irritating to the skin.

AMPD and TRIS AMINO, normally crystalline solids, are of less concern in terms of irritancy to skin and eyes.

The 40% aqueous solution of TRIS AMINO is nonirritating to the eyes and skin. In general, the toxicology of the alkanolamines is typical of alkaline materials, ie, the greater the base strength, the greater the effect. Neutralized alkanolamines are much less toxic; their stearate soaps, eg, have been found to be nonhazardous.

TRIS AMINO, the least toxic of this series of alkanolamines, has been studied extensively as a buffer (12). It is used in a number of pharmaceutical applications (13).

Environmentally, these alkanolamines present little problem. Only AMP has been studied in great detail, but it was found to be degradable according to OECD guidelines, to be of low toxicity to fish and microorganisms, and to be nonaccumulative. TRIS AMINO is actually added to the water used for shipment of living fish, in order to improve their viability (14).

7. Uses

Because they are closely related, the alkanolamines can sometimes be used interchangeably. However, cost/performance considerations generally dictate a best choice for specific applications.

Functional Fluids. The fatty acid soaps of alkanolamines are excellent emulsification agents for use in functional fluids such as hydraulic and metalworking fluids. For example, improved hardwater stability of a hydraulic fluid emulsion is obtained using AMP in the formulation (15). AMP has also been shown to enhance the chemical and biostability of metalworking fluids (16).

Pigment Dispersion. AMP is used widely as a pigment co-dispersant in water-based paints and paper coatings. In small amounts, it efficiently disperses pigments and improves pH and viscosity stability, corrosion inhibition, odor and dried-film properties (17). When AMP is used in conjunction with other surfactants, enhanced performance is obtained with lower levels of these ingredients in the dispersion.

TiO_2 and clay slurries which utilize AMP as part of the dispersant system are available in bulk for the paint and paper industries (see PIGMENTS). AMP has been used as a particle-surface treatment for production of TiO_2 with improved luster and dispersibility (18).

Resin Solubilizers. In general, water-soluble resins are amine salts of acidic polymers. Water-soluble coatings formulated with AMP-95 or DMAMP-80 exhibit superior performance (19,20) (see WATER-SOLUBLE POLYMERS). AMP-95, used in conjunction with associative thickeners (21) or derivatized cellulose, provides for the most efficient utilization of such thickeners.

AMP is also the neutralizer/solubilizer of choice for use with acid-functional hair-fixative resins (22). AMP and TRIS AMINO are efficient neutralizers of polyacrylate (carbomer) resins used for thickening or gelling of topical pharmaceutical and personal-care products (23).

Catalysts. The alkanolamines continue to find use in blocked-catalyst systems for textile resins, coatings resins, adhesives, etc. Of particular utility in curing durable-press textiles is AMP·HCl. Other salts, such as those of the benzoin tosylate or *p*-toluenesulfonic acid, find utility in melamine- or urea-based coatings (24) (see AMINO RESINS AND PLASTICS).

Boiler Water Treatment. Alkanolamines in general provide corrosion protection for ferrous metals in many applications. When used in boiler water treatment, AMP provides excellent protection to condensate-return lines through efficient absorption of CO_2, effective distribution ratio for transport throughout the system, and minimum amine loss in the deaerating heater (25–27).

Formaldehyde Scavenging. The formation of oxazolidines from alkanolamines and formaldehyde is rapid at room temperature and provides a method for the elimination of excess free formaldehyde from products such as urea–formaldehyde resins. AEPD and TRIS AMINO are the products of choice for this purpose because 1 mol of each will react with 2 mol of formaldehyde (28).

Applications in Oil and Gas Production. AMP, as a hindered amine, is useful for removing acid-gas contaminants such as CO_2 or H_2S from gas streams (29). It also has been utilized in tertiary oil recovery to enhance removal of petroleum from marginal wells (30).

AB has also shown value in purification systems for fluid streams (31).

Biomedical Applications. TRIS AMINO is used for a number of biomedical purposes. In its pure form, it is an acidimetric standard; TRIS AMINO also is useful in biotechnology as a buffering agent for enzyme systems, industrial protein purification, and electrophoretic separations (32).

TRIS AMINO, together with ethylenediaminetetraacetic acid (EDTA), can enhance the effectiveness of antimicrobial agents and antibiotics (33).

AB, the only optically active member of this alkanolamine series, is used as a raw material for production of the anti-tuberculosis drug ethambutol[74-55-5], and in chiral synthesis/resolution of drug optical isomers (34).

Synthetic Applications. Oxazolines, which are synthesized as indicated above, have been utilized in many different applications (35). When used in resin formulations, AMP, AEPD, and TRIS AMINO can incorporate the oxazoline structure into the polymer structure (36). Because they are polyols, both AEPD and TRIS AMINO can be used in polyester-resin modification. Oxazoline alkyd films are characterized by improved performance, particularly salt-spray resistance and gloss (see ALKYD RESINS; COATINGS, SPECIAL PURPOSE, HIGH PERFORMANCE).

Other oxazolines produced from alkanolamines are useful as oil-soluble surface-active agents and corrosion inhibitors. Synthetic oxazoline waxes promote lubricity and mar-resistance of coatings.

Oxazolidines, formed by reaction of alkanolamines with aldehydes, are useful as leather tanning agents (37) and are effective curing agents for proteins, phenolic resins, moisture-cure urethanes, etc. They also find use as antimicrobial agents (38).

BIBLIOGRAPHY

"Alkanolamines from Nitro Alcohols" in *ECT* 2nd ed., Vol. 1, pp. 824–831, by R. H. Dewey, Commercial Solvents Corporation; in *ECT* 3rd ed., Vol. 1, pp. 961–967, by Robert H.

Dewey and Allen F. Bollmeier, Jr., International Minerals & Chemical Corporation; in *ECT* 4th ed., Vol. 2, pp. 26–34, by Allen F. Bollmeier, Jr., ANGUS Chemical Company; "Alkanolamines from Nitro Alcohols" in *ECT* (online), posting date: December 4, 2000, by Allen F. Bollmeier, Jr., ANGUS Chemical Corporation.

CITED PUBLICATIONS

1. H. B. Haas and E. F. Riley, *Chem. Rev.* **32**, 373 (1943).
2. J. Y. Zhang and co-workers, *Biomed. Environ. Mass Spectrom.* **15**(1), 33 (1988).
3. A. I. Meyers and co-workers, *J. Org. Chem.* **39**(18), 2787 (1974).
4. A. I. Meyers and K. A. Lutomski, *Synthesis* **2**, 105 (1983).
5. U.S. Pat. 2,559,440 (July 3, 1951), W. A. Jordan and S. H. Shapiro (to Armour and Company).
6. H. G. Johnson, *J. Am. Chem. Soc.* **68**, 14 (1946).
7. J. L. Riebsomer, *J. Am. Chem. Soc.* **70**, 1629 (1948).
8. J. H. Billman and E. E. Parker, *J. Am. Chem. Soc.* **66**, 538 (1944).
9. M. Senkus, *J. Am. Chem. Soc.* **67**, 1515 (1945).
10. M. Senkus, *Ind. Eng. Chem.* **40**, 506 (1948).
11. U.S. Pat. 3,564,057 (Feb. 16, 1971), J. B. Tindall (to Commercial Solvents Corp.).
12. G. G. Nahas and co-workers, *Ann. NY Acad. Sci.* **92**, 333 (1961).
13. A. C. Eachus, *Chimica Oggi*, **12**(1–2), 24 (1994).
14. W. N. McFarland and K. S. Norris, *Calif. Fish Game*, **44**(4), 291 (1958).
15. U.S. Pat. 4,428,855 (Jan. 31, 1984), D. A. Law and J. Shim (to Mobil Oil Corp.).
16. U. Aumann and co-workers, *Lubricants World*, **9**(5), 20 (1999).
17. G. N. Robinson, *Paint Coatings Ind.*, **16**(10), 160 (2000).
18. Ger. Pat. 2,924,849 (Jan. 22, 1981), K. Koehler and co-workers (to A. G. Bayer).
19. M. E. Woods, *Mod. Paint. Coat.* **65**, 21 (1975).
20. Can. Pat. 1,242,182 (Sep. 20, 1988), M. A. Fentrup and co-workers (to BASF Corp.).
21. L. W. Hill and Z. W. Wicks, *Prog. Org. Coat.* **8**(2), 161 (1980).
22. U. Aumann and A. C. Eachus, *Euro Cosmetics*, **9**(3), 52 (2001).
23. U. Aumann and A. C. Eachus, *SOFW-Journal*, **125**(10), 3 (1999).
24. L. R. Gatechair, *Polym. Mater. Sci. Eng.* **59**, 289 (1988).
25. D. G. Dionisio, *Mater. Perform.* **16**(5), 21 (1977).
26. M. J. Fountain and co-workers, *Proceedings of the International Water Conference, Engineering Society of Western Pennsylvania*, 45th, 1984, p. 393.
27. C. Fernandez Palomero and J. Perez Pallares, *Rev. Soc. Nucl. Esp.* **1**(56), 27 (1987).
28. Jpn. Pat. 53/37794 (Apr. 7, 1978), K. Ebisawa and O. Matsudaira (to Hitachi Chemical Co., Ltd.).
29. G. Sartori and co-workers, *Sep. Purif. Methods,* **16**, 171 (1987).
30. U.S. Pat. 4,485,021 (Nov. 27, 1984), R. F. Purcell and R. B. Kayser (to ANGUS Chemical Company).
31. International Pat. Appl. WO 00/18492 (Apr. 6, 2000), P. C. Rooney (to The Dow Chemical Company).
32. *Biotechnology Applications of TRIS AMINO (Tris Buffer), Technology Review No. 10,* ANGUS Chemical Company, Buffalo Grove, Ill., March 1994.
33. A. C. Eachus and U. Aumann, *Pharma. Chem.* **1**(9), 56 (2002).
34. *Applications of ANGUS AB in Chiral-Drug Synthesis, Benchmark No. 54,* ANGUS Chemical Company, Buffalo Grove, Ill., August 1997.

35. J. A. Frump, *Chem. Rev.* **71**, 483 (1971).
36. U.S. Pat. 3,382,197 (May 7, 1968), R. F. Purcell (to Commercial Solvents Corp.).
37. S. das Gupta, *J. Sol. Leather Technol. Chem.* **61**, 97 (1977).
38. A. C. Eachus and co-workers, *S.D.F.W.*, (9), 337 (1991).

ALAN C. EACHUS
ALLEN F. BOLLMEIER
ANGUS Chemical Co.

ALKANOLAMINES FROM OLEFIN OXIDES AND AMMONIA

1. Introduction

Ethylene oxide [75-21-8], propylene oxide [75-56-9], or butylene oxide [106-88-7] react with ammonia to produce alkanolamines (Table 1). Ethanolamines, $NH_{3-n}(C_2H_4OH)_n$ ($n = 1, 2, 3$, mono-, di-, and tri-), are derived from the reaction of ammonia with ethylene oxide. Isopropanolamines, $NH_{3-n}(CH_2CHOHCH_3)_n$ (mono-, di-, and tri-), result from the reaction of ammonia with propylene oxide. Secondary butanolamines, $NH_{3-n}(CH_2CHOHCH_2CH_3)_n$ (mono-, di-, and tri-), are the result of the reaction of ammonia with butylene oxide. Mixed alkanolamines can be produced from a mixture of oxides reacting with ammonia.

Ethanolamines have been commercially available for over 50 years and isopropanolamines, for over 40 years. *sec*-Butanolamines have been prepared in research quantities, but are not available commercially. Primary butanolamines, eg, 2-amino-1-butanol [96-20-8] are made by a different chemical route (see ALKANOLAMINES FROM NITROALCOHOLS).

A variety of substituted alkanolamines are also available commercially, but have not reached the volume popularity of the ethanolamines and isopropanolamines (see Table 2).

2. Physical Properties

The freezing points of alkanolamines are moderately high as shown in Tables 1 and 2. The ethanolamines, monoisopropanolamine and mono-*sec*-butanolamine, are colorless liquids at or near room temperature. Di- and triisopropanolamine and di- and tri-*sec*-butanolamine are white solids at room temperature.

All the ethanolamines and isopropanolamines except monoisopropanolamine are available in low freezing grades, to provide liquid handling at room temperature.

Alkanolamines have a mild ammoniacal odor and are extremely hygroscopic (1). The mono- and dialkanolamines have a basicity similar to aqueous ammonia; the trialkanolamines are slightly weaker bases.

Table 1. **Physical Properties of Alkanolamines Prepared from Ammonia and Olefin Oxides**

Common name	Molecular formula	CAS Registry Number	Chemical Abstracts name	Structural formula	Freezing point, °C	Boiling point[a], °C	Water solubility[b], g/100 g	n-Heptane solubility[b], g/100 g	Viscosity[b], mPa·s (= cP)
monoethanolamine (MEA)	C_2H_7NO	[141-43-5]	2-aminoethanol	$NH_2C_2H_4OH$	10	171	∞	0.06	19
diethanolamine (DEA)	$C_4H_{11}NO_2$	[111-42-2]	2,2'-iminobisethanol	$NH(C_2H_4OH)_2$	28	268	∞	0.01	54 (60°C)
triethanolamine (TEA)	$C_6H_{15}NO_3$	[102-71-6]	2,2',2''-nitrilotrisethanol	$N(C_2H_4OH)_3$	21	340	∞	0.02	600
monoisopropanolamine (MIPA)	C_3H_9NO	[78-96-6]	1-amino-2-propanol	$NH_2CH_2CHOHCH_3$	3c	159	∞	0.4	23
diisopropanolamine (DIPA)	$C_6H_{15}NO_2$	[110-97-4]	1,1'-iminodi-2-propanol	$NH(CH_2CHOHCH_3)_2$	44c	249	1200	0.1	86 (54°C)
triisopropanolamine (TIPA)	$C_9H_{21}NO_3$	[122-20-3]	1,1',1''-nitrilotris-2-propanol	$N(CH_2CHOHCH_3)_3$	44c	306	>500	3.4	100 (60°C)
mono-sec-butanolamine	$C_4H_{11}NO$	[13552-21-1]	1-amino-2-butanol	$NH_2CH_2CHOHC_2H_5$	3	169	∞	0.04	29
di-sec-butanolamine	$C_8H_{19}NO_2$	[21838-75-5]	1,1'-iminodi-2-butanol	$NH(CH_2CHOHC_2H_5)_2$	68–70	256	∞	4.7	890
tri-sec-butanolamine	$C_{12}H_{27}NO_3$	[2421-02-5]	1,1',1''-nitrilotris-2-butanol	$N(CH_2CHOHC_2H_5)_3$	41–47	310	ca 7	>100	ca 6000

[a] At 101.3 kPa = 1 atm.
[b] Approximate, at 25°C unless otherwise noted.
[c] Supercools; freezing points may show variation.

Table 2. Physical Properties of Substituted Alkanolamines

Common name	Molecular formula	CAS Registry Number	Chemical Abstracts name	Freezing point, °C	Boiling point[a], °C	Water solubility[b], g/100 g[a]	Viscosity[c], mPa·s (= cP)
dimethylethanolamine	$C_4H_{11}NO$	[108-01-0]	2-dimethylaminoethanol	−59	135	∞	3
diethylethanolamine	$C_6H_{15}NO$	[100-37-8]	2-diethylaminoethanol		162	∞	4
aminoethylethanolamine (AEEA)	$C_4H_{12}N_2O$	[111-41-1]	2-(2-aminoethylamino)ethanol	−38[d]	244	∞	141
methylethanolamine	C_3H_9NO	[109-83-1]	2-methylaminoethanol	−4.5	160	∞	13
butylethanolamine	$C_6H_{15}NO$	[111-75-1]	2-butylaminoethanol	−2	199	∞	20
N-acetylethanolamine	$C_4H_9NO_2$	[142-26-7]	N-2-hydroxyethyl acetamide	16	decompn	∞	203
phenylethanolamine	$C_8H_{11}NO$	[122-98-5]	2-anilinoethanol	11	285	4.6	101
dibutylethanolamine	$C_{10}H_{23}NO$	[102-81-8]	2-dibutylaminoethanol	−75[e]	229	0.4	8
diisopropylethanolamine	$C_8H_{19}NO$	[96-80-0]	2-diisopropylaminoethanol	−39	191	1.2	8
phenylethylethanolamine	$C_{10}H_{15}NO$	[92-50-2]	2-N-ethylanilinoethanol	37[f]	decompn	0.2	
methyldiethanolamine	$C_5H_{13}NO_2$	[105-59-9]	2,2'-(methylimino)diethanol	−21	247	∞	101
ethyldiethanolamine	$C_6H_{15}NO_2$	[139-87-7]	2,2'-(ethylimino)diethanol	44e	253	∞	87
phenyldiethanolamine	$C_{10}H_{15}NO_2$	[120-07-0]	2,2'-(phenylimino)diethanol	57d		2.8	119 (60°C)
dimethylisopropanolamine	$C_5H_{13}NO$	[108-61-7]	1-dimethylamino-2-propanol	−85[e]	126		
N-(2-hydroxypropyl)ethylenediamine	$C_5H_{14}N_2O$	[123-84-2]	1-(2-aminoethylamino)-2-propanol	−50[e]	155[g]	∞	112 (25°C)

[a] At 101.3 kPa = 1 atm unless otherwise noted.
[b] Approximate, at 25°C.
[c] At 20°C unless otherwise noted.
[d] Pour point.
[e] Sets to a glasslike solid below this temperature.
[f] Melting point.
[g] At 8 kPa (60 mm Hg).

All the alkanolamines, except tri-*sec*-butanolamine, are completely miscible in water and polar solvents. Solubility in nonpolar solvents varies, as noted in Tables 1 and 2.

3. Chemical Reactions

Alkanolamines are bifunctional molecules because of the alcohol and the amine functional groups in the same compound. This allows them to react in a wide variety of ways, with similarities to primary, secondary, and tertiary amines, and primary and secondary alcohols.

3.1. Reaction with Acids. Under anhydrous conditions, mono- and diethanolamines and isopropanolamines form carbamates with carbon dioxide (2,3).

$$2HOCHRCH_2NH_2 + CO_2 \longrightarrow HOCHRCH_2NHCOOH \cdot H_2NCH_2CHROH \tag{1}$$

Trialkanolamines, lacking an amine hydrogen, do not undergo the carbamate reaction.

Alkanolamines in aqueous solution react with carbon dioxide and hydrogen sulfide to yield salts, important to gas conditioning reactions. The dissociation of the salts upon heating results in recovery of the original starting material. These reactions form the basis of an important industrial application, ie, the "sweetening" of natural gas.

$$HOCHRCH_2NH_2 + CO_2 + H_2O \overset{\Delta}{\rightleftharpoons} HOCHRCH_2NH_2 \cdot H_2CO_3 \tag{2}$$

$$HOCHRCH_2NH_2 + H_2S \overset{\Delta}{\rightleftharpoons} HOCHRCH_2NH_2 \cdot H_2S\} \tag{3}$$

Halogen acids and strong organic acids, such as *p*-nitrobenzoic acid and trichloroacetic acid, form crystalline salts (4), with alkanolamines.

$$HOCHRCH_2NH_2 + HCl \longrightarrow HOCHRCH_2NH^+{}_3Cl^- \tag{4}$$

Heating with sulfuric acid gives bisulfates (5).

$$H_2NCH_2CHROH + H_2SO_4 \overset{\Delta}{\longrightarrow} H_2NCH_2CHROSO_3H + H_2O \tag{5}$$

These can be cyclized by heating in the presence of sodium hydroxide to give a dehydration product. Thus monoalkanolamines form 2-alkylaziridines.

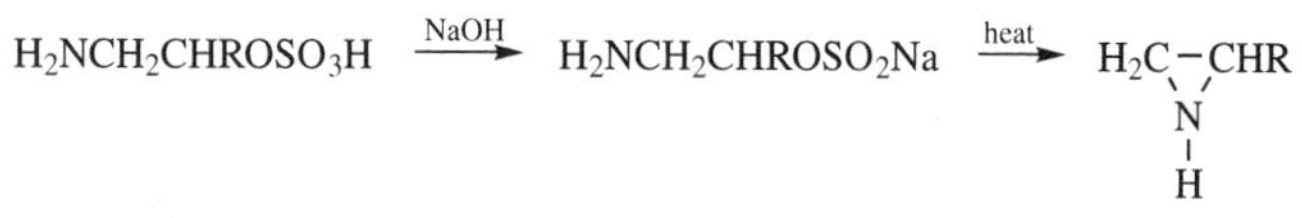

$$H_2NCH_2CHROSO_3H \xrightarrow{NaOH} H_2NCH_2CHROSO_2Na \xrightarrow{heat} \underset{\underset{H}{\overset{|}{N}}}{H_2C{-}CHR} \tag{6}$$

In a similar manner, diethanolamine and diisopropanolamine can be cyclized to give morpholines (6).

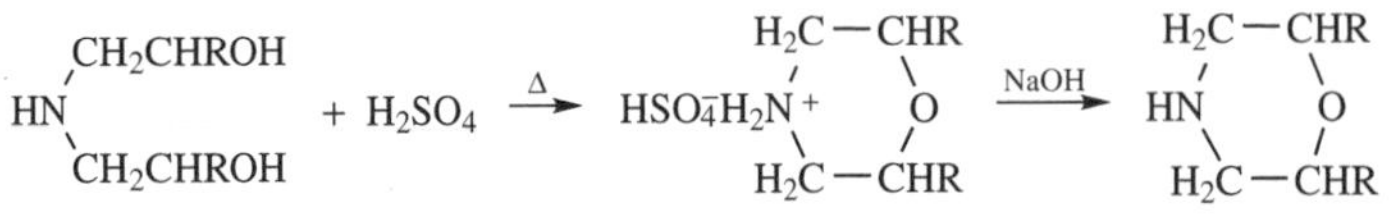

$$\underset{\substack{\diagup CH_2CHROH \\ HN \\ \diagdown CH_2CHROH}}{} + H_2SO_4 \xrightarrow{\Delta} HSO_4^- H_2N^+ \underset{\substack{H_2C-CHR \\ \diagup \qquad \diagdown \\ O \\ \diagdown \qquad \diagup \\ H_2C-CHR}}{} \xrightarrow{NaOH} HN \underset{\substack{H_2C-CHR \\ \diagup \qquad \diagdown \\ O \\ \diagdown \qquad \diagup \\ H_2C-CHR}}{} \qquad (7)$$

Nitric acid gives nitrates and concentrated (48%) hydrobromic acid gives olefins, upon reaction with alkanolamines.

$$R_2NCH_2CHROH \xrightarrow{HBr} R_2NCH\!=\!CHR + H_2O \qquad (8)$$

3.2. Reaction with Fatty Acids and Esters.

Alkanolamines and long-chain fatty acids react at room temperature to give neutral alkanolamine soaps, which are waxy, noncrystalline materials with widespread commercial applications as emulsifiers. At elevated temperatures, 140–160°C, N-alkanolamides are the main products, at a 1:1 reaction ratio (7,8).

$$HOCHRCH_2NH_2 + C_{17}H_{35}COOH \longrightarrow HOCHRCH_2NH_2 \cdot HOOCC_{17}H_{35} \qquad (9)$$

$$HOCHRCH_2NH_2 + R\,COOH \xrightarrow{\Delta} HOCHRCH_2NH-\overset{\overset{\textstyle O}{\|}}{C}-R + H_2O \qquad (10)$$

Significant quantities of amine and amide esters are formed by side reactions (9). In addition, with dialkanolamines, amide diesters, morpholines, and piperazines can be obtained, depending on the starting material. Reaction of dialkanolamines with fatty acids in a 2:1 ratio, at 140–160°C, produces a second major type of alkanolamide. These products, in contrast to the 1:1 alkanolamides, are water soluble; they are complex mixtures of N-alkanolamides, amine esters, and diesters, and still contain a considerable amount of unreacted dialkanolamine, accounting for the water solubility of the product. Both the 1:1 and the 2:1 alkanolamides are of commercial importance in detergents.

Trialkanolamines cannot form amides, but they do give esters at temperatures sufficiently high to eliminate water.

$$(HOCHRCH_2)_3N + 3R'COOH \longrightarrow N(CH_2CHROOCR')_3 + 3H_2O \qquad (11)$$

3.3. Reaction with Acyl Halides.

Acyl halides react at room temperature with mono- and dialkanolamines to give amides (10,11).

$$2\,HOCHRCH_2NH_2 + R'\overset{\overset{\textstyle O}{\|}}{C}Cl \longrightarrow HOCHRCH_2NH\overset{\overset{\textstyle O}{\|}}{C}R' + HOCHRCH_2NH_3^+Cl^- \qquad (12)$$

At elevated temperatures, in the presence of alkali, mono-, di-, and trialkanolamines produce esters.

$$(HOCHRCH_2)_3N + 2\,R'\overset{\overset{\textstyle O}{\|}}{C}Cl \xrightarrow{NaOH} HOCHRCH_2N(CH_2CHROOCR')_2 \qquad (13)$$

3.4. Reaction with Acid Anhydrides. Below room temperature, acid anhydrides react with alkanolamines to produce amides, which partially rearrange to esters on warming to room temperature or slightly above.

$$\text{HOCHRCH}_2\text{NH}_2 \ + \ \text{O}(\overset{\overset{\text{O}}{\|}}{\text{C}}\text{R}')_2 \ \longrightarrow \ \text{HOCHRCH}_2\text{NH} - \overset{\overset{\text{O}}{\|}}{\text{C}} - \text{R}' \ \overset{\Delta}{\longrightarrow} \ \text{R}'\text{COOCHRCH}_2\text{NH}_2 \qquad (14)$$

A 2:1 molar ratio of alkanolamine and fatty acid anhydride, at room temperature, gives a mixture of amide and alkanolamine soap.

$$2\,\text{HOCHRCH}_2\text{NH}_2 \ + \ \text{R}\overset{\overset{\text{O}}{\|}}{\text{C}} - \text{O} - \overset{\overset{\text{O}}{\|}}{\text{C}}\text{R} \ \longrightarrow \ \text{HOCHRCH}_2\text{NH}\overset{\overset{\text{O}}{\|}}{\text{C}}\text{R} \ + \ \text{HOCHRCH}_2\text{NH}_2 \cdot \text{HOOCR} \qquad (15)$$

3.5. Reaction with Aldehydes and Ketones. Formaldehyde combines with primary and secondary alkanolamines in the presence of alkali to give methylol derivatives. For the reaction of monoethanolamine with formaldehyde (12), the reaction scheme shown in Figure 1 occurs.

Primary alkanolamines react with aliphatic and aromatic aldehydes or ketones (other than formaldehyde) to give Schiff bases.

$$\text{HOCHRCH}_2\text{NH}_2 + \text{R}'\text{R}''\text{C}{=}\text{O} \longrightarrow \text{HOCHRCH}_2\text{N}{=}\text{CR}'\text{R}'' + \text{H}_2\text{O} \qquad (16)$$

The Schiff bases can be catalytically hydrogenated to the corresponding saturated derivatives. More severe conditions cause cyclization to oxazolidines.

With secondary alkanolamines, aldehydes in the presence of K_2CO_3 yield di-tertiary amines, which, on distillation, break down into α,β-unsaturated

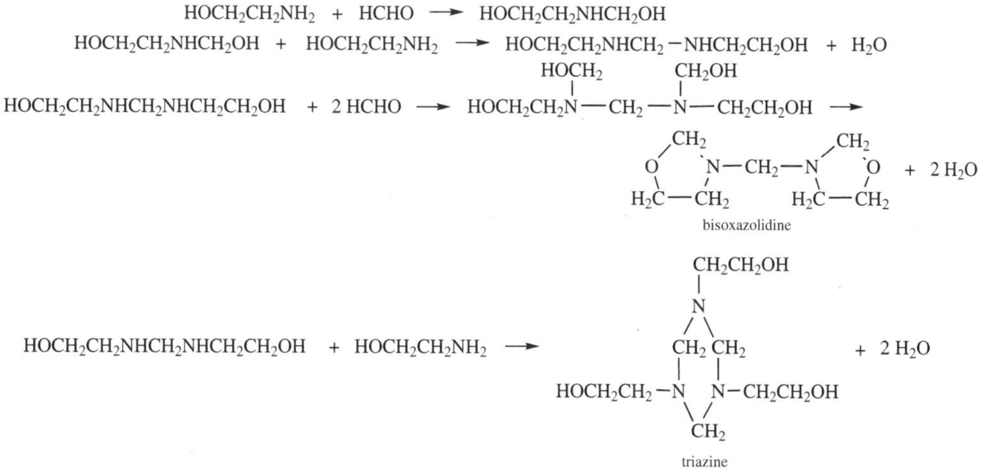

Fig. 1. Reaction of formaldehyde with monoethanolamine. If two moles of monoethanolamine react with three moles of formaldehyde, the bisoxazolidine is the only product isolated. A 1:1 mole ratio of monoethanolamine to formaldehyde produces a mixture of the triazine and the bisoxazolidine.

amines and secondary amines. With a mono- or dialkanolamine, an alkali metal cyanide, and an aldehyde or ketone, aminoacetonitriles are formed.

$$HOCHRCH_2NH_3^+Cl^- + R'CHO + KCN \longrightarrow HOCHRCH_2NHCHCN + KCl + H_2O \qquad (17)$$

$$\overset{R'}{\underset{}{|}}$$

These nitriles can be saponified to the corresponding acids.

3.6. Reaction with Alkyl and Aralkyl Halides.

Alkyl halides form *N*-alkyl derivatives of alkanolamines.

$$HOCHRCH_2NH_2 + R'Cl \longrightarrow HOCHRCH_2NHR' + HCl \qquad (18)$$

Aralkyl halides (R' either phenyl or naphthyl) give disubstituted products.

$$HOCHRCH_2NH_2 + 2R'CH_2Cl \longrightarrow HOCHRCH_2N(CH_2R')_2 \cdot HCl + HCl \qquad (19)$$

Mono- and dialkyl derivatives can also be prepared using alkyl sulfates. Aryl chlorides are usually inert, unless activated by an electron-withdrawing group. Conversion to alkoxides allows formation of ethers.

$$HOCH_2CH_2NH_2 \xrightarrow{NaOH} NaOCH_2CH_2NH_2 \xrightarrow{RCl} ROCH_2CH_2NH_2 + NaCl \qquad (20)$$

3.7. Other Reactions.

Polyester polyamides can be formed from dicarboxylic acids (13).

$$HOOC-R'-COOH + H_2NCH_2CHOH \longrightarrow \underset{R}{\underset{|}{}} \left(\underset{O}{\overset{}{\underset{\|}{C}}} -R'-\underset{O}{\overset{}{\underset{\|}{C}}} -NH-CH_2CHO \underset{R}{\underset{|}{} } \right)_{\!n} \qquad (21)$$

Monoalkanolamines are converted by carbon disulfide into 2-mercaptothiazolines. Ethyleneurea (2-imidazolidinone) can be prepared by heating a mixture of monoethanolamine and urea for several hours.

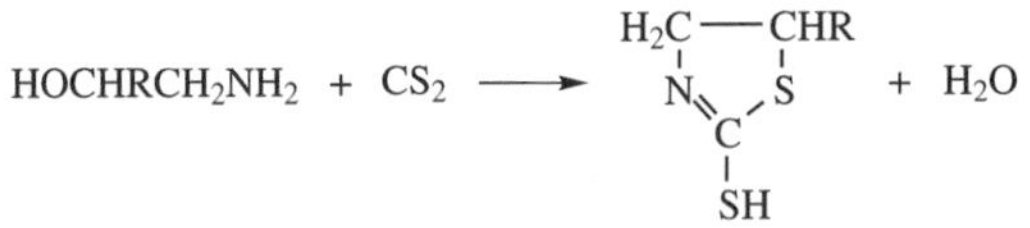

$$HOCHRCH_2NH_2 + CS_2 \longrightarrow \begin{array}{c} H_2C \!-\! CHR \\ | \qquad | \\ N \qquad S \\ \diagdown C \diagup \\ | \\ SH \end{array} + H_2O \qquad (22)$$

$$HOCHRCH_2NH_2 + NH_2CNH_2 \longrightarrow \begin{array}{c} H_2C \!-\! CHR \\ | \qquad | \\ HN \qquad NH \\ \diagdown C \diagup \\ \| \\ O \end{array} + NH_3 + H_2O \qquad (23)$$

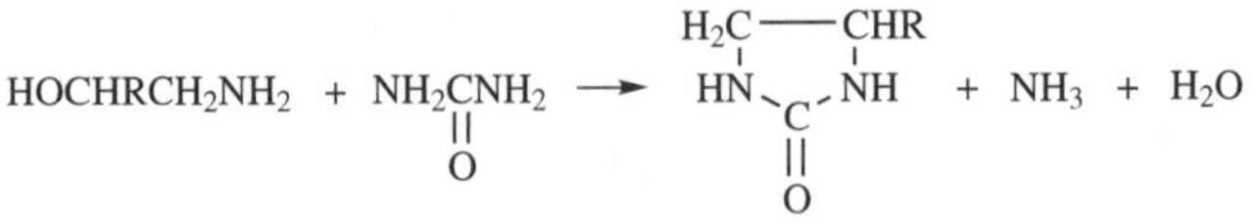

$$\overset{\|}{O}$$

With acrylonitrile, mono- and dialkanolamines undergo a Michael addition to give the β-aminonitrile.

$$\underset{NH(CH_2CHOH)_2}{\overset{CH_3}{\overset{|}{}}} + CH_2\!=\!CHCN \longrightarrow \underset{(HOCHCH_2)_2NCH_2CH_2CN}{\overset{CH_3}{\overset{|}{}}} \qquad (24)$$

Mono- and dialkanolamines react readily with ethylene or propylene carbonates to yield carbamates.

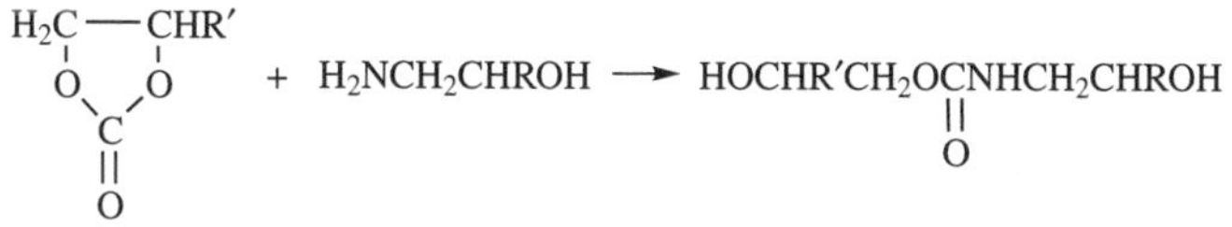

$$H_2C\!-\!CHR' \quad + \ H_2NCH_2CHROH \ \longrightarrow \ HOCHR'CH_2OCNHCH_2CHROH \tag{25}$$

Alkanolamines can be oxidized with various oxidizing agents. With acidic potassium permanganate or excess potassium hydroxide, the potassium salts of the corresponding amino acid are obtained.

$$NH(CH_2CH_2OH)_2 \xrightarrow[200-300°C]{KOH} NH(CH_2COOK)_2 + 3H_2 \tag{26}$$

Mono- and diethanolamine are converted to formaldehyde and ammonia by acidic periodates.

Numerous patents exist for the production ofnitrilotriacetic acid [139-13-9] and its salts from triethanolamine (14–16).

$$N(CH_2CH_2OH)_3 \xrightarrow{O_2} N(CH_2COOH)_3 \tag{27}$$

In certain cases, alkanolamines function as reducing agents. For example, monoethanolamine reduces anthraquinone to anthranols, acetone to 2-propanol, and azobenzene to aniline (17). The reduction reaction depends on the decomposition of the alkanolamine into ammonia and an aldehyde. Similarly, diethanolamine converts o-chloronitrobenzene to 2,2'-dichloroazobenzene and m-dinitrobenzene to 3,3'-diaminoazobenzene.

Monoethanolamine can also be reduced catalytically with hydrogen and ammonia over Raney nickel at 200°C and 20.7 MPa (3000 psig) to produce ethylenediamine [107-15-3] (18,19).

$$NH_2CH_2OHOH + NH_3 \xrightarrow[Raney\ Ni]{H_2} NH_2CH_2CH_2NH_2 + H_2O \tag{28}$$

4. Manufacture

Alkanolamines are manufactured from the corresponding oxide and ammonia. Anhydrous or aqueous ammonia may be used, although anhydrous ammonia is typically used to favor monoalkanolamine production and requires high temperature and pressure (20). Mono-, di-, and trialkanolamines are produced in the reactor and sent to downstream columns for separation (Fig. 2).

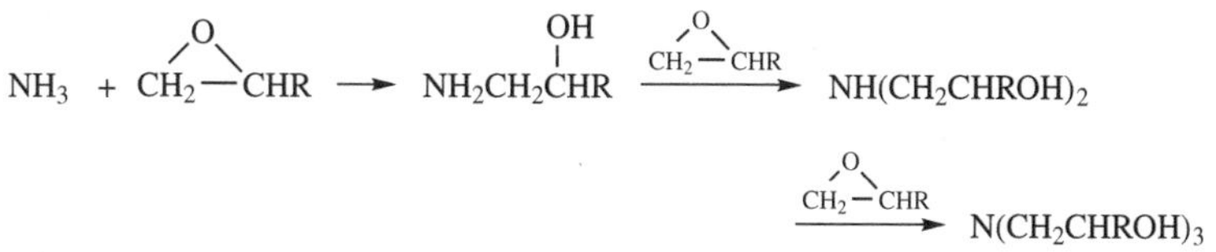

$$NH_3 + CH_2\!-\!CHR \ \longrightarrow \ NH_2CH_2CHR \xrightarrow{CH_2-CHR} NH(CH_2CHROH)_2 \xrightarrow{CH_2-CHR} N(CH_2CHROH)_3$$

where R = H, CH₃, C₂H₅

The reaction is exothermic; reaction rates decrease with increased carbon number of the oxide (ethylene oxide > propylene oxide > butylene oxide). The

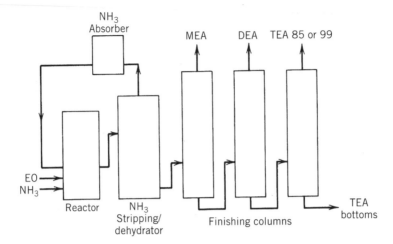

Fig. 2. Flow sheet for ethanolamine production. EO = ethylene oxide; MEA, DEA, and TEA are defined in Table 1.

ammonia–oxide ratio determines the product split among the mono-, di-, and trialkanolamines. A high ammonia to oxide ratio favors mono- production; a low ammonia to oxide ratio favors trialkanolamine production. Mono- and dialkanolamines can also be recycled to the reactor to increase di- or trialkanolamine production. Mono- and dialkanolamines can also be converted to trialkanolamines by reaction of the mono- and di- with oxide in batch reactors. In all cases, the reaction is run with excess ammonia to prevent unreacted oxide from leaving the reactor.

A variety of substituted alkanolamines (Table 2) can also be made by reaction of oxide with the appropriate amine. Aminoethylethanolamine is made from the reaction of ethylenediamine [107-15-3] and ethylene oxide. Methyldiethanolamine is made from the reaction of ethylene oxide andmethylamine [74-89-5]. Diethylethanolamine is made by the reaction of diethylamine [109-87-7] and ethylene oxide.

5. Economic Aspects

U.S. capacity of the ethanolamines in 1989, almost one-half of global capacity, was estimated to be 379,000 t. Global capacity for 1989 was estimated at 692,000 t. Estimated annual U.S. production figures are listed in Table 3 (21). U.S. consumption of ethanolamines for various applications is shown in Table 4.

Isopropanolamine global capacity for 1989 was estimated at 37,000 t. U.S. capacities for isopropanolamine (IPA), aminoethylethanolamine (AEEA), and methyldiethanolamine (MDEA) are as follows:

Akanolamine	U.S. capacity, 10^3 t
IPA	23,000
AEEA	9,000
MDEA	20,000

Table 3. **Estimated Annual U.S. Production of Ethanolamines, 10^3 t**

Year	Total	MEA	DEA	TEA
1955	36	12	14	10
1960	57	20	24	13
1965	91	31	35	25
1970	120	40	42	38
1975	118	38	39	41
1980	171	59	56	56
1985	245	98	76	71
1988	278	103	86	89
1990	302	113	91	98
1995	362	137	106	119

Consumption of ethanolamines in the United States has changed dramatically since the 1960s. Consumption in gas conditioning applications has peaked and chemical processing intermediates (captive use for ethyleneamine and surfactant applications) has increased significantly.

U.S. production of ethanolamines grew at just under 8% during the 1980s. Export of ethanolamines has contributed significantly to the increased production. Exports have averaged 11% growth per year during the 1980s. U.S. sales consumption have grown an average of 6% a year since 1970.

List pricing (22) has ranged from ~45/kg in the 1960s, to a low of 33¢/kg in the early 1970s. In the mid-1970s, the price increased to over 65¢/kg to peak in the early 1980s at over $1.10/kg. Prices decreased again in the mid-1980s to peak again at $1.43/kg in 1989 for bulk quantities.

6. Specifications

Specifications for the most commonly used alkanolamines are listed in Table 5. Special grades of products have been developed as needed by the customer. *National Formulary* grade meets the specifications in the *U.S. Pharmacopeia*. Low iron and electronic grades have particular specifications for iron and other metals. Low freezing grades were developed because of the high freezing points of most of the product. Generally, the addition of 15% water, by weight, significantly reduces the freezing point to a manageable level.

Table 4. **U.S. Consumption by Use of Ethanolamines, 10^3 t**

Use	1950	1960	1970	1980	1990
surfactants, detergents, and personal care products	5.9	25.4	43	60	95
gas purification	1.8	13.2	23	35	30
textiles	1.8	9.1	6	4	5
agricultural products	2.7	1.1	2	4	4
cement and concrete			3	5	7
metal working	1.8	3.6	5	6	7
chemical processing intermediate			6	8	42
other	1.4	4.5	4	5	5

Table 5. **Typical Specifications of Alkanolamines**[a]

Compound	Assay[b], min %	Apparent[c] equivalent weight	APHA[d] color max	Water, max %
MEA, CG[e]	99	61–62	15	0.3
DEA, CG[e]	99	104–106	15	0.15
TEA 85[f]	85	140–145	40	0.20
TEA 99	99	148–150	50	0.2
MIPA	99	75–76	20	0.5
DIPA, CG[e]	98	132–134	40	0.5
TIPA 99	99	190–192	50	0.3
AEEA	99.6		25	0.2

[a] Specification values and items may vary with different manufacturers. Analytical methods may be obtained from manufacturers.
[b] Assay determined by gas chromatography.
[c] Apparent equivalent weight determined by titration with hydrochloric acid.
[d] APHA color determined by ASTM D1209, platinum–cobalt method.
[e] CG = commercial grade.
[f] Contains approximately 15% DEA.

7. Analytical Test Methods

Generally, alkanolamines are analyzed by gas chromatography or wet test methods. Details on gas chromatography conditions are available in the literature (1) for packed or glass capillary columns.

Apparent equivalent weight can be determined by titration with hydrochloric acid using a bromocresol green indicator. Calculations give the equivalent weight of total amines and are not specific for the mono-, di- or trialkanolamines.

APHA color is determined using ASTM D1209; percent water is determined by Karl Fischer titration following ASTM E203. Detailed analytical procedures are available in the literature (1) or from producers.

8. Storage and Handling

Stainless steel, 316L and 304L, is the preferred material of construction for shipment and storage of alkanolamines, if product quality is of importance. Aluminum can be used for short term transport, at temperatures below 60°C, for the pure amines. Plastic liners have not been found acceptable for all alkanolamines. Phosphatized-lined steel drums are used for transport of ethanlomaines and isopropanolamines, except MEA, which is shipped in high density polyethylene drums to maintain low product color. Monoethanolamine and monoisopropanolamine are classified as DOT corrosive liquids.

Polyethylene, polypropylene, and Teflon resins have been found to be acceptable in contact with all of the alkanolamines at low temperatures. Lined tanks have not been found to successfully withstand monoethanolamine and monoisopropanolamine. These two alkanolamines degrade all linings tested. Extensive testing of any lined tank or piping should be initiated before being used in any alkanolamine service. Some producers use only stainless steel in handling, storage, and shipping of their alkanolamines.

Mild steel can be used for transport and storage if product discoloration is not a problem, such as in gas conditioning applications. Contact with copper, brass, and other copper alloys may cause corrosion of the metal.

Storage tanks, lines, and pumps should be heat traced and insulated to enable product handling. Temperature control is required to prevent product degradation because of color; alkanolamines have poor heat transfer properties. Exposure to air will also cause product discoloration. Storage tanks should be nitrogen-padded if low color product is required.

Normally, centrifugal pumps, made of stainless steel, are used to transfer amines to maintain good product quality. Carbon steel or black iron pumps can be used if quality is unimportant. Heat tracing the pumps may also be required if exposure to cold weather is a possibility. Teflon gaskets and Teflon tape can be used for applications under 200°C. Graphite-filled gaskets can also be used. Graphite packing or mechanical seals with Teflon resins are also suggested. Alkanolamines can leach conventional pipe dope.

Reactions of monoethanolamine with mild steel are referenced in the literature (23). The complex formed, identified as trisethanolamino–iron, can decompose in air to pyrophoric iron, with the potential to cause a fire, if contacted with combustible materials.

9. Health and Safety Factors

A brief summary of safety and health hazards follows; detailed health hazards, however, should be obtained from producers by requesting Material Safety Data Sheets. Proper protective equipment and exposure hazards should be noted before handling any alkanolamine. Detailed toxicological testing is found in the CTFA Chemical Ingredient Review Board Reports on ethanolamines and isopropanolamines (24).

9.1. Oral Toxicity. Alkanolamines generally have low acute oral toxicity, but swallowing substantial quantities could have serious toxic effects, including injury to mouth, throat, and digestive tract.

Concentrated monoethanolamine and monoisopropanolamine can cause severe local irritation or even burns to the mouth, throat, and digestive tract. If monoethanolamine and monoisopropanolamine are swallowed, large volumes of milk or water should be administered immediately. If diethanolamine, triethanolamine, diisopropanolamine, or triisopropanolamine are swallowed, vomiting should be induced after drinking two glasses of water.

9.2. Vapor Toxicity. Laboratory exposure data indicate that vapor inhalation of alkanolamines presents low hazards at ordinary temperatures (generally, alkanolamines have low vapor pressures). Heated material may cause generation of sufficient vapors to cause adverse effects, including eye and nose irritation. If inhalation exposure is likely, approved respirators are suggested. Monoethanolamine and diethanolamine have OSHA TLVs of 3 ppm.

9.3. Eye Irritation. Exposure of the eye to undiluted alkanolamines can cause serious injury. Solutions as dilute as 1% of monoethanolamine and monoisopropanolamine can cause some eye irritation.

Diethanolamine, diisopropanolamine, and isopropanolamine mixtures are also irritating to the eyes, even diluted, but less so than monoethanolamine and monoisopropanolamine. Undiluted triethanolamine and triisopropanolamine and concentrated solutions have an irritating action on the eyes, but only slight transient or no corneal injury would be expected. Proper protective equipment, as detailed in the Material Safety Data Sheets should be used. Chemical workers' goggles or the equivalent, and suitable facilities for washing the eyes should be readily available. If any of the amines contact the eyes, the eyes should be flushed thoroughly with flowing water for 30 min for monoethanolamines and isopropanolamines, and 15 min for di- and triethanolamines, and di- and triisopropanolamines.

9.4. Skin Irritation. Monoethanolamine and monoisopropanolamine, being strongly alkaline, are skin irritants, capable of producing serious injury in concentrations of 10% or higher upon repeated or prolonged contact. Occasional short contact, assuming the material is thoroughly washed off, should have little adverse effect.

Diethanolamine, diisopropanolamine, and isopropanolamine mixtures are less irritating to the skin than MEA and MIPA; however, any one of them may produce severe skin irritation, even mild burns, if contact is prolonged or frequently repeated. Occasional short contact should not result in more than slight irritation. Undiluted triethanolamine and triisopropanolamine are slightly to moderately irritating to the skin. A burn may result from prolonged and repeated contact. Short occasional contact and solutions of less than 10% concentration are unlikely to cause more than very slight irritation, if any.

Monoethanolamine and monoisopropanolamine may be moderately toxic by absorption through the skin. The other amines are low in toxicity by this route and are not likely to be absorbed in acutely toxic amounts. In the event of skin contact, clothing and shoes should be removed promptly, and the skin thoroughly washed with water. Contaminated clothing should be thoroughly cleaned before reuse; shoes and leather products should be discarded.

9.5. Special Precautions. Use of sodium nitrite or other nitrosating agents in formulations containing alkanolamines could lead to formation of suspected cancer-causing nitrosamines.

Strong oxidizers and strong acids are incompatible with alkanolamines. Reactions, generating temperature and/or pressure increases, may occur with halogenated organic compounds. Alkanolamines are corrosive to copper and brass and may react. Contact with aluminum by alkanolamines, particularly when wet or at elevated temperatures (60°C), should be avoided.

Spills and leaks should be cleaned up with alkanolamine-compatible absorbents or sands. Local, state, and federal requirements should be followed for disposal. Incineration in an approved facility is suggested for final disposal.

10. Uses

Alkanolamines and their derivatives are used in a wide variety of household and industrial applications. Nonionic surfactants (alkanolamides) can be formed by the reaction of alkanolamines with fatty acids, at elevated temperatures (eq. 10). The amides can be liquid, water-soluble materials as produced from a 2:1 ratio, or solid, poorly water-soluble materials, or "super" amides as produced from a 1:1 ratio of reactants. These products are useful as foam stabilizers, and aid cleaning in laundry detergents, dishwashing liquids, shampoos, and cosmetics. They are also used as antistatic agents, glass coatings, fuel gelling agents, drilling mud stabilizers, demulsifiers, and in mining flotation. Reaction of alkanolamines with a fatty acid at room temperature produces neutral alkanolamine soaps (eq. 9). Alkanolamine soaps are found in cosmetics, polishes, metalworking fluids, textile applications, agricultural products, household cleaners, and pharmaceuticals.

Alkanolamine salts are anionic surfactants (qv) formed from the reaction of alkanolamines and the acids of synthetic detergents, such as alkylarylsulfonates, alcohol sulfates, and alcohol ether sulfates. These add to the surfactants line used in detergents, cosmetics, textiles, polishes, agricultural sprays, household cleaners, pharmaceutical ointments, and metalworking compounds. Salts of alkanolamines and inorganic acids are useful chemical intermediates (eq. 4), and are also used in corrosion inhibitors, antistatic agents, glass coatings, electroplating, high octane fuels, inks, metalworking, dust control in mining, and in textiles.

10.1. Adhesives. Alkanolamines are used in hot melt adhesives for binding nonpervious materials to polyolefins (25). Diethanolamine and triethanolamine have been used in the area of phenol formaldehyde adhesives to improve bond strengths (26,27), improve storage stability (28–30), and to increase water dispersibility (31). Starch-based adhesives have increased stability, viscosity, and gel temperature with the addition of monoethanolamine or diethanolamine to Mannich reaction products (32,33). Asphalt compositions have increased bonding to the aggregate in the emulsion when monoethanolamine is added (34,35). Refractory binder formulations, ceramics, and molds are improved by alkanolamine usage for stimulating gel formation (36–38). Anaerobic adhesive sheets, useful in adhering to ABS resin sheets, are made from the combination of diethanolamine, epoxidized polybutadiene, and peroxide (39).

10.2. Cement and Concrete. Low concentrations of triethanolamine or its salts are added to cement clinkers to increase the efficiency of the grinding mill by reducing particle agglomeration (40). The resulting cement is more free-flowing and pack-set is reduced. The mechanism is presumed to be due to the dispersing capability of the triethanolamine, creating a uniform particle dispersion throughout the mixture.

In concrete, triethanolamine accelerates set time and increases early set strength (41–43). These are often formulated as admixtures (44), for later addition to the concrete mixtures. Compared to calcium chloride, another common set accelerator, triethanolamine is less corrosive to steel-reinforcing materials, and gives a concrete that is more resistant to creep under stress (45). Triethanolamine can also neutralize any acid in the concrete and forms a salt with chlorides. Improvement of mechanical properties, whiteness, and more even distribution of iron impurities in the mixture of portland cements, can be effected by addition of 2% triethanolamine (46). Triethanolamine bottoms and alkanolamine soaps can also be used in these type applications. Waterproofing or sealing concrete can be accomplished by using formulations containing triethanolamine (47,48).

10.3. Cleaners. Properties, such as foaming and detergency (qv), make alkanolamines useful in cleaning formulations. Monoethanolamine is particularly effective in wax removal formulations because of its ability to penetrate films. Cleaners that involve skin contact use triethanolamine because of its mildness. Derivatives of the amines (49,50) as well as the free alkanolamines (51–53), may be formulated into cleaning products.

10.4. Coatings. A wide variety of applications have been reported in coatings technology. The alkanolamines are added to neutralize polyester, epoxy (54–56), polyamide, and urethane (57) resin coatings to enhance dispersibility (58–60). In metal-coating preparations, the alkanolamines serve as complexing agents, neutralizers, promoters, modifiers, corrosion inhibitors, and electrophoretic bath components (61,62). Alkanolamines assist in improving adhesion, curing resins, complexing metals, improving storage stability, and improving both fresh and salt water resistance for some types of coatings (63–67).

Alkanolamines are used in urethane coatings for glass shatter proofing (68) and have been utilized as amides, salts, or free amines in providing antifrosting, antifogging, and dirt-resistant films on glass and plastics (69–72).

Triethanolamine and diethanolamine are accelerators for photopolymerization coatings (73–75). Diethanolamine is the most commonly used alkanolamine in the preparation of cationic polymers for electrophoretic coatings. The use of acrylate–amine–isocyanate–anhydride copolymers provides electrodeposited coatings that can be cured with uv light (76,77). All of the ethanolamines are effective in improving thermal properties and reducing cracking in prepared wire coatings (78–82).

10.5. Corrosion Inhibitors. Alkanolamines inhibit corrosion of ferrous metals (83). These can be used in a wide variety of applications, such as coolant systems, lubricating oils (84,85), metal working fluids, petroleum antifouling (86), and drilling needs (87). Monoethanolamine is most often used in applications requiring free base, and triethanolamine as a salt with organic and inorganic acids. Fatty acid salts are used in lubricating and metal working fluids (88), whereas inorganic salts, the phosphate or borate, are used in aqueous or mixed-solvent systems such as glycol antifreeze (89). Corrosion inhibitors for aluminum, containing alkanolamines, are also detailed in the literature (90).

Alkanolamines are also used in formulations for corrosion prevention. Triethanolamine-containing formulations can be used to coat stainless steel to prevent high temperature oxidation (91). Short term anticorrosion protection for metal parts during quenching can be accomplished by using triethanolamine salts of phosphoric acid esters (92,93). Rusted steel surfaces can be stabilized before painting, using triethanolamine-containing formulations in the prime paint (94).

It should be noted that corrosion inhibitors and protection systems are generally designed for specific conditions, and the effectiveness of the inhibitor can change with conditions.

10.6. Cosmetics and Personal Care Products. Alkanolamines are important raw materials in the manufacture of creams (95–97), lotions, shampoos, soaps, and cosmetics. Soaps (98) formed from triethanolamine and fatty acids are mild, with low alkalinity and excellent detergency. Triethanolamine lauryl sulfate is a common base for shampoos (99–101) and offers significant mildness over sodium lauryl sulfate. Diethanolamine lauryl sulfate and fatty acid soaps of mono- and triethanolamine can also be used in shampoos and bubble bath formulations. Chemistry similar to that used in soluble oils and other emulsifiers is applicable to cleansing creams and lotions (102, 103). Alkanolamides or salts are added to the shampoo base to give a smooth, dense foam (104).

10.7. Detergents. Alkanolamines, alkanolamine fatty acid soaps, and alkanolamides, with their mild alkalinity and excellent detergency characteristics, are used extensively in soaps and detergents. An alkanolamine salt of LAS (linear alkylbenzene sulfonate) forms the anionic surfactant component of many liquid laundry detergents (105). Monoalkanolamides have very limited water solubility and excellent alkaline resistance, resulting in widespread use in heavy-duty powder detergents as foam stabilizers and rinse improvers (106). Dialkanolamides are used in light-duty liquid detergents, where high alkalinity is not needed and high solubility is required.

The new liquid laundry detergents, with no phosphates, have developed a use for alkanolamines. In nonenzyme formulations, they contribute alkalinity, pH control, and enhanced product stability. In enzyme products, alkanolamines contribute to the stability of the enzyme in water solutions (107).

10.8. Electroplating, Electroless Plating and Stripping. Alkanolamines form complexes with many metal ions, and are consequently useful in electrolytic and chemical plating to improve coating properties. Diethanolamine is used in baths for electroplating palladium (108), triethanolamine in electroless baths for coating copper (109), and monoethanolamine in electrodeposition baths for zinc (110). Various coating formulations also contain alkanolamines, such as diethanolamine in cationic electrophoretic coatings (111), monoethanolamine in photoresist developers (112), and triethanolamine in formulations to protect electric switches and contact elements (113).

Alkanolamines are also used in formulations for removing photoresists (114,115) and cleaning printed-circuit boards (116). Formulations containing diethanolamine are used for electrodip coating of substrates (117).

Various documents relate to the use of formulations containing triethanolamine as flux for a variety of metals (118,119), for solders (120), for soldering pastes (121, 122), and for low corrosion solder pastes (123).

10.9. Gas Purification. In the "sweetening" of natural gas, aqueous ethanolamine reacts with the hydrogen sulfide, carbon dioxide, or other acid constituents of the gas to give a water-soluble salt. The amine is then regenerated by steam stripping. Generally, monoethanolamine is used because of its low equivalent weight. Diethanolamine must be used, however, if there is a significant quantity of carbonyl sulfide in the gas, since this forms an unregenerable complex with monoethanolamine. The advantages of diethanolamine and monoethanolamine in treating refinery gases include low vapor pressure and hydrocarbon solubility (124).

Work continues on improving the efficiency of this process, such as for freeing the alkanolamine from heat-stable salts that can form (125). Formulations have been developed which inhibit degradation of mono- and diethanolamine in processing (126). Models (127), computer programs (128), and kinetics and enthalpies (129–136) have been developed to help determine equilibria of the acid gas–alkanolamine–water system. Additional references relate to the use of tertiary alkanolamines, such as triethanolamine, for gas conditioning (137–139).

Extensive work has been done on corrosion inhibitors (140), activated carbon use (141–144), multiple absorption zones and packed columns (145,146), and selective absorption and desorption of gas components (147,148). Alkanolamines can also be used for acid gas removal in ammonia plants (149).

Methyldiethanolamine (MDEA) and solutions of MDEA have increased in use for gas treating (150,151). Additional gas treating capacity can often be obtained with the same working equipment, because of the higher amine concentrations that can be used.

The Sulfinol gas conditioning process of Shell uses diisopropanolamine in a sulfolane solvent system. This system also increases gas capacity with improved efficiencies (152).

10.10. Metal Working, Cleaning, and Lubricating. Alkanolamines find wide use in the metal working industry, in formulations used for cutting fluids, lubricating oils, and cleaning solutions. The literature cites diethanolamine and triethanolamine as contributing biostatic effectiveness and high corrosion resistance (153). Triethanolamine derivatives, reacted with fatty acids, are formulated in emulsifiable oils, cutting fluids, and water-soluble cutting fluids (154–157). Triethanolamine is an excellent chelating agent in basic solutions. This chelating ability makes triethanolamine particularly useful in metal cleaning (158,159). Lubricants and coolants containing triethanolamine can increase the life of the cutting tool (160), increase the efficiency of the process (161), reduce surface roughness of the cutting tool (162), and reduce bactericidal concentrations (163).

10.11. Mining. Numerous patents have advocated the use of alkanolamines in mining applications. Triethanolamine has been used as a depressent in the flotation of copper (164), in the electrotwinning of gold (165), and as an aid in the froth flotation of nickel ores. Phosphate ore flotation has been improved through the use of a fatty acid condensate with ethanolamine (166). Beneficiation of tin ore has been accomplished using fatty acid alkanolamides (167).

10.12. Petroleum and Coal. The alkanolamines have found wide use in the petroleum industry. The ethanolamines are used as lubricants and stabilizers in drilling muds. Reaction products of the ethanolamines and fatty acids are used as emulsion stabilizers, chemical washes, and bore cleaners (168). Oil recovery has been enhanced through the use of ethanolamine petroleum sulfonates (169–174). Oil–water emulsions pumped from wells have been demulsified through the addition of triethanolamine derivatives. Alkanolamines have been used in recovering coal in aqueous slurries and as coal–oil mix stabilizers (175–177).

10.13. Polymers. Because of their polyfunctionality, alkanolamines are used as additives in polyurethanes, polycarbonates, epoxy resins, polyesters, and other various resins and rubbers. Triethanolamine and triisopropanolamine are cited in the literature as cross-linking agents and curing agents in injection-molding formulations, contributing stability, strength, good impact, heat, and flame resistance (178–182) for urethanes. In polyester formulations, triethanolamine-based formulations give good physical and mechanical properties (183), form good coatings for concrete floors (184), and decrease molding pressure (185).

Monoethanolamine-containing formulations can be used to make polycarbonate molds with good dyeability and crack resistance (186). Monoethanolamine, diethanolamine, and triethanolamine have been cited in formulations for epoxy resins compounds (187–192).

Triisopropanolamine is used in natural rubber cross-linking and as a color stabilizer for polyethylene formulations. Chain termination of polybutadiene with triisopropanolamine gives improved cold-flow properties.

Alkanolamines can also be used to improve thermal stability in PVC copolymers and polystyrene.

10.14. Textiles. Alkanolamines are used in many stages in the production of textiles, from fiber manufacture, fiber lubrication, and fabric bleaching to fiber dyeing, and finishing. The alkanolamine fatty acid soaps and salts are used to lend antistatic properties to the fibers (193–195). Soluble oils can be formulated for easy removal from the fabric. Fabric lubrication can be accomplished with alkanolamine derivatives or alkanolamine-containing formulations (196, 197). Ethanolamines can improve dyeing in a number of fabrics, including cotton (198), triacetate textiles (199), polyester-cellulosic fibers (200), acrylic fibers (201), polycarbonates (202), and leather (203). Quarternary alkanolamines can improve the dye stability (204), and act as fabric softeners as part of a formulation (205,206). Treatment of wool with a dilute solution of monoethanolamine, followed by steaming, imparts a durable crease (207) and sizing (208).

10.15. Pigment Dispersion. The alkanolamines and their derivatives are useful in dispersing titanium dioxide and other pigments (209). Monoisopropanolamine and triethanolamine are particularly effective in aiding titanium dioxide dispersion in the production of TiO_2 and in water-based paints (210). The alkanolamines are also an aid in the grinding of titanium dioxide (211).

10.16. Wood Pulping. Numerous patents have appeared in recent years regarding the use of alkanolamines in wood pulping. Pretreatment of wood chips with monoethanolamine has resulted in increasing yields from the alkaline pulping process (212). Monoethanolamine has also been used as an aid to increase pulp (qv) strength and brightness (213). Monoethanolamine has been used to impregnate wood to a more dense configuration (214). Combinations of an aqueous lignin slurry with monoethanolamine have produced a lignin salt that is suitable for printing compositions (215).

10.17. Chemical Processing Intermediates and Other Applications. Monoethanolamine can be used as a raw material to produce ethylenediamine. This technology has some advantages over the ethylene dichloride process in that salts are not a by-product. Additional reactions are required to produce the higher ethyleneamines that are normally produced in the ethylene dichloride process.

Alkanolamines are used in the manufacture of a variety of pharmaceutical compounds. Some of these products include antitumor agents, anti-inflammatory and allergy agents, and anticonvulsants. The literature reports ethanolamine derivatives in the treatment of Alzheimer's disease (216), the treatment of cerebral psychoorganic syndromes (217), and veterinary drugs (218).

The major use of alkanolamines in agricultural products is as a neutralizer for acidic herbicides. They also contribute increased water solubility, reduced volatility, and more uniform solutions. Various ethanolamines are reported in formulations to improve potato tuber size (219) and enhance the resistance to salt of some crops (220).

A variety of applications, including photography, employ alkanolamines for pH control. Reports have described formulations including monoethanolamine and triethanolamine in films and processing (221–224).

Diethanolamine can be used in the manufacture of morpholine (eq. 7).

10.18. Alkylalkanolamines.   Aminoethylethanolamine and its derivatives are used in textiles, detergents, fabric softeners, chelating agents, water treating, petroleum, oil field and gas conditioning products, agricultural and pharmaceutical products, emulsifiers, mining chemicals, corrosion inhibitors, and surfactants for cosmetics (225).

Dimethylethanolamine, diethylethanolamine, and their derivatives are used in pesticides, corrosion inhibitors, drugs and pharmaceuticals, emulsification, paints and coatings, metal fabrication and finishing, petroleum and petroleum products, and plastics and resins (226).

BIBLIOGRAPHY

"Ethanolamines" in *ECT* 1st ed., Vol. 5, pp. 851–858, by J. Conway, Carbide and Carbon Chemicals Company, Division of Union Carbide and Carbon Corporation; "Propanolamines" in *ECT* 1st ed., Vol. 11, pp. 163–168, by T. Houtman, Jr., The Dow Chemical Company; "Alkanolamines from Olefin Oxides and Ammonia" in *ECT* 2nd ed., Vol. 1, pp. 810–824, by A. Hart, The Dow Chemical Company; in *ECT* 3rd ed., Vol. 1, pp. 944–960, by R. M. Mullins, The Dow Chemical Company.

CITED PUBLICATIONS

1. *The Alkanolamines Handbook*, The Dow Chemical Company, Midland, Mich., 1988, p. 32.
2. M. B. Jensen, E. Jorgensen, and C. Fourholt, *Acta Chem. Scand.* **8**, 1137–1140 (1954).
3. A. K. Holliday, G. Hughes, and S. M. Walker, *Comprehensive Inorganic Chemistry*, Vol. 1, Pergamon Press, New York, 1973, p. 123.
4. H. Imanishi and co-workers, *Aichi-Ken Kogyo Shidosho Hokoku (Japan)* **11**, 56–60 (1975).
5. P. A. Leighton, W. A. Perkins, and M. L. Renquist, *J. Am. Chem. Soc.* **69**, 1540 (1947).
6. Czech. Pat. 146,401 (1972), V. Patek and co-workers.
7. A. Davidson and B. Milwidsky, *Synthetic Detergents*, 4th ed., C.R.C. Press, Cleveland, Ohio, 1968, 131–136.
8. E. Jungerman and D. Tabor, *Nonionic Surfactants*, Marcel Dekker, New York, 1967, 226–239.
9. H. Kroll and H. Nadeau, *J. Am. Oil Chem. Soc.* **34**, 323 (1957).
10. J. R. Trowbridge, R. A. Falk, and I. J. Krems, *J. Org. Chem.* **20**, 990 (1955).
11. H. Brintzinger and H. Koddebusch, *Chem. Ber.* **82**, 201 (1949).

12. A. N. Gafanov and co-workers, *Izv. Akad. Nauk. SSSR Ser. Khim.* (9), 2189 (1978).
13. U.S. Pat. 2,440,516 (Aug. 27, 1948), E. L. Kropa (to American Cyanamid Corp.).
14. U.S. Pat. 3,535,373 (Oct. 20, 1970), P. F. Jackisole (to Ethyl Corp.).
15. U.S. Pat. 3,708,533 (Jan. 2, 1973), I. S. Backman and co-workers (to Air Products & Chemicals Corp.).
16. U.S. Pat. 3,833,650 (Sept. 3, 1974), H. Schulze and E. T. Marquis (to Jefferson Chemical Co.).
17. Meltsner and co-workers, *J. Am. Chem. Soc.* **57**, 2554 (1935).
18. U.S. Pat. 3,270,059 (Aug. 30, 1966), S. Winderl and co-workers (to BASF).
19. U.S. Pat. 4,123,462 (Oct. 31, 1978), D. C. Best (to Union Carbide Corp.).
20. U.S. Pat. 4,845,296 (July 1989), M. Ahmed and co-workers (to Union Carbide Corp.).
21. U.S. International Trade Commission, 1955–1988.
22. *Chem. Mark. Rep.*, various issues from 1970–1990.
23. B. E. Dixon and R. A. Williams, *Chem. Ind.* **69**, 69 (1950).
24. Expert Panel of the Cosmetic Ingredient Review, *Final Report on the Safety Assessment for Diisopropanolamine, Triisopropanolamine, Isopropanolamine, Mixed Isopropanolamines*, Sept. 26, 1986; *Final Report for the Safety Assessment for Triethanolamine, Diethanolamine, Monoethanolamine*, May 19, 1983.
25. U.S. Pat. 3,922,469 (1975), J. W. Bayer (to Owens-Illinois Corp.).
26. Ger. Offen. 2,401,544 (1975), Siegerl and co-workers (to BASF, A.G.).
27. USSR Pat. 544,663 (1977) Sklyarskii (to State Scientific Research and Design Institute of Polymeric Adhesives).
28. Fr. Demande 2,163,578 (1973), (to BASF A.G.).
29. U.S. Pat. 3,962,166 (1976) (to United Merchants and Manufacturers Corp.).
30. U.S. Pat. 4,182,839 (1980), G. Tesson (to Manufacture de Produits Chemiques Protex S.A.).
31. U.S. Pat. 3,624,246 (1971), H. Hendrick and N. Lumley (to Fiberglas Canada Ltd.).
32. U.S. Pat. 4,033,914 (1977), E. Bovier and J. Voight (to Anheuser-Busch Corp.).
33. U.S. Pat. 4,009,311 (1977), J. Schoenberg (to National Starch and Chemical Co.).
34. Ger. Offen. 2,452,454 (1975), L. Houizot and R. Smadja (to Mobil Oil Corp.).
35. Swed. Pat. 354,478 (1973), K. Hellsten and co-workers (to Mo Och Domojo AB).
36. Ger. Offen. 2,513,740 (1975), I. Walters and H. Emblem (to Zirconal Processes Ltd.).
37. USSR Pat. 538,808 (1976), G. Temnova and co-workers.
38. Jpn. Kokai Tokkyo Koho 75 01,031 (1975), Y. Shimada and T. Shimizu (to Hitachi Ltd.).
39. Jpn. Kokai Tokkyo Koho 74 55,733 (1974), T. Nakenishi (to Matsushita Electric Works Ltd.).
40. W. Scheibe, W. Dallman, and A. Rosenbaum, *Silikattechnik* **21**, 11 (1970).
41. V. S. Ramachadran, *Cem. Conr. Res.* **6**, 623–632 (1976).
42. Czech Pat. 254,243 (Nov. 15, 1989) R. Fedorid, M. Potancok, and L. Novak.
43. USSR Pat. 1,470,709 (Apr. 7, 1989), S. Makhudov and co-workers.
44. Jpn. Kokai Tokkyo Koho 89 03,039 (Jan. 6, 1989), S. Kobayaski and A. Fukazawa.
45. S. M. Royak, V. S. Klement'eva, and G. M. Tarnarutskii, *Zh. Prikl. Khim. (Leningrad)* **43**, 82 (1970).
46. I. Teoreanu and co-workers, *Mater. Constr.* **19**(1), 15–19 (1989).
47. Jpn. Kokai Tokkyo Koho 88 260,845 (Oct. 27, 1988), H. Owada and co-workers (to Miyoshi Oil and Fat Co., Ltd.).
48. Jpn. Kokai Tokkyo Koho 89 45,788 (Feb. 20, 1989), Y. Watanabe and co-workers (to Denki Kagaku Kogyo K.K.).
49. U.S. Pat. 4,822,514 (Apr. 18, 1989), B. D. Becker (to Murphy Phoenix Co.).
50. PRC Pat. 87,100,025 (Mar. 16, 1988), B. Tang (to Faming Zhuanli Shenging Gongkai Shuomingshu); *Chem. Abstr.* **111**, 60037n (1989).

51. PRC Pat. 86,107,086 (May 4, 1988), W. Zhang (to Faming Zhuanli Shenging Gongkai Shuomingshu); *Chem. Abstr.* **111**, 80389g (1989).
52. Brit. Pat. Appl. 2,205,851 (Dec. 21, 1988), R. J. Corring and L. A. S. Mallows (to Unilever PLC).
53. Pol. Pat. 141,759 (May 30, 1988), W. Barwinski and co-workers (to Wojskowy Instytut Chemii Radiometrii).
54. Eur. Pat. Appl. 298,034 (Jan. 4, 1989), F. H. Howell and co-workers (to CIBA-GEIGY A.G.).
55. Eur. Pat. Appl. 301,433 (Feb. 1, 1989), M. Kitabatake (to Kansai Paint Co., Ltd.).
56. Jpn. Kokai Tokkyo Koho 89 85,262 (Mar. 30, 1989), K. Kamikado (to Kansai Paint Co., Ltd.).
57. USSR Pat. 1,479,481 (May 15, 1989), V. P. Shaboldin and co-workers (to All-Union Correspondence Institute of Mechanical Engineering).
58. U.S. Pat. Appl. 803,192 (1977), W. Schneider and L. Gast (to USAD).
59. Ger. Offen. 2,627,697 (1977), K. Shen and T. Pickering (to CIBA-GEIGY A.G.).
60. U.S. Pat. 3,962,522 (1976), W. Chang and M. Hartman (to PPG Industries).
61. Ger. Offen. 2,223,850 (1972), F. Wehrmann (to "Isovolta" Oersterrechisches Isolierstoffwerk K.G.).
62. Jpn. Kokai Tokkyo Koho 79 132,638 (1979), S. Obana and N. Miyagawa (to Mitsubishi Rayon Co., Ltd.).
63. Jpn. Kokai Tokkyo Koho 77 73,932 (1977), H. Ito and H. Kigure (to Kansai Paint Co., Ltd.).
64. U.S. Pat. 3,986,993 (1976), E. Vassiliou (to E. I. du Pont de Nemours & Company, Inc.).
65. Jpn. Kokai Tokkyo Koho 74 08,497 (1974), T. Takahashi and co-workers (to Riken Light Metal Industries Corporation).
66. Ger. Offen. 2,751,761 (1978), R. Butler (to AKZO GmbH).
67. Jpn. Kokai Tokkyo Koho 80 67,368 (1980) (to Nissen Kagaku Kogyo K.K.).
68. Jpn. Kokai Tokkyo Koho 78 139,662 (1978), T. Kawahata and co-workers (to Mitsui Nisso Urethane K.K.).
69. Jpn. Kokai Tokkyo Koho 74 116,180 (1974), H. Toshima and M. Ishihara (to Sohai Chemical Industry Co., Ltd.).
70. Jpn. Kokai Tokkyo Koho 77 141,873 (1977), A. Okhara (to Teigin Chemical, Ltd.).
71. Jpn. Kokai Tokkyo Koho 75 35,035 (1975), H. Ukihahi and A. Hara (to Asahi Glass Co., Ltd.).
72. Jpn. Kokai Tokkyo Koho 79 118,404 (1979), Y. Inoue and co-workers (to Shin-Etsu Chemical Industry Company, Limited; Nitto Chemical Industry Co., Ltd.).
73. U.S. Pat. 3,551,311 (1970) (to Sun Chemical Corp.).
74. U.S. Pat. 3,876,518 (1975), G. Borden and co-workers (to Union Carbide Corp.).
75. Ger. Offen. 2,833,825 (1979), G. Gruber and co-workers (to PPG Industries).
76. U.S. Pat. 4,039,414 (1977), V. McGuiness (to SCM Corp.).
77. U.S. Pat. 4,793,864 (Dec. 27, 1988), P. J. Neumiller and R. M. Etter (to S. C. Johnson and Son Corp.).
78. Ger. Offen. 2,221,834 (1972), B. Durif-Varambon (to Institut Francais du Petrole, des Carburants et Lubrifiants).
79. Jpn. Kokai Tokkyo Koho 75 109,249 (1975), N. Nahamura and co-workers (to Furukawa Electric Co., Ltd.).
80. Jpn. Kokai Tokkyo Koho 75 35,932 (1975), K. Hanaoka (to Osaka Henatsuki, K. K.).
81. U.S. Pat. 4,070,524 (1978), R. Keske (to Standard Oil Co., Indiana).
82. Jpn. Kokai Tokkyo Koho 79 00,094 (1979), E. Kosokawa and co-workers (to Showa Electric Wire and Cable Co., Ltd.).
83. Fr. Pat. 1,582,591 (1969), Brangs and Heinrich.

84. Eur. Pat. Appl. 311,166 (Apr. 12, 1989), F. De Jong and co-workers (to Shell International Research Maatschappij B.V.).

85. Pol. Pat. 144,233 (Apr. 30, 1988), W. Stanik and co-workers (to Instytut Technologii Nafty).

86. U.S. Pat. 4,804,456 (Feb. 14, 1989), D. R. Forester (to Betz Laboratories).

87. USSR Pat. 1,484,825 (June 7, 1989), L. K. Mukhin and co-workers (to Moscow Institute of the Petrochemical and Gas Industry).

88. Fr. Pat. 1,484,815 (1967), H. W. Peabody.

89. Fr. Pat. Add. 92,119 to Fr. Pat. 1,311,943 (Sept. 27, 1968), H. Brunel.

90. Jpn. Kokai Tokkyo Koho 88 210,196 (Aug. 31, 1988), T. Imai and co-workers (to Sanyo Chemical Industries, Ltd.).

91. Jpn. Kokai Tokkyo Koho 88 293,178 (Nov. 30, 1988), T. Yazaki and F. Uchida (to Hakusui Chemical Industry, Ltd.).

92. Czech. Pat. 259,590 (Mar. 15, 1989), J. Boxa and co-workers.

93. T. Szauer, *Powloki Ochr.* **16**(4), 14–17 (1988); *Chem. Abstr.* **111**, 26785v (1989).

94. Pol. Pat. 146,131 (Dec. 31, 1988), G. Wieczorek and co-workers (to Instytut Techniki Budowlanej).

95. M. S. Balsam and E. Sagarin, eds., *Cosmetics: Science and Technology*, 2nd ed., Wiley-Interscience, New York, 1972.

96. J. Jellinek, ed., *Formulation and Function of Cosmetics*, 2nd ed., Wiley-Interscience, New York, 1970.

97. M. G. Navarre, ed., *The Chemistry and Manufacture of Cosmetics*, 2nd ed., Continental Press, Orlando, Fla., 1975.

98. Jpn. Kokai Tokkyo Koho 88 275,700 (Nov. 14, 1988) (to Neutrogena Corp.).

99. Eur. Pat. Appl. 253,489 (Jan. 20, 1988), D. M. Lapetina and C. Patel (to Helene Curtis Corp.).

100. Eur. Pat. Appl. 304,846 (Mar. 1, 1989), V. Manohar Deshpande and co-workers (to Sterling Drug Corp.).

101. Jpn. Kokai Tokkyo Koho 89 09,913 (Jan. 13, 1988), Y. Yamamoto (to Kao Corp.).

102. Jpn. Kokai Tokkyo Koho 89 09,908 (Jan. 13, 1989), I. Sukai and co-workers (to Kao Corp.).

103. Jpn. Kokai Tokkyo Koho 89 04,236 (Jan. 9, 1989), T. Otomo and co-workers (to Kao Corporation).

104. USSR Pat. 1,456,393 (Feb. 7, 1989), M. A. Podustov and co-workers (to Kharkov Polytechnic Institute).

105. H. A. Segalas, *Hydrocarbon Process.* 71, (March 1975).

106. Jpn. Kokai Tokkyo Koho 89 90,295 (Apr. 6, 1989), K. Yahagi (to Kao Corp.).

107. U.S. Pat. 4,507,219 (Mar. 26, 1985), L. J. Hughes (to Procter & Gamble Corp.); U.S. Pat. 4,318,818 (Mar. 9, 1982), J. C. Letton (Procter & Gamble Corp.); U.S. Pat. 4,261,868 (April 14, 1981), J. Hora and co-workers (to Lever Brothers Corp.); U.S. Pat. 4,142,999 (Mar. 6, 1979), H. Blocking (to Henkel Corp.); U.S. Pat. 4,238,345 (Dec. 9, 1980), C. C. Guilbert (to Economics Laboratory Corp.); U.S. Pat. 4,243,546 (Jan. 6, 1981), E. H. Shaer (to The Drackett Co.); U.S. Pat. 4,243,543 (Jan. 6, 1981), C. C. Guilbert and co-workers (to Economics Laboratory Corp.).

108. U.S. Pat. 4,778,574 (Oct. 18, 1988), Z. F. Mathe and A. Fletcher (to American Chemical and Refining Co.).

109. U.S. Pat. 4,818,286 (Apr. 4, 1989), R. Jagannathan and co-workers (to International Business Machines Corp.).

110. Pol. Pat. 140,547 (May 30, 1987), S. Szczepaniak.

111. Jpn. Kokai Tokkyo Koho 89 85,262 (Mar. 30, 1989), K. Kamikado (to Kansai Paint Co., Ltd.)

112. Eur. Pat. Appl. 286,272 (Oct. 12, 1988), R. M. Lazarus and co-workers (to Morton Thiokol Corp.).

113. Ger. Pat. 266,935 (Apr. 19, 1989), K. Nitsche and co-workers (to VEB Carl Zeiss Jena).

114. Jpn. Kokai Tokkyo Koho 89 42,653 (Feb. 14, 1989), M. Kobayashi and co-workers (to Tokyo Ohka Kogyo Co., Ltd.).

115. Jpn. Kokai Tokkyo Koho 89 88,548 (Apr. 3, 1989), S. Shiozu and M. Sugita (to Nagase Denshi Kagaku K.K.).

116. USSR Pat. 1,461,757 (Feb. 28, 1989), A. N. Pichugin and co-workers.

117. Ger. Offen. 3,735,600 (May 3, 1989), G. Ott and co-workers (to BASF Lacke and Farben A.G.).

118. Pol. Pat. 146,096 (Dec. 31, 1988), K. Mloczeke and co-workers (to Glowny Instytut Gornictwa).

119. Pol. Pat. 145,184 (Aug. 31, 1988), S. Drzewiecka and co-workers (to Instytut Tele-i-Radiotechniczny).

120. USSR Pat. 1,488,168 (June 23, 1989), O. Kallas and J. Kuslapuu (to Institute of Cybernetics, Academy of Sciences, Estonian S.S.R.).

121. Ger. Pat. 264,175 (Jan. 25, 1989), J. Schulze (to VEB Elektronische Bauelmente, Teltow).

122. USSR Pat. 1,493,430 (July 15, 1989), Yu. A. Dinaev and S. G. Radkovskii (to Kabardino-Balkar State University).

123. Czech. Pat. 259,554 (Mar. 15, 1989), G. Racz and co-workers.

124. *Gas Conditioning Fact Book*, The Dow Chemical Company, Midland, Mich., 1962.

125. U.S. Pat. 4,814,051 (Mar. 21, 1989), S. A. Bedell and co-workers (to The Dow Chemical Company).

126. U.S. Pat. 4,840,777 (June 20, 1989), J. A. Faucher (to UOP Corp.).

127. D. M. Austgen and co-workers, *Ind. Eng. Chem. Res.* **28**(7), 1060–73 (1989).

128. V. D. Pitsinigos and co-workers, *Hydrocarbon Process., Int. Ed.* **68**(4), 43–44 (1989).

129. J. E. Crooks and J. P. Donnellan, *J. Chem. Soc., Perkin Trans. 2* (4), 331–333 (1989).

130. J. L. Oscarson and co-workers, *Thermochim. Acta* **146**, 107–114 (1989).

131. H. Bosch and co-workers, *Gas Sep. Purif.* **3**(2), 75–83 (1989).

132. M. S. DuPart and B. D. Marchant, *Proc. Laurance Reid Gas Cond. Conf.*, 1989, 279–292.

133. O. Dawodu and co-workers in Ref. 132, 9–71.

134. C. L. Kimtantes, *Oil Gas J.* **87**(21), 58–64 (1989).

135. D. A. Glasscock and G. T. Rochelle, *AIChE J.* **35**(8), 1271–1281 (1989).

136. T. R. Bacon in Ref. 132, 1–8.

137. Eur. Pat. Appl. 322,924 (July 5, 1989), J. C. Thomas and co-workers (to Union Carbide Corp.); U.S. Pat. Appl. 140,127 (Dec. 31, 1987).

138. U.S. Pat. 4,814,104 (Mar. 21, 1989), D. J. Kubek and D. S. Kovach (to UOP Corp.).

139. A. Chakma and co-workers, *Gas. Sep. Purif.* **3**(2), 65–70 (1989).

140. A. P. Mitina and co-workers, *Zh. Prikl. Khim. (Leningrad)* **61**(8), 1780–1784 (1988).

141. M. J. Bourke and A. F. Mazzoni in Ref. 132, 137–158.

142. USSR Pat. 1,466,781 (Mar. 23, 1989), F. R. Ismagilov and co-workers (to Volga-Ural Scientific Research Institute for the Extracting and Processing of Hydrogen Sulfide Gas).

143. USSR Pat. 1,437,351 (Nov. 15, 1988), F. R. Ismagilov and co-workers (to Volga-Ural Scientific Research Institute for the Extracting and Processing of Hydrogen Sulfide Gas).

144. USSR Pat. 1,494,947 (July 23, 1989), N. S. Chernozemov and co-workers (to All-Union Scientific Research of Hydrocarbon Raw Materials).

145. G. Y. Polishchuk, *Vestn. L'vov. Politekh Inst.* **221**, 65–66 (1988).

146. S. Zhu and co-workers, *Huagong Xuebao* **40**(1), 96–103 (1989); *Chem. Abstr.* **111**, 25475g (1989).

147. USSR Pat. 1,490,124 (June 30, 1989), V. R. Granval'd and co-workers (to Volga-Ural Scientific Research Institute for the Extracting and Processing of Hydrogen Sulfide Gas).

148. USSR Pat. 1,477,674 (May 7, 1989), K. A. Bogatin and co-workers.

149. C. R. Gagliardi and co-workers, *Oil Gas J.* **87**(10), 44–49 (1989).

150. U.S. Pat. Appl. 36,486 (Oct. 12, 1988) and Eur. Pat. Appl. 286,143 (Oct. 12, 1988), R. Gregory, Jr. and M. F. Cohen (to Union Carbide Corp.).

151. H. A. Al-Ghawas and co-workers, *J. Chem. Eng. Data* **34**(4), 385–391 (1989).

152. U.S. Pat. 4,808,765 (Feb. 28, 1989), R. L. Pearce and R. A. Wolcott (to The Dow Chemical Company).

153. Pol. Pat. 144,799 (July 30, 1988), M. Marcinski and co-workers (to Instytut Ciezkiej Syntezy Organicznej "Blachownia").

154. Pol. Pat. 144,774 (Apr. 30, 1988), A. Pasierb and co-workers (to Akademia Gorniczo-Hutnicza; and Rafineria Nafty "Glimar").

155. S. Watanabe and co-workers, *Ind. Eng. Chem. Res.* **28**, 1264–1266 (1989).

156. Jpn. Kokai Tokkyo Koho 88 291,995 (Nov. 29, 1988), H. Tomihari and co-workers (to Yushiro Chemical Industry Co., Ltd.).

157. USSR Pat. 1,488,361 (June 23, 1989), A. N. Tarasov and co-workers.

158. Jpn. Kokai Tokkyo Koho 88 247,384 (Oct. 14, 1988), N. Tada and co-workers (to Ricoh Kyosan Co., Ltd.).

159. Hung. Pat. 3,238 (Jan. 30, 1989), M. Major.

160. USSR 1,456,460 (Feb. 7, 1989), A. M. Tikhontsov and co-workers (to Dnepropetrovsk Industrial Institute).

161. USSR Pat. 1,447,845 (Dec. 30, 1988), V. E. Ivanov and co-workers (Belorussian Institute of Railroad Transport Engineers).

162. USSR Pat. 1,490,146 (June 30, 1980), E. M. Berliner and co-workers (to Moscow Automobile Paint).

163. Pol. Pat. 131,755 (Dec. 30, 1985), A. Klopotek and co-workers (to Instytut Chemmi Przemyslowej).

164. U.S. Pat. 4,139,455 (Feb. 13, 1979), R. M. Griffith and co-workers (to Allied Colloids, Ltd.).

165. U.S. Pat. 3,836,443 (Sept. 17, 1974), D. MacGregor.

166. U.S. Pat. 4,059,509 (Nov. 22, 1977), F. B. Eisenhardt and S. F. Muchlberger (to Mobil Oil Corp.).

167. U.S. Pat. 3,286,837 (Nov. 22, 1966), V. Mercade and J. B. Duke (to Minerals and Chemicals Philipp Corp.).

168. U.S. Pat. 3,387,402 (1974), C. Stringer (to Radon Development Corp.).

169. U.S. Pat. 3,799,263 (1974), M. Prillieux and R. Tirtiaux (to Esso Product Research Co.).

170. U.S. Pat. 3,861,466 (1975), W. Gale (to Exxon Product Research Co.).

171. U.S. Pat. 3,994,342 (1976), R. Healy and W. Gale (to Exxon Product Research Co.).

172. U.S. Pat. 3,983,940 (1976), C. Carpenter, Jr. and W. Gale (to Exxon Research and Engineering Co.).

173. U.S. Pat. 3,811,504 (1974), K. Flournoy and co-workers (to Texaco Corp.).

174. Brit. Pat. 1,504,789 (1978), D. Grist (to British Petroleum Co., Ltd.).

175. U.S. Pat. 4,191,425 (1980), B. Davis (to Chevron Research Co.).

176. Jpn. Pat. Kokai Tokkyo Koho 79 68,805 (1979), S. Honjo (to Daiichi Kogyo Seiyaku Co., Ltd.).

177. U.S. Pat. 4,101,293 (1978), A. Krause and J. Korose (to Reichhold Chemicals Corp.).

178. U.S. Pat. 4,780,482 (Oct. 25, 1988), D. C. Kreuger (to BASF Corp.).

179. Eur. Pat. Appl. 302,591 (Feb. 8, 1989), J. M. Larkin and M. Cuscurida (to Texaco Development Corp.).
180. U.S. Pat. 4,822,518 (Apr. 18, 1989), A. B. Goel and H. J. Richards (to Ashland Oil Corp.).
181. USSR Pat. 1,479,475 (May 15, 1989), S. S. Pesetskii and co-workers (to Institute of the Mechanics of Metal-Polymer Systems, Academy of Sciences, Belorussian S.S.R.; Institute of Colloidal and Water Chemistry, Academy of Sciences, Ukrainian S.S.R.).
182. V. M. Shimakskii and co-workers, *Plast. Massy.* (3), 19–22 (1989).
183. E. Ceausescu and co-workers *Rev. Roum. Chim* **34**, 437–443 (1989).
184. Czech Pat. 254,031 (Nov. 15, 1988), E. Pavlacka and co-workers.
185. USSR Pat. 1,442,525 (Dec. 7, 1988), B. K. Kupchiov and co-workers (to Institute of the Mechanics of Metal-Polymer Systems, Academy of Sciences, Belorussian S.S.R.).
186. Eur. Pat. Appl. 300,672 (Jan. 25, 1989), R. Kobashi and co-workers (to Nippon Oils and Fats Co., Ltd. and PPG Industries Corp.).
187. Pol. Pat. 141,291 (July 31, 1987), H. Staniak and co-workers (to Instytut Chemii Przemyslowej).
188. USSR Pat. 1,428,732 (Oct. 7, 1988), V. P. Selyaev and co-workers (to Mordovian State University).
189. Pol. Pat. 133,789 (June 29, 1985), B. Szczepaniak and co-workers (to Instytut Chemii Przemyslowej).
190. Jpn. Pat. Kokai Tokkyo Koho 88 295,628 (Dec. 2, 1988), I. Yamashita (to Toyo Electric Manufacturing Co., Ltd.).
191. U.S. Pat. 4,800,222 (Jan. 24, 1989), H. G. Waddill (to Texaco Corp.).
192. PRC Pat. 87 101,669 (Sept. 7, 1988), X. Xiang and X. Zheng (to Faming Zhuanli Shenqing Gongkai Shuomingshu; *Chem. Abstr.* **110**, 174467n (1989).
193. USSR Pat. 1,484,850 (June 7, 1989), L. P. Loseva (to Ivanovo Chemical-Technological Institute; and Ivanovo Worsted Fabric Combine).
194. Pol. Pat. 140,720 (May 30, 1987), J. Berger and co-workers (to Nadodrzanskie Zaklady Przemyslu Organiczanego "Organika-Rokita").
195. Pol. Pat. 142,213 (Mar. 30, 1988), A. Pasternak and co-workers (to Politechnika Wroclawska; and Paczkowskie Zaklady hemii Gospodarczej "Pollena").
196. USSR Pat. 1,432,054 (Oct. 23, 1988), D. A. Zhukov and co-workers (to All-Union Scientific Research Institute of the Chemical Industry, Tula).
197. USSR Pat. 1,484,847 (June 7, 1989), Y. N. Tsibizov and co-workers (to Central Scientific Research Institute of the Wool Industry, Nevinnomyssk).
198. E. J. Beanchard and R. M. Reinhardt, *Int. Conf. Exhib. of AATCC*, American Association of Textile Chemists and Colonists, 1988, 261–266.
199. USSR Pat. 1,432,118 (Oct. 23, 1988), I. Saladene and co-workers (to P. Zilbertus Silk Combine).
200. K. V. Lunyaka and L. I. Logacheva, *Izv. Vyssh. Uchebn. Zaved., Technol. Tekst. Promsti* (6), 76–79 (1988).
201. Eur. Pat. Appl. 318,294 (May 31, 1989), K. Taniguchi (to Nippon Chemical Works Co., Ltd.).
202. Eur. Pat. Appl. 300,672 (Jan. 25, 1988), R. Kobashi and co-workers (to Nippon Oils and Fats Co., Ltd. and PPG Industries, Inc.).
203. Eur. Pat. Appl. 290,384 (Nov. 9, 1988), J. M. Adam and A. Kaeser (to CIBA-GEIGY A.G.).
204. Jpn. Kokai Tokkyo Koho 89 20,268 (Jan. 24, 1989), N. Yamanaka and co-workers (to Nippon Kayaku Co., Ltd.).
205. W. Ruback, *Commun. Jorn. Com. Esp. Deterg.* **20**, 181–192 (1989).
206. Eur. Pat. Appl. 295,385 (Dec. 21, 1988), W. Ruback and J. Schut (to Huels A. G.).

207. Belg. Pat. 617,966 (Sept. 14, 1962), D. M. Gagarine and co-workers (to Deering Milliken Research Corp.); and Belg. Pat. 648,426 (Sept. 15, 1964), J. H. Dusenbury and co-workers (to Deering Milliken Research Corp.).

208. Eur. Pat. Appl. 307,778 (Mar. 22, 1989), G. Sackman and co-workers (to Bayer A. G.).

209. U.S. Pat. 3,722,046 (Nov. 13, 1973), D. E. Knapp and L. P. Nageroni (to American Cyanamid Corp.).

210. U.S. Pat. 3,808,023 (Apr. 30, 1974), J. Whitehead and C. L. Denton (to British Titan, Limited).

211. U.S. Pat. 4,165,239 (Aug. 21, 1979), H. Linden and H. Bormann (to Henkel Co.).

212. U.S. Pat. 4,045,280 (Aug. 30, 1977), D. M. Mackie (to MacMillan Bloedd, Ltd.).

213. U.S. Pat. 4,298,428 (Nov. 3, 1981), M. D. Breslin and D. R. Cosper (to Nalco Chemical Co.).

214. U.S. Pat. 4,397,712 (Aug. 9, 1983), J. Gordy (to New Fibers International Corp.).

215. U.S. Pat. 4,740,591 (Apr. 26, 1988), P. Dilling and M. S. Demitri (to Westvaco Corp.).

216. Int. Pat. Appl. 88 09,171 (Dec. 1, 1988), S. H. Appel (to Baylor College of Medicine).

217. Int. Pat. Appl. 88 08,860 (Oct. 20, 1988), R. Lodi (to Riace Establishment).

218. Jpn. Kokai Tokkyo Koho 87 255,461 (Nov. 7, 1987), H. Horada and F. Hanzawa (to Chisso Corp.).

219. Ger. Pat. 265,313 (Mar. 1, 1989), G. Meisgeier and co-workers (to Akademie der Landwirtschaftswissenschaften der DDR, Forschungszenstrum fur Bodenfruchtbarkeit).

220. Ger. Pat. 255,873 (Apr. 20, 1988), H. Bergmann and co-workers (to Akademie der Landwirtschaftswissenschaften der DDR, Forschungszenstrum fur Bodenfruchtbarkeit).

221. U.S. Pat. 4,786,583 (Nov. 22, 1988), P. A. Schwartz (to Eastman Kodak Co.).

222. USSR Pat. 1,474,579 (Apr. 23, 1989), B. A. Shashlov and co-workers (to Moscow Printing Institute).

223. V. Weiss and co-workers, *Proc. Spie. Int. Soc. Opt. Eng.* **1038**, 110–114 (1989).

224. I. I. Breido, *Zh. Nauchn. Prikl. Fotogr. Kinematogr.* **33**, 441–445 (1988); *Chem. Abstr.* **110**, 104698g (1989).

225. *Aminoethylethanoleamine (AEEA): A Versatile Intermediate from Dow*, Form No. 118–1102–R88, The Dow Chemical Company, Midland, Mich., 1988.

226. *Pennsalt Amines*, Pennsalt Chemicals Corp., Philadelphia, Pa., 1967.

MARTHA R. EDENS
J. FRED LOCHARY
The Dow Chemical Company

ALKYD RESINS

1. Introduction

While no longer the largest volume vehicles in coatings, alkyds still are of major importance. Alkyds are prepared from polyols, dibasic acids, and fatty acids. They are polyesters, but in the coatings field the term polyester is reserved for "oil-free polyesters". The term *alkyd* is derived from *alcohol* and *acid*. Alkyds tend to be lower in cost than most other vehicles and tend to give coatings that exhibit fewer film defects during application. However, durability of alkyd films, especially outdoors, tends to be poorer than films from acrylics, polyesters, and polyurethanes. In a comparison of resistance to acid rain among coconut alkyd-MF, polyester-MF, and silicone-modified polyester-MF (MF = melamine–formaldehyde) coatings at five locations in the United States and Canada, the alkyd coating showed the poorest resistance (1).

There are several types of alkyds. One classification is into *oxidizing* and *nonoxidizing* types. Oxidizing alkyds cross-link by the same mechanism as drying oils. Nonoxidizing alkyds are used as polymeric plasticizers or as hydroxy-functional resins, which are cross-linked by MF resins, by urea–formaldehyde (UF) resins, or by isocyanate cross-linkers. A second classification is based on the ratio of monobasic fatty acids to dibasic acids utilized in their preparation. The terminology used was adapted from terminology used to classify varnishes. Varnishes with high ratios of oil to resin were called long oil varnishes; those with a lower ratio, medium oil varnishes; and those with an even lower ratio, short oil varnishes. Oil length of an alkyd is calculated by dividing the amount of "oil" in the final alkyd by the total weight of the alkyd solids, expressed as a percentage, as shown in equation 1. The amount of oil is defined as the triglyceride equivalent to the amount of fatty acids in the alkyd. The 1.04 factor in equation 2 converts the weight of fatty acids to the corresponding weight of triglyceride oil. Alkyds with oil lengths >60 are *long oil alkyds*; those with oil lengths from 40 to 60, *medium oil alkyds*, and those with oil lengths <40, *short oil alkyds*. There is some variation in the dividing lines between these classes in the literature.

$$\text{Oil length} = \frac{\text{Weight of oil}}{\text{Weight of alkyd} - \text{water evolved}} \times 100 \tag{1}$$

$$\text{Oil length} = \frac{1.04 \times \text{Weight of fatty acids}}{\text{Weight of alkyd} - \text{water evolved}} \times 100 \tag{2}$$

Another classification is *unmodified* or *modified alkyds*. Modified alkyds contain other monomers in addition to polyols, polybasic acids, and fatty acids. Examples are *styrenated alkyds* and *silicone alkyds*. Since they are closely related to alkyd resins, uralkyds and epoxy esters are also discussed.

2. Oxidizing Alkyds

Oxidizing alkyds can be considered as synthetic drying oils. They are polyesters of one or more polyols, one or more dibasic acids, and fatty acids from one or more drying or semidrying oils.

2.1. Film Formation. Most of the studies of the chemistry of cross-linking have been with drying oils and not the alkyds derived from them but the mechanisms are applicable to both (see DRYING OILS). Films exposed to air undergo autoxidative cross-linking. In nonconjugated unsaturated oils, the active group initiating drying is the diallylic group ($-CH=CHCH_2CH=CH-$) from esters of (Z,Z)-9,12-octadecadienoic acid [60-33-3] (linoleic acid) and

(Z,Z,Z)-9,12,15-octadecatrienoic acid [463-40-1] (linolenic acid). They have one and two diallylic groups per molecule, respectively. Drying is related to the average number of diallylic groups per molecule. If this number is greater than ~2.2, the oil is a drying oil and if it is moderately <2.2, the oil is a semidrying oil; there is no sharp dividing line between semidrying oils and nondrying oils. Since diallylic groups are the sites for cross-linking, it is convenient to relate the average number of such groups per molecule to the number average functionality f_n of the triglyceride or synthetic drying oil. It is probable that some of the sites are involved in more than one cross-linking reaction.

When a film is applied, initially naturally present hydroperoxides decompose to form free radicals. Hydrogens on methylene groups between double bonds are particularly susceptible to abstraction, yielding a resonance stabilized free radical that reacts with oxygen to give predominantly a conjugated peroxy free radicals. The peroxy free radicals can abstract hydrogens from other methylene groups between double bonds to form additional hydroperoxides and generate free radicals like. Thus, a chain reaction is established, resulting in autoxidation. At least part of the cross-linking occurs by radical–radical combination reactions forming C–C, ether, and peroxide bonds. These reactions correspond to termination by combination reactions in free-radical chain-growth polymerization. Reactions analogous to the addition step in chain-growth polymerization could also produce cross-links (see DRYING OILS for further discussion).

Rearrangement and cleavage of hydroperoxides to aldehydes and ketones, among other products, lead to low molecular weight byproducts. The characteristic odor of oil and alkyd paints during drying is attributable to such volatile byproducts, as well as to the odor of organic solvents. Undesirable odor has been a factor motivating replacement of oil and alkyds in paints with latex, particularly for interior applications.

Dried films, especially of alkyds with three double-bond fatty acids, yellow with aging. The yellow color bleaches significantly when exposed to light; hence, yellowing is most severe when films are covered, such as by a picture hanging on a wall. The reactions leading to color are complex and are not fully understood. Yellowing has been shown to result from incorporation of nitrogen compounds and is markedly increased by exposure to ammonia. It has been proposed that ammonia reacts with 1,4-diketones formed in autoxidation to yield pyrroles, which oxidize to yield highly colored products (2).

The autoxidation rates of uncatalyzed nonconjugated oxidizing alkyds dry are slow. Many years ago, it was found that metal salts (*driers*) catalyze drying. The most widely used driers are oil-soluble cobalt, manganese, lead, zirconium, and calcium salts of octanoic or naphthenic acids. Salts of other metals, including rare earths, are also used. Cobalt and manganese salts, so-called *top driers* or *surface driers*, primarily catalyze drying at the film surface. Lead and zirconium salts catalyze drying throughout the film and are called *through driers*. The surface-drying catalysis by cobalt and manganese salts results from the catalysis of hydroperoxide decomposition. The cobalt or manganese cycle between the two oxidation states. The activity of through driers has not been adequately explained.

Combinations of metal salts are almost always used. Mixtures of lead with cobalt and/or manganese are particularly effective, but as a result of toxicity control regulations, lead driers can no longer be used in consumer paints sold in interstate commerce in the United States. Combinations of cobalt and/or manganese with zirconium, frequently with calcium, are commonly used. Calcium does not undergo redox reactions; it has been suggested that it may promote drying by preferentially adsorbing on pigment surfaces, minimizing adsorption of active driers. The amounts of driers needed are system specific. Their use should be kept to the minimum possible level, since they not only catalyze drying but also catalyze the reactions that cause postdrying embrittlement, discoloration, and cleavage. For further discussion see section on catalysis in DRYING OILS.)

Oils containing conjugated double bonds, such as tung oil, dry more rapidly than any nonconjugated drying oil. Free-radical polymerization of the conjugated diene systems can lead to chain-growth polymerization, rather than just a combination of free radicals to form cross-links. High degrees of polymerization are unlikely because of the high concentration of abstractable hydrogens acting as chain-transfer agents. However, the free radicals formed by chain transfer also yield cross-links. In general, the water and alkali resistance of films derived from conjugated oils are superior, presumably because more of the cross-links are stable carbon–carbon bonds. However, since the (E,Z,E)-9,11,13-octadecatrienoic acid [506-23-0] (α-eleostearic acid) in tung oil has three double bonds, discoloration on baking and aging is severe.

The most commonly used polyol in preparing alkyds is glycerol (1,2,3-propanetriol) [56-81-5], the most commonly used dibasic acid is phthalic anhydride (PA) (1,3-isobenzofurandione) [85-44-9], and a widely used oil is soybean oil. Let us consider a simple, idealized example of the alkyd prepared from 1 mol of PA, 2 mol of glycerol, and 4 mol of soybean fatty acids. A typical fatty acid composition data for soybean oil is as follows saturated fatty acids, palmitic acid (hexadecanoic acid) [57-10-3] and stearic acid (octadecanoic acid) [57-11-4], 15%; oleic acid (Z)-9-octadecenoic acid [112-80-1], 25%; linoleic acid, 51%; and linolenic acid, 9%. Any oil with an $\bar{f}_n$ higher than 2.2 is a drying oil. Although soybean oil is a semidrying oil, this alkyd would have an $\bar{f}_n$ of 2.76 per molecule and, therefore, would dry to a solid film. The alkyd would form a solvent-resistant film in about the same time as a pentaerythritol (PE) (2,2-bis(hydroxymethyl)-1,3-pentanediol) [115-77-5] ester of soybean fatty acids, since they have the same $\bar{f}_n$. However, the alkyd would form a tack-free film faster because the rigid aromatic rings from PA increase the T_g of the film.

If the mole ratio of PA to glycerol were 2–3, 5 mol of soybean fatty acid could be esterified to yield an alkyd with an $\bar{f}_n$ of 3.45. This alkyd would cross-link more rapidly than the 1:2:4 mol ratio alkyd and would also form tack-free films even faster because the ratio of aromatic rings to long aliphatic chains would be 2:5 instead of 1:4. As the ratio of PA to glycerol is increased further, the average functionality for autoxidation increases and the T_g after solvent evaporation increases because of the increasing ratio of aromatic to long aliphatic chains. For both reasons, films dry faster.

A theoretical alkyd prepared from 1 mol each of glycerol, PA, and fatty acid would have an oil length of ~60. However, if one were to try to prepare such an alkyd, the resin would gel prior to complete reaction. Gelation would result from reaction of a sufficient number of trifunctional glycerol molecules with three difunctional PA molecules to form cross-linked polymer molecules, swollen with partially reacted components. Gelation can be avoided by using a sufficient excess of glycerol to reduce the extent of cross-linking. When the reaction is carried to near completion with excess glycerol, there are few unreacted carboxylic acid groups, but many unreacted hydroxyl groups.

There have been many attempts, none fully successful, to calculate the ratios of functional groups and the extent of reaction that can be reached without encountering gelation. The problem is complex. The reactivity of the hydroxyl groups can be different; for example, glycerol contains both primary and secondary alcohol groups. Under esterification conditions, polyol molecules can self-condense to form ethers and, in some cases, can dehydrate to form volatile aldehydes. Reactivity of the carboxylic acids also varies. The rate of formation of the first ester from a cyclic anhydride is more rapid than formation of the second ester. Aliphatic acids esterify more rapidly than aromatic acids. Polyunsaturated fatty acids and their esters can dimerize or oligomerize to form cross-links. Of the many papers in the field, that by Blackinton recognizes the complexities best (3). In addition to the above complexities, particular emphasis is placed on the extent of formation of cyclic compounds by intramolecular esterification reactions. Equations have been developed that permit calculation of ratios of ingredients theoretically needed to prepare an alkyd of any desired oil length, number average molecular weight, and hydroxy content (4). Just like in other equations, the important effect of dimerization of fatty acids is not included as a factor in these equations. In practice, alkyd resin formulators have found that the mole ratio of dibasic acid to polyol should be <1 to avoid gelation. How much <1 depends on many variables.

For medium oil alkyds, the ratio of dibasic acid to polyol is not generally changed much relative to alkyds with an oil length of ~60, but the fatty acid content is reduced to the extent desired. This results in a larger excess of hydroxyl groups in the final alkyd. It is commonly said that as the oil length of an oxidizing alkyd is reduced <60, the drying time decreases to a minimum at an oil length of ~50. However, this conventional wisdom must be viewed cautiously. The ratio of aromatic rings to aliphatic chains continues to increase, increasing T_g after the solvent evaporates from the film tending to shorten the time to form a tack-free film. However, at the same molecular weight, the number of fatty acid ester groups per molecule decreases as the oil length decreases <60, since more hydroxyl groups are left unesterified. Therefore, the time required to achieve sufficient cross-linking for solvent resistance increases.

Long oil alkyds are soluble in aliphatic hydrocarbon solvents. As the oil length decreases, mixtures of aliphatic and aromatic solvents are required, and oil lengths below ~50 require aromatic solvents, which are more expensive than aliphatics. The viscosity of solutions of long oil alkyds, especially of those with oil lengths <65, is higher in aliphatic than in aromatic solvents; in medium oil alkyds, which require mixtures of aliphatic and aromatic solvents, viscosity decreases as the proportion of aromatic solvents increases. In former days, and to some extent still today, it was considered desirable to use a solvent mixture that gave the highest possible viscosity; then, at application viscosity, the solids were lower and the raw material cost per unit volume was less. Accordingly, alkyds were designed to have high dilutability with aliphatic solvents. This was false economy, but it was a common practice and is still being practiced to some extent. Increasingly, the emphasis is on reducing volatile organic compound (VOC) emissions and so the question becomes how to design alkyds with low solvent requirements rather than high dilutability potential. Furthermore, the aromatic solvents are on the hazardous air pollutants (HAP) list. High solids alkyds are discussed in a later section.

2.2. Monobasic Acid Selection.   Drying alkyds can be made with fatty acids from semidrying oils, since the $\bar{f}_n$ can be well >2.2. For alkyds made by the monoglyceride process, soybean oil is used in the largest volume. Soybean oil is economical and supplies are dependable because it is a large scale agricultural commodity; alkyd production takes only a few percent of the world supply. For alkyds made by the fatty acid process, tall oil fatty acids (TOFA) are more economical than soybean fatty acids. Both soybean oil and TOFA contain roughly 40–60% linoleic acid and significant amounts of linolenic acid. White coatings containing linolenic acid esters gradually turn yellow. Premium cost "nonyellowing" alkyds are made with safflower or sunflower oils, which are high in linoleic acid but contain very little linolenic acid.

Applications in which fast drying and high cross-link density are important require alkyds made with drying oils. The rate of oxidative cross-linking is affected by the functionality of the drying oils used. At the same oil length and molecular weight, the time required to achieve a specific degree of cross-linking decreases as the average number of diallylic groups ($\bar{f}_n$) increases. Linseed long oil alkyds therefore cross-link more rapidly than soybean long oil alkyds. The effect is especially large in very long oil alkyds and less noticeable in alkyds with oil lengths ~60, where $\bar{f}_n$ is very high even with soybean oil and the effect of further increase in functionality by using linseed oil is small. Because of the large fraction of esters of fatty acids with three double bonds in linseed alkyds, their color and color retention are poorer than that of soybean alkyds. Tung oil based alkyds, because of the high proportion of esters with three conjugated double bonds, dry still faster. Tung oil alkyds also exhibit a high degree of yellowing. Dehydrated castor alkyds have fairly good color retention and cure more rapidly than those made with nonconjugated fatty acids. Since they contain only a small proportion of esters of fatty acids with three double bonds; they are used primarily in baking coatings. Conjugated acids made by isomerizing tall oil acids give similar results to those obtained with dehydrated castor oil acids.

Drying oils and drying oil fatty acids undergo dimerization at elevated temperatures. Dimerization occurs concurrently with esterification during alkyd synthesis; it generates difunctional acids, increasing the mole ratio of dibasic acids to polyol. The rate of dimerization is faster with drying oils having a higher average number of diallylic groups per molecule and with those having conjugated double bonds. Thus, the molecular weight, and therefore the viscosity of an alkyd, made with the same ratio of reactants depends on the fatty acid composition. The higher the degree of unsaturation, the higher the viscosity because of the greater extent of dimerization. Linseed alkyds have higher viscosities than soybean alkyds made with the same monomer ratios under the same conditions. The effect is particularly marked with tung oil. It is difficult to prepare straight tung alkyds because of the risk of gelation; commonly, mixed linseed–tung alkyds are used when high oxidative cross-linking functionality is desired.

A critical factor involved in the choice of fatty acid is cost. Drying oils are agricultural products and, hence, tend to be volatile in price. By far, the major use of vegetable oils is for foods. Depending on relative prices, one drying oil is often substituted for another in certain alkyds. By adjusting for functionality differences, substitutions can frequently be made without significant changes in properties.

Fatty acids are not the only monobasic acids used in making alkyds. Benzoic acid is also used, especially to esterify some of the excess hydroxyl groups remaining in the preparation of medium oil alkyds. The benzoic acid [65-85-0] increases the ratio of aromatic to aliphatic chains in the alkyd, thus contributing to a higher T_g of the solvent-free alkyd and more rapid formation of a tack-free film. At the same time, the reduction in the free hydroxyl content may somewhat reduce water sensitivity of the dried films. Rosin can also be used in the same fashion. Although rosin is not an aromatic acid, its polynuclear ring structure is rigid enough to increase T_g. If the critical requirement in drying is rapid development of solvent resistance, such benzoic acid and rosin modifications do not serve the purpose; they only reduce tack-free time. Frequently, benzoic acid modified alkyds are called *chain-stopped* alkyds. The implication of the terminology is that the benzoic acid stops chain growth, which is not the case; the benzoic acid simply esterifies hydroxyl groups that would not have been esterified if the benzoic acid were absent. The effect on degree of polymerization is negligible.

2.3. Polyol Selection. Glycerol is the most widely used polyol because it is present in naturally occurring oils from which alkyds are commonly synthesized. The next most widely used polyol is PE. In order to avoid gelation, the tetrafunctionality of PE must be taken into account when replacing glycerol with PE. If the substitution is made on a mole basis, rather than an equivalent basis, chances for gelation are minimized. As mentioned earlier, the ratio of moles of dibasic acid to polyol should be <1, and generally, a slightly lower mole ratio is required with PE than with glycerol. At the same mole ratio of dibasic acid to polyol, more moles of fatty acid can be esterified with PE. Hence, in long oil alkyds, the average functionality for cross-linking is higher, and the time to reach a given degree of solvent resistance is shorter for a PE alkyd as compared to a glycerol alkyd. Because of this difference, one must be careful in comparing oil lengths of glycerol and PE alkyds. Films from PE based alkyds generally are superior to their glycerol counterparts in drying, hardness, and humidity resistance.

When PE is synthesized, dipentaerythritol (2,2′-[oxybis(methylene)]-bis(2-hydroxymethyl)-1,3-propanediol) [126-58-9] and tripentaerythritol (2,2-bis([3-hydroxy-2,2-bis(hydroxymethyl)propoxy]methyl}1,3-propanediol) [78-24-0] are byproducts, and commercial PE contains some of these higher polyols. Consequently, care must be exercised in changing sources of PE, since the amount of the higher polyols may differ. Because of the very high functionality, diPE and triPE ($F = 6$ and 8, respectively) are useful in making fast drying low molecular weight alkyds.

To reduce cost, it is sometimes desirable to use mixtures of PE and ethylene glycol (1,2-ethanediol) [107-21-1] or propylene glycol (1,3-propanediol). A 1:1 mole ratio of tetrafunctional and difunctional polyols gives an average functionality of 3, corresponding to glycerol. The corresponding alkyds can be expected to be similar, but not identical. Trimethylolpropane (TMP) (2-ethyl-2-(hydroxy-methyl)-1,3-propanediol) [77-99-6] can also be used, but the rate of esterification is slower than with glycerol. Although all of TMPs alcohol groups are primary, they are somewhat sterically hindered by the neopentyl structure (5). Trimethyl-olpropane, however, gives a narrower molecular weight distribution, which provides alkyds with a somewhat lower viscosity than the comparable glycerol-based alkyd. A kinetic study demonstrated that esterification of one or two of the hydroxyl groups of TMP has little effect on the rate constant for esterification of the third hydroxyl group (6). It can be speculated that PE behaves similarly.

2.4. Dibasic Acid Selection.   Dibasic acids used to prepare alkyds are usually aromatic. Their rigid aromatic rings increase the T_g of the resin. Cyclo-aliphatic anhydrides, such as hexahydrophthalic anhydride, are also used. While they are not as rigid as aromatic rings, the cycloaliphatic rings also increase T_g.

By far, the most widely used dibasic acid is PA. It has the advantage that the first esterification reaction proceeds rapidly by opening the anhydride ring. The amount of water evolved is lower, which also reduces reaction time. The relatively low melting point (the pure compound melts at 131°C) is desirable, since the crystals melt and dissolve readily in the reaction mixture. In large-scale manufacturing, molten PA is used, which reduces packaging, ship-ping, and handling costs.

The next most widely used dibasic acid is isophthalic acid (IPA) (1,3-benze-nedicarboxylic acid) [121-91-5]. Esters of IPA are more resistant to hydrolysis than are those of PA in the pH range of 4–8, the most important range for exter-ior durability. On the other hand, under alkaline conditions esters of phthalic acid are more resistant to hydrolysis than isophthalic esters. The raw material cost for IPA is not particularly different from PA (even after adjusting for the extra mole of water that is lost), but the manufacturing cost is higher. The high melting point of IPA (330°C) leads to problems getting it to dissolve in the reaction mixture so that it can react. High temperatures are required for longer times than with PA; hence more dimerization of fatty acids occurs with IPA resulting in higher viscosity. The longer time at higher temperature also leads to greater extents of side reactions of the polyol components (7). Thus, when substituting IPA for PA, one must use a lower mole ratio of IPA to polyol in order to make an alkyd of similar viscosity.

Maleic anhydride (2,5-furandione) [108-31-6] is sometimes used with PA to give faster drying with improved adhesion and water resistance alkyds. Aliphatic acids, such as adipic acid (1,6-hexanedioic acid) [124-04-9], are sometimes used as partial replacements for PA to give more flexible alkyds.

Chlorinated dibasic acids, such as chlorendic anhydride (3,4,5,6,7,7-hexa-chloroendomethylene-1,2,3,6-tetrahydrophthalic anhydride), are used in making alkyds for fire retardant coatings (8).

2.5. High Solids Oxidizing Alkyds.   The need to minimize VOC emis-sions has led to efforts to increase solids content of alkyd resin coatings. Since xylene is on the HAP list, its use is being reduced. Some increase in solids can be realized by a change of solvents. Aliphatic (and to a somewhat lesser degree, aromatic) hydrocarbon solvents promote intermolecular hydrogen bonding, espe-cially between carboxylic acids, but also between hydroxyl groups, thereby increasing viscosity. Use of at least some hydrogen-bond acceptor solvent, such as an ester or ketone, or hydrogen-bond acceptor–donor solvent such as an alcohol, gives a significant reduction in viscosity at equal solids.

The molecular weight of conventional alkyds is usually >50,000. Solids can be increased by decreasing molecular weight, which is easily accomplished by decreasing the dibasic acid to polyol ratio. Alkyds with solids in the range of 60–80% are commercially available with molecular weights in the range of 12,000–20,000 (9). High solids alkyds tend to have lower functionality for cross-linking and a lower ratio of aromatic to aliphatic chains. Both changes increase the time for drying. There is also a decrease in branching with the higher hydroxyl excess.

The effect of longer oil length on functionality can be minimized by using drying oils with higher average functionality. Use of oils containing linolenic or α-eleostearic acid is limited by their tendency to discolor. One can use safflower oil, which has a higher linoleic acid content and less linolenic acid than soybean oil. Proprietary fatty acids with 78% linoleic acid are commercially available. Early hardness of the films can be improved by using some benzoic acid to esterify part of the free hydroxy groups. As noted earlier, the rigid rings of benzoic acid increase T_g to increase hardness after solvent evaporation.

Different drier combinations are recommended for use with high solids alkyds. A study of a variety of driers and drier combinations with high solids coatings has been published (10). Cobalt, neodymium, aluminum, and barium carboxylic acid salts were of particular interest. Performance was enhanced by adding bipyridyl as an accelerator. The author reports that the best drier system was 0.04% Co, 0.3% Nd with 0.07% bipyridyl (percentages based on the vehicle solids). Reference (11) reports studies of mechanisms of action of cobalt and mixed cobalt–zirconium driers.

By using optimized resins, good quality air dry and baking alkyd coatings can be formulated with VOC levels of 280–350 g/L of coating. A 250-g/L level is attainable only with some sacrifice of application and film properties; still lower limits of permissible VOC are projected.

Solids can be increased by making resins with narrower molecular weight distributions. For example, one can add a transesterification catalyst near the end of the alkyd cook; this gives more uniform molecular weight and a lower viscosity product. To study the effect of molecular weight distribution, model alkyds with very narrow molecular weight distribution were synthesized by using dicyclohexylcarbodiimide, which allows low temperature esterification (5). With the same ratio of reactants, the $\bar{M}_n$ and polydispersity were lower than that of the conventional alkyd control. These differences resulted from less dimerization through reactions of the double-bond systems of the fatty acids and avoidance of self-etherification of polyol in the low temperature preparation. It was found that the solids could be 2–10% higher than with the conventionally prepared alkyd of the same raw material composition. The model alkyds dried more rapidly, but their film properties, especially impact resistance, were inferior to those obtained with control resins with the usual broad molecular weight distribution (12). Conventionally prepared TMP alkyds had lower molecular weights and viscosities than the glycerol alkyds. This difference may result from less self-etherification of TMP as compared to glycerol.

High solids alkyds for baking applications have been made using tripentaerythritol. The high functionality obtained using this polyol ($F = 8$) gives alkyds that cross-link as rapidly as shorter oil length, higher viscosity glycerol alkyds (13). However, for air dry applications, the lower aromatic to aliphatic ratio lengthens the tack-free time. Presumably, progress could be made using a high functionality polyol with some combination of phthalic and benzoic acid, together with fatty acids with as high functionality fatty acids as possible. The cost of such an alkyd would be high.

Another approach to high solids alkyds is to use *reactive diluents* in place of part of the solvent. The idea is to have a component of lower molecular weight and that lower viscosity than the alkyd resin, which reacts with the alkyd during drying, so that it is not part of the VOC emissions. This permits the use of somewhat higher molecular weight alkyds that improves performance. The use of reactive diluents is reviewed in (14), the authors give a list of the key properties of a reactive diluent: low viscosity, good compatibility with alkyds, low volatility (bp $>300°C$), nontoxic, low color, and economic replacement for solvent. A variety of possible reactive diluents were studied. A combination of 2,7-octadienyl maleate and fumarate and 2-(2,7-octadienyloxy)succinate was reported to be particularly effective. The mechanism of reactions between this combination with alkyds has been studied (15). The use of esters prepared from ricinoic acid (9,11-octadecdienoic acid) and polyols gave faster drying reactive diluents.

Several other types of reactive diluents have been used to formulate high solids alkyd coatings. Polyfunctional acrylate monomers (eg, trimethylolpropane trimethacrylate) have been used in force dry coatings (coatings designed to be cured in the range of 60–80°C) (16). Another example is use of dicyclopentadienyloxyethyl methacrylate [70191-60-5] as a reactive diluent (17). It is difunctional, because of the easily abstractable allylic hydrogen on the dicyclopentadiene ring structure and the methacrylate double bond. The compound coreacts with drying oil groups in the alkyd. Mixed acrylic and drying oil fatty acid amides of hexa(aminomethoxymethyl)melamine have been recommended as reactive diluents (18,19). They contain high functionalities of $>NCH_2NHCOCH=CH_2$ and $>NCH_2NHCOC_{17}H_x$ moieties and promote fast drying. A recent patent discloses use of a reactive diluent prepared by reacting drying oil fatty acids with excess dipentaerythritol and then with isophorone diisocyanate (20).

3. Waterborne Alkyds

As with almost all other resin classes, work has been done to make alkyd resins for coatings that can be reduced with water. One approach that has been more extensively used in Europe than in the United States is the use of alkyd emulsions (21–23). The emulsions are stabilized with surfactants and can be prepared with little, if any, volatile solvent. Some problems limit use of alkyd emulsions (24). Coatings prepared using alkyd emulsion loose dry time on storage because of absorption of cobalt drier on the surface of pigments and precipitation of cobalt hydroxide. Best results were obtained with a combination of cobalt neodecanoate and 2,2′-bipyridyl (bpy). Incorporation of driers in the alkyd resin can adversely affect emulsion stability. It has been recommended to emulsify the driers separately and mix the emulsions so that the drier and the alkyd are in separate phases (25). It was shown that the surfactant tends to bloom to the surface of films formed from emulsions of long oil alkyds, washing a dry film tends to leave pits in the film showing a hexagonal pattern.

It is common to add a few percent of an alkyd–surfactant blend to latex paints to improve adhesion to chalky surfaces and, in some cases, to improve adhesion to metals. It is important to use alkyds that are as resistant as possible to hydrolysis. Hybrid alkyd–acrylic latexes have been prepared by dissolving an oxidizing alkyd in the monomers used in emulsion polymerization, yielding a latex with an alkyd grafted on the acrylic polymer (26,27). Nonyellowing waterborne alkyds based on rosin–fatty acid modified acrylic latexes have been reported (28). Hybrid alkyd–acrylic latexes have been prepared by emulsion polymerization (29). Hydroperoxidized sunflower oil was used as the initiator to polymerize a combination of a long oil alkyd and ethyl methacrylate. The use of the hydroperoxidized sunflower led to homogeneous polymerization in contrast to use of *tert*-butyl hydroperoxide. Films formed from the latex gave to fast dry expected from a latex followed by autoxidation to give cross-linked films.

Another approach has been to make alkyds with an acid number in the range of 50, using secondary alcohols or ether alcohols as solvents. The acid groups are neutralized with ammonia or an amine. The resultant solution can be diluted with water to form a dispersion of solvent swollen aggregates in water. Molecular weight can be higher than in the case of high solids alkyd because the major factor affecting viscosity at application solids is the volume fraction of internal phase of the dispersion rather than the molecular weight of the polymer. Use of primary alcohol solvents should be avoided because they can more readily transesterify with the alkyd during resin production and storage, leading to reduction in molecular weight and $\bar{f}_n$ (30). Hydrolytic stability can be a problem with water-reducible alkyds. If the carboxylic acid groups are half esters from PA or trimellitic acid anhydride (1,3-dihydro-1,3-dioxo-5-iso-benzofurancarboxylic acid) [552-30-7], the hydrolytic stability will be poor and probably inadequate for paints that require a shelf life of more than a few months. Because of the anchimeric effect of the neighboring carboxylic acid group, such esters are relatively easily hydrolyzed. As hydrolysis occurs, the solubilizing acid salt is detached from the resin molecules, and the aqueous dispersion loses stability. A more satisfactory way to introduce free carboxylic acid groups is by reacting a completed alkyd with maleic anhydride. Part of the maleic anhydride adds to the unsaturated fatty acid esters. The anhydride groups are then hydrolyzed with amine and water to give the desired carboxylate salt groups, which are attached to resin molecules with C–C bonds and cannot be hydrolyzed off. There is still a hydrolytic stability problem with the alkyd backbone, but hydrolysis does not result in destabilization of the dispersion. Similarly acrylated fatty acids can be used to synthesize water-reducible alkyds with improved hydrolytic stability (31). Another approach to improving package stability is to react some of the free hydroxyl groups of an alkyd with isophorone diisocyanate (IPDI) [(1-isocyanato-3-isocyanatomethyl)-3,5,5-trimethylcyclohexane) [4098-71-9] and dimethylolpropionic acid, (DMPA) (3-hydroxy-2-(hydroxymethyl)-2-methylpropanoic acid) [4767-03-7] (32) or with α,α-tetramethylxylylene diisocyanate (TMXDI) (1,3-bis(2,2-dimethyl-2-isocyanato)benzene] [2778-42-9] and DMPA (33).

After the film is applied, the water, solvent, and amine evaporate, and the film cross-links by autoxidation. Since there are a fairly large number of residual carboxylic acid groups left in the cross-linked binder, the water resistance and particularly the alkali resistance of the films are reduced, but are still satisfactory for some applications (34). *Early water resistance* can be a problem if, for example, a freshly painted surface is rained on before all the amine has evaporated from the film. Commonly, ammonia is used as the neutralizing amine because it is assumed that ammonia volatilizes faster than any other amine. This assumption is not necessarily valid; if the T_g of the alkyd film is sufficiently high before all of the amine has volatilized, loss of amine becomes controlled by diffusion rate. The rate of diffusion of amine through the carboxylic acid-functional film is affected by the base strength of the amine. A less basic amine, such as morpholine, may leave the film before ammonia even though its volatility is considerably lower.

4. Modified Alkyds

Oxidizing alkyds have been modified by reacting with a variety of other components; vinyl-, silicone-, phenolic-, and polyamide-modified alkyds are the most common examples.

Oxidizing alkyds can be modified by reaction with vinyl monomers. The most widely used monomers are styrene (ethenylbenzene) [100-42-5], vinyl toluene (1-ethenyl-2-methylbenzene), and methyl methacrylate (2-methyl-2-propenoic acid methyl ester), but essentially any vinyl monomer can be reacted in the presence of an alkyd to give a modified alkyd. Methyl methacrylate imparts better heat resistance than styrene but at higher cost.

In making styrenated alkyds, an oxidizing alkyd is prepared in the usual way and cooled to about 130°C in the reactor; then styrene and a free-radical initiator such as dibenzoyl peroxide [94-36-0] are added. The resulting free-radical chain process leads to a variety of reactions, including formation of low molecular weight homopolymer of styrene, grafting of polystyrene onto the alkyd, and dimerization of alkyd molecules. The reaction is generally carried out at ~130°C, which favors decomposition of benzoyl peroxide to form phenyl free radicals; phenyl radicals have a greater tendency to abstract hydrogen, which favors grafting. After the reaction is complete, the resin is diluted with solvent. Alkyds made with some maleic anhydride as one of the dibasic acids give a higher ratio of grafting. The ratio of alkyd to styrene can be varied over a wide range; commonly 50% alkyd and 50% styrene is used. The ratio of aromatic rings to aliphatic chains is greatly increased, and as a result, the T_g of styrenated alkyd films is higher and tack-free time is shorter. Styrenated alkyds give a "dry" film in 1 h or less versus 4–6 h for the counterpart nonstyrenated alkyd. However, the average functionality for oxidative cross-linking is reduced, not just by dilution with styrene, but also because the free-radical reactions involved in the styrenation consume some activated methylene groups. As a result, the time required to develop solvent resistance is longer than for the counterpart alkyd. The fast drying and low cost make styrenated alkyds very attractive for some applications, but in other cases, the longer time required for cross-linking is more critical, in which case styrenated alkyds are not appropriate.

Styrenated alkyd vehicles are often used for air dry primers. One must be careful to apply top coat almost immediately or not until after the film has had ample time to cross-link. During the intermediate time interval, application of top coat is likely to cause nonuniform swelling of the primer, leading to what is called *lifting* of the primer. The result of lifting is the development of wrinkled areas in the surface of the dried film. End users who are accustomed to using alkyd primers, which do not give a hard film until a significant degree of cross-linking has occurred, are particularly likely to encounter problems of lifting if they switch to styrenated alkyd primers.

Graft copolymers prepared from anhydride-functional acrylic resins and hydroxyamides of soybean–conjugated tall oil acids are another approach to making acrylic resins that undergo oxidative curing (35).

Silicone resins have exceptional exterior durability but are expensive. Silicone modification of alkyd resins improves their exterior curability. The earliest approach was simply to add a silicone resin to an alkyd resin in the reactor at the end of the alkyd cook. While some covalent bonds between silicone resin and alkyd might form, probably most of the silicone resin simply dissolves in the alkyd. Exterior durability of silicone-modified alkyd coatings is significantly better than unmodified alkyd coatings. The improvement in durability is roughly proportional to the amount of added silicone resin; 30% silicone resin is a common degree of modification. Further improvements in exterior durability are obtained by coreacting a silicone intermediate during synthesis of the alkyd. Such intermediates react readily with free hydroxyl groups of the alkyd resin. Silicone resins designed for this purpose may contain higher alkyl, as well as methyl and phenyl, groups to improve compatibility. Alkyd coatings modified with high-phenyl silicone resins are reported to have greater thermoplasticity, faster air drying, and higher solubility than high methyl silicone-modified alkyds. These differences result from the higher rigidity of the aromatic rings, which leads to a "solid" film at an earlier stage of cross-linking. Less cross-linking in the phenylsilicone-modified coatings makes them more thermoplastic and soluble. Since IPA based alkyd resins have better exterior durability than PA alkyds, they are generally used as the alkyd component. Silicone-modified alkyds are used mainly in outdoor air dry coatings for which application is expensive (eg, in a topcoat for steel petroleum storage tanks).

Phenolic-modified alkyds are made by heating the alkyd with a low molecular weight resole phenolic resin based on *p*-alkylphenols. Presumably, the methylol groups on the phenols react with some of the unsaturated groups of the alkyd to form chroman structures. The resins give harder films with improved water and chemical resistance as compared to the unmodified alkyd.

Ceramer (organic–inorganic hybrid) coatings prepared with long oil linseed alkyds and titanium tetraisopropoxide gave films with excellent hardness, tensile strength, flexibility, and impact resistance (36).

Polyamide-modified alkyds are used as thixotropic agents to increase the low shear viscosity of alkyd resin based paints. Typically, ~10% of a polyamide resin made from diamines such as ethylenediamine (1,2-ethanediamine) with dimer acids is reacted with an alkyd resin. High solids thixotropic alkyds based on polyamides made with aromatic diamines have been developed, which give superior performance in high solids alkyd coatings (37).

5. Nonoxidizing Alkyds

Certain low molecular weight short-medium and short oil alkyds are compatible with such polymers as nitrocellulose and thermoplastic polyacrylates. Therefore, such alkyds can be used as plasticizers for these polymers. They have the advantage over monomeric plasticizers (eg, dibutyl or dioctyl phthalate) in that they do not volatilize appreciably when films are baked. It is generally not desirable to use oxidizing alkyds, which would cross-link and lead to embrittlement of the films, especially on exterior exposure. Therefore, nondrying oil fatty acids (or oils) are used in the preparation of alkyds for such applications. For exterior acrylic lacquers, pelargonic acid (nonanoic acid) alkyds combine excellent resistance to photodegradation with good compatibility with the thermoplastic acrylic resins. An interesting sidelight on terminology is that these pelargonic alkyds have been called polyesters rather than alkyds because the word polyester connotes higher quality than the word alkyd. Castor oil derived alkyds are particularly appropriate for nitrocellulose lacquers for interior applications, since the hydroxyl groups on the ricinoleic acid (12-hydroxy-(Z)-9-octadecenoic acid [141-22-0] promote compatibility.

All alkyds, particularly short-medium oil and short oil alkyds, are made with a large excess of hydroxyl groups to avoid gelation. These hydroxyl groups can be cross-linked with MF resins or with polyisocyanates. In some cases, relatively small amounts of MF resin are used to supplement the cross-linking during baking of medium oil oxidizing alkyds. To achieve compatibility, butylated MF resins are used. Such coatings provide somewhat better durability and faster curing than alkyd resins alone, with little increase in cost. The important advantage of relative freedom from film defects common to alkyd coatings can be retained. However, the high levels of unsaturation remaining in the cured films reduce resistance to discoloration on overbake and exterior exposure and cause loss of gloss and embrittlement on exterior exposure. These difficulties can be reduced by using nondrying oils with minimal levels of unsaturated fatty acids. Coconut oil has been widely used; its performance can be further enhanced by hydrogenation of the small amount of unsaturated acids present in it.

Since isophthalic (IPA) esters are more stable to hydrolysis in the pH range of 4–8 than phthalate esters, the highest performance exterior alkyd-MF enamels use nonoxidizing IPA alkyds. Exterior durability of such coatings is satisfactory for automobile topcoats with opaque pigmentation. The films have an appearance of greater "depth" than acrylic-MF coatings. The films are perceived to be thicker than films of acrylic-MF coatings of comparable thickness and pigmentation. However, for many applications, alkyd-MF coatings have been replaced with acrylic-MF or polyester-MF coatings to improve the overall balance of film properties.

6. Synthesis of Alkyd Resins

Various synthetic procedures, each with many variables, are used to produce alkyd resins. The general reference and (38) and (39) provide useful reviews of manufacturing procedures. Alkyds can be made directly from oils or by using free fatty acids as raw materials.

6.1. Synthesis from Oils or Fatty Acids. *Monoglyceride Process.*

In the case of glycerol alkyds, it would be absurd to first saponify an oil to obtain fatty acids and glycerol, and then reesterify the same groups in a different combination. Rather, the oil is first reacted with sufficient glycerol to give the total desired glycerol content, including the glycerol in the oil. Since PA is not soluble in the oil, but is soluble in the glycerol, transesterification of oil with glycerol must be carried out as a separate step before the PA is added; otherwise, glyceryl phthalate gel particles would form early in the process. This two-stage procedure is often called the monoglyceride process. The transesterification reaction is run at 230–250°C in the presence of a catalyst; many catalysts have been used. Before the strict regulation of lead in coatings, litharge (PbO) was widely used; the residual transesterification catalyst also acted as a drier. Examples of catalysts now used in the United States are tetraisopropyl titanate, lithium hydroxide, and lithium ricinoleate. The reaction is run under an inert atmosphere such as CO_2 or N_2 to minimize discoloration and dimerization of drying oils.

While the process is called the monoglyceride process, the transesterification reaction actually results in a mixture of unreacted glycerol, monoglycerides, diglycerides, and unconverted drying oil. The composition depends on the ratio of glycerol to oil and on catalyst, time, and temperature. In general, the reaction is not taken to equilibrium. At some relatively arbitrary point, the PA is added, beginning the second stage. The viscosity and properties of the alkyd can be affected by the extent of reaction before the PA addition. While many tests have been devised to evaluate the extent of transesterification, none is very general because the starting ratio of glycerol to oil varies over a considerable range, depending on the oil length of the alkyd being made. (In calculating the mole ratio of dibasic acid to polyol, the glycerol already esterified in the oil must also be counted.) A useful empirical test is to follow the solubility of molten PA in the reaction mixture. This test has the advantage that it is directly related to a major requirement that must be met. In the first stage, it is common to transesterify the oil with PE to obtain mixed partial esters. The second stage, esterification of the "monoglyceride" with PA, is carried out at a temperature of 220–255°C.

Fatty Acid Process. It is often desirable to base an alkyd on a polyol (eg, PE) other than glycerol. In this case, fatty acids must be used instead of oils, and the process can be performed in a single step with reduced time in the reactor. Any drying, semidrying, or nondrying oil can be saponified to yield fatty acids, but the cost of separating fatty acids from the reaction mixture increases the cost of the alkyd. A more economical alternative is to use TOFA, which have the advantage that they are produced as fatty acids. Tall oil fatty acid composition is fairly similar to that of soybean fatty acids. Specially refined tall oils with higher linoleic acid content are available, as are other grades that have been treated with alkaline catalysts to isomerize the double bonds partially to conjugated structures. Generally, when fatty acids are used, the polyol, fatty acids, and dibasic acid are all added at the start of the reaction, and the esterification of both aliphatic and aromatic acids is carried out simultaneously in the range of 220–255°C.

6.2. Process Variations.   Esterification is a reversible reaction; therefore, an important factor affecting the rate of esterification is the rate of removal of water from the reactor. Most alkyds are produced using a reflux solvent, such as xylene, to promote the removal of water by azeotroping. Since the reaction is run at a temperature far above the boiling point of xylene, <5% of xylene is used. The amount is dependent on the reactor and is set empirically such that there is enough to reflux vigorously, but not so much as to cause flooding of the condenser. Some of the xylene is distilled off along with the water; water is separated and xylene is returned to the reactor. The presence of solvent is desirable for other reasons: vapor serves as an inert atmosphere, reducing the amount of inert gas needed, and the solvent serves to avoid accumulation of sublimed solid monomers, mainly PA, in the reflux condenser.

Reaction time is affected by reaction temperature. Higher temperatures obviously accelerate the reaction. If the reaction is carried too far, there is a major risk of gelation. There are economic advantages to short reaction times. Operating costs are reduced, and the shorter times permit more batches of alkyd to be produced in a year, increasing capacity without capital investment in more reactors. Therefore, it is desirable to operate at as high a temperature as possible without risking gelation.

A critical aspect of alkyd synthesis is deciding when the reaction is completed. Disappearance of carboxylic acid is followed by titration, and increase in molecular weight is followed by viscosity. Determination of acid number and viscosity both take some time. Meanwhile, in the reactor, the reaction is continuing. After it is decided that the extent of reaction is sufficient, the reaction mixture must be "dropped" into a larger tank containing solvent. When a 40,000-L batch of alkyd is being made, a significant time is required to get the resin out of the reactor into the reducing tank; meanwhile, the reaction is continuing. The decision to start dropping the batch must be made so that the acid number and viscosity of the batch will be right after the continuing reaction that occurs between the time of sampling, determination of acid number and viscosity, and discharging of the reactor. The time for these determinations becomes the rate-controlling step in production. If they can be done rapidly enough, the reaction can be carried out at 240°C or even higher without overshooting the target acid number and viscosity. On the other hand, if the control tests are done slowly, it may be necessary to run the reaction at only 220°C, which may require 2 h or more of additional reaction time. Automatic titration instruments permit rapid determination of acid number, so the usual limit on time required is viscosity determination. While attempts have been made to use viscosity of the resin at reaction temperature to monitor change in molecular weight, the dependence of viscosity on molecular weight at that high temperature is not sensitive enough to be very useful. The viscosity must be determined on a solution at some lower standard temperature. Since viscosity depends strongly on solution concentration and temperature, these variables must be carefully controlled.

In alkyd production, viscosity is commonly determined using Gardner bubble tubes. The cook is continued until the viscosity is high enough so that by the time the resin batch is dropped into the solvent and the batch cooled, its viscosity will be what is called for in the specification. This means starting to discharge the reactor when the test sample is at some lower viscosity. It is not possible to generalize how large this difference should be; it depends on the specific alkyd composition, the temperature at which the reaction is being run, the time required to do the determination, the time required to empty the reactor, and so on. Viscosities can be determined more rapidly using a cone and plate viscometer than with bubble tubes; the very small sample required for a cone and plate viscometer can be cooled and equilibrated at the measurement temperature more quickly.

Many variables affect the acid number and viscosity of alkyds. One is the ratio of reactants: The closer the ratio of moles of dibasic acid to polyol approaches 1, the higher the molecular weight of the backbone of the resin, but also the greater the likelihood of gelation. A useful rule of thumb for a starting point is to use a mole ratio of 0.95. The final ratio is determined by adjustments such that the combination of acid number and solution viscosity come out at the desired levels. The greater the ratio of hydroxyl groups to carboxylic acid groups, the faster the acid groups are reduced to a low level. The degree of completion of the reaction is an important factor controlling the viscosity, as well as the acid number. It is usually desirable to have a low acid number, typically in the range of 5–10.

The composition of the fatty acids is a major factor affecting the viscosity, and compositions of an oil or grade of TOFA can be expected to vary somewhat from lot to lot. Dimerization and oligomerization of the unsaturated fatty acids occur in the same temperature range at which the esterification is carried out. Fatty acids with conjugated double bonds dimerize more rapidly than those with nonconjugated bonds, and dimerization rates increase with the level of unsaturation. At the same ratio of phthalic to polyol to fatty acids, alkyds of the same acid number and solution concentration will increase in viscosity in the order soybean < linseed < tung.

Some volatilization of polyol, PA, and fatty acids out of the reactor will occur depending on the design of the reactor, the rate of reflux of the azeotroping solvent, the rate of inert gas flow, and the reaction temperature, among other variables; the amount and ratio of these losses affect the viscosity at the standard acid number. The exact ratio of reactants must be established in the reactor that is actually used for synthesis. Since gelation can occur if the ratio of dibasic acid to polyol is too high, it is better not to put all the PA into the reactor in the beginning. If the viscosity is too low when the acid number is getting down near the standard, more PA can easily be added. The amount of PA held back can be reduced as experience is gained cooking a particular alkyd in a particular reactor.

Side reactions can affect the viscosity–acid number relationship. Glycerol and other polyols form ethers to some degree during the reaction. Glycerol can also form acrolein by successive dehydrations. When these reactions occur, the mole ratio of dibasic acid to polyol increases and the number of hydroxyl groups decreases; therefore, at the same acid number, the molecular weight will be higher. Excessively high viscosity and even gelation can result. Ether formation is catalyzed by strong protonic acids; therefore, it is desirable to avoid them as catalysts for the esterification. Monobutyltin oxide has been used as an esterification catalyst; presumably, it does not significantly catalyze ether formation. As noted earlier, PE and TMP seem less vulnerable than glycerol to undesirable side reactions such as ether formation, and glycerol is the only polyol that can decompose to form acrolein. A hydroxyl group on one end of a growing polyester chain can react with a carboxylic acid group on another end of the same molecule, leading to ring formation. Transesterification of chain linkages can have the same result. Since cyclization reactions reduce chain length, their net effect is to reduce viscosity.

Many alkyd resins have broad, uneven molecular weight distributions. It has been shown that even modest changes in reaction conditions can cause large differences in molecular weight distribution, which can have significant effects on final film properties (40). In many alkyds, very small gel particles (microgels) are formed. It has been shown that these microgels play an important role in giving greater strength properties to final films (40). Process changes that may make the alkyd more uniform may be undesirable. For example, allowing glycerolysis to approach equilibrium before addition of PA and using transesterification catalysts in the final stages of esterification both favor narrower molecular weight distributions and lower viscosities, but films made from the more uniform alkyds may exhibit inferior mechanical properties.

7. Urethane Derivatives

Uralkyds are also called *urethane alkyds* or *urethane oils*. They are alkyd resins in which a diisocyanate, usually 2,4(6)-toluene diisocyanate [584-84-9] (TDI) or bis(4-isocyanatophenyl)methane (MDI), has fully or partly replaced the PA usually used in the preparation of alkyds. One transesterifies a drying oil with a polyol such as glycerol or PE to make a "monoglyceride" and reacts it with some PA (if desired) and then with somewhat less diisocyanate than the equivalent amount of N=C=O based on the free OH content. To assure that no N=C=O groups remain unreacted, methanol is added at the end of the process. Just like alkyds, uralkyds dry faster than the drying oil from which they were made, since they have a higher average functionality (more activated diallylic groups per average molecule). The rigidity of the aromatic rings also speeds up the drying by increasing the T_g of the resin.

Two principal advantages of uralkyd over alkyd coatings are superior abrasion resistance and resistance to hydrolysis. Disadvantages are inferior color retention (when aromatic isocyanates used) of the films, higher viscosity of resin solutions at the same percent solids, and higher cost. Uralkyds made with aliphatic diisocyanates have better color retention, but are more expensive and have lower T_g. The largest use of uralkyds is in architectural coatings. Many so-called varnishes sold to the consumer today are based on uralkyds; they are not really varnishes in the original sense of the word. They are used as transparent coatings for furniture, woodwork, and floors: applications in which good abrasion resistance is important. Since they are generally made with aromatic isocyanates, they tend to turn yellow and then light brown with age; yellowing is acceptable in clear varnishes, but would be a substantial drawback in light colored pigmented paints.

Water-reducible polyunsaturated acid substituted aqueous polyurethane dispersions are also being used (41). They can be made by reacting an diisocyanate with a polyol, monoglyceride of a drying oil, and dimethylolpropionic acid. The carboxylic acid groups are neutralized with a tertiary amine and dispersed in water. If aliphatic isocyanates are used, good color retention can be obtained. They are much more resistant to hydrolysis than conventional alkyd resins. Films also have excellent abrasion resistance. Cost can be reduced by blending in 10–20% of acrylic latex.

8. Epoxy Esters

Bisphenol A (BPA) [4,4'-(1-methylethylidene)bisphenol] [80-05-7] epoxy resins can be converted to what are commonly called *epoxy esters* by reacting with fatty acids. Drying or semidrying oil fatty acids are used so that the products cross-link by autoxidation. The epoxy groups undergo a ring-opening reaction with carboxylic acids to generate an ester and a hydroxyl group. These hydroxyl groups, as well as the hydroxyl groups originally present on the epoxy resin, can esterify with fatty acids. They are generally made by starting with a low molecular weight epoxy resin (ie, the standard liquid resin, $n = 0.13$) and extending with BPA by the advancement process to the desired molecular weight. Off-specification epoxy resin is often used to reduce cost. The fatty acids are added to the molten, hot resin, and the esterification reaction is continued until the acid number is low, usually <7 mg of KOH/g of resin. In the esterification reaction with fatty acids, the average number of sites for reaction is the n value, corresponding to the number of hydroxyl groups on the resin, plus twice the number of epoxy groups. The esterification is carried out at high temperatures (220–240°C). The rate of esterification slows as the concentration of hydroxyl groups diminishes, and side reactions occur, especially dimerization of the drying oil fatty acids (or their esters). It is not practical to esterify more than ~90% of the potential hydroxyl groups, including those from ring opening the epoxy groups. The lower useful limit of the extent of esterification is ~50%. This is required to ensure sufficient fatty acid groups for oxidative cross-linking.

Tall oil fatty acids are commonly used because of their low cost. Linseed fatty acids give faster cross-linking coatings because of higher average functionality. However, their viscosity is higher because of the greater extent of dimerization during esterification, and their cost is higher. For still faster cross-linking, part of the linseed fatty acids can be replaced with tung fatty acids, but the viscosity and cost are still higher. The color of epoxy esters from linseed and linseed–tung fatty acids is darker than the tall oil esters. Dehydrated castor oil fatty acids give faster curing epoxy esters for baked coatings. The rate of formation of a dry film from epoxy esters depends on two factors: the average number of diallylic groups $\bar{f}_n$ and the ratio of aromatic rings to long aliphatic chains. The $\bar{f}_n$ can be maximized by using higher molecular weight BPA epoxy resin and by using enough fatty acid to react with a large fraction of the epoxy and hydroxyl groups. The ratio of aromatic rings to fatty acids can be maximized by using high molecular weight epoxy resin and esterifying a smaller fraction of epoxy and hydroxyl groups.

Epoxy esters are used in coatings in which adhesion to metal is important. While the reasons are not completely understood, it is common for epoxy coatings, including epoxy esters, to have good adhesion to metals and to retain adhesion after exposure of the coated metal to high humidity, a critical factor in corrosion protection. A distinct advantage of epoxy esters over alkyd resins is their greater resistance to hydrolysis and saponification. The backbone of alkyds is held together with esters from PA and the polyol, whereas in epoxy esters, the backbone is held together with C–C and ether bonds. Of course, the fatty acids are bonded to the backbone with ester groups in both cases, but the fraction of polymer bonds in a dry film subject to hydrolysis is substantially lower in the case of epoxy esters. On the other hand, exterior durability of epoxy ester coatings is poor, as is the case with all films made with BPA epoxy resins. As a result of these advantages and disadvantages, the major uses for epoxy resins are in primers for metal and in can coatings, such as for crowns (bottle caps), in which the important requirements are adhesion and hydrolytic stability. In baking primers, it is sometimes desirable to supplement the cross-linking through oxidation by including a small amount of MF resin in the formulation to cross-link with part of the free hydroxyl groups on the epoxy ester.

Epoxy ester resins with good exterior durability (better than alkyds) can be prepared by reacting epoxy-functional acrylic copolymers (made with glycidyl methacrylate) with fatty acids. The product is an acrylic resin with multiple fatty acid ester side chains. By appropriate selection of acrylate ester comonomers and molecular weight, the T_g of the resin can be designed so that a tack-free film is obtained by solvent evaporation; then the coating cross-links by autoxidation. For an application like repainting an automobile at ambient temperatures, the cross-linking can proceed relatively slowly and need not be catalyzed by metal salt driers. The rate of cross-linking is slower without driers, but exterior durability is better.

Epoxy esters can also be made water reducible. The most widely used water-reducible epoxy esters have been made by reacting maleic anhydride with epoxy esters prepared from dehydrated castor oil fatty acids. Subsequent addition of a tertiary amine, such as 2-(dimethylamino)ethanol [108-01-0], in water results in ring opening of the anhydride to give amine salts. Like other water-reducible resins, these resins are not soluble in water but form a dispersion of resin aggregates swollen with water and solvent in an aqueous continuous phase. The hydrolytic stability of these epoxy esters is better than corresponding alkyds and sufficient for use in electrodeposition primers until anionic primers were replaced by cationic primers. Water-reducible epoxy esters are still used in spray applied baking primers and primer surfacers. They are also used in dip coating primers in which nonflammability is an advantage. Their performance equals that of solvent–soluble epoxy ester primers.

9. Uses

In 1997, the U.S. consumption of alkyds was ∼310 metric tons (t) and projected use in 5 years is estimated to be 280–290 t (42). Coatings are the largest market with use in 1997 of ~250 t (43). European consumption of alkyd coating resins in 1996 has been reported to be 360 t (43). Use of alkyds has been declining at ~2%/ year and is projected to decline further as they are replaced with resins with higher performance and lower volatile emissions. Higher solids alkyds have been replacing conventional solids alkyds. In 1997, ~81,000 t with solids of 50–60% and 16,000 t of >60% were used in the United States, in comparison with 150,000 t of alkyds with <50% solids. 10,000 t of waterborne alkyds were used (42).

The principal advantages of alkyds are low cost, low toxicity, and low surface tension. The low surface tension permits wetting of most surfaces including oily steel. Also, the low surface tension minimizes application defects such as cratering. The principal limitations are generally poorer exterior durability and corrosion protection than alternative coating resins. While high solids and waterborne alkyd resins are manufactured, their properties are generally somewhat inferior to conventional solventborne alkyds.

The largest use for alkyds in coatings is in architectural paints, particularly in gloss enamels for application by contractors. Contractors tend to prefer alkyd enamels over latex enamels because coverage can be achieved with a single coat. Also alkyd paints can be applied at low temperatures whereas latex paints can only be applied at temperatures above ~5°C. The do-it-yourself market is served primarily with latex paints because of ease of cleanup and lower odor. While initial gloss of alkyd enamels is higher than of latex enamels, the latex enamels exhibit far superior gloss retention especially in exterior applications. Alkyd primers provide better adhesion to chalky surfaces than most latex paints.

The largest uses of alkyds in industrial applications is in general industrial coatings for such applications as machinery and metal furniture. Significant amounts are used with UF resins in coatings for wood furniture. Alkyd resin–chlorinated rubber based coatings are used in traffic paints, but use is decreasing because of high VOC content. An approach to overcoming this problem is the use of solvent free alkyds in hot melt traffic paints (44). Some alkyds are still used in refinish paints for automobiles since they give high gloss coatings with a minimum of polishing. Soy alkyds are used in topcoats for complete refinishing of cheap old cars, where the primary requirement is for low cost. For intermediate performance and cost refinish enamels for painting whole cars, drying oil functional acrylic resins are still used. Higher grades of coatings are urethane coatings. Some nitrocellulose primers with nonoxidizing alkyd plasticizers and some alkyd underbody sealers are still used in the United States but will undoubtedly be phased out completely in the next few years. Alkyds are still fairly widely used for refinish coatings in some underdeveloped countries. An example of recent work in formulating refinish coatings is preparing an alkyd by reacting tris(hydroxyethyl)isocyanurate with drying oil fatty acids and formulating with trimethylolpropane trimethacrylate as a reactive diluent (45).

About 39,000 t of uralkyds were used in the United States in 1997 (42). The largest use for uralkyds is as the vehicle for so-called urethane varnishes for the do-it-yourself market. The abrasion resistance of such coatings is greatly superior to that obtained with conventional varnishes or alkyd resins. Epoxy esters give coatings with markedly superior corrosion protection as compared with alkyd resins while retaining the advantage of low surface tension. However, as with any BPA epoxy system, exterior durability is poor. They are used primarily in primers for steel and in flexible coatings such as for metal crowns. Maleated epoxy esters give primers with equivalent properties of solvent borne epoxy ester coatings and are widely used in formulating waterborne primers for steel.

Noncoatings applications include foundry core binders and printing inks, especially lithographic inks.

BIBLIOGRAPHY

"Alkyd Resins" in *ECT* 1st ed., Vol. 1, pp. 517–532, by W. Howlett Gardner, National Aniline Division, Allied Chemical & Dye Corporation; in *ECT* 2nd ed., Vol. 1, pp. 851–882, by Richard G. Mraz and Raymond P. Silver, Hercules Powder Company; in *ECT* 3rd ed., Vol. 2, pp. 18–48, by H. J. Lanson, Poly-Chem Resin Corporation; "Alkyd Resins" in *ECT* 4th ed., Vol. 2, pp. 53–85, by K. F. Lin Hercules Incorporated; "Alkyd Resins" in *ECT* (online), posting date: December 4, 2000, by K. F. Lin, Hercules Incorporated.

CITED PUBLICATIONS

1. Northwest and Montreal Sections of FSCT, *J. Coat. Technol.*, **67**(850) 19 (1995).
2. T. L. T. Robey and S. M. Rybicka, *Paint Research Station Technical Papers No. 217*, Vol. 13, No. 1, 1962, p. 2.
3. R. J. Blackinton, *J. Paint Technol.*, **39**(513) 606 (1967).
4. T. A. Misev, *Prog. Org. Coat.*, **21**, 79 (1992).
5. S. L. Kangas and F. N. Jones, *J. Coat. Technol.*, **59**(744) 89 (1987).
6. R. Bacaloglu and co-workers, *Angew. Makromol. Chem.*, **164**, 1 (1988).
7. R. Brown, H. Ashjian, and W. Levine, *Off. Digest*, **33**, 539 (1961).
8. *Tech. Bull. No. 524-5*, Velsicol Chemical Corp., Chicago Ill.
9. D. Ryer, *Paint Coat. Ind.*, **14**(1) 76 (1998).
10. R. W. Hein, *J. Coat. Technol.*, **71**(898) 21 (1999).
11. J. Mallegol, J. Lemaire, and J.-L. Gardette, *Prog. Org. Coat.*, **39**, 107 (2000).
12. S. L. Kangas and F. N. Jones, *J. Coat. Technol.*, **59**(744) 99 (1987).
13. U.S. Patent 2,577,770 (Dec. 11, 1951), P. Kass and Z. W. Wicks, Jr. (to Interchemical Corporation).
14. K. H. Zabel and co-workers, *Prog. Org. Coat.*, **35**, 255 (1999).
15. W. J. Muizebelt and co-workers, *Prog. Org. Coat.*, **40**, 121 (2000).
16. E. Levine, *Proc. Water-Borne Higher-Solids Coat. Symp.*, New Orleans, La., 1977, p. 155.
17. D. B. Larson and W. D. Emmons, *J. Coat. Technol.*, **55**(702) 49 (1983).
18. *Tech. Bull., Resimene AM-300 and AM-325*, Monsanto Chemical Co., (now Solutia, Inc.), January 1986.
19. U.S. Patent 4,293,461 (Oct. 6, 1981), W. F. Strazik, J. O. Santer, and J. R. LeBlanc (to Monsanto Company).
20. U.S. Patent 6,075,088 (June 13, 2000), J. Braeken, (to Fina Research, S. A.).
21. G. Osterberg, M. Hulden, B. Bergenstahl, and K. Holmberg, *Prog. Org. Coat.*, **24**, 281 (1994); G. Ostberg and B. Bergenstahl, *J. Coat. Technol.*, **68**(858), 39 (1996).
22. A. Hofland, in J. E. Glass, ed., *Technology for Waterborne Coatings*, American Chemical Society, Washington, D.C. 1997, p. 183.
23. E. Makarewicz, *Prog. Org. Coat.*, **28**, 125 (1996).
24. P. K. Weissenborn and A. Motiejauskaite, *Prog. Org. Coat.*, **40**, 253 (2000).
25. V. Verkholantsev, *Eur. Coat. J.*, (1–2) 120 (2000).

26. T. Nabuurs, R. A. Baijards, and A. L. Germna, *Prog. Org. Coat.*, **27**, 163 (1996).
27. J. W. Gooch, S. T. Wang, F. J. Schork, and G. W. Poehlein, *Proc. Waterborne, High Solids, Powder Coat. Symp.*, New Orleans, La., 1997, p. 366.
28. W. S. Sisson and R. J. Shah, *Proc. Waterborne, High Solids, Powder Coat. Symp.*, New Orleans, La., 2001, pp. 329–336.
29. E. M. S. Van Hamersfeld and co-workers, *Prog. Org. Coat.*, **35**, 235 (1999).
30. C. J. Bouboulis, *Proc. Water-Borne Higher-Solids Coat. Symp.*, New Orleans, La., 1982, p. 18.
31. B. Zuchert and H. Biemann, *Farg och Lack Scandinavia*, (2) 9 (1993); W. Weger, *Fitture e Vernici*, **B66**(9), 25 (1990).
32. U.S. Patent 5,004,779 (Apr. 2, 1991), H. Blum and co-workers, (to Bayer Aktiengesellschaft).
33. U.S. Patent 6,187,384 (Feb. 13, 2001), G. Wilke, D. Grapatin, and H.-P. Rink, (to BASF Coatings A. G).
34. R. Hurley and F. Buona, *J. Coat. Technol.*, **54**(694), 55 (1982).
35. U.S. Patent 5,223,582 (June 29, 1993), H. Blum and L. Fleiter, (to Bayer Aktiengesellschaft).
36. R. A. Sailer and co-workers, *Prog. Org. Coat.*, **33**, 117 (1998).
37. P. J. Bakker and co-workers, *Water-borne High-Solids, Powder Coating Symposium*, New Orleans, La, 2001, pp. 439–453.
38. J. Kaska and F. Lesek, *Prog. Org. Coat.*, **19**, 283 (1991).
39. Anonymous, *The Chemistry and Processing of Alkyd Resins*, Monsanto Chemical Co. (now Solutia, Inc.), 1962.
40. J. Kumanotani, H. Hironori, and H. Masuda, *Adv. Org. Coat. Sci. Tech. Ser.*, **6**, 35 (1984).
41. W. Liu, S. Wang, and T. Rende, *Western Coat. Symp.*, Reno, Nv. (1999).
42. Skeist Report VI, Skeist Inc., Whippany, NJ, 1998, pp. 805–826.
43. E. Connolly, E. Anderson, and Y. Sakuma, *Alkyd/Polyester Surface Coatings*, SRI International, Pasadena, Calif., 1998.
44. U.S. Patent 6,011,085 (Jan. 4, 2000), B. A. Maxwell, M. A. Weaver, G. R. Robe, and R. A. Miller, (to Eastman Chemical Co.).
45. U.S. Patent 6,083,312 (July 4, 2000), G. L. Bajc, (to BASF Corporation).
46. R. A. Sailer and M. D. Soucek, *Prog. Org. Coat.*, **33**, 36 (1998).

GENERAL REFERENCE

T. C. Patton, *Alkyd Resin Technology*, John Wiley & Sons, New York, 1962.

ZENO W. WICKS, JR.
Consultant

ALKYLATION

1. Introduction

The alkylation described in this article is the substitution of a hydrogen atom bonded to the carbon atom of a paraffin or aromatic ring by an alkyl group.

The alkylations of nitrogen, oxygen, and sulfur are described in separate articles (see AMINES; ETHERS).

Significant technological development has been made in the area of alkylation in recent years. Environmental concerns associated with mineral acid catalysts have encouraged process changes and the development of solid-bed alkylation processes. The application of heterogenous catalysts, especially zeolite catalysts, has led to new alkylation technologies. Research efforts to develop environmentally acceptable, economical technologies by applying new materials as alkylation catalysts will continue, and more new technologies are expected to be commercialized in the 1990s.

This article covers important industrial technologies and the direction of future technological development. The description of alkylation chemistry and conventional alkylation technologies covered in the earlier editions of this *Encyclopedia* and other references is minimized (1,2) (see also FRIEDEL-CRAFTS REACTIONS).

2. Nomenclature

Open-chain saturated hydrocarbons have the generic names alkanes and paraffins. In this article, terms such as hexanes, heptanes, and octanes are synonymous with C_6, C_7, and C_8 alkanes, respectively, and do not refer to the straight chains of six carbons, seven carbons, and eight carbons, as defined in the IUPAC system.

The ending *ene* is adopted for straight-chain monounsaturated hydrocarbons. Thus, butenes refer to 1-butene and 2-butene. The ending *ylene* denotes a monounsaturated hydrocarbon that consists of the same number of carbons as expressed by the name; ie, butylenes are 1-butene, 2-butene, and isobutylene (methylpropene). The generic names alkenes and olefins refer to monounsaturated hydrocarbons.

The prefix *iso* is used loosely to denote branched alkanes or alkenes that have one or more methyl groups only as side chains.

3. Alkylation of Paraffinic Hydrocarbons

Paraffin alkylation as discussed here refers to the addition reaction of an isoparaffin and an olefin. The desired product is a higher molecular weight paraffin that exhibits a greater degree of branching than either of the reactants.

The principal industrial application of paraffin alkylation is in the production of premium-quality fuels for spark-ignition engines. Originally developed in the late 1930s to meet the fuel requirements of high performance aviation engines, alkylation is now primarily used to provide a high octane blending component for automotive fuels. Future gasoline specifications will continue to favor the clean-burning characteristics and the low emissions typical of alkylate. These specifications will include reductions in total aromatics, benzene, methyl-*tert*-butyl ether (MTBE), vapor pressure, olefins, sulfur, and distillation endpoint that will reduce the demand for every major gasoline blending component except alkylate (3). Alkylate is an ideal gasoline blend stock because of its high octane and paraffinic nature. Alkylate production capacity as of 2001 reached $\sim$75 million tons per year compared to 58 million tons in 1990 (4) and is expected to grow as worldwide gasoline specifications become more stringent. In addition to this demand for cleaner gasoline, there is a growing demand for gasoline in the United States as well as in many other areas of the world.

3.1. Catalysts and Reactions. Although the alkylation of paraffins can be carried out thermally (5), catalytic alkylation is the basis of all processes in commercial use. Early studies of catalytic alkylation led to the formulation of a proposed mechanism based on a chain of ionic reactions (6–8). The reaction steps include the formation of a light tertiary cation, the addition of the cation to an olefin to form a heavier cation, and the production of a heavier paraffin (alkylate) by a hydride transfer from a light isoparaffin. This last step generates another light tertiary cation to continue the chain.

In practice, the alkylate is a complex mixture of branched paraffins that cannot be explained solely by the chain mechanism. Since the 1960s, studies using more sophisticated experiments and analytical techniques have shown that a complex combination of parallel and sequential ionic reactions must be involved as well (9–13). Oligomerization to C_{12}^{+} cations followed by scission and/or hydride transfer can produce light and heavy ends as well as alkylate of the expected molecular weight. An alternative route to alkylate, especially with isobutylene feeds, is dimerization of feed olefins followed by hydride transfer. Isomerization of the feed olefin prior to alkylation is significant for the n-butenes. Hydrogen transfer, or self-alkylation, results in the production of iso-octane and a light paraffin from two moles of isobutane and 1 mol of a light olefin. The relative extent of the various reactions depends on the catalyst as well as the feed olefin and operating conditions.

The catalysts used in the industrial alkylation processes are strong liquid acids, either sulfuric acid [7664-93-9] (H_2SO_4) or hydrofluoric acid [7664-39-3] (HF). Other strong acids have been shown to be capable of alkylation in the laboratory but have not been used commercially. Aluminum chloride [7446-70-0] ($AlCl_3$) is suitable for the alkylation of isobutane with ethylene (14). Superacids, such as trifluoromethanesulfonic acid [1493-13-6], also produce alkylate (15). Solid strong acid catalysts, such as Y-type zeolite or BF_3-promoted acidic ion-exchange resin, have also been investigated (16–18). Currently, there is not a commercial operation utilizing a heterogeneous acid catalyst for the production of motor fuel alkylate.

Sulfuric Acid Alkylation. The H_2SO_4 alkylation process was developed during the late 1930s. In the late 1980s, the H_2SO_4 process accounted for $\sim$50% of the motor fuel alkylate produced worldwide.

The modern H_2SO_4 processes are differentiated primarily by the type of reactor system that is used. The reactor must generate a high degree of mixing of the two-phase system (hydrocarbons and H_2SO_4), provide efficient heat removal via refrigeration to keep temperatures in the range of 5–10°C, and provide sufficient time for completion of the reaction. Two reactor systems, the Stratco Contactor (19) and the Kellogg Cascade Reactor (20), account for most of the licensed operating capacity.

A simplified flow diagram of a modern H_2SO_4 alkylation unit is shown in Figure 1. Excess isobutane is supplied as recycle to the reactor section to suppress

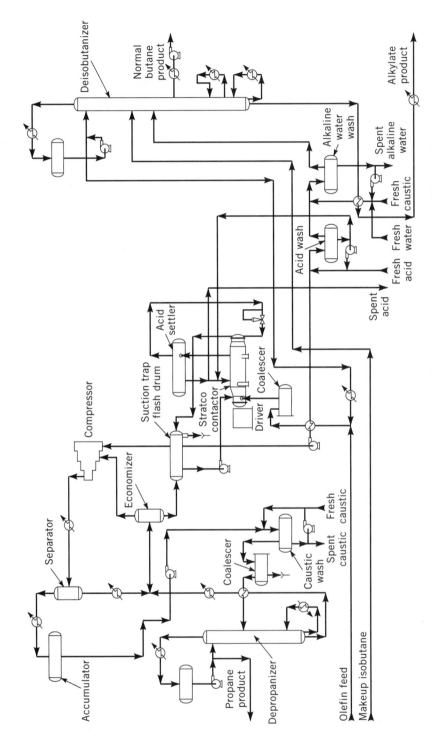

Fig. 1. The H_2SO_4 alkylation unit with effluent refrigeration. Courtesy of Stratco Inc.

polymerization and other undesirable side reactions. The isobutane is supplied both by fractionation and by return of flashed reactor effluent from the refrigeration cycle.

Propane and light ends are rejected by routing a portion of the compressor discharge to the depropanizer column. The reactor effluent is treated prior to debutanization to remove residual esters by means of acid and alkaline water washes. The deisobutanizer is designed to provide a high purity isobutane stream for recycle to the reactor, a sidecut normal butane stream, and a low vapor pressure alkylate product.

The H_2SO_4 concentration is controlled above 90% to provide the optimum activity and selectivity. Purity is maintained by the withdrawal of system acid and replacement with fresh 98% acid. The spent acid is returned to an acid manufacturing plant for reprocessing.

Continuing efforts to reduce residence time and acid level in the acid settler has led to a settler design that incorporates two stages of coalescing for hydrocarbon product separation from the acid phase (21). This new settler design reduces the acid settler size by ~10% as well as reducing residence time and acid level.

A tube insert technology is currently being implemented to provide a cost effective incremental increase in alkylation capacity (22). The inserts are placed in the Contact Reactor and optimize the overall heat transfer coefficient of the bundle by ~20% and minimize corrosion. Other process benefits include lower reaction temperature, higher alkylate octane, lower acid consumption, and increased Contact Reactor capacity.

HF Alkylation.   The HF alkylation process was developed in the late 1930s and commercialized in the 1940s. Initially, the growth rate of capacity was lower than for H_2SO_4, but by the 1980s, the capacity was approximately equivalent.

The modern HF alkylation processes are also differentiated primarily by the reactor system that is used. The Phillips process employs a gravity acid circulation system and a riser reactor (23). The UOP process uses a pumped acid circulation system and an exchanger reactor (24).

A simplified flow diagram of a modern HF alkylation unit is shown in Figure 2. Olefin feed and recycle isobutane are combined prior to contacting the acid catalyst in the reactor. Cooling water maintains the reactor temperature in the range of 20–40°C. Acid is settled from the reactor effluent and is returned to the reactor. The hydrocarbon phase is routed to the isostripper for fractionation into an isobutane recycle stream, a sidecut normal butane product, and an alkylate bottoms product. Propane is removed from the system by routing an overhead stream from the isostripper to the HF stripper. If the unit processes any significant quantity of propylene, a depropanizer is included to produce a high purity propane product.

Before leaving the unit, the products are treated with potassium hydroxide to remove any trace acidity. In addition, any product streams that are used for liquefied petroleum gas (LPG) are processed over alumina at elevated temperatures to remove residual organic fluorides.

The HF concentration of the acid catalyst is maintained in the range of 85–95% by regeneration within the unit's fractionation facilities. A separate acid regeneration column (not shown in Fig. 2) is also included to provide a

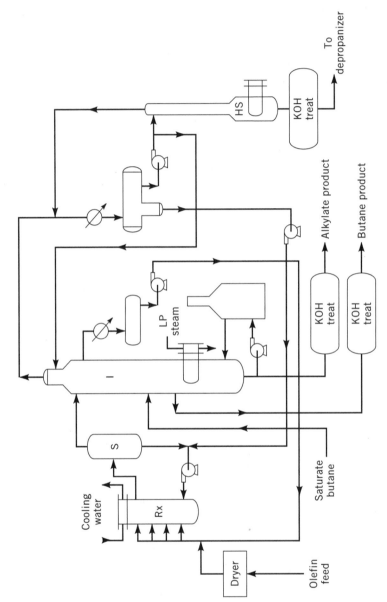

Fig. 2. The UOP HF alkylation process with butylene feed: Rx = reactor; S = settler; I = isostripper; HS = HF stripper.

means to remove excess acid-soluble oils and water. The regeneration of acid in the unit accounts for the low consumption of fresh acid by the HF process.

New and existing unit modifications have been developed and incorporated over the years to increase operational safety to reduce risk of HF acid release or dramatically minimize its impact if an accidental release occurs. Some of these risk reduction features include HF detectors, water sprays, isolation valves, reactor acid compartmentalization, rapid gravity acid transfer, and reduced acid inventory requirements (25,26). Quantitative risk analyses have been conducted and have shown dramatic decreases in risk with the addition of these mechanical modifications to both new and existing operating units.

A further advancement in the HF alkylation technology in the 1990s included the development of additives to reduce the volatility of the HF acid. Two separate but parallel HF additive technologies were developed. Chevron Texaco–UOP introduced the Alkad Process (27) and Phillips–Mobil developed the ReVap Process (25,28). In both of these technologies, the additive is handled as a "drop in" to the existing equipment. The reactor section and HF acid and hydrocarbon fractionation towers are utilized normally. In both cases, however, additional equipment is required to recover and recirculate the proprietary additives. The additive in the ReVap Process is recovered from both the acid phase as well as from the hydrocarbon phase. In this case, the additive is only recovered from the acid phase in an additional additive stripper column. In both cases, additive can be recovered with essentially no loss.

The additive technologies have demonstrated several advantages. A significant aerosol reduction is expected with both additives. Large-scale test releases of these modified acids were conducted at the Quest Consultants test site in Oklahoma. The results showed a 60–83% reduction in aerosol with the Alkad additive samples and 60–90% reduction for the ReVap additive depending on additive level and release conditions. This reduction in aerosol indicates that these additives may also be used in the transport of HF to minimize hazards upon accidental release. Additional additive could be added upon transport in order to achieve ~100% aerosol reduction and adjusted to appropriate level upon delivery. Materials of construction used in conventional HF processes are acceptable, alkylate can be produced over a wide range of processing and feedstock conditions, and alkylate quality is similar or slightly better with the additives.

3.2. Feedstock and Products. *Isobutane.* Although other isoparaffins can be alkylated, isobutane [75-28-5] is the only paraffin commonly used as a commercial feedstock. The hydrocarbon cracking operations that generate feed olefins generally do not produce sufficient isobutane to satisfy the reaction requirements. Additional isobutane must be recovered from crude oil, natural gas liquids, or generated by other refinery operations. A growing quantity of isobutane is produced by the isomerization of *n*-butane [106-97-8].

Butylenes. Butylenes are the primary olefin feedstock to alkylation and produce a product high in trimethylpentanes. The research octane number, which is typically in the range of 94–98, depends on isomer distribution, catalyst, and operating conditions.

The effect of butene isomer distribution on alkylate composition produced with HF catalyst (29) is shown in Table 1. The alkylate product octane is highest

Table 1. HF Alkylation Products from Pure Butene Isomers[a]

Alkylate product	Feed isomer			
	1-Butene [106-98-9]	trans-2-butene [590-18-1]	cis-2-butene [624-64-6]	Isobutylene [115-11-7]
carbon number distribution, wt%				
C_5	3.3	1.9	1.8	5.5
C_6	1.7	1.5	1.5	3.1
C_7	2.4	2.3	2.1	3.7
C_8	85.0	91.4	91.8	80.1
C_{9+}	7.6	2.9	2.8	7.6
total	*100.0*	*100.0*	*100.0*	*100.0*
C_8H_{18} structural distribution, wt%				
trimethylpentanes	77.5	92.1	91.4	89.0
dimethylhexanes	22.1	7.9	7.8	11.0
methylheptanes	0.4		0.8	
estimated octane				
research—clear	94.4	97.8	97.6	95.4
motor—clear	91.6	94.6	94.4	93.4

[a] Ref. 30.

for 2-butene feedstock and lowest for 1-butene; isobutylene is intermediate. The fact that the major product from 1-butene is trimethylpentane and not the expected primary product dimethylhexane indicates that significant isomerization of 1-butene has occurred before alkylation.

The H_2SO_4 catalyst produces a high octane product of similar composition from either 2-butene or 1-butene. This fact suggests that the isomerization of 1-butene to 2-butene is more complete than in the HF system. Isobutylene produces a slightly lower product octane than do the *n*-butenes. The location of a MTBE [1634-04-4] process upstream of the H_2SO_4 alkylation unit has a favorable effect on performance because isobutylene is selectively removed from the alkylation feed.

Propylene. Propylene alkylation produces a product that is rich in dimethylpentane and has a research octane typically in the range of 89–92. The HF catalyst tends to produce somewhat higher octane alkylate than the H_2SO_4 catalyst because of the hydrogen-transfer reaction, which consumes additional isobutane and results in the production of trimethylpentane and propane.

Amylenes. Amylenes (C_5 monoolefins) produce alkylates with a research octane in the range of 90–93. In the past, amylenes have not been used widely as an industrial alkylation charge, although in specific instances, alkylation with amylenes has been practiced (31). In the future, alkylation with amylenes will become more important as limits are placed on the vapor pressure and light olefin content of gasolines. J. Peterson and his collegues in a recent paper have shown economics and product quality of alkylate obtained from amylenes (32).

3.3. Future Technology Trends. As previously discussed, the future technology developments in paraffin alkylation will be greatly influenced by environmental considerations. The demand for alkylate product will continue to increase because alkylate is one of the most desirable components in modern low-emission gasoline formulations. Increased attention will be focused on improving process safety, reducing waste disposal requirements, and limiting the environmental consequences of any process emissions.

Hydrofluoric acid has long been recognized as a hazardous material that must be handled with care. However, in recent years concerns have increased over the possible consequences of an accidental release of HF. The results of a 1986 spill test showed that a large portion of the released HF can form a vapor cloud (33). In response to this information, the refining industry has acted to further tighten the already rigorous operating and design standards for HF plants. Periodic hazard reviews are being conducted for all operating units to ensure that the proper systems and procedures are in place (34,35).

Improved feedstock pretreatment is important to minimize catalyst consumption and reduce subsequent spent-catalyst handling requirements. Selective hydrogenation of dienes can be used to reduce acid consumption, both in HF and H_2SO_4 alkylation (36). More effective adsorptive treating systems have been applied to remove oxygen-containing contaminants that are frequently introduced in upstream processing steps.

Because solid acid catalyst systems offer advantages with respect to their handling and noncorrosive nature, research on the development of a commercially practical solid acid system to replace the liquid acids will continue. A major hurdle for solid systems is the relatively rapid catalyst deactivation caused by fouling of the acid sites by heavy reaction intermediates and by-products. At this time there are two technologies which are being offered for commercialization, but have not yet been demonstrated on a large scale. These technologies are the ABB Lummus Global ALKYCLEAN technology (37) and the UOP Alkylene technology (38). Both of these technologies require frequent catalyst regeneration, which is accomplished via hydrogen stripping. A moving bed process design is used in the UOP Alkylene technology, while a number of fixed bed reactors operating in a cyclical regeneration mode is used in the ABB Lummus Global ALKYCLEAN process design.

4. Alkylation of Aromatic Hydrocarbons

Most of the industrially important alkyl aromatics used for petrochemical intermediates are produced by alkylating benzene [71-43-2] with monoolefins. The most important monoolefins for the production of ethylbenzene, cumene, and detergent alkylate are ethylene, propylene, and olefins with 10–18 carbons, respectively. This section focuses primarily on these alkylation technologies.

4.1. Acid Catalysts and Reaction Mechanism. Acid catalysts promote the addition of alkyl groups to aromatic rings. Olefins, alcohols, ethers, halides, and other olefin-producing compounds can be used as alkylating reagents. In addition to traditional protonic acid catalysts (H_2SO_4, HF, phosphoric acid) and Friedel-Crafts-type catalysts ($AlCl_3$, boron fluoride), any solid acid catalyst having a comparable acid strength is effective for aromatic alkylation. Typical solid acid catalysts are amorphous and crystalline alumino-silicates, clays, ion-exchange resins, mixed oxides, and supported acids (39). Among these solid acid catalysts, ZSM-5, Y-type zeolites, and more recently MCM-22 (40) and beta-zeolite (41) have become the new commercial catalysts for aromatic alkylation. A new catalytic function, shape selectivity, was found in the application of zeolite catalysts as represented by selective formation of p-xylene in toluene alkylation with methanol over a ZSM-5 catalyst (42). The specific catalysts used in the commercial alkylation processes are described in discussions of specific products, eg, for ethylbenzene production.

The first step in the catalytic alkylation of aromatics is the conversion of an olefin or olefin-producing reagent into a carbonium ion or polarized complex. Then, this carbonium ion or complex, which is a powerful electrophile, attacks the aromatic ring (43).

A tertiary carbonium ion is more stable than a secondary carbonium ion, which is in turn more stable than a primary carbonium ion. Therefore, the alkylation of benzene with isobutylene is much easier than is alkylation with ethylene. The reactivity of substituted aromatics for electrophilic substitution is affected by the inductive and resonance effects of a substituent. An electron-donating group, such as the hydroxyl and methyl groups, activates the alkylation; and an electron-withdrawing group, such as chloride, deactivates it.

The rearrangement of carbonium ions that readily occurs according to the thermodynamic stability of cations sometimes limits synthetic utility of aromatic alkylation. For example, the alkylation of benzene with n-propyl bromide gives mostly isopropylbenzene (cumene) C_9H_{12} and much less n-propylbenzene. However, the selectivity to n-propylbenzene [103-65-1] versus isopropylbenzene [98-82-8] changes depending on alkylating reagents, conditions, and catalysts; eg, the alkylation of benzene with n-propyl chloride at room temperature gives mostly n-propylbenzene (44).

4.2. Base Catalysts and Reaction Mechanism.

Alkali metals and their derivatives can catalyze the alkylation of aromatics with olefins (45). In contrast to acid-catalyzed alkylation, in which the aromatic ring is alkylated, an olefin is added to the alkyl group of aromatics over a base catalyst through a carbanion intermediate. The carbanion intermediate is produced from an aromatic compound by the abstraction of benzylic hydrogen as a proton by a base. The carbanion reacts with an olefin to grow the side chain of the aromatic compound (46).

The side-chain alkylation of toluene with methanol to produce a mixture of styrene and ethylbenzene can be catalyzed by alkali-cation-exchanged X- and Y-type zeolites (47), magnesium oxide [1309-48-4] (MgO), titanium oxide [13463-67-7] (TiO$_2$), and mixtures of MgO and TiO$_2$ and calcium oxide [1305-78-8] (CaO) and TiO$_2$ (48). Toluene is activated on a basic site and reacts with formaldehyde, which is produced from methanol. The coexistence of weak acid sites promotes the reaction (49). The conversion of relatively low cost toluene into more valuable ethylbenzene and styrene is attractive. However, the ethylbenzene or styrene process based on side-chain alkylation has not been developed for commercial applications.

4.3. Industrial Application. *Ethylbenzene.* This alkylbenzene is almost exclusively used as an intermediate for the manufacture of styrene monomer [100-42-5]. A small amount ($<1\%$) is used as a solvent and as an intermediate in dye manufacture (1,50,51). The ethylbenzene growth rate projections for 1990–1995 range from 3.0 to 3.5% per year (50).

Ethylbenzene [100-41-4] is primarily produced by the alkylation of benzene with ethylene [74-85-1], although a small percentage of the world's ethylbenzene capacity is based on the superfractionation of ethylbenzene from mixed xylene streams (52). A wide variety of different alkylation processes have been developed and commercialized since the 1940s. These processes can generally be divided into liquid- and vapor-phase processes.

Liquid-Phase Processes. Prior to 1980, commercial liquid-phase processes were based primarily on an $AlCl_3$ catalyst. $AlCl_3$ systems have been developed since the 1930s by a number of companies, including Dow, BASF, Shell Chemical, Monsanto, Société Chimique des Charbonnages, and Union Carbide–Badger. These processes generally involve ethyl chloride or occasionally hydrogen chloride as a catalyst promoter. Recycled alkylated benzenes are combined with the $AlCl_3$ and ethyl chloride to form a separate catalyst–complex phase that is heavier than the hydrocarbon phase and can be separated and recycled.

In 1974, Monsanto brought on-stream an improved liquid-phase $AlCl_3$ alkylation process that significantly reduced the $AlCl_3$ catalyst used by operating the reactor at a higher temperature (53–55). In this process, the separate heavy catalyst–complex phase previously mentioned was eliminated. Eliminating the catalyst–complex phase increases selectivities and overall yields in addition to lessening the problem of waste catalyst disposal. The ethylbenzene yields exceed 99%.

In the 1980s, environmental pressures associated with the problem of disposal of the waste $AlCl_3$ catalyst led to the development of two new liquid-phase processes based on zeolite catalysts, which are considered environmentally inert. The first of these processes is a conventional fixed-bed catalyst system (Fig. 3). The catalyst was developed by Unocal, and the process is jointly licensed by ABB Lummus Global and UOP. The operating conditions with regard to temperature and pressure are mild, and carbon steel can be used throughout the process. The technology has no corrosive elements, and the ethylbenzene yields exceed 99% (56–58). The first commercial unit successfully started in Japan in August, 1990.

The second new zeolite-based liquid-phase process was developed by Chemical Research & Licensing Company (CR&L). The process is based on the concept of catalytic distillation, ie, reaction and separation in the same vessel. The concept has been applied commercially for the production of MTBE (59–62) and for the production of ethylbenzene.

Current state-of-the-art technology from the mid-1990s to present involves the use of liquid phase technologies offered by the partnerships of ABB Lummus Global/UOP based on a proprietary beta-zeolite catalyst and ExxonMobil/ Washington Group International, Inc. process based on MCM-22 catalyst.

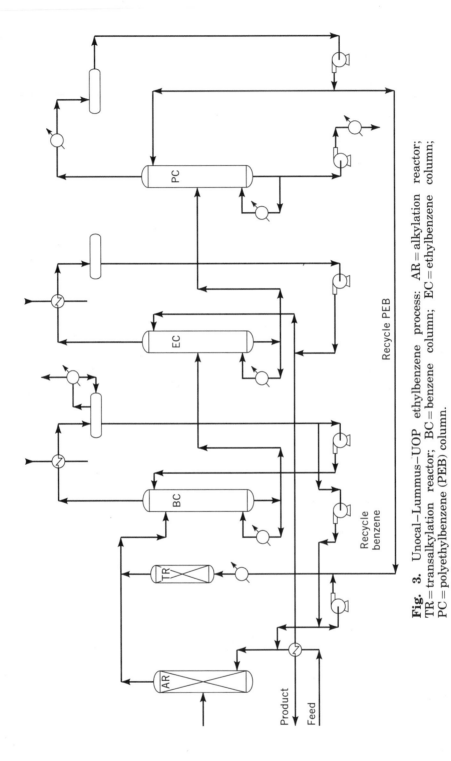

Fig. 3. Unocal–Lummus–UOP ethylbenzene process: AR = alkylation reactor; TR = transalkylation reactor; BC = benzene column; EC = ethylbenzene column; PC = polyethylbenzene (PEB) column.

Beta-zeolite is quickly becoming the catalyst of choice for commercial production of ethylbenzene and cumene. Mobil invented the basic beta-zeolite composition of matter in 1967 (63). Since that time, catalysts utilizing beta-zeolite have undergone a series of evolutionary steps leading to the development of state-of-the-art catalysts such as the UOP EBZ-500 and QZ-2000 for ethylbenzene and cumene alkylation service, respectively.

Much of the effort between 1967 and the early 1980s involved characterization of beta's perplexing structure. It was quickly recognized that beta (BEA) had a large three dimensional pore structure and had a high acidity capable of catalyzing many reactions. It wasn't until early 1988, however, that scientists at Exxon finally solved the chiral nature of the BEA structure.

At the same time that the structure of beta was being investigated, extensive research was being conducted to identify new uses for this zeolite. A major breakthrough came in late 1988 with the invention by workers at Chevron of a liquid phase alkylation process using beta-zeolite catalyst. While Chevron had significant commercial experience with the use of Y (FAU) zeolite in liquid phase aromatic alkylation service, they were quick to recognize the benefits of BEA over Y as well as the other acidic zeolites used at the time, such as mordenite (MOR) or ZSM-5 (MFI). Chevron discovered that the open 12-membered ring structure characteristic of beta coupled with the high acidity of the material made it an ideal catalyst for aromatic alkylation. These properties were shown to be key in the production of aromatic derivative products such as ethylbenzene and cumene with extremely high yields and product purities approaching 100%. Moreover, the combination of high activity and porous structure imparted a high degree of tolerance to many of the contaminants ordinarily found in the feedstocks to these processes. A liquid-phase process was developed by Chevron in 1990 and the rights were acquired by UOP in 1995 as a basis for the Lummus/UOP EB*One* process for ethylbenzene and Q-Max process for cumene production.

The superior performance of the new liquid-phase process, however, provided the incentive for the development of a new manufacturing technology to make this catalyst a commercial reality. In 1991, a new cost-effective synthesis route invented by UOP paved the way for the successful commercialization of the process. The new synthesis route involved the substitution of alkanolamines as a low cost replacement for the tetraethylammonium hydroxide that had to be used heretofore as the templating agent. Finally, the new synthesis route enabled the practical synthesis of beta-zeolite over a wider range of silica to alumina ratios, a factor that has a profound effect on the catalyst's performance.

In contrast to UOP, ExxonMobil uses MCM-22 catalyst in its EBMax liquid phase EB process (43). MCM-22 consists of two nonintersecting 10- and 12-ring pore systems. It is believed that the primary alkylation reactions take place in the 12-ring "pockets" giving rise to somewhat higher EB selectivity. However, there remains some debate in the literature as to how much of an effect this structure has on enhancing selectivity. Separate studies conducted at Enichem (64) show very little difference in monoalkylate selectivity for MCM-22 catalyst vs. well-optimized low Si/Al2 beta-zeolite catalyst.

Vapor-Phase Processes. Although vapor-phase alkylation has been practiced since the early 1940s, it could not compete with liquid-phase processes until the 1970s when the Mobil-Badger vapor-phase ethylbenzene process was introduced (Fig. 4). The process is based on Mobil's ZSM-5 zeolite catalyst (49,65,66). The nonpolluting and noncorrosive nature of the process is one of its major advantages over the AlCl₃ liquid-phase system. Unlike the liquid-phase system, the reactors operate at high temperature (400–450°C) and low pressure (2–3 MPa). The high temperature allows the net process heat input and exothermic heat of reaction to be recovered as steam. However, the high temperature vapor-phase operation causes catalytic deactivation by fouling as a result of the deposition of carbonaceous materials, and so the catalyst requires periodic regeneration. Two reactors are required so that processing and regeneration can proceed alternately without interrupting production. Ethylbenzene yields are ~98%.

A modified ZSM-5 catalyst has a unique shape-selective property for producing p-ethyltoluene [622-96-8] selectively by the alkylation of toluene [108-88-3] with ethylene (67). p-Ethyltoluene is an intermediate in the production of poly (p-methylstyrene) [24936-41-2] (PPMS), which is reported to have physical advantages, such as higher flash point and glass-transition temperatures and lower specific gravity, over polystyrene (68,69).

Cumene.   Cumene processes were originally developed between 1939 and 1945 to meet the demand for high octane aviation gasoline during World War II (1,2). In 1989, ~95% of cumene demand was as an intermediate for the production of phenol [108-95-2] and acetone [67-64-1]. A small percentage is used for the production of α-methylstyrene. The demand for cumene [98-82-8] has risen at an average rate of 2–3% per year since 1970 (70,71), and this trend continued throughout the 1990s.

Currently, almost all cumene is produced commercially by two processes: (1) a fixed-bed, kieselguhr-supported phosphoric acid catalyst system developed by UOP and (2) a homogeneous AlCl₃ and hydrogen chloride catalyst system developed by Monsanto.

Two new processes using zeolite-based catalyst systems were developed in the late 1980s. Unocal's technology is based on a conventional fixed-bed system. CR&L has developed a catalytic distillation system based on an extension of the CR&L MTBE technology (59–62).

SPA Catalyst.   The solid phosphoric acid (SPA) catalyst process has been the dominant source of cumene since the 1930s. This process accounts for >90% of cumene operating capacity (72). A simplified process flow diagram is given in Figure 5.

Propylene feed, fresh benzene feed, and recycle benzene are charged to the upflow reactor, which operates at 3–4 MPa and at 200–260°C. The SPA catalyst provides an essentially complete conversion of propylene [115-07-1] on a one-pass basis. A typical reactor effluent yield contains 94.8 wt% cumene and 3.1 wt% diisopropylbenzene [25321-09-9] (DIPB). The remaining 2.1% is primarily heavy aromatics. This high yield of cumene is achieved without transalkylation of DIPB and is unique to the SPA catalyst process.

The cumene product is 99.9 wt% pure, and the heavy aromatics, which have a research octane number (RON) of 109, can either be used as high octane gasoline-blending components or combined with additional benzene and sent to a transalkylation section of the plant where DIPB is converted to cumene. The

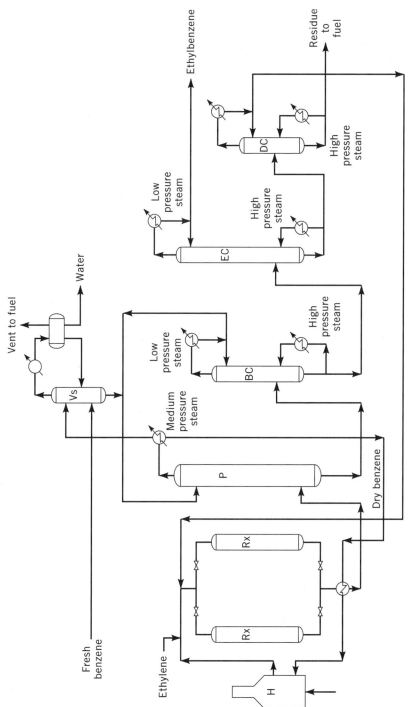

Fig. 4. Mobil-Badger process for ethylbenzene production: H = heater; Rx = reactor; P = prefractionator; BC = benzene recovery column; VS = vent gas scrubber; EC = ethylbenzene recovery column; DC = diethylbenzene recovery column; Courtesy of VCH Publishers, Inc.

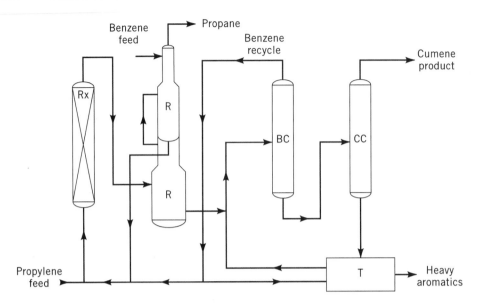

Fig. 5. UOP Cumene process: Rx = reactor; R = rectifier; BC = benzene column; CC = cu-cumene column; T = transalkylation.

overall yields of cumene for this process are typically 97–98 wt% with trans-alkylation and 94–96 wt% without transalkylation.

AlCl$_3$ and Hydrogen Chloride Catalyst. Historically, AlCl$_3$ processes have been used more extensively for the production of ethylbenzene than for the production of cumene. In 1976, Monsanto developed an improved cumene process that uses an AlCl$_3$ catalyst, and by the mid-1980s, the technology had been successfully commercialized. The overall yields of cumene for this process can be as high as 99 wt% based on benzene and 98 wt% based on propylene (73).

A simplified process flow diagram is shown in Figure 6 (74). Dry benzene, fresh and recycle, and propylene are mixed in the alkylation reaction zone with the AlCl$_3$ and hydrogen chloride catalyst at a temperature of <135°C and a pressure of <0.4 MPa (74). The effluent from the alkylation zone is combined with recycle polyisopropylbenzene and fed to the transalkylation zone, where polyiso-propylbenzenes are transalkylated to cumene. The strongly acidic catalyst is separated from the organic phase by washing the reactor effluent with water and caustic.

The distillation system is designed to recover a high-purity cumene product. The unconverted benzene and polyisopropylbenzenes are separated and recycled to the reaction system. Propane in the propylene feed is recovered as liquid petroleum gas (LPG).

Zeolite Catalysts. Unocal introduced a fixed-bed liquid-phase reactor system based on a Y-type zeolite catalyst (75) in the early 1980s. The selectivity to cumene is generally between 70 and 90 wt%. The remaining components are primarily polyisopropylbenzenes, which are transalkylated to cumene in a separate reaction zone to give an overall yield of cumene of ~99 wt%. The distillation requirements involve the separation of propane for LPG use, the recycle of excess

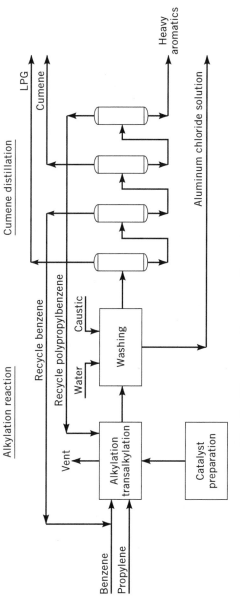

Fig. 6. Monsanto–Lummus Crest cumene process.

185

benzene to the reaction zones, the separation of polyisopropylbenzene for trans-alkylation to cumene, and the production of a purified cumene product.

The second zeolite process was developed by CR&L and is based on the concept of catalytic distillation (59–62), which is a combination of catalytic reaction and distillation in a single column. The basic principle is to use the heat of reaction directly to supply heat for fractionation. This concept has been applied commercially for the production of MTBE and cumene.

Current state-of-the-art processes for cumene are similar to ethylbenzene and consist of liquid-phase technologies offered by UOP and ExxonMobil based on beta-zeolite and MCM-22 catalysts, respectively. Over the past decade, great progress has been made in improving and optimizing catalyst formulations for use in both the EB and cumene alkylation applications. For example, the ability to synthesize beta-zeolite in a wide range of Si/Al_2 ratios has given catalyst designers the ability to tailor the zeolite into a form that optimizes activity and selectivity. A parametric study on the effects of Si/Al_2 ratio on activity and selectivity was published by Bellusi (76). In this work, it was found that as the silica to alumina ratio was increased from 28 to 70, there was a decrease in both activity and selectivity toward IPBs. Additionally, the less active catalysts had a greater tendency toward oligomerization and were more prone toward coking. An analogous trend was observed for ethylene, as well.

This study parallels work performed at UOP, where, through the use of nonconventional synthesis techniques, samples have also been prepared with Si/Al_2 ratios down to 10. Through this work it has been found that with a Si/Al_2 ratio of 25, the catalyst maintains sufficient activity to achieve polyalkylate equilibrium (eg, diisopropylbenzene equilibrium) and, at the same time, minimizes formation of heavier diphenyl compounds (and hence maximizes yield) in cumene service.

Perhaps the most critical understanding was developed with regard to the need to minimize the Lewis acidity of the catalyst and at the same time maintain high Brønsted acidity. Studies at UOP demonstrated that olefin oligomerization was directly related to the Lewis acid function of the catalyst. Olefin oligomerization reactions can lead to the formation of heavy compounds (coke-type precursors), which have a negative effect on catalyst stability. Thus, minimization of the Lewis character of the beta leads to a catalyst with high stability. Generally, Lewis acidity in beta-zeolite has been attributed to the existence of nonframework aluminum atoms. The most common mechanism for the formation of non-framework alumina is through steam dealumination during the catalyst calcination step of the manufacturing process. By careful control of the temperature, time, and steam levels during the manufacturing process, it is possible to produce a catalyst that is extremely stable at typical alkylation conditions.

The feature of complete regenerability is another attribute that distinguishes beta-zeolite catalysts from other commercially practiced technology, where selectivity can be lost upon regeneration (77).The ability to regenerate catalyst is essential in a commercial environment to provide additional flexibility to cope with a wide range of feedstock sources, feedstock contaminants, and potential operational upsets.

The historical development of beta-zeolite showed that early versions of beta catalyst demonstrated less than optimal performance when compared to

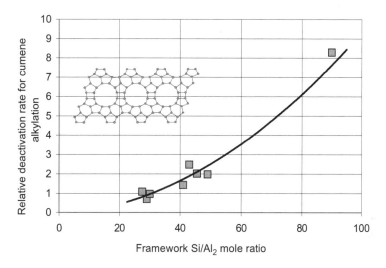

Fig. 7. Relative stability of beta-zeolite as a function the Si/Al ratio.

today's state-of-the-art formulation. Figure 7 is a plot of the relative stability of beta-zeolite as a function of the Si/Al_2 ratio of the beta-zeolite structure in which the dominating influence of this parameter is evident. Stabilizing the zeolitic structure through careful process and chemical means results in a catalyst system that is extremely robust, highly regenerable, and tolerant of most common feedstock impurities. Additional studies of beta-zeolite have come to similar conclusions. For example, Enichem finds that beta-zeolite is the most effective catalyst for cumene alkylation among others tested including Y, mordenite and an isostructural synthesis of MCM-22.

The principles described above also led to the development of a new generation cumene alkylation catalyst, QZ-2001. In Figure 8, results from accelerated stability testing of QZ-2000 and QZ-2001 catalyst demonstrates the superior stability of the latest catalyst version.

Since new high activity beta-zeolite catalysts such as QZ-2000 are such strong acids, they can be used at lower temperatures than SPA catalyst or competing lower activity zeolites such as MCM-22 (43,78). The lower reaction temperature reduces the rate of competing olefin oligomerization reactions that is particularly high in SPA based processes. The result is higher selectivity to cumene and lower production of non-aromatics that distill with cumene (including olefins, which are analyzed as Bromine Index, and saturates) as well as lower heavy by-products production. For example, although butylbenzene is typically produced from traces of butylene in the propylene feed, there is always the potential for butylbenzene formation through the oligomerization of propylene to nonene, followed by cracking and alkylation to produce butylbenzenes and amylbenzenes. As a result of the high-activity and low-operating temperature of the beta-zeolite catalyst system, the Q-Max process essentially eliminates oligomerization. This results in almost no butylbenzene formation beyond that from butylenes in the feed. The cumene product from a Q-Max unit processing a butylene-free propylene feedstock typically contains <15 wt-ppm butylbenzenes.

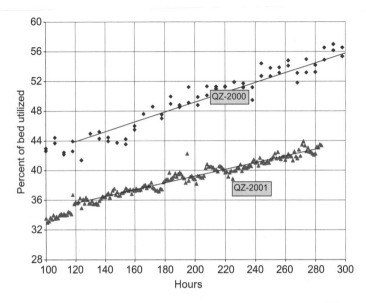

Fig. 8. Comparison of QZ-2000 and QZ-2001 catalyst stability.

The Q-Max process typically produces equilibrium levels of cumene (between 85 and 95 mol.%) and DIPB (between 5 and 15 mol%). The DIPB is fractionated from the cumene and reacted with recycle benzene at optimal conditions for transalkylation to produce additional cumene. Beta-zeolite catalyst is also an extremely effective catalyst for the transalkylation of DIPB to produce cumene. Due to the high activity of beta-zeolite, transalkylation in the Q-Max process can be accomplished at very low temperatures to achieve high conversion and minimum side products such as heavy aromatics and additional n-propylbenzene. As a result of the high activity and selectivity properties of beta-zeolite, the same catalyst (eg, QZ-2000) is specified for both the alkylation and transalkylation sections of the process. With both of these reactors working together to take full advantage of the QZ-2000 catalyst, the overall yield of cumene is increased to at least 99.7 wt%.

The improvement in beta-zeolite catalyst quality has progressed to the point that any significant impurities in the cumene product are governed largely by trace impurities in the feeds. The selectivity of the catalyst typically reduces by-products to a level resulting in production of ultra-high cumene product purities of up to 99.97 wt%. At this level, the only significant by-product is n-propylbenzene with the catalyst producing essentially no ethylbenzene, butylbenzene, or cymene beyond precursors in the feed.

Cymene. Methylisopropylbenzene [25155-15-1] can be produced over a number of different acid catalysts by alkylation of toluene with propylene (79–82). Although the demand for cymene is much lower than for cumene, one commercial plant was started up in 1987 at the Yan Shan Petrochemical Company in the People's Republic of China. The operation of this plant is based on SPA technology offered by UOP for cumene. The cymene is an intermediate for the production of m-cresol (3-methylphenol) [108-39-4].

Detergent Alkylate. In the 1940s, sodium dodecylbenzene sulfonate (DDBS) [25155-30-0] produced by the alkylation of benzene with propylene tetramer ($C_{12}H_{24}$) [6842-15-5] followed by sulfonation with oleum [8014-95-7] (H_2SO_4 mixture with sulfur trioxide) or sulfur trioxide and then neutralization was found to have detergent characteristics superior to those of natural soaps. Because of its price stability and effectiveness, DDBS became the standard synthetic surfactant in the industry. By 1955, these efficient surfactants were also leading to environmental problems, such as buildup of foam in downstream discharge sites. This buildup was attributed to the poor biodegradability of the highly branched structure of the propylene tetramer side chain (83–85).

During the early 1960s, linear alkylbenzene sulfonates (LABS), prepared by the sulfonation of linear alkylbenzenes (LAB), began to replace DDBS in industrialized countries due to its superior biodegradability. LAB is produced by the alkylation of benzene with linear aliphatic olefins; such as alpha olefins produced via ethylene oligomerization or linear internal olefins produced via catalytic dehydrogenation of linear paraffins. In the 1970s, LABS capacity increased rapidly with facilities being installed around the world. Except in a few parts of the world, the use of DDBS was phased out by 1980.

The synthetic detergent industry has become one of the largest chemical process industries. The worldwide annual production of LAB has increased from 1.1 million tons in 1980 to 1.8 million tons in 1990 and 2.4 million tons in 2000 (87). Paraffin dehydrogenation followed by alkylation accounts for ~88% of the current world production.

Industrial Processes. A variety of acid catalysts have been used for the production of alkylbenzenes by the alkylation of benzene with higher olefins (C_{10}–C_{15} detergent-range olefins). HF and $AlCl_3$ have been used since the 1960s and H_2SO_4 was used in some earlier units. In 1995, the first detergent alkylation unit, using a solid acid catalyst developed by UOP and CEPSA, was started-up. The Detal process offers superior LAB product quality and lower capital costs due to simplified catalyst handling and downstream product clean up compared with either HF or $AlCl_3$. The main reaction in detergent alkylation is the alkylation of benzene with the straight-chain olefins to yield a linear alkylbenzene:

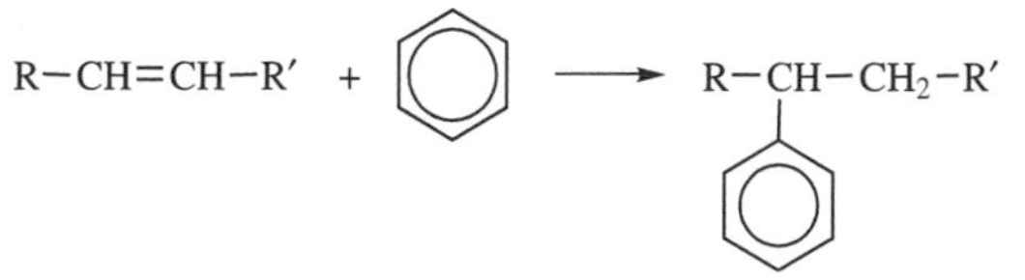

$$R-CH=CH-R' \; + \; \bigcirc \longrightarrow \; R-\underset{\underset{\displaystyle \bigcirc}{|}}{CH}-CH_2-R'$$

At the conditions used, some side reactions, such as the formation of dialkylbenzenes, take place:

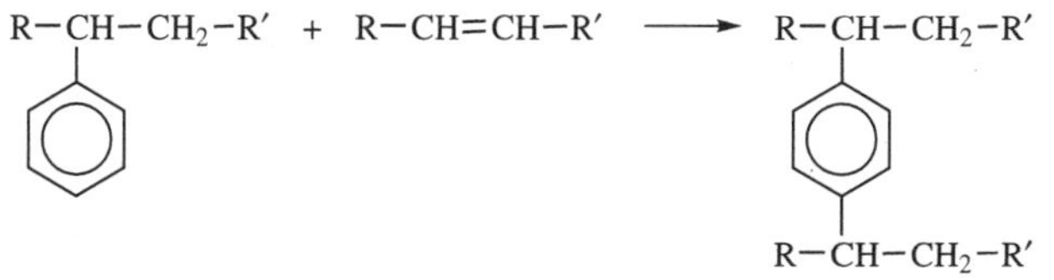

$$R-\underset{\underset{\displaystyle \bigcirc}{|}}{CH}-CH_2-R' \; + \; R-CH=CH-R' \longrightarrow \; R-\underset{\underset{\displaystyle \underset{\displaystyle R-CH-CH_2-R'}{|}}{\bigcirc}}{CH}-CH_2-R'$$

Table 2. **Isomer Distribution, wt%, of Dodecylbenzene from 1-Dodecene and Benzene**

Phenyl position	Catalyst system		
	HF	$AlCl_3$	H_2SO_4
1	0	0	0
2	20	32	41
3	17	22	20
4	16	16	13
5	23	15	13
6	24	15	13

Any diolefins present in the olefin stream can also react to form diphenylalkanes:

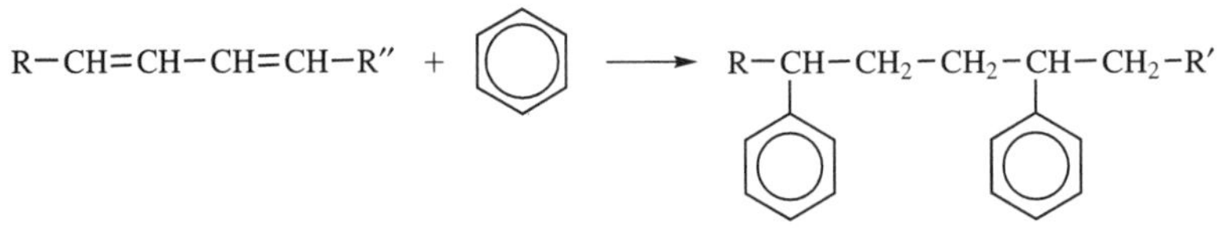

$$R-CH=CH-CH=CH-R'' \; + \; \bigcirc \longrightarrow R-CH-CH_2-CH_2-CH-CH_2-R'$$

Some heavier compounds are also formed by a combination of these reactions. In addition to the alkylation activity, all of these acid catalysts possess, in varying degrees, activity to shift the olefinic double bond along the chain. Thus, regardless of the position of the double bond in the olefin feed, the position of the phenyl group in the final product, as shown in Table 2 for the reaction of 1-dodecene [112-41-4] with benzene, is specific to the catalyst system used (87,88).

AlCl$_3$ Alkylation Process. The first step in the AlCl$_3$ process is the chlorination of *n*-paraffins to form primary monochloroparaffins. Then in the second step, the monochloroparaffin is alkylated with benzene in the presence of AlCl$_3$ catalyst (89,90). Considerable amounts of indane (2,3-dihydro-1H-indene [496-11-7]) and tetralin (1,2,3,4-tetrahydronaphthalene [119-64-2]) derivatives are formed as by-products because of the dichlorination of paraffins in the first step (91). Only a few industrial plants built during the early 1960s use this technology to produce LAB from linear paraffins. The $C_{10}-C_{15}$ alpha olefins also can be alkylated with benzene using this catalyst system.

HF Alkylation Process. The most widely used technology today is based on the HF catalyst system (92). During the mid-1960s, commercial processes were developed to selectively dehydrogenate linear paraffins to linear internal olefins (93–95). Although these linear internal olefins are of lower purity than are alpha olefins, they are more cost-effective because of their lower cost of production. Furthermore, with improvement over the years in dehydrogenation catalysts and processes, such as selective hydrogenation of diolefins to monoolefins (96,99), the quality of linear internal olefins has improved.

A simplified flow diagram for a typical UOP Detergent Alkylate Process is shown in Figure 9. A necessary feature of the reaction section of early alkylation units was the use of two reactors: the first-stage reactor completes the major part

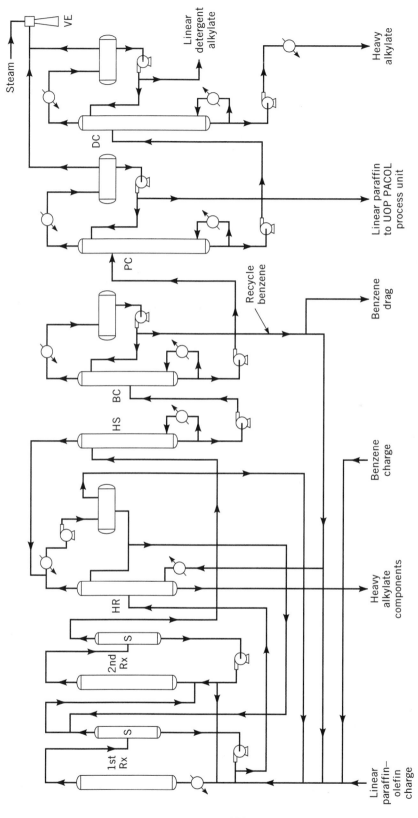

Fig. 9. UOP Detergent alkylate process: Rx = reactor; S = settler; HR = HF regenerator; HS = HF stripper; BC = benzene column; PC = paraffin column; DC = detergent alkylate column; VE = vacuum ejector.

191

of the alkylation reaction, and the second-stage reactor the last traces of unsaturated hydrocarbons react, and a sizable portion of the soluble polyaromatics is removed. Modern units with lower diene-containing feeds employ a single alkylation reactor (93).

High-purity olefin feed or an olefin–paraffin mixture from a dehydrogenation unit is combined with makeup and recycle benzene and cooled prior to mixing with HF acid. The reaction section consists of a mixer reactor and an acid settler. Because of dilution by excess benzene and paraffin, the temperature rise resulting from the exothermic reaction is relatively small. A portion of the HF phase from the settler is sent to the HF regenerator, where heavy byproducts are removed to maintain the required purity of the HF acid. The hydrocarbon phase from the acid settler proceeds to the fractionation section, where the remaining HF catalyst, excess benzene, unreacted n-paraffins, heavy alkylate, and the LAB product are separated by means of sequential fractionation columns. The HF acid and benzene are recycled to the alkylation reactor. The unreacted n-paraffins are passed through an alumina treater to remove combined fluorides and are then recycled back to the dehydrogenation unit. Not shown in Figure 9 is the HF acid handling and neutralization section. This section is not basic to the process but is required for the safe operation of the unit (98).

Detal Process. The most recent advance in detergent alkylation is the development of a solid catalyst system. UOP and Compania Espanola de Petroleos SA (CEPSA) have jointly developed the Detal process, which uses a fixed-bed heterogeneous aromatic alkylation catalyst system for the production of LAB (99). Petresa, a subsidiary of CEPSA, started up the first unit utilizing this process in Quebec, Canada in 1995 (100). Two additional Detal units are currently in operation. In contrast to HF and $AlCl_3$, the Detal catalyst is non-corrosive and eliminates problems associated with the handling and disposal of the previous catalysts. A variety of other solid acid catalysts for detergent alkylation have been described in the literature (101–109), but none have been used commercially at the present time.

The flow scheme of the UOP/CEPSA Detal process is presented in Figure 10. The process is operated in conjunction with UOPs dehydrogenation technology to produce linear olefins. The olefin feed and recycle benzene are combined with make up benzene before introduction to the fixed-bed reactor containing the solid acid catalyst. The reaction occurs in the liquid phase under mild conditions to achieve optimal product quality. The reactor effluent flows directly to the fractionation system that is identical to that for the hydrofluoric acid process. The hydrofluoric acid stripper column, settlers, other hydrofluoric acid related piping and equipment as well as the product alumina treater are eliminated. Carbon steel metallurgy can now be used due to the elimination of the liquid acid. In order to improve product yield and quality, there are two additional process units included in the overall Detal process scheme. First, a DeFine unit selectively hydrogenates diolefins to monoolefins (same as in hydrofluoric acid technology) to increase alkylate yield. Second, a PEP Process is added to eliminate aromatics from the olefin feed stream. These aromatics would alkylate in the Detal unit leading to faster catalyst deactivation and lower quality product.

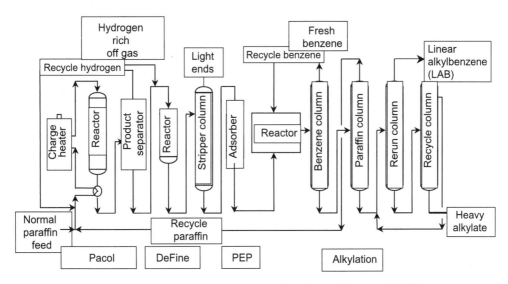

Fig. 10. UOP/Cepsa detal process for the production of LAB.

Table 3 compares linear alkylbenzene product properties for the HF and Detal catalyst systems. Bromine index and sulfonatability are key measures of product quality because they affect final product cost. High bromine index product also produces a highly colored sulfonate that requires further treatment. Recently, reduction in non-alkylbenzene components, in particular reduction of tetralins, and improved linearity has also become important. Both of these parameters are related to the rate of biodegradation of the ultimate LAS product. (105).

As can be seen in Table 3, the Detal linear alkylbenzene product is made in higher yield, with higher linearity, improved sulfonate color, and less tetralin by-product compared to the HF alkylation process. It also has higher 2-phenylalkane content that gives improved solubility in many detergent formulations. For example, the cloud point of a liquid detergent formulation prepared with LABs

Table 3. **Typical LAB Product Properties**

Property	AlCl$_3$ alkylate	HF alkylate	Detal alkylate
specific gravity	0.860	0.860	0.860
bromine number	0.015	0.015	0.015
Saybolt color	+30	+30	+30
Doctur test	negative	negative	negative
water, wt%	0.01	0.01	0.01
sulfonation, wt%	98.5	98.5	98.5
biodegradability, wt%	95	95	95
paraffins, wt%	0.3	0.3	0.3
indanes or tetralins, wt%	5–15	1–3	1–3
2-phenylalkanes, wt.%	30	15	25
n-alkylbenzene, wt%	90	94	94
average molecular weight	235–260	235–260	235–260

derived from Detal LAB is lower than that of the same formulation produced from HF LAB over a wide range of surfactant concentrations (106). All of these properties demonstrate that the current Detal technology produces a superior product than the hydrofluoric acid technology. Due to the higher tetralin content from aluminum chloride based alkylation process it is being phased out.

Economics of the current Detal and hydrofluoric acid technologies have been summarized in detail (106). For an 80,000 MTA linear alkylbenzene unit, the estimated erected costs for the Detal and HF systems are $67 and 72 million, respectively. Thus, fixed plant investment has been reduced by ~15% by the use of a solid acid catalyst. The absence of the hydrofluoric liquid acid and required neutralization facilities for the acid wastes is also reflected in lower operating costs.

Xylenes. The main application of xylene isomers, primarily *p*- and *o*-xylenes, is in the manufacture of polyester fibers, films, resins and plasticizers. Demands for xylene isomers and other aromatics such as benzene have steadily been increasing over the last two decades. Food packaging applications, which use polyester blends derived from *m*-xylene, have been increasing 10–15% per year during the 1990s. This has led to a singificant increase in the capacity for the *m*-xylene isomer. The major source of xylenes is catalytic reforming of naphtha and pyrolysis of naphtha and gas oils. A significant amount of toluene and $C_{9}+$ aromatics, which have lower petrochemical value, is also produced by these processes. More valuable pure xylene isomers can be manufactured from these low value aromatics by transalkylation, eg, the UOP Tatoray process (109), the Toray TAC9 process. (110) and the ExxonMobil TransPlus technology (111).

It is also possible to produce *p*-xylene in concentrations that are significantly above equilibrium by selective toluene disproportionation processes such as the UOP PX-Plus (112) process and the ExxonMobil MSTDP or MTPX technologies (113). These latter coproduce a significant amount of benzene from the toluene feed. The Xylene isomers are recovered by technologies such as the UOP Parex process (114) for *p*-xylene and the UOP MX Sorbex process (115) for *m*-xylene. The *o*-xylene isomer can be recovered by fractionation. The "raffinate" product from the recovery processes is sent to an isomerization unit, which reestablishes the equilibrium among the C_{8} aromatics. Typical technologies are the UOP Isomar process (116–118) and ExxonMobil Xymax process (119–120). The Tatoray process can be applied to the production of xylenes and benzene from feedstock that consists typically of toluene [108-88-3], either alone or blended with C_{9} aromatics (particularly trimethylbenzenes and ethyl-toluenes) and C_{10} alkyl aromatics. The main reactions are transalkylation (or disproportionation) of toluene to xylene and benzene or of toluene and trimethylbenzenes to xylenes in the vapor phase over a highly selective fixed-bed catalyst in a hydrogen atmosphere at 350–500°C and 1–5 MPa. Ethyl groups are dealkylated or transalkylated (121,122). The TAC9 process, developed by Toray and licensed by UOP, converts $C_{9}+$ aromatics primarily to xylenes, at conditions that are similar to those of the Tatoray process.

The PX-Plus, MSTDP and MTPX processes produce *p*-xylene [106-42-3] (*p*-dimethylbenzene) over a pretreated modified MFI catalyst at concentrations in the xylene products that range from 80 to 93% depending on catalyst pretreatment and operating conditions. A p-xylene concentration higher than an equilibrium concentration of 24% results from the much greater diffusivity of p-xylene in the selectivated MFI pore structure than that of ortho and meta isomers. Further isomerization of the p-xylene is minimized by the passivation of active sites on the exterior surface of the MFI. Toluene conversions of 25–32 wt% are achieved at 2–6 h^{-1} weight hourly space velocity, 400–420°C, 2.3–3.5 MPa and 1–4 H$_2$ to hydrocarbon molar feed ratio. The selective alkylation of toluene with methanol to produce p-xylene as a predominant isomer can be achieved over shape-selective catalysts (123–125). With a modified MFI zeolite catalyst, >99% p-xylene in xylene isomers can be produced at 550°C. This p-xylene concentration exceeds the equilibrium concentration of 23% (123). The primary by-product is water, making this technology economical only under very particular circumstances.

Polynuclear Aromatics. The alkylation of polynuclear aromatics with olefins and olefin-producing reagents is effected by acid catalysts. The alkylated products are more complicated than are those produced by the alkylation of benzene because polynuclear aromatics have more than one position for substitution. For example, the alkylation of naphthalene [91-20-3] with methanol over mordenite and Y-type zeolites at 400–450°C produces 1-methylnaphthalene [90-12-0] and 2-methylnaphthalene at a two to one ratio of ~1.8. The selectivity to 2-methylnaphthalene [91-57-6] is increased by applying a ZSM-5 catalyst to give a 2:1 ratio of ~8 (126).

2,6-Dimethylnaphthalene [581-42-0] (2,6-DMN) can be a precursor for 2,6-naphthalenedicarboxylic acid [1141-38-4], which is a starting material for high performance polyesters (polyethylene naphthalate or PEN) as well as polyamides.

2,6-DMN can be produced by alkylating naphthalene or 2-methylnaphthalene at 250–450°C over zeolite catalysts (126,127). However, no commercial technology by this synthetic route had been developed as of 1991, primarily because of low catalytic selectivity. A multistep synthetic route to 2,6-DMN starting with butadiene and o-xylene has been used on a commercial scale. However, the market growth for PEN has been very slow due to its relatively high cost.

4.4. Future Technology Trends. Over the years, improvements in aromatic alkylation technology have come in the form of both improved catalysts and improved processes. This trend is expected to continue into the future.

Catalysts. Nearly all of the industrially significant aromatic alkylation processes of the past have been carried out in the liquid phase with unsupported acid catalysts. For example, AlCl$_3$ and HF have been used commercially for at least one of the benzene alkylation processes to produce ethylbenzene (128,129), cumene (130), and detergent alkylates (92–95). Exceptions to this historical trend have been the use of a supported boron trifluoride for the production of ethylbenzene and of a solid phosphoric acid (SPA) catalyst for the production of cumene (72,131).

Since 1976, these forms of acids have become a significant environmental concern from both a physical handling and disposal perspective. This concern has fueled much development work toward solid acid catalysts, including zeolites, silica–aluminas, and clays (132,133).

A liquid-phase ethylbenzene process jointly licensed by ABB Lummus and UOP initially used a Y-type zeolite catalyst developed by Unocal. During the 1990s, significant advances took place with numerous new zeoletic and other solid acid catalysts that were introduced for the production of ethylbenzene (43,63), cumene (75–77) and detergent alkylate (99). Because of their initial commercial success and the industry's growing awareness of environmental issues, solid acid catalysts are expected to ultimately replace liquid acid catalysts (134).

Process. As solid acid catalysts have replaced liquid acid catalysts, they have typically been placed in conventional fixed-bed reactors. An extension of fixed-bed reactor technology is the concept of catalytic distillation being offered by CR&L (59). In catalytic distillation, the catalytic reaction and separation of products occurs in the same vessel. The concept has been applied commercially for the production of MTBE and is also being offered for the production of ethylbenzene and cumene.

A new alkylation process, the Alkymax process was introduced by UOP in 1990. This process was developed in response to proposed legislation requiring the reduction of benzene (a known carcinogen) in gasoline (135). Refinery propylene, typically from a fluid catalytic cracker, is used to alkylate the benzene in light reformate (136). In addition to lowering the benzene content, the alkylate formed has a high octane value and can typically boost the octane of the gasoline pool by 0.5 RON.

5. Other Alkylations

Alkylation of Phenol. The hydroxyl group activates the alkylation of the benzene ring because it is a strong electron-donating group; therefore, the alkylation of phenol [108-95-2] can be achieved with olefins and olefin-producing reagents under milder conditions than the alkylation of aromatic hydrocarbons. The alkylation of phenol with olefins and other alkylating reagents is discussed in other publications (137 and 138) (see also ALKYLPHENOLS).

Alkylated phenol derivatives are used as raw materials for the production of resins, novolaks (alcohol-soluble resins of the phenol–formaldehyde type), herbicides, insecticides, antioxidants, and other chemicals. The synthesis of 2,6-xylenol [576-26-1] has become commercially important since PPO resin, poly (2,6-dimethyl phenylene oxide), an engineering thermoplastic, was developed (139,140). The demand for *o*-cresol and 2,6-xylenol (2,6-dimethylphenol) increased further in the 1980s along with the growing use of epoxy cresol novolak (ECN) in the electronics industries and poly(phenylene ether) resin in the automobile industries. The ECN is derived from *o*-cresol, and poly(phenylene ether) resin is derived from 2,6-xylenol.

o-cresol   2,6-xylenol

Cresol and xylenol can be prepared by the methylation of phenol with methanol over both acid and base catalysts. It is postulated that phenol methylation on acid catalysts proceeds through the initial formation of anisole (methoxybenzene [100-66-3]) followed by intramolecular rearrangement of the methyl group to form o-cresol. The methyl group in the ortho position can further undergo isomerization to form meta and para isomers. The formation of m- and p-cresols is accelerated at higher temperatures and with stronger acid catalysts (141,142). Xylenol isomers are produced by the consecutive methylation of cresol isomers. The methylation of phenol is more selective to o-cresol with base catalysts than with acid catalysts. On base catalysts, phenol adsorbs dissociatively on a pair of basic and acidic sites to form a protonic site and an adsorbed phenolate species, respectively. This proton site activates methanol to produce a carbonium ion, which reacts with the benzene ring of an adjacently adsorbed phenolate species at the ortho position (138).

The commercial process for the selective synthesis of o-cresol [95-48-7] and 2,6-xylenol by the alkylation of phenol with methanol in a fixed-bed reactor was developed by the General Electric Company. The high selectivity is effected by using a magnesium oxide catalyst at high temperatures (475–600°C). The alkylation occurs at the positions ortho to the hydroxyl group. Because the catalyst does not have isomerization activity, the products are o-cresol, 2,6-xylenol, and minor amounts of 2,4,6-trimethylphenol [527-60-6] and anisole [100-66-3] (143). Similar commercial processes using a fixed-bed reactor system have been commercialized by BASF, Groda, and Mitsubishi Gas Chemicals (118). A new phenol methylation process technology for the production of o-cresol and 2,6-xylenol has recently been developed by Asashi Chemical Industry (144). This new process uses a new catalyst and a fluidized-bed reactor. The catalyst is a silica-supported iron–vanadium mixed oxide modified by metal promoters. This catalyst is active at 300–350°C and is selective for o-cresol and 2,6-xylenol formation. The catalyst-bed temperature can be maintained uniformly at relatively low temperatures because of the high heat-transfer efficiency of the fluidized-bed reactor system; therefore, side reactions caused at high temperatures are eliminated and the catalyst life becomes long. Because no significant amount of meta and para isomers is produced in this process, high purity o-cresol (99.95%) and 2,6-xylenol (99.85%) can be produced.

Alkylation of Aromatic Amines and Pyridines. Commercially important aromatic amines are aniline [62-53-3], toluidine [26915-12-8], phenylenediamines [25265-76-3], and toluenediamines [25376-45-8] (see AMINES, AROMATIC). The ortho alkylation of these aromatic amines with olefins, alcohols, and dienes to produce more valuable derivatives can be achieved with solid acid catalysts. For example, 5-tert-butyl-2,4-toluenediamine ($C_{11}H_{18}N_2$), which is used for performance polymer applications, is produced at 85% selectivity and 84%

2,4-toluenediamine [95-80-7] (2,4-TDA) conversion by alkylation of 2,4-TDA with isobutylene over a Y-type zeolite catalyst at 180–200°C (145).

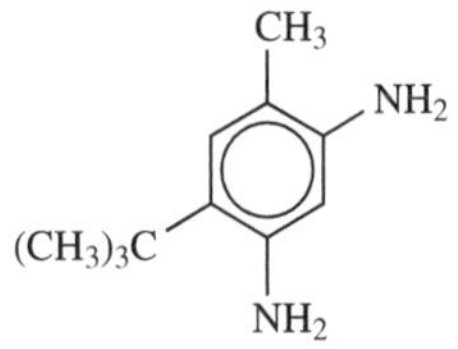

The alkylation of pyridine [110-86-1] takes place through nucleophilic or homolytic substitution because the π-electron-deficient pyridine nucleus does not allow electrophilic substitution, eg, Friedel-Crafts alkylation. Nucleophilic substitution, which occurs with alkali or alkaline metal compounds, and free-radical processes are not attractive for commercial applications. Commercially, catalytic alkylation processes via homolytic substitution of pyridine rings are important. The catalysts effective for this reaction include boron phosphate, alumina, silica–alumina, and Raney nickel (146).

6. Health and Safety Factors

Generally, specific health and safety factors relating to feedstock and products must be addressed for each particular industrial alkylation process. The reader is referred to the sections of the *Encyclopedia* that describe specific chemical compounds. In addition, the properties of the catalyst systems employed in alkylation must be considered. The hazardous properties of the homogeneous acid catalysts, HF, H_2SO_4, and $AlCl_3$, are documented in other sections of this encyclopedia (see ALUMINUM COMPOUNDS; FLUORINE COMPOUNDS, INORGANIC; SULFURIC ACID AND SULFUR TRIOXIDE). In industrial applications, specialized procedures are required to ensure the safe handling of these materials. Replacing these materials with solid acid catalysts will become more important in the future. The solid acid catalysts themselves present a disposal problem that favors the development of regenerable catalysts or the implementation of recycling procedures.

BIBLIOGRAPHY

"Alkylation" in *ECT* 1st ed., Vol. 1, pp. 532–550, by R. Norris Shreve, Purdue University; in *ECT* 2nd ed., Vol. 1, pp. 822–901, by R. H. Rosenwald, Universal Oil Products Company; in *ECT* 3rd ed., Vol. 2, pp. 50–72, by R. H. Rosenwald, Universal Oil Products Company; in *ECT* 4th ed., Vol. 2, pp. 85–112, by H. U. Hammershaimb, T. Imai, G. J. Thompson, B. V. Vora, UOP LLC; "Alkylation" in *ECT* (online), posting date: December 4, 2000, by H. U. Hammershaimb, T. Imai, G. J. Thompson, B. V. Vora, UOP LLC.

CITED PUBLICATIONS

1. G. Stefanidakis and J. E. Gwyn, in J. J. McKetta and W. A. Cunningham, eds., *Encyclopedia of Chemical Processing and Design*, Vol. 2, Marcel Dekker, New York, 1977, p. 357.
2. W. Keim and M. Roper, in W. Gerhartz, ed., *Ullmann's* Encyclopedia of Industrial Chemistry, Vol. A1, VCH Verlagsgesellschaft, Weinheim, 1985, p. 185.
3. M. Graham, P. Pryor, and M. Sarna, NPRA Meeting March, 2000, AM-00-53.
4. Stratco Web Site http://www.stratco.com/alkylation.html
5. F. E. Frey and H. J. Hepp, *Ind. Eng. Chem.* **28**, 1439 (1936).
6. L. Schmerling, *J. Am. Chem. Soc.* **68**, 275 (1946).
7. F. G. Ciapetta, *Ind. Eng. Chem.* **37**, 1210 (1945).
8. Schmerling, *Ind. Eng. Chem.* **45**, 1447 (1953).
9. J. E. Hoffman and A. J. Schriescheim, *J. Am. Chem. Soc.* **84**, 953 (1962).
10. T. Hutson and G. E. Hays, in L. F. Albright and A. R. Goldsby, eds., *Industrial and Laboratory Alkylations* (ACS Symposium Series) American Chemical Society, Washington, D.C., 1977, pp. 27–56.
11. L. F. Albright, in Ref. 8, pp. 128–146.
12. L. F. Albright, M. A. Spalding, J. Faunce, and R. E. Eckert, *Ind. Eng. Chem. Res.* **27** (3), 391 (1988).
13. L. F. Albright, *Oil Gas J.* **88** (46), 79 (1990).
14. R. B. Thompson and J. A. Chenicek, *Ind. Eng. Chem.* **40**, 1265 (1948).
15. R. A. Innes, in Ref. 8, p. 57.
16. F. W. Kirsh, J. D. Potts, and D. S. Barmby *J. Catal.* **27**, 142 (1972).
17. U.S. Pat. 3,893,942 (July 8, 1975), C. L. Yang (to Union Carbide Corp.).
18. T. J. Huang and S. Yurchak, in Ref. 8, p. 75.
19. L. E. Chapin, G. C. Liolios, and T. M. Robertson, *Hydrocarbon Process* **64** (9), 67 (1985).
20. *Hydrocarbon Process* **67** (9), 84 (1988).
21. J. Branzaru, Introduction to Sulfuric Acid Alkylation Unit Process Design, Stratco Technology Conference, November, 2001.
22. P. Pryor, R. Peterson, T. Godry, and Y. Lin, NPRA Meeting March, 2002, AM-02-50.
23. Ref. 18, p. 85.
24. H. U. Hammershaimb and B. R. Shah, *Hydrocarbon Process* **64** (6), 73 (1985).
25. L. Shoemaker, K. Hovis, K. Hoover, B. Randolph, and M. Pfile, NPRA Meeting March, 1997, AM-97-44.
26. P. Pryor, Alkylation Current Events, Stateo Technology Conference, November, 2001.
27. J. Sheckler, H. Hammershaimb, L. Ross, and K. Comey, NPRA Meeting March, 1994, AM-94-14.
28. L. Shoemaker, B. Randolph, and K. Hovis, NPRA Meeting March, 1998, AM-98-35.
29. T. Hutson and R. S. Logan, *Hydrocarbon Process* **54** (9), 107 (1975).
30. Ref. 21, Tables 3 and 4.
31. D. H. Vahlsing, *Hydrocarbon Process* **56** (9), 125 (1977).
32. J. Peterson, D. Graves, K. Kkranz, and D. Buckler, NPRA Meeting March 1999, AM-99-28.
33. D. N. Blewitt, J. F. Yohn, R. P. Koopman, and T. C. Brown, International Conference on Vapor Cloud Modeling, American Institute of Chemical Engineers, New York, 1987, pp. 1–38.
34. D. K. Whittle, D. K. Lorenzo, and J. Q. Kirkham, *Oil Gas J.* **87** (28), 96 (1989).
35. R. L. Van Zele and R. Diener, *Hydrocarbon Process* **69** (6), 92 (1990); **69** (7), 77 (1990).

36. B. V. Vora and C. P. Luebke, *Oil Gas J.* **86** (49), 40 (1988).
37. V. D'Amico, NPRA Meeting March, 2002, AM-02-19.
38. C. Roeseler, NPRA Meeting March, 2002, AM-02-17.
39. K. Tanabe, *Solid Acids and Bases*, Kodansha, Tokyo, 1976, p. 1.
40. J. C. Cheng and co-workers *Sci. Technol. Catalysis* **6** 52 (1998).
41. R. J. Schmidt, A. Zarchy, and G. Peterson, Paper 124b AIChE Spring Meeting, March 10–14, 2002.
42. W. W. Kaeding, G. C. Barile, and M. M. Wu, *Catal. Rev. Sci. Eng.* **26** (3–4), 597 (1984).
43. N. L. Allinger and co-workers, *Organic Chemistry*, 2nd ed., Worth Publishers, New York,1976, p. 339.
44. J. March, *Advanced Organic Chemistry*, 2nd ed., McGraw-Hill, Inc., New York,1977, p. 487.
45. H. Pines, in Ref. 8, p. 205.
46. H. Pines, J. A. Vesely, and V. N. Ipatieff, *J. Am. Chem. Soc.* **77**, 554 (1955).
47. Y. N. Sidorenko, P. N. Galich, V. S. Gutyrya, V. G. Ilin, and I. E. Neimark, *Dokl. Akad. Nauk. SSSR* **173**, 132 (1967); T. Yashima, H. Ahmad, K. Yamazaki, M. Katsuta, and N. Hara, *J. Catal.* **26**, 303 (1972); J. J. Freeman and M. L. Unland, *J. Catal.* **54**, 183 (1978); S. T. King and J. M. Garces, *J. Catal.* **104**, 59 (1987); J. Engelhardt, J. Szanyi, and J. Valyon, *J. Catal.* **107**, 296 (1987).
48. K. Tanabe, O. Takahashi, and H. Hattori, *React. Kinet. Catal. Lett.* **7**, 347 (1977).
49. H. Itoh, A. Miyamoto, and Y. Murakami, *J. Catal.* **64**, 284 (1980); H. Itoh, T. Hattori, K. Suzuki, and Y. Murakami, *J. Catal.* **79**, 21 (1983).
50. R. R. Cody, V. A. Welch, S. Ram, and J. Singh, Ref. 2, Vol. A10, p. 35.
51. *Chem. Mark. Rep.* **236** (8), 50 (1989).
52. M. Fox, "Ethylbenzene," *Chemical Economics Handbook*, SRI International, Menlo Park, Calif., June 1988, p. 645.3000A.
53. T. Wett, *Oil Gas J.* **79** (29), 76 (1981).
54. A. C. MacFarlane, *Oil Gas J.* **74** (6), 99 (1976).
55. *Chem. Week* **116** (23), 29 (1975).
56. *Chem. Week* **143** (22), 9 (1988).
57. *Oil Gas J.* **86** (49), 27 (1988).
58. *Chem. Week* **143** (21), 38 (1988).
59. J. D. Shoemaker and E. M. Jones, Jr., "Cumene by Catalytic Distillation," 1987 NPRA Annual Meeting, March 29–31, 1987.
60. E. M. Jones, Jr., and J. Mawer, "Cumene by Catalytic Distillation," AICHE Meeting, New Orleans, La., April 6–10, 1986.
61. W. P. Stadig, *Chem. Process.* **50** (2), 27 (1987).
62. J. D. Shoemaker and E. M. Jones, Jr., *Hydrocarbon Process* **66** (6), 57 (1987).
63. U.S. Pat. 3,308,069 (March 6, 1967) R. L. Wadlinger and G. T. Kerr (to Mobil Oil Corp.).
64. C. Perego, and co-workers *Microporous Mater.* **6**, 395 (1996).
65. F. G. Dwyer, P. J. Lewis, and F. H. Schneider, "The Mobil-Badger Ethylbenzene Process," Mexico City, Mexies, November 30, 1975.
66. R. Cody and S. Ram, "Use of Dilute Ethylbenzene Streams for Ethylbenzene Production," AICHE, Session 65, Paper No. 65c, New Orleans, La., Mar. 9, 1988.
67. U.S. Pat. 4,673,767 (June 16, 1987), T. S. Nimry and R. E. DeSimone (to Amoco Corp.); U.S. Pat. 4,812,536 (Mar. 14, 1989), R. E. DeSimone and L. B. Lane (to Amoco Corp.).
68. U.S. Pat. 4,575,573 (Mar. 11, 1986), R. M. Dessan and G. T. Kerr (to Mobil Oil Corp.).
69. U.S. Pat. 4,670,617 (June 2 1987), R. E. DeSimone and M. S. Haddad (to Amoco Corp.).

70. Z. Sedaglat-Pour, *Cumene*, CEH Data Summary, SRI International, Menlo Park, Calif., March 1989, p. 638.5000A.
71. *Chem. Mark. Rep.* **232** (10), 54 (1987).
72. R. C. Schulz, G. J. Thompson, and H. C. Ward, "Cumene Technology Improvements," AICHE Summer National Meeting, Denver, Colo., Aug. 21–24, 1988.
73. R. C. Canfield and T. L. Unruh, *Chem. Eng.* **90** (6), 32 (1983).
74. R. C. Canfield, R. C. Cox, and D. M. McCarthy, "Monsanto/Lummus Crest Process Produces Lowest Cost Cumene," AICHE 1988 Spring Meeting, New Orleans, La., April 6–10, 1986.
75. U.S. Pat. 4,459,426 (July 10, 1984), T. V. Inwood, C. G. Wight, and J. W. Ward (to Union Oil).
76. G. Bellussi, and co-workers *J. Catal* **157**, 227 (1995).
77. World Pat. 01/83408 (Nov. 2001), A. B. Dandekar, and co-workers (to Mobil Oil Corp.).
78. U.S. Pat. 5,600,048 (February 4, 1997) J. C. Cheng, Smith, C. M. Smith, C. R. Venkat, and D. E. Walsh (to Mobil Oil Corp.).
79. W. W. Kaeding, L. B. Young, and A. G. Prapas, *CHEMTECH* **12** (9), 556 (1982).
80. *Chem. Week* **130** (7), 42 (1982).
81. W. W. Kaeding, L. B. Young, and C. J. Chu, *J. Catal.* **89**, 267 (1984).
82. D. Fraenkel and M. Levy, *J. Catal.* **118**, 10 (1989).
83. R. D. Swisher, *Surfactant Biodegration*, 2nd ed., Marcel Dekker, New York, 1970.
84. A. S. Davidsohn and B. M. Milwidsky, *Synthetic Detergents*, 7th ed., John Wiley & Sons, Inc., New York, 1987.
85. L. Huber, *Soap Cosmet. Chem. Spec.* **65** (5), 44 (1989).
86. T. Imai, J. A. Kocal, and B. V. Vora, *Sci. Technol. Catalysis* 339 (1994).
87. W. M. Linfield, ed., *Anionic Surfactants*, Vol. 7, Part 1, Marcel Dekker, New York, 1976, p. 258.
88. Ref. 2, p. 195.
89. ARCO Technology, Inc., *Hydrocarbon Process* **64** (11), 127 (1985).
90. Eteco Impianti SPA, *Hydrocarbon Process* **60** (11), 175 (1981).
91. L. Cavalli, A. Landone, and T. Pellizzan, *Linear Alkylation for Detergency— Characterization of Secondary Components*, XIX Jornadas Del Comite Espanola De la Detergeneia, Barcelona, Spain, 1988, pp. 41–52.
92. B. V. Vora, P. R. Pujado, M. A. Allawala, and T. R. Fritsch, "Production of Biodegradable Detergent Intermediates," Second World Surfactants Congress, Paris, France, May 24–27, 1988.
93. H. S. Bloch, "A New Route to Linear Alkylbenzenes," Symposium on *n*-Paraffins, Institute of Chemical Engineers (NW Branch).
94. *Eur. Chem. News*, Manchester, England (Nov. 1966).
95. U.S. Pat. 3,356,757 (1967), J. F. Roth and A. R. Schaefer (to Monsanto).
96. U.S. Pat. 4,523,048 (June 11, 1985), B. V. Vora (to UOP).
97. U.S. Pat. 4,761,509 (Aug. 2, 1988), b B. V. Vora and D. L. Ellig (to UOP).
98. R. C. Berg and B. V. Vora, in Ref. 1, Vol. 15, pp. 266–284.
99. J. A. Kocal, B. V. Vor and T. Imai, *Appl. Catal. A:Gen.* **221**, 295 (2001).
100. B. V. Vora, P. R. Pujado, T. Imai, and T. R. Fritsch, *Recent Advances in the Production of Detergent Olefins and Linear Alkylbenzenes*, Society of Chemical Industry, University of Cambridge, England, March 26–28, 1990.
101. *Eur. Chem. News* **54** (1428), 26 (1990).
102. P. B. Venuto, A. L. Hamilton, P. S. Landis, and J. J. Wise, *J. Catal.* **5** 272 (1996).
103. Y. Cao, R. Kessas, C. Naccache, and Y. Ben Taarit, *Appl. Catal. A:Gen.* **184**, 231 (1999).
104. P. M. Price, J. H. Clark, K. Martin, T. W. Macquarrie and T. W. Bastock, *Org. Process Res. Dev.* **2** 221 (1998).

105. J. L. Brena Tejero and A. Moreno Danvilla, Eur. Pat. 353,813 (1990).

106. Ref. 71, pp. 25–26.

107. L. Cavalli, R. Clerici, P. Radici, and L. Valtorta, *Tenside Surf. Det.* **36** 254 (1999).

108. P. R. Pujado, *Handbook of Petroleum Refining Processes*, R. A. Meyers, ed., McGraw-Hill, New York, 1997, 1.53–1.66.

109. A. Negiz, T. Stoodt, C. H. Tan, and J Noe, Paper 123c, AIChE Spring Meeting, March 10–14, 2002.

110. U.S. Pat. 5,847,256 (Dec. 8, 1998), Ryoji Ichioka, Shinobu Yamakawa, Hirohito Okino (to Toray Industries Inc.).

111. S. Ramsey, Paper 122b, AIChE Spring Meeting, March 10–14, 2002.

112. J. A. Johnson, C. M. Roeseler, and T. J. Stoodt, *PX-Plus Process: Innovation for Para Xylene*, DeWitt Petrochemical Review, Houston, Tx, March 18–20, 1997.

113. D. Stern, Paper 122e, AIChE Spring Meeting, March 10–14, 2002.

114. J. Jeanneret in R. A. Myers, ed., *Petroleum Refining Processes*, McGraw-Hill, New York, 1996, pp. 2.45–2.53.

115. Kirkpatrick Awards, *Chem. Eng.*, **106**, 12, 96 (1999).

116. J. R. Mowry, in R. A. Meyers, ed., *Petroleum Refining Processes*, McGraw-Hill, New York, 1986, pp. 5–67.

117. J. Jeanneret, in R. A. Myers, ed., *Petroleum Refining Processes*, McGraw-Hill, New York, 1996, pp. 2.37–2.44.

118. T. A. Ebner, K. M. Oneil, and P. J. Silady, Paper 123b, AIChE Spring Meeting, March 10–14, 2002.

119. J. R. Green, "The Mobil High Temperature Xylene Isomerization (MHTI) Process," 1988 Petrochemical Review, DeWitt Company, Houston, Tex., March 23–25, 1988.

120. G. D. Mohr, Paper 123e, AIChE Spring Meeting, March 10–14, 2002.

121. M. Sato and M. Kanaoka, *Kagaku Kogyo* **37** (11), 1075 (1973).

122. Ref. 81, pp. 5–61.

123. L. B. Young, S. A. Butter, and W. W. Kaeding, *J. Catal.* **76**, 418 (1982).

124. U.S. Pat. 4,283,306 (Aug. 11, 1981), F. E. Herkes (to E. I. du Pont de Nemours & Co., Inc.).

125. U.S. Pat. 4,444,989 (Apr. 24, 1984), F. E. Herkes (to E. I. du Pont de Nemours & Co., Inc.).

126. D. Fraenkel, M. Cherniavsky, B. Ittah, and M. Levy, *J. Catal.* **101**, 273 (1986).

127. U.S. Pat. 4,795,847 (Jan. 3, 1989), J. Weikamp, M. Neuber, W. Holtmann, and H. Spengler (to Rutgerswerke Aktiengesellschaft).

128. A. C. MacFarlane, in Ref. 8, p. 371.

129. R. C. Canfield, R. P. Cox, and D. M. McCarthy, *Chem. Eng. Prog.* **36** (Aug. 1986).

130. Ref. 87, pp. 1–29.

131. P. J. Lewis and F. G. Dwyer, *Oil Gas J.* **75** (40), 55 (1977).

132. U.S. Pat. 4,185,040 (Jan. 22, 1980), J. W. Ward, and T. V. Inwood (to Union Oil); F. Figueras, *Catal. Rev. Sci. Eng.* **30**, 457 (1988).

133. W. W. Kaeding and R. E. Holland, *J. Catal.* **109**, 212 (1988).

134. M. Dewey, *Oil Daily* (Dec. 1989); G. Parkinson, *Chem. Eng.* **97** (1), 30 (1990).

135. B. M. Wood, M. E. Reno, and G. J. Thompson, "Alkylate Aromatics in Gasoline via the UOP Alkylmax Process," UOP Technology Conference, April 1990.

136. O. N. Tsvetkow, K. D. Korenev, N. M. Karavaev, and S. A. Dmitriev, *Int. Chem. Eng.* **7** (1), 104 (1967).

137. T. Kotanigawa, *Sekiyu Gakkai Shi* **17** (4), 286 (1974).

138. A. S. Hay, *J. Polym. Sci.* **58**, 581 (1962).

139. A. S. Hay, G. F. Endres, and J. W. Eustance, *J. Am. Chem. Soc.* **81**, 6335 (1959).

140. T. Nishizaki, H. Hattori, and K. Tanabe, *Shokubai* **14**, 138 (1972).

141. M. Inoue and S. Enomoto, *Chem. Pharm. Bull.* **29**, 232 (1972).

142. U.S. Pat. 3,446,856 (May 27, 1969), S. B. Hamilton (to General Electric Comp.).
143. T. Dozono, *Petrotech* **11** (9), 776 (1988).
144. T. Katsumata and T. Dozono, *AICHE J.* **83** (255), 86 (1987).
145. W. F. Burgoyne, D. D. Dixon, and J. P. Casey, *CHEMTECH* **19** 690 (1989).
146. C. V. Digiovanna, P. J. Cislak, and G. N. Cislak, in Ref. 8, p. 397.

Bipin V. Vora
Joseph A. Kocal
Paul T. Barger
Robert J. Schmidt
James A. Johnson
UOP LLC

ALKYLPHENOLS

1. Introduction

Alkylphenols of greatest commercial importance have alkyl groups ranging in size from one to twelve carbons. The direct use of alkylphenols is limited to a few minor applications such as epoxy-curing catalysts and biocides. The vast majority of alkylphenols are used to synthesize derivatives which have applications ranging from surfactants to pharmaceuticals. The four principal markets are nonionic surfactants, phenolic resins, polymer additives, and agrochemicals.

Nonionic surfactants and phenolic resins based on alkylphenols are mature markets and only moderate growth in these derivatives is expected. Concerns over the biodegradability and toxicity of these alkylphenol derivatives to aquatic species may limit their use in the future. The use of alkylphenols in the production of both polymer additives and monomers for engineering plastics is expected to show above average growth as plastics continue to replace traditional building materials.

Alkylphenols containing 3–12-carbon alkyl groups are produced from the corresponding alkenes under acid catalysis. Alkylphenols containing the methyl group were traditionally extracted from coal tar. Today they are produced by the alkylation of phenol with methanol.

2. Nomenclature

An alkylphenol is a phenol derivative wherein one or more of the ring hydrogens has been replaced by an alkyl group(s). Phenol is a heading parent in the CAS indexing system. Appropriate names of alkylphenols for abstract citations can be derived by using the appropriate aids (1). The names generated in this manner are unambiguous and refer to a specific compound, but are lengthy and cumbersome to use. Common names are used on a daily basis and are especially prevalent for alkylphenols that have gained commercial importance.

For monosubstituted alkylphenols, the position of the alkyl radical relative to the hydroxyl function is designated either with a numerical locant or ortho, meta, or para. The alkyl side chain typically retains a trivial name. Thus 4-(1,1,3,3-tetramethylbutyl)phenol, 4-*tert*-octylphenol, and *para-tert*-octylphenol (PTOP) all refer to structure (**1**).

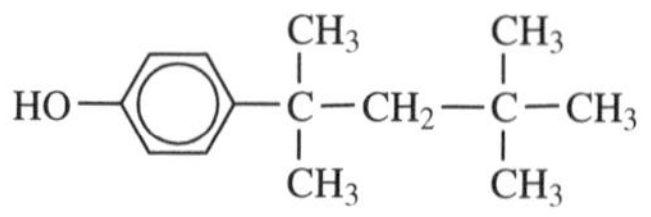

$$HO-\text{C}_6\text{H}_4-\underset{\underset{\displaystyle CH_3}{|}}{\overset{\overset{\displaystyle CH_3}{|}}{C}}-CH_2-\underset{\underset{\displaystyle CH_3}{|}}{\overset{\overset{\displaystyle CH_3}{|}}{C}}-CH_3$$

(**1**)

Dialkylphenols employ locants to designate the position of the alkyl groups on the ring. Thus, 2,6-bis(1,1-dimethylethyl)phenol, 2,6-di-*tert*-butylphenol, and 2,6-DTBP each refer to structure (**2**).

Other common names are cresol and xylenol for methyl- and dimethylphenols respectively, eg, *o*-cresol is 2-methylphenol, and 2,5-xylenol is 2,5-dimethylphenol.

For phenols with three or more alkyl substituents, trade names, abbreviations, and associative names predominate, eg, BHT and 2,6-di-*tert*-butyl-4-methylphenol refer to structure (**3**) and mesitol and 2,4,6-trimethylphenol refer to structure (**4**).

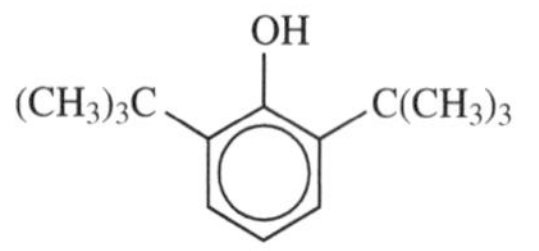

(**2**) 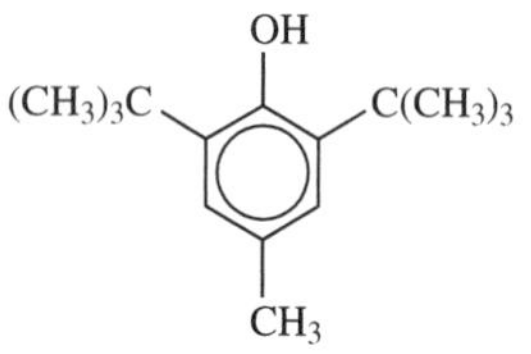   (**3**) 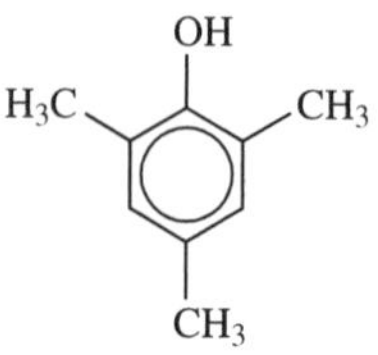   (**4**)

3. Physical Properties

Of course, the physical properties of alkylphenols are comparable to phenol. The properties are strongly influenced by the type of alkyl substituent and its position on the ring. Alkylphenols, like phenol, are typically solids at 25°C. Their form is affected by the size and configuration of the alkyl group, its position on the ring, and purity. They appear colorless, or white, to a pale yellow when pure (Table 1).

Para-alkylphenols have higher melting points and boiling points than the corresponding ortho-isomers. The melting points of para-alkylphenols go through a maximum for *tert*-butyl and then decrease. An alkene stream consisting of a mixture of isomers produces an alkylphenol that has a depressed melting point. As the carbon chain of the alkyl group surpasses 20, the resulting phenols take on a waxy form. Alkylphenols, especially when di- and trisubstituted, tend to supercool. Alkylphenols show the same sensitivity to oxidation that phenol

Table 1. **Commercially Important Alkylphenols**

Name	CAS Registry Number	Molecular formula	Molecular weight	Physical form at 25°C	Boiling point, °C[a]	Freezing point, °C	Density[b], g/mL	Typical Assay	Flash point, °C	Molten color APHA
4-*tert*-amylphenol	[80-46-6]	$C_{11}H_{16}O$	164.0	solid	249	90.0	0.915^{107}	99	121	200
4-*tert*-butylphenol	[98-54-4]	$C_{10}H_{14}O$	150.2	solid	237	97.5	0.890^{107}	98–99	117	100
2-*sec*-butylphenol	[89-72-5]	$C_{10}H_{14}O$	150.2	liquid	224	20.0	0.938^{43}	98	>93	100
4-cumylphenol	[599-64-4]	$C_{15}H_{16}O$	212.0	solid	335	70.0	1.029^{93}	99	188	100
4-dodecylphenol	[27193-86-8]	$C_{18}H_{30}O$	262.0	liquid	334		0.914^{20}	89–95	>100	500
4-nonylphenol	[84852-15-3][c]	$C_{15}H_{24}O$	220.3	liquid	310		0.933^{43}	90–95	146	100
4-*tert*-octylphenol	[140-66-9]	$C_{14}H_{22}O$	220.3	solid	290	81.0	0.940^{25}	90–98	132	200
2,4-di-*tert*-amylphenol	[25231-47-4]	$C_{16}H_{26}O$	234.4	liquid	275	23.0	0.900^{49}	99	104	100
2,4-di-*tert*-butylphenol	[96-76-4]	$C_{14}H_{22}O$	206.3	solid	263	52.0	0.867^{82}	99	115	100
2,6-di-*tert*-butylphenol	[128-39-2]	$C_{14}H_{22}O$	206.3	solid	253	36.0	0.898^{43}	99	>99	100
di-*sec*-butylphenol	[31291-60-8]	$C_{14}H_{22}O$	206.3	liquid			0.902^{66}	90	127	500
2,4-dicumylphenol	[2772-45-4]	$C_{24}H_{26}O$	330.0	solid		65.0	1.030^{66}	99	462	100
2-methylphenol	[95-48-7]	C_7H_8O	108.1	solid	191	30.0	$1.049^{15.5}$	99	81	25
3-methylphenol	[108-39-4]	C_7H_8O	108.1	liquid	202	10.0	$1.042^{15.5}$	97	86	
4-methylphenol	[106-44-5]	C_7H_8O	108.1	solid	202	34.0	1.022^{25}	99	86	25
2,6-dimethylphenol	[576-26-1]	$C_8H_{10}O$	122.1	solid	203	48.0	1.020^{25}	99	88	

[a] At 101.3 kPa = 1 atm.
[b] At the temperature indicated by the superscript, °C.
[c] Mixture, branched chains.

does. The presence of trace amounts of metals or alkaline impurities accelerates oxidation. Oxidation products cause discoloration.

The solubility of alkylphenols in water falls off precipitously as the number of carbons attached to the ring increases. They are generally soluble in common organic solvents: acetone, alcohols, hydrocarbons, toluene. Solubility in alcohols or heptane follows the generalization that "like dissolves like." The more polar the alkylphenol, the greater its solubility in alcohols, but not in aliphatic hydrocarbons; likewise with cresols and xylenols. The solubility of an alkylphenol in a hydrocarbon solvent increases as the number of carbon atoms in the alkyl chain increases. High purity para substituted phenols, C_3 through C_8, can be obtained by crystallization from heptane.

The aromatic ring of alkylphenols imparts an acidic character to the hydroxyl group; the pK_a of unhindered alkylphenols is 10–11 (2). Alkylphenols unsubstituted in the ortho position dissolve in aqueous caustic. As the carbon number of the alkyl chain increases, the solubility of the alkali phenolate salt in water decreases, but aqueous caustic extractions of alkylphenols from an organic solution can be accomplished at elevated temperatures. Bulky ortho substituents reduce the solubility of the alkali phenolate in water. The term cryptophenol has been used to describe this phenomenon. A 35% solution of potassium hydroxide in methanol (Claisen's alkali) dissolves such hindered phenols (3).

Alkyl groups in the ortho position affect the environment about the hydroxyl group; the larger the group the greater the effect. Intermolecular hydrogen bonding decreases with the introduction of a *tert*-butyl group ortho to the hydroxyl, reducing the atmospheric boiling point of the ortho isomer by about 20°C. Substitution of the second ortho position with a *tert*-butyl group effectively precludes any hydrogen bonding as shown by the infrared spectrum of 2,6-DTBP; a sharp absorbance is found at 2.75 μm, characteristic of an unassociated hydroxyl stretching (4). The impact of an ortho alkyl substituent on hydrogen bonding can be advantageously applied to the analysis and separation of alkylphenols. A mixture of ortho and para isomers is separated effectively using normal phase chromatography (flash or hplc).

There is a health benefit associated with hindering hydrogen bonding. Alkylphenols as a class are generally regarded as corrosive health hazards, but this corrosivity is eliminated when the hydroxyl group is flanked by bulky substituents in the ortho positions. In fact, hindered phenols as a class of compounds are utilized as antioxidants in plastics with FDA approval for indirect food contact.

4. Chemical Properties

Alkylphenols undergo a variety of chemical transformations, involving the hydroxyl group or the aromatic nucleus that convert them to value-added products.

4.1. The Hydroxyl Group. The unshared pairs of electrons on hydroxyl oxygens seek electron deficient centers. Alkylphenols tend to be less nucleophilic than aliphatic alcohols as a direct result of the attraction of the electron density by the aromatic nucleus. The reactivity of the hydroxyl group can be enhanced in spite of the attraction of the ring current by use of a basic catalyst which removes the acidic proton from the hydroxyl group leaving the more nucleophilic alkyl-phenoxide.

Esterification. Alkylphenols react with acid chlorides and acids to produce commercially important esters. Three equivalents of *p*-nonylphenol (**5**) react with phosphorus trichloride or tributyl phosphite to produce tris(4-nonylphenyl) phosphite (TNPP) (**6**).

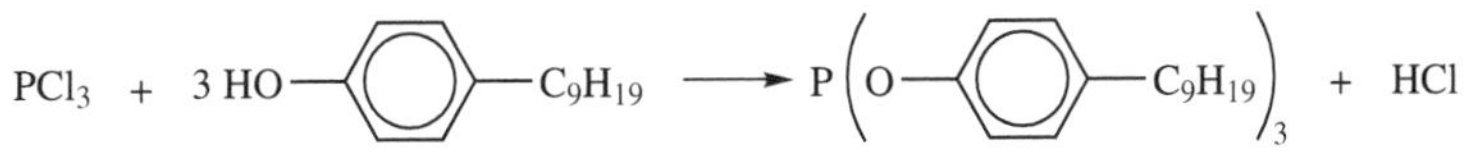

(5)   (6)

When the phenol reactant is 2,4-di-*tert*-butylphenol, a phosphite ester with greater hydrolytic stability is produced.

Alkylphenyl esters (**7**) of aromatic carboxylic acids can be made from appropriate acid chlorides or by transesterification of benzoate esters. The resulting alkylphenyl benzoates undergo a transformation called the Fries rearrangement which involves cleavage of the ester linkage followed by the migration of the benzoyl group to the ortho or para position of the phenol. The reaction can be catalyzed by metal halides in a typical Friedel-Crafts reaction or the process can be photochemical, ie, the photo-Fries. The resulting product is a hydroxybenzophenone (**8**). These benzophenones have found applicability as uv absorbers in thermoplastics.

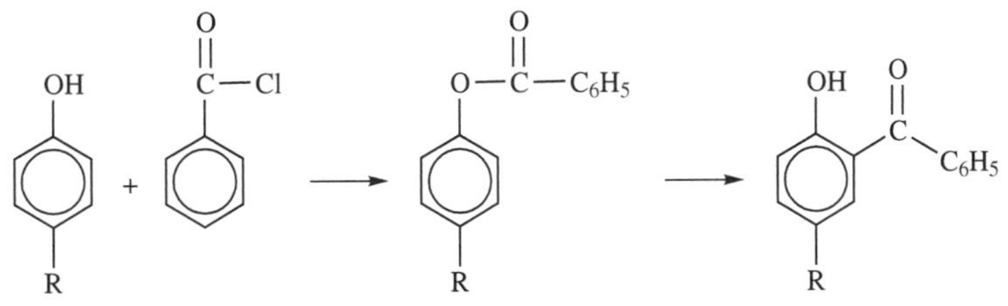

(7)   (8)

Alkylphenols have been substituted for phenol as chain terminators in polycarbonates. In this role, PTBP (**9**) competes with the diol monomer for reactive chlorocarbonate sites. The ratio of butylphenol to diol controls the molecular weight of the polymer.

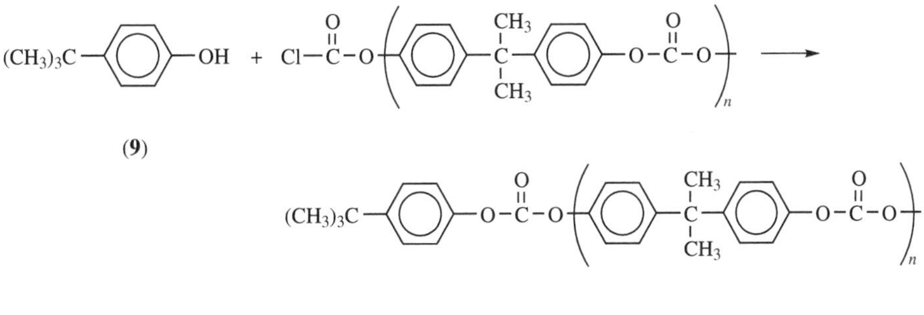

(9)

Etherification. Many of the monoalkylphenols and some of the dialkylphenols are converted into ethoxylates which find commercial application as nonionic surfactants (5). For example, *p*-nonylphenol reacts with ethylene oxide under mild basic conditions.

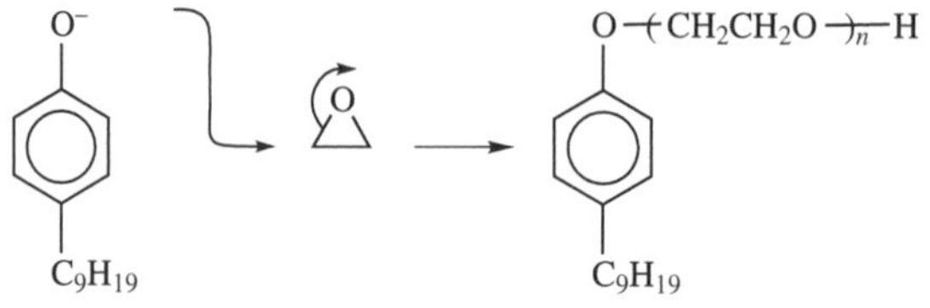

(10)

The number of ethylene oxide units added to the phenoxide depends on the application of the ethoxylate. This chemistry is closely related to the reaction between an alkylphenol and epichlorohydrin which is used in epoxy resins (qv).

4.2. Reactions Involving the Ring. The aromatic nucleus of alkylphenols can undergo a variety of aromatic electrophilic substitutions. Electron density from the hydroxyl group is fed into the ring. Besides activating the aromatic nucleus, the hydroxyl group controls the orientation of the incoming electrophile.

Alkylphenols undergo a carboxylation reaction known as the Kolbe Schmidt reaction. In the following example, the phenolate anion of *p*-nonylphenol (10) reacts with carbon dioxide under pressure. Neutralization generates a salicylic acid (11) (6).

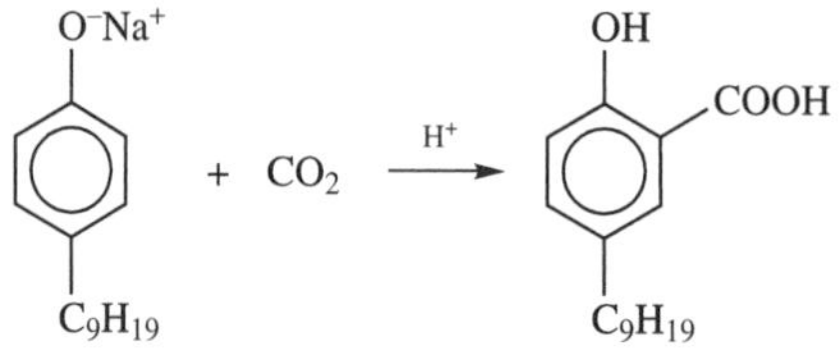

(10)   (11)

Reactions with Aldehydes and Ketones. An important use for alkylphenols is in phenol–formaldehyde resins. These resins are classified as resoles or novolaks (see PHENOLIC RESINS). Resoles are produced when one or more moles of formaldehyde react with one mole of phenol under basic catalysis. These resins are thermosets. Novolaks are thermoplastic resins formed when an excess of phenol reacts with formaldehyde under acidic conditions. The acid protonates formaldehyde to generate the alkylating electrophile (12).

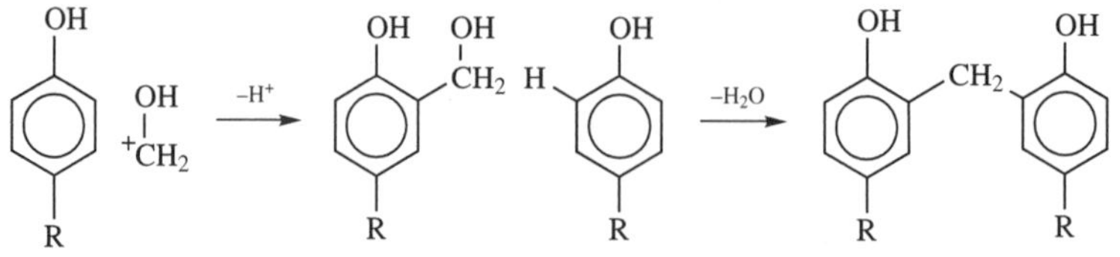

(12)   (13)   (14)

The intermediate methylol derivative (**13**) is unstable under the reaction conditions and generates an electrophile that undergoes another substitution reaction resulting in the methylene bridge between alkylphenol groups (**14**). This reaction is repeated to propagate the polymer chain. The choice of alkylphenol as well as the alkylphenol:formaldehyde ratio allows a good deal of control over the properties of the resulting resin.

A newer development in alkylphenol–formaldehyde resins has been the application of this condensation reaction to produce calixarenes, cyclic oligomers of methylene-bridged alkylphenols (**15**). The smallest cyclic oligomer that has been isolated to date contains four alkylphenol groups.

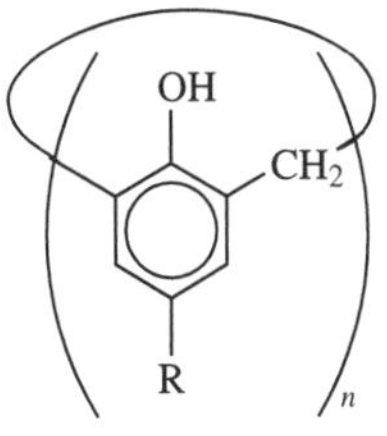

(**15**)

These molecules are significant in the field of research devoted to host–guest complexation. Synthetic routes to a number of calixarenes have been developed (7).

2,4-Dialkylphenols react with aldehydes similarly. A bisphenol (**16**) is formed by the condensation.

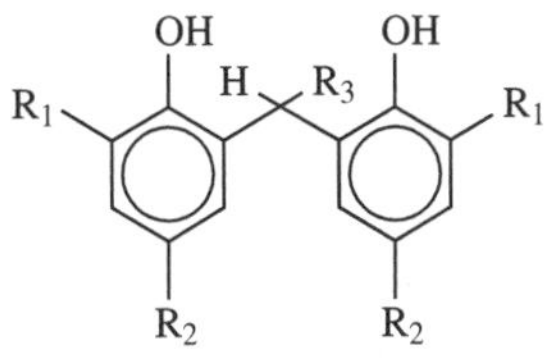

(**16**)

Commercial application of this type of reaction is used to produce 2,2′-methylene-bis(6-*tert*-butyl-4-methylphenol) ($R^1 = tert-$ butyl; $R^2 =$ methyl; $R^3 =$ H) and 2,2′-ethylidenebis(4,6-di-*tert*-butylphenol) $R^1 = R^2 = tert -$ butyl; $R^3 = CH_3$).

Quinone Methides.   The reaction between aldehydes and alkylphenols can also be base-catalyzed. Under mild conditions, 2,6-DTBP reacts with formaldehyde in the presence of a base to produce the methylol derivative (**17**) which reacts further with base to eliminate a molecule of water and form a reactive intermediate, the quinone methide (**18**). Quinone methides undergo a broad array of transformations by way of addition reactions. These molecules are conjugated homologues of vinyl ketones, but are more reactive because of the driving force associated with rearomatization after addition. An example of this type of addition is between the quinone methide and methanol to produce the substituted benzyl methyl ether (**19**).

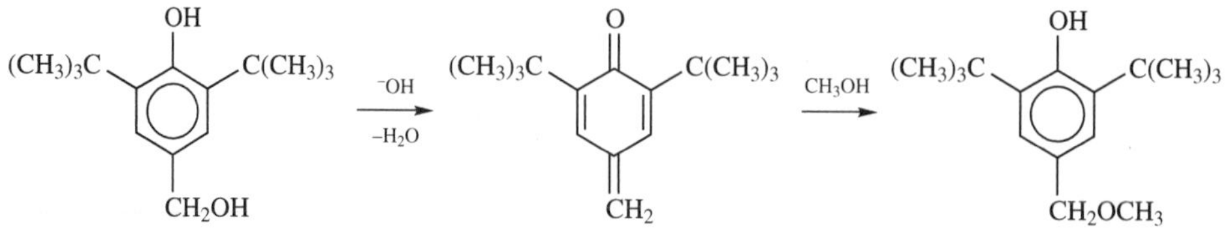

(17)   (18)   (19)

This addition is general, extending to nitrogen, oxygen, carbon, and sulfur nucleophiles. This reactivity of the quinone methide (**18**) is applied in the synthesis of a variety of stabilizers for plastics. The presence of two *tert*-butyl groups ortho to the hydroxyl group, is the structural feature responsible for the antioxidant activity that these molecules exhibit (see ANTIOXIDANTS).

4,4′-Methylenebis(2,6-di-*tert*-butylphenol) (**20**) (R = H) [118-82-1], the reaction product of two molecules of 2,6-DTBP with formaldehyde under basic conditions, is a bisphenolic antioxidant. The quinone methide in this case is generated *in situ*. The product results from the addition of 2,6-di-*tert*-butylphenolate to (**18**) (8).

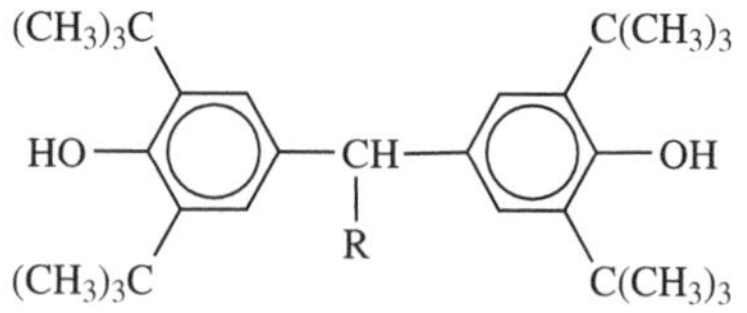

(20)

The versatility of this reaction is extended to a variety of aldehydes. The bisphenol derived from 2,6-di-*tert*-butylphenol and furfural, (**20**) where R = furfuryl (9), is also used as an antioxidant. The utility of the 3,5-di-*tert*-butyl-4-hydroxybenzyl moiety is evident in stabilizers of all types (16), and its effectiveness has spurred investigations of derivatives of hindered alkylphenols to achieve better stabilizing qualities. Another example is the Michael addition of 2,6-di-*tert*-butyl phenol to methyl acrylate. This reaction is carried out under basic conditions and yields methyl 3-(3,5-di-*tert*-butyl-4-hydroxyphenyl)propionate [6386-38-5] (**21**) (11).

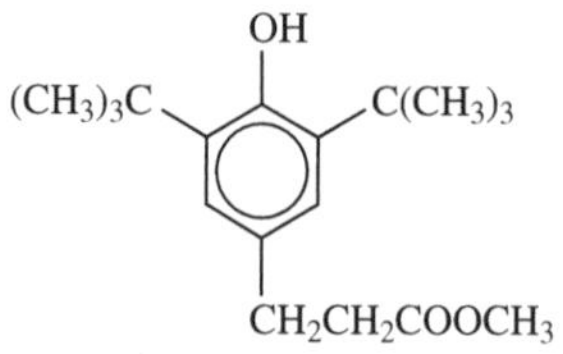

(21)

Transesterification reactions between the methyl propionate and various alcohols produce another family of stabilizers. Stearyl alcohol yields octadecyl 3-(3,5-di-*tert*-butyl-4-hydroxyphenyl)propionate (**22**) (12), pentaerythritol gives the tetrakis ester (**23**) (13), and trishydroxyethyl isocyanurate gives (**24**) (14).

R—OC$_{18}$H$_{37}$

(**22**)

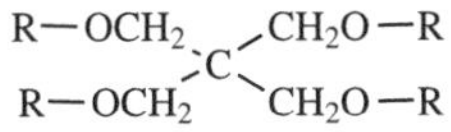

R—OCH$_2$ ⟍ ⟋ CH$_2$O—R
⟍C⟋
R—OCH$_2$ ⟋ ⟍ CH$_2$O—R

(**23**)

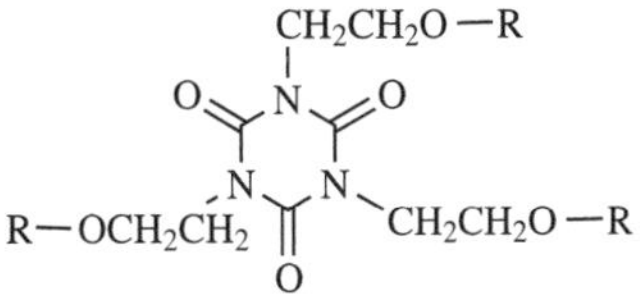

CH$_2$CH$_2$O—R
|
O ⟍ N ⟋ O
⟍ ⟋
N   N
R—OCH$_2$CH$_2$ ‖ ⟍CH$_2$CH$_2$O—R
O

(**24**)

where R =

OH
|
(CH$_3$)$_3$C ⟍ ⟋ C(CH$_3$)$_3$

O
‖
CH$_2$CH$_2$C—

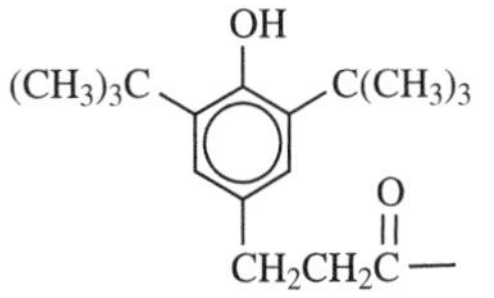

Diazo Coupling Reactions. Alkylphenols undergo a coupling reaction with diazonium salts which is the basis for the preparation of a class of uv light stabilizers for polymers. The interaction of *ortho*-nitrobenzenediazonium chloride with 2,4-di-*tert*-butylphenol results in an azo-coupled product (**25**). Reduction of the nitro group followed by *in situ* cyclization affords the benzotriazole (**26**) (15).

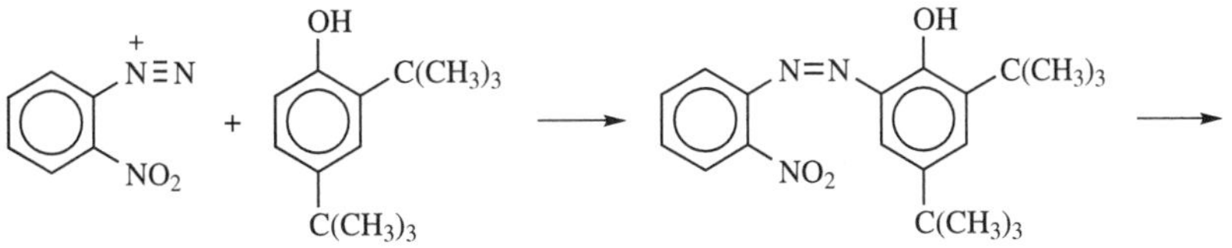

(**25**)

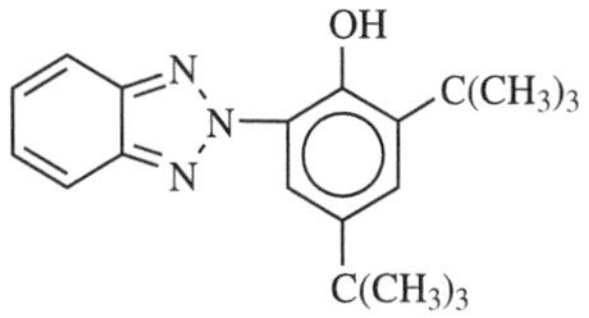

(**26**)

Benzotriazoles stabilize resins based on their ability to absorb uv radiation and re-emit the energy as thermal energy through molecular vibrations. Nickel-containing light stabilizers can act as both absorbers and quenchers of excited states of carbonyl impurities (16). One such nickel-containing stabilizer is made by coupling 4-*tert*-octylphenol with sulfur dichloride. The resulting bisphenol-sulfide is used to complex nickel(II) and form (**27**) (17).

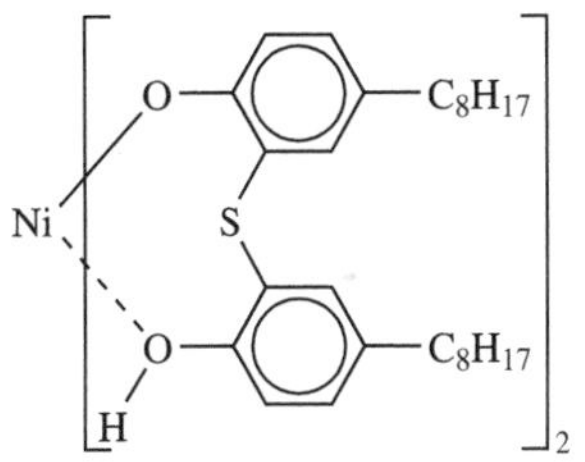

(**27**)

Removal of tert-Alkyl Groups.   *tert*-Alkyl groups on a phenol nucleus can be removed selectively to produce a desired synthetic result. The oxidative coupling of phenol offers a good example. 2,6-Di-*tert*-butylphenol can be coupled under oxidative conditions to 3,3′,5,5′-tetra-*tert*-butyl-4,4′-dihydroxybiphenyl (**28**). The *tert*-butyl groups can then be removed under acid conditions and recovered as isobutylene. The net result is the formation of 4,4′-dihydroxybiphenyl (**29**) from phenol (18).

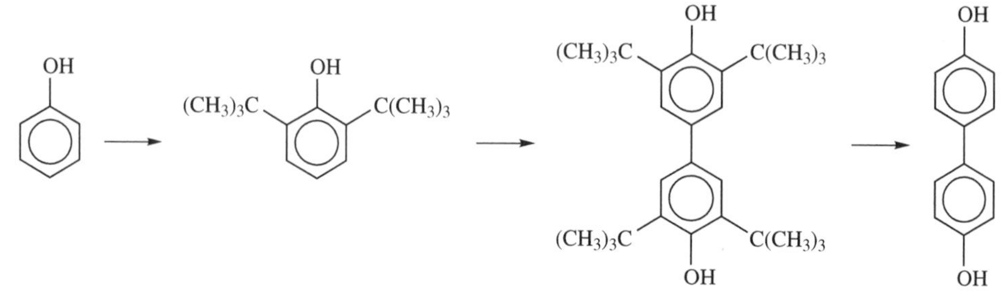

5. Manufacture and Processing

Alkylphenols of commercial importance are generally manufactured by the reaction of an alkene with phenol in the presence of an acid catalyst. The alkenes used vary from single species, such as isobutylene, to complicated mixtures, such as propylene tetramer (dodecene). The alkene reacts with phenol to produce monoalkylphenols, dialkylphenols, and trialkylphenols. The monoalkylphenols comprise ~85% of all alkylphenol production.

The choice of catalyst is based primarily on economic effects and product purity requirements. More recently, the handling of waste associated with the choice of catalyst has become an important factor in the economic evaluation. Catalysts that produce less waste and more easily handled waste by-products are strongly preferred by alkylphenol producers. Some commonly used catalysts are sulfuric acid, boron trifluoride, aluminum phenoxide, methanesulfonic acid, toluene–xylene sulfonic acid, cationic-exchange resin, acidic clays, and modified zeolites.

To describe the varied processes by which alkylphenols are produced, it is convenient to consider the reaction and recovery separately. In some cases this distinction is artificial because the operations are intimately linked, but in many processes the break is operationally significant.

5.1. Alkylation of Phenols.   The approach used to synthesize commercially available alkylphenols is Friedel-Crafts alkylation. The specific procedure typically uses an alkene as the alkylating agent and an acid catalyst, generally a sulfonic acid. Alkene and catalyst interact to form a carbocation and counter ion (**30**) which interacts with phenol to form a π complex (**31**). This complex is held together by the overlap of the filled π-orbital of the aromatic ring with the empty orbital of the carbocation (19). The π-complex can rearrange to the ς complex (**32**).

Loss of R^+ from complex (**32**) results in no reaction whereas loss of H^+ results in the alkylation of phenol.

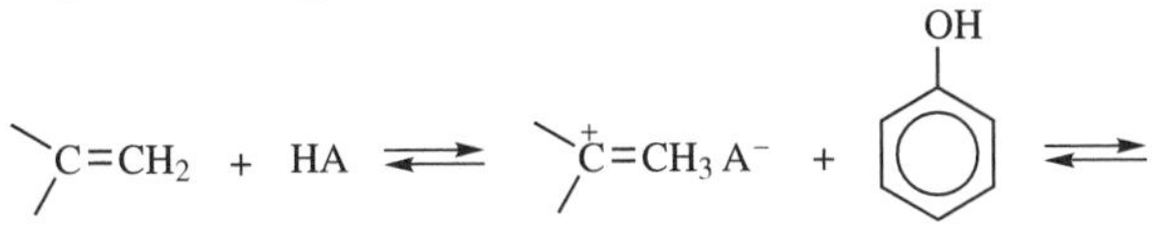

(**30**)

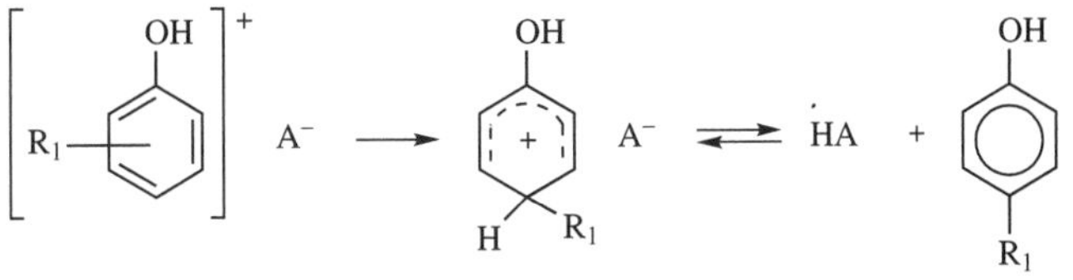

(**31**)   (**32**)

The driving force for the formation of (**31**) can be viewed as the result of electrostatic interaction, an electron deficient species being attracted to an electron rich species. This type of interaction explains some of the by-products that are formed during these alkylation reactions. Phenylalkyl ethers are the result of an interaction between the unshared electron pairs of the phenol oxygen with the carbocation. Alkene oligomers form when the carbocation complexes with the filled π-orbital of another alkene. These alkene oligomers can in turn react with phenol to produce correspondingly higher molecular weight alkylphenols.

Alkylations of phenol can give rise to three positional isomers, ortho, meta, and para, and multisubstitution products as well. A reaction may yield some of each potential product and this product mix presents a separation problem. Fortunately, judicious choice of catalyst and/or reaction conditions allows some control over selectivity. The reactivity of the alkylating agent also controls selectivity. The reaction between phenol and isobutylene [115-11-7], wherein the alkylating agent is the *tert*-butyl carbocation, shows a very high para selectivity using acid catalysis. Tertiary carbocations are the most stable, hence least reactive species and consequently are the most selective. Isomer selectivity in Friedel-Crafts reactions (qv) has also been shown to rely on steric effects, charge distribution, and relative stability of the quinonoid intermediates (20).

In 1957 a procedure was described that selectively alkylated phenol in the ortho position (21). This approach, using aluminum catalysis, made a variety of 2,6-dialkylphenols accessible. The mechanism proposed for this ortho alkylation is outlined as follows:

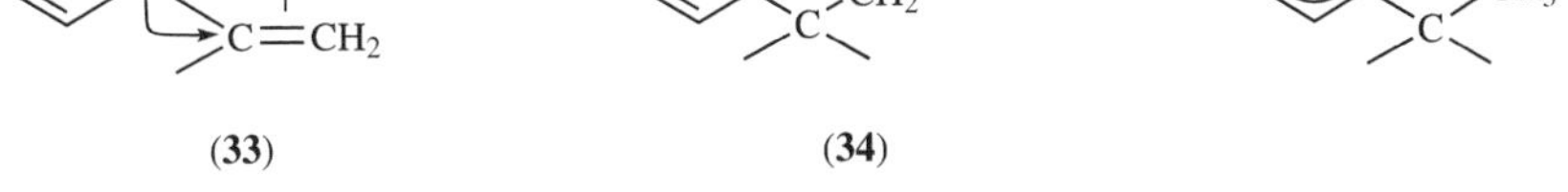

(**33**)   (**34**)   (**8**)

An aluminum trisphenoxide (**33**) is generated from phenol and aluminum or a trialkyl aluminum. The attraction between the electropositive aluminum and the electron-rich alkene places the latter in close proximity to the ortho position of the phenol yielding (**34**). An unreacted phenol molecule can then displace the ortho alkylated phenol. Using an alkene to phenol ratio of 2:1 yields proportionately more 2,6-dialkylphenol. The schematic representation of this reaction is a gross simplification. Some intermolecular alkylation does not take place. The aluminum catalyzed synthesis of 2,6-dialkylphenols generates several isomeric products, albeit in low yields. This procedure works well for making dialkylphenols from alkenes containing up to six carbon atoms. Thereafter, the nonpolar nature of the products and intermediates causes the aluminum trisphenoxide to fall out of solution. Given the bulkiness of the aluminum complex (**34**), terminal alkenes react more readily than internal alkenes. The use of forcing conditions with internal alkenes presents its own set of problems. At high temperatures the aluminum phenoxide produces a polymeric aluminum species accompanied by the formation of phenyl ethers (22).

The alkylation of phenol with an alkene using either acid or aluminum catalysis probably accounts for 95% of the commercially produced alkylphenols with alkyl groups of three carbons or larger. The alkenes are commercially available and environmentally kind. They do not produce by-products as do alkylations which use alcohols or alkyl halides. Together with an acid catalyst and the appropriate amount of phenol, mono-, di-, and trialkylphenols can be produced.

Several methods are available to supplement the phenol alkylations described above. Primary alkylphenols can be produced using the more traditional Friedel-Crafts reaction. Thus an *n*-butylphenol can be synthesized directly from a butyl halide, phenol, and mild Lewis acid catalyst. Alternatively, butyryl chloride can be used to acylate phenol producing a butyrophenone. Reduction with hydrazine (a Wolff-Kishner reduction) generates butylphenol.

5.2. Reactors. Reactors used to produce alkylphenols are simple batch reactors, complex batch reactors, and continuous reactors. All of these reactors have good mixing and heat removal capability. Good mixing is required for contacting the alkene and catalyst with the phenol. Typically, alkene–alkene reactions compete with phenol–alkene reactions at operating conditions. Good mixing minimizes locally high alkene concentrations and thus favors the desired reactions relative to the undesired ones. Good heat removal capability is needed to maintain controlled temperatures because of the highly exothermic nature of these reactions. The selectivity of alkylation is greatly affected by temperature (23).

The simple batch reactor generally consists of a cooled, agitated mixing tank. Figure 1 shows one type of simple batch reactor arrangement. There are four basic operating steps for this type of reactor in alkylphenol service: (*1*) the phenol is loaded; (*2*) the catalyst is loaded; (This step may be avoided if an appropriate heterogeneous catalyst is available.) (*3*) the alkene is loaded at such a rate that the reactor's heat removal capability is not exceeded and the desired reaction temperature is maintained; and (*4*) the alkylate is removed. This final step is trivial if a homogeneous catalyst is used, but it may require special equipment design if a heterogeneous catalyst is used.

The relatively low capital cost of the simple batch reactor is its most enticing feature. The inability to operate under pressure typically limits the simple batch reactor to use with the higher alkenes; ie, octenes, nonenes, and dodecenes.

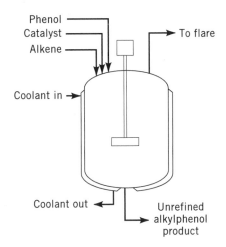

Fig. 1. Flow sheet of simple batch reactor.

For mainly economic reasons, these reactors are usually run at phenol to alkene mole ratios of between 0.9 and 1.1 to 1.

The complex batch reactor is a specialized pressure vessel with excellent heat transfer and gas liquid contacting capability. These reactors are becoming more common in alkylphenol production, mainly due to their high efficiency and flexibility of operation. Figure 2 shows one arrangement for a complex batch reactor. Complex batch reactors produce the more difficult to make alkylphenols; they also produce some conventional alkylphenols through improved processes.

The same four operating steps are used with the complex batch reactor as with the simple batch reactor. The powerful capabilities of the complex batch reactor offset their relatively high capital cost. These reactors can operate at phenol to alkene mole ratios from 0.3 to 1 and up. This ability is achieved by designing for positive pressure operation, typically 200 to 2000 kPa (30 to 300 psig), and

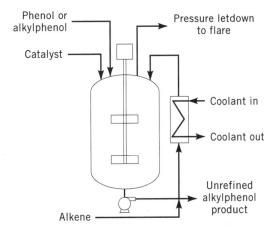

Fig. 2. Flow sheet of complex batch reactor.

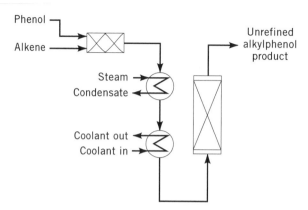

Fig. 3.   Flow sheet of continuous reactor.

for the use of highly selective catalysts. Because these reactors can operate at low phenol to alkene mole ratios, they are ideal for production of di- and trialkylphenols.

Among continuous reactors, the dominant system used to produce parasubstituted alkylphenols is a fixed-bed reactor holding a solid acid catalyst. Figure 3 shows an example of this type of reactor. The phenol and alkene are premixed and heated or cooled to the desired feed temperature. This mix is fed to the reactor where it contacts the porous solid, acid-impregnated catalyst. A key design consideration for this type of reactor is the removal of the heat of reaction.

Phenol and alkenes react quite exothermically. The reaction between 1 mole of phenol and 1 mole of isobutylene to yield 1 mole of *p-tert*-butylphenol PTBP liberates approximately 79.8 kJ/mol (19.1 kcal/mol) (24). In an adiabatic system, this reaction, if started at 40°C, would result in a reaction product at about 250°C. Temperatures above 200°C are considered unacceptably high in the reactor so design measures are employed to keep the temperature down.

The most common approach to maintaining the desired reaction temperature is to operate with a significant excess of phenol in the reactor. An adiabatic reactor fed with 2 moles of phenol and 1 mole of isobutylene at 40°C would reach about 180°C if all the isobutylene formed PTBP. The selectivity towards the desired monoalkylphenol product almost always improves as the phenol to alkene mole ratio increases. These gains must be weighed against the higher costs associated with higher mole ratio operation. Both the capital cost and the operating costs increase as the mole ratio is increased, since larger equipment that consumes more energy is needed to separate the excess phenol from the unrefined alkylphenol stream.

5.3. Purification.   The method used to recover the desired alkylphenol product from the reactor output is highly dependent on the downstream use of the product and the physical properties of the alkylphenol. The downstream uses vary enormously; some require no refining of the alkylphenol feedstock; others require very high purity materials. Physical property differences affect both the basic type of process used for recovery and the operating conditions used within that process.

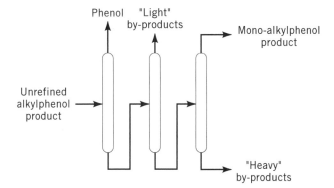

Fig. 4. Monoalkylphenol distillation train.

Some alkylphenol applications can tolerate "as is" reactor products, most significantly in the production of alkylphenol–formaldehyde resins. These resins can tolerate some of the reactant and by-product from the alkylphenol reactor because they undergo purification steps. This resin production route has both capital and operating cost advantages over using purer alkylphenol streams as feedstock. For these savings, the resin producer must operate the process in such a way as to tolerate a more widely varying feedstock and assume the burden of waste disposal of some unreactive materials from the alkylphenol process.

Most alkylphenols sold today require refinement. Distillation is by far the most common separation route. Multiple distillation tower separations are used to recover over 80% of the alkylphenol products in North America. Figure 4 shows a basic alkylphenol distillation train. Excess phenol is removed from the unrefined alkylphenol stream in the first tower. The by-products, which are less volatile than phenol but more volatile than the product, are removed in the second tower. The product comes off the third tower overhead while the heavy by-products come out the bottom.

The design of these distillation systems and the operating conditions used depend on the physical properties of the alkylphenols involved and on the product requirements. Essentially all alkylphenol distillation systems operate under vacuum, but the actual pressures maintained vary considerably. Vacuum operation allows reasonable reboiler temperatures (200–350°C) so that thermal dealkylation reactions of the alkylphenols are slow.

Some alkylphenols in commercial production have low vapor pressures and/ or low thermal decomposition temperatures. For these products, the economics of distillation are poor and other recovery processes are used. Crystallization from a solvent is the most common nondistillation method for the purification of these alkylphenols.

6. Shipment

Most commercially important alkylphenol production is of three types, unrefined alkylphenols, monoalkylphenols, and dialkylphenols. Together, these processes comprise over 95% of all alkylphenol production in the United States. The boundaries between types of production are not rigid; and some commercially important production is through a combination of these processes.

Unrefined alkylphenols are generally produced in the simple batch reactors described earlier. An alkene with between 8 and 12 carbon atoms reacts with phenol to produce a mixture of reactants, monoalkylphenols, and dialkylphenols. These mixtures usually do not freeze above 25°C and so are liquid at production and storage conditions. The product is generally used in the same factory or complex in which it is produced so shipment typically consists of pumping the material from the reactor to a storage tank.

Monoalkylphenols are generally produced in specialized plants that have both continuous reactors and continuous vacuum distillation trains. Alkenes with between 4 and 24 carbon atoms react with phenol to produce an unrefined phenol–alkylphenol mixture. This mixture is fed to the distillation train where the phenol is removed for recycle and the product is isolated. The product is then stored in heated tanks made of stainless steel or phenolic resin lined carbon steel. These tanks are blanketed with inert gas to avoid product discoloration associated with oxidation.

Large volumes of monoalkylphenols are shipped in liquid form by railcar, tank wagon, or export container. These shipping vessels must be stainless steel or phenolic resin lined carbon steel. For smaller volumes, drums and tote-tanks are used. For high freezing point alkylphenols, such as PTBP, the product is flaked and shipped in either bags or supersacs. For low freezing point products, such as p-nonylphenol (PNP) (fp < 20°C), the product is shipped in drums or tote-tanks.

Dialkylphenols are also produced in specialized plants. These plants combine complex batch reactors with vacuum distillation trains or other recovery systems. Alkenes with carbon numbers between 4 and 9 react with phenol to make an unrefined alkylphenol mixture, which is fed into the recovery section where very high purity product is isolated. The product is stored, handled, and shipped just as are the monoalkylphenols.

7. Economic Aspects

Among the key variables in strategic alkylphenol planning are feedstock quality and availability, equipment capability, environmental needs, and product quality. In the past decade, environmental needs have grown enormously in their effect on economic decisions. The manufacturing cost of alkylphenols includes raw-material cost, nonraw-material variable cost, fixed cost, and depreciation.

Raw-material costs are the largest cost items over the lifetime of a plant and typically make up between 40 and 90% of the total manufacturing cost. The placement of plants near production facilities making alkenes and/or phenol is important to producers of alkylphenols. The raw-material costs are so important that a large fluctuation in a raw material price can drive a product from a reasonably profitable situation to a clearly unprofitable one.

Nonraw-material variable costs consist largely of utility costs, but other costs can be significant in this area. For example, operating costs for additional waste treatment are in this category. Nonraw-material variable costs account for 5–20% of the total manufacturing cost of an alkylphenol operation.

Fixed costs include corporate overhead and administration costs as well as those plant-related costs that do not vary with production and contribute 5 to 30% of the total manufacturing cost.

Depreciation costs can drop from as high as 50% of the total manufacturing cost to less than 10% as the plant ages. The effect of the changing depreciation rates is tempered if after-tax analysis is used.

Schenectady International Classes is the leading global supplier of alkylphenols. Plants are located in Rotterdam Junction, N.Y, and Freeport, Texas (25).

8. Health and Safety Factors

As a class of compounds the toxicity of alkylphenols range from moderately toxic (oral rat LD_{50} 50–500 mg/kg) to practically nontoxic (oral rat LD_{50} 5,000–15,000 mg/kg) and most are irritants or corrosive toward skin (Table 2). Sensitization to alkylphenols has been reported. 4-tert-Butylphenol is known to cause leucoderma (depigmentation of the skin) in sensitive individuals. In general, precautions should be taken when handling alkylphenols to avoid contact with the skin by wearing appropriate protective gloves, clothing, and a face shield or goggles. Most of the alkylphenols are combustible when heated and emit irritating vapors upon decomposition the Dot label for octyl phenol is 6.1. KEEP AWAY FROM FOOD (27). Alkylphenols should only be stored and handled in well-ventilated areas and appropriate respiration equipment with carbon filters should be worn if PEL or TLV limits are exceeded.

9. Uses and Derivatives of Alkylphenols

Typical physical properties and assays of the commercially most important alkylphenols in terms of worldwide volume were given in Table 1.

9.1. 4-*tert*-Amylphenol. *p-tert*-Amylphenol (PTAP) or 4-(1,1-dimethyl-propyl)phenol is commercially produced by the alkylation of phenol with isoamylene under acidic catalysis. Isoamylene is a mixture of 2-methyl-1-butene and 2-methyl-2-butene, which is produced by dehydration of the corresponding alcohol or backcracking of the corresponding methyl ether. The highest purity isoamylene is available from the backcracking of *tert*-amyl methyl ether (TAME), produced from a C_5 raffinate stream by reaction with methanol under acid catalysis.

The crude product formed from the alkylation of phenol with isoamylene contains principally 2-*tert*-amylphenol, 4-*tert*-amylphenol, and 2,4-di-*tert*-amylphenol. 4-*tert*-Amylphenol is purified to its typical assay of 99+% by fractional distillation. 4-*tert*-Amylphenol [80-46-6] is commercially available as a solid, flaked material packaged in paper or plastic bags (25 kg net weight) or as a molten material in tank wagon or railcar quantities.

4-*tert*-Amylphenol is employed as a germicide in cleaning solutions (28), but it is being replaced by environmentally safer quaternary ammonium salts.

Table 2. Health and Safety Data[a]

Name	CAS Registry Number	RTECS NO.[b]	Toxicity data				
			Skin (rabbit)	Oral (rat) LD$_{50}$, mg/kg	Skin (rabbit) LD$_{50}$, mg/kg	Dot skin	Eye (rabbit)
4-tert-amylphenol	[80-46-6]	SM6825000	open, 100 µg/24 h	severe, 500 mg	1830	2000	yes
4-tert-butylphenol	[98-54-4]	SJ8925000	mild, 500 mg/24 h	severe, 50 µg/24 h	2951	2288	yes
2-sec-butylphenol	[89-72-5]	SJ8920000	severe (2 mg/24 h)	severe, 50 µg/24 h	340	5560	yes
4-cumylphenol	[599-64-4]	SL1942450			1770		yes
4-dodecylphenol	[27193-86-8]	SL367500	severe, 8.0–8.0/24 h	moderate, 33.3/110	2140	5000	yes
4-nonylphenol	[84852-15-3]	SM5630000	severe, 10 mg/24 h	severe, 50 µg/24 h	1620	2140	yes
4-tert-octylphenol	[140-66-9]	SM9625000	moderate, 20 mg/24 h	severe, 50 µg/24 h	2160		yes
2,4-di-tert-amylphenol	[25231-47-4]	SL3500000		moderate, 100 mg	330		yes
2,4-d-tert-butylphenol	[96-76-4]	SK8260000			1500		yes
2,6-di-tert-butylphenol	[128-39-2]	SK8265000	mild, 0.5 g/24 h	minimal, 100 mg/24 h	>5000		no
2,6-di-sec-butylphenol	[5510-99-6]	SK8225000	severe, 500 mg/24 h	severe, 50 µg/24 h	1320		yes
2,4-di-cumylphenol	[2772-45-4]						yes
2-methylphenol	[95-48-7]	G06300000	severe, 524 mg/24 h	severe, 105 mg	121	620	yes
3-methylphenol	[108-39-4]	G06125000	severe, 517 mg/24 h	severe, 103 mg	242	1100	yes
4-methylphenol	[106-44-5]	G06475000	severe, 517 mg/24 h	severe, 103 mg	207	750	yes
2,6-dimethylphenol	[576-26-1]	ZE6125000		severe, 100 mg	296	1000	yes

[a] Ref. 26.
[b] RTECS = Registry of Toxic Effects of Chemical Substances.

Another commercial application of 4-*tert*-amylphenol is in phenolic resins (qv) (novolaks and resoles). These resins are used in paints and varnishes and as printing ink resins. The ethoxylated novolaks are used as oil field demulsifiers. Because of the high cost of 4-*tert*-amylphenol, which is related to the high cost of isoamylene, many of the resin applications have been reformulated using 4-*tert*-butylphenol. 4-*tert*-Amylphenol is also used as a vulcanizing agent as the disulfide derivative for the curing of rubber.

9.2. 4-*tert*-Butylphenol. *p-tert*-Butylphenol (PTBP) or 4-(1,1-dimethylethyl)phenol is produced from the alkylation of phenol with isobutylene under acid catalysis. Isobutylene [115-11-7] is commercially produced mostly from the dehydration of *tert*-butyl alcohol or from the cracking of methyl *tert*-butyl ether (MTBE). The principal products of this alkylation are 2-*tert*-butylphenol, 4-*tert*-butylphenol, 2,4-di-*tert*-butylphenol, and minor amounts of 2,4,6-tri-*tert*-butylphenol and 4-*tert*-octylphenol. 4-*tert*-Butylphenol is available in a technical grade which is used in the production of phenolic resins. A high purity grade is available for the production of glycidyl ethers and for the chain termination of polycarbonates. 4-*tert*-Butylphenol [98-54-4] is available as a flaked solid packaged in paper or plastic bags (25 kg net weight), and as a molten material in tankwagon and railcar quantities.

Phenolic resin applications account for 60–70% of all 4-*tert*-butylphenol consumed worldwide. These resins are used in a wide range of applications which include paints, coating resins, and printing inks (29). 4-*tert*-Butylphenol novolak resins react with ethylene oxide to form oil field demulsifiers or are converted into phosphate esters for use as hydraulic fluids and synthetic lubricants. 4-*tert*-Butylphenol resoles react with alkaline-earth metal hydroxides to produce metal resinates. The resinate is combined with a rubber component to give an adhesive. Other uses of 4-*tert*-butylphenol include the chain termination of polycarbonates where its use in the place of phenol gives a polycarbonate with better heat distortion characteristics and improved processability for injection grade material. 4-*tert*-Butylphenol is converted to its corresponding glycidyl ether by the reaction with epichlorohydrin followed by dehydrohalogenation. The glycidyl ether is used as a hardner in epoxy resins (qv). 4-*tert*-Butylphenol can be reduced with hydrogen to a mixture of *cis*-4-tert-butylcyclohexanol and *trans*-4-*tert*-butylcyclohexanol under nickel catalysis. The corresponding acetate derivative is widely used as a perfume in soaps and detergents. The sodium salt of the phosphoric ester of 4-*tert*-butylphenol, sodium bis(4-*tert*-butylphenyl) phosphate, is used as a nucleating agent for polypropylene (30). The use of a nucleator provides a polypropylene with improved thermal properties and increased clarity. 4-*tert*-Butylphenol is used in the production of pesticides such as 2-(4-*tert*-butylphenoxy)cyclohexyl-2-propynyl sulfide [2312-35-8]2-(4-*tert*-Butylphenoxy), which is used on a variety of fruits and vegetables (31).

The consumption of 4-*tert*-butylphenol in the production of phenolic resins represents an application in a mature market and little growth is projected. Its use in end-capping polycarbonates, in the production of glycidyl ethers, and in the production of nucleation agents for polypropylene was expected to grow.

9.3. 2-*sec*-Butylphenol. *o-sec*-Butylphenol (OSBP) or 2-(1-methylpropyl) phenol is produced by the alkylation of phenol with butene under aluminum or acid catalysis. The aluminum catalysis route selectively yields 2-*sec*-butylphenol whereas the acid catalysis route yields a 2:1 mixture of 2-*sec*-butylphenol and 4-*sec*-butylphenol. 2-*sec*-Butylphenol [89-72-5] is a liquid available in 55-gal drums (208-L) and in bulk quantities in tank wagons and railcars.

Up until 1986 the major use for 2-*sec*-butylphenol was in the production of the herbicide, 2-*sec*-butyl-4,6-dinitrophenol [88-85-7], which was used as a pre- and postemergent herbicide and as a defoliant for potatoes (32). The EPA banned its use in October 1986 based on a European study which showed that workers who came in contact with 2-*sec*-butyl-4,6-dinitrophenol experienced an abnormally high rate of reproduction problems. France and the Netherlands followed with a ban in 1991. A significant volume of 2-*sec*-butyl-4,6-dinitrophenol is used worldwide as a polymerization inhibitor in the production of styrene where it is added to the reboiler of the styrene distillation tower to prevent the formation of polystyrene (33). OSBP is used in the Far East as the carbamate derivative, 2-*sec*-butylphenyl-*N*-methylcarbamate [3766-81-2] (BPMC) (34). BPMC is an insecticide used against leaf hoppers which affect the rice fields.

Because of environmental concerns about 2-*sec*-butylphenol-based derivatives, the market growth was expected to be negative in the future, with the exception of possible significant growth in the use of the carbamate insecticide.

9.4. 4-Cumylphenol. *p*-Cumylphenol (PCP) or 4-(1-methyl-1-phenyl-ethyl)phenol is produced by the alkylation of phenol with α-methylstyrene under acid catalysis. An inexpensive and selective method has been reported (35). α-Methylstyrene is a by-product from the production of phenol via the cumene oxidation process. The principal by-products from the production of 4-cumylphenol result from the dimerization and intramolecular alkylation of α-methylstyrene to yield substituted indanes. 4-Cumylphenol [599-64-4] is purified by either fractional distillation or crystallization from a suitable solvent. Purification by crystallization results in the easy separation of the substituted indanes from the product and yields a solid material which is packaged in plastic or paper bags (20 kg net weight). Purification of 4-cumylphenol by fractional distillation yields a product which is almost totally free of any dicumylphenol. The molten product resulting from purification by distillation can be flaked to yield a solid form; however, the solid form of 4-cumylphenol sinters severely over time. PCP is best stored and transported as a molten material.

The major use of 4-cumylphenol is as a chain terminator for polycarbonates. Its use in place of phenol gives a polycarbonate with superior properties (36). For a low molecular weight polycarbonate used for injection-molding applications, the use of 4-cumylphenol as a chain terminator significantly lowers the volatility of the resin. Other uses of 4-cumylphenol include the production of phenolic resins, some of which have applications in the electronics industry (37). Another application of 4-cumylphenol involves its reaction with ethylene oxide to form a specialty surfactant.

The growth rate of 4-cumylphenol is expected to parallel the growth rate of polycarbonates, particularly the grades used to produce compact discs.

9.5. 4-Dodecylphenol. *p*-Dodecylphenol (PDDP) may be produced by the reaction of phenol with dodecene under acid catalysis, but commercial 4-dodecylphenol is produced from propylene tetramer. Dodecene differs significantly from the idealized tetramer of propylene because of skeletal rearrangements which occur during the oligomerization process, producing a complex mixture of tri- and tetrasubstituted monoolefins. Although dodecene is purified by fractional distillation, it contains olefins with carbon numbers ranging from C_{10} to C_{14}; the C_{12} content is typically 60–70%. Two grades of 4-dodecylphenol [27193-86-8] are commercially available. A technical grade of 4-dodecylphenol is a nondistilled product and contains approximately 10% 2-dodecylphenol, 85% 4-dodecylphenol, and 5% 2,4-didodecylphenol. A high purity grade is fractionally distilled and contains approximately 5% 2-dodecylphenol, 95% 4-dodecylphenol, and only a trace of 2,4-didodecylphenol. The technical grade of 4-dodecylphenol is amber in color, whereas the high purity grade is colorless. 4-Dodecylphenol is available in 55-gal drums (208-L) and as bulk shipments in tank wagons and railcars.

The major use of technical grade 4-dodecylphenol is in lube oil additives. 4-Dodecylphenol is converted to a calcium phenolate [50910-68-4] and used as a detergent in lubricating oils (38). The phenolate combines with combustion debris to prevent the accumulation of the debris on engine parts and neutralizes the strong acids formed during combustion and oxidation. The reaction product of 4-dodecylphenol with ethylene oxide and propylene oxide is used as a corrosion inhibitor in oil. It coats the inner surfaces of an engine to prevent the corrosion of moving parts. 4-Dodecylphenol is used in the production of a zinc dithiophosphate ester [54261-67-5], which is used as an antioxidant and antiwear additive for high temperature applications (39).

High purity 4-dodecylphenol is used to produce specialty surfactants by its reaction with ethylene oxide. The low color of high purity 4-dodecylphenol is important in this application from a standpoint of aesthetics. 4-Dodecylphenol is also used to produce phenolic resins which are used in adhesive applications and printing inks. 4-Dodecylphenol is also used as an epoxy curing catalyst where the addition of 4-dodecylphenol accelerates curing of the epoxy resin to a hard, nontacky solid.

The worldwide consumption of 4-dodecylphenol is difficult to estimate since the majority of 4-dodecylphenol produced is captively used.

9.6. 2-Methylphenol. This phenol, commonly known as *o*-cresol, is produced synthetically by the gas phase alkylation of phenol with methanol using modified alumina catalysis or it may be recovered from naturally occurring petroleum streams and coal tars. Most is produced synthetically. Reaction of phenol with methanol using modified zeolite catalysts is a concerted dehydration of the methanol and alkylation of the aromatic ring. 2-Methylphenol [95-48-7] is available in 55-gal drums (208-L) and in bulk quantities in tank wagons and railcars.

The majority of 2-methylphenol is used in the production of novolak phenolic resins. High purity novolaks based on 2-methylphenol are used in photoresist applications (40). Novolaks based on 2-methylphenol are also epoxidized with epichlorohydrin, yielding epoxy resins after dehydrohalogenation, which are used as encapsulating resins in the electronics industry. Other uses of 2-methylphenol include its conversion to a dinitro compound, 4,6-dinitro-2-methylphenol [534-52-1] (DNOC), which is used as a herbicide (41). DNOC is also used to a limited extent as a polymerization inhibitor in the production of styrene, but this use is expected to decline because of concerns about the toxicity of the dinitro derivative.

2-Methylphenol is converted to 6-*tert*-butyl-2-methylphenol [2219-82-1] by alkylation with isobutylene under aluminum catalysis. A number of phenolic anti-oxidants used to stabilize rubber and plastics against thermal oxidative degradation are based on this compound. The condensation of 6-*tert*-butyl-2-methylphenol with formaldehyde yields 4,4'-methylenebis(2-methyl-6-*tert*-butylphenol) [96-65-1], reaction with sulfur dichloride yields 4,4'-thiobis(2-methyl-6-*tert*-butylphenol) [96-66-2], and reaction with methyl acrylate under base catalysis yields the corresponding hydrocinnamate. Transesterification of the hydrocinnamate with triethylene glycol yields triethylene glycol-bis[3-(3-*tert*-butyl-5-methyl-4-hydroxyphenyl)propionate] [36443-68-2] (42). 2-Methylphenol is also a component of cresylic acids, blends of phenol, cresols, and xylenols. Cresylic acids are used as solvents in a number of coating applications. An oxidative hair dye precursor composition containing 5-amino-2- methyl phenol has been reported (43).

9.7. 3-Methylphenol. *m*-Cresol is produced synthetically from toluene. Toluene is chlorinated and the resulting chlorotoluene is hydrolyzed to a mixture of methylphenols. Purification by distillation gives a mixture of 3-methylphenol and 4-methylphenol since they have nearly identical boiling points. Reaction of this mixture with isobutylene under acid catalysis forms 2,6-di-*tert*-butyl-4-methylphenol and 2,4-di-*tert*-butyl-5-methylphenol, which can then be separated by fractional distillation and debutylated to give the corresponding 3- and 4-methylphenols. A mixture of 3- and 4-methylphenols is also derived from petroleum crude and coal tars.

A major use of 3-methylphenol [108-39-4] is in the production of phenolic based antioxidants which are particularly good at stabilizing polymers in contact with copper against thermal oxidative degradation. The alkylation of 3-methylphenol with isobutylene under acid or aluminum catalysis yields 2-*tert*-butyl-5-methylphenol [88-60-8]. Condensation of 2-*tert*-butyl-5-methylphenol with butyraldehyde yields the corresponding 4,4'-butylidenebis(6-*tert*-butyl-3-methylphenol) [85-60-9] which is used in the stabilization of rubber and latex (44). Condensation of 2-*tert*-butyl-5-methylphenol with crotonaldehyde yields a corresponding trisphenol derivative, 1,1,3-tris(5-*tert*-butyl-2-methyl-4-hydroxyphenyl)butane [1843-03-4], which is used to stabilize polymers for high temperature applications (45). The reaction of 2-*tert*-butyl-5-methylphenol with sulfur dichloride yields 4,4'-thiobis(6-*tert*-butyl-3-methylphenol) [96-69-5] which is widely used in curable rubber.

Another significant use of 3-methylphenol is in the production of herbicides and insecticides. 2-*tert*-Butyl-5-methylphenol is converted to the dinitro acetate derivative, 2-*tert*-butyl-5-methyl-4,6-dinitrophenyl acetate [2487-01-6] which is used as both a pre- and postemergent herbicide to control broad leaf weeds (46). Carbamate derivatives of 3-methylphenol based compounds are used as insecticides. The condensation of 3-methylphenol with formaldehyde yields a curable phenolic resin. Since 3-methylphenol is trifunctional with respect to its reaction with formaldehyde, it is possible to form a thermosetting resin by the reaction of a prepolymer with paraformaldehyde or other suitable formaldehyde sources. 3-Methylphenol is also used in the production of fragrances and flavors. It is reduced with hydrogen under nickel catalysis and the corresponding esters are used as synthetic musk.

9.8. 4-Methylphenol. *p*-Cresol is produced synthetically from toluene. Toluene is sulfonated to yield *para*-toluenesulfonic acid, which is then converted to 4-methylphenol via the caustic fusion route. A minor amount of 4-methylphenol is also derived from petroleum crude and coal tars. 4-Methylphenol [106-44-5] is available in 55-gal drums (208-L) and in bulk quantities as a molten material.

The bulk of 4-methylphenol is used in the production of phenolic antioxidants. The alkylation of 4-methylphenol with isobutylene under acid catalysis yields 2-*tert*-butyl-4-methylphenol [2409-55-4] and 2,6-di-*tert*-butyl-4-methylphenol [128-37-0]. The former condenses with formaldehyde under acid catalysis to yield 2,2'-methylene bis(6-*tert*-butyl-4-methylphenol) [119-47-1],which is widely used in the stabilization of natural and synthetic rubber (47). The reaction of 2-*tert*-butyl-4-methylphenol with sulfur dichloride yields 2,2'-thiobis(6-*tert*-butyl-4-methylphenol) [90-66-4]. 2,6-Di-*tert*-butyl-4-methylphenol, which is commonly known as BHT (butylated hydroxy toluene), is a widely used phenolic antioxidant in the stabilization of oils, rubber, and polyolefins (48). BHT is also one of the few *phenolic antioxidants* approved by the FDA as a direct food additive where it is used to retard the oxidation of naturally occurring oils in food.

Other uses of 4-methylphenol include its conversion to a benzotriazole uv stabilizer, 2-(2'-hydroxy-5'-methylphenyl)benzotriazole [2440-22-4] (49). The benzotriazole-based uv stabilizer makes possible the extended use of thermoplastics in outdoor applications. Other minor applications for 4-methylphenol include its use in the production of novolak or resole phenolic resins. It is also used in the production of certain dyes and fragrances.

9.9. 4-Nonylphenol. *p*-Nonylphenol (PNP) is produced by the alkylation of phenol with nonene under acid catalysis. All commercially produced PNP is made from nonene based on the trimerization of propylene. Because of the skeletal rearrangements which occur during the oligomerization of propylene, commercial grade nonene does not have a high percentage of the idealized structure, 4,6-dimethyl-2-heptene. Rather it is a complex mixture of olefins, mostly tri- and tetrasubstituted monoolefins. Nonene is fractionally distilled from other oligomers and contains approximately 90% C_9 olefins; the remaining 10% is C_8 and C_{10} olefins. The two commercial purity grades of 4-nonylphenol are a technical grade which is composed of 10–12% 2-nonylphenol, 85–90% 4-nonylphenol, and up to 5% 2,4-dinonylphenol, and a high purity grade which contains 5% maximum 2-nonylphenol, 95% minimum 4-nonylphenol, and only a trace of 2,4-dinonylphenol. 4-Nonylphenol [84852-15-3] is available in 55-gal drums (280-L) and in bulk quantities in tank wagons and railcars.

The major use for 4-nonylphenol is in the production of nonionic surfactants (10) and constitutes 80% of usage (50). 4-Nonylphenol reacts with ethylene oxide in a mole ratio that varies from 1–40 under base catalysis. 4-Nonylphenol based nonionic surfactants are the largest volume alkylphenol based ethoxylate. The most common nonionic surfactant based on 4-nonylphenol is the nine-mole ethoxylate. Higher mole ratios in the range of 12–15 moles of ethylene oxide per mole of 4-nonylphenol are used as emulsifiers for agrochemicals. Other major applications for alkylphenol ethoxylates are as cleaners of metal surfaces; detergents for car washes, commercial laundries, and the rug cleaning industry; as a dispersant for wood pulp in the production of paper, and as an emulsifier in the production of latex paint (51). 4-Nonylphenol based ethoxylates can be converted to phosphate esters or sulfonated to the corresponding sulfate to yield higher performing surfactants. Phosphate esters of 4-nonylphenol are also reported to be excellent flame retardants (52). Nonylphenol nonionic geminic surfactants are extremely effective in improving detergency (53).

Another significant use of 4-nonylphenol is in the production of tris(4-nonylphenyl) phosphite [3050-88-2] (TNPP) (54), a secondary antioxidant which protects organic materials against oxidative degradation by decomposing hydroperoxides. In this process, the hydroperoxide is converted to the corresponding alcohol and the phosphite is converted to the corresponding phosphate. Phosphites form synergistic mixtures with phenolic-based antioxidants. Tris(4-nonylphenyl) phosphite is widely used in the stabilization of natural and synthetic rubber, vinyl polymers, and polyolefins and styrenics; and it has the distinction of being one of the few commercially available liquid triaryl phosphites. Other uses of 4-nonylphenol include its conversion to barium [41157-58-8] and calcium [100842-25-9] phenolates, which are used as heat stabilizers in poly(vinyl chloride) (55). 4-Nonylphenol also has application as a catalyst in the curing of epoxy resins (56,57). Hydrogen bonding of the acidic proton on the hydroxyl group of 4-nonylphenol with the oxygen of the oxirane ring may assist in the opening of the epoxide.

The worldwide consumption of 4-nonylphenol is somewhat difficult to ascertain because of the captive consumption by some producers. Future growth in the consumption of 4-nonylphenol is predicted to be at the rate of 2%, but some market share in surfactants has been lost to the linear alcohol surfactants because of environmental concerns over aquatic toxicity and biodegradability of alkylphenol based ethoxylates (50). The use of alkylphenol ethoxylates was banned in Switzerland in 1986 and Germany is undergoing a voluntary phase out by the early 1990s. An industry group of 4-nonylphenol producers and surfactant producers is currently operating under a consent order from EPA to carry out a wide range study on the biodegradability of 4-nonylphenol-based ethoxylates and toxicity studies on aquatic life forms of the resulting break-down products. Early results from testing of sediments from river beds and waste plant sludge show that nonylphenol-based ethoxylates are readily biodegraded and the residual level of 4-nonylphenol is very low. Tests on endocrine interference potential are incouraging, but research continues (50).

The use of 4-nonylphenol in the production of tris(4-nonylphenyl) phosphite is also not expected to show much growth because of its replacement in many polymers by higher performing and more hydrolytically stable phosphites.

9.10. 4-*tert*-Octylphenol.

p-tert-Octylphenol (PTOP) or 4-(1,1,3,3-tetramethylbutyl)phenol is produced by the alkylation of phenol with diisobutylene under acid catalysis. Diisobutylene is a mixture of 2,4,4-trimethyl-1-pentene and 2,4,4-trimethyl-2-pentene. A small amount of skeletal rearrangement during alkylation leads to a second 4-*tert*-octylphenol isomer which has the proposed structure of 1,1,2,3-tetramethylbutyl as the alkyl radical. The crude alkylation product typically contains 4-*tert*-butylphenol (from the backcracking of the diisobutylene), 2-*tert*-octylphenol, 4-*tert*-octylphenol, butyloctylphenol, and 2,4-dioctylphenol. 4-*tert*-Octylphenol is purified by fractional distillation under reduced pressure; it is available as a technical grade which contains 5–8% 2-*tert*-octylphenol,

90–95% 4-*tert*-octylphenol, and 1–2% butyloctylphenol. A high purity grade is also available which contains less than 2% 2-*tert*-octylphenol, 98–99% 4-*tert*-octylphenol, and only a trace of higher boiling alkylphenols. 4-*tert*-Octylphenol [140-66-9] is packaged as a flaked solid in paper or plastic bags (25 kg net weight) or as a molten product for bulk shipments in tank wagons and railcars.

4-*tert*-Octylphenol reacts with ethylene oxide under base catalysis and the resulting ethoxylates are used in many of the same applications as the 4-nonyl-phenol-based surfactants. A major use area for ethoxylates based on 4-*tert*-octyl-phenol is as a surfactant in the emulsion polymerization of acrylic and vinyl monomers (58). The 4-*tert*-octylphenol ethoxylate aids in the dispersion of the monomers in the aqueous medium and stabilizes the latex formed as a result of polymerization.

Another important application for 4-*tert*-octylphenol is in the production of phenolic resins. Novolak resins based on 4-*tert*-octylphenol are widely used in the tire industry as tackifiers. The tackiness of these resins binds the many parts of an automobile tire prior to final vulcanization. A specialty use for novolak resins based on 4-*tert*-octylphenol is the production of a zincated resin, which is formulated as a dispersion in water and coated onto paper in combination with encapsulated leuco dyes to yield carbonless copy paper (see MICROENCAPSULATION). Pressure from writing bursts the encapsulated leuco dye, which is converted from its colorless form to its colored form by the *zincated resin* (59). Novolak resins based on 4-*tert*-octylphenol are also used in the production of specialty printing inks.

Resoles based on 4-*tert*-octylphenol react with alkaline-earth metal hydroxides to yield metal resinates. The resinate is combined with natural or synthetic rubber to produce an adhesive. The 4-*tert*-octylphenol-based resin gives the adhesive its initial tack which holds the two surfaces undergoing bonding together while the other components of the adhesive undergo final curing.

Other applications for 4-*tert*-octylphenol include chain termination of polycarbonates (60). The properties of low molecular weight polycarbonates used in injection-molding applications to form compact disks are enhanced when the polymer is terminated using 4-*tert*-octylphenol.

Another use of 4-*tert*-octylphenol is in the production of uv stabilizers. 4-*tert*-Octylphenol reacts with sulfur dichloride to yield the thio-bisphenol derivative, which then reacts with nickel acetate to form 2,2'-thiobis(4-*tert*-octyl-phenolate)-*N*-butylamine nickel [14516-71-3]. This type of stabilizer is widely used in the production of outdoor carpeting based on polypropylene fibers. Nickel compounds give a green discoloration which limits their applications. A second class of uv stabilizers based on the benzotriazole structure. 2-(2'-hydroxy-5'-*tert*-octylphenyl)benzotriazole [3147-75-9] is produced from 4-*tert*-octylphenol (61).

It is difficult to estimate the world consumption of 4-*tert*-octylphenol because a significant volume is captively consumed as a crude alkylphenol by producers of phenolic resins. The overall growth rate of 4-*tert*-octylphenol is expected to track the growth in the GNP. As with 4-nonylphenol-based surfactant, the surfactants based on 4-*tert*-octylphenol have also lost market share to the linear alcohol ethoxylates because of concerns over biodegradability and product aquatic toxicity. The use of 4-*tert*-octylphenol in the production of phenolic resins is a mature market and little growth is forecasted. However, the use of 4-*tert*-octylphenol in the chain termination of polycarbonates and in the production of uv stabilizers is expected to have an above average growth rate.

9.11. Dialkylated Phenols.   *2,4-Di-tert-amylphenol (2,4-DTAP)* or 2,4-bis(1,1-dimethylpropyl)phenol is produced by the alkylation of phenol with isoamylene under acid catalysis in a mole ratio of 2:1 (isoamylene to phenol). The crude alkylation product contains 4-*tert*-amylphenol, 2,4-di-*tert*-amylphenol, and 2,4,6-tri-*tert*-amylphenol. The 2,4-di-*tert*-amylphenol is purified via fractional distillation under reduced pressure. 2,4-Di-*tert*-amylphenol [25231-47-4] is available in 55-gal drums (208-L) and bulk quantities in tank wagon and railcar shipments.

A major use for 2,4-di-*tert*-amylphenol is in the production of uv stabilizers; the principal one is a benzotriazole-based uv absorber, 2-(2′-hydroxy-3′,5′-di-*tert*-amylphenyl)-5-chlorobenzotriazole [25973-55-1], which is widely used in polyolefin films, outdoor furniture, and clear coat automotive finishes (62). Another significant use for 2,4-di-*tert*-amylphenol is in the photographic industry. A number of phenoxyacetic acid derivatives of 2,4-di-*tert*-amylphenol are used as developing agents in color photography (qv) (63). Other uses include its reaction with ethylene oxide under base catalysis to form a specialty surfactant used to treat cotton fibers to prevent the redeposition of dirt onto the fibers during the production of cotton-based fabric. A phenoxy poly(ethylene oxide) based on 2,4-di-*tert*-amylphenol also has an application as a fuel additive where it is used as a corrosion inhibitor.

The growth rate for 2,4-di-*tert*-amylphenol is predicted to be above GNP growth rate, mainly driven by the use of the benzotriazole derivative as a uv stabilizer for clear-coat applications in auto finishes.

2,4-Di-tert-butylphenol (2,4-DTBP) or 2,4-bis(1,1-dimethylethyl)phenol is produced by the alkylation of phenol with isobutylene under acid catalysis using a mole ratio of 2:1 (isobutylene to phenol). The crude product contains 4-*tert*-butylphenol, 2,4-di-*tert*-butylphenol, and 2,4,6-tri-*tert*-butylphenol. The 2,4-*tert*-butylphenol is purified by fractional distillation under reduced pressure. 2,4-Di-*tert*-butylphenol [96-76-4] is available in 55-gal drums (208-L) and in bulk as a molten material.

The primary use for 2,4-di-*tert*-butylphenol is in the production of substituted triaryl phosphites. 2,4-Di-*tert*-butylphenol reacts with phosphorus trichloride typically using a trialkylamine or quaternary ammonium salt as the catalyst. Hydrogen chloride is formed and either complexed with the amine or liberated as free hydrogen chloride gas forming the phosphite ester, tris(2,4-di-*tert*-butylphenyl)phosphite [31570-04-4] (64). The phosphite-based on 2,4-di-*tert*-butylphenol is a solid and very hydrolytically stable. Because of this hydrolytic stability it has replaced the use of tris(4-nonylphenyl) phosphite and other less stable phosphites in polyolefins and engineering resins. Another secondary antioxidant based on 2,4-di-*tert*-butylphenol is the diphosphite derived from pentaerythritol, bis(2,4-di-*tert*-butylphenyl)pentaerythrityl diphosphite [26741-53-7] (65). It too is widely used in polyolefins, and in engineering resins where high performance is required. 2,4-Di-*tert*-butylphenol is also used in the production of a benzotriazole-based uv stabilizer, 2-(2′-hydroxy-3′,5′-di-*tert*-butylphenyl)-5-

chlorobenzotriazole [3864-99-1]. Another uv stabilizer is the phenyl ester based on the carboxylic acid of 2,6-di-*tert*-butylphenol, 2,4-di-*tert*-butylphenyl 3′,5′-di-*tert*-butyl-4′-hydroxybenzoate [4221-80-1] (66). 2,4-Di-*tert*-butylphenol condenses with acetaldehyde under acid catalysis to produce a phenolic-based primary antioxidant, 2,2′-ethylidenebis(4,6-di-*tert*-butylphenol) [35958-30-6] which is used in the stabilization of polyolefins, styrenics, synthetic and natural rubber against thermal oxidative degradation.

The growth rate for 2,4-di-*tert*-butylphenol is estimated to be above average as the market share for high performance phosphites increase. *2,6-Di-tert-butylphenol (2,6-DTBP)* or 2,6-bis(1,1-dimethylethyl)phenol is produced from phenol by alkylation with isobutylene under aluminum catalysis. The crude alkylated phenol contains 2-*tert*-butylphenol, 2,6-di-*tert*-butylphenol, and 2,4,6-tri-*tert*-butylphenol. Pure 2,6-di-*tert*-butylphenol is produced by vacuum fractional distillation. Aluminum trisphenoxide is very selective in the formation of 2,6-di-*tert*-butylphenol and yields in excess of 80% are possible. 2,6-Di-*tert*-butylphenol [128-39-2] is packaged in 55-gal drums (208-L) and sold as a molten material in bulk quantity shipments in tank wagons and railcars.

The principal use for 2,6-di-*tert*-butylphenol is in the production of hindered phenolic antioxidants and this application accounts for 80−90% of all of this compound produced. Reaction of 2,6-DTBP with formaldehyde under base catalysis forms the methylene bisphenolic, 4,4′-methylenebis(2,6-di-*tert*-butylphenol) [118-82-1] which is used as a primary antioxidant in the stabilization of greases (67). In an excess of formaldehyde the hydroxymethyl compound is obtained. This hydroxymethyl derivative condenses with trimethylbenzene on a 3:1 mole ratio to yield 1,3,5-trimethyl-2,4,6,-tris(3′,5′-di-*tert*-butyl-4′-hydroxybenzyl)benzene [1709-70-2], which is widely used in the stabilization of polypropylene, particularly polypropylene fibers (68).

A large number of hindered phenolic antioxidants are based on the Michael addition of 2,6-di-*tert*-butylphenol and methyl acrylate under basic catalysis to yield the hydrocinnamate which is a basic building block used in the production of octadecyl 3-(3,5-di-*tert*-butyl-4-hydroxyphenyl)propionate, [2082-79-3], tetrakis(methylene-3(3,5-di-*tert*-butyl-4-hydroxylphenyl)propionate)methane [6683-19-8], and many others (69,70). These hindered phenolic antioxidants are the most widely used primary stabilizers in the world and are used in polyolefins, synthetic and natural rubber, styrenics, vinyl polymers, and engineering resins. 2,6-Di-*tert*-butylphenol is converted to a methylene isocyanate which is trimerized to a triazine derivative 1,3,5-tris(3′,5′-di-*tert*-butyl-4′-hydroxybenzyl)isocyanurate [27676-62-6]; this molecule has a special application in the protection of articles made from polypropylene fibers, such as surgical gowns, that are sterilized by ionizing radiation (71). Other uses for 2,6-di-*tert*-butylphenol include the production of a diisopropylidene sulfide derivative which is a cholesterol-lowering drug. Another sulfur derivative is useful as an antiinflammatory agent (72). 2,6-Di-*tert*-butylphenol is also used in the production of 4,4′-biphenol [92-88-6] which is used as a replacement for bisphenol-A in the production of liquid crystal polyesters, polysulfones, and polyetherimides. In these applications biphenol imparts an increased rigidity and higher heat distortion temperature to the polymer. Pure 2,6-di-*tert*-butylphenol is used as is as an antioxidant in oils and greases and the crude alkylate from the alkylation of phenol with isobutylene is commonly used in gasoline, diesel fuels, and jet fuels (73).

With the growth in thermoplastic materials replacing more traditional materials such as glass, wood, paper, and metal, the growth rate for 2,6-di-tert-butylphenol is estimated to be above average.

Di-sec-butylphenol (DSBP) is produced by the alkylation of phenol with 1-butene or 2-butene under acidic or aluminum catalysis. The production of di-sec-butylphenol under acid catalysis gives a 2 to 1 mixture of 2,4-di-sec-butylphenol and 2,6-di-sec-butylphenol [2,4-bis(1-methylpropyl)phenol and 2,6-bis(1-methylpropyl)phenol]. Under aluminum catalysis 2,6-di-sec-butylphenol is produced in greater than 90% yield. Di-sec-butylphenol [31291-60-8] is available in 55-gal drums (208-L) and in bulk shipments in tank wagons and railcars.

The only significant use for di-sec-butylphenol is a specialty nonionic surfactant produced by reaction with ethylene oxide under base catalysis. This surfactant is registered with EPA for use in emulsifying agrochemicals.

2,4-Dicumylphenol (2,4-DCP) or 2,4-bis(1-methyl-1-phenylethyl)phenol is produced by the alkylation of phenol with α-methylstyrene under acidic catalysis. The crude alkylation product contains 4-cumylphenol, 2,4-dicumylphenol, and 2,4,6-tricumylphenol along with some olefin oligomers. Pure 2,4-dicumylphenol can be obtained either by vacuum fractional distillation or crystallization from a suitable solvent. 2,4-Dicumylphenol [2772-45-4] is packaged in 55-gal drums (208-L) and sold as a bulk material in molten form.

The largest use for 2,4-dicumylphenol is in a production of a uv stabilizer of the benzotriazole class, 2-(2'-hydroxy-3',5'-dicumylphenyl)benzotriazole [70321-86-7] which is used in engineering thermoplastics where high molding temperatures are encountered (74). The high molecular weight of 2,4-dicumylphenol makes this uv stabilizer the highest molecular weight benzotriazole based uv stabilizer in commercial production.

2,6-Dimethylphenol (2,6-xylenol) is produced by the gas phase alkylation of phenol with methanol using modified alumina catalysis. The crude product contains 2-methylphenol, 2,6-dimethylphenol, a minor amount of 2,4-dimethylphenol, and a mixture of trimethylphenols. The 2,6-dimethylphenol is purified by fractional distillation. The mixture of di- and trimethylphenols is sold as cresylic acid for use as a solvent. 2,6-Dimethylphenol [576-26-1] is available in 55-gal drums (208-L) and in bulk shipments in tank wagons and railcars.

The oxidative coupling of 2,6-dimethylphenol to yield poly(phenylene oxide) represents 90–95% of the consumption of 2,6-dimethylphenol (75). The oxidation with air is catalyzed by a copper–amine complex. The poly(phenylene oxide) derived from 2,6-dimethylphenol is blended with other polymers, primarily high impact polystyrene, and the resulting alloy is widely used in housings for business machines, electronic equipment and in the manufacture of automobiles (see POLYETHERS, AROMATIC). A minor use of 2,6-dimethylphenol involves its oxidative coupling to 3,3',5,5'-tetramethyl-4,4'-biphenol [2417-40-1] (76). Tetramethyl-biphenol is used as a monomer in the production of specialty polycarbonates and reacts with epichlorohydrin to produce an epoxy resin.

The worldwide consumption of 2,6-dimethylphenol is difficult to estimate accurately because the majority is captively consumed. Growth rate for

2,6-dimethylphenol is directly related to the growth of engineering resins, which is generally predicted to be above average.

BIBLIOGRAPHY

"Alkylphenols" in *ECT* 2nd ed., Vol. 1, pp. 901–906, by R. W. G. Preston and H. W. B. Reed, Imperial Chemical Industries Limited; Supplement, pp. 27–31, by Gerd Leston, Koppers Company; "Alkylphenols" in *ECT* 3rd ed. Vol. 2, pp. 72–96, by H. W. B. Reed, Imperial Chemical Industries Limited; in *ECT* 4th ed., Vol. 2, pp. 113–143 by John F. Lorenc, Gregory Lambeth, and William Scheffer, Schenectady Chemicals Inc.

CITED PUBLICATIONS

1. *CAS Index Guide*, Chemical Abstract Service, Columbus, Ohio, 1982, Appendix 4.
2. Z. Rappoport, *Handbook of Tables for Organic Compound Identification*, 3rd ed., The Chemical Rubber Co., Cleveland, Ohio, 1967, p. 434.
3. R. L. Shriner, R. C. Fuson, D. Y. Curtin, and T. C. Morrill, *The Systematic Identification of Organic Compounds*, 6th ed., John Wiley & Sons, Inc., New York, 1980, p. 102.
4. C. J. Pouchert, ed., *The Aldrich Library of Infrared Spectra*, Aldrich Chemical Company, Inc., Milwaukee, Wis., 1981.
5. *SURFONIC N-Series Surface Active Agents*, Technical Product Bulletin, Texaco Chemical Company, Houston, Tex., 1987.
6. A. S. Lindsey and H. Jeskey, *Chem. Rev.* **57**, 583–620 (1957).
7. C. D. Gutsche in J. F. Stoddart, ed., *Monographs in Supramolecular Chemistry, Calixarenes*, Royal Society of Chemistry, Cambridge, England, 1989.
8. *Ethyl 702, Technical Bulletin*, Ethyl Corporation, Baton Rouge, La.
9. U.S. Pat. 4,222,883 (Sept. 16, 1980), E. Clinton (to Ethyl Corporation).
10. P. R. Dean, *Index of Commercial Antioxidants and Antiozonants*, Technical Bulletin, Goodyear Chemicals, Akron, Ohio, 1983.
11. Fr. Pat. 1,343,301 (Nov. 15, 1963), E. A. Meier and M. Dexter (to J. R. Geigy A-G).
12. U.S. Pat. 3,330,859 (July 11, 1967), M. Dexter, J. D. Spivack, and D. H. Steinberg (to Geigy Chemical Corporation).
13. U.S. Pat. 3,644,482 (Feb. 22, 1972), M. Dexter, J. D. Spivack, and D. H. Steinberg (to CIBA-GEIGY Corporation).
14. U.S. Pat. 3,678,047 (July 18, 1972), G. Kletecka and P. D. Smith (to B. F. Goodrich Company).
15. U.S. Pat. 4,226,763 (Oct. 7, 1980), M. Dexter and R. A. E. Winter (to CIBA-GEIGY Corporation).
16. J. W. Chien, W. P. Connor, *J. Am. Chem. Soc.* **90**, 1001 (1968); D. J. Harper, J. F. McKellar, and P. H. Turner, *J. Appl. Polym. Sci.* **18**, 2805 (1974); N. S. Allen, J. Homer, and J. F. McKellar, *Macromol. Chem.* **179**, 1575 (1978).
17. U.S. Pat. 2,971,968 (Feb. 14, 1961), A. M. Nicholson and B. Zurensky (to Ferro Corporation); U.S. Pat. 2,971,941 (Feb. 14, 1961), S. B. Elliott, C. H. Furchman, and A. M. Nicholson (to Ferro Corporation).
18. U.S. Pat. 4,447,656 (May 8, 1984), L. D. Kershner, L. R. Thompson, and R. M. Strom (to Dow Chemical Company); U.S. Pat. 4,205,187 (May 27, 1980), J. N. Cardenes and W. T. Reichle (to Union Carbide Corporation).

19. J. March, *Advanced Organic Chemistry Reactions, Mechanisms, and Structure*, 3rd ed., John Wiley & Sons, Inc., New York, 1985, p. 448.

20. *Ibid*, p. 463; S. H. Patinkin and B. S. Friedman in G. A. Olah, ed., *Friedel-Crafts and Related Reactions*, Vol. 2, Wiley-Interscience, New York, 1964, p. 7.

21. A. J. Kolka and co-workers, *J. Org. Chem.* **22**, 642 (1957).

22. U.S. Pat. 4,360,699 (Nov. 23, 1982). W. E. Wright (to Ethyl Corp.).

23. C. B. Campbell, A. Onopchenko, and D. C. Young, *Ind. Eng. Chem. Res.* **29**, 642–647 (1990).

24. W. A. Pardee and W. Weinrich, *Ind. Eng. Chem.* 36, Vol. 7, 602 (1944).

25. B. Schmeth, *Chem. Week* (Jan 2, 2002).

26. *Registry of Toxic Effects of Chemical Substances*, National Institute for Occupational Safety and Health, Cincinnati, Ohio, July 1990.

27. R. J. Lewis, Sr., *Sax's Properties of Industrial Materials*, 10th ed.,Vols. 2 and 3, John Wiley & Sons, Inc., New York, 2000.

28. *Pentaphen, Technical Bulletin*, Pennwalt Chemicals, Philadelphia, Pa., March 1981.

29. Jpn. Pat. 62,050,186 (March 4, 1987), K. Ogawa (to Matsushita Electric Industrial Company).

30. U.S. Pat. 4,258,142 (Mar. 24, 1981), M. H. Fisch, S. S. Ahluwalia, and B. A. Hegranes (to Witco Corporation).

31. Neth. Pat. Appl. 6,406,854 (Jan. 19, 1965) (to United States Rubber Company).

32. U.S. Pat. 3,694,513 (Sept. 26, 1972), S. W. Tobey and M. Z. Lourandos (to The Dow Chemical Company).

33. D.E. Pat. 3,539,776 (May 21, 1987), D. Lansberg, M. Lieb, H. Moeckel, and H. Uho (to BASF A.G.).

34. D.E. Pat. 2,545,389 (April 29, 1976), I. Yamamoto, Y. Takahashi, and N. Kyomura (to Mitsubishi Chemical Industries Company).

35. U. S. Pat. 6, 448,453 (Sept. 10, 2002), J. Oberholtzer and co-workers (to General Electric Company).

36. Eur. Pat. 305,214 (Mar. 1, 1989), M. Okamoto (to Idemitsu Petrochemical Company).

37. U.S. Pat. 4,837,086 (June 6, 1989), T. K. Hiroshi, M. H. Naohiro, and T. S. Shir (to Hitachi Chemical Company).

38. U.S. Pat. 3,761,414 (Sept. 25, 1973), H. Hangen and H. Chafetz (to Texaco, Inc.).

39. U.S. Pat. 4,089,791 (May 16, 1978), H. Hangen and D. G. Weetman (to Texaco, Inc.).

40. E.P. Pat. 204,659 (Dec. 10, 1986), C. E. Monnico and S. A. C. Sheik (to CIBA-GEIGY A.G.).

41. U.S. Pat. 3,737,480 (June 5, 1973), E. E. Stahly and E. W. Lard (to W. R. Grace Company).

42. Jpn. Pat. 62,053,942 (Mar. 9, 1987), M. Sasaki and co-workers (to Sumitomo Chemical Company, Ltd.).

43. U.S. Pat. 6,554,871 (April 29, 2003) H. J. Braun (to Wella A6).

44. U.S. Pat. 3,170,893 (Feb. 23, 1965), S. Steinguseo and I. D. Salyer(to Monsanto Company).

45. Brit. Pat. 951,935 (Mar. 11, 1964), D. Ranson (to Imperial Chemical Industries Ltd.).

46. U.S. Pat. 3,679,736 (July 25, 1972), I. C. Popuff and K. I. H. Williams (to Pennwalt Corporation).

47. U.S. Pat. 2,762,787 (Sept. 11, 1956), I. Goodman and A. Lambert (to Imperial Chemical Industries, Ltd.).

48. U.S. Pat. 4,433,181 (Feb. 21, 1984), S. N. Holter (to Koppers Company, Inc.).

49. D.E. Pat. 2,621,006 (Dec. 2, 1976), C. E. Ziegler and H. J. Peterli (to CIBA-GEIGY A.G.).

50. "Nonylphenol, Chemical Profile, "*Chemical Market Reporter* (July 2, 2001).

51. D.E. Pat. 270,726 (Aug. 9, 1989), A. Kirchhof and A. Strauch (to Veb Galvanotechnik Leipzig).
52. Eur. Pat. 324,716 (July 19, 1989), B. Holt (to CIBA-GEIGY Corp.).
53. U.S. Pat. 5,900,397 (May 4, 1999), D. J. Tracy and R. Li (Rhodes, Inc.).
54. U.S. Pat. 2,220,845 (Nov. 5, 1940), C. L. Moylet to The Dow Chemical Company).
55. B.R. Pat. 7,803,418 (May 8, 1979), J. M. Bohen (to Pennwalt Corporation).
56. U.S. Pat. 4,668,757 (May 26, 1987), G. Nichols.
57. U.S. Pat. 4,828,879 (July 30, 1987), K. B. Sellstrom and H. G. Waddill (to Texaco, Inc.).
58. Jpn. Pat. 01,138,275 (May 31, 1989), Y. Masuda (to Agency of Industrial Science and Technology).
59. Jpn. Pat. 1,042,288 (Feb. 14, 1989), H. Hasegawa (to Richo Company).
60. D.E. Pat. 3,506,680 (Aug. 28, 1986), W. Paul, U. Grigo, P. R. Mueller and W. Nouvertne (to Bayer A.G.).
61. U.S. Pat. 4,224,451 (Sept. 23, 1980), R. D. Roberts and W. B. Hardy (to American Cyanamid Company).
62. D.E. Pat. 2,620,897 (Nov. 25, 1976), C. E. Tiegler (to CIBA-GEIGY A.G.).
63. U.S. Pat. 4,443,536 (April 17, 1984), G. J. Lestina (to Eastman Kodak Company).
64. U.S. Pat. 4,312,818 (Jan. 26, 1982), R. M. Larsch, E. O. Linderfels, and H. Z. Ernsthofin (to CIBA-GEIGY Corporation).
65. Ger. Offen. 2,709,528 (Sept. 15, 1977), J. F. York (to Borg-Warner Corporation).
66. U.S. Pat. 4,128,726 (Dec. 5, 1978), V. G. Grosso and R. L. Hillard (to American Cyanamid Company).
67. U.S. Pat. 3,211,652 (Oct. 12, 1965), J. B. Hinkamp (to Ethyl Corporation).
68. U.S. Pat. 3,026,264 (Sept. 12, 1962), A. L. Rocklin and J. L. Van Winkle (to Shell Oil Company).
69. U.S. Pat. 3,330,859 (July 11, 1967), M. Dexter, J. D. Spivack, and D. H. Steinberg (to Geigy Chemical Corporation).
70. U.S. Pat. 3,642,868 (Feb. 15, 1972), M. Dexter and co-workers, (to CIBA-GEIGY Corp.).
71. U.S. Pat. 3,531,483 (Sept. 29, 1970), J. C. Gilles (to B.F. Goodrich Company).
72. U. S. Pat. 5,684,204 (Nov. 4, 1997), R. S. Matthew (to Procter and Gamble).
73. *Ethyl Antioxidants in Fuels, Technical Bulletin*, Ethyl Corporation, Baton Rouge, La., 1977.
74. U.S. Pat. 4,275,004 (June 23, 1981), R. A. E. Winter and M. Dexter (to CIBA-GEIGY Corp.).
75. U.S. Pat. 3,306,874 (Feb. 28, 1967), A. S. Hay (to General Electric Company).
76. U.S. Pat. 3,247,262 (April 19, 1966), W. W. Kneding (to The Dow Chemical Company).

JOHN F. LORENC
GREGORY LAMBETH
WILLIAM SCHEFFER
Schenectady Chemicals, Inc.

ALLYL ALCOHOL AND MONOALLYL DERIVATIVES

1. Introduction

Allyl alcohol, $CH_2=CH-CH_2OH$ (2-propen-1-ol) [107-18-6] is the simplest unsaturated alcohol. One hydrogen atom can easily be abstracted from the allylic methylene ($-CH_2-$) to form a radical. Since the radical is stabilized by resonance with the $C=C$ double bond, it is very difficult to get high molecular weight polymers by radical polymerization. In spite of the fact that allyl alcohol has been produced commercially for some years (1), it has not found use as a monomer in large volumes as have other vinyl monomers.

More recently, however, the technology of introducing a new functional group to the double bond of allyl alcohol has been developed. Allyl alcohol is accordingly used as an intermediate compound for synthesizing raw materials such as epichlorohydrin and 1,4-butanediol, and this development is bringing about expansion of the range of uses of allyl alcohol.

2. Physical Properties

Allyl alcohol is a colorless liquid having a pungent odor; its vapor may cause severe irritation and injury to eyes, nose, throat, and lungs. It is also corrosive. Allyl alcohol is freely miscible with water and miscible with many polar organic solvents and aromatic hydrocarbons, but is not miscible with n-hexane. It forms an azeotropic mixture with water and a ternary azeotropic mixture with water and organic solvents (Table 1). Allyl alcohol has both bacterial and fungicidal effects. Properties of allyl alcohol are shown in Table 2.

3. Chemical Properties

3.1. Addition Reactions. The $C=C$ double bond of allyl alcohol undergoes addition reactions typical of olefinic double bonds. For example, when

Table 1. **Azeotropic Boiling Points of Allyl Alcohol–Water–Organic Solvent Systems**

Organic solvent	Boiling point, °C	Component, wt %		
		Allyl alcohol	Water	Solvent
none	88.9	72	28	
benzene	68.2	9.1	7.3	83.6
diallyl ether	77.8	8.7	12.4	78.9
allyl acetate	82.6	9	20	71
cyclohexane	66.2	10.9	8.1	81.0
toluene	80.6	31.4	15.2	53.4

Table 2. **Properties of Allyl Alcohol**[a]

Property	Value
molecular formula	C_3H_6O
molecular weight	58.08
boiling point, °C	96.90
freezing point, °C	−129.00
density, d_4^{20}	0.8520
refractive index, n_D^{20}	1.413
viscosity at 20°C, mPa·s(=cP)	1.37
flash point[b], °C	25
solubility in water at 20°C, wt %	infinity

[a] Ref. 2.
[b] Closed cup.

bromine is added, a good yield of 2,3-dibromopropanol is obtained although 1,2,3-tribromopropane is obtained as a by-product. 1,2,3-tribromopropane is formed from substitution of the hydroxyl group of allyl alcohol by bromide and further addition of bromine to the C=C double bond. When this addition reaction is carried out in 2,3-dibromopropanol solvent, the substitution reaction is reduced and the yield of 2,3-dibromo adduct is increased (3). The addition of chlorine is different from that of bromine; the yield of 2,3-dichloro adduct is low (4), and much intermolecularly condensed ether by-product is formed. When hydrogen chloride dissolved in a solvent, such as a low boiling point ether, is used, the yield of 2,3-dichloro adduct can be increased (5). Furthermore, when an aqueous solution of hydrogen chloride above 45 wt % is used, the formation of ether by-product can be reduced, as can the formation of chlorohydrin which normally occurs in aqueous chlorine solution. Thus, a high yield of 2,3-dichloropropanol [616-23-9] can be obtained. For example, when chlorination is done continuously in a 50–60 wt % aqueous solution of hydrogen chloride at 0°C, allyl alcohol reacts completely with chlorine and 2,3-dichloropropanol is obtained in 95% yield (6). Ephichlorohydrin [106-89-8] is obtained by saponifying 2,3-dichloropropanol with calcium hydroxide.

$$CH_2{=}CHCH_2OH + Cl_2 \xrightarrow{HCl/H_2O} CH_2ClCHClCH_2OH$$

$$CH_2ClCHClCH_2OH + 1/2\ Ca(OH)_2 \longrightarrow \underset{\underset{Cl}{|}}{CH_2CH}{-}\overset{O}{CH_2} + 1/2\ CaCl_2 + H_2O$$

In fact, epichlorohydrin is being industrially manufactured by this method (7). The merit of this method is that it consumes only half the amount of chlorine and $Ca(OH)_2$ compared to that of the method via allyl chloride.

In the reaction of allyl alcohol with an aqueous chlorine solution, addition of hypochlorous acid to the double bond of allyl alcohol yields glycerol monochlorohydrin and as a by-product, glycerol dichlorohydrin. Thus, a poor yield of glycerol monochlorohydrin is obtained (8). To improve the yield of glycerol monochlorohydrin, addition of sodium carbonate in an amount equivalent to that of the hydrogen chloride in the aqueous chlorine solution, has been proposed (9).

When thiol is added to the double bond of allyl alcohol under radical form-ing conditions, Markovnikov reaction selectivity takes place. Mercury com-pounds, light, and oxygen accelerate the addition reaction. In the presence of $(CH_3S)_2Hg$, light, and oxygen, $CH_3S(CH_2)_3OH$ can be obtained in 93% yield. In the presence of light and oxygen only, the yield decreases to 61%; the reaction cannot occur in the presence of light only (10). On the other hand, under ionic reaction conditions, an anti-Markovnikov reaction takes place (11).

Under alkaline conditions, an amine addition reaction can occur. For exam-ple, in the reaction of $C_6H_5CH_2CH_2NH_2$ and allyl alcohol in the presence of sodium alcoholate at 108°C for 80 h, 43.4% N-(3-hydroxypropyl)phenylethyl-amine is formed (12).

3.2. Hydroformylation.   Hydroformylation of allyl alcohol is a synthetic route for producing 1,4-butanediol [110-63-4], a raw material for poly(butylene terephthalate), an engineering plastic (qv); many studies on the process have been carried out.

$$CH_2\!=\!CHCH_2OH \;+\; H_2 \;+\; CO$$

CHO
|
$CH_2CH_2CH_2OH$

(1)

(3)

CH_3CHCH_2OH
|
CHO

(2)

After it was found that the rhodium carbonyl–triphenylphospine–complex, $HRh(CO)[P(C_6H_5)_3]_3$, is very effective in increasing the yield of hydroformylated product, many studies were done in greater detail with rhodium catalysts (13). The reactions are generally performed under the following conditions: 60–100°C, 0.69–3.4 MPa (7–35 kg/cm^2), H_2 to CO molar ratio more than one, and excess ligand phosphine. By-products are branched aldehyde **(2)**, 2-hydroxymethylpro-pionaldehyde [38433-80-6], propionaldehyde produced by isomerization of allyl alcohol and n-propanol produced by hydrogenation of allyl alcohol. The types of by-products and the yield are affected by the reaction temperature, the molar ratio between phosphine and rhodium, the kind of phosphine, and the molar ratio between H_2 and CO. The yield of linear aldehyde **(1)**, 4-hydroxybu-tyraldehyde [25714-71-0], depends on the kind of ligand. A yield of 60–70% is obtained even with excess triphenylphosphine, but a yield of more than 80% is obtained with 1-bis(diphenylphosphino)ferrocene (14). Addition of a great excess of triphenylphosphine causes a gradual decrease in catalytic activity. On the other hand, addition of excess triphenylphosphine and bidentate phos-phine, especially, 1,4-bis(diphenylphosphino) butane to the rhodium complex in equimolar amounts enables the catalytic activity to be maintained, and the molar ratio of linear aldehyde to branched aldehyde is 9:1 and the selectivity to 2-hydroxytetrahydrofuran [5371-52-8] **(3)** is 80% (13). In the case of homogeneous reaction, separation and recovery of catalyst poses a problem. To solve this pro-blem, gas-phase reaction was attempted and it was found that linear aldehyde is selectively produced in 99% yield, using silica containing a small amount of alumina as the catalyst carrier. From this experimental result, the gas-phase reaction was found to be regiospecific. Catalytic activity is maintained for at least 250 hours by using tris-*p*-tolylphosphine (15).

In the reaction of allyl alcohol with carbon monoxide using cobalt carbonyl, $Co(CO)_8$ as the catalyst, in the presence of a small amount of hydrogen and carbon monoxide under pressure, 9.8 MPa (1420 psi), at 100°C, intramolecular hydroesterification takes place, yielding γ-butyrolactone [96-48-0] (16).

$$CH_2{=}CHCH_2OH \; + \; CO \; + \; H_2 \; \longrightarrow \; + \; CH_3CH_2CHO$$

With solvents having a nitrile group like acetonitrile, the selectivity of γ-butyro-lactone is increased, resulting in a yield of 60%.

3.3. Substitution of Hydroxyl Group.   The substitution activity of the hydroxyl group of allyl alcohol is lower than that of the chloride group of allyl chloride and the acetate group of allyl acetate. However, allyl alcohol undergoes substitution reactions under conditions in which saturated alcohols do not react. Reactions proceed in catalytic systems in which a π-allyl complex is considered as an intermediate. It can thus be said that this substitution reaction is a specific reaction of allyl alcohol. The reaction of allyl alcohol with diethylamine, using palladium acetyl acetonate [14024-61-4] and triphenylphosphine [603-35-0] as the catalyst, at 50°C for 30 min yields 95% diethylallylamine (17). However, in this reaction, the catalyst is deactivated gradually during the reaction due to oxidation of phosphine by allyl alcohol. The reactivity of ammonia is lower than that of dialkylamine and even under the above-mentioned conditions, a substitution reaction cannot take place. Catalytic activity decreases markedly in the presence of ammonia; however, using diphosphine as the ligand, improves catalytic activity and stability (18). For instance, in the reaction of allyl alcohol and ammonia with palladium acetyl acetonate and 1,3-bis(diphenylphosphono)propane [6737-42-4] as the catalyst and propylene glycol as the solvent at 110°C for 4 h, a mixture of monoallylamine, diallylamine, and triallylamine is obtained with a 73.5% conversion of allyl alcohol and a selectivity of 98.9% to amines. In the conventional process for synthetizing allylamine from allyl chloride and amine, the reaction vessel becomes badly corroded (19). Moreover, it has the disadvantage of forming sodium chloride as a by-product. The new process is, therefore, economically attractive.

With active methylene compounds, the carbanion substitutes for the hydroxyl group of allyl alcohol (17,20). Reaction of allyl alcohol with acetylacetone at 85°C for 3 h yields 70% monoallyl compound and 26% diallyl compound. Malonic acid ester in which the hydrogen atom of its active methylene is substituted by *N*-acetyl, undergoes the same substitution reaction with allyl alcohol and subsequently yields α-amino acid by decarboxylation (21).

In the reaction of allyl alcohol and Grignard reagent with $[P(C_6H_5)_3]_2$-$NiCl_2$ as the catalyst, formation of the carbon–carbon bond proceeds at a high yield (22).

$$CH_2{=}CHCH_2OH + RMgBr \longrightarrow CH_2{=}CHCH_2R + MgBrOH$$

For reaction with hydrogen halides, the substitution reaction with halide ion easily occurs when a cuprous or cupric compound is used as the catalyst

(23) and yields a halogenated allyl compound. With a cuprous compound as the catalyst at 18°C, the reaction is completed in 6 h. Zinc chloride is also a good catalyst (24), but a by-product, diallyl ether, is formed.

$$CH_2{=}CHCH_2OH + HCl \longrightarrow CH_2{=}CHCH_2Cl + H_2O$$

3.4. Oxidation. The C=C double bond of allyl alcohol undergoes epoxidation by peroxide, yielding glycidol [556-52-5]. This epoxidation reaction is applied in manufacturing glycidol as an intermediate for industrial production of glycerol [56-8-5], using a typical epoxidation agent such as peracetic acid.

$$CH_2{=}CHCH_2OH \ + \ CH_3COOOH \ \longrightarrow \ \underset{\underset{O}{\diagdown\diagup}}{CH_2{-}CHCH_2OH} \ + \ CH_3COOH$$

In the past, FMC Corporation industrially produced glycerol in two steps: production of glycidol from allyl alcohol in high boiling point ketone solvent, followed by hydrolysis of glycidol to glycerol. Daicel Chemical Industries produces glycerol with a reaction–distillation system to prevent a decrease in yield caused by intermolecular reaction of glycidol, by feeding water to the distillation column and converting the main part of the glycidol to glycerol *in situ* and not isolating glycidol as an intermediate (25). Shell Chemical Company produced glycerol, using hydrogen peroxide as the oxidant (25). The reaction, which is carried out with tungstic acid, H_2WO_4, as the catalyst, enables glycerol to be obtained in one step. When the pH (4–6), concentration and other conditions are controlled and the reaction temperature is at 45°C, yield of glycidol can reach 82–87% (26). Degussa A.G. produced glycidol by using sodium tungstate, $NaHWO_4$, as the catalyst. It is necessary to use a catalyst in the epoxidation reaction of olefinic double bonds (such as hydrogen peroxide or alkyl hydroperoxide). However, epoxidation of allyl alcohol is different from that of typical olefins in reactivity considerations such as the reaction rate and selectivity, because of interaction between the hydroxyl group of allyl alcohol and the catalyst. When tungstenic acid is used as the catalyst, the reaction rate of epoxidation of allyl alcohol by hydroperoxide as the oxidation reagent is 30 times faster than that of allyl chloride (27). Further, in the case of epoxidation by $(CH_3)_3COOH$ with a vanadium catalyst, the epoxidation rate of allyl alcohol is 1000 times faster than that of methyl allyl ether. It is postulated that a covalent alkoxide intermediate is formed between the metal and the hydroxyl group (28).

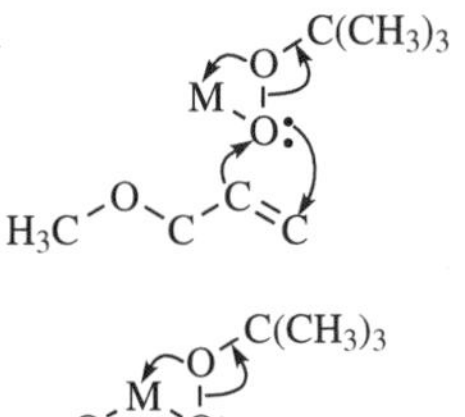

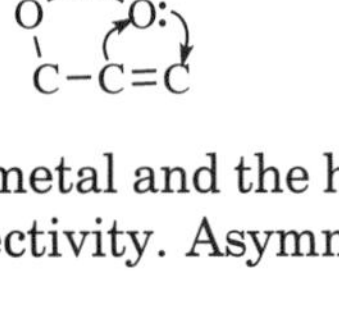

This chemical bond between the metal and the hydroxyl group of allyl alcohol has an important effect on stereoselectivity. Asymmetric epoxidation is well-known.

The most stereoselective catalyst is $Ti(OR)_4$, which is one of the early transition metal compounds and has no oxo group (28). Epoxidation of isopropylvinylcarbinol [4798-45-2] (1-isopropylallyl alcohol) using a combined chiral catalyst of $Ti(OR)_4$ and L-(+)-diethyl tartrate and $(CH_3)_3COOH$ as the oxidant, stops at 50% conversion, and the erythro:threo ratio of the product is 97:3. The reason for the reaction stopping at 50% conversion is that only one enantiomer can react and the unreacted enantiomer is recovered in optically pure form (28).

Allyl alcohol can be easily oxidized to yield acrolein [107-02-8] and acrylic acid [79-10-7]. In an aqueous potassium hydroxide solution of $RuCl_3$, allyl alcohol is oxidized by a persulfate such as $K_2S_2O_8$ at room temperature, yielding acrylic acid in 45% yield (29). There are also examples of gas-phase oxidation reactions of allyl alcohol, such as that with Pd–Cu or Pd–Ag as the catalyst at 150–200°C, in which allyl alcohol is converted by 80% and acrolein and acrylic acid are selectively produced in 83% yield (30).

3.5. Miscellaneous Reactions.   Allyl alcohol can be isomerized to propionaldehyde [123-38-6] in the presence of solid acid catalyst at 200–300°C. When copper or alumina is used as the catalyst, only propionaldehyde is obtained, because of intramolecular hydrogen transfer. On the other hand, acrolein and hydrogen are produced by a zinc oxide catalyst. In this case, it is considered that propionaldehyde is obtained mainly by intermolecular hydrogen transfer between allyl alcohol and acrolein (31).

$$CH_2{=}CHCH_2OH \longrightarrow CH_2{=}CHCHO + H_2$$

$$CH_2{=}CHCH_2OH + CH_2{=}CHCHO \longrightarrow CH_2{=}CHCHO + CH_3CH_2CHO$$

Friedel-Crafts reaction of allyl alcohol with benzene or alkylbenzene yields many kinds of products, in which the reaction species and the product ratio depend on the type of catalyst. Zinc chloride is the most effective catalyst for producing allyl compounds by this reaction (32).

Allyl alcohol undergoes reactions typical of saturated, aliphatic alcohols. Allyl compounds derived from allyl alcohol and used industrially, are widely manufactured by these reactions. For example, reactions of allyl alcohol with acid anhydrides, esters, and acid chlorides yield allyl esters, such as diallyl phthalates and allyl methacrylate; reaction with chloroformate yields carbonates, such as diethylene glycol bis(allyl carbonate); addition of allyl alcohol to epoxy groups yields products used to produce allyl glycidyl ether (33,34).

4. Industrial Manufacturing Processes for Allyl Alcohol

There are four processes for industrial production of allyl alcohol. One is alkaline hydrolysis of allyl chloride (1). In this process, the amount of allyl chloride, 20 wt % aqueous NaOH solution, water, and steam are controlled as they are added to the reactor and the hydrolysis is carried out at 150°C, 1.4 MPa (203 psi) and pH 10–12. Under these conditions, conversion of allyl chloride is 97–98%, and allyl alcohol is selectively produced in 92–93% yield. The main by-products are diallyl ether and a small amount of high boiling point substance. The alkali concentration and pH value are important factors. At high alkali concentrations, the amount of by-product, diallyl ether, increases and at low concentrations, conversion of allyl chloride does not increase.

A second process has two steps. The first step is oxidation of propylene [115-07-1] to acrolein and the second step is reduction of acrolein to allyl alcohol by a hydrogen transfer reaction, using isopropyl alcohol (25).

$$CH_2{=}CHCH_3 + O_2 \longrightarrow CH_2{=}CHCHO + H_2O$$

$$CH_2{=}CHCHO + (CH_3)_2CHOH \longrightarrow CH_2{=}CHCH_2OH + CH_3COCH_3$$

This process has defects such as co-production of acetone and a low yield of allyl alcohol.

At present, neither of these two processes are being used industrially. Another process is isomerization of propylene oxide [75-56-9].

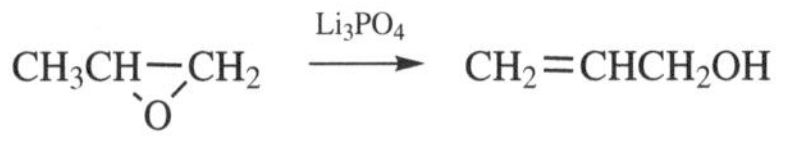

$$CH_3CH{-}CH_2 \xrightarrow{\text{Li}_3\text{PO}_4} CH_2{=}CHCH_2OH$$

In this process, the fine powder of lithium phosphate used as catalyst is dispersed, and propylene oxide is fed at 300°C to the reactor, and the product, allyl alcohol, together with unreacted propylene oxide is removed by distillation (25). By-products such as acetone and propionaldehyde, which are isomers of propylene oxide, are formed, but the conversion of propylene oxide is 40% and the selectivity to allyl alcohol reaches more than 90% (25). However, allyl alcohol obtained by this process contains approximately 0.6% of propanol. Until 1984, all allyl alcohol manufacturers were using this process. Since 1985 Showa Denko K.K. has produced allyl alcohol industrially by a new process which they developed (6,7). This process, which was developed partly for the purpose of producing epichlorohydrin via allyl alcohol as the intermediate, has the potential to be the main process for production of allyl alcohol. The reaction scheme is as follows:

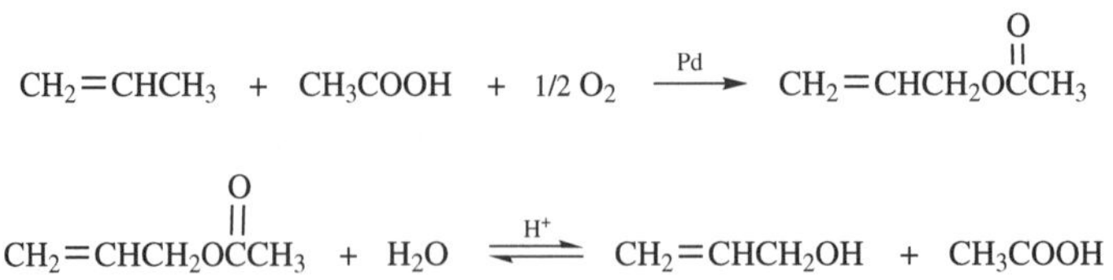

$$CH_2{=}CHCH_3 + CH_3COOH + 1/2\,O_2 \xrightarrow{\text{Pd}} CH_2{=}CHCH_2O\overset{\displaystyle O}{\overset{\|}{C}}CH_3$$

$$CH_2{=}CHCH_2O\overset{\displaystyle O}{\overset{\|}{C}}CH_3 + H_2O \underset{}{\overset{H^+}{\rightleftharpoons}} CH_2{=}CHCH_2OH + CH_3COOH$$

In the first step of the reaction, the acetoxylation of propylene is carried out in the gas phase, using solid catalyst containing palladium as the main catalyst at 160–180°C and 0.49–0.98 MPa (70–140 psi). Components from the reactor are separated into liquid components and gas components. The liquid components containing the product, allyl acetate, are sent to the hydrolysis process. The gas components contain unreacted gases and CO_2. After removal of CO_2, the unreacted gases, are recycled to the reactor. In the second step, the hydrolysis, which is an equilibrium reaction of allyl acetate, an acid catalyst is used. To simplify the process, a solid acid catalyst such as ion-exchange resin is used, and the

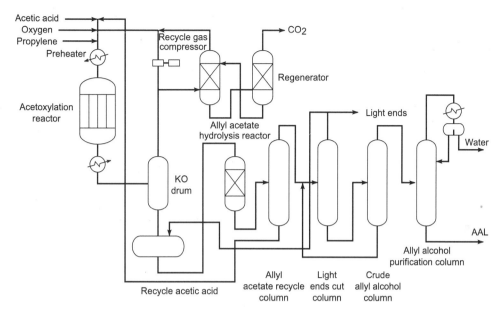

Fig. 1. The process for allyl alcohol (AAL) production via allyl acetate.

reaction is carried out at the fixed-bed liquid phase. The reaction takes place under the mild condition of 60–80°C and allyl alcohol is selectively produced in almost 100% yield. Acetic acid recovered from the hydrolysis process, is reused in the first step. As a result, it can be said that allyl alcohol is produced from oxidation of propylene by oxygen. Allyl alcohol forms an azeotropic mixture with water, and the mixture is a homogeneous liquid. Therefore, to obtain dry allyl alcohol, ternary azeotropic distillation and dehydration are required. This process for allyl alcohol production is shown in Figure 1.

The world's manufacturers of allyl alcohol are ARCO Chemical Company, Showa Denko K.K., Daicel Chemical Industries, and Rhône-Poulenc Chimie; total production is approximately 70,000 tons per year.

5. Uses of Allyl Alcohol

Recently, the uses of allyl alcohol have been greatly changing and increasing. Before 1985, the two principal uses of allyl alcohol were as a raw material for glycerol, which is industrially produced by the Daicel Chemical Company, and a monomer, diethylene glycol bis(allyl carbonate), for plastic optical lens (see ALLYL MONOMERS AND POLYMERS). It is estimated that each use is consuming several thousand tons of allyl alcohol per year. In 1985, Showa Denko K.K. started producing about 12,000 tons per year of epichlorohydrin using allyl alcohol as the raw material (7). Further in 1990, its production of epichlorohydrin increased to 24,000 tons per year, consuming 17,000–18,000 tons per year of allyl alcohol. Some epichlorohydrin manufacturers are planning to switch their production process to the allyl alcohol process. The consumption of allyl alcohol may therefore be expected to expand. In 1990, ARCO Chemical Company put on stream the world's largest allyl alcohol production plant, using the propylene oxide isomerization process, and at the same time, started producing 1,4-butanediol, consuming most of the allyl alcohol they produced, as the raw material. Since their production of 1,4-butanediol is said to be 35,000 tons per year, approximately 30,000 tons per year of allyl alcohol are needed. This is the first example of industrially producing 1,4-butanediol using allyl alcohol as the raw material (see ACETYLENE-DERIVED CHEMICALS). With ARCO Chemical Company starting up the allyl alcohol production plant, FMC Corporation stopped their allyl alcohol production plant. In addition to these applications, allyl alcohol is used as the raw material for producing allyl esters (diallyl phthalates and allyl methacrylate), allyl ether (allyl glycidyl ether), and a styrene–allyl alcohol copolymer. The styrene–allyl alcohol copolymer is produced by Monsanto Chemical Company (35) and is used in water-soluble paints, alkyd resins (qv), and urethanes as polyols.

6. Monoallyl Derivatives

In this article, mainly monoallyl compounds are described. Diallyl and triallyl compounds used as monomers are covered in the article entitled ALLYL MONOMERS AND POLYMERS and also in the literature (36,37).

6.1. Reactivity of Allyl Compounds. *Hydrosilylation.* The addition reaction of silane

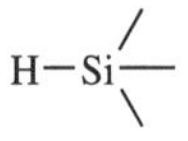

$$H-Si\overset{\diagup}{\underset{\diagdown}{\rule{0pt}{1.2em}}}$$

to the C=C double bond of allyl compounds is applied in the industrial synthesis of silane coupling agents.

$$CH_2=CHCH_2Y + HSiX_3 \longrightarrow X_3SiCH_2CH_2CH_2Y$$

Usually, trichlorosilane, trialkoxysilane, methyl dichlorosilane, and methyl dialkoxysilane are used. For example, the reaction of trichlorosilane with allyl methacrylate is as follows:

$$CH_2=C(CH_3)COOCH_2CH=CH_2 + HSiCl_3 \longrightarrow CH_2=C(CH_3)COO(CH_2)_3SiCl_3 \quad (4)$$

$$CH_2=C(CH_3)COO(CH_2)_3SiCl_3 + CH_3OH \longrightarrow CH_2=C(CH_3)COO(CH_2)_3Si(OCH_3)_3 \quad (5)$$

Platinum compounds are the most active catalysts for hydrosilylation. Compounds such as H_2PtCl_6 and $PtCl_2(CH_3CHCOCH_3)_2$ are effective. In the reaction with allyl methacrylate, which has two C=C double bonds, the allylic double bond selectively reacts. For example, while refluxing 10.5 moles (1422 g) of $HSiCl_3$ with 1 mL of 0.01 M $PtCl_2(CH^3COCHCOCH_3)$ in an acetone solution, as the catalyst, allyl methacrylate is added dropwise over 2 h and the reaction solution is agitated for 2 h at room temperature, whereupon 9.94 moles (2600 g) of product (4) is obtained after removal of excess $HSiCl_3$ (38). Also, in the case of allyl glycidyl ether, the allylic double bond is more reactive than the glycidyl group to silane. In the presence of mesityl oxide dichloro platinum complex as the catalyst, trimethoxysilane reacts with allyl glycidyl ether of equivalent moles at 130–140°C yielding 91.5% (3-glycidoxypropyl)-trimethoxysilane [2530-83-8] (39).

These examples show that silane reacts selectively with the γ-position of allyl compounds. However, in its reaction with allyl amine, a side reaction in which silane binds to the β-position takes place (40).

$$(CH_3CH_2O)_3SiH \; + \; CH_2{=}CHCH_2NH_2 \xrightarrow{H_2PtCl_6} \underset{\underset{Si(OCH_2CH_3)_3}{|}}{CH_2CH_2CH_2NH_2} \; + \; \underset{\underset{Si(OCH_2CH_3)_3}{|}}{CH_3CHCH_2NH_2}$$

Use of $HRh(CO)[P(C_6H_5)_3]_3$ as the catalyst and an excess of triphenylphosphine improves the $\gamma{:}\beta$ ratio. For example, reaction of triethoxysilane with allylamine of equivalent moles at 150°C for 10 h, yields the γ-form product in more than 70% and the $\gamma{:}\beta$ ratio is 26. Compared with this, when H_2PtCl_6 is used as the catalyst, the $\gamma{:}\beta$ ratio is 4 (41). Furthermore, when $Rh[(\mu\text{-}P(C_6H_5)_2\text{-}(cyclooctadiene)]_2$ is used as the catalyst, the yield of γ-form product is selectively increased to 92% and that of β-form product is decreased to 1.1% (42).

π-*Allyl Complex Formation.*   Allyl halide, allyl ester, and other allyl compounds undergo oxidative addition reactions with low atomic valent metal complexes to form π-allyl complexes. This is a specific reaction of allyl compounds.

$$R\underset{CH}{\overset{CH}{\diagdown}}\underset{CH_2}{\overset{X}{\diagup}} \; + \; M^n \longrightarrow R\underset{CH}{\overset{CH}{\diagdown}}\underset{\underset{M^{n+2}}{CH_2}}{\overset{CH}{\diagup}} \xrightarrow[Y^-]{E^+} \begin{array}{l} R\underset{CH}{\overset{CH}{\diagdown}}\underset{CH_2}{\overset{E}{\diagup}} \; + \; M^{n+2} \\[10pt] R\underset{CH}{\overset{CH}{\diagdown}}\underset{CH_2}{\overset{Y}{\diagup}} \; + \; M^n \end{array}$$

This π-allyl complex does not react with electrophilic reagent, E^+, in a catalytic way because the central metal remains in an oxidized state even after reaction. On the other hand, in the reaction with nucleophilic reagent, Y^-; the central metal is easily reduced to M^n, and the oxidative addition of allyl compound with M^n again takes place. Thus, this reaction continues in a catalytic way. Allyl compounds that carry out oxidative addition have as their functional group, X, halogen, RCOO, ROCOO, $(RO)_2COO$, RO, RNH, R_3N, NO_2, RSO_2, R_2S, and others. Metals that are effective as catalysts for the reaction of allyl compounds with nucleophilic reagents are Pd, Pt, Rh, Ru, Ni, Fe, Co, W, Mo, and others (43). Of these metals, Pd catalysts are the ones most studied. In the presence of nucleophilic reagent and carbon monoxide, a CO insertion can be done (44).

$$\underset{CH_2}{\overset{CH_2}{HC{\Big(}}}Pd\underset{2}{\overset{Cl}{\diagdown}} \xrightarrow{CO,\; CH_3CH_2OH} CH_2{=}CHCH_2COOCH_2CH_3$$

Table 3. **Properties of Important Allyl Compounds**

Property	Allylchloride	Allylacetate	Allylmethacrylate	AGE[a]	Allylamine	DMAA[b]
molecular formula	C_3H_5Cl	$C_5H_8O_2$	$C_7H_{10}O_2$	$C_6H_{10}O_2$	C_3H_7N	$C_5H_{11}N$
CAS Registry Number	[107-05-1]	[591-87-7]	[96-05-9]	[106-92-3]	[107-11-9]	[2155-94-4]
molecular weight	76.53	100.12	126.16	114.14	57.10	85.15
boiling point, °C	44.69	104	150	153.9	52.9	64.5
freezing point, °C	−134.5	−96	−60	−100	−88.2	
density, d_4^{20}	0.9382	0.9276	0.934	0.9698	0.7627	0.72
refractive index, n_D^{20}	1.416	1.404	1.436	1.435	1.420	
viscosity at 20°C, mPa·s(=cP)	3.36	0.52	13	1.20		0.44
flash point, °C	−31.7	6	33	57.2	−29	−23
solubility in water at 20°C, %	0.36	2.8			infinity	
limits of inflammability, %	11.30	13.0			2.2	
	2.9	2.1			22	

[a] Allyl glycidyl ether.
[b] Dimethylallylamine.

6.2. Physical Properties of Derivatives. The physical properties of some important monoallyl compounds are summarized in Table 3.

6.3. Allyl Chloride. This derivative, abbreviated AC, is a transparent, mobile, and irritative liquid. It can be easily synthesized from allyl alcohol and hydrogen chloride (23). However, it is industrially produced by chlorination of propylene at high temperature.

$$CH_2{=}CHCH_3 + Cl_2 \longrightarrow CH_2{=}CHCH_2Cl + HCl$$

The process involving allyl alcohol has not been industrially adopted because of the high production cost of this alcohol. However, if the allyl alcohol production cost can be markedly reduced, and also if the evaluated cost of hydrogen chloride, which is obtained as a by-product from the substitutive chlorination reaction, is cheap, then this process would have commercial potential. The high temperature propylene–chlorination process was started by Shell Chemical Corporation in 1945 as an industrial process (1). The reaction conditions are a temperature of 500°C, residence time 2–3 s, pressure 1.5 MPa (218 psi), and an excess of propylene to chlorine. The yield of allyl chloride is 75–80% and the main by-product is dichloropropane, which is obtained as a result of addition of chlorine. Other by-products include monochloropropenes, dichloropropenes, 1,5-hexadiene. At low temperatures, the amount of by-product dichloropropane increases and above 550°C, the amount of by-product benzene increases. Excess propylene is recovered and recycled to the reactor after washing and dehydration treatment, and hydrogen chloride, the by-product is recovered as concentrated hydrochloric acid (1). The purity of allyl chloride in the market is 99–99.5%; the main impurities are 1,5-hexadiene and monochlorinated compounds.

Uses. Allyl chloride is industrially the most important allyl compound among all the allyl compounds (see CHLOROCARBONS AND CHLOROHYDROCARBONS, ALLYL CHLORIDE). It is used mostly as an intermediate compound for producing epichlorohydrin, which is consumed as a raw material for epoxy resins (qv). World production of AC is approximately 700,000 tons per year, the same as that of epichlorohydrin. Epichlorohydrin is produced in two steps: reaction of AC with an aqueous chlorine solution to yield dichloropropanol (mixture of 1,3-dichloropropanol and 2,3-dichloropropanol) by chlorohydrination, and then saponification with a calcium hydroxide slurry to yield epichlorohydrin.

$$CH_2{=}CHCH_2Cl + Cl_2 + H_2O \longrightarrow (CH_2ClCHOHCH_2Cl, CH_2ClCHClCH_2OH) + HCl$$

$$(CH_2ClCHOHCH_2Cl, CH_2ClCHClCH_2OH) + HCl + Ca(OH)_2 \longrightarrow \underset{\underset{Cl}{|}}{CH_2CH}{-}\underset{\diagdown O \diagup}{CH_2} + CaCl_2 + 2\,H_2O$$

In the second step, a distillation-reaction system is applied to prevent hydrolysis of epichlorohydrin, by removing epichlorohydrin and water as an azeotropic mixture from the top of the distillation column. This operation is known as steam-stripping. In addition to being used in the synthesis of epichlorohydrin, AC is also used as a raw material for synthesizing other allyl compounds such as allyl esters, allyl ethers, and allylamines by nucleophilic substitution, utilizing the easily substituting property of its chloride group.

6.4. Allyl Esters. *Allyl Acetate.* Industrial production of allyl acetate started only rather recently. Nevertheless, among the allyl compounds, its production is second to that of allyl chloride. It is produced mostly for manufacturing allyl alcohol and its manufacture by acetoxylation of propylene has been described previously. The allyl acetate obtained may be separated and purified by distillation.

Allyl Methacrylate. At present, allyl methacrylate, AMA, is used mostly as a raw material for silane coupling agents. Utilizing the difference in the polymerizing ability of the allyl double bond and that of the methacrylate double bond, polymerization at the methacrylate double bond only is done by an anionic initiated reaction (45), yielding linear and soluble polymers with an allylic group attached to the pendant side chain. There are various methods for synthesizing AMA. For example, transesterification between allyl acetate and methyl methacrylate (46), esterification of methacrylic acid [79-41-4] with an excess of allyl alcohol (47), transesterification between methyl methacrylate [80-62-6] and allyl alcohol (48). This last method gives the highest yield of AMA. With an excess of methyl methacrylate and a combined catalyst of CaO and LiCl, allyl alcohol is converted by 97.7% and AMA is selectively produced in 95% yield (see also METHACRYLIC ACID AND DERIVATIVES).

Other monoallyl esters are esters of caproic acid and amyl glycolic acid, which are used as perfumes.

6.5. Allyl Ethers. The C—H bond of the allyl position easily undergoes radical fission, especially in the case of allyl ethers, reacting with the oxygen in the air to form peroxide compounds.

$$CH_2\!=\!CHCH_2OR \ + \ O_2 \ \longrightarrow \ CH_2\!=\!CHCHOR$$
$$\underset{\displaystyle OOH}{|}$$

Therefore, in order to keep allyl ether for a long time, it must be stored in an air-tight container under nitrogen. Utilizing the peroxidation property, allyl glycidyl ether, glycerol monoallyl ether [25136-53-2], ethylene glycol monoallyl ether [111-45-45], and others are employed in unsaturated polyesters for "air-drying" coatings, but in this application, usually polyfunctional allyl ethers are used (36).

Allyl Glycidyl Ether. This ether is used mainly as a raw material for silane coupling agents and epichlorohydrin rubber. Epichlorohydrin rubber is synthesized by polymerizing the epoxy group of epichlorohydrin, ethylene oxide, propylene oxide, and allyl glycidyl ether, AGE, with an aluminum alkyl catalyst (36). This rubber has high cold-resistance.

In the synthesis of AGE with an acid as the catalyst, allyl alcohol is added to the epoxy group of epichlorohydrin, yielding 3-allyloxy-1-chloro-2-propanol [4638-03-3], which then undergoes cyclization with alkali to yield AGE. Catalysts such as H_2SO_4, $SnCl_4$, BF_4^-; $(C_2H_5)_2O$ (33), heteropolyacids, $HClO_4$, and p-$CH_3C_6H_4SO_3H$ (34) are used.

6.6. Allyl Amines. *Allylamine.* This amine can be synthesized by reaction of allyl chloride with ammonia at the comparatively high temperature of 50–100°C (49), or at lower temperatures using $CuCl_2$ (50) or CuCl (51) as the catalyst. In all such methods, a mixture of monoallyl, diallyl, and triallyl amines is obtained. For selectively obtaining monoallylamine, AAm, hydrolysis of by hydrochloric acid is used (52). The degree of polymerization of monoAAm is low. However, its hydrogen chloride salt is polymerized readily (53) and an industrial process for manufacturing polyAAm has been developed (54). AAm polymers are used as fixing agents for reactive dyes on fibers. If the production of silane coupling agents from monoAAm is established, industrial consumption of monoAAm may increase.

Organotin compounds derived from diallylamine and alkylthiophosphine, have low phytotoxicity and are useful as relatively stable insecticides (55).

Dimethylallylamine. When 1-dimethylamino-2,3-dichloropropane [5443-48-1], which is obtained by addition of chlorine to dimethylallylamine, DMAA, reacts with NaSCN, the dimethylamino group is transferred, yielding 1,3-dithio-cyano-2-dimethylaminopropane, which is used to synthesize the carbamothioic ester, $(CH_3)_2NCH(CH_2SCONH_2)_2$ [15263-53-3] for pesticide use (56). Similarly, when 1-dimethylamino-2,3-dichloropropane reacts with Na_2S_x, 1,3-trithia-2-dimethylaminocyclopropane [31895-21-3], which is used as an insecticide (57), is produced. Furthermore, diallyldimethylammonium chloride (DADMAC) [7398-69-8], which is used as a monomer for synthesizing water-soluble polymers, is obtained from the reaction between DMAA and allyl chloride. DMAA is obtained in 95% yield by reaction of allyl chloride with two equivalent moles of dimethylamine at 23°C (58).

7. Health and Safety Factors

Most allyl compounds are toxic and many are irritants. Those with a low boiling point are lachrymators. Precautions should be taken at all times to ensure safe handling (59). Allyl compounds are harmful and may be fatal if inhaled, swallowed, or absorbed through skin. They are destructive to the tissues of the mucous membranes and upper respiratory tract, eyes, and skin (Table 4).

7.1. Handling and Storage.   Workers should be provided with appropriate respirators (NIOSH/MSHA approved), chemical resistant gloves, safety goggles, and other protective equipment. Work areas should be well-ventilated and be equipped with a safety shower and an eye bath. Care must be taken not to inhale any vapor and prevent it from getting into eyes, on skin, or on clothing. Prolonged or repeated exposure should be avoided. Thorough washing is

Table 4. **Toxicity of Important Monoallyl Compounds**[a]

Compound	LD_{50} rat, mg/kg
allyl alcohol	64
allyl chloride	64
allyl acetate	130
allyl methacrylate	430
allyl glycidyl ether	922
allylamine	102

[a] Ref. 59.

required after handling. The compounds should be kept in a tightly closed container away from heat, sparks, and open flames and be stored in a cool dry place. Allyl ethers must be stored under nitrogen, but allyl (meth)acrylate must not be stored under an inert atmosphere in order to inhibit polymerization.

BIBLIOGRAPHY

"Allyl Alcohol" in *ECT* 1st ed., Vol. 1, pp. 584–589, by H. G. Vesper, Shell Development Co.; "Allyl Compounds" in *ECT* 2nd ed., Vol. 1, pp. 916–928, by N. M. Bikales and N. G. Gaylord, Gaylord Associates, Inc.; "Allyl Compounds" in *ECT* 3rd ed., Vol. 2, pp. 97–108, by H. H. Beacham, FMC Corporation; in *ECT* 4th ed., Vol. 2, pp. 144–160, by Nobuyuki Nagato, Showa Denko K.K.

CITED PUBLICATIONS

1. A. W. Fairbain and H. A. Cheney, *Chem. Eng. Prog.* **43**(6), 280 (1947).
2. *Allyl Alcohol, Technical Publication SC: 46–32*, Shell Chemical Corp., San Francisco, Calif., Nov. 1, 1946, p. 95.
3. U.S. Pat. 3,268,597 (Aug. 23, 1966), C. W. Clemons and D. E. Overbeek (to Michigan Chemicals Corp.).
4. H. King and F. L. Pyman, *J. Chem. Soc.* **105**, 1238 (1914).
5. Ger. Pat. 2,007,867 (Aug. 26, 1971), D. Freudenberger and H. Fernholz (to Farbwerke Hoechst A.G.).
6. U.S. Pat. 4,634,784 (Jan. 6, 1987), N. Nagato (to Showa Denko K.K.).
7. N. Nagato, *Nikkakyo Geppo* **40**, 13 (Nov. 1987).
8. J. Read and E. Hurst, *J. Chem. Soc.* 989 (1922).
9. U.S. Pat. 2,311,023 (Dec. 16, 1939), B. T. Brooks (to Standard Alcohol Co.).
10. T. Kaneko, *J. Chem. Soc. Jpn.* **59**, 1139 (1938).
11. M. K. Gadzhiev and Kh. I. Areshidze, *Soobshch. Akad. Nauk Gruz, USSR*, **95**(1), 93 (1979).
12. O. Hromatka, *Ber.* **75B**, 379 (1942).
13. C. Botteghi and R. Ganzerla, *J. Mol. Catal.* **40**, 129 (1987).
14. C. U. Pittman, Jr. and W. D. Honnick, *J. Org. Chem.* **45**(11), 2132 (1980).
15. N. A. De Munk and J. P. Nootenboon, *J. Mol. Catal.* **11**, 233 (1981).
16. A. Matsuda, *Bull. Chem. Soc. Jpn.* **41**(8), 1876 (Aug. 1968).
17. K. E. Adkins and W. E. Walker, *Tetrahedron Lett.* **43**, 3821 (1970).
18. U.S. Pat. 4,942,261 (June 17, 1990), Y. Ishimura and N. Nagato (to Showa Denko K.K.).
19. R. S. Treseder and R. F. Miller, *Corrosion* **7**, 225 (1951).
20. X. Lu, L. Lu, and J. Sun, *J. Mol. Catal.* **41**, 245 (1987).
21. J. P. Haudegond and Y. Chauvin, *J. Org. Chem.* **44**(17), 3063 (1979).
22. C. Chuit and H. Felkin, *Chem. Commun.* (24), 1604 (1968).
23. J. Jacques, *Bull. Soc. Chim.* **12**, 843 (1945).
24. S. Coffey and C. F. Ward, *J. Chem. Soc.* **119**, 1301 (1921).
25. K. Yamagishi, *Chem. Econ. Eng. Rev.* **6**(7), 40 (July 1974).
26. A. Kleemann and R. Wagner, *Glycidol: Properties, Reactions, Applications*, Dr. Alfred Huthig Verlag, Heidelberg, Basel, New York, 1981.
27. V. N. Sapunov and N. N. Lebedev, *Zh. Org. Khim. (Engl.)* **2**(2), 263 (Feb. 1966).

28. K. B. Sharpless, *CHEMTECH*. 692 (Nov. 1985).
29. G. Green and W. P. Griffith, *J. Chem. Soc. Perkin Trans. I* 681 (1984).
30. U.S. Pat. 4,051,181 (Sept. 27, 1977), J. H. Murib (to National Distillers and Chemicals Corp.).
31. P. E. Weston and H. Adkins, *J. Am. Chem. Soc.* **51**, 2430 (Aug. 1929).
32. Y. M. Paushkin and I. H. Galal, *J. Chem. U.A.R. (Engl.)* **9**(2), 145 (1966).
33. Jpn. Pat. 77 3924 (Jan. 31, 1977), S. Sato and K. Arakida (to Nippon Oils and Fats Co., Ltd.).
34. Y. Izumi and K. Hayashi, *Chem. Lett.* (7), 787 (1980).
35. U.S. Pat. 2,894,938 (July 14, 1959), E. C. Chapin (to Monsanto Chemical Co.).
36. C. E. Schildknecht, *Allylic Compounds and Their Polymers*, John Wiley & Sons, Inc., New York, 1973.
37. H. Raech, Jr., *Allylic Resins and Monomers*, Reinhold Publishing Corporation, New York, 1965.
38. Ger. Offen. 1,271,712 (July 4, 1968), H. Knorre and W. Rothe (to Deutsche Gold-und-silber-Scheideanstalt).
39. Ger. Offen. 1,937,904 (Feb. 25, 1971), H. J. Vahlensieck and C. D. Seiler (to Dyamit Nobel A.G.).
40. Z. V. Belyakova and V. N. Bochkarev, *J. Gen. Chem. USSR* **42**(4), 848 (Apr. 1972).
41. U.S. Pat. 4,556,722 (Dec. 3, 1985), J. M. Quirk and S. Turner (to Union Carbide Corp.).
42. Eur. Pat. Appli. 302,672 (Feb. 8, 1989), K. Takatsuna and M. Tachikawa (to Toa Nenryo Kogyo K.K.).
43. J. Tsuji, *Kagaku Zokan* (109), 123 (Sept. 1986).
44. J. Tsuji and S. Imamura, *J. Am. Chem. Soc.* **86**, 4491 (1964).
45. G. F. Dalelio and T. R. Hoffend, *J. Polym. Sci. Part A* **1**(5), 323 (1967).
46. Brit. Pat. 1,059,875 (Feb. 22, 1967), D. K. V. Steel (to Imperial Chemicals Industries, Ltd.).
47. B. N. Rutovskii and A. M. Shur, *Zh. Prikl. Khim.* **24**, 851 (1951).
48. Ger. Pat. 3,423,441 A1 (Jan. 2, 1986), F. Schlosser and P. J. Arndt (to Roehm Gesellschaft mit beschrankten Hoftung).
49. U.S. Pat. 2,216,548 (Dec. 2, 1938), W. Converse (to Shell Development Co.).
50. Rom. Pat. 76,823 (Nov. 30, 1981), C. Benedek and C. Fagarasan (to Combinatul Chimic, Rimnicu-Vilcea).
51. USSR Pat. 578,301 (Oct. 30, 1977), A. M. Mezheritsukii and M. M. Krivenko.
52. M. T. Leefler, *Org. Synth.* **18**, 5 (1938).
53. S. Harada and S. Hasegawa, *Makromol. Chem., Rapid Commun.* **5**, 27 (1984).
54. Jpn. Kokai Tokkyo Koho 83-201811 (Nov. 24, 1983), S. Harada (to Nitto Boseki Co., Ltd.).
55. U.S. Pat. 3,984,542 (Oct. 5, 1976), D. R. Baker and O. Calif (to Stauffer Chemical Co.).
56. C. Xing and Y. Sun, *Shandong Haiyang Xueyuan Xuebao* **15**(1), 19 (1985).
57. RC Pat. CN 85,102,251 (Sept. 24, 1986), W. Guo and H. Peng.
58. RC Pat. CN 85,102,989 (Jan. 10, 1987), S. Cao and W. Zeng.
59. R. E. Lenga and S. Aldrich, *The Sigma-Aldrich Library of Chemical Safety Data Edition* II, 1988.

NOBUYUKI NAGATO
Showa Denko K.K.

ALLYL MONOMERS AND POLYMERS

1. Introduction

Allyl compounds comprise a large group of ethylenic compounds having unique reactivities and uses often contrasting with those of typical vinyl-type compounds (styrenes, acrylics, vinyl esters and ethers, and related compounds). In allyl compounds the double bond is not substituted by a strong activating group to promote polymerization but is attached to a carbon which generally bears one or more reactive hydrogen atoms, eg,

$$CH_2{=}CHCH_3 \qquad CH_2{=}CHCH_2OOCR \qquad CH_2{=}CHCHX_2$$

propylene   allyl ester   3,3-dihalo-1-propenes

The allylic 1-alkenes, which yield useful polymers by Ziegler-type catalysts, are not discussed here. Unlike monovinyl compounds, monoallyl compounds do not form homopolymers of high molecular weight by free-radical or conventional ionic mechanisms; in general, only viscous liquid homopolymers of limited use have been obtained. This is explained by the low reactivity of the ethylenic double bond together with the high reactivity of hydrogen atoms on the allylic carbon in reducing the molecular weight by degradative chain transfer (1). However, numerous monoallyl compounds, including some that occur in nature, are known outside the polymer field. Examples are allyl sulfur compounds of onions, mustard, and other food flavors; allyl esters used in perfumes; allylic drugs and other compounds of biologic activity; as well as important intermediates for organic syntheses (see ALLYL ALCOHOL AND DERIVATIVES).

In contrast, many allyl compounds containing two or more reactive double bonds yield solid, high molecular weight polymers by initiation with suitable free-radical catalysts. A number of polyfunctional allyl esters have achieved importance in polymerization and copolymerization especially to obtain heat-resistant cast sheets and thermoset moldings. The reactivities of these monomers often permit polymerization in two stages: a solid prepolymer containing reactive double bonds can be molded by heating; then completion of polymerization gives cross-linked articles of superior heat resistance. The most important examples of allyl polymers are the CR-39 or diallyl diglycol carbonate polymers and molding materials based on diallyl phthalates.

Another use is of minor proportions of polyfunctional allyl esters, eg, diallyl maleate, triallyl cyanurate, and triallyl isocyanurate, for cross-linking or curing preformed vinyl-type polymers such as polyethylene and vinyl chloride copolymers. These reactions are examples of graft copolymerization in which specific added peroxides or high energy radiation achieve optimum cross-linking (see COPOLYMERS).

Small proportions of mono- or polyfunctional allylic monomers also may be added as regulators or modifiers of vinyl polymerization for controlling molecular weight and polymer properties. Polyfunctional allylic compounds of high boiling point and compatibility are employed as stabilizers against oxidative degradation and heat discoloration of polymers. Diallyl ammonium salt copolymers are used in water purification and flocculation. Compounds containing one or more methallyl groups

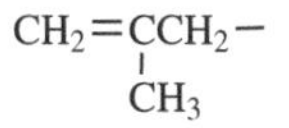

$$CH_2=CCH_2-$$
$$\overset{|}{CH_3}$$

find relatively little commercial utility.

2. Reactivity of Allyl Compounds

Whereas vinyl acetate [108-05-4] ($C_4H_6O_2$), upon heating with benzoyl peroxide or other free-radical initators, forms solid polymers of high molecular weight, similar treatment of allyl acetate [591-87-7] ($C_5H_8O_2$) gives only viscous liquid polymers.

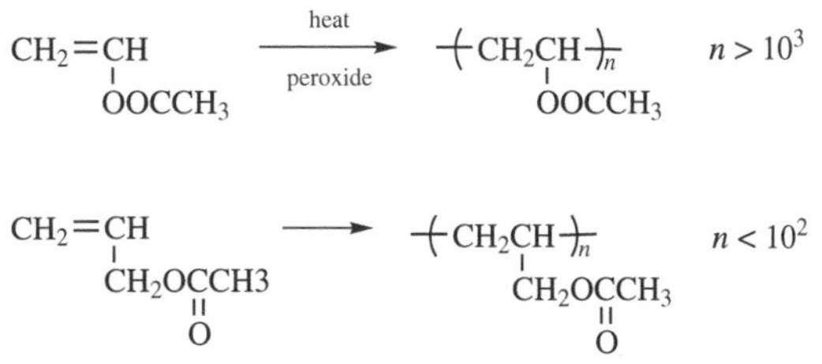

$$CH_2=CH \xrightarrow[\text{peroxide}]{\text{heat}} \left(CH_2CH\right)_n \quad n > 10^3$$
$$\overset{|}{OOCCH_3} \qquad\qquad \overset{|}{OOCCH_3}$$

$$CH_2=CH \longrightarrow \left(CH_2CH\right)_n \quad n < 10^2$$
$$\overset{|}{CH_2OCCH3} \qquad \overset{|}{CH_2OCCH_3}$$
$$\overset{\|}{O} \qquad\qquad \overset{\|}{O}$$

This is explained by the low reactivity of the double bond of the allyl compound together with prevalence of chain termination through reaction of allylic H atoms as shown (2).

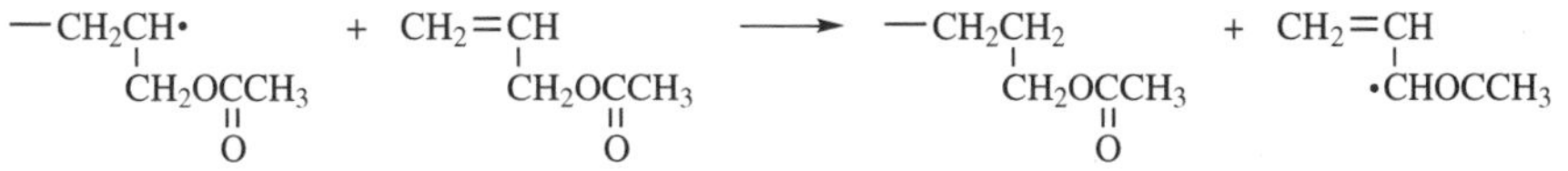

$$-CH_2CH\cdot \quad + \quad CH_2=CH \longrightarrow -CH_2CH_2 \quad + \quad CH_2=CH$$
$$\overset{|}{CH_2OCCH_3} \qquad \overset{|}{CH_2OCCH_3} \qquad\qquad \overset{|}{CH_2OCCH_3} \qquad \overset{|}{\cdot CHOCCH_3}$$
$$\overset{\|}{O} \qquad\qquad \overset{\|}{O} \qquad\qquad\qquad \overset{\|}{O}$$

When vinyl and allyl monomers undergo chain transfer with solvents, so-called telomers may form, generally of rather low molecular weight (1).

Because of the low reactivity and tendency to undergo chain transfer, small additions of most allyl compounds retard polymerization of typical vinyl monomers in free-radical systems (1,3) and may be useful in controlling molecular weight and structure in polymers.

Many polyallyl compounds, upon heating with radical initiators, form solid high polymers in spite of chain transfer and loss of some double bonds by cyclization. With diallyl esters, such as diallyl phthalates, these slower polymerizations can be controlled more readily than in the polymerization of poly-functional vinyl compounds to give soluble prepolymers containing reactive double bonds (1,4). Cyclization in polymerization of diallyl compounds also can occur depending on the reaction conditions (5). In general, more cyclization occurs at lower monomer concentrations in solution and cross-linking is thereby reduced.

Few allyl monomers have been polymerized to useful, well-characterized products of high molecular weight by ionic methods, eg, by Lewis acid or base catalysts. Polymerization of the 1-alkenes by Ziegler catalysts is an exception. However, addition of acidic substances, at room temperature or upon heating, often gives viscous liquid low mol wt polymers, frequently along with by-products of uncertain structure.

In special cases allyl compounds, such as diallyl [592-42-7], (C_6H_{10}), (1,5-hexadiene) and diallyl ether [557-40-4] ($C_6H_{10}O$), can form high polymers by addition of active hydrogen atoms. The best known case is the polymerization of diallyl with dimercaptans using free-radical initiation (6). These reactions are related to chain transfer and what has been called telocopolymerization (1,7).

Allyl compounds, depending on the structure of substituents, can undergo other reactions such as rearrangements, hydrolysis, and additions. Such reactions that may affect polymerization and have health considerations have been closely studied only with important industrial compounds (1).

3. Diallyl Carbonate Cast Plastics

From a number of diallyl esters investigated, diallyl diglycol carbonate or diethylene glycol bis(allyl carbonate), DADC, was developed to produce by bulk polymerization cast sheets, lenses, and other shapes of outstanding scratch resistance, and optical and mechanical properties. CR-39, a trademark of PPG Industries, is used to describe this material. DADC is the CR-39 monomer [142-22-3] from which CR-39 homopolymer [25656-90-0] ($C_{12}H_{18}O_7$)$_x$ is made. CR-39 polymers have greater impact resistance and lower density than glass. These polymers are the most important clear, organic optical materials shaped by the casting process and by machining. Many polyfunctional ethylenic monomers have been patented (8,9).

3.1. DADC Monomers.   Reaction of allyl alcohol in the presence of alkali with diethylene glycol bis(chloroformate), obtained from the glycol and phosgene, gives the monomer

$$O(CH_2CH_2O\overset{\overset{\displaystyle O}{\|}}{C}Cl)_2 \ + \ 2\,CH_2{=}CHCH_2OH \ + \ 2\,NaOH \ \longrightarrow$$

$$O(CH_2CH_2O\overset{\overset{\displaystyle O}{\|}}{C}OCH_2CH{=}CH_2)_2 \ + \ 2\,NaCl \ + \ 2\,H_2O$$

In another method, phosgene is gradually passed into 1,2-propylene glycol (9). The chloroformate is washed, dried, and distilled at 266 Pa (2 mm Hg) and added slowly to a mixture of allyl alcohol and pyridine below 15°C. The purified monomer 1,2-propylene glycol bis(allyl carbonate) ($C_{11}H_{16}O_6$) heated with lauroyl peroxide at 70°C gives a hard clear, polymer.

Reaction of allyl chloroformate and diethylene glycol in the presence of alkali with cooling is another method of preparing the diallyl carbonate ester DADC. The properties of diallyl carbonate monomers are given in Table 1.

DADC monomer is a colorless liquid of mild odor and a viscosity of 9 mPa s(=cP) at 25°C. It is low in toxicity, but can produce skin irritation. It is fairly resistant to saponification by dilute alkali. Contact with strong alkali at

Table 1. **Properties of Diallyl Glycol Carbonate Monomers**

Monomer	Molecular formula	CAS Registry Number	Boiling point, °C$_{Pa}$[a]	n_D^{20}	d_4^{20}
ethylene glycol bis(allyl carbonate)	$C_{10}H_{14}O_6$	[4074-91-3]	122$_{133}$	1.444	1.114
diethylene glycol bis(allyl carbonate)[b]	$C_{12}H_{18}O_7$	[142-22-3]	160$_{266}$	1.452	1.143
triethylene glycol bis(allyl carbonate)	$C_{14}H_{22}O_8$		polymerized	1.452	1.135
tetraethylene glycol bis(allyl carbonate)	$C_{16}H_{26}O_9$			1.454	1.133
glycerol tris (allyl carbonate)	$C_{15}H_{20}O_9$			1.456	1.194
ethylene glycol bis(methallyl carbonate)	$C_{12}H_{18}O_6$	[64653-60-7]	142$_{266}$	1.449	1.110

[a] To convert Pa to mm Hg, multiply by 0.0075.
[b] DADC (CR-39 monomer).

higher temperature produces the more toxic allyl alcohol. Properties are given in Table 2 and the trade literature (10). DADC is soluble in common organic solvents and in methyl methacrylate, styrene, and vinyl acetate. It is partially soluble in amyl alcohol, gasoline, and ligroin. It is insoluble in ethylene glycol, glycerol, and water.

3.2. DADC Homopolymerization.

Bulk polymerization of CR-39 monomer gives clear, colorless, abrasion-resistant polymer castings that offer advantages over glass and acrylic plastics in optical applications. Free-radical initiators are required for thermal or photochemical polymerization.

Table 2. **Typical Properties of Commercial DADC Monomer**[a]

Property	Value
appearance	clear, colorless liquid
color, APHA	10
odor	none to slight
specific gravity	1.15$^{20}_4$
refractive index, n_D^{20}	1.452
boiling point at 266 Pa[b], °C	166
melting point (supercooled), °C	−4 to 0
viscosity at 25°C, mm^2/s (=cSt)	15
flash point	
Seta closed cup, °C	173
Cleveland open cup, °C	186
water content, slightly hygroscopic, %	0.1

[a] Ref. 10.
[b] To convert Pa to mm Hg, multiply by 0.0075.

Relatively high concentrations of organic peroxide or azo initiators are needed to obtain complete polymerization. After the reaction peak exotherm, polymerization slows down. Initiator concentrations must be high enough to complete conversion. Polymerization is inhibited by oxygen and copper, lead, and sulfur compounds (11).

Bulk polymerization has been studied at relatively low temperatures and in toluene and carbon tetrachloride solutions carried to low conversions (12). The effects of temperature and different organic peroxide initiators have been observed. The molecular weight of soluble polymer after 3% conversion is ca $M_n = 19,000$ and is somewhat dependent on initiator concentration or temperature between 35 and 65°C. With di-2-methylpentanoyl peroxide, polymerization can be carried out at temperatures as low as 13°C. Nuclear magnetic resonance studies of the unsaturated prepolymer show that less than 10% of monomer units undergo cyclopolymerization.

3.3. Casting of DADC.

Sheets, rods, and lens preforms are cast from CR-39 or prepolymer syrup by methods similar to those used for methacrylate ester syrups (13). Casting in glass cells with flexible gaskets is described in reference 4. Horizontal cells may be heated at 60–70°C to reach the gel state, and later at 125°C for completing the polymerization of thin sheets. In some cases polymerization of gel shapes can be completed by heating the outside of the mold with additional shaping. Shrinkage between monomer and polymer is about 14%. Under controlled casting conditions, density increases and initiator concentration decreases, as shown in Figure 1.

Many proprietary methods have been developed for casting and shaping DADC, especially for lenses. In one method DADC containing 3.5% diisopropyl percarbonate is prepolymerized by warming to a syrup of viscosity 40–60 mm²/s

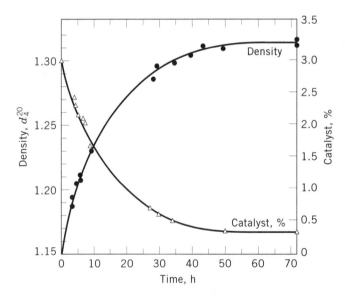

Fig. 1.   Bulk polymerization of diethylene glycol bis(allylcarbonate) at 45°C with initial addition of 3.0% diisopropyl percarbonate. Rates of polymerization as measured by density and catalyst consumption decrease with time at a given temperature (14).

(=cSt) (15). Polymerization is continued in a lens for 18 h at 90°C followed by annealing at 120°C.

Scratch resistance of polymer from DADC is improved by novel mixtures of peroxide initiators such as 5% isopropyl percarbonate with 3.5% benzoyl peroxide (16). In order to force completion of polymerization and attain the best scratch resistance in lenses, uv radiation is applied (17). Eyeglass lenses can be made by prepolymerization in molds followed by removal for final thermal cross-linking (18).

Initially, DADC polymers were used in military aircraft for windows of fuel and deicer-fluid gauges and in glass-fiber laminates for wing reinforcements of B-17 bombers. Usage in impact-resistant, lightweight eyewear lenses has grown rapidly and is now the principal application. Other uses include safety shields, filters for photographic and electronic equipment, transparent enclosures, equipment for office, laboratory, and hospital use, and for detection of nuclear radiation.

Many lens casters use the term hard-resin lenses for DADC products. Sheet castings for covers for welding-mask lenses are produced. These covers are resistant to hot metal fragments.

Typical properties of homopolymers from DADC are given in Tables 3 and 4, and in manufacturers' literature (20).

3.4. Coatings. In recent years methods have been developed to improve abrasion resistance of DADC polymer surfaces in optical devices and glazings by means of special coatings. Hard or glasslike coatings may be applied by near-vacuum vapor deposition of quartz (silica) or by hydrolysis of alkoxysilanes. Vapor deposition or sputtering processes of SiO_2 may be facilitated by electron beams or glow discharge. A second class of coatings are elastomeric polymers,

Table 3. **Optical and Electrical Properties of DADC Homopolymer** [a,b]

Property	Value	ASTM method
refractive index at 20°C 589.3 nm, n_D	1.4980	D542
dispersion factor f	0.0084	
Abbe number	59.3	
uv transmission[c], %		
280 nm	6	
300 nm	27	
340 nm	78	
380 nm	88	
visible transmission[c], %		
400–700 nm	89–91	
volume resistivity, $M\Omega$ cm	10.4×10^{14}	D257
dielectric strength, V/μm	13.9	D149
dielectric constant		
10^3 Hz	4.2	D150
10^6 Hz	3.6	D150

[a] Ref. 19.
[b] CAS Registry Number = [25656-90-].
[c] 2.7-mm thickness.

Table 4. Physical and Mechanical Properties of DADC Homopolymer[a,b]

Property	Value	ASTM method
density at 20°C, g/cm^3	1.31	D792
tensile strength, MPa[c]	35–41	D638
flexural yield strength, MPa[c]	52–58.6	D790
compressive strength, MPa[c]	155	D695
Izod impact, notched 25°C, J/m[d]	10.7–21.4	D256
hardness		
Barcol, 15 s	25–28	
abrasion		
Taber (X PMMA)[e]	15–20	D1044
thermal conductivity, W/(m · K)	0.21	C177
specific heat, kJ/(kg · K)[f]	2.3	C351
linear coefficient of expansion/°C		
– 40 to 25°C	8.1×10^{-5}	D696
25 to 75°C	11.4×10^{-5}	D696
75 to 125°C	14.3×10^{-5}	D696
heat distortion at 1.8 MPac, °C/10 mL	55–65	D648
glass transition, °C	85	
burn rate, 1.3–2.9-mm thickness, mm/min	0.04	D635

[a] Ref. 19.
[b] CAS Registry Number = [25656-90-0].
[c] To convert MPa to psi, multiply by 145.
[d] To convert J/m to ftlb/in., divide by 53.38.
[e] Reference of poly(methyl methacrylate) = 1.
[f] To convert J to cal, divide by 4.184.

which appear to be scratch-resistant because they heal by slow flow into the scratch depression. According to the Bayer abrader test, some coated DADC lenses are 10 times more resistant than the unmodified homopolymer surfaces. Subcoats and pretreatments of cast polymer surfaces have been patented. Antireflection, antistatic, and antifogging properties may also be improved by such coatings. Photochromic agents and color tints may be incorporated or added in coatings. A photochromic lens called Transitions is are designed to surface and edge in the same manner as lenses made from CR-39 under normal processing procedures (21). Many features such as blue color, uv absorption, and a scratch-resistant coating are inherent in the Transitions (trademark of PPG Industries) lenses (see CHROMOGENIC MATERIALS, PHOTOCHROMIC).

3.5. Modified Polymers and Copolymers. DADC pure monomer and mixtures with small amounts of comonomers or other additions are commercially available for casting. Monomer formulations are available including agents for protecting the eyes against uv light. Another grade is designed to absorb infrared radiation, and several modified monomers give copolymers of increased heat resistance and hardness. Heat-resistant castings have special advantages in high temperature vacuum deposition of scratch-resistant and antireflective coatings and metallization.

DADC may be polymerized industrially with small amounts of other miscible liquid monomers. Some acrylic ester monomers and maleic anhydride may accelerate polymerization. Copolymerization with methacrylates, diallyl phthalates, triallyl isocyanurate, maleates, maleimides, and unsaturated polyesters are among the examples in the early literature. Copolymers of DADC with poly-functional unsaturated esters give castings of high clarity for eyeglass lenses and other optical applications (20).

Various methacrylate esters have been disclosed as modifiers of DADC. Thus methyl methacrylate polymer may be dissolved in DADC and the sheets cast (22). When DADC is copolymerized with methyl methacrylate, a silane derivative may be added to control the release from the mold (23). CR-39 has been copolymerized with benzyl methacrylate and triallyl cyanurate, also with benzyl methacrylate [2495-37-6] and diallyl phthalate (24), and with trifluoroethyl methacrylate by a two-step process (25).

The DADC monomer has been copolymerized with small amounts of polyfunctional methacrylic or acrylic monomers. For example, 3% triethylene glycol dimethacrylate was used as a flexibilizing, cross-linking agent with a percarbonate as initiator (26). CR-39 and diethylene glycol diacrylate containing isopropyl percarbonate were irradiated with a mercury lamp to a 92% conversion and then cured at 150°C (27). By a similar two-step process DADC was copolymerized with methyl methacrylate and tetraethylene glycol dimethacrylate (28).

Light-focusing plastic rods and other optical devices with graduated refractive indexes may use DADC and other monomers (29). Preparation and properties of plastic lenses from CR-39 are reviewed in reference 30.

3.6. Polymeric Nuclear-Track Detectors. DADC polymer is used in solid-state track detectors (SSTD) of nuclear particles, including alpha-particles, fast neutrons, cosmic rays, and ions of elements of atomic number 10 and above. The outstanding sensitivity of the cast polymers in thin sheets and films to diverse types of ionic radiations with a wide range of mass-to-charge ratio has increased applications in nuclear and space sciences, medicine, mining, and ecological research (1). The tracks formed in the polymer are made visible by partial saponification of the polymer with warm aqueous alkali. The resulting surface holes or spots are counted under a light microscope or a scanning device. Polymerization has been optimized to improve sensitivity, resolution, and measurement. Electrons, x rays, and gamma rays are not recorded, but ions of energies above 0.5 MeV are. As little as 1 ppm uranium in river waters can be detected, autoradiographs from alpha particles or other radioactivity in body tissues can be made, and personnel radiation exposures as low as 100 kSv (10 Mrem) can be monitored. SSTD badges of DADC polymer for monitoring exposure to high energy radiations are used in sensitive areas. Commercial SSTD films from CR-39 and modifiers of 5 μm and thicker are available.

Radiation sensitive cast polymers from DADC are also used in resists for microelectronic circuitry. Relief images result from differential rates of solution in alkali induced by exposure to high energy radiations.

4. Other Allyl Carbonate Polymers

In bulk polymerization triallyl carbonates show less than the 13% shrinkage of CR-39 (31). For example, a trimethylolpropane derivative of average molecular weight 300 was treated with phosgene, and the resulting chloroformate, treated with allyl alcohol, gave a polyfunctional allyl carbonate monomer. The purified monomer was heated with a percarbonate initiator to form a polymer lens.

Similarly, a propoxylated glycerol derivative of molecular weight ca 700 was treated with phosgene and then with allyl alcohol in the presence of a base (32). A polyfunctional allyl carbonate was prepared from Uvithane 893, an oligomer of molecular weight ca 1300 (33). A copolymer of this monomer with CR-39 had better impact strength than that of the homopolymer. Polymers with 1,4-cyclohexane dimethanol bis(allyl carbonate) have been disclosed (34). A lens prepared from equal parts of this monomer and phenyl methacrylate has a refractive index $n_D = 1.565$.

5. Diallyl Phthalates and Their Polymerization

The three isomeric diallyl phthalates are colorless liquids of mild odor, low volatility, and relatively slow polymerization in the early stages. At ca 25% conversion, the viscous liquid undergoes gelation and polymerization accelerates; however, the last monomer disappears at a slow rate.

The monomers are prepared by conventional esterification. Diallyl phthalate (DAP) [131-17-9] is prepared from phthalic anhydride and allyl alcohol:

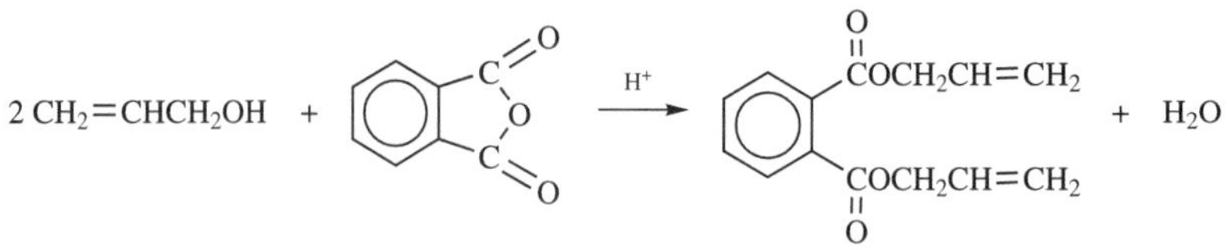

Properties of two diallyl phthalate monomers, $C_{14}H_{14}O_4$, are given in Table 5. The liquids are soluble in common organic solvents but insoluble in water.

If DAP is partially polymerized by heating with peroxide initiator to give a viscous solution and the polymerization is terminated with methanol and branched soluble prepolymer precipitated (35), the dried prepolymer melts near 90°C and exhibits about one-third of the unsaturation of the monomer. When bulk polymerization is allowed to continue, gelation occurs at about 25%

Table 5. **Properties of Commercial Diallyl Esters**[a]

Property	DAP[b]	DAIP[c]
CAS Registry Number	[131-17-9]	[1087-21-4]
boiling point, °C at 0.53 kPa[d]	161	181
density, g/mL	1.117^{25}	1.124^{20}
refractive index, n_D^{25}	1.518	1.5212
surface tension at 20°C, Pa[e]	3.9	3.54
viscosity at 20°C, mPa s($=$cP)	12	17
freezing point, °C	below −70	−3
flash point, °C		171
solubility in gasoline at 25°C, %	24	miscible

[a] Sources: Osaka Soda Company, Hardwick Chemical Company, and FMC Corporation.
[b] Diallyl phthalate.
[c] Diallyl isophthalate.
[d] To convert kPa to mm Hg, multiply by 7.5.
[e] To convert Pa to dyn/cm^2, multiply by 10.

conversion with very rapid polymerization which, however, stops at 65–93% conversion, depending on the initial benzoyl peroxide concentrations. At higher temperatures cross-linking is less complete. Adding diallyl maleate to DAP delays gelation and gives copolymers.

For all three diallyl phthalate isomers, gelation occurs at nearly the same conversion; DAP prepolymer contains fewer reactive allyl groups than the other isomeric prepolymers (36). More double bonds are lost by cyclization in DAP polymerization, but this does not affect gelation. The heat-distortion temperature of cross-linked DAP polymer is influenced by the initiator chosen and its concentration (37). Heat resistance is increased by electron beam irradiation.

Films from prepolymer solutions can be cured by heating at 150°C. Heating the prepolymer in molds gives clear, insoluble moldings (38). The bulk polymerization of DAP at 80°C has been studied (35). In conversions to ca 25% soluble prepolymer, rates were nearly linear with time and concentrations of benzoyl peroxide. A higher initiator concentration is required than in typical vinyl-type polymerizations.

The bulk polymerization of DAP has been studied at 60°C with azobisisobutyronitrile as initiator (39). Branching of the polymer chains is confirmed by enhanced broadening of the molecular weight distribution until gelation occurred at about 25% conversion. In copolymerizations with styrene at 80°C with benzoyl peroxide as initiator the gel time increases with fraction comonomer in the feed. Both the yield of gel and the styrene units in the gel increase with copolymerization time. Heating DAP prepolymer with styrene in benzene solution at 60–100°C with the initiators gives no gelation, but slow formation of polystyrene and copolymer.

Theoretical calculations to predict the conversions at which gelation of polyfunctional monomers occur are reviewed in reference 40. The gelations of DAP, DAIP, and diallyl terephthalate (DATP) near 25% conversion are little affected by conditions and are much higher than predicted.

5.1. DAP Copolymerization. The diallyl phthalates copolymerize readily with monomers bearing strong electron-attracting groups attached to the ethylenic group. These include maleic anhydride, maleate and fumarate esters, and unsaturated polyesters. For example, maleic anhydride copolymerizes with DAP or DAIP in the presence of free-radical initiators at practical rates at ca 50°C; rates are higher than those of the copolymerizations. Additions of styrene, methyl methacrylate, acrylic esters, or acrylonitrile reduce rates of reaction in a linear manner (see Fig. 2). Vinyl chloride, vinyl acetate, or alkenes impede polymerization even more; products are low in comonomer units. Some conjugated comonomers are believed to retard less because they lead to less degradative chain transfer.

The Q and e values of the allyl group in DAP have been estimated as 0.029 and 0.04, respectively, suggesting that DAP acts as a fairly typical unconjugated, bifunctional monomer (42). Cyclization affects copolymerization, since cyclized radicals are less reactive in chain propagation. Thus DAP is less reactive in copolymerization than DAIP or DATP where cyclization is sterically hindered. Particular comonomers affect cyclization, chain transfer, and residual unsaturation in the copolymer products. Diallyl tetrachloro- and tetrabromophthalates are low in reactivity.

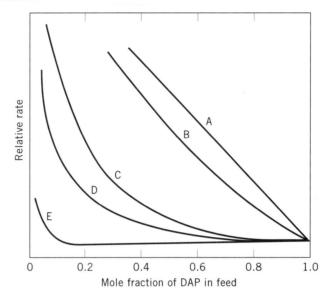

Fig. 2. Relationship between relative rate and monomer composition in the copolymerization of DAP with vinyl monomers: A, styrene or methyl methacrylate; B, methyl acrylate or acrylonitrile; C, vinyl chloride; D, vinyl acetate, and E, ethylene (41).

Diallyl phthalate copolymerizes at 80°C with peroxide catalyst and small
amounts of long chain vinyl monomers including vinyl laurate, dioctyl fumarate,
lauryl methacrylate, and stearyl methacrylate (43). The products show increased
elongations but reduced tensile strengths.

In copolymerization with unsaturated polyesters in fiber-reinforced thermosetting high impact materials, DAP offers advantages over styrene. Glass roving
can be immersed in a solution of unsaturated polyester–DAP containing dicumyl
peroxide (44). A resin, $CaCO_3$, and lubricant are added and the moldings are
cured by heating. A solution of DAP prepolymer, benzoyl peroxide, and polyester
(from maleic anhydride, phthalic acid, and propylene glycol) in acetone–toluene
solution can be impregnated into a nonwoven glass fabric (45), which is cured in
decorative panels by heating under pressure. In electronic moldings, polyester,
DAP, peroxide, glass fibers, and alumina hydrate are molded by heating at
140°C under pressure (46). In compositions for carbon fiber-reinforced fishing
rods DAP is cured with a polyester to give flexural strength as high as 713 MPa
(103,000 psi) (47). Prepregs containing DAP, polyester, and mica filler have been
applied to electrical coils (48), and other prepregs, containing less DAP, used for
decorative panels (49).

Other examples of DAP copolymerizations of industrial interest include
copolymerization with MMA in emulsion (50) and for light focusing rods (51);
with vinylnaphthalene for lenses (52); with epoxy acrylates and glass fibers
(53); epoxy acrylates and coatings (54); with diacetone acrylamide (55); with
aliphatic diepoxide compounds (56); triallyl cyanurate in lacquers for printed
circuits (57); and DAIP with MMA (58).

5.2. Diallyl Isophthalate. DAIP polymerizes faster than DAP, undergoes less cyclization, and yields cured polymers of better heat resistance, eg, up to ca 200°C. Prepolymer molding materials such as Dapon M, are not sticky. Maleic anhydride accelerates polymerization, whereas vinyl isobutyl ether retards it and delays gelation in castings. Copolymers with maleic anhydride are exceptionally hard and tough and may scratch homopolymer surfaces.

Besides application as heat-resistant molding powders for electronic and other applications, DAIP copolymers have been proposed for optical applications. Lenses of high impact resistance contain 50% DAIP, 20% benzyl methacrylate, and larger amounts of CR-39 (59). A lens of refractive index $n_D = 1.569$ and low dispersion can be cast from phenyl methacrylate, DAIP, and isopropyl peroxide (60). Lenses of better impact properties can be obtained by modifying DAIP with allyl benzoate (61).

Diallyl terephthalate [1026-92-2] is utilized less, but lenses made of copolymers with triallyl cyanurate and methacrylates have been suggested (62). Diallyl tetrabromophthalate and tetrachlorophthalate polymers have been proposed for electronic circuit boards of low flammability (63). They are uv-curable and solder-resistant. Copolymers with unsaturated polyester, vinyl acetate and DAP have been studied (64).

5.3. Telomerization. Polymerization of DAP is accelerated by telogens such as CBr_4, which are more effective chain-transfer agents than the monomer itself (65); gelation is delayed. The telomers are more readily cured in uv than DAP prepolymers. In telomerizations with CCl_4 with peroxide initiator, at a DAP/CCl_4 ratio of 20, the polymer recovered at low conversion has a DP of 12 (66).

5.4. Uses. The largest use of diallyl phthalate thermoset polymers is in moldings and coatings for electronic devices requiring high reliability under long-term adverse environmental conditions. A photocurable diallyl phthalate resin composition for use on printed circuit boards has been developed (67). These devices include electrical connectors and insulators in communication, computer, and aerospace systems. The flow in molding and curing of these prepolymer formulations must be carefully controlled by selection of initiator, monomer or comonomer content, and heating and radiation. Proprietary compositions may contain about equal weights of prepolymer and filler, eg, fine glass fibers or calcium silicate, 10% antimony oxide, 1–2% DAP or DAIP monomer, silane derivative, peroxide, lubricant, and polymerization-control agents. Formulations are designed for injection, compression, and transfer molding.

For best impact strength, glass fibers should be pretreated with a silane derivative. Fillers may include clays, calcium carbonate, silicates, various silicas, glass, and carbon, preferably as short fibers. Small amounts of metallic stearates or long chain organic acids are added as molding lubricants. In addition to antimony oxide, chlorine- and bromine-containing monomers may be added to reduce flammability. For encapsulation of fragile electronic components, prepolymer of lower molecular weight along with more monomer may be used for flow at low mold pressures. Potting with monomer-polymer syrup is normally too slow. Properties of DAP thermoset moldings are given in Table 6.

Diallyl phthalates are used with glass cloth and roving in tubular ducts, radomes, aircraft, and missile parts of high heat resistance. They offer the advantages of low volatility, little odor, and high heat resistance, replacing

Table 6. **Properties of DAP Thermoset Moldings**[a]

Property	ASTM method	Unfilled	Glass-fiber reinforced	Mineral filled
molding temperature, °C		140–190	144–193	133–193
mold shrinkage, cm/cm	D955		0.0005–0.005	0.002–0.007
tensile strength, MPa[b]	D638	27.6	41–76	34–55
flexural strength, MPa[b]	D790	62.0	62–138	55–76
compressive strength, MPa[b]	D695	165	172–241	138–220
elongation, %	D638		3–5	3–5
Izod impact, N per notch[c]	D256A	1.1–1.6	2.1–80	1.6–4.3
hardness, Rockwell	D785	115	E80–87	E61
coefficient of linear expansion, 10^{-6} cm/cm per °C	D696		10–36	10–42
deflection temp under 18.2 MPa, °C	D648	155	166–290	162–290
thermal conductivity, W/(m · K)	C177		0.21–0.62	0.29–1.04
refractive index, n_{D}^{25}		1.571		
specific gravity	D792		1.70–1.96	1.65–1.85
water absorption[d], 24 h at 25°C, %	D570	0.20	0.12–0.35	0.2–0.5
dielectric strength[d], V/μm	D149		16–18	16–18
dielectric constant at 25°C				
10^3 Hz		3.4	4.4	~4.8
10^6 Hz		3.4	4.4	~4.4
arc resistance, s		118	125	140

[a] Ref. 68.

[b] To convert MPa to psi, multiply by 145.

[c] To convert N per notch to lbf/in. per notch, multiply by 0.187.

[d] 3-mm thickness.

styrene with unsaturated polyesters in fiber-reinforced plastic structures. High curing temperatures, however, and high costs limit structures. Glass cloth, textiles, and papers may be impregnated by prepolymer-monomer mixtures in solution along with peroxide initiator and lubricant. Molding and curing gives decorative stain- and heat-resistant overlays for wall panels, tables, and furniture (1).

Favorable rates and yields of DAP prepolymer are obtained by solution polymerization in CCl4–benzene mixtures (69). Bulk polymerization at 80°C with benzoyl peroxide is advanced to a certain viscosity before addition of ethanol to precipitate the prepolymer that is then dried (70).

Dapon 35 of FMC and a similar Japanese product have been studied by gel permeation chromatography. Hydrogen peroxide acts as a regulator as well as initiator, and gives relatively large fractions of oligomers. In polymerization between 80 and 220°C gelation occurs at 25–45% conversion (71).

Tableware is molded from DAP polymer and prepolymer, cellulose, pigment, dicumyl peroxide, and a silane coupling agent (72). Nontoxic DAP-based polymers have been suggested as being agents of low viscosity and long pot life for glass-reinforced plastics and electroluminescent coatings of high transparency (73). Abrasion-resistant CR-39-DAP copolymer recording disks have low moisture permeability and good heat resistance (74).

Diallyl esters find little application in lenses. However, DAP and DAIP can be polymerized by high energy radiation in lens molds (75). Coatings of silica and alumina by vaporization give antiglare, scratch-resistant lenses.

Table 7. **Properties of Some Diallyl Esters**

Diallyl ester	Molecular formula	CAS Registry Number	$Bp_{Pa}{}^a$	n_D^{20}	d_4^{20}
oxalate	$C_8H_{10}O_4$	[615-99-6]	$107_{1.9}$	1.4460	1.0081
malonate	$C_9H_{12}O_4$	[1797-75-7]	$119_{1.9}$	1.4489	1.060
succinate	$C_{10}H_{14}O_4$	[925-16-6]	$94_{0.13}$	1.4507	1.056
adipate	$C_{12}H_{18}O_4$	[2998-04-1]	$115_{0.13}$	1.4542	1.023
sebacate	$C_{16}H_{26}O_4$	[3137-00-6]	$164_{0.26}$	1.4550	0.978
tartrate	$C_{10}H_{14}O_6$	[57833-54-2]	$171_{1.3}$	1.187	

^{a} To convert Pa to mm Hg, multiply by 0.0075.

6. Other Diallyl Esters

Tables 7 and 8 give properties of some diallyl esters. Dimethallyl phthalate [5085-00-7] has been copolymerized with vinyl acetate and benzoyl peroxide, and reactivity ratios have been reported (76).

Copolymers of diallyl succinate and unsaturated polyesters cured by x rays provide wear-resistant coatings of MMA dental polymers (77). Copolymers of diallyl itaconate [2767-99-9] with N-vinylpyrrolidinone and styrene have been proposed as oxygen-permeable contact lenses (qv) (78). Reactivity ratios have been studied in the copolymerization of diallyl tartrate (79). A lens of a high refractive index ($n_D = 1.63$) and a heat distortion above 280°C has been reported for diallyl 2,6-naphthalene dicarboxylate [51223-57-5] (80). Diallyl chlorendate [3232-62-0] polymerized in the presence of di-t-butyl peroxide gives a lens with a refractive index of $n = 1.57$ (81). Hardness as high as Rockwell 150 is obtained by polymerization of triallyl trimellitate [2694-54-4] initiated by benzoyl peroxide (82).

Among the preformed polymers cured by minor additions of allyl ester monomers and catalysts followed by heat or irradiation are PVC cured by diallyl fumarate (83), PVC cured by diallyl sebacate (84), fluoropolymers cured by triallyl trimellitate (85), and ABS copolymers cured by triallyl trimellitate (86).

Table 8. **Properties of Allyl–Vinyl Monomers**

Property	Allyl methacrylate, AMA	Diallyl maleate, DAM	Diallyl fumarate, DAF
structure	$CH_2{=}\overset{\overset{\displaystyle CH_3}{\vert}}{C}{-}COOA^a$	$\overset{\displaystyle HC-COOA}{\underset{\displaystyle HC-COOA^a}{\parallel}}$	$\overset{\displaystyle AOOCCH}{\underset{\displaystyle HC-COOA^a}{\parallel}}$
molecular formula	$C_7H_{10}O_2$	$C_{10}H_{12}O_4$	$C_{10}H_{12}O_4$
CAS Registry Number	[96-05-9]	[999-21-3]	[2807-54-7]
boiling point, $°C_{kPa}{}^b$	55_4	$112_{0.53}$	$140_{0.40}$
density at 25°C, g/cm^3	0.930	1.070	1.0516
refractive index, n_D^{25}	1.453	1.4664^c	1.4669
viscosity mPa·s(=cP)	13	4.3	3.0
flash point, open cup, °C		123	74

^{a} A = allyl = $—CH_2CH{=}CH_2$
^{b} To convert kPa to mm Hg, multiply by 7.5.
^{c} At 20°C.

In studies of the polymerization kinetics of triallyl citrate [6299-73-6], the cyclization constant was found to be intermediate between that of diallyl succinate and DAP (86). Copolymerization reactivity ratios with vinyl monomers have been reported (87). At 60°C with benzoyl peroxide as initiator, triallyl citrate retards polymerization of styrene, acrylonitrile, vinyl choloride, and vinyl acetate. Properties of polyfunctional allyl esters are given in Table 7; some of these esters have sharp odors and cause skin irritation.

A series of glycol bis(allyl phthalates) and bis(allyl succinates) and their properties are reported in reference 88. In homopolymerizations, cyclization increases in the order: diallyl aliphatic carboxylates < glycol bis(allyl succinates) < glycol bis(allyl phthalates). Copolymerizations with small amounts of DAP can give thermoset moldings of improved impact (89).

7. Allyl–Vinyl Compounds

Monomers such as allyl methacrylate and diallyl maleate have applications as cross-linking and branching agents selected especially for the different reactivities of their double bonds (91); some physical properties are given in Table 8. These esters are colorless liquids soluble in most organic liquids but little soluble in water; DAM and DAF have pungent odors and are skin irritants.

Addition of dialkyl fumarates to DAP accelerates polymerization; maximum rates are obtained for 1:1 molar feeds (41). Methyl allyl fumarate [74856-71-6] (MAF), $C_8H_{10}O_4$, homopolymerizes much faster than methyl allyl maleate [51304-28-0] (MAM) and gelation occurs at low conversion; more cyclization occurs with MAM. The greater reactivity of the fumarate double bond is shown in copolymerization of MAF with styrene in bulk. The maximum rate of copolymerization occurs from monomer ratios, almost 1:1 molar, but no maximum is observed from MAM and styrene. Styrene hinders cyclization of both MAF and MAM.

Diallyl fumarate polymerizes much more rapidly than diallyl maleate. Because of its moderate reactivity, DAM is favored as a cross-linking and branching agent with some vinyl-type monomers (1). Cyclization from homopolymerizations in different concentrations in benzene has been investigated (91). Diallyl itaconate and several other polyfunctional allyl–vinyl monomers are available.

7.1. Allyl Methacrylate (AMA).   Of the compounds containing both allyl and vinyl-type double bonds, allyl methacrylate is the most important. The lower reactivity of the allyl group permits controlled, second-stage cross-linking when used in low concentrations. A copolymer with MMA in cured moldings has a heat distortion temperature of 96°C compared to 70°C for MMA homopolymer (92). AMA graft copolymers with acrylate and methacrylate esters blended with PVC and other molding plastics have high impact strength (93). AMA is used as cross-linking agent with methacrylate esters in contact lenses (94). AMA is also used in low concentrations in curable acrylic coatings (95). For cross-linked pressure-sensitive adhesives, a small amount of AMA is used with alkyl acrylates and cured by electron beam irradiation (96). AMA may be added as branching comonomer with acrylic acid for preparing mucilages and thickening agents for aqueous systems (97).

8. Polyfunctional Allyl Nitrogen Monomers

8.1. Triallyl Cyanurate as Cross-linking Agent.
Triallyl cyanurate (TAC), 2,4,6-tris(allyloxy)-s-triazine [101-37-1], and its isomer triallyl isocyanurate (TAIC) are used as cross-linking agents with comonomers and for aftercuring preformed polymers such as olefin copolymers in electrical insulations. TAC monomer melts at 20–25°C. It is prepared by gradual addition of cyanuric chloride to an excess of allyl alcohol in the presence of aqueous alkali (98).

Properties of TAC and TAIC are given in Table 9.

Crystalline TAC monomer can be stored at room temperature with only slow change, but heating may cause polymerization with violence. In storage, the liquid monomer slowly forms a viscous syrup of prepolymer solution. In homopolymerizations at elevated temperatures with free-radical initiators, two exotherms are observed. An initial release of heat indicates reaction of two allyl groups, a second larger exotherm apparently results from the remaining allyl groups together with rearrangement of the polymer to the more stable isocyanurate structure. Because of brittleness, the homopolymers of TAC and TAIC have had little application.

Triallyl cyanurate is used as a comonomer in small amounts with methacrylate esters and unsaturated polyesters. The addition of 5% or more of TAC to MMA in castings improves heat and solvent resistance as well as thermooxidative stability (99). For optical applications, up to 20% TAC has been suggested. Reactivity ratios for TAC and methacrylate esters have been reported (100).

Small amounts of TAC and TAIC are copolymerized with unsaturated polyesters in glass cloth or fibers in high strength laminates, and electrical sealing and potting applications. In one case, a glass fiber mat impregnated with a

Table 9. **Properties of TAC and TAIC**

Property	TAC	TAIC
molecular formula	$C_{12}H_{15}N_3O_3$	$C_{12}H_{15}N_3O_3$
CAS Registry Number	[101-37-1]	[1025-15-6]
melting point, °C	31	24
boiling point, °C	$140_{67\ Pa}$[a]	$126_{40\ Pa}$[a]
density at 30°C, g/cm^3	1.1133	1.1720
refractive index n_D^{25}	1.5049	1.5115

[a] To convert Pa to mm Hg, multiply by 0.0075.

mixture of a maleate polyester, DAP, and TAC was cured with *t*-butyl peroxide at 150°C (101). Solid unsaturated polyesters with 4% TAC have been used in powder coating formulations cured by heating with *t*-butyl perbenzoate and cobalt octanoate (102). Allyl cyanoacrylate modified by TAC has been used in dental adhesives (103).

8.2. Triallyl Cyanurate Cure of Preformed Polymers. TAC and TAIC are often used in small amounts with vinyl-type and condensation polymers for cured plastics, rubber and adhesive products of high strength, and heat and solvent resistance. In some cases, chemical stability is also improved. These effects are the result of grafting at active H atoms as well as penetrating the polymer network. Improvement in strength of PVC by addition of TAC is discussed in reference 104. The modification of PVC by increasing proportions of TAC and diallyl sebacate (DAS) has been studied (105). Up to about 15% TAC, graft polymerization predominates, whereas with higher concentrations polymer networks are observed. DAS is less effective in forming such networks. Both comonomers have a stabilizing effect against loss of HCl under exposure to gamma radiation (106); decomposition products have been studied (107). TAC may act also as sensitizer and stabilizer in electron radiation curing (qv) of PE, PP, and PVC. TAC has been applied to curing PVC elastomers (108).

The use of TAC as a curing agent continues to grow for polyolefins and olefin copolymer plastics and rubbers. Examples include polyethylene (109), chlorosulfonated polyethylene (110), polypropylene (111), ethylene–vinyl acetate (112), ethylene–propylene copolymer (113), acrylonitrile copolymers (114), and methylstyrene polymers (115). In ethylene–propylene copolymer rubber compositions. TAC has been used for injection molding of fenders (116). Unsaturated elastomers, such as EPDM, cross link with TAC by hydrogen abstraction and addition to double bonds in the presence of peroxyketal catalysts (117).

For curing copolymers of tetrafluoroethylene and perfluorovinyl ether, addition of ca 4% TAC has been proposed (118). TFE–propylene copolymers have been cured by TAC and organic peroxide (119). Copolymers of TFE–propylene–vinylidene fluoride are cured with TAC by heating at 200°C.

Nonvinyl polymers cured by TAC include polyamides (120), polyamide–polyurethane blends (121), caprolactone polymers (122), terephthalate polymers (123), epoxy resins (124), and acrylic epoxies (125).

8.3. TAIC as Curing Agent. Triallyl isocyanurate is prepared from an alkali cyanate with allyl chloride. Homopolymers are brittle, intractable, and of little use. A viscous solution of prepolymer first forms under mild polymerization conditions. Addition of alcohol to benzoyl peroxide-catalyzed syrup gives a solid prepolymer of molecular weight 5800 (126). TAIC prepolymer of molecular weight 6000 cures faster than DAP prepolymer (127). TAIC with methacrylic acid in aqueous solution along with azobisisobutyronitrile gives, at first, soluble polymers which cross link on further conversion (128). Zinc chloride accelerates polymerization rates of TAIC (129). Copolymers have been prepared with diallylamine and chloroprene (130).

Small amounts of TAIC together with DAP have been used to cure unsaturated polyesters in glass-reinforced thermosets (131). It has been used with polyfunctional methacrylate esters in anaerobic adhesives (132). TAIC and vinyl acetate are copolymerized in aqueous suspension, and vinyl alcohol copolymer gels are made from the products (133). Electron cure of poly(ethylene terephthalate) moldings containing TAIC improves heat resistance and transparency (134).

Publications on curing polymers with TAIC include TFE–propylene copolymer (135), TFE–propylene–perfluoroallyl ether (136), ethylene–chlorotrifluoroethylene copolymers (137), polyethylene (138), ethylene–vinyl acetate copolymers (139), polybutadienes (140), PVC (141), polyamide (142), polyester (143), poly(ethylene terephthalate) (144), siloxane elastomers (145), maleimide polymers (146), and polyimide esters (147). Compositions containing allyl compounds and processes for forming and curing polymer compositions have been described (148).

8.4. Diallyl Ammonium Polymers. *N,N*-Diallyldimethyl(DADM)ammonium salts are used for the preparation of polyelectrolytes. Polymerization from concentrated water solution can be initiated by *t*-butyl hydroperoxide (149). The polymers are used as flocculating agents, and in water purification (1). The polyelectrolytes are used for aqueous coal flotation (150). Copolymers of DADM ammonium chloride with acrylamide have also been proposed for water purification (151). DADM ammonium polyelectrolyte is used in coatings for copy paper (152). Electrical conductivity of the polyelectrolytes plasticized with poly(ethylene glycol) has been studied (153), and microencapsulation of polyelectrolytes such as DADM ammonium chloride has been reported (154). DADM ammonium compounds can be used as cross-linking agents for temporary gel-blocking in petroleum and natural gas wells (155). Diallyl diammonium polymers have been reported to control snails (156).

Copolymers of diallyldimethylammonium chloride [7398-69-8] with acrylamide have been used in electroconductive coatings (157). Copolymers with acrylamide made in activated aqueous persulfate solution have flocculating activity increasing with molecular weight (158). DADM ammonium chloride can be grafted with cellulose from concentrated aqueous solution; catalysis is by ammonium persulfate (159). Diallyl didodecylammonium bromide [96499-24-0] has been used for preparation of polymerized vesicles (160).

Molded polyamide surfaces can be hardened by grafting with *N,N*-diallylacrylamide [3085-68-5] monomer under exposure to electron beam (161). *N,N*-Diallyltartardiamide [58477-85-3] is a cross-linking agent for acrylamide reversible gels in electrophoresis. Such gels can be dissolved by a dilute periodic acid solution in order to recover protein fractions.

9. Other Allyl Compounds

Although much research has been carried out with allyl ethers there has been only limited commercial use (1). Multifunctional allyl glycidyl ether [106-92-3],

$$CH_2{=}CHCH_2OCH_2CH{-}CH_2,$$
$$\underset{O}{\diagdown\diagup}$$

a toxic liquid, has been used as an additive to epoxy resins, in copolymers with ethylene oxide and derivatives, and in copolymers with vinyl comonomers. The

Table 10. **Physical Properties of Some Diallyl Ethers**

Ether	Molecular formula	CAS Registry Number	Bp, °C[a]	Sp gr$^{20}_{20}$	n^{20}
diallyl ether	$C_6H_{10}O$	[557-40-4]	96	0.805	1.4165
allyl methallyl ether	$C_7H_{12}O$	[14289-96-4]	115		1.4236
1,2-diallyloxyethane	$C_8H_{14}O_2$	[7529-27-3]	$37_{133\ Pa}$	0.894	1.4340
1,2-dimethallyloxyethane	$C_{10}H_{18}O_2$	[79719-27-0]	$50_{33.2\ Pa}$	0.8779	1.4383
trimethylolpropane diallyl ether	$C_{12}H_{22}O_3$	[682-09-7]	258	0.957	1.4560
pentaerythritol diallyl ether	$C_{11}H_{20}O_4$	[2590-16-1]	$174_{1.3kPa}$	1.037	1.4695

[a] To convert Pa to mm Hg, multiply by 0.0075.

second reactive group permits final cross-linking with radiation or heating with peroxides.

A great number of other allyl compounds have been prepared, especially allyl ethers and allyl ether derivatives of carbohydrates and other polymers. They are made by the reaction of hydroxyl groups with allyl chloride in the presence of alkali (1). Polymerizations and copolymerizations are generally slow and incomplete. Products have only limited use in coatings, inks, and specialties. Properties of a few allyl ethers are given in Table 10. An allyl ether-based resin used as a scale inhibitor has been described (162).

These compounds are miscible with most organic solvents. Much research has been directed toward the preparation of air-drying prepolymers or oligomers. Thus allyl groups have been introduced into low molecular weight polyesters, polyurethanes, and formaldehyde condensates (1), but curing rates have been low. Some examples include copolymers of trimethylolpropane diallyl ether with unsaturated polyesters in uv-curable coatings (163). A bis(allyloxy)sulfolane can be used to cross-link unsaturated polyesters in coatings (164). Pentaerythritol triallyl ether [1471-17-6] serves as a curing agent in acrylic photolacquers (165). Diallyl ether reacts with SO_2 to give soluble copolymers containing cyclic structures (166).

BIBLIOGRAPHY

"Allyl Monomers and Polymers" in *ECT* 3, Vol. 2, pp. 109–129, by C. E. Schildknecht, Gettysburg College.; in *ECT* 4th ed., Vol. 2, pp. 161–184, by R. Dowbenko, PPG Industries; "Allyl Monomers and Polymers" in *ECT* (online), posting date: December 4, 2000, by R. Dowbenko, PPG Industries.

CITED PUBLICATIONS

1. C. E. Schildknecht, *Allyl Compounds and Their Polymers*, Wiley-Interscience, New York, 1973.
2. P. D. Bartlett and R. Altschul, *J. Am. Chem. Soc.* **67**, 812 (1945); P. D. Bartlett and F. O. Tate, *J. Am. Chem. Soc.* **75**, 91 (1953).
3. E. F. Jordan and co-workers, *J. Polym. Sci. Chem. Ed.* **11**, 1475 (1973).

4. W. Simpson and co-workers, *J. Polym. Sci.* **10**, 489 (1953); **16**, 440 (1955); R. N. Haward, *J. Polym. Sci.* **14**, 535 (1954).
5. G. B. Butler and R. J. Angelo, *J. Am. Chem. Soc.* **79**, 3128 (1957).
6. C. S. Marvel and co-workers, *J. Am. Chem. Soc.* **70**, 993 (1948); **75**, 6318 (1953).
7. C. E. Schildknecht and co-workers, *Polym. Prepr. Am. Chem. Soc. Div. Polym. Chem.* **12**(2), 117 (Sept. 1971).
8. U.S. Pats. 2,370,567; 2,370,569; and 2,370,571 (Feb. 27, 1945), I. E. Muskat and F. Strain (to Pittsburgh Plate Glass Co.).
9. U.S. Pat. 2,403,113 (July 2, 1946), I. E. Muskat and F. Strain (to PPG Industries, Inc.).
10. *CR-39 Allyl Diglycol Carbonate Monomer Bulletin 45A* and *Casting and Material Safety Data Sheet*, PPG Industries, Inc., Pittsburgh, Pa., 1984.
11. F. Strain, *ASTM Symposium on Plastics*, (Feb. 22, 1944), American Society for Testing and Materials, Philadelphia, Pa., 1944, 152–164.
12. E. Schnarr and K. E. Russell, *J. Polym. Sci. Chem. Ed.* **18**, 913 (1980).
13. L. S. Luskin, *Modern Plastics Encyclopedia*, McGraw-Hill, New York, 1984–1985, p. 196.
14. U.S. Pat. 2,379,218 (June 26, 1948), W. R. Dial and C. Gould (to PPG Industries, Inc.).
15. Jpn. Kokai Tokkyo Koho 58 167,125 (Oct. 3, 1983) (to Nippon Oils and Fats Co.).
16. U.S. Pat. 4,311,462 (Jan. 19, 1982), A. A. Spycher and D. J. Damico (to Corning Glass Corp.).
17. Jpn. Kokai Tokkyo Koho 57 158,240 (Sept. 30, 1982) (Toray Industries).
18. Jpn. Kokai Tokkyo Koho 1 82,221 (July 4, 1982) (to Swa Seikosla).
19. *Bulletin A-691-45*, PPG Industries, Inc., Pittsburgh, Pa., 1983.
20. U.S. Pat. 4,139,578 (Feb. 13, 1979), G. L. Baughman and H. C. Stevens (to PPG Industries, Inc.).
21. *Technical Bulletin Transitions TM*, PPG Industries, Inc., Pittsburgh, Pa., 1990.
22. Ger. Offen. (June 24, 1982), J. C. Crano and R. L. Haynes (to PPG Industries, Inc.).
23. U.S. Pat. 4,146,696 (Mar. 27, 1979), H. M. Bond and co-workers (to Buckbee-Mears Co.).
24. Jpn. Kokai Tokkyo Koho 81 61,412 (May 26, 1981) (to Hoya Lens Co.).
25. Y. Otsuka and co-workers, *Organic Coat. Plast. Chem.* **40**, 382 (1979).
26. Ger. Offen. 2,938,098 (Mar. 26, 1981), H. Fricke and H. Schiller (to Deutsche Special-glas A. G.).
27. Jpn. Kokai Tokkyo Koho 76,125,486 (Nov. 1, 1976), I. Kaetsu and co-workers (to Japan Atomic Energy Research Institute).
28. H. Okubo and co-workers, *J. Appl. Polym. Sci.* **22**(1), 27 (1978).
29. U.S. Pat. 4,022,855 (May 10, 1977), D. P. Hamblen (to Eastman Kodak Corp.).
30. W. Koeppen, *Proc. SPIE-Int. Soc. Opt. Eng.* **381**, 78 (1983).
31. U.S. Pat. 4,144,262 (Mar. 13, 1979), H. C. Stevens (to PPG Industries, Inc.).
32. U.S. Pat. 4,205,154 (May 27, 1980), H. C. Stevens (to PPG Industries, Inc.).
33. U.S. Pat. 4,360,653 (Nov. 23, 1982), H. C. Stevens (to PPG Industries, Inc.).
34. Jpn. Kokai Tokkyo Koho 59 08,710 (Jan. 18, 1984) (to Mitsui Toatsu Chemical Company).
35. W. Simpson, *J. Soc. Chem. Ind. London* **65**, 107 (1946).
36. W. Simpson and T. Holt, *J. Polym. Sci.* **18**, 335 (1955).
37. *Ibid.* **28**, 445 (1958).
38. U.S. Pat. 2,273,891 (Feb. 24, 1942), M. A. Pollak, I. E. Muskat, and F. Strain (to PPG Industries, Inc.).
39. K. Ito and co-workers, *J. Polym. Sci. Chem. Ed.* **13**, 87 (1975).
40. A. Matsumoto and co-workers, *J. Polym. Sci. Phys. Ed.* **15**, 127 (1977).

41. A. Matsumoto and co-workers, *J. Polym. Sci. Chem. Ed.* **20**, 2611, 3207 (1982).
42. M. Oiwa and A. Matsumoto, *Prog. Polym. Sci. Jpn.* **17**, 128 (1974).
43. A. Matsumoto and co-workers, *Polym. Bull.* **10**, 438 (1983).
44. Jpn. Kokai Tokkyo Koho 77 10,374 (July 16, 1975), H. Takamoto and N. Tanedo (to Teijin, Ltd.).
45. Jpn. Kokai Tokkyo Koho 78 145,890 (Dec. 19, 1978), K. Satomo and Y. Kono.
46. Jpn. Kokai Tokkyo Koho 82 16,018 (Jan. 27, 1982) (to Matsushita Electric Works, Ltd.).
47. Jpn. Kokai Tokkyo Koho 58 208,316 (Dec. 5, 1983) (to Toray Industries, Inc.).
48. Jpn. Kokai Tokkyo Koho 82 18,731 (Jan. 30, 1982) (to Hitachi Ltd.).
49. Jpn. Kokai Tokkyo Koho 81 65,010 (June 2, 1981) (to Osaka Soda Co., Ltd.).
50. A. E. Kulikova, V. D. Mal'kov, L. V. Mal'kova, E. G. Pomerantseva, I. I. Yatchishin, E. F. Samarin, I. N. Vishnevskaya, and I. M. Monich, *Vysokomol. Soedin. Ser. B* **22**(2), 131 (1980).
51. Y. Ohtsuka and A. Hirosawa, *Kobunshi Ronbunshu* **359**, 535 (1978).
52. J. J. Mauer in E. A. Turi, ed., *Thermal Characteristics of Polymeric Materials*, Academic Press, Inc., New York, 1981, 571–708.
53. Jpn. Kokai Tokkyo Koho 82 41,543 (Mar. 8, 1982) (to Osaka Soda Co., Ltd.).
54. Jpn. Kokai Tokkyo Koho 57 133,107 (Aug. 17, 1982) (to Osaka Soda Co., Ltd.).
55. Jpn. Kokai Tokkyo Koho 81 67,318 (June 6, 1981) (to Osaka Soda Co., Ltd.).
56. Jpn. Kokai Tokkyo Koho 58 164,617 (Sept. 29, 1983) (to Toray Industries, Inc.).
57. Jpn. Kokai Tokkyo Koho 58 212,945 (Dec. 10, 1983) (to Sumitomo Bakelite Co., Ltd.).
58. Y. Ohtsuka and T. Senga, *Kobunshi Ronbunshu* **35**, 721 (1978).
59. Jpn. Kokai Tokkyo Koho 81 61,411 (May 26, 1981) (to Hoya Lens Co.).
60. Ger. Offen. 3,146,075 (June 24, 1982), N. Tarumi and co-workers (to Hoya Lens Co.).
61. Jpn. Kokai Tokkyo Koho 59 81,318 (May 11, 1984).
62. Jpn. Kokai Tokkyo Koho 58 179,210 (Oct. 20, 1983) (to Olympia Optical Co.).
63. Ger. Offen. 3,323,222 (Jan. 12, 1984), U. G. Kang and co-workers (to W. R. Grace & Co.).
64. D. P. Braksmayer and co-workers, *Plast. Compd.* **6**(7), 59 (1983).
65. A. Matsumoto and co-workers, *J. Appl. Polym. Sci.* **28**, 1105 (1983).
66. A. Matsumoto, T. Nakane, and M. Oiwa, *J. Polym. Sci. Lett. Ed.* **21**, 699 (1983).
67. U.S. Pat. 5,091,283 (Feb. 25, 1992), I. Tanaka and co-workers (to Hitachi, Ltd.)
68. *Modern Plastics Encyclopedia*, McGraw-Hill, New York, 1984–1985, p. 454.
69. USSR Pat. 563,424 (Mar. 15, 1976), L. L. Nikogosyan and co-workers.
70. Czech. Pat. 169,373 (May 15, 1977), F. Kaspar and F. Lesek.
71. M. I. Prusinska and W. Kolikowski, *Angew. Macromol. Chem.* **64**(1), 29 (1977).
72. Jpn. Kokai Tokkyo Koho 58 157,846 (Sept. 20, 1983) (to Osaka Soda Co., Ltd.).
73. N. H. Alexeev and co-workers, *Plast. Massy* (6), 6 (1984).
74. Jpn. Kokai Tokkyo Koho 58 137,150 (Aug. 15, 1983) (to Suwa Seikosha Co.).
75. Jpn. Kokai Tokkyo Koho 58 721,030 (Apr. 30, 1983) (to Suwa Seikosha Co.).
76. A. Matsumoto and M. Oiwa, *J. Polym. Sci. Chem. Ed.* **19**, 3607 (1971).
77. H. Kimura, *J. Osaka Univ. Dental School* **20**, 43 (1980).
78. Ger. Offen. 864,275 (Dec. 27, 1977) (to Wesley-Jessen, Inc.).
79. T. Ohata and co-workers, *J. Polym. Sci. Chem. Ed.* **18**, 1011 (1980).
80. Jpn. Kokai Tokkyo Koho 82 08,806 (Feb. 18, 1982) (to Teijin Chemical Co.).
81. Jpn. Kokai Tokkyo Koho 58 168,609 (Oct. 5, 1983) (to Toray Industries, Inc.).
82. Jpn. Kokai Tokkyo Koho 57 185,306 (Nov. 15, 1982) (to Wako Pure Chemical Industries, Ltd.).
83. Jpn. Kokai Tokkyo Koho 76,579 (May 1, 1984) (to Sumitomo Electric Industries, Ltd.).
84. V. I. Kakin and V. L. Karpov, *Khim. Vys. Energ.* **18**(1), 20 (1984).

85. U.S. Pat. 4,031,167 (June 21, 1977) (to International Telephone and Telegraph Corp.).
86. Jpn. Kokai Tokkyo Koho 58 208,312 (Dec. 5, 1983) (to Nisshin Electric Co., Ltd.).
87. A. Matsumoto and co-workers, *Bull. Chem. Soc. Jpn.* **47**, 673, 928 (1974).
88. A. Matsumoto and co-workers, *J. Polym. Sci. Chem. Ed.* **21**, 3493 (1983).
89. A. Matsumoto and co-workers, *J. Polym. Sci. Lett.* **21**, 837 (1983).
90. A. Matsumoto and co-workers, *J. Polym. Sci. Chem. Ed.* **16**, 1081, 2695 (1978); **19**, 245 (1981).
91. A. Matsumoto and co-workers, *J. Polym. Sci. Chem. Ed.* **17**, 4089 (1979).
92. U.S. Pat. 4,213,417 (Apr. 21, 1981), C. J. Dyball (to Pennwalt Corp.).
93. U.S. Pat. 4,180,529 (Dec. 25, 1979), G. H. Hoffman (to E. I. du Pont de Nemours & Co., Inc.).
94. Brit. Pat. 1,467,416 (Mar. 16, 1977), J. D. Frankland; U.S. Pat. 4,139,692 (Feb. 13, 1979), Y. Tanaka and co-workers (to Toyo Contact Lens Co.).
95. Ger. Offen. 2,536,312 (Feb. 24, 1977), J. Fock and E. Schamberg (to T. Goldschmidt A.G.).
96. Ger. Offen. 3,015,463 (Oct. 30, 1980), S. D. Pastor and S. H. Ganslaw (to National Starch and Chemical Co.).
97. Ger. Offen. 3,2212,284 (Dec. 8, 1983), A. Koschik and co-workers (to Roehm G.m.b.H.).
98. U.S. Pat. 2,510,564 (June 6, 1950), J. R. Dudley (to American Cyanamid Co.).
99. M. Kucharski and A. Ryttel, *Polimery Warsaw* **22**, 412 (1977).
100. M. Kucharski and A. Ryttel, *J. Polym. Sci. Chem. Ed.* **16**, 3011 (1978).
101. Neth. Pat. Appl. 79 03,427 (Nov. 4, 1980), A. N. J. Verwer (to Stamicarbon Co.).
102. Eur. Pat. Appl. 106,399 (Apr. 25, 1984), A. N. J. Verwer and co-workers (to DSM Resins B. V.).
103. U.S. Pat. 4,134,929 (Jan. 16, 1979), D. M. Stoakley and J. R. Dombroski (to Eastman Kodak Corp.).
104. S. H. Pinner, *J. Appl. Polym. Sci.* **3**, 338 (1960).
105. V. I. Dakin, *Plast. Massy* (6), 58 (1984).
106. V. I. Dakin and co-workers, *Khim. Vys. Energ.* **11**(4), 378 (1977).
107. V. I. Dakin and V. L. Karpov, *Plast. Massy* (1), 17 (1981).
108. Jpn. Kokai Tokkyo Koho 76 117,742 (Oct. 16, 1976), Y. Masumoto and H. Eguchi (to Dainippon Ink and Chemical Co.).
109. A. Zyball, *Kunststoffe* **67**(8), 461 (1977).
110. W. Honsberg, *Rubber World*, **190**(3), 34, 36 (1984).
111. Jpn. Kokai Tokkyo Koho 80 115,438 (Sept. 5, 1980) (to Denki Kagaku Kogyo K.K.).
112. Ger. Offen. 2,602,689 (July 28, 1977), E. Koenlein, R. Glaser, and L. Koessler (to BASF A.G.).
113. Jpn. Kokai Tokkyo Koho 79 157,152 (Dec. 11, 1979), T. Kawada, M. Nagasawa, and Y. Yaeda (to Synthetic Rubber Co., Ltd.).
114. Jpn. Kokai Tokkyo Koho 58 208,309 (Dec. 5, 1983) (to Nisshin Electric Co., Ltd.).
115. Eur. Pat. Appl. 47,050 (Mar. 10, 1982), A. B. Robertson (to Mobil Oil Corp.).
116. U.S. Pat. 4,244,861 (Jan. 13, 1981), L. Spenadel and co-workers (to Exxon Research and Engineering Co.).
117. V. R. Kamath and S. E. Stromberg in Ref. 13, p. 160.
118. U.S. Pat. 3,987,126 (Oct. 19, 1976), N. Brodoway (to E. I. du Pont de Nemours & Co., Inc.).
119. G. Kojima and co-workers, *Rubber Chem. Technol.* **50**(2), 403 (1977).
120. Jpn. Kokai Tokkyo Koho 58 163,107 (Sept. 27, 1983) (to Furukawa Electric, Ltd.).
121. U.S. Pat 4,419,499 (Dec. 6, 1983), A. Y. Coran, R. Patel, and D. Williams (to Monsanto Co.).

122. Jpn. Kokai Tokkyo Koho 59 11,315 (Jan. 20, 1984) (to Sumitomo Electric Industries, Ltd.).

123. Jpn. Kokai Tokkyo Koho 80 161,827 (Dec. 16, 1980) (to Teijin Ltd.).

124. Eur. Pat. Appl. 35,072 (Sept. 9, 1981), Z. Kovacs and R. Schuler (to BBC A.G.).

125. W. J. Miller, *Soc. Vac. Coaters Proc. Ann. Tech. Conf.* **26**, 20 (1983).

126. Jpn. Kokai Tokkyo Koho 78 77,294 (Dec. 20, 1976), A. Hamano and co-workers (to Nippon Kasei Co.).

127. A. Kameyama and co-workers, *Kagaku to Kogyo (Osaka)* **56**(7), 254 (1983).

128. Yu. V. Baranov and co-workers, *Deposited Doc.* 9 (1980).

129. Jpn. Kokai Tokkyo Koho 79 77,794 (June 21, 1979), T. Takata, M. Kamiyoshi, and S. Tanaka (to Teijin Ltd.).

130. Jpn. Kokai Tokkyo Koho 59 49,217 (Sept. 14, 1982) (to Matsushita Electric Works).

131. Jpn. Kokai Tokkyo Koho 80 02,346 (Jan. 19, 1980) (to Fuji Polymers).

132. U.S. Pat. 4,126,134 (Aug. 5, 1980), W. Brenner.

133. Jpn. Kokai Tokkyo Koho 81 151,703 (Nov. 24, 1981) (to Kureha Chemical Industries Co.).

134. Jpn. Kokai Tokkyo Koho 59 109,507 (June 24, 1984) (to Nisshin Electric Co.).

135. Jpn. Kokai Tokkyo Koho 79 129,072 (Oct. 6, 1979), G. Kojima and H. Wachi (to Asahi Glass Co., Ltd.).

136. Jpn. Kokai Tokkyo Koho 81 84,711 (July 10, 1981) (to Asahi Chemical Industry Co., Ltd.).

137. U.S. Pat. 4,264,650 (Apr. 28, 1981), S. R. Schulze, E. C. Lupton, W. A. Miller, and R. H. Hutzler (to Allied Chemical Corp.).

138. V. I. Dakin, Z. S. Egorova, and V. L. Karpov, *Vysokomol. Soedin. Ser. A* **23**, 2727 (1981).

139. Ger. Offen. 2,825,995 (June 15, 1977), S. C. Zingheim (to Raychem Corp.).

140. Jpn. Kokai Tokkyo Koho 81 115,313 (Sept. 10, 1981) (to Hitachi, Ltd.).

141. Z. Mrazek and co-workers, *Plast. Kau.* **29**, 566 (1982).

142. Jpn. Kokai Tokkyo Koho 58 164,110 (Sept. 29, 1983) (to Furukawa Electric Co., Ltd.).

143. Jpn. Kokai Tokkyo Koho 58 160,373 (Sept. 22, 1983) (to Showa Electric Wire and Cable Co., Ltd.).

144. Ger. Offen. 2,745,906 (Apr. 13, 1978), H. Inato, T. Nishihari, and T. Arakawa (to Teijin Ltd.).

145. Jpn. Kokai Tokkyo Koho 80 108,456 (Aug. 20, 1980) (to General Electric Co.).

146. U.S. Pat. 4,393,177 (July 12, 1983), T. Ishii and co-workers (to Hitachi, Ltd.).

147. Jpn. Kokai Tokkyo Koho 57 133,108 (Aug. 17, 1982) (to Nitto Electric Industrial Co., Ltd.).

148. U.S. Pat. 6,277,925 (Aug. 21, 2001), A. Biswas and co-workers (to Hercules Inc., and Plastics Technology Corporation).

149. D. Sheehan and co-workers, *J. Appl. Polym. Sci.* **6**, 47 (1962).

150. U.S. Pat. 4,141,691 (Feb. 27, 1979), J. M. Antonette and G. F. Snow (to Calgon Corp.).

151. U.S. Pat. 4,160,731 (July 10, 1979), C. L. Doyle (to American Cyanamid Co.).

152. Eur. Pat. Appl. 12,517 (June 25, 1980), R. H. Windhager and M. H. Hwang (to Calgon Corp.).

153. L. C. Hardy and D. F. Shriver, *Macromolecules* **17**(4), 975 (1984).

154. Ger. Pat. DD 160,393 (July 27, 1983), H. Dautzenberg and co-workers.

155. Ger. Pat. DD 160,779 (Mar. 7, 1984), G. Pacholke and co-workers.

156. U.S. Pat. 6,315,910 (Nov. 13, 2001), J. E. Farmerie and J. K. Dicksa (to Calgon Corp.).

157. Eur. Pat. Appl. 11,486 (May 20, 1989), G. D. Sinkovitz (to Calgon Corp.).

158. A. Wyroba, *Przem. Chem.* **62**(12), 681 (1983).

159. I. S. Shkurnikova and co-workers, *Izv. Akad. Nauk SSSR Ser. Khim.* (4), 928 (1984).

160. D. Babilis and co-workers, *Polym. Prepr.* **26**(1), 204 (1984).
161. Eur. Pat. Appl. 81,334 (June 15, 1983), W. M. Prest, Jr. and R. J. Roberts, Jr.
162. U.S. Pat. 6,451,952 (Sept. 17, 2002), S. Yamaguchi and T. Fujisawa (to Nippon Sho-kubai Co., Ltd).
163. Jpn. Kokai Tokkyo Koho 58 176,209 (Oct. 15, 1983) (to Dainippon Ink and Chemical Co.); Ger. Offen. 3,218,200 (Nov. 17, 1983), K. Reuter and co-workers (to Bayer AG).
164. N. G. Videnina and co-workers, *Lakokras. Mater.* (2), 28 (1982).
165. Ger. Offen. 3,233,912 (Mar. 15, 1984), R. Klug and co-workers (to Merck G.m.b.H.).
166. K. Fujimori, *J. Polym. Sci. Chem. Ed.* **21**, 25 (1983).

R. Dowbenko
PPG Industries

ALUMINATES

1. Introduction

Sodium aluminate [1302-42-7], $NaAlO_2$, has been used as a commercial product since about 1925 when it was introduced as an effective water treatment chemical. Among industrial users of sodium aluminate are producers of paper (qv), paint pigments (qv), silica–alumina or alumina-based catalysts, dishwasher detergents (qv), molecular sieves (qv), concrete, antacids, and others. Sodium aluminate is used in removal of phosphates from municipal and industrial waste waters and for clarification of industrial process and potable water. Commercial sodium aluminate products are available as liquids, and to a lesser degree, in solid form. The formula of anhydrous sodium aluminate is variously given as $NaAlO_2$ (aluminum sodium oxide [1302-42-7]), $Na_2O \cdot Al_2O_3$, or $Na_2Al_2O_4$. Commercial sodium aluminates are not accurately represented by these formulas because the products contain more than the stoichiometric amount of sodium oxide [1313-59-3], Na_2O. The amount of excess caustic in commercial products is indicated by ratios of Na_2O/Al_2O_3 that are typically between 1.05 and 1.15 for dry products, and 1.26 and 1.5 for liquids.

2. Physical and Chemical Properties

Commercial grades of sodium aluminate contain both waters of hydration and excess sodium hydroxide. In solution, a high pH retards the reversion of sodium aluminate to insoluble aluminum hydroxide. The chemical identity of the soluble species in sodium aluminate solutions has been the focus of much work (1). Solutions of sodium aluminate appear to be totally ionic. The aluminate ion is monovalent and the predominant species present is determined by the Na_2O concentration. The tetrahydroxyaluminate ion [14485-39-3], $Al(OH)^-_4$, exists in lower concentrations of caustic; dehydration of $Al(OH)^-_4$, to the *meta*-aluminate ion [20653-98-9], AlO^-_2, is postulated at concentrations of Na_2O above 25%. The

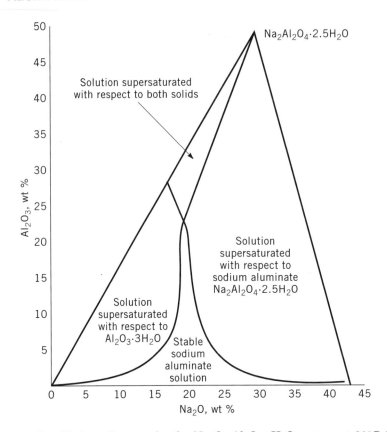

Fig. 1. Equilibrium diagram for the $Na_2O-Al_2O_3-H_2O$ system at 30°C (3).

formation of polymeric aluminate ions similar to the positively charged polymeric ions formed by hydrolysis of aluminum at low pH does not seem to
occur. $Al(OH)^-{}_4$ has been identified as the predominant ion in dilute aluminate
solutions (2).

Figure 1 shows a phase equilibrium diagram for soda–alumina–water (3).
The point of maximum solubility at 30°C for alumina trihydrate [12252-70-9],
$Al_2O_3 \cdot 3H_2O$, and sodium aluminate is near 23% Al_2O_3 and 20% Na_2O. At higher
Na_2O concentrations, the solutions are supersaturated with respect to sodium
aluminate; at lower concentrations of Na_2O, the solutions are supersaturated
with respect to the trihydrate. At concentrations greater than 23% Al_2O_3 and
20% Na_2O, the solution is supersaturated with respect to both solids. Most commercial solutions are supersaturated with regard to one or both of the solids.

3. Manufacture

Small amounts of sodium aluminate are prepared in the laboratory by fusion
of equimolar quantities of sodium carbonate [497-19-8] and aluminum acetate
[139-12-8], $Al(C_2H_3O_2)_3$, at 800°C (4). Other methods involve reaction of sodium
hydroxide with amorphous alumina or aluminum [7429-90-5] metal. Commercial

quantities of sodium aluminate are made from hydrated alumina, in the form of aluminum hydroxy oxide [24623-77-6],AlO(OH), or aluminum hydroxide [21645-51-2], Al(OH)$_3$, a product of the Bayer process (5,6) which is used to refine bauxite [1318-16-7], the principal aluminum ore.

Commercial grades of sodium aluminate are obtained by digestion of aluminum trihydroxide in aqueous caustic at atmospheric pressure and near the boiling temperature (7). Digestion of the aluminum hydroxy oxide in aqueous sodium hydroxide requires pressures of up to 1.38 MPa (13.6 atm) and temperatures of about 200°C. Dry sodium aluminate is obtained by evaporation of water. Several processes for the production of sodium aluminate are known that do not require the addition of water. In one process, bauxite reacts with molten sodium hydroxide at approximately 400°C (8); in another, aluminum trihydroxide reacts with solid sodium hydroxide (9); additionally, sodium carbonate has been treated with alumina or bauxite in rotary kilns at about 1000°C (10, 11). Other similar methods have also been described.

The form of sodium aluminate produced depends upon the manufacturing process. In high temperature, nonaqueous processes, the sodium aluminate product normally contains less than 1% moisture. If a dry product is desired from aqueous digestion processes, some sort of drying operation is required. In practice about half of the digestion liquor is sold as liquid sodium aluminate after filtration and dilution or concentration to conform with product specifications. Some products contain small amounts of powdered insoluble solids that act as precipitation or coagulation aids in water-treatment processes (12). Concentrated solutions of sodium aluminate tend to decompose by precipitation of alumina. Such decomposition is controlled in commercial products by addition of excess caustic and small quantities of certain organic compounds, or stabilizers (7,13–15).

Liquid sodium aluminate is available in steel drums having an approximate capacity of 210 L and bulk shipments are available in either tank trucks or railroad tank cars. The density of liquid sodium aluminate is usually from 1450 to 1510 kg/m^3. Solid products are available in moisture-proof paper bags or fiber drums containing approximately 23 and 150 kg, respectively.

4. Economic Aspects

Demand for aluminates for the treatment of industrial water should grow annually at the rate of 9% to $40 × 10^6 or 84 × 10^6 kg (185 × 10^6 lb) in 2004 and 110 × 10^6 kg (242 × 10^6 lb) in 2009 (16).

The primary U. S. producers of aluminates for water treatment are General Chemical, Geo Speciality Chemicals, and Kemwater North America.

Price range history 1989–2000 was $0.07–0.10/kg ($0.15–0.22/lb). Expected price in 2004 should be $0.11/kg ($0.25/lb) (16).

5. Analytical Methods

Commercial liquid sodium aluminates are normally analyzed for total alumina and for sodium oxide by titration with ethylene diaminetetraacetic acid

[60-00-4] (EDTA) or hydrochloric acid. Further analysis includes the determination of soluble alumina, soluble silica, total insoluble material, sodium oxide content, and carbon dioxide. Aluminum and sodium can also be determined by emission spectroscopy. The total insoluble material is determined by weighing the ignited residue after extraction of the soluble material with sodium hydroxide. The sodium oxide content is determined in a flame photometer by comparison to proper standards. Carbon dioxide is usually determined by the amount evolved, as in the Underwood method.

The determination of the soluble or available sodium aluminate presents difficulties because sodium aluminate begins to hydrolyze to the insoluble alumina trihydrate in water. The degree of hydrolysis depends on concentration, temperature, and time. It is therefore necessary to use a method of analysis that simultaneously affords control of the hydrolysis and gives the amount of available sodium aluminate encountered. This is best done by extracting the soluble alumina using sodium hydroxide solutions and subsequently determining the alumina content by gravimetric methods or titration with EDTA or hydrochloric acid (17).

6. Health and Safety Aspects

Sodium aluminate is highly alkaline and should be treated as a corrosive.

ACGIH has assigned a maximum concentration for TLV TWA OF 2.0 mg/m^3. It is considered a soluble salt of aluminum because of its relatively high solubility (18).

7. Uses

Sodium aluminate is used in the treatment of industrial and municipal water supplies and the use of sodium aluminate is approved in the clarification of drinking water (19). The FDA approves the use of sodium aluminate in steam generation systems where the steam contacts food. One early use of sodium aluminate was in lime softening processes, where it increases the precipitation of ions contributing to hardness and improves suspended solids removal from the treated water (20). Sodium aluminate reacts with silica to leave very low residual concentrations of silica in hot process water softeners. Sodium aluminate is often used with other chemicals such as alum, ferric salts, clays, and polyelectrolytes, as a coagulant aid (21,22).

Sodium aluminate is an effective precipitant for soluble phosphate in sewage and is especially useful in wastewater having low alkalinity (23,24). Sodium aluminate hydrolyzes in water to $Al(OH)_3$ and Al^{+3} which precipitate soluble phosphate as aluminum phosphate [7784-30-7], $AlPO_4$. Sodium aluminate has also been described as an effective aid for the removal of fluorides from some industrial waste waters (25). Combinations of sodium aluminate and other chemicals are being used to improve the detackification of paint particles in water from spray-painting operations (26).

Large quantities of sodium aluminate are used in papermaking where it improves sizing, filler retention, and pitch deposition (19,27–31). The addition of sodium aluminate to titanium dioxide paint pigment improves the nonchalking performance of outdoor paints (19,32,33). The etching of glass (qv) and ceramics (qv) by alkaline washing solutions is inhibited by inclusion of sodium aluminate in the formulations (34,35). The use of sodium aluminate solutions in the processing of acrylic and polyester synthetic fibers improves drying, antipiling, and antistatic properties of the fibers (36,37).

The recovery of sand from foundry molds and cores is much easier when binders made water soluble by use of sodium aluminate are used in place of insoluble resin binders (38,39). Sodium aluminate acts as a setting accelerator for Portland cement (qv) (40). In similar application, addition to concrete provides a longer gel time before fully curing (41).

One application patented in 1989 is the injection of sodium aluminate into silica-containing formations for enhanced petroleum recovery (42). Additionally, the pharmaceutical industry uses sodium aluminate as an alkaline source of aluminum for the production of certain antacids (43).

Sodium aluminate is widely used in the preparation of alumina-based catalysts. Aluminosilicate [1327-36-2] can be prepared by impregnating silica gel with alumina obtained from sodium aluminate and aluminum sulfate (44,45). Reaction of sodium aluminate with silica or silicates has produced porous crystalline aluminosilicates which are useful as adsorbents and catalyst support materials, ie, molecular sieves (qv) (46,47).

One method for using sodium aluminate to desulfurize flue gas containing sulfur dioxide is described (48). This procedure led to a process where aluminum sulfate [10043-01-3] could be generated as a by-product of flue gas desulfurization (49).

BIBLIOGRAPHY

"Aluminates" in *ECT* 1st ed., Vol. 1, pp. 649–653, by J. W. Ryznar, National Aluminate Corporation; in *ECT* 2nd ed., Vol. 2, pp. 6–11, by J. W. Ryznar and A. C. Thompson, Nalco Chemical Company; in *ECT* 3rd ed., Vol. 2, pp. 197–202, by W. R. Busler (Nalco Chemical Company); in *ECT* 4th ed., Vol. 2, pp. 267–273, by Robert J. Keller and Joseph J. Len, Vinings Industries, Inc.

CITED PUBLICATIONS

1. J. R. Glastonbury, *Chem. Ind.* (London), **121** (Feb. 1969).
2. P. L. Hayden and A. J. Rubin, in A. J. Rubin, ed., *Aqueous-Environmental Chemistry of Metals*, Ann Arbor Science, Ann Arbor, Mich. 1976, 317–381.
3. R. Fricke and P. Jucaistis, *Z. Anorg. Allg. Chem.*, 129–149 (1930).
4. J. Thery, A. M. Lejus, D. Briancon, and R. Collongues, *Bull. Soc. Chim. Fr.* 973 (1961).
5. U.S. Pat. 382,505 (May 8, 1888), K. J. Bayer.
6. U.S. Pat. 515,859 (Mar. 6, 1894), K. J. Bayer.
7. U.S. Pat. 2,345,134 (Mar. 28, 1944), F. K. Lindsay and B. F. Willey (to National Aluminate Corp.).

8. U.S. Pat. 2,159,843 (May 23, 1939), R. L. Davies (to Pennsylvania Salt Mfg. Co.).
9. U.S. Pat. 2,018,607 (Oct. 22, 1935), R. E. Cushing and C. W. Burkhardt (to Pennsylvania Salt Mfg. Co.).
10. K. Kammermeyer and A. B. Peck, *J. Am. Ceram. Soc.* **16**, 363 (1933).
11. C. Matignon, *Compt. Rend.* **177**, 1290 (1923).
12. U.S. Pat. 3,342,742 (Sept. 19, 1967), T. G. Cooks (to Nalco Chemical Co.).
13. Fr. Pat. 1,356,638 (Apr. 7, 1964), W. H. Brown and B. F. Armburst, Jr. (to Reynolds Metals Co.).
14. Fr. Pat. 1,476,257 (Apr. 7, 1967), R. Beverini.
15. U.S. Pat. 3,656,889 (Apr. 18, 1972), E. W. Olewinski (to Nalco Chemical Co.).
16. *Industrial Water Management Chemicals to 2001*, Freedonia Group, Inc., March 2000.
17. H. L. Watts and D. W. Utley, *Anal. Chem.* **25**, 864 (1953).
18. B. D. Dinman, in E. Bingham, B. Cohrssen, and C. H. Powell, eds., *Patty's* Toxicology, 5th ed., Vol. 2, John Wiley & Sons, Inc., New York, 2001, p. 403.
19. B. Suresh, A. Kishi, and S. Schlag, *Chemical Economics Handbook*, SRI, Menlo Park, Calif., Nov. 2001.
20. U.S. Pat. 1,620,332 (Mar. 8, 1927), W. Evans (to National Aluminate Corp.).
21. M. Adhikari, S. K. Gupta, and B. Banerjee, *J. Indian Chem. Soc.* **51**(10), 891 (1974).
22. Can. Pat. 964,808 (Mar. 25, 1975), J. Kane and L. O. Boots (to Nalco Chemical Co.).
23. U.S. Pat. 3,617,542 (Nov. 2, 1971), R. A. Boehler and M. R. Purvis, Jr., (to Nalco Chemical Co.).
24. Ger. Offen. 2,016,758 (Nov. 19, 1979), R. D. Sawyer and J. D. Tnsley (to Nalco Chemical Co.).
25. Jpn. Pat. 74 32,472 (Mar. 25, 1974), M. Watanabe, T. Okamoto, and K. Iida (to Sumitomo Chemical Co.).
26. Ger. Offen. 2,247,164 (Mar. 28, 1974), H. Wirth (to Chemische Werke Kluthe K. G.).
27. P. E. Barr, *Pulp Pap. Ind.* **21**(3), 54 (1947).
28. H. E. Berg, *Tappi* **39**(1), 153A (1956).
29. W. S. Wilson, *Paper Trade J.* **121**(8), 39 (1945).
30. E. Hechler, *Papier (Darmstadt)* **28**(11), 473 (1974).
31. J. V. Lamarre and J. R. Nelson, *TAPPI Monogr. Ser.* (33), 47 (1971).
32. U.S. Pat. 2,671,031 (Mar. 2, 1954), W. R. Whately (to American Cyanamid Co.).
33. U.S. Pat. 3,086,877 (Apr. 23, 1963), G. M. Sheehan and E. R. Lawhorne (to American Cyanamid Co.).
34. U.S. Pat. 2,575,576 (Nov. 20, 1951), L. R. Bacon and J. V. Otrhalek (to Wyandotte Chemicals Corp.).
35. U.S. Pat. 3,250,318 (Oct. 31, 1967), R. L. Green (to FMC Corp.).
36. Jpn. Pat. 71 09,869 (Mar. 12, 1971), H. Sugimoto, H. Sahara, and K. Nimura (to Mitsubishi Rayon Co.).
37. Jpn. Pat. 71 27, 783 (Aug. 21, 1971), T. Ito and M. Sotomurs (to Kanogafuchi Spinning Co.).
38. Czech. Pat. 147,361 (Feb. 2, 1973), J. Ornst and A. Burian; *Chem. Abstr.* **79**, 6961d (1973).
39. Jpn. Pat. 73 39,695 (Nov. 26, 1973), K. Kobayashi (Nihon Shell-Mold Assn.), and K. Ohtani (to Chisso Corp.).
40. U.S. Pat. 3,656,985 (Apr. 18, 1972), B. Bonnel and C. Houasse (to Progil, France).
41. Jpn. Pat. 01 111,761 (Apr. 28, 1989), O. Imamura (to Sanko Colloid Kagaku K. K.).
42. Fr. Pat. 3,822,734 (Jan. 19, 1989), J. Labrid (to Institute Francais du Petrole).
43. Jpn. Pat. 60,161,915 (Aug. 23, 1985), T. Kajino (to Sato Pharmaceutical Co.).
44. U.S. Pat. 2,929,973 (Jan. 5, 1960), W. D. Stillwell, L. Bakker, and J R. Lytle.

45. U.S. Pat. 2,996,460 (Aug. 15, 1960), D. G. Braithwaite (to Nalco Chemical Co.).
46. U.S. Pat. 2,882,243 (Apr. 14, 1959), R. M. Milton (to Union Carbide Corp.).
47. U.S. Pat. 2,882,244 (Apr. 14, 1959), R. M. Milton (to Union Carbide Corp.).
48. U.S. Pat. 4,134,961 (Jan. 16, 1979), D. Lurie.
49. U.S. Pat. 4,296,079 (Oct. 20, 1981), H. W. Hauser (to Vinings Chemical Co.).

ROBERT J. KELLER
JOSEPH J. LEN
Vinings Industries Inc.

ALUMINUM AND ALUMINUM ALLOYS

1. Introduction

Aluminum [7429-90-5], Al, is a silver-white metallic element in group III of the periodic table having an electronic configuration of $1s^2 2s^2 2p^6 3s^2 3p^1$. Aluminum exhibits a valence of +3 in all compounds except for a few high temperature gaseous species in which the aluminum may be monovalent or divalent. Aluminum is the most abundant metallic element on the surfaces of the earth and moon, comprising 8.8% by weight (6.6 atomic %) of the earth's crust. However, it is rarely found free in nature. Nearly all rocks, particularly igneous rocks, contain aluminum as aluminosilicate minerals.

Although impure metallic aluminum was isolated in 1824 by reducing aluminum chloride [7446-70-0], $AlCl_3$, using potassium amalgam, Wöhler, who produced higher purity aluminum by a similar method in 1827 and determined its properties, is often given the credit for the discovery. The metal remained a laboratory curiosity until 1854 when the method of preparation was improved, using sodium as a reductant, and commercial quantities of relatively pure aluminum were produced. The present industrial production method was developed independently in 1886 by both Paul-Louis Héroult in France and Charles Martin Hall in Oberlin, Ohio.

Aluminum reflects radiant energy throughout the spectrum. It is odorless, tasteless, nontoxic, and nonmagnetic. Because of its many desirable physical, chemical, and metallurgical properties, aluminum is the most widely used nonferrous metal. The utility of the metal is enhanced by the formation of a stable adherent oxide surface that resists corrosion. Because of high electrical conductivity and lightness, aluminum is used extensively in electrical transmission lines. High purity aluminum is soft and lacks strength but its alloys, containing small amounts of other elements, have high strength-to-weight ratios. Alloys of aluminum are readily formable by many metalworking processes; they can be joined, cast, or machined and accept a wide variety of finishes. Aluminum, having a density about one-third that of ferrous alloys, is used in transportation and structural applications where weight saving is important.

2. Physical Properties

The properties of aluminum vary significantly according to purity and alloying (1). Physical properties for aluminum of a minimum of 99.99% purity are summarized in Table 1. Although a number of radioactive isotopes have been artificially produced (see RADIOISOTOPES), naturally occurring aluminum consists of a single stable isotope, ^{27}Al, having a cross section for thermal neutrons of 0.215×10^{-28} m^2. This low cross section and the short half-life of the radioactive product from neutron irradiation make aluminum an attractive material for use within nuclear reactors. Aluminum crystallizes in the face-centered cubic system having a unit cell of 0.40496 nm at 20°C. The unit cell contains four atoms and has a coordination number of 12. The distance of closest approach of atoms is 0.2863 nm. The effects of temperature on the density of aluminum are shown in Table 2. The change in density at the solid–liquid transformation corresponds to a volume increase of ca 6.5%. The heat capacities of solid and liquid aluminum are shown in Table 3. Other thermodynamic data are available in reference 3.

The linear expansion of aluminum over several temperature ranges can be calculated from equations 1–3

$$L_t(-200 \text{ to } 0°C) = L_0\left[1 + C\left(21.57t + 0.00443t^2 - 0.000124t^3\right)10^{-6}\right] \qquad (1)$$

$$L_t(-60 \text{ to } 100°C) = L_0\left[1 + C\left(22.17t + 0.012t^2\right)10^{-6}\right] \qquad (2)$$

$$L_t(0 \text{ to } 500°C) = L_0\left[1 + C\left(22.34t + 0.00997t^2\right)10^{-6}\right] \qquad (3)$$

where L_0 = length at 0°C, L_t = length at temperature t, and C = alloy constant, which is unity for pure aluminum. The thermal conductivity of aluminum at

Table 1. **Physical Properties of Aluminum**

Property	Value
atomic number	13
atomic weight	26.9815
density at 25°C, kg/m^3	2698
melting point, °C	660.2
boiling point, °C	2494
thermal conductivity at 25°C, W/m · K)	234.3
latent heat of fusion, J/g[a]	395
latent heat of vaporization at bp ΔH_v, kJ/g[a]	10,777
electrical conductivity	65% IACS[b]
electrical resistivity at 20°C, Ω·m	2.6548×10^{-8}
temperature coefficient of electrical resistivity, Ω·m/°C	0.0043
electrochemical equivalent, mg/°C	0.0932
electrode potential, V	−1.66
magnetic susceptibility, g^{-1}	0.6276×10^{-6}
Young's modulus, MPa[c]	65,000
tensile strength, MPa[c]	50

[a] To convert J to cal, divide by 4.184.
[b] International Annealed Copper Standard.
[c] To convert MPa to psi, multiply by 145.

Table 2. **Density of 99.996% Aluminum**

Temperature, °C	Density, kg/m^3
Solid	
25	2698
100	2680[a]
300	2660[a]
500	2620[a]
660	2550[a]
Liquid[b]	
660	2368
700	2357
750	2345
800	2332
850	2319
900	2304

[a] Calculated from density at 25°C and volume expansion at elevated temperatures.
[b] Ref. 2.

various temperatures is shown in Table 4. The thermal and electrical conductivities are related by

$$K = 2.1 \, \lambda T \cdot 10^{-6} + 12.55 \qquad (4)$$

where K = thermal conductivity in W/m · K), λ = electrical conductivity in S/m (or reciprocal Ωm), and T = temperature K. The electrical resistivity of aluminum of 99.996% purity is 2.6548×10^{-8} Ω · m at 20°C and the temperature coefficient of electrical resistivity is 0.0043 Ω°C. The effects of temperature on electrical resistivity for solid and liquid aluminum are shown in Table 5. Aluminum becomes superconductive at temperatures near absolute zero and has a

Table 3. **Heat Capacity of Aluminum**[a]

Temperature, K	Heat capacity, J/(kg · K)[b]
Crystalline solid	
0	0.00
100	481.7
200	790.8
298	897.2
300	898.7
400	955.6
500	994.8
600	1033.5
700	1078.5
800	1132.7
900	1197.4
Liquid	
1000 and higher	1273.5

[a] Ref.3.
[b] To convert J to cal, divide by 4.184.

Table 4. **Thermal Conductivity of Aluminum**

Temperature, K	Thermal conductivity, $W/(m \cdot K)$	
	99.996% Al[a]	99.95% Al[b]
4	3150.6	
6	4196.6	
8	5196.5	
10	5995.7	
15	6794.8	
20	5497.8	
50	949.8	
100	301.2	
150	244.7	
200	234.3	
300	234.3	
298		225.1
523		203.3
723		189.9
923		186.2
1013		59.8[c]
1173		75.3[c]

[a] Ref.4.
[b] Ref.5.
[c] Liquid aluminum.

transition temperature of 1.187 K. Resistivity at very low temperatures is strongly increased by impurities.

2.1. Optical Properties. The index of refraction and extinction coefficient of vacuum-deposited aluminum films have been reported (8,9) as have the total reflectance at various wavelengths and emissivity at various temperatures

Table 5. **Electrical Resistivity of Aluminum**

Temperature, °C	Resistivity, $\Omega \cdot m \times 10^{-8}$
Solid[a]	
0	2.42
100	3.50
200	4.63
300	5.81
400	7.05
500	8.36
600	9.77
650	10.56
Liquid[b]	
660	24.20
700	24.75
800	26.25
1000	29.20
1200	32.15

[a] Ref. 6.
[b] Ref. 7.

Table 6. **Solubility of Hydrogen in Aluminum**

| | Solubility[a], mL H_2/100 g Al | |
| | Reference 11 | Reference 12 |
Temperature, °C		
Solid		
400	0.004	0.003
500	0.01	0.01
600	0.025	0.03
660	0.04	0.05
Liquid		
660	0.69	0.46
700	0.91	0.63
800	1.68	1.23
850	2.18	1.66

[a] Values at 20°C and 202.3 kPa (14.7 psi); pressure of hydrogen over sample is 101.3 kPa (1 atm).

(10). Emissivity increases significantly as the thickness of the oxide film on aluminum increases and can be 70–80% for oxide films of 100 nm.

2.2. Solution Potential. The standard electrode potential of aluminum ($Al \longrightarrow Al^{3+} + 3e$) is −1.66 V on the standard hydrogen scale and −1.99 V on the 0.1 N calomel scale at 25°C. In the electromotive force series this places aluminum cathodic to magnesium and anodic to zinc, cadmium, iron, nickel, and copper. Aluminum electrode potentials are irreversible in aqueous solutions and vary significantly with pH. Potentials measured in highly basic solutions are anodic to those measured in acid solutions.

2.3. Surface Tension and Viscosity. The surface tension of molten aluminum, determined by the method of maximum bubble pressure, is 0.86 ± 0.2 N/m (860 dyn/cm) in the range of 700–750°C. The viscosity of 99.996% aluminum at these same temperatures is $1 - 1.2$ mPa $\cdot$ s($=$ cP).

2.4. Aluminum and Hydrogen. Hydrogen is the only gas known to be appreciably soluble in solid or molten aluminum. Hydrogen can be introduced into liquid aluminum from reaction with moisture present in the furnace atmosphere or the refractories, or with moisture entrapped in the oxide film of the solid aluminum before melting. The solubility of hydrogen in molten and solid aluminum is shown in Table 6.

3. Chemical Properties

3.1. Reactions with Elements and Inorganic Compounds. Aluminum reacts with oxygen O_2, having a heat of reaction of −1675.7 kJ/mol (−400.5 kcal/mol) Al_2O_3 produced.

$$2\,Al + 3/2\,O_2 \longrightarrow Al_2O_3 \tag{5}$$

In dry air at room temperature this reaction is self-limiting, producing a highly impervious film of oxide ca 5 nm in thickness. The film provides both stability at ambient temperature and resistance to corrosion by seawater and other aqueous and chemical solutions. Thicker oxide films are formed at elevated temperatures and other conditions of exposure. Molten aluminum is also protected by an oxide film and oxidation of the liquid proceeds very slowly in the absence of agitation.

At high temperatures, aluminum reduces many oxygen-containing compounds, particularly metal oxides. These reactions, of the type shown in equation 6, are used in the manufacture of certain metals and alloys, as well as in the thermite welding process.

$$3\,MO + 2\,Al \longrightarrow Al_2O_3 + 3\,M \tag{6}$$

Molten aluminum reacts violently with water and the molten metal should not be allowed to touch damp tools or containers. In finely divided powder form, aluminum also reacts with boiling water to form hydrogen and aluminum hydroxide [21645-51-2]; this reaction proceeds slowly in cold water.

Aluminum does not combine directly with hydrogen, but it does react with nitrogen sulfur and carbon in oxygen-free atmospheres at high temperatures. To form aluminum carbide [1299-86-1], Al_4C_3, temperatures above $1000°C$ are required. Aluminum nitride [24304-00-5], AlN, is produced by arcing high purity aluminum electrodes in a nitrogen atmosphere. Aluminum sulfide [1302-81-4], Al_2S_3, and aluminum phosphide [20859-73-8], AlP, result from reaction at high temperatures with sulfur and phosphorus respectively.

Very high purity aluminum, resistant to attack by most acids, is used in the storage of nitric acid, concentrated sulfuric acid, organic acids, and other chemical reagents. Aluminum is, however, dissolved by aqua regia. Because of its amphoteric nature, aluminum is attacked rapidly by solutions of alkali hydroxides evolving hydrogen and forming soluble aluminates. Aluminum reacts vigorously with fluorine chlorine, bromine, and iodine to form trihalides, and with chlorinated hydrocarbons in the presence of water. It also reacts to form a volatile aluminum chloride when heated in a current of dry oxygen-free chlorine or hydrogen chloride. Gaseous monohalides can be formed by passing the trihalides over aluminum at temperatures above $800°C$

$$AlX_3 + 2\,Al \longrightarrow 3\,AlX \tag{7}$$

Aluminum hydroxide and aluminum chloride do not ionize appreciably in solution but behave in some respects as covalent compounds. The aluminum ion has a coordination number of six and in solution binds six molecules of water existing as $[Al(H_2O)_6]^{3+}$. On addition of a base, substitution of the hydroxyl ion for the water molecule proceeds until the normal hydroxide results and precipitation is observed. Dehydration is essentially complete at pH 7.

Aluminum is attacked by salts of more noble metals. In particular, aluminum and its alloys should not be used in contact with mercury or mercury compounds.

3.2. Reaction with Organic Compounds. Aluminum is not attacked by saturated or unsaturated, aliphatic or aromatic hydrocarbons. Halogenated derivatives of hydrocarbons do not generally react with aluminum except in the presence of water, which leads to the formation of halogen acids. The chemical stability of aluminum in the presence of alcohols is very good and stability is excellent in the presence of aldehydes, ketones, and quinones.

Organic compounds that form with aluminum other than through a direct metal-to-carbon bond include the metallo-organics, represented as $Al-X-R$ where X may be oxygen, nitrogen, or sulfur, and R is a suitable organic radical. The alcoholates or alkoxides are compounds of this type where R is an alcohol (see ALKOXIDES, METAL). Compounds having an aluminum-to-carbon bond include polymers that may be linear or cross-linked. They are best described as vinyl, divinyl, and trivinyl aluminum halides that polymerize rapidly to compounds the structures of which are not completely understood. Aluminum alkyls are also aluminum–carbon bond compounds. These are not polymeric but are viewed as being bridge compounds of varying complexity. Alkylaluminum compounds are used as catalysts in the preparation of oriented crystalline polyolefins leading to the production of elastomers that are essentially identical in structure to natural rubber (see ELASTOMERS, SYNTHETIC).

4. Manufacture and Processing

4.1. Raw Materials. Aluminum, the third most abundant element in the earth's crust, is usually combined with silicon and oxygen in rock. When aluminum silicate minerals are subjected to tropical weathering, aluminum hydroxide may be formed. Rock that contains high concentrations of aluminum hydroxide minerals is called bauxite [1318-16-7]. Although bauxite is, with rare exception, the starting material for the production of aluminum, the industry generally refers to metallurgical grade alumina [1344-28-1], Al_2O_3, extracted from bauxite by the Bayer Process (see ALUMINUM COMPOUNDS), as the ore. Aluminum is obtained by electrolysis of this purified ore. The specification of metallurgical alumina is given in Table 7.

4.2. Cryolite. Cryolite [15096-52-3], Na_3AlF_6, is the primary constituent of the Hall-Héroult cell electrolyte. High purity, natural cryolite is found in Greenland, but its rarity and cost have caused the aluminum industry to substitute synthetic cryolite. The latter is produced by the reaction of hydrofluoric acid HF, with sodium aluminate [1302-42-7], $NaAlO_2$, from the Bayer process

$$6\,HF + 3\,NaAlO_2 \longrightarrow Na_3AlF_6 + 3\,H_2O + Al_2O_3 \qquad (8)$$

Gaseous hydrofluoric acid is generally made by the reaction of acid-grade fluorspar CaF_2, with sulfuric acid (see FLUORINE COMPOUNDS, INORGANIC)

$$CaF_2 + H_2SO_4 \longrightarrow 2\,HF + CaSO_4 \qquad (9)$$

No cryolite is actually needed once the smelting process is in operation because cryolite is produced in the reduction cells by neutralizing the Na_2O brought into the cell as an impurity in the alumina using aluminum fluoride.

$$4\,AlF_3 + 3\,Na_2O \longrightarrow 2\,Na_3AlF_6 + Al_2O_3 \qquad (10)$$

Table 7. **Properties of Metallurgical Grade Alumina**

Item	Influencing operations[a]		Normal range	Specifications	
	Major	Minor		Typical	Desirable
particle size, µm	P	Ca	10–200	>10% >150% 10% 44	>2% >150% 2%20
chemical composition					
Na_2O	P	Ca	0.3–0.7%	0.5%	0.35%
Fe_2O_3	D	Cl	0.01–0.04%	0.03%	0.015%
SiO_2	D		0.01–0.03%	0.025%	0.020%
CaO	D,Cl		0.02–0.08%	0.06%	0.030%
TiO_2			0.002–0.005%	0.005%	0.002
CuO			0.001–0.01%	0.01%	
K_2O			0.000–0.05%	0.005%	
MgO				0.002%	
P_2O_5				0.001%	
NiO				0.005%	
Cr_2O_3				0.002%	
LOI[b]			0.3–1.5%		
total water			1.5–4.0%	3.5%	
surface area, m²/g			20–100	60	80
α-Al_2O_3	Ca		5–90%	40%	10%
attrition index	P,Ca		4–15		8
crystallite size, µm	P		10–200		20
angle of repose, deg	Ca		30–45		35
bulk density, kg/m³					
loose			800–1100		
packed			950–1300		

[a] Designations are as follows: Ca = calcination, Cl = clarification, D = digestion, and P = precipitation.
[b] LOI = loss on ignition.

Thus operating cells need aluminum fluoride [7784-18-1], AlF_3, rather than cryolite. Much aluminum fluoride is produced in a fluidized bed by the reaction of hydrofluoric acid gas and activated alumina made by partially calcining the alumina hydrate from the Bayer process

$$Al_2O_3 \cdot XH_2O + 6\ HF \longrightarrow 2\ AlF_3 + (X+3)\ H_2O(g) \tag{11}$$

Aluminum fluoride is also made by the reaction of fluosilicic acid H_2SiF_6, a byproduct from phosphoric acid production (see PHOSPHORIC ACID AND THE PHOSPHATES), and aluminum hydroxide from the Bayer process.

$$H_2SiF_6 + 2\ Al(OH)_3 \longrightarrow 2\ AlF_3(aq) + SiO_{2(s)} + 4\ H_2O \tag{12}$$

The AlF_3 solution is filtered, AlF_3 precipitated by heating, flash dried, and calcined.

The equivalent of 3–4 kg of F per metric ton of aluminum produced is adsorbed from the bath into the cell lining over the lining's life (3–10 yr). The recoverable fluoride, 70–75%, requires an expensive recovery plant justified largely on an ecological basis. The most common method of recovery treats the crushed lining using dilute NaOH to dissolve the cryolite and other fluorides. The solution is filtered and the NaF:AlF$_3$ mol ratio adjusted to 3:1 whereupon-Na$_3$AlF$_6$ is precipitated by neutralizing the NaOH using CO$_2$.

Fluoride Availability. The aluminum industry in the United States uses about 15 kg of fluoride ion per metric ton aluminum, 10–25% of which is lost. The remainder, consisting of cryolite generated in reduction cells and of bath in scrap cell linings, is stored for future use. New fluoride for the aluminum industry comes largely from fluorspar, the world reserves of which are widespread and abundant (see FLUORINE COMPOUNDS, INORGANIC). Fluoride is also recovered from phosphate rock, generally as fluosilicic acid, in producing phosphoric acid. Mexico, the world's largest producer of fluorspar, supplies over 20% of the world's demand and 75% of U.S. usage. The aluminum industry uses about 21% of the fluoride consumed in the United States. Improved emission control (see EXHAUST CONTROL, INDUSTRIAL) and recycling of fluorides continue to reduce the fluoride requirements of the aluminum industry.

4.3. Electrolysis of Alumina. Since the discovery of the process by Hall and Héroult, nearly all aluminum has been produced by electrolysis of alumina dissolved in a molten cryolite based bath. The aluminum is deposited molten on a carbon cathode, which serves also as the melt container. Simultaneously, oxygen is deposited on and consumes the cell's carbon carbon anode(s) (13). Pure cryolite melts at 1012°C, but alumina and additives, namely 4–8% calcium fluoride CaF$_2$, 5–13% aluminum fluoride, 0–7% lithium fluoride LiF, and 0–5% magnesium fluoride, MgF$_2$, lower the melting point, allowing operation at 920–980°C. The system Na$_3$AlF$_6$–Al$_2$O$_3$ (Fig. 1) has a eutectic at 10.5 wt % Al$_2$O$_3$ at 960°C (14). Figure 2 gives liquidus temperatures in the system Na$_3$AlF$_6$–AlF$_3$–Al$_2$O$_3$

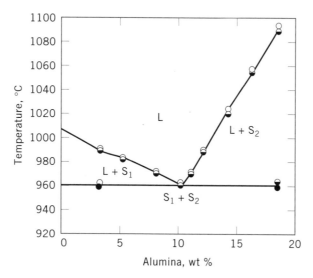

Fig. 1. Cryolite–alumina phase diagram from 0 to 18.5% alumina. L, liquid; S$_1$, cryolite; S$_2$, corundum; ○, liquid; ◐, liquid and solid; ●, solid (14).

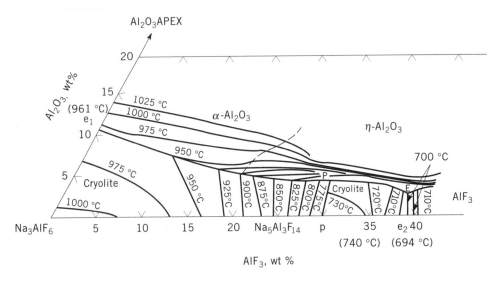

Fig. 2. The system Na_3AlF_6–AlF_3–Al_2O_3: P, ternary peritectic point (28.3% AlF_3–4.4%Al_2O_3–67.3%Na_3AlF_6, 723°C); E, ternary eutectic point (37.3% AlF_3–3.2%Al_2O_3–59.5%Na_3AlF_6, 684°C); p, binary peritectic point; e, e_1, binary points (15).

(15). Equations for the liquidus temperature and alumina solubility have also been worked out for the Na_3AlF_3–LiF–CaF_2–AlF_3–Al_2O_3 system (16). Joule heat from the flow of electric current is more than adequate to maintain the melt temperature.

Although the mechanism of electrolysis is still imperfectly understood, most investigators (17) agree that cryolite ionizes as

$$Na_3AlF_6 \longrightarrow 3\,Na^+ + AlF_6^{3-} \tag{13}$$

$$AlF_6^{3-} \rightleftharpoons AlF_4^- + 2\,F^- \tag{14}$$

Alumina dissolves at low concentrations by forming oxyfluoride ions having a 2:1 ratio of aluminum to oxygen ($Al_2OF_{2n}^{4-2n}$),

$$Al_2O_3 + 4\,AlF_6^{3-} \longrightarrow 3\,Al_2OF_6^{2-} + 6\,F^- \tag{15}$$

for example, at higher alumina concentrations, oxyfluoride ions with a 1:1 ratio of aluminum to oxygen ($Al_2O_2F_{2n}^{2-2n}$) are formed (18)

$$2\,Al_2O_3 + 2\,AlF_6^{3-} \longrightarrow 3\,Al_2O_2F_4^{2-} \tag{16}$$

Cells are generally operated using 1.5–6 wt % Al_2O_3 in the electrolyte. Saturation ranges between 6–12% Al_2O_3 depending upon composition and temperature.

Ion transport measurements indicate that Na^+ ions carry most of the current, yet aluminum is deposited. A charge transfer probably occurs at the cathode interface and hexafluoroaluminate ions are discharged, forming aluminum and F^- ions to neutralize the charge of the current carrying Na^+

$$12\,Na^+ + 4\,AlF_6^{3-} + 12\,e^- \longrightarrow 12\,(Na^+ + F^-) + 4\,Al + 12\,F^- \tag{17}$$

Oxyfluoride ions are discharged at the anode, forming carbon dioxideand aluminum fluoride. The first oxygen can be removed more readily from $Al_2O_2F_4^{2-}$ than either the second or the oxygen from $Al_2OF_6^{2-}$

$$2\,Al_2O_2F_4^{2-} + C \longrightarrow CO_2 + 2\,Al_2OF_4 + 4\,e^- \tag{18}$$

$$Al_2OF_4 + Al_2OF_6^{2-} \rightleftharpoons Al_2O_2F_4^{2-} + 2\,AlF_3 \tag{19}$$

Combining alumina dissolution, equation 16, and the anode and cathode reactions, equations 17–19, gives the overall reaction

$$2\,Al_2O_3 + 3\,C \longrightarrow 4\,Al + 3\,CO_2 \tag{20}$$

According to Faraday's law, one Faraday (26.80 A · h) should deposit one gram equivalent (8.994 g) of aluminum. In practice only 85–95% of this amount is obtained. Loss of Faraday efficiency is caused mainly by reduced species (Al, Na, or AlF) dissolving or dispersing in the electrolyte (bath) at the cathode and being transported toward the anode where these species are reoxidized by carbon dioxide forming carbon monoxide and metal oxide, which then dissolves in the electrolyte. Certain bath additives, particularly aluminum fluoride, lower the content of reduced species in the electrolyte and thereby improve current efficiency.

Equipment. A modern alumina smelting cell consists of a rectangular steel shell typically $9 - 16\,m \times 3 - 4\,m \times 1 - 1.3\,m$. It is lined with refractory insulation that surrounds an inner lining of baked carbon (see CARBON-ARTIFICIAL GRAPHITE). Few materials other than carbon are able to withstand the combined corrosive action of molten fluorides and molten aluminum. Thermal insulation is adjusted to provide sufficient heat loss to freeze a protective coating of electrolyte on the inner walls but not on the bottom, which must remain substantially bare for electrical contact to the molten aluminum cathode. Steel (collector) bars are joined to the carbon cathode at the bottom to conduct electric current from the cell. Current enters the cell either through prebaked carbon anodes (Fig. 3) or through a continuous self-baking Soderberg anode (Fig. 4).

Prebaked anodes are produced by molding petroleum coke and coal tar pitch binder into blocks typically $70\,cm \times 125\,cm \times 50\,cm$, and baking to 1000–1200°C. Petroleum coke is used because of its low impurity (ash) content. The more noble impurities, such as iron and silicon, deposit in the aluminum whereas less noble ones such as calcium and magnesium, accumulate as fluorides in the bath. Coal-based coke could be used, but extensive and expensive prepurification would be required. Steel stubs seated in the anode using cast iron support the anodes (via anode rods) in the electrolyte and conduct electric current into the anodes (Fig. 3). Electrical resistivity of prebaked anodes ranges from $5 - 6\,\Omega \cdot m$; anode current density ranges from 0.65 to 1.3 A/cm^2.

A Soderberg anode is formed continuously from a paste of petroleum coke and coal tar pitch added to the top of a rectangular steel casing typically $6 - 8\,m \times 2\,m \times 1\,m$ (Fig. 4). While passing through the casing, the paste bakes forming carbon to replace the anode being consumed. The baked portion extends past the casing and into the molten electrolyte. Electric current enters

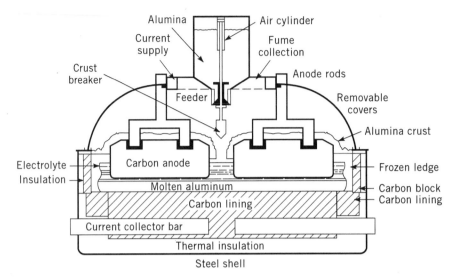

Fig. 3. Aluminum electrolyzing cell with prebaked anode.

the anode through vertical or sloping steel spikes (also called pins). Periodically, the lowest spikes are reset to a higher level. Resistivity of Soderberg anodes is about 30% higher than prebaked anodes. Current density is lower, ranging from 0.65 to 0.9 A/cm^2.

Molten aluminum is removed from the cells by siphoning, generally daily, into a crucible. Normally the metal is 99.6–99.9% pure. The principal impurities

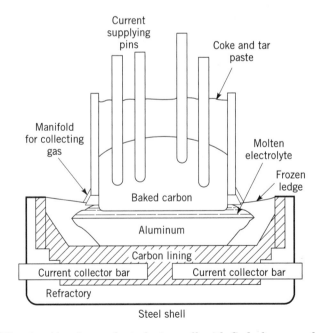

Fig. 4. Aluminum electrolyzing cell with Soderberg anode.

Table 8. **Energy Consumption Per Metric Ton of Aluminum Produced**[a]

Operation	Thermal, MJ[b]	Electric, kW · h	Total energy	
			Fossil and hydro, MJ[b]	If all fossil, MJ[b]
mining and refining	30,000	480	35,200	35,200
smelting	19,000[c]	15,000	146,400	182,500
mill processing	19,000	1,830	33,700	38,900
Total	*68,000*	*17,310*	*215,300*	*256,600*

[a] Values are approximate. Actual energy consumption depends upon the particular plant, alloy produced, and product formed.
[b] To convert MJ to Mcal, divide by 4.184.
[c] Includes forming, baking, and fuel value of anodes.

are Fe, Si, Ti, V, and Mn, and come largely from the anode, but also from the alumina.

4.4. Energy Considerations. Table 8 gives a breakdown of the energy required to produce aluminum. Note that smelting consumes about 65% of the required energy. In the United States most of this energy comes from fossil fuels. In other parts of the world, hydropower is a significant source of power for melting aluminum.

Some steps in aluminum production are quite energy efficient. Bayer plants make extensive use of heat exchangers to recover heat from material leaving high temperature operations for use in lower temperature operations (see HEAT EXCHANGE TECHNOLOGY). Fluid-bed calciners reduce energy consumption by 30% over the older rotary kiln calciners of alumina. On the other hand, melting and holding furnaces are only about 30% efficient in use of energy. Efficiency has been increased by improved firing practice and in some cases further improved by using heat from stack gases to preheat incoming air.

Furnaces used to bake anodes for prebake cells use the cooling anodes to preheat combustion air. Hot combustion gases from the baking zone are used to preheat incoming anodes. Using these techniques, about 4.2 MJ/kg (1004 kcal/kg) of anode carbon or only 2520 MJ/t (6.03×10^5 kcal/t) of aluminum is required to produce anodes.

Alternating current is converted to direct current (dc) for the smelting cells by silicon rectifiers. High conversion efficiency (over 99%) and minimum capital costs are achieved when the rectified voltage is 600–900 V dc. Because aluminum smelting cells operate at 4.5–5.0 V, 130 or more cells are connected in series, forming what the industry calls a potline, which may operate at 50–360 kA.

The individual components of smelting-cell voltage and energy are shown in Figure 5. The electrical energy required to decompose alumina $E_1 = 2.233$ V in a cryolite bath 65% saturated with alumina is given by

$$E_1 = (-\Delta G_1/nF) - (RT/nF)\ln a(\text{Al}_2\text{O}_3) \tag{21}$$

where ΔG_1, the free energy of formation of α-alumina [12252-63-0], is -1280.02 kJ/mol(-305.93 kcal/mol), (although the feed to the cell is largely

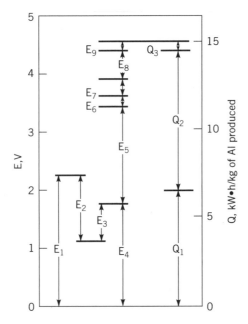

Fig. 5. Energy requirements of the Hall-Héroult cell (25–27). E_1, decomposition of alumina; E_2, depolarization by carbon; E_3, anode overvoltage; E_4, counter electromotive force; E_5, bath voltage drop; E_6, bath bubble voltage; E_7, anode voltage drop; E_8, cathode voltage drop; E_9, external voltage drop; Q_1, enthalpy to produce aluminum; Q_2, cell heat losses; Q_3 bus heat loss.

γ-Al_2O_3, α-Al_2O_3 is the phase in equilibrium with the bath); R is the gas constant (8.314 J/(mol · K)); T is bath temperature (1245 K); $n = 6$ electrons transferred; F is the Faraday constant (96.491 kJ/g · equiv); and $a(Al_2O_3)$ is the activity of alumina; a is 0.3 at 65% saturation. Oxygen deposition on the anode is depolarized by carbon, subtracting $E_2 = 1.026$ V given by

$$E_2 = (-\Delta G_2/nF) - (RT/\ln F)\ln K \tag{22}$$

where ΔG_2 is the free energy of formation of carbon dioxide and K the oxygen plus carbon reaction equilibrium constant. There is an anode overvoltage slightly over 0.5 V given by

$$E_3 = (RT/\alpha nF)\ln(i/i_o) \tag{23}$$

where i is current density, $n = 2$ electrons per oxygen, and i_o and α are experimental constants typically 50 A/m^2 and 0.52, respectively. Cathode overvoltage is minimal. Combining E_1, E_2, and E_3 gives E_4 the counter electromotive force (CEMF) of the cell of about 1.73 V. Linear extrapolation of cell volts vs current to zero current gives an apparent CEMF about 0.1 V lower because of the change in overvoltage with current. Below 2% alumina concentration, overvoltage becomes significant, the CEMF rises and fluoride is codeposited with oxygen. This causes the bath no longer to wet the anode and a gas envelope to form around it. In order for the current to arc across this poorly conductive layer the voltage must rise (>30 V) and this high voltage, called anode effect, generates some carbon tetrafluoride gas [75-73-0], CF_4, wastes power, and may overheat the cell. Anode effects can be minimized or eliminated by careful control of alumina additions.

$$E_5 = IL/KA \qquad\qquad (24)$$

Bath voltage E_5 is about 1.7 V, where I is current, L the anode-to-cathode spacing (4–5 cm), K the electrical conductivity of the bath, 2.0–2.4 $(\Omega \cdot m)^{-1}$), and A the effective bath area. Anode gas bubbles raise the bath's apparent resistivity, increasing its voltage about 0.17 V as indicated by E_6. Anode voltage E_7 represents the potential drop through the anode, its connecting stub, and the contact between them. It ranges from 0.25 to 0.3 V using prebake anodes and 0.45–0.55 V using Soderberg anodes. Cathode voltage E_8 includes the voltage loss in the lining, collector bars, and the contact between the two. It ranges from 0.35 to 0.55 V. There are also voltage losses in external conductors in the range 0.1–0.2 V, E_9, that do not contribute heat to the cell.

The right side of Figure 5, depicting heat, shows where the electrical power, in volts, is expended. The two scales correspond at 90% current efficiency. Reduction of alumina consumes carbon producing aluminum, carbon dioxide, and carbon monoxide. Assuming a typical $CO{:}CO_2$ ratio of 0.265, the energy (enthalpy) Q_1 required to reduce alumina (γ-Al_2O_3) using carbon at 970°C is 6.419 kW $\cdot$ h/kg of aluminum. Q_1 (19) includes the energy required to heat the alumina and carbon. The free energy (ΔG) part of the enthalpy (ΔH) required for this reduction is supplied by voltage, $E_1 - E_2$. The heat Q_2 that must be dissipated is 8.12 kW $\cdot$ h/kg aluminum. Heating external to this cell is Q_3, ca 0.53 kW $\cdot$ h/kg aluminum. Dividing the productive energy Q_1 by the total electrical energy input $Q_1 + Q_2 + Q_3$ gives an electrical power efficiency of 42.6% for this typical cell.

Carbon consumption also represents energy consumption. Carbon reacting stoichiometrically with alumina to produce aluminum and carbon dioxide would require 0.33 kg carbon per kg aluminum; however, about 0.43 kg carbon per kg aluminum is consumed in the system, making the efficiency of carbon consumption 77%. Both carbon and electrical power can be saved by lowering bath temperature, but decreased alumina solubility and freezing limit this option. Bath additives allow temperature lowering but they also lower alumina solubility. Increasing aluminum fluoride concentration improves current efficiency but also increases the emission of fluoride fumes into the atmosphere. Power can be saved by lowering cell voltage, but when this is accomplished by lowering the anode–cathode spacing, current efficiency is lowered, often nullifying the power saving. Considerable voltage is lost in the anode and cathode. Soderberg cell efficiency has been increased through improved anode material and design, better insulation of the cathode, and automatic alumina feed systems. It is difficult to obtain large gains in prebaked anode cells because they already have higher power efficiency. However, use of cements that improve the electrical contact to the carbon cathode block and substituting semigraphitic carbon for carbon can save 2–5% of electric power.

Energy Conservation.   The U.S. Department of Energy (DOE) has sponsored research on inert anodes and refractory hard metal (RHM) composite cathodes. Success in these developments could significantly lower the energy required to reduce alumina. Inert anodes would eliminate the consumption of anode carbon and allow improved sealing of the cell for reduced heat loss and reduced fluoride emission. RHM cathodes would be wetted with a thin film of aluminum which would drain to a sump and provide stable cathodic surface rather than the present aluminum pool that sloshes about owing to electrohydrodynamic effects. The stable cathode should improve current efficiency and also allow closer interelectrode spacing for reduced power consumption.

A patent describes the use of ceramic inert anodes for electrolytic production of high purity aluminum (20). The most promising route to reduced energy requirement, however, is through recycling of scrap aluminum. Recycling scrap requires less than 5% of the energy required to produce new metal. The recycling rate for aluminum beverage cans in the United States has typically been in the 62–67% range for the decade of the 1990's. Where possible, these recycling approaches should be extended to other aluminum scrap (see RECYCLING).

Comparisons of production energies per metric ton for different metals strongly prejudice the case against light metals. A fairer comparison is energy per unit volume, which more closely approximates the basis for substituting metals. Additionally, the lower energy requirement of recycled metal is often not considered. The energy savings in the application of light metals frequently exceeds the energy for their production. In fact, saving gasoline by substituting aluminum for ferrous materials in automobiles may be one of the nation's lowest cost options for extending fuel supplies (21). One of the goals of the Partnership for a New Generation of Vehicles is the development of family car prototype that will get nearly three times the corporate average fuel economy federal regulation requirement of 27.5 mi/gal (22). At least 10% of the electrical energy generated in the U.S. is lost through heating of transmission and distribution lines. Larger aluminum conductors often give lower lifetime costs (cost of energy lost plus investment costs) and would save energy. Other energy-saving applications include aluminum storm doors and windows, aluminum-pigmented heat-reflecting paints, and aluminum heat exchangers.

4.5. Alternative Processes for Aluminum Production.   In spite of its industrial dominance, the Hall-Héroult process has several inherent disadvantages. The most serious is the large capital investment required resulting from: the multiplicity of units (250–1000 cells in a typical plant), the cost of the Bayer alumina-purification plant, and the cost of the carbon–anode plant (or paste plant for Soderberg anodes). Additionally, Hall-Héroult cells require expensive electrical power rather than thermal energy, most producing countries must import alumina or bauxite, and petroleum coke for anodes is in limited supply.

Aluminum can be produced by metallothermic, carbothermic, or electrolytic reduction processes. The earliest commercial process for producing aluminum (1855–1893) was sodiothermic reduction of aluminum halides. Once the Hall-Héroult process became commercial, however, sodiothermic reduction was not competitive.

Most attempts at direct carbothermic reduction of alumina have been electrically heated. Low yields of aluminum have been obtained owing to the formation of solid aluminum carbide and aluminum suboxide [12004-36-3], Al_2O, and aluminum vapors that react with carbon monoxide as they leave the furnace (23). However, yields approaching 100% at high energy efficiency can be obtained by staging the reactions as shown below and recovering the volatiles, the back-reaction products, and the heat on incoming reactants (24):

$$2\,Al_2O_3 + 9\,C \longrightarrow Al_4C_3 + 6\,CO\;(\text{at } 1930 - 2030°C) \qquad (25)$$

$$Al_4C_3 + Al_2O_3 \longrightarrow 6\,Al + 3\,CO\;(\text{at } 2030 - 2130°C) \qquad (26)$$

Vaporization and carbide formation can be reduced, however, by adding to the electric furnace a metal (or metal oxide that is subsequently reduced to the metal), such as iron, silicon, copper, or tin to alloy with the aluminum produced and lower its vapor pressure. However this may require a separate purification step to remove the other metal. The greater volume of gas passing through the charge in an aluminum blast furnace generally causes very low yields. Nevertheless, a 33% Al, 42% Fe, 21% Si, 4% others alloy has been successfully produced in a pilot oxygen blast furnace at moderately high yield (25). It is then necessary to extract the aluminum from the alloy in a separate operation. Extraction of pure aluminum from the molten alloy has been accomplished electrolytically by the three-layer process. Attempts have also been made to electro-refine the solid alloy using a low melting chloride electrolyte. Neither method has proved to be economic.

Selective solution of the aluminum from the alloy using a volatile metal, such as mercury, lead, bismuth, cadmium, magnesium, or zinc, has been investigated. After extracting the aluminum from the original alloy into the volatile metal, the volatile metal is distilled, leaving pure aluminum. Neither electrolysis nor volatile metal extraction can extract aluminum from iron aluminide [12004-62-5], $FeAl_3$, titanium aluminide [12004-78-3], $TiAl_3$, or Al_4C_3.

A third technique employs monovalent aluminum. By bringing vapors of aluminum fluoride or aluminum chloride into contact with carbothermically reduced aluminum alloy at 1000–1400°C, the following reaction occurs

$$AlX_3(g) + 2\,Al(\text{alloy}) \longrightarrow 3\,AlX(g) \qquad (27)$$

The monohalide vapors are conveyed to a slightly cooler zone (700–800°C) where the reaction reverses, resulting in the condensation of pure aluminum. The monochloride process was carried to the demonstration plant stage but was abandoned because of corrosion problems (26).

A fourth alloy separation technique is fractional crystallization. If silica is co-reduced with alumina, nearly pure silicon and an aluminum silicon eutectic can be obtained by fractional crystallization. Tin can be removed to low levels in aluminum by fractional crystallization and a carbothermic reduction process using tin to alloy the aluminum produced, followed by fractional crystallization and sodium treatment to obtain pure aluminum, has been developed (27). This method looked very promising in the laboratory, but has not been tested on an industrial scale.

The vapor pressure of aluminum can be lowered by alloying it with aluminum carbide: over 40% aluminum carbide is soluble in aluminum at 2200°C (25).

Thus an alloy of aluminum and aluminum carbide was produced at 2400°C by controlling the stoichiometry of the charge (28). When the alloy was tapped from the furnace and allowed to cool slowly, the aluminum carbide crystallized in an open lattice and the interstices filled with pure aluminum. The aluminum was removed by leaching with molten chlorides or by vacuum distillation of the aluminum from the alloy. The aluminum carbide residue was recycled to the arc furnace. This process has been further improved (29), going to three stages to decarbonize the aluminum—two in the hearth and one in an external holding furnace. A liquid aluminum and a slag which was recycled were obtained. The process gave good yields and produced aluminum at less than 11 kW · h/kg, but was not developed past the pilot stage.

In 1976 the first section of a new smelting process that required 30% less electric power than the best Hall-Héroult cells came on stream. In this process, alumina, carbon, and chlorine reacted to produce aluminum chloride and carbon dioxide. The aluminum chloride was electrolyzed in bipolar electrode cells to produce aluminum and chlorine and the chlorine was recycled to make more aluminum chloride. After six years of operation, the plant was shut down for economic reasons: repair and maintenance were excessive. The process was designed to use heavy fuel oil (bunker C) as the source of carbon. The chemical reactors never reached design capacity. Increasing their size or number would have been too expensive. The chemical reactor produced a trace of polychlorinated biphenyl which, rather than being destroyed in the cell, continued to accumulate within the system. Its removal and decomposition to environmentally safe products was expensive.

4.6. High Purity Aluminum. The Hall-Héroult process cannot ensure aluminum purity higher than 99.9%. Techniques such as electrolytic refining and fractional crystallization are required to produce metal of higher purity. Development of an electrolytic refining process was begun in 1901 and made workable by 1919. The Hoopes process is based upon the use of a cell containing three liquid layers, as depicted in Figure 6. Aluminum is electrochemically transported from the bottom alloy layer (anode) through an intermediate electrolyte

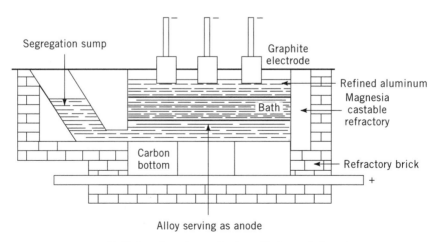

Fig. 6. Cell for the electrolytic refining of aluminum.

Table 9. **Electrolytic Purification Processes**[a]

Characteristic	Hoopes	Pechiney	AIAG Neuhausen
	Cathode layer		
Al purity	99.98%	99.99+%	99.99+%
density, g/mL	2.29	2.30	2.30
	Electrolyte		
composition, %			
NaF	25–30	17	18
AlF$_3$	30–38	23	48
CaF$_2$			16
BaF$_2$	30–38		18
BaCl$_2$		60	
Al$_2$O$_3$	0.5–7.0		
density, g/mL	2.5	2.7	2.6
resistivity, $\Omega \cdot$ cm	0.3	0.75–0.85	1.1
	Anode layer		
composition, %			
Al	75	67	70
Cu	25	33	30
density, g/cm^3	2.8	3.14	3.05
	Operating characteristic		
cell amperage, A	20,000	25,000	14,000
voltage, V	5–7	6.9	5.5–5.3
current density, A/cm^2	0.95	0.40	0.36
current efficiency, %	90–98	96–98	92–95
operating temperature, °C	950–1000	750	740

[a] Refs. 14 and 31.

layer to the high purity top layer (cathode). The bottom phase consists of impure aluminum plus an alloying agent (usually copper) to increase the density above that of the fused salt electrolyte. The composition of the electrolyte is selected to have a density less than the alloy layer but greater than pure aluminum. Electrical connection is made to the alloy through carbon or graphite blocks and graphite is used for the cathode connection. Aluminum is purified because metals more noble than aluminum are not oxidized and remain in the anodic layer. Metals less noble than aluminum are oxidized at the anode, but not reduced at the cathode. Hence impurities accumulate as chlorides or fluorides in the electrolyte.

This process used an all-fluoride electrolyte, a portion of which was frozen on the carbon sidewalls to prevent short circuiting through the walls. One version of the cell operated at 20,000 A and 950–1000°C. The highest purity aluminum produced was 99.98%. A summary of the cell characteristics is given in Table 9.

In 1932 the Pechiney Company in France perfected an electrolyte that consists of a chloride–fluoride mixture (Table 9) and melts at 720°C. This allowed low temperature operation (750°C) and the use of a nonconducting magnesia brick lining. Purities as high as 99.995% have been reported. In 1937 the Société Suisse de l'Aluminum Industrie (AIAG) at Neuhausen patented a low melting (720°C) all-fluoride electrolyte. This electrolyte also allows low temperature operation and the use of a magnesia brick lining. Operating characteristics are similar to the Pechiney process (Table 9).

An important design feature of modern purification cells is the segregation sump pictured in Figure 6 (30). It is normally operated at least 30°C cooler than the main cell and serves as the charging port of the cell. As impurities concentrate on the bottom layer, saturation is reached and crystals preferentially form in the cooler sump where they can be removed. This procedure greatly extends the operating life of the cell and allows use of the process to recover certain types of scrap (31), as well as purifying smelting-grade aluminum. Electrolytic purification, using either of the low melting electrolytes, is a principal source of commercially produced high purity aluminum.

Fractional crystallization processes are also used commercially to produce high purity metal from lower grade aluminum. These processes rely on the fact that most impurities preferentially concentrate either in the liquid or the solid as aluminum freezes, eg, zone melting (see ZONE REFINING) (32) where a molten zone is moved along an aluminum bar. Impurities that lower the melting point of aluminum, such as silicon and iron, concentrate in one end of the bar while impurities that raise the melting point, such as titanium and vanadium, concentrate in the other end. After several passes the ends are removed leaving the middle region as the high purity product.

In one process (33) high purity crystals are formed by cooling a molten aluminum surface using air. After the furnace is nearly filled with crystals, the remaining liquid is drained and the crystals remelted to yield the high purity product. Starting with 99.9% smelting-grade metal, aluminum of purity higher than 99.995% has been produced by fractional crystallization.

Table 10. **World Smelter Production and Capacity**[a,b]

	Production		Year end capacity	
Country	2000	2001[c]	2000	2001[c]
United States	3,668	2,600	4,270	4,280
Australia	1,770	1,800	1,770	1,770
Brazil	1,280	1,200	1,260	1,260
Canada	2,370	2,500	2,370	2,550
China	2,550	2,700	2,640	2,640
France	441	450	450	450
Norway	1,030	1,000	1,020	1,020
Russia	3,240	3,200	3,200	3,200
South Africa	671	680	676	676
Venezuela	570	570	640	640
other countries	6,440	6,680	7,500	7,670
World total (rounded)	*24,000*	*23,400*	*25,800*	*26,200*

[a] Ref. 34.
[b] Data in thousand metric tons of metal.
[c] Estimated.

5. Production

World smelter production and capacity are given in Table 10 (34). Primary U.S. production by company is listed in Table 11 (35).

In 2000, U.S. production of aluminum was 3.7×10^6 t. U.S. metal recovered from new and old scrap decreased by 7% to 3.45×10^6 t.

The tremendous growth of aluminum production as compared to other metals is shown in Table 12 (36).

6. Economic Aspects

Aluminum prices have historically been more stable than other nonferrous metals. Beginning in the 1970s, however, aluminum prices have fluctuated as shown in Table 13 (31). These fluctuations reflect increased energy costs as well as increased costs of raw materials. Improvements in production processes as well as a rebalancing of demand and supply are expected to stabilize aluminum prices in the future.

U.S. imports for consumption decreased in 2000 compared to those in 1999 reversing a trend starting in 1997. Total exports increased 7% in 2000. Canada remains the largest shipper to the United States. Exports from the U.S. are listed in Table 14 and imports to the U.S. are listed in Table 15 (35).

7. Analysis

Aluminum is best detected qualitatively by optical emission spectroscopy. Solids can be vaporized directly in a d-c arc and solutions can be dried on a carbon electrode. Alternatively, aluminum can be detected by plasma emission spectroscopy using an inductively coupled argon plasma or a d-c plasma. Atomic absorption using an aluminum hollow cathode lamp is also an unambiguous and sensitive qualitative method for determining aluminum.

Quantitative aluminum determinations in aluminum and aluminum base alloys is rarely done. The aluminum content is generally inferred as the balance after determining alloying additions and tramp elements. When aluminum is present as an alloying component in alternative alloy systems it is commonly determined by some form of spectroscopy (qv): spark source emission, x-ray fluorescence, plasma emission (both inductively coupled and d-c plasmas), or atomic absorption using a nitrous oxide acetylene flame.

The predominant method for the analysis of aluminum-base alloys is spark source emission spectroscopy. Solid metal samples are sparked directly, simultaneously eroding the metal surface, vaporizing the metal, and exciting the atomic vapor to emit light in proportion to the amount of material present. Standard spark emission analytical techniques are described in ASTM E101, E607, E1251 and E716 (37). A wide variety of well-characterized solid reference materials are available from major aluminum producers for instrument calibration.

Table 11. **Primary Annual Aluminum Production Capacity in the United States, by Company**[a]

Company	Year end capacity (thousand metric tons)		2000 ownership
	1999	2000	
Alcan Aluminum Corp.:			
Sebree, KY	186	186	Alcan Aluminium Ltd., 100%
Alcoa Inc.:[b]			
Alcoa, TN	210	210	Alcoa Inc., 100%
Badin, NC	115	115	Do.
Evansville, IN (Warrick)	300	300	Do.
Femdale, WA (Intalco)	272	272	Alcoa Inc., 61%; Mitsui & Co. Ltd., 32%; YKK Corp., 7%
Frederick, MD (Eastalco)	174	174	Do.
Longview, WA[d]	204	204	Alcoa Inc., 100%
Massena, NY[d]	123	123	Do.
Massena, NY	125	125	Do.
Mount Holly, SC	215[c]	215	Alcoa Inc., 50.3%; Century Aluminum Co., 49.7%.
Rockdale, TX	315	315	Alcoa Inc., 100%
Troutdale, OR [d]	121	121	Do.
Wenatchee, WA	220	220	Do.
Total	*2,390[c]*	*2,390*	
Century Aluminum Co.:			
Ravenswood, WV	168[c]	168	Century Aluminum Co., 100%
Columbia Falls Aluminum Co.:			
Columbia Falls, MT	168	168	Glencore AG, 100%
Goldendale Aluminum Co.:			
Goldendale, WA	168	168	Glencore AG, 100%
Goldendale, WA	168	168	Private interest, 60%; employees, 40%
Kaiser Aluminum & Chemical Corp.:			
Mead, WA (Spokane)	200	200	MAXXAM Inc., 100%
Tacoma, WA	73	73	Do.
Total	*273*	*273*	
NSA:			
Hawesville, KY	237	237	Southwire Co., 100%
Noranda Aluminum Inc.:			
New Madrid, MO	222[c]	222	Noranda Mines Ltd., 100%
Northwest Aluminum Corp.:			
The Dalles, OR	82	82	Private interests, 100%
Ormet Primary Aluminum Corp.:			
Hannibal, OH	255	257	Ormet Corp., 100%
Vanalco Inc.:			
Vancouver, WA	116	116	Vanalco Inc., 100%
Grand total	*4,270*	*4,270*	

[a] Data are reounded to no more the significant digits; may not add to totals shown.
[b] Individual plant capacities are U.S. Geological Survey estimates based on company reported total.
[c] Revised.
[d] Alcoa and Reynolds in June 2000.

Table 12. **Growth of Aluminum Production Compared to Other Metals,** $\times 10^3$t/yr[a]

Year	Al[b]	Cu[c]	Mg[d]	Pb[d]	Zn
1900	5.7	449	0.01	877	479
1950	1,516	2,791	21	1,752	1,985
1960	4,732	4,631	93	2,436	3,019
1970	9,780	6,885	223	3,660	5,022
1980	16,043	7,984	317	5,456	6,115
1990	18,174	9,668	368	5,699	7,086
1999	23,074	11,337	393	6,120	8,406

[a] Ref. 36.
[b] Al - primary production only.
[c] Cu - smelter production only.
[d] Mg, Pb - include both primary and secondary.

Table 13. **Aluminum**[a] **Price Averages from 1966 to 2001**[a]

Year	Price, $/kg
1966	0.540
1967	0.551
1968	0.564
1969	0.599
1970	0.633
1971	0.639
1972	0.583
1973	0.559
1974	0.752
1975	0.877
1976	0.977
1977	1.132
1978	1.170
1979	1.309
1980	1.534
1981	1.676
1982	1.676
1983	1.712
1984	1.786
1985	1.786
1986	1.232
1987	1.594
1988	2.427
1989	1.937
1990	1.631
1991	1.311
1992	1.267
1993	1.175
1994	1.569
1995	1.893
1996	1.571
1997	1.699
1998	1.443
1999	1.448
2000	1.64
2001	1.54

[a] From Ref. 35, Pure, 99.5%, ingot aluminum at New York.

Table 14. **U.S. Exports of Aluminum, by Country**[a]

	Total, 1999		Total, 2000	
Country or territory	Quantity (metric tons)	Value (thousands, $)	Quantity (metric tons)	Value (thousands, $)
Azerbaijan	1	3		
Brazil	60,300	173,000	32,900	99,800
Canada	795,000	1,560,000	854,000	1,740,000
France	7,220	31,500	7,350	32,800
Germany	7,340	39,400	6,900	38,100
Hong Kong	29,400	49,900	33,000	49,900
Italy	2,710	12,900	2,110	12,100
Japan	117,000	241,000	81,900	184,000
Korea, Republic of	46,800	84,600	58,200	127,000
Mexico	306,000	670,000	321,000	768,000
Netherlands	2.150	7,700	3,940	14,500
Philippines	847	2,620	1,080	4,020
Russia	14	158	258	1,250
Saudi Arabia	8,210	22,500	8,380	20,300
Singapore	3,250	15,500	3,390	17,300
Slovakia	9	54	11	35
Slovenia	58	140	(3/)	4
South Africa	328	1,940	118	1,210
Taiwan	42,900	64,800	40,900	66,000
Thailand	4,100	14,700	8,700	27,400
Ukraine	(3/)	4	1	25
United Kingdom	18,100	78,800	26,900	104,000
Venezuela	18,100	40,700	17,300	40,500
other	175,000	418,000	248,000	532,000
Total	*1,640,000*	*3,530,000*	*1,760,000*	*3,880,000*

[a] Source: U.S. Census Bureau.

In addition to the spark emission methods, quantitative analysis directly on solids can be accomplished using x-ray fluorescence, or, after sample dissolution, accurate analyses can be made using plasma emission or atomic absorption spectroscopy (38).

8. Environmental Considerations

Fluoride emission from aluminum smelting cells has long been an area of great concern (39). One source of fluoride evolution is melt entrained in the gas generated at the anodes. Additionally, sodium tetrafluoroaluminate [13821-15-3], $NaAlF_4$, has a significant vapor pressure, 190–330 Pa (1.4–2.5 mm Hg), over the melt. The melt vapor pressure increases as melt temperature and aluminum fluoride concentration increase, but decreases as alumina, calcium fluoride, and sodium fluoride concentration increase. Both entrained liquid and $NaAlF_4$ gas solidify to form fluoride particulates. Moisture entering the melt adsorbed on alumina, and from the hydrogen content of the anodes, results in hydrolysis of AlF_3 in the melt forming hydrogen fluoride (HF) gas. Historically, treatment consisted of passing the exhausted gases through electrostatic precipitators to remove

Table 15. **U.S. Imports for Consumption of Aluminum, by Country**[a]

Country	Total, 1999		Total, 2000	
	Quantity (metric tons)	Value (thousands, $)	Quantity (metric tons)	Value (thousands, $)
Argentina	27,700	$39,700	61,700	105,000
Australia	63,000	94,000	25,800	45,100
Bahrain	49,200	80,700	44,100	82,500
Belgium	4,260	12,600	3,810	11,800
Brazil	91,500	115,000	85,500	136,000
Canada	2,210,000	3,460,000	2,150,000	3,780,000
Croatia	273	933	163	547
Czech Republic	714	1,890	328	1,070
France	23,900	68,900	22,700	61,300
Germany	46,500	161,000	49,100	176,000
Italy	5,670	12,900	3,220	11,000
Japan	24,800	82,000	22,500	82,700
Korea, Republic of	21,400	42,300	13,900	35,200
Mexico	106,000	151,000	102,000	171,000
Netherlands	15,400	25,600	6,930	16,300
Norway	5,030	9,240	7,110	14,000
Panama	7,420	10,400	8,690	13,800
Russia	831,000	1,100,000	837,000	1,280,000
Slovakia	204	218	257	282
Slovenia	4,000	11,800	4,350	14,000
South Africa	22,500	34,500	47,900	90,600
Spain	5,600	10,200	905	2,470
Tajikistan	17,200	20,600		
Ukraine	32,200	36,900	17,000	19,800
United Arab Emirates	21,000	28,300	58,300	101,000
United Kingdom	45,500	91,300	26,300	75,200
Venezuela	169,000	236,000	165,000	236,000
Other	151,000	258,000	148,000	293,000
Total	*4,000,000*	*6,200,000*	*3,910,000*	*6,860,000*

[a] Ref. 35

particulates, followed by wet scrubbers to remove HF. This technique has largely been replaced by highly (over 99%) efficient dry scrubbers that catch particulates and adsorb HF on alumina (40) that is subsequently fed to the cells. Hence, nearly all the fluoride evolved is fed back into the cell. A monolayer of HF is chemisorbed in the Al_2O_3 and converted to AlF_3 on heating when the alumina is fed to the cell. HF in excess of the monolayer is desorbed and becomes a rapidly increasing circulating load in the system if the alumina used does not have adequate surface area. A surface area of 45 m^2/g chemisorbs 1.35% HF and is generally adequate.

Hydrocarbon fumes evolved during anode baking are generally disposed of by burning. Additional fuel is required to support combustion because the hydrocarbon concentration is low. In some plants these products are now absorbed in a fluid bed of alumina for burning in a concentrated form. This treatment also catches the fluoride evolved during anode baking.

Handling of alumina and coke presents dusting problems. Hoods and exhaust systems collect the dust, which is then separated from the exhaust air either by cyclones, electrostatic precipitators, filter bags, or a combination of these methods, and recycled to the process (see AIR POLLUTION CONTROL METHODS).

Chlorine fluxing of aluminum to remove hydrogen and undesirable metallic impurities has largely been supplanted by fumeless fluxing procedures, which generally employ a low vapor pressure melt of alkali chlorides containing a small amount of aluminum chloride as the active ingredient.

The linings of aluminum reduction cells must be replaced periodically. These spent linings represent the largest volume of waste associated with the smelting process. Because they contain fluorides and cyanide, they must be either stored under roof or buried in landfills lined with impervious materials to prevent leaching and contamination of the environment. The large volumes involved make these procedures costly and research is being directed toward destroying the cyanide and recovering valuable components from the lining. Some potential uses for recoverables are as a flux in the steel industry, in making rock wool, as supplemented fuel for cement manufacture, and as fuel in fluidized-bed boilers.

Table 16. **Representative Guides and Regulatory Standards**[a,b]

Standards	Concentration (mg/m^3)
United States	
ACGIH TLV TWA	
metal dust, as Ai	10
pyropowder, as Ai	5
welding fume, as Ai	5
OSHA PEL	
total dust	15
respirable dust	5
welding fumes, pyropowders, as Ai	5
NIOSH REL	
total dust, as Ai	10
respirable dust	5
pyropowders, welding fumes, as Ai	5
UK, metal, as Ai	10
10-min STEL	20
Germany, as metal, fine dust	6
Sweden, total dust	10
respirable	4
Australia, for metal	10
welding fumes, pyropowder	5

[a] Ref. 42.
[b] ACGIH = The American Conference of Governmental Industrial Hygienists. NIOSH = National Institute of Occupational Safety and Health. OSHA = Occupational Safety and Health Agency. PEL = Permissible exposure level. REL = Relative exposure level. STEL = Short-term exposure limit. TLV = Threshold limit values. TWA = Time weighted average.

9. Recycling

Aluminum recovered in 2001 from purchased scrap was about 3.2×10^6 t, of which 60% came from new (manufacturing) scrap and 40% from old scrap (discarded aluminum products). Aluminum recovered from old scap was equivalent to approximately 20% apparent consumption (35).

According to figures released by the Aluminum Association Inc, the Can Manufacturer's Institute, and the Institute of Scrap Recycling Industries, 62.6×10^9 used beverage containers were recycled in the United States in 2000. The recycling rate was 62.1% in 2000, a decrease from 62.5% in 1999. However, 2000 was the twelfth consecutive year that aluminum can recycling rate was greater than 60%. According to the organizations, aluminum beverage cans produced in the United State in 2000 had an average of 51% postconsumer recycled content, the highest percentage of all packaging materials (41).

10. Health and Safety Factors

Table 16 lists some representative guides and regulatory standards presently extant for aluminum metal dusts or powders (42). The major difference in the U.S. standards is the discordance between the higher OSHA PEL as compared with ACGIH and NIOSH. The lower recommendations of the latter two organizations are based on their adherence to a generic standard recommended for all dusts not otherwise classified (NOC).

11. Aluminum Alloys

Aluminum obtained by electrolysis of cryolite baths contains iron [7439-89-6] and silicon [7440-21-3] as impurities. Iron content may vary from 0.05 to 0.4% and silicon from 0.05 to 0.1% depending on the raw materials and the age and condition of the reduction cell. Primary aluminum metal also contains small, usually not to exceed 0.05% in total, amounts of many other elements. Some of these trace impurities are Cu, Mn, Ni, Zn, V, Na, Ti, Mg, and Ga, most of which are present in quantities substantially below 100 ppm.

Many of the properties of aluminum alloy products depend on metallurgical structure which is controlled both by the chemical composition and by processing (43). In addition to features such as voids, inclusions, grains, subgrains, dislocations, and vacancies which are present in virtually all metallic products, the structure of aluminum alloys is characterized by three types of intermetallic particles. Aluminum metallurgists refer to these as constituent particles, dispersoid particles, and precipitate particles. Constituent particles are formed during solidification generally as a byproduct of a divorced eutectic reaction and range in size in the final product from about 1–20 micrometers, Figure 7**a**. These negatively affect toughness of high strength alloy products. Dispersoids and precipitates both form by a solid-state reaction. The particles known as dispersoids characteristically form during thermal treatment of an ingot by precipitation of solid solution which exceeds maximum solid solubility because of nonequilibrium

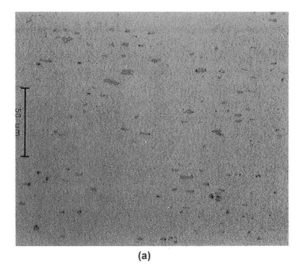

(a)

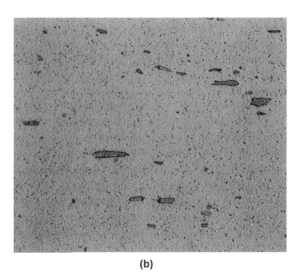

(b)

Fig. 7. Alloy 3004 sheet. (**a**) Al$_6$(Fe, Mn) andAl$_{12}$(Fe, Mn) Si constituent particles. Magnification is ×500. (**b**) Etched to reveal the Al$_6$(Fe, Mn) and Al$_{12}$(Fe, Mn) dispersoids. Magnification is ×1000.

conditions during ingot solidification. Dispersoids, about 10-200 nm in the largest dimension, are present in most aluminum alloy products. See Figures 7**b** and 8. Their primary function is to control grain size, grain orientation (texture), and degree of recrystallization. Particles classified as precipitates from during heat treatment of the final mill product by precipitation from a supersaturated solid solution that does not exceed the maximum equilibrium solid solubility. In the final product, their size may range from disks a few atoms thick by a few nm in diameter up to needlelike or platelike particles which may exceed 1 micrometer

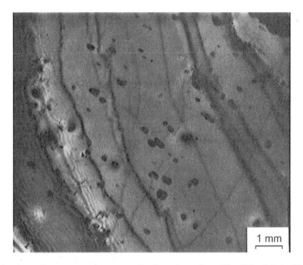

Fig. 8.   Transmission electron micrograph showing the $Al_{12}(Fe, Mn)_3Si$ and $Al_6(Fe, Mn)$ dispersoid particles in alloy 3004 sheet.

in the largest dimension (Fig. 9). Precipitates may confer high strength. The nature of the constituent, dispersoid, and precipitate particles depends strongly on the phase diagrams of the particular alloy.

11.1. Binary Alloys.   Aluminum-rich binary phase diagrams show three types of reaction between liquid alloy, aluminum solid solution, and other

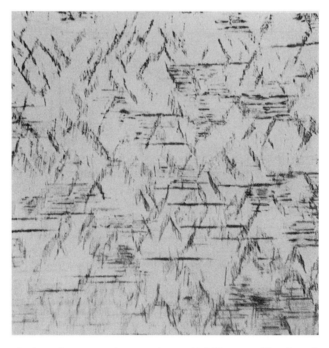

Fig. 9.   Transmission electron micrograph at 40,000 magnification showing S′-phase precipitates in Al–Cu–Mg alloy sheet aged 12 h at 100°C.

Table 17. **Phase Transformations in Binary Aluminum Alloys**

Al–M system	Reaction		Solubility of M %		Other phase
	Type	Temperature, °C	Solid	Liquid	
Al–Si	eutectic	577	1.65	12.60	Si
Al–Cr	peritectic	651	0.77	0.41	CrAl$_7$
Al–Pb	monotectic	326.8			Pb

phases: eutectic, peritectic, and monotectic. Table 17 gives representative data for reactions in the systems Al-M. Diagrams are shown in figures 10–19. Compilations of phase diagrams may be found in reference (44).

Al–Fe. TheAl–Fe system (Fig. 10), is important because virtually all commercial aluminum alloys contain some iron [7439-89-6], Fe. The system has a eutectic at 1.9% Fe, but solid solubility of only 0.05% Fe. Consider an alloy containing 0.3% Fe. During solidification, most of the Fe remains in the liquid phase until a eutectic of solid solution plus Al$_3$Fe constituent particles freezes. Alternatively, constituents of the metastable Al$_6$Fe phase [12005-28-6] may form during solidification. These transform to equilibrium Al$_3$Fe by heating

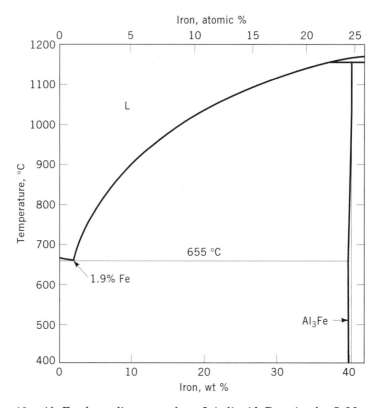

Fig. 10. Al–Fe phase diagram, where L is liquid. Drawing by J. Murray.

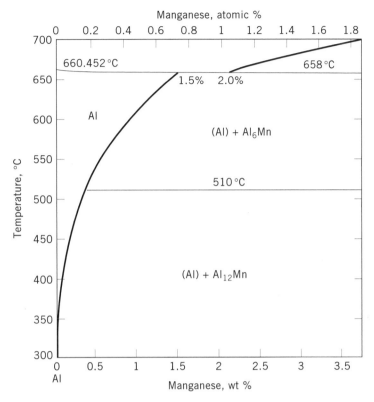

Fig. 11. Al–Mn phase diagram, where L is liquid. Drawing by J. Murray.

above 500°C. If the solidification rate is rapid enough, more than 0.05% Fe remains in supersaturated solid solution. This excess amount can precipitate as Al_3Fe dispersoids during subsequent thermal treatment. In more complex alloys, such as Al–Fe–Si, Al–Fe–Mn, and Al–Cu–Fe, most of the iron is contained in constituent particles formed by eutectic decomposition. In all cases the maximum solid solubility of iron in aluminum is 0.05% or less.

Al–Mn.   The Al–Mn system (Fig. 11), the basis for the oldest yet most widely used aluminum alloys, is characterized by a eutectic at 1.95% Mn and 658°C. Maximum solid solubility is 1.76% manganese, Mn, and the intermetallic phase in aluminum-rich alloys is Al_6Mn [12043-69-5]. The $Al_{12}Mn$ phase [12446-45-6] is difficult to nucleate in binary Al–Mn alloys, but $Al_{12}(Mn,Cr)$ forms readily in the Al–Mn–Cr ternary alloy. Most of the manganese added to commercial alloys is usually retained in supersaturated solution in the ingot. Commercial alloys based on the Al–Mn system always contain some iron and silicon, and in the presence of these elements manganese precipitates as dispersoids which may be $Al_6(Fe,Mn)$ or $Al_{12}(Fe,Mn)_3Si$. The dispersoid particles provide a modicum of dispersion strengthening, increase the degree of strain hardening, refine the recrystallized grain structure and influence its crystallographic orientation, and minimize strain localization during plastic deformation.

Al–Cu.   Many structural aluminum alloys contain significant amounts of copper Cu. There is a eutectic in the Al–Cu system at 33.2% Cu and 548°C, but

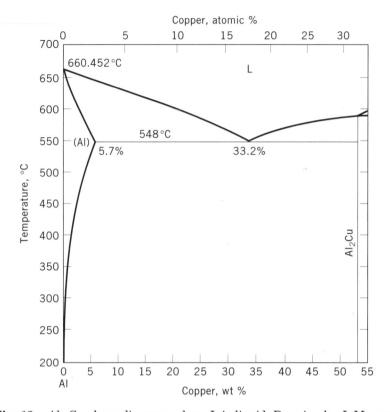

Fig. 12.   Al–Cu phase diagram, where L is liquid. Drawing by J. Murray.

the important feature (Fig. 12) is the maximum solubility of 5.7% Cu at 548°C which decreases drastically at lower temperatures. This decreasing solubility with decreasing temperature is necessary for the phenomenon known as age hardening or precipitation strengthening. Consider an alloy containing 4% Cu. During solidification, it separates into a solid solution of copper in aluminum and a liquid containing copper in aluminum until a temperature of 548°C is reached. Then, the remaining liquid freezes as a solid solution of 5.7% Cu in Al plus a eutectic of this solid solution and Al₂Cu constituent particles. The alloy product may be subsequently heated in the solid-state above about 425°C to dissolve the Al₂Cu constituents (solution heat treatment). Rapid cooling to room temperature (quenching) retains the Cu in supersaturated solution. This unstable solution decomposes at room temperature (natural aging) by the formation of clusters of Cu atoms (G–P or Guinier–Preston zones) on preferred crystallographic planes. The G–P zones strengthen the alloy product by resisting the passage of dislocations. Heating to 100–200°C (artificial aging or elevated temperature precipitation treatment) provides enough energy for the formation of metastable forms of Al₂Cu precipitates (θ' or θ'') which provide even more strengthening. Thermal treatment at 250°C and higher (annealing) converts the metastable precipitates to the equilibrium Al₂Cu precipitates (θ), and the product softens.

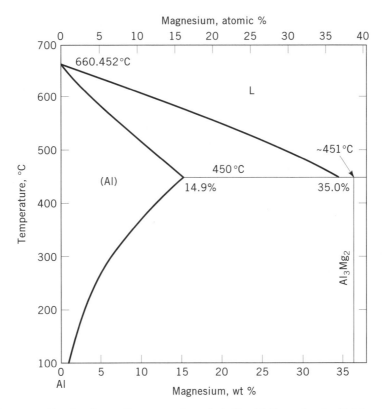

Fig. 13. Al–Mg phase diagram, where L is liquid. Drawing by J. Murray.

Al–Mg. Almost every commercial structural aluminum alloy contains magnesium as an alloying element. The Al–Mg system (Fig. 13) has a eutectic at 35% magnesium Mg, and 451°C. Maximum solid solubility is 14.9% Mg, and solubility decreases to about 0.8% Mg at room temperature. Despite this decreased solubility, precipitation strengthening by the mestable β′-phase precursor to the equilibrium β-phase Al_3Mg_2 precipitates is observed only at very high magnesium levels. Commercial alloy products based on the Al–Mg system are strengthened by the effect of the magnesium in solid solution (solid solution hardening). Strength is enhanced by cold-working the products (strain-hardening). Strength levels are not as high as in most precipitation strengthened alloys, but the products have good ductility and corrosion resistance.

Al–Si. Al–Si alloys (Fig. 14) possess high fluidity and castability and are consequently used for weld wire, brazing, and as casting alloys. The range of silicon [7440-21-3], Si, is about 5 to 20% in commercial casting alloys. This system has a eutectic at 12.6% Si and 577°C; maximum solid solubility of Si is 1.65%. The constituent which forms during eutectic decomposition is essentially pure Si. Hypereutectic alloys are used for engine blocks because the hard Si particles are wear-resistant. The Si particles, which form during solidification, are coarse and acicular. Micro additions of sodium [7440-23-5], Na, antimony [7440-36-0], Sb, and strontium [7440-24-6], Sr, modify the faceted structure of primary silicon

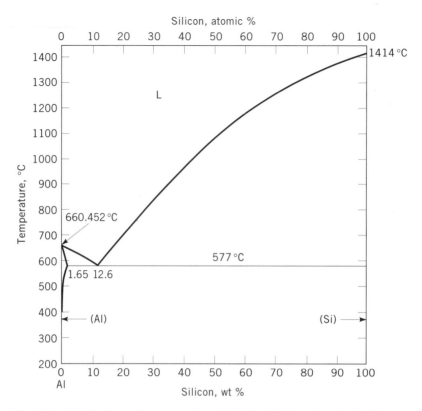

Fig. 14. Al–Si phase diagram, where L is liquid. Drawing by J. Murray.

particles in hypereutectic alloys to spherulitic and the flake structure of eutectic silicon to a finer fibrous morphology.

Al–Li. Alloys containing about two to three percent lithium Li, (Fig. 15) received much attention in the 1980s because of their low density and high elastic modulus. Each weight percent of lithium in aluminum alloys decreases density by about three percent and increases elastic modulus by about six percent. The system is characterized by a eutectic reaction at 8.1% Li at 579°C. The maximum solid solubility is 4.7% Li. The strengthening precipitate in binary Al–Li alloys is metastable Al_3Li [12359-85-2], δ', having the cubic $L1_2$ crystal structure, and the equilibrium precipitate is complex cubic Al–Li [12042-37-4], δ. The nature of the phase relationships involving δ' has been the subject of much discussion. Portions of the metastable phase boundaries have not yet been agreed upon.

Al–Cr. Although no commercial alloys are based on this system, chromium, Cr, is an ingredient of several complex and commercially significant alloys (Fig. 16). The Cr is added for control of grain structure. The Al–Cr system has a peritectic portion at 661°C where solid solubility is 0.7% Cr and liquid solubility is 0.4% Cr. Freezing of binary alloys containing >0.4% Cr produces coarse primary crystals (metallic inclusions) of Al_7Cr [12005-37-7] which may adversely affect mechanical properties. In complex alloys, the liquid solubility can be reduced such that formation of primary Al_7Cr crystals may occur at significantly lower Cr contents.

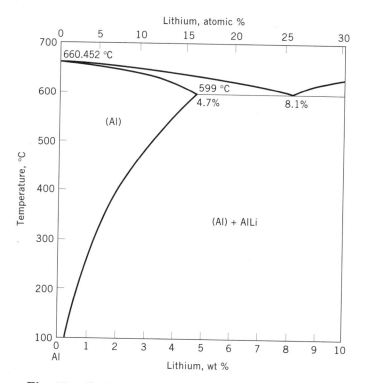

Fig. 15.   Al–Li phase diagram. Drawing by J. Murray.

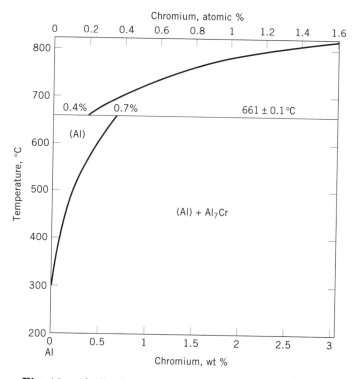

Fig. 16.   Al–Cr phase diagram. Drawing by J. Murray.

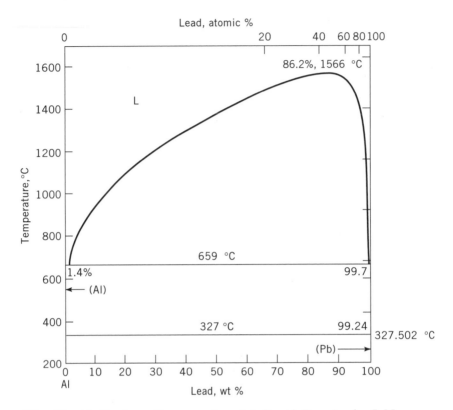

Fig. 17. Al–Pb phase diagram, where L is liquid. Drawing by J. Murray.

Al–Pb. Both lead Pb, and bismuth, Bi, which form similar systems (Fig. 17), are added to aluminum alloys to promote machinability by providing particles to act as chip breakers. The Al–Pb system has a monotectic reaction in which Al-rich liquid freezes partially to solid aluminum plus a Pb-rich liquid. This Pb-rich liquid does not freeze until the temperature has fallen to the eutectic temperature of 327°C. Solid solubility of lead in aluminum is negligible; the products contain small spherical particles of lead which melt if they are heated above 327°C.

Al–Zn. Aluminum-rich binary alloys (Fig. 18) are not age hardenable to any commercial significance, and zinc, Zn, additions do not significantly increase the ability of aluminum to strain harden. Al–Zn alloys find commercial use as sacrificial claddings on high strength aircraft sheet or as sacrificial components in heat exchangers.

Al–Zr. This system (Fig. 19) has a peritectic reaction at 660.8°C at which solubility is 0.28% zirconium Zr, solid and 0.11% Zr liquid. The equilibrium phase on the aluminum-rich end of the phase diagram is tetragonal Al_3Zr [12004-83-0], β. Coarse primary particles of β-phase have a tendency to form during solidification when the zirconium content is much above 0.12%. A metastable form of Al_3Zr having a cubic $L1_2$ structure, β′, is formed when supersaturated

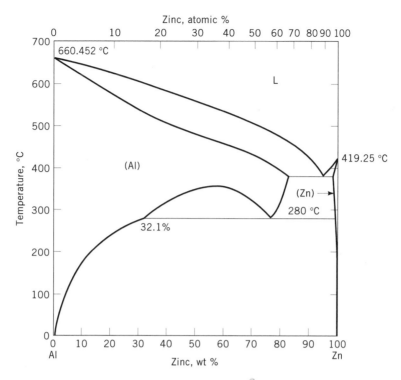

Fig. 18. Al–Zn phase diagram, where L is liquid. Drawing by J. Murray.

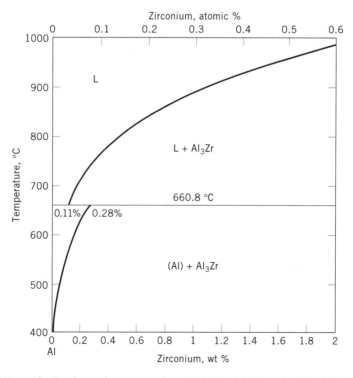

Fig. 19. Al–Zr phase diagram, where L is liquid. Drawing by J. Murray.

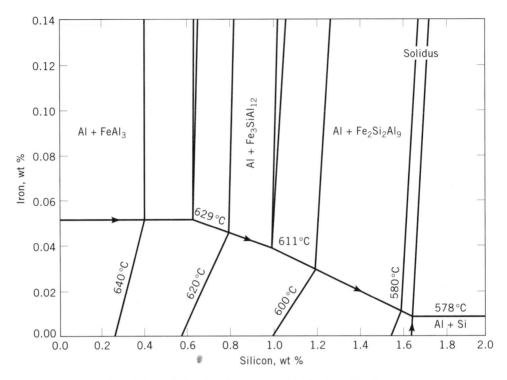

Fig. 20.   Solidus for aluminum-rich Al–Fe–Si alloys.

zirconium precipitates as a dispersoid. Although this phase is nonequilibrium, it is extremely resistant to transformation to the equilibrium β-phase.

11.2. Ternary Alloys.   Almost all commercial alloys are of ternary or higher complexity. Alloy type is defined by the nature of the principal alloying additions, and phase reactions in several classes of alloys can be described by reference to ternary phase diagrams. Minor alloying additions may have a powerful influence on properties of the product because of the influence on the morphology and distribution of constituents, dispersoids, and precipitates. Phase diagrams, which represent equilibrium, may not be indicative of these effects.

Al–Fe–Si.   Iron and silicon, present in primary aluminum, may also be added to produce enriched alloys for specific purposes. The equilibrium phase fields in the Al–Fe–Si system are shown in Figure 20 and Table 18. The inter-metallic phases have a limited range of composition when in equilibrium with

Table 18. **Intermetallic Phases in Al–Mg–Zn Alloys**

Designation	Formula	CAS Registry Number	Fe, %	Si, %
	Al_3Fe	[12004-62-5]	41	negligible
α-(Al–Fe–Si)	$Al_{12}Fe_3Si$	[12397-58-9]	32	5.5
β-(Al–Fe–Si)	$Al_9Fe_2Si_2$	[12397-57-8]	27	14
	Si	[7440-21-3]	negligible	~100

the aluminum solid solution. The amount of iron in solid solution in the matrix is small, so almost all of the iron is in the intermetallic compounds. At low silicon contents the iron is present as Al_3Fe except for about 0.01% Fe in solid solution. As silicon content increases, the ternary intermetallic compound $Al_{12}Fe_3Si$, α-(Al–Fe–Si), appears. At higher silicon contents $Al_9Fe_2Si_2$, β-(Al–Fe–Si), appears in which a higher amount of silicon is combined with iron. The positions of the phase fields move to lower silicon content with decreasing temperature.

The phases present in products can differ from those predicted from equilibrium diagrams. Nonequilibrium metastable phases form at solidification rates experienced in commercial ingots. Because of the low rate of diffusion of iron in aluminum, equilibrium conditions can only be established by long heat treatments and are very slowly approached at temperatures below about 550°C. Small additions of other elements, particularly manganese, can also modify the phase relations.

Al–Mg–Si.   An important class of commercial alloys is based on the Al–Mg–Si system because of its precipitation hardening capabilities and good corrosion resistance. The precipitation hardening results from precipitation of a metastable precursor of magnesium silicide Mg_2Si, from solid solution. The phase relations are as shown in Figure 21. At low magnesium contents the alloys may contain silicon as a second phase. As magnesium increases, the ternary phase field Al–Mg_2Si–Si is entered, and at higher magnesium contents the phase field Al–Mg_2Si. The solubility of Mg_2Si decreases at higher magnesium contents. Heat treatment at high temperatures can put as much as 1.8%

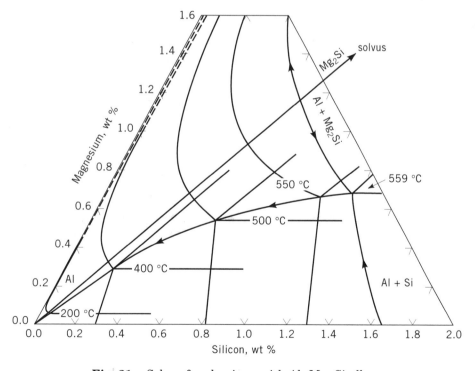

Fig. 21.   Solvus for aluminum-rich Al–Mg–Si alloys.

Mg_2Si into solid solution, and cooling at a sufficient rate can retain it therein. The cooling rate required depends on the amount of Mg_2Si dissolved. Compositions that have magnesium and silicon in the exact ratio to form Mg_2Si are not widely employed commercially, because alloys containing silicon in excess of that required to form Mg_2Si develop higher strength. Silicon, however, has a tendency to precipitate on grain boundaries. Consequently, excess-silicon alloy products must be quenched at a higher rate to prevent brittle intergranular fracture.

Al–Mg–Mn. The basis for the alloys used as bodies, ends, and tabs of the cans used for beer and carbonated beverages is the Al–Mg–Mn alloy system. It is also used in other applications that require excellent weldability and corrosion resistance. These alloys have the unique ability to be highly strain hardened yet retain a high degree of ductility. Some of the manganese combines with the iron to form $Al_6(Fe,Mn)$ or $Al_{12}(Fe,Mn)_3Si$ constituent particles during solidification, but most remains in supersaturated solid solution until it precipitates as $Al_6(Fe,Mn)$ or $Al_{12}(Fe,Mn)_3Si$ dispersoids during ingot thermal treatment. No ternary Al–Mg–Mn phases are present in commercial alloys; mechanical properties reflect the enhanced strain hardening characteristics of the matrix with magnesium in solution and the effects of the Mn-bearing dispersoids in stabilizing the deformation substructure.

Most alloys contain no more than 5% Mg and because of the slow precipitation kinetics of Al_3Mg_2 at these levels, the magnesium is easily retained in solution during processing. Natural aging is insignificant at these low magnesium levels, but with levels of 4% Mg or more, small amounts of β may precipitate at grain boundaries after long times, and this may cause susceptibility to intergranular corrosion. Special thermal treatments (tempers) have been developed to prevent this.

Al–Cu–Mg. The first precipitation hardenable alloy was an Al–Cu–Mg alloy. There is a ternary eutectic at 508°C, and there are nine binary and five ternary intermetallic phases. For aluminum-rich alloys, only four phases are encountered in addition to the aluminum solid solution (Table 19). Several commercial alloys are based on the age hardening characteristics of the metastable precursors of θ or S-phase, principally θ′ or S′. Hardening by T- and β-phases is not very effective. Alloys of greatest age hardenability have compositions near the Cu:Mg ratio of the S-phase. Additions of about 0.12% Mg to alloys containing as much as 6% Cu, however, significantly increase strength by refining the θ′ precipitate.

Solid-state reactions in these alloys can be understood with reference to Figures 22 and 23. Heat-treatable Al–Cu–Mg alloys are solution treated above the solvus temperatures for S-phase or θ-phase and quenched. At age hardening

Table 19. **Intermetallic Phases in Al–Cu–Mg Alloys**

Designation	Formula	CAS Registry Number	Cu, %	Mg, %
θ	Al_2Cu	[12004-15-8]	54	negligible
S	Al_2CuMg	[12004-18-1]	46	17
T	Al_6CuMg_4	[12253-80-7]	25	28
β	Al_3Mg_2	[12004-68-1]	negligible	38

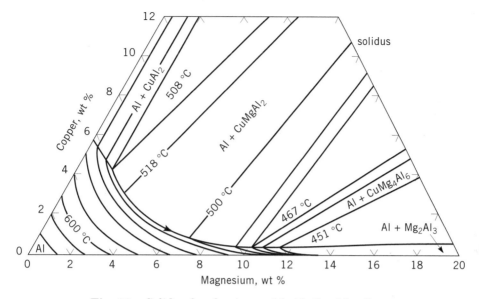

Fig. 22. Solidus for aluminum-rich Al–Cu–Mg alloys.

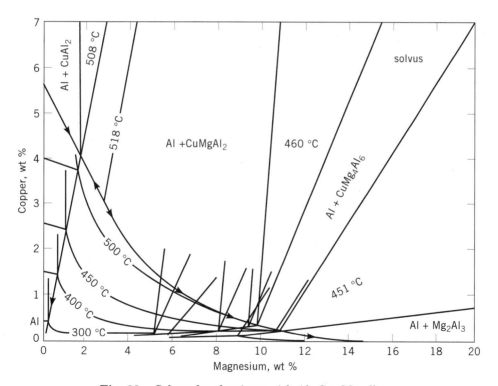

Fig. 23. Solvus for aluminum-rich Al–Cu–Mg alloys.

Table 20. Intermetallic Phases in Al–Mg–Zn Alloys

Designation	Formula	CAS Registry Number	Mg, %	Zn, %
M or η	$MgZn_2$	[12032-47-2]	17	79
T	$Al_2Mg_3Zn_3$	[12004-33-0]	30–20	25–70
β	Al_3Mg_2	[12004-68-1]	38	negligible

temperatures the solvus curves are at very low copper and magnesium contents, and appreciable supersaturation is achieved. As magnesium increases above the ratio in S-phase the amount of this phase that can be formed decreases, so the amount of age hardening decreases. The excess magnesium contributes to solid solution strengthening. Excess copper is restricted by the solubility of θ-phase. The precursors to both θ- and S-phases are heterogeneously nucleated during elevated temperature precipitation heat treatments. Cold-working after quenching and prior to artificial aging increases the density of dislocations that nucleate precipitation, thus refining the precipitate structure and increasing strength.

Al–Mg–Zn. Although neither aluminum-rich binary Al–Zn nor Al–Mg alloys are precipitation hardenable, the ternary system is a source of alloys strengthened in this manner. The intermetallic phases encountered in aluminum-rich alloys are shown in Table 20. Alloy compositions are selected for precipitation of an M- or η-phase precursor because T-phase is less effective as a strengthener. Commercially important alloys always contain more zinc than magnesium to provide attractive combinations of strength, extrudability, and

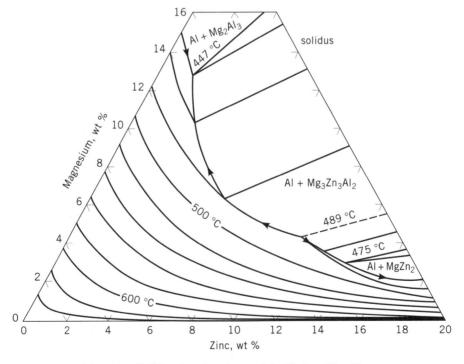

Fig. 24. Solidus for aluminum-rich Al–Zn–Mg alloys.

weldability. As a generality, resistance to stress-corrosion cracking increases as the Zn:Mg ratio increases.

The Al–Mg–Zn solidus in Figure 24 illustrates the high solid solubility of both magnesium and zinc. Melting temperatures of the eutectics are lower than in the Al–Mg–Si and Al–Cu–Mg systems. Solid-state reactions can be understood with reference to Figure 25. Substantial amounts of magnesium and zinc can be dissolved well below melting temperatures, and appreciable supersaturation and age hardening can be achieved.

Al–Cu–Li. Although the addition of Cu to Al–Li alloys increases density, the boost in strength more than offsets the density increase so that Al–Cu–Li alloy products develop higher specific strengths (strength/density) than do binary Al–Li alloy products. Furthermore, the fracture toughness and corrosion resistance of products manufactured from Al–Cu–Li alloys are higher than these properties in binary Al–Li alloy products. The phases in equilibrium at 327–350°C are presented in Figure 26 (45). Designations for equilibrium and metastable phases in Al–Cu–Li alloys and their crystal structures are presented in Table 21 in order of increasing lithium content of the phase. The strengthening precipitates in most Al–Cu–Li alloys that have commercial significance are δ' and T_1. The T_2-phase decreases toughness when it precipitates on grain boundaries.

Al–Li–Mg. In aluminum-rich alloys the ternary phase, T, sometimes designated as Al_2LiMg, is encountered in addition to AlLi (δ), Al_3Mg_2 (β), and $Al_{12}Mg_{17}$ [12254-22-7] (γ), Figure 27. Assessment of the composition of the

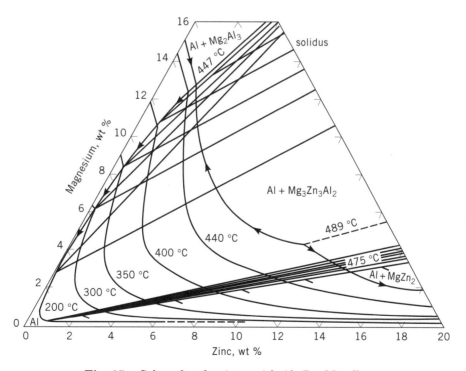

Fig. 25. Solvus for aluminum-rich Al–Zn–Mg alloys.

Table 21. **Equilibrium and Metastable Phases in Al–Cu–Li Alloys**

Designation	Formula	Crystal structure
θ	Al_2Cu	tetragonal
θ'	Al–Cu	tetragonal
θ''	Al–Cu	tetragonal
Ω	Al–Cu	tetragonal
T_B	Al_6Cu_4Li	CaF_2
T_1	Al_2CuLi	hexagonal
R	Al_5CuLi_3	Im3
T_2	Al_6CuLi_3	icosohedral
T_2	Al_6CuLi_3	rhombohedral
δ'	Al_3Li	$L1_2$
δ	AlLi	B32

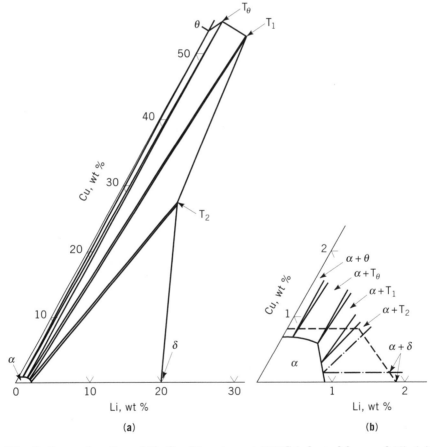

Fig. 26. Isothermal section of Al–Cu–Li system at 350°C (adapted from ref.46); **(a)**, isothermal section; **(b)**, limits of solid solubility for α-phase. Solid lines correspond to proposed limits for 327°C (47); dashed lines correspond to 350°C section.

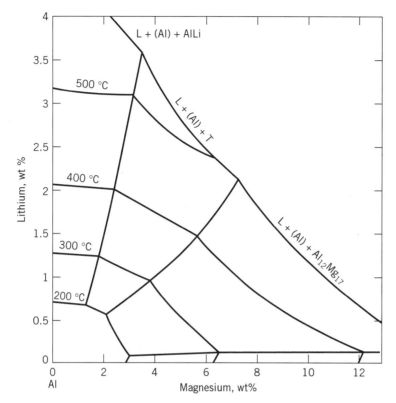

Fig. 27. Projection of solvus surface of Al–Li–Mg system, where L is liquid and T is temperature.

T-phase indicates that it contains 15.5 atomic % Mg and 32 atomic % Li. The high solubility of magnesium is unaffected by lithium, and no metastable phase in addition to Al_3Li (δ') is formed. Alloys in this system are solid solution strengthened by magnesium and age harden by precipitation of δ'.

11.3. Quaternary and Higher Alloys. Further additions to commercial aluminum alloys usually are made either to modify the metastable strengthening precipitates or to produce dispersoids.

Modifications to Precipitates. Silicon is sometimes added to Al–Cu–Mg alloys to help nucleate S′ precipitates without the need for cold work prior to the elevated temperature aging treatments. Additions of elements such as tin cadmium, and indium I, to Al–Cu alloys serve a similar purpose for θ' precipitates. Copper is often added to Al–Mg–Si alloys in the range of about 0.25% to 1.0% Cu to modify the metastable precursor to Mg_2Si. The copper additions provide a substantial strength increase. When the copper addition is high, the quaternary $Al_4CuMg_5Si_4$ Q-phase must be considered and dissolved during solution heat treatment.

The highest strength aluminum alloy products are based on the Al–Cu–Mg–Zn system and all are strengthened by precursors to the η-phase. Copper and aluminum can substitute for zinc in $MgZn_2$, so η-phase may be represented as Mg (Al,Cu,Zn). No quaternary phases are a factor in aluminum-rich Al–Cu–Mg–Zn

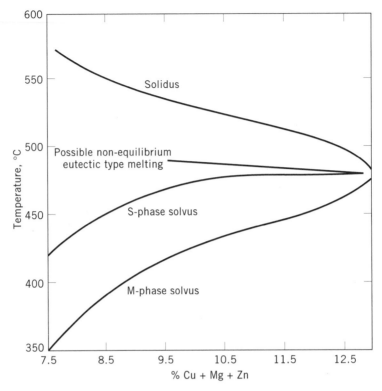

Fig. 28. Isopleth in Al–Cu–Mg–Zn system where the appropriate ratio is Cu:Mg:Zn = 1 : 1.4 : 3.3. Courtesy of E. DaPos.

alloys, but Al$_2$CuMg, S-phase, may be present in higher solute alloys. The isopleth in Figure 28 for alloys in the ratio of 1.6% Cu, 2.5% Mg, and 5.6% Zn illustrates that the S-phase solvus is always higher than that for η-phase for such alloys. The addition of copper to Al–Mg–Zn alloys above about 0.25% Cu increases strength and reduces the weldability and resistance to general corrosion of products, whereas additions above about 1.5% Cu significantly increase strength and resistance to stress-corrosion cracking.

Several of the Al–Li alloys developed in the 1980s contain both magnesium and copper. No quaternary Al–Cu–Li–Mg phase has been found in the alloys. The S′-phase in addition to δ′ and T$_1$ provides precipitation hardening.

When combined with magnesium, silver has found commercial use as an alloying element in several aluminum alloys for specialized applications. It was added to an Al–Cu–Mg–Zn forging alloy to increase the resistance to stress-corrosion cracking and to an Al–Cu–Mg casting alloy to increase strength. In the late 1980s, combined Ag and Mg additions were shown to increase strength of an Al–Cu alloy (48). Additions of lithium to this basic composition further increased strength (49). The effect of the combined usage of Ag and Mg was attributed to promoting the precipitation of a phase designated Ω that was believed to contain Mg and Ag atoms, but has recently been identified in a binary Al–Cu alloy (50).

Dispersoid Formers. The three elements commonly added to precipitation hardenable alloys to form dispersoids are manganese, chromium, and zirconium. The amounts customarily used (0.5% Mn, 0.2% Cr, and 0.1% Zr) remain in supersaturated solid solution during ingot casting and precipitate as dispersoids during thermal treatment of the ingot. These dispersoids serve to minimize recrystallization during solution heat treatment of products such as plate, forgings, and extrusions which are hot-worked, and to maintain a fine recrystallized grain size in sheet and tubing which is cold-worked.

The manganese in commercial Al–Cu–Mg alloys precipitates as $Al_{20}Cu_2$ Mn_3 [12355-72-5] dispersoid particles. These dispersoids also help strengthen the products because, during quenching, they promote the formation of dislocations that serve as sites for heterogeneous nucleation during artificial aging. The chromium in commercial Al–Cu–Mg–Zn alloys precipitates as $Al_{12}Mg_2Cr$ dispersoids. These particles are more effective than $Al_{20}Cu_2Mn_3$ dispersoids in controlling recrystallization and in providing fine grain size in cold-worked products. They do not contribute to age hardening, but serve to nucleate η-phase particles heterogeneously during a slow quench, decreasing the strength of slowly quenched products because the solute is unavailable for controlled precipitation during artificial aging treatments. The zirconium in commercial Al–Cu–Mg–Zn,

Table 22. **Compositions of Aluminum Foundry Alloys**

Aluminum Association designation	Casting process[a]	Alloying elements, %				Applications
		Si	Cu	Mg	Others[b]	
208.0	S	3.0	4.0			general purpose
213.0	P	2.0	7.0			cylinder heads, timing gears
242.0	S,P		4.0	1.5	2.0 Ni	cylinder heads, pistons
295.0	S	1.1	4.5			general purpose
B295.0	P	2.5	4.5			general purpose
308.0	P	5.5	4.5			general purpose
319.0	S,P	6.0	3.5			engine parts, piano plates
A332.0	P	12.0	1.0	1.0	2.5 Ni	pistons, sheaves
F332.0	P	9.5	3.0	1.0		pistons, elevated temperatures
333.0	P	9.0	3.5	0.3		engine parts, meter housings
355.0	S,P	5.0	1.3	0.5		general: high strength, pressure tightness
356.0	S,P	7.0		0.3		intricate castings: good strength, ductility
360.0	D	9.5		0.5	2.0 Fe max	marine parts, general purpose
380.0	D	8.5	3.5		2.5 Fe max	general purpose
A413.0	D	12.0				large intricate parts
443.0	D	5.3			2.0 Fe max	carburetors, fittings, cooking utensils
B443.0	S,P	5.3			0.8 Fe max	general purpose
514.0	S			4.0		hardware, tire molds, cooking utensils
520.0	S			10.0		aircraft fittings
A712.0	S		0.5	0.7	6.5 Zn	general purpose

[a] S, sand cast; P, permanent mold cast; D, pressure die cast.
[b] Aluminum and impurities constitute remainder.

Al–Cu–Li, and Al–Cu–Mg alloys precipitates as metastable Al_3Zr, β', dispersoids. Products which contain β'-phase dispersoids are not as quench sensitive as those which contain $Al_{12}Mg_2Cr$. Alloys in the Al–Mg–Si system may contain either manganese, chromium, or zirconium. Some Al–Mg–Zn alloys contain both chromium and manganese, and at least one contains all three dispersoid forming elements.

11.4. Foundry Alloys and Their Characteristics.

Unalloyed aluminum does not have either mechanical properties or casting characteristics suitable for general foundry use yet both can be greatly improved by the addition of other elements. The most common addition is silicon which enhances fluidity, increases resistance to hot-cracking, and improves pressure tightness. Because binary Al–Si alloys have relatively low strengths and ductility, other elements such as copper and magnesium are added to obtain higher strengths through heat treatment. The compositions of representative foundry alloys are shown in Table 22, and typical mechanical properties are presented in Table 23.

The alloys are identified by a series of three-digit numbers where the first digit indicates the alloy group as shown. The aluminum is at least 99% pure.

Table 23. **Mechanical Properties of Aluminum Foundry Alloys**

Alloy and temper	Tensile strength, MPa[a]	Yield strength, MPa[a]	Elongation, %	Brinell hardness
Sand castings				
208.0–F	145	95	2.5	55
242.0–T21	185	125	1	70
242.0–T77	205	160	2	75
295.0–T4	220	110	9	60
295.0–T6	250	165	5	75
319.0–F	185	125	2	70
319.0–T6	250	165	2	80
355.0–T6	240	170	3	80
256.0–T6	225	165	3	70
B443.0–F	130	55	8	40
514.0–F	170	85	9	50
520.0–T4	330	180	16	75
A712.0–T5	240	170	5	75
Permanent mold castings				
242.0–T61	325	290	0.5	110
308.0–F	195	110	2	70
319.0–F	235	130	2.5	85
319.0–T6	275	185	3	95
F332.0–T5	250	195	1	105
355.0–T6	290	185	4	90
356.0–T6	260	185	5	80
B443.0–F	160	60	10	45
Die castings				
360.0	305	170	3	75
380.0	315	160	3	80
A413.0	290	130	3	80
443.0	225	95	9	50

[a] To convert MPa to psi, multiply by 145.

First digit	1	2	3	4	5	7	8	9
Element	Al	Cu	Si–Cu–Mg	Si	Mg	Zn	Zn	Other

In addition to these elements, foundry alloys may contain a small amount of titanium, for grain refinement, as well as small additions of manganese, chromium, or nickel. A high strength Al–Cu–Mg alloy for aircraft use contains silver for added strength by modifying the precipitate phase. Alloys intended for pressure die casting may have high iron contents to resist welding to the dies. Lower grades of the same alloys may be used in sand casting and permanent mold castings with some benefits to mechanical properties. Castings are produced in an F temper (as-cast) and a T temper (heat treated) where higher strengths or other special characteristics are desired.

Die castings produced in a cavity which has been evacuated contain significantly less porosity and hence, are more ductile. Such castings were produced on an experimental basis in the 1980s. This process can product some parts which compete with either forged or sheet metal components.

11.5. Wrought Alloys and Their Characteristics. Alloys for the production of wrought products are selected for fabricability as well as their physical, chemical, and mechanical properties. Usually, these alloys are less highly alloyed than those for foundry use and contain less iron and silicon. A series of alloys based on the eutectic Al–Fe–Si composition, however, has been developed to provide good combinations of strength and formability in thin sheet products. The compositions of some of the more common alloys are shown in Table 24. The alloys are identified by four-digit numbers, where the value of the first digit identifies the alloy type and principal alloying addition. The aluminum is at least 99% pure.

First digit	1	2	3	4	5	6	7	8
Element	Al	Cu	Mn	Si	Mg	Mg+Si	Zn	Other

In addition to these principal alloying elements, which provide solid solution strengthening and/or precipitation strengthening, wrought alloys may contain small amounts of titanium and boron [7440-42-8], B, for control of ingot grain size, and ancillary additions of chromium, manganese, and zirconium to provide dispersoids. All commercial alloys also contain iron and silicon.

Mechanical properties of wrought aluminum alloy mill products depend on temper as well as on chemical composition. The letter O following the alloy designation indicates a final annealing operation to achieve lowest strength and highest ductility. The letter H indicates that the product is in a strain-hardened condition or temper as a result of cold rolling, drawing, or other types of working. The first digit following the H indicates whether the product received some thermal treatment after cold working; the second digit implies property level. The letter T signifies a heat treatment to obtain solute precipitation and

Table 24. Typical Compositions of Wrought Aluminum Alloys (Major Alloying Elements Only)

Aluminum Association designation	Alloy elements, %								Applications
	Si	Fe	Cu	Mn	Mg	Cr	Zn	Others[a]	
1100			0.12					1.0 Si + Fe max	cooking utensils, heat exchanger
1350								0.5 max	electrical conductor
2014	0.8		4.4	0.8	0.5				aircraft structures, and forced wheels
2024			4.4	0.6	1.5				aircraft structures,
2195			4.0		0.6			1.1 Li, 0.12 Zr, 0.4 Ag	aerospace structure, welded
2219			6.3	0.3				0.1 V, 0.18 Zr, 0.06 Ti	aerospace structures, welded
2519			5.9	0.3	0.2			0.1 V, 0.18 Zr	armor and aerospace structures, welded
3003			0.12	1.1					general purpose, cooking utensils
3104			0.15	1.0	1.1				general purpose, can body sheet
3105				0.6	0.5				building products, bottle caps (sheet)
5052					2.5	0.25			general purpose sheet
5083				0.7	4.4	0.15			pressure vessels, marine applications
5182				0.35	4.5				can lids, automotive sheet
5657					0.8				bright trim and lighting sheet
6022	1.0		0.05		0.55				automotive body sheet
6111	0.8		0.70		0.7				automotive body sheet
6013	0.8		0.8		1.0				tubing and aerospace structures, welded
6061	0.6		0.25	0.5	1.0	0.20			general purpose, structures, forged truck wheels
6063	0.4				0.7				general purpose, building products (extrusions)
6201	0.7				0.8				electrical conductor
7005				0.45	1.4	0.13	4.5	0.14 Zr	truck and, railroad cars (extrusions)
7075			1.6		2.5	0.25	5.6		aircraft structures
7150			2.2		2.3		6.4	0.12 Zr	aerospace structures
8079	0.2	1.0							packaging and household foil
8011	0.7	0.8							packaging foil, formed containers

[a] Aluminum and impurities constitute remainder.

hardening. The numbers following the T further define the condition of the product by indicating the type of thermal or working operation employed in producing the material.

Mechanical properties also depend on product form and test direction because the manufacturing operations used to produce the mill products may significantly affect the final crystallographic texture and extent of mechanical fibering. For example, the longitudinal yield strength of alloy 7075–T6 rod produced by axisymmetric extrusion may be as much as 140 MPa (20,000 psi) higher than that of 7075–T6 sheet. The typical long-transverse mechanical properties of some wrought aluminum alloys in the form of sheet or plate are presented in Table 25.

12. Thermal Treatment of Alloys

Aluminum alloys are subjected during manufacture to a variety of thermal treatments that range from heating to assist fabrication, to heating for control of final properties. Although the natural oxide film on aluminum provides good protection against surface oxidation and deterioration during such treatments, controlled atmospheres are sometimes employed for products requiring minimum surface oxide such as foil and sheet for reflectors. The atmosphere may be nitrogen or burned producer gas that is dried prior to use in the furnace. The introduction of sulfur compounds into furnace atmospheres should be avoided because in the presence of moisture sulfur compounds tend to break down the protective aluminum oxide film permitting further oxidation to occur. Nascent hydrogen is a product of such a reaction and may diffuse into the metal, forming voids or blisters, especially in the high strength Al–Cu–Mg or Al–Cu–Mg–Zn alloys. Such problems can be minimized by using low dew point atmospheres, eliminating sulfur compounds, and paying careful attention to recommended times and temperatures of heat treatment. Additional protection can be obtained by adding small quantities of boron trifluoride BF_3, or one of several fluoroborate compounds to the furnace atmospheres. These fluorides tend to form protective films that resist breakdown by moisture and sulfur compounds.

12.1. Homogenization. Ingots are usually preheated prior to rolling, forging, or extrusion to increase workability. The process is commonly referred to as homogenization because chemical segregation of the major alloying elements that are completely soluble in the solid-state is reduced. In addition, however, supersaturation of elements such as manganese, chromium, and zirconium is relieved during the preheat operation, so these elements become more segregated when they precipitate as dispersoids. Also during the homogenization treatment, metastable compounds may transform into the more stable equilibrium form. Incipient melting is usually avoided during preheating, but homogenization treatments are often close to eutectic or solidus temperatures, so care must be taken.

12.2. Annealing. The resistance to further deformation of aluminum alloy products at elevated temperatures reaches a steady value after a modest strain when the rate of formation of fresh dislocations is balanced by the rate

Table 25. **Long Transverse Properties of Wrought Aluminum Alloy Products**

Alloy and temper	Tensile strength, MPa[a]	Yield strength, MPa[a]	Elongation,%	Brinell hardness	Elastic modulus, MPa[a] × 10³
1100–O	90	35	35	23	69
1100–H14	125	115	9	32	69
1100–H18	165	150	5	44	69
1350–O	85	30			69
1350–H19	185	165			69
2014–T6	485	415	13	135	73
2024–T3	485	345	18	120	73
2024–T81	485	450	7	125	73
2195–T8	585	545	8		
2219–T62	415	290	10	110	73
2219–T81	455	350	10	123	73
2519–T87	500	450	11		72
3003–O	110	40	30	28	69
3003–H14	150	145	8	40	69
3003–H18	200	185	4	55	69
3004–O	180	70	20	45	69
3004–H34	240	200	9	63	69
3004–H38	285	250	5	77	69
3105–O	115	55	24		69
3105–H25	180	160	8		69
5052–O	195	90	25	47	70
5052–H34	260	215	10	68	70
5052–H38	290	255	7	77	70
5083–O	290	145	22	65	71
5083–H321	315	225	16	82	71
5182–O	270	130	23	64	71
5654–H25	185	145	9		69
6013–T6	395	350	11		69
6022–T4	230	120	28		68
6061–O	125	55	25	30	69
6061–T4	240	145	22	65	69
6061–T6	310	275	12	95	69
6063–T6	240	215	12	73	69
6111–T4	265	130	25		69
7005–T53	395	345	15	105	70
7075–T6	570	505	11	150	72
7075–T76	525	450	11		72
7075–T73	505	415	11		72
7150–T77	600	565	10		71
8011-O	45	95	20		—

[a] To convert MPa to psi, multiply by 145.

of annihilation. This process is known as dynamic recovery. During deformation processing by hot rolling, the metal temperature has a tendency to decrease. When this happens the rate of dynamic recovery decreases, so the resistance to further deformation increases because the dislocation density increases. In the rolling of sheet to thicknesses below about 4 mm, the final 25 to 50% reduction is usually done cold to provide close dimensional control. In-process annealing is employed to decrease the dislocation density thereby increasing the plasticity of the hot-rolled metal prior to cold rolling. This thermal process also modifies the crystallographic texture, a very important consideration in producing products requiring control of anisotropy. During cold rolling, the dislocation density progressively increases with increasing strain, so for the thinnest sheet or foil, more than one in-process anneal may be employed. The temperatures employed for annealing are usually in the range of 350–400°C.

Annealing is also employed as a final mill operation to produce a material having high formability for subsequent customer shaping or forming operations. For precipitation hardenable alloys, the cooling rate must be controlled so that the equilibrium phases precipitate as coarse, nonstrengthening particles. Some tempers of strain-hardenable alloy products are produced by partial annealing or recovery treatments to provide attractive combinations of strength and formability. Highly cold-worked products are heated for combinations of time and temperature sufficient to provide enough energy to rearrange and eliminate some of the dislocations introduced during cold working but insufficient to allow the product to recrystallize by the motion of high angle grain boundaries.

12.3. Solution Heat Treatment.   Solution heat treatment is the first stage of a series of operations to achieve precipitation hardening. The product is heated to a high temperature to dissolve as many soluble particles as practicable. Times for adequate dissolution depend on the size of the particles to be dissolved. Thick castings generally require many hours because the particles are a combination of products of eutectic decomposition and of precipitates which form and coarsen during the slow cool of the casting. Thin sheet may require less than a minute provided that the process was controlled so that the precipitate particles did not coarsen during fabrication. Heating is usually performed in an electrical resistance furnace, but molten salt is used for higher heating rates to produce fine grain size in sheet. Much sheet is solution heat treated in a continuous web, fed from a coil into a horizontal furnace.

12.4. Quenching.   After solution treatment, the product is generally cooled to room temperature at such a rate to retain essentially all of the solute in solution. The central portions of thicker products cannot be cooled at a sufficient rate to prevent extensive precipitation in some alloys. Moreover, some forgings and castings are deliberately cooled slowly to minimize distortion and residual stress produced by differential cooling in different portions of the products. Cold water, either by immersion or by sprays, is the most commonly used cooling medium. Hot water or a solution of a polymer in cold water is used when the highest rates are not desired. Dilute Al–Mg–Si and Al–Mg–Zn extrusions can be effectively solution heat treated by the extrusion process; therefore, they may be quenched at the extrusion press by either air or water.

12.5. Precipitation Heat Treatment.   The supersaturated solution produced by the quench from the solution temperature is unstable, and the alloys tend to approach equilibrium by precipitation of solute. Because the activation energies required to form equilibrium precipitate phases are higher than those to form metastable phases, the solid solution decomposes to form G-P zones at room temperature (natural aging). Metastable precursors to the equilibrium phases are formed at the temperatures employed for commercial precipitation heat treatments (artificial aging).

Table 26. **Age Hardening of Aluminum Alloys 6061 and 2024 at Room Temperature after Heat Treatment and Quench**

Aging time	Alloy 6061			Alloy 2024		
	Tensile strength, MPa[a]	Yield strength, MPa[a]	Elongation, %	Tensile strength, MPa[a]	Yield strength, MPa[a]	Elongation, %
5 min	165	55	28	365	155	23
30 min	170	60	28	370	170	23
1 h	175	65	28	385	195	23
2 h	185	70	27	415	225	23
4 h	195	80	27	440	265	23
8 h	205	90	26	455	275	23
1 day	220	105	25	460	285	22
1 wk	235	125	24	465	290	22
1 mo	250	130	24	470	290	21
1 yr	260	140	23	470	290	21

[a] To convert MPa to psi, multiply by 145.

Natural Aging (Room Temperature Precipitation Heat Treatments)
Certain alloys are capable of developing commercially useful properties when the products are allowed to age at room temperature. In general, the properties of 2xxx alloy products stabilize after about a week, and those of 6xxx alloy products attain about 90% of their peak properties after the same time period (Table 26). Products of these alloys are referred to as being in the T4 temper after about one week natural aging. For several Al–Cu–Mg alloys, cold working a few percent before natural aging is complete significantly increases strength relative to material in a T4 temper. The T3 temper designation is used to describe such products. In contrast to the behavior of 2xxx and 6xxx alloy products, properties of 7xxx alloy products continue to change at a logarithmic rate for decades. Consequently, naturally aged 7xxx alloy products are not used for

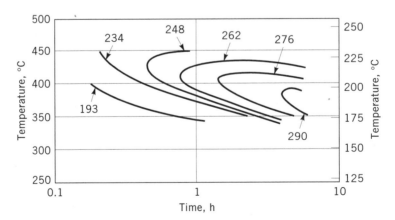

Fig. 29. Effects of time and temperature of precipitation treatments on yield strength of alloy 2036–T4 sheet. Numbers represent strength in MPa. To convert MPa to psi, multiply by 145.

commercial applications. The W temper designation accompanied by a time is used to describe 7xxx alloy products in naturally aged conditions.

Artificial Aging (Elevated Temperature Precipitation Heat Treatments). Most precipitation hardenable alloy products are heated at slightly elevated temperatures after solution treatment and quenching to accelerate age hardening. Effects of time and temperature on yield strength of alloy 2036 sheet are illustrated in Figure 29 (51). The T6 temper is generally used to designate material aged by the practice which produces the highest practicable strength. When cold work prior to artificial aging produces a significant positive effect on strength, the T8 temper designation is employed for material worked and aged to the highest practicable strength. Multistep aging treatments are sometimes employed to produce T7 tempers which enhance the combination of strength and resistance to exfoliation and stress-corrosion cracking of 7xxx alloy products.

13. Shaping and Fabricating of Alloys

Aluminum alloys are commercially available in a wide variety of cast forms and in wrought mill products produced by rolling, extrusion, drawing, or forging. The mill products may be further shaped by a variety of metal working and forming processes and assembled by conventional joining procedures into more complex components and structures.

13.1. Melting and Casting. For energy savings and economy, some aluminum is alloyed and cast into ingot at smelting plants. Alloy ingot is then shipped to fabricating plants for conversion into mill products such as plate, forgings, extrusions, and tubing. Considerable scrap is generated in such mill operations, and plants remelt the scrap and cast ingots to salvage the metal. In remelting aluminum for the production of new ingot, selected mill scrap and purchased scrap are combined with sufficient pure metal and additions in the form of alloys rich in the alloying elements. Melting temperatures are in the range of 700–750°C; casting temperatures vary with alloy but typically are about 50°C above the temperature of initial solidification.

In remelting for the production of ingot, dross is formed in substantial quantities and is collected by skimming. Dross may consist of up to 5% of the melt where clean mill scrap is used but is much higher for operations using painted foil or post-consumer scrap. The primary components of dross are aluminum oxide and entrained aluminum plus small amounts of magnesium oxide, aluminum carbide, nitride, and extraneous matter. Such dross is generally treated in rotary furnaces using fluxes of sodium chloride and potassium chloride to recover the metal values.

Scrap that is unsuitable for recycling into products by the primary aluminum producers is used in the secondary aluminum industry for castings that have modest property requirements. Oxide formation and dross buildup are encountered in the secondary aluminum industry, and fluxes are employed to assist in the collection of dross and removal of inclusions and gas. Such fluxes are usually mixtures of sodium and potassium chlorides. Fumes and residues from these fluxes and treatment of dross are problems of environmental and economic importance, and efforts are made to reclaim both flux and metal values in the dross.

Concern for the environment has prompted legislation which encourages recycling (qv) of aluminum beverage containers. Billions of used beverage containers are returned every year to the primary aluminum companies (about 62% of the aluminum cans produced in 2000 were recycled), and technology has been developed to remelt this thin scrap metal with acceptable loss from oxidation. The use of recycled metal has obviated the need for additional smelting capacity, thus saving energy and minimizing cost increases of aluminum products.

After remelting and skimming, the molten metal is treated to remove dissolved hydrogen (the only gas soluble in aluminum), inclusions, and undesirable trace elements such as sodium. Generally this is accomplished by bubbling chlorine or a mixture of chlorine and nitrogen through the molten metal by means of graphite tubes. In large-scale modern operations use is made of in-line filtering and fluxing systems between the holding furnace and the casting station. Choice between deep-bed filters having a counter flow of inert gas and disposable foam ceramic filters depends on quality and economic considerations (52). In-line processes are more efficient than furnace fluxing, and their use reduces hydrogen, inclusion content, and undesirable trace elements to very low levels without significant fume generation and air pollution.

Solidification of molten aluminum containing equilibrium volumes of hydrogen (Table 6) can result in significant porosity because of the greatly reduced solubility of hydrogen in solid aluminum. Commercial filtering and fluxing practices reduce hydrogen contents of the molten metal to an operating level where this is usually not a problem. Hydrogen contents of between about 0.06 and 0.2 ppm are usually obtained in the final product depending on the alloy. The solid solubility of hydrogen is considerably higher in Al–Li alloys, hence they can tolerate more hydrogen without inducing porosity (52).

Most ingot for fabrication into rolled products, forgings, or extrusions is produced by the direct chill (DC) process or a modification. The process employs a short mold having a moveable bottom. During casting the bottom of the mold is moved downward through a spray of water that cools the mold and chills the ingot. Ingot length is limited only by the supply of metal and the depth of the casting pit. Rectangular cross-section ingot for rolling flat products is typically 40 to 60 cm thick and may be as much as 120 cm in width. Ingot having a round cross section is routinely cast in diameters to 90 cm.

Since the early 1980s, electromagnetic (EM) casting, where an electromagnetic field is used to contain the molten metal during casting, has been gaining increased acceptance (53). Because the EM process provides the closest approach to continuous heat removal, the surface defects associated with the DC processes are largely eliminated.

Some aluminum alloys are cast as strip or rod for fabrication into sheet, foil, or wire by introducing the molten metal directly between rolls or between moving plates or steel bands that form the desired shape. Casting operations of this type have the advantage of eliminating capital investment in large break-down mills and extrusion presses. The more rapid solidification also reduces the constituent size and increases the amount of supersaturated solute. The process is limited to 1xxx, 3xxx, and certain 8xxx alloys which have a small freezing range.

13.2. Fabrication. After the preheat or homogenization step, the ingots may be fabricated directly. Often, however, the preheated ingots are reheated in a separate operation before the first metal working operation. Bulk deformation temperatures usually range from about 350 to 500°C.

Rectangular ingots are generally broken down using a four-high reversing mill. Some gauges may be rolled to completion on a reversing mill, but thinner plate gauges are usually transferred to a series of rolling mills for fabricating to final gauge. Thicker sheet gauges, above about 5 mm, are generally hot-rolled to flat sheet on such a continuous hot line. Thinner gauges are usually coiled at an intermediate gauge as they exit the continuous hot line, annealed, then cold-rolled about 25% or more to final gauge. Sheet may be annealed and cold-rolled to foil. Thin sheet wider than the equipment used for continuous solution heat treatment must be produced by a batch process as flat sheet.

In extrusion, a heated length of ingot (billet) is forced under pressure to flow from a container through a die opening to form an elongated shape or tube. In the direct extrusion process the metal is pushed through the die. In the indirect extrusion process the die is forced into the metal. Productivity is higher using the indirect process because friction of the billet with the container walls is eliminated. The surface layers of the ingot used for the indirect process must be machined (scalped) because the surface of the billet becomes the surface of the extruded product in this process. Extrusion can economically produce many shapes, including hollows, that would be difficult or expensive to produce by alternative fabricating processes. Extruded rod or tube is frequently employed as stock in the production of drawn wire or tubing. Extrusion is sometimes used to produce starting billet for forging.

Forgings are produced by pressing or hammering ingots or billets either into simple shapes on flat dies or into complex shapes in cavity dies. Several sets of dies may be employed in arriving at the final shape. Equipment varies from simple drop hammers and mechanical presses to large hydraulic presses.

Most aluminum mill products are used in the manufacture of more complex shapes. This can be accomplished by conventional metal removal techniques and by common fabricating and joining procedures. The latter include welding, brazing, soldering, and adhesive bonding, as well as mechanical methods of joining. Methods of fabricating include bending, deep drawing, stamping, stretch forming, impact extrusion, ironing, and swaging. The bodies of two-piece carbonated beverage cans, the largest market for aluminum sheet, are produced at high rates (about 200–300 per minute per bodymaker) by stamping a disk, drawing the disk into a cup, ironing to elongate the cup into a can, shearing to trim the can, and swaging to provide the neck.

Intricate shapes which once had to be made by joining a large number of parts are made by a process known as superplastic forming. In this process, aluminum alloy sheet that has been processed to develop an extremely fine grain size (typically about 10 micrometers in diameter) is slowly deformed either into a female die or over a male die. For those components that require metal removal, techniques include shearing, grinding, and a variety of common machining procedures. Aluminum is also being successfully processed using high speed machining techniques that employ ceramic tools and special equipment.

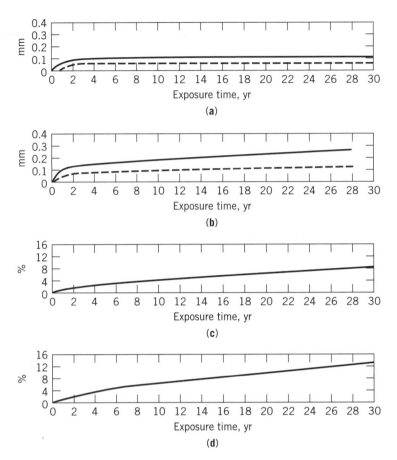

Fig. 30. Weathering of wrought aluminum alloys 1100, 3003, and 3004: (**a**) and (**b**), ——
maximum and – – – average depth of pitting-type corrosion; (**c**) and (**d**), loss in tensile
strength; in industrial (**a** and **c**) and seacoast (**b** and **d**) environments.

14. Corrosion

Aluminum and aluminum alloys are employed in many applications because of
the ability to resist corrosion. Corrosion resistance is attributable to the tightly
adherent, protective oxide film present on the surface of the products. This film is
5–10 nm thick when formed in air; if disrupted it begins to form immediately in
most environments. The weathering characteristics of several common aluminum alloy sheet products used for architectural applications are shown in Figure
30. The loss in strength as a result of atmospheric weathering and corrosion is
small, and the rate decreases with time. The amount of corrosion that occurs is a
function of the alloy as well as the severity of the corrosive environment.
Wrought alloys of the Al, Al–Mn, Al–Mg, and Al–Mg–Si types have excellent
corrosion resistance in most weathering exposures including industrial and seacoast atmospheres. Alloys based on additions of copper, or copper, magnesium,
and zinc, have significantly lower resistance to corrosion.

Table 27. **Electrode Potentials of Aluminum Solid Solutions and Intermetallic Particles**

Solution	Potential, V^a
Al_3Mg_2	−1.24
$Al + 4\ MgZn_2$	−1.07
$Al + 4\ Zn^b$	−1.05
$MgZn_2$	−1.05
Al_2CuMg	−1.00
$Al + 1\ Zn^b$	−0.96
$Al + 7\ Mg^b$	−0.89
$Al + 5\ Mg^b$	−0.87
Al_6Mn	−0.85
99.95 Al	−0.85
$Al + 1\ Mg_2Si$	−0.83
$al + 1\ Si^b$	−0.81
$Al + 2\ Cu^b$	−0.75
Al_2Cu	−0.73
$Al + 4\ Cu^b$	−0.69
Al_3Fe	−0.56
Al_3Ni	−0.52
Si	−0.26

[a] 0.1 N calomel scale, measured in an aqueous solution of 53 g/L $NaCl + 3$ g/L H_2O_2 at 25°C.
[b] Solid solution.

Atmospheric corrosion is electrochemical in nature and depends on the flow of current between anodic and cathodic areas. The resulting attack is generally localized to particular features of the metallurgical structure. Features that contribute to differences in potential include the intermetallic particles and the electrode potentials of the matrix. The electrode potentials of some solid solutions and intermetallic particles are shown in Table 27. Iron and silicon impurities in commercially pure aluminum form intermetallic constituent particles that are cathodic to aluminum. Because the oxide film over these constituents may be weak, they can promote electrochemical attack of the surrounding aluminum matrix. The superior resistance to corrosion of high purity aluminum is attributed to the small number of these constituents.

Intermetallic particles formed by interaction of aluminum and alloying elements such as copper, magnesium, lithium, and zinc also have electrode potentials that differ from those of aluminum or the solid solutions in which they exist. If such particles are finely divided and uniformly distributed, their effect on corrosion is minimal. During quenching or aging treatments, however, precipitates may localize at grain boundaries, and render them susceptible to intergranular exfoliation corrosion or stress-corrosion cracking (SCC). In the 1960s, a series of tempers was developed for commercial Al–Cu–Mg–Zn alloys which modify the precipitate structure so that the products are highly resistant to these phenomena (54) even in the short-transverse direction. These tempers are designated T73, highly resistant to both SCC and exfoliation corrosion; T76, resistant to exfoliation corrosion and more resistant to SCC than products in T6 tempers, and T74, corrosion characteristics intermediate to T73 and T76. Because the strength of products in these T7 tempers was as much as 20% less than that possible using the T6 temper, in the late 1980s a process was developed that provided T6 strength and T76 corrosion characteristics (52). The process was first commercialized on alloy 7150 in and later alloy 7055 as the T77 temper. These materials have been specified for upper wing skins and other aircraft structure that is dominated by compressive loading.

The resistance to corrosion of some alloy sheet is improved by cladding the sheet with a thin layer of aluminum or aluminum alloy that is anodic to the base alloy. These anodic layers are typically 5–10% of the sheet thickness. Under corrosive conditions, the cladding provides electrochemical protection to the core at cut edges, abrasions, and fastener holes by corroding preferentially. Aircraft skin sheet is an example of such a clad product.

Aluminum alloys are used widely for processing, handling, and storing a variety of chemicals. Aluminum generally has a high resistance to corrosion by chemical solutions in the pH range of 4.5–8.5 because the protective oxide film is relatively stable within this range. In strongly acidic or alkaline solutions, the film is less stable and resistance to corrosion is impaired. Dry inorganic salts are usually not corrosive to aluminum. Organic acids and alcohols can be handled in aluminum equipment if a trace of water is present to maintain the oxide film.

15. Finishes for Aluminum

Finishes for aluminum products can be both decorative and useful. Processes in use include anodic oxidation, chemical conversion coating, electrochemical graining, electroplating (qv), thin film deposition, porcelain enameling, organic coating, and painting. Some alloys respond better than others to such treatments.

Anodic oxidation is an electrochemical process in which the oxide film is increased in thickness by making the product the anode in a cell containing a suitable electrolyte and an inert cathode (see ELECTROCHEMICAL PROCESSING). Sulfuric acid solutions are widely employed as electrolytes for anodizing; they produce clear oxide coatings on pure aluminum and high purity Al–Mg and Al–Mg–Si alloys. As the purity of the metal decreases, iron and silicon and other impurities and their reaction products form in the oxide film and decrease its transparency. This imparts a gray appearance to the finish. Sheet for reflectors and for automotive and appliance trim is made from high purity alloys to obtain maximum brightness and specularity. Oxide thickness generally does not exceed 0.01 mm for such products. Colored oxide films are produced by anodizing in mixed acid electrolytes the principal components of which are organic sulfonic acids and a small amount of sulfuric acid. Colors range from pale yellow to dark bronze and black. The color results from modification of the oxide and inclusion of intermetallic particles, metal, and reaction products in the oxide. Such coatings are widely employed in architectural applications for aluminum and are usually 0.02–0.03 mm thick.

Thin, porous anodic films provide an excellent base for paint (qv) coatings (qv), as do certain chemical and mechanical treatments. These are employed to disrupt the natural oxide film on aluminum and eliminate contaminants that would interfere with wetting of the surface by organic coatings. Chemical conversion coatings convert a portion of the base alloy to one of the components of a chemical film which is then integral with the metal surface. Carbonate, chromate, and phosphate films are produced in this manner and have excellent adhesion qualities. Chemically produced metallic coatings (qv) are also used extensively to provide a base for organic coatings. The zinc immersion coating is the most widely employed of the chemical displacement films.

Zinc, tin, and nickel immersion coatings are used as a base for electroplating. Anodizing and mechanical roughening treatments are also employed. Good surface preparation is required for good adherence of electrodeposits, which can be of any commonly electroplated metal. Electroplating conditions and solutions are the same as those for steel and other metals (see METAL SURFACE TREATMENTS). Deposition of thin films of oxides or other ceramics by vapor deposition or other means is beginning to be commercially applied to aluminum products (see FILM DEPOSITION TECHNIQUES).

Porcelain enameling requires the use of frits and melting temperatures of 550°C or below. Enamels are applied over chemical conversion coatings that are compatible with the frit. Alloy selection is important to obtain good spall resistance. Alloys 1100, 3003, and 6061 are employed most extensively among wrought products and alloy 356 for castings.

16. Uses

Transportation accounted for 35% of U.S. consumption in 2001, packaging, 25% building 15%, consumer durables, 8% electrical, 7%, and other, 10% (34). Table 28 gives data on U.S. shipments by market.

16.1. Containers and Packaging. Aluminum packages exploit the barrier properties of aluminum (impervious to light and gas transmission) and corrosion resistance of selected alloys. By the end of 1999, the steel beverage can had been virtually eliminated from the market in the United States. Worldwide approximately 180 billion aluminum cans are produced each year, with more than 1/2 of that consumption in North America. The two-piece aluminum can has been continuously improved to reduce metal content and maintain performance and reliability in the demanding beer and soft drink market. Alloys 3004 (can body) and 5182 (can lid) have also been optimized to improve can making efficiency with thinner sheet. Aluminum convenience food cans for items

Table 28. **U.S. Shipments of Aluminum by Market,** $\times 10^3$ [a]

market	1975	1980	1985	1990	1995	1999
building and construction	1,019	1,189	1,375	1,208	1,215	1,468
transportation	775	1,012	1,383	1,454	2,608	3,601
containers and packaging	914	1,512	1,865	2,165	2,308	2,316
consumer durables	342	398	484	509	621	˙760
electrical	552	633	632	595	630	739

[a] Aluminum Statical Review, 1980–1999

such as meat, pudding and fish are also produced in large quantities. Other aluminum rigid containers are the aerosol can, produced from an impact extrusion process and formed containers for frozen foods, pie plates, and baked goods.

Aluminum foil continues to be an important portion of the packaging business. Foil may be produced as household wrap ("Reynolds Wrap") or used in a laminate with paper or plastic for packaging of cereal, frozen foods or pharmaceutical products.

16.2. Building and Construction. Despite the growth of other markets for aluminum products, the worldwide building and construction industry continues to consume vast quantities of aluminum in a wide variety of applications. Good resistance to weather, adequate strength, light weight, and the ability to accept a wide range of finishes are advantages for aluminum in these applications. Alloys such as 3003, 3004, and 3105 produced as sheet are used for residential siding, gutters, industrial and farm roofing, and siding. Alloy 6063 and 6060 extrusions are typically used for both structural and decorative door and window frame structures. Aluminum wire and rod are used to fabricate nails, screen wire, and many types of fasteners. Anodized and/or painted panels of many colors and textures are used for building panels and store fronts.

16.3. Transportation. Aluminum serves the worldwide transportation market for aerospace, automotive and truck, railroad, and marine transportation. Aluminum's low density, moderate cost, high damage tolerance, and wide range of product forms and tempers in comparison to other materials makes it ideal for most types of vehicles where life cycle costs are an issue.

Aerospace. The high strength/weight ratio and relatively moderate cost of aluminum alloys still make them attractive for aircraft construction. The heat treatable 2xxx and 7xxx alloys are used bor both supporting structure and skins for wings, fuselages and other components of military and commercial airplanes. Weldable alloys are used for construction of large fuel tanks for the Space Shuttle. The choice of alloy, temper, and final product depend upon the mechanical properties required for the initial design and the expected service conditions over the life of the aircraft. Static strength and fracture toughness, corrosion resistance, and response to cyclic (fatigue) loads and corrosion are all considered in selecting the appropriate aluminum product for the application.

Ground Transportation. Aluminum's biggest growth has been in the use of cast components (engines, wheels, cylinder heads) for cars and trucks. Side panels of aluminum sheet and floors made from extrusions help to save weight in the truck trailer market. The use of aluminum in the typical family vehicle grew from 75 kg in 1991 to more than 11 kg in 2000 (56). New alloys with higher strength after paint baking and improved formability have made significant inroads into the sheet metal applications such as hoods and deck lids of automobiles. Brazed heat exchangers are replacing copper radiators in large quantities for automotive applications. Aluminum radiators and other heat exchangers may be produced in a variety of ways, but typically use a special clad brazing sheet and formed aluminum foil (fin stock). On the high end of the market, high performance luxury and sports vehicles have been produced which exploit the advantages of various aluminum product forms: sheet extrusions, and castings. When designed and joined properly the weight savings and performance of these all aluminum vehicles are truly impressive. Most important, however, is the experience gained in producing these vehicles that will be put to use in higher volume models in future years.

Railroad and Marine.   Aluminum continues to be designed into new railroad vehicles either for hauling commodities such as coal or high performance cars for passenger trains. The design and construction of new high speed "fast ferries" for use in the Pacific regions have raised the demand for higher performance welded 5xxx aluminum sheet and plate structures. Large 5xxx plates up to 200 mm thick are used for the construction of LNG (Liquefied Natural Gas) tanker ships. Aluminum canoes, pleasure craft, and deck houses for naval vehicles also employ significant amounts of aluminum sheet, plate, and extrusions.

Other.   Aluminum is used in the home as cooking utensils (the first commercial use of aluminum), refrigerators, air conditioners, satellite dishes, appliances, insect screening, and hardware. It is also for toys, sporting equipment, lawn furniture, lawn mowers, and portable tools. In many of these applications, anodized or colored coatings are employed for decorative purposes.

Aluminum is an excellent conductor of electricity, having a volume conductivity 62% of that of copper. Because of the difference in densities of the two metals, an aluminum conductor weighs only half as much as a copper conductor of equal current carrying capacity. Because of its lightness, aluminum has been used extensively for overhead transmission lines. Both standed aluminum conductors and steel core stranded conductors are employed. Alloys include conductor grades 1350 and 6201 along with newer compositions. Aluminum is also used as cable sheathing, replacing the neutral in three-phase systems. Aluminum is used in insulated cable, for bus bars, and for transformer windings, magnet wire, building wire, and 99.99% aluminum is used to connect integrated circuits. An aluminum-air fuel cell has exhibited a high energy output, which results from the characteristic energy density of aluminum. These cells can be used in electric vehicles, emergency power sources, and portable electronic devices (57).

Aluminum has many applications in the chemical and petrochemical industries such as for piping and tanks in alloys 1100, 3003, 6061, 6063, and the Al-Mg alloys. Aluminum is chosen for its resistance to corrosion by the chemicals involved as well as for its mechanical properties, light weight, and thermal conductivity. Aluminum is also used in the storage and packaging of chemicals. Applications range from large Al-Mg alloy tanks for the storage of ammonium nitrate and liquefied natural gas to collapsible tubes for dispensing pharmaceuticals and toiletries. The absence of any harmful reaction with the microorganisms involved in the manufacture of pharmaceuticals is obviously important in this application of aluminum.

BIBLIOGRAPHY

"Aluminum and Aluminum Alloys" in *ECT* 1st ed., Vol. 1, pp. 591–623, by J. D. Edwards and F. Keller, Aluminum Research Laboratories, Aluminum Company of America; in *ECT* 2nd ed., Vol. 1, pp. 929–990, by P. Vachet, Cie de Produits Chimiques et Électrométallurgiques, Péchiney; in *ECT* 3rd ed., Vol. 2, pp. 129–188, by W. A. Anderson and W. E. Haupin, Aluminum Company of America; in *ECT* 4th ed., Vol. 2, pp. 184–251 by James T. Stally and Warren Haupin, Aluminum Company of America; "Aluminum and

Aluminum Alloys" in *ECT* (online), posting date: December 4, 2000, by James T. Stally and Warren Haupin, Aluminum Company of America.

CITED PUBLICATIONS

1. J. Hatch, ed., *Aluminum: Properties and Physical Metallurgy*, American Society for Metals, Metals Park, Ohio, 1984.
2. E. Gebhardt, M. Becker, and A. Dorner, *Z. Metallkd.* **44**, 573 (1953).
3. M. W. Chase, Jr. and co-workers, eds., *JANAF* Thermochemical Tables, 3rd ed., *J. Phys. Chem. Ref. Data.* **14** (1985).
4. R. A. Andrews, R. T. Webber, and P. A. Spohr, *Phys.Rev.* **84**, 994 (1951).
5. C. C. Bidwell and C. L. Hogan, *J. Appl. Phys.* **18**, 776 (1947).
6. T. E. Pochapsky, *Acta Metall.* **1**, 747 (1953).
7. A. Roll and H. Motz, *Z.Metallkd.* **48**, 272 (1957).
8. G. Hass and J. E. Waylonis, *J. Opt. Soc. Am.* **51**, 719 (1961).
9. R. P. Madden, L. R. Canfield, and G. Hass, *J. Opt. Soc. Am.* **53**, 620 (1963).
10. H. E. Bennett, M. Silver, and E. J. Ashley, *J. Opt. Soc. Am.* **53**, 1089 (1963).
11. C. E. Ransley and H. Neufield, *J. Inst. Met.* **74**, 599 (1947–1948).
12. W. Eichenauer, K. Hattenbach, and Z. Pebles, *Z. Metallkd.* **52**, 682 (1961).
13. T. G. Pearson, *The Chemical Background of the Aluminum Industry,* Monogr. 3, The Royal Institute of Chemistry, 1955.
14. P. A. Foster, Jr. *J. Am. Ceram. Soc.* **43(8)**, 437 (1960).
15. *Ibid.*, **58(7–8)**, 288 (1975).
16. A. Rostum, A. Solheim, and A. Sterten, *Light Metals 1990,* Minerals, Metals, and Materials Society, Warrendale, Pa., 1990, 311–323.
17. E. W. Dewing, *Can Metall. Q.* **13(4)**, 607 (1974).
18. S. K. Ratkje and T. Forland, *Light Met.* **1**, 223 (1976).
19. *Joint Army-Navy-Air Force Thermochemical Tables,*2nd ed., National Bureau of Standards, Washington, D.C., 1971.
20. U.S. Pat. Appl. 2002/0056650 (May 16, 2002), S. P. Ray (to Alcoa, Inc.).
21. C. N. Cochran and co-workers, *Use of Aluminum in Automobiles—Effect on the Energy Dilemma,* SAE Paper 750421, Society of Automotive Engineers, Warrendale, Pa., Feb. 1975.
22. A. Wrigley, *Am. Metal Market*, **108**(7), 6 (Jan. 12, 2000).
23. P. T. Stroup, *Trans. Metall. Soc. AIME* **230**, 356 (Apr. 1964).
24. A. F. Saavedra, C. J. McMinn, and N. E. Richards, in H. Y. Sohn and E. S. Geskin, eds., *Metallurgical Processes for the Year 2000 and Beyond*, Minerals, Metals, and Materials Society, Warrendale, Pa., 1988, 517–534.
25. A. Mote and co-workers, *Denki Kagaku* **55(9)**, 676 (1987).
26. *Can. Chem. Process.* **51(2)**, 45; **(3)**, 75 (1967).
27. J. F. Elliot, *Phys. Chem. of Carbothermic Reduction of Aluminum,* DOE/ID/12467-3 (DE89015201), June **16**, 1989.
28. L. M. Foster, G. Long, and M. S. Hunter, *J. Am. Chem. Soc.* **21**, 1 (1956).
29. A. F. Saavedra and R. M. Kibby, *J. Met.* **40**, 32–36 (Nov. 1988).
30. H. Ginsberg, *Aluminum Düsseldorf* **23**, 131 (1941).
31. L. Evans and W. B. C. Perrycoste, *B. I. O. S. Final Report No 1757, Item. No. 21* Manufacture of Super-Purity Aluminum at the Vereinigte Aluminum Werke, Erftwerk, Grevenbroich, UK, March and July 1946.
32. W. G. Pfann, *J.Met.* **4**, 747 (July 1952).
33. U.S. Pat. 3,303,019 (Feb. 7, 1967), S. C. Jacobs (to Aluminum Company of America).

34. P. A. Plunkert, "Aluminum," *Mineral Commodity Summaries*, U.S. Geological Survey, Reston, Va., Jan. 2002.
35. P. A. Plunkert, M. George, and L. Roberts, "Aluminum" *Minerals Yearbook*, U.S. Geological Survey, Reston, Va., 2000.
36. *Nonferrous Metal Data 1960–1999* American Bureau of Metal Statistics, New York, 1960–1999.
37. *ASTM Annual Book of Standards*, Vol. 03.06, Sect. 3, American Society for Testing and Materials, Philadelphia, Pa., 1989.
38. "Methods of Chemical and Spectrochemical Analysis of Aluminum" in *Light Metals and Their Alloys,* Technical Committee ISO/TC 79, International Organization for Standardization, Geneva, Switzerland, 1979.
39. K. Grjotheim, H. Kvande, K. Motzfeldt, and B. J. Welch, *Can. Metall. Q.* **11(4)**, 585 (1972).
40. C. N. Cochran, W. C. Sleppy, and W. B. Frank, *J. Met.* **22(9)**, 54 (1970).
41. Aluminum Association of America, Washington, D. C., press release, April 19, 2001.
42. B. Dinman, "Aluminum," in E. Bingham, B. Cohrssen, and C. H. Powell, eds., *Patty's Toxicology*, 5th ed., John Wiley & Sons, Inc., New York, 2001 Chapt. 31.
43. J. T. Staley, in A. K. Vasudévan and R. D. Doherty, eds., *Aluminum Alloys: Contemporary Research and Applications,* Academic Press, San Diego, Calif., 1989, pp. 3–31.
44. T. Massalski, ed., *Binary Alloy Phase Diagrams,* Vol. 1, American Society for Metals, Metals Park, Ohio, 1986.
45. H. M. Flower and P. J. Gregson, *Mater. Sci. Tech.* **33**, 81 (1987).
46. H. K. Hardy and J. M. Silcock, *J. Inst. Met.* **84**, 423 (1955).
47. L. F. Mondolfo, *Aluminum Alloys: Structure and Properties*, Butterworths, London, 1976.
48. I. J. Polmear, in E. A. Starke, Jr. and T. H. Sanders, Jr., eds., *Aluminum Alloys: Their Physical and Mechanical Properties,* 1986, pp. 661–674.
49. J. R. Pickens, F. H. Heubaum, T. J. Langan, and L. S. Kramer in T. H. Sanders, Jr. and E. A. Starke, Jr., eds., *Proc. 5th Int. Conf. on Al–Li Alloys,* Vol. 3, Williamsburg, Va., 1989, 1397–1414.
50. A. Garg, Y. C. Chang, and J. Howe, *Scrip. Met.* **24**, 677 (1990).
51. R. F. Ashton, I. Broveman, P. R. Sperry, and J. T. Staley in Ref. 1, p. 180.
52. P. N. Anyalebechi, E. J. Talbot, and D. A. Granger in E. W. Lee, E. H. Chia, and N. J. Kim, eds., *Light Weight Alloys for Aerospace Applications,* The Minerals, Metals, and Materials Society, Warrendale, Pa., 1989.
53. D. A. Granger in A. K. Vasudévan and R. D. Doherty, eds., *Aluminum Alloys: Contemporary Research and Applications,* Academic Press, San Diego, Calif., 1989, pp. 109–135.
54. J. T. Staley, R. J. Rioja, R. K. Wyss, and J. Liu, *Processing to Improve High Strength Aluminum Alloy Products*, 9th Int. Conf. on Production Research, Cincinnati, Ohio, 1987.
55. R. J. Bucci, L. N. Mueller, L. B. Vogelsang, and J. W. Gunnick in Ref. 51, pp. 295–322.
56. A. Wrigley, *Am. Metal Market*, **108**(38), 6 (Feb. 28, 2000).
57. *Am. Metal Market*, **108**(133), 4 (July 12, 2000).

Robert E. Sanders, Jr.
Alcoa Technical Center

ALUMINUM COMPOUNDS, SURVEY

1. Introduction

Aluminum [7429-90-5], atomic number 13, atomic weight 26.981, is, at 8.8 wt%, the third most abundant element in the earth's crust. It is usually found in silicate minerals such as feldspar [68476-25-5], clays, and mica [12001-26-2]. Aluminum also occurs in hydroxide, oxide–hydroxide, fluoride, sulfate, or phosphate compounds in a large variety of minerals and ores.

The CAS registry lists 5,037 aluminum-containing compounds exclusive of alloys and intermetallics. Some of these are listed in Table 1. Except for nepheline and alunite in the former USSR and Poland, bauxite is the raw material for all manufactured aluminum compounds. The term bauxite is used for ores that contain economically recoverable quantities of the aluminum hydroxide mineral gibbsite or the oxide–hydroxide forms boehmite and diaspore.

World bauxite production in 2000 totaled about 136×10^6 t, approximately 85% of which was refined to aluminum hydroxide by the Bayer process. Most of the hydroxide was then calcined to alumina and consumed in making aluminum metal. An additional 10% was used in nonmetallurgical applications in the form of speciality aluminum, 5% was used for nonmetallurgical bauxite applications such as abrasives, refractories, cement additives, and aluminum chemicals (1). Other uses included catalysts used in petrochemical processes and automobile catalytic converter systems (see PETROLEUM; EXHAUST CONTROL, AUTOMOTIVE); ceramics that insulate electronic components such as semiconductors and spark plugs; chemicals such as alum, aluminum halides, and zeolite; countertop materials for kitchens and baths; cultured marble; fire-retardant filler for acrylic and plastic materials used in automobile seats, carpet backing, and insulation wrap for wire and cable (see FLAME RETARDANTS); paper (qv); cosmetics (qv); toothpaste manufacture; refractory linings for furnaces and kilns; and separation systems that remove impurities from liquids and gases.

Aluminum compounds, particularly the hydroxides and oxides are very versatile. Properties range from a hardness indicative of sapphire and corundum to a softness similar to that of talc [14807-96-6] and from inertness to marked reactivity. Aluminas that flow and filter like sand may be used for chromatography (qv); others are viscous, thick, unfilterable, and even thixotropic (2).

2. Bauxite Occurence

The term bauxite originates from the location of the deposit discovered in 1821 near the village of Les Baux in Provence, France. Bauxite is weathered rock consisting mainly of aluminum hydroxide minerals but having small and variable amounts of silica [7631-86-9], the iron oxides hematite [1309-37-1], Fe_2O_3, and magnetite [1317-61-9], Fe_3O_4, rutile or titanium oxide [1317-80-2], and alumina silicate clays. Deposits were formed over many geologic time periods in a tropical or subtropical climate having enough rainfall and circulating groundwater to

Table 1. **Aluminum Compounds Referred to in Text**

Compounds	CAS Registry Number	Molecular formula
alum	[7784-24-9]	$KAl(SO_4)_2 \cdot 12H_2O$
alumina	[1344-28-1]	Al_2O_3
aluminum bromide	[77727-15-3]	$AlBr_3$
aluminum chlorhydroxide (ACH)	[12042-91-0]	$Al_2Cl(OH)_5$
aluminum(I) chloride	[13595-81-8]	$AlCl$
aluminum(III) chloride	[7446-70-6]	$AlCl_3$
aluminumchloride hexahydrate	[7784-13-6]	$AlCl_3 \cdot 6H_2O$
aluminum(I) fluoride	[13595-82-9]	AlF
aluminum(III) fluoride	[7784-18-1]	AlF_3
aluminum hydroxide	[21645-51-2]	$Al(OH)_3$
aluminum iodide	[7784-23-8]	AlI_3
aluminum(II) oxide	[14457-64-8]	AlO
aluminum silicate	[12141-46-7]	$Al_2(SiO_3)_3$
aluminum sulfate	[10043-01-3]	$Al_2(SO_4)_3$
aluminum sulfate octadecahydrate	[7784-31-8]	$Al_2(SO_4)_3 \cdot 18H_2O$
alunite	[12588-67-9]	$K_2Al_6(SO_4)_4(OH)_{12}$
anorthite	[1302-54-1]	$CaO \cdot Al_2O_3 \cdot 2SiO_2$
bauxite	[1318-16-7]	
boehmite	[1318-23-6]	$AlO(OH)$
calcium aluminate	[12042-78-3]	$Al_2O_3 \cdot 3CaO$
corundum	[1302-74-5]	$\alpha\text{-}Al_2O_3$
diaspore	[14457-84-2]	$\alpha\text{-}AlO(OH)$
gibbsite	[14762-49-3]	$\alpha\text{-}Al(OH)_3$
halloysite	[12244-16-5]	$Al_2Si_2O_5(OH)_4 \cdot 2H_2O$
kaolin	[1332-58-7]	$H_2Al_2Si_2O_8 \cdot H_2O$
kaolinite	[1318-74-7]	$Al_2O_3 \cdot 2SiO^2 \cdot 2H_2O$
kyanite	[1302-76-7]	$H_6O_5Si \cdot 2Al$
montmorillonite	[1318-93-0]	
nepheline	[12251-27-3]	$NaAl(OH)SiO_3$
nepheline	[12251-28-4]	$NaAl_2(OH)_2(SiO_3)_2 \cdot H_2O$
nepheline	[14797-52-5]	AlH_4O_4Si
sapphire	[1317-82-4]	Al_2O_3
sodium aluminate	[1302-42-7]	$NaAlO_2$
triethylaluminum	[97-93-8]	$(C_2H_5)_3Al$
triisobutylaluminum	[100-99-2]	$(C_4H_9)_3Al$
zeolite A	[1318-02-1]	$Na_{12}[(Al_{12}Si_{12})O_{48}] \cdot 27H_2O$
zeolite Y		$Na_{56}[(AlO_2)_{56}$ $(SiO_2)_{136}] \cdot 250H_2O$
zeolite X		$Na_{86}[(AlO_2)_{86}$ $(SiO_2)_{106}] \cdot 264H_2O$

dissolve constituents of the parent rock and carry them away. Weathering is intensified by good drainage and by decaying vegetation, which makes the process acidic. Table 2 gives typical compositions for bauxites found throughout the world.

Uses of bauxite other than for aluminum production are in refractories, abrasives, chemicals, and aluminous cements. Bauxites for these markets must meet more rigid compositional requirements with respect to Fe_2O_3, SiO_2, and TiO_2 content that those used for alumina production.

Table 2. Composition of Bauxite Used for Alumina Production

Location	Constituents, wt %				Loss on ignition (LOI), %	Mineral components
	Al_2O_3	SiO_2	Fe_2O_3	TiO_2		
Australia						
Weipa	54.8	5.3	5.20			gibbsite, some boehmite
Gove	50	3.4–4.2	17.1	3.4	26.4	gibbsite, some boehmite
Darling Range	30–35	0.3–2.0	10–25			gibbsite, trace boehmite
Brazil						
Trombetas	55.9	4.8	9.4	1.3	28.6	gibbsite, boehmite
France	53.0	7.8	21.4	2.6	13.3	boehmite
Greece	57.6	3.0	22.8	2.75	12.17	boehmite, some diaspore
Guinea						
Boke-Sangaredi	59.1–59.6	0.7–0.9	4.9–5.9	3.3–3.5	30.6–31.0	83–86% gibbsite, 3.5–5.5% boehmite
Fria	45.5	4.0	23.6	2.5	23.9	gibbsite
Guyana	55–61	1–10	0.8–5	2–5	30–35	nearly pure gibbsite
Hungary	50–60	1–5				boehmite, some gibbsite
India						
Ranchi	51–60	0.1–5	4–10	0.3–17	22–28	gibbsite, some boehmite
East Coast province	43.7–56.5	0.5–4.2	8.6–38.4	2.1–3.5	24.6–30.5	gibbsite, boehmite
Jamaica	49.1–50.6	0.7–6.1	18.9–20	2.5–2.7	24.6–27.3	gibbsite, 7–10% boehmite
Surinam	58.5–60	3.4–4.3	2.7–4.4	2.4–2.7	30.7–31.4	nearly pure gibbsite
United States	45–50	13	8	2.5–3	25±	gibbsite, SiO_2 is mainly in kaolinite
Former USSR	45–55	2–10	5–15			boehmite; somemixed with diaspore; a few are gibbsite
Yugoslavia	56–58	3–5	20–22	2.5–2.7	13.0	boehmite, SiO_2 in kaolinite

Table 3. **World Bauxite Mine Production, Reserves, and Reserve Base, 10^3 t**[a]

	Mine production			Reserve base 10^3 t[a]
	2000	2001[b]	Reverves	
United States	NA	NA	20,000	40,000
Australia	53,800	53,500	3,800,000	7,400,000
Brazil	14,000	14,000	3,900,000	4,900,000
China	9,000	9,200	720,000	2,000,000
Guinea	15,000	15,000	7,400,000	8,600,000
Guyana	2,400	2,000	700,000	900,000
India	7,370	8,000	770,000	1,400,000
Jamaica	11,100	13,000	2,000,000	2,500,000
Russia	4,200	4,000	200,000	250,000
Suriname	3,610	4,000	580,000	600,000
Venezuela	4,200	4,400	320,000	350,000
other countries	10,800	10,200	4,100,000	4,700,000
World total (rounded)	135,000	137,000	24,000,000	34,000,000

[a] Ref. 3.
[b] Estimated

The largest world bauxite producers in Australia, Brazil, Guinea, and Jamaica. They accounted for 70% of the total bauxite mined in 2000 and 2001. See Table 3 for world bauxite mine production, reserves, and reserve base (3).

Bauxite reserves are estimated to be $55-75 \times 10^9$ t, located in South America (33%), Africa (27%), Asia (17%), Oceania (13%), and elsewhere (10%) (3).

Bauxite exists in many varieties. The physical appearance ranges from earthy dark brown ferruginous material to cream or light pink-colored layers of hard, crystalline, gibbsitic bauxite. Ores composed chiefly of gibbsite, α-Al(OH)$_3$, are commonly termed trihydrate bauxites or the Surinam type; those composed of boehmite, AlO(OH), are called monohydrate bauxite, or the European type; and those composed of a mixture of gibbsite and boehmite are referred to as mixed bauxites. The term Jamaica type is applied to very fine-grained gibbsitic bauxite ore containing <10% boehmite. Diaspore, α-AlO(OH), is the major constitutent of bauxites in Greece, Romania, Russia, and China.

Iron and titanium minerals commonly found in bauxites are hematite, goethite [1310-14-1], FeO(OH), siderite [14476-16-5], FeCO$_3$, magnetite, ilmenite [12168-52-4], FeTiO$_3$, anatase [1317-70-0], TiO$_2$, and rutile. Silicon dioxide may occur as quartz [14808-60-7], but it is silica associated with the clay (qv) minerals kaolite, halloysite, or montmorillonite that is important in processing bauxite for alumina production. These aluminosilicates react during the extraction stage of alumina production to form insoluble sodium aluminum silicates resulting in losses of NaOH and Al$_2$O$_3$. The amount of this reactive silica is one of the bauxite quality and price-determining factors (4). Some ores, Brazilian Trombetas, for example, are upgraded by washing some of the fine kaolin away.

Hardness, texture, and the amount of overburden determine the methods for mining bauxite. Tropical gibbsitic bauxites are usually located so close to the earth's surface that strip mining can be used. However, European bauxites frequently require mining to depths of several hundred meters. The Bayer process, patented in 1888 (5), involved hot leaching of bauxite withNaOH solution in pressure vessels to obtain a supersaturated sodium aluminate solution from which $Al(OH)_3$ was precipitated by seeding. This process achieved immediate industrial success, replacing the soda ash roasting of bauxite used to form leachable sodium aluminate. Development of the Bayer process also coincided with the rapid increase in demand for pure alumina to be used in metal production by the Hall-Héroult process discovered in 1886 (see ALUMINUM AND ALUMINUM ALLOYS).

3. Chemical Properties

The ground state distribution of electrons in the aluminum atom is $1s^2 2s^2 2p^6 3s^2 3p^1$. The oxidation state of aluminum is +3, except at high temperatures where monovalent species such as AlCl, AlF, and Al_2O have been spectrally identified. At lower temperatures, these compounds disproportionate

$$3\,Al_2O \longrightarrow Al_2O_3 + 4\,Al$$

Aluminum, although highly electropositive, does not react with water under ordinary conditions because it is protected by a thin (2–3 nm) impervious oxide film that rapidly forms even at room temperature on nascent aluminum surfaces exposed to oxygen. If the protective film is overcome by amalgamation or scratching, water rapidly attacks to form hydrous aluminum oxide. Because of the tendency to amalgamate, aluminum and its alloys should not be used in contact with mercury or its compounds. Molten aluminum (mp 660°C) is known to react explosively with water (16). Thus the molten metal should not be allowed to touch damp tools or containers.

There is much discussion on the nature of the aluminum species present in slightly acidic and basic solutions. There is general agreement that in solutions below pH 4, the mononuclear Al^{3+} exists coordinated by six water molecules, ie, $[Al(H_2O)_6]^{3+}$. The strong positive charge of the Al^{3+} ion polarizes each water molecule and as the pH is increased, a proton is eventually released, forming the monomeric complex ion $[Al(OH)(H_2O)_5]^{2+}$. At about pH 5, this complex ion and the hexahydrated Al^{3+} are in equal abundance. The pentahydrate complex ion may dimerize by losing two water molecules

$$2\left[Al(OH)(H_2O)_5\right]^{2+} \longrightarrow \left[Al_2(OH)_2(H_2O)_8\right]^{4+} + 2\,H_2O$$

Further deprotonation, dehydration, and polymerization of monomers and dimers may yield ringlike structures of hydroxy–aluminum complexes(17). Coalescence of ring compounds into layers by further growth results in the formation of crystalline aluminum hydroxide at pH 6, the point of minimum aqueous solubility.

Highly purified aluminum, 99.999% Al, is resistant to attack by most acids but can be dissolved in aqua regia or in hydrochloric acid containing a trace of

$CuCl_2$. Addition of hydrogen peroxide [7722-84-1], H_2O_2, during dissolution in HCl speeds the process (18). Typical commercial aluminum, 99.85 to 99.95% Al, is soluble in dilute mineral acids but is passivated by concentrated nitric acid. Dissolution in acids is accompanied by evolution of hydrogen and hydration of the aluminum ion

$$2\,Al + 6\,H_3O^+ + 6\,H_2O \longrightarrow 2\,\left[Al(H_2O)_6\right]^{3+} + 3\,H_2$$

Because of its amphoteric nature, aluminum is also readily attacked by solutions of strong bases. At a pH >8.5, the solubility of aluminum increases sharply because of the formation and hydration of $Al(OH)_4^-$ ions

$$2\,Al + 2\,OH^- + 10\,H_2O \longrightarrow 2\,\left[Al(OH)_4(H_2O)_2\right]^- + 3\,H_2$$

In high caustic Bayer liquor, $Al(OH)_4^-$ ions exist because there is not enough water to hydrate them.

Aluminum hydroxide is capable of reacting as either an acid or a base (18).

$$Al(OH)_3(s) \longrightarrow Al^{3+} + 3\,OH^-$$

$$Al(OH)_3(s) \longrightarrow AlO_2^- + H^+ + H_2O$$

and the hydroxide is readily soluble in both acids and strong bases

$$Al(OH)_3 + 3\,H^+ + 3\,H_2O \longrightarrow \left[Al(H_2O)_6\right]^{3+}$$

$$Al(OH)_3 + OH^- + 2\,H_2O \longrightarrow \left[Al(OH)_4(H_2O)_2\right]^-$$

Calcined alumina, α-Al_2O_3, and naturally occurring corundum are practically insoluble in acids and bases, but partially calcined and low temperature amorphous oxide, such as that which forms on nacent commercial aluminum surfaces, is soluble

$$Al_2O_3 + 6\,H_3O^+ + 3\,H_2O \longrightarrow 2\,\left[Al(H_2O)_6\right]^{3+}$$

$$Al_2O_3 + 2\,OH^- + 7\,H_2O \longrightarrow 2\,\left[Al(OH)_4(H_2O)_2\right]^-$$

The amphoteric nature of the oxide is illustrated by its ability to form silicates and aluminates in the dry state at elevated temperatures

$$Al_2O_3 + 3\,SiO_2 \longrightarrow Al_2(SiO_3)_3$$

$$Al_2O_3 + CaO \longrightarrow Ca(AlO_2)_2$$

Aluminum also reacts vigorously with the halogens to form trihalides

$$2\,Al + 3\,X_2 \longrightarrow 2\,AlX_3$$

4. Bauxite Production

Figure 1 illustrates the Bayer process as it is practiced. The primary purpose of a Bayer plant is to process bauxite to provide pure alumina for the production of aluminum. World production of alumina totaled ca 49×10^6 t in 2000 (1). Practically all of the hydroxide used was obtained by Bayer processing and 85% of it was calcined to metallurgical grade alumina (Al_2O_3). However, about 15% of the bauxite processed serves as feedstock to the growing aluminum chemicals industry.

Bauxite grade, mineralogy, composition, uniformity and location of the ore deposit, site infrastructure, availability of low cost energy, and designer bias, as well as legislative and environmental restrictions all affect Bayer plant design.

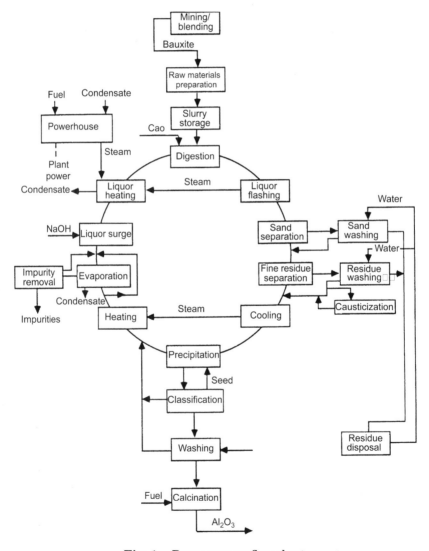

Fig. 1. Bayer process flow sheet.

All bauxite refineries share six common process steps: bauxite mining; raw material preparation; bauxite digestion; separation, washing, and disposal of insoluble bauxite residue; aluminum hydroxide (trihydrate) precipitation; and calcination to anhydrous alumina.

Additional operations essential to commercial bauxite processing are steam and power generation, heat recovery to minimize energy consumption, process liquor evaporation to maintain a water balance, impurity removal from process liquor streams, classification and washing of trihydrate, lime caustication of sodium carbonate to sodium hydroxide, repair and maintenance of equipment, rehabilitation of mine and residue disposal sites, and quality and process control. Each operation in the process can be carried out in a variety of ways depending upon bauxite properties and optimum economic tradeoffs.

4.1. Mining and Ore Preparation. Bauxite mining practice is dictated by the nature of the ore body. Blending operations, physical beneficiation (washing out clay slurries), and bauxite drying are used if the ore body is not uniform, contains an excessive amount of kaolin (clay) that would consume caustic, or is difficult to handle because of its moisture content (see CLAYS). Grinding is designed to produce feed material small enough to ensure easy alumina extraction yet coarse enough to avoid clarification problems with bauxite residue. Uniform, consistent, easily digested bauxite slurry is formed by blending properly ground bauxite slurry in slurry storage "surge" tanks prior to digestion.

4.2. Digestion. Digestion extracts and solubilizes the available aluminum minerals from the bauxite. Digest conditions and equipment vary widely, depending on bauxite mineralogy, ie, the solution rates of the ore, and on achieving high productivity from a given liquor flow rate. In digestion, which is performed in steel vessels, autoclaves, or tubular reactors, hot spent liquor reacts with the aluminum minerals in the bauxite to form soluble sodium aluminate. Virtually all the other constituents are rejected as undissolved solids.

$$\mathrm{Al(OH)_3 + Na^+ + OH^- \longrightarrow Na^+ + Al(OH)_4^-} \tag{1}$$

$$\mathrm{AlO(OH) + Na^+ + OH^- + H_2O \longrightarrow Na^+ + Al(OH)_4^-} \tag{2}$$

The term spent liquor is used to distinguish recycled sodium hydroxide solution containing a low amount of dissolved sodium aluminate from the "green" or "pregnant" liquor leaving digestion that has a high sodium aluminate content. Liquor concentrations are expressed as g/L of a given constituent and depending upon the refinery location, the total caustic content, including that tied up as sodium aluminate, is expressed as g/L NaOH, Na_2CO_3, or sodium oxide Na_2O.

Gibbsitic bauxite, Equation1, is the most economical to process because of gibbsite's high solubility in Bayer process liquor at moderate temperature and pressure. Liquor containing 120–135 g/L Na_2O is used at about 140°C. Although boehmite and diaspore have the same chemical composition (AlOOH), diaspore is denser and harder than boehmite. Boehmitic bauxite requires temperatures from 200–250°C and pressures of 3.45 MPa (34 atm) or higher to obtain complete extraction and form high ratio green liquors. This ratio, which ultimately determines liquor productivity, is the g/L Al_2O_3 dissolved divided by the g/L Na_2O in solution. Processing diasporic bauxite requires stronger caustic solutions,

200–300 g/L Na_2O, in addition to higher temperatures and pressures. Whereas some bauxites may be composed entirely of one aluminous mineral phase, others may contain all three aluminum hydroxide minerals and the mineral most difficult to extract sets the digest conditions.

Other important reactions that occur in digestion are desilication, causticization of liquor, and precipitation of impurities. The reactive silica in bauxite, for example that in kaolin, reacts with caustic to form soluble sodium silicate [1344-09-8], Na_2SiO_3

$$Al_2O_3 \cdot 2\,SiO_2 \cdot 2\,H_2O + 6\,NaOH \longrightarrow 2\,NaAlO_2 + 2\,Na_2SiO_3 + 5\,H_2O \qquad (3)$$

which then reacts at digest temperature to form an insoluble sodium aluminum silicate known as desilication product (DSP), the most probable formula of which is $3\,Na_2O \cdot 3\,Al_2O_3 \cdot 6\,SiO_2 \cdot 2\,Na_2 X \cdot YH_2O$, where X can be any of a number of anions, including CO_3^{2-}, SO_4^{2-}, Cl^-, OH^-, and $Al(OH)^-_4$, all of which are present in Bayer liquors. Good desilication is effected by high temperatures and holding times and is essential for product purity.

Causticization, the reaction of hydrated lime [1305-62-0], $Ca(OH)_2$, with sodium carbonate to regenerate sodium hydroxide and precipitate calcium carbonate, is an important part of the Bayer process chemistry.

$$Na_2CO_3 + Ca(OH)_2 \longrightarrow CaCO_3 + 2\,NaOH \qquad (4)$$

Na_2CO_3 is formed in Bayer liquors by caustic degradation of the organics (humic acids) in bauxite and by absorption of CO_2 during exposure of process liquors to the atmosphere. Although poor lime efficiency and alumina losses during digestion as calcium aluminates have led to the practice of "outside" causticization of dilute green liquor flows in the residue washing area of the plant, digestion lime additions are still made to control impurities such as phosphorous pentoxide [1314-56-3], P_2O_5.

4.3. Clarification. Clarification is the term used to describe separation of bauxite residue solids from the supersaturated green liquor near its boiling point. Coarse particles, called sand because of the high silica content, are usually removed by cycloning followed by washing on sand classifiers prior to disposal. In most plants, the fine fraction of residue is settled in raking thickeners with the addition of flocculants to improve the clarity of thickener overflow. The concentrated thickener underflow is washed before disposal in countercurrent decantation washers (similar to the raking thickeners) or on vacuum drum-type filters, or a combination of both. Thickener overflow is filtered to remove the final traces of solids and ensure product purity. Kelly-type pressure filters are most widely used, but some plants use sand filters in which the liquor is filtered by gravity through a bed of properly sized sand. Filtered solids are removed from filter press cloth by hosing and are elutriated from the sand by backwashing.

4.4. Precipitation. Precipitation is the heart of the Bayer plant where recovery of the $Al(OH)_3$ from process liquors occurs in high yield and product quality is controlled. The dominant use for Bayer $Al(OH)_3$ is to calcine it into smelting grade alumina.

Cooling after digestion and clarification enhances the supersaturation of dissolved $Al(OH)_3$, but the solution remains metastable and autoprecipitation, which would result in losses during clarification, still is not rapid. The liquor is usually seeded with fine gibbsite seed from previous cycles to initiate precipitation. This reaction is the reverse of Equation 1, except that seed is present so that agglomeration and creation of new particles occur simultaneously. Maintaining a seed balance around precipitation and classification is important in achieving good yield and proper particle size, ie, the number of particles created in precipitation should equal the number leaving the system as product. Thus, nucleation, agglomeration, particle growth, and particle breakage need to be balanced. Problems arise because of the lack of a complete understanding of the interaction among variables such as specific precipitation rate (SPR), residence time, liquor concentrations, inorganic and organic impurities, mixed and split seeding, active vs poisoned seed area, seed retention, temperature profiles, and the effect on yield and product quality.

Precipitation can be continuous or batch. Modern plants use the continuous system where 10 to 14 flat-bottom, internally agitated tanks, approximately 30 m in height and 10–12 m in diameter, are placed in series so that flow of the liquor–seed slurry moves by gravity through launders connecting the tank tops.

Classification. Slurry leaving precipitation is classified into a coarse and one or more fine fractions, usually by elutriation in hydroclassifiers. Cyclones and combinations of hydroclassifiers and cyclones are gaining popularity. In smelting grade alumina plants, the coarse fraction, called primary product, is sent to calcination; the fine fractions, called secondary and tertiary seed, are recycled to be grown to product size.

4.5. Calcination. Calcination, the final operation in production of metallurgical grade alumina, is done either in rotary kilns or fluid bed stationary calciners at about 1100°C. Prior to calcination, the process liquor is washed from the $Al(OH)_3$ using storage tanks and horizontal vacuum filters. During heating, the trihydroxide undergoes a series of changes in composition and crystal structure but essentially no change in particle shape. The product is a white powder consisting of aggregates ranging in size from 20 μm to about 200 μm. Details of the calcination process are given in the literature (7,9,10).

4.6. Evaporation and Impurity Removal. Evaporation over and above that obtained in the cooling areas from flashed steam is usually required to maintain a water balance by removing dilution arising from residue and $Al(OH)_3$ washing, free moisture in the ore, injected steam, purge water, and uncontrolled dilutions. Evaporation also serves to concentrate impurities in the liquor stream such as sodium oxalate [62-76-0], $Na_2C_2O_4$, a product of organics degradation, making impurity removal easier.

4.7. Energy Conservation. Cogeneration of steam and power is usually an integral part of a Bayer plant, unless there is abundant low cost (usually hydroelectric) power available from an existing grid. The cogeneration feature enables fuel efficiencies of $\geq 85\%$ compared to ca 35% in public utilities power plants where turbine exhaust steam is merely condensed and recycled rather than being used for process heating.

Minimizing energy consumption per ton of alumina while maintaining a steam-power balance is an industry-wide, ongoing effort. Reduction of steam consumption has been limited by the cost of purchased power to compensate for loss of power generation.

In most plants heat recovery from the hot slurry leaving digestion is accomplished by flashing the slurry through a series of pressure vessels (flash tanks) to its atmospheric boiling point. The steam evolved accounts for a significant amount of the evaporation needed to maintain a water balance. It is used to heat the liquor returning to digestion before being recycled to the powerhouse boilers. Heat recovery from clarified liquor to precipitation is done in an analogous system, but because the liquor is cooled below its boiling point, flash vessels and heat exchangers must operate under vacuum.

4.8. Residue Disposal.   The major environmental problem in the Bayer process is disposal of bauxite residue which is effected by marine disposal, lagooning, use of underdrain lakes, or semidry disposal. Marine disposal in oceans or rivers, diluting the alkaline residue by large quantities of water, is environmentally unacceptable. Lagooning behind retaining dikes built around clay-sealed ground is commonly used, but there have been isolated leaks into aquifers. This has motivated installation of underdrains between the residue and clay-sealed, plastic-lined, lake bottom. This design removes the hydraulic head from the lake bottom and improves consolidation of the residue.

This is semidry disposal, sometimes referred to as dry-stacking, or the drying field method of disposal, takes advantage of the thixotropic nature of the residue. In this method, the residue is concentrated by vacuum filtration or other means to 35–50% solids. The percent solids required depends on residue rheology which varies as the source of the bauxite. Using agitation and/or additives, the viscosity of the concentrated slurry is reduced so it can be pumped to the disposal area where it flows like lava, establishing a gentle slope away from the discharge point. The slurry is called nonsegregating because neither water nor sand separate from it. In the absence of shear, the viscosity increases and flow stops. Any rainwater rapidly drains to the perimeter of the deposit. There is no free water on the surface of the impoundment, so the deposited residue dries and cracks whenever it is not raining. When the percent solids approaches 70–75%, bulldozers can work on the deposit. This method of disposal provides maximum storage for a given area and easy recovery if, in the future, any economical use for the residue is found. The disadvantages include a need for dust control and a separate cooling pond for the plant.

To return land once devoted to bauxite residue disposal to productive use, abandoned semidry disposal areas can be contoured and landscaped into surroundings capable of supporting homes and/or light industrial buildings. In the case of lagoons and lakes, the land has been returned to agricultural use. This has been accomplished, particularly in Australia, by dewatering (qv) abandoned lakes, covering them with several meters of neutral sand, and adding fertilizer to support crop growth.

4.9. Alternative Processes.   Bayer processing of bauxite is the most economical method for $Al(OH)_3$ and Al_2O_3 production; however, the lime–soda sintering process (9,11) is available to treat the high silica bauxites in the United States and nepheline in the former USSR. In the lime-soda process, leachable sodium aluminate is formed and silica is insolubilized by reaction with calcium. Research has shown that alumina production from nonbauxitic raw materials such as clay, anorthosite, alunite, coal waste, and oil shale (qv) has been technically feasible (12–14), but these technologies cannot compete with the Bayer process economically. The one that comes closest is acid (HCl) processing of clay (see CLAYS).

5. Bauxite Economics

Alumina, Al_2O_3, production capacity, is summarized in Table 4. The smaller plants produce products solely for chemical applications; the larger refineries

Table 4. **Alumina World Production, by Country** $\times 10^3$t [a,b,c,d]

Country	1996	1997	1998	1999	2000 (est)
Australia	13,348	13,385	13,853	14,532	15,681[e]
Azerbaijan (est)	5	10	(5)[f]	50 r[e]	200
Bosnia and Herzegovina (est)	50	50	50	50	50
Brazil	2,752	3,088	3,322	3,515 r	3,500
Canada	1,060	1,165	1,229	1,233	1,200
China (est)	2,550	2,940	3,330	3,840	4,330
France	440	454	450 (est)	400 (est)	400
Germany	755	738	600 r/ (est)	583 r/	700
Greece (est)	602[e]	602	600	600	600
Guinea (est)	640	650 [e]	480	500	550
Hungary	208	76	138	145 r/	150
India (est)	1,780	1,860	1,890	1,900	2,000
Ireland	1,234	1,273	1,200 (est)	1,200 (est)	1,200
Italy	881	913	930	973	950
Jamaica	3,200	3,394	3,440	3,570	3,600
Japan[g]	337	368	359	335 r/	340
Kazakhstan	1,083	1,095	1,085	1,152	1,200
Romania	261	282	250	277	417[e]
Russia	2,105	2,400 (est)	2,465	2,657	2,850
Serbia and Montenergro	186	160 (est)	153	156	250
Slovakia (est)	100	100	100	100	100
Slovenia	88	85	70 (est)	70 (est)	70
Spain[h]	1,095	1,110	1,100 (est)	1,200 (est)	1,200
Suriname (est)	1,600	1,600	1,600	– r/	–[e]
Turkey	159	164	157	159 r/	155[e]
Ukraine	1,000 (est)	1,080 (est)	1,291	1,230	1,360[e]
United Kingdom	99	100 (est)	96	90 r/	100
United States	4,700	5,090	5,650 r/	5,140 r/	4,780[e]
Venezuela	1,701	1,730	1,553	1,335	1,400
Total	*44,000*	*46,000*	*47,400 r/*	*47,000 r/*	*49,300*

[a] Ref. 3, est = estimated, r = revised,– = zero

[b] Figures represent calcined alumina or the total of calcined alumina plus the calcined equivalent of hydrate when available; exceptions, if known, are noted.

[c] World totals, U.S. data and estimated data are rounded to no more than three significant digits; may not add to totals shown.

[d] Table includes data available through July 25, 2001.

[e] Reported figure.

[f] Production sharply curtailed or ceased.

[g] Data presented are for alumina used principally for specialty applications. Information on aluminum hydrate for all uses is not adequate to formulate estimates of production levels.

[h] Hydrate.

Table 5. **Bauxite Salient Statistics—United States**[a]

	1997	1998	1999	2000	2001[b]
production, bauxite, mine	NA	NA	NA	NA	NA
imports of bauxite for consumption[c]	11,300	11,600	10,400	9,030	9,500
exports of bauxite[c]	97	108	168	147	100
shipments of bauxite from government stockpile excesses[c]	1,430	3,300	4,180	1,100	200
consumption, apparent, bauxite (and alumina) (in aluminum equivalents)[d]	4,210	5,000	4,870	3,870	3,200
price, bauxite, average value U.S. imports (f.a.s.) dollars per ton	25	23	22	23	24
stocks, bauxite, industry, yearend[c]	2,260	1,860	1,440	1,300	1,200
net import reliance,[e] bauxite (and alumina) as a percentage of apparent consumption	100	100	100	100	100

[a] Ref. 1, NA = not available, also includes Virgin Islands.
[b] Estimated.
[c] Includes all forms of bauxite, expressed as dry equivalent weights.
[d] The sum of U. S. bauxite production and net import reliance.
[e] Defined as imports-exports + adjustments for Government and industry stock changes (all in aluminum equivalents). Treated as separate commodities, the net import reliance equaled 100% for bauxite and 31% for alumina in 2001. For the years 1997–2000, the net import reliance was 100% for bauxite and ranged from 33% to 37% for alumina.

are almost fully devoted to production of metallurgical grade alumina. Production facilities are located adjacent to a bauxite reserve or in countries having the largest markets for the product.

The aluminum industry is a cyclic industry responding to global economic activity. The price of metal grade alumina ranged from about $165–440/t in 2000. Prices had started to decline in May 2000 (1). The largest cost elements in alumina production are bauxite, caustic, energy, and sustaining capital for production equipment. Some U.S. statistics for bauxite are given in Table 5 (1).

6. Commercially Significant Compounds

The aluminum containing compound having the largest worldwide market, estimated to be over 49×10^6 t in 2000, is metal grade alumina (15). Second, is aluminum hydroxide. The split between additive and feedstock applications for $Al(OH)_3$ (16) is roughly 50:50. Additive applications include those as flame retardants (qv) in products such as carpets, and to enhance the properties of paper (qv), plastic, polymer, and rubber products. Significant quantities are also used in pharmaceuticals (qv), cosmetics (qv), adhesives (qv), polishes (qv), dentifrices (qv), and glass (qv).

Feedstock applications of $Al(OH)_3$ for production of other chemicals include almost all of the 5000 plus compounds listed in the CAS registry.

6.1. Aluminum Sulfate (Alum). Alum, a double sulfate of potassium and aluminum having twelve waters of crystallization, $KAl(SO_4)_2 \cdot 12H_2O$, is the earliest referenced aluminum containing compound. It was mentioned by Herodotus in the fifth century BC. The Egyptians used alum as a mordant and as a medicine; the Romans used it for fireproofing. Some alums contain sodium or ammonium ions in place of potassium.

Aluminum sulfate, $Al_2(SO_4)_3 \cdot 18H_2O$, also known as alum cake, is industrially produced by reaction of $Al(OH)_3$ and sulfuric acid [7664-93-9], H_2SO_4, in agitated pressure vessels at about $170°C$. The commercial product has about 10% less water of hydration than the theoretical amount. Aluminum sulfate has largely replaced alums for the major applications as a sizing agent in the paper industry and as a coagulant to clarify municipal and industrial water supplies. In terms of worldwide production, it ranks third behind alumina and aluminum hydroxide, The U.S. exported 7,690 t of aluminum sulfate and imported 23,500 t of aluminum sulfate in 2000 (1).

6.2. Aluminum Halides. All the halogens form covalent aluminum compounds having the formula AlX_3. The commercially most important are the anhydrous chloride and fluoride, and aluminum chloride hexahydrate.

Anhydrous aluminum chloride, $AlCl_3$, is manufactured primarily by reaction of chlorine [7782-50-5] vapor with molten aluminum and used mainly as a catalyst in organic chemistry; ie, in Friedel-Crafts reactions (qv) and in proprietary steps in the production of titanium dioxide [13463-67-7], TiO_2, pigment. Its manufacture by carbochlorination of alumina or clay is less energy-intensive and is the preferred route for a few producers (20). Aqueous $AlCl_3$ is used in water treatment (15).

Aluminum chloride hexahydrate, $AlCl_3 \cdot 6H_2O$, manufactured from aluminum hydroxide andhydrochloric acid [7647-01-0], HCl, is used in pharmaceuticals and cosmetics as a flocculant and for impregnating textiles. Conversion of solutions of hydrated aluminum chloride with aluminum to the aluminum chlorohydroxy complexes serve as the basis of the most widely used antiperspirant ingredients (18). The U.S. exported 19,600 t of aluminum chloride and imported 1,700 t in 2000 (1).

Another cosmetic application of aluminum compounds is as lakes for lipstick manufacture (19). A water-soluble dye can become a lipstick ingredient if combined with compounds that are colorless and insoluble. The result, called a lake, is insoluble in both oil and water. Some dyes are laked with alumina; others are dissolved in water and treated with solutions that precipitate $Al(OH)_3$ with the dye molecules occluded in the precipitate. These lakes are mixed with castor oil [8001-79-4] (qv), finely ground, and used as lipstick ingredients.

Aluminum fluoride is important as an additive to the electrolyte of aluminum smelting cells. World production, is by the hydrofluoric acid [7664-39-3] HF process and the fluorosilicic acid [16961-83-4], H_2SiF_6, process. In the HF process, acid grade feldspar [7789-75-5], CaF_2, reacts in a rotary kiln with fuming sulfuric acid (oleum) [8014-95-7], $H_2SO_4 \cdot SO_3$, to produce anhydrous HF gas and gypsum (see FLUORINE COMPOUNDS, INORGANIC, HYDROGEN)

$$CaF_2 + H_2SO_4 \longrightarrow 2\,HF + CaSO_4$$

The HF is fed into the bottom stage of a three-stage fluid bed reactor and $Al(OH)_3$ is fed to the top stage where it is converted to activated alumina at 300–400°C. In the middle stage, rising HF gas contacts downcoming alumina and forms AlF_3

$$6\,HF + Al_2O_3 \xrightarrow{400-600°C} 2\,AlF_3 + 3\,H_2O$$

The AlF_3 goes to the bottom stage of the reactor and is removed as product.

In the fluorosilicic acid process, H_2SiF_6 solution, obtained from scrubbing stack gases from phosphate rock fertilizer plants, is reacted with $Al(OH)_3$ at about 100°C, whereupon silica precipitates and AlF_3 is dissolved.

$$H_2SiF_6 + 2\,Al(OH)_3 \xrightarrow{100°C} 2\,AlF_3{\cdot}3H_2O(aq) + SiO_2 + H_2O$$

After the SiO_2 is filtered off, the aluminum fluoride is crystallized as the trihydrate which is then calcined at 500–550°C to yield anhydrous AlF_3.

6.3. Organoaluminum Compounds. Application of aluminum compounds in organic chemistry came of age in the 1950s when the direct synthesis of trialkylaluminum compounds, particularly triethylaluminum and triisobutylaluminum from metallic aluminum, hydrogen, and the olefins ethylene and isobutylene, made available economic organoaluminum raw materials for a wide variety of chemical reactions (see ORGANOMETALLICS, Σ-BONDED ALKYLS AND ARYLS).

The alkyls and aryls, R_3Al (in monomer form), are colorless liquids or low melting solids easily oxidized and hydrolyzed when exposed to the atmosphere. Triethylaluminum (TEA), one of the most commercially important members of this family of chemicals, is so reactive it bursts into flame on contact with air, ie, it is pyrophoric, and it reacts violently with water. This behavior is typical and special techniques are necessary for the safe handling and use of organoaluminum compounds.

The alkylaluminum halides, R_nAlX_{3-n}, where X is Cl, Br, I, and R is methyl, ethyl, propyl, iso-butyl, etc, in monomer form, and $n = 1$ or 2, are less easily oxidized and hydrolyzed than the trialkyls. Organoaluminum hydrides such as diisobutylaluminum hydride [1191-15-7], $(iso\text{-}C_4H_9)_2AlH$, are also available. Organoaluminum compounds are used commercially in multimillion kg/yr quantities as catalysts or starting materials for the manufacture of organic compounds such as plastics, elastomers (qv) biodegradable detergents, and organometallics containing zinc, phosphorus, or tin (8,20–22).

6.4. Sodium Aluminate. Sodium aluminate is manufactured by dissolving high purity $Al(OH)_3$ in 50% sodium hydroxide solution

$$Al(OH)_3 + NaOH \longrightarrow Na^+ + Al(OH)_4^-$$

The resulting solutions contain high dissolved solids content in the range of 30 wt% or more. Special surfactant technology (23) is sometimes used to avoid precipitation of $Al(OH)_3$ or at least to extend the shelf life of the caustic liquor.

Sodium aluminate is used in water purification, in the paper industry, for the after treatment of TiO_2 pigment, and in the manufacture of aluminum containing catalysts and zeolite.

6.5. Zeolites. A large and growing industrial use of aluminum hydroxide and sodium aluminate is the manufacture of synthetic zeolites (see MOLECULAR SIEVES). Zeolites are aluminosilicates with Si/Al ratios between 1 and infinity. There are 40 natural, and over 100 synthetic, zeolites. All the synthetic structures are made by relatively low (100–150°C) temperature, high pH hydrothermal synthesis. For example the manufacture of the industrially important zeolites A, X, and Y is generally carried out by mixing sodium aluminate and sodium silicate solutions to form a sodium aluminosilicate gel. Gel-aging under hydrothermal conditions crystallizes the final product. In special cases, a small amount of seed crystal is used to control the synthesis.

Zeolite-based materials are extremely versatile: uses include detergent manufacture, ion-exchange resins (ie, water softeners), catalytic applications in the petroleum industry, separation processes (ie, molecular sieves), and as an adsorbent for water, carbon dioxide, mercaptans, and hydrogen sulfide.

BIBLIOGRAPHY

"Aluminum Compounds" in *ECT* 1st ed., Vol. 1, pp. 623–630, by F. J. Mann, Mann Fine Chemicals; in *ECT* 2nd ed., Vol. 2, pp. 1–5, by C. L. Rollinson, University of Maryland; in *ECT* 3rd ed., Vol. 2, pp. 188–197, by C. I. Rollinson, University of Maryland; in *ECT* 4th ed., Vol. 1, pp. 252–267, by William C. Sleppy, Aluminum Company of America; "Aluminum Compounds, Introduction", in *ECT* (online), posting date: December 4, 2000, by William C. Sleppy, Aluminum Company of America.

CITED PUBLICATIONS

1. P. A. Plunkett, Bauxite and Alumina, *Mineral Commodity Summaries*, U.S. Geological Survey 2002.
2. F. C. Frary, *Ind. Eng. Chem.* **38**(2) (Feb. 1946).
3. P. A. Plunkett, "Bauxite and Alumina" *Minerals Yearbook*, U.S. Geological Survey, 2000.
4. C. Misra, *Industrial Alumina Chemicals, ACS Monogr. Ser. No. 184*, American Chemical Society, Washington, D.C., 1986.
5. Ger. Pat. 43977 (Aug. 3, 1888) to Karl Josef Bayer, the inventor of the Bauxite process.
6. A. W. Lemmon, Jr., *Explosions of Molten Aluminum and Water*, Light Metals, New York, 1980, pp. 817–836.
7. K. Wefers and C. Misra, *Oxides and Hydroxides of Aluminum, Alcoa Technical Paper No. 19*, revised, Alcoa Laboratories, Alcoa Division, Aluminum Company of America, Pittsburgh, Pa., 1987.
8. J. J. Eisch, *Comprehensive Organometallic Chemistry*, Vol. 1, Pergamon Press, Oxford, UK, 1982, Chapt. 6.
9. W. M. Fish, *Alumina Calcination in the Fluid-Flash Calciner, TMS Paper No. A74-63*, AIME, New York, 1974.

10. T. A. Wheat, *J. Can. Ceram. Soc.* **40**, 43 (1971).
11. W. Gerhartz, ed., *Ullmann's* Encyclopedia of Industrial Chemistry, 5th ed., VCH, Weinheim, Germany, 1985.
12. C. A. Hamer, *Acid Extraction Processes for Non-Bauxitic Alumina Materials, Canmet Report 77–54*, Canada Center for Mineral and Energy Technology, August 1977.
13. F. A. Peters, P. W. Johnson, *Cost Estimates for Producing Alumina from Domestic Raw Materials*, Information Circular 8648, U.S. Bureau of Mines, Washington, D.C., 1974.
14. K. S. Bengston, *A Technical Comparison of Six Processes for the Production of Reduction Grade Alumina from Non-Bauxite Raw Materials*, Light Metals, AIME Annual Meeting, New Orleans, 1979.
15. B. Suresh A. Kiski, and S. Schlag, *Chemical Economics Handbook*, SRI, Menlo Park, Calif., 2001.
16. L. D. Hart, ed., *Aluminum Chemicals: Science and Technology Handbook*, American Ceramics Society, Columbus, Ohio, 1990.
17. W. Buchner, R. Schliebs, G. Winter, and K. H. Buchel, *Industrial Inorganic Chemistry*, VCH Publishers, New York, 1989, 247–255.
18. *Chlorhydrol*, Reheis Chemical Company, division of Armour Pharmaceutical Company, Chicago, Ill., 1970.
19. L. K. Sibley, *Today's* Chemist **2**(4), (Aug. 1989).
20. *The Use of Aluminum Alkyls in Organic Synthesis*, Industrial Chemicals Division, Ethyl Corporation, Baton Rouge, La., March 1977.
21. *The Use of Aluminum Alkyls in Organic Synthesis*, Industrial Chemicals Division, Ethyl Corporation, Baton Rouge, La., 1972–1978 Supplement.
22. J. R. Zietz, Jr., and co-workers, in J. J. Eisch, *Comprehensive Organometallic Chemistry*, Vol. 7, Pergamon Press, Oxford, UK, 1982, Chapt. 6.
23. U.S. Pat. 4,252,735 (Feb. 24, 1981), W. O. Layer and S. A. Khan.

WILLIAM C. SLEPPY
Aluminum Company of America

ALUMINUM HALIDES AND ALUMINUM NITRATE

1. Introduction

Both the binary and complex fluorides of aluminum have played a significant role in the aluminum industry. Aluminum trifluoride [7784-18-1], AlF_3, and its trihydrate [15098-87-0], $AlF_3 \cdot 3H_2O$, have thus far remained to be the only binary fluorides of industrial interest. The nonahydrate [15098-89-2], $AlF_3 \cdot 9H_2O$, and the monohydrate [12252-28-7, 15621-55-3], $AlF_3 \cdot H_2O$, are of only academic curiosity. The monofluoride [13595-82-9], AlF, and the difluoride [13569-23-8], AlF_2, have been observed as transient species at high temperatures.

Of the fluoroaluminates known, cryolite, ie, sodium hexafluoroaluminate [15096-52-2], Na_3AlF_6, has been an integral part of the process for production of aluminum. The mixtures of potassium tetrafluoroaluminate [14484-69-6],

KAlF$_4$, and potassium hexafluoroaluminate [13575-52-5], K$_3$AlF$_6$, have been employed as brazing fluxes in the manufacture of aluminum parts.

Two new types of aluminates, with far-ranging commercial potential, have been prepared and characterized in the past few years. The first is the tetrafluoroaluminate anion (AlF$_4^-$) for which a reproducible, high-yield synthesis has been developed. This anion is able to stimulate various guanosine nucleotide binding proteins (G-proteins), and inhibit P-type ATPases by serving as a non-hydrolyzing phosphate mimic. Additionally, tetrafluoroaluminate complexes serve as precursors to aluminum trifluoride, which is used as a catalyst for chlorofluorocarbon isomerizations and fluorinations. The aluminum halides and aluminum nitrates have similar properties with the exception of the fluorides. In this group the chlorides are the most commercially important.

2. Aluminum Monofluoride and Aluminum Difluoride

Significant vapor pressure of aluminum monofluoride [13595-82-9], AlF, has been observed when aluminum trifluoride [7784-18-1] is heated in the presence of reducing agents such as aluminum or magnesium metal, or is in contact with the cathode in the electrolysis of fused salt mixtures. AlF disproportionates into AlF$_3$ and aluminum at lower temperatures. The heat of formation at 25°C is −264 kJ/mol (−63.1 kcal/mol) and the free energy of formation is −290 kJ/mol (−69.3 kcal/mol) (1). Aluminum difluoride [13569-23-8] has been detected in the high temperature equilibrium between aluminum and its fluorides (2).

3. Aluminum Trifluoride

Aluminum trifluoride trihydrate [15098-87-0], AlF$_3$·3H$_2$O, appears to exist in a soluble metastable α-form as well as a less soluble β-form (3). The α-form can be obtained only when the heat of the reaction between alumina and hydrofluoric acid is controlled and the temperature of the reaction is kept below 25°C. Upon warming the α-form changes into a irreversible β-form which is insoluble in water and is much more stable. The β-form is commercially available.

Aluminum trifluoride trihydrate is prepared by reacting alumina trihydrate and aqueous hydrofluoric acid. The concentration of acid can vary between 15 to 60% (4). In the beginning of the reaction, addition of Al(OH)$_3$ to hydrofluoric acid produces a clear solution which results from the formation of the soluble α-form of AlF$_3$·3H$_2$O. As the addition of Al(OH)$_3$ is continued and the reaction temperature increases, irreversible change takes place and the α-form of AlF$_3$·3H$_2$O gets converted to the β-form and precipitation is observed. After all the alumina is added, the reaction mixture is continuously agitated for several hours at 90–95°C. After the precipitate settles down, the supernatant liquid is removed using rotary or table vacuum filters and the slurry is centrifuged. The cake is washed with cold water, dried, and calcined in rotating horizontal kilns (5), flash dryers, or fluid-bed calciners to produce anhydrous AlF$_3$ for aluminum reduction cells. This process is known as a wet process.

Aluminum trifluoride can also be advantageously made by a dry process in which dried $Al(OH)_3$ is treated at elevated temperatures with gaseous hydrogen fluoride. High temperature corrosion-resistant alloys, such as Monel, Inconel, and titanium are used in the construction of fluidized-bed reactors. In one instance, an Inconel reactor is divided into three superimposed compartments by two horizontal fluidizing grid sieve plates. Aluminum hydroxide is fed into the top zone where it is dried by the existing gases. The gases such as HF and SiF_4 are scrubbed from stack gases with water. These gases are recycled or used in the manufacture of cryolite [15096-52-3]. Solids are transported from top to bottom by downcomers while HF enters at the bottom zone getting preheated by heat exchange from the departing AlF_3. The bulk of the reaction occurs in the middle compartment which is maintained at 590°C.

The third process involves careful addition of aluminum hydroxide to fluor-osilicic acid (6) which is generated by fertilizer and phosphoric acid-producing plants. The addition of $Al(OH)_3$ is critical. It must be added gradually and slowly so that the silica produced as by-product remains filterable and the $AlF_3 \cdot 3H_2O$ formed is in the soluble α-form. If the addition of $Al(OH)_3 \cdot 3H_2O$ is too slow, the α-form after some time changes into the insoluble β-form. Then separation of silica from insoluble β-$AlF_3 \cdot 3H_2O$ becomes difficult.

$$H_2SiF_6 + 2\,Al(OH)_3 \longrightarrow 2\,AlF_3 \cdot 3H_2O + SiO_2 + H_2O$$

Environmentally sound phosphate fertilizer plants recover as much of the fluoride value as H_2SiF_6 as possible. Sales for production of $AlF_3 \cdot 3H_2O$ is one of the most important markets (see FERTILIZERS; PHOSPHORIC ACID AND THE PHOSPHATES).

Dehydration of $AlF_3 \cdot 3H_2O$ above 300°C leads to a partial pyrohydrolysis forming HF and Al_2O_3 which can be avoided by heating the trihydrate gradually to 200°C to remove 2.5 moles of water and then rapidly removing the remainder at 700°C. This latter procedure yields a product having less than 3.5% water content and Al_2O_3 content below 8% (7). This product is a typical material used in aluminum reduction cells. The presence of alumina does not interfere in the process of aluminum reduction because it replaces part of the alumina that is fed to the cells.

The principal use of AlF_3 is as a makeup ingredient in the molten cryolite, $Na_3AlF_6 \cdot Al_2O_3$, bath used in aluminum reduction cells in the Hall-Haroult process and in the electrolytic process for refining of aluminum metal in the Hoopes cell. A typical composition of the molten salt bath is 80–85% Na_3AlF_6, 5–7% AlF_3, 5–7% CaF_2, 2–6% Al_2O_3, and 0–7% LiF with an operating temperature of 950°C. Ideally fluorine is not consumed in the process, but substantial quantities of fluorine are absorbed by the cell lining and fluorine is lost to the atmosphere. Modern aluminum industry plants efficiently recycle the fluorine values.

Minor uses of aluminum fluoride include flux compositions for casting, welding (qv), brazing, and soldering (see SOLDERS AND BRAZING ALLOYS) (8,9); passivation of stainless steel (qv) surfaces (10); low melting glazes and enamels (see ENAMELS, PORCELAIN OR VITREOUS); and catalyst compositions as inhibitors in fermentation (qv) processes. Table 1 gives typical specifications for a commercial sample of AlF_3.

Table 1. **Specification for Commercial Aluminum Trifluoride**

Parameter	Specification
assay as AlF_3, %	90–92
Al_2O_3, typical, %	8–9
SiO_2, max, %	0.1
iron as Fe_2O_3, %	0.1
sulfur as SO_2, %	0.32
bulk density, g/cm^3	
loose	1.3
packed	1.6
screen analysis, % retained	
105 μm (140 mesh)	20
74 μm (200 mesh)	60
44 μm (325 mesh)	90

Other hydrates of aluminum trifluoride are the nonahydrate [15098-89-2], $AlF_3 \cdot 9H_2O$, which is stable only below 8°C, and aluminum trifluoride monohydrate [12252-28-7], [15621-55-3], $AlF_3 \cdot H_2O$, which occurs naturally as a rare mineral, fluellite found in Stenna-Gwyn Cornwall, U.K. (11).

3.1. High Purity Aluminum Trifluoride. High purity anhydrous aluminum trifluoride that is free from oxide impurities can be prepared by reaction of gaseous anhydrous HF and $AlCl_3$ at 100°C, gradually raising the temperature to 400°C. It can also be prepared by the action of elemental fluorine on metal/metal oxide and subsequent sublimation (12) or the decomposition of ammonium fluoroaluminate at 700°C.

Relatively smaller amounts of very high purity AlF_3 are used in ultra low loss optical fiber–fluoride glass compositions, the most common of which is ZBLAN containing zirconium, barium, lanthanum, aluminum, and sodium (see Fiber Optics). High purity AlF_3 is also used in the manufacture of aluminum silicate fiber and in ceramics for electrical resistors (see Ceramics as electrical materials; Refractory fibers).

Anhydrous aluminum trifluoride, AlF_3, is a white crystalline solid. Physical properties are listed in Table 2. Aluminum fluoride is sparingly soluble in water (0.4%) and insoluble in dilute mineral acids as well as organic acids at ambient temperatures, but when heated with concentrated sulfuric acid, HF is liberated, and with strong alkali solutions, aluminates are formed. AlF_3 is slowly attacked by fused alkalies with the formation of soluble metal fluorides and aluminate. A series of double salts with the fluorides of many metals and with ammonium ion can be made by precipitation or by solid-state reactions.

3.2. Health and Safety Factors. Owing to very low solubility in water and body fluids, AlF_3 is relatively less toxic than many inorganic fluorides. The toxicity values are oral LD_{LO}, 600 mg/kg; subcutaneous, 3000 mg/kg. The ACGIH adopted (1992–1993) TLV for fluorides as F^- is TWA 2.5 mg/m^3. Pyrohydrolysis and strong acidic conditions can be a source of toxicity owing to liberated HF.

Table 2. **Physical Properties of Anhydrous Aluminum Trifluoride**

Property	Value
mol wt	83.977
mp, °C	1278^a
transition point, °C	455
density, g/cm^3	3.10
dielectric constant	6
heat of transition at 455°C, kJ/mol^b	0.677
heat of sublimation for crystals at 25°C, kJ/mol^b	300
ΔH_f at 25°C, kJ/mol^b	-1505
ΔG_f at 25°C, kJ/mol^b	-1426
S at 25°C, J/(mol·K)^b	66.23
C_p at 25°C J/(mol·k)^b	
α-crystals	74.85
β-crystals	100.5

^{a}Sublimes.
^{b}To convert J to cal, divide by 4.184.

4. Fluoroaluminates

The naturally occurring fluoroaluminates are listed in Table 3.

The common structural element in the crystal lattice of fluoroaluminates is the hexafluoroaluminate octahedron, AlF_6^{3-}. The differing structural features of the fluoroaluminates confer distinct physical properties to the species as compared to aluminum trifluoride. For example, in AlF_3 all corners are shared and the crystal becomes a giant molecule of very high melting point (13). In $KAlF_4$, all four equatorial atoms of each octahedron are shared and a layer lattice results. When the ratio of fluorine to aluminum is 6, as in cryolite, Na_3AlF_6, the AlF_6^{3-} ions are separate and bound in position by the balancing metal ions. Fluorine atoms may be shared between octahedrons. When opposite corners of each octahedron are shared with a corner of each neighboring octahedron, an infinite

Table 3. **Naturally Occurring Fluoroaluminates**

Name	Cas Registry Number	Molecular formula
cryolite	[15096-52-2]	Na_3AlF_6
chiolite	[1302-84-7]	$Na_5Al_3F_{14}$
cryolithionate	[15491-07-3]	$Na_3Li_3(AlF_6)_2$
thomsenolite, hagemannite	[16970-11-9]	$NaCaAlF_6 \cdot H_2O$
ralstonite	[12199-10-9]	$Na_{2x}(Al_{2x},Na_x)\,(F,OH)_6 \cdot yH_2O$
prosopite	[12420-95-0]	$CaAl_2(F,OH)_8$
jarlite, *meta*-jarlite	[12004-61-4]	$NaSr_3Al_3F_{16}$
weberite	[12423-93-7]	Na_2MgAlF_7
gearksutite	[12415-96-2]	$CaAl(F,OH)_5 \cdot H_2O$
pachnolite	[15489-46-0]	$NaCaAlF_6 \cdot H_2O$

chain is formed as, for example, in Tl_2AlF_5 [33897-68-6]. More complex relations exist in chiolite, wherein one-third of the hexafluoroaluminate octahedra share four corners each and two-thirds share only two corners (14).

4.1. Cryolite. Cryolite constitutes an important raw material for aluminum manufacturing. The natural mineral is accurately depicted as $3NaF \cdot AlF_3$, but synthetic cryolite is often deficient in sodium fluoride. Physical properties are given in Table 4.

Table 4. **Physical Properties of Cryolite**

Property	Value
mol wt	209.94
mp, °C	1012
transition temperature, °C	
monoclinic-to-rhombic	565
second-order	880
dimensions of unit cell, nm	
a	0.546
b	0.561
c	0.780
vapor pressure of liquid at 1012°C, Pa[a]	253
heat of fusion at 1012°C, kJ/mol[b]	107
heat of vaporization at 1012°C, kJ/mol[b]	225
heat of transition, kJ/mol[b]	
monoclinic-to-rhombic at 565°C	8.21
second-order at 880°C	0.4
heat capacity, J/(mol·K)[b]	
monoclinic crystal at 25°C	215
cubic crystal at 560°C	281
liquid at 1012°C	395
S, J(mol·K)[b,c]	238
ΔH_f^0 at 25°C,[c] kJ/mol[b]	−3297
ΔG_f^0 at 25°C,[c] kJ/mol[b]	−3133
density, g/cm^3	
monoclinic crystal at 25°C	2.97
cubic crystal from x-ray	2.77
solid at 1012°C	2.62
liquid at 1012°C	2.087
hardness, Mohs'	2.5
refractive index	
α-fom	1.3385
β-fom	1.3389
τ-fom	1.3396
electrical conductivity, $(\Omega \cdot cm)^{-1}$	
solid at 400°C	4.0×10^{-6}
liquid at 1012°C	2.82
viscosity, liquid at 1012°C, mPa·s(= cP)[a]	6.7
surface tension, liquid in air, mN/m(= dyn/cm)	125
activity product constant in water at 25°C	1.46×10^{-34}
solubility in water, g/100 g	
at 25°C	0.0042
at 100°C	0.0135

[a] To convert Pa to mm Hg, multiply by 7.

[b] To convert J to cal, divide by 4.184.

[c] Monoclinic crystal.

Cryolite derives its name from its resemblance to ice when immersed in water as a result of the closely matched refractive indexes. The only commercially viable source of cryolite deposits has been found in the south of Greenland at Ivigtut (15). Minor localities, not all authenticated, are in the Ilmen Mountains in the former USSR; Sallent, in the Pyrenees, Spain; and Pikes Peak, Colorado (16). For the most part the ore from Ivigtut is a coarse-grained aggregate carrying 10–30% of admixtures, including siderite, quartz, sphalerite, galena, chalcopyrite, and pyrite, in descending order of frequency.

The mineral cryolite is usually white, but may also be black, purple, or violet, and occasionally brownish or reddish. The lustre is vitreous to greasy, sometimes pearly, and the streak is white. The crystals are monoclinic, differing only slightly from orthorhombic symmetry, and have an axial angle of 90°11. The space group is $P2_1/m$. The [001] and [110] axes are usually dominant, giving the crystals a cubic appearance. Twinning is ubiquitous, and because the lamellae tend to be perpendicular, cleavage appears to be cubic. The fracture of individual crystals, however, is uneven. Because its refractive indexes are close to that of water, powdered cryolite becomes nearly invisible when immersed in water, but because the optical dispersion is different for the two materials the suspension shows Christiansen colors.

Upon heating the crystallographic angles approach 90° and the transition to the cubic form at 565°C is accompanied by a small heat change. The transition also involves a substantial change in density as evidenced by a characteristic decrepitation (17). The second transformation occurs at 880°C as indicated by the slope of the heating curve. It is also accompanied by a sharp rise in electrical conductivity. The heat change is very small and the transitions with rising temperatures probably mark the onset of a lattice disorder. The more plastic character of the solid near the melting point seems to corroborate this view (18).

Liquid cryolite is an equilibrium mixture of the products of the dissociation:

$$Na_3AlF_6 \longrightarrow 2\,NaF + NaAlF_4$$

The composition to the melting point is estimated to be 65% Na_3AlF_6, 14% NaF, and 21% $NaAlF_4$ [1382-15-3]. The ions Na^+ and F^-; are the principal current carrying species in molten cryolite whereas the AlF^{4-}; is less mobile. The structural evidences are provided by electrical conductivity, density, thermodynamic data, cryoscopic behavior, and the presence of $NaAlF_4$ in the equilibrium vapor (19,20).

Molten cryolite dissolves many salts and oxides, forming solutions of melting point lower than the components. Figure 1 combines the melting point diagrams for cryolite–AlF_3 and for cryolite-NaF. Cryolite systems are of great importance in the Hall-Heroult electrolysis process for the manufacture of aluminum (see ALUMINUM AND ALUMINUM ALLOYS). Table 5 lists the additional examples of cryolite as a component in minimum melting compositions.

The vapor from molten cryolite is largely $NaAlF_4$, the vapor pressures of Na_3AlF_6, NaF, and $NaAlF_4$ near the melting point are about in the ratios 5:1:30. Therefore, the liquid tends to become depleted in AlF_3, and the composition of the aluminum cell electrolyte has to be regularly adjusted by the addition of AlF_3 (20,22).

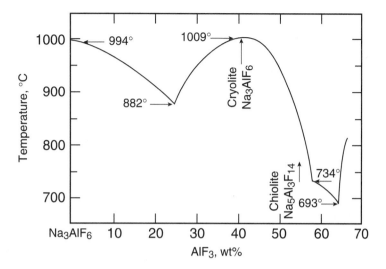

Fig. 1. Liquidus curves in the system NaF–AlF$_3$ (21).

In contact with moist air, molten cryolite loses HF and is depleted in AlF$_3$.

$$2\ Na_3AlF_6 + 3\ H_2O \longrightarrow 6\ NaF + 6\ HF\ +\ Al_2O_3$$

The more electropositive metals react with cryolite, liberating aluminum or aluminum monofluoride (22,23). The reduction of cryolite by magnesium is a current

Table 5. **Minimum Melting Compositions Containing Cryolite**

NaF	24.5	882
AlF$_3$	64	693
Al$_2$O$_3$	10.5	962
Li$_3$AlF$_6$	62	710
CaF$_2$	25.8	945
ZrO$_2$	14	969
MgO	7.5	902
CaO	11.3	896
ZnO	2.4	974
CdO	6.0	971
TiO$_2$	4.0	970
BaF$_2$	62.5	835
PbF$_2$	40	730
feldspar	70	830
NaF	34.0	870
Al$_2$O$_3$	12.0	
CaF$_2$	23.0	867
Al$_2$O$_3$	17.7	
CaF$_2$	37.8	675
AlF$_3$	6.2	
SiO$_2$	17	ca 800
Al$_2$O$_3$	50	

method for removal of magnesium in the refining of aluminum. Upon contact with strong acids cryolite liberates hydrogen fluoride.

4.2. Synthetic Cryolite.   The supply of cryolite is almost entirely met by synthetic material which possesses the same properties and composition with a minor difference in that it is deficient in NaF. Millions of tons of cryolite are used per year. Synthetic cryolite also commonly contains oxygen, hydroxyl group, and/ or sulfate groups. The NaF deficiency does not interfere for most applications but the presence of moisture leads to the fluorine losses as HF on heating. Because synthetic cryolite is lighter than the natural mineral, losses by dusting are also higher.

There are several processes available for the manufacture of cryolite. The choice is mainly dictated by the cost and quality of the available sources of soda, alumina, and fluorine. Starting materials include sodium aluminate from Bayer's alumina process; hydrogen fluoride from kiln gases or aqueous hydro-fluoric acid; sodium fluoride; ammonium bifluoride, fluorosilicic acid, fluoroboric acid, sodium fluosilicate, and aluminum fluorosilicate; aluminum oxide, aluminum sulfate, aluminum chloride, alumina hydrate; and sodium hydroxide, sodium carbonate, sodium chloride, and sodium aluminate.

The manufacture of cryolite is commonly integrated with the production of alumina hydrate and aluminum trifluoride. The intermediate stream of sodium aluminate from the Bayer alumina hydrate process can be used along with aqueous hydrofluoric acid, hydrogen fluoride kiln gases, or hydrogen fluoride-rich effluent from dry-process aluminum trifluoride manufacture.

$$NaAlO_2 + Na_2CO_3 + 6\,HF \longrightarrow Na_3AlF_6 + 3\,H_2O + CO_2$$

The HF and Na_2CO_3 give a sodium fluoride solution. Bayer sodium aluminate solution is added in the stoichiometric ratio. Cryolite is precipitated at 30–70°C by bubbling CO_2, until the pH reaches 8.5–10.0. Seed crystals are desirable. The slurry is thickened and filtered, or settled and decanted, or centrifuged. The resulting product is calcined at 500–700°C. The weight ratio of fluorine to aluminum in the product should exceed 3.9. The calculated value is 4.2 (24). Cryolite can also be made by passing gaseous HF over briquettes of alumina hydrate, sodium chloride, and sodium carbonate at 400–700°C, followed by sintering at 720°C (25).

In addition, there are other methods of manufacture of cryolite from low fluorine value sources, eg, the effluent gases from phosphate plants or from low grade fluorspar. In the former case, making use of the fluorosilicic acid, the silica is separated by precipitation with ammonia, and the ammonium fluoride solution is added to a solution of sodium sulfate and aluminum sulfate at 60–90°C to precipitate cryolite (26,27):

$$12\,NH_4F + 3\,Na_2SO_4 + Al_2(SO_4)_3 \longrightarrow 2\,Na_3AlF_6 + 6\,(NH_4)_2SO_4$$

The ammonia values can be recycled or sold for fertilizer use. The most important consideration in this process is the efficient elimination of the phosphorus from the product, because as little as 0.01% P_2O_5 in the electrolyte causes a 1–1.5% reduction in current efficiency for aluminum production (28).

Significant amounts of cryolite are also recovered from waste material in the manufacture of aluminum. The carbon lining of the electrolysis cells, which may contain 10–30% by weight of cryolite, is extracted with sodium hydroxide or sodium carbonate solution and the cryolite precipitated with carbon dioxide (28). Gases from operating cells containing HF, CO_2, and fluorine-containing dusts may be used for the carbonation (29).

The specifications for natural cryolite include 95% content of sodium aluminum fluorides as Na_3AlF_6, 4% of other fluorides calculated as CaF_2, and 88% of the product passing through 44 µm sieve (325 mesh). Product for the ceramic industry contains a small amount of selected lump especially low in iron. The following is a typical analysis for commercial-grade cryolite: cryolite as Na_3AlF_6, 91%; fluorine, 48–52%; sodium, 31–34%; aluminum, 13–15%; alumina, 6.0%; silica (max), 0.70%; calcium fluoride, 0.04–0.06%; iron as Fe_2O_3, 0.10%; with moisture at 0.05–0.15%, bulk density at 1.4–1.5 g/cm^3, and screen analysis passing through 74 µm (200 mesh) at 65–75%.

In spite of the fact that cryolite is relatively less soluble, its fluoride toxicity by oral routes are reported to be about the same as for soluble fluorides: LD_{50} = 200 mg/kg; for NaF, 180 mg/kg; KF, 245 mg/kg (30). Apparently, stomach fluids are acid enough to bring the solubility of cryolite up to values comparable with other fluorides. Chronic exposure may eventually lead to symptoms of fluorosis. The toxicity to insects is in many cases high enough for control. Because of its variable composition, synthetic cryolite may show physiological activity greater than the natural mineral (31).

The effective dissolution of Al_2O_3 by molten cryolite to provide a conducting bath has spurred the need for its use in manufacture of aluminum. Additives enhance the physical and electrical properties of the electrolyte, for example the lowering of melting point by AlF_3 (Fig. 1). Figure 2 illustrates the effect of various additives on the electrical conductivity of liquid cryolite. AlF_3 has the

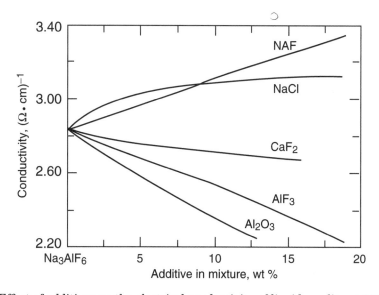

Fig. 2.   Effect of additives on the electrical conductivity of liquid cryolite at 1009°C (32).

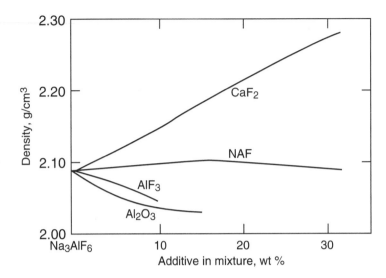

Fig. 3. Effect of additives on the density of liquid cryolite at 1009°C.

adverse effect of decreasing the electrical conductivity. Calcium fluoride is better in this regard but again too much of it can lead to rise in density of the melt close to that of aluminum (ca 2.28 g/cm^3), inhibiting the separation of metal and electrolyte as indicated in Figure 3. Sodium fluoride has the disadvantage of reducing the current efficiency while increasing density and conductivity. Small amounts of lithium fluoride may also improve density and conductivity. Compromises on all of these factors have led to the following composition of the electrolyte: 80–85% cryolite, 5–7% AlF$_3$, 5–7% CaF$_2$, 0–7% LiF, and 2–8% Al$_2$O$_3$.

Another use for cryolite is in the production of pure metal by electrolytic refining. A high density electrolyte capable of floating liquid aluminum is needed, and compositions are used containing cryolite with barium fluoride to raise the density, and aluminum fluoride to raise the current efficiency.

Other applications of cryolite include use in reworking of scrap aluminum as flux component to remove magnesium by electrochemical displacement; as a flux in aluminizing steel as well as in processing a variety of metals; in the compounding of welding-rod coatings; as a flux in glass manufacture owing to its ability to dissolve the oxides of aluminum, silicon, and calcium, and also because of the low melting compositions formed with the components; for lowering the surface tension in enamels and thereby improving spreading (33); as a filler for resin-bonded grinding wheels for longer wheel life, reducing metal buildup on the wheel, and faster and cooler grinding action; and in insecticide preparations making use of the fines residue from the refining operation of the cryolite.

4.3. Potassium Tetrafluoroaluminate. Potassium tetrafluoroaluminate, KAlF$_4$, an important fluoroaluminates, mainly because of developments in the automotive industry involving attempts to replace the copper and solder employed in the manufacture of heat exchangers. The source mineral for aluminum radiator manufacture, bauxite, is highly abundant and also available in steady supply. Research and developmental work on the aluminum radiators started in the 1960s using chloride salt mixtures for brazing. The resulting products and the process itself could not compete with conventional radiators because these processes were comparatively uneconomical. This led to the development of an all fluoride-based flux which confers corrosion-resistant features to the product as well as to the process. Potassium tetrafluoroaluminate in mixtures with other fluoroaluminates, potassium hexafluoroaluminate [13775-52-5], K_3AlF_6, and potassium pentafluoroaluminate monohydrate [41627-26-3], $K_2AlF_5 \cdot H_2O$, has emerged as a highly efficient, noncorrosive, and nonhazardous flux for brazing aluminum parts of heat exchangers. Nocolok 100 Flux (Alcan Aluminum Corp.) developed by Alcan (Aluminum Co. of Canada) has been the first commercial product. Its use and mechanistic aspects of the associated brazing process have been well documented (33–37).

The important task performed by all brazing processes is the removal of oxide films lying on the surfaces of metals to be joined. The process should also permit wetting and flow of the molten filler metal at the brazing temperature (38). The fluxes employed should melt and become active for a successful brazing action. Thus if the flux melts at a temperature higher than that of the filler metal, it leads to the development of thick oxide films on the liquid filler metal inhibiting the flux action. The system $KF \cdot AlF_3$ (Fig. 4) (39) provides the most suitable flux for this applications. The system presents a eutectic mixture of $KAlf_4$ and K_3AlF_6 which melts at $559 \pm 2°C$ (40). This is just below the eutectic temperature of the Al-Si filler metal, which is $577°C$. The melting point of pure $KAlF_4$ is $574 \pm 1°C$ and that of K_3AlF_6 is $990°C$ (40).

Both $KAlF_4$ and K_3AlF_6 are white solids. The former is less soluble (0.22%) in water than the latter (1.4%). The generally cubic form of $KAlF_4$ inverts to the orthorhombic modification between -23 and $50°C$. On heating the cubic form is stable to its congruent melting temperature. The materials are generally inert and infinitely stable under ambient conditions. At melting temperatures and more significantly at temperatures above $730°C$ they react with water releasing hydrogen fluoride (41). Dissolution in strong acids is also slow but is enhanced at higher temperatures leading to the evolution of HF. Several possible interactions of $KAlF_4$ and the metal oxides in the brazing processes have been proposed as part of the mechanism for the latter (34).

An early method of preparation of $KAlF_4$ (42) involved combining aqueous solutions of HF, AlF_3, and KHF_2 in stoichiometric proportions and evaporating the suspension to a dry mixture. The product was subsequently melted and recrystallized. Some of the other conventional technical methods comprise reacting hydrated alumina, hydrofluoric acid, and potassium hydroxide followed by separation of the product from the mother liquor; concentrating by evaporation, a suspension obtained by combining stoichiometric amounts of components; and melting together comminuted potassium fluoride and aluminum fluoride at $600°C$ and grinding the resulting solidified melt.

Several other proprietary methods have been reported, which in general have the aim of producing lower melting products thereby aiming more at the preparation of a eutectic mixtures of the fluoroaluminates as discussed in the beginning of this section. One process (42) describes the making of $KAlF_4$, melting below $575°C$, by addition of potassium hydroxide to the aqueous solution of fluoroaluminum acid. The fluoroaluminum acid is prepared from a reaction of

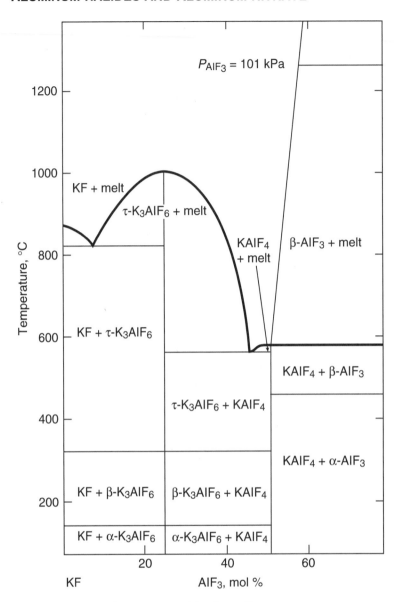

Fig. 4. KF–AlF$_3$ phase diagram.

hydrofluoric acid and hydrated alumina. A fairly similar method has been reported in making a flux mixture comprising of K$_2$AlF$_5$ or K$_2$AlF$_5$·H$_2$O and KAlF$_4$, wherein a potassium compound is added to the mixed aqueous fluoroaluminic acid (HAlF$_4$, H$_2$AlF$_5$, and H$_2$AlF$_6$) solution (43).

4.4. Tetrafluoroaluminate. *Synthesis. Aqueous.* Historically, many methods have been used to synthesize the tetrafluoroaluminate (TFA) anion (44). Indeed, aluminofluoride complexes are formed spontaneously in water containing fluoride and trace amounts of aluminum. These methods, however, produce mixtures of aluminates. This partially explains why most applications employing this anion have utilized *in situ* syntheses. Not surprisingly, there was some controversy over the actual products obtained from early syntheses of TFA (45). The first of these methods involved the neutralization of H_3 at 70–80°C (eq. 1) (46). The product was found to be soluble in water. A second proposed method involved the combination of AlF_3 and NaF at 80°C (eq.2) (47). This reaction did not produce TFA but two other products, $AlF_3(H_2O)_3$ and 1.5 NaF·AlF_3 (chiolite). In an effort to explain this earlier work, and to determine the water solubility of NaTFA, a more tightly controlled synthesis was devised (48). Thus, HTFA was prepared in situ at 80°C (higher temperatures drive the equilibrium towards more HFTA)(49) and then neutralized to a residual acidity of ~0.2% HF (eq. 3). The resulting NaTFA precipitate was isolated by filtration, washed with alcohol, and dried at 105°C. It was subsequently determined that NaTFA decomposes to chiolite and other species when dissolved in water at 25°C. However, at 75°C, the equilibrium is shifted toward the maintenance of TFA and decomposition does not occur. The solubility in water at this temperature is 0.133%.

$$H_3[AlF_6] \ (aq) + NaOH(aq) \longrightarrow Na \ [AlF_4] \cdot H_2O \tag{1}$$

$$NaF \ (aq) + AlF_3 \ (aq) \longrightarrow AlF_3(H_2O)_3 \ + \ 1.5NaF \cdot AlF_3 \tag{2}$$

$$HF \ + \ Al(OH)_3 \ + \ Na_2CO_3 \longrightarrow Na \ [AlF_4](s) \tag{3}$$

Ammonium TFA has been prepared in a number of ways. In one method an inorganic acid such as H_2SO_4 is added to an aqueous slurry of $(NH_4)_3AlF_6$ and either $Al(OH)_3$ or Al_2O_3 leading to the precipitation of $[NH_4^+] \ [AlF_4^-]$ (50). The reaction system was kept under atmospheric pressure and at 70 to 100°C. To neutralize the free ammonia in the slurry and maintain the solubility of the ammonium tetrafluoroaluminate, the inorganic acid was added in such an amount that the pH of the slurry after completion of the reaction was 4 to 7. Varying the method in which the inorganic acid is added to the slurry can control the particle size of the NH_4AlF_4. When the total amount of the acid is poured into the slurry at one time, the particle size of the crystalline NH_4AlF_4 is in the range from about 10 to 20 µm. The particle size increases to the range 20–50 µm when the acid is added intermittently.

[($CH_3)_4N$]TFA has also been prepared by neutralization of an aqueous solution of hydrated aluminum trifluoride in 40% HF with $N(CH_3)_4OH$ (51). Dehydration of the resulting precipitate at <120°C produced hygroscopic $[(CH_3)_4N^+]$ $[AlF_4^-]$. $KAlF_4$ can be prepared from aluminum hydroxide, hydrogen fluoride, and potassium hydroxide. The salt could be isolated as a precipitate and dried under vacuum at 80°C. It was found to have a melting point of 546–550°C by differential scanning calorimetry.

$$4 \ (NH_4)_3AlF_6 \ + \ 2 \ Al(OH)_3 \ + \ 3 \ H_2SO_4 \longrightarrow 6 \ NH_4AlF_4 \ + \ 3 \ (NH_4)_2SO_4 \ + \ 6H_2O \tag{4}$$

$$AlCl_3 \cdot N(C_2H_5)_3 \ + \ 2 \ NH_4HF_2 \longrightarrow NH_4AlF_4 \ + \ (C_2H_5)_3N \cdot HCl \ + \ HCl \tag{5}$$

$$Al(OH)_3 \ + \ 5 \ HF \ + \ 2 \ KOH \longrightarrow [K^+] \ [AlF_4^-] \ + \ 5 \ H_2O \ + \ KF \tag{6}$$

Fluoroaluminates with organic base counter-cations can be prepared by evaporating a mixture of $Al(OH)_3$ dissolved in HF and the organic base (52). The organic bases used were hydroxylamine, pyridine, quinoline, morpholine, 2-aminopyridine, and α,α'-bipyridyl. The fluoroaluminates were highly hygroscopic solids. Thermal decomposition produced a mixture of aluminum fluoride and oxide, indicating the presence of either hydrated [baseH$^+$] [AlF$_4^-$] or a mixed fluoro/aquo/hydroxo compound. TFA can be extracted from mixtures of aluminum and sodium fluoride into dimethylsulfoxide (DMSO) and acetonitrile (AN) solutions containing benzo-15-crown-5 (B15C5) (53). This solubilizes, and apparently stabilizes, salts having the general formula [M(B15C5)]+[TFA]– (with M=Na and 1,8-bis(dimethylamino)napthalene). Varying the concentration of B15C5 from 0.2 to 0.7 M both in DMSO and AN solutions, and the ratio of NaF/Al from 4-50, had no significant effect on the extraction. However in DMSO, six-coordinate aluminum fluoride complexes were also present. The (27) Al NMR data was consistent with quintets ~49 ppm, and sextets for the ^{19}F NMR in the range, -188.1 to -194.2 ppm.

$$2\,NH_4F \cdot HF + Al(NO_3)_3 \longrightarrow NH_4AlF_4 + NH_4^+ + 2\,H^+ + 3\,NO_3^- \qquad (7)$$

Preparation of NH_4AlF_4 can be simplified by using equation 7. Mixing of the solutions to form a precipitate follows solvation of each reactant in water. The precipitate is then filtered and dried with acetone. There is no need to heat the precipitate as it dries out in a short amount of time at room temperature.

Cation exchange has not proven to be a useful method to vary the cation paired with TFA. For example, it is not possible to replace a univalent cation with a divalent cation (for two AlF$_4^-$ units) since the divalent anion (AlF$_5^{2-}$) is readily formed. This is demonstrated in the synthesis of $Mg[AlF_5] \cdot 2.2\,H_2O$ (54).

$$NaF\,(aq) + AlF_3\,(aq) \longrightarrow [M^+]\,[AlF_4^-] \text{ where M = group I metal, Tl, NH}_4 \qquad (8)$$

$$Al(NO_3)_3\,(aq) + 4\,NaF\,(aq) \longrightarrow [Na^+]\,[AlF_4^-]\,(aq) + 3\,NaNO_3\,(aq) \qquad (9)$$

Anhydrous. The addition of trimethylaluminum to pyridinium fluoride produced the first anhydrous TFA product (eq. 10) (55). This compound could subsequently be used to prepare other anhydrous salts through cation exchange. Thus, the compound [PS]+ [AlF$_4$]– (PS = 1,8-bis(dimethylamino)naphthalene) was prepared in a glove box by slurring [pyridineH$^+$] [AlF$_4^-$] into a solution containing excess PS dissolved in dry acetomtrile (56). Likewise, slurring the pyridinium derivative in neat collidine (collidine = 2,4,6-trimethylpyridine) in a glove box, and heating to 120°C for 30 minutes produced [collidineH$^+$] [AlF$_4^-$].

$$(CH_3)_3Al + 4\,HF \cdot pyridine \longrightarrow 4\,CH_4 + [pyridineH^+]\,[AlF_4^-] \qquad (10)$$

The compound, [tetraphenylphosphonium$^+$] [AlF$_4^-$], was prepared by cation exchange with the collidine salt dissolved in methanol and $(C_6H_5)_4PBr$. The product was recrystallized from hot acetone or acetonitrile. The arsonium derivative [$(C_6H_5)_4As^+$] [AlF$_4^-$] was prepared (using $(C_6H_5)_4AsCl$) and characterized in a similar manner. Another synthesis using the collidine salt involved mixing with $N(CH_3)_4Cl$ in dry methanol (57). The by-product collidine HCl was sublimed away in flowing nitrogen at <200°C, and the $[(CH_3)_4N^+]$ $[AlF_4^-]$ left behind.

The ordinarily aqueous salt, NH_4AlF_4, could be prepared anhydrous by adding an alkylamine·aluminum trichloride complex to a bifluoride (eg, NH_4HF_2 or $NaHF_2$) in toluene (58). After completion of the reaction the toluene was distilled off and the product was purified by washing with water (to remove NH_4Cl) and then dried in an oven. The beta phase of NH_4AlF_4 could be obtained by heating pyridineHAlF$_4$ (under N_2) to about 180°C in formamide (59). Pyridine is evolved from the solution and HAlF$_4$ remains behind. The HAlF$_4$ reacts with the formamide solvent eliminating CO gas (eq. 11).

$$C_6H_5NH\ AlF_4\ +\ NH_2CHO\ \longrightarrow\ NH_4AlF_4\ +\ C_6H_5N\ +\ CO \qquad (11)$$

Characterization. *Spectroscopic.* The IR spectra for these compounds showed a sharp band at 785 cm^{-1} attributed to the Al-F stretching frequencies. The Raman spectra showed a sharp band at 635 cm^{-1}. Mixtures, presumably also containing some TFA have IR values in the range of 410–675 cm^{-1}. In the author's laboratory IR values of 567–825 cm^{-1} for hydrated tetrafluoroaluminate anions with inorganic cations (Na^+, Li^+, K^+, Rb^+, Cs^+, and Ti^+) have been obtained.

NMR peaks at −187 to −194 ppm (^{19}F NMR) and 49 to 52 ppm (^{27}Al NMR) were observed for the PSH, collidineH, (17), $(CH_3)_4N$, $(CH_3)_4P$, $(CH_3)_4As$, and $(CH_3CH_2)_4P$ tetrafluoroaluminate species. Similar NMR data was observed for the AlF_4^- anion stabilized by benzo-15-crown-5 in solutions of donor solvents. The coupling for $[PSH^+][AlF^{4-}]$ in CD_3CN was observed as a sextet from ^{19}F-^{27}Al in the ^{19}F NMR, and a quintet from ^{27}Al- ^{19}F in the ^{27}Al NMR.

In aqueous solutions, there is a rapid, pH dependant exchange between H_2O, OH^-, and F^- ligands binding to the aluminum cation. The solution behavior of fluoroaluminate complexes in aqueous solutions has been studied using ^{27}Al and ^{19}F NMR.

The question of coordination of the fluoroaluminate species in aqueous solutions has also been investigated. The coordination of the fluoroaluminate species is an important one with regards to its ability to interfere with the activities of nucleoside-binding proteins. As noted before, the tetrafluoroaluminate anion has been proposed to stimulate 6-proteins and P-type ATPases by assuming a geometry that is similar to a γ-phosphate. The AlF_4^--nucleoside diphosphate (NDP) complex is thought to mimic the size and shape of a nucleoside triphosphate (60). However, theoretical studies have ruled out any tetracoordination for AlF_x in aqueous solutions (61), although it is possible that ternary species such as $AlOH_yF_x$ may be tetrahedra. A reversible equilibria exist between the different fluoroaluminate species. The proportions of multifluorinated species, such as $AlF_x(H_2O)_{6-x}^{(3-x)+}$ (where $x = 3-6$) or $AlF_xOH(H_2O)_{5-x}^{(2-x)+}$ (where $x = 3-5$), depend on the excess concentration of free fluoride ions and on the pH of the solution (62).

Structures. There are several examples of tetrahedral $[AlX_4^-]$ species when X=Cl, Br, I. The $[AlF_4^-]$ species has however been controversial due to lack of structural proof, malthough there is indirect evidence, including IR Raman (in molten salts) and NMR data. Tetrahedrally coordinated AlF_4^- compounds had been proposed to exist in a hot melt or in vapor phase, but upon cooling reassembled into six-coordinate forms. In the structure of $[PSH^+][AlF_4^-]$ (where PS = Proton Sponge) the expected tetrahedral anion was confirmed. Accordingly, the F-Al-F angles were $\sim109°$ (with Al-F distances of ~1.62 Å). The closest contact between the coordinated fluoride and the chelated proton of the cation was 2.77 Å, which could be considered a long hydrogen bond. (check H bonds to F, add sum of the Van der Waals for H and F=). The crystal structure of [tetraphenylphosphonium$^+$] $[AlF_4^-]$ and [tetraphenylarsonium$^+$] $[AlF_4^-]$ has also been reported. Both structures are similar to that of $[PSH]$ $[AlF_4^-]$ with discrete cations and anions.

The structure of [collidineH$^+$] $[AlF_4^-]$ contains $[AlF_4^-]\infty$ chains, with six-coordinate aluminum. The collidinium cations form strong hydrogen bonds to the terminal fluoride ions of the chain, effectively forming a sheath around these chains. In between the chains and residing in a hydrophobic region defined by the collidinium ions are two independent, discrete, tetrahedral $[AlF_4^-]$ species. As before, the $[AlF_4^-]$ anions have no contact, other than Van der Waals, with other species in the lattice. In the structure of a related compound, $AlF_3(NH_3)_2$ (bridging F), only octahedral aluminum is observed (63). Indeed, there is evidence for octahedral coordination in the tetrafluoroaluminate complex in the active sites of proteins such as the G-protein Giα_1, (64) transducin (65) NDP kinase (66) nitrogenase (67) the Ras-RasGAP complex (68) and the G-protein RhoA (69).

Commercial Production.   The commercially utilized Al-F compounds are inorganic and synthetic in origin. Tetrafluoroaluminate salts, $M[AlF_4]$ can be prepared by many different methods. These preparations are mostly *in situ* due to the small amounts of the material needed for the various applications. AlF_3 is a catalyst for various reactions and is prepared by the thermolysis of the $[AlF_4]^-$ anions (70). This must take place at elevated temperatures (between 700 and 900 K) since the enthalpy of this reaction is $+66.9$ kJ. As an example, the beta phase of $[NH_4^+][AlF_4^-]$ can be thermolyzed at 550°C to form the kappa phase of AlF_3 (59). Fluoroaluminum catalysts can also be prepared by pyrolysis of precursors obtained from aqueous solution, treatment of Al_2O_3 with HF at elevated temperatures, and treatment of $AlCl_3$ with HF or chlorofluorocarbons. One advantage of the $[AlF_4]$- route, however, is that the AlF_3 is produced without any oxide or hydroxide contamination.

Health and Safety Factors.   In view of the ubiquity of phosphate in cell metabolism together with the dramatic increase in the amount of aluminum and fluoride now found in our ecosystem, aluminofluoride complexes represent a strong potential danger for living organisms, including humans (71). One area of important research will be the investigation of the long-term pharmacological and toxicological effects of exposure to tetrafluoroaluminate complexes on animals and plants. Another area of future research will be the determination of the relationship, if any, between aluminum in everyday products (cooking utensils, deodorants, antacids, food and beverage packaging), the increasing use of fluoride (water fluoridation, dental products, industrial fertilizers), and the health of humans.

The toxicity of these fluoroaluminates is mainly as inorganic fluorides. The ACGIH adopted (1992–1993) values for fluorides as F^- is TLV 2.5 mg/m^3. The oral toxicity in laboratory animal tests is reported to be LD$_{50}$ rat 2.15 mg/kg (41). Because of the fine nature of the products they can also be sources of chronic toxicity effects as dusts.

Uses. *Catalysis.* Tetrafluoroaluminates are used in the preparation of AlF$_3$. Aluminum trifluoride is important in the industrial production of aluminum metal, as it increases the conductivity of electrolytes in the electrolysis process. Aluminum trifluoride is also used as a catalyst for chlorofluorocarbon isomerization and fluorination. High surface area AlF$_3$ dispersed onto carbon, organic, or inorganic supports may be a useful catalyst for these or other reactions (72). For example, a fluoroaluminum species (obtained from ammonium bifluoride in anhydrous methanol slurried with calcined alumina) generated on a support of alumina was treated with chromium to obtain active olefin polymerization catalysts (73).

$$5 \, NaAlF_4 \, (s) \longrightarrow Na_5Al_3F_{14} \, (s) \, + \, 2 \, AlF_3 \, (s) \tag{12}$$

Biological Activity. Aluminum fluoride complexes, especially tetrafluoroaluminates [AlF$_4{}^-$], are currently receiving intense scrutiny because of their ability to act as phosphate analogues and thereby stimulate various guanosine nucleotide binding proteins (G-proteins) (74) and inhibit P-type ATPases (75) (See Fig. 5). G-proteins take part in an enormous variety of biological signaling systems, helping to control almost all important life processes. As a result of the ubiquitous nature of G-proteins, tetrafluoroaluminates are used in laboratory studies to investigate the physiological and biochemical changes caused in cellular systems by aluminofluoride complexes.

For most of the biochemical and physiological studies involving the putative AlF$_4{}^-$ anion, the fluoride source is usually sodium fluoride, and the aluminum source is aluminum nitrate or aluminum chloride. Solutions are usually prepared with millimolar concentrations of sodium fluoride, and micromolar concentrations of the aluminum source.

Transfer of phosphate groups is the basic mechanism in the regulation of the activity of numerous enzymes, including energy metabolism, cell signaling, movement, and regulation of cell growth. Phosphate is an important component of phospholipid in the cell membranes. AlF$_4{}^-$ acts as a high affinity analog of the γ-phosphate (76,77). AlF$_4{}^-$ mimics the role of γ-phosphate only if the β-phosphate

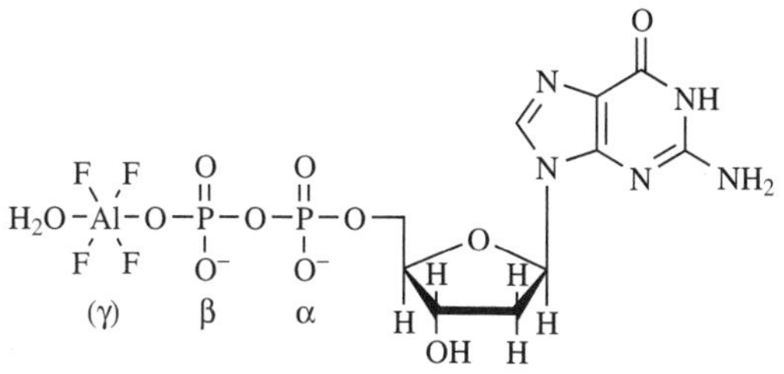

Fig. 5. AlF$_4{}^-$-GDP binding. AlF$_4{}^-$ binds strongly to the β-phosphate.

is present and remains unsubstituted. The effect is more readily seen with G proteins because guanosine diphosphate (GDP) is always tightly bound at the site afier the hydrolysis of guanosine triphosphate (GTP).

The tetrafluoroaluminate complex was proposed to act as an analogue of the terminal phosphate of GTP because the Al-F bond length is close to the P-O phosphate bond length, and the AlF_4^- and PO_4^{3-} structures are both tetrahedral. Fluorine and oxygen have nearly the same size and the same valence orbitals. Aluminum and phosphorus have their valence electrons in the same third shell. However, the two bonding schemes differ in that the former is more ionic while the latter is more covalent. In phosphate, oxygen is covalently bound to the phosphorus and does not exchange with oxygen from the solvent. In $[AlF_4^-]$ the bonding between the electropositive aluminum and the highly electronegative fluorine is more ionic in character. The reaction of a bound phosphate compound with orthophosphate is endergonic and slow, whereas the corresponding reaction with $[AlF_4^-]$ is rapid and spontaneous. Fluorides in the bound complex can also exchange with free fluoride ions in solution.

G protein-mediated cell responses are of key importance in the processes of neurotransmission and intercellular signaling in the brain (78) and AlF_4^- acts as an active stimulatory species (79). Aluminofluoride complexes mimic the action of many neurotransmitters, harmones, and growth factors. Exposure of osteoclasts to AlF_4^- resulted in a marked inhibition of bone resorption (80). Brief exposure to aluminum fluoride complexes induced prolonged enhancement of synaptic transmission (81) and can affect the activity of many other ion channels and enzymes in the kidney (82). Rapid and dynamic changes of the actin network are of vital importance for the motility of human neutrophils. AlF_4^- induction expressed a pronounced and sustained increase in a filamentous form of actin in intact human neutrophils (83).

It should be noted that the human body does posses natural barrier systems to aluminum intake. There are various physiological ligands, such as transferrin, citrate, and silicilic acid, which are efficient buffers in preventing the intake of aluminum under natural conditions (84). However, the formation of AlF_4^- only requires trace amounts of aluminum, and the increased bioavailability of aluminum in the environment will certainly lead to increased absorption of aluminum by living organisms.

It should be noted that there is some uncertainty over the identity of the biologically active aluminum compound. Species with four fluorines, six fluorines, and with the fluorines replaced by hydroxide appear possible. For example, multinuclear NMR spectroscopy was used to study the ternary system Al^{3+}, F^-, and NDP in aqueous solutions (pH = 6) without protein (85,86). Ternary complexes (NDP) AlF_x ($x = 1-3$) were found, but no (NDP)AlF_4^- was detected. Further multinuclear NMR studies of fluoroaluminate species in aqueous solutions over a wider pH range (2–8) with varying [F]/[Al] concluded that all the fluoroaluminate complexes observed in aqueous solution are hexacoordinated with an octahedral geometry (87). Further research specifically targeted to elucidate the nature of the aluminum phosphate interaction will certainly provide the answer in the near future.

As Flux Material. Potassium tetrafluoroaluminate can be used as a flux when soldering aluminum. A flux is added in order to remove oxides and other disruptive covering films on the metal surface (88). A mix of fluoroaluminates (including $KAlF_4$ and $KAlF_6$) can be used as a flux in the brazing of aluminum and aluminum alloy parts. The flux containing the mix of fluoroaluminates allows brazing of aluminum at temperatures lower than what could be accomplished with fluoride fluxes, thus saving heating energy (89). Another method utilizes a flux-coated soldering rod that is used to deliver a solder alloy and a flux compound to the region to be repaired. The flux compound has a higher melting temperature than the solder alloy, and is present as a coating that thermally insulates the alloy to cause the flux compound and the solder alloy to melt nearly simultaneously during the soldering operation. The solder alloy is preferably a zinc-aluminum alloy, while the flux coating preferably contains a cesium-aluminum flux compound such as potassium cesium tetrafluoroaluminate, dispersed in an adhesive binder that will readily volatilize or cleanly bum off during the soldering operation. The flux compound and binder form a hard coating that adheres to and thermally insulates the solder alloy until the flux compound melts. By controlling the relative amounts of flux compounds and binders, the flux coating remains protective and insulating on the alloy until melting of the flux compound begins (90).

Other. $LiAlF_4$ can be incorporated into the carbon cathode current collector of non-aqueous lithium batteries. $LiAlF_4$ is also used as the non-aqueous electrolyte, which can be dispersed throughout the cathode collector (91).

Future Considerations.   Industrially, continued AlF_4^- research will be seen in the development of better AlF_4^- precursors, which provide for cleaner decomposition at lower temperatures to AlF_3. The deposition of AlF_4^- onto organic or inorganic supports for subsequent decomposition to AlF_3 continues to be investigated. Future research will also be conducted with the view of preparing more pure $M^+ AlF_4^-$ (M = metal or organic cation) with less dangerous starting materials, lower temperatures, and better cost effectiveness. Preparation of better fluxes containing AlF_4^- for the soldering of aluminum and aluminum alloys will also be a continuing area of research and development.

All of these applications will benifit from fundamental synthetic and structural results. Thus, there will be a clear connection between laboratory research and commercial applications.

5. Aluminum Chloride

The chemistry of aluminum chloride is influenced significantly by hydration. Aluminum chloride hexahydrate [7784-13-6], $AlCl_3 \cdot 6H_2O$, is a crystalline solid that dissolves easily in water forming ionic species. Heating the hydrate results in the loss of hydrogen chloride [7647-01-0], HCl, and formation of aluminum oxide [1344-28-1], Al_2O_3. On the other hand, anhydrous aluminum chloride [7446-70-0] reacts violently with water evolving heat, a gas consisting of hydrogen chloride and steam, and aluminum oxide particulates. Anhydrous aluminum chloride sublimes at 180°C leaving no residue. The uses of anhydrous aluminum chloride and the hydrated form are also very different. The anhydrous material is a Lewis acid used as an alkylation catalyst. The hydrate is used principally as a flocculating aid.

Table 6. **Physical Properties of Anhydrous Aluminum Chloride**[a]

Property	Value
molecular weight	133.34054
density at 25°C[b], g/mL	2.46
sublimation temperature[b], °C	180.2
triple point, °C, 233 kPa[c]	192.5 ± 0.2
heat of formation, 25°C, kJ/mol[d]	-705.63 ± 0.84
heat of sublimation of dimer, 25°C, kJ/mol[d]	115.52 ± 2.3
heat of solution, 20°C, kJ/mol[d]	-329.1
heat of fusion, kJ/mol[d]	35.35 ± 0.84
entropy, 25°C, J/(K·mol)[d]	109.29 ± 0.42
heat capacity, 25°C, J/(K·mol)[d]	91.128

[a] Ref. 92.
[b] Ref. 93.
[c] To convert kPa to psi, multiply by 0.145.
[d] To convert J to cal, divide by 4.184.

Commercially, aluminum chloride is available as the anhydrous $AlCl_3$, as the hexahydrate, $AlCl_3 \cdot 6H_2O$, or as a 28% aqueous solution designated 32°Be′. Polyaluminum chloride, or poly(aluminum hydroxy) chloride [1327-41-9] is a member of the family of basic aluminum chlorides. These are partially neutralized hydrates having the formula $Al_2Cl_{6-x}(OH)_x \cdot 6H_2O$ where $x = 1 - 5$.

5.1. Anhydrous Aluminum Chloride. *Properties.* Anhydrous aluminum chloride is a hygroscopic, white solid that reacts with moisture in air. Properties are shown in Table 6. Commercial grades vary in color from light yellow to light gray as a result of impurities. Crystal size is dependent upon method of manufacture. At atmospheric pressure, anhydrous aluminum chloride sublimes at 180°C as the dimer [13845-12-0], Al_2Cl_6, which dissociates to the monomer beginning at approximately 300°C. Dissociation is essentially complete at 1100°C. As can be seen from Figure 6, the liquid form of aluminum chloride exists only at elevated temperatures and pressures.

Aluminum chloride dissolves readily in chlorinated solvents such as chloroform, methylene chloride, and carbon tetrachloride. In polar aprotic solvents, such as acetonitrile, ethyl ether, anisole, nitromethane, and nitrobenzene, it dissolves forming a complex with the solvent. The catalytic activity of aluminum chloride is moderated by these complexes. Anhydrous aluminum chloride reacts vigorously with most protic solvents, such as water and alcohols. The ability to catalyze alkylation reactions is lost by complexing aluminum chloride with these protic solvents. However, small amounts of these "procatalysts" can promote the formation of catalytically active aluminum chloride complexes.

Manufacture. In the United States anhydrous aluminum chloride is manufactured by the exothermic reaction of chlorine [7782-50-5], Cl_2, vapor with molten aluminum [7429-90-5]. The aluminum may be scrap, secondary ingot of varying purity, or prime ingot. Melting of additional metal feed, external cooling of the reactor, and regulation of the chlorine feed rate, control the reactor temperature between 600–750°C. Chlorine is fed into the molten aluminum pool below the pool's surface. Aluminum chloride sublimes out of the pool and into a condensing vessel where the product solidifies on the condenser walls.

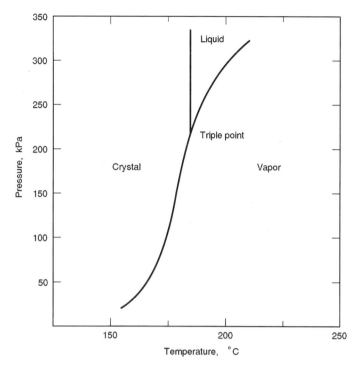

Fig. 6. Aluminum chloride phase diagram. The triple point occurs at 192.5°C at 33 kPa. To convert to psi, multiply by 0.145.

Condensers are normally air-cooled, thin-walled steel vertical cylinders having cone-shaped bottoms. The aluminum chloride grows into teardrop-shaped crystals which are periodically removed, crushed, screened, and packaged under a dry air or nitrogen atmosphere. The product may be colored yellow because of the presence of excess chlorine or ferric chloride [7705-08-0]. A gray or greenish coloration indicates the presence of condensed aluminum vapor in the product.

The chlorination of aluminous materials in the production of aluminum chloride has been thoroughly investigated (93). The Gulf Oil Company produced aluminum chloride from calcined bauxite [1318-16-7] and coke from 1920 to 1960 (94).

In 1948 BASF commercialized the process of catalytic chlorination of gamma-alumina in a fluidized bed (95). This process is still used in Germany. A mixture of chlorine and carbon monoxide passes over a beechwood charcoal catalyst in a preliminary step which partially converts the gases to phosgene. The hot gas mixture enters the bottom of a lined reactor and as the gas flows upward through a bed of finely divided particles of gamma-alumina, aluminum chloride gas forms. Catalytic amounts of sodium chloride are added to maintain a liquid phase within the fluidized bed of alumina and promote the conversion of the solid gamma-alumina to aluminum chloride vapor. Finely divided droplets of molten sodium aluminum chloride develop under these reaction conditions. The hot gas which exits the top of the reactor is filtered to remove entrained liquids and solids. The anhydrous aluminum chloride solidifies in a condenser, is collected, and packaged. Off-gases are purified before recycling.

The hot (400–500°C) inlet gases warm the alumina particles and the mildly exothermic reaction serves to maintain the heat of the furnace between 500 and 600°C. The alumina particle size is critical for maintaining a good reaction rate and a fluidized bed.

The main impurities of the reaction product are iron (0.02 to 0.03%) and small quantities of sodium (<0.01%) as chlorides. No chlorine-consuming side reactions occur and the product requires only one step to obtain a product free of significant impurities. The lower process temperatures also reduce demands on the materials of plant construction and maintenance.

Both the Toth and Alcoa processes provide aluminum chloride for subsequent reduction to aluminum. Pilot-plant tests of these processes have shown difficulties exist in producing aluminum chloride of the purity needed. In the Toth process for the production of aluminum chloride, kaolin [1332-58-7] clay is used as the source of alumina (96). The clay is mixed with sulfur and carbon, and the mixture is ground together, pelletized, and calcined at 700°C. The calcined mixture is chlorinated at 800°C and gaseous aluminum chloride is evolved. The clay used contains considerable amounts of silica, titania, and iron oxides, which chlorinate and must be separated. Silicon tetrachloride and titanium tetrachloride are separated by distillation. Resublimation of aluminum chloride is required to reduce contamination from iron chloride.

In the Alcoa process, high purity aluminum chloride is prepared for electrolytic reduction to aluminum (97). The starting material is Bayer process alumina. A small amount of sodium compounds (2.0–0.5% as Na_2O) is added to facilitate alumina chlorination. This alumina is coked or impregnated with carbon to yield a product containing 15–24 wt % carbon and the coked alumina, contaminated with sodium but otherwise substantially free of elements yielding volatile chlorides, reacts with chlorine in a fluidized-bed reactor. Reaction temperature is controlled preferably at about 600°C. Effluent gases from the reactor are cooled below the reaction temperature, but above the condensation temperature of aluminum chloride, and filtered. Solids consisting of aluminum oxide, carbon dust, aluminum oxychloride [13596-11-7], AlOCl, and a liquid composed of equimolar quantities of sodium and aluminum chlorides are removed from the gas stream. Material retained on the filter is returned in controlled amounts to the reactor to improve the efficiency of operation. After filtration, the gas stream containing essentially pure aluminum chloride and oxides of carbon is conducted to a desublimer (98). The aluminum chloride vapor can be condensed on solid particles of aluminum chloride maintained in a temperature-controlled fluidized bed (99). Product condensed in this manner is a lobular, free-flowing powder suitable for electrolytic reduction to aluminum. Traces of chlorinated biphenyls and phosgene contaminate aluminum chloride produced in this manner, making this product unsuitable for most commercial aluminum chloride applications. An electrolytic method for the production of sodium and aluminum chloride has been reported (100).

Economic Aspects.   The U.S. exported 15,700 t of aluminum chloride and imported 1,160 t of aluminum chloride in 2001 (101). Current price of aluminum chloride anhydrous is in the range of $ 0.85–90/lb (102).

Table 7. **Properties of Commercial Anhydrous Aluminum Chloride**[a]

Property	Value
mesh size, U.S. sieve series	-20^b
nonvolatile material, %	0.15
aluminum chloride, %	99.6
water insolubles, ppm	150
free aluminum, ppm	30
iron, ppm	30
magnesium, ppm	5

[a] Ref. 11.
[b] Corresponds to pore size of 840 μm.

Specifications and Packaging. Aluminum chloride's catalytic activity depends on its purity and particle size. Moisture contamination is an important concern and exposure to humid air must be prevented to preserve product integrity. Moisture contamination can be determined by a sample's nonvolatile material content. After subliming, the material remaining is principally nonvolatile aluminum oxide. Water contamination leads to a higher content of nonvolatile material.

In many chemical processes the catalyst particle size is important. The smaller the aluminum chloride particles, the faster it dissolves in reaction solvents. Particle-size distribution is controlled in the manufacturer's screening process. Typical properties of a commercial powder are shown in Table 7.

Aluminum chloride is available in a wide variety of moisture-free packages. Pails and drums are often used when fixed amounts of aluminum chloride are required for batch operations. For small operations, bags having a specially designed liner to maintain moisture-free product are available. For shipments from 200 to 1200 kilograms net, suppliers offer 37.8, 75.7, 113.6, and 208-L drums. Semibulk bins hold up to 11,000 kilograms net. These returnable containers are constructed of fiberglass to make shipping, storage, and handling of aluminum chloride more convenient. Aluminum chloride can also be purchased in bulk truck trailers in quantity up to 90,000 kg net (103).

Safety and Handling. Anhydrous aluminum chloride reacts with water or moisture, generating heat, steam, and hydrochloric acid vapors. Product containers should be stored inside a cool, dry, well-ventilated area and bulk handling systems must be waterproofed; the product transferred only in a nitrogen or dry air system. Although aluminum chloride is nonflammable, it should also be stored away from combustible materials. In storage, some reaction with moisture may occur and over time can lead to a pressure build-up from HCl in the container. Containers should be carefully vented before being opened (103). Safety goggles or face shields, rubber gloves, rubber shoes, and coveralls made of acid-resistant material should be used in handling. A NIOSH/OSHA-certified respirator is also required to prevent breathing fumes and dust (103). Aluminum chloride reacts with moisture in the skin, in the eyes, ears, nose, and throat (103).

The ACGIH TLV TWA is 2.0 mg/m^3 soluble salts, as aluminum. The TLV is based on the amount of hydrolyzed acid and the corresponding TLV (104).

Environmental Protection.   Fumes resulting from exposure of anhydrous aluminum chloride to moisture are corrosive and acidic. Collection systems should be provided to conduct aluminum chloride dusts or gases to a scrubbing device. The choice of equipment, usually one of economics, ranges from simple packed-tower scrubbers to sophisticated high energy devices such as those of a Venturi design (103).

Spills should be picked up before flushing thoroughly with water and neutralizing with soda ash or lime. The introduction of aluminum chloride into any drainage system results in the reduction of effluent pH, which can be adjusted using caustic soda or lime (103).

Uses.   Aluminum chloride is used as a catalyst in a wide variety of manufacturing processes, such as the polymerization of light molecular weight hydrocarbons in the manufacture of hydrocarbon resins. Friedel-Crafts reactions (qv) which employ this catalyst are used extensively in the synthesis of agricultural chemicals, pharmaceuticals (qv), detergents, and dyes (105).

Aluminum chloride is a nucleating agent in the production of titanium dioxide [13463-67-7] (rutile) used as a white pigment in a variety of paints, paper, and plastics. In the manufacture of titanium dioxide (106), aluminum chloride is mixed with titanium tetrachloride to ensure the formation of the rutile crystalline structure during the reaction with oxygen at 1300–1450°C (see TITANIUM COMPOUNDS). Sufficient aluminum chloride is used to produce TiO_2 containing 1% Al_2O_3. The pigment is wet treated, filtered, washed, dried, and fluid-energy-milled to form a dry TiO_2 for plastic pigmentation and for paints. Environmental and cost problems have favored use of this chloride process. Aluminum chloride is also used in water treatment (107).

5.2. Aluminum Chloride Hexahydrate.   The hexahydrate of aluminum chloride is a deliquescent, crystalline solid soluble in water and alcohol and usually made by dissolving aluminum hydroxide [21645-51-2], $Al(OH)_3$, in concentrated hydrochloric acid. When the acid is depleted, the solution is cooled to 0°C and gaseous hydrogen chloride is introduced. Crystalline aluminum chloride hexahydrate, $AlCl_3 \cdot 6H_2O$, is precipitated, filtered from the liquor, washed with ethyl ether, and dried. Alternatively, anhydrous aluminum chloride may be hydrolyzed in chilled dilute hydrochloric acid. Briquetting of the anhydrous material slows the reaction and the hydrogen chloride evolved may be recycled to aid precipitation of the hexahydrate.

Aluminum chloride hexahydrate is available in a 28% by weight (32° Be′) aqueous solution shipped in glass carboys, tank cars, or trucks. Crystalline hexahydrate is shipped in glass containers or plastic-lined drums. In 1980, 5200 metric tons of aluminum chloride hexahydrate on a 100% $AlCl_3$ basis was produced in the United States (106).

Roofing granules and mineral aggregate for bituminous products are treated with aluminum chloride solution to improve adhesion of the asphalt (108) (see BUILDING MATERIALS, SURVEY). Pigmented coatings (qv), containing sodium silicate, Na_2SiO_3, and used to color roofing granules, are insolubilized by spraying with aluminum chloride solution and then heating. Aluminum chloride hydrates are the alumina sources used in the manufacture of special forms of alumina and alumina-silica refractories (qv) such as alumina fibers (109), finely dispersed alumina for pesticide carriers (110), and catalyst substrates. Certain casting molds are hardened by spraying with a solution of aluminum chloride before firing.

Aluminum chloride hydrate is used in textile finishing to impart crease recovery and nonyellowing properties to cotton (qv) fabrics, antistatic characteristics to polyester, polymide, and acrylic fabrics, and to improve the flammability rating of nylon (see TEXTILES). Dye-bleeding of printed textile may be blocked (111) by treatment with aluminum chloride and zinc acetate, $Zn(O_2CCH_3)_2$, followed by solubilizing with ethylenediamine tetraacetic acid, and washing from the fabric. Aluminum chloride hexahydrate is used in cosmetics (antiperspirants) and the pharmaceutical industry.

5.3. Basic Aluminum Chlorides. The class of compounds identified as basic aluminum chlorides [1327-41-9] is used primarily in deodorant, antiperspirant, and fungicidal preparations. They have the formula $Al_2(OH)_{6-x}Cl_x$, where $x = 1$–5, and are prepared by the reaction of an excess of aluminum with 5–15% hydrochloric acid at a temperature of 67–97°C (112). The same compounds are obtained by hydrolyzing aluminum alkoxides with hydrochloric acid (113,114) (see ALKOXIDES, METAL). Basic aluminum chloride has also been prepared by the reaction of an equivalent or less of hydrochloric acid with aluminum hydroxide at 117–980 kPa (17–143 psi) (114).

Aluminum chloride solutions used in antiperspirants and deodorant preparations must be buffered for the protection of skin and clothing (see COSMETICS). Lactic acid [598-82-3] is usually employed for neutralizing these formulations. Hydrates of aluminum chloride and basic aluminum chlorides are also effective in a number of difficult water treatment problems (see FLOCCULATING AGENTS). A polymeric form, called polyaluminum chloride, is added in amounts of 50–500 ppm and the pH adjusted to about 6.5 in the presence of treatment aids such as emulsion breakers, anionic surfactants, diatomaceous earth, or a high molecular weight flocculant. The resulting floc may be separated by aeration and flotation, settling and decantation, electrophoresis, or filtration. Latex, acrylic paint and oil emulsions, dyes, clay suspensions, and effluent from sanitary waste digestion respond to this treatment (see ALUMINUM COMPOUNDS, POLYALUMINUMS).

6. Aluminum Bromide

Anhydrous aluminum bromide, $AlBr_3$, forms colorless trigonal crystals and exists in dimeric form, Al_2Br_6, in the crystal and liquid phases (1). Dissociation of the dimer to the monomer occurs in the gas phase. The bromide is produced commercially only in small quantities. This product melts at 97.45°C, boils at 256°C, and has a specific gravity at 25°C of 3.01.

Aluminum halides change from ionic to covalent character as the electronegativity of the halogen decreases (F > Cl > Br > I). Aluminum bromide, because of its covalent nature, is more soluble in many organic solvents than anhydrous aluminum chloride. Although its catalytic activity is moderate, it can be used in Friedel-Crafts reactions (qv) where selectivity is important (105). Anhydrous aluminum bromide, prepared from bromine [7726-95-6] and metallic aluminum, decomposes upon heating in air to bromine and alumina. Caution should be exercised in handling this hazardous compound because of its reactivity with water. Aluminum bromide may cause tissue burns, and both the anhydrous and the hydrate forms may be toxic upon ingestion.

Aluminum bromide hexahydrate [7784-11-4], $AlBr_3 \cdot 6H_2O$, may be made by dissolving aluminum or aluminum hydroxide in hydrobromic acid [10035-10-6], HBr. This white, crystalline solid is precipitated from aqueous solution.

7. Aluminum Iodide

Aluminum iodide [7884-23-8], AlI_3, is a crystalline solid with a melting point of 191°C. The presence of free iodine in the anhydrous form causes the platelets to be yellow or brown. The specific gravity of this solid is 3.98 at 25°C. Aluminum iodide hexahydrate [10090-53-6], $AlI_3 \cdot 6H_2O$, and aluminum iodide pentadecahydrate [65016-30-0], $AlI_3 \cdot 15H_2O$, are precipitated from aqueous solution. They may be prepared by the reaction of hydroiodic acid [10034-85-2], HI, with aluminum or aluminum hydroxide.

8. Aluminum Nitrate

Aluminum nitrate is available commercially as aluminum nitrate nonahydrate [7784-27-2], $Al(NO_3)_3 \cdot 9H_2O$. It is a white, crystalline material with a melting point of 73.5°C, that is soluble in cold water, alcohols, and acetone. Decomposition to nitric acid [7699-37-2], HNO_3, and basic aluminum nitrates [13473-90-0], $Al(OH)_x(NO_3)_y$ where $x + y = 3$, begins at 130°C, and dissociation to aluminum oxide and oxides of nitrogen occurs above 500°C. Aluminum nitrate nonahydrate is prepared by dissolving aluminum or aluminum hydroxide in dilute nitric acid, and crystallizing the product from the resulting aqueous solution. It is made commercially from aluminous materials such as bauxite. Iron compounds may be extracted from the solution with naphthenic acids (115) before hydrate precipitation. In the laboratory it is prepared from aluminum sulfate and barium nitrate.

Anhydrous aluminum nitrate [13473-90-0] is covalent in character, easily volatilized, and decomposes on heating (116). Hydrated aluminum nitrate is used in the preparation of insulating papers, on transformer core laminates, and in cathode-ray tube heating elements. Solution of aluminum hydroxide through digestion of core materials in nitric acid has been proposed in aluminum extractive metallurgy. The resulting solution of aluminum nitrate is separated from other metal ions, and the aluminum oxide is recovered by thermal decomposition of the aluminum nitrate solution (117,118). A process for producing the nitrate from gibbsite has been reported (119).

BIBLIOGRAPHY

"Aluminum Fluoride" under "Fluorine Compounds, Inorganic," in *ECT* 1st ed., Vol. 6, pp. 668–671 by R. G. Danehower, Pennsylvania Salt Manufacturing Co.; "Aluminum Fluorides" under "Fluorine Compounds, Inorganic," in *ECT* 2nd ed., Vol. 9, pp. 529–533, by J. F.

Gall, Pennsalt Chemicals Corp.; "Fluoroaluminates" under "Fluorine Compounds, Inorganic," in *ECT* 1st ed., Vol. 6. pp. 671–675, by I. Mockrin, Pennsylvania Salt Manufacturing Co.; "Fluoroaluminates" under "Fluorine Compounds, Inorganic," in *ECT* 2nd ed., Vol. 9, pp. 534–548, by J. F. Gall, Pennsalt Chemicals Corp.; "Aluminum" under "Fluorine Compounds, Inorganic," in *ECT* 3rd ed., Vol. 10, pp. 660–675, by J. F. Gall, Philadelphia College of Textiles and Science; in *ECT* 4th ed., Vol. 11, pp. 273–287, by Dayal T. Meshri, Advance Research Chemicals, Inc.; "Fluorine Compounds, Inorganic, Aluminum" in *ECT* (online), posting date: December 4, 2000, by Dayal T. Meshri, Advance Research Chemicals, Inc. "Aluminum Halides" in *ECT* 1st ed., Vol. 1, pp. 632–639, by H. E. Morris and C. L. Rollinson; "Aluminum Nitrate" in *ECT* 1st ed., Vol. 1, p. 640, by W. H. Schliffer, Jr.; "Aluminum Halides" in *ECT* 2nd ed., Vol. 2, pp. 17–25, by R. Gottlieb, Stauffer Chemical Co.; "Aluminum Nitrate" in *ECT* 2nd ed., Vol. 2, pp. 25–41, by A. R. Anderson, Anderson Chemical Division, Stauffer Chemical Co.; "Aluminum Halides and Aluminum Nitrate" in *ECT* 3rd ed., Vol. 2, pp. 209–218, by C. M. Marstiller, Aluminum Company of America; in *ECT* 4th ed., Vol. 2, pp. 281–290, by G. W. Grams, Witco Corporation; "Aluminum Halides and Aluminum Nitrate" in *ECT* (online), posting date: December 4, 2000, by G. W. Grams, Witco Corporation. "Aluminum Halides and Aluminum Nitrate" in *ECT* (online), posting date: February 14, 2000, by G. W. Grams, Witco Corporation; Fluorine Compounds, Inorganic, Aluminum" in *ECT* (online), posting date: December 4, 2000, by B. Conley, T. Shaikh, D. A. Atwood, University of Kentucky.

CITED PUBLICATIONS

1. *JANAF Thermochemical Tables*, 2nd ed., NSR DS-NBS 37, National Bureau of Standards, Washington, D.C., 1985.
2. T. C. Ehlert and J. L. Margrave, *J. Am. Chem. Soc.* **86**, 3901 (1964).
3. W. F. Ehret and F. J. Frere, *J. Am. Chem. Soc.* **67**, 64 (1945).
4. U.S. Pat. 2,958,575 (Nov. 1, 1960), D. R. Allen (to The Dow Chemical Co.).
5. J. K. Callaham, *Chem. Met. End.* **52**(3), 94 (1945).
6. F. Weinratter, *Chem. Eng.* **71**, 132 (Apr. 27, 1964).
7. J. K. Bradley, *Chem. Ind., London*, 1027 (1960).
8. Jpn. Kokai Tokkyo Koho 04 04,991 [92 04,991] (Jan. 9, 1992), T. Usui and S. Kagoshige (to Showa Aluminum Corp.).
9. Jpn. Kokai Tokkyo Koho 04 09,274 [92 09,274] (Jan. 14, 1992), K. Toma and co-workers (to Mitsubishi Aluminum Co. Ltd.).
10. Jpn. Kokai Tokkyo Koho 03 215,656 [91 215,656] (Sept. 20, 1991), T. Omi and co-workers (to Hashimoto Industries Co. Ltd.).
11. J. D. Dana and co-workers, *The System of Mineralogy*, 7th ed., Vol. 2, John Wiley & Sons, Inc., New York, 1951, pp. 124–125.
12. U.S. Pat. 4,983,373 (Jan. 8, 1991), H. P. Withers Jr. and co-workers (to Air Products & Chemicals, Inc.).
13. P. J. Durrant and B. Durrant, *Introduction to Advanced Inorganic Chemistry*, John Wiley & Sons, Inc., New York, 1970, p. 570.
14. A. F. Wells, *Structural Inorganic Chemistry*, 4th ed., Clarendon Press, Oxford, U.K., 1975, 388–390.
15. H. Pauly, *Met. Assoc. Acid. Magmat.* **I**, 393 (1974).
16. C. Palache and co-workers, in Ref. 4, pp. 110–113.
17. P. P. Fedotiev and V. Hyinskii, *Z. Anorg. Chem.* **80**, 113 (1913).
18. G. G. Landon and A. R. Ubbelohde, *Trans. Faraday Soc.* **52**, 647 (1955).
19. P. A. Foster, Jr. and W. B. Frank, *J. Electrochem. Soc.* **107**, 997 (1960).

20. L. M. Foster, *Ann. N.Y. Acad. Sci.* **79**, 919 (1960).

21. N. W. F. Philips and co-workers, *J. Electrochem. Soc.* **102**, 648–690 (1955).

22. K. Grjotheim and co-workers, *Light. Met.* **1**, 125 (1975).

23. M. Feinleib and B. Porter, *J. Electrochem. Soc.* **103**, 231 (1956); W. E. Haupin, *J. Electrochem. Soc.* **107**, 232 (1960).

24. U.S. Pat. 3,061,411 (Oct. 30, 1962), D.C. Gernes (to Kaiser Aluminum & Chemicals Corp.).

25. U.S. Pat. 3,104,156 (Sept. 17, 1963), P. Saccardo and F. Gozzo (to Sicedoison SpA).

26. G. Tarbutton and co-workers *Ind. Eng. Chem.* **50**, 1525 (1958).

27. U.S. Pat. 2,687,341 (Aug. 24, 1954), I. Mockrin (to Pennsylvania Salt Manufacturing Co.).

28. E. Elchardus, *Compt. Rend.* **206**, 1460 (1938).

29. U.S. Pat. 3,065,051 (Nov. 20, 1962), H. Mader (to Vereinigte Metallwerke Ranshofen-Berndorf A.G.).

30. *The Toxic Substances List*, 1974 ed., U.S. Dept. of Health, Education & Welfare, National Institute for Occupational Safety & Health, Rockville, Md., June 1974.

31. E. J. Largent, *J. Ind. Hyg. Toxicol.* **30**, 92 (1948).

32. J. D. Edwards and co-workers, *J. Electrochem. Soc.* **100**, 508 (1953); K. Matiasovsky and co-workers, *J. Electrochem. Soc.* **111**, 973 (1964).

33. R. Marker, *Glas Email Keramo Tech.* **4**, 117 (1957); **5**, 178 (1957).

34. Y. Ando and co-workers, *SAE Technical Paper Series, International Congress and Exposition*, paper no. 870180, Detroit, Mich., Feb. 23–27, 1987; D. J. Field and N. I. Steward, *ibid.*, paper no. 870186.

35. D. G. W. Claydon and A. Sugihara, in Ref. 34, paper no. 830021.

36. W. E. Cooke and H. Bowman, *Welding J.* (Oct. 1980).

37. W. E. Cooke and co-workers, *SAE Technical Paper Series, Congress and Exposition* paper no. 780300, Detroit, Mich., Feb. 27–Mar. 3, 1978.

38. J. R. Terril and co-workers, *Welding J.* **50**(12), 833–839 (1971).

39. B. Jensen, *Phase and Structure Determination of a New Complex Alkali Aluminum Fluoride*, Institute of Inorganic Chemistry, Norwegian Technical University, Trandheim, 1969.

40. B. Philips and co-workers, *J. Am. Ceram. Soc.* **49**(2), 631–634 (1966).

41. *Nocolok 100 Flux, Material Safety Data Sheet*, Alcan Aluminum Corp., Apr. 1986.

42. U.S. Pat. 4,428,920 (Jan. 31, 1984), H. Willenberg and co-workers (to Kali-Chemie Aktiengesellachaft).

43. U.S. Pat. 4,579,605 (Apr. 1, 1986), H. Kawase and co-workers (to Furukawa Aluminum Co., Ltd.).

44. For a brief overview of the syntheses see: U. Dutta and co-workers, *ACS Symposium Series*, American Chemical Society, Washington, D.C., 2001.

45. A. S. Korobitsyn and co-workers, *Zh. Prikl. Khim.* **56**, 887 (1983).

46. V. S. Yatlov, *Zh. Obshch. Khim.* **7**, 2439 (1937).

47. Yu. A. Kozlov, N. V. Belova, I. A. Leont'eva, and G. N. Bogachov, in M. E. Pozin, ed., *Studies in the Chemistry and Technology of Mineral Salts and Oxides*, Izd. Nauka, Leningrad, 1965, p. 119.

48. V. M. Masalovich and co-workers, *J. Inorg. Chem.* **35**, 968 (1990).

49. V. M. Masalovich, A. S. Korobitsyn, and T. A. Permyakova, *Russ. J. Inorg. Chem.* **33**, 264 (1988).

50. U.S. Pat. 4,034,068 (July 5, 1977), M. Aramaki and U. Etsuo (to Central Glass Co. Ltd.).

51. U.S. Pat. 5,985,233 (Nov. 16, 1999), H. J. Belt, R. Sander, and W. Rudolph (to Solvay Fluor und Derivative GmbH).

52. A. K. Sengupta and K. Sen, *Indian J. Chem.* **17A**, 107 (1979).

53. S. P. Petrosyants, M. A. Maliarik, E. O. Tolkacheva, and A. Y. Tsivadze, *Main Group Chemistry* **2**, 183 (1998).
54. K. W. Riley and A. Horne, *Anal. Chim. Acta.* **182**, 257 (1986).
55. U.S. Pat. 5,986,023 (Nov. 16, 1999), R. L. Harlow, N. Herron, and D. L. Thorn (to DuPont).
56. N. Herron, D. L. Thorn, R. L. Harlow, and F. Davidson, *J. Am. Chem. Soc.* **115**, 3028 (1993).
57. N. Herron, R. L. Harlow, and D. L. Thorn, *Inorg. Chem.* **32**, 2985 (1993).
58. U.S. Pat. 5,045,300 (Sept. 3, 1991), E. Marlett (to Ethyl Corporation).
59. U.S. Pat. 5,417,954 (May 23, 1995), R. L. Harlow and N. Herron (to DuPont).
60. E. J. Martinez, J. L. Girardet, and C. Morat, *Inorg. Chem.* **35**, 706 (1996).
61. R. B. Martin, *Biochem. Biophys. Res. Commun.* **155**, 1194 (1998).
62. G. Goldstein, *Anal. Chem.* **36**, 36, 243 (1964).
63. D. R. Ketchum, G. L. Schimek, W. T. Pennington, and J. W. Kolis, *Inorg. Chim. Acta.* **294**, 200 (1999).
64. D. E. Coleman and co-workers, *Science* **265**, 1405 (1994).
65. D. G. J. Sondek and co-workers, *Nature* **372**, 276 (1994).
66. Y. W. Xu, S. Moréra, J. Janin, and J. Cherfils, *Proc. Natl. Acad. Sci. U.S.A.* **94**, 3579 (1997).
67. H. Schindelin and co-workers, *Nature* **387**, 370 (1997).
68. K. Scheffzek and co-workers, *Science* **277**, 333 (1997).
69. K. Rittinger and co-workers, *Nature* **389**, 758 (1997).
70. D. B. Shinn, D. S. Crocket, and H. M. Haendler, *Inorg. Chem.* **5**, 1927 (1966).
71. A. Strunecka and J. Patocka, *Fluoride* **32**, 230 (1999).
72. U.S. Pat. 5,986,023 (Nov. 16, 1999), R. L. Harlow, N. Herron, D. L. Thorn (to DuPont).
73. U.S. Pat. 5,171,798 (Dec. 15, 1992), M. P. McDaniel, D. D. Klendworth, and M. M. Johnson (to Philips Petroleum).
74. P. C. Sternweis and A. G. Gilman, *Proc. Natl. Acad. Sci. U.S.A.* **78**, 4888 (1982). A. G. Gilman, *Annu. Rev. Biochem.* **56**, 615 (1987).
75. A. Troullier, J. L. Girardet, and Y. Dupont, *J. Biol. Chem.* **267**, 22821 (1992) and references therein.
76. J. Bigay, P. Deterre, C. Pfister, and M. Chabre, *EMBO J.* **6**, 2907 (1987).
77. M. Chabre, *TIBS* **15**, 6–10 (1990).
78. R. S. Rana and L. E. Hokin, *Physiol. Rev.* **70**, 115–164 (1990).
79. S. M. Candura, A. F. Castoldi, L. Manzo, and L. G. Costa, *Life Sci.* **49**, 1245 (1991).
80. B. S. Moonga and co-workers, *Biochem. Biphys. Res. Comun.* **190**, 496 (1993).
81. S. J. Publicover, *Exp. Brain Res.* **84**, 680 (1991).
82. J. Zhou and co-workers, *Proc. Natl. Acad. Sci. USA* **7**, 7532 (1990).
83. T. Bengtsson, E. Sarndahl, O. Stendahl, and T. Andersson, *Proc. Natl. Acad. Sci. USA* **87**, 2921 (1990).
84. M. Wilhelm, D. E. Jager, and F. K. Ohnesorge, *Pharmacol. Toxicol.* **66**, 4 (1990).
85. D. J. Nelson and R. B. Martin, *J. Inorg. Biochem.* **43**, 37 (1991).
86. X. Wang, J. H. Simpson, and D. J. Nelson, *J. Inorg. Biochem.* **58**, 29 (1995).
87. E. J. Martinez, J. L. Girardet, and C. Morat, *Inorg. Chem.* **35**, 706 (1996).
88. U.S. Pat. 6,244,497 (June 12, 2001), P. J. Conn and J. H. Bowling Jr. (to S. A. Day Manufacturing Co.).
89. U.S. Pat. 6,010,578 (Jan. 4, 2000), M. Ono, M. Hattori, E. Itaya, and Y. Yanagawa (to Morita Chemical Industry Co., Ltd.).
90. U.S. Pat. 5,985,233 (Nov. 16, 1999), H. J. Belt, R. Sander, and W. Rudolph (to Solvay fluor und Derivative GmbH), and references therein.

91. U.S. Pat. 5,714,279 (Feb. 3, 1998) W. V. Zajack Jr., and co-workers (to the United States Government as represented by the Secretary of the Navy).

92. *JANAF Thermochemical Tables*, 3rd ed., American Chemical Society, Washington, D.C., and American Institute of Physics, New York, for the National Bureau of Standards, 1985.

93. C. A. Thomas, *Anhydrous Aluminum Chloride in Organic Chemistry*, ACS Monogr. Ser. #87, Reinhold Publishing Corp., New York, 1941. Out of print, but available in facsimile by the University Microfilms, a Xerox Company, Ann Arbor, Mich.

94. U.S. Pat. 2,832,668 (Apr. 29, 1958), O. L. Culbertson and W. A. Pardee (to Gulf Oil Company).

95. J. Hille and W. Durrwachter, *Angew. Chem.* **72**(22), 850–855 (1960).

96. U.S. Pat. 4,695,436 (Sept. 22, 1987), R. Wyndham, G. M. Chaplin, and W. M. Swanson (to Toth Aluminum Corporation).

97. U.S. Pat. 3,929,975 (Dec. 30, 1975), L. K. King and N. Jarrett (to Aluminum Company of America).

98. U.S. Pat. 3,930,800 (Jan. 6, 1976), R. C. Schoener, L. K. King, L. L. Knapp, and N. A. Kloap (to Aluminum Company of America).

99. U.S. Pat. 3,725,222 (Apr. 3, 1973), A. S. Russell, L. L. Knapp, and W. E. Haupin (to Aluminum Company of America).

100. U.S. Pat. 6,235,183 (May 22, 2001), H. Putter and co-workers (to BASF).

101. P. A. Plunkert, "Bauxite and Alumina," *Minerals Yearbook*, U.S. Geological Survey, Reston, Va, 2001.

102. *Chemical Market Reporter*, **263**(6), 16 (Feb 10, 2003).

103. *Publication Release*, Witco Corporation, New York, N.Y., 1990.

104. B. D. Dinman in E. Bingham, B. Cohrssen, and C. H. Powell, eds., *Patty's Toxicology*, 5th ed., Vol. 2, John Wiley & Sons, Inc., New York, 2001, p. 396.

105. G. A. Olah, *Friedel-Crafts Chemistry*, John Wiley & Sons, Inc., New York, 1973.

106. U.S. Pat. 4,214,913 (July 29, 1980), H. H. Glaeser (to E. I. du Pont de Nemours & Co., Inc.).

107. "Aluminum Chemicals," *Chemical Economics Handbook*, SRI International, Menlo Park, Calif., 2001.

108. U.S. Pat. 2,313,759 (Mar. 16, 1943), P. E. McCoy (to American Bitumuls Co.).

109. Ger. Offen. 2,511,650 (Sept. 25, 1975), R. W. Grimshaw and J. G. Peacey (to Hepworth and Grandage, Ltd.).

110. Ind. Pat. 130,124 (May 4, 1974), H. Bhavanagary and S. K. Mijumdev (to Council of Science and Indian Research, India).

111. Jpn. Pat. 74,45,751 (Dec. 5, 1974), H. Amabe (to Toybo Co., Ltd.).

112. U.S. Pat. 3,891,745 (June 24, 1975), A. Bellan and K. Deneke (to Dynamit-Nobel, A.G.).

113. U.S. Pat. 3,887,691 (Jan. 3, 1975), P. Kobetz (to Ethyl Corporation).

114. Jpn. Pat. 75,00,839 (Jan. 11, 1975), S. Ban, S. Hatano, and T. Mijazawa (to Daimei Chemical Industries).

115. USSR Pat. 513,006 (May 5, 1976), Kh. R. Ismatov, T. P. Rasulov, T. Kh. Klycher, and N. D. Prabova (to Institute of Chemistry, Academy of Sciences Uzbe, USSR).

116. USSR Pat. 456,785 (Jan. 15, 1975), G. N. Sherokova, V. Ya. Rosolovski, and S. Ya. Zhuka (to Institute of New Chemical Problems, Academy of Sciences, USSR).

117. U.S. Pat. 3,869,543 (Mar. 4, 1975), A. H. Schutte and J. T. Stevens (to Arthur D. Little, Inc.).

118. U.S. Pat. 3,864,462 (Feb. 4, 1975), C. P. Bruen and D. H. Kelly (to Reynolds Metals Co.).

119. U.S. Pat. Appl. 2001/0046469 (Nov. 29, 2001), K. Johansen (to Haldertopsoe).

GENERAL REFERENCES

Refs. 93 and 105 are good general references.

Ki Wade and A. J. Banister, *Comprehensive Inorganic Chemistry*, 1st ed., Pergamon Press, 1973.

A. I. Vogel and co-workers, *Vogel's* Textbook of Practical Organic Chemistry, 4th ed., John Wiley & Sons, Inc., New York, 1978.

E. W. Post and J. C. Kotz, "Aluminum Halides," in M. F. Lappert, ed., *Int. Rev. Sci., Inorganic Chemistry, Series 2*, Vol. 1, Butterworth, London, Chapt. 7, p. 219, A review of 252 references.

G. W. Grams
Witco Corporation
B. Conley
T. Shaikh
D. A. Atwood
University of Kentucky

ALUMINUM OXIDE (ALUMINA), ACTIVATED

1. Activated

The activated aluminas comprise a series of nonequilibrium forms of partially hydroxylated aluminum oxide [1344-28-1], Al_2O_3. The chemical composition can be represented by $Al_2O_{(3-x)}(OH)_{2x}$ where x ranges from about 0 to 0.8. They are porous solids made by thermal treatment ofaluminum hydroxide [21645-57-2] precursors and find application mainly as adsorbents, catalysts, and catalyst supports. Activated alumina, for purposes of this discussion, refers to thermal decomposition products (excluding α-alumina [12252-63-0]) of aluminum trihydroxides, oxide hydroxides, and nonstoichiometric gelatinous hydroxides. The term "activation" is used in this article to indicate a change in properties resulting from heating (calcining). Other names for these products are active alumina, gamma alumina, catalytic alumina, and transition alumina. Transition alumina is probably the most accurate because the various phases identified by x-ray diffraction are really stages in a continuous transition between the disordered structures immediately following decomposition of the hydrous precursors and the stable α-alumina which is the product of high temperature calcination.

2. Physical and Chemical Properties

In general, as a hydrous alumina precursor is heated, hydroxyl groups are driven off leaving a porous solid structure of activated alumina. The transformation is

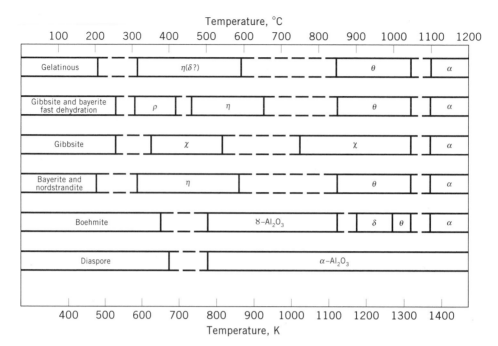

Fig. 1. Decomposition sequence of hydrous aluminas.

topotactic and little change in size or shape of the material is observed at low magnifications. At magnifications higher than about 10,000, changes in texture resulting from recrystallization can be seen. The physical properties of the material are set by the choice of precursor, the forming process, and the activation conditions.

Figure 1 shows the decomposition sequence for several hydrous precursors and indicates approximate temperatures at which the activated forms occur (1). As activation temperature is increased, the crystal structures become more ordered as can be seen by the x-ray diffraction patterns of Figure 2 (2). The similarity of these patterns combined with subtle effects of precursor crystal size, trace impurities, and details of sample preparation have led to some confusion in the literature (3). The crystal structures of the activated aluminas have, however, been well-documented by x-ray diffraction (4) and by nmr techniques (5).

2.1. Decomposition of Boehmite. Boehmite [1318-23-6], AlO(OH), can be synthesized having surface areas ranging from about 1 to over 800 m^2/g depending upon the method of preparation (6). The properties of activated boehmite products are strongly influenced by the crystallite size of the precursor material (7). When a crystallized, low surface area boehmite is heated, conversion to gamma alumina occurs at about 725 K yielding a low surface area product having a well-defined x-ray pattern. On further heating, the transition follows the gamma-delta-theta-alpha sequence shown in Figure 1. In contrast, a very high surface area boehmite (also referred to as pseudoboehmite or gelatinous boehmite) decomposes at 575–625 K yielding a high surface area product having

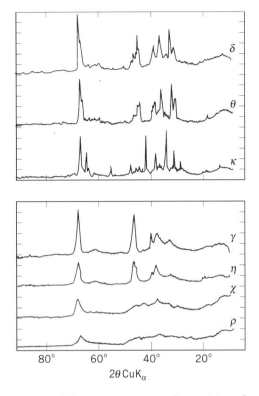

Fig. 2. X-ray diffraction patterns of transition aluminas.

a poorly defined gamma alumina x-ray pattern. On further heating, the transformation of this material to delta and theta phases may be retarded or may not occur at all, depending upon trace impurities (8). This sequence is also indicated in Figure 1. Figure 3 shows surface areas of three boehmite samples having different initial surface areas after heating for 16 hours at various temperatures. In

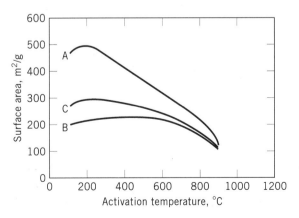

Fig. 3. Surface area of boehmite samples after activation, where A and B are experimental boehmite and C is Catapal SB.

the low temperature region, the highest surface area starting material maintains a higher surface area than the others, but in the range where delta-theta phases would be expected, the curves converge. This loss of surface area is a manifestation of coarsening of the activated alumina crystals and is also accompanied by a shift toward larger pore sizes on heating (9). In the high surface area materials, the surface area is mostly attributable to the external surface area of the precursor crystallites. Evolution of internal pore structure has been well-documented for activation of well-crystallized boehmite (10), but this is only a minor contribution to surface areas of the commercially important activated boehmites, typically in the 200–350 m^2/g range (11,12). Loss of surface area on heating is inevitable, but can be either retarded or accelerated by various additives (13).

2.2. Activation Products of Aluminum Hydroxide.

Figure 4 shows the effect of temperature on some properties of activated alumina produced by heating gibbsite [14762-49-3], α-Al(OH)$_3$ (14). As the material is heated, surface area reaches a maximum of 300 m^2/g or more at about 650 K. As temperature is increased further, surface area decreases and the skeletal structure becomes more dense reflecting increased ordering of the crystalline structure during the progression from chi to kappa to alpha. At about 1450 K conversion to alpha alumina occurs with a major rearrangement of crystal structure and corresponding decrease in surface area to about 5 m^2/g. These trend vary somewhat according to precursor crystal size, purity, and the atmosphere of heating.

Unlike the case for boehmite, the surface area of activated gibbsite results mainly from internal porosity rather than external surface of the precursor. There is a minor effect of initial crystal size, however. Some (10% or less) boehmite tends to form when coarse crystals of gibbsite are activated because of hydrothermal conditions which are generated in the particles. On further activation, this boehmite decomposes, but contributes little to the overall surface area. For this reason, activation of fine gibbsite tends to give somewhat higher maximum surface areas than coarser material (15). High humidity in the activating gas can also promote boehmite formation and lower surface area.

Activation of bayerite [20257-20-9] and nordstrandite [13840-05-6] qualitatively follows the pattern shown in Figure 4, but the transition sequence is eta-theta-alpha. The structures of these transition phases are somewhat different from those obtained from gibbsite, reflecting the differences in crystal structure of the hydroxides (16).

If aluminum hydroxide is decomposed by heating at low temperature under vacuum or by rapidly heating at high temperature, a nearly amorphous (x-ray indifferent) phase known as rho alumina is produced which has the interesting property of recrystallizing (rehydrating) to boehmite or bayerite when mixed with water (17). This behavior is known as rehydration bonding and occurs to a significant degree in hot water at atmospheric pressure (18). Rho alumina can be formed from any of the aluminum hydroxides. The crystal structures are probably somewhat different depending upon which precursor (gibbsite or bayerite) is used, but this cannot be detected by x-ray and rehydration properties are similar. Some recrystallization to boehmite occurs in all of the activated aluminas under severe hydrothermal conditions and generally the degree to which this occurs decreases with increased crystalline order (higher activation

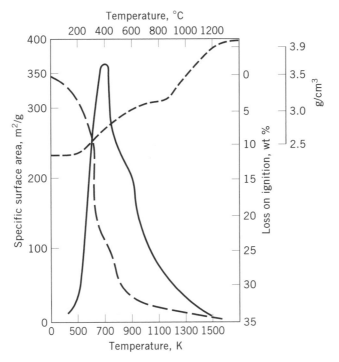

Fig. 4. Properties of activated gibbsite, where --- represents density, ––– is the loss of ignition, and — is specific surface area.

temperature). The ease with which rho alumina rehydrates sets it apart from the other activated alumina phases.

Except for rho alumina, the activated aluminas are quite stable when mixed with water in the pH range from about 4 to 10. Below pH 2 and above pH 12 they degrade rapidly.

The surface of activated alumina is a complex mixture of aluminum, oxygen, and hydroxyl ions which combine in specific ways to produce both acid and base sites. These sites are the cause of surface activity and so are important in adsorption, chromatographic, and catalytic applications. Models have been developed to help explain the evolution of these sites on activation (19). Other ions present on the surface can alter the surface chemistry and this approach is commonly used to manipulate properties for various applications.

3. Manufacturing Processes

The large majority of activated alumina products are derived from activation of aluminum hydroxide, rehydrated alumina, or pseudoboehmite gel. Other commercial methods to produce specialty activated aluminas are roasting of aluminum chloride [7446-70-0], $AlCl_3$, and calcination of precursors such as ammonium alum [7784-25-0], $AlH_7NO_8S_2$. Processing is tailored to optimize one or more of the product properties such as surface area, purity, pore size distribution, particle size, shape, or strength. A process for the production of β-alumina solid electrolyte without calcination has been reported (20).

3.1. Activated Aluminum Hydroxide.

The principal precursor for this class of products is gibbsite derived from the Bayer process although a small amount of bayerite is also used for some specialty catalytic applications. Bayer process gibbsite is available in very large tonnages as an intermediate product in aluminum production. It is 99+% pure: the main impurity is sodium oxide [1313-59-3], Na_2O, at 0.2–0.3% on a $Al(OH)_3$ basis. Low cost, relatively high purity, and availability make gibbsite the raw material of choice for many activated alumina products.

Gibbsite is a free-flowing powder having a median particle diameter of about 100 micrometers. Traditionally, activated alumina powders were produced by passing this material through a rotary kiln at an appropriate temperature. However, the aluminum industry has developed fluid bed and flash heating technology for calcination of metallurgical alumina and Bayer alumina-based activated aluminas are often produced in this type of equipment (21,22). The particle sizes of the activated aluminas are essentially the same as that of the precursor powder and in the unground form are amenable to applications involving fluid bed handling. These powders can also be ground to sizes of 10 micrometers or less by standard techniques. Conventional activated aluminas are produced by slow heat treatment whereas "rehydratable" powders are activated within a few seconds.

Larger particle size products can be produced by direct activation of agglomerated gibbsite. The oldest product of this class is made from "scale" which forms on the walls of Bayer process precipitators and is periodically removed in massive chunks, then crushed and activated. This gives a granular product, having particle sizes up to 1 cm or more, which has good mechanical properties and is relatively inexpensive. A similar product is manufactured by high pressure compaction of gibbsite powder. Once the gibbsite agglomerates are formed they are converted to the desired activated alumina product by a combination of crushing, screening, and heat treatment.

Unrefined bauxite [1318-16-7] is also used as a precursor to low cost activated alumina because bauxites can contain as much as 90% gibbsite (dry basis). These materials represent the low cost, low performance end of the activated alumina spectrum of products, but they are still used in significant quantities.

A method for processing bauxite rich in alumina monohydrate had been reported. The monohydrate is more difficult to digest using the Bayer process and normally require higher temperatures (23).

3.2. Rehydration Bonded Alumina.

Rehydration bonded aluminas are agglomerates of activated alumina, which derive their strength from the rehydration bonding mechanism. Because more processing steps are involved in the manufacture, they are generally more expensive than activated aluminum hydroxides. On the other hand, rehydration bonded aluminas can be produced in a wider range of particle shape, surface area, and pore size distribution.

The generic process used to manufacture these materials begins with an activated powder produced by rapid (flash) activation and grinding of Bayer process gibbsite. This powder is generally composed of a mixture of chi and rho forms, the proportion depending upon the specifics of the activation process. The powder is then mixed with water and formed into a shape by tumbling (24), extrusion (25), or dropping a slurry into an immiscible fluid (oil-drop) (26,27). During or after the forming process, the shapes are heated without drying from several minutes to several hours at temperatures between about 330 and 370 K. This allows partial rehydration to occur, rigidizing the structure. At this point, the alumina is a mixture of pseudoboehmite, bayerite, chi, and amorphous phases (18, 24). The shapes are then given a second heat treatment or calcination to establish the desired degree of activation. The surface area, which depends upon the micropores, is largely determined by the second heat treatment whereas the total pore volume is set by specifics of the forming operation. Activated aluminas prepared by this method are mixtures of phases reflecting the composition after rehydration.

3.3. Gel-Based Activated Aluminas. Alumina gels can be formed by wet chemical reaction of soluble aluminum compounds. An example is rapid mixing of aluminum sulfate [17927-65-0],$Al_2(SO_4)_3 \cdot XH_2O$, and sodium aluminate [1302-42-7], $NaAlO_2$, solutions to form pseudoboehmite and a near neutral sodium sulfate [7757-82-6] solution (28). After extensive washing to remove the sodium sulfate, the resulting gel can be dried or partially dewatered and formed directly by extrusion or oil drop. If the gel is dried prior to forming, it can be ground to fine particle size, then mixed with water and tumbled or formed by the general processes described above (29). The shapes are then activated to produce relatively pure phase gamma alumina. Gels made from aluminates and aluminum salts must be carefully washed to remove undesirable anions and cations which would be detrimental to the final application.

Hydrolysis of aluminum alkoxides is also used commercially to produce precursor gels. This approach avoids the introduction of undesirable anions or cations so that the need for extensive washing is reduced. Although gels having surface area over 800 m^2/g can be produced by this approach, the commercial products are mostly pseudoboehmite powders in the 200–300 m^2/g range (30). The forming processes already described are used to convert these powders into activated alumina shapes.

"Oil-drop" covers an interesting group of processes to produce small activated alumina spheres or beads by dispersing an aqueous alumina sol or solution in an immiscible liquid. The surface tension effects cause the aqueous droplets to attain a spherical shape and, while in this condition, they gel. Gellation can be accomplished by neutralization with ammonia (28, 30), dehydration with alcohol, or through rehydration bonding (26). Beads of very uniform size have been generated by forming droplets of aluminum nitrate [13473-90-0], $Al(NO_3)_3$, solution using a specially designed nozzle, allowing them to be partially gelled with ammonia [7664-41-7] vapor then falling into an oil– ammonia solution for final gellation (31). A new process for producing aluminum oxide beads is discussed in Ref. 32.

The gel-based products have traditionally been the most expensive and highest performance activated alumina products. They have very good mechanical properties, high surface area, and their purity and gamma-alumina structure make them somewhat resistant to thermal degradation. On the other hand, they are the most difficult to manufacture; and disposal of by-product salts can present an environmental problem.

4. Economic Aspects

The least expensive products are those derived directly from Bayer-process gibbsite and powders are generally less expensive than formed products. The soda content (0.2–0.3% Na_2O) of Bayer gibbsite makes it unattractive for many catalytic applications. Gel-based products are normally used where low soda level is required. Soda content of gels prepared from inorganic salts or aluminate solutions is typically about 0.03% whereas soda in alkoxide-based gels is much lower. Specialty activated aluminas having purity as high as 99.99% are also available at a much higher price.

Shaped products used for adsorbent purposes are generally less sophisticated and therefore less expensive than catalytic products.

5. Safety and Handling

Activated alumina is a relatively innocuous material from a health and safety standpoint. It is nonflammable and nontoxic. Fine dusts can cause eye irritation and there is some record of lung damage because of inhalation of activated alumina dust mixed with silica [7631-86-9] and iron oxide [1317-61-9] (33). Normal precautions associated with handling of nuisance dusts should be taken. Activated alumina is normally shipped in moisture-proof containers (bags, drums, sling bins) because of its strong desiccating action.

U.S. and international exposure standards for aluminum oxide are given in Table 1 (35).

Table 1. **U.S. and International Exposure Limits for Aluminum Oxide**

U.S. OSHA PEL	TWA 15 mg/m^3, total dust
	TWA 5 mg/m^3, respirable fraction
U.S. NIOSH REL	No REL
U.S. ACGIH TLV	TWA 10 mg/m^3, A4
Australia	TWA 10 mg/m^3
Belgium	TWA 10 mg/m^3
Denmark	TWA 10 mg/m^3
France	TWA 10 mg/m^3
Germany (DFG MAK)	TWA mg/m^3 1.5, respirable fraction of the aerosol, (fume)
Ireland	TWA 10 mg/m^3, total inhalable dust
	TWA 5 mg/m^3, respirable dust
Japan (JSOH)	TWA 05 mg/m^3, respirable dust
	TWA 2 mg/m^3, total dust
The Netherlands	TWA 10 mg/m^3, inhalable
Poland	TWA 2 mg/m^3, STEL 16 mg/m^2
Russia	STEL 4 mg/m^3
Switzerland	TWA 6 mg/m^3

6. Uses

6.1. Catalytic Applications.   Activated alumina is used commercially in catalytic processes as a catalyst, catalyst substrate, or as a modifying additive. Activated alumina serves as the catalyst in the Claus process for recovering sulfur from H_2S that originates from natural gas processing or petroleum refinery operations (34). The alumina is generally in the form of spheres, about 5 mm in diameter. This size has evolved as a good compromise between high activity and low pressure drop for fixed bed application (36). Promoting the alumina using a small amount of alkali has been claimed to enhance performance in certain Claus operations (37) (see SULFUR REMOVAL AND RECOVERY).

The largest application for activated alumina as a catalyst substrate is in hydrotreating of petroleum feedstocks (qv) (38). The purpose of hydrotreating is threefold: to increase the H/C ratio; to remove O, S, and N impurities; and to remove V, Ni, and other tramp contaminants, especially from residuum of heavier feedstocks. Specialized alumina-based catalysts typically promoted with compounds of Co, Mo, W, and Ni have been developed for these operations (34). The catalysts are usually in the form of extrudates having variously shaped cross sections (39) (circular, lobed, wagon-wheel) about one millimeter in diameter. Spherical catalysts 1 mm or less in diameter have also been described for hydrotreating (26). Much attention has been given to optimizing pore volume and pore size distribution of the activated alumina substrates used in these operations and the "optimum" properties vary with operating conditions and the petroleum feedstock being processed (40,41).

Another catalytic application for promoted alumina is in automotive exhaust catalysts which enhance oxidation of hydrocarbons, carbon monoxide, and nitrogen oxide in exhaust gas (see EXHAUST CONTROL, AUTOMOTIVE). There are two general configurations of catalyst currently in use: beads and monoliths. In both systems, the catalytic component is precious metal (platinum, palladium, and rhodium) on alumina (42). The bead system consists of a bed of 3-mm diameter alumina spheres which act as both catalyst substrate and mechanical support. In the monolith system, the mechanical support is provided by a porous ceramic (cordierite [12182-53-5]) multichannel "honeycomb" having a thin alumina coating (washcoat) as the catalytic substrate. Automotive exhaust catalyst was the largest volume application (about 18,000 t/yr) for promoted alumina in the mid-1970s and was a significant driving force in improving the technology of low density alumina sphere manufacture. Since that time, however, monoliths have become the system of choice for this application. Beads have maintained only a small share of the market and consequently the volume of activated alumina used in automotive catalysts has dropped significantly (43).

The largest tonnage single application for catalyst particles is in fluid cracking (FCC). These materials are typically made from zeolite having a clay or alumina–silica binder system to provide the necessary mechanical strength for fluid bed handling. Addition of alumina (aluminum hydroxide, pseudoboehmite, or rehydratable alumina) to these formulations has been reported to improve various properties (44,45). Any alumina powder would be activated under FCC processing conditions, thus the activated alumina is used as a modifying additive in the catalytic process. The actual usage of activated alumina in FCC catalysts is unclear because of the highly competitive and proprietary nature of this market.

A number of smaller but nevertheless important applications in which activated alumina is used as the catalyst substrate include: alcohol dehydration, olefin isomerization, hydrogenation, oxidation, and polymerization (46). A new method for synthesizing hydrocarbons using a silica–alumina support for group VIII metal has been reported (47).

6.2. Chromatographic Applications. Activated alumina has been used for many years in the separation of various organic compounds by normal phase chromatography (qv) because of its natural hydrophilic surface characteristics. More recently, stable surface coatings have been developed which impart hydrophobic properties to the particle surface (48). These coatings, coupled with improved technology to produce closely sized particles have allowed alumina to compete in reverse-phase chromatographic markets. Compared to silica, the dominant reverse-phase packing material, alumina has better chemical stability at moderately high pH levels, which gives it a natural advantage for separations in this pH range (49).

6.3. Membranes. Membranes comprised of activated alumina films less than 20 μm thick have been reported (50). These films are initially deposited via sol–gel technology (qv) from pseudoboehmite sols and are subsequently calcined to produce controlled pore sizes in the 2 to 10-nm range. Inorganic membrane systems based on this type of film and supported on solid porous substrates have been introduced commercially. They are said to have better mechanical and thermal stability than organic membranes (51). The activated alumina film comprises only a miniscule part of the total system (see MEMBRANE TECHNOLOGY).

6.4. Adsorbent Applications. One of the earliest uses for activated alumina was removal of water vapor from gases and this remains an important application. Under equilibrium conditions alumina adsorbs an increasing amount of water as the relative humidity of the contacting gas increases. At 50% rh, for example, a good quality activated alumina can adsorb water at levels of 15 to 20% or more of its own weight (52). By heating the activated alumina to about 525 K under low rh conditions, essentially all of the adsorbed water is removed and the alumina is returned to its original state. This adsorption–regeneration cycle can be repeated hundreds of times with little deterioration of the adsorbent. Industrial scale countercurrent drying systems use this cyclic approach, but generally are designed for lower water loadings because economic cycle times are too short for equilibrium to be established. Another consideration in the large beds of industrial dryers is the amount of heat generated which amounts to about 46 kJ/mol (11 kcal/mol) of water adsorbed. A common strategy is to operate the drying systems under pressure. This partially reduces the water content in the gas before it enters the bed by condensation and also increases the heat capacity of the gas, facilitating heat removal from the bed (53). Under proper operating conditions, moisture in the dried gas can be as low as 11 ppmv. The usual forms of activated alumina used in desiccant applications are granules or spheres about 3 to 12 mm in size. Besides air, a partial list of gases that can be dried includes Ar, He, H_2, CH_4, C_2H_6, C_3H_8, C_2H_2, C_2H_4, C_3H_6, Cl_2, HCl, SO_2, NH_3, and fluorochloroalkanes. Activated alumina is also used to remove water from organic liquids including gasoline, kerosene, oils, aromatic hydrocarbons, cyclohexane, butane and heavier alkanes, butylenes, and many chlorinated hydrocarbons.

Besides drying applications, activated alumina is used to selectively remove various species from gas and liquid systems. In refining and petrochemical operations, activated alumina is used to remove trace HCl from reformer hydrogen, fluorides from hydrocarbons produced by HF alkylation, and a variety of other stream cleanup applications (52). An important example in air pollution abatement is adsorption of fluoride vapors emanating from the aluminum smelting operation (see ALUMINUM AND ALUMINUM ALLOYS). Fluoride-contaminated air from the smelting cell area is countercurrently passed through fluid beds of alumina cell feed which adsorbs the fluorides for recycle. Fluorides are also effectively removed from potable water by activated alumina (54). Fluoride tends to adsorb on alumina at low pH levels and desorb (regenerate) when the pH is increased so in these systems, regeneration of the alumina adsorbent is accomplished by pH control rather than thermal treatment. Arsenic has also been effectively removed from potable water by activated alumina (55). Activated alumina is receiving renewed attention as an adsorbent and a wealth of information has been published on adsorption characteristics of many chemical species (56, 57). Decontamination of chemical warfare agents using activated aluminum oxide has been patented (58).

BIBLIOGRAPHY

"Aluminum Oxide (Alumina)" under "Aluminum Compounds" in *ECT* 1st ed., Vol. 1, pp. 640–649, by J. D. Edwards, Aluminum Company of America, and A. J. Abbott, Shawinigan Chemicals Limited; in *ECT* 2nd ed., Vol. 2, pp. 41–58, by D. Papée and R. Tertian, Cie de Produits Chimiques et Electrométallurgiques, Péchiney; in *ECT* 3rd ed., Vol. 2, pp. 218–244, by G. MacZura, K. P. Goodboy, and J. J. Koenig, Aluminum Company of America; "Aluminum oxide (alumina), activated" on *ECT* 4th ed., Vol. 2, pp. 291–302 by Alan Pearson, Aluminum Company of America; "Aluminum Oxide (Alumina), Activated" on *ECT* (online), posting date: December 4, 2000, Alan Pearson, Aluminum Company of America.

CITED PUBLICATIONS

1. C. Misra, *Industrial Alumina Chemicals, ACS Monogr. 184*, American Chemical Society, Washington, D.C., 1986, p. 78.
2. K. Wefers and C. Misra, *Oxides and Hydroxides of Aluminum, Alcoa Technical Paper 19*, revised, Alcoa Laboratories, Aluminum Company of America, Pittsburgh, Pa., 1987, p. 52.
3. B. C. Lippens and J. J. Steggerda in B. G. Linsen, ed., *Physical and Chemical Aspects of Absorbents and Catalysts,* Academic Press, New York, 1970, pp. 188–194.
4. Ref. 2, 51–54.
5. C.S. John, N.C.M. Alma, and G.R. Hays, *Appl. Catal.* **6**, 341–346 (1983).
6. M. Astier and K. S. W. Sing, *J. Chem. Tech. Biotechnol.* **30**, 691–698 (1980).
7. D. Aldroft, G. C. Bye, J. G. Robinson, and K. S. W. Sing, *J. Appl. Chem.* **18**, 301–306 (1968).
8. Ref. 3, pp. 189, 190.

9. R. K. Oberlander in B. E. Leach, ed., *Aluminas for Catalysis: Their Preparation and Properties,* Vol. 3, *Applied Industrial Catalysis*, Academic Press, New York, 1984, pp. 98–102.

10. S. J. Wilson and M. H. Stacey, *J. Colloid and Interface Sci.* **82**(2), 507–517 (1981).

11. *Alumina Products and Technology*, Product Data, Kaiser Chemicals Corporation, Baton Rouge, La., 1984. Note: Kaiser is now owned by La Roche Chemicals Corporation.

12. *Calcination of Catapal Aluminas*, Technical Information, Vista Chemical Company, Houston, Tex., 1986.

13. P. Burtin, J. P. Brunelle, M. Pijolat, and M. Soustelle, *Appl. Catal.* **34**, 225–238 (1987).

14. Ref. 2, p. 49.

15. U.S. Pat. 2,876,068 (Mar. 3, 1959), R. Tertian and D. Papée (to Péchiney Co., France).

16. Ref. 3, p. 186.

17. U.S. Pat. 3,226,191 (Dec. 18, 1965), H. E. Osment and R. L. Jones (to Kaiser Aluminum and Chemicals Corporation).

18. K. Yamada, *Nippon Kagaku Kaishi* **9**, 1486–1492 (1981).

19. J. B. Peri, *J. Phys. Chem.* **69**(1), 220–230 (1965).

20. U.S. Pat. Appl.2002/Q113344 (Aug. 22, 2002), T. Kitagama and M. Kajiita (to Burr and Brown).

21. W. M. Fish, *Light Met. 1974* **3**, 673 (1974).

22. U.S. Pat. 2,915,365 (Dec. 1, 1959), F. Saussol (to Pechiney Company, France).

23. U.S. Pat. Appl.2001/0012498 (Aug. 9, 2001), J.-M. Lamerant.

24. U.S. Pat. 3,392,125 (July 9, 1968), A. C. Kelly, H. J. Ducote, and L. R. Barsotti (to Kaiser Aluminum and Chemicals Corporation).

25. U.S. Pat. 3,856,708 (Dec. 14, 1974), V. G. Carithers (to Reynolds Metals Company).

26. U.S. Pat. 4,411,771 (Oct. 25, 1983), W. E. Bambrick and M. S. Goldstein (to American Cyanamid Company).

27. U.S. Pat. 4,579,839 (Apr. 1, 1986), A. Pearson (to Aluminum Company of America).

28. U.S. Pat. 4,390,456 (June 28, 1983), M. G. Sanchez, M. V. Earnest, and N. R. Laine (to W. R. Grace & Company).

29. U.S. Pat. 3,714,313 (Jan. 30, 1973), W. Belding and co-workers (to Kaiser Aluminum and Chemicals Corporation).

30. U.S. Pat. 2,620,314 (Dec. 2, 1952), J. Hoekstra (to Universal Oil Products Company).

31. U.S. Pat. 3,933,679 (Jan. 20, 1976), W. H. Weitzel and L. D. LaGrange (to General Atomics Corporation).

32. U.S. Pat. 6,197,073 (March 6, 2001), M. Kadner and co-workers (to Egbert Brandau).

33. R. J. Lewis, Jr., *Dangerous Properties of Industrial Materials*, 10th ed., John Wiley & Sons, Inc., New York, 2000, p. 133.

34. Z. M. George, *Sulfur Removal and Recovery from Industrial Processes*, American Chemical Society, Washington, D.C., 1975, pp. 75–92.

35. E. Bingham, B. Cohrrsen, and C. H. Powell, eds, *Patty's* Toxicology, John Wiley & Sons, Inc., New York, Vol. 8, 2001, p. 1120.

36. *S-100 Activated Alumina for Claus Catalysis*, Case Histories, Alcoa Chemicals Division, Aluminum Company of America, Pittsburgh, Pa., 1985.

37. *SP-100 Promoted Activated Alumina for Claus Catalysis,* Product Data, Alcoa Chemicals Division, Aluminum Company of America, Pittsburgh, Pa., 1984.

38. B. C. Gates, J. R. Katzer, and G. C. A. Schuit, *Chemistry of Catalytic Processes*, McGraw-Hill, New York, 1979, p. 393.

39. U.S. Pat. 3,674,680 (July 4, 1972), G. B. Hoekstra and R. B. Jacobs (to Standard Oil Company).

40. J. C. Downing and K. P. Goodboy, "Claus Catalysts and Alumina Catalyst Materials and Their Application," in L. D. Hart, ed., *Alumina Chemicals Handbook*, American Ceramic Society, Westerville, Ohio, 1990.

41. *Hydrocarbon Process.*, 19 (Apr. 1985).
42. C. J. Pereira, G. Kim, and L. L. Hegedus, *Catal. Rev. Sci. Eng.* **26**, 503–623 (1984).
43. "Catalysis '85", Special Advertising Supplement, *Chem. Week*, 32 (June 26, 1985).
44. U.S. Pat. 4,010,116 (Mar. 1, 1977), R. B. Secor, R. A. Van Nordstrand, and D. R. Pegg (to Filtrol Corporation).
45. U.S. Pat. 4,606,813 (Aug. 19, 1986), J. W. Byrne and B. K. Speronello (to Engelhard Corporation).
46. Ref. 9, pp. 73–76.
47. U.S. Pat. Appl.2002/0012629 (Jan. 31, 2002), M. Roy Auberger and co-workers.
48. U. Bien-Vogelsang and co-workers, *Chromatographia* **19**, 170–199 (1984).
49. P. R. Brown and R. A. Hartwick, eds., *High Performance Liquid Chromatography*, John Wiley & Sons, Inc., New York, 1988, pp. 165–168.
50. A. F. M. Leenars, K. Keizer, and A. J. Burggraaf, *J. Mater. Sci.* **19**, 1077–1088 (1984).
51. *Membralox Ceramic Multichannel Membrane Modules*, Technical Brochure, Alcoa/SCT, Aluminum Company of America, Pittsburgh, Pa., 1987.
52. *F-200 Activated Alumina for Adsorption Applications*, Product Data, Alcoa Chemicals Division, Aluminum Company of America, Pittsburgh, Pa., 1985.
53. R. D. Woosley, "Activated Alumina Desiccants," in L. D. Hart, ed., *Alumina Chemicals: Science and Technology Handbook*, American Ceramic Society, Westerville, Ohio, 1990.
54. R. Rubel, Jr. and R. D. Woosley, *J. Am. Water Works Assoc.* **1**, 24–49 (1979).
55. F. Rubel, Jr. and F. S. Williams, *Pilot Study of Fluoride and Arsenic Removal from Potable Water, EPA-600/2-80-100,* U.S Environmental Protection Agency, Research Triangle Park, N. C., 1980.
56. J. W. Novak, Jr., R. R. Burr, and R. Bednarik, "Mechanisms of Metal Ion Adsorption of Activated Alumina," Vol. 35, *Proc. Int. Symp. on Metals Speciation, Separation, and Recovery,* Chicago, Ill., July 27–Aug. 1, 1986, Industrial Waste Elimination Research Center of the Illinois Institute of Technology, Chicago, Ill.
57. C. P. Huang and co-workers, "Chemical Interactions Between Heavy Metal Ions and Hydrous Solids," Vol. 1, in Ref. 56.
58. U. S. Pat. 5,689,038 (Nov. 11, 1997), P. W. Bartram and G. Wagner (to the United States of America/Secretary of the Army).

ALAN PEARSON
Aluminum Company of America

ALUMINUM OXIDE (ALUMINA), CALCINED, TABULAR, AND ALUMINATE CEMENTS

1. Calcined Alumina

Calcined aluminas are generally obtained from Bayer process gibbsite [14762-49-3], α-Al(OH)$_3$(1–8), thermal decomposition of which follows the transition through the generic gamma alumina phases to α-alumina [1302-74-5] (corundum), α-Al$_2$O$_3$. Nonmineralized metal-grade or smelter-grade alumina (SGA) for

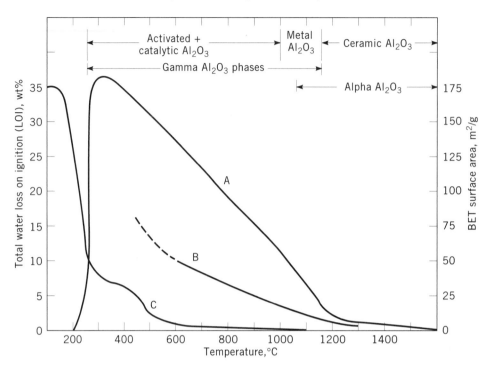

Fig. 1. Normal soda Bayer hydrate heated for one hour. A, change in surface area; B, total water, LOI plus sorbed water after exposure to 44% rh; and C, water loss on ignition when heated to 1100°C.

aluminum production is calcined at lower temperatures and usually contains about 20 to 50% α-Al_2O_3. The remainder consists of higher temperature transition aluminas, usually theta, kappa, delta, and gamma, depending upon the consolidation of the original gibbsite structure, impurities, heating rate, and furnace atmosphere. Figure 1 (9) shows the changes in gibbsite upon heating for one hour where no mineralizers are added. Loss on ignition (LOI), total water, and surface area may be used to define the degree of calcination of Bayer hydrates.

Nonmineralized SGA flows freely, and is often known as sandy alumina because it easily covers the cryolite bath of aluminum electrolysis cells (see ALUMINUM COMPOUNDS, INTRODUCTION). Properties typical of a sandy SGA are shown in Table 1. Aluminum smelting technology in the United States is primarily based upon sandy alumina. Older European smelting technology, however, is based upon a poor flowing, low bulk density, highly mineralized SGA called floury alumina, composed principally of α-Al_2O_3.

Specialty alumina derived from the Bayer process resemble SGA, except for a higher α-Al_2O_3 content, which is usually >80%, and a lower surface area, typically <20 m^2/g. The remaining material is generallysodium β-alumina (Table 2) formed as a result of the soda contamination common to the process (2–7). The mineralogical and structural properties of calcined aluminas are given in Table 2 (3).

α-Alumina, the stable form of anhydrous alumina, occurs in nature as corundum, some varieties of which are natural abrasives (qv). It is common to

Table 1. **Properties of Sandy SGA Metallurgical Alumina**[a]

Property	Value
α-Al$_2$O content, wt %[b]	20
specific surface area, m^2/g	50–80
particle size distribution, wt %	
+100 mesh[c]	5
+325[d]	92
−325[e]	8
bulk density, kg/L	
loose	0.95–1.00
packed	1.05–1.10
attrition index, wt %[f]	4–15
moisture at 573 K, wt %	1.0
loss on ignition, 573 to 1473 K, wt %	1.0
chemical composition, wt %	
Fe$_2$O$_3$	0.020
SiO$_2$	0.020
TiO$_2$	0.004
CaO	0.040
Na$_2$O	0.500

[a] Ref. 6.
[b] Determined by optical or x-ray method.
[c] Retained on 100 mesh = 149 μm Tyler screen.
[d] Retained on 325 mesh = 44 μm Tyler screen.
[e] Passes through 325 mesh = 44 μm Tyler screen.
[f] A 44 μm increase is observed by the modified Forsythe-Hertwig fluidizing test method.

igneous and metamorphic rocks and the red and blue varieties of gem quality are called ruby [12174-49-1] and sapphire [1317-82-4], respectively (6). The corundum structure consists of alternating layers of Al and O (10).

Commercial Bayer calcined aluminas, other than for aluminum production, contain substantial amounts of α-Al$_2$O$_3$. α-Al$_2$O$_3$ also results industrially from solidification of molten alumina to form artificial sapphires, abrasives, or other corundum varieties (see GEMS, SYNTHETIC) or else by sintering processes (tabular aluminas). Properties include extreme hardness, resistance to wear and abrasion, chemical inertness, outstanding electrical and electronic properties, good thermal shock resistance and dimensional stability, and high mechanical strength at elevated temperatures (see ADVANCED CERAMICS; CERAMICS; GLASS; ENAMELS).

Shapes fabricated from ground, calcined α-Al$_2$O$_3$ powders have reached 98% theoretical density at temperatures below 1723 K even though α-Al$_2$O$_3$ melts at 2326 K. The fine (submicrometer) hexagonal crystals obtained during calcination permit sintering to occur at much lower temperatures after supergrinding to dislodge the individual crystals from the Bayer agglomerate. These thermally reactive, fine-crystalline, fully ground α-Al$_2$O$_3$ products have been classified as reactive aluminas by the alumina ceramic industry, because they densify at lower sintering temperatures. The increased reactivity corresponds to decreasing sintering temperatures achieved by decreasing crystal size.

Table 2. Structural Properties of Transition Aluminas[a]

Phase	CAS Registry Number	Formula	Crystal system	Space group	Molecules per unit cell	Unit cell length, nm			Density, g/cm³
						a	b	c	
gamma		γ-Al_2O_3	tetragonal			0.562	0.780		3.2[b]
delta		δ-Al_2O_3	orthorhombic		12	0.425	1.275	1.021	3.2b
			tetragonal			0.796		2.34	
theta[c]		θ-Al_2O_3	monoclinic	C^3_{2h}	4	1.124	0.572	1.174	3.56
kappa		κ-Al_2O_3	hexagonal		28	0.971		1.786	3.1–3.3
			hexagonal			0.970		1.786	
			hexagonal			1.678		1.786	
α-alumina (corundum)	[1302-74-5]	α-Al_2O_3	hexagonal (rhombic)	D^6_{3d}	2	0.475		1.299	3.98
sodium β-aluminate	[1302-42-7]	$NaAlO_2$	orthorhombic	C^9_{2v}	4	0.537	0.521	0.707	2.693
			tetragonal			0.532		0.705	
sodium β-alumina	[11138-49-1]	$Na_2O \cdot 11Al_2O_3$	hexagonal	D^4_{6h}	1	0.558		2.245	3.24
sodium β-alumina	[12005-48-0]	$Na_2O \cdot 5Al_2O_3$	hexagonal	D^4_{6h}		0.561		3.395	
potassium β-alumina	[12005-47-9]	$K_2O \cdot 11Al_2O_3$	hexagonal	D^4_{6h}	1	0.558		2.267	3.30
magnesium β-alumina	[12428-93-2]	$MgO \cdot 11Al_2O_3$	hexagonal	D^4_{6h}	1	0.556		2.255	
calcium β-alumina	[12005-50-4]	$CaO \cdot 6Al_2O_3$	hexagonal	D^4_{6h}	2	0.554		2.183	
strontium β-alumina	[12254-24-9]	$SrO \cdot 6Al_2O_3$	hexagonal	D^4_{6h}	2	0.556		2.195	
barium β-alumina	[12254-17-0]	$BaO \cdot 6Al_2O_3$	hexagonal	D^4_{6h}	2	0.558		2.267	3.69
lithium δ-alumina	[12005-14-0]	$Li_2O \cdot 5Al_2O_3$	cubic	O^7_h	2	0.790		2.267	3.61

[a] Ref.3.
[b] Estimated.
[c] Unit cell angle is 103°20′.

2. Preparation

Calcination of gibbsite has been done in rotary kilns for many years. Specialty calcined aluminas can also be prepared in stationary or fluid bed calciners (see FLUIDIZATION) similar to those used for producing SGA. Fluid-flash calciners producing up to 1500 t/d provide a 30–40% fuel savings compared to the rotary kiln (11). Only a few specialty calcines have sufficient commercial requirements to justify manufacture in these high production capacity fluid-flash calciners. The resulting calcined products have degrees of crystallization that vary according to the temperature, the duration of the calcination, and whether or not mineralizers are used (1–4,7,9). Mineralizer usage reduces the temperature and enhances α-Al_2O_3 crystal growth. Data for a wide variety of calcined alumina is available (1,12,13).

Rotary kiln production is preferred for coarse crystalline products which are easily attrited. Particle breakdown causes dust and materials handling problems but electrostatic precipitators are used to remove particulates from combustion products (see POWDER HANDLING). The particle size of unground calcined Bayer alumina is primarily controlled during precipitation: very little particle shrinkage occurs during calcination, even though 3 moles of water of crystallization is lost. The resulting unground porous agglomerates are nominally 100 to 325 mesh (149 to 44 µm). Scanning electron microscope (sem) data differentiates between unground Bayer agglomerates (Fig. 2) which can have crystals as fine as 0.5 µm

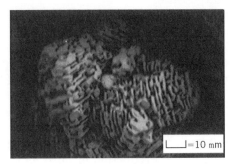

Fig. 2.   Sem photographs of controlled crystal size aluminas.

and as large as 15 μm. Physical properties begin to correlate with crystal size above about 2–3 μm. As crystal size increases, bulk density decreases, angle of repose increases, attrition rate increases, and thus, bulk handling characteristics worsen.

Soft-burned calcined products having α-Al_2O_3 crystals no larger than 1 μm may include a considerable portion of transition aluminas (1–7) and exhibit specific surfaces as high as 50 m^2/g. Most special calcined products are 80–100% α-Al_2O_3 and have a specific surface area from <0.5 to ~20 m^2/g. A considerable portion of the normal 0.3–0.6% Na_2O contained in Bayer alumina is in the form of β-alumina, $Na_2O·11Al_2O_3$.

Special calcined alumina can be broadly categorized according to soda content in addition to crystal size. Besides the normal soda products, special grades are made at the intermediate (0.15–0.25% Na_2O) and low (<0.1% Na_2O) soda levels. Extra high purity 0.5 μm calcines, typically 99.95% Al_2O_3 having less than 0.01% Na_2O, are also available in ton quantities.

2.1. Ground, Calcined, and Reactive Aluminas. Most ceramic grade aluminas are supplied dry ground to about 95% −325 mesh (44 μm) using 85–90% Al_2O_3 ceramic ball, attrition, vibro-energy, or fluid-energy milling. Particles larger than 44 μm can be removed by air classification during continuous milling to produce 99+% −325 mesh product. More fully ground, or superground, calcined aluminas having particle size distributions that approximate the natural or ultimate crystal size of the Bayer grain as calcined are often desired (8).

Thermally reactive aluminas contain submicrometer crystals. These must be separated from the Bayer agglomerate during grinding to permit dense compaction upon ceramic forming, and thus, enhance densification upon sintering at lower temperatures (14–21). Usually from 10 to greater than 30 hours dry batch ball milling is required for this separation. Such superground, thermally reactive aluminas exhibit higher densification rates when compacted and sintered into ceramic products and complete densification is obtained about 200°C lower than using the coarser, continuously ground aluminas (1).

Severe mill packing problems can occur in dry batch-milling aluminas and dry grinding aids may be used (14,15). These sorb on the surface of the aluminas, reduce the energy required to separate the individual crystals from the Bayer agglomerates and prevent mill packing by developing repelling surface charges on the alumina crystals. Mill packing tendencies also decrease with increasing mill size and ball-to charge ratio.

Reactive aluminas have enabled 85, 90, and 95% Al_2O_3 ceramics to be upgraded (16,17), because they could be sintered without fluxes in the temperature range of about 1723–2023 K, rather than 2073–2123 K. Advances in microminiaturization of components for the electronic, computer, and aerospace industries have been directly related to the development of low soda and reactive aluminas (18).

2.2. Specialty Aluminas. Process control (qv) techniques permit production of calcined specialty aluminas having controlled median particle sizes differentiated by about 0.5 μm. This broad selection enables closer shrinkage control of high tech ceramic parts. Production of pure 99.99% Al_2O_3 powder from alkoxide precursors (see ALKOXIDES, METAL), apparently in spherical form, offers the potential of satisfying the most advanced applications for calcined aluminas requiring tolerances of ±0.1% shrinkage.

The difficulty of dispersing superground aluminas which contain dense, repelletized agglomerates is illustrated in Figure 3**a**. Deflocculated 0.5 μm material is only partially dispersed when magnetically stirred, in comparison to the completely dispersed sample produced by using an ultrasonic probe. In contrast, the dispersible alumina is essentially fully dispersed as shown in Figure 3**b** (8). This product, with agglomerates mechanically disintegrated, also disperses fully

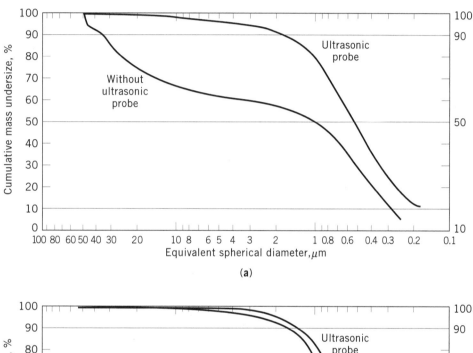

(**a**)

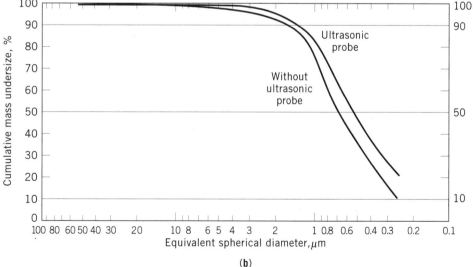

(**b**)

Fig. 3. Sedigraph particle size distribution for superground submicrometer alumina. (**a**) Partially dispersed; (**b**) fully dispersed.

in 3 to 5 min where low shear mixers used to homogenize refractories are employed. Improved predictability of the rheology and forming characteristics of advanced refractory mixes results.

3. Economic Aspects

Calcined alumina markets consume slightly less than 50% of the specialty alumina chemicals production (1–8,20,22–115). Worldwide usage is estimated to be

Table 3. **World Production of Alumina by Country** $\times 10^3$ t ^{a,b}

Country	1996	1997	1998	1999	2000^c
Australia	13,348	13,385	13,853	14,532	15,681^e
Azerbaijan^c	5	10	^{f}	50^{d,e}	200
Bosnia and Herzegovina^c	50	50	50	50^d	50
Brazil	2,752	3,088	3,322	3,515^d	3,500
Canada	1,060	1,165	1,229	1,233	1,200
China^c	2,550	2,940	3,330	3,840	4,330^e
France	440	454	450^c	400^c	400
Germany	755	738	600^{c,d}	583^d	700
Greece^c	602^e	602	600	600	600
Guinea^c	640	650^e	480	500	550
Hungary	208	76	138	145^d	150
India^c	1,780	1,860	1,890	1,900	2,000
Ireland	1,234	1,273	1,200^c	1,200^c	1,200
Italy	881	913	930	973	950
Jamaica	3,200	3,394	3,440	3,570	3,600
Japan^g	337	368	359	335^d	340
Kazakhstan	1,083	1,095	1,085	1,152	1,200
Romania	261	282	250	277	417^e
Russia	2,105	2,400^c	2,465	2,657	2,850
Serbia and Montenegro	186	160^c	153	156	250
Slovakia^c	100	100	100	100	100
Slovenia	88	85	70^c	70^c	70
Spain^h	1,095	1,110	1,100^c	1,200^c	1,200
Suriname^c	1,600	1,600	1,600	0^d	0
Turkey	159	164	157	159^d	155^e
Ukraine	1,000^e	1,080^c	1,291	1,230	1,360
United Kingdom	99	100^c	96	90^d	100
United States	4,700	5,090	5,650^d	5,140^d	4,780^e
Venezuela	1,701	1,730	1,553	1,335	1,400
Total	*44,000*	*46,000*	*47,400*^{d}	*47,000*^{d}	*49,300*

^{a} From Ref. 116.

^{b} Figures represent calcined alumina or the total of calcined alumina plus the calcined equivalent of hydrate when available; exceptions, if known, are noted. World totals, U.S. data, and estimated data are rounded to more than three significant digits; may not add to totals shown. Table includes data available through July 25, 2001.

^{c} Estimated.

^{d} Revised.

^{e} Reported figure.

^{f} Production shaprly curtailed or ceased.

^{g} Data presented are for alumina used principally for specialty appalications. Information on aluminum hydrate for all uses is not adequate to formulate estimates of production levels.

^{h} Hydrate.

for refractories (qv), for abrasives, and for ceramics (qv). Calcined aluminas are also used in the manufacture of tabular alumina and calcium aluminate cements (CAC).

World output of alumina increased 5% in 2000. Principal producing countries were Australia, the United States, China, and Jamaica. They accounted for 60% of world production. See Table 3 for world production data (116).

The United States depends on imports for one half of its metallurgical requirements despite the fact that it is one of the top four producers. Table 4 shows export/import data for the United States (117). Average annual value of U.S. imports of calcined alumina was $226/t fas port shipment and $238/t cif U.S. (116).

Capacities of U.S. plants are given in Table 5. Annual capacity was 6.35 × 10^6 t with four Bayer refineries in operation. Of the total alumina used, 90% went to primary aluminum smelters and the remainder went to nonmetallurgical uses (116).

Metallurgical-grade alumina spot prices began the year 2000 at $375–385/t, in mid-year prices rose to $420–440/t, and at year-end, prices dropped to $165–175/t. The alumina market ended in 2000 in an oversupply of product. This continued into the first half of 2001 (116).

Table 4. **U.S. Imports and Exports of Alumina, t, calcined equivalent**[a,b] (metric tons, calcined equivalent)

Country	2001	2002 First quarter	Second quarter	Year to Date
Imports (for consumption)[c]				
Australia	1,810,000	421,000	451,000	872,000
Canada	83,800	23,900	21,300	45,200
India	2,830	54	21	75
Jamaica	278,000	93,300	45,900	139,000
Suriname	654,000	195,000	140,000	335,000
Other	272,000	31,200	42,600	73,800
Total	*3,100,000*	*764,000*	*701,000*	*1,460,000*
Exports[d]				
Brazil	1,420	276	331	607
Canada	1,100,000	292,000	207,000	498,000
Mexico	35,000	16,200	9,670	25,900
Other	107,000	13,400	48,500	61,900
Total	*1,250,000*	*322,000*	*265,000*	*587,000*

[a] Ref. 117. U.S. Census Bureau; data adjusted by the U.S. Geological Survey.
[b] Data are rounded to no more than three significant digits; may not add to totals shown.
[c] Includes imports of aluminum hydroxide: 159,000 tons in 2001; 31,800 tons in the first quarter, and 45,300 tons in the second quarter of 2002.
[d] Includes exports of alumina from the U.S. Virgin Islands to foreign countries. Includes exports of aluminum hydroxide: 36,600 tons in 2001; 16,500 tons in the first quarter, and 10,300 tons in the second quarter of 2002.

Table 5. **Capacities of Domestic Alumina Plants,**
$\times 10^3$ **t/yr** [a,b]

Company and Plant	1999	2000
Alcoa Inc.		
Point Comfort, Texas	2,300	2,300
St. Croix, VI[c]	600	600
Total	*2,900*	*2,900*
BPU Reynolds, Inc., Corpus Christi, Texas	1,600	1,600
Kaiser Aluminum & Chemical Corp., Gramercy, La.	(4/)[d]	1,250[d]
Ormet Corp., Burnside, La.	600	600
Grand total	*5,100*	*6,350*

[a] From Ref. 116, as of Dec. 31.

[b] Data are rounded to no more than three significant digits; may not add to totals shown. Capacity may vary depending on the bauxite used.

[c] Temporarily shutdown.

[d] Damaged in an explosion, partial restart in December 2000.

4. Uses

4.1. Ceramics. Calcined aluminas are used in both electronic and structural ceramics (see ADVANCED CERAMICS). Electronic applications are dominant in the United States and Japan whereas mechanical applications are predominant in Europe (1). Specialty electronic integrated circuit packages generally use the low soda and thermally reactive aluminas.

Enamels, glass, chinaware glazes, china and hotel ware, and electrical porcelain insulators usually contain 5 to 25% alumina additions to increase strength and chip resistance, whereas electronic and mechanical alumina ceramics contain greater than 85% Al_2O_3 as calcined alumina. Coarse crystalline (2 to 10 µm) aluminas having 0.05 to 0.20% Na_2O are used in spark plug insulators which is the largest use of alumina in the electronics field. High purity 99.99% Al_2O_3 (2,3) is used to make translucent polycrystalline alumina tubes for sodium vapor lamps. Traditional glass tubes allow significant sodium diffusion at the operating temperatures of the sodium vapor lamps. All varieties of calcined aluminas are used in mechanical and technical applications. But when optimum hardness, density, and wear resistance are required, the thermally reactive aluminas are used in 95% and higher Al_2O_3 compositions. Lower price, normal soda calcined aluminas are used in compositions as low as 85% Al_2O_3 whenever lower performance can be tolerated.

Cutting tools of thermally reactive, high purity aluminas in combination withzirconia, titanium carbide ortitanium nitride, the SIALONS, and boron, nitride have high mechanical strength, fracture toughness, and cutting behavior for high speed cutting of hard steel and cast iron. High mechanical strength, fine surface finish, high density, and high purity are also the requirements for alumina ceramics used in prosthetics such as hip joints and dental implants.

Other alumina ceramic applications include ceramic armor for bullet-proof vests, balls and rods for grinding media, abrasion-resistant tiles for lining coal and ash transfer lines in power stations, electrical high tension insulators, bioceramics, integrated electronic circuits, vacuum tube envelopes, r-f windows, rectifier housing, integrated circuit packages, and thick and thin film substrates.

4.2. Abrasives. Special unground calcined aluminas are used both as feedstock in the manufacture of white fused alumina and as abrasives themselves. Fused alumina production processes and applications have been documented (2,47,48). Graphite electrode, arc-resistance type furnaces are used to melt the calcined alumina in large water-cooled steel shells/pots. The optimum unground calcined alumina contains less than 1 µm crystals having a low water adsorptive capacity of about 1% LOI at 1100°C after exposure to 44% rh. This nominally 100 to 325 mesh (149 to 44 µm) free-flowing, sandy alumina provides good coverage of the fused alumina melt. Dust losses are reduced by minimizing the −325 mesh fraction. Controlled water and fluoride additions are claimed to improve whiteness or discoloration by increasing conductivity at the electrode−melt interface (114,115).

A method for using recycled aluminum oxide ceramics in industrial application has been reported (118). Alumina ceramics products which 90–97.5 wt% aluminum oxide content have been made (119).

Calcined aluminas are also used for polishing applications by mixing into polishing compounds in the form of paste or suspensions. Polishing aluminas are used to alter the surfaces of metals, plastics, glass, and stones in the manufacture of cutlery, automobiles, computers, furniture, eyewear, semiconductors, and jewelry. Polishing aluminas are also used to coat surfaces, such as video tapes (1).

4.3. Refractories. Calcined alumina is used in the bond matrix to improve the refractoriness, high temperature strength/creep resistance, and abrasion/corrosion resistance of refractories (1,2,4,7). The normal, coarse (2 to 5 µm median) crystalline, nominally 100% α-Al_2O_3, calcined aluminas ground to 95% −325 mesh are used to extend the particle size distribution of refractory mixes, for alumina enrichment, and for reaction with chemical binders and/or clays for reaction bonding. One or more of the calcined aluminas are utilized in amounts to 10% for special refractories requiring optimum density (4,55).

Unground calcined aluminas are also used as feedstock in the production of fused refractory fibers, bubbles, aggregate, and fused-cast alumina refractories. Abrasive grade calcined alumina is satisfactory for production of all, except for the fused-cast refractories which require a hard-burned Bayer alumina calcined to a low (less than 0.2% ignition loss after exposure to 44% rh) water adsorptive capacity to minimize porosity. Low sulfur requirements for both abrasive and fusion-cast grades of calcined alumina limit the type of fuels that can be used for calcining. The alumina bubbles and $Al_2O_3 \cdot SiO_2$ ceramic fibers are used to produce insulating refractories having high porosity and low density structures. A process for producing aluminum oxide fibers has been patented (120).

Refractories made using aluminas are used in the iron and steel, chemical and petroleum, ceramics and glass manufacture, minerals processing (cement, lime, etc), public utilities, waste incineration, and power generation industries.

Porous high-alumina fused cost refractory having corrosion resistance and methods of production are discussed in Ref. 121. Aluminum oxide is also used in semiconductors (122–124).

5. Tabular Alumina

Tabular alumina is a high density, high strength form of α-Al$_2$O$_3$ made by sintering an agglomerated shape of ground, calcined alumina. It is available in the form of smooth balls having diameters from 3 to 25 mm and imperfect 19 mm diameter spheres, which are crushed, screened, and ground to obtain a wide variety of graded, granular, and powdered products having various particle size distributions ranging from a top size of 12.7 mm to −325 mesh (44 μm).

Tabular alumina is a recrystallized, sintered α-Al$_2$O$_3$ that gets its name from the large, elongated, flat, tablet like corundum crystals, typically 50 to greater than 400 μm, that develop upon rapid heating of briquettes or balls. It is also characterized by closed spherical porosity (about 5–8%) entrapped in the large crystals during rapid sintering of the less than 1 μm α-Al$_2$O$_3$ crystals. Open porosity is characteristically low, being less than 5% and typically 2 to 3% (55).

Tabular alumina processing is similar to that required for making sintered alumina ceramics. Ground calcined alumina is shaped by agglomerating, extruding or pressing, and sometimes using organic binders. The compacted pieces are recrystallized upon sintering at 1873–2123 K, producing a 3.40–3.65 t/m^3 bulk-specific gravity product (99.5% Al$_2$O$_3$) having closed spherical porosity, typical of a fully sintered ceramic with secondary crystallization. Impurities typically include about 0.05% SiO$_2$, 0.05% Fe$_2$O$_3$, and less than 0.1–0.4% Na$_2$O. The lower soda levels are evident in the U.S. products; the higher level in Japanese and European products. Extensive magnetic separation is required to remove iron from the crushed grain to minimize discoloration in ceramic and refractory products. Some processes have been patented (55).

Not all sintered aluminas are tabular Al$_2$O$_3$, primarily because sintering aids, such as MgO compounds, are used to achieve densification at lower temperatures and the resultant α-Al$_2$O$_3$ crystal size is not large enough to warrant the distinction of tabular Al$_2$O$_3$. Such dense sintered alumina grain exhibits poor thermal shock characteristics in comparison with tabular alumina.

6. Uses

The large α-Al$_2$O$_3$ crystals containing closed round pores make tabular alumina an excellent refractory raw material. Advantages include: high refractoriness; high fusion point; good abrasion resistance; excellent hot load strength; low creep; good resistance to chemical attack; high density; low permeability; low reheat shrinkage; and good thermal shock resistance; and high purity, minimizing system contamination.

Tabular alumina is the ideal base material for high alumina brick and monolith liners in the metal, ceramic, and petrochemical industries. Applications in the steel industry include alumina slide-grade valves, nozzles, shrouds, weir plates, impact pads, runners, troughs, torpedo car linings, tap holes, high alumina brick, ladle linings, snorkles, and lances. Tabular alumina is universally accepted as the most effective alumina aggregate for manufacture of mullite-bonded and carbon-bonded alumina slide-gate plates used to control the flow of steel from 250 tons and larger ladles.

Tabular alumina also offers advantages over other materials as an aggregate in castables made from calcium aluminate cement as the binder and in phosphate-bonded monolithic furnace linings in all thermal processing industries. Other applications include their use in electrical insulators, electronic components, and kiln furniture. Applications other than refractories and high Al_2O_3 ceramics include molten metal filter media (125), ground filler for epoxy and polyester resins (see FILLERS), inert supporting beds for adsorbents or catalysts, and heat exchange media, among others.

7. Aluminate Cement

Refined calcined alumina is commonly used in combination with high purity limestone [1317-65-3] to produce high purity calcium aluminate cement (CAC). The manufacture, properties, and applications of CAC from bauxite limestone, as well as high purity CAC, has been described (104). High purity CAC sinters readily in gas-fired rotary kiln calcinations at 1600–1700 K. CAC reactions are considered practically complete when content of free CaO is less than 0.15% and loss on ignition is less than 0.5% at 1373 K.

Table 6. **Characteristics of Calcium Aluminate Cement (CAC) Mineral Constituents**[a]

Mineral[b]	Chemical composition, wt %		Melting point, K	Molecular weight	Density, g/cm^3	Crystal system
	CaO	Al_2O_3				
C	99.8		2843	56.1	3.25–3.38	cubic
$C_{12}A_7$	48.6	51.4	1633–1663	1387	2.69	cubic
CA	35.4	64.6	1873 2023–2038	158	2.98	monoclinic (triclinic)
CA_2	21.7	78.3	decomposes	260	2.91	monoclinic
C_2S^c	65.1		2339	172	3.27	monoclinic
C_4AF^d	46.2	20.9	1688	486	3.77	orthorhombic
C_2AS^e	40.9	37.2	1863	274	3.04	tetragonal
CA_6	8.4	91.6	2103	668	3.38	hexagonal
α-A		99.8	2324[a]	102	3.98	hexagonal

[a] Ref. 2.
[b] A = Al_2O_3, C = CaO, F = Fe_2O_3, and S = SiO_2.
[c] Contains 34.9% SiO_2.
[d] Contains 32.9% Fe_2O_3.
[e] Contains 21.9% SiO_2.

Table 6 lists the characteristics of CAC mineral constituents that can occur in CAC of varying purities. The primary hydraulic setting cement phase in all CAC grades is CA [12042-68-1], $CaO \cdot Al_2O_3$; the main secondary phase in 70% Al_2O_3 CAC is CA_2, [12004-88-5], $CaO \cdot 2Al_2O_3$. CA_2 and $C_{12}A_7$ [12005-57-1], $12CaO \cdot 7Al_2O_3$, occur in minor amounts in the 80% Al_2O_3 CAC products. The $C_{12}A_7$ phase reacts rapidly with water to initiate hydraulic bonding and early strength development for rapid mold removal when manufacturing precast shapes. A significant advantage of CAC over Portland cement is its rapid strength development, developing in 24 hours of moist curing, the same proportional strength as Portland cement after 28 days. CA_6 [12005-50-4], $CaO \cdot 6Al_2O_3$, is the only nonhydrating CAC phase.

Emplaced CAC concrete should be moist cured at temperatures greater than 294 K to avoid explosive steam spalling upon heating after the curing cycle. Below 294 K, an alumina gel forms reducing the permeability of the refractory lining. Low permeability has resulted in severe explosions during heatup of thick (>10 mm) linings. At temperatures above 300 K, the gel crystallizes into $Al(OH)_3$ resulting in increased permeability.

CAC castables have rather short working times for placement in comparison to Portland cement concretes. Thus, preconditioning, mixing, and placement of CAC refractory concrete at lower (about 285–290 K) temperatures is favored. This provides sufficiently long placement times to minimize formation defects, which enlarge on drying and firing restricting strength development. Immediately after placement, the temperature of the lining should be increased to above 300 K while being covered with polyethylene plastic sheet or a curing compound to prevent water loss during the moist curing. Hydraulic, dried, and fired strengths increase in proportion to the moist curing time used, but strength development becomes asymptotic after 48–96 hours.

8. Uses

High purity CA cements are primarily used as binders for high strength refractory castables to form linings up to about 1.0 m thick, as, for example, in iron blast furnaces. Since the 1970s, large monolithic precast CAC castable shapes have found increased usage in a variety of specialty fired shapes that are too expensive to be inventoried.

The high purity CAC finds extensive use as an efficient binder for other aggregates such as fire clays, kaolin, and alusite, kyanite, pyrophyllite, sillimanite, mullite, and refractory grade bauxite, having the added advantage of increasing the refractoriness of some of these aggregates. The many applications cited for tabular alumina in refractories are also common for high purity CAC.

High purity CAC is also used as a steel slag conditioner during ladle refining of steel. CAC clinker in the unground form is added to the steel ladle either prior to or after filling the ladle with steel to form a clean slag, after first removing most of the dirty slag used to protect the steel from oxidation in the transfer ladles. The nominal 50 to 65% Al_2O_3 CAC clinker melts rapidly to provide a protective insulating layer over the molten steel, prevent oxidation, entrap and remove sulfur and metal oxide inclusions from the steel during mixing.

BIBLIOGRAPHY

"Aluminum Oxide (Alumina)" under "Aluminum Compounds" in *ECT* 1st ed., Vol. 1, pp. 640–649, by J. D. Edwards, Aluminum Research Laboratories, Aluminum Company of America, and A. J. Abbott, Shawinigan Chemicals Limited; in *ECT* 2nd ed., Vol. 2, pp. 41–58, by D. Papée and R. Tertian, Cie de Produits Chimiques et Electrométallurgiques, Péchiney; in *ECT* 3rd ed., pp. 233–244, by G. MacZura, K. P. Goodboy, and J. J. Koenig, Aluminum Company of America. "Aluminum oxide, Calcined, Tabular, and Aluminate Cements" in *ECT* 4th ed., Vol. 2, pp. 302–317, by George MacZura, Aluminum Company of America; "Aluminum Oxide (Alumina), Calcined, Tabular, and Aluminate Cements" in *ECT* (online) posting date: December 4, 2000, by George MacZura, Aluminum Company of America.

CITED PUBLICATIONS

1. T. J. Carbone, "Production Processes, Properties, and Applications for Calcined in High-Purity Aluminas," in L. D. Hart, ed., *Alumina Chemicals: Science and Technology Handbook*, The American Ceramic Society, Columbus, Ohio, 1990.
2. W. H. Gitzen, ed., *Alumina as a Ceramic Material*, The American Ceramic Society, Columbus, Ohio, 1970, 1–253.
3. K. Wefers and G. M. Bell, *Oxides and Hydroxides of Aluminum, Technical Paper No. 19*, Aluminum Company of America, Pittsburgh, Pa., 1972, 1–51.
4. G. MacZura, T. L. Francis, and R. E. Roesel, *Interceram.* **25**(3), 200 (1976).
5. C. Misra, *Industrial Alumina Chemicals, ACS Monogr. Ser. No. 184*, The American Chemical Society, Washington, D. C., 1986.
6. L. Hudson, C. Misra, and K. Wefers, "Aluminum Oxide," in W. Gerhartz, ed., *Ullmann's* Encyclopedia of Industrial Chemistry, 5th ed., Vol. A1, VCH Publishers, Deerfield Beach, Fla., 1985.
7. J. A. Everts and G. MacZura, *Industrial Minerals Refractory Supplement*, Apr. 1983.
8. G. MacZura, K. J. Moody, and J. T. Kennedy, *Am. Ceram. Soc. Bull.* **69**(5), 844–846 (1990).
9. G. MacZura and R. J. Getty, "Bayer Process Aluminas for Ceramics," *24th Pacific Coast Regional Meeting of the ACerS*, Anaheim, Calif., Oct. 1971.
10. W. H. Bragg, *J. Chem. Soc.* **121**, 2766 (1922).
11. W. H. Fish, *TMS-AIME Paper A74-63*, 1974, p. 673.
12. D. J. De Renzo, *Ceramic Raw Materials*, Noyes Publications, Noyes Data Corporation, Park Ridge, N.J., 1987.
13. L. D. Hart and L. K. Hudson, *Am. Ceram. Soc. Bull.* **43**(1), 3 (1964).
14. U.S. Pat. 3,358,937 (Dec. 19, 1967), A. Pearson and G. MacZura (to Aluminum Company of America).
15. *Characteristics of Alumina Powders*, Alumina Ceramic Manufacturers Association, New York, 1976, pp. 1–4.
16. *Standards of the Aluminum Ceramic Manufacturers Association for High Alumina Ceramics*, 3rd ed., Alumina Ceramic Manufacturers Association, New York, 1969, pp. 1–16.
17. *Standards for High Alumina Ceramic Substrates for Microelectronic Applications*, Alumina Ceramic Manufacturers Association, New York, 1971, pp. 1–12.
18. J. C. Williams, *Am. Ceram. Soc. Bull.* **56**(7), (1977).
19. ASTM Standard Methods for Particle Size Analysis: C371, C678, C690, C721, C925, and C958, *Annual Book of Standards*, ASTM, Philadelphia, Pa., 1990.

20. A. Pearson, J. E. Marhanka, G. MacZura, and L. D. Hart, *Am. Ceram. Soc. Bull.* **47**(7), 654 (1968).

21. L. D. Hart, *Am. Ceram. Soc. Bull.* **64**(10), 968–971 (1985).

22. A. M. Houston, *Mater. Eng.* **83**(6), 51 (1976).

23. R. E. Birch, *Iron Steel Eng.* **43**, 143 (1966).

24. F. H. Norton, *Refractories*, 4th ed., McGraw-Hill Book Company, New York, 1968, pp. 331–392.

25. B. L. Bryson, Jr., *Refract. J.* **46**, 6, 9 (1971).

26. D. H. Houseman, *Steel Times Annual Review*, 1971.

27. J. A. Keitch and R. L. Stanford, *J. Met.* **25**(7), 38 (1973).

28. R. L. Shultz, *Am. Ceram. Soc. Bull.* **52**(11), 833 (1973).

29. H. Ohba and K. Sugito, *Nippon Steel Technical Report Overseas No. 5*, Tokyo, Japan, 1974, 12–22.

30. K. K. Kappemeyer, C. K. Russell, and D. H. Hubble, *Am. Ceram. Soc. Bull.* **53**(7), 519, 527 (1974).

31. *Refractories: Firebrick-Specialties, Uses and Industrial Importance*, The Refractories Institute, Pittsburgh, Pa., 1–43.

32. R. H. Herron and K. K. Baab, *Am. Ceram. Soc. Bull.* **54**(7), 654, 661 (1975).

33. K. K. Kappmeyer and D. H. Hubble, *Ironmaking Steelmaking* **3**, 113 (1976).

34. R. A. Ayers and co-workers, *Am. Ceram. Soc. Bull.* **53**(3), 220 (1974).

35. J. C. Hicks, *Am. Ceram. Soc. Bull.* **54**(7), 644 (1975).

36. W. T. Hogan and D. C. Martin, *Refractory* **1**(2), 21 (1976).

37. P. D. Hess and M. J. Caprio, "Refractories for Aluminum Melting Operations," *22nd Pacific Coast Regional Meeting*, Seattle, Wash., American Ceramic Society, Columbus, Ohio, 1969.

38. E. R. Broadfield, *Refract. J.* **48**(4), 11, 21, 48 (1973); **48**(5), 13, 15 (1973).

39. D. J. Whittemore, Jr., *Am. Ceram. Soc. Bull.* **53**(5), 456 (1974).

40. R. I. Jaffee, *Am. Ceram. Soc. Bull.* **54**(7), 657 (1975).

41. M. S. Crowley and J. F. Wygant, *Am. Ceram. Soc. Bull.* **52**(11), 828, 837 (1973).

42. R. E. Dial, *Am. Ceram. Soc. Bull.* **54**(7), 640 (1975).

43. M. S. Crowley, *Am. Ceram. Soc. Bull.* **54**(12), 1072 (1975).

44. D. Hale, *Pipeline Gas J.* **203**, 22 (Mar. 1976).

45. "Refractories in the Supply and Use of Energy," *12th Annual Symp. on Refractories*, American Ceramic Society, St. Louis Section, St. Louis, Mo., 1976.

46. P. Cichy, *Electr. Furn. Conf. Proc.* **29**, 162 (1971).

47. P. Cichy, *Electr. Furn. Conf. Proc.* **31**, 71 (1973).

48. IIT Research Institute, *Ceramics Bull. No. 4*, Chicago, Ill., 1966, pp. 1–3.

49. *Alcoa Alumina in Glass Fibers*, Aluminum Company of America, Pittsburgh, Pa., 1968, pp. 1–6.

50. U.S. Pat. 3,705,223 (Dec. 5, 1972), A. Pearson and J. E. Marhanka (to Aluminum Company of America).

51. *Chem. Eng. News* **52**(18), 24 (1974).

52. B. D. Wakefield, *Iron Age* **213**(33), 43 (1974).

53. A. E. Pickle and E. Norcross, *J. Br. Ceram. Soc.* **73**(7), 239 (1974).

54. U. S. Pat. 3,953,561 (Feb. 18, 1975), H. Shin (to E. I. du Pont de Nemours & Co., Inc.).

55. G. MacZura, "Production Processes, Properties, and Applications for Tabular Alumina Refractory Aggregates," in L. D. Hart, ed., *Alumina Chemicals: Science and Technology Handbook*, The American Ceramic Society, Columbus, Ohio, 1990.

56. A. Nishikawa, *Technology of Monolithic Refractories*, Plibrico Japan Company, Tokyo, Japan, 1984.

57. *Refractory Sp*, American Concrete Institute, Detroit, Mich., 1988.

58. *Proc. of 25th International Colloquium on Refractories*, Aachen, Germany, Oct. 14–15, 1982.

59. D. H. Hubble and K. K. Kappmeyer, *Workshop on Critical Materials*, Vanderbilt University, Nashville, Tenn., Oct. 4–7, 1982, to be published in *U.S. Bureau of Standards Bulletin*, 1983.

60. H. D. Leigh, "Refractories," in M. Grayson, ed., *Encyclopedia of Glass, Ceramics, and Cement*, John Wiley & Sons, Inc., New York, 1985.

61. H. E. McGannon, *Making, Shaping and Treating of Steel*, U.S. Steel Corporation, Pittsburgh, Pa., Dec. 1970.

62. *Refractories, Fire Brick-Specialties, Uses in Industrial Importance*, The Refractories Institute, Pittsburgh, Pa., 1975.

63. Harbison-Walker Refractories Company Technical Staff, "Refractories for Iron and Steel Plants," *Watkins Cyclopedia of the Steel Industry*, 1969.

64. *Refractories, No. 7901*, The Refractories Institute, Pittsburgh, Pa., 1979.

65. Y. Sakano and H. Takahashi, *Am. Ceram. Soc. Bull.* **67**(7), 1164–1175 (1988).

66. *International Iron and Steel Institute Statistics*, Iron Steel Maker, AIME-ISS, **15**(13), 2–16 (1988).

67. A. Kadano, *Am. Ceram. Soc. Bull.* **63**(9), 1124–1127 (1984).

68. *Modern Refractories Practice*, 4th ed., Harbison-Walker Refractories Company, Pittsburgh, Pa., 1961.

69. F. H. Norton, *Refractories*, 4th ed., McGraw-Hill, New York, 1968.

70. *Interceram* **32**, 3–146 (1983).

71. "Refractories in the Cement Industry," *Interceram* **33**, 3–82 (1984).

72. *Proc. of 1st International Conference on Refractories*, Nov. 15–18, 1983, The Technical Association of Refractories—Japan, Tokyo, Japan, 1983.

73. K. Shaw, *Refractories and Their Uses*, Halsted Press Division, John Wiley & Sons, Inc., New York, 1972.

74. W. T. Lankford and co-workers, eds., *The Making, Shaping and Treating of Steel*, 10th ed., United States Steel Company, New York, 1985.

75. C. W. Hardy and co-workers, *Committee on Technology—Special Study Team on Refractories, Refractory Materials for Steelmaking*, International Iron and Steel Institute, Brussels, Belgium, 1985.

76. R. J. Fruehan, *Ladle Metallurgy Principles and Practices*, AIME-ISS, Pittsburgh, Pa., 1985.

77. *Advances in Ceramics*, Vol. 13, The American Ceramic Society, Columbus, Ohio, 1985.

78. C. R. Beechan and co-workers, eds., "Applications of Refractories" *Ceram. Sci. Eng. Proc.* **7**(1–2) (1986).

79. *Proc. of 29th International Colloquium on Refractories*, Aachen, Germany, Oct. 9–10, 1986.

80. *Foundry Industry Scoping Study, Report No. 86–5*, Center for Metals Production, Pittsburgh, Pa., Nov. 1986.

81. J. E. Kopanda and co-workers, *Ceram. Sci. Eng. Proc.* **8**(1–2) (1987).

82. *Proc. of 2nd International Conference on Refractories*, Vols. 1–2, The Technical Association of Refractories—Japan, Tokyo, Japan, 1987.

83. M. A. J. Rigaud and co-workers, *Proc. of International Symposium on Advances in Refractories for the Metallurgical Industries*, Vol. 4, Pergamon Press, New York, 1988.

84. J. Benzel and co-workers, *Ceram. Sci. Eng. Proc.* **9**(1–2) (1988).

85. *Shinagawa Technical Report 31*, Shinagawa Refractories Company, Tokyo, Japan, 1988.

86. *Refractories*, The Refractories Institute, Pittsburgh, Pa., 1987.

87. K. Hiragushi, *Proc. of ALAFAR Congress XIV*, Canela, Brazil, Nov. 4–7, 1984.
88. P. L. Smith, J. White, and P. G. Whiteley, in *Proc. of the 2nd International Conference on Refractories*, Vol. 1, The Technical Association of Refractories—Japan, Tokyo, Japan, 1987, 101–117.
89. W. Ishikawa, T. Yamamoto, Y. Abe, and K. Okuda, *Proc. European Iron Institute Congress, 1986, Paper V-2*, Vol. 3.
90. J. M. Bauer and J. P. Kiehl, *Proc. Brit. Ceram. Soc.* **29**(10), 191 (1980).
91. R. Eschenberg and co-workers, *Interceram.* **32**, 19–24 (1983).
92. J. M. Bauer and co-workers, *Interceram.* **32**, 25–32 (1983).
93. R. L. Wessel, *ACI Sp-57 Refractory Concrete*, American Concrete Institute, Detroit, Mich., 1978, pp. 179–222.
94. A. Egami, *Taikabutsu Overseas-Special Topics: Refractories for Iron Making* **2**(1), 71–77 (1982).
95. L. P. Krietz, R. Woodhead, S. Chadhuri, and A. Egami, *Advances in Ceramics*, Vol. 13, American Ceramic Society, Columbus, Ohio, 1985, 323–330.
96. A. Watanabe and co-workers, *Proc. of 2nd International Conference on Refractories*, Vol. 1, The Technical Association of Refractories—Japan, Tokyo, Japan, 1987, pp. 118–132.
97. Y. Naruse, *Proc. of 2nd International Conference on Refractories*, Vol. 1, The Technical Association of Refractories—Japan, Tokyo, Japan, 1987, 3–60.
98. W. Kroenert, *Advances in Ceramics*, Vol. 13, American Ceramic Society, Columbus, Ohio, 1985, pp. 21–45.
99. T. Nishina, S. Takehara, and M. Terao, "Special Topics: Refractories for Iron Making," *Taikabutsu Overseas* **2**(1), 98–109 (1982).
100. K. Sugita and Y. Shinohara, *Interceram.* **32**, 111–118 (1983).
101. *Ironmaking, Proc. of Symposium on Taphole Mixes*, AIME 35, St. Louis, Mo., 1976, pp. 79–96.
102. J. A. Cummins, S. A. Nightingale, and I. N. Mackay, *Proc. 2nd International Conference on Refractories*, Vol. 1, Technical Association of Refractories—Japan, Tokyo, Japan, 1987, pp. 133–146.
103. J. Kopanda and G. MacZura, "Production Processes, Properties, and Applications for Calcium Aluminate Cements," in L. D. Hart, ed., *Alumina Chemicals: Science and Technology Handbook*, American Ceramic Society, Columbus, Ohio, 1990.
104. T. D. Robson, *High-Alumina Cements and Concretes*, John Wiley & Sons, Inc., New York, 1962.
105. T. D. Robson, *Refractory Concrete, SP-57*, American Concrete Institute, Detroit, Mich., 1978.
106. *Refractory Concrete, State-of-the-Art Report, ACI Committee 547, Report ACI 547R-79*, American Concrete Institute, Detroit, Mich., 1979.
107. R. E. Fisher, ed., *Advances in Ceramics*, Vol. 13, American Ceramic Society, Columbus, Ohio, 1985.
108. R. E. Fisher, ed., *Ceramic Transactions*, Vol. 4 of *Advances in Refractories Technology*, American Ceramic Society, Columbus, Ohio, 1989.
109. *UNITECR '89*, Vols. 1 and 2, American Ceramic Society, Columbus, Ohio, 1989.
110. *Proc. of 18th ALAFAR Congress*, San Juan, Puerto Rico, Oct. 23–27, 1988.
111. *Proc. of 19th ALAFAR Congress*, Caracas, Venezuela, Nov. 7–10, 1989.
112. *Proc. 31st International Colloquium on Refractories*, German Refractories Association, Aachen, Germany, Oct. 10–11, 1988.
113. B. McMichael, *Ind. Min. (London)* **267**(12), 19–37 (1989).
114. U.S. Pat. 3,397,952 (Aug. 20, 1968), G. MacZura and W. H. Gitzen (to Aluminum Company of America).

115. U.S. Pat. 3,409,396 (Nov. 5, 1968), H. E. Osment, R. B. Emerson, and R. L. Jones (to Kaiser Aluminum and Chemical Corporation).
116. P. Plunkert, "Aluminum and Bauxite," *Minerals Yearbook*, U.S. Geological Survey, Reston, Va., 2000.
117. P. Plunkert, "Aluminum and Bauxite," *Mineral Industry Surveys*, U.S. Geological Survey, Reston, Va., March 2002.
118. U.S. Pat. 6,203,405 (March 20, 2001), R. W. Hansen (to Idaho Powder Product LLC).
119. U.S. Pat. Appl. 2002/0010071 (Jan. 24, 2002), M. Cohen.
120. U.S. Pat. 6,036,930 (March 14, 2000), Y. Shinatani and Y. Okochi (to Toyoya).
121. U.S. Pat. Appl. 2002/010370 (Aug. 1, 2002), I. Toshihiro (to Asahi Glass Co., Ltd.).
122. U.S. Pat. Appl. 2001/10053615 (Aug. 22, 2002), J. S. Kim and co-workers (to Jones Valentine LLC).
123. U.S. Pat. 6,426,307 (July 30, 2002), C. Lim (to Hyundai Electronic Industries Co. Inc.).
124. U.S. Pat. 6,436,817 (Aug. 20, 2002), S.-J. Lee (to Hyundai Electronics Industries Co., Inc.).
125. U.S. Pat. 3,737,303 (June 5, 1973), L. C. Blayden, K. J. Brondyke, and R. E. Spear (to Aluminum Company of America).

GENERAL REFERENCES

Alumina Product Data Bulletins: No. CHE920, CHE 922, Alcoa, Bauxite, Arkansas; Alcoa Chemie, Lausanne, Switzerland; Alcan Chemicals, Cleveland, Ohio; Kaiser Chemicals, Cleveland, Ohio; Malakoff Industries (Reynolds), Malakoff, Texas; Baikowski, Charlotte, North Carolina; Criceram (Pechiney), Cedex, France; Union Carbide Specialty Powders, Indianapolis, Indiana; Sumitomo Aluminum Smelting, Tokyo, Japan; Showa Alumina Industries SAL, Tokyo, Japan; VAW Aluminum, Germany; Martinswerke, Germany; Pechiney, France; Arco Specialty Chemicals, Newton Square, Pennsylvania.

GEORGE MACZURA
Aluminum Company of America

ALUMINUM OXIDE (ALUMINA), HYDRATED

1. Introduction

The term alumina hydrates or hydrated aluminas is used in industry and commerce to designate aluminum hydroxides. These compounds are true hydroxides and do not contain water of hydration. Several forms are known; a general classification is shown in Figure 1. The most well-defined *crystalline forms* are the trihydroxides, $Al(OH)_3$: gibbsite [14762-49-3], bayerite [20257-20-9], and nordstrandite [13840-05-6]. In addition, two aluminum oxide–hydroxides, $AlO(OH)$, boelimite [1318-23-6] and diaspore [14457-84-2], have been clearly defined. The

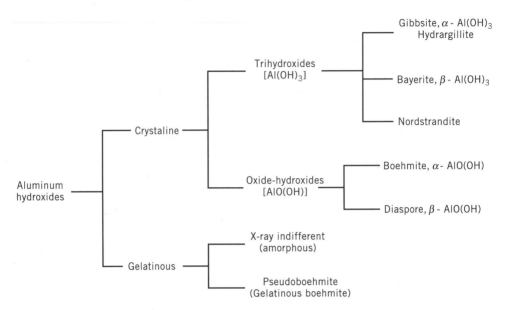

Fig. 1. Classification of aluminum hydroxides. Courtesy of Aluminum Company of America.

existence of several other forms of aluminum hydroxides have been claimed. However, there is controversy as to whether they are truly new phases or structures having distorted lattices containing adsorbed or interlamellar water and impurities.

The terms gelatinous alumina or alumina gel cover a range of products in which colloidal hydrated alumina is the predominant solid phase. Structural order varies from x-ray indifferent (amorphous) to some degree of crystallinity. The latter product has been named pseudoboehmite or gelatinous boehmite. Its x-ray diffraction pattern shows broad bands that coincide with the strong reflections of the well-crystallized boehmite.

2. Crystalline Alumina Hydrates

The mineralogical, structural, physical, and thermodynamic properties of the various crystalline alumina hydrates are listed in Tables 1, 2, and 3, respectively. X-ray diffraction methods are commonly used to differentiate between materials. Density, refractive index, tga, and dta measurements may also be used.

2.1. Gibbsite (α-Aluminum Trihydroxide, Hydrargillite). Gibbsite, commonly associated with bauxite [1318-16-7] deposits of tropical regions, is the most important aluminum compound. The gibbsite lattice consists of double layers of hydroxide ions, and aluminum occupies two-thirds of the interstices within the layers. The hydroxyls of adjacent layers are situated directly opposite each other. The layers are somewhat displaced in the direction of the a-axis and the hexagonal symmetry (brucite type) is lowered to monoclinic. The particle size of gibbsite varies from 0.5 to nearly 200 μm depending on the method of prepara-

Table 1. **Mineralogical Properties of Aluminum Hydroxides**

Material	Index of refraction[a]			Cleavage	Brittleness	Mohs' hardness	Luster
	α	β	γ				
gibbsite	1.568	1.568	1.587	(001) perfect	tough	$2\frac{1}{2} - 3\frac{1}{2}$	pearly vitreous
boehmite	1.649	1.659	1.665	(010)		$3\frac{1}{2} - 4$	
diaspore	1.702	1.722	1.750	(010) perfect	brittle	$6\frac{1}{2} - 7$	brilliant pearly

[a] The average index of refraction for bayerite is 1.583.

Table 2. **Structural Properties of Aluminum Hydroxides**

Material	Crystal system[a]	Space group	Unit axis length, nm			Angle	Density, g/cm³
			a	b	c		
$Al(OH)_3$							
gibbsite	monoclinic[b]	C_{2h}^5	0.8684	0.5078	0.9136	94°34′	2.42
bayerite	monoclinic	C_{2h}^5	0.5062	0.8671	0.4713	90°27′	2.53
nordstrandite	triclinic	C_1^1	0.5114	0.5082	0.5127	70°16′ 74°0′ 58°28′	
$AlO(OH)$							
boehmite	orthorhombic	D_{2h}^{17}	0.2868	0.1223	0.3692		3.01
diaspore	orthorhombic	D_{2h}^{16}	0.4396	0.9426	0.2844		3.44

[a] Unit cell contains two molecules unless otherwise indicated.
[b] Unit cell contains four molecules.

tion. The smaller crystals are composed of plates and prisms whereas the larger particles appear as agglomerates of tabular and prismatic crystals. The basic crystal habit is pseudo-hexagonal tabular.

The usual commercial method of preparation of gibbsite is by crystallization from a supersaturated caustic aluminate, $NaAlO_2$, solution. Seed gibbsite crystals are used.

$$NaAlO_2 + 2\,H_2O \longrightarrow Al(OH)_3 + NaOH$$

Table 3. **Thermodynamic Data for Crystalline Aluminum Hydroxides at 298.15 K and 0.1 MPa[a]**

Substance	Molecular weight	Molar vol, cm³/mol	ΔH°_f, kJ/mol[b]	ΔG°_f, kJ/mol[b]	S°, J/(mol·K)[b]	C_p, J/(mol·K)[b]
gibbsite	78.004	31.956	−1293.2	−1155.0	68.44	91.7
bayerite	78.004		−1288.2	−1153.0		
boehmite	59.989	19.55	−990.4	−915.9	48.43	65.6
diaspore	59.989	17.76	−999.8	−921.0	35.33	53.3

[a] To convert MPa to psi, multiply by 145.
[b] To convert J to cal, divide by 4.184.

Fig. 2. Aluminum trihydroxides: (**a**) coarse gibbsite from the Bayer process, ×100; (**b**) Schmäh bayerite, ×10,000. Courtesy of Aluminum Company of America.

Alternatively, neutralization of sodium aluminate [1302-42-7] by CO_2 can be employed.

$$2\,NaAlO_2 + CO_2 + 3\,H_2O \longrightarrow 2\,Al(OH)_3 + Na_2CO_3$$

Crystallization at temperatures of about 40°C results in heavy nucleation and a fine product. At temperatures above 75°C, only crystal growth occurs, giving rise to large, well-crystallized aggregates composed of hexagonal rods and prisms (Fig. 2a).

Gibbsite usually contains several tenths of a percent of alkali metal ions; the technical product, precipitated from a sodium aluminate solution, contains up to 0.3% Na_2O which cannot be washed out even using dilute HCl. Several authors (1,2) suggest that these alkali ions are an essential component of gibbsite structure.

Gibbsite is an important technical product and world production, predominantly by the Bayer process. Most (90%) is calcined to alumina [1344-28-1], Al_2O_3, to be used for aluminum production. The remainder is used by the chemical industry as filler for paper, plastics, rubber, and as the starting material for the preparation of various aluminum compounds, alumina ceramics, refractories, polishing products, catalysts, and catalyst supports.

2.2. Bayerite (β-Aluminum Trihydroxide).   Bayerite is rarely found in nature. It has been synthesized by several methods: A pure product is prepared by the Schmäh method (3) in which amalgamated aluminum reacts with water at room temperature. Other methods include rapid precipitation from sodium aluminate solution by CO_2 gassing, aging of gels produced by neutralization of aluminum salts with NH_4OH, and rehydration of transition rho alumina.

Unlike gibbsite, pure bayerite can be prepared without any alkali ions and there is evidence that bayerite converts irreversibly to gibbsite in the presence of alkali (Na and K) metal ions. Bayerite does not form well-defined single crystals for proper structural analysis. The most commonly observed growth forms are spindle or hourglass shapes formed by stacking of $Al(OH)_3$ layers in a direction perpendicular to the basal plane (Fig. **2b**). The bayerite lattice is also composed of double layers of OH, but hydroxyl groups of one layer lie in the depressions between the OH positions of the second. This approximately hexagonal close packing results in the higher density of bayerite compared to that of gibbsite.

Bayerite is a commercially available technical product that is produced in small quantities mainly for alumina catalyst manufacture. High purity aluminum [7429-90-5] metal has been converted to bayerite to produce very high purity aluminum oxides.

2.3. Nordstrandite.   The x-ray diffraction pattern of an aluminum trihydroxide which differed from the patterns of gibbsite and bayerite was published (4) prior to the material, named nordstrandite, being found in nature. The nordstrandite structure is also assumed to consist of double layers of hydroxyl ions and aluminum occupies two-thirds of the octahedral interstices. Two double layers are stacked with gibbsite sequence followed by two double layers in bayerite sequence.

Pure nordstrandite has been prepared (5) by reaction of aluminum, aluminum hydroxide gel, or hydrolyzable aluminum compounds with aqueous ethylenediamine [107-15-3]. However, no commercial production or uses have been reported.

2.4. Boehmite (α-Aluminum Oxide-Hydroxide).   Boehmite, the main constituent of bauxite deposits in Europe, is also found associated with gibbsite in tropical bauxites in Africa, Asia, and Australia. Hydrothermal transformation of gibbsite at temperatures above 150°C is a common method for the synthesis of well-crystallized boehmite. Higher temperatures and the presence of alkali increase the rate of transformation. Boehmite crystals of 5–10 μm size (Fig. 3) are produced by this method. Fibrous (acicular) boehmite is obtained under acidic hydrothermal conditions (6). Excess water, about 1% to 2% higher than the stoichiometric 15%, is usually found in hydrothermally produced boehmite.

The structure of boehmite consists of double layers in which the oxygen ions exhibit cubic packing. Hydroxyl ions of one double layer are located over the depression between OH ions in the adjacent layer such that the double layers

5.0 mm

Fig. 3.   Aluminum oxide–hydroxide: hydrothermally prepared boehmite, ×2,000. Courtesy of Aluminum Company of America.

are linked by hydrogen bonds between hydroxyls in neighboring planes. There is some technical production and use of synthetically produced boehmite.

2.5. Diaspore (β-Aluminum Oxide Hydroxide). Diaspore, found in bauxites of Greece, China, and the former USSR, can also be obtained by hydrothermal transformation of gibbsite and boehmite. Higher (>200°) temperatures and pressure (>15 MPa–150bar) are needed for synthesis and the presence of diaspore seed crystals helps to avoid boehmite formation.

In the diaspore structure, the oxygen ions are nearly equivalent, each being joined to another by way of a hydrogen ion and arranged in hexagonal close packing. This arrangement accounts for the higher density of diaspore as compared to boehmite. Although diaspore-containing bauxites and clays have been used for the production of high alumina refractories, no commercial use or large-scale synthesis of diaspore has been reported.

3. Gelatinous Aluminum Hydroxides

Apart from the crystalline forms, aluminum hydroxide often forms a gel. Fresh gels are usually amorphous, but crystallize on aging and gel composition and properties depend largely on the method of preparation. Gel products have considerable technical use.

The amphoteric behavior of aluminum hydroxide, which dissolves readily in strong acids and bases, is shown in Figure 4. In the pH range of 4 to 9, a small change in pH towards the neutral value causes rapid and voluminous precipitation of colloidal hydroxide which readily forms a gel. Gels are also formed by the hydrolysis of organoaluminum compounds such as aluminum alkoxides (see ALKOXIDES, METAL).

Aluminum hydroxide gels contain considerable excess water and variable amounts of anions. Even after prolonged drying at 100–110°C, the water content

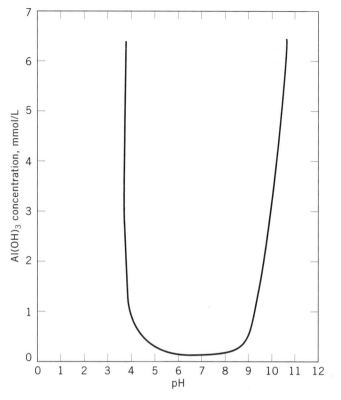

Fig. 4. Amphoteric behavior of Al(OH)$_3$. Courtesy of Aluminum Company of America.

can be as high as 5 mols H$_2$O/mol Al$_2$O$_3$. The initial precipitated product is usually amorphous (x-ray indifferent), except for material prepared at a pH above 7 or at high temperatures. Gradual transformation to crystalline hydroxide (aging) occurs; the rate is dependent on the OH ion concentration and temperature. Gelatinous boehmite is the first x-ray identifiable crystalline phase and the diffraction pattern shows broad bands that coincide with strong reflections of well-crystallized boehmite (7). In contact with a mother liquor having a pH greater than 7, gelatinous boehmite transforms into crystalline trihydroxides; the rate of transformation increases with pH and temperature. The transformation sequence may be represented by (8)

$$\text{Al}^{3+} + 3\,\text{OH}^- \xrightarrow{\text{pH}<8} \text{basic Al}^{3+} \text{ salt} \xrightarrow{\text{pH 8}} \text{xray amorphouse gel} \xrightarrow{\text{pH 9}}$$

$$\text{gelantinous boehmite} \xrightarrow{\text{pH 10}} \text{bayerite (or nordstrandite)} \xrightarrow{+\text{Na}^+ \text{ or K}^+} \text{gibbsite}$$

Gelatinous boehmite, called alumina gel in commercial use, is used in the preparation of adsorbents, desiccants (qv), catalysts, and catalyst supports (see CATALYSTS, SUPPORTED). A significant amount is used in pharmaceutical preparations.

4. Phase Relations in the $Al_2O_3-H_2O$ System

Under equilibrium vapor pressure of water, the crystalline trihydroxides, $Al(OH)_3$ convert to oxide–hydroxides at above 100°C (9,10). Below 280°–300°C, boehmite is the prevailing phase, unless diaspore seed is present. Although spontaneous nucleation of diaspore requires temperatures in excess of 300°C and 20 MPa (200 bar) pressure, growth on seed crystals occurs at temperatures as low as 180°C. For this reason it has been suggested that boehmite is the metastable phase although its formation is kinetically favored at lower temperatures and pressures. The ultimate conversion of the hydroxides to corundum [1302-74-5], Al_2O_3, the final oxide form, occurs above 360°C and 20 MPa.

Several nonequilibrium forms of aluminum oxides have been observed (11,12) in hydrothermal experiments at low water vapor pressures in the temperature region of 300–500°C. The $KI-Al_2O_3$ form, also known as tohdite [12043-15-1], $Al_2O_3 \cdot 1/5H_2O$, is characterized by a distinct x-ray diffraction pattern.

5. Production

Aluminum hydroxides are technically the most widely used members of the alumina chemicals family. The most important source of aluminum hydroxides is the bauxite refining plant for alumina production. A small amount of somewhat purer aluminum hydroxide is produced by the Sinter process.

The operating conditions and practices of alumina refining plants are generally geared to the production of a single grade of metallurgical alumina and the hydroxide has rather limited chemical uses in the form produced for this conversion. Variations in particle size, purity, and other properties are generally needed before the hydroxide can be utilized for chemical applications. Thus a part of the aluminate liquor is diverted to special purpose crystallization vessels. The crystallized product is filtered, washed thoroughly using clean hot water to remove adhering caustic liquor, and dried. Care has to be exercised in the drying operation to avoid exposure to high (>150°C) temperatures which could cause dehydration and thus affect such properties as surface activity and dissolution rate in acids and alkali. Several commercial grades of aluminum hydroxide are produced. The properties of some grades are given in Table 4.

Hydroxide grades can be surface-treated to modify dispersion behavior and rheological properties. The most widely used surface coating agents are stearic acid [57-11-4] and stearates. Additionally, compounds from the silane group have been used as coupling agents to give improved adhesion to polymers when the hydroxide is used as a filler.

5.1. Normal Coarse Grade Bayer Hydrate. The normal coarse grade Bayer hydrate has the lowest cost. It corresponds to the usual Bayer plant product produced for conversion to metallurgical alumina. Traditionally, the American Bayer product is coarser than the European product; chemical compositions are nearly identical. The hydroxide is approximately 99.5% pure. Principal impurities are soda, Fe_2O_3, SiO_2, and small amounts of TiO_2. Organic impurities originating from bauxite impart an off-white or yellowish color to the hydroxide. This low cost commodity grade material is widely used for the manufacture of alum and other aluminum chemicals.

Table 4. **Properties of Commercial Grade Aluminum Hydroxides**

Property	Normal coarse grade[a]	Normal white grade[b]	Ground[c]	Fine precipitated[d]
Al_2O_3, wt %	65.0	65.0	65.0	64.7
SiO_2, wt %	0.012	0.01	0.02	0.04
Fe_2O_3, wt %	0.015	0.004	0.03	0.01
Na_2O (total), wt %	0.40	0.15	0.30	0.45
Na_2O (soluble), wt %	0.05	0.05	0.05	0.1–0.25
LOI at 1200°C, wt %[e]	34.5	34.5	34.5	34.5
moisture at 100°C, wt %	0.1	0.1	0.4	0.3–1.0
specific gravity	2.42	2.42	2.42	2.42
bulk density (loose), g/cm^3	1.2–1.4	1.0–1.1	0.7–1.25	0.13–0.22
surface area, m^2/g	0.1	0.15	2–4	6–8
color	off-white	white	off-white	white
refractive index	1.57	1.57	1.57	1.57
Mohs' hardness	2.5–3.5	2.5–3.5	2.5–3.5	2.5–3.5
Particle size, cumulative wt %				
retained 100 mesh = 149 μm	5–20	0–1		
retained 200 mesh = 74 μm	65–90	5–15		
retained 325 mesh = 44 μm	90–98	30–65	1–2	0.1–0.2
passing 325 mesh = 44 μm	2–10	35–70	98–99	99.8
median particle size, μm			6.5–9.5	0.6

[a] Alcoa C-30.
[b] Alcoa C-31.
[c] Alcoa C-330.
[d] Alcoa Aydral 710.
[e] Loss on ignition.

5.2. White Hydroxide. The soda sinter process applied to bauxite or bauxite residue produces a hydroxide that is completely free from organic coloring matter and is very white. A value of more than 95% is obtained on the GE brightness scale relative to TiO_2 as followed in the paper (qv) industry. This compares to about 70% on the same scale for the normal Bayer product. The white hydroxide is preferred in the paper, toothpaste, and artificial marble industries.

5.3. Ground Hydroxides. Many applications of aluminum hydroxides require smaller particle sizes obtained mostly by grinding the dry hydroxide from the refining plant, followed, if necessary, by size classification. Types of grinding equipment include: ball and jar mills (sometimes ceramic lined and using ceramic balls to avoid iron contamination), vibrating mills, disc mills, and air or steam jet pulverizers. Aluminum hydroxide is relatively easy to grind; consistency of particle size distribution is desirable in the production and applications of ground hydroxides. Direct precipitation of a fine hydroxide is possible, but is not widely practiced because filtration, washing, and drying of fine products are relatively more difficult. Ground hydroxides are used extensively as fire retardant fillers in plastics (see FLAME RETARDANTS).

5.4. Low Soda Hydroxide. The Na_2O content of normal Bayer hydroxide is around 0.2–0.4%, 0.1% of which can be removed by thorough washing. The remaining soda is trapped within the hydroxide crystal. Experience shows that the occluded soda content is reduced when crystallization is carried out

Fig. 5. Fine precipitated aluminum hydroxide, ×10,000. Courtesy of Aluminum Company of America.

under low alumina-supersaturation conditions and at relatively higher temperatures (80–95°C). Soda contents as low as 0.05% Na_2O can be obtained by this procedure. However, these conditions also reduce hydroxide yield and thus increase the production cost. Low soda aluminum hydroxide is generally employed in the production of aluminas for the ceramics industries.

5.5. Extra-Fine Precipitated Hydroxide. Very fine (<1 μm-diameter) particle size hydroxide is produced by precipitation under carefully controlled conditions using specially prepared hydroxide seed. Precipitation is usually carried out at low (30–40°C) temperatures causing massive nucleation of fine, uniform hydroxide particles (Fig. 5). Tray or tunnel type dryers are used to dry the thoroughly washed filter cake to a granular product which is easily pulverized to obtain the fine hydroxide. Alternatively, the washed product is spray dried. Precipitation from an organic-free aluminate liquor, such as that obtained from the soda–sinter process, yields a very white product. The fine precipitated hydroxide is used by the paper and plastic industries as fillers.

5.6. Production of Other Aluminum Hydroxides. Commercial production of gelatinous aluminum hydroxide involves the neutralization of aluminum sulfate [10043-01-3] or ammonium alum [7784-25-0] by ammonia or sodium hydroxide. Neutralization of a sodium aluminate solution by acids, $NaHCO_3$, or CO_2 has also been used. The gelatinous product is difficult to filter and wash, and the product is often aged to improve washability. The filter cake is either dried and pulverized, or directly extruded and cut to the desired shape and size, and then dried. Spray drying has also been used. The dried powder may be agglomerated to yield a spherical product or pressed into pellets. Drying and dehydration in a hot oil bath also has been used to produce spherical particles. Rate, temperature, pH variation, agitation during neutralization, and the subsequent aging all play significant roles in the development of reproducible properties of the final product. Gelatinous aluminum hydroxides are extensively used in the preparation of adsorbent and catalytic aluminas. They are also used in the pharmaceutical industry and in the manufacture of printing inks.

The Ziegler process involves the formation of aluminum alkoxides at an intermediate stage (see ALCOHOLS, HIGHER ALIPHATIC). Hydrolysis of the alkoxides produces aluminum hydroxide having pseudoboehmite structure. The hydroxide product is further processed to remove residual alcohols and then dried. The chemical purity is generally high; the most significant contaminant is sometimes TiO_2. Catalytic applications have been the principal market for this product because of its higher purity.

Commercial production of bayerite is relatively small and employs CO_2 neutralization of caustic aluminate liquor obtained from either Bayer or sinter processes. The product obtained is about 90% crystalline bayerite having small amounts of gibbsite, pseudoboehmite, and amorphous aluminum hydroxides.

6. Shipping and Analysis

Shipping of aluminum hydroxide powders is usually in paper bags of 10 to 25 kg size. Bulk shipment by road or rail wagons is also common. Aluminum hydroxides are not hygroscopic, but could be dusty and precautions against dust inhalation should be taken during handling.

Many of the procedures used for technical analysis of aluminum hydroxides are readily available from the major producers of aluminum hydroxides.

Phase Composition. Weight loss on ignition $(110° – 1200°C)$ can differentiate between pure (34.5% $Al(OH)_3$) trihydroxides and oxide–hydroxides (15% $Al(OH)_3$). However, distinction between individual trihydroxides and oxide–hydroxides is not possible and the method is not useful when several phases are present together. X-ray powder diffraction is the most useful method for identifying and roughly quantifying the phase composition of hydroxide products.

Thermal Analyses. Thermal analysis often complements x-ray data in providing information on phase composition. The thermal behavior of aluminum hydroxides is particularly important in filler type applications.

Particle Size and Shape. Particle size information is important for many aluminum hydroxide applications. Sieve analysis, both dry and wet, is commonly used in the 20 μm (using micro-sieves) to 200 μm size range. Sizes in the 2–200 μm range have been analyzed by light microscopy, sedimentation (eg, Andreasen pipette, sedimentation balances, Sedigraph instrument), Coulter Counter, and MicroTrac instruments. Electron microscopy is the only reliable procedure for characterizing particles below 2-μm size. Scanning electron microscopy has proved useful in obtaining information on particle shape and the relation to behavior in many applications.

Alkalinity (Soluble Soda) Determination. The surface alkalinity or soluble or leachable soda is determined by making a fixed weight percent slurry in water and determining the alkalinity of the solution by pH measurement or acid titration. Sodium ion-sensitive electrodes have been investigated.

Whiteness and Brightness. Photometric instruments, originally developed by the paper industry, are used for these measurements. Values are compared against standard white pigments such as $BaSO_4$, TiO_2, or MgO.

Chemical Analysis. Chemical impurities commonly analyzed include Na_2O, Fe_2O_3, and SiO_2. The hydroxide is first dissolved in boiling concentrated HCl. Atomic absorption methods have replaced older colorimetric procedures.

Other Measurements.   Other tests include free moisture content, rate of dissolution and undissolved residue in acids and alkali, resin and plasticizer absorption, suspension viscosity, and specific surface area. Test procedures for these properties are developed to satisfy application-related specifications.

7. Economic Aspects

U.S aluminum hydroxide production capacity was nearly 940×10^3t in 1997 (13). About 90% of world production came from the Bayer process; the remaining came from Sinter, Ziegler, and gel processes. Although many alumina plants possess the capability to make the normal Bayer-grade hydroxide, specialty grade aluminum hydroxides for the chemical industry are generally produced in alumina refining plants dedicated to nonmetallurgical alumina products.

In 2000, 1×10^3t of aluminum hydroxide were consumed in the U.S.; 960×10^3t were consumed in Western Europe, and 430×10^3t were used in Japan (14).

8. Health and Safety Factors

Aluminum hydroxides are minimally absorbed by the body and LD_{50} values for ingestion are unavailable. Death upon ingestion occurs from intestinal blockage rather than systemic aluminum toxicity (15). It is only as a fine particulate suspended in air that aluminum hydroxides may gain entry (via the lungs) into the body in amounts of physiological significance. Evidence collected among workers in the alumina refining industry has failed to show any effect of aluminum hydroxide dust on the lungs. However, in recognition of the possible adverse effects of long term exposure to alumina dusts, threshold limit values have been established by the ACGIH (16) as follows: 10 mg/m^3 TLV–TWA and 20 mg/m^3 TLV–STEL. Aluminum hydroxide [21645-51-2] and aluminum hydroxide oxide [24623-77-6] are reported in EPA TSCA inventory.

9. Uses

In 2000, aluminum hydrate was used in flame retardants, reinforcement fillers in plastics, elastomers, and adhesives, filler pigments, coatings in papermaking, precursors for the production of activated alumina and other specialty aluminas, and as a raw material for the production of aluminum (14).

BIBLIOGRAPHY

"Aluminum Oxide (Alumina)" under "Aluminum Compounds" in *ECT* 1st ed., Vol. 1, pp. 640–649, by J. D. Edwards, Aluminum Research Laboratories, Aluminum Company of America, and A. J. Abbott, Shawinigan Chemicals Limited; in *ECT* 2nd ed., Vol. 2, pp. 41–58, by D. Papée and R. Tertian, Cie de Produits Chimiques et Electrométallurgiques, Péchiney; in *ECT* 3rd ed., Vol. 2, pp. 218–244, by G. MacZura, K. P. Goodboy, and J. J. Koenig, Aluminum Company of America.

CITED PUBLICATIONS

1. H. Ginsberg and M. Köster, *Z. Anorg. Allg. Chem.* **271**, 41 (1952).
2. K. Wefers, *Erzmetall* **18**, 459 (1965).
3. H. Schmäh, *Z. Naturforsch.* **1**, 323 (1946).
4. R. A. Van Nordstrand, W. P. Hettinger, and C. D. Keith, *Nature (London)* **177**, 713 (1956).
5. U. Hauschild, *Z. Anorg. Allg. Chem.* **324**, 15 (1963).
6. U. S. Pat. 2,915,475 (Dec. 1, 1959), J. Bugosh (to the E. I. du Pont de Nemour & Co., Inc.).
7. B. C. Lippens, *The Texture of Aluminas*, Poestschrift (Thesis), University of Delft, Netherlands, 1961.
8. H. Ginsberg, W. Hüttig, and H. Stiehl, *Z. Anorg. Allg. Chem.* **318**, 238 (1962).
9. G. C. Kennedy, *Am. J. Sci.* **257**, 563 (1959).
10. A. Neuhaus and H. Heide, *Ber. Dtsch. Keram. Ges.* **42**, 167 (1965).
11. K. Torkar and H. Krischner, *Monatsh. Chem.* **91**, 764 (1960).
12. G. Yamaguchi, H. Yanagida, and S. Ono, *Bull. Chem. Soc. Jpn.* **37**, 752 (1964).
13. "Inorganic Chemicals", *1997 Current Industrial Reports*, Series MA-28A, U.S. Bureau of the Census.
14. B. Suresh, A. Kishi, and S. Schlage, *Chemical Economics Handbook*, SRI, Menlo Park, Nov. 2001.
15. R. L. Bertholf, M. R. Wills, and J. Savory, "Aluminum" in H. G. Seiter and H. Sigel, eds., *Handbook on Toxicity of Inorganic Compounds*, Marcel Dekker, New York, 1988, Chapt. 6.
16. *Threshold Limit Values and Biological Exposure Indices for 1985–86*, American Conference of Governmental Industrial Hygienists, Cincinnati, Ohio, 1985.

GENERAL REFERENCES

C. Misra, *Industrial Alumina Chemicals, ACS Monogr. 184*, American Chemical Society, Washington, D.C., 1986.

K. Wefers and C. Misra, *Oxides and Hydroxides of Aluminum, Technical Paper No. 19*, Revised, Aluminum Company of America, Pittsburgh, Pa., 1987.

H. Ginsberg and K. Wefers, *Aluminum and Magnesium*, Vol. 15, Die Metallischen Rohstoffe, Enke Verlag, Stuttgart, Germany, 1971.

L. D. Hart, ed., *Alumina Chemicals Science and Technology Handbook*, The American Ceramic Society, Westerville, Ohio, 1990.

CHANAKYA MISRA
Aluminum Company of America

ALUMINUM SULFATE
AND ALUMS

1. Introduction

Aluminum sulfate octadecahydrate [7784-31-8], $Al_2(SO_4)_3 \cdot 18H_2O$, and its aqueous solutions are used primarily in the paper (qv) industry for sizing and as a flocculating agent in water (qv) and wastewater treatment. This material is often called papermakers' alum or alum. Because this salt is precipitated from aqueous solution, aluminum sulfate hydrate [17927-65-0], $Al_2(SO_4)_3 \cdot nH_2O$, can have variable composition and is sometimes referred to as cake alum or patent alum. The solid commercial hydrate, generally written as the 18-hydrate, is typically dehydrated to correspond to from 17.0–17.5% Al_2O_3 where $n = 13 - 14$ (1,2). This dehydrated form is called dry alum, ground or lump. Aluminum sulfate solutions are typically 7.5–8.5% Al_2O_3 and are known as liquid alum.

Confusion arises in the nomenclature of alum because double salt compounds, $M(I)Al(SO_4)_2$, where M is in the +1 oxidation state, have also traditionally been called alums. In particular, potassium aluminum sulfate [15007-61-1] $KAl(SO_4)_2 \cdot nH_2O$, is referred to as ordinary alum or potash alum.

Anhydrous aluminum sulfate [10043-01-3], $Al_2(SO_4)_3$, is a specialty item used in food applications.

2. Properties

Over 50 acidic, basic, and neutral aluminum sulfate hydrates have been reported. Only a few of these are well characterized because the exact compositions depend on conditions of precipitation from solution. Variables such as supersaturation, nucleation and crystal growth rates, occlusion, nonequilibrium conditions, and hydrolysis can each play a role in the final composition. Commercial dry alum is likely not a single crystalline hydrate, but rather it contains significant amounts of amorphous material.

2.1. Hydrates. Aluminum sulfate hydrates, $Al_2(SO_4)_3 \cdot nH_2O$, where n ranges from 0 to 27 have been reported (3–6). Relative decreasing vapor pressure studies indicate the presence of an octadecahydrate, hexadecahydrate, dodecahydrate, dihydrate, and the anhydrous salt, assuming that basic aluminum sulfates are not formed during the dehydration (3).

Thermal analysis of the dehydration of $Al_2(SO_4)_3 \cdot 18H_2O$ shows loss of 15 moles of water between 40°C and 250°C and 3 moles between 250°C and 420°C (7). Heating rate can affect the product. Although rapid heating can fuse hydrated aluminum sulfate containing >35% water and puff material containing 25–35% water, these problems do not occur during slow heating in an agitated bed. An aluminum sulfate hydrate can dissolve in its water of crystallization during heating.

2.2. Solubility. The aqueous solubility of a typical commercial aluminum sulfate, $Al_2(SO_4)_3 \cdot nH_2O$, where $n = 14 - 18$, is shown compared to the solubility of the octadecahydrate in Table 1 (2,8). Differences in solubilities probably

Table 1. **Aqueous Solubility of Aluminum Sulfate Hydrates, $Al_2(SO_4)_3 \cdot nH_2O$**

Temperature, °C	Solubility, $\dfrac{\text{g anhydrous salt}}{\text{100 g satd soln}}$	
	$n = 18^a$	$n = 14-18^b$
0	27.5	23.9
10	27.6	25.1
20		26.6
25	27.8	
30	28.0	28.8
35	28.4	
40	28.8	31.4
50	29.9	34.2
60	31.0	37.2
70	32.8	39.8
80		42.2
82	36.6	
90		44.7
95	41.9	
100		47.1
103	46.9	

[a] Ref. 8.
[b] Commercial hydrate data from ref. 2.

result from small amounts of impurities such as iron, and the slight excess of Al_2O_3 base present in technical grade commercial aluminum sulfate hydrate.

Aqueous solutions of aluminum sulfate can hydrolyze at temperatures of about 150°C (9) resulting in products having a formula such as $3Al_2O_3 \cdot 4SO_3 \cdot 9H_2O$, which is structurally related to alunite [12588-67-9], $K_2Al_6(SO_4)_4(OH)_{12}$ (10).

2.3. Crystallization. Acidified aluminum sulfate solutions can be supercooled 10°C or more below the saturation point. However, once nucleation begins, the crystallization rate is rapid and the supersaturated solution sets up. The onset of nucleation in a gently stirred supersaturated solution is marked by the appearance of silky, curling streamers of microscopic nuclei resulting from orientation effects of hydraulic currents on the thin, platelike crystals. Without agitation, nucleation in an acidified solution, in glass tubes, can yield extended crystalline membranes of such thinness to exhibit colors resulting from optical interference.

2.4. Other Properties. The formula weight for the octadecahydrate is 666.45 and its specific gravity is 1.69 at 17°C (11). Other physical properties such as percentage of Al_2O_3 in alum liquor vs specific gravity at 15.6°C, pH as a function of alum concentration, heat evolved on mixing $Al_2(SO_4)_3 \cdot 18H_2O$ and water, viscosity of aqueous alum solutions as a function of temperature, crystallization temperature of $Al_2(SO_4)_3 \cdot 18H_2O$ as a function of the alum concentration, and the boiling point of alum solutions as a function of the alum concentration have been reported (2,11). Representative data are given in Tables 2 and 3.

Table 2. **Aluminum Sulfate Solution Properties**

Composition, % Al_2O_3	pH^a	Specific gravity at $15.6°C^b$	Crystallization temperature, $°C^c$
2.0	2.9		
4.0	2.4	—	−4
5.0		1.182	−6
5.5		1.204	
6.0	2.1	1.222	−8
6.5		1.246	
7.0		1.269	−11
7.5		1.291	
8.0	1.9	1.314	−15
8.2			−16
8.4			−14
8.5		1.337	
8.6			8
8.8			29

^{a} Values are for neutral alum solutions; commercial alum liquor contains excess H_2SO_4, and values run about 0.5 pH units lower; slightly basic alum solutions, some excess Al_2O_3, have values of about 0.5 pH units higher.
^{b} Values depend on alum producing point; for example, at 8.0% Al_2O_3, range can be 1.311 to 1.334.
^{c} Values are for the 18-hydrate, $Al_2(SO_4)_3 \cdot 18H_2O$.

3. Manufacture

In the United States, aluminum sulfate is usually produced by the reaction of bauxite or clay (qv) with sulfuric acid (see SULFURIC ACID AND SULFUR TRIOXIDE). Bauxite is imported and more expensive than local clay, generally kaolin, which is more often used. Clay is first roasted to remove organics and break down the crystalline structure in order to make it more reactive. This is an energy intensive process. The purity of the starting clay or bauxite ore, especially the iron and potassium contents, are reflected in the assay of the final product. Thus the selection of the raw material is governed by the overall economics of producing a satisfying product.

The optimum conditions for roasting the clay and the optimum strength (30–60%) of the sulfuric acid used depend on the particular raw material. Finely ground bauxite or roasted clay is digested with sulfuric acid near the boiling

Table 3. **Dissolution Time of Dry Alum, $Al_2(SO_4)_3 \cdot 14.3H_2O$, and Heat Evolved on Mixing**

Dry alum added to 378.5 L H_2O, kg	Dissolving time, min at			Heat evolved on mixing at 18°C, kJ^a
	4°C	16°C	38°C	
91	9	6	2	8
181	15	9	3	18
272	25	15	5	26

^{a} To convert J to cal, divide by 4.184.

point of the solution (100–120°C). The clay or bauxite-to-acid ratio is adjusted to produce either acidic or basic alum as desired and solids are removed by sedimentation. If necessary, the solution can be treated to remove iron. However, few, if any, of the many methods claimed to be useful for iron removal have been used industrially (12). Instead, most alum producers prefer to use raw materials that are naturally low in iron and potassium.

The clear supernatant solution is decanted and sold in liquid form or concentrated to approximately 61.5° Bé and then allowed to solidify to form blocks that are crushed, ground, and graded. A typical analysis for the dry product is: total Al_2O_3, 17.0–17.5%; Fe_2O_3 <0.5%; water of composition 42–43%; insoluble <1.0%. Liquid alum contains 7.5–8.5% Al_2O_3. At concentrations >8.5% Al_2O_3, crystallization of the solution may occur.

The iron-free grade of aluminum sulfate hydrate contains less than 0.005% iron as Fe_2O_3. It is manufactured from pure alumina trihydrate [12252-70-9] $Al_2O_3 \cdot 3H_2O$, rather than from bauxite or clay. Although a technical or commercial grade alum can be treated to remove iron, such processes are not sufficiently economical to meet the requirements for the iron-free grade (12). The presence of iron can cause discoloration or staining of the product employing the aluminum sulfate.

A process to produce aluminum sulfate from coal ash has been reported (13).

4. Economic Aspects

In the United States 79 companies produced aluminum sulfate in 1997 (14). The total U.S. production was 1161×10^3/t on a 17% Al_2O_3 basis.

The United States exported 7690 t of aluminum sulfate and imported 23,500 t in 2000 (15).

Aluminum sulfate hydrate is marketed as material commercial or technical grade containing a maximum of 0.5% Fe_2O_3. The commercial grade is available as a dry ground product containing from 17.0 to 17.5% Al_2O_3, 57–59% $Al_2(SO_4)_3$, and as an aqueous solution containing from 7.5 to 8.5% Al_2O_3. The price of aluminum sulfate from \$152 to 292 /t in October 2002 (16).

5. Health and Safety Factors

Aluminum sulfate is orderless. When it is heated or burned, irrirant sulfur oxide fumes are produced.

The ACGIH threshold limit (TWA) is 2.0 mg/m^3 OSHA PEL is 2.0 mg/m^3 (17).

6. Uses

The pulp and paper industry and potable and wastewater treatment industry are the principal markets for aluminum sulfate. Over half of the U.S. aluminum sulfate produced is employed by the pulp and paper industry. About 37% is used to precipitate and fix rosin size on paper fibers, set dyes, and control slurry pH. Another 16% is utilized to clarify process waters. The alum sold for these purposes is usually liquid alum. It is frequently acidic as a result of a slight excess of H_2SO_4. Aluminum sulfate consumption by the pulp and paper industry is projected to remain constant or decline slightly in the near term because of more efficient use of the alum and an increased use of alkaline sizing processes (18).

Aluminum sulfate is used as a flocculating agent or coagulant in water and wastewater treatment (see FLOCCULATING AGENTS). This use, excluding the 16% employed in process water treatment by the pulp and paper industry, accounts for 44% of U.S. consumption. Alum sold to municipalities in the United States is required by the American Water Works Association (AWWA) to be basic as a result of a slight excess of Al_2O_3. When alum is added to water, positively charged hydrated or hydroxylated aluminum ions neutralize and adsorb negatively charged colloidal and suspended matter in the water causing the material to floc and settle out (19). Increased use of organic polymeric coagulants and combinations of organic and inorganic coagulants has decreased somewhat the aluminum sulfate market. Polyaluminum chloride, basic aluminum sulfate (20–22), and a more recent basic aluminum chlorosulfate (23,24) are all effective coagulants. The near term market for aluminum sulfate in water treatment is projected to remain flat.

Other uses for aluminum sulfate include tanning (see LEATHER), and as a pH stabilizer, cement (qv) hardening accelerator, mordant in dyeing (see DYES AND DYE INTERMEDIATES), anticaking agent, fire extinguisher, and as a flame retardant additive, a use which dates to the early Egyptians (see FLAME RETARDANTS). Wood (qv) and cellulose (qv) treatment, catalysis (qv), and pharmaceutical and cosmetic applications are known (see COSMETICS AND PHARMACEUTICALS). Japanese literature suggests uses as an absorbent for odor and gases (25). Alum is effective in reducing the phosphate content of sewage and wastewater and in combating the eutrophication of lakes (26,27). Russian literature cites uses in drilling muds (28).

Aluminum sulfate is a starting material in the manufacture of many other aluminum compounds. Aluminum sulfate from clay could potentially provide local sourcing of raw materials for aluminum production. Processes have been studied (29) and the relative economics of using clay versus bauxite have been reviewed (30). It is, however, difficult to remove impurities economically by precipitation, and purification of aluminum sulfate by crystallization is not practiced commercially because the resulting crystals are soft, microscopic, and difficult to wash effectively on a production scale (31–33).

Patents have been issued for the production of animal litter from ammonium sulfate (34,35).

7. Other Alums

The word alum is derived from the Latin *alumen,* which was applied to several astringent substances, most of which contained aluminum sulfate (25). Unfortunately, the term alum is now used for several different materials. Papermakers' alum or simply alum refers to commercial aluminum sulfate. Common alum or ordinary alum usually refers to potash alum which can be written in the form $K_2SO_4 \cdot Al_2(SO_4)_3 \cdot 24H_2O$, or it can refer to ammonium alum, ammonium aluminum sulfate. The term is also applied to a whole series of crystallized double sulfates $[M(I)M'(III)(SO_4)_2 \cdot 12H_2O]$ having the same crystal structure as the common alums, in which sodium and other univalent metals may replace the potassium or ammonium, and other metals may replace the aluminum. Even the sulfate radical may be replaced, by selenate, for example. Some examples of alums are cesium alum [7784-17-0], $CsAl(SO_4)_2 \cdot 12H_2O$; iron alum [13463-29-1], $KFe(SO_4)_2 \cdot 12H_2O$; chrome alum [7788-99-0], $KCr(SO_4)_2 \cdot 12H_2O$; and chromoselenic alum [17855-06-0], $KCr(SeO_4)_2 \cdot 12H_2O$.

Pseudoalums are a series of double sulfates, such as iron(II) aluminum sulfate [22429-82-9], $FeSO_4 \cdot Al_2(SO_4)_3 \cdot 24H_2O$, containing a bivalent metal ion in place of the univalent element of ordinary alums. These pseudoalums have different crystal structures from those of the ordinary alums.

In industrial practice it is generally the aluminum content of alums that is important. Because aluminum sulfate is widely available, other alums are more specialty items and are no longer produced in quantities comparable to those of aluminum sulfate (19).

7.1. Ammonium Aluminum Sulfate.

Ammonium aluminum sulfate [7784-26-1], $NH_4Al(SO_4)_2 \cdot 12H_2O$, is also known as ammonium alum or ammonia alum. It is a colorless crystal having a strong, astringent taste; formula weight 453.33; mp, 94.5°C; sp gr 1.64; and solubility of 12.0 g per 100 mL H_2O at 20°C (8). It is soluble in dilute acid and insoluble in alcohol. The material dehydrates at about 250°C (36) to porous, dry ammonium alum [7784-25-0], $NH_4Al(SO_4)_2$, formula weight 237.14, which is also called dried or burnt alum. Decomposition of the anhydrous material begins at 280°C and γ-Al_2O_3 is formed between 1000 and 1250°C.

Ammonium alum is manufactured by crystallization from a mixture of ammonium sulfate and aluminum sulfate or by the treatment of aluminum sulfate and sulfuric acid with ammonia gas. Ammonium alum is used in medicine, as a mordant in dyeing, dressing furs in tanning, paper sizing, and water purification. It can be used in baking powders (see BAKERY PROCESSES AND LEAVENING AGENTS). It has also been used as the starting material to produce finely powdered aluminum oxide for the manufacture of synthetic corundum gems (see GEMS, SYNTHETIC) (19).

7.2. Potassium Aluminum Sulfate.

Potassium aluminum sulfate [7784-24-9]. $KAl(SO_4)_2 \cdot 12H_2O$, is a white, astringent crystal known as potassium alum, ordinary alum, or potash alum. Its formula weight is 474.39; mp 92.5°C; sp gr 1.75; and solubility 11.4 g per 100 mL H_2O at 20°C (8). It is soluble in dilute acid and insoluble in alcohol. It dehydrates at about 200°C to porous desiccated potassium alum [10043-67-1], $KAl(SO_4)_2$, a dried or burnt alum, which has a formula weight of 258.20.

Potassium alum is manufactured by treating bauxite with sulfuric acid and then potassium sulfate. Alternatively, aluminum sulfate is reacted with potassium sulfate, or the mineral alum stone, alunite, can be calcined and leached with sulfuric acid. Alunite is a basic potassium aluminum sulfate [1302-91-6], $K_2Al_6(SO_4)_4(OH)_{12}$, sp gr 2.58–2.75.

Potassium alum, which also occurs naturally as the mineral kalinite [7784-24-9], $KAl(SO_4)_2 \cdot 12H_2O$, sp gr 1.75, is used in tanning skins, as a mordant in dyeing, and in the pharmaceutical and cosmetic industries (see PHARMACEUTICALS; COSMETICS). It is used as a styptic pencil and as a hardening agent and set accelerator for cement and plaster. The ACGIH threshold limit value TWA is 2 mgAl/m^3 (12).

7.3. Sodium Aluminum Sulfate. Sodium aluminum sulfate [7784-28-3], $NaAl(SO_4)_2 \cdot 12H_2O$, known as sodium alum, is a colorless crystal having an astringent taste; mp 61°C; and sp gr 1.675. It is the most water soluble alum: 75 g per 100 mL H_2O at 20°C (8). It is soluble in dilute acid and insoluble in alcohol.

Sodium alum occurs naturally as the mineral mendozite. Commercially, it is produced by the addition of a sodium sulfate solution to aluminum sulfate. Small amounts of potassium sulfate, sodium silicate, and soda ash can be added to improve product handling and performance. After adjustment of the ratio of aluminum sulfate to sodium sulfate, water is evaporated to give a hard cake in the cooling pans. This cake is further heated in roasters and ground to a fineness of 99% through a 100-mesh (~150 μm) sieve.

In the United States, sodium aluminum sulfate is used as a leavening agent in baking applications.

BIBLIOGRAPHY

"Aluminum Sulfate and Alums," under "Aluminum Compounds," in *ECT* 1st ed., Vol. 1, pp. 653–656, by W. F. Phillips, American Cyanamid Company; in *ECT* 2nd ed., Vol. 2, pp. 58–65, by W. C. Saeman, Olin Matheson Chemical Corporation; in *ECT* 3rd ed., Vol. 2, pp. 244–251, by K. V. Darragh, Stauffer Chemical Company; in *ECT* 4th ed., Vol. 2, pp. 330–338, by K. V. Darragh and C. A. Ertell, Rhône-Poulenc, Inc.; "Aluminum Sulfate and Alums" in *ECT* (online), posting date: December 4, 2000, by K. V. Darragh and C. A. Ertell, Rhône-Poulenc, Inc.

CITED PUBLICATIONS

1. *Merck Index*, 11th ed., Merck Corp., Rahway, N.J., 1989.
2. *Aluminum Sulfate*, Technical Brochure, Rhône-Poulenc Basic Chemicals Company, Shelton, Conn., 1982.
3. *A Comprehensive Treatise on Inorganic and Theoretical Chemistry*, Vol. 5, Longmans, Green and Co. Ltd., London, UK, pp. 332–357.
4. D. Taylor and H. Bassett, *J. Chem. Soc.* **1952**, 4431 (1952).
5. N. O. Smith, *J. Am. Chem. Soc.* **64**, 41 (1942).
6. N. O. Smith and P. N. Walsh, *J. Am. Chem. Soc.* **72**, 1282 (1950).
7. E. B. Gitis and co-workers, *Zh. Prikl. Khim. (Leningrad)* **46**, 1838 (1973); *Chem. Abstr.* **79**, 132465w (1973).
8. W. F. Linke, ed., *Solubilities of Inorganic and Metal-Organic Compounds*, Vol. 1, 4th ed., American Chemical Society, Washington, D.C., 1958.
9. N. F. Dyson and T. R. Scott, *Nature (London)* **205**, 358 (1965).
10. P. T. Davey, G. M. Lukaszewski, and T. R. Scott, *Aust. J. Appl. Sci.* **14**, 137 (1963).

11. P. Domaige and O. Brouhier, *Trib. CEBEDEAU* **25**(340), 114 (1972).
12. R. Siebert, G. Scholze, and S. Ziegenbalg, *Proc. 2nd Int. Symp. ICSOBA 1969* **3**, 403 (1971).
13. U.S. Pat. 6,214,302 (April 10, 2001), G. Malybaeva and D. Partovi.
14. "Inorganic Chemicals," *1997 Current Industrial Reports*, Series MA-28A, U.S. Bureau of the Census.
15. P. A. Plunkert, "Bauxite and Alumina," *Minerals Yearbook*, U.S. Geological Survey, Reston, Va., 2000.
16. *Chem. Market Rep.* **262**(15), 28 (2002).
17. B. D. Dinman, "Aluminum," in E. Bingham, B. Cohrssen, and C. H. Powell, eds., *Patty's* Toxicology, 5th ed., Vol. 2, John Wiley & Sons, Inc., New York, 2001, p. 397.
18. B. Suresh, A. Kishi, S. Schlag and J. W. Mellor, *Chemical Economics Handbook*, Stanford Research Institute, Menlo Park, Calif., Nov. 2001.
19. W. Gerhartz, *Ullmann's* Encyclopedia of Industrial Chemistry, 5th ed., vol. A1, VCH, Deerfield Beach, Fla., 1985, pp. 527–534.
20. Eur. Pat. Appl. 247,987 (Dec. 2, 1987), (to Bolinden AB).
21. U.S. Pat. 4,826,606 (May 2, 1989) (to General Chemical Corp.).
22. Brit. Pat. Appl. 2,190,074 (Nov. 11, 1987) (to Laporte Industries Ltd.).
23. Fr. Demande, 2,584,699 (Jan. 16, 1987) (to Rhône-poulenc).
24. Eur. Pat. Appl. 317,393 (May 24, 1989) (to Rhône-Poulenc).
25. Jpn. Kokai Tokkyo Koho 63 300,768 (Dec. 7, 1988) (to Daicel Chemical Industries, Ltd.).
26. J. N. Connor and G. N. Smith, *Water Resour. Bull.* **22**, 661 (1986).
27. A. Jernelov, *Proc. 5th Int. Conf. Adv. Water Pollut. Res.* **1**, 1–15, 16 (1970).
28. USSR Pat. 1,239,143 (June 23, 1986) (to Moscow Inst. Petrochem. Gas Industry).
29. H. W. St. Clair, S. F. Ravitz, A. Y. Sweet, and C. E. Plummer, *Trans. Am. Inst. Min. Metall. Pet. Eng.* **159**, 255 (1944).
30. F. A. Peters and P. W. Johnson, *Revised and Updated Cost Estimates for Producing Alumina from Domestic Raw Materials*, Circular 8648, Bureau of Mines Information, Washington, D.C., 1974.
31. E. A. Gee, W. K. Cunningham, and R. A. Heindl, *Ind. Eng. Chem.* **39**, 1178 (1947).
32. U.S. Pat. 2,951,743 (Sept. 6, 1960) (to Vereinigte Aluminum Werke, Aktiengesellschaft).
33. *Chem. Eng. N.Y.* **68**, 36 (May 1, 1961).
34. U.S. Pat. 6,206,947 (March 27, 2001), D. F. Evans, R. B. Steele, and M. E. Summer (to Waste Reduction Products Corp.)
35. U.S. Pat. 6,029,603 (Feb. 29, 2000), D. F. Evans, R. B. Steele, and M. E. Summer (to Waste Reduction Products Corp.).
36. V. V. Volodin and E. I. Khazanov, *Izv. Nauchno-Issled. Inst. Nefte Uglekhim. Sint. Irkutsk. Univ.* **1970**(12), 121 (1970); *Chem. Abstr.* **75**, 44416t (1971).

K. V. DARRAGH
C. A. ERTELL
Rhône-Poulenc, Inc.

AMIDES, FATTY ACID

1. Introduction

Fatty acid amides are of the general formula

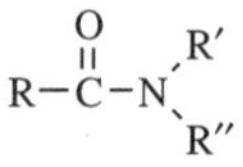

$$R-\overset{\overset{\textstyle O}{\|}}{C}-N\overset{\textstyle R'}{\underset{\textstyle R''}{\diagdown}}$$

in which R may be a saturated or unsaturated alkyl chain derived from a fatty acid. They can be divided into three categories (1). The first is primary mono-amides in which R is a fatty alkyl or alkenyl chain of C_5–C_{23} and $R' = R'' = H$. The second, and by far the largest category, is substituted monoamides, including secondary, tertiary, and alkanolamides in which R is a fatty alkyl or alkenyl chain of C_5–C_{23}; R' and R'' may be a hydrogen, fatty alkyl, aryl, or alkylene oxide condensation groups with at least one alkyl, aryl, or alkylene oxide group. The third category is bis(amides) of the general formula:

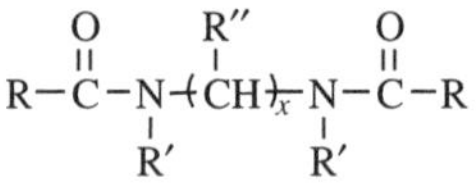

$$R-\overset{\overset{\textstyle O}{\|}}{\underset{\underset{\textstyle R'}{|}}{C}}-N\overset{\overset{\textstyle R''}{|}}{\underset{\underset{\textstyle R'}{|}}{}}(CH)_{\overline{x}}N-\overset{\overset{\textstyle O}{\|}}{C}-R$$

where R groups are fatty alkyl or alkenyl chains. R' and R'' may be hydrogen, fatty alkyl, aryl, or alkylene oxide condensation groups. Other amides include halogenated amides and multifunctional amides such as amidoamines and poly-amides. A more detailed description of the synthesis, properties, and reactions of amides can be found in (2). Examples of fatty acid amides and common nomen-clature for primary amides can be seen in Table 1.

Table 1. **Primary Fatty Amide ($RCONH_2$) Nomenclature**

Carbon atoms[a]	Common name	Molecular formula	IUPAC name	CAS Registry Number
		Alkyl		
12	lauramide	$C_{12}H_{25}NO$	dodecylamide	[1120-16-7]
14	myristamide	$C_{14}H_{29}NO$	tetradecylamide	[638-58-54]
16	palmitamide	$C_{16}H_{33}NO$	hexadecylamide	[629-54-9]
18	stearamide	$C_{18}H_{37}NO$	octadecylamide	[124-26-5]
		Alkenyl		
16(1)	palmitoleamide	$C_{16}H_{31}NO$	hexadecena-mide	[b]
18(1)	oleamide	$C_{18}H_{35}NO$	9-octadecena-mide	[301-02-0]
18(2)	linoleamide	$C_{18}H_{33}NO$	9,12-octadeca-dienamide	[3999-01-7]

[a] The number in parentheses designates double bonds in the carbon chain.
[b] Not available.

Table 2. **Fatty Amide Melting Points**

Alkyl(R) length	RCONH$_2$ Amide		RCONHCH$_2$CH$_2$OH MEA		RCON(CH$_2$CH$_2$OH)$_2$ DEA	
	CAS Registry Number	mp, °C	CAS Registry Number	mp, °C	CAS Registry Number	mp, °C
12	[1120-16-7]	103	[142-78-9]	78.2	[120-40-1]	38.7
14	[638-58-4]	104	[142-58-5]	87.4	[7545-23-5]	47.9
16	[629-54-9]	108	[544-31-0]	94.4	[7545-24-6]	65.1
18	[124-26-5]	109	[111-57-9]	96.1	[93-82-3]	69.7

For alkanolamides the abbreviation MEA and DEA are used after the common name as seen in Table 2 to signify monoethanolamide (MEA) and diethanolamide (DEA) according to INCI (The International Nomenclature of Cosmetic Ingredients) as wellas in this article.

2. Physical Properties

Many of the physical properties of fatty acid amides have been explained on the basis of the tautomeric structures:

$$R-\overset{\overset{O}{\|}}{C}-\underset{\underset{H}{|}}{N}-H \quad \rightleftharpoons \quad R-\overset{\overset{O^-}{|}}{C}-\underset{\underset{H}{|}}{N^+}-H$$

Primary and secondary amides show strong hydrogen bonding that accounts for their high melting points and low solubilities in most solvents. With tertiary amides (disubstituted amides), hydrogen bonding is not possible as exhibited by their increased solubility and lower melting points as shown in Table 2 (3).

Many fatty acid amides are essentially insoluble in water. Polar solvent solubilities decrease with longer alkyl chain lengths. Amides with alkyl chain lengths $>C_{12}$, with the exception of alkanolamides, have low solubility in all solvents. In nonpolar solvents, solubility is low and varies irregularly with chain length (4).

Solubility is influenced by the production procedure used resulting in different by-product composition. As an example the following can be mentioned. Equimolar quantities of diethanolamine with a fatty acid results in water insoluble products and high levels of ester amide. Wolff Kritchevsky in 1937 (5,6) observed that reaction between 2 mol of diethanolamine and 1 mol of fatty acid led to a water soluble product with good surface activity. These were found to be complex mixtures of free diethanolamine, fatty acid amine soap, fatty amide (50–60%), ester amide, diester amide, amine ester, amine diester, and N,N'-bis(2-hydroxyethyl) piperazine. Later, the so-called superamides were developed by Meade in 1949 (7), which contain fatty acid amide up to 95%. It is a two-stage process via triglyceride reacted with 3 mol methanol to form the methyl ester and then the methyl ester is reacted with diethanolamine to give fatty acid diethanolamide

Table 3. **Physicochemical Properties of Selected Alkyl Amide Ethoxylates**

Chemical substance[a]	HLB (Griffin)	Cloud point, °C	Surface tension mN/m 0.1%	Foam, Ross Miles (mm), 50°C, 0 and 5 min 0.05%	Draves wetting (s) 0.1%
OMA 2	6.4		28	14/12	580
OMA 4	8.8	69[b]	30	20/18	280
OMA 7	10.5		30	26/23	210
OMA 13	15.1	68[c]	37		
CMA 2	7.7		27	113/108	22
CMA 5	11.1	80[b]	30	125/122	20
CMA 8	13	60–70[c]	33		
CMA 12	14.6	72[c]	34		
HTMA 5	8.8		31	39/33	220
HTMA 13	13.4		37	65/66	560
HTMA 50	17.7		39	84/32	>600
Cocamide DEA	8.5		27	90/85	27

[a] OMA = Oleylmonoethanolamide, CMA = Cocomonoethanolamide, HTMA = Hydrogenated tallow monoethanolamide.
[b] A 5-g product in 25 mL 25% butyl glycol.
[c] 1 % in 10% NaCl, pH < 3 (HCl).

Source: Akzo Nobel internal reports and product data sheets. The figures show the number of ethylene oxide.

and methanol. The superamides are high foamers and foam stabilizers and exhibit synergistic effects with other surfactants. They are easily solubilized by other surfactants and soap. They gradually have taken over from Kritchevski amides in personal care products, hard surface cleaners, and higher quality liquid detergents Table 3.

Amides have a strong tendency to reduce friction by adsorption on surfaces. This coating action may be attributed to their hydrophobic character and strong hydrogen bonding. Hydrogen bonding between amide groups lead to higher surfactant crystal stability and higher Krafft point (8,9).

The question arises whether hydrogen bonding occurs in aqueous solution? The role of the amide linkage in the molecular structure when it comes to physicochemical properties in aqueous solution has been investigated lately by Folmer (10,11) and Kjellin (12). Their focus is on alkyl amide ethoxylates that are compared to alkyl ethoxylates. The discussion concerns hydrogen-bond formation on one hand and increase in hydrophilicity and hydrophilic head group size on the other. These two functional properties may be expected to influence the physicochemical behavior in opposite directions, with hydrogen bonding causing denser packing and hydrophilicity resulting in less tendency to aggregate.

The general conclusion is that an amide linkage renders the surfactant more hydrophilic and increases the CMC and the cloud point. (13). However, the surface tension at CMC is higher. Attractive interactions could be found between amide groups in an adsorbed layer (13,14), which is all explained in terms of changes in the packing in several ways and not necessarily depending on direct hydrogen bonding between the amide groups.

The amide group also changes the form of the micelle to a more prolate shaped structure (15), which would mean a tendency to form rodlike structures at higher concentrations. This in turn points at the uses of alkyl amides for rheology control. The existence of rodlike micelles in a typical alkyl amide ethoxylate has been shown by Khan and co-workers (16) for an octadecylamide, poly(oxyethylene)ether, made from rapeseed oil fatty acid according to the superamide route described above. A special application for the threadlike micelles formed is as drag reducing agents in cold water distribution systems (17).

3. Chemical Properties

Amides in general are stable to elevated processing temperatures, air oxidation, and dilute acids and bases. Stability is reduced in amides containing unsaturated alkyl chains; unsaturation offers reactive sites for many reactions. Hydrolysis of primary amides catalyzed by acids or bases is rather slow compared to other fatty acid derivatives. Even more difficult is the hydrolysis of substituted amides. More information can be found in Table D under Applications. The dehydration of amides that produce nitriles is of great commercial value (18). Amides can also be reduced to primary and secondary amines using a copper chromite catalyst (19) or metallic hydrides (20).

4. Synthesis and Manufacture

4.1. Amides. Fatty amides are prepared from fatty acids by reaction with ammonia at elevated temperature and pressure followed by dehydration, or from fatty esters by ammonolysis. Catalysts for the amidation of fatty acids include boric acid, alumina, titanium and zinc alkoxides, and various metallic oxides (21,22). Partially dehydrated metal hydroxides of zirconium, titanium, and tin have also been described (23). A low-pressure process from alkyl esters utilizes a soluble tetravalent tin catalyst (24). Purification of the crude amides to remove residual fatty acid and nitrile is achieved by neutralization of the amide with an alkali metal hydroxide followed by distillation using a thin-film evaporator. Antioxidants can be included in the purification process to stabilize the color and odor of the product, particularly for unsaturated amides (25,26).

Substituted amides can be made directly from triglycerides by the base-catalyzed reaction with amines (27). The process may also be run under high pressure in the absence of catalyst (28). Phenyl substituted amides can be made using boric acid or boronic acid catalyst in the presence of a chelating agent (29), and alkenyl substituted amides by dehydration of 2-hydroxyalkyl substituted amides (30).

4.2. Alkanolamides. The methods for manufacture of alkanolamides are well known and comprise the reaction of an alkanolamine, such as ethanolamine, diethanolamine, or monoisopropanolamine, with a fatty acid, fatty acid ester, or triglyceride (17,21). The reaction of fatty acid with diethanolamine yields a mixture of DEA and diethanolamine esteramide. The milder conditions used with fatty acid esters leads to improved selectivity and better color.

Recent developments in alkanolamides have focused on improving the product quality in terms of color, odor, and ester content. Avoiding excess fatty acid in the system by a controlled addition of fatty acid with efficient mixing has been claimed to improve the purity of alkanolamides (31). However, others believe that a stepwise addition of the alkanolamine to fatty acid is beneficial in improving purity and color while reducing reaction times (32). Conducting the reaction in the presence of an inert gas or steam (33), or using a reducing agent comprising a mixture of alkali metal borohydride and alkali metal hydride (34) are claimed to improve both quality and odor. A product with reduced esteramide content is obtained when the reaction is run at 70–150°C at 10–400 Torr with a very specific amount of an alkali metal catalyst (35). Monoalkanolamides with low color, odor, and ester content can be made from fatty acid ester and monoethanolamine with an alkali metal catalyst and posttreating the product with water (36). Pretreatment of the alkanolamines with alkali metal hydroxide or alkoxide has been claimed to improve color (37). Novel alkanolamides, useful as thickeners, are obtained from the reaction of branched fatty acids with monoisopropanolamine (38).

Polyoxyalkylene amides are made either by alkoxylation of alkanolamides or by direct alkoxylation of amides using well-documented procedures (21). Monoethanolamides are typically solids but by alkoxylating with branched alkylene oxides such as propylene oxide liquid products with surfactant properties similar to DEA are obtained (39,40). The product distribution obtained when ethoxylating alkanolamides is impacted by esteramide impurities, increased levels of esteramide giving rise to increased amounts of polyglycol ethers in the product (17). Other by-product formation during the alkoxylation can be controlled by reducing the water content of the alkanolamide prior to alkoxylation, either by using a dehydrating agent such as a metal alkoxide (41), or by thermal processes (42).

4.3. Sugar Amides.
The increased demand for biodegradable surfactants, particularly in the detergent industry, promoted the development of amide surfactants derived from sugars, notably glucamides and lactobionamides. Of these, glucamides are the most studied (17). Glucamides are prepared by the reductive amination of glucose, typically with methylamine to give the N-methylglucamine, followed by amidation with a fatty acid ester. The amidation step may be catalyzed by a variety of metal salts or aluminosilicates. Residual fatty acid can be reduced by reaction with a more reactive amine such as ethanolamine (43). Improvements in yield and quality are claimed if the glucamide is recovered from solution by crystallization (44). Amidation with akoxylated triglycerides has been claimed to provide a useful surfactant composition of N-methylglucamide and alkoxylated mono- and diglycerides (45), while a synergistic blend of N-methylglucamide and N-propylglucamide was prepared by blending the corresponding amines prior to the base-catalyzed amidation with a fatty acid ester (46). Lactobionamides are made by oxidation of lactose to lactobionic acid followed by amidation with an amine (47).

4.4. Amidoamines.
The established process for making amidoamines is to react a fatty acid, ester, or triglyceride with a polyamine, usually without a catalyst, although several have been suggested (48). Common polyamines include ethylenediamine (EDA), diethylenetriamine (DETA), dipropylenetriamine (DPTA), aminoethylethanolamine (AEEA), and dimethylaminopropylamine

(DMAPA). The addition of fatty acid to DETA has the potential to proceed to give the 1,2-diamide or the 1,3-diamide, however, in the absence of solvent the reaction proceeds to give the 1,3-adduct in high yield. Amidoamines derived from DETA and AEEA cyclize at elevated temperatures, 200–300°C, to form imidazolines. The imidazolines tend to be unstable and care has to be exercised during

Table 4. **Fatty Amides, Producers, and Trade Names**

Product category, trade name	Company	Web site
Alkyl amides		
Armid	Akzo Nobel Surface Chemistry AB	www.surface.akzonobel.com
Petrac Vyn-Eze Addit.	Ferro Corp./Polymer Additives Divison	www.ferro.com
Ethylene bis(stearamide)		
Alkamide STEDA	Rhodia Home, Personal Care, Industrial Ingreds. (HPCII),	www.rpsurfactants.com
Glycowax 765	Lonza Inc.	www.lonza.com
Advawax	Rohm and Haas Co.	www.rohmhaas.com
Kemamide W-39	Crompton Corp./Olefins & Styrenics	www.uniroyalchemical.com
Amido amines		
Indulin QTS	Westvaco Corp., Chemical Division	www.westvaco.com
Ethoxylated amides		
Schercoterge 140	Scher Chemicals, Inc.	
Bermodol Amadol	Akzo Nobel Surface Chemistry AB	www.bermodol.com www.surface.akzonobel.com
Alkanolamides		
Ablumide	Taiwan Surf.	www.taiwansurfactant.com.tw
Amidex	Chemron Corp.	www.chemron.com
Alkamide	Rhodia Home, Personal Care, Industrial Ingreds. (HPCII)	www.rpsurfactants.com
Mackamide	McIntyre	www.mcintyregroup.com
NINOL Manromid STEPANOL	Stepan Co Stepan UK Ltd.	www.stepan.com
Monamid	Uniqema	www.uniqema.com
Surfonamide Empilan	Huntsman Corp.	www.huntsman.com
Aminol Amidet	Kao Corp.	www.kao.co.jp
Calamide	Pilot Chemical Co.	www.pilotchemical.com
Chimipal Rolamid	Cesalpinia Chemicals SpA	www.cesalpinia.com
Comperlan	Cognis Deutschland GmbH	www.es.cognis.com
Witcamide	Akzo Nobel Surface Chemistry AB	www.surface.akzonobel.com
Foamid	Alzo International Inc.	www.alzointernational.com
Incromide	Croda Chemicals (SA) (Pty) Ltd	www.croda.com
Mazamide.	BASF Corp./Performance Chemicals	www.basf.com/businesses/chemicals/performance

Source: Industrial Surfactants Electronic Handbook, 2002 ed.

manufacture and storage to minimize color degradation and hydrolysis. Hydrolysis of imidazolines can give a mixture of 1,2- and 1,3-amidoamines (49). The hydrolysis of imidazolines in the presence of metal salts and sodium borohydride has been described for making amidoamines of good color (50). Amidoamines with reduced odor and irritation properties can be obtained by using thin-film distillation or reverse osmosis [] to remove impurities (51,52). The manufacture of amidoamines from volatile amines like DMAPA often requires the use of an excess of the amine. Conducting the reaction under pressure with periodic venting to remove the water reduces the need for excess reagent (53). Alternatively, the reactor can be fitted with a membrane device to allow for continuous removal of the water (54).

5. Economic Aspects

Several large chemical producers make alkyl amides and derivatives today. In Table 4, typical brand names for alkyl amides, alkanol amides, amido amines, and alkyl bis(stearamides) and the companies producing them are listed together with the web addresses. Extensive product information can be found on the web sites and it is unneccessary to give more detailed information in this article. Approximate volumes for alkanolamides are estimated to be 90,000 tons/year for 2003 by Colin A. Houston & Associates, Inc.

6. Analytical Methods

Amides can be titrated directly by perchloric acid in a nonaqueous solvent (52,53) and by potentiometric titration (54), which gives the sum of amide and amine salts. Infrared (ir) spectroscopy has been used to characterize fatty acid amides (55). Mass spectroscopy (ms) has been able to indicate the position of the unsaturation in unsaturated fatty amides (56).

Alkyl glucamides have been analyzed by liquid chromatography coupled with ms with electrospray ionization (I) and fatty acid amides were separated and determined as trimethylsilyl derivatives by gas chromatography (gc) with mass spectrometric detection (II). Configurational and conformational nuclear magnetic resonance (nmr) analysis of sugar amides have been described (III), which also is a generally applicable technique to characterize fatty acid amides.

Typical specifications of some primary fatty acid amides and properties of bis(amides) are shown in Tables 5 and 6.

7. Health and Safety Factors

7.1. Environmental Properties. Products contributing to sustainable development are becoming more and more important. The alkyl amides are derived directly from fatty acids that come from renewable sources and are thus seen as an attractive surfactant type (17). The biodegradability of some representative structures is shown in Table 7. Most of the amide-based

Table 5. **Amide Specifications**[a]

Primary amide	CAS Registry Number	Iodine number[b]	Fatty acid max%	Mp, °C	Gardner color[b]
coco fatty amide	[61789-19-3]	10	4.0	85 min	10
hydrogenated tallow amide	[61790-31-6]	5	5.0	98–103	7
stearamide	[124-26-5]	95[c]	3.5	68 min	7
oleamide	[361-02-0]	2	5.0	99–109	7

[a] Amide, 90% minimum.
[b] Maximum.
[c] Minimum = 80.

products that are sold in this area are considered to be easily biodegradable. Their aquatic toxicity is also rather low, with a fish toxicity LC_{50} value of 1–10 mg/L or above.

The positive environmental profile makes them a good alternative to the nonyl phenol ethoxylates as emulsifiers or dispersants. Their structure with delocalized π-electrons over the amide function is similar to the phenol ring with its aromatic structure. This in combination with an alkyl chain-containing conjugated unsaturation affords effective dispersants for pigments, see, eg, the Bermodol series from Akzo Nobel.

7.2. Nitrosamines in Alkanol Amides. Cosmetics are controlled in the EU by Council Directive 76/768/EEC and its subsequent amendments and adaptations. The 15th adaptation, Council Directive 92/86/EEC was implemented in many EU states in the mid 1990s, eg, in the United Kingdom by the Cosmetic Products (Safety) (Amendment) Regulations of 1993. The 15th adaptation includes certain requirements in respect of fatty acid dialkanolamides, the most commonly used in cosmetics being diethanolamides. The Directive specifies a maximum 50 ppb level of *N*-nitrosodiethanolamine, commonly referred to as "NDELA", a maximum of 5% free diethanolamine in fatty acid diethanolamides and a maximum 0.5% free diethanolamine in formulated products containing them. The directive did not provide or refer to any analytical procedures for the determination of ppb levels of NDELA in fatty acid diethanolamides,

Table 6. **Substituted Amide Properties**

Substituted amides	CAS Registry Number	Molecular formula	Melting range, °C	Flash point[a], °C	Gardner color
ethylenebis-stearamide (EBS)	[110-30-5]	$C_{38}H_{76}N_2O_2$	140–145	304	3
methylenebis-stearamide (MBS)	[109-23-9]	$C_{37}H_{74}N_2O_2$	135–140	260	5
oleyl stearamide	[b]	$C_{36}H_{71}NO$	70–75	260	3
stearyl stearamide	[13276-08-9]	$C_{36}H_{73}NO$	92–95	246	4

[a] Tag closed cup.
[b] Not available.

Table 7. CAS Registry Numbers, EINECS/ELINCS, Physical, and Environmental Data

Compounds, INCI	CAS Registry Number	EINECS/ELINCS	Physical form	Melting point, °C	Biodegradable	Fish toxicity 96 h LC$_{50}$; mg/L
lauramide MEA	[142-78-9]	205-560-1	cream wax	85		
cocamide MEA	[68140-0-1]	268-770-2	flakes to solid	70–74	readily	10–100
lauramide DEA	[120-40-1]	204-393-1	solid	39		
cocamide DEA	[8051-30-7]	263-163-9	liquid		readily	1–10
oleamide DEA	[93-83-4]	202-281-7	liquid			1–10
tallamide DEA	[68153-57-1]	268-949-5	liquid			
soyamide DEA	[68425-47-8]	270-355-6	liquid			
erucamide	[112-84-5]	204-009-2	waxy beads	75–80	Not readily	
oleamide	[301-20-0]	206-103-9	waxy beads or powder	72	readily	
stearamide	[124-26-5]	204-693-2	flakes	98–109		
hydrogenated tallow amide	[61790-31-6]	263-123-0	flakes	98–103	readily	
coco monoethanol amide ethoxylate	[68425-44-5]				readily	1–10
cocamidopropyl dimethylamine	[68140-01-2]	268-771-8	soft solid to liquid			
ethylene distearamide	[110-30-5]	203-755-6	solid	140–145		

Source: Industrial Surfactants Electronic Handbook, 2002 ed., environmental data from company product information.

diethanolamine raw materials or formulated products containing fatty acid diethanolamides.

N-nitroso derivatives, including NDELA, $(HO-CH_2-CH_2)_2N-NO$, are well-known carcinogens (57,58). In the case of NDELA liver and nasal tumors are produced. The formation of potentially carcinogenic N-nitroso products has been recorded for virtually every kind of amino substrate, but the most extensive reactions relate to secondary amines and secondary N-acyl compounds (59). The nitrosation of secondary amines leads, of course, directly to N-nitrosamines (eq. 1) and under appropriate conditions these reactions can be both rapid and extensive (59).

$$R_2NH + NOX \rightarrow R_2NNO + HX \qquad (2)$$

Further, the N-nitrosamine products as well as being carcinogenic are usually stable and difficult to destroy (60). The principle sources of nitrosating agents are either HNO_2(nitrite salts) or nitrogen oxide (NO_x) pollutants (59). Exhaust fumes from a stacker truck passing an open drum of a fatty acid diethanolamide is sufficient to produce ppb levels of NDELA. The formation of N-nitrosamine in surfactants and related products has been extensively studied and reported by Challis and co-workers at the United Kingdom's Open University (61,62).

It is recommended that fatty acid DEA producers ensure DEA raw material suppliers provide DEA with a minimum level of NDELA, inorganic nitrite and organic nitrite esters and that DEA is carefully stored and transported to avoid NDELA formation. Fatty acid DEAs should not be stored long term at elevated temperature in the presence of air, likewise during blending by a formulator.

Commercial test houses in the United Kingdom, Germany, and the United States were found to be using variations based on the Eisenbrand-Preussmar cleavage reaction (63) and utilising headspace analysis. The levels of NDELA being measured are close to the limit of detection of these methods and explain the variations in results obtained from different test houses.

The anticipated difficulties with DEA-based amides has also initiated work to substitute them with combinations of other amides. Since MEA amides are less soluble they can only partly substitute the DEA based ones. A suggestion has been to combine them with suitable betaines (64).

8. Uses

The many applications of fatty acid amides and their ethoxylated derivatives are determined by their functional properties that include

> Foam boosting and stabilization
> Emulsification
> Detergency
> Viscosity modification
> Lubricity

Antistatic properties

Corrosion inhibition

Wetting, etc

Alkanolamides enhance the foaming properties of detergents, personal care products, hard surface cleaners, hand cleaning gels and many general purpose cleaners by providing a more stable, finer foam. This is often accompanied by an increase in viscosity of the formulation by inclusion of alkanolamide; the nonethoxylated derivatives being the most effective. Enhanced detergency may also be observed. Their ability to emulsify a range of oils, waxes, solvents, and fuels affords application in metal working fluids/cutting oils, lubricants, fuels, textile auxiliaries, rust inhibitors, and plant protection formulations. Their antistatic properties are utilized to advantage in a range of plastics and polymers as well as in additives for fabric softener formulations.

The performance of alkanolamides will also depend on the choice of hydrophobe and their overall composition. Ethanolamines, such as DEA or monoisopropanol amine (MIPA) may be reacted 1:1 or 2:1 with fatty acids. Thus in a 2:1 fatty acid diethanolamide there will be an excess of DEA which may optionally be neutralized with another fatty acid. Wolff Kritchevsky in 1937 observed that reaction between 2 mol of DEA and 1 mol of fatty acid led to a water soluble product with good surface activity, whereas the 1:1 DEA/fatty acid reaction led to water insoluble products and high ester amide levels. In the 2:1 reaction of, eg, DEA and lauric acid, at 150–180°C with removal of water of condensation, <5% free fatty acid remains. In later years, analysis showed the 2:1 reaction products to be complex mixtures (65), including fatty amide (50–60%), free diethanolamine (20–30%), fatty amine soap (3–5%), ester amide (5–10%), and amine ester (5–10%).

In 1949, E. M. Meade (7) described the preparation of the so-called "superamides" with fatty alkanolamide contents up to 95%. They are prepared in a two-stage reaction, first by producing the fatty acid methyl ester from reaction of 3 mol of methanol with a triglyceride (at 80–95°C) with separation and removal of glycerol followed by reaction of the methyl ester with equimolar quantities of alkanolamine at 105°C using sodium methylate catalyst. The lower temperature reaction than the Kritchevsky process ensures far less by-products and improved product colors. Superamides may also be prepared directly from fats and oils, but in such cases the glycerol (8–10%) will remain in the product. Typical composition for a fatty acid DEA would be

	% Component	
	Kritchevsky	"Superamide"
fatty acid diethanolamide	60	90
diethanolamine	23	7
fatty acid	5	0.5
ester amines/amides	12	2.5

The higher fatty amide content of the superamides means that they are not as water soluble, but are readily solubilized by other surfactants and fatty acid additions to produce soap. The Kritchevsky amides are generally better wetters and detergents but poorer foam stabilisers. The superamides have excellent foam stabilizing properties and increase the viscosity of many anionic-based cleaning and personal care compositions. The superamides are also preferred in personal care products because of their higher purity and lower by product content.

Alkanolamides being derived from natural oils and fats, can vary in price depending on climatic conditions and harvest yields. Severe weather conditions in South East Asia, eg, can dramatically reduce the supply of coconut oil. Hence, compared to many petrochemical derived nonionic surfactants, their long-term price is less predictable. In some more price sensitive areas, such as detergent and personal care products, usage has diminished as a consequence. Their continued use is in areas where combinations of their unique properties (as listed above) are required and alternatives not readily available.

The following application sectors are considered in more detail.

8.1. Detergents and Personal Care Products. It is reasonable to consider the use of alkanolamides together in all aspects of cleaning, whether in household detergents, personal care or industrial and institutional cleaning products, as their inherent properties of detergency, enhancement of foam properties and viscosity modification apply to all these sectors (66,67). Anionic surfactants including alkylbenzene sulfonates, fatty alcohol sulfates, and alcohol ether (2 or 3 mol EO) sulfates are frequently the prime component of a formulation selected to impart detergency. Hence, it is the interaction between these surfactants and the alkanolamides that is the key to enhancement of formulation properties. Table 8 illustrates the effect of a coconut diethanolamide on the foam properties of various anionic surfactants.

Ether sulfates are already high foamers and addition of alkanolamide is most marked in the case of alkyl sulfates, particularly when used in personal care products.

Table 8. Effect of Coconut Diethanolamide on the Foaming of Anionic Surfactants[a]

Anionic surfactant	Active concentration, %	Foam height (cm) at 45°C in hard water			
		100% Anionic		80% Anionic/20% diethanolamide	
		0 min	5 min	0 min	5 min
triethanolamine lauryl sulfate	0.02	38	15	110	90
monoethanolamine lauryl sulfate	0.02	30	12	105	100
sodium lauryl ether (3 EO) sulfate	0.02	160	150	150	150
sodium lauryl sulfate	0.02	55	50	135	125

[a] See Ref. (67).

Table 9. **Plate Washing Test for Alkylaryl Sulfonate/Coconut Diethanolamide Blends**[a]

Alkyl aryl sulfonate, % active	Coconut, diethanolamide %	Number of plates washed to foam end-point		
		Distilled water	10 water hardness	30 water hardness
100	0	0	8	11
80	20	4	11	12
70	30	5	12	15
60	40	6	12	15

[a] See Ref. (67).

Alkyl aryl sulfonates are the base detergent used in washing up liquids and many general purpose cleaners. Although high foaming they generate a rather coarse, open structured foam and secondary surfactants, such as alkanolamides, fatty alcohol ether sulfates, or betaines are used to enrich the foam and enhance detergency. Table 9 illustrates partial substitution of an alkyl aryl sulfonate by a coconut DEA in a standard plate washing test for the evaluation of washing-up liquids. Here plates soiled with a standard fat-based soil are washed in 0.25-g/L active solutions of surfactant at 40°C until the surface foam disappears.

Both the Kritchevsky and superamides exhibit detergency in their own right, but they also exhibit a synergistic effect with other surfactants. This finding is illustrated in Table 10, where scouring tests on wool and cotton with a sodium alkyl aryl sulfonate alone and blended 3:1 in an alkyl aryl sulfonate/coconut DEA blend are compared, the latter affording up to 30% improvement in fabric brightness.

Similar improvements in detergency can also be shown with nonionic/alkanolamide blends.

Formulation of alkanolamides in both highly alkaline and acidic cleaning systems can be restricted by their limited hydrolytic stability. However, as illustrated in Table 11, some derivatives are more stable than others.

Table 11 clearly shows that the hydrolytic stability is in decreasing order lauric monoethanolamide ethoxylates > lauric monoisopropanolamide > lauric monoethanolamide > lauric diethanolamide. The monoalkanolamides have

Table 10. **Fabric Brightness Results from Alkyl Aryl Sulfonate/Coconut Diethanolamide Scouring Systems**[a]

Concentration, g/L	Brightness			
	Sodium alkyl aryl sulfonate		75 parts alkyl aryl sulfonate/25 parts high active coconut DEA	
	Wool	Cotton	Wool	Cotton
0.75	38	39	51	39
1.0	44	42	58	43
1.25	48	45	61	51
1.5	50	47	64	52

[a] See Ref. 67.

Table 11. **Hydrolytic Stability of Alkanolamide Derivatives (% Breakdown after 2-h reflux at 100°C)**[a]

Alkanolamide	Sulfuric acid hydrolysis		Caustic soda saponification	
	0.1 N, pH 1.2	1.0 N, pH 0.4	0.1 N, pH 12.4	1.0 N, pH 12.9
lauric diethanolamide	3.2	88.2	33.6	94.1
lauric isopropanolamide	4.3	83.7	0.6	1.8
lauric monoethanol- amide	9.5	71.9	2.7	22.2
lauric monoethanol- amide + 2 EO	2.8	22.5	0.0	6.2
lauric monoethanol- amide + 5 EO	0.0	6.8	0.0	0.0

[a] See Ref. 67.

better hydrolytic stability than the corresponding DEAs, with monoisopropanolamides being more stable than the monoethanolamides, most likely due to steric hindrance. Hydrolytic stability is further increased when the monalkanolamides are ethoxylated. The 5-mol ethoxylate of lauric monoethanolamide is extremely stable and hence suitable for both heavy duty acidic and alkaline cleaners.

The degree of ethoxylation, of, eg, lauric monoethanolamide, also influences the physical form, solubility, foaming, and wetting properties. Transition from 3 to 5 mol EO changes physical form from solid through paste to a liquid, with a minimum melting point at 5-6EO, optimum wetting at 4EO and highest foaming (Ross & Miles test) at 4-5EO.

8.2. Personal Care. The prime use of alkanolamides, most commonly diethanolamide superamides, is as foam boosters and stabilisers and to increase the viscosity of shampoo and bath products.

Table 12 illustrates the ability of alkanolamides to boost the viscosity of typical anionic surfactants used to formulate personal care products, namely, lauryl and lauryl ether sulfates. Not all anionics respond in this way, as illustrated by the alpha-olefin sulfonate. However, in the latter case small additions of sodium chloride will increase viscosity.

Table 12. **Viscosity Enhancement of Anionic Surfactant Solutions**

Anionic surfactant, 15% solution	Viscosity[a]			
	Oleamide MIPA (1:1)			
	1%	3%	5%	5% + 0.75% NaCl
sodium lauryl ether (3 EO) sulfate	3.5	34	2650	
sodium lauryl sulfate	3	18	2600	
sodium $C_{14,16}$-alpha olefin sulfonate	2	4.5	45	2500

[a] Brookfield viscometer model DVIII+, spindle #21 (1 rpm) at 25°C.

Table 13. **Ross & Miles Foam Heights for Anionic Surfactants Containing Alkanolamide**

Blend		Foam height (mm)	
anionic	:oleamide MIPA	initial	after 5 min
sodium lauryl ether sulfate	:oleamide MIPA	155	154
sodium lauryl sulfate	:oleamide MIPA	150	150
sodium $C_{14,16}$ alpha olefin sulfonate	:oleamide MIPA	168	166

Ross and Miles foaming tests of 0.5% solutions of the blends in Table 12 afforded excellent foam stability, even at the 1% addition level of the oleamide MIPA, as shown in Table 13.

Anionic derivatives such as sulfated monoalkanolamide ethoxylates and monoester sulfosuccinates may also be used as major components in such products, the latter in part to being milder to skin and eyes compared to the frequently used alcohol ether sulfates. More specialized applications include lipsticks (68), medicated face washes (69) and a range of applications for the polyoxyethylated amides (70–73). Propoxylated C_{10}–C_{18} fatty acid alkanolamides have also found application as foam stabilizers and thickeners in shampoos (74). Permanent wave setting lotions are frequently based on quaternary ammonium salts and alkanolamides (75).

8.3. Detergents and Cleaners. In the household sector, laundry detergents have been formulated based on combinations of polyoxyethylated amides with soaps (76), alkyl sulfates (77), fatty alcohol ethoxylates (78) and in combination with anionics and enzyme lipase (79–80).

N-Methyl lauric acid ethanolamide and N-Methyl cocoethanolamides have been found to enhance the stability of enzymes in liquid detergents (81) and may be incorporated in detergent tables without loss of strength and disintegration rate (82). They also enhance the foam profile of many anionic surfactants (C_{12} methyl ester sulfonates, C_{12} alkyl and alkyl ether phosphates, etc) (83) as well as mixtures of primary and secondary alcohol ethoxylates (84).

Polyoxyethylated rapeseed oil monoethanolamide + 5-7EO has also been formulated into low-foam laundry detergents (85). Ethoxylated oleic acid amides find application as textile softeners (86) and in fabric conditioners to improve antistatic properties (87–88).

Industrial cleaning and degreasing compositions containing various alkanolamides are used in many applications, including degreasing of metallic surfaces prior to painting or coating (89–92), degreasing to reduce corrosion (93), removal of deposits (94) or hydrocarbons from metallic surfaces (95), and removal and emulsification of many oily soils (96).

Ethoxylated amides in combination with glycol ethers are effective cleaners for glass and ceramic surfaces (97). These derivatives may be used in disinfectant-cleaner compositions in combination with quaternary ammonium compounds (eg, alkyltrimethylammonium chloride) for use in, say, the dairy industry (98–99). The polyoxyethylated coconut oil analogue when used in disinfectant-cleaners affords reduced eye irritation (100). Combinations of alkyl polyglucoside and polyoxyalkylated amide may be used in a variety of cleaning compositions for wool, dishwashing, etc.

Alkanolamides may be used as cosurfactants with quaternary ammonium compounds in the formulation of fabric conditioners and softeners (101,102). More recently, N-alkylglucamides have been used as cosurfactants in the formulation of detergents, syndet bars, and cosmetics because of their mildness and, like APGs, their synergistic effects with other surfactants (103–106, 111).

Alkanolamides continue to be used as minor components (typically 5% by weight or less) or as optional cosurfactants in numerous detergent and personal care product patents, too many to list here.

8.4. Textile Processing (67). Outside the disinfectant and personal care sector, textile auxiliaries and fiber lubricants represents a major application sector for fatty acid amides. Coconut diethanolamide, eg, may be used as an antimigration agent in the dyeing of polyesters with dispersed dyes and as a setting agent (sometimes referred to as an antifrosting agent) in the dyeing of polyamides and cellulosics.

Alkanolamides are excellent emulsifiers for mineral oils and are used as coemulsifiers in coning oils to improve the shock emulsion and the scourability of the coning oil, as well as imparting antistatic properties. A typical formulation for a mineral oil-based coning oil would be

200 s mineral oil	90 parts by weight (%)
fatty alcohol ethoxylate	4
PEG oleate	4
coconut diethanolamide	2

Another example of textile auxiliary use is application in textile oversprays, where the alkanolamide assists wetting and spreading, as well as affording antistatic protection. A typical overspray would consist of

coconut diethanolamide	5–10 parts by weight (5)
PEG 400 monolaurate	90–95

Fatty amides, mainly stearamide, are used in the production of water repellant fabrics (Zelan, Velan types) by reaction with formaldehyde, pyridine, and hydrochloric acid to produce a quaternary salt.

8.5. Antistatic Agents for Plastics and Polymers. Coconut diethanolamide is used widely as an antistatic agent in plastics, eg, in polyethylene film for food packaging and rigid Poly vinyl chloride (PVC). It has been employed in combination with metallic salts as an antistat for polystyrene and in impact resistant rubber–polystyrene blends. Lauric monoethanolamide is employed as an antistat for polyolefins, polystyrene, and acrylonitrile–butadiene–styrene (ABS) (107).

Alkylene bis (saturated higher fatty acid amides) are effective antistatic agents for films, sheets or moldings based on ethylene–styrene copolymers, as well as imparting antiblocking properties (108).

8.6. Metal Working Fluids / Cutting Oils.   Alkanolamides are used in the metal working industry as emulsifiers for mineral oils, but they also provide lubricant properties and good corrosion resistance. Oil soluble types such as groundnut oil derived alkanolamides are particularly effective both in mineral oil-based and semisynthetic cutting oils (109).

8.7. Lubricants and Antiblock Agents for Plastics and Polymers. Alkanolamides are used as antiblock and mould release agents, lubricants, and slip agents for plastics and polymers (110). Applications include low density polyethylene (LDPE), polypropylene and linear low density polyethylene (LLDPE) polymers, especially in polymer films, incorporated at 0.05–0.4% levels. They have also been used in polyolefins, vinyl, and acrylic polymers and polystyrene as antiblock additives. Oleamide, and to a lesser extent, stearamide and erucamide are used as antislip and antiblock additives (via masterbatches) for polyethylene film used for packaging. Stearamide is also used as rubber mould release agent and, along with oleamide, as an ink additive to prevent printed pages sticking together after printing and stacking.

Ethylene bis(stearamide) is used as an internal lubricant in ABS and PVC plastics and is used as a lubricant in the manufacture of extrusion grade polyethylene.

8.8. Intermediates.   Ethoxylated fatty acid alkanolamides are used as intermediates for the production of anionic surfactants including amide ether sulfates and monoester sulfosuccinates.

Continuous, partial SO_3 sulfation of lauryl monoethanolamide +2EO, eg, can yield varying ratios of sodium lauryl monoethanolamide ether (2EO) sulfate / lauryl monethanolamide + 2EO blends suitable for use as a shampoo or bath product base. Here in a single product there is a combination of prime detergent component, foam booster–stabilizer, and viscosity modifier combined with low skin irritancy.

Likewise, conversion of this same ethoxylated alkanolamide intermediate to a monoester sulfosuccinate, by reaction first with maleic anhydride to produce a half ester maleate followed by reaction with sodium bisulfite, affords a mild anionic component for personal care applications. The lauric monoethanolamide ether (2EO) sulfosuccinate exhibits milder skin and eye irritation characteristics when compared to C_{12}-C_{14} lauryl ether (2 or 3 mol EO) sulfates widely used in shampoos, shower, and bath products. This half ester also finds application in carpet shampoos.

The monoester sulfosuccinate derived from an undecylenic fatty acid amide is unique, in as much as it possesses antidandruff properties, and hence may be used in "medicated" shampoos.

Bis(amides), eg, N,N'-ethylenebis(stearamide), may be ethoxylated and then acrylated to produce acrylate intermediates for radiation ultraviolet(uv) or (EB) curable coatings.

Fatty amide ethoxylate phosphate esters may be used in conveyor lubricants (111) for the beverage industry.

Quaternary ammonium compounds containing fatty acid amido moieties are used in fabric finishing compositions (112) cosmetic formulations (113) and in nonwovens to improve long-term hydrophilicity (114).

8.9. Miscellaneous Applications. The claimed uses of alkanolamides and derivatives are many and varied but not necessarily all fully commercialized. Some of those applications where alkanolamide usage is established include the following:

Alkyd resin emulsions are unsaturated fatty acid amide ethoxylates, based on soybean- and linseed-based hydrophobes (115) or N-alkyl glucamides (116) may be used to prepare alkyd emulsions.

Asphalt compositions to prevent phase separation (117) or as release agents for paving equipment (118). Fat liquoring agents for leather (119).

"Spreader-stickers" in crop protection formulations (120).

Defoamer components, especially N, N'-ethylene bis(stearamide).

Demulsifiers for water-in-oil emulsions.

9. Acknowledgments

The author wishes to thank the following Akzo Nobel Surface Chemistry colleagues for their contributions to this article: Ralph Franklin (Synthesis and Manufacture of Fatty Amides), David Karsa (Uses and Nitrosoamines in alkanolamides) and Bo Nilsson (Analysis).

BIBLIOGRAPHY

"Amides, Fatty Acid" in *ECT* 2nd ed., Vol. 2, pp. 72–76, by H. J. Harwood, Durkee Famous Foods; in *ECT* 3rd ed., Vol. 2, pp. 252–259, by H. B. Bathina and R. A. Reck, Armak Company; in *ECT* 4th ed. By R. Opsahl from Akzo Chemicals Corp.

1. A. L. McKenna, *Fatty Amides*, Witco Chemical Corporation, Tenn., 1982, p. 1.
2. Reference 1, pp. 11–31, 69–87, 111–127, 170–173.
3. G. G. D'Alelio and E. E. Reid, *J. Am. Chem. Soc.* **59**, 111 (1937).
4. S. H. Shapiro, in E. S. Pattison, ed., *Fatty Acids and Their Industrial Application*, Vol. 5, Marcel Dekker, New York, 1968, p. 77.
5. U.S. Pat. 2,089,212 (1937), W. Kritchevsky (transferred to Ninol Inc.).
6. U.S. Pat. 2,173,058 (1939), W. Kritchevsky (transferred to Ninol Inc.).
7. U.S. Pat. 2,464,094 (1949), E. M. Meade (to Lankro Chemicals Ltd.).
8. L. Syper, K. A. Wilk, A. Solowski, and B. Burczyk, *Prog. Colloid Polym Sci.*, **110**, 199 (1998).
9. H. Mizushima, T. Matsuo, N. Satoh, H. Hoffmann, and D. Graebner, *Langmuir* **15**, 6664 (1994).
10. B. M. Folmer, M. Nyden, and K. Holmberg, *J. Colloid Interface Sci.* **24**, 404 (2001).
11. B. M. Folmer, K. Holmberg, E. Gottberg-Klingskog, and K. Bergström, *J. Surf. Det.* **4**(2), 175 (2001).
12. U. R. Kjellin, Ph.D. Thesis, *Department of Surface Chemistry*, Royal Institute of Technology, Stockholm 2002.
13. U. R. M. Kjellin, P. M. Claesson, and P. Linse, *Langmuir* **18**(20), 6745 (2002).
14. U. R. M. Kjellin and P. M. Claesson, *Langmuir* **18**(20), 6754 (2002).

15. U. R. M. Kjellin, J. Reimer, and P. Hansson, *J. Colloid Interface Sci.* **262**(2), 506 (2003).
16. A. Khan, A. Kaplun, Y. Talmon, and M. Hellsten, *J. Colloid Interface Sci.* **181**, 191 (1996).
17. A. Lif and M. Hellsten, *Nonionic Surfactants*, Vol. 72, Marcel Dekker, New York, 177 (1998).
18. U.S. Pat. 2,546,521 (Mar. 27, 1951), R. H. Potts (to Akzo Chemicals).
19. H. Adkins, *J. Am. Chem. Soc.* **56**, 247 (1934).
20. F. Wessely and W. Swoboda, *Monatsh. Chem.* **82**, 621 (1951).
21. R. Opsahl, ECT.
22. K. Matsumoto, S. Hashimoto, and S. Otani, *Angew. Chem. Int. Ed. Engl.* **25**(6), 565 (1986).
23. Eur. Pat. 0239954B1 (1992), H. Matsushita, K. Takahashi, and M. Shibagaki (Japan Tobacco Inc.).
24. WO Pat. 0119781A1 (2001), B. Gutsche, C. Sicre, R. Armengaud, J. Rigal, and G. Wollman (Cognis Deutschland GmbH).
25. U.S. Pat. 5,419,815 (1993), N. Doerpinghaus and S. Rittner (Hoechst AG).
26. U.S. Pat. 4,897,492 (1990), B. R. Bailey, III, and J. M. Richmond (Akzo America Inc.).
27. U.S. Pat. 6,034,257 (2000), A. Oftring, G. Oetter, R. Baur, O. Borzyk, B. Burkhart, C. Ott, and M. aus dem Kahmen (BASF Aktiengesellschaft).
28. U.S. Pat. 5,681,971 (1997), J. J. Scheibel, and R. E. Schumate (The Proctor and Gamble Company).
29. U.S. Pat. 6,384,278 (2002), T. Pingwah, and Y. Feng (Emisphere Technologies, Inc.).
30. U.S. Pat. 5,625,076 (1997), Y. Shimasaki, Y. Hitoshi, and K. Ariyoshi (Nippon Shokubai Co., Ltd.).
31. Jpn. Pat. 8301827A2 (1996), K. Nishikawa, K. Miyoshi, and M. Yamanishi (Kao Corp.).
32. Jpn. Pat. 9157234A2 (1997), A. Utsunomiya, F. Hayakawa, M. Dobashi, and Y. Watabe (Mitsui Toatsu Chem. Inc.).
33. Jpn. Pat. 2002037765A2 (2002), T. Sakai, M. Kubo, and M. Tetsu (Kao Corp.).
34. Jpn. Pat. 2002037766A2 (2002), T. Sakai, M. Kubo, and M. Tetsu (Kao Corp.).
35. Jpn. Pat. 9143133A2 (1997), Y. Oshima, H. Imoto, and A. Fujio (Kao Corp.).
36. Jpn. Pat. 2,001,055,365 (2001), K. Kado (Kawaken Fine Chem.Co Ltd.).
37. Jpn. Pat. 2001302601 (2001), T. Murayama and Kawaken (Fine Chem. Co. Ltd.).
38. U.S. Pat. 5,688,978 (1997), G. Lefebvre and L. Fiquet (Witco Corp.).
39. U.S. Pat. 6531443B2 (2003), J. E. Perella, J. A. Komor, D. L. Frost, and R. D. Katstra (Mono Industries, Inc.).
40. Jpn. Pat. 8,337,560 (1996), T. Fujii and A. Shiroichi (Kawaken Fine Chem. Co. Ltd.).
41. Jpn. Pat. 9255773A2 (1997), H. Sekido (Takefu Fine Chem. Co. Ltd.).
42. Jpn. Pat. 9087379A2 (1997), H. Imoto, Y. Oshima, and A. Fujio (Kao Corp.).
43. U.S. Pat. 5,188,769 (1993), D. S. Connor, J. J. Scheibel, B. P. Murch, M. H. Mao, E. P. Gosselink, and R. G. Severson, Jr. (The Proctor and Gamble Company).
44. U.S. Pat. 5,646,318 (1997), M. Dery and N. Brolund (Akzo Nobel Nev.).
45. U.S. Pat. 5,750,749, U. Weerasooriya and J. Lin (Condea Vista Company).
46. U.S. Pat. 5,965,516 (1999), J.-P. Boutique, and P. F. A. Delplancke (The Proctor and Gamble Company).
47. U.S. Pat. 5,401,426 (1995), K-G. Gerling, S. Joisten, K. Wendler, and C. Schreer (Solvay Deutschland GmbH).
48. F. Freidli, in, *Cationic Surfactants, Surfactant Science Series*, Vol. 34, Marcel Dekker, New York, 1990, pp. 51–99.
49. Y. Wu and P. R. Herrington, *J. Am. Oil Chem. Soc.* **74**(1), 61 (1997).
50. Jpn. Pat. 4026663A2 (1992), K. Sotodani and M. Kubo (Kao Corp.).

51. Jpn. Pat. 8176084A2 (1996), S. Tanahashi, T. Maeda, M. Morishita, M. Kubo, and T. Sakai (Kao Corp.).
52. Jpn. Pat. 7048592A2 (1995), S. Tanahashi, T. Kawai, and M. Morishita (Kao Corp.).
53. U.S. Pat. 6,107,498 (2000), B. Maisonneuve, D. Steichen, R. Franklin, and K. Over-kempe, Akzo Nobel, Nev.
54. Jpn. Pat. 6,279,375 (1994), N. Ueno, Y. Yamaji, and K. Kida (Kao Corp.).
55. D. C. Wimer, *Talanta* **13**(10) (1967).
56. D. C. Wimer, *Anal. Chem.* **30**, 77 (1958).
57. Ecetoc. Human Exposure to *N*-Nitrosamines, their Effects, and a Risk Assessment for *N*-Nitrosdiethanolamine in Personal Care Products. *Technical Report No 41*, Ecetoc, Brussels (1991).
58. H. Druckrei and co-workers, *Z. Krebsforsch* **69**, 103 (1967).
59. B. C. Challis, in G. G. Gibson and C. Ioannides, eds., *Safety Evaluation of Nitrosatable Drugs and Chemicals*, Taylor & Francis, London, (1981) p. 16.
60. IARC "Laboratory Decontamination and Destruction of Carcinogensin Laboratory Wastes: some *N*-Nitrosamines," in M. Castegnaro and co-workers, *IARC Scientific Publication No. 43*, IARC, Lyon, Franc, 1982, pp. 143.
61. B. C. Challis, '*N*-Nitrosamine contamination in Surfactants', 3rd CESIO International Surfactants Congress & Exhibition − A World Market, Barbican Centre, London, Proceedings Section F, 1992, pp. 138–154.
62. B. C. Challis, D. F. Trew, W. G. Guthrie, and D. V. Roper, *Inter. J. Cosmetic Sci.* **17**, 119 (1995).
63. G. Eisenbrand and co-workers, *Arzneim. Forsch.* **20**, 1513 (1970).
64. T. Schoenberg, Happi, 76 (July, 1998).
65. S. A. Zelenaya, T. I. Pantelei, and O. F. Kostenko, *Zarodsk Lab.* **30**(9) (1964).
66. W. E. Link and K. M. Buswell, *J. Am. Oil Chem. Soc.* **39**, 39 (1962).
67. P. Eichhorn and T. P. Knepper, *J. Mass Spectrom.* **35**(3), 468 (2000).
68. A. J. Gee, L. A. Groen, and M. E. Mitchell, *J. Chromatog.* **A849**(2), 541 (1999).
69. M. Avalos, R. Babiano, C. J. Duran, J. L. Jimenez, and J. C. Palacios, *J. Chem. Soc., Perkin Trans. 2, Phys. Org. Chem. (1972–1999)* **12**, 2205 (1992).
70. A. T. Pugh, *Lankro Chemicals Ltd.*, Manuf. Chemist, Vol. 28, 1957, p. 557.
71. Martin J. Schick, ed., *Nonionic Surfactants, Surfactant Science Series*, Vol. 1, Marcel Dekker, New York, 1967, pp. 401–403.
72. The Preparation and Application of Alkanolamides and their Derivatives, B. Shelmerdine, Y. Garner, and P. Nelson, in D. R. Karsa, ed. Industrial Applications of Surfactants II (1989)/ Royal Society of Chemistry Special Publication No. 97, pp. 132–149.
73. DE Pat. 1,930,954 (1969), (to Yardley of London Inc.).
74. SU Pat. 1,046,280 (1983), V. A. Yushchenko, E. A. Melnik, and V. A. Drashchink (to Chem. Ind. Res. Des. and Biotech. Res. Inst.).
75. DE Pat. 4,409,189 (1995) (to Chem-Y Chem. Fab. GmbH).
76. Jpn. Pat. 8,041,489 (1994) (to Ajinomoto KK).
77. DE Pat. 2,846,639 (1980) (to P Jurgensen).
78. Eur. Pat. 552,032 (1993), A. R. Naik (to Unilever).
79. Eur. Pat. 1,179,335 (2002), D. Heinz (to Goldwell GmbH).
80. Jpn. Pat. 2,001,322,916 (2001), K. Sugimoto (to Kanebo Ltd.).
81. Jpn. Pat. 56,038,399 (1981), (to Asahi Denka Kogyo and Cope Clean KK).
82. Jpn. Pat. 60,060,196 (1985), (to Kanebo KK).
83. SU Pat. 1,011,682 (1983), B. K. Daurov, Z. H. E. Golovina, and L. K. Dzhelmach (to Che. Ind. Res. Des.).
84. Jpn. Pat. 2,002,294,282 (2002), Y. Yamaguchi and H. Nishimura (to Kao Corp.).
85. Eur. Pat. 341,999 (1989) J. Klugkist (to Unilever).

86. Jpn. Pat. 2,002,294,284 (2002), A. Ishikawa, T. Sakai, and H. Nishimura (to Kao Corp.).
87. Jpn. Pat. 2,002,294,297 (2002), Y. Yamaguchi, Y. Kaneko, and H. Nishimura (to Kao Corp.).
88. Jpn. Pat. 2,002,285,193 (2002), T. Sakai, M. Kubo, and M. Tetsu (to Kao Corp.).
89. Jpn. Pat. 2,002,294,280 (2002), A. Ishikawa, K. Ide, and H. Nishimura (to Kao Corp.).
90. J. Przondo, Z. Kot, and Z. Kossinski, Pollena: Tluszcze, *Srodki Piorace, Kosmet*, **26**: 6 (1982).
91. Eur. Pat. 415,279 (1991), R. Brueckmann and T. Simenc (to BASF AG).
92. Eur. Pat. 43,547 (1982), S. Billenstei, A. May, and H. W. Buecking (to Hoechst AG).
93. Jpn. Pat. 60,096,695 (1985) (to Sanyo Chem Ind Ltd.).
94. SU Pat. 732,370 (1980), A. Y. A. Nagina, A. I. Mikhalska, and E. I. Nechesova (to Atom Eng Reactor).
95. SU Pat. 732,350 (1980), A. Y. A. Nagina, A. I. Mikhalska, and E. I. Nechesova (to Atom Eng Reactor).
96. SU Pat. 745,925 (1980), M. P. Shemelyuk, M. M. Oleinik, and L. A. Gnutenko (to Car Ind Cons Techn).
97. SU Pat. 749,888 (1980), M. M. Oleinik, V. T. Protsishin, and A. V. Galkin (Auto Ind Tech. Inst).
98. SU Pat. 662,578 (1979), I. K. Getmanskii, M. M. Bebko, and V. D. Yakovlev (to I K Getmanskii).
99. Eur. Pat. (1989), R. Baur, H. H. Goertz, H. W. Neumann, D. Stoeckigt, and N. Wagner (to BASF AG).
100. Eur. Pat. 84,411 (1983), M. Blezard and W. H. McAllister (to Albright & Wilson Ltd.).
101. Jpn. Pat. 2,002,285,190 (2002), T. Sakai, M. Kubo, M. Tetsu, and Y. Kaneko (to Kao Corp.).
102. Eur. Pat. 288,856 (1988), R. Osberghaus, K. H. Rogmann, and B. Frohlich (to Henkel KgaA).
103. Eur. Pat. 621,335 (1994), C. A. Boronio, B. T. G. Graubart, E. J. Sachs, and A. L. Streit (to Eastman Kodak Co and Reckitt & Coleman Inc.).
104. SU Pat. 735,630 (1980), R. G. Alagezyan, B. V. Andriasyan, and A. G. Nikitenko (to Erev Zool Veter Ins.).
105. U.S. Pat. 4,336,151 (1982), B. M. Like, D. Smialowicz, and E. Brandli (to American Cyanamid Co.).
106. U.S. Pat. 6,191,101 (2001), A. Jacques and L. Laitem (to Colgate-Palmolive Co.).
107. WO Pat. 9,927,046 (1999), E. S. Baker and R. G. Baker (to The Procter & Gamble Company).
108. Starch derived products in detergents, R. Beck, in D. R. Karsa, ed., *Industrial Applications of Surfactants IV*, Royal Society of Chemistry Special Publication No. 230, 1999, pp. 115–129.
109. New carbohydrate surfactants, C. Schmidt, R. R. Schmidt, G. Oetter, and A. Oftring, *Tenside; Surfactants & Detergents* **36**(4), 244 (1999).
110. R. Beck, *Chimica Oggi* **20**(5), 45 (2002).
111. Jpn. Pat. 2,001,214,196 (2001), A. Ishikawa, Y. Fujii, H. Nishimura, and K. Ide (to Kao Corp.).
112. Additives for Non-vinyl Polymers, Akzo Nobel, Akcros Chemicals SBU, PO Box 1, Eccles, Manchester, M30 0BH, 1998.
113. Jpn. Pat. 2,000,248,135 (2000), Y. Nishitoba, T. Oda, and A. Arai (to Denki Kagaku Kogyo KK).
114. *Chemistry & Technology of Lubricants*, Blackie & Son Ltd., in R. M. Mortier and S. T. Orszulik, eds., 1992, pp. 213–219.
115. R. A. Rech, *J. Am. Oil. Chem. Soc.* **61**, 176 (1984).

116. WO Pat. 2,000,022,073 (2000), D. D. McSherry and G-J. J. Wei (to Ecolab Inc.).

117. DE Pat. 10,021,169 (2001), R. Jeschke and K-H. Scheffler (to Henkel KgaA).

118. WO Pat. 2,002,074,729 (2002), I. Bigorra Llosas, K. H. Schmid, R. Pi Subirana, and G. Bonastre Nuria (to Cognis Iberia S L).

119. Jpn. Pat. 2,002,235,285 (2002), S. Tanaka and H. Nishinaka (to Toyobo Co Ltd.).

120. Eur. Pat. 593,487 (1995), K. Holmberg (to Berol Nobel AB).

121. WO Pat. 2,000,075,243 (2000), A. Bouvy and H. S. Bovinakatti (to Imperial Chemical Industries plc).

122. Jpn. Pat. 2,003,020,406 (2003), H. Yanagi (to Kao Corp.).

123. U.S. Pat. 6,126,757 (2000), M. G. Kinnaird (to Chemtek Inc.).

124. CN Pat. 1,330,156 (2002), L. Liu (to Chengdu Inst of Organic Chemistry, Chinese Academy of Sciences).

125. U.S. Pat. 5,906,961 (1999), J. R. Roberts and G. Volgar (to Helena Chemical Company).

INGEGÄRD JOHANSSON
Akzo Nobel Surface Chemistry AB

AMINE OXIDES

1. Introduction

Amine oxides, known as *N*-oxides of tertiary amines, are classified as aromatic or aliphatic, depending on whether the nitrogen is part of an aromatic ring system or not. This structural difference accounts for the difference in chemical and physical properties between the two types.

The higher aliphatic amine oxides are commercially important because of their surfactant properties and are used extensively in detergents. Amine oxides that have surface-acting properties can be further categorized as nonionic surfactants; however, because under acidic conditions they become protonated and show cationic properties, they have also been called cationic surfactants. Typical commercial amine oxides include the types shown in Table 1.

Aromatic amine oxides, produced on a much smaller scale and having some pharmaceutical importance, do not demonstrate the surface-acting properties that the aliphatic amine oxides do.

2. Physical Properties

The physical properties of amine oxides are attributed to the semipolar or coordinate bond between the oxygen and nitrogen atoms with high electron density residing on oxygen.

$$(CH_3)_3 \overset{+}{N} - O^-$$

Table 1. **Commercial Amine Oxides**

Name	Molecular formula	CAS Registry Number	Structural formula
dimethyldodecyl-amine oxide	$C_{14}H_{31}NO$	[1643-20-5]	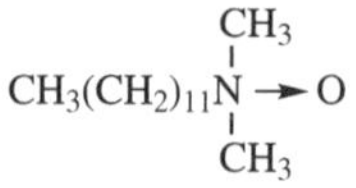 $CH_3(CH_2)_{11}\overset{\displaystyle CH_3}{\underset{\displaystyle CH_3}{N}} \rightarrow O$
dihydroxyethyl-dodecylamine oxide	$C_{16}H_{35}NO_3$	[2530-44-1]	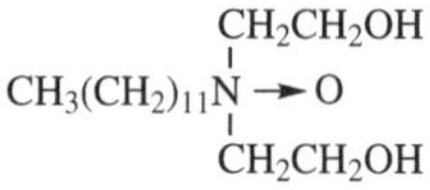 $CH_3(CH_2)_{11}\overset{\displaystyle CH_2CH_2OH}{\underset{\displaystyle CH_2CH_2OH}{N}} \rightarrow O$
dimethyltetradecyl-amidopropyl amine oxide	$C_{20}H_{40}NO_2$		 $CH_3(CH_2)_{13}\overset{\displaystyle O}{\overset{\|}{C}}NHCH_2CH_2CH_2\overset{\displaystyle CH_3}{\underset{\displaystyle CH_3}{N}} \rightarrow O$
N-dodecylmorpho-line *N*-oxide	$C_{16}H_{33}NO_2$	[2530-46-3]	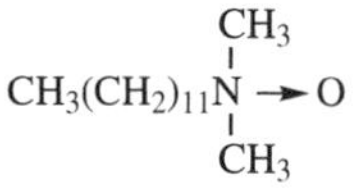 $CH_3(CH_2)_{11}\overset{\displaystyle CH_3}{\underset{\displaystyle CH_3}{N}} \rightarrow O$
1-hydroxyethyl-2-octa-decyl imida-zoline oxide	$C_{23}H_{46}N_2O_2$		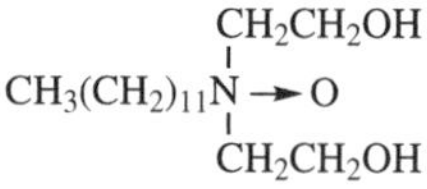 $CH_3(CH_2)_{11}\overset{\displaystyle CH_2CH_2OH}{\underset{\displaystyle CH_2CH_2OH}{N}} \rightarrow O$
N,N′,N′-hydroxy ethyl-*N*-octadecyl-1,3-propylene-diamine oxide	$C_{27}H_{58}N_2O_5$		

The N–O bond distances, found to be 0.133 to 0.139 nm for trimethylamine oxide (1), are somewhat shorter than the single N–C bond distance of 0.147 nm in methylamine. The N–C bond distance of 0.154 nm in trimethylamine oxide approaches that of the C–C bond. This is in agreement with the respective absorptions in the infrared region; valence vibrations of N–O bonds of aliphatic amine oxides are found between $970 - 920$ cm^{-1} (2).

A dipole moment of 1.46×10^{-29} C·m (4.38 D) for the nitrogen–oxygen bond is larger than the moments of other semipolar bonds (3). The spatial arrangement around nitrogen in amine oxides is tetrahedral as in quaternary ammonium salts. Tetrahedral configuration was demonstrated by Meisenheimer, who separated *N*-ethyl-*N*-methylaniline *N*-oxide [825-19-4] into its optical isomers (4), and later confirmed by electron diffraction studies (1).

For the aromatic amine oxides, the trigonal nitrogen forces the oxygen into the same plane as the aromatic ring and permits resonance structures involving the nonbonded electrons on oxygen. This is largely responsible for the distinction between aliphatic and aromatic amine oxides and accounts for the added stability and special properties of aromatic amine oxides. Although amine oxides are

Table 2. pK_a Values of Protonated Amines and Their *N*-Oxides

Parent amine	Amine Registry Number	pK_a amine	Oxide molecular formula	Oxide Registry Number	pK_a *N*-oxide
$(CH_3)_3N$	[75-50-3]	9.74	C_3H_6NO	[1184-78-7]	4.65
$(C_2H_5)_3N$	[121-44-8]	10.76	$C_6H_{15}NO$	[2687-45-8]	5.13
$C_6H_5N(CH_3)_2$	[121-69-7]	5.06	$C_8H_{11}NO$	[874-52-2]	4.21
$C_6H_5N(C_2H_5)_2$	[91-66-7]	6.56	$C_{10}H_{15}NO$	[826-42-6]	4.53
o-$CH_3C_6H_4N(CH_3)_2$	[609-72-3]	5.86	$C_9H_{13}NO$	[6852-47-7]	4.78
p-$CH_3C_6H_4N(CH_3)_2$	[99-97-8]	5.50	$C_9H_{13}NO$	[825-85-4]	4.32

weaker bases than the amines from which they are derived, there is a base leveling effect in the oxides that is not found in the amines. This leveling effect is probably caused by the decreased effect of substituents which are further removed from the basic oxygen center (Table 2). Aromatic amine oxides are generally weaker bases than aliphatic amine oxides. However there are exceptions, eg, 1,10-phenanthroline monoxide [1891-19-6] is a stronger base than most aliphatic amine oxides because of stabilization of the conjugate acid through intramolecular hydrogen bonding (5).

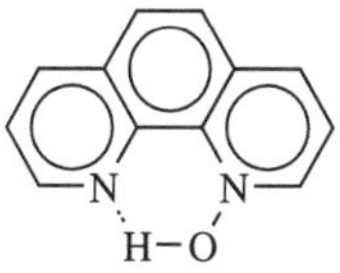

Amine oxides show either nonionic or cationic behavior in aqueous solution depending on pH. In acid solution the cationic form (R_3N^+OH) is observed (2) while in neutral and alkaline solution the nonionic form predominates as the hydrate $R_3NO \cdot H_2O$. The formation of an ionic species in the acidic pH range stabilizes the form generated by the most studied commercial amine oxide, dimethyldodecylamine oxide (6).

Aliphatic amine oxides behave as typical surfactants in aqueous solutions. Below the critical micelle concentration (CMC), dimethyldodecylamine oxide exists as single molecules. Above this concentration micellar (spherical) aggregates predominate in solution. Aliphatic amine oxides are similar to other typical nonionic surfactants in that their CMC decreases with increasing temperature.

Temperature, °C	CMC, mol/L
1	0.0028
27	0.0021
40	0.0018
50	0.0017

Wetting times of *N,N*-dimethyl-*n*-alkylamine oxides as a function of the alkyl chain length show a minimum with dimethyldodecylamine oxide

Table 3. **Surfactant Properties of *N,N*-Dimethylalkylamine Oxides**

Amine oxide	CAS Registry Number	Alkyl chain length	Wetting time, s	Foam height[a], mm
dimethyloctylamine oxide	[2605-78-9]	8	900	
dimethyldecylamine oxide	[2605-79-0]	10	150	138
dimethyldodecylamine oxide	[1643-20-5]	12	37	175
dimethyltetradecylamine oxide	[3332-27-2]	14	225	185
dimethylhexadecylamine oxide	[7128-91-8]	16	900	30
dimethyloctadecylamine oxide	[2571-88-2]	18		17

[a] Concentration = 3 g/L at 20°C.

(Table 3). Foam generation of dimethyl-*n*-alkylamine oxides solutions show a maximum when the alkyl group contains 14 carbons.

In the presence of an anionic surfactant such as sodium dodecyl-benzene-sulfonate [25155-30-0] any protonated amine salt present forms an insoluble salt (4). Salt formation results in an increase in the pH of the solution.

$$R_3NO + H_2O \rightleftharpoons R_3\overset{+}{N}OH + {}^-OH$$

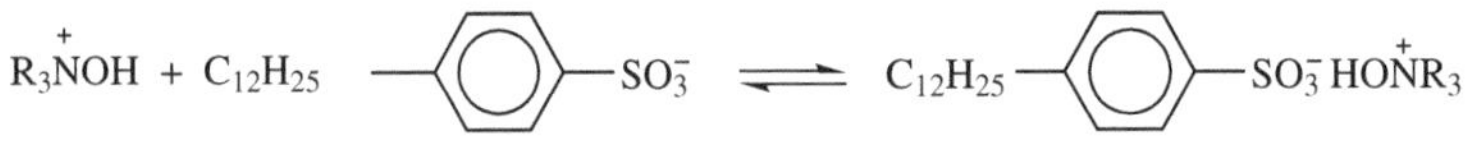

$$R_3\overset{+}{N}OH + C_{12}H_{25} - \langle \bigcirc \rangle - SO_3^- \rightleftharpoons C_{12}H_{25} - \langle \bigcirc \rangle - SO_3^- \overset{+}{HONR_3}$$

The effect of added inorganic salts on the micellar properties of the nonionic and cationic forms of dimethyldodecylamine oxide has been determined (2).

Aliphatic amine oxides form charge-transfer complexes with iodine (7) because of the asymmetric electron distribution in the N–O bond with oxygen being the electron donor. Complexes of aliphatic amine oxides are stronger than those of aromatic ones (8). Amine oxides form hydrogen bonds with strong acids (9) and weak acids such as phenols (10). Trimethylamine oxide dihydrate contains strongly bound tricoordinated water and the hydrate water forms with the oxygen of the amine oxide is a structure similar to the water–fluoride cluster in tetraethylammonium fluoride dihydrate [63123-01-3] (11). Aliphatic amine oxides form complexes with metals or metallic salts (12).

3. Chemical Properties

3.1. Decomposition.
Most amine oxides undergo thermal decomposition between 90 and 200°C. Aromatic amine oxides generally decompose at higher temperatures than aliphatic amine oxides and yield the parent amine.

Rearrangement. Aliphatic amine oxides without an aliphatic hydrogen atom β to the nitrogen undergo Meisenheimer's rearrangement when heated to give trisubstituted hydroxylamines.

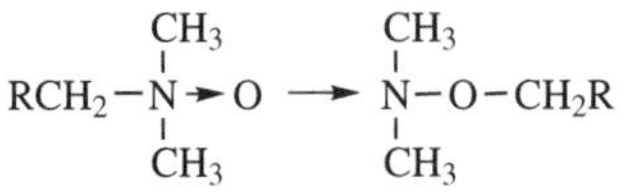

$$\underset{\underset{CH_3}{|}}{\overset{\overset{CH_3}{|}}{RCH_2-N \rightarrow O}} \longrightarrow \underset{\underset{CH_3}{|}}{\overset{\overset{CH_3}{|}}{N-O-CH_2R}}$$

Allyl or benzyl groups on the nitrogen facilitate the process. The rearrangement appears to be intramolecular (13), proceeding by a cyclic mechanism as in the case of N-2-butenyl-N-methylaniline oxide giving N-methyl-O-1-methylallyl-N-phenyl-hydroxylamine.

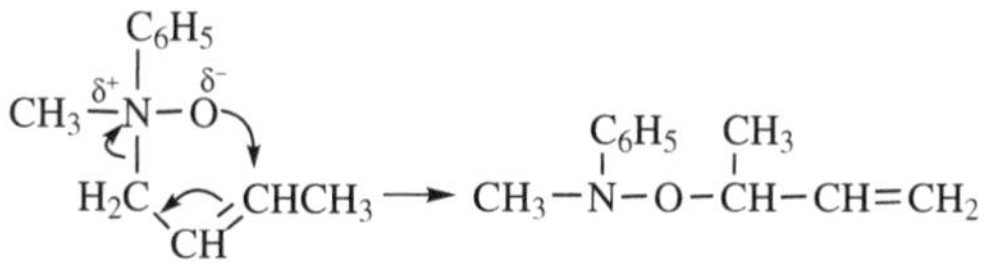

$$
\underset{\underset{\displaystyle H_2C\diagdown CHCH_3}{\overset{\displaystyle C_6H_5}{\underset{CH}{\mid}}}}{CH_3\overset{\delta+}{-}N\overset{\delta-}{-}O} \longrightarrow CH_3-\underset{\underset{\displaystyle C_6H_5}{\mid}}{N}-O-\underset{\underset{\displaystyle CH_3}{\mid}}{CH}-CH=CH_2
$$

The rate of rearrangement increases as the basicity of the parent tertiary amine decreases (14). Strong support for a free-radical mechanism has been demonstrated (15,16).

Elimination. Aliphatic amine oxides having an aliphatic hydrogen β to the nitrogen form olefins and dialkyl hydroxylamines when heated. This reaction is known as the Cope elimination (17)

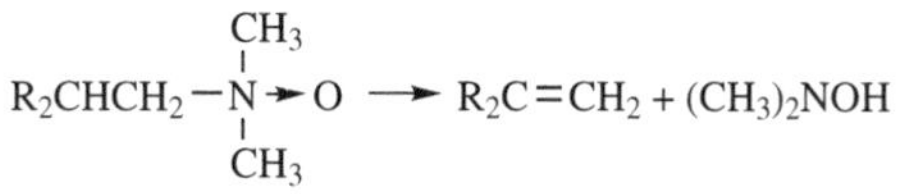

$$
R_2CHCH_2-\underset{\underset{\displaystyle CH_3}{\mid}}{\overset{\overset{\displaystyle CH_3}{\mid}}{N}}\rightarrow O \longrightarrow R_2C=CH_2 + (CH_3)_2NOH
$$

and proceeds through a planer five-center intermolecular mechanism (18,19):

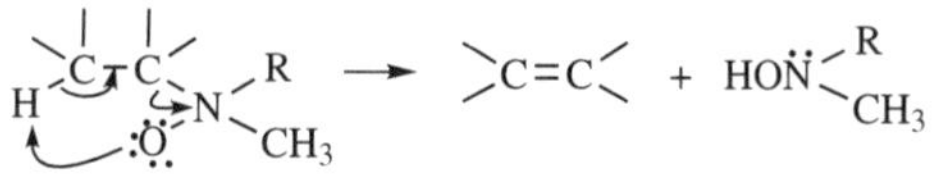

$$
\underset{H}{\overset{\mid\ \ \mid}{C\text{-}C}}\diagup\overset{R}{\underset{CH_3}{N}} \longrightarrow \ \diagup C=C\diagdown + HO\ddot{N}\diagup\overset{R}{\underset{CH_3}{}}
$$

N-methylpiperidine oxide [17206-00-7] does not undergo the reaction because of its inability to achieve the highly strained transition configuration, whereas the

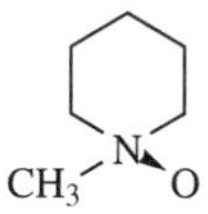

$$
\underset{CH_3 \diagup \overset{\displaystyle N}{\diagdown} O}{\bigcirc}
$$

corresponding seven- and eight-membered ring compounds yield respectively 57% and 78% of the elimination product (18). When more than one possibility for elimination exists, it occurs predominantly toward the alkyl group having the most β-hydrogens, or one having an electron withdrawing group attached to the β-carbon (20). Stereoselectivity for this elimination reaction is predominantly cis (21) and allows for the preparation of certain olefins that could not otherwise be made using other elimination reactions that follow either the Hoffmann rule or the Saytzeff rule (19).

In the pyrolysis of pure amine oxides, temperature has a significant effect on the ratio of products obtained (22). The principal reaction during thermal decomposition of *N,N*-dimethyllaurylamine oxide [1643-20-5] at 80–100°C is deoxygenation to *N,N*-dimethyllaurylamine [112-18-5] (lauryl = dodecyl).

$$\underset{\underset{CH_3}{|}}{\overset{\overset{CH_3}{|}}{C_{12}H_{25}N}} \rightarrow O \cdot H_2O \longrightarrow C_{12}H_{25}N(CH_3)_2 + \tfrac{1}{2}O_2 + H_2O$$

However, when the temperature is increased to 120°C, the principal reaction is the elimination to olefin. The thermal decomposition of dimethyldodecylamine oxide at 125°C in a sealed system, as opposed to a vacuum used by Cope and others, produces 2-methyl-5-decylisoxazolidine, dimethyldodecylamine, and olefin (23). The amine oxide oxidizes *N,N*-dialkylhydroxylamine to the nitrone during the pyrolysis and is reduced to a tertiary amine in the process.

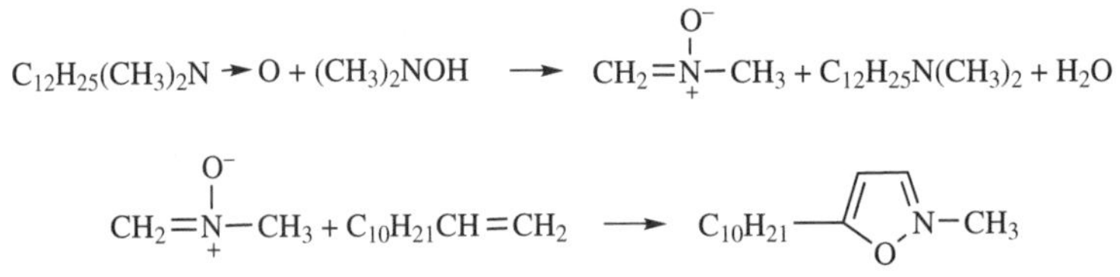

$$C_{12}H_{25}(CH_3)_2N \rightarrow O + (CH_3)_2NOH \longrightarrow \overset{\overset{O^-}{|}}{CH_2=\underset{+}{N}-CH_3} + C_{12}H_{25}N(CH_3)_2 + H_2O$$

$$\overset{\overset{O^-}{|}}{CH_2=\underset{+}{N}-CH_3} + C_{10}H_{21}CH=CH_2 \longrightarrow C_{10}H_{21}-\underset{O}{\overset{\diagdown}{\diagup}}N-CH_3$$

Metal Catalysis. Aqueous solutions of amine oxides are unstable in the presence of mild steel and thermal decomposition to secondary amines and aldehydes under acidic conditions occurs (24,25). The reaction proceeds by a free-radical mechanism (26). The decomposition is also catalyzed by V(III) and Cu(I).

3.2. Reduction. Just as aromatic amine oxides are resistant to the foregoing decomposition reactions, they are more resistant than aliphatic amine oxides to reduction. Aliphatic amine oxides are readily reduced to tertiary amines by sulfurous acid at room temperature; in contrast, few aromatic amine oxides can be reduced under these conditions. The aliphatic amine oxides can also be reduced by catalytic hydrogenation (27), with zinc in acid, or with stannous chloride (28). For the aromatic amine oxides, catalytic hydrogenation with Raney nickel is a fairly general means of deoxygenation (29). Iron in acetic acid (30), phosphorus trichloride (31), and titanium trichloride (32) are also widely used systems for deoxygenation of aromatic amine oxides.

3.3. Alkylation. Alkylating agents such as dialkyl sulfates and alkyl halides react with aliphatic amine oxides to form trialkylalkoxyammonium quaternaries. For example (33), methyl iodide reacts with trimethylamine oxide to form trimethylmethoxyammonium iodide

$$(CH_3)_3N \rightarrow O + CH_3I \longrightarrow (CH_3)_3\overset{+}{N}OCH_3I^-$$

3.4. Acylation. Aliphatic amine oxides react with acylating agents such as acetic anhydride and acetyl chloride to form either *N,N*-dialkylamides and aldehyde (34), the Polonovski reaction, or an ester, depending upon the polarity of the solvent used (35,36). Along with a polar mechanism (37), a metal-complex-induced mechanism involving a free-radical intermediate has been proposed.

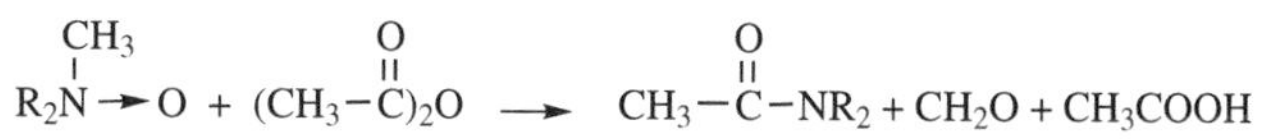

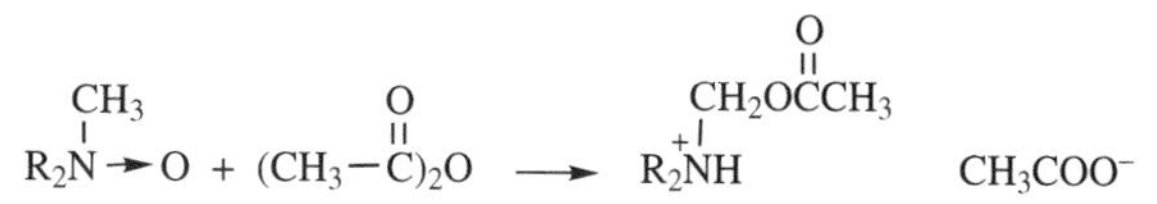

$$R_2\overset{\underset{\displaystyle CH_3}{|}}{N}\!\!\rightarrow\!O + (CH_3-\overset{\overset{\displaystyle O}{\|}}{C})_2O \longrightarrow CH_3-\overset{\overset{\displaystyle O}{\|}}{C}-NR_2 + CH_2O + CH_3COOH$$

$$R_2\overset{\underset{\displaystyle CH_3}{|}}{N}\!\!\rightarrow\!O + (CH_3-\overset{\overset{\displaystyle O}{\|}}{C})_2O \longrightarrow R_2\overset{\underset{\displaystyle CH_2O\overset{\overset{\displaystyle O}{\|}}{C}CH_3}{+|}}{N}H \qquad CH_3COO^-$$

3.5. Substitution Reactions. Aromatic heterocyclic N-oxides undergo both electrophilic and nucleophilic substitution because the dipolar N-oxide group is both an electron donor and an electron acceptor, giving rise to the resonance structures:

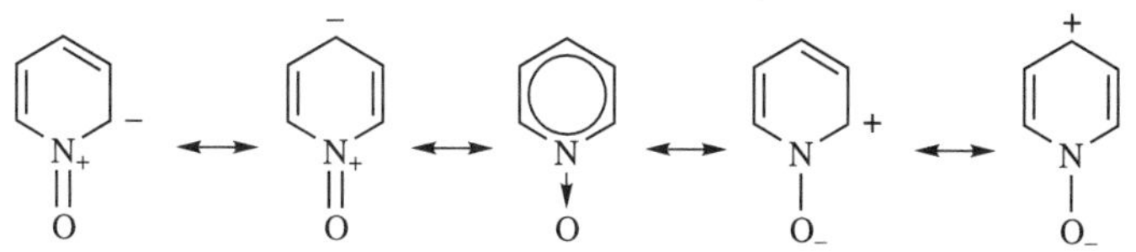

Pyridine oxide [694-59-7] is converted to 4-nitropyridine oxide in 80–90% yield on heating with concentrated sulfuric acid and fuming nitric acid at 100°C (38).

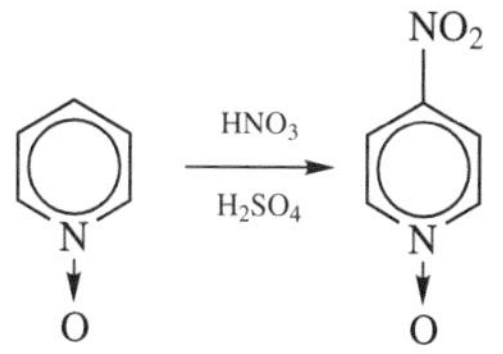

Nucleophilic substitution occurs in positions α and γ to the N-oxide group. In nearly all these reactions deoxygenation occurs giving the substituted heterocyclic amine.

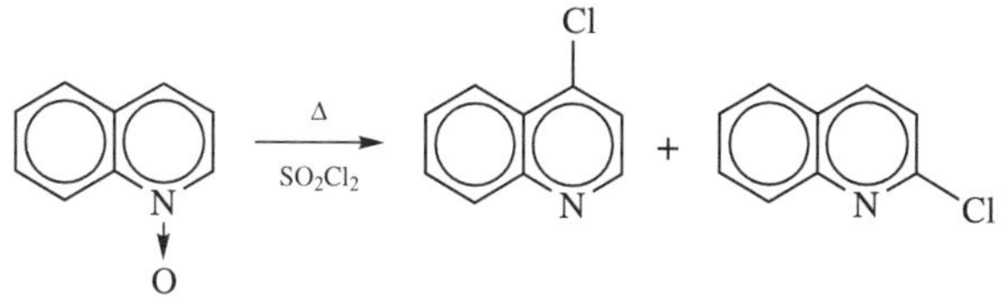

Heterocyclic N-oxides can react at the oxygen atom with a variety of electrophilic reagents to give adducts which, according to the reagent and reaction conditions, may be stable or react further (39). Heterocyclic N-oxides are reduced by reaction of nucleophiles at the N-oxide oxygen.

4. Manufacturing and Processing

Linear alpha-olefins are the source of the largest volume of aliphatic amine oxides. The olefin reacts with hydrogen bromide in the presence of peroxide catalyst, to yield primary alkyl bromide, which then reacts with dimethylamine to yield the corresponding alkyldimethylamine. Fatty alcohols and fatty acids are also used to produce amine oxides (Fig. 1).

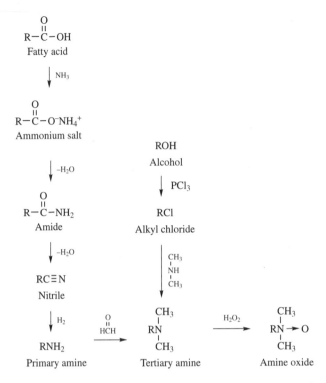

Fig. 1.   Routes to tertiary amines from fatty acids or fatty alcohols.

Amine oxides used in industry are prepared by oxidation of tertiary amines with hydrogen peroxide solution using either water or water and alcohol solution as a solvent. A typical industrial formulation is as follows:

N,N-dimethyltetradecylamine [112-75-4]	1475 kg
hydrogen peroxide (35%)	640 kg
EDTA	1.5 kg
water	3225 kg

A process for preparation of high quality amine oxides using secondary and tertiary amines with hydrogen peroxide has been reported (40).

EDTA (ethylenediaminetetraacetic acid, [60-00-4]) chelates any trace metals that would otherwise decompose the hydrogen peroxide [7722-84-1]. The amine is preheated to 55–65°C and the hydrogen peroxide is added over one hour with agitation; the temperature is maintained between 60–70°C. The reaction is exothermic and cooling must be applied to maintain the temperature below 70°C. After all the peroxide has been added, the temperature of the reaction mixture is raised to 75°C and held there from three to four hours until the unreacted amine is less than 2.0%. The solution is cooled and the unreacted hydrogen peroxide can be destroyed by addition of a stoichiometric amount of sodium bisulfite. This may not be desirable if a low colored product is desired, in which case residual amounts of hydrogen peroxide enhance long-term color stability.

Primary and secondary amines are oxidized to the respective hydroxyl amines, and further oxidation to the nitro compound occurs in the case of primary amines.

$$RNH_2 + H_2O_2 \longrightarrow RNHOH + H_2O$$

$$RNHOH + 2\,H_2O_2 \longrightarrow RNO_2 + 3\,H_2O$$

Owing to the lower basicity of the parent amines, aromatic amine oxides cannot be formed directly by hydrogen peroxide oxidation. These compounds may be obtained by oxidation of the corresponding amine with a peracid; perbenzoic, monoperphthalic, and monopermaleic acids have been employed.

5. Economic Aspects

Demand for amines in the United States is expected to grow to $ 1.9 \times 10^9$ in 2004. Specialty amines, the group in which amine oxides are categorized, will lead the demand because of strong performance characteristics. A major use for amine oxides is as surfactants in a variety of soaps, detergents and personal care products, Multifunctionality and mildness of ingredients are reasons for the demand (41).

Global demand for cationic and amphoteric surfactants is projected to grow at a rate of 5.4% to 1.97×10^6 t for a value of $\$3.26 \times 10^9$ in 2005. Europe is the largest consumer of cationic surfactants (about 36% of total), the United States is the second largest at approximately 27% of total. Asia (25%), South America (8%), and Mexico (4%) follow (42).

6. Specifications

Industrial specifications for aliphatic tertiary amine oxides generally require an amine oxide content of 20–50%. These products may contain as much as 5% unreacted amine, although normally less than 2% is present. Residual hydrogen peroxide content is usually less than 0.5%. The most common solvent systems employed are water and aqueous isopropyl alcohol, although some amine oxides are available in nonpolar solvents. Specifications for individual products are available from the producers.

7. Analytical Methods

Analytical methods include thin-layer chromatography (43), gas chromatography (44), and specific methods for determining amine oxides in detergents (45) and foods (46). Nuclear magnetic resonance (47–49) and mass spectrometry (50) have also been used. A frequently used procedure for industrial amine oxides (51) involves titration with hydrochloric acid before and after conversion of the amine to the quaternary ammonium salt by reaction with methyl iodide. A simple, rapid quality control procedure has been developed for the determination of amine oxide and unreacted tertiary amine (52).

8. Health and Safety Factors

Aliphatic amine oxides such as alkyldimethylamine oxides and alkylbis (2-hydroxylethyl)amine oxides range from practically nontoxic to slightly toxic (53). Reported LD_{50}s range from 1.77 g/kg to 6.50 g/kg. The commercial concentrated products are primary skin and eye irritants. At concentrations of 2%, these products may be considered as nonirritating to the skin or eye.

Test	Result
bacterial toxicity, Bringmann-Kuhn	$EC_{10} = 80$ ppm
algae toxicity, growth inhibition 72 h	$EC_{50} = 0.66$ ppm
	$NOEC = 0.25$ ppm
daphnia toxicity, acute 48 h	$EC_{50} = 9.5$ ppm
	$NOEC = 4.6$ ppm
fish toxicity (zebra fish), acute 96 h	$LC_{50} = 42.0$ ppm
	$NOEC = 33.5$ ppm
biodegradation, Closed Bottle, 28 d	readily $= 93\%$

Among the aromatics, it was found that 4-nitroquinoline N-oxide [56-57-5] is a powerful carcinogen producing malignant tumors when painted on the skin of mice (54). It was further established that the 2-methyl, 2-ethyl, and 6-chloro derivatives of 4-nitroquinoline oxide are also carcinogens (55).

9. Uses

9.1. Detergents.   Aliphatic amine oxides find wide use in the detergent and personal care industries. A comparison of detergents and admixtures with other surfactants in a light duty liquid detergent revealed that the most effective performance came from a blend of ammonium ether sulfate and an alkyldimethylamine oxide where the alkyl group contained 14 carbon atoms (56). Amine oxides improve the stability and amount of foam generated in light liquid detergents and because of this they are used extensively in shampoos, dishwashing liquids, and liquid soaps. A high foaming, grease cutting, light duty liquid detergent has been described (57). Other uses for amine oxides are found in paper and textile production, electroplating, oil and petroleum, plastics and rubber, metal and mining, polymerization, and photographic industries.

Amine oxides compete with alkanolamides as foam boosters in the detergent and personal care industry. Although amine oxides are more expensive than alkanolamides they have the advantage of being milder to the skin and eyes and are more effective surfactants, so that on a cost performance basis they are a better buy than alkanolamides in many cases (see ALKANOLAMINES FROM OLEFIN OXIDES AND AMMONIA). As well as being excellent foam stabilizers in liquid detergents, alkyl amine oxides also increase viscosity, emolliency, detergency, and antistatic properties in many detergent and cosmetic formulas (58).

The surface active properties of aliphatic amine oxides were discovered in the 1930s and the wetting, detergent, emulsion, and foam stabilizing properties were published shortly thereafter (59). However, the use of amine oxides was not significant until Procter and Gamble started using them in household products around 1960 (60–63).

9.2. Organic Reagents. Amine oxides are used in synthetic organic chemistry in the preparation of olefins, or phase-transfer catalysts (64), in alkoxylation reactions (65), in polymerization, and as oxidizing agents (66,67).

9.3. Textiles. In the area of textile and synthetic fiber processing, amine oxides have been used as dyeing auxiliaries as well as wetting agents (68,69), as antistatic agents (qv) (70–72), and as bleaching agents (73,74).

Selected amine oxides in textile technology as dye receptors and for aesthetic purposes has been described (75).

9.4. Pharmaceutical Uses. The biochemistry of heteroaromatic amine oxides has been extensively explored and has led to the synthesis of many biochemically and pharmaceutically important compounds (76). Aromatic amine oxides are useful as analgesics, antihistamines, antitussives, diuretics, tranquilizers, and drug potentiators. In many cases, the N-oxides of pharmacologically active tertiary amines have added benefits, ranging from lower toxicity and better solubility to enhanced therapeutic behavior. The biological activity of these materials has led to patented uses as bactericides, fungicides, insecticides, nematocides, filaricides, amoebicides, anthelmintics, antiparasitics, and disinfectants. The pharmacology and biochemistry of amine oxides has been reviewed (77). Earlier references covering these and other applications were reported in a survey on amine oxides by the Du Pont Company (78).

9.5. Other. Other uses of aliphatic amine oxides are as corrosion inhibitors for nonferrous metals (79) and in aqueous systems (80), as fuel oil antiicing and pourpoint additives that also depress combustion chamber fouling (81–83), in the plastic industry as molecular weight regulators in ethylene and propylene copolymerization (84), in photography to prevent waterspots in drying photographic films (85), as complexing developers and dyes (86), and as asphalt emulsifiers (87).

BIBLIOGRAPHY

"Amine Oxides" in *ECT* 2nd ed., Supplement Volume, pp. 32–50, by S. H. Shapiro, Armour Industrial Chemical Co.; in *ECT* 3rd ed., Vol. 2, pp. 259–271, by R. J. Nadolsky, Armak Co.; in *ECT* 4th ed., Vol. 2, pp. 357–368, by B. Maisonneuve, Akzo Chemicals, Inc.; "Amine Oxides" in *ECT* (online), posting date: December 4, 2000, by B. Maisonneuve, Akzo Chemicals, Inc.

CITED PUBLICATIONS

1. M. W. Lister and I. E. Sutton, *Trans. Faraday Soc.* **35**, 495 (1939).
2. K. W. Herrmann, *J. Phys. Chem.* **66**, 295 (1962).

3. E. P. Linton, *J. Am. Chem. Soc.* **62**, 1945 (1940).

4. J. Meisenheimer, *Ber.* **41**, 3966 (1908).

5. E. J. Corey, A. L. Borror, and T. Foglia, *J. Org. Chem.* **30**, 288 (1965).

6. K. Tsuji and H. Arai, *J. Am. Chem. Soc.* **55**, 558 (1978).

7. T. Kubota, *J. Am. Chem. Soc.* **88**, 211 (1966).

8. T. Kubota, *J. Am. Chem. Soc.* **87**, 458 (1965).

9. D. Hadzi, *J. Chem. Soc.* 5128 (1962).

10. W. A. Bueno and N. M. Mazzaro, *Can. J. Chem.* **54**, 1579 (1978).

11. K. M. Harmon and J. Harmon, *J. Mol. Struct.* **78**, 43 (1982).

12. J. R. Shapley, G. A. Pearson, M. Tachikawa, G. E. Schmidt, M. R. Churchill, and F. J. Hollander, *J. Am. Chem. Soc.* **99**, 8064 (1977).

13. Cope and co-workers, *J. Am. Chem. Soc.* **66**, 1929 (1944); **71**, 3423, 3929 (1949).

14. A. H. Wragg, T. S. Stevens, and D. M. Ostle, *J. Chem. Soc.* 4057 (1958).

15. R. A. W. Johnstone, *Mech. Mol. Migr.* **2**, 249 (1969).

16. G. P. Shulman, P. Ellgen, and M. Conner, *Can. J. Chem.* **43**, 3459 (1965).

17. A. C. Cope, T. T. Foster, and P. H. Towle, *J. Am. Chem. Soc.* **71**, 3929 (1949).

18. A. C. Cope and N. A. Lebel, *J. Chem. Soc.* **82**, 4656 (1960).

19. A. C. Cope, N. A. Lebel, H. H. Lee, and W. R. Moore, *J. Am. Chem. Soc.* **79**, 4720 (1957).

20. J. Zavada, M. Pankova, and M. Svoboda, *Collect. Czech. Chem. Commun.* **38**(7), 2102 (1973).

21. J. Cram and J. E. McCarty, *J. Am. Chem. Soc.* **76**, 5740 (1954).

22. G. P. Shulman and W. E. Link, *J. Am. Oil Chem. Soc.* **41**, 329 (1964).

23. R. G. Lauglin, *J. Am. Chem. Soc.* **95**, 3295 (1973).

24. J. P. Ferris, R. D. Gerwe, and G. R. Gapski, *J. Am. Chem. Soc.* **89**, 5270 (1967).

25. J. P. Ferris, R. D. Gerwe, and G. R. Gapski, *J. Org. Chem.* **33**(9), 3493 (1968).

26. F. Devinski, *Acta. Fac. Pharm.* **XXXIX**, 189 (1985).

27. K. Bodendorf and B. Binder, *Arch. Pharm.* **287**, 326 (1954).

28. E. Glynn, *Analyst* **72**, 248 (1947).

29. E. Hayashi, H. Yamanaka, and K. Shimizu, *Chem. Pharm. Bull. (Tokyo)* **6**, 323 (1958).

30. H. J. den Hertog and J. Overhoff, *Rec. Trav. Chim. Pays-Bas* **69**, 468 (1950).

31. E. Ochiai, *J. Org. Chem.* **18**, 534 (1953); M. Hamana, *J. Pharm. Soc. Japan* **71**, 263 (1951).

32. R. T. Brooks and P. D. Sternglanz, *Anal. Chem.* **31**, 561 (1959).

33. W. R. Dunstun and E. Goulding, *Trans. Chem. Soc.* **75**, 792 (1899).

34. M. Polonovski and M. Polonovski, *Bull. Soc. Chim. Fr.* **41**, 1190 (1927).

35. R. Huisgen, F. Bayerlein, and W. Heydkarp, *Chem. Ber.* **92**, 3223 (1959).

36. S. Oae, T. Kitao, and Y. Kitaoka, *J. Chem. Soc.* **84**, 3366 (1962).

37. J. C. Craig, F. P. Dwyer, A. N. Glazer, and E. C. Horning, *J. Am. Chem. Soc.* **83**, 1871 (1961).

38. E. Ochiai, K. Arimu, and M. Ishikawa, *J. Pharm. Soc. Japan* **63**, 79 (1943).

39. A. R. Katritzky and J. M. Lagowski, *Chemistry of the Heterocyclic N-oxides*, Academic Press, London, 1971.

40. U.S. Pat. 6,455,735 (Sept. 24, 2002), B. M. Choudary and co-workers (to Council of Scientific and Industrial Research, India).

41. *Chemical Market Reporter* (March 24, 2003).

42. D. Sheraga, *Chemical Market Reporter* (Jan. 26, 1998).

43. J. R. Pelka and L. D. Metcalfe, *Anal. Chem.* **37**(4), 603 (1965).

44. T. H. Liddicoet and L. H. Smithson, *J. Am. Oil Chem. Soc.* **42**(12), 1097 (1965).

45. H. Y. Lew, *J. Am. Oil Chem. Soc.* **41**(40), 297 (1964); M. E. Turney and D. W. Cannel, *J. Am. Oil Chem. Soc.* **42**(6), 544 (1965).

46. A. Ruiter, M. B. Krol, and B. J. Tinbergen, eds., *Proceedings of the International Symposium on Nitrite Meat Production*, 1973, 37–43.

47. D. L. Chang, H. L. Rosano, and A. E. Woodward, *Langmuir* **1**(6), 669 (1985).

48. G. J. T. Tiddy, K. Rendall, and M. A. Trevethan, *Commun. J. Com. Esp. Deterg.* **15**, 51 (1984).

49. K. Rendall, G. J. T. Tiddy, and M. A. Trevethan, *J. Colloid Interface Sci.* **98**(2), 565 (1984).

50. N. Bild and M. Hesse, *Helv. Chim. Acta* **50**(70), 1885 (1967).

51. L. D. Metcalfe, *Anal. Chem.* **34**, 1849 (1962).

52. C. N. Wang and L. D. Metcalfe, *J. Am. Oil Chem. Soc.* **62**(3), 558 (1985).

53. *Toxicity Data for Aromox Amine Oxides*, Bull. 68, Armak Co.

54. W. Nakahara, *Prog. Exp. Tumor Res.* **2**, 158 (1961).

55. W. Nakahara, F. Fukuoka, and S. Sakai, *Gann* **49**, 33 (1958).

56. H. Stupel, *Soap Chem. Specialties* **42**(9), 55–7, 135 (1966).

57. U.S. Pat. 6,180,579 (Jan. 30, 2001), R. Erill and C. Gallant (to Colgate Palmolive).

58. E. Jungermann and M. E. Gium, *Soap Chem. Spec.* **40**, 59 (1964).

59. U.S. Pat. 2,159,967 (1939), M. Engleman (to E. I. du Pont de Nemours & Co., Inc.).

60. U.S. Pat. 3,159,581 (1964), F. L. Diehi (to Procter and Gamble).

61. U.S. Pat. 3,001,945 (1961), H. F. Drew and R. E. Zimmer (to Procter and Gamble).

62. U.S. Pat. 3,192,166 (1965), H. F. Drew (to Procter and Gamble).

63. U.S. Pat. 3,346,504 (1967), K. W. Herrmann (to Procter and Gamble).

64. U.S. Pat. 4,307,249 (1981), E. L. Derrenbacker.

65. W. Umbach and W. Stein, *Tenside* **7**(3), 132 (1970).

66. K. B. Sharpless, K. Akushi, and K. Oshima, *Tetrahedron Lett.* 2503 (1976).

67. U.S. Pat. 4,186,077 (1980) D. D. Carlos.

68. Brit. Pat. 1,125,259 (1968), W. Langman, H. Pantke, V. W. Hendricks, and M. Quadvlieg.

69. U.S. Pat. 3,309,319 (1967), T. L. Coward and N. R. Smith.

70. T. P. Matson, *J. Am. Oil Chem. Soc.* **40**, 640 (1963).

71. U.S. Pat. 3,468,869 (1969), E. C. Sherburne.

72. U.S. Pat. 4,395,373 (1983), R. B. Login.

73. U.S. Pat. 3,876,551 (1975), R. J. Laufer and J. H. Geiger.

74. U.S. Pat. 4,390,448 (1983), R. M. Boden, M. Licciordello, J. J. Maisano, and M. R. Hanna.

75. U.S. Pat. 6,500,215 (Dec. 31, 2002), R. Blogin and co-workers (to Sybron Chemicals).

76. S. Oae and K. Ogino, *Heterocycles* **6**(5) (1977).

77. M. H. Bickel, *Pharmacol. Rev.* **21**(4), 325 (1969).

78. *Amine Oxides,* Electrochemicals Dept., E. I. du Pont de Nemours & Co. Inc., Wilmington, Del., 1963.

79. Brit. Pat. 1,185,865 (1970), R. J. Betty and R. E. Malec.

80. U.S. Pat. 5,167,866 (Dec. 1, 1992), C. Hwa and co-workers (to W. R. Grace).

81. U.S. Pat. 3,007,784 (1961), H. G. Ebner.

82. U.S. Pat. 3,387,953 (1968), R. A. Bouford (to Esso).

83. U.S. Pat. 3,594,139 (1971), R. A. Bouford (to Esso).

84. U.S. Pat. 3,405,107 (1968), D. N. Mathews and R. J. Kelly (to Uniroyal).

85. U.S. Pat. 2,490,760 (1950) (to Eastman Kodak Co.).

86. U.S. Pat. 3,68,901 (1972), H. S. Flins and J. P. Van Meter (to Eastman Kodak Co.).

87. U.S. Pat. 6,494,944 (Dec. 17, 2002), J. M. Wates, B. A. Thorstensson, and A. James (to Akzo Nobel NV).

GENERAL REFERENCES

P. A. S. Smith, *The Chemistry of Open-Chain Organic Nitrogen Compounds*, Vol. II, W. A. Benjamin, Inc., New York, 1966, pp. 21–28.

C. C. J. Culvenor, *Rev. Pure Appl. Chem. (Australia)* **3**, 83 (1953).

L. W. Burnette, in M. J. Shick, ed., *Nonionic Surfactants*, Vol. I, Marcel Dekker, Inc., New York, 1967, pp. 403–410.

E. Ochiai, *Aromatic Amine Oxides*, Elsevier Publishing Co., Amsterdam, 1967.

A. R. Katritzky, *Q. Rev. (London)* **10**, 395 (1956).

J. D. Sauer, in J. M. Richmond, ed., *Surfactant Science Series,* Vol. 34, Marcel Dekker, New York, 1990, pp. 275–295.

B. MAISONNEUVE
Akzo Chemicals, Inc.

AMINES BY REDUCTION

1. Introduction

Amines are derivatives of ammonia in which one or more of the hydrogens is replaced with an alkyl, aryl, cycloalkyl, or heterocyclic group. When more than one hydrogen has been replaced, the substituents can either be the same or different. Amines are classified as primary, secondary, or tertiary depending on the number of hydrogens which have been replaced. It is important to note that the designations primary, secondary, and tertiary refer only to the number of substituents and not to the nature of the substituents as in some classes of compounds.

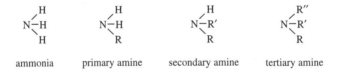

ammonia   primary amine   secondary amine   tertiary amine

Amines can be prepared by a variety of methods including substitution reactions, rearrangements, ammonolysis, and reductions. On a large scale, however, ammonolysis and reductions are usually the most efficient and are generally used in commercial processes. In reductive methods, the nitrogen is already incorporated in the molecule, and the amine is formed by reducing the oxidation state of the compound with the addition of hydrogen. In theory, many different types of nitrogen-containing compounds can be reduced to amines. In practice, however, nitriles or nitro compounds are usually used because they are the most easily obtained starting materials.

There are several commercial processes for reducing nitro or nitrile groups to amines. Most large volume aromatic and aliphatic amines are made by continuous high pressure catalytic hydrogenation. Nitro compounds can also be reduced in good yields with iron and hydrochloric acid in the Béchamp process.

The importance of the Béchamp process has declined over the last few decades, but it is still used in the pigment and dyestuff industry and to make iron oxide pigments; aniline is produced as a by-product. Other more specialized methods used for making amines by reduction include the Zinin reduction, in which sulfides in alkaline media are used to reduce aromatic nitro compounds, bisulfite reductions, electrochemical reductions, and reductions using metal amalgams or hydrides. The special case involving preparation of aliphatic amines by hydrogenation of aromatic amines is also included.

2. Catalytic Hydrogenation

In catalytic hydrogenation, a compound is reduced with molecular hydrogen in the presence of a catalyst. This reaction has found applications in many areas of chemistry including the preparation of amines. Nitro, nitroso, hydroxylamino, azoxy, azo, and hydrazo compounds can all be reduced to amines by catalytic hydrogenation under the right conditions. Nitriles, amides, thioamides, and oximes can also be hydrogenated to give amines (1). Some examples of these reactions follow:

Nitro $R-NO_2 + 3 H_2 \longrightarrow R-NH_2 + 2 H_2O$

Nitrile $R-CN + 2 H_2 \longrightarrow R-CH_2NH_2$

Amide $R-\overset{\overset{\text{O}}{\|}}{C}-NH_2 + 2 H_2 \longrightarrow R-CH_2NH_2 + H_2O$

Thioamide $R-\overset{\overset{\text{S}}{\|}}{C}-NH_2 + 2 H_2 \longrightarrow R-CH_2NH_2 + H_2S$

Azo $R-N{=}N-R' + 2 H_2 \longrightarrow R-NH_2 + R'-NH_2$

Catalytic hydrogenation is the most efficient method for the large scale manufacture of many aromatic and aliphatic amines. Some of the commercially important amines produced by catalytic hydrogenation include aniline (from nitrobenzene), 1,6-hexanediamine (from adiponitrile), isophoronediamine (from 3-nitro-1,5,5-trimethylcyclohexanecarbonitrile), phenylenediamine (from dinitrobenzene), toluenediamine (from dinitrotoluene), toluidine (from nitrotoluene), and xylidine (from nitroxylene). As these examples suggest, aromatic amines are usually made by hydrogenating the corresponding nitro compound, whereas the aliphatic amines generally start with the corresponding nitrile. The main reason for this difference is the availability of the necessary raw materials. Many aromatic hydrocarbons can be easily nitrated to give a variety of aromatic nitro compounds. For aliphatic amines, however, the nitrile precursor is generally easier to obtain than the corresponding aliphatic nitro compound. Nitriles, however, can only yield amines which are located at a primary carbon.

Catalytic hydrogenations can be carried out in the vapor phase or in the liquid phase, either with or without the use of a solvent. The vapor phase

reaction is limited to compounds which are thermally stable and relatively volatile. High boiling compounds and those which are thermally unstable must be
hydrogenated in the liquid phase.

3. Mechanism and Kinetics of Hydrogenation.

3.1. Reduction of Nitro Compounds.
The mechanism for catalytic
hydrogenation of nitro compounds has been the subject of many investigations
and there is much evidence that this reaction proceeds through several intermediate species. The most widely accepted mechanism for the hydrogenation
of nitro compounds was proposed by Haber in 1898 (2) (see Fig. 1).

Haber based this mechanism on the electrochemical reduction of nitrobenzene, but it has since been used by many researchers to explain the results of
hydrogenation studies. For example, a mechanistic and kinetic study of the
hydrogenation of nitrobenzene to aniline concluded that the reaction proceeds
through a number of intermediates including nitrosobenzene, phenylhydroxylamine, azoxybenzene, azobenzene, and hydrazobenzene, even though not all of
these intermediates were detected (3). Further evidence of the existence of
these intermediates is provided by reports that under certain conditions, catalytic hydrogenation of nitro compounds can yield either hydroxylamines (4,5) or
azoxy compounds (6) as the major product.

Many kinetic studies on hydrogenation reactions of nitro compounds have
been aimed at determining not only rate equations, but also the effect of various
factors on the reaction rate. In one study, the vapor-phase hydrogenation of
nitrobenzene to aniline using a copper oxide–chromium oxide catalyst was
found to be half order with respect to hydrogen concentration and first order
in nitrobenzene (7). Similarly, in a liquid-phase hydrogenation using a homogeneous catalyst, the reaction rate was found to be first order in nitrobenzene concentration and dependent on the hydrogen pressure (8). However, other workers
have concluded that the order of the reaction with respect to the nitro compound

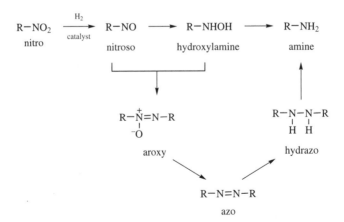

Fig. 1. Mechanism for the hydrogenation of nitro compounds. Functionalities are
labeled.

can vary between 0 and 1, depending on factors such as the amount of catalyst and the nature of the solvent (9).

3.2. Reductions of Nitriles. In the reduction of nitriles, hydrogen is added progressively across the carbon–nitrogen triple bond, forming first the imine and then the amine.

$$R{-}CN \xrightarrow{H_2} R{-}CH{=}NH \xrightarrow{H_2} R{-}CH_2{-}NH_2$$

One characteristic of this reaction that can cause problems is that secondary and tertiary amines are produced in addition to the primary amine. It has been proposed that these side reactions occur through reaction of the imine intermediate with the product amine, followed by the loss of ammonia and further hydrogenation (10).

$$RCH{=}NH \;+\; RCH_2NH_2 \;\rightleftharpoons\; \underset{NHCH_2R}{R\,CH{-}NH_2} \;\overset{2NH_3}{\rightleftharpoons}\; RCH{=}NHCH_2R \;\xrightarrow{H_2}\; RCH_2NHCH_2R$$

The tertiary amine is formed in a similar manner from the imine and a secondary amine. This side reaction can be minimized by carrying out the hydrogenation in the presence of ammonia, which tends to shift the equilibrium back towards the imine. When a compound with two or more nitrile groups is hydrogenated, the formation of both cyclic and acyclic secondary and tertiary amines is possible, depending on whether the side reaction is intramolecular or intermolecular. For example, for the hydrogenation of adiponitrile:

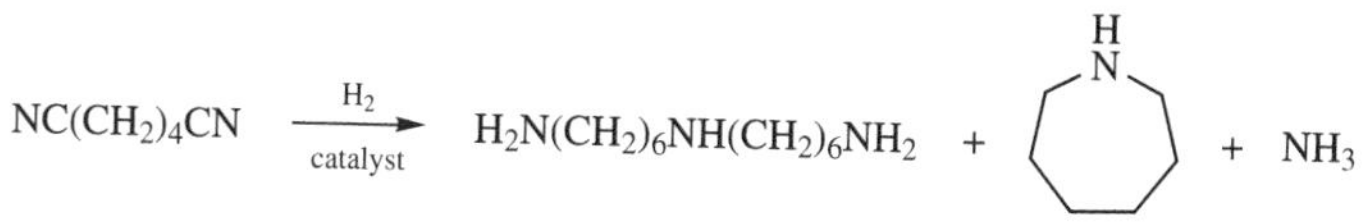

$$NC(CH_2)_4CN \;\xrightarrow[\text{catalyst}]{H_2}\; H_2N(CH_2)_6NH(CH_2)_6NH_2 \;+\; \text{(cyclic amine)} \;+\; NH_3$$

4. Hydrogenation Raw Materials

4.1. Substrates. Many different types of nitrogen-containing compounds can be hydrogenated to amines, but nitro compounds and nitriles are the most commonly used starting materials.

Nitro Compounds. Many aromatic hydrocarbons react with nitric acid in the presence of sulfuric acid to form aromatic nitro compounds. The nitration reaction is an electrophilic aromatic substitution in which the orientation of the nitro group being introduced is controlled by the substituents which are already present (11). Because the nitro group is deactivating, each nitro group becomes progressively more difficult to add, and the reaction conditions can be controlled to give almost exclusively mono-, di- or trinitro compounds. For general reviews of nitration reactions see also references 12 and 13.

Nitriles. Nitriles can be prepared by a number of methods, including (1) the reaction of alkyl halides with alkali metal cyanides, (2) addition of hydrogen cyanide to a carbon–carbon, carbon–oxygen, or carbon–nitrogen multiple bond,

(*3*) reaction of hydrogen cyanide with a carboxylic acid over a dehydration catalyst, and (*4*) ammoxidation of hydrocarbons containing an activated methyl
group. For reviews on the preparation of nitriles see references 14 and 15.

4.2. Hydrogenation Catalysts.   The key to catalytic hydrogenation is
the catalyst, which promotes a reaction which otherwise would occur too slowly
to be useful. Catalysts for the hydrogenation of nitro compounds and nitriles are
generally based on one or more of the group VIII metals. The metals most commonly used are cobalt, nickel, palladium, platinum, rhodium, and ruthenium,
but others, including copper (16), iron (17), and tellurium (18) have been used.
Despite this relatively small list, a wide variety of catalysts and catalyst modifications have been reported in the literature.

4.3. Physical Characteristics.   Heterogeneous catalysts are generally
used for both large and small scale hydrogenations, but many homogeneous catalysts have also been used. Homogeneous and heterogeneous catalysts generally
contain the same type of metals, but in homogenous catalysts the metals are present in the form of complexes or clusters with various organic and inorganic
ligands. Some of the homogeneous catalysts which have been reported include
copper triphenylphosphine complexes (19), ruthenium phosphine complexes
(20,21), rhodium carbonyl clusters (8), and complexes of palladium with quinoline (22), pyridine (23,24), and phenylisocyano (25) ligands. Despite the high
activity (20) and selectivity (21) sometimes claimed for homogeneous catalysts,
heterogeneous catalysts are usually preferred because they are easy to use and
recover for reuse.

Heterogeneous hydrogenation catalysts can be used in either a supported or
an unsupported form. The most common supports are based on alumina, carbon,
and silica. Supports are usually used with the more expensive metals and serve
several purposes. Most importantly, they increase the efficiency of the catalyst
based on the weight of metal used and they aid in the recovery of the catalyst,
both of which help to keep costs low. When supported catalysts are employed,
they can be used as a fixed bed or as a slurry (liquid phase) or a fluidized bed
(vapor phase). In a fixed-bed process, the amine or amine solution flows over
the immobile catalyst. This eliminates the need for an elaborate catalyst recovery system and minimizes catalyst loss. When a slurry or fluidized bed is used,
the catalyst must be separated from the amine by gravity (settling), filtration, or
other means.

The available surface area of the catalyst greatly affects the rate of a hydrogenation reaction. The surface area is dependent on both the amount of catalyst
used and the surface characteristics of the catalyst. Generally, a large surface
area is desired to minimize the amount of catalyst needed. This can be accomplished by using either a catalyst with a small particle size or one with a porous
surface. Catalysts with a small particle size, however, can be difficult to recover
from the material being reduced. Therefore, larger particle size catalyst with a
porous surface is often preferred. A common example of such a catalyst is
Raney nickel.

Because of its relatively low cost, high activity, and long life, Raney nickel is
one of the most commonly used hydrogenation catalysts. Raney nickel is composed primarily of nickel and aluminum which has been processed to give it a
large surface area. This is done by activating an alloy of nickel and aluminum

with aqueous sodium hydroxide under controlled conditions to dissolve most of the aluminum, leaving behind the nickel. The resulting catalyst is extremely porous, having a surface area of up to 100 m^2/g (26). There are several standard types of Raney nickel which are prepared using different activation conditions. These differ both in their physical characteristics and in their activity. Similar activation processes have also been applied to other metals to increase their surface area and catalytic activity.

4.4. Activity and Selectivity. The activity of a catalyst refers to its ability to promote the desired reaction whereas the selectivity relates to how effective a catalyst is at promoting only a specific reaction. The selectivity which is required in a particular hydrogenation depends on the functional groups present in the material being hydrogenated. Many common functional groups can be reduced by catalytic hydrogenation with varying degrees of difficulty, and often one functional group must be reduced selectively in the presence of other groups which are to be left unchanged. For example, in the reduction of aromatic nitro compounds the catalyst and reaction conditions must be selected so that the nitro group is completely reduced while the aromatic ring is left intact. Since aromatic rings are generally much more difficult to reduce than nitro groups, this reduction can be carried out very selectively. However, other cases exist where it is much harder to selectively reduce one functional group in the presence of another. An impressive example of the selectivity which can be achieved is in the reduction of 2,4-dinitroaniline [97-02-9], (C$_6$H$_5$N$_3$O$_4$). Not only is it possible to reduce one of the nitro groups to an amine while leaving the other unchanged, but through proper choice of catalyst and reaction conditions, either 4-nitro-1,2-benzenediamine [99-56-9] (**1**) (C$_6$H$_7$N$_3$O$_2$) (27) or 2-nitro-1,4-benzenediamine [5307-14-2] (**2**) (28) can be made in good yield.

(**1**)　　　(**2**)

The subject of catalyst selection for hydrogenation reactions has been summarized in several books (29,30).

Poisoning and Deactivation. Many catalysts can be inhibited or poisoned by the presence of certain materials. Sulfur, phosphorus, arsenic, and bismuth compounds that have unshared electrons are common catalyst poisons. They act by binding to the catalyst surface, thereby preventing the desired reaction from occurring. Catalysts can also be deactivated by other means including pore plugging and physical degradation. However, the effects of poisoning or deactivation are not always permanent, and the catalyst can sometimes be regenerated. For a review of catalyst poisoning, deactivation and regeneration, see reference 31.

4.5. Hydrogen. Most large scale industrial processes operate with essentially pure hydrogen, but hydrogen mixed with carbon monoxide or inert gases can also be used. Hydrogen can be obtained by any of several known processes (32), but those which simultaneously produce carbon monoxide as a by-product are often favored. The carbon monoxide can react with chlorine to form phosgene, the principal reagent required to convert amines to isocyanates. Methane gas is a convenient starting material for hydrogen generation, and the hydrogen and carbon monoxide produced from methane are generally of a high purity. One common route which produces hydrogen and carbon monoxide from methane is steam reforming, in which methane gas and steam are passed over a nickel–magnesia catalyst at about 800°C under pressure:

$$CH_4 + H_2O \xrightarrow{\text{catalyst}} CO + 3\ H_2$$

4.6. Solvents. A solvent is not always required in catalytic hydrogenations, but the use of a solvent can have several benefits. Using a solvent can make it simpler to hydrogenate solids or materials which are otherwise difficult to handle. A solvent can also act to moderate the heat of reaction, which can be quite large. Finally, a solvent can affect the course of a reaction by influencing the selectivity of the hydrogenation (33). Since hydrogenation processes usually operate under high pressure, the reaction temperature can be significantly above the normal boiling point of the solvent. In practice, low molecular weight alcohols, particularly methanol or ethanol, are the most commonly used hydrogenation solvents. Other solvents which have been used include acetic acid, acetone, ammonia, benzene, glycerol, ethylene glycol, hydrochloric acid, sulfuric acid, and water. When applicable the hydrogenation product itself can act as a solvent (34).

4.7. Additives. Several literature references and patents deal with the use of additives to promote or suppress particular reactions during the preparation of amines by catalytic hydrogenation. For example, the use of triethyl phosphite [122-52-1] (35), phosphoric acid [7664-38-2] (36), and 2,2′-thiodiethanol [111-48-8] $C_4H_{10}O_2S$ (37) has been reported to suppress dehalogenation during the reduction of halogen containing aromatic nitro compounds. Additives have also been reported to be useful both in suppressing the formation of amidines during the hydrogenation of nitriles (38) and promoting the reduction of aromatic nitrosulfonic acids to the corresponding amino derivatives (39). The presence of ammonia as an additive or solvent during the hydrogenation of nitriles helps to inhibit the formation of secondary and tertiary amine by-products.

5. Hydrogenation Reaction Conditions

5.1. Heat of Reaction and Heat Transfer. Catalytic hydrogenation is a very exothermic reaction and, therefore, removal of the heat generated is an important consideration in any hydrogenation process. For example, the heat of reaction for the catalytic hydrogenation of nitroxylene to xylidine is about 488 kJ/mol (117 kcal/mol) at 200°C (40). This value is comparable to the 544 kJ/mol (130 kcal/mol), which is often quoted for the hydrogenation of nitrobenzene to aniline. By comparison, the heat of reaction for the hydrogenation of nitriles is much smaller, about 310 kJ/mol (74 kcal/mol) for the hydrogenation of adiponitrile to 1,6-hexanediamine (41). This difference is, in part, because water is formed in the hydrogenation of nitro compounds, but not in the hydrogenation of nitriles. Nevertheless, any hydrogenation process must be able to remove sufficient heat to keep the reaction mixture at the desired temperature.

5.2. Temperature and Pressure.   Temperature and pressure both have large effects on the course of a hydrogenation process. Higher temperatures lead to faster reactions, but excessive temperatures can result in undesirable by-products and unsafe conditions. A study of the hydrogenation of 3,4-dichloroaniline showed that this reaction involved the initial formation of the hydroxylamine, which at low temperatures disproportionates to give the amine and the nitroso compound. At high temperatures, however, a highly exothermic reaction occurs in which the hydroxylamine and nitroso compounds form the azoxy compound (42).

Increasing hydrogen pressure also increases the rate of reaction by increasing the contact between the hydrogen gas, the catalyst, and the substrate which is needed for the reaction to take place. Usually higher pressures do not cause the problems associated with high temperatures. In fact, increasing the hydrogen pressure may actually serve to suppress side reactions which can occur under hydrogen deficient conditions (43).

5.3. Mixing.   Agitation also plays a large role in optimizing a hydrogenation process. The metal catalyst is often much heavier than the rest of the reaction mixture and therefore vigorous mixing is needed to keep the catalyst in contact with the material being reduced and the hydrogen. Various methods have been used to achieve this mixing. With a fixed-bed catalyst system, the reactants flow through the stationary catalyst. In a fluidized-bed reactor the upward vapor flow keeps the catalyst suspended, ensuring good contact. A similar effect is possible with a liquid slurry, where the flow of liquid and gas keeps the catalyst and reactants mixed. It has also been reported that pulsation at frequencies of 30 to 3000 Hz can provide the necessary mixing when the liquid phase hydrogenation is carried out in tube reactors (44). More conventional agitators can also be used in stirred reactors, and in the laboratory the entire reactor is sometimes shaken to provide the necessary mixing.

6. Industrial Hydrogenation Processes

6.1. Liquid-Phase Hydrogenation of Nitro Compounds.   Most high boiling aromatic amines are prepared by liquid-phase hydrogenation of the corresponding nitro compound, either with or without the use of a solvent. Early catalytic hydrogenations were performed in stirred batch reactors. More recently these have been replaced by continuous processes which provide better control in terms of heat transfer and mixing of the reactants. A continuous liquid-phase hydrogenation process is illustrated by the process for toluenediamine. Toluene can be readily nitrated by conventional means to give dinitrotoluene (DNT) (methyldinitrobenzene [25321-14-6], $C_7H_6N_2O_4$), consisting primarily of the 2,4- and 2,6-isomers in a ratio of about 4:1 (45). The DNT can be reduced by high pressure liquid-phase catalytic hydrogenation to the corresponding toluenediamine (TDA) (ar-methylbenzenediamine [25376-45-8], $C_7H_{10}N_2$) (see AMINES,

AROMATIC, DIAMINOTOLUENES). Most of the TDA produced is phosgenated to give toluenediisocyanate (TDI) (diisocyanatomethylbenzene [26471-62-5], $C_9H_6N_2O_2$), which is used in the manufacture of polyurethanes. A typical industrial process for the reduction of DNT to TDA, as described by patents, is shown in Figure 2 and described in the following (46,47).

A solution containing about 25% DNT in methanol is pumped along with a Raney nickel slurry and enough hydrogen gas to complete the reaction through a series of reactors. There are three high pressure reactors plus an auxiliary reactor, each with a volume of approximately 450 L. The reactors are about 6 m long and 35 cm in diameter and are equipped with water cooling for temperature control. This reaction mixture is fed to the first reactor at a rate of approximately 2000 kg/h. The material from the first reactor is split and fed to the second and third reactors, which are run in parallel. The temperature of the reactors is maintained at about 100°C and the hydrogen pressure at between 15,000 and 20,000 kPa (150 and 200 atm). After the reaction is completed, the pressure is reduced and the excess hydrogen removed in a liquid–gas separator. The hydrogen is recycled to the beginning of the process and the product stream continues on through a series of catalyst removal steps. The recovered catalyst is also recycled with only a small amount, generally 0.1 to 0.3%, being lost in the process. After the catalyst is removed, the product is sent to a methanol column where the solvent is removed and recycled. The water is then removed in the dehydration column. The recovered water from this step contains volatile amines and other by-products and must be processed further before it can be recycled or discharged to the waste treatment plant. The product from the dehydration column can be used directly or taken to a final TDA column where it is distilled to give a product which is more than 99% pure. The residue from the final column can be disposed of by incineration or processed to recover some of the amine, thereby improving the yield and reducing waste-disposal problems.

Numerous modifications to the above process are possible and many variations have been suggested. Inert solvents other than methanol can be used; however, low molecular weight alcohols are usually considered preferable. Part of the reaction product can be recycled back to the front of the process to reduce the amount of solvent required and to eliminate problems associated with DNT solidification. A 76:24 mixture of DNT:TDA has been found to exhibit a minimum freezing point of 26°C, as compared to 50°C for pure DNT (46,47). The temperature at which the reaction is carried out can also be varied. Higher temperatures not only reduce the reaction time needed, but also result in less residue being formed (46). A temperature of 115 to 130°C is considered ideal, whereas temperatures above 170°C are considered unsafe.

6.2. Vapor Phase Hydrogenations of Nitro Compounds.

Catalytic hydrogenation of nitro compounds can be carried out in the vapor phase, provided the boiling point of the compound is low enough and the material is thermally stable. These two conditions effectively limit this process to aliphatic and relatively simple aromatic nitro compounds such as nitrobenzene or nitroxylene. Early vapor-phase hydrogenation processes used fixed-bed catalysts. However, fluidized-bed catalytic vapor-phase hydrogenations, such as the one illustrated by the process for aniline, have become more common (see AMINES, AROMATIC, ANILINE AND DERIVATIVES).

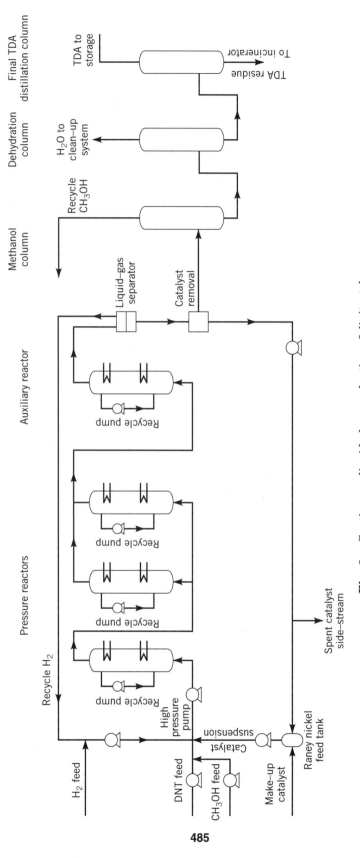

Fig. 2. Continuous liquid phase reduction of dinitrotoluene.

Nitrobenzene [98-95-3], $C_6H_5NO_2$ is produced by the nitration of benzene with a mixture of nitric and sulfuric acid. A process for the manufacture of aniline [62-53-3], C_6H_7N, from nitrobenzene is shown in Figure 3 (16). Nitrobenzene, which contains less than 10 ppm nitrothiophene, a catalyst poison, is fed to a vaporizer where it is vaporized. As the gaseous nitrobenzene leaves the vaporizer, it is mixed with a 200% excess of hydrogen gas. The gaseous mixture then passes upward through a porous distributor plate into the reduction chamber of the fluidized-bed reactor which contains the silica-supported copper catalyst. The catalyst powder is carried upward through the reactor by the vapor flow. The vapor velocity through the reactor is about 30 cm/s. The reaction occurs at 270°C and 234 kPa (2.3 atm) with a very short contact time. The excess heat of reaction is removed by circulating heat transfer fluid through cooling tubes located in the catalyst bed. The upper portion of the reactor is large enough to allow most of the catalyst to fall back into the main catalyst bed. Any catalyst which escapes from the reactor is removed from the product by stainless steel filters.

After leaving the reactor, the reaction mixture consisting of aniline, water, and excess hydrogen is cooled and condensed prior to the purification steps. First, the excess hydrogen is removed and recycled back to the reactor. The rest of the mixture is sent to the decanter where the water and aniline are separated. The crude aniline, which contains less than 0.5% of unreacted nitrobenzene and about 5% water, is distilled in the crude aniline column. The aniline is further dehydrated in the finishing column to yield the purified aniline. Meanwhile, the aqueous layer from the decanter, which contains about 3.5% aniline, is extracted to recover the aniline and clean up the water before it is sent to the waste-water treatment plant.

This process produces aniline with a yield of greater than 99% of theory. The appearance of nitrobenzene in the product is a sign of catalyst deactivation and the catalyst must be regenerated. This is done by stopping the nitrobenzene and hydrogen flow, and passing air through the catalyst at temperatures between 250 and 350°C. With regeneration, each gram of catalyst can produce a minimum of 600 grams of aniline before it must be replaced.

6.3. Aliphatic Amines from the Hydrogenation of Nitriles. One of the most common and economical means of producing aliphatic amines is by the catalytic hydrogenation of nitriles, eg, the process for 1,6-hexanediamine. Adiponitrile (hexanedinitrile [111-69-3], $C_6H_8N_2$), is produced commercially by several routes which are based on different raw materials. It can be made by reaction of adipic acid (qv) with ammonia over a catalyst or by the dimerization of acrylonitrile (qv) at the cathode in an electrolytic cell. Adiponitrile is also made from butadiene, either by direct reaction with hydrogen cyanide or by first chlorinating the butadiene to give 1,4-dichlorobutane, which then reacts with sodium cyanide. A continuous process for the reduction of adiponitrile to 1,6-hexanediamine ([124-09-4], $C_6H_{16}N_2$), is illustrated in Figure 4 and described below (41). This process uses a fixed-bed catalyst with liquid ammonia as a solvent to suppress formation of secondary and tertiary amines.

Adiponitrile, liquid ammonia, and hydrogen gas are introduced together in a molar ratio of approximately 1:25:38 under a pressure of about 60 MPa (600 atm). The mixture passes through a preheater where the temperature is

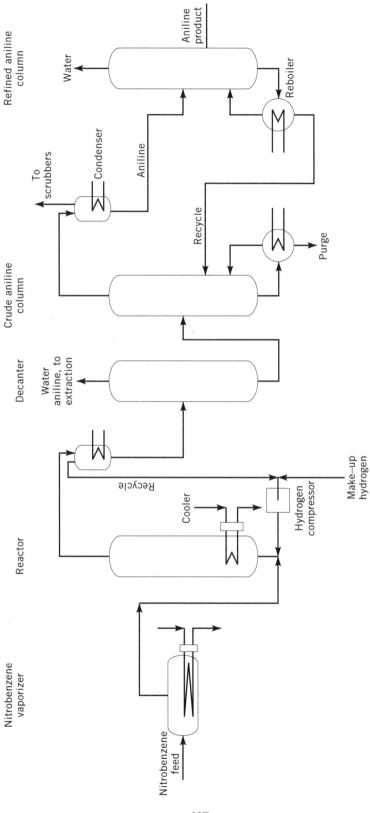

Fig. 3. Continuous fluidized-bed vapor phase reduction of nitrobenzene.

487

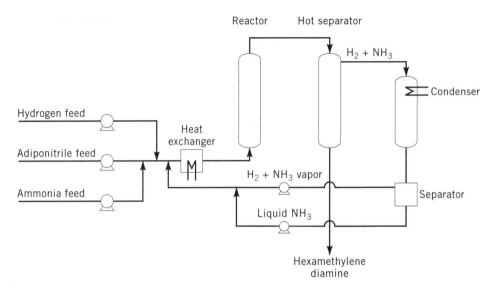

Fig. 4. Continuous fixed-bed reduction of adiponitrile to 1,6-hexanediamine.

raised to about 30°C below the desired reaction temperature. The mixture then enters the reactor which contains the catalyst in the form of a fixed bed. Several different catalysts can be used, including copper–cobalt, cobalt–aluminum, cobalt oxide, and chromium oxide–cobalt oxide. The preferred temperature range is from 110 to 135°C, but any temperature between 100 and 150°C can be used. The adiponitrile is converted to 1,6-hexanediamine in the reactor and transferred to the hot separator where the gaseous hydrogen and ammonia are removed. The hydrogen and ammonia are then taken to a condenser for recycling while the diamine is taken forward for purification.

About 310 kJ/mol (75 kcal/mol) of heat is released during this reaction and the exotherm is controlled by adjusting the recycle streams. Part of the heat is used to raise the temperature of the reaction mixture the final 30°C while a large amount of heat is expended in vaporizing the ammonia.

7. Béchamp Process

In the Béchamp process, nitro compounds are reduced to amines in the presence of iron and an acid. This is the oldest commercial process for preparing amines, but in more recent years it has been largely replaced by catalytic hydrogenation. Nevertheless, the Béchamp reduction is still used in the dyestuff industry for the production of small volume amines and for the manufacture of iron oxide pigments; aniline is produced as a by-product. The Béchamp reduction is generally run as a batch process; however, it can also be run as a continuous (48) or semi-continuous process (49).

7.1. Reaction Mechanism. The overall reaction in the Béchamp process is as follows

$$4\,RNO_2 + 9\,Fe + 4\,H_2O \xrightarrow{FeCl_2} 4\,RNH_2 + 3\,Fe_3O_4$$

However, this is an oversimplification, and the Béchamp reduction is a complex reaction which can be better represented by the following stepwise mechanism (50).

$$2\ R\text{---}NO_2 + FeCl_2 + 6\ Fe + 10\ H_2O \longrightarrow 2\ R\text{---}NH_3Cl + 7\ Fe(OH)_2$$

$$R\text{---}NO_2 + 6\ Fe(OH)_2 + 4\ H_2O \longrightarrow R\text{---}NH_2 + 6\ Fe(OH)_3$$

$$Fe(OH)_2 + 2\ Fe(OH)_3 \longrightarrow Fe_3O_4 + 4\ H_2O$$

In this representation the $FeCl_2$ which takes part in the first step of the reaction is not a true catalyst, but is continuously formed from HCl and iron. This is a highly exothermic process with a heat of reaction of 546 kJ/mol (130 kcal/mol) for the combined charging and reaction steps (50). Despite the complexity of the Béchamp process, yields of 90–98% are often obtained. One of the major advantages of the Béchamp process over catalytic hydrogenation is that it can be run at atmospheric pressure. This eliminates the need for expensive high pressure equipment and makes it practical for use in small batch operations. The Béchamp process can also be used in the laboratory for the synthesis of amines when catalytic hydrogenation cannot be used (51).

Some of the important parameters in the Béchamp process are the physical state of the iron, the amount of water used, the amount and type of acid used, agitation efficiency, reaction temperature, and the use of various catalysts or additives. When these variables are properly controlled, the amine can be obtained in high yields while controlling the color and physical characteristics of the iron oxide pigment which is produced.

7.2. Raw Materials. *Iron.* Clean, finely divided, soft, gray cast iron yields the best results. In practice, the iron is often etched by heating it in an acid solution prior to addition of the nitro compound. This ensures a good iron surface and prevents the possibility of a violent delayed reaction. Since the rate of reaction depends in part on the surface area of the iron, finely divided iron leads to a smoother reaction requiring less time. When coarse iron is used, the reaction is slower and a larger excess of iron is required to complete the reaction. For these reasons, iron turnings, shavings, or borings are generally preferred. It has been reported that impurities in the iron can lead to undesirable side reactions in certain cases. For example, when ring-chlorinated aromatic nitro compounds are reduced, dechlorination is possible. Dechlorination is promoted by the presence of nickel in the iron (52), but can be retarded by addition of certain inhibitors including potassium thiocyanate and dicyandiamide (53,54).

Water. Based on the overall balanced equation for this reaction, a minimum of one mole of water per mole of nitro compound is required for the reduction to take place. In practice, however, 4 to 5 moles of water per mole of nitro compound are used to ensure that enough water is present to convert all of the iron to the intermediate ferrous and ferric hydroxides. In some cases, much larger amounts of water are used to dissolve the amino compound and help separate it from the iron oxide sludge after the reaction is complete.

Acid. The reaction requires only enough acid to generate the ferrous ion which is needed to participate in the first step. Alternatively, a ferrous salt can be added directly. Generally 0.05 to 0.2 equivalents of either hydrochloric or sulfuric acid is used, but both acids have their drawbacks. Hydrochloric acid can cause the formation of chlorinated amines and sulfuric acid can cause the rearrangement of intermediate arylhydroxylamines to form hydroxyaryl amines. Occasionally an organic carboxylic acid such as acetic or formic acid is used when there is a danger of hydrolysis products being formed.

7.3. Reaction Conditions. *Mixing.* Because of the heterogeneous nature of this system, efficient mixing is essential to ensure the intimate contact of the iron, nitro compound, and water soluble catalyst. An agitator which allows the iron to settle to the bottom and the other materials to separate into layers does not function efficiently. On the other hand, a reaction whose rate is limited by the quality of the iron will not be significantly improved by better mixing.

Amine Recovery. Once the reaction is completed, the amine must be separated from the iron oxide sludge. First, the iron oxide is allowed to settle and the liquid amine layer is removed by siphoning or decanting. A significant amount of amine remains in the iron oxide and this must also be recovered. Several methods have been used to accomplish this (55). Either steam distillation or vacuum distillation of the amine from the iron oxide is possible. In steam distillation, steam is fed into the iron oxide sludge and the condensate is collected. This is an effective but expensive method of recovering the amine. Vacuum distillation is less expensive than steam distillation, but it is difficult to recover all of the amine from the iron oxide using vacuum distillation. Filtration is another way of recovering the amine from the iron oxide. In this method, the iron oxide is first washed with water recovered from the amine purification. This is followed by fresh boiling water and finally, the iron oxide is blown with hot air or steam to remove the remaining liquid.

7.4. Preparation of Aniline by the Béchamp Process. About 1500 L of a ferrous chloride solution and 1300–1500 L of aniline water are charged to a 20,000-L reactor. Because of the abrasive nature of this reaction mixture, the reactor is often lined with tile or contains a replaceable liner. To this mixture are added 1000 kg of iron filings and 300 L of nitrobenzene. Once the reaction is underway, an additional 4700 kg of nitrobenzene and 5300 kg of iron are added together over a period of 6 to 9 h. After about 12 h, the reaction is complete and the mixture is neutralized. Then the iron sludge is allowed to settle and the aniline is decanted off through an adjustable dip tube. The iron oxide sludge is steam distilled by passing steam through it for 4–5 h and the condensate collected. The condensate is added to the aniline which was decanted and the material is further neutralized and purified by distillation. The aniline water from the distillation is returned to the process or extracted with nitrobenzene to recover the residual aniline (56).

Several modifications to this process are possible (55). Instead of adding ferrous chloride directly, it is more common to generate it by using iron and hydrochloric acid. The order in which the reactants are added can also be altered, and it is even possible to add all of the iron or aniline at the beginning of the reaction. There are also other ways to recover the aniline from the iron oxide sludge.

8. Miscellaneous Reductions

8.1. Zinin Reduction. The method of reducing aromatic nitro compounds with divalent sulfur is known as the Zinin reduction (57). This reaction can be carried out in a basic media using sulfides, polysulfides, or hydrosulfides as the reducing agent. These reactions can be represented as follows when the counter ion is sodium:

$$4\,ArNO_2 + 6\,Na_2S + 7\,H_2O \longrightarrow 4\,ArNH_2 + 3\,Na_2S_2O_3 + 6\,NaOH$$

$$ArNO_2 + Na_2S_2 + H_2O \longrightarrow ArNH_2 + Na_2S_2O_3$$

$$4\,ArNO_2 + 6\,NaSH + H_2O \longrightarrow 4\,ArNH_2 + 3\,Na_2S_2O_3$$

Although this reduction is more expensive than the Béchamp reduction, it is used to manufacture aromatic amines which are too sensitive to be made by other methods. Such processes are used extensively where selectivity is required such as in the preparation of nitro amines from dinitro compounds, the reduction of nitrophenol and nitroanthraquinones, and the preparation of aminoazo compounds from the corresponding nitro derivatives. Amines are also formed under the conditions of the Zinin reduction from aromatic nitroso and azo compounds.

The Zinin reduction is also useful for the reduction of aromatic nitro compounds to amines in the laboratory. It requires no special equipment, as is the case with catalytic hydrogenations, and is milder than reductions with iron and acid. Usually ammonium or alkali sulfides, hydrosulfides or polysulfides are used as the reactant with methanol or ethanol as the solvent.

The reduction of *m*-dinitrobenzene [99-65-0] to *m*-nitroaniline [99-09-2] on a laboratory scale offers an example. A solution containing 18 g of sodium sulfide and 6 g of sodium bicarbonate in 50 mL of water is prepared, mixed with 50 mL of methanol, and filtered to remove precipitated sodium carbonate. Next, 6.7 g of *m*-dinitrobenzene, $C_6H_4N_2O_4$, in 50 mL of methanol is added and the mixture refluxed for 20 min. The methanol is removed by distillation and the residue poured into cold water where it solidifies. The material can then be recrystallized from 75% aqueous methanol to give a 69% yield of *m*-nitroaniline, $C_6H_6N_2O_2$, with a melting point of 114°C (58).

8.2. Sodium Bisulfite. Sodium bisulfite [7631-90-5], $NaHSO_3$, is occasionally used to perform simultaneous reduction of a nitro group to an amine and the addition of a sulfonic acid group. For example, 4-amino-3-hydroxyl-1-naphthalenesulfonic acid [116-63-2], $C_{10}H_9NO_4S$, is manufactured from 2-naphthol in a process which uses sodium bisulfite (59). The process involves nitrosation of 2-naphthol in aqueous medium, followed by addition of sodium bisulfite and acidification with sulfuric acid.

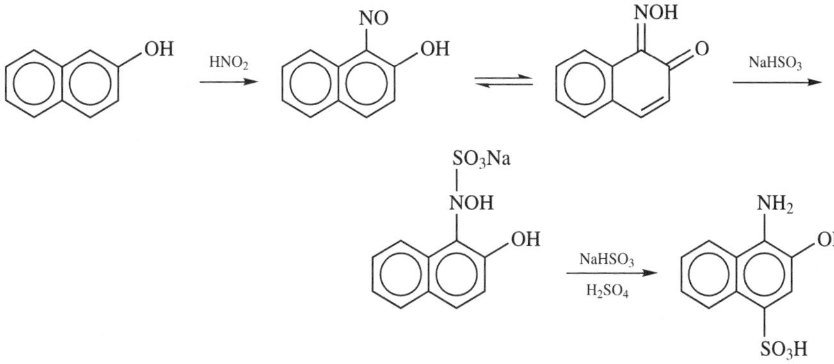

8.3. Electrolytic Reductions.   Both nitro compounds and nitriles can be reduced electrochemically. One advantage of electrochemical reduction is the cleanness of the operation. Since there are a minimum of by-products, both waste disposal and purification of the product are greatly simplified. However, unless very cheap electricity is available, these processes are generally too expensive to compete with the traditional chemical methods.

Nitro Compounds.   When nitro compounds are reduced by electrochemical methods a number of products are possible depending on such factors as the nature of the electrode, the electrode potential, and the reaction media. For the reduction of nitrobenzene these products include aniline, *p*-aminophenol, *p*-chloroaniline, phenylhydroxylamine, azoxybenzene, azobenzene, and hydrazo-benzene (60).

Electrolytic reductions generally cannot compete economically with chemical reductions of nitro compounds to amines, but they have been applied in some specific reactions, such as the preparation of aminophenols (qv) from aromatic nitro compounds. For example, in the presence of sulfuric acid, cathodic reduction of aromatic nitro compounds with a free para-position leads to *p*-aminophenol [123-30-8] by rearrangement of the intermediate *N*-phenyl-hydroxylamine [100-65-2] (61).

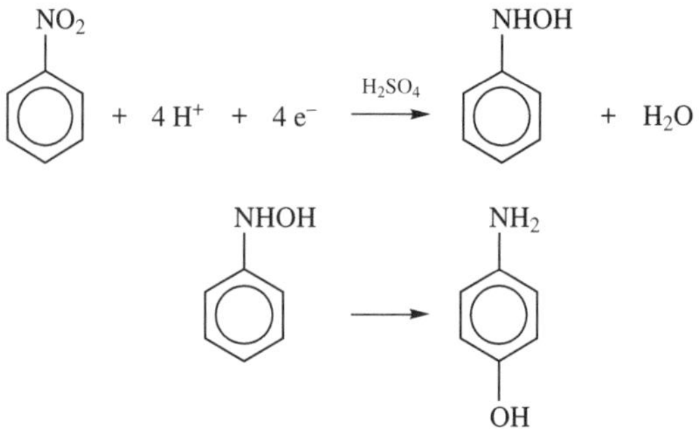

Nitriles.   The electrolytic reduction of nitriles requires a high negative potential, but can lead to amines in good yields under the right conditions. This reaction occurs in acidic media according to the following equation (62).

$$R\text{---}CN + 4\,H^+ + 4\,e^- \longrightarrow R\text{---}CH_2NH_2$$

In general, however, the electrochemical reduction of nitriles offers no significant advantages over traditional chemical methods and has not been widely used.

8.4. Metal Amalgams and Hydrides.   Metal hydrides and amalgams are sometimes the preferred method of reducing various functional groups in the laboratory, especially when the necessary equipment for catalytic hydrogenations is unavailable. However, these reagents are usually too expensive to make their use on a large commercial scale feasible.

Metal Amalgams.   Alkali metal amalgams function in a manner similar to a mercury cathode in an electrochemical reaction (63). However, it is more difficult to control the reducing power of an amalgam. In the reduction of nitro compounds with an $NH_4(Hg)$ amalgam, a variety of products are possible. Aliphatic nitro compounds are reduced to the hydroxylamines, whereas aromatic nitro compounds can give amino, hydrazo, azo, or azoxy compounds.

Metal Hydrides.   Metal hydrides can sometimes be used to prepare amines by reduction of various functional groups, but they are seldom the preferred method. Most metal hydrides do not reduce nitro compounds at all (64), although aliphatic nitro compounds can be reduced to amines with lithium aluminum hydride. When aromatic amines are reduced with this reagent, azo compounds are produced. Nitriles, on the other hand, can be reduced to amines with lithium aluminum hydride or sodium borohydride under certain conditions. Other functional groups which can be reduced to amines using metal hydrides include amides, oximes, isocyanates, isothiocyanates, and azides (64).

8.5. Aliphatic Amines from the Ring Reduction of Aromatic Amines.

Certain aliphatic amines can be prepared by reduction of aromatic amines by catalytic hydrogenation. This method is applicable only when a corresponding aromatic amine is available. Nevertheless, it is used for the production of several important amines including cyclohexylamine and bis(4-aminocyclohexyl)-methane. Reduction of an aromatic ring can be carried out using the same types of catalysts that are used for the hydrogenation of nitro compounds and nitriles. The conditions required are much harsher, however, and higher temperatures and pressures are generally required.

Bis(4-aminocyclohexyl)methane.   Aniline can react with formaldehyde in the presence of an acid catalyst to give a mixture of polymeric amines. The principal component of this mixture is 4,4'-diaminodiphenylmethane (4,4'-methylenebisbenzenamine [101-77-9], $C_{13}H_{14}N_2$). Most of this material is used in the production of isocyanates for polyurethanes. However, part of the 4,4'-diaminodiphenylmethane is converted to bis(4-aminocyclohexyl)methane (4,4'-methylenebiscyclohexanamine [1761-71-3], $C_{13}H_{26}N_2$) by catalytic hydrogenation, which results in a mixture of cis,cis-, cis,trans-, and trans,trans-stereoisomers. Typically a ruthenium catalyst is used at relatively high temperatures and pressures. For example, one patent describes a process for preparation of bis(4-aminocyclohexyl)methane from 4,4'-diaminodiphenylmethane with excellent yields in short times (65). The reaction is carried out using any of several ruthenium catalysts in the presence of a solvent and ammonia. The preferred temperature range is 225 to 250°C with an initial hydrogen pressure of about 14 to 24 MPa (140–240 atm). Under these conditions, the reaction is completed in less than 30 min with yields of 93 to 97% or higher.

mixture of isomers

9. Intermediates from the Reduction of Nitro Compounds

In the reduction of nitro compounds to amines, several of the intermediate species are stable and under the right conditions, it is possible to stop the reduction at these intermediate stages and isolate the products (see Figure 1, where

$R = C_6H_5$). Nitrosobenzene [586-96-9], C_6H_5NO, can be obtained by electrochemical reduction of nitrobenzene [98-95-3]. Phenylhydroxylamine, C_6H_5NHOH, is obtained when nitrobenzene reacts with zinc dust and calcium chloride in an alcoholic solution. When a similar reaction is carried out with iron or zinc in an acidic solution, aniline is the reduction product. Hydrazobenzene [122-66-7], $C_{12}H_{12}N_2$, is formed when nitrobenzene reacts with zinc dust in an alkaline solution. Azoxybenzene [495-48-7], $C_{12}H_{10}N_2O$, is formed from two molecules of nitrobenzene when heated in an alcoholic solution containing a strong base such as sodium hydroxide. Azobenzene [103-33-3], $C_{12}H_{10}N_2$, can be obtained by distilling azoxybenzene in the presence of iron powder or by reaction of nitrobenzene with sodium stannite [12214-41-7], Na_2SnO_2.

10. Environmental and Safety Aspects

Amines, nitro compounds, nitriles, and the various solvents and reagents used in the preparation of amines by reduction vary widely in the hazards they may pose. Some of these materials are acutely toxic by ingestion, inhalation, or absorption through the skin. Others are skin irritants or sensitizers. Still others may cause damage by chronic exposure to organs, such as the liver, or may be carcinogenic. Since amines vary so widely in their potential danger, no general rules can govern their safe use in all cases. The Material Safety Data Sheet (MSDS) for the material in question should be consulted before the chemicals are used.

10.1. Waste Treatment and Effluent Monitoring. Modern chemical plants are closed systems to prevent the emission of pollutants into the environment. Vents are protected by fume scrubbers and process waste water is cleaned up before being discharged. Any unavoidable by-products are collected and disposed of in a manner consistent with government regulations. In the water treatment plant diagrammed in Figure 5, the incoming water is first treated to neutralize the normally acidic stream and the solids are removed in a 4 million liter clarifier. Biological treatment then removes many of the the organic compounds. Aniline and methanol are readily removed at this stage; nitro compounds are much more difficult for the bacteria to digest. An activated carbon system removes final traces of organics and much of the remaining color, providing an effluent which meets state and federal regulations. The carbon is routinely regenerated in a system that incinerates the organics absorbed from the waste stream. Finally, the water leaving the waste water treatment plant is monitored by chemical and biological tests to ensure its quality.

10.2. Air Monitoring. The atmosphere in work areas is monitored for worker safety. Volatile amines and related compounds can be detected at low concentrations in the air by a number of methods. Suitable methods include chemical, chromatographic, and spectroscopic techniques. For example, the NIOSH Manual of Analytical Methods has methods based on gas chromatography which are suitable for common aromatic and aliphatic amines as well as ethanolamines (67). Aromatic amines which diazotize readily can also be detected photometrically using a treated paper which changes color (68). Other methods based on infrared spectroscopy (69) and mass spectroscopy (70) have also been reported.

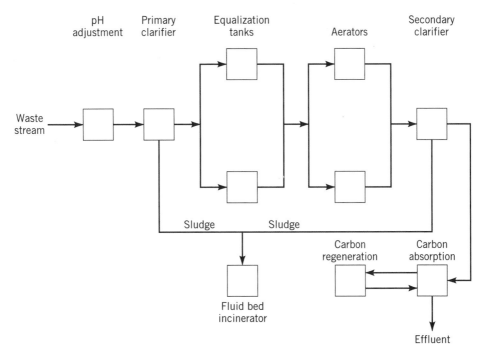

Fig. 5. Typical industrial waste-water treatment plant, eg, 25×10^6 L/day, Mobay Corporation plant at New Martinsville, W.V. (66).

10.3. Handling of Amines. Regulations governing the safe handling and shipping of amines in interstate commerce are given in U.S. Department of Transportation publications (71). Specific information on the safe handling and hazards associated with a particular amine can be found in the Material Safety Data Sheet for that material. For further information on the safety of industrial material see also references 72 and 73.

BIBLIOGRAPHY

"Amination by Reduction" in *ECT* 1st ed., Vol. 1, pp. 673–702, by P. H. Groggins and A. J. Stirton, U.S. Department of Agriculture; in *ECT* 2nd ed., Vol. 2, pp. 76–99, by R. Norris Shreve, Purdue University; "Amines by Reduction," in *ECT* 3rd ed., Vol. 2, pp. 355–375, by R. L. Sandridge and H. B. Staley, Mobay Chemical Corporation; in *ECT* 4th ed., Vol. 2, pp. 482–503, by Steven L. Schilling, Mobay Corporation; "Amines by Reduction" in *ECT* (online), posting date: December 4, 2000, by Steven L. Schilling, Mobay Corporation.

CITED PUBLICATIONS

1. P. H. Groggins, ed., *Unit Processes in Organic Synthesis,* 5th ed., McGraw-Hill Book Co., Inc., New York, 1958, 170–171.
2. F. Haber, *Z. Elektrochem.* **22**, 506 (1898).

3. H. D. Burge, D. J. Collins, and B. H. Davis, *Ind. Eng. Chem. Prod. Res. Dev.* **19**, 389–391 (1980).

4. T. A. Bolshinskova, E. V. Şelyaeva, and V. N. Kravtosva, *Issled. Obl. Sint. Katal. Org. Soedin.* 40–42 (1975).

5. Eur. Pat. Appl. 147,879 A1 (July 10, 1985), A. H. Sharma and P. Hope (to Akzo N.V.).

6. A. D. Shebadova and co-workers, *Izv. Akad. Nauk. SSSR Ser. Khim.* **7**, 1665 (1975).

7. D. Data and co-workers, *Fert. Technol.* **13**, 125–127 (1976).

8. R. C. Ryan, G. M. Wilemon, M. P. Dalsanto, and C. U. Pittman, Jr., *J. Mol. Catal.* **5**, 319–330 (1979).

9. H. C. Yao and P. H. Emmett, *J. Am. Chem. Soc.* **81**, 4125–4132 (1959).

10. H. Greenfield, *Ind. Eng. Chem. Prod. Res. Dev.* **6**, 142–144 (1967).

11. R. T. Morrison and R. N. Boyd, *Organic Chemistry,* 3rd. ed., Allen & Bacon, Inc., Boston, 1973, Chapt. 11.

12. T. Urbanski, *Chemistry and Technology of Explosives,* Vol. 1, MacMillan Company, New York, 1964.

13. K. Schofield, *Aromatic Nitration,* Cambridge University Press, Cambridge, UK, 1980.

14. D. T. Mowby, *Chem. Rev.* **42**, 189–283 (1948).

15. K. Friedrich and K. Wallenfels in Z. Rappoport, ed., *The Chemistry of the Cyano Group,* Wiley-Interscience, New York, 1970, Chapt. 2.

16. U.S. Pat. 2,891,094 (June 16, 1959), O. C. Karkalits, Jr., C. M. Vanderwaart, and F. H. Megson (to American Cyanamid Co.).

17. U.S. Pat. 4,480,051 (Oct. 30, 1984), J. C. Wu (to E. I. du Pont de Nemours & Co., Inc.).

18. I. Susuki and Y. Hanazaki, *Chem. Lett.* 459–550 (1986).

19. M. C. Datta, C. R. Saha, and D. Sen, *J. Chem. Technol. Biotechnol.* **29**, 88–99 (1979).

20. P. Kvintovics, B. Heil, and L. Marko, *Adv. Chem. Ser.* **173**, 26–30 (1979).

21. U.S. Pat. 4,169,853 (Oct. 2, 1979), J. F. Knifton (to Texaco Development Corp.).

22. P. Khandual, P. K. Santra, and C. R. Saha, *Proc. Indian Natl. Sci. Acad. Part A* **51**, 538–549 (1985).

23. S. Bhattacharya, P. Khandual, and C. R. Saha, *Chem. Ind.,* 600–601 (1982).

24. A. Bose and C. R. Saha, *Chem. Ind.* 199–201 (1987).

25. T. K. Mondal, T. K. Banerjee, and D. Sen, *Indian J. Chem. Sect A* **19A**, 846–848 (1980).

26. C. N. Satterfield, *Heterogeneous Catalysis in Practice,* McGraw-Hill Book Company, New York, 1980, p. 85.

27. R. J. Alaimo and R. J. Storrin, *Chem. Ind.,* 473–478 (1981).

28. E. S. Lazer, J. S. Anderson, J. E. Kijek, and K. C. Brown, *Synth. Commun.* **12**, 691–694 (1982).

29. R. L. Augustine, *Catalytic Hydrogenation,* Marcel Dekker, Inc., New York, 1965.

30. P. N. Rylander, *Hydrogenation Methods,* Academic Press, London, 1985.

31. C. H. Bartholomew, *Chem. Eng.* **91**(23), (*NY*), 96–112 (Nov. 12, 1984).

32. J. G. Santangel and W. M. Smith, eds., *Hydrogen: Production and Marketing, ACS Symp. Ser. 116,* American Chemical Society, Washington, D.C., 1980, 4–16.

33. P. Rylander, *Chemical Catalyst News,* the Engelhard Corporation, Iselin N.J., Oct. 1989.

34. P. N. Rylander, *Catalytic Hydrogenation in Organic Synthesis,* Academic Press, New York, 1979, pp. 3, 114.

35. Brit. Pat. 1,498,722 (Jan. 25, 1978), R. J. Gait (to Imperial Chemical Industries, Ltd.).

36. Ger. Offen. 2,615,079 (Oct. 28, 1976), J. R. Kosak (to E. I. du Pont de Nemours & Co., Inc.).

37. Ger. Offen. 2,549,900 (May 12, 1977), H. Kritzler, W. Boehm, W. Kiel, and U. Birkenstock (to Bayer AG).
38. U.S. Pat. 4,313,005 (Jan. 26, 1982), M. E. Ford and R. J. Daughenbaugh (to Air Products and Chemicals, Inc.).
39. Ger. Offen. 3,347,452 (July 11, 1985), A. Hackenberger and M. Patsch (to BASF AG).
40. A. Voorhies, Jr., W. M. Smith, and R. B. Mason, *Ind. Eng. Chem.* **40**, 1543–1548 (1948).
41. U.S. Pat. 2,284,525 (May 26, 1942), A. W. Larchar and H. S. Young (to E. I. du Pont de Nemours & Co., Inc.).
42. P. Cardillo, A. Quattrini, E. Vajna de Pava, and A. Girelli, *J. Calorim. Anal. Therm. Thermodyn. Chim.* **17**, 394–397 (1986).
43. Ref. 34, p. 7.
44. Ger. Pat. 226,872 (Sept. 4, 1985), G. Wolter and co-workers (to VEB Chemiekombinat Bitterfeld).
45. Ref. 12, 281–289.
46. U.S. Pat. 3,032,586 (Apr. 18, 1957), H. Dierichs and H. Holzrichter (to Bayer AG and Mobay Corporation).
47. Brit. Pat. 768,111 (Feb. 13, 1957), H. Dierichs and H. Holzrichter (to Bayer AG).
48. USSR Pat. 118,506 (Mar. 10, 1959), M. I. Gol'dfarb and co-workers.
49. Ger. Offen. 2,534,176 (Feb. 10, 1977), G. Franz, G. Halfter, W. Jaeckle, and F. Mindermann (to Ciba-Geigy AG).
50. Ref. 1, p. 143.
51. D. C. Owsley and J. J. Bloomfield, *Synthesis,* 118–120 (1977).
52. Czech. Pat. 202,124 (July 27, 1979), J. Prachensky, J. Teri, and J. Vokal; *Chem. Abstr.* **98**, 34225 (1983).
53. J. Terc, J. Vokal, and J. Prachensky, *Chem. Prum.* **31**, 20–23 (1981).
54. Czech. Pat. 196,476 (Mar. 1, 1982), J. Prachensky, J. Terc, and J. Vokal; *Chem. Abstr.* **97**, 72059 (1982).
55. Ref. 1, 150–161.
56. *Report of the British Intelligence Objectives Subcommittee on (1) The Manufacture of Nitration Products of Benzene, Toluene, and Chlorobenzene at Griesheim and Leverkusen, (2) The Manufacture of Aniline and Iron Oxide Pigments at Uerdigen, PB 77729 (also issued as BIOS Report No. 1144) and BIOS Trip Report No. 2526,* Sept.–Oct. 1946, 25–32.
57. H. K. Porter in W. G. Dauben, ed., *Organic Reactions,* John Wiley & Sons, Inc., New York, Vol. 20, 1973, 455–481.
58. B. S. Furniss and co-workers, *Vogel's Textbook of Practical Organic Chemistry,* 4th ed., Longman, Inc., New York, 1978, 662–663.
59. Eur. Pat. 262,093 (Mar. 30, 1988), B. Albrecht and J. Beyrich (to Ciba-Geigy AG).
60. M. R. Rifi and F. H. Covitz, *Introduction to Organic Electrochemistry,* Marcel Dekker, Inc., New York, 1974, 182–183.
61. M. M. Baizer, ed., *Organic Electrochemistry,* Marcel Dekker, Inc., New York, 1973, 326–330.
62. Ref. 61, 423–425.
63. Ref. 61, 805–812.
64. J. March, *Advanced Organic Chemistry: Reactions, Mechanism and Structure,* 2nd ed., McGraw-Hill Book Co., New York, 1977, 1125–1128.
65. U.S. Pat. 3,347,917 (Oct. 17, 1967), W. J. Arthur (to E. I. du Pont de Nemours & Co., Inc.).
66. J. Myers, *Chem. Process.* **39**(9), 18 (1976).
67. P. M. Eller, ed. *NIOSH Manual of Analytical Methods,* Pub. No. 84–100, U.S. Dept. of Health and Human Services, Washington, D.C., 1984, method nos. 2003, 2007, and 2010.

68. *Miniature Continuous Monitor, Tech. Bull.*, MDA Scientific, Inc. Park Ridge, Ill., Dec. 1974.
69. *Miron II, Tech, Bull.*, Wilks Scientific, South Norwalk, Conn., 1975.
70. J. E. Evans and J. T. Arnold, *Environ. Sci. Technol.* **9**, 1134 (1975).
71. *Code of Federal Regulations,* Title 46, Parts 150–165, U.S. Government Printing Office, Washington, D.C., 1975.
72. *Toxic and Hazardous Industrial Chemicals Safety Manual,* The International Technical Information Institute, Tokyo, Japan, 1975.
73. I. N. Sax, *Dangerous Properties of Industrial Materials,* 6th ed., Van Nostrand-Reinhold, New York, 1984.

STEVEN L. SCHILLING
Mobay Corporation

AMINES, CYCLOALIPHATIC

1. Introduction

Cycloaliphatic amines are comprised of a cyclic hydrocarbon structural component and an amine functional group external to that ring. Included in an extended cycloaliphatic amine definition are aminomethyl cycloaliphatics. Although some cycloaliphatic amine and diamine products have direct end use applications, their major function is as low cost organic intermediates sold as moderate volume specification products.

2. Physical Properties

For simple primary amines directly bonded to a cycloalkane by a single C–N bond to a secondary carbon the homologous series is given in Table 1. Up through C_8 each is a colorless liquid at room temperature. The ammoniacal or fishy odor

Table 1. Properties of Primary Aminocycloalkanes

Cycloaliphatic amine	CAS Registry Number	Molecular formula	Boiling point, °C	Flash point, °C	Specific gravity, g/mL	Refractive index, n_D
cyclopropylamine	[765-30-0]	C_3H_7N	49	−26	0.824	1.4210
cyclobutylamine	[2516-34-9]	C_4H_9N	82	−4	0.833	1.4363
cyclopentylamine	[1003-03-8]	$C_5H_{11}N$	108	17	0.863	1.4478
cyclohexylamine	[108-91-8]	$C_6H_{13}N$	134	32	0.868	1.4565
cycloheptylamine	[5452-35-7]	$C_7H_{15}N$	169	42		1.4724
cyclooctylamine	[5452-37-9]	$C_8H_{17}N$	190	80	0.928	1.4804
cyclododecylamine	[1502-03-0]	$C_{12}H_{25}N$	280^a	121		

[a] Melting point 27°C.

Table 2. **Properties of Substituted Aminocycloalkanes**

Cycloaliphatic amine	CAS Registry Number	Molecular formula	Boiling point, °C	Flash point, °C
1-methylcyclohexylamine	[6526-78-9]	$C_7H_{15}N$	140	
2-methylcyclohexylamine	[7003-32-9]	$C_7H_{15}N$		22
(±)cis-2-methylcyclohexylamine	[2164-19-4]	$C_7H_{15}N$	154	
(±)trans-2-methylcyclohexylamine	[931-10-2]	$C_7H_{15}N$	150	
(+)t-2-methylcyclohexylamine	[29569-76-4]	$C_7H_{15}N$		
(−)t-2-methylcyclohexylamine	[931-11-3]	$C_7H_{15}N$		
3-methylcyclohexylamine	[6850-35-7]	$C_7H_{15}N$		24
(±)cis-3-methylcyclohexylamine	[1193-16-4]	$C_7H_{15}N$	153	
(±)trans-3-methylcyclo-hexylamine	[1193-17-5]	$C_7H_{15}N$	152	
	[6321-23-9]			
4-methylcyclohexylamine	[17746-6]	$C_7H_{15}N$		27
cis-4-methylcyclohexylamine	[2523-56-0]	$C_7H_{15}N$	154	
trans-4-methylcyclohexylamine	[2523-55-9]	$C_7H_{15}N$	152	
3,3,5-trimethylcyclohexylamine	[15901-42-5]	$C_9H_{19}N$	180	60
4-tert-butylcyclohexylamine	[5400-88-4]	$C_{10}H_{21}N$	213	79
N-methylcyclohexylamine	[100-60-7]	$C_7H_{15}N$	149	30
N-ethylcyclohexylamine	[5459-93-8]	$C_8H_{17}N$	165	44
N,N-dimethylcyclohexylamine	[98-94-2]	$C_8H_{17}N$	159	42
N,N-diethylcyclohexylamine	[91-65-6]	$C_{10}H_{21}N$	194	58
dicyclohexylamine	[101-83-7]	$C_{12}H_{23}N$	256	96
N-methyldicyclohexylamine	[7560-83-0]	$C_{13}H_{25}N$	265	101
1-adamantylamine	[768-94-5]	$C_{10}H_{17}N$	[a]	

[a]Melting point 207°C.

and high degree of water solubility decrease with increased molecular weight and boiling point for these corrosive, hygroscopic mobile fluids.

When additional substituents are bonded to other alicyclic carbons, geometric isomers result. Table 2 lists primary (1°), secondary (2°), and tertiary (3°) amine derivatives of cyclohexane and includes CAS Registry Numbers for cis and trans isomers of the 2-, 3-, and 4-methylcyclohexylamines in addition to identification of the isomer mixtures usually sold commercially. For the 1,2- and 1,3-isomers, the racemic mixture of optical isomers is specified; ultimate identification by CAS Registry Number is listed for the (+) and (−) enantiomers of trans-2-methylcyclohexylamine. The 1,4-isomer has a plane of symmetry and hence no chiral centers and no stereoisomers. The methylcyclohexylamine geometric isomers have different physical properties and are interconvertible by dehydrogenation–hydrogenation through the imine.

cis-1,4   trans-1,4

Table 3 lists cycloaliphatic diamines. Specific registry numbers are assigned to the optical isomers of trans-1,2-cyclohexanediamine; the cis isomer is achiral at ambient temperatures because of rapid interconversion of ring

Table 3. Properties of Cycloaliphatic Diamines

Diamine	CAS Registry Number	Molecular formula	Boiling point[a], °C	Flash point, °C
cis,trans-1,2-cyclohexanediamine	[694-83-7]	$C_6H_{14}N_2$	183	75
cis-1,2-cyclohexanediamine	[1436-59-5]	$C_6H_{14}N_2$	182	72
(±)trans-1,2-cyclohexanediamine	[1121-22-8]	$C_6H_{14}N_2$		
(+)trans-1,2-cyclohexanediamine	[21436-03-3]	$C_6H_{14}N_2$		
(−)trans-1,2-cyclohexanediamine	[20439-47-8]	$C_6H_{14}N_2$		
cis,trans-1,3-cyclohexanediamine	[3385-21-5]	$C_6H_{14}N_2$		91
cis-1,3-cyclohexanediamine	[26772-34-9]	$C_6H_{14}N_2$	198	
trans-1,3-cyclohexanediamine	[26883-70-5]	$C_6H_{14}N_2$	203	
methylcyclohexanediamine	[28282-16-0]	$C_7H_{16}N_2$	99 (1.66)	83
cis,trans-1,3-cyclohexanediamine,2-methyl	[13897-56-8]			
cis,trans-1,3-cyclohexanediamine,4-methyl	[13897-55-7]			
cis,trans-1,4-cyclohexanediamine	[1436-59-5]	$C_6H_{14}N_2$	181	80
cis-1,4-cyclohexanediamine	[15827-56-2]	$C_6H_{14}N_2$		
trans-1,4-cyclohexanediamine	[2615-25-0]	$C_6H_{14}N_2$	197	71
cis-1,8-menthanediamine	[80-52-4]	$C_{10}H_{22}N_2$	210	102
cis,trans-1,3-di(aminomethyl)cyclohexane	[2579-20-6]	$C_8H_{18}N_2$	114 (1.07)	106
cis-1,3-di(aminomethyl)cyclohexane	[10304-00-8]	$C_8H_{18}N_2$	117 (1.33)	
trans-1,3-di(aminomethyl)cyclohexane	[10339-97-6]			
cis,trans-1,4-di(aminomethyl)cyclohexane	[2549-93-1]	$C_8H_{18}N_2$	245	107
cis-1,4-di(aminomethyl)cyclohexane	[10029-09-9]			
trans-1,4-di(aminomethyl)cyclohexane	[10029-07-9]			
cis,trans-isophoronediamine	[2855-13-2]	$C_{10}H_{22}N_2$	252	112
methylenedi(cyclohexylamine)	[1761-71-3]	$C_{13}H_{26}N_2$	162 (2.40)	>110
isopropylidenedi(cyclohexylamine)	[3377-24-0]	$C_{15}H_{30}N_2$	182 (1.32)	>110
3,3'-dimethylmethylene-di(cyclohexylamine)	[6864-37-5]	$C_{15}H_{30}N_2$	160 (0.27)	174
cis,trans-tricyclodecanediamine[b]	[68889-71-4]	$C_{12}H_{22}N_2$	~314	165

[a] At 101.3 kPa unless otherwise indicated by the value (in kPa) in parentheses. To convert kPa to mm Hg, multiply by 7.5.
[b] (4,7-Methano-1H-indene-dimethaneamine, octahydro).

conformers. Commercial products are most often marketed as geometric isomer mixtures, though large differences in symmetry may lead to such wide variations in physical properties that separations by classical unit operations are practicable, as in Du Pont's fractional crystallization of *trans*-1,4-cyclohexanediamine (mp 72°C) from the low melting (5°C) cis–trans mixture.

Two-ring cycloaliphatic diamines such as methylenedi(cyclohexylamine) (MDCHA), historically misnamed bis(para-amino cyclohexylmethane), or PACM, also exhibit critically dependent fundamental physical properties as a function of configurational isomerism, the simplest and most important being melting point.

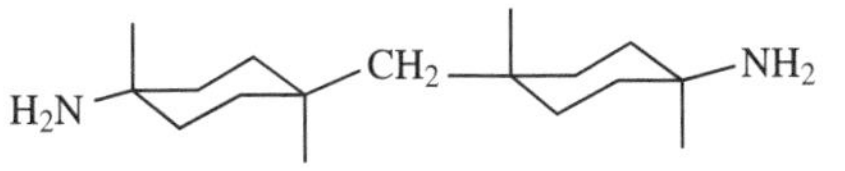

trans, trans	*cis, trans*
[6693-29-4]	[6693-30-7]
mp 65°C	mp 36°C
(**1**)	(**2**)

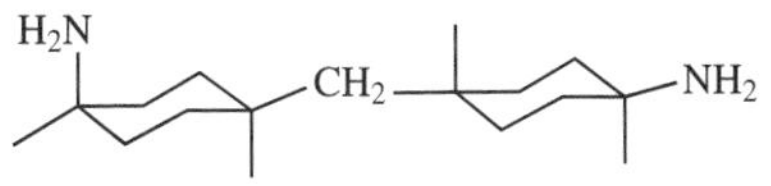

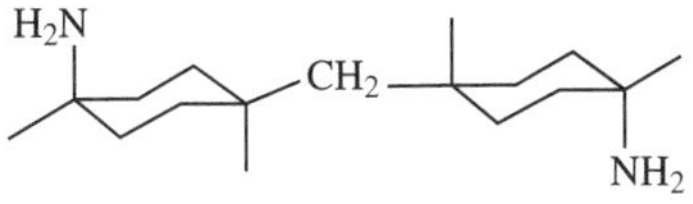

cis, cis
[6693-31-8]
mp 61°C
(**3**)

3. Chemical Properties

Cycloaliphatic amines are strong bases with chemistry similar to that of simpler primary, secondary, or tertiary amines. Upon reaction with nitrous acid, primary amines evolve nitrogen and generate alcohols; secondary amines form mutagenic nitrosamines. Substituted amides are formed under forcing Schotten-Baumann alkaline conditions from primary and secondary amines using acid chlorides; benzamides from benzoyl chloride distinguish 1° and 2° from 3° amines. The Hinsberg test using benzenesulfonyl chloride differentiates water soluble primary amine sulfonamide derivatives from insoluble secondary amine derivatives; tertiary amines are unreactive. Oxidation of secondary carbon primary amines proceeds through hydroxylamine (CH–NHOH) to oxime (C=NOH) and ultimately to the nitroalkane (CH–NO$_2$). Hydrogen peroxide generates amine oxides from tertiary cycloaliphatic amines.

Salt formation with Brønsted and Lewis acids and exhaustive alkylation to form quaternary ammonium cations are part of the rich derivatization chemistry of these amines. Carbamates and thiocarbamates are formed with CO$_2$ and CS$_2$, respectively; the former precipitate from neat amine as carbamate salts but are highly water soluble.

Primary cycloaliphatic amines react with phosgene to form isocyanates. Reaction of isocyanates with primary and secondary amines forms ureas.

Dehydration of ureas or dehydrosulfurization of thioureas results in carbodiimides. The nucleophilicity that determines rapid amine reactivity with acid chlorides and isocyanates also promotes epoxide ring opening to form hydroxyalkyl- and dihydroxyalkylamines. Michael addition to acrylonitrile yields stable cyanoethylcycloalkylamines.

Cycloaliphatic diamines react with dicarboxylic acids or their chlorides, dianhydrides, diisocyanates and di- (or poly-)epoxides as comonomers to form high molecular weight polyamides, polyimides, polyureas, and epoxies. Polymer property dependence on diamine structure is greater in the linear amorphous thermoplastic polyamides and elastomeric polyureas than in the highly crosslinked thermoset epoxies (2–4).

4. Manufacture and Processing

Cycloaliphatic amine synthesis routes may be described as distinct synthetic methods, though practice often combines, or hybridizes, the steps that occur: amination of cycloalkanols, reductive amination of cyclic ketones, ring reduction of cycloalkenylamines, nitrile addition to alicyclic carbocations, reduction of cyanocycloalkanes to aminomethylcycloalkanes, and reduction of nitrocycloalkanes or cyclic ketoximes.

Secondary alcohols are aminated to secondary amines by dehydration catalysts or under H_2 pressure using metal dehydrogenation catalysts such as Ni or Co. The latter process becomes mechanistically equivalent to reductive alkylation of ammonia, though no hydrogen is consumed. Cyclohexylamine (CHA) is commercially produced from cyclohexanol [108-93-0] by reaction in the vapor phase with NH_3 and H_2. Controlled alkyl:ammonia, hydrogen ratios over metal dehydrogenation catalysts on solid supports at 160–200°C and 1350–2000 kPa (196–290 psi) at gas hourly space velocities of 1000–2500 vol/vol are analogous conditions to those of the preferred manufacturing process for other secondary aliphatic amines. Reduction of ammonia to cyclohexanol feed ratios in the fixed bed vapor phase process promotes dicyclohexylamine (DCHA) coproduction.

Reductive amination of cyclic ketones, or reductive alkylation of ammonia, is a general route to cycloaliphatic amines (5). Use of pressurized hydrogen and metal catalyst is the process technology of choice commercially; alternative (6) hydrogen sources include formic acid. Batch liquid-phase reaction technology predominates because cyclic ketone volatilization in the presence of ammonia leads to by-product-forming aldol condensations; higher molecular weight alicyclic ketones such as 2-adamantanone [700-58-3] are insufficently volatile. Short contact time (1–30 s), high temperature (to 275°C), and atmospheric vapor-phase reaction conditions for production of cyclohexylamine from mixtures of cyclohexanol and cyclohexanone [108-94-1] over heated copper chromite–nickel catalyst with an ammonia:alkyl ratio of 3.3:1 and hydrogen:alkyl ratio of 6.5:1 have, however, been claimed (8).

Aniline [62-53-3] ring reduction produces cyclohexylamine. Alternative historical synthetic routes and early (1905–1931) metal-catalyzed hydrogen additions under hydrogen pressure have been well reviewed (9). Increased efficiencies compared to those with Ni and Co catalysts are available from the more precious elements of the same subgroup, Ru and Rh. Batch reaction giving >90% selectivity to CHA with <5% DCHA may use 1–3% of supported precious metal catalyst and neat substrate. Representative reaction conditions are 100–150°C with 1400–3500 kPa (200–500 psi) H_2 requiring 4–20 hours. Subsequent distillation may be batch or continuous. CHA fractionation from trace reaction by-product lights, then from recoverable DCHA and distillate heavies is done under reduced pressure. CHA has been made directly from phenol at low H_2 plus NH_3 pressure using rhodium catalyst in batch reactions (10) and in the vapor phase over nickel (11). Reaction selectivity to CHA is but 56% with 37% DCHA in the latter case.

Reductive amination of cyclohexanone using primary and secondary aliphatic amines provides *N*-alkylated cyclohexylamines. Dehydration to imine for the primary amines, to endocyclic enamine for the secondary amines is usually performed *in situ* prior to hydrogenation in batch processing. Alternatively, reduction of the *N*-alkylanilines may be performed, as for *N,N*-dimethylcyclohexylamine from *N,N*-dimethylaniline [121-69-7] (12,13). One-step routes from phenol and the alkylamine (14) have also been practiced.

Dicyclohexylamine may be selectively generated by reductive alkylation of cyclohexylamine by cyclohexanone (15). Stated batch reaction conditions are specifically 0.05–2.0% Pd or Pt catalyst, which is reusable, pressures of 400–700 kPa (55–100 psi), and temperatures of 75–100°C to give complete reduction in 4 h. Continuous vapor-phase amination selective to dicyclohexylamine is claimed for cyclohexanone (16) or mixed cyclohexanone plus cyclohexanol (17) feeds. Conditions are 5–15 s contact time of <1:1 ammonia:ketone, ~3:1 hydrogen:ketone at 260°C over nickel on kieselguhr. With mixed feed the preferred conditions over a mixed copper chromite plus nickel catalyst are 18-s contact time at 250°C with ammonia:alkyl = 0.6:1 and hydrogen:alkyl = 1:1.

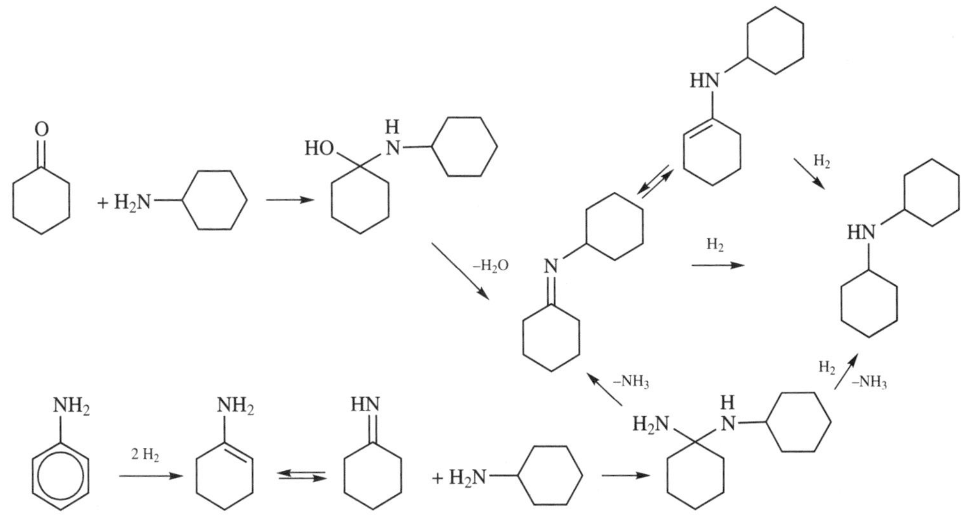

Fig. 1.   Routes to dicyclohexylamine (DCHA).

DCHA is normally obtained in low yields as a coproduct of aniline hydrogenation. The proposed mechanism of secondary amine formation in either reductive amination of cyclohexanone or arene hydrogenation illuminates specific steps (Fig. 1) on which catalyst, solvents, and additives moderating catalyst supports all have effects.

Alkali moderation of supported precious metal catalysts reduces secondary amine formation and generation of ammonia (18). Ammonia in the reaction medium inhibits Rh, but not Ru precious metal catalyst. More secondary amine results from use of more polar protic solvents, $CH_3OH > C_2H_5OH > t\text{-}C_4H_9OH$. Lithium hydroxide is the most effective alkali promoter (19), reducing secondary amine formation and hydrogenolysis. The general order of catalyst proclivity toward secondary amine formation is $Pt > Pd \gg Ru > Rh$ (20). Rhodium's catalyst support contribution to secondary amine formation decreases in the order carbon > alumina > barium carbonate > barium sulfate > calcium carbonate.

Methylenedianiline (4) (MDA) [101-77-9] hydrogenation to methylenedi (cyclohexylamine) generates first the cis-(6) and trans-(7) isomers of half-reduced 4-(p-aminobenzyl)-cyclohexylamine (5) [28480-77-5], a differentially reactive diamine offered in developmental quantities by Air Products and Chemicals.

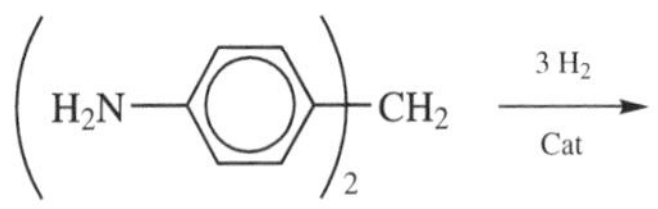

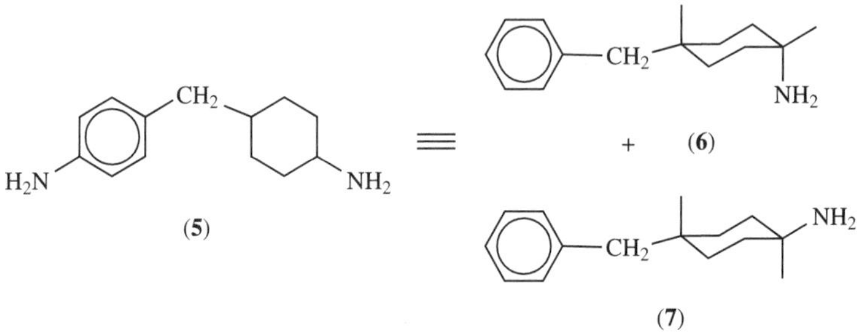

Addition of H_2 to the aromatic ring occurs cis, yielding a kinetic product subject to isomerization to the more thermodynamically stable trans isomer. Subsequent hydrogen addition to the remaining aromatic ring then produces the three fully reduced isomers (1–3). Catalyst systems were first optimized for efficient maximum trans isomer production (21–23). Batch reaction conditions using Ru on alumina catalyst for obtaining the thermodynamic mixture of product isomers were 200°C and 28–35 MPa (4000–5000 psi). Improved yields, including isomerization to a 50/40/10 mixture of (1,2,3), are enhanced by Ru alkali moderation (13,24,25).

Conditions cited for Rh on alumina hydrogenation of MDA are much less severe, 117°C and 760 kPA (110 psi) (26). With 550 kPa (80 psi) ammonia partial pressure present in the hydrogenation of twice-distilled MDA employing

2-propanol solvent at 121°C and 1.3 MPa (190 psi) total pressure, the supported Rh catalyst could be extensively reused (27). Medium pressure (3.9 MPa = 566 psi) and temperature (80°C) hydrogenation using iridium yields low *trans/trans* isomer MDCHA (28). Improved selectivity to alicyclic diamine from MDA has been claimed (29) for alumina-supported iridium and rhodium by introducing the tertiary amines 1,4-diazabicyclo[2.2.2]octane [280-57-9] and quinuclidine [100-76-5].

Direct production of select MDCHA isomer mixtures has been accomplished using ruthenium dioxide (30), ruthenium on alumina (31), alkali-moderated ruthenium (32) and rhodium (33). Specific isomer mixtures are commercially available from an improved 5–7 MPa (700–1000 psi) medium pressure process tolerant of oligomer-containing MDA feeds (34). Dimethylenetri(cyclohexylamine) **(8)** [25131-42-4] is a coproduct.

(8)

Continuous solvent-free hydrogenation of MDA over alkaline-earth-supported metals at 240°C and 25 MPa (3600 psi) has been described (35) as well as a similar high pressure process employing diluent solvent (36). MDA with isomeric impurities and oligomeric contaminants has been hydrogenated in a continuous flow system to advantage by first pretreating with Ni catalyst (150°C), then sequentially performing Ru hydrogenation at 185–200°C and 18 MPa (2600 psi) (37) (see also METHYLENEDIANILINE).

Batch syntheses comparable to those used for MDA produce 3,3′-dimethylmethylenedi(cyclohexylamine) marketed under the trade name Laromin C-260. The starting aromatic diamine, 3,3′-dimethylmethylenedianiline [838-88-0], is prepared from o-toluidine [95-53-4] condensation with formaldehyde. Similarly 3,3′-dimethyldicyclohexylamine [24066-10-2] may be produced (38) from o-tolidine [119-93-7] derived from o-nitrotoluene [88-72-2]. The resultant isomer mixtures are dependent on reduction conditions as in MDA hydrogenation.

Isopropylidenedi(cyclohexylamine) (PDCHA) (10) may be made by precious metal hydrogenation of isopropylidenedianiline **(9)** [2479-47-2], the condensation product of acetone [67-64-1] and aniline, commonly termed bisaniline A. A number of metal catalysts have been shown effective in an alternative route, the amination of isopropylidenedi(cyclohexanol) **(12)** [80-04-6] (39,40), the ring reduction product of bisphenol A **(11)** [80-05-7]. Ruthenium has been used for both the bisphenol ring reduction (210°C, 24 MPa H_2 = 3500 psi) and then a subsequent amination following ammonia addition. Batch amination of the cycloaliphatic diol **(12)** over a Co–Mn phosphoric acid–modified catalyst is only accomplished by initially pressurizing to 5 MPa (700 psi) with H_2, then adding NH_3 to a new autoclave pressure of 30 MPa (4350 psi) at 170°C, removing the water of reaction by venting, and cycling H_2 and NH_3 anew.

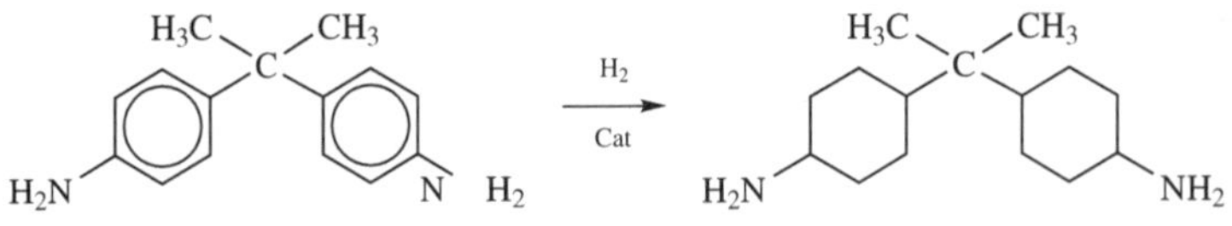

(9)   (10)

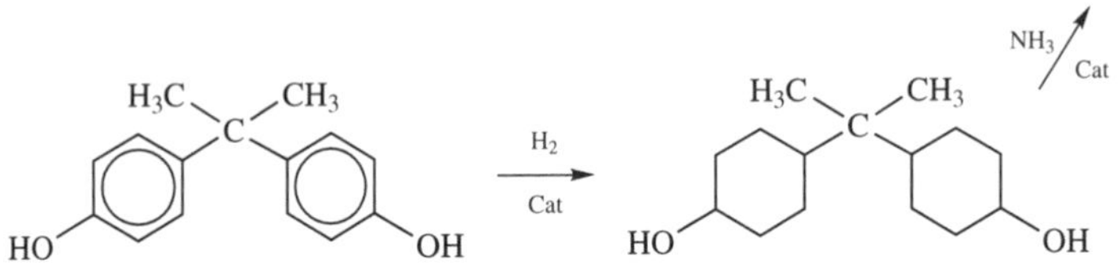

(11)   (12)

Trickle bed reaction of diol **(12)** using amine solvents (41) has been found effective for producing PDCHA, and heavy hydrocarbon codistillation may be used to enhance diamine purification from contaminant monoamines (42). Continuous flow amination of the cycloaliphatic diol in a liquid ammonia mixed feed gives >90% yields of cycloaliphatic diamine over reduced Co/Ni/Cu catalyst on phosphoric acid-treated alumina at 220°C with H_2 to yield a system pressure of 30 MPa (4350 psi) (43).

Cycloaliphatic diamines such as **(13)** [115172-12-8] which retain some aromatic character have been made from end-ring hydrogenation (44) of 1,3-bis (*p*-aminocumyl)benzene [2687-27-6], the double alkylation adduct of aniline to *m*-diisopropenylbenzene [3748-13-8] (45) using Ru catalysts (46).

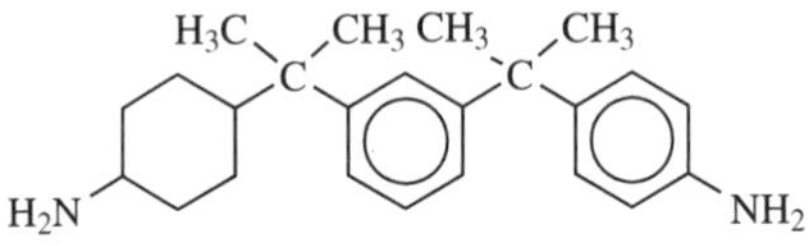

(13)

Cycloaliphatics capable of tertiary carbocation formation are candidates for nucleophilic addition of nitriles. HCN in strong sulfuric acid transforms 1-methyl-1-cyclohexanol to 1-methyl-1-cyclohexylamine through the formamide (47). The terpenes pinene **(14)** [2437-95-8] and limonene [5989-27-5] **(15)** each undergo a double addition of HCN to provide, after hydrolysis, the cycloaliphatic diamine 1,8-menthanediamine **(16)** (48).

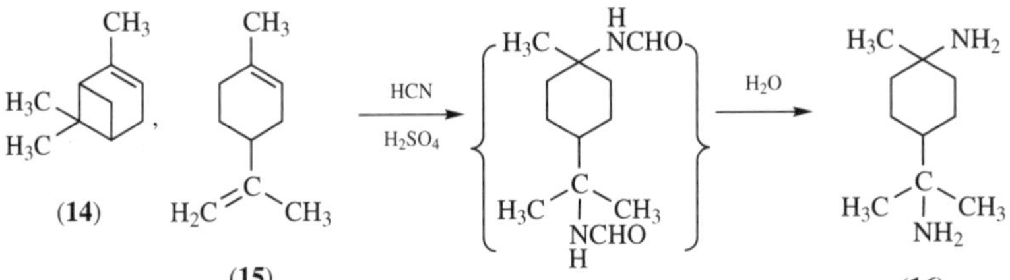

(14)

(15)   (16)

1-Adamantylamine is prepared from the corresponding alcohol or bromide by bridgehead cation generation in the presence of acetonitrile (49). Selective hydrolysis of the resultant acetamide to the rigid cycloaliphatic amine by acid or base is difficult.

Acetone's cyclic trimer, isophorone (17) [78-59-1], has reacted directly with ammonia (Ra Ni, 120°C), methylamine, (Pt/C, 100°C) and dimethylamine (Pd/C, 95°C) at 2400–3450 kPa H_2 pressure to form 3,3,5-trimethylcyclohexylamines (18), where R_1, R_2 = H, alkyl (50). The double bond is hydrogenated simultaneously with the imine:

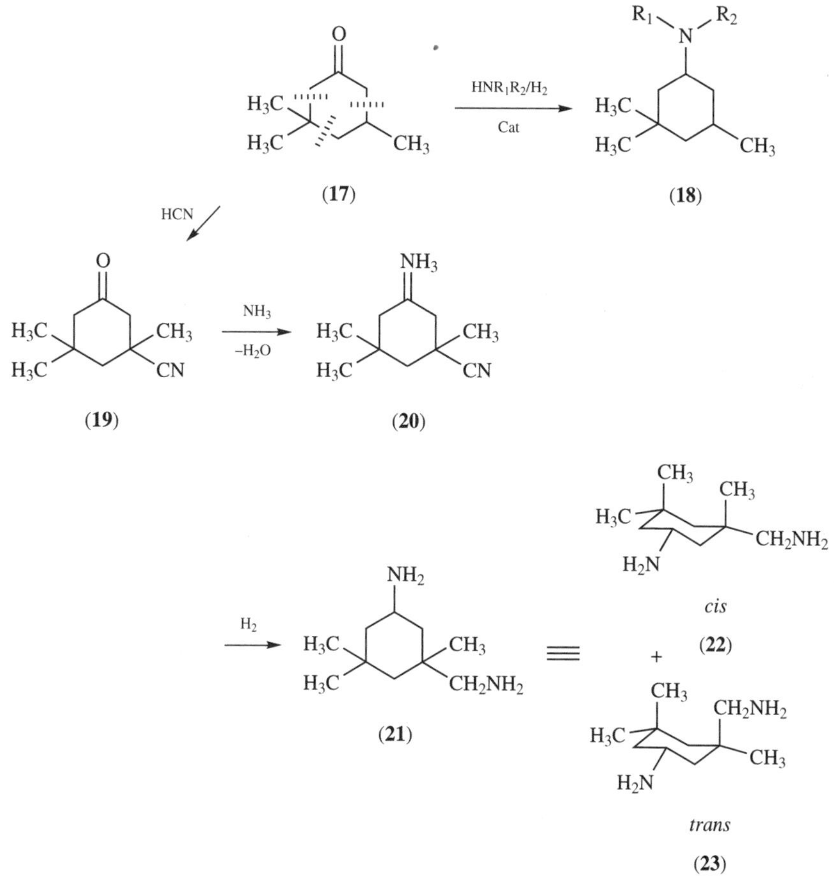

3-Aminomethyl-3,5,5-trimethylcyclohexylamine (21), commonly called isophoronediamine (IPD) (51), is made by hydrocyanation of (17) (52), (53) followed by transformation of the ketone (19) to an imine (20) by dehydrative condensation of ammonia (54), then concomitant hydrogenation of the imine and nitrile functions at 15–16 MPa (∼2200 psi) system pressure and 120°C using methanol diluent in addition to H_2 and NH_3. Integrated imine formation and nitrile reduction by reductive amination of the ketone leads to alcohol by-product. There are two geometric isomers of IPD; the major product is cis-(22) [71954-30-5] and the minor, trans-(23) [71954-29-5] (55).

1,2-Cyclohexanediamine's commercial origin is its presence as a minor $0.1 - <1\%$ coproduct of hexamethylenediamine [124-09-4] produced by hydrogenation of adiponitrile [111-69-3]. Fractional distillation by up to four columns in a series is routine commercial practice to purify nylon grade acyclic diamine; the crude cycloaliphatic diamine requires further refining before use as a specification intermediate.

Dicyclopentadiene **(24)** [77-73-6] is an inexpensive raw material for hydrocyanation to **(25)**, a mixture of 1,5-dicarbonitrile [70874-28-1] and 2,5-dicarbonitrile [70874-29-2], then subsequent hydrogenation to produce tricyclodecanediamine, TCD diamine **(26)**. This developmental product, a mixture of endo and exo, cis and trans isomers, is offered by Hoechst.

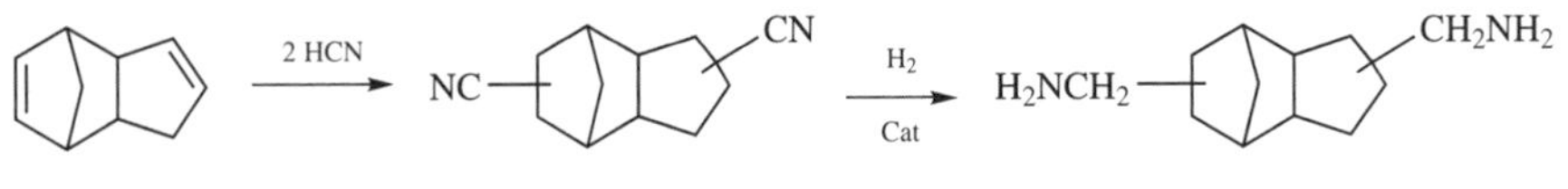

(24)   (25)   (26)

Di(aminomethyl)cyclohexanes are potentially available from large volume starting materials. The 1,3-isomer **(29)** may be produced by reduction of *m*-xylylenediamine **(28)** [1477-55-0] or directly upon exhaustive hydrogenation of isophthalonitrile **(27)** [626-17-5].

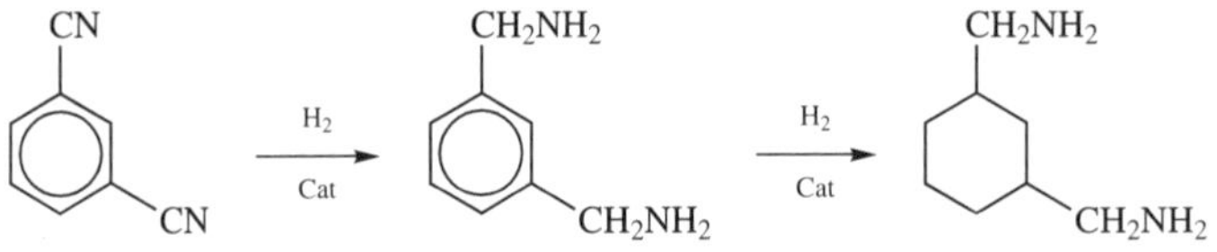

(27)   (28)   (29)

The 1,4-isomer has been similarly generated from terephthalonitrile [623-26-7] (56) using a mixed Pd/Ru catalyst and ammonia plus solvent at 125°C and 10 MPa (100 atm). It is also potentially derived (57) from terephthalic acid [100-21-0] by amination of 1,4-cyclohexanedimethanol **(30)** [105-08-8]. Endocyclization, however, competes favorably and results in formation of the secondary amine **(31)** 3-azabicyclo[3.2.2]nonane [283-24-9] upon diol reaction with ammonia over dehydration and dehydrogenation catalysts (58):

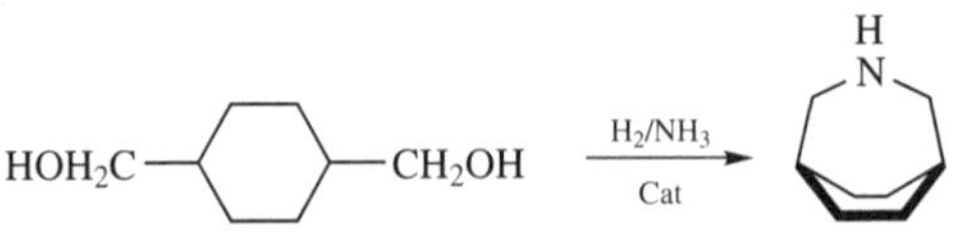

(30)   (31)

5. Shipment

Shipment of these liquid products is by nitrogen-blanketed tank truck or tank car. Drum shipments are usually in carbon steel, DOT-17E.

6. Economic Aspects

Cycloaliphatic amine production economics are dominated by raw material charges and process equipment capital costs. Acetone (isophorone), adiponitrile, aniline, and MDA are all large-volume specification organic intermediates bordering on commodity chemicals. They are each cost-effective precursors.

Reductive alkylations and aminations require pressure-rated reaction vessels and fully contained and blanketed support equipment. Nitrile hydrogenations are similar in their requirements. Arylamine hydrogenations have historically required very high pressure vessel materials of construction. A nominal breakpoint of 8 MPa (~1200 psi) requires yet heavier wall construction and correspondingly more expensive hydrogen pressurization. Heat transfer must be adequate, for the heat of reaction in arylamine ring reduction is ~50 kJ/mol (12 kcal/mol) (59). Solvents employed to maintain catalyst activity and improve heat-transfer efficiency reduce effective hydrogen partial pressures and require fractionation from product and recycle to prove cost-effective.

Production of cyclohexylamine reflects this balance of raw material versus operating cost structure. When aniline cost and availability are reasonable, the preferred route is aniline ring reduction; alternatively the cyclohexanol amination route is chosen.

Demand for cyclohexylamine should remain the same as current demand at 7030×10^3 t (15.5×10^6 lb) (60).

Price history for 1995–2000 was a high of \$0.69/kg (\$1.52/lb) and a low of \$0.64/kg (\$1.40/lb). Both on the same basis: list, tech., tanks delivered (60).

Cyclohexylamine's market has become stagnant as its major applications are losing ground. Since the ban in the U.S. on cyclamate sweeteners in the 1970s (cyclohexylamine's biggest market at that time), there has been an overcapacity in the market (60).

7. Specifications, Standards, and Quality Control

Liquid cycloaliphatic amines and diamines have exacting purity and color standards. Almost all are sold to specification, not performance standards. Use as isocyanate precursors requires low water content criteria for these hygroscopic fluids, hence nitrogen blanketing is often specified for product sampling as well as storage and transport.

Contaminant by-products depend upon process routes to the product, so maximum impurity specifications may vary, eg, for CHA produced by aniline hydrogenation versus that made by cyclohexanol amination. Capillary column chromatography has improved resolution and quantitation of contaminants beyond the more fully described packed column methods (61) used historically to define specification standards. Wet chemical titrimetry for water by Karl Fisher or amine number by acid titration have changed little except for their automation. Colorimetric methods remain based on APHA standards.

8. Analytical Methods

Isomer separation beyond physical fractional crystallization has been accomplished by derivatization using methyl formate to make N-formyl derivatives and acetic anhydride to prepare the corresponding acetamides (1). Alkaline hydrolysis regenerates the analytically pure amine configurational isomers.

Amine chromatographic analyses suffer poor resolution in many gas–liquid column separations because of strong interactions of the basic functional group and surface active components of even glass column supports. Tailing results. For close-eluting isomers, the problem is magnified. Derivatization of isomeric cycloaliphatic diamines has been reported using N,N'-trifluoroacetyl derivatives from reaction with trifluoroacetic anhydride (62). N,N,N',N'-Tetramethyl derivatives from formaldehyde–formic acid methylation, the Eschweiler–Clarke procedural variant of the Leuckart reaction (63), avoid amide hydrogen bonding column interactions (64). A more efficient procedure has been detailed for N,N-dimethylformamide dimethyacetal reaction with methylene- and isopropylidene-di(cyclohexylamine) geometric isomers to form chromatographically resolvable N-dimethylaminomethylene derivatives (65). Improved capillary column technology allows geometric isomer resolution directly, without derivatization (34).

Wet chemical methods determining titratable amine are reported for products entering urethane (amine number as meq/g) or epoxy (AHEW = amine hydrogen equivalent weight) trade applications. For secondary amines N-nitrosamine contaminants are reportable down to ppb using Thermoelectron Corporation thermal energy analyzer techniques.

9. Health and Safety Factors

Cycloaliphatic amines and diamines are extreme lung, skin, and eye irritants. MSD sheets universally carry severe personal protective equipment use warnings due to the risk of irreversible eye damage. These compounds are generally not mutagenic in the Ames test, and are highly (50–500 mg/kg) to moderately (500–5000 mg/kg) toxic as graded by the Hodge–Sterner scale by acute animal testing (66) (Table 4).Use of dry chemical, alcohol foam, or carbon dioxide is recommended for cycloaliphatic amine fire fighting. Water spray is recommended only to flush spills away to prevent exposures. In the aquatic environment, cyclohexylamine has a high (420 mg/L) toxicity threshold for bacteria (*Pseudomonas putida*) (68), and is considered biodegradable, that is, mineralizable to CO_2 and H_2O, by acclimatized bacteria.

The OSHA PEL time-weighted average for cyclohexylamine is 10 ppm.

The ACGIH TLV time-weighted average is 10 ppm. Cyclohexylamine is not classified as a human carcinogen (69).

10. Uses

Cyclohexylamine is miscible with water, with which it forms an azeotrope (55.8% H_2O) at 96.4°C, making it especially suitable for low pressure steam systems in

Table 4. **Acute Toxicity of Cycloaliphatic Amines**

Cycloaliphatic amine	Rat oral LD_{50}, mg/kg
cyclohexylamine	360
dicyclohexylamine	370
N-methylcyclohexylamine	400
dimethylcyclohexylamine	348
N-ethylcyclohexylamine	590
cis,trans-1,3-cyclohexanediamine	390
methyl-1,3-cyclohexanediamine	1060
4-methyl-1,3-cyclohexanediamine	1410[a]
1,3-diaminomethylcyclohexane	880
1,4-diaminomethylcyclohexane	530
1,8-menthanediamine	700
1-adamantylamine	900
methylenedi(cyclohexylamine)	450
3,3′-dimethylmethylenedi(cyclohexyla- mine)	550
tricyclodecanediamine	502

[a] Ref. 67.

which it acts as a protective film-former in addition to being a neutralizing amine.

Nearly 55% of the 2000 U.S. production of 22,680 t/yr (50×10^6 lb/yr) cyclohexylamine serviced this application (60). Carbon dioxide corrosion is inhibited by deposition of nonwettable film on metal (70). In high pressure systems CHA is chemically more stable than morpholine [110-91-8] (71). A primary amine, CHA does not directly generate nitrosamine upon nitrite exposure as does morpholine. CHA is used for corrosion inhibitor radiator alcohol solutions, also in paper- and metal-coating industries for moisture and oxidation protection.

Monofunctional, cyclohexylamine is used as a polyamide polymerization chain terminator to control polymer molecular weight. 3,3,5-Trimethylcyclohexylamines are useful fuel additives, corrosion inhibitors, and biocides (50). Dicyclohexylamine has direct uses as a solvent for cephalosporin antibiotic production, as a corrosion inhibitor, and as a fuel oil additive, in addition to serving as an organic intermediate. Cycloaliphatic tertiary amines are used as urethane catalysts (72). Dimethylcyclohexylamine (DMCHA) is marketed by Air Products as POLYCAT 8 for pour-in-place rigid insulating foam. Methyldicyclohexylamine is POLYCAT 12 used for flexible slabstock and molded foam. DMCHA is also sold as a fuel oil additive, which acts as an antioxidant. Sterically hindered secondary cycloaliphatic amines, specifically dicyclohexylamine, effectively catalyze polycarbonate polymerization (73).

Cycloaliphatic diamines which have reacted with diacids to form polyamides generate performance polymers whose physical properties are dependent on the diamine geometric isomers (58, 74). Proprietary transparent thermoplastic polyadipamides have been optimized by selecting the proper mixtures of PDCHA geometric isomers (32–34) for incorporation (75):

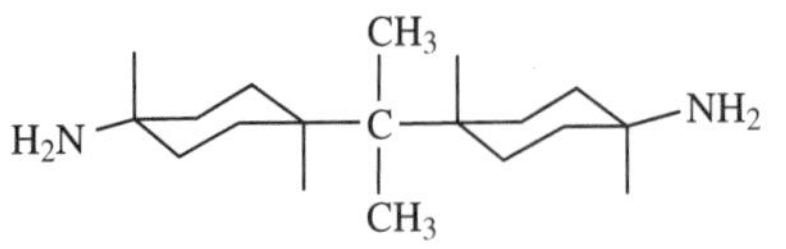

trans, trans
[28465-04-5]
(**32**)

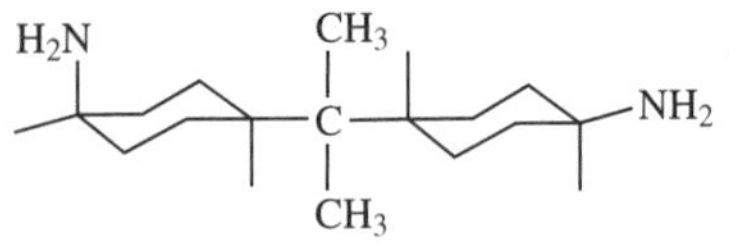

cis, trans
[28465-03-4]
(**33**)

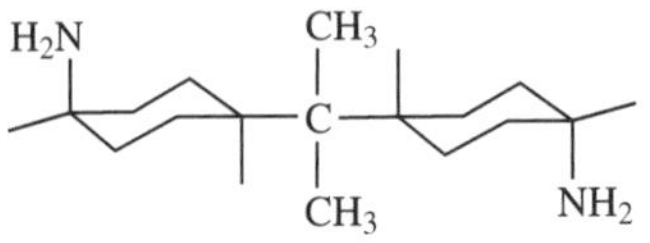

cis, cis
[28465-02-3]
(**34**)

The polyamide copolymer of dodecanoic acid with methylenedi(cyclohexylamine) (MDCHA, PACM) was sold as continuous filament yarn fiber under the tradename QIANA. The low melting raffinate coproduct left after trans, trans isomer separation by fractional crystallization was phosgenated to produce a liquid aliphatic diisocyanate marketed by Du Pont as Hylene W. Upon termination of their QIANA commitment, Du Pont sold the urethane intermediate product rights to Mobay, who now markets the 20% trans, trans–50% cis, trans–30% cis, cis diisocyanate isomer mixture as Desmodur W. In addition to its use in polyamides and as an isocyanate precursor, methylenedi(cyclohexylamine) is used directly as an epoxy curative.

1,2-Cyclohexanediamine is used as an epoxy curative, (Millamine 5260). It may be adducted with epichlorohydrin to generate solventless low viscosity curatives for varnishes and surface coatings (76). Other cycloaliphatic diamines have long been modified as epoxy curatives to modify their reactivity profile (77).

MCHD from ring reduction of TDA (78,79) has been cited as an epoxy curative (80) and is available as a developmental cycloaliphatic diamine. Ring reduction of sterically hindered arylenediamines such as diethyltoluenediamine [68479-98-1] provides slower-reacting alkylated 1,3-cyclohexanediamines for polyurethane, polyurea, and epoxy use (81).

Use of 1,3-cycloaliphatic diamines as organic intermediates appears limited because of cis isomer endocyclization reactions. Ring hydrogenation of the low cost 80/20 2,4-toluenediamine/2,6-toluenediamine isomer mixture results in 4 geometric isomers of 4-methyl-1,3-cyclohexanediamine, 3 isomers of 2-methyl-1,3-cyclohexanediamine; the overall sum of methyl *cis*-1,3-diamine is ~50%. Phosgenation of the free-base or dicarbamate of hydrogenated TDA to produce methylcyclohexanediisocyanate results in low yields (82,83), possibly because of endocyclic urea formation. Diequatorial 1,3-cyclohexanediamine (**35**) is conformationally labile, and in the alternative 1,3-diaxial diamine conformation (**36**) allows facile condensation to urea (**37**). Phosgenation of the methylcyclohexanediamine dihydrochloride, however, is efficient, giving ~90% yields of methylcyclohexanediisocyanate in 4–14 hours at 125–185°C and, depending on solvent, pressure to 1 MPa (145 psi) (84).

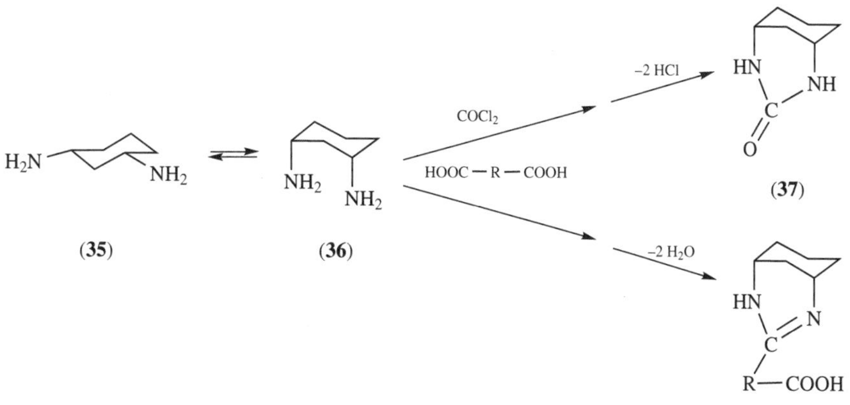

Use of 1,3 cycloaliphatic diamines in polyamides may be similarly limited by internal amide dehydration of the conformationally labile cis isomers to form a tetrahydropyrimidine (38) rather than high molecular weight polyamide. 1,3-Cyclohexanediamine is, however, a component of Spandex polyureas; Du Pont uses the hydrogenation product of *m*-phenylenediamine [108-45-2] (24) captively to produce Lycra (see FIBERS, ELASTOMERIC).

1,8-Menthanediamine has been effectively reacted to form polyamides (85) and is sold in metric tons per year volume as a premium epoxy curative (86). 1-Adamantylamine hydrochloride [665-66-7] is a prophylactic against type A viral infections sold by Du Pont under the trade name Symmetrel. Cyclohexylamine derivatives as subtype selective *N*-methyl-D-aspartate antagonists useful for treating cerebral vascular disorders have been described (87).

11. Derivatives

Before a 1/1/70 FDA ban, cyclamate noncaloric sweeteners were the major derivatives driving cyclohexylamine production. The cyclohexylsulfamic acid sodium salt (39) [139-05-9] and more thermally stable calcium cyclohexylsulfamic acid (40) [139-06-1] salts were prepared from high purity cyclohexylamine by, among other routes, a reaction cycle with sulfamic acid.

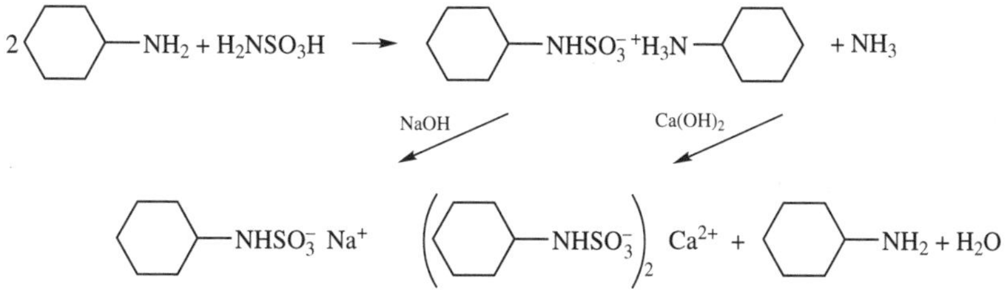

Cyclohexylamine condensed with mercaptobenzothiazole produces the large volume moderated rubber accelerator N-cyclohexyl-2-benzothiazolesulfenamide (41) [95-33-0] (see RUBBER COMPOUNDING). DCHA similarly is used in preparing N,N-dicyclohexyl-2-benzothiazolesulfenamide (42) [4979-32-2]. The cyclohexylamine derivative is preferred over *tert*-butylamine [75-64-9] and morpholine sulfenamide analogues because of lower amine volatility and less nitrosamine risk respectively.

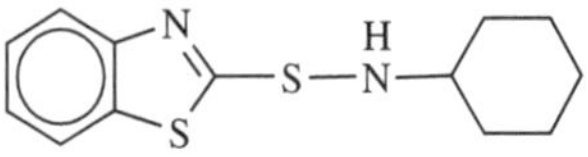

(41) 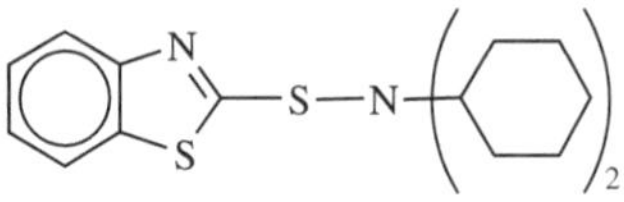   (42)

1,3-Dicyclohexylcarbodiimide (43) [538-75-0] is an important peptide-condensing agent and analytical reagent (88).

trans-1,2-Cyclohexanediamine is derivatized by Mannich reaction of formaldehyde and HCN, then hydrolyzed to the tetraacetate (44) [13291-61-7] and sold as a chelating agent.

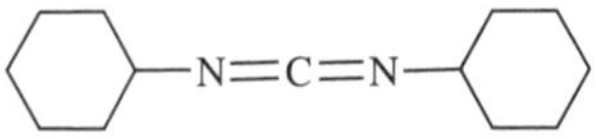

(43) 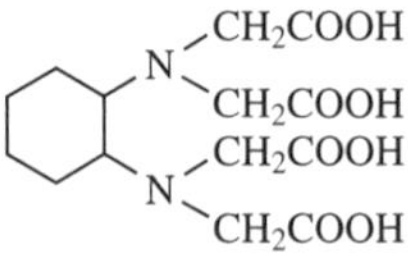   (44)

Methylenedi(cyclohexylisocyanate) (45) [5124-30-1] (MDCHI, Desmodur W) is the dominant derivative of MDCHA and is used in light-stable urethanes. Polyurethane physical properties are dependent on the diamine geometric isomer composition used for the derivative diisocyanate which reacts with diol (89).

Isophoronediisocyanate (46) [4098-71-9] made by phosgenation of IPD (90) competes effectively in this same polyurethane market, predominantly coatings, and is the major commercial application of isophoronediamine.

1,4-Cyclohexanediamine from hydrogenation of p-phenylenediamine [106-50-3] may be easily phosgenated, unlike the corresponding 1,2- and 1,3- isomers to produce a useful (92) diisocyanate for performance polyurethanes efficiently (91), particularly *trans*-1,4-cyclohexanediisocyanate [2556-36-47] (47) (CHDI). This diamine organic intermediate use competes with an Akzo route to *t*-CHDI from the corresponding diacid without intermediacy of the diamine.

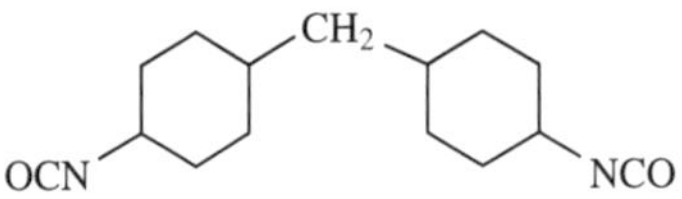

(45) 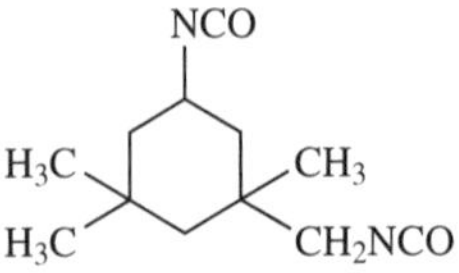   (46) 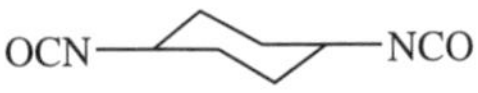   (47)

A representative agrochemical application of cycloaliphatic amines is the reaction of the commercial 30/70 cis/trans isomer mixture of 2-methylcyclohexylamine with phenylisocyanate to give the crabgrass and weed control agent Siduron (1-(2-methylcyclohexyl)-3-phenylurea **(48)** [1982-49-6] (93). The preplant herbicide Cycloate used for sugar beets, vegetable beets, and spinach, (S-ethyl-N-ethyl-N-cyclohexylthiocarbamate **(49)** [1134-23-2], incorporates N-ethylcyclohexylamine. The herbicide Hexazinone, (3-cyclohexyl-6-dimethylamino-1-methyl-1,3,5-triazine-2,4-dione **(50)** [51235-04-2]) is prepared from cyclohexylisocyanate [3173-53-3] (94).

         (48)   **(49)**   **(50)**

BIBLIOGRAPHY

"Amines" in *ECT* 1st ed., Vol. 1, pp. 702–717, by E. F. Landau, Celanese Corporation of America; "Amines, Survey" in *ECT* 2nd ed., Vol. 2, pp. 99–116, by Richard L. Bent, Eastman Kodak Company; "Amines, Cyclic" in *ECT* 3rd ed., Vol. 2, pp. 295–308, by Kenneth Mjos, Jefferson Chemical Company; in *ECT* 4th ed., Vol. 2, pp. 386–405, by Jeremiah P. Casey, Air Products and Chemicals.

CITED PUBLICATIONS

1. A. E. Barkdoll, H. W. Gray, and W. Kirk, *J. Am. Chem. Soc.* **73**, 741–746 (1951).
2. T. F. Mika and R. S. Bauer, in C. A. May, ed., *Epoxy Resins Chemistry and Technology*, 2nd ed., Marcel Dekker, New York, 1988, 465–550.
3. H. Lee and K. Neville, in *Handbook of Epoxy Resins*, McGraw-Hill, 1967, (1982 reissue), Chapt. 7.
4. *EPON Resin Structural Reference Manual*, Shell Chemical Company.
5. W. S. Emerson, in R. Adams, ed., *Organic Reactions*, Vol. 4, John Wiley & Sons, Inc., New York, 1948, Chapt. 3, 174–255.
6. J. March, *Advanced Organic Chemistry, Reactions, Mechanisms and Structure*, 3rd ed., John Wiley & Sons, Inc., New York, 1985.
7. U.S. Pat. 3,532,748 (Oct. 6, 1970), G. W. Smith (to E. I. du Pont de Nemours & Co.).
8. U.S. Pat. 3,551,487 (Dec. 29, 1970), B. R. Bluestein, J. M. Solomon, and L. B. Nelson (to Witco).
9. T. S. Carswell and H. L. Morrill, *Ind. Eng. Chem.* **29**, 1247–1251 (1937).
10. U.S. Pat. 3,364,261 (Jan. 16, 1968), F. H. Van Munster (to Abbott).
11. Jpn. Kokai 85 239,444 (Nov. 28, 1985), H. Matsumoto, K. Hiraska, M. Hasiguchi, and H. Arimatsu (to Honshu).
12. U.S. Pat. 3,376,341 (Apr. 2, 1965), C. R. Bauer (to E. I. du Pont de Nemours & Co., Inc.).

13. U.S. Pat. 3,644,542 (Feb. 22, 1972), L. D. Brake and A. B. Stiles (to E. I. du Pont de Nemours & Co., Inc.).
14. U.S. Pat. 3,355,490 (Nov. 28, 1967), F. H. Van Munster (to Abbott).
15. U.S. Pat. 3,154,580 (Oct. 27, 1964), R. M. Robinson and W. C. Braaten (to Abbott).
16. U.S. Pat. 3,551,486 (Dec. 29, 1970), J. M. Solomon, B. Ishitsky, and B. R. Bluestein (to Witco).
17. U.S. Pat. 3,551,488 (Dec. 29, 1970), B. R. Bluestein and J. M. Solomon (to Witco).
18. S. Nishimura, T. Shu, T. Hara, and Y. Takagi, *Bull. Chem. Soc. Japan* **39**, 329–333 (1966).
19. S. Nishimura, Y. Kono, Y. Otsuki, and Y. Fukaya, *Bull. Chem. Soc. Japan* **44**, 240–243 (1971).
20. P. N. Rylander, L. Hasbrouck, and I. Karpenko, *Ann. N.Y. Acad. Sci.* **214**, 100–109 (1973).
21. A. E. Bardoll, and co-workers, *J. Am. Chem. Soc.* **75**, 1156–1159 (1953).
22. U.S. Pat. 3,347,917 (Oct. 17, 1967), W. J. Arthur (to E. I. du Pont de Nemours & Co., Inc.).
23. U.S. Pat. 3,697,499 (Oct. 10, 1972), L. D. Brake (to E. I. du Pont de Nemours & Co., Inc.).
24. U.S. Pat. 3,636,108 (Jan. 18, 1972), L. D. Brake (to E. I. du Pont de Nemours & Co., Inc.).
25. U.S. Pat. 3,697,449 (Oct. 10, 1972), L. D. Brake (to E. I. du Pont de Nemours & Co., Inc.).
26. U.S. Pat. 3,591,635 (July 6, 1971), W. J. Farrissey and F. F. Frulla (to Upjohn).
27. U.S. Pat. 3,856,862 (Dec. 24, 1974), T. H. Chung, M. L. Dillon, and G. L. Lines (to Upjohn).
28. Jpn. Kokai 75 37,758, (Apr. 8, 1975), K. Watanabe, Y. Hirai, and K. Miyata (to Mitsui Toatsu).
29. U.S. Pat. 3,914,307 (Oct. 21, 1975), S. N. Massie (to Universal Oil Products).
30. U.S. Pat. 4,394,522 (July 19, 1983), G. F. Allen (to Mobay).
31. U.S. Pat. 4,394,523 (July 19, 1983), G. F. Allen (to Mobay).
32. U.S. Pat. 4,448,995 (May 15, 1984), G. F. Allen (to Mobay).
33. Eur. Pat. Appl. 0 066 212 (Dec. 8, 1982), G. F. Allen (to Mobay).
34. U.S. Pat. 4,754,070 (June 28, 1988), J. P. Casey and M. J. Fasolka (to Air Products).
35. U.S. Pat. 3,634,512 (Jan. 11, 1972), L. Wolf, H. Corr, and K. Pilch (to BASF).
36. U.S. Pat. 3,743,677 (July 3, 1973), O. Grosskinsky and K. Merkel (to BASF).
37. U.S. Pat. 3,959,374 (May 25, 1976), M. E. Brennan and E. L. Yeakey (to Texaco).
38. E. V. Genkina, and co-workers. *Zh. Vses. Khim. Obshchest.* **14**, 475–476 (1969).
39. U.S. Pat. 3,551,485 (Dec. 29, 1970), P. Raff, H. G. Peine, L. Schuster, and K. Adam (to BASF).
40. U.S. Pat. 3,283,002 (Nov. 1, 1966), L. D. Brake (to E. I. du Pont de Nemours & Co., Inc.).
41. U.S. Pat. 4,479,009 (Oct. 23, 1984), T. K. Shioyama (to Phillips).
42. U.S. Pat. 4,399,307 (Aug. 13, 1983), T. K. Shioyama (to Phillips).
43. U.S. Pat. 4,014,933 (Mar. 29, 1977), G. Boettger and co-workers (to BASF).
44. U.S. Pat. 4,161,492 (July 17, 1979), O. Weissel (to Bayer).
45. H. A. Colvin and J. Muse, *CHEMTECH* **16**, 500–504 (1986).
46. U.S. Pat. 4,186,145 (Jan. 29, 1980), O. Weissel (to Bayer).
47. L. I. Krimen and D. J. Cota, in W. G. Dauben, ed., *Organic Reactions*, Vol. 17, John Wiley & Sons, Inc., New York, 1948, Chapt. 3, 213–235.
48. U.S. Pat. 2,632,022 (Mar. 17, 1963), M. N. Bortnik (to Rohm and Haas).
49. R. C. Fort and P. v. R. Schleyer, *Chem. Rev.* **64**, 277–300 (1964).

50. U.S. Pat. 3,994,975 (Nov. 30, 1976), B. A. Oude Alink and N. E. S. Thompson (to Petrolite Corporation).

51. U.S. Pat. 3,352,913 (Nov. 14, 1967), K. Schmitt, J. Disteldorf, and W. Hubel (to Schloven-Chemie).

52. U.S. Pat. 3,270,044 (Aug. 30, 1966), K. Schmitt, J. Disteldorf, W. Hubel, and K. Rindtorff (to Hibernia-Chemie Gesellschaft).

53. U.S. Pat. 4,299,775 (Nov. 10, 1981), B. Dubreux (to PCUK).

54. U.S. Pat. 4,429,157 (Jan. 31, 1984), J. Disteldorf, W. Hubel, and L. Broschinski.

55. L. Born, D. Wendisch, H. Reiff, and D. Dieterich, *Angew. Makrol. Chem.* **171**, 213–231 (1989).

56. U.S. Pat. 4,070,399, (Jan. 24, 1978), W. A. Butte (to Suntech).

57. U.S. Pat. 3,334,149 (Aug. 1, 1967), G. A. Akin, H. J. Lewis, and T. F. Reid (to Eastman Kodak).

58. Fr. Pat. 1575505, (May 21, 1962), V. L. R. Brown, J. G. Smith and T. E. Stanin (to Eastman Kodak).

59. D. V. Sokol'skii, A. E. Temirbulatova, and A. Ualikhanova, *Zh. Prik. Khim.* **56**, 1610–1614 (1983).

60. "Cyclohexylamine, Chemical Profile", *Chemical Market Reporter*, (May 28, 2001).

61. J. G. Theivagt, P. F. Helgren, and D. R. Luebke, "Cyclohexylamine," *Encyclopedia of Industrial Chemical Analysis*, Vol. 11, John Wiley & Sons, Inc., New York, 1971, 209–219.

62. F. R. Prince and E. M. Pearce, *Macromolecules* **4**, 347–350 (1971).

63. M. L. Moore, in R. Adams, ed., *Organic Reactions*, Vol. 5, John Wiley & Sons, Inc., New York, 1949, Chapt. 7, 301–330.

64. M. W. Scroggins, L. Skurcenski, and D. S. Weinberg, *J. Chrom. Sci.* **10**, 678–681 (1972).

65. M. W. Scroggins, *J. Chrom. Sci.* **13**, 146–148 (1975).

66. T. J. Haley in N. L. Sax, ed., *Dangerous Properties of Industrial Materials*, 6th ed., sect. 1, Van Nostrand-Reinhold, New York, 1984.

67. Reference 66, p. 887 (misprinted as 1410 μg/kg).

68. K. Verschueren, ed., *Handbook of Environmental Data on Organic Chemicals*, 2nd ed., Van Nostrand-Reinhold, New York, 1983.

69. R. J. Lewis, Sr., *Sax's Dangerous Properties of Industrial Materials*, 10th ed., Vol. 2, John Wiley & Sons, Inc., New York, 2000.

70. H. L. Kahler and J. K. Brown, *Combustion*, 55–58 (1954).

71. B. Tuck and E. M. Osborn, *Chem. Ind.*, 326–331 (1960).

72. U.S. Pat. 4,473,666 (Sept. 25, 1984), F. A. Casati, D. S. Raden, and F. W. Arbir (to Abbott).

73. U.S. Pat. 4,286,086 (Aug. 25, 1981), V. Mark (to General Electric).

74. U.S. Pat. 4,794,158 (Dec. 27, 1988), M. Hasuo, H. Urabe, M. Kawai, and T. Ohsako (to Mitsubishi Kasei).

75. U.S. Pat. 3,703,595 (Nov. 21, 1972), G. Falkenstein and co-workers (to BASF).

76. U.S. Pat. 4,525,571 (June 25, 1985), C. Burba and H. Franz (to Schering AG).

77. U.S. Pat. 3,629,181 (Dec. 21, 1971), A. Heer, W. Schneider, and B. Dreher (to CIBA Limited).

78. Can. Pat. 892,636 (Feb. 8, 1972), J. M. Cross and R. V. Norton (to Mobay).

79. U.S. Pat. 3,450,759 (June 17, 1969), J. M. Cross, C. D. Campbell, and S. H. Metzger (to Mobay).

80. U.S. Pat. 2,817,644 (Dec. 24, 1957), E. C. Shokal and H. A. Newey (to Shell).

81. U.S. Pat. 4,849,544 (July 18, 1989), S. A. Culley, K. A. Keblys, and C. J. Nalepa (to Ethyl).

82. U.S. Pat. 3,351,650 (Nov. 7, 1967), J. M. Cross, S. H. Metzger, and C. D. Campbell (to Mobay).
83. U.S. Pat. 3,419,612 (Dec. 31, 1968), J. M. Cross, S. H. Metzger, and C. D. Campbell (to Mobay).
84. U.S. Pat. 3,651,118 (Mar. 21, 1972), M. Cenker and P. T. Kan (to Mobay).
85. D. L. Trumbo, *J. Poly. Sci., part A, Poly. Chem.* **26**, 1859–2862 (1988).
86. A. Sabra, J. P. Pascault, and G. Seytre, *J. Appl. Polym. Sci.* **32**, 5147–5160 (1986).
87. U.S. Pat. Appl. 2003000 4212 (Jan. 2, 2003), R. J. De Orazio and co-workers (to Warner-Lambert Company).
88. H. G. Khorana, *Chem. Ind.* 1087–1088 (1955).
89. C. A. Byrne, D. P. Mack, and J. M. Sloan, *Rubber Chem. Tech.* **58**, 985–996 (1985).
90. U.S. Pat. 3,401,190 (Sept. 10, 1968), K. Schmitt, F. Gude, K. Rindtorff, and J. Disteldorf (to Schloven-Chemie).
91. U.S. Pat. 4,256,869 (Mar. 17, 1981), H. Schulze, and co-workers (to Akzo).
92. U.S. Pat. 4,892,920 (Jan. 16, 1990), J. R. Quay and J. P. Casey (to Air Products).
93. U.S. Pat. 3,309,192 (Mar. 14, 1967), (to E. I. du Pont de Nemours & Co., Inc.).
94. U.S. Pat. 3,902,887 (Sept. 2, 1975), K. Lin (to E. I. du Pont de Nemours & Co., Inc.).

JEREMIAH P. CASEY
Air Products and Chemicals

AMINES, FATTY

1. Introduction

Fatty amines are nitrogen derivatives of fatty acids, olefins, or alcohols prepared from natural sources, fats and oils, or petrochemical raw materials. Commercially available fatty amines consist of either a mixture of carbon chains or a specific chain length from C_8–C_{22}. The amines are classified as primary, secondary, or tertiary depending on the number of hydrogen atoms of an ammonia molecule replaced by fatty alkyl or methyl groups (Fig. 1). The amino nitrogen is most frequently found on a primary carbon atom, but secondary and tertiary carbon substitution derivatives have been made and are commercially available. Fatty

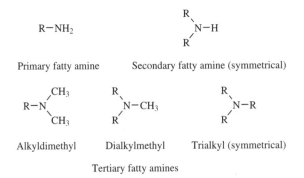

Fig. 1. Types of commercially available fatty amines. $R = C_8 - C_{22}$.

Table 1. **Typical Fatty Acid Composition**

Fatty acid[a], %	Fat		Fatty oils		
	Tallow	Coconut	Soya	Palm kernel	Palm
caproic (C6:0)		1.2		<0.5	
caprylic (C8:0)		3.4–15.0		2.4–6.2	
capric (C10:0)		3.2–15.0		2.6–6.2	
lauric (C12:0)		41–56		41–55	
myristic (C14:0)	3.0	13–23	0.9	14–20	1
palmitic (C16:0)	29.2	4.2–12.0	7–12	6.5–11.0	43.5
stearic (C18:0)	19.1	1.0–4.7	2.0–5.5	1.3–3.5	4.5
oleic (C18:1)	43.6	3.4–12.0	20–50	10–23	40
linoleic (C18:2)	2.1	0.9–3.7	35–60	0.7–5.4	
linolenic (C18:3)	0.5		2–13		
other	2.5				11

[a] The number of carbon atoms and number of double bonds are designated in parentheses.

amines are cationic surface-active compounds (see SURFACTANTS), which strongly adhere to surfaces by either physical or chemical bonding, thus modifying surface properties. Important commercial products are prepared using fatty amines as reactive intermediates.

Commercially available fatty amines are most frequently prepared from naturally occurring materials (see FATS AND FATTY OILS) by hydrogenation of a fatty nitrile intermediate using a variety of catalysts (1–3). Naturally occurring fats and oils (triglycerides) are continuously hydrolyzed at 200–280°C to yield saturated and unsaturated fatty acids and glycerol. Fatty nitriles are prepared from fatty acid mixtures by batch or continuous processes. The alkyl chain-length composition of the amines varies depending on the type of fat or oil. Other factors influencing alkyl chain composition include location of source, time of harvest, and, more recently, hybridization. Typical compositions for various fats and oils used to prepare commercially available fatty amines are given in Table 1 (4–8).

Fatty amines derived from fats and oils, containing several carbon-chain-length moieties, are designated as such by common names which describe these mixtures: tallowalkylamines [61790-33-8], cocoalkylamines [61788-46-3], and soyaalkylamines [61970-18-9], for example. High purity fatty amines are also commercially available. These amines are prepared by distillation of either the precursor fatty acid or amine product mixture. There are common names for single chain-length fatty amines in addition to IUPAC nomenclature, which uses the alkyl chain length in naming the amine. Secondary and tertiary amines have been named as primary amine derivatives, for example, N-hexadecyl-1-hexadecylamine instead of di-n-hexadecylamine (IUPAC). Examples include:

Common name	IUPAC name	Molecular formula	CAS Registry Number
laurylamine	1-dodecylamine	$C_{12}H_{27}N$	[124-22-1]
palmitylamine	1-hexadecylamine	$C_{16}H_{35}N$	[143-27-1]
stearylamine	1-octadecylamine	$C_{18}H_{39}N$	[124-30-1]
oleylamine	1-octadecen-9-ylamine	$C_{18}H_{37}N$ (unsaturated)	[112-90-3]

Trade names are commonly used for commercial products.

In the former USSR and Europe, synthetic fatty acids, prepared via hydrocarbon oxidation, have been used to prepare fatty amines (2,9).

Fatty alcohols, prepared from fatty acids or via petrochemical processes, aldol or hydroformylation reactions, or the Ziegler process, react with ammonia or a primary or secondary amine in the presence of a catalyst to form amines (10–12).

In addition to the nitrile and alcohol routes just described, many other methods of preparation of fatty amines are available. Some commercially available tertiary fatty amines are prepared via a petrochemical route (13). The amines have been prepared by reaction of an olefin with ammonia or a primary or secondary amine in the presence of a catalyst prepared from a Group 8–10 metal or an ammonium halide (14–16). Nitration of paraffins having from 6 to 30 carbon atoms with nitrogen dioxide at elevated temperatures followed by hydrogenation in the presence of a nickel or palladium catalyst produces secondary alkyl primary amines (17–20). Long-chain, unbranched, aliphatic tertiary amines can be prepared by reaction of an alkyl chloride with an alkyl secondary amine at 100–250°C (21,22). Other methods of producing amines include reaction of a carboxylic acid ester with a secondary amine in the presence of hydrogen at high pressure using a metal oxide catalyst, zinc oxide–chromium oxide or zinc oxide–aluminum oxide (23), or by catalytic hydroammonolysis of carboxylic acids at high pressure and temperature in the presence of a mixture of sulfides of metals of Groups 6 and 8–10 (24). The Hofmann rearrangement, preparation of an amine from an unsubstituted amide using solutions of chlorine in sodium hydroxide, has been useful in preparing long-chain primary amines (25). Amines prepared using the Hofmann rearrangement contain one less carbon atom than the amide. Thus, odd-chain-length fatty amines, not available from natural sources, can be prepared. The Ritter reaction of olefins with hydrogen cyanide in the presence of a strong acid, after hydrolysis, produces tertiary-alkyl primary amines. Rohm and Haas Primene amines are prepared using the Ritter reaction.

2. Physical Properties

Data on physical properties of fatty amines have been well documented and summarized in many reference works on fatty acids and nitrogen derivatives (3,8,13,26–29). Table 2 lists melting point data of some commercially available primary, secondary, and tertiary fatty amines, and it is evident that: (1) melting points within a homologous series of single-chain-length fatty amines increase with molecular weight, (2) symmetrical secondary amines have a higher melting point than the primary amine of the same alkyl group, but are lower melting than a primary amine with the same number of carbon atoms (hydrogen bonding), (3) symmetrical tertiary amines are lower melting than a symmetrical secondary amine of the same alkyl group, (4) symmetrical tertiary amines are lower melting than a primary or secondary amine containing the same number

Table 2. **Melting Points of Fatty Amines**

Amine	Molecular formula	CAS Registry Number	Mp, °C
Primary amines			
cocoalkylamines		[61788-46-3]	16.0
1-dodecylamine	$C_{12}H_{27}N$	[124-22-1]	28.0
1-hexadecylamine	$C_{16}H_{35}N$	[143-27-1]	46.2
1-octadecylamine	$C_{18}H_{39}N$	[124-30-1]	53.0
oleylamine	$C_{18}H_{37}N$	[112-90-3]	21.0
soyaalkylamines		[61970-18-9]	29.0
tallowalkylamines		[61790-33-8]	40.0
hydrogenated tallowalkylamines		[61788-45-2]	55.0
Secondary amines			
dicocoalkylamines		[61789-76-2]	43.0
di-*n*-dodecylamine	$C_{24}H_{51}N$	[3007-31-6]	47.0
di-*n*-hexadecylamine	$C_{32}H_{67}N$	[16724-63-3]	67.0
di-*n*-octadecylamine	$C_{36}H_{75}N$	[112-99-2]	72.3
ditallowalkylamines		[68783-24-4]	55.0
dihydrogenated tallowalkylamines		[61789-79-5]	62
Tertiary amines			
Alkyldimethyl			
cocoalkyldimethylamines		[61788-93-0]	−22
dimethyl-*n*-octylamine	$C_{10}H_{23}N$	[7378-99-6]	−57
dimethyl-*n*-decylamine	$C_{12}H_{27}N$	[1120-24-7]	−35
dimethyl-*n*-dodecylamine	$C_{14}H_{31}N$	[112-18-5]	−15
dimethyl-*n*-tetradecylamine	$C_{16}H_{35}N$	[112-75-4]	−6
dimethyl-*n*-hexadecylamine	$C_{18}H_{39}N$	[112-69-6]	8
dimethyl-*n*-octadecylamine	$C_{20}H_{43}N$	[124-28-7]	21
dimethyloleylamine	$C_{20}H_{41}N$	[28061-69-0]	−10
Dialkylmethyl			
di-*n*-decylmethylamine	$C_{21}H_{45}N$	[7396-58-9]	−6.3
dicocoalkylmethylamines		[61788-62-3]	−2
dihydrogenated tallowalkyl-methyl-amines		[61788-63-4]	38
Trialkyl			
tri-*n*-octylamine	$C_{24}H_{51}N$	[1116-76-3]	−34.6
tri-*n*-dodecylamine	$C_{36}H_{75}N$	[102-87-4]	−9
tri-*n*-hexadecylamines		[67701-00-2]	3

of carbon atoms, and (5) unsaturation lowers the melting point of the fatty amine, eg, oleylamine versus 1-octadecylamine and ditallowalkylamines versus dihydrogenated tallowalkylamines.

Boiling points of fatty amines have been reported (13,27–29). A direct correlation between molecular weight and boiling point is observed. Mixtures of primary fatty amines prepared from fats and oils can be separated into component amines by fractional distillation; an approximately 10°C increment in boiling point per carbon in the chain length is maintained throughout the series. Symmetrical secondary and tertiary fatty amines have a tendency to decompose during distillation and boiling point data are not reliable. The presence of residual catalyst from the preparation of these amines can lead to decomposition products during distillation.

Fatty amines are insoluble in water, but soluble in organic solvents to varying degrees (26–29). Water, however, is soluble in the amines, and hydrates are formed. Solubility in organic solvents is dependent on solvent polarity and temperature. The solubilities of primary amine acetates and hydrochlorides have been documented (29). Fatty amine acetates, available commercially, are very soluble in 95% ethanol.

The unshared pair of electrons on the nitrogen atom provides the basic character to the fatty amines. Basicity of amines has been determined as (29):

secondary amines > primary amines > tertiary amines > ammonia

Water and carbon dioxide from the atmosphere can be absorbed by the amines to form hydrates and carbamates, from primary and secondary amines, respectively.

3. Chemical Properties

General amine chemistry is applicable to fatty amines. Many chemical reactions using fatty amines as reactive intermediates are run on an industrial scale to produce a wide range of important products. Important industrial reactions are as follows.

3.1. Salt Formation.   Amines react with inorganic and organic acids.

$$RNH_2 + HCl \longrightarrow RNH_3^+ Cl^-$$

$$RNH_2 + R'COOH \longrightarrow RNH_3^+ R'COO^-$$

3.2. Methylation of Primary and Secondary Fatty Amines.   This is done by the Leuckart reaction (1,30) or reductive methylation (1,29,31,32).

Leuckart reaction

$$RNH_2 + 2\, CH_2O + 2\, HCOOH \longrightarrow RN(CH_3)_2 + 2\, H_2O + 2\, CO_2$$

Reductive methylation

$$RNH_2 + 2\, CH_2O + 2\, H_2 \longrightarrow RN(CH_3)_2 + 2\, H_2O$$

The Leuckart reaction uses formic acid as reducing agent. Reductive alkylation using formaldehyde, hydrogen, and catalyst, usually nickel, is used commercially to prepare methylated amines. These tertiary amines are used to prepare quaternary ammonium salts.

3.3. Quaternization.   Quaternary ammonium compounds are formed by alkylation of alkyl, alkyldimethyl, dialkyl, and dialkylmethyl fatty amines with methyl chloride, dimethyl sulfate, or benzyl chloride (1,3,7,12,29).

$$RNH_2 + 3\ CH_3X + 2\ NaOH \longrightarrow RN(CH_3)_3^+X^- + 2\ NaX + 2\ H_2O$$

$$R_2NCH_3 + CH_3X \longrightarrow R_2N(CH_3)_2^+X^-$$

A wide variety of quaternaries can be prepared. Alkylation with benzyl chloride may produce quaternaries that are biologically active, namely, bactericides, germicides, or algaecides. Reaction of a tertiary amine with chloroacetic acid produces an amphoteric compound, a betaine.

3.4. Ethoxylation and Propoxylation. Ethylene oxide [75-21-8] or propylene oxide [75-56-9] add readily to primary fatty amines to form bis(2-hydroxyethyl) or bis(2-hydroxypropyl) tertiary amines; secondary amines also react with ethylene or propylene oxide to form 2-hydroxyalkyl tertiary amines (1,3,7,33–36). The initial addition is completed at approximately 170°C. Additional ethylene or propylene oxide can be added by using a basic catalyst, usually sodium or potassium hydroxide.

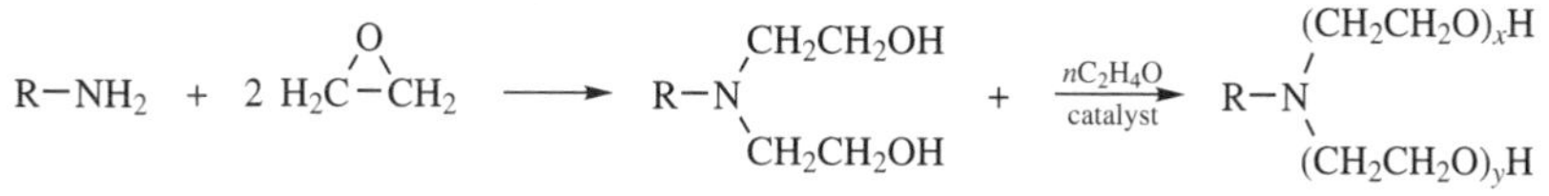

Ethylene oxide adds to the bis(2-hydroxyethyl) tertiary amine in a random fashion where $x + y = n + 2$. Ethoxylated amines, varying from strongly cationic to very weakly cationic in character, are available containing up to 50 mol of ethylene oxide/mol of amine. Ethyoxylated fatty amine quaternaries, cationic surfactants (both chloride from methyl chloride and acetate from acetic acid), are also available.

3.5. Oxidation by Hydrogen Peroxide. This reaction produces amine oxides (qv) (1,7,33,34,36).

Fatty amine oxides are most frequently prepared from alkyldimethylamines by reaction with hydrogen peroxide. Aqueous 2-propanol is used as solvent to prepare amine oxides at concentrations of 50–60%. With water only as a solvent, amine oxides can only be prepared at lower concentrations because aqueous solutions are very viscous. Fatty amine oxides are weak cationic surfactants.

3.6. Cyanoethylation. The reaction of primary fatty amines with acrylonitrile followed by hydrogenation produces diamines and triamines (4,7,31, 32,37,38).

Cyanoethylation (Michael addition)

$$RNH_2 + CH_2{=}CHC{\equiv}N \xrightarrow{70\ to\ 80°C} RNHCH_2CH_2C{\equiv}N$$

$$RNHCH_2CH_2C{\equiv}N + CH_2{=}CHC{\equiv}N \xrightarrow{>150°C} RN(CH_2CH_2C{\equiv}N)_2$$

Hydrogenation

$$RNHCH_2CH_2C{\equiv}N + 2\,H_2 \longrightarrow RNHCH_2CH_2CH_2NH_2$$

$$RN(CH_2CH_2C{\equiv}N)_2 + 4\,H_2 \longrightarrow RN(CH_2CH_2CH_2NH_2)_2$$

The addition of 1 mol of acrylonitrile (Michael addition) to the primary amine is an exothermic reaction, carried out at moderate temperature (70°C), either neat or using a polar solvent (water or low molecular weight alcohol). A fatty diamine, containing one primary and one secondary nitrogen, is formed when the nitrile adduct is catalytically hydrogenated. It is possible to add a second mole of acrylonitrile to the initial mononitrile adduct. However, the second addition does not occur as readily. Reduction of a dinitrile adduct forms a triamine, which contains one tertiary and two primary amino groups. Reduction of the nitrile can be problematic in that varying degrees of decyanoethylation occur depending on catalyst and temperature used during reduction.

Primary fatty amines also add (Michael addition) to esters of acrylic acid, $H_2C{=}CHCOOH$, methacrylic acid, $H_2C{=}C(CH_3)COOH$, or crotonic acid, $CH_3CH{=}CHCOOH$. Hydrolysis of the Michael ester forms an amphoteric surfactant. Crotonic acid can be used to form the amphoteric compound directly.

4. Manufacture

The principal industrial production route used to prepare fatty amines is the hydrogenation of nitriles, a route which has been used since the 1940s. Commercial preparation of fatty amines from fatty alcohols is a fairly new process, created around 1970, which utilizes petrochemical technology, Ziegler or Oxo processes, and feedstock.

In addition to the nitrile and alcohol routes for fatty amine preparation, processes have been described by Unocal and Pennwalt Corporation, using an olefin and secondary amine (14–16); by Texaco Inc., hydrogenation of nitroparaffins (17–20); by Onyx Corporation, reaction of an alkyl halide with secondary amines (21,22); by Henkel & Cie, GmbH, reduction of an ester in the presence of a secondary amine (23); by catalytic hydroammonolysis of carboxylic acids (24); and by the Hofmann rearrangement (25).

4.1. Nitrile Process. Fatty nitriles are readily prepared via batch, liquid-phase, or continuous gas-phase processes from fatty acids and ammonia. Nitrile formation is carried out at an elevated temperature (usually >250°C) with catalyst. An ammonia soap which initially forms, readily dehydrates at temperatures above 150°C to form an amide. In the presence of catalyst, zinc (ZnO) for batch and bauxite for continuous processes, and temperatures >250°C, dehydration of the amide occurs to produce nitrile. Removal of water drives the reaction to completion.

$$RCOOH + NH_3 \longrightarrow RCOO^-NH^+_4 \longrightarrow RCONH_2 + H_2O$$

$$\overset{\overset{\textstyle O}{\|}}{RCNH_2} \longrightarrow R{-}C{\equiv}N + H_2O$$

Hydrogenation of fatty nitrile to amine can be either a batch or continuous process (39–42). For preparing primary amines, ammonia is used to suppress secondary amine formation, at a minimum partial pressure of ~1 MPa (150 psig). Batch hydrogenation of a nitrile produces primary amine at 96% purity with low secondary (1.5%) and tertiary (<1%) amine levels (39). Reduction of the nitrile is carried out at elevated temperature (from 50 to 200°C), using hydrogen gas at high pressure, 3.5 MPa (500 psig) and higher total pressure of ammonia and hydrogen gases, in the presence of various catalysts. Catalysts useful for nitrile reduction include (1) aluminum–nickel Raney alloy (39), (2) Raney cobalt (40,43,44), (3) zinc–chromium or zinc–aluminum (41), (4) Raney nickel (42,45), (5) cobalt, copper, and chromium pellets (46), and (6) various nickel-supported catalysts (47–49).

A mixture of primary and secondary amines is formed when ammonia is not used during the nitrile reduction. It is possible to prepare high purity secondary amines by carrying the reduction out at low pressure and passing hydrogen through the reaction in a batch process (47,48), 2 $RNH_2 \longrightarrow R_2NH + NH_3$. A nickel–diatomaceous earth catalyst has been used at 220°C at low pressure and hydrogen purge to prepare secondary amines in high yield. Ammonia is removed continuously to drive the reaction to completion. The selectivity to secondary amine can be further enhanced by adding small amounts of an aliphatic carboxylic acid amide (48). Cobalt or copper chromite catalysts are also used to prepare secondary amines (1). Imines ($RCH=NH$ and $RCH=NCH_2R$) are suggested as intermediates in the process (3).

Symmetrical tertiary amines can be prepared in an analogous manner to secondary amines (1). Catalytic hydrogenation at elevated temperature and low pressure with a hydrogen purge produces a mixture of primary, secondary, and tertiary amines.

$$6\,R\!-\!C\!\equiv\!N + 12\,H_2 \longrightarrow RCH_2\!-\!NH_2 + (RCH_2)_2N\!-\!H + (RCH_2)_3N + 3\,NH_3$$

The rate of reaction slows down as the conversion to tertiary amine increases and primary amine concentrations drop below 1%. Conversion to 100% tertiary amine is difficult.

Alkyldimethyl and dialkylmethyl tertiary amines are commercially available. These amines are prepared by reductive methylation of primary and secondary amines using formaldehyde and nickel catalysts (1,3,47,48). The asymmetrical tertiary amines are used as reactive intermediates for preparing many commercial products.

Catalytic hydrogenation of raw materials prepared from natural fats and oils can be difficult because impurities are present which can affect the catalyst performance. Tallow fatty acid can contain the following catalyst poisons in differing amounts: sulfur, phosphorus, chlorine, and nitrogen (49). These catalyst poisons are removed to varying degrees during the refining process. Thus, catalyst loading levels can change from batch to batch.

Homogeneous and heterogenous catalysts which selectively or partially hydrogenate fatty amines have been developed (50). Selective hydrogenation of cis and trans isomers, and partial hydrogenation of polyunsaturated moieties, such as linoleic and linolenic to oleic, is possible.

4.2. Alcohol Process.   Fatty alcohols react with ammonia, or a low molecular weight primary or secondary amine, to form fatty amines. The fatty alcohols can be prepared from natural sources, fats and oils, or may be petroleum derived. Processes for manufacturing the fatty alcohol vary and depend on the raw material source, natural or petrochemical. Natural fatty acids, or preferably methyl esters of fatty acids, are catalytically reduced to alcohols by high pressure >20 MPa (3000 psig) and temperature (250–300°C) hydrogenation using copper chromite catalyst (11,51). Synthetic fatty alcohols are prepared using the Oxo (hydroformylation, reaction of carbon monoxide and hydrogen) or the Ziegler (ethylene and triethylaluminum) processes (10).

Primary amines can be prepared from alcohols and an excess of ammonia (52–55). Either a batch or continuous process can be used. The reaction is run at elevated temperature (50–340°C) and high pressure, 3.5 MPa (500 psig), with an ammonia-to-alcohol ratio of 5:1 to 30:1.

$$R—OH + NH_3 \longrightarrow R—NH_2 + H_2O$$

For example, a secondary alcohol of an average molecular weight of 202, containing 12–14 carbon atoms, is introduced into a reaction tube packed with a cobalt promoted zirconium catalyst on alumina at a rate of 60 mL/h along with liquid ammonia (90 mL/h) and hydrogen (5 L/h) to produce an amine mixture composed of 92.2% primary amine, 2.15% secondary and tertiary amines, and 5.7% unreacted alcohol (52). The reaction, at 180–190°C and 24 MPa ($\approx$3500 psig), is run by withdrawing the liquid reaction product from the bottom of the reactor.

Secondary amines can be prepared from an alcohol and either ammonia (56) or primary amine (57).

$$6\,R—OH + 3\,NH_3 \longrightarrow R—NH_2 + R_2N—H + R_3N + 6\,H_2O$$

$$R—OH + R'—NH_2 \longrightarrow R—NH—R' + H_2O$$

When a mixture of long-chain alcohols, Neodol 25 (Shell Chemical Company) and Alfol 1618 [Conoco Inc. (E. I. du Pont de Nemours & Co.)], containing a nickel catalyst (Girdler G49B) is heated at 190°C for six hours with a hydrogen and ammonia sparge, a mixture of amines is formed: 3.7% primary, 76.7% secondary, and 16.9% tertiary (56). This is similar to what is observed when fatty nitriles are reduced at low pressure and high temperature. High purity secondary amines have been prepared by reaction of an alcohol with a primary amine at elevated temperature (to 250°C) and low pressure (atmospheric) using a selective catalyst; use of copper, nickel, and a noble metal catalyst such as platinum, palladium, or ruthenium is recommended (57). For example, dilaurylamine was prepared at 92% purity by reaction of lauryl alcohol with laurylamine in the presence of a ruthenium catalyst.

Batch or continuous processes can be used to prepare tertiary amines from alcohols and ammonia or a secondary amine, such as, dimethylamine.

$$3\,C_8H_{17}OH + NH_3 \longrightarrow (C_8H_{17})_3 + 3\,H_2O$$

Trioctylamine has been prepared, in a continuous process, using 5,200 kg of n-octanol, 100 kg of copper formate catalyst, 500 kg of n-octylamine, 10 kg of calcium hydroxide, and 240 kg of ammonia (58). Ammonia was added over a 10-h period while 10 m^3 of hydrogen/h was passed through the reactor at a reaction temperature of 180–200°C. The final product was composed of 94% trioctylamine, 2% dioctylamine, 1% octylamine, and 0.5% n-octanol. A tertiary amine was prepared from dodecanol and dimethylamine using a nickel promoted-catalyst.

$$C_{12}H_{25}OH + (CH_3)_2N\!\!-\!\!H \longrightarrow C_{12}H_{25}N(CH_3)_2 + H_2O$$

Thus, a 94% conversion to dimethyldodecylamine was obtained in five hours using nickel catalyst promoted with chromium and iron at 180°C and 1.1 MPa (160 psig) (59).

Catalysts used for preparing amines from alcohols include: cobalt promoted with zirconium, lanthanum, cerium, or uranium (52); the metals and oxides of nickel, cobalt, and/or copper (53,54,56,60,61); metal oxides of antimony, tin, and manganese on alumina support (55); copper, nickel, and a metal belonging to the platinum group 8–10 (57); copper formate (58); nickel promoted with chromium and/or iron on alumina support (53,59); and cobalt, copper, and either iron, zinc, or zirconium (62).

5. Shipment

Fatty amine products are normally shipped in 55-gal (208 L), lined and unlined, steel drums or in tank cars or tank trucks for bulk shipments. High melting amines can be flaked and shipped in cardboard cartons or paper bags. The amines are corrosive to skin and eyes. Protective splash goggles and gloves should be worn when handling these materials.

6. Economic Aspects

Demand for amines in the United States will grow to nearly 2×10^9 in 2004. Advances will be led by the specialty amines. The fatty amines will show more modest gains since it is a mature market. Growth will exist in water treatment and plastics. The set markets, detergents, cleaning products, personal care products, and agricultural products will advance moderately. Fatty amines and derivatives are used in fabric softeners, dishwashing liquids, car wash detergents, and carpet cleaners (63).

Fatty amines and ethanolamines are the largest amine types used in the detergent and cleaner markets. The amine demand for 2004 is expected to be 70×10^6 kg up from 62×10^6 kg in 1999 (64).

The top U.S. producers of amines are Union Carbide, Air Products and Chemicals, Huntsman, and Dow Chemical.

The major source of raw materials for the preparation of fatty amines is fats and oils such as tallow, and coconut, soya, and palm oils. Commercially available high purity fatty amines are listed in Table 3. Cost of the amines can vary owing to supply of raw materials.

7. Specifications

Specifications of commercially available fatty amines are listed in product bulletins which are available from the manufacturers (31,65–72). For primary

Table 3. **Commercially Available Fatty Amines**

Fatty amine (common name)	Molecular formula	CAS Registry Number	Trade name	Manufacturer	Appearance
Primary amines					
2-ethylhexylamine	$C_8H_{19}N$	[104-75-6]	Armeen L8D	Akzo	liquid
cocoalkylamines		[61788-46-3]	Armeen CD	Akzo	liquid
			Alamine 21D	Henkel	
			Kemamine P-650	Humko	liquid
			Jet Amine PCD	Jetco	liquid
			Adogen 160-D	Sherex	liquid
dodecylamine	$C_{12}H_{27}N$	[124-22-1]	Armeen 12D	Akzo	semisolid
			Alamine 4D	Henkel	
			Adogen 163-D	Sherex	liquid
hexadecylamine	$C_{16}H_{35}N$	[143-27-1]	Armeen 16D	Akzo	solid
			Kemamine P-880D	Humko	solid
octadecylamine	$C_{18}H_{39}N$	[124-30-1]	Armeen 18D	Akzo	solid
			Alamine 7D	Henkel	
			Kemamine P-990D	Humko	solid
			Adogen 142-D	Sherex	solid
oleylamine	$C_{18}H_{37}N$	[112-90-3]	Armeen OD	Akzo	semisolid
			Alamine 11D	Henkel	
			Kemamine P-989D	Humko	liquid
			Jet Amine POD	Jetco	liquid
			Adogen 172-D	Sherex	liquid
soyaalkylamines		[61970-18-9]	Armeen SD	Akzo	paste
			Jet Amine PSD	Jetco	paste
			Adogen 115-D	Sherex	liquid
tallowalkylamines		[61790-33-8]	Armeen TD	Akzo	solid
			Alamine 26D	Henkel	
			Kemamine P-974D	Humko	paste
			Jet Amine PTD	Jetco	solid
			Adogen 170-D	Sherex	paste
			Tomah P-2B	Tomah	
hydrogenated tallow-alkylamines		[61788-45-2]	Armeen HTD	Akzo	solid
			Alamine H26D	Henkel	
			Kemamine P-970D	Humko	solid
			Jet Amine PHTD	Jetco	solid
			Adogen 140-D	Sherex	solid

Table 3 (*Continued*)

Fatty amine (common name)	Molecular formula	CAS Registry Number	Trade name	Manufacturer	Appearance
eicosylamine/docosylamine	(C_{20}/C_{22})	[60800-66-0]	Kemamine P-190D and P-150D	Humko	solid
			Adogen 101	Sherex	solid
Secondary amines					
didecylamine	$C_{20}H_{43}N$	[1120-49-6]	Armeen 2-10	Akzo	liquid
dicocoalkylamines		[61789-76-2]	Armeen 2C	Akzo	solid
			Jet Amine 2C	Jetco	solid
diisotridecylamine	$C_{26}H_{55}N$	[57157-80-9]	Adogen 283	Sherex	liquid
dioctadecylamine	$C_{36}H_{75}N$	[112-99-2]	Armeen 2-18	Akzo	solid
ditallowalkylamines		[68783-24-4]	Armeen 2T	Akzo	solid
dihydrogenated tallowalkylamines		[61789-79-5]	Armeen 2HT	Akzo	solid
			Kemamine S-970	Humko	solid
			Jet Amine 2HT	Jetco	solid
			Adogen 240	Sherex	solid
dieicosylamine/didocosylamines		[53529-35-4]	Kemamine S-190	Humko	solid
Tertiary amines					
Alkyldimethylamines					
dimethyloctylamine	$C_{10}H_{23}N$	[7378-99-6]	ADMA-8	Ethyl	liquid
decyldimethylamine	$C_{12}H_{27}N$	[1120-24-7]	ADMA-10	Ethyl	liquid
dimethyldodecylamine	$C_{14}H_{31}N$	[112-18-5]	Armeen DM12D	Akzo	liquid
			ADMA-12	Ethyl	liquid
cocoalkyldimethylamines		[61788-93-0]	Armeen DMCD	Akzo	liquid
			Kemamine T-6502D	Humko	liquid
			Jet Amine DMCD	Jetco	liquid
dimethyltetradecylamine	$C_{16}H_{35}N$	[112-75-4]	ADMA-14	Ethyl	liquid
dimethylhexadecylamine	$C_{18}H_{39}N$	[112-69-6]	Armeen DM16D	Akzo	liquid
			ADMA-16	Ethyl	liquid
dimethyloctadecylamine	$C_{20}H_{43}N$	[124-28-7]	Armeen DM18D	Akzo	liquid
			ADMA-18	Ethyl	liquid
			Kemamine T-9902D	Humko	liquid
dimethyloleylamine	$C_{20}H_{41}N$	[28061-69-0]	Armeen DMOD	Akzo	liquid
			Jet Amine DMOD	Jetco	liquid
dimethylsoyaalkylamines		[61788-91-8]	Armeen DMSD	Akzo	liquid
			Kemamine T-9972D	Humko	liquid
			Jet Amine DMSD	Jetco	liquid

Table 3   (*Continued*)

Fatty amine (common name)	Molecular formula	CAS Registry Number	Trade name	Manufacturer	Appearance
dimethyltallow-alkylamines		[68814-69-7]	Armeen DMTD	Akzo	liquid
			Kemamine T-9742D	Humko	liquid
			Jet Amine DMTD	Jetco	liquid
dimethylhydro-genated tallow-alkylamines		[61788-95-2]	Armeen DMHTD	Akzo	liquid
			Kemamine T-9702D	Humko	liquid
			Jet Amine DMHTD	Jetco	liquid
dimethyleicosyl-amine/-dimethyl-docosylamines		[93164-85-3]	Kemamine T-1902D	Humko	solid
Dialkylmethyl-amines					
di(C$_8$–C$_{10}$)-methylamine	C$_{17}$H$_{37}$N	[4455-26-9]	DAMA-810	Ethyl	liquid
	C$_{19}$H$_{41}$N	[22020-14-0]			
	C$_{21}$H$_{45}$N	[7396-58-9]			
didecylmethyl-amine	C$_{21}$H$_{45}$N	[7396-58-9]	Armeen M2-10D	Akzo	liquid
			DAMA-1010	Ethyl	liquid
dicocoalkyl-methylamines		[61788-62-3]	Armeen M2C	Akzo	liquid
			Kemamine T-6501	Humko	liquid
			Jet Amine M2C	Jetco	solid
			Adogen 369	Sherex	liquid
dioctadecyl-methylamine	C$_{37}$H$_{77}$N	[4088-22-6]	Adogen 349	Sherex	solid
dihydrogenated tallowalkyl-methylamines		[61788-63-4]	Armeen M2HT	Akzo	solid
			Kemamine T-9701	Humko	solid
			Jet Amine M2HT	Jetco	solid
			Adogen 343	Sherex	solid
Trialkylamines					
trioctylamine	C$_{24}$H$_{51}$N	[1116-76-3]	Alamine 336	Henkel	liquid
tri(isooctyl)-amine	C$_{24}$H$_{51}$N	[25549-16-0]	Adogen 381	Sherex	liquid
tri(C$_8$–C$_{10}$)-amines		[68814-95-9]	Adogen 364	Sherex	liquid
tri(isodecyl)amine	C$_{30}$H$_{63}$N	[35723-89-8]	Adogen 382	Sherex	liquid
trilaurylamine	C$_{36}$H$_{75}$N	[102-87-4]	Armeen 3-12	Akzo	liquid
			Alamine 304	Henkel	liquid
tri(isotridecyl)-amine	C$_{39}$H$_{81}$N	[35723-83-2]	Adogen 383	Sherex	liquid

Table 3　(*Continued*)

Fatty amine (common name)	Molecular formula	CAS Registry Number	Trade name	Manufacturer	Appearance
trihexadecyl- amines		[67701-00-2]	Armeen 3-16	Akzo	solid
trioctadecylamine	$C_{54}H_{111}N$	[102-88-5]	Adogen 340	Sherex	solid
Diamines					
N-cocoalkyltrimethy- lenedi- amines		[61791-63-7]	Duomeen CD	Akzo	semisolid
			Diam 21D Kemamine D-650	Henkel Humko	liquid
N-tallowalkyltri- methylenedi- amines		[61791-55-7]	Jet Amine DC Adogen 560 Duomeen T	Jetco Sherex Akzo	liquid liquid solid
			Diam 26 Kemamine D-974	Henkel Humko	solid
			Jet Amine DT Adogen 570-S Tallow Diamine	Jetco Sherex Tomah	paste solid
N-hydrogenated tallowalkyltri- methylenedia- mines		[37231-11-1]	Kemamine D-970	Humko	solid
N-oleyl-1,3-diamino- propane	$C_{21}H_{44}N_2$	[7173-62-8]	Adogen 540 Duomeen O	Sherex Akzo	flakes liquid
			Diam 11 Kemamine D-989	Henkel Humko	liquid
N-eicosyl-1,3-di- aminopropane/ N-docosyl-1,3-di- aminopropanes		[89234-29-7]	Jet Amine DO Adogen 572 Kemamine D-150	Jetco Sherex Humko	liquid liquid solid
N,N,N'-trimethyl-N'- tallowalkyltri- methylenedi- amines		[68783-25-5]	Duomeen TTM	Akzo	liquid
Triamines					
N-tallowalkyl dipropylenetri- amines		[61791-57-9]	Triameen T	Akzo	paste
			Tallow tria- mine	Tomah	
Tetramines					
N-tallowalkyl tripro- pylenetetramine			Tallow tetra- mine	Tomah	

amines, specifications can include minimum and maximum values for the following: the percentage of primary and apparent secondary amines, color (Gardner), equivalent weight, amine number, moisture, and iodine value (IV). Secondary amine specifications, in addition to the above, have a minimum value for apparent secondary amine content, while tertiary amines would have specifications ona minimum percentage of tertiary and maximum percentage of primary and secondary amine content. Specifications for diamines can include a minimum percentage of diamine content. Some product bulletins in addition to listing specifications give typical analytical values.

8. Analytical Methods

To analyze fatty amines, both wet and instrumental methods of analysis are used. Wet methods routinely used are: total amine value (ASTM Method D2073); combining weight or neutralization equivalent; primary, secondary, and tertiary amine content (ASTM Method D2083); moisture, Karl-Fischer (ASTM Method D2072); and iodine value, measure of unsaturation (ASTM Method D2075). These provide important information on physical and chemical characteristics of the amine products used in various application areas (8,68,73). In addition to the ASTM methods available, the American Oil Chemists' Society has developed methods of analysis for fatty amines (74).

Instrumental methods of analysis provide information about the specific composition and purity of the amines. Qualitative information about the identity of the product (functional groups present) and quantitative analysis (amount of various components such as nitrile, amide, acid, and determination of unsaturation) can be obtained by infrared analysis. Gas chromatography (gc), with a liquid phase of either Apiezon grease or Carbowax, and high performance liquid chromatography (hplc), using silica columns and solvent systems such as isooctane, methyl *tert*-butyl ether, tetrahydrofuran, and methanol, are used for quantitative analysis of fatty amine mixtures. Nuclear magnetic resonance spectroscopy (nmr), both proton (^{1}H) and carbon-13 (^{13}C), which can be used for qualitative and quantitative analysis, is an important method used to analyze fatty amines (8,73).

9. Health and Safety Factors

9.1. Skin and Eye Irritation.   Fatty alkylamines are generally considered to be irritating to both the skin and eyes (75). The severity or degree of irritation is usually dependent on the type of alkylamine, concentration of the chemical, time of exposure to the chemical, and sensitivity to the chemical. A small percentage of the population who come into contact with fatty amines may develop a skin hypersensitivity to certain amines and diamines.

Alkylamines and diamines are generally classified as corrosive to the skin based on results from laboratory animal (rabbit) studies performed in accordance with the Department of Transportation (DOT) test method (76); rabbits are considered to be especially sensitive to alkylamines which even at low concentrations can induce skin redness and swelling. Oleylamine has been shown to induce mild to moderate skin irritation in laboratory rats when applied at a concentration of 0.3% in mineral oil (Chemical Manufacturer's Association, 1985). Fatty amines which contain alkyl chains of 10–14 carbons are considered more irritating than related products which contain alkyl chains of 14–18 carbon atoms. Ethoxylation generally decreases the irritation potential of alkylamines.

9.2. Oral Toxicity. Depending on the chemical class, most fatty amines range from moderately toxic to practically nontoxic by acute oral ingestion. Laboratory animal testing has revealed that the amines and diamines can induce irritation and damage to the gastrointestinal tract. The acute oral LD_{50} of cocoalkyldiamine in rats is less than 200 mg/kg body weight (Armak Test Data, 1981), whereas the corresponding value for dihydrogenated tallowalkyl-methylamine (Akzo Chemie Data, 1984) is greater than 5,000 mg/kg. Products which have LD_{50} values greater than 5000 mg/kg are considered practically nontoxic by accidental ingestion.

9.3. Dermal Toxicity. Fatty alkylamines are not considered especially toxic with regard to skin penetration and systemic absorption into the body; certain polyamines may be absorbed through the skin to a much greater degree. The acute dermal LD_{50} of decylamine in rabbits has been reported to be 350 mg/kg [RTECS, 1982 (77)]; dialkyl(C_{14}–C_{18})methylamine has been shown to have an acute dermal LD_{50} value of greater than 2000 mg/kg in rabbits (Dynamac Corporation, 1988). Products with acute dermal LD_{50} values greater than 2000 mg/kg are considered to exhibit a low degree of hazard by acute dermal exposure.

9.4. Inhalation. Long-chain amines are not considered an inhalation hazard at ambient conditions because of their relatively low volatility. Inhalation of aerosols or heated vapors may result in irritation of the nose, throat, and upper respiratory system. Lower molecular weight and branched-chain amines are more volatile and can cause irritation if inhaled. Volatile amines are easily recognized by their unpleasant, fishy odor.

10. Uses

Fatty amines and chemical products derived from the amines are used in many industries.

Amine salts, especially acetate salts prepared by neutralization of a fatty amine with acetic acid, are useful as flotation agents (collectors), corrosion inhibitors, and lubricants (3,8).

The single largest market use for quaternary fatty amines is in fabric softeners. Monoalkyl quaternaries (chloride) have been used in liquid detergent softener antistat formulations (LDSA), dialkyldimethyl quaternaries (chloride) in the rinse cycle, and dialkyldimethyl quaternaries (sulfate) as dryer softeners.

Another significant use for dialkyldimethyl quaternary ammonium salts and alkylbenzyldimethylammonium salts is in preparing organoclays for use as drilling muds, paint thickeners, and lubricants.

Betaines, or specialty quaternaries, are used in the personal care industry in shampoos, conditioners, foaming, and wetting agents.

A significant use of ethoxylated and propoxylated amines is as antistatic agents (qv) in the textile and plastics industry (78). Ethoxylates are also used in the agricultural area as adjuvants.

Examples of uses for amine oxides include: detergent and personal care areas as a foam booster and stabilizer, as a dispersant for glass fibers, and as a foaming component in gas recovery systems.

Important uses for the diamines include: corrosion inhibitors, gasoline and fuel oil additives, flotation agents, pigment wetting agents, epoxy curing agents, herbicides, and asphalt emulsifiers (3,75,77).

Fatty amines and derivatives are widely used in the oil field, as corrosion inhibitors, surfactants, emulsifying/deemulsifying and gelling agents (79). In the mining industry, amines and diamines are used in the recovery and purification of minerals, flotation, and benefication. A significant use of fatty diamines is as asphalt emulsifiers for preparing asphalt emulsions. Diamines have also been used as epoxy curing agents, corrosion inhibitors, gasoline and fuel oil additives, and pigment wetting agents. Oleylamine is a petroleum additive useful as a detergent in gasoline (8). In addition, derivatives of the amines, amphoterics, and long-chain alkylamines are used as anionic and cationic surfactants in the personal care industry (80).

BIBLIOGRAPHY

"Amines, Fatty" in *ECT* 2nd ed., Vol. 2, pp. 127–138, by H. J. Harwood, Durkee Famous Foods; in *ECT* 3rd ed., Vol. 2, pp. 283–295, by Harinath B. Bathina and Richard A. Reck, Armak Company; in *ECT* 4th ed., Vol. 2, pp. 405–425, by K. Visek, Akzo Chemicals, Inc.

CITED PUBLICATIONS

1. S. Billenstein and G. Blaschke, *J. Am. Oil Chem. Soc.* **61**, 353 (1984).
2. N. O. V. Sonntag, *J. Am. Oil Chem. Soc.* **56**, 861A (1979).
3. R. A. Reck, in R. W. Johnson and E. Fritz, ed., *Fatty Acids in Industry*, Marcel Dekker, Inc., New York, 1989, 177–199, 201–215.
4. R. A. Reck, *J. Am. Oil Chem. Soc.* **39**, 461 (1962).
5. D. R. Erickson, *J. Am. Oil Chem. Soc.* **60**, 351 (1983).
6. F. V. K. Young, *J. Am. Oil Chem. Soc.* **60**, 374 (1983).
7. R. A. Reck, *J. Am. Oil Chem. Soc.* **62**, 355 (1985).
8. *Oleochemicals: Fatty Acids, Fatty Alcohols, Fatty Amines*, Course sponsored by the Education Committee of the American Oil Chemists' Society, Kings Island, Ohio, Sept. 13–16, 1987.
9. H. Fineberg, *J. Am. Oil Chem. Soc.* **56**, 805A (1979).
10. J. A. Monick, *J. Am. Oil Chem. Soc.* **56**, 853A (1979).
11. T. Voeste and H. Buchold, *J. Am. Oil Chem. Soc.* **61**, 350 (1984).
12. R. Puchta, *J. Am. Oil Chem. Soc.* **61**, 367 (1984).
13. *Fatty Tertiary Amines, Product Bulletin*, CG-180R(288), Ethyl Corporation, Baton Rouge, La., Feb. 1988.

14. U.S. Pat. 3,513,200 (May 19, 1970), G. Biale (to Union Oil Company of California).
15. U.S. Pat. 4,483,757 (Nov. 20, 1984), D. M. Gardner and P. J. McElligott (to Pennwalt Corporation).
16. U.S. Pat. 4,827,031 (May 2, 1989), D. M. Gardner, P. J. McElligott and R. T. Clark (to Pennwalt Corporation).
17. U.S. Pat. 3,739,027 (June 12, 1973), W. C. Gates (to Texaco Inc.).
18. U.S. Pat. 3,917,705 (Nov. 4, 1975), R. W. Swanson and H. K. Zang (to Texaco Inc.).
19. U.S. Pat. 3,917,706 (Nov. 4, 1975), R. B. Hudson and W. C. Gates (to Texaco Inc.).
20. U.S. Pat. 3,920,744 (Nov. 18, 1975), R. M. Suggitt and W. C. Gates (to Texaco Inc.).
21. U.S. Pat. 3,385,893 (May 28, 1968), R. L. Wakeman (to Millmaster Onyx Corporation).
22. U.S. Pat. 3,548,001 (Dec. 15, 1970), Z. J. Dudzinski (to Millmaster Onyx Corporation).
23. U.S. Pat. 3,579,584 (May 18, 1971), H. Rutzen and R. Brockmann (to Henkel & Cie, GmbH).
24. Brit. Pat. 1,135,915 (Dec. 11, 1968), D. Zalmanovich Zavelsky and co-workers, (to Gosudarstvenny Ordena Trudovogo Krasnogo Znameni Institut Prikladnoi Khimii).
25. U.S. Pat. 4,198,348 (Apr. 15, 1980), F. Bertini and C. A. Pauri (to SNIA VISCOSA Societa Nazionale Industria Applicazioni Viscosa SpA).
26. *Armeen, Duomeen and Triamine Aliphatic Amines, Product Bulletin*, Bulletin 89–134, Akzo Chemicals Inc., Chicago, Ill., 1989.
27. N. O. V. Sonntag, "Nitrogen Derivatives," in K. S. Markley, ed., *Fatty Acids*, Part 3, John Wiley & Sons, Inc., New York, 1964, 1551–1715.
28. S. H. Shapiro, "Commercial Nitrogen Derivatives of Fatty Acids," in E. Pattison, ed., *Fatty Acids and Their Industrial Applications*, Marcel Dekker, Inc., New York, 1968, 77–154.
29. W. M. Linfield, "Straight-Chain Alkylammonium Compounds," in E. Jungermann, ed., *Cationic Surfactants*, Vol. 4, Marcel Dekker, Inc., New York, 1970, 9–70.
30. S. H. Pine and B. L. Sanchez, *J. Org. Chem.* **56**, 829 (1971).
31. *Kemamine Fatty Amines, Product Bulletin AMN:901/MI*, Humko Sheffield Chemical, Memphis, Tenn., 1978.
32. C. W. Glankler, *J. Am. Oil Chem. Soc.* **56**, 802A (1979).
33. H. Maag, *J. Am. Oil Chem. Soc.* **61**, 259 (1984).
34. R. A. Reck, *J. Am. Oil Chem. Soc.* **56**, 796A (1979).
35. R. A. Reck, "Polyoxyethylene Alkylamines," in M. Schick, ed., *Nonionic Surfactants, Surfactant Science Series*, Vol. 1, Marcel Dekker, Inc., New York, 1967, 187–207.
36. B. H. Babu, P. K. S. Amma, and S. V. Rao, *Indian J. Technol.* **5**, p. 262 (Aug. 1967).
37. U.S. Pat. 3,222,402 (Dec. 7, 1965), M. C. Cooperman (to Armour and Company).
38. R. W. Fulmer, *J. Org. Chem.* **27**, 4115 (1962).
39. Brit. Pat. 1,321,981 (July 4, 1973), N. Waddleton.
40. Brit. Pat. 1,388,053 (Mar. 19, 1975) (W. R. Grace & Co.).
41. Brit. Pat. 1,153,919 (June 4, 1969) (Henkel & Cie, GmbH).
42. U.S. Pat. 3,574,754 (Apr. 13, 1971), G. A. Specken.
43. U.S. Pat. 4,375,003 (Feb. 22, 1983), R. J. Allain and G. D. Smith (to Nalco Chemical Company).
44. U.S. Pat. 4,140,720 (Feb. 20, 1979), C. A. Drake (to Phillips Petroleum Company).
45. U.S. Pat. 4,359,585 (Nov. 16, 1982), C. R. Campbell and C. E. Cutchens (to Monsanto Company).
46. U.S. Pat. 4,552,862 (Nov. 12, 1985), J. M. Larkin (to Texaco Inc.).

47. U.S. Pat. 4,248,801 (Feb. 3, 1981), S. Tomidokoro, M. Sato, and D. Saika (to The Lion Fat & Oil Co., Ltd.).
48. U.S. Pat. 4,845,298 (July 4, 1989), T. Inagaki, A. Fukasawa, and H. Yamagishi (to Lion Akzo Company Limited).
49. H. Klimmek, *J. Am. Oil Chem. Soc.* **61**, 200 (1984).
50. E. Draguez De Hault and A. Demoulin, *J. Am. Oil Chem. Soc.* **61**, 195 (1984).
51. U. R. Kreutzer, *J. Am. Oil Chem. Soc.* **61**, 343 (1984).
52. Brit. Pat. 1,361,363 (July 24, 1974), H. Koike, T. Sawano, N. Kurata, and Y. Okuda (to Nippon Shokubai Kagaku Kogyo Co., Ltd.).
53. Brit. Pat. 1,074,603 (July 5, 1967) (to Jefferson Chemical Company, Inc.).
54. U.S. Pat. 4,418,214 (Nov. 29, 1983), M. G. Turcotte (to Air Products and Chemicals, Inc.).
55. U.S. Pat. 4,654,440 (Mar. 31, 1987), R. J. Card and J. L. Schmitt (to American Cyanamid Company).
56. U.S. Pat. 3,803,137 (Apr. 9, 1974), R. R. Egan, G. K. Hughs, and J. W. Sigan (to Ashland Oil, Inc.).
57. U.S. Pat. 4,792,622 (Dec. 20, 1988), Y. Yokota and co-workers (to Kao Corporation).
58. U.S. Pat. 4,827,035 (May 2, 1989), H. Mueller and H. Axel (to BASF Aktiengesellschaft).
59. Brit. Pat. 1,553,285 (Sept. 26, 1979), F. Wattimena and C. Borstlap (to Shell Internationale Research Maatschappij B.V.).
60. U.S. Pat. 4,851,580 (July 25, 1989), H. Mueller, R. Fischer, G. Jeschek, and W. Schoenleben (to BASF Aktiengesellschaft).
61. U.S. Pat. 4,014,933 (Mar. 29, 1977), G. Boettger and co-workers (to BASF Aktiengesellschaft).
62. U.S. Pat. 4,153,581 (May 8, 1979), C. E. Habermann (to The Dow Chemical Company).
63. http://www.healthcare-information.com/R154-306.html, accessed April 2003.
64. Amines to 2004, Amine Markets, Freedonia Group, June 2000.
65. *Fine and Functional Chemicals, Nitrogen Derivatives, General Catalog Bulletin 89–74*, Akzo Chemicals, Inc., Chicago, Ill., 1989.
66. *Industrial Chemicals / Fine Chemicals / Industrial Gums, Product Catalog, G-30*, General Mills Chemicals, Inc., Minneapolis, Minn., 1976.
67. *Jetco Chemicals, Product Guide to Jet Amines—Jet Quats*, Jetco Chemicals Inc., Corsicana, Tex., 1989.
68. *Adogen Fatty Amines, Diamines & Amides, Product Bulletin*, Sherex Chemical Company, Inc., Dublin, Ohio, 1989.
69. *Tomah Products, Product Bulletin, 1989 Formulary*, Exxon Chemical Company, Miltom, Wis., Aug. 1, 1989.
70. *Innovative Specialty Surfactants Product Bulletin*, Jordan Chemical Company, Folcroft, Pa.
71. *Specialty Chemical Products, Product Bulletin, 5M 682 TGT*, Lonza Inc., Fair Lawn, N.J., 1987.
72. *Ethyl Chemicals Group, Product Bulletin, CG-220R (3/89), Industrial Chemicals Division Products*, Ethyl Corporation, Baton Rouge, La., Mar. 1989.
73. L. D. Metcalfe, *J. Am. Oil Chem. Soc.* **61**, 363 (1984).
74. D. Firestone, ed., *Official Methods and Recommended Practices of the American Oil Chemists' Society*, 3rd ed., The American Oil Chemists' Society, Champaign, Ill., 1988.
75. N. Irving Sax, ed., *Dangerous Properties of Industrial Materials*, 5th ed., Van Nostrand Reinhold Company, New York, 1979, pp. 357, 683.
76. A. McRae and L. Whelchel, eds., *Toxic Substances Control Sourcebook*, Aspen Systems Corporation, 1978, p. 124.

77. R. L. Tatken and R. J. Lewis, eds., *Registry of Toxic Effects of Chemical Substances*, 1981–1982 ed., Vol. 2, U.S. Department of Health and Human Services, 1983, p. 8.

78. R. A. Reck, *J. Am. Oil Chem. Soc.* **61**, 187 (1984).

79. C. D. LaSusa, *J. Am. Oil Chem. Soc.* **61**, 184 (1984).

80. "Cosmetics & Toiletries," *Surfactant Encyclopedia* **104**, 67 (1989).

K. Visek
Akzo Chemicals Inc.

AMINES, LOWER ALIPHATIC AMINES

1. Introduction

Lower aliphatic amines are derivatives of ammonia with one, two, or all three of the hydrogen atoms replaced by alkyl groups of five carbons or fewer. Amines with higher alkyl groups are known as fatty amines. The names, chemical formulas, molecular weights, CAS Registry Numbers, and common names or abbreviations of commercially important amines are given in Table 1. Amines are toxic,

Table 1. **Commercial Lower Aliphatic Amines**

Alkylamine	CAS Registry Number	Molecular Formula	Molecular Weight	Synonym or common abbreviation
Ethylamines				
ethylamine	[75-04-7]	C_2H_7N	45.08	monomethylamine, aminomethane, MEA
diethylamine	[109-89-7]	$C_4H_{11}N$	73.14	N-ethylethanamine, DEA
triethylamine	[121-44-8]	$C_6H_{15}N$	101.19	N,N-diethylethanamine, TEA
n-Propylamines				
n-propylamine	[107-10-8]	C_3H_9N	59.11	mono-n-propylamine, 1-aminopropane,1-propanamine, MNPA
di-n-propyl-amine	[142-84-7]	$C_6H_{15}N$	101.19	N-propyl-1-propanamine,DNPA
tri-n-propyl-amine	[102-69-2]	$C_9H_{21}N$	143.27	N,N-dipropyl-1-propanamine, TNPA
iso-Propylamines				
isopropylamine	[75-31-0]	C_3H_9N	59.11	2-aminopropane, 2-propanamine-MIPA
diisopropyl-amine	[108-18-9]	$C_6H_{15}N$	101.19	N-(1-methylethyl)-2-propanamine, DIPA
Allylamines				
allylamine	[107-11-9]	C_3H_7N	57.10	monoallylamine, 2-propenamine,3-aminopropene

Table 1 (*Continued*)

Alkylamine	CAS Registry Number	Molecular Formula	Molecular Weight	Synonym or common abbreviation
diallylamine	[124-02-7]	$C_6H_{11}N$	97.16	N-2-propenyl-2-propenamine, di-2-propenylamine
triallylamine	[102-70-5]	$C_9H_{15}N$	137.23	N,N-di-2-propenyl-2-propenamine,tris(2-propenyl)amine
*n*Butylamines				
n-butylamine	[109-73-9]	$C_4H_{11}N$	73.14	mono-n-butylamine, 1-aminobutane, MNBA
di-n-butyl-amine	[111-92-2]	$C_8H_{19}N$	129.25	N-butyl-1-butanamine, DNBA
tri-n-butyl-amine	[102-82-9]	$C_{12}H_{27}N$	185.36	N,N-dibutyl-1-butanamine, TNBA
Isobutylamines				
isobutylamine	[78-81-9]	$C_4H_{11}N$	73.14	monoisobutylamine, 2-methyl-1-propanamine, 1-amino-2-methylpropane,1-aminobutane, MIBA
diisobutyl-amine	[110-96-3]	$C_8H_{19}N$	129.25	2-methyl-N-(2-methylpropyl)-1-propanamine, DIBA
triisobutyl-amine	[1116-40-1]	$C_{12}H_{27}N$	185.36	2-methyl-N,N-bis(2-methylpropyl)-1-propanamine, TIBA
sec-Butylamine				
sec-butylamine	[13952-84-6]	$C_4H_{11}N$	73.14	2-aminobutane, 2-butanamine, 1-methylpropanamine
tert-Butylamine				
tert-butylamine	[75-64-9]	$C_4H_{11}N$	73.14	2-methyl-2-propanamine, 2-aminoisobutane, 1,1-dimethylethanamine, trimethylaminomethane
Amylamines				
amylamine		$C_5H_{13}N$	87.17	mixture of 1-pentylamine and 2-methyl-1-butylamine
diamylamine		$C_{10}H_{23}N$	157.30	mixture of linear and branched isomers
triamylamine		$C_{15}H_{33}N$	227.44	mixture of linear and branched isomers
Mixed Amines				
dimethylethyl-amine	[598-56-1]	$C_4H_{11}N$	73.14	N,N-dimethylethanamine, N-ethyldimethylamine
dimethyl-n-propylamine	[926-63-6]	$C_5H_{13}N$	87.17	N,N-dimethyl-1-propanamine, propyldimethylamine
ethyl-nbutyl-amine	[13360-63-9]	$C_6H_{15}N$	101.19	N-ethyl-1-butanamine, butylethylamine, EBA
dimethyl-n-butylamine	[927-62-8]	$C_6H_{15}N$	101.19	N,N-dimethyl-1-butanamine, butyldimethylamine, DMBA

colorless gases or liquids, highly flammable, and have strong odors. Lower-molecular-weight amines are water soluble and are sold as aqueous solutions and in pure form. Amines react with water and acids to form alkylammonium compounds analogous to ammonia. The base strengths in water of the primary, secondary, and tertiary amines and ammonia are essentially the same, as shown by the equilibrium constants. Values of pK_a for some individual amines are given in Table 2.

$$RNH_2 + H_2O \rightleftharpoons RNH_3^+ + OH^-$$

Primary and secondary amines can also act as very weak acids ($K_a \sim 10^{-33}$). They react with acyl halides, anhydrides, and esters with rates depending on the size of the alkyl group(s). These amines react with carbon disulfide and carbon dioxide, to form alkyl amminium salts of dithiocarbamic and carbamic acid, respectively. Substituted ureas are obtained upon reaction with isocyanic acid and alkyl or aryl isocyanates. The corresponding thioureas are prepared by reaction with isothiocyanate. Primary amines react with nitrous acid to give highly toxic nitrosamines (see *N*-NITROSAMINES). The lower aliphatic amines are widely used in the manufacture of pharmaceutical, agricultural, textile, rubber, and plastic chemicals.

2. Properties

2.1. Physical Properties.   Table 2 lists the physical and chemical properties of the commercially important alkylamines (1–5). Thermodynamic data are available only for the lower alkylamines and are mainly estimates based on a few experimental determinations (6–7). Recently, quantum mechanics-based computational methods have been used to calculate equilibrium constants for amines systems (8–9). This methodology may find practical application as the product selectivities from many manufacturing processes appear to be limited by thermodynamic equilibria.

2.2. Chemical Properties.   The chemistry of the lower aliphatic amines is dominated by their basicity and their nucleophilic character, which result from the presence of an unshared pair of electrons on the nitrogen atom. Due to their basicity, amines are often used as neutralization agents or pH adjusters. For example, tertiary amines, such as triethylamine may be used to neutralize inorganic acids liberated during alkylation reactions. The resulting trialkylaminium salt is water-soluble and therefore, is easily separated from the organic phase (10).

Amines react with a variety of substrates such as epoxides, aldehydes and ketones, alkyl halides, carboxylic acids/halides/esters/anhydrides, and carbon disulfide to produce products used in agricultural, pharmaceutical, textile, polymer, and rubber chemical applications. As shown in the following, respective products of these reactions are amines containing additional functionality (alkanolamines), amines containing additional alkyl groups, quaternary

Table 2. **Alkylamine Physical and Chemical Properties**

Alkylamine	Mp, °C	Bp, °C	Vapor pressure at 20°C, kPa	Density, gm/ml at 20°C	Refractive index, 20°C	Water solubility, g/100 g H₂O at 20°C	p-K_a at 25°C
Ethylamines							
ethylamine	−81	16.6	116	0.68	1.3663	complete[a]	10.63
diethylamine	−50	55.9	25.9	0.700	1.3864	complete	11.09
triethylamine	−114	88.2	7.2	0.73	1.3980^{25}	11.24	10.72
n-Propylamines							
n-propylamine	−83	47.4	33.9	0.72	1.3882	complete	10.57
di-n-propylamine	−47	109.1	2.8	0.74	1.4043	5.30	10.91
tri-n-propylamine	−93	156.9	0.3	0.76	1.4160	0.022	10.66
Isopropylamines							
isopropylamine	−101	32.0	63.7	0.69	1.3771^{25}	complete	10.64
diisopropylamine	−96.3	83.0	8	0.72	1.3924	10.32 (30°C)	11.21
Allylamines							
allylamine	−88.2	55–58		0.76	1.4185	complete	9.52
diallylamine	−88	111–112		0.79	1.4405	11.64	9.29
triallylamine		155		0.81 (14°C)	1.4486	0.29	8.31
n-Butylamines							
n-butylamine	−47	77.1	9.6	0.74	1.3992^{25}	complete	10.64
di-n-butylamine	−61.9	156.2	0.3	0.76	1.4177	0.57	11.25
tri-n-butylamine	−70	212.6		0.78	1.4283	0.096	
Isobutylamines							
isobutylamine	−84.6	67.6	13.3	0.73	1.3972	complete	10.41
diisobutylamine	−77	140	1.3	0.74	1.4081	0.27	10.65
isobutylamine	<−24	183		0.77	1.4252	trace1	
sec-Butylamine	−72	63	20	0.73	1.3928	complete	10.56
tert-Butylamine	−70.0	44.4		0.69	1.3788	complete	10.68
Amylamines							
amylamine	−55	104		0.75	1.4110		
diamylamine				0.77	1.4258		
triamylamine				0.78	1.4360		
Mixed Amines							
dimethylethylamine	−140	36–38	8.1	0.67	1.3720	complete	9.25
dimethyl-n-propylamine		65		0.71	1.3869		
ethyl-n-butylamine		111	2.4	0.74	1.4050	5.06	10.96
dimethyl-n-butylamine		95		0.72	1.3970	3.52	10.1

[a] Under pressure.

ammonium salts, amides, and dithiocarbamates (11).

$$R_2NH \ + \ \overset{O}{\underset{\triangle}{\bigtriangleup}} \ \longrightarrow \ R_2NCH_2CH_2OH$$

$$R_2NH \ + \ R'-\overset{O}{\overset{\|}{C}}-R'' \ \xrightarrow[\text{Catalyst}]{H_2} \ R'-\overset{R''}{\overset{|}{CH}}-NR_2 \ + \ H_2O$$

$$R_3N \ + \ R'Cl \ \longrightarrow \ R'R_3N^+Cl^-$$

$$R_2NH \ + \ R'-\overset{O}{\overset{\|}{C}}-X \ \longrightarrow \ R'-\overset{O}{\overset{\|}{C}}-NR_2 \ + \ HX$$

$$2\,R_2NH \ + \ CS_2 \ \xrightarrow{M^{2+}} \ \left[R_2N-\overset{S}{\overset{\|}{C}}-S^- \right]_2 M^{2+}$$

where R, R′= H or alkyl, X=RCO$_2$, OR, Cl, or OH. Reaction of primary or secondary amines with carbon dioxide produces amine carbamate salts under anhydrous conditions or amine carbonate salts in the presence of water:

$$2\,R_2NH \ + \ CO_2 \ \longrightarrow \ R_2N-\overset{O}{\overset{\|}{C}}-O^- \ \overset{+}{N}R_2H_2$$

$$2\,R_2NH \ + \ CO_2 \ \xrightarrow{H_2O} \ R_2H_2\overset{+}{N} \ ^-O-\overset{O}{\overset{\|}{C}}-O^- \ \overset{+}{N}R_2H_2$$

The white solid material often observed on the caps of bottles containing amines typically results from amine carbamate formation due to reaction between the amine and carbon dioxide in the air. While tertiary amines do not form carbamates, they can react with carbon dioxide in aqueous solution to produce carbonate salts.

Reaction of secondary amines with nitrous acid or with nitrogen oxides present in air produces nitrosamines (11, 12), which are probable carcinogens. Nitrosamine formation is a concern in several applications that use secondary amines as raw materials. In the rubber industry, use of accelerators made from primary amines or from hindered secondary amines has been investigated as a means of eliminating or reducing formation of nitrosamines (13).

$$R_2NH \ \xrightarrow[\text{or nitrogen oxides}]{HNO_2} \ R_2N-N=O$$

Nitrosamines also can be formed from tertiary amines in the presence of excess nitrous acid. An alkyl group is cleaved from the amine, and the secondary nitrosamine is obtained along with aldehyde or ketone. However, formation of nitrosamines as a result of contact of tertiary amines with nitrogen oxides contained in air is unlikely to occur.

$$R_3N \ \xrightarrow{HNO_2} \ R_2N-N=O \ + \ R_2C=O \ + \ N_2O$$

Primary amines do not form nitrosamines. Reaction of primary amines with nitrous acid leads to formation of unstable diazonium salts which decompose to olefins, alcohols (in the presence of water), and nitrogen.

$$RNH_2 \xrightarrow[H_2O/HX]{HNO_2} \left[R\overset{+}{N}\equiv N \right] X^- \longrightarrow N_2 + \text{olefins} + \text{alcohols}$$

Nitrosamines typically are not formed as by-products during manufacture of amines. Storage and use of secondary amines under a nitrogen atmosphere is an effective method of preventing nitrosamine formation.

Oxidation of amines with hydrogen peroxide or peracids gives a mixture of products from primary amines including hydroxamic acids and oximes, hydroxylamines from secondary amines via amine oxide intermediates, and amine oxides from tertiary amines (12).

$$R_2C-NH_2 \xrightarrow{H_2O_2} R_2C=NOH + R-\overset{\overset{O}{\|}}{C}-NHOH$$

$$R_2NH \xrightarrow{H_2O_2} R_2NOH$$

$$R_3N \xrightarrow{H_2O_2} R_3\overset{+}{N}-O^-$$

Use of other oxidizing agents such as permanganate or manganese dioxide results in removal of hydrogen to give imines, enamines, or nitriles.

Alkylamines are corrosive to copper, copper-containing alloys (brass), aluminum, zinc, zinc alloys, and galvanized surfaces. Aqueous solutions of alkylamines slowly etch glass as a consequence of the basic properties of the amines in water. Carbon or stainless steel vessels and piping have been used satisfactorily for handling alkylamines.

3. Manufacture

Lower aliphatic amines can be prepared by a variety of methods, and from many different types of raw materials. By far the largest commercial applications involve the reaction of alcohol with ammonia to form the corresponding amines. Other methods are employed depending on the particular amine desired, raw material availability, plant economics, and the ability to sell coproducts. The following manufacturing methods are used commercially to produce the lower alkylamines:

Method 1. Alcohol amination: amination of an alcohol over a metal catalyst under reducing conditions or over a solid acid catalyst at high temperature.

Method 2. Reductive alkylation: reaction of an amine or ammonia and hydrogen with an aldehyde or ketone over a hydrogenation catalyst.

Method 3. Ritter reaction: reaction of hydrogen cyanide with an olefin in an acidic medium to produce a primary amine.

Table 3. **Manufacturing Data for Aliphatic Amines**

Company and Plant Location	Products	Capacity, t/yr	Method
United States Producers			
Air Products and Chemicals, Inc., Pace, Fla. & St. Gabriel, La.	C_2–C_4	137,000	1
Celanese, Bucks, Ala. & Portsmouth, Va.	C_2–C_4	25,500	1,4
Elf Atochem, Riverview, MIch.	C_2–C_4	13,500	1
Sterling Chemicals, Texas City, Tex.	*t*-butyl	10,000	3
		186,000	
Other American Producers			
Air Products and Chemicals, Inc., Bahia, Brazil	C_2–C_4	41,000	1
Petramin, Iraputo, Mexico	C_2–C_3	3,000	1
		44,000	
West European Producers			
BASF, Antwerp, Belgium & Ludwigshafen, Germany	C_2–C_4, *t*-butyl	75,000	1,5
Celanese, Oberhausen, Germany	C_3–C_4	20,000	1,4
Elf Atochem, La Chambre, France	C_2–C_3	25,000	1
		120,000	
Far East			
Mitsubishi Gas Chemical, Niigata, Japan	ethyl	3,000	1
Mitsubishi Rayon, Otake, Japan	*t*-butyl	2,300	3
Daicel Chemical Industries, Japan	C_2–C_3	12,000	1
Koei Chemical, Sodegaura, Japan	C_3–C_4	2,600	1
Sumitomo Chemical, Niihama, Japan	*t*-butyl	500	3
Shandong Zibo, P.R.C.	C_2	3,000	1
Xuanhau, P.R C.	C_2	3,000	1
Various, P.R.C.	C_2–C_3	10,000	1
		36,400	
Rest of World			
African Amines, South Africa	C_2–C_3	5,000	1
Alkylamines, Kurkumbh, India	CN	10,000	1
Chemicoplex, Borzesti, Rumania	C_2–C_3	6,000	1
Moravske Chemicke, Zavody, Czech.	C_3-C_4	3,300	1
		24,300	

Method 4. Nitrile reduction: reaction of a nitrile with hydrogen over a hydrogenation catalyst.

Method 5. Olefin amination: reaction of an olefin with ammonia.

Method 6. Alkyl halide amination: reaction of ammonia or alkylamine with an alkyl halide.

Table 3 gives plant and capacity information for these methods (14).

3.1. Alcohol Amination (Method 1). Amination of alcohols with ammonia typically is conducted over a fixed catalyst bed at elevated temperature and pressure, and produces a mixture of mono-, di-, and tri-alkylamines (15). The reaction section consists of feed systems, vaporizers, and/or preheaters, which pass a liquid or gaseous feed mixture over the catalyst bed in the desired ratio, temperature, and pressure. The amination reaction may be catalyzed either by solid acid catalysts, such as metal oxides and zeolites (16–18) or by supported metal catalysts. Operating conditions for the acid-catalyzed reaction are maintained in the range from 300–500°C and 790–3550 kPa (100–500 psig) at a gas hourly space velocity between 500–1500 vol/vol per hour. The ammonia to alcohol mole ratio varies from 2:1 to 6:1 depending on the amine desired as shown in Figure 1.

The metal catalysts generally contain cobalt, nickel, copper, chromium, zirconium, or mixtures thereof (19–25). Metal-catalyzed alcohol amination reactions are conducted in the presence of hydrogen to maintain catalyst activity (26). The mole ratio of hydrogen is typically in the range of 1 to 2.5:1 with respect to the alcohol. Operating conditions are maintained in the range of 130–250°C and 790–22,000 kPa (100–3200 psig). Yields are usually in excess of 95%. Most commercial processes employ metal catalysts as increased yields are obtained relative to acid-catalyzed processes. As noted, the acid catalysts require higher reaction temperatures for amination, and side reactions, such as alcohol dehydration (27), are likely to occur.

Alcohol amination reactions are described by a network of two general types of reaction:

1. Sequential substitution reactions, which convert alcohols into a family of primary, secondary, and tertiary amines.

$$ROH + NH_3 \longrightarrow RNH_2 + H_2O \qquad \text{primary amine}$$

$$ROH + RNH_2 \longrightarrow R_2NH + H_2O \qquad \text{secondary amine}$$

$$ROH + R_2NH \longrightarrow R_3N + H_2O \qquad \text{tertiary amine}$$

2. Reforming reactions which equilibrate the alkylamines.

$$2\,RNH_2 \rightleftharpoons R_2NH + NH_3$$

$$RNH_2 + R_2NH \rightleftharpoons R_3N + NH_3$$

$$2\,R_2NH \rightleftharpoons RNH_2 + R_3N$$

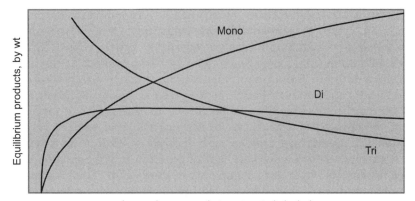

Fig. 1. Alkylamine products vs increasing ammonia.

As the equilibrium product selectivities may not reflect market requirements, process schemes have been developed to increase the flexibility of the process (28) and to selectively produce one or more of the amines (29,30).

Water is a by-product of the amination reactions from alcohols. The mixture of ammonia, water, unconverted alcohol, and amines is continuously separated by distillation. Ammonia, unconverted alcohol, and any amines produced in excess of anticipated sales are recycled to the reactor section. A schematic depicting the typical process flows is presented as Figure 2. Note that the separation requirements vary from process to process depending on the physical properties of the specific alkylamines involved, their product specifications, and the by-products involved. Processing variations might include extractive distillation and/or liquid–liquid separation, particularly for secondary and tertiary amines.

Reforming of the amines produced in excess of anticipated sales may be practiced commercially using both types of catalyst.

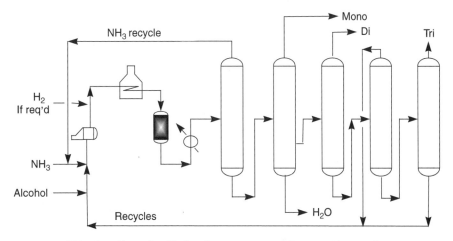

Fig. 2. Generic alkylamines reactor and separation train.

3.2. Reductive Alkylation (Method 2).   Ammonia or a primary or secondary amine, hydrogen, and an aldehyde orketone form aliphatic amines when they react over a hydrogenation catalyst, eg, nickel, cobalt, copper chromite, platinum, or palladium (31). These reactions may be carried out at conditions similar to Method 1 and in similar equipment. Liquid-phase stirred-tank equipment operated either continuous or batch mode is also an option. A larger exothermic heat of reaction limits fixed-bed applications to those with a large excess of ammonia, which acts as a heat sink to restrict the total bed temperature rise. Alternatively, a multistage reactor with interstage cooling or a multitubular reactor with cooling to remove the heat of reaction may be employed, but this increases the cost and operating complexity of the reactor. The heat removal problem is more readily controlled in stirred tank systems with cooling coils or systems that pump the reaction mass through external heat exchangers, eg, loop reactors. Typically these reactions take place at lower temperatures than alcohol amination, and reforming reactions compete less favorably with the reductive alkylation. Consequently, it is quite often possible to control the selectivity to a single product through the judicious choice of catalyst and of the mole ratio of the starting aldehyde or ketone to the amine or ammonia 32, 33. Further it is possible to produce amines with mixed alkyl groups by this method.

$$
\begin{matrix} R_1 \\ \diagdown \\ C=O \\ \diagup \\ R_2 \end{matrix} \ + \ \begin{matrix} R_3 \\ \diagup \\ HN \\ \diagdown \\ R_4 \end{matrix} \ + \ H_2 \ \longrightarrow \ \begin{matrix} R_1 \quad R_3 \\ \diagdown \ \diagup \\ H-C-N \\ \diagup \ \diagdown \\ R_2 \quad R_4 \end{matrix} \ + \ H_2O
$$

R_n = alkyl or H. Operating conditions vary radically depending on the type of equipment selected but typically temperatures used are in the range of 50 to 180°C and pressures of 446–3550 kPa (50–500 psig) are sufficient (34).

3.3. Ritter Reaction. (Method 3).   A small but important class of amines are manufactured by the Ritter reaction (35). These are the amines in which the nitrogen atom is adjacent to a tertiary alkyl group. In the Ritter reaction a substituted olefin such as isobutylene reacts with hydrogen cyanide under acidic conditions. The resulting formamide is then hydrolyzed to the parent primary amine. Typically sulfuric acid is used in this transformation of an olefin to an amine. Stoichiometric quantities of sulfate and formate salts are produced along with the desired amine.

$$
\begin{matrix} H_3C \\ \diagdown \\ C=CH_2 \\ \diagup \\ H_3C \end{matrix} \ + \ HCN \ + \ H_2SO_4 \ + \ 3\,NaOH \ \longrightarrow \ \begin{matrix} H_3C \\ \diagdown \\ H_3C-C-NH_2 \\ \diagup \\ H_3C \end{matrix} \ + \ Na_2SO_4 \ + \ H_2O \ + \ HCO_2Na
$$

The only low-molecular-weight alkylamine produced by this method commercially is *t*-butylamine.

3.4. Nitrile Reduction (Method 4).   The reduction of nitriles with hydrogen to simple alkylamines is another technology that is practiced commercially (36,37). As with Method 2, both continuous packed-bed reactor systems designed for removal of the heat of reaction or batch stirred tank or loop reactor systems may be used. Catalysts for this transformation are nickel,cobalt, platinum, palladium, and rhodium. Again the operating conditions vary widely, depending on the type of equipment; but temperatures and pressures are generally in the range, 50–150°C, and 446–73,900 kPa (50–2000 psig), respectively. Selectivity to primary amine is normally controlled by introducing ammonia as a diluent and nickel or cobalt as catalyst (38). Use of a lithium hydroxide modified Raney cobalt catalyst to favor primary amine formation recently has been described (39)

$$R-C\equiv N \xrightarrow{\text{H}_2/\text{catalyst}} RCH_2NH_2 \;+\; (RCH_2)_2NH \;+\; (RCH_2)_3N \;+\; NH_3$$

Secondary and tertiary amines are preferentially produced when rhodium or palladium are chosen as catalyst (40). As in Method 2, reforming reactions do not normally compete with the hydrogenation reaction, and high selectivities to the desired product are possible.

3.5. Olefin Amination (Method 5). The most recent technology developed for the production of lower alkylamines is olefin amination (41,42). This zeolite-catalyzed reaction of ammonia with an olefin is practiced in a packed-bed reactor system. t-Butylamine currently is the only amine manufactured on a commercial scale via this process (43).

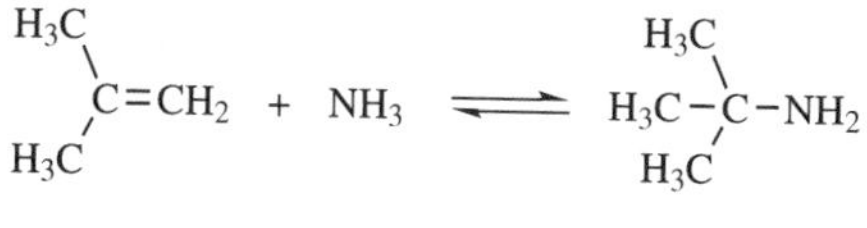

$$\begin{array}{c} H_3C \\ \diagdown \\ \diagup \quad C=CH_2 \;+\; NH_3 \;\rightleftharpoons\; H_3C-\overset{\textstyle H_3C}{\underset{\textstyle H_3C}{C}}-NH_2 \\ H_3C \end{array}$$

A number of zeolitic materials have been claimed to catalyze this reaction, and reaction temperatures are on the order of 200–350°C with pressures as high as 30000 kPa (4350 psi) reported. Conversion of the olefin to the amine typically is low, and recycle of the unconverted starting materials is necessary to provide an economical process.

3.6. Alkyl Halide Amination (Method 6). The oldest technology for producing amines is the reaction of ammonia with an alkyl halide (11,12). This method is still of commercial importance for the manufacture of allylamines (44) and select mixed alkylamines. Allylamines are not readily made by the other methods mentioned here primarily because either the double bond in the allylamine becomes saturated in methods involving hydrogen or the severe reaction conditions result in cyclization or polymerization by-products (45). Alkyl halide amination occurs under very mild conditions, typically with the addition of aqueous alkali to neutralize the hydrogen halide coproduced by the reaction. As in the Ritter reaction, one mole of salt is produced for every mole of alkyl halide consumed. The resulting waste disposal problem detracts problem detracts from the general utility of both this method and the Ritter reaction.

$$RX + NH_3 \xrightarrow{\text{NaOH}} RNH_2 + R_2NH + R_3N + NaCl$$

4. Shipment and Handling

All the lower alkylamines are classified as either flammable or combustible liquids at normal temperatures and pressure with the exception of

Table 4. **ACGIH Threshold Limit Values**

Alkylamine	TWA, ppm	STEL, ppm
ethylamine	5	15
diethylamine	5	15
triethylamine	1	3
isopropylamine	5	10
diisopropylamine	5	

monoethylamine, which is a flammable gas under these conditions. Anhydrous monoethylamine therefore is shipped under pressure in bulk tank trucks and railcars. Both monoethylamine and monoisopropylamine are available as 70% solutions in water and are shipped in this form as flammable liquids. The liquid amines are available in drums and isocontainers as well as tank trucks and railcars.

The lower alkylamines are toxic and have strong odors. Labeling and packaging of amines must conform with Department of Transportation (DOT) requirements. Amine shipments are regulated by the Coast Guard, the DOT, the International Air Transport Association (IATA), and in some cases, the Drug Enforcement Administration (DEA).

5. Health and Safety Factors

The lower alkylamines all have strong fishy or ammoniacal odors. The American Conference of Governmental Industrial Hygienists has established exposure limits for some of these substances. The values are shown in Table 4 (46). As a general practice, exposure to all alkylamines should be limited, and therefore, they should be handled only in well-ventilated areas. A full face shield with goggles underneath, neoprene, nitrile, or butyl rubber gloves and impervious clothing should be worn when working with alkylamines.

The lower alkylamines are toxic by ingestion, inhalation, and/or skin absorption. Alkylamine vapors in low concentrations can cause lacrimation, conjunctivitis, and corneal edema when absorbed into the tissue of the eye from the atmosphere. Inhalation of vapors may cause irritation in the respiratory tract. Contact of undiluted product with the eyes or skin quickly causes severe irritation and pain and may cause burns, necrosis, and permanent injury. Repeated exposure may result in adverse respiratory, eye, and/or skin effects. If contact with the eyes or skin occurs, the affected area should be washed with water for at least 15 min. If these products are inhaled, the patient should be moved to fresh air and assisted with respiration if required.

6. Economic Aspects, Specifications, and Uses

Table 5 provides a list of major applications for some of the lower alkylamines. As shown in the table, these products are primarily used as intermediates for pesticides, rubber chemicals, and catalysts. Total U.S. consumption of C_2-C_5 amines

Table 5. **Products Manufactured Using Alkylamines**

Chemical name	CAS Registry Number	Trade Name/Common Name	Use
From Monomethylamine			
2-chloro-4-ethylamino-6-isopropylamino-1,3,5-triazine	[1912-24-9]	Atrazine	herbicide
2-chloro-4-(1-cyano-1-methylethylamino)-6-ethylamino-1,3,5-triazine	[21725-46-2]	Bladex/Cyanazine	herbicide
2-(*tert*-butylamino)-4-chloro-6-ethylamino-1,3,5-triazine	[5915-41-3]	Terbuthylazine	herbicide
From Diethylamine			
2-diethylaminoethanol	[100-37-8]		corrosion inhibitor
N,N-diethyldithiocarbamate salts			
triethylamine salt	[2391-78-8]		rubber accelerator
diethylamine salt	[1518-58-7]		
sodium salt	[148-18-5]		
zinc salt	[14324-55-1]		
tetraethylthiuram disulfide	[97-77-8]		rubber accelerator
N,N-diethyl-*m*-toluamide	[134-62-3]	DEET	insect repellent
From Triethylamine			
triethylamine	[121-44-8]		catalyst for foundry molds, phenol-formaldehyde resins; extraction solvent; pH adjuster
diethylhydroxylamine	[3710-84-7]	DEHA	radical scavenger
tetraethylammonium bromide	[71-91-0]		phase transfer catalyst
From Monoisopropylamine			
N-phosphonomethylglycine, monoisopropylamine salt	[38641-94-0]	Roundup/Glyphosate	herbicide
2-chloro-4-ethylamino-6-isopropylamino-1,3,5-triazine	[1912-24-9]	Atrazine	herbicide
dodecylbenzenesulfonate, monoisopropylamine salt	[26264-05-1]		dry-cleaning detergent, fabric finish
From Diisopropylamine			
S-2,3,3-trichloroallyl diisopropyl-(thiocarbamate)	[2303-17-5]	Far-Go/Triallate	herbicide
N,N-diisopropylethylamine	[7087-68-5]	Hünig's Base	proton scavenger, catalyst

Table 5. **Products Manufactured Using Alkylamines**

Chemical name	CAS Registry Number	Trade Name/Common Name	Use
From Di-n-propylamine			
α,α,α-trifluoro-2,6-dinitro-N,N-dipropyl-p-tolui-dine	[1582-09-8]	Treflan/Trifluralin	herbicide
From Monobutylamine			
N-butylbenzenesul-fonamide	[3622-84-2]		plasticizer
butyl isocyanate	[111-36-4]		intermediate in polyurethane manufacture
From Dibutylamine			
dibutylamine	[111-92-2]		corrosion inhibitor
dibutyldithiocarbamate salts			rubber accelerators
sodium salt	[136-30-1]	NaDBC	
zinc salt	[136-23-2]	ZDBC	
2,3-dihydro-2,2-dimethyl-7-benzofuranyl[(dibutyl-amino)thio]methylcar-bamate	[55285-14-8]	Carbosulfan	insecticide
tetrabutylurea	[4559-86-8]		processing aid
From Tributylamine			
tributylamine	[102-82-9]		catalyst, stabilizer
tetrabutylammonium bromide	[1643-19-2]		phase transfer catalyst
From tert-Butylamine			
N-tert-butyl-2-benzothia-zolesulfenamide	[95-31-8]	TBTS	rubber accelerator
2-(tert-butylamino)-4-chloro-6-ethylamino-s-triazine	[5915-41-3]	Terbuthylazine	herbicide
From Diamylamine			
diamyldithiocarbamic acid, zinc salt	[15337-18-5]	ZDAC	rubber accelerator

in 1995 was estimated to be 88,000 mt (47). The percent consumption by carbon chain length and by end use are depicted in Figures 3 and 4. As shown in Figure 3, C_2 and C_3 amines have the greatest market demand of the lower alkylamines. The high demand for C_3 amines is driven by the use of monoisopropylamine in Monsanto's Roundup® herbicide. Worldwide manufacturers of alkylamines, their plant locations, and plant capacities are listed in Table 3. The information provided in this table is based on market studies and capacity announcements published in the trade literature (14). Sales specifications (48) and list prices (49) of some alkylamines products are shown in Table 6. Detailed product specifications can be obtained from the manufacturers.

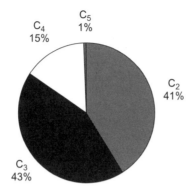

Fig. 3. Percent U.S. consumption of C_2-C_5 alkylamines, 1995.

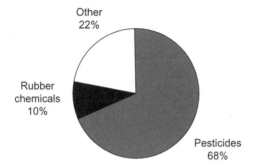

Fig. 4. Percent U.S. consumption of C_2-C_5 alkylamines by end use, 1995.

Table 6. **Alkylamines Specifications and Economic Data**

Alkylamine	Assay, wt %	Other Amines, wt %	Water, wt %	2000 U.S. price, $/kg
ethylamine	99.5	0.4	0.1	3.13
ethylamine, 70% solution	70.0–71.0	0.4	balance	3.31
diethylamine	99.2	0.4	0.3	3.22
triethylamine	99.5	0.4	0.1	3.31
n-propylamine	99.0	0.5	0.5	3.02
di-n-propylamine	99.0	0.8	0.2	3.26
tri-n-propylamine	98.5	1.2	0.3	4.17
isopropylamine	99.5	0.4	0.1	2.51
isopropylamine, 70% solution	70.0–72.0	0.3	balance	2.67
diisopropylamine	99.5	0.2	0.1	3.55
n-butylamine	99.5	0.4	0.1	3.42
di-n-butylamine	99.0	0.6	0.2	3.55
tri-n-butylamine	98.0	1.7	0.3	4.14
isobutylamine	99.0	0.5	0.5	8.33
diisobutylamine	98.5	1.3	0.2	2.89

BIBLIOGRAPHY

"Ethylamines" in Vol. 5, pp. 876–879, and "Propylamines" in Vol. 11, pp. 190–192, by R. H. Goshorn, Sharples Chemicals, Inc.: "Butylamines" in Vol. 2, pp. 680–683, and "Amylamines" in Vol. 1, pp. 849–851, by C. K. Hunt, Sharples Chemicals, Inc.; "Amines, Lower Aliphatic" in ECt 2nd ed., Vol. 2, pp. 116–127; "Amines, Lower Aliphatic" in ECT 3rd ed., Vol. 2, pp. 272–283, by A. Schweizer, R. Fowlkes, J. McMakin, and T. Whyte, Air Products and Chemicals, Inc., "Amines (Lower Aliphatic)" in ECT 4th ed., Vol. 2, pp. 369–386, by M. Turcotte and T. Johnson, Air Products and Chemicals, Inc.; "Amines, Lower Aliphatic Amines" in *ECT* (online), posting date: December 4, 2000, by Michael G. Turcotte, and Thomas A. Johnson, Air Products and Chemical, Inc.

CITED PUBLICATIONS

1. Air Products and Chemicals, Inc., "Amines from Air Products", 1996.
2. J. A. Dean, ed. *Lange's Handbook of Chemistry* McGraw-Hill, Inc., New York, 1992.
3. R. M. Stephenson, *J. Chem. Eng. Data* **38**, 625 (1993).
4. D. D. Perrin, *Dissociation Constants of Organic Bases in Aqueous Solution*, Butterworth, Washington, D.C., 1965.
5. K. N. Campbell, F. C. Fatora, Jr., and B. K. Campbell, *J. Org. Chem.* **17**, 1141 (1952).
6. D. R. Stull, E. F. Westrum, and G. C. Sinke, *The Chemical Thermodynamics of Organic Compounds*, John Wiley & Sons, Inc., New York, 1969, pp. 462, 466–467.
7. J. Pasek, P. Kondelik, and P. Richter, *Ind. Eng. Chem. Prod. Res. Dev.* **11**, 333 (1972).
8. H. Cheng, V. S. Parekh, J. W. Mitchell, and K. S. Hayes, *AIChE J.* **43**, 2153 (1997).
9. H. Cheng, V. S. Parekh, J. W. Mitchell, and K. S. Hayes, *J. Phys. Chem. A* **102**, 1568 (1998).
10. G. Solomons and C. Fryhle, *Organic Chemistry*, 7th ed., John Wiley & Sons, Inc., New York, 2000, pp. 951952.
11. B. C. Challis and A. R. Butler in S. Patai, ed., *The Chemistry of the Amino Group*, Interscience, London, 1968, pp. 277–347.
12. P. A. S. Smith, *The Chemistry of Open-Chain Organic Nitrogen Compounds: Vol. 1*, W. A. Benjamin, Inc., New York, 1965.
13. B. Spiegelhalder and C.-D. Wacker in R. N. Loeppky and C. J. Michejda, ed., *Nitrosamines and Related N-Nitroso Compounds: Chemistry and Biochemistry*, American Chemical Society, Washington, 1994, pp. 42–51.
14. Capacities are based on announcements published in the trade literature, including: *Chemical Marketing Reporter, Chemical Week, China Chemical Reporter, Financial Express, Frankfurter Allgemeine, Manufacturing Chemist, Japan Chemical Week*, and the Chem Expo website at http://www.chemexpo.com/news
15. T. Mallat and A. Baiker in G. Ertl, H. Knözinger, and J. Weitkamp, ed., *Handbook of Heterogeneous Catalysis*, Vol. **5**, VCH, Weinheim, 1997, pp. 2334–2348.
16. U.S. Pat. 3,384,667 (May 21, 1968), L. A. Hamilton (to Mobil Oil Corporation).
17. U.S. Pat. 4,082,805 (Apr. 4, 1978), W. Kaeding (to Mobil Oil Corporation).
18. U.S. Pat. 4,913,234 (Apr. 17, 1990), M. Deeba (to Air Products and Chemicals, Inc.).
19. U.S. Pat. 4,314,084 (Feb. 2, 1982), J. V. Martinez de Pinillos and R. L. Fowlkes (to Air Products and Chemicals, Inc.).
20. U.S. Pat. 5,002,922 (Mar. 26, 1991), M. Irgana, J. Schossig, W. Schroeder, and S. Winderl (to BASF Aktiengesellschaft).
21. U.S. Pat. 5,530,127 (Jun. 25, 1996), W. Reif, L. Franz, P. Stops, V. Menger, R. Becker, R. Kummer, and S. Winderl (to BASF Aktiengesellschaft).

22. U.S. Pat. 5,932,679 (Aug. 3, 1999), G. A. Vedage, K. S. Hayes, M. Leeaphon, and J. N. Armor (to Air Products and Chemicals, Inc.).
23. U.S. Pat. 4,255,357 (Mar. 10, 1981), D. A. Gardner and R. T. Clark (to Pennwalt Corporation).
24. G. Sewell, C. O'Connor, and E. van Steen, *Appl. Catal. A* **125**, 99 (1995).
25. G. S. Sewell, C. T. O'Connor, and E. van Steen, *J. Catal.* **167**, 513 (1997).
26. A. Baiker and J. Kijenski, *Catal. Rev. Sci. Eng.* **27**, 653 (1985).
27. V. A. Veefkind and J. A. Lercher, *J. Catal.* **180**, 258 (1998).
28. Eur. Pat. EP 379,939 (Aug. 1, 1990), B. K. Heft, C. A. Cooper, R. L. Fowlkes, and L. S. Forester, Jr. (to Air Products and Chemicals, Inc.).
29. U.S. Pat. 4,851,578 (Jul. 25, 1989), R. Fischer, and H. Mueller (to BASF Aktiengesesellschaft).
30. U.S. Pat. 4,851,580 (Jul. 25, 1989), H. Mueller, R. Fischer, G. Jeschek, and W. Schoenleben (to BASF Aktiengesellschaft).
31. M. Freifelder, *Catalytic Hydrogenation in Organic Synthesis*, John Wiley & Sons, New York, 1978, pp. 90–106.
32. U.S. Pat. 3,346,640 (Oct. 10, 1967), A. Guyer and P. Guyer (to Lonza Ltd.).
33. Jap. Pat. 54039006 (Mar. 24, 1979), K. Yasuda, T. Itokazu, and Y. Kurata (to Daicel Ltd.).
34. W. S. Emerson, in R. Adams, ed., *Organic Reactions*, Vol. **4**, John Wiley & Sons, Inc., New York, 1948, pp. 174–255.
35. L. I. Krimen and D. J. Cota, in W. G. Dauben, ed., *Organic Reactions*, Vol. **17**, John Wiley & Sons, Inc., New York, 1969, pp. 213–325.
36. J. Volf and J. Pasek in L. Cerveny, ed., *Studies in Surface Science and Catalysis, Vol. 27: Catalytic Hydrogenation*, Elsevier, Amsterdam, 1986, pp. 105–144.
37. C. De Bellefon and P. Fouilloux, *Catal. Rev. Sci. Eng.* **36**, 459 (1994).
38. U.S. Pat. 4,248,799 (Feb. 3, 1981), C. A. Drake (to Phillips Petroleum Company).
39. U.S. Pat. 5,869,653 (Feb. 9, 1999), T. A. Johnson (to Air Products and Chemicals, Inc.).
40. P. N. Rylander, *Catalytic Hydrogenation over Platinum Metals*, Academic Press, New York, 1967, 209–211.
41. U.S. Pat. 4,929,758 (May 29, 1990), V. Taglieber, W. Hoelderich, R. Kummer, W. D. Mross, and G. Saladin (to BASF A-G).
42. U.S. Pat. 4,929,759 (May 29, 1990), V. Taglieber, W. Hoelderich, R. Kummer, W. D. Mross, and G. Saladin (to BASF A-G).
43. *Chemical Week* **154** (1), 23 (1994).
44. U.S. Pat. 2,216,548 (Oct. 1, 1940), W. Converse (to Shell Development Company).
45. W. E. Carroll, R. J. Daughenbaugh, D. D. Dixon, M. E. Ford, and M. Deeba, *J. Mol. Catalysis*, **44**, 213 (1988).
46. ACGIH, 1999 TLVs and BEIs, American Conference of Governmental Industrial Hygienists, Cincinnati, OH, 1999.
47. W. K. Johnson, A. Leder, and Y. Yoshida, CEH Marketing Research Report 611.5030A, Alkylamines (C1-C6), 1997.
48. http://www.airproducts.com/amines
49. Air Products and Chemicals, Inc., List Prices, 1 June 2000.

MICHAEL G. TURCOTTE
Air Products and Chemicals, Inc.
KATHRYN S. HAYES
Air Products and Chemicals, Inc.

AMINO ACIDS

1. Introduction

Amino acids are the main components of proteins. Approximately twenty amino acids are common constituents of proteins (1) and are called protein amino acids, or primary protein amino acids because they are found in proteins as they emerge from the ribosome in the translation process of protein synthesis (2), or natural amino acids. In 1820 the simplest amino acid, glycine, was isolated from gelatin (3); the most recently isolated, of nutritional importance, is L-threonine which was found (4) in 1935 to be a growth factor of rats. The history of the discoveries of the amino acids has been reviewed (5,6).

Hydroxylated amino acids (eg, 4-hydroxyproline, 5-hydroxylysine) and N-methylated amino acids (eg, N-methylhistidine) are obtained by the acid hydrolysis of proteins. γ-Carboxyglutamic acid occurs as a component of some sections of protein molecules; it decarboxylates spontaneously to L-glutamate at low pH. These examples are formed upon the nontranslational modification of protein and are often called secondary protein amino acids (1,6).

The presence of many nonprotein amino acids has been reported in various living metabolites, such as in antibiotics, some other microbial products, and in nonproteinaceous substances of animals and plants (7). Plant amino acids (8) and seleno amino acids (9) have been reviewed.

The general formula of an α-amino acid may be written:

$$R-\overset{*}{C}H-COOH$$
$$\underset{NH_2}{|}$$

The asterisk signifies an asymmetric carbon. All of the amino acids, except glycine, have two optically active isomers designated D- or L-. Isoleucine and threonine also have centers of asymmetry at their β-carbon atoms (1,10). Protein amino acids are of the L-α-form (1,10) as illustrated in Table 1.

Amino acids are important components of the elementary nutrients of living organisms. For humans, ten amino acids are essential for existence and must be ingested in food. The nutritional value of proteins is governed by the quantitative and qualitative balance of individual essential amino acids.

The nutritional value of a protein can be improved by the addition of amino acids of low abundance in that protein. Thus the fortification of plant proteins such as wheat, corn, and soybean with L-lysine, DL-methionine, or other essential amino acids (L-tryptophan and L-threonine) is expected to alleviate some food problems (11). Such fortification has been widespread in the feedstuff of domestic animals.

Proteins are metabolized continuously by all living organisms, and are in dynamic equilibrium in living cells (6,12). The role of amino acids in protein biosynthesis has been described (2). Most of the amino acids absorbed through the digestion of proteins are used to replace body proteins. The remaining portion is metabolized into various bioactive substances such as hormones and purine and pyrimidine nucleotides, (the precursors of DNA and RNA) or is consumed as an energy source (6,13).

Table 1. α-Amino Acids

Common name	CAS Registry Number	Abbreviation	Systematic name	Formula	Molecular weight
Aliphatic			*Monocarboxylic*		
glycine	[56-40-6]	Gly	aminoacetic acid	H_2NCH_2COOH	75.07
alanine	[6898-94-8]	Ala	2-amino-propanoic acid	$CH_3CHCOOH$ $\overset{\|}{NH_2}$	89.09
L-alanine	[56-41-7]				
D-alanine	[338-69-2]				
DL-alanine	[302-72-7]				
valine[a]	[7004-03-7]	Val	2-amino-3-methyl-buta-noic acid	$(CH_3)_2CHCHCOOH$ $\overset{\|}{NH_2}$	117.15
L-valine	[72-18-4]				
D-valine	[640-68-6]				
DL-valine	[516-06-3]				
leucine[a]	[7005-03-0]	Leu	2-amino-4-methyl-penta-noic acid	$(CH_3)_2CHCH_2CHCOOH$ $\overset{\|}{NH_2}$	131.17
L-leucine	[61-90-5]				
D-leucine	[328-38-1]				
DL-leucine	[328-39-2]				
isoleucine[a]	[7004-09-3]	Ileu	2-amino-3-methyl-penta-noic acid	$CH_3CH_2CH—CHCOOH$ $\overset{\|}{CH_3}\ \overset{\|}{NH_2}$	131.17
L-isoleucine	[73-32-5]				
D-isoleucine	[319-78-8]				
DL-isoleucine	[443-79-8]				
Aliphatic containing —OH, —S—, —NH— *group*					
serine	[6898-95-9]	Ser	2-amino-3-hydroxy-pro-panoic acid	$HOCH_2CHCOOH$ $\overset{\|}{NH_2}$	105.09
L-serine	[56-45-1]				
D-serine	[312-84-5]				
DL-serine	[302-84-1]				

Table 1 (*Continued*)

Common name	CAS Registry Number	Abbreviation	Systematic name	Formula	Molecular weight
threonine[a]	[36676-50-3]	Thr	2-amino-3-hydroxy-buta-noic acid	CH_3CH—$CHCOOH$ OH NH_2	119.12
L-threonine D-threonine DL-threonine	[72-19-5] [632-20-2] [80-68-2]				
cysteine	[4371-52-2]	Cys	2-amino-3-mercapto-pro-panoic acid	$HSCH_2CHCOOH$ NH_2	121.16
L-cysteine D-cysteine DL-cysteine	[52-90-4] [921-01-7] [3374-22-9]				

cystine	[24645-67-8]	$(Cys)_2$	3,3′-dithio-bis-(2-amino-propanoic acid)	$SCH_2CHCOOH$
				    | NH_2
				$SCH_2CHCOOH$
				        NH_2
					240.30

L-cystine D-cystine DL-cystine	[56-893] [349-46-2] [923-32-0]				
methionine[a]	[7005-18-7]	Met	2-amino-4-methyl-thio-butanoic acid	$CH_3SCH_2CH_2CHCOOH$ NH_2	149.21
L-methionine D-methionine DL-methionine	[63-68-3] [348-67-4] [59-51-8]				
lysine[a]	[6899-06-5]	Lys	2,6-diamino-hexanoic acid	$H_2N(CH_2)_4CHCOOH$ NH_2	146.19
L-lysine D-lysine DL-lysine	[56-87-1] [923-27-3] [70-54-2]				
arginine[b]	[7004-12-8]	Arg	2-amino-5-guani-dopen-tanoic acid	$HN{=}CNH(CH_2)_3CHCOOH$ H_2N NH_2	174.20

Name / [CAS]	Abbrev.	IUPAC name	Structure	M.W.
L-arginine [74-79-3] D-arginine [157-06-2] DL-arginine [7200-25-1]				
Aromatic				
phenylalanine[a] [3617-44-5]	Phe	2-amino-3-phenyl-propanoic acid	$C_6H_5CH_2CHCOOH$, NH_2	165.19
L-phenylalanine [63-91-2] D-phenylalanine [673-06-3] DL-phenylalanine [150-30-1]				
tyrosine [55520-40-6]	Tyr	2-amino-3-(4-hydroxy-phenyl)-propanoic acid	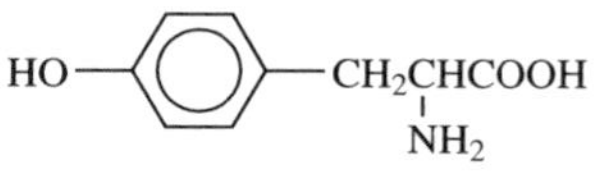 HO—C$_6$H$_4$—$CH_2CHCOOH$, NH_2	181.19
L-tyrosine [60-18-4] D-tyrosine [556-02-5] DL-tyrosine [556-03-6]				
Heterocyclic				
proline [7005-20-1]	Pro	2-pyrrolidine-carboxylic acid	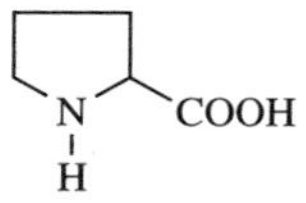 (pyrrolidine ring, N—H, COOH)	115.13
L-proline [147-85-3] D-proline [344-25-2] DL-proline [609-36-9] (cis)				
hydroxyproline	Hypro	4-hydroxy-2-pyrrolidine-carboxylic acid	(HO—pyrrolidine ring, N—H, COOH)	131.13
L-hydroxyproline [618-27-9] D-hydroxyproline [2584-71-6] DL-hydroxyproline [49761-17-3] (trans) L-hydroxyproline (trans) [51-35-4] D-hydroxyproline (trans) [3398-22-9] DL-hydroxyproline (trans) [618-28-0]				
histidine[b] [7006-35-1]	His	2-amino-3-imidazole-pro-panoic acid	(imidazole ring, N, NH)—$CH_2CHCOOH$, NH_2	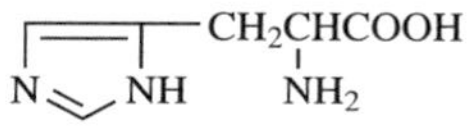 155.16
L-histidine [71-00-1] D-histidine [351-50-8]				

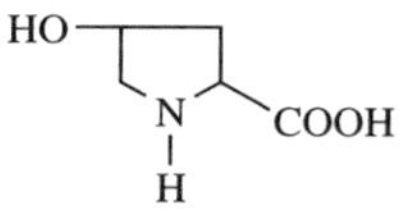

Table 1 (Continued)

Common name	CAS Registry Number	Abbreviation	Systematic name	Formula	Molecular weight
DL-histidine tryptophan[a]	[4998-57-6] [6912-86-3]	Trp	2-amino-3-indoyl-propanoic acid		204.22
L-tryptophan D-tryptophan DL-tryptophan	[73-22-3] [153-94-6] [54-12-6]				
			Dicarboxylic		
aspartic acid	[6899-03-2]	Asp	2-amino-butane-dioic acid	$HOOCCH_2CHCOOH$ NH_2	133.10
L-aspartic acid D-aspartic acid DL-aspartic acid	[56-84-8] [1783-96-6] [617-45-8]				
glutamic acid	[6899-05-4]	Glu	2-amino-pentane-dioic acid	$HOOCCH_2CH_2CHCOOH$ NH_2	147.13
L-glutamic acid D-glutamic acid DL-glutamic acid	[56-86-0] [6893-26-1] [617-65-2]				
asparagine	[7006-34-0]	Asn	2-amino-3-carbamoyl-propanoic acid	$H_2NCOCH_2CHCOOH$ NH_2	132.12
L-asparagine D-asparagine DL-asparagine	[70-47-3] [2058-58-4] [3130-87-8]				
glutamine	[6899-04-3]	Gln	2-amino-4-carbamoyl-butanoic acid	$H_2NCOCH_2CH_2CHCOOH$ NH_2	146.15
L-glutamine D-glutamine DL-glutamine	[56-85-9] [5959-95-5] [585-21-7]				

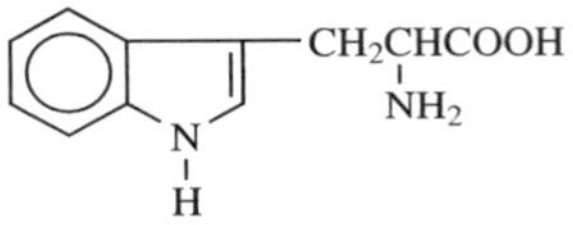

[a] Essential amino acid.
[b] Arginine and histidine are also essential for children (6).

The history of the discovery of amino acids is closely related to advances in analytical methods. Initially, quantitative and qualitative analysis depended exclusively upon crystallization from protein hydrolysates. The quantitative precipitation of several basic amino acids including phosphotungstates, the separation of amino acid esters by vacuum distillation, and precipitation by sulfonic acid derivatives were developed successively during the last century.

After World War II, analytical methods for amino acids were improved and new methods were introduced. The first was microbial assay using lactic acid bacteria which require all of the regular amino acids for growth. Manometric determination (by use of a Warburg manometer) of CO_2 liberated by the action of amino acid decarboxylase is now also classical. However, these methods are still used for the microdetermination of amino acids. Later, chromatographic separations using filter paper, ion-exchange resins, and other adsorbants were rapidly developed. Automatic analyzers of amino acids using ion-exchange resin chromatography (14) are now used widely in routine analyses of amino acid mixtures. More recently, high performance liquid chromatography (hplc) has been extensively developed and the separate determination of L- and D-amino acids has been possible by this method. Amino acid sensors have been studied (15,16). The contribution of various biosynthetic pathways for amino acids has been analyzed by ^{13}C-nmr with glucose, labeled with carbon-13 at C-1 or C-6, as substrate (17).

The determination of amino acids in proteins requires pretreatment by either acid or alkaline hydrolysis. However, L-tryptophan is decomposed by acid, and the racemization of several amino acids takes place during alkaline hydrolysis. Moreover, it is very difficult to confirm the presence of cysteine in either case. The use of methanesulfonic acid (18) and mercaptoethanesulfonic acid (19) as the protein hydrolyzing reagent to prevent decomposition of L-tryptophan and L-cysteine is recommended. Enzymatic hydrolysis of proteins has been studied (20).

In 1950 all L-amino acids were manufactured by isolation from protein hydrolyzates or by separation of L-amino acids from the synthesized racemic mixtures. Since the mid-1950s, methods of production of L-amino acids have changed extensively. The first important change was made by the invention of a new fermentation process using so-called glutamic acid bacteria (eg, *Corynebacterium glutamicum, Brevibacterium flavum*) to produce L-glutamic acid (21). Thereafter, fermentation processes were developed to produce many other amino acids. Most amino acids (except for glycine, L-methionine, L-cysteine, and L-serine) are now economically produced by fermentation (22,23). Subsequently, enzymatic processes were developed to produce L-aspartic acid, L-alanine, L-tryptophan, L-cysteine, L-serine, L-lysine, L-phenylalanine, and some D-amino acids from chemically synthesized substrates (24). Glycine, DL-alanine, DL-methionine, and DL-cysteine, and some other amino acids are still produced by chemical synthesis. Chemical manufacturing procedures for amino acids have been discussed (25).

All of the protein amino acids are currently available commercially and their uses are growing. Amino acids and their analogues have their own characteristic effects in flavoring, nutrition, and pharmacology.

In the food industries a number of amino acids have been widely used as flavor enhancers and flavor modifiers (see FLAVORS AND SPICES). For example, monosodium L-glutamate is well-known as a meat flavor-enhancer and an enormous quantity of it is now used in various food applications (see L-MONOSODIUM GLUTAMATE (MSG)). Protein, hydrolyzed by acid or enzyme to be palatable, has been used for a long time in flavoring agents. The addition of L-glutamate, L-aspartate, glycine, DL-alanine, and other palatable amino acids can improve flavoring by these protein hydrolyzates. In addition, some nucleotides, such as 5′-inosinic acid [131-99-7] and 5′-guanylic acid [85-32-5], have a synergistic effect on the meat flavor enhancing effects of L-glutamate and L-aspartate. Tricholomic acid [2644-49-7] (**1**) and ibotenic acid [2552-55-8] (**2**), nonprotein amino acids found in mushrooms, have 4 to 25 times stronger umami taste than L-glutamic acid (26). However, they have not been used in food.

$$\begin{array}{cc}
\mathrm{H_2C-CH-CH-COOH} & \mathrm{HC=C-CH-COOH} \\
\mathrm{O=C\quad O\quad NH_2} & \mathrm{O=C\quad O\quad NH_2} \\
\mathrm{N} & \mathrm{N} \\
\mathrm{H} & \mathrm{H} \\
(1) & (2)
\end{array}$$

Some peptides have special tastes. L-Aspartyl phenylalanine methyl ester is very sweet and is used as an artificial sweetener (see SWEETENERS). In contrast, some oligopeptides (such as L-ornithinyltaurine·HCl and L-ornithinyl-β-alanine·HCl), and glycine methyl or ethyl ester·HCl have been found to have a very salty taste (27).

Amino acids are also used in medicine. Amino acid infusions prepared from crystalline amino acids are used as nutritional supplements for patients before and after surgery. Some amino acids and their analogues are used for treatment of major diseases. L-DOPA, L-3-(3,4-dihydroxyphenyl)alanine, [59-92-7] is an important drug in the treatment of Parkinson's disease, and L-glutamine and its derivatives are used for treatment of stomach ulcers. α-Methyl-DOPA [555-30-6] is an effective antidepressant (see PSYCHOPHARMACOLOGICAL AGENTS). Some peptides, eg, oxytocin [50-56-6], angiotensin [1407-47-2], gastrin, and cerulein, have hormonal effects which have medical utility. The physiological effect of glutathione [70-18-8] (L-glutamyl-L-cysteinyl glycine) has been reviewed (28).

Amino acid polymers like poly(γ-methyl-L-glutamate) [29967-97-3] have been developed as raw materials for artificial leathers (see LEATHERLIKE MATERIALS). Derivatives of amino acids are now finding new applications in industry and agriculture.

2. Physical Properties

2.1. Melting Point. Amino acids are solids, even the lower carbon-number amino acids such as glycine and alanine. The melting points of amino acids generally lie between 200 and 300°C. Frequently amino acids decompose before reaching their melting points (Table 2).

2.2. Crystalline Structures. Crystal shape of amino acids varies widely, for example, monoclinic prisms in glycine and orthorhombic needles in L-alanine. X-ray crystallographic analyses of 23 amino acids have been described

Table 2. Physical Constants of Amino Acids[a]

Amino acid		Melting point, °C	Density, d_{t1}^{t2}	Specific rotation			
				$[\alpha]_D$	t, °C	c, %	Solvent
Ala	L-	297 (dec)	1.401	+2.8	25	6	H_2O
		314 (dec)	1.432^{23}	+2.8	25	6	H_2O
	L-·HCl	204 (dec)		+8.5	26	9.3	
	D-	314 (dec)	1.424	−13.6	25	1	6 N HCl
	DL-	264 (dec)	1.424				
		295 (dec)					
Arg	L-	244 (dec)		+12.5	20	3.5	H_2O
	L-·HCl	235 (dec)		+12.0	20	4	
	DL-	217–218					
Asn	L-·H_2O	234–235	1.543^{15}_{4}	−5.42	20	1.3	
	D-·H_2O	215		+5.41	20	1.3	
		234.5	1.543^{15}				
	DL-·H_2O	182–183	1.4540^{15}_{4}	+5.41	20	1.3	
Asp	L-	270–271	$1.661^{12.5}_{4}$	+25.0	20	1.97	6 N HCl
			1.6613^{13}_{13}	+24.6	24	2	6 N HCl
	D-	324 (dec)		−23.0	27	2.30	6 N HCl
				−25.5	20		HCl
	DL-	269–271	1.6613^{13}_{13}				
		338–339	1.6632^{13}_{13}				
Cys	L-	240 (dec)		+6.5	25		5 N HCl
				+9.8	30	1.3	H_2O
	L-·HCl	175–178		+5.0	25		5 N HCl
(Cys)₂	L-	260–261 (dec)	1.677	−223.4	20	1	1 N HCl
	D-	247–249		+223	20		1 N HCl
	DL-	260		+224	20	1	1 N HCl
Glu	L-	247–249 (dec)	1.538^{20}	+31.4	22.4		6 N HCl
	L-·HCl	224–225 (dec)	1.538^{20}_{4}	+31.4	22	1	6 N HCl
	D-	214		+24.4	22	6	
	DL-	213 (dec)	1.538^{20}_{4}	−30.5	20	1.0	6 N HCl
		225–227 (dec)	1.4601^{20}	−31.7	25		1.7 N HCl

Table 2 (Continued)

Amino acid		Melting point, °C	Density, d_{t1}^{t2}	Specific rotation			
				$[\alpha]_D$	t, °C	c, %	Solvent
Gly		199 (dec)	$1.4601^{20}_{\ 4}$				
		233 (dec)	1.1607^{17}				
		262 (dec)	0.828^{17}				
His	L-	287 (dec)		−39.74	20	1.13	H_2O
		287 (dec)		−39.7	20	1.13	$3\,N$ HCl
	L-·HCl·H_2O	259 (dec)		+8.0	26	2	H_2O
	D-	287 (dec)		+40.2	20		
	DL-	285 (dec)					
Ileu	L-	284 (dec)		+11.29	20	3	$6.1\,N$ HCl
				+40.61	20	4.6	H_2O
		285–286 (dec)		+12.2	25	3.2	$1\,N$ HCl
				+36.7		4	H_2O
	D-	283–284 (dec)		−12.2	20	3.2	$5\,N$ HCl
				−40.7		1	
	DL-	280 (dec)					
Leu	L-	293–295 (dec)	$1.293^{18}_{\ 4}$	−10.8	25	2.2	H_2O
		293–295		−10.42	25	22	
	D-	293		+10.34	20		
	DL-	332 (dec)	$1.293^{18}_{\ 4}$				
		293–295					
Lys	L-	224.5 (dec)		+25.9	23	2	$6\,N$ HCl
		224–245 (dec)		+14.6	20	6	H_2O
	L-·HCl	263–264		+14.6	25	2	$0.6\,N$ HCl
	L-·2HCl	193		+15.3	20	2	
	DL-·HCl	201–202		+15.29	20		
		260–263					
	DL-·2HCl	187–189					
Met	L-	280–282 (dec)		−8.2	25		H_2O
		283 (dec)		−8.2	25	1	
	DL-	281 (dec)	1.340				
Phe	L-	283 (dec)		−35.1	20	1.94	H_2O
	D-	285 (dec)		+35.0	20	2.04	

		MP		$[\alpha]$	Temp	Conc	Solvent
Pro	DL-	271–273 (dec)					
	L-	220–222 (dec)		−52.6	20	0.58	0.5 N HCl
		220–222 (dec)		−80.9	20	1	H_2O
Hyp	DL-	205 (dec)					
	L-	274		−76.5	20	2.5	H_2O
Ser	L-	228 (dec)		−6.83	20	10	H_2O
	D-	228 (dec)		+6.87	20	10	H_2O
	DL-	246 (dec)	1.537				
		246 (dec)	$1.603^{22.5}$				
Thr	L-	255–257 (dec)		−28.3	26	1.1	
	DL-	229–230 (dec)					
Trp	L-	289 (dec)		−31.5	23	1	0.5 N HCl
				+2.4	20		0.5 N NaOH
				+0.15	20	2.43	H_2O
		290–292 (dec)		−31.5	20	0.5	1 N NaOH
				+6.1	20	11	H_2O
	D-	281–282		+33	20		
	DL-	282					
Tyr	L-	342–344 (dec)	1.456	−10.6	22	4	1 N HCl
				−13.2	18	4	3 N NaOH
	D-	310–314 (dec)		+10.3	25	4	1 N HCl
	DL-	316 (dec)					
		340 (dec)					
Val	L-	315	1.230	+22.9	23	0.8	20% HCl
		93–96(?)	1.230	+22.9	23	0.8	20% alc
	D-	156–157.5	1.310	−29.4	20		20% alc
	DL-	298 (dec)					

[a] From refs. 29 and 30.

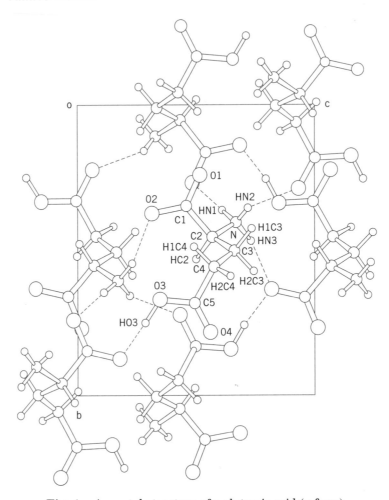

Fig. 1. A crystal structure of L-glutamic acid (α-form).

(31). L-Glutamic acid crystallizes in two polymorphic forms (α and β) (32), and the α-form is more facilely handled in industrial processes. The crystal structure has been determined (33) and is shown in Figure 1.

    2.3. Dipole. Every amino acid molecule has two equal electric charges of opposite sign caused by the amino and carboxyl groups on the α-carbon atom.

$$R-CH-COO^-$$
$$|$$
$$NH_3^+$$

    The dielectric constants of amino acid solutions are very high. Their ionic dipolar structures confer special vibrational spectra (Raman, ir), as well as characteristic properties (specific volumes, specific heats, electrostriction) (34).

    2.4. Optical Configuration. With the exception of glycine, all α-amino acids contain at least one asymmetric carbon atom and may be characterized by their ability to rotate light to the right (+) or to the left (−), depending on the solvent and the degree of ionization. Specific rotations are given in Table 2. They are also characterized by the stereochemical configuration of the asymmetric carbon based on the configuration of glyceraldehyde; D,L-notation is popular for amino acids, but R,S-notation is a more precise designation of chirality.

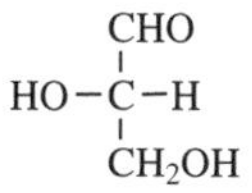

$$
\begin{array}{c}
\text{CHO} \\
| \\
\text{HO} - \text{C} - \text{H} \\
| \\
\text{CH}_2\text{OH}
\end{array}
\qquad
\begin{array}{c}
\text{COOH} \\
| \\
\text{HO} - \text{C} - \text{H} \\
| \\
\text{CH}_3
\end{array}
\qquad
\begin{array}{c}
\text{COOH} \\
| \\
\text{H}_2\text{N} - \text{C} - \text{H} \\
| \\
\text{CH}_3
\end{array}
$$

L-glyceraldehyde;   L-lactic acid;   L-alanine;

(S)-2,3-dihydroxypropanal   (S)-2-hydroxypropanoic acid   (S)-2-aminopropanoic acid

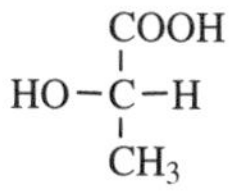

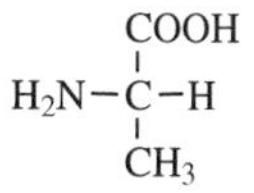

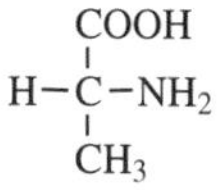

$$
\begin{array}{c}
\text{COOH} \\
| \\
\text{H} - \text{C} - \text{NH}_2 \\
| \\
\text{CH}_3
\end{array}
$$

L-alanine;

(R)-2-aminopropanoic acid

2.5. Solubility. Solubility data of amino acids are given in Table 3. In all instances there are at least two polar groups, acting synergistically on the solubility in water. The solubility of amino acids having additional polar groups, eg, −OH, −SH, −COOH, −NH$_2$, is even more enhanced.

2.6. Dissociation. In aqueous solution, amino acids undergo a pH-dependent dissociation (37):

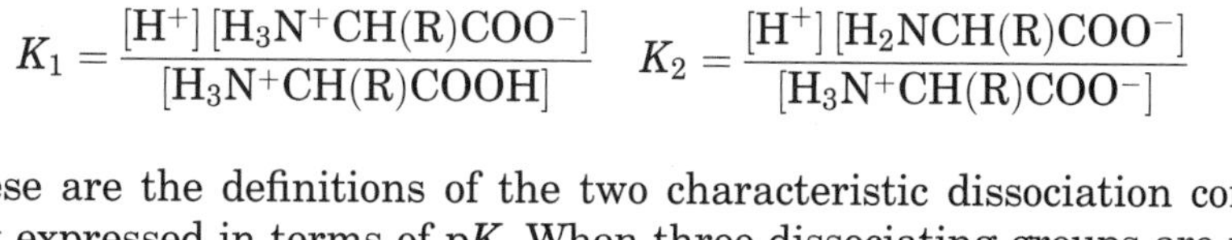

at pH = 1   at pH = 6   at pH = 11

$$
\underset{\substack{|\\ \text{R}}}{\text{H}_3\text{N}^+ - \text{CH} - \text{COOH}}
\;\underset{+\text{H}^+}{\overset{-\text{H}^+}{\rightleftharpoons}}\;
\underset{\substack{|\\ \text{R}}}{\text{H}_3\text{N}^+ - \text{CH} - \text{COO}^-}
\;\underset{+\text{H}^+}{\overset{-\text{H}^+}{\rightleftharpoons}}\;
\underset{\substack{|\\ \text{R}}}{\text{H}_2\text{N} - \text{CH} - \text{COO}^-}
$$

cationic form   reaction 1, K_1 ampholyte   reaction 2, K_2 anionic form

where

$$K_1 = \frac{[\text{H}^+]\,[\text{H}_3\text{N}^+\text{CH(R)COO}^-]}{[\text{H}_3\text{N}^+\text{CH(R)COOH}]} \qquad K_2 = \frac{[\text{H}^+]\,[\text{H}_2\text{NCH(R)COO}^-]}{[\text{H}_3\text{N}^+\text{CH(R)COO}^-]}$$

These are the definitions of the two characteristic dissociation constants normally expressed in terms of pK. When three dissociating groups are present in a molecule there are three pK values, ie, pK$_1$, pK$_2$, pK$_3$. A knowledge of these pK values is important in the separation or isolation of each amino acid by ion-exchange chromatography.

A large part of the dissolved amino acid exists as the ampholyte (zwitterion). The isoelectric point (pI) is the pH at which the net electric charge of a dissolved amino acid molecule is zero. pI is expressed as

$$pI = \frac{pK_1 + pK_2}{2}$$

Table 3. **pK and pI at 25 °C and Solubility of Amino Acids**[a]

Amino acid		pK_1 (COOH)	pK_2 (NH_3^+)[b]	pK_3 (NH_3^+)	pI	Solubility in water, g/L				
						0°C	25°C	50°C	75°C	100°C
Divalent acids										
Gly	L-	2.34	9.60		5.97		250	391	544	672
Ala	L-	2.34	9.69[c]		6.00	127.3	166.5	217.9	285.1	373.0
	DL-	2.35	9.87			121	167	231	319	440
Val	L-	2.32	9.62[c]		5.96	83.4	88.5	96.2	102.4 (65°C)	188.1
	DL-	2.29	9.72			59.8	70.9	91.1	126.1	
Leu	L-	2.36	9.60		5.98	22.7	24.26	28.87	38.23	56.38
	DL-	2.26	9.62[c]			7.97	9.91	14.06	22.76	42.06
Ileu	L-	2.32	9.76		5.94	37.9	41.2	48.2	60.8	82.6
	DL-					18.3	22.3	30.3	46.1	78.0
Ser	L-	2.21	9.15[c]		5.68	22.04	50.23	soluble	192	322
	DL-	2.21	9.15					103		
Thr	L-	2.15	9.12		5.64			freely soluble		
Pro	L-	1.99	10.60		6.30	1272	1623	2067	2509 (70°C)	99.0
Hyp	L-	1.82	9.65		5.74	288.6	361.1	451.8	516.7 (65°C)	
Phe	L-	1.83	9.13[c]		5.48	19.8	29.6	44.3	66.2	68.9
	DL-	2.58	9.24			9.97	14.11	21.87	37.08	49.9
Trp	L-	2.38	9.39		5.89	0.23	11.4	17.1	27.95	
Met	L-	2.28	9.21		5.74	18.18	33.81	60.70	105.2	176.0
Trivalent acids										
Asp	L-	1.88	3.65 (COOH)	9.60	2.77	2.1	5.0	12.0	28.8	68.9
Glu	L-	2.19	4.25 (COOH)	9.67	3.22		8.64	21.86	55.32	140.0
	DL-						20.54	49.34	118.6	284.9
Tyr	L-	2.20	9.11	10.07 (OH)	5.66	0.196	0.453	1.052	2.438	5.65
	DL-					0.147	0.351	0.836		
Cys	L-	1.71	8.33	10.78				freely soluble		
His	L-	1.78	5.97	8.97	7.47		41.9			
Arg	L-	2.18	9.09	13.2	11.15			the satd aq soln contains 15% (w/w) at 21°C		
Lys	L-	2.20	8.90	10.28 (38°C)	9.59			freely soluble		
Tetravalent acids										
$(Cys)_2$	L-	<1	2.1 (COOH)	8.02[d]	5.03	0.05	0.112	0.239	0.523	1.142

[a] Refs. 29 and 35.
[b] Unless indicated (COOH).
[c] Ref. 36.
[d] pK_4 (NH_3^+) = 8.71.

The solubility of each amino acid is minimal at its pI. pK and pI values are given in Table 3.

3. Chemical Properties

α-Amino acids are ampholytic compounds. The chemical reactions of amino acids can be classified according to their carboxyl, amino, and side-chain groups. Most of the reactions have been well known for a long time; the details of these reactions have been reviewed (38).

3.1. Reactions of the Amino Group

N-Acylation. N-Acylation and related reactions are brought about in straightforward ways with acyl chloride or acid anhydride, although the proximity of the carboxyl group may produce other reactions, eg, oxazolinone formation, under some conditions.

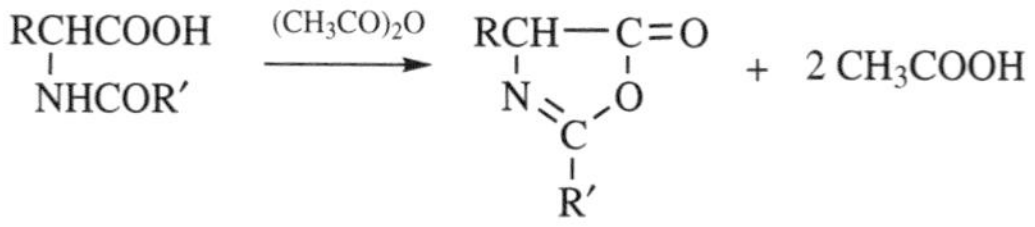

$$\underset{\underset{\text{NHCOR}'}{|}}{\text{RCHCOOH}} \xrightarrow{(CH_3CO)_2O} \underset{\underset{N\diagdown_{\underset{|}{C}\diagup}O}{|}}{\text{RCH}-\text{C}=\text{O}} \;+\; 2\,CH_3COOH$$
$$R'$$

In these cases, it is better to protect the carboxyl group. Optimized conditions for N-acetylation have been studied (39). N-Acylation can be utilized for protecting the amino group in the reaction of amino acids, for example in peptide synthesis.

Reaction with Phosgene. This reaction of amino acid esters is used for preparing the corresponding isocyanates, especially lysine diisocyanate [4460-02-0] (LDI). LDI is a valuable nonyellowing isocyanate with a functional side group for incorporation in polyurethanes.

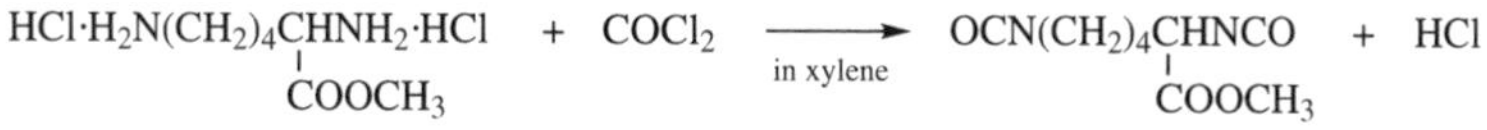

$$\underset{\underset{\text{COOCH}_3}{|}}{\text{HCl·H}_2\text{N(CH}_2)_4\text{CHNH}_2\text{·HCl}} \;+\; COCl_2 \xrightarrow[\text{in xylene}]{} \underset{\underset{\text{COOCH}_3}{|}}{\text{OCN(CH}_2)_4\text{CHNCO}} \;+\; HCl$$

$$LDI$$

In the case of β-hydroxy-α-amino acids, oxazolidinone derivatives are formed with retention of configuration.

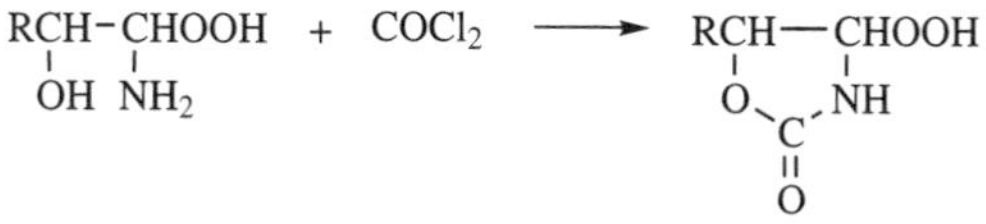

$$\underset{\underset{\text{OH \ NH}_2}{|\ \ \ |}}{\text{RCH}-\text{CHOOH}} \;+\; COCl_2 \longrightarrow \underset{\underset{O\diagdown_{\underset{\overset{||}{O}}{C}\diagup}\text{NH}}{|\qquad|}}{\text{RCH}-\text{CHOOH}} \;+\; HCl$$

Formation of Schiff-Bases. Reaction of an amino acid and an aldehyde or ketone gives a Schiff-base in neutral or alkaline solution, and following reduction gives the corresponding N-alkylamino acid.

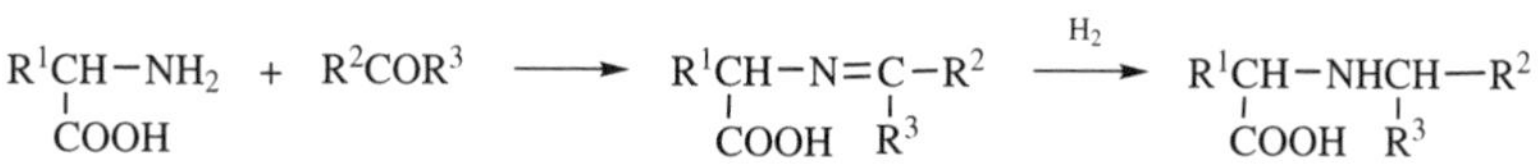

$$\underset{\underset{\text{COOH}}{|}}{\text{R}^1\text{CH}-\text{NH}_2} \;+\; R^2COR^3 \longrightarrow \underset{\underset{\text{COOH \ R}^3}{|\qquad|}}{\text{R}^1\text{CH}-\text{N}=\text{C}-\text{R}^2} \xrightarrow{H_2} \underset{\underset{\text{COOH \ R}^3}{|\qquad\quad|}}{\text{R}^1\text{CH}-\text{NHCH}-\text{R}^2}$$

Schiff-base

Maillard Reaction (Nonenzymatic Glycation). Browned reaction products are formed by heating amino acid and simple sugar. This reaction is important in food science relating to coloring, taste, and flavor enhancement (40), and is illustrated as follows:

$$\text{sugar} + \text{amino acid (or protein)} \longrightarrow \underset{(\text{aldimine})}{\text{Schiff base}} \longrightarrow \underset{(\text{ketoamine})}{\text{Amadori compound}} \longrightarrow \text{browned material}$$

Substitution Reactions. Reaction with nitrous acid in dilute aqueous solutions yields the corresponding hydroxy acid or in solution containing a hydrohalic acid, the corresponding α-halo acid, with inversion in many cases.

3.2. Reactions of the Carboxyl Group

Esterification, Amidation, and Acid Chloride Formation. Amino acids undergo these common reactions of the carboxyl group with due regard for the need for *N*-protection.

Reduction to Amino Alcohols. Reduction can be brought about using diborane–dimethyl sulfide in THF (41). $NaBH_4$ in ethanol is also effective, but requires that the carboxyl group be esterified first (42). $LiAlH_4$ is inferior in terms of yield and practical convenience to $LiAlH_2(OCH_3)_2$ for the reduction of amino acid esters (43). Solvent effects have been reported in the case of $LiAlH_4$ (44).

Anhydride Formation. The carboxyl group in *N*-protected amino acids is converted into the symmetrical anhydride on treatment with the carbodiimide (45).

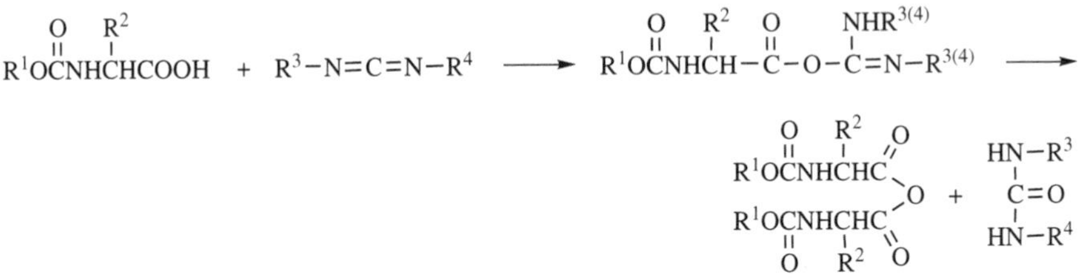

Cyclic anhydrides are formed readily from *N*-protected aspartic and glutamic acids.

3.3. Reactions Depending on Both Amino and Carboxyl Groups.

Formation of Diketopiperazines. Esters of α-amino acids can be readily prepared by refluxing anhydrous alcoholic suspensions of α-amino acids saturated with dry HCl. Diketopiperazines are formed by heating the alcoholic solution of the α-amino acid ester.

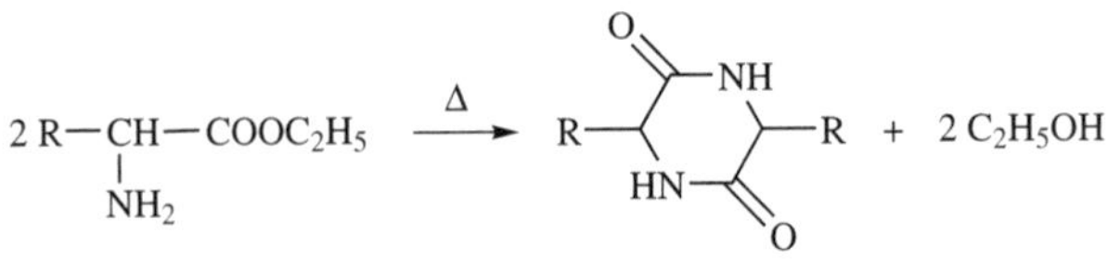

Formation of Hydantoin

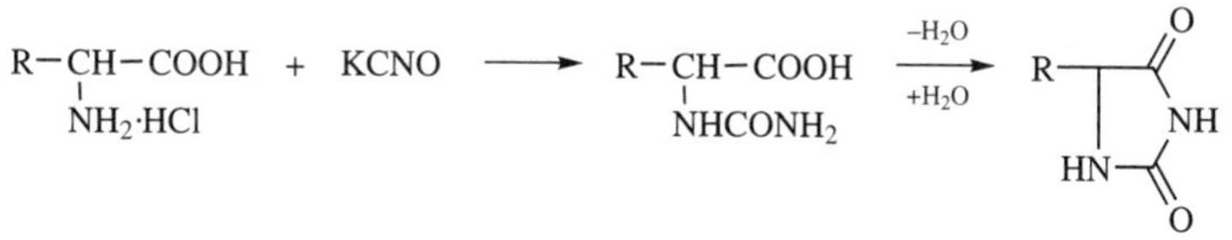

The use of an organic isocyanate instead of potassium isocyanate gives a *N*-substituted hydantoin.

Strecker Degradation (Oxidative Deamination). Mild oxidizing agents such as aqueous sodium hypochlorite or aqueous *N*-bromosuccinimide, cause decarboxylation and concurrent deamination of amino acids to give aldehydes.

$$H_3N^+CHRCO_2^- \xrightarrow{[O]} H_2N^+ = CR + CO_2 + H_2O \xrightarrow{H_2O} RCHO + NH_3$$

Similarly, silver(II) picolinate and lead tetraacetate can be used to produce carbonyl compounds.

Formation of N-Carboxy-α-Amino Acid Anhydride (NCA). NCAs are important as starting materials for amino acid polymers. They are prepared by the reaction of amino acids with phosgene in an aprotic solvent (46).

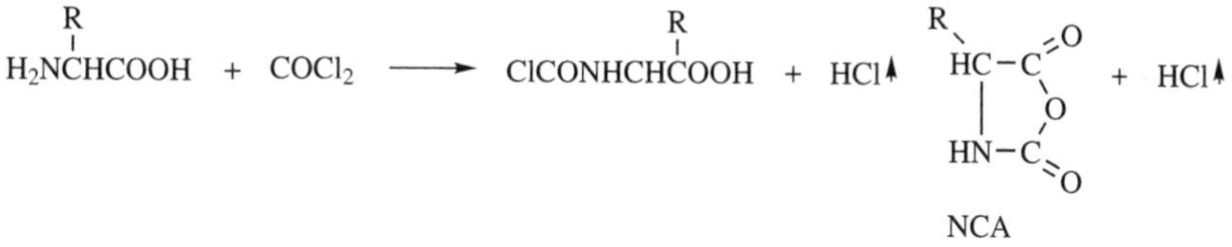

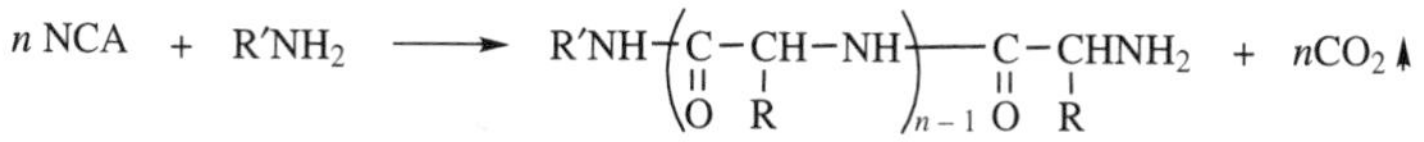

Although polymerization of NCA is popular, other types of amino acid polymerization have also been reported, for example (47,48):

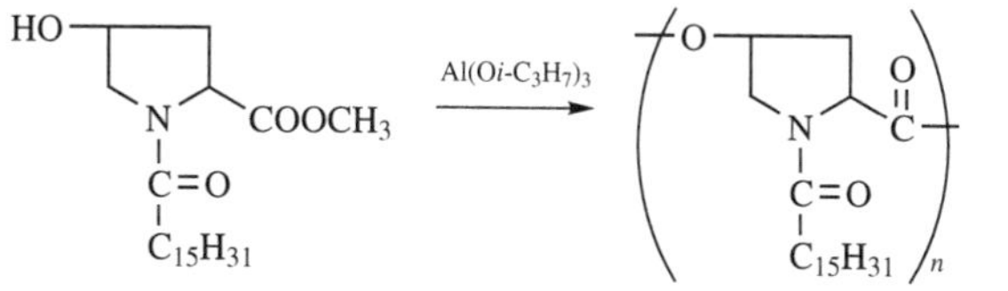

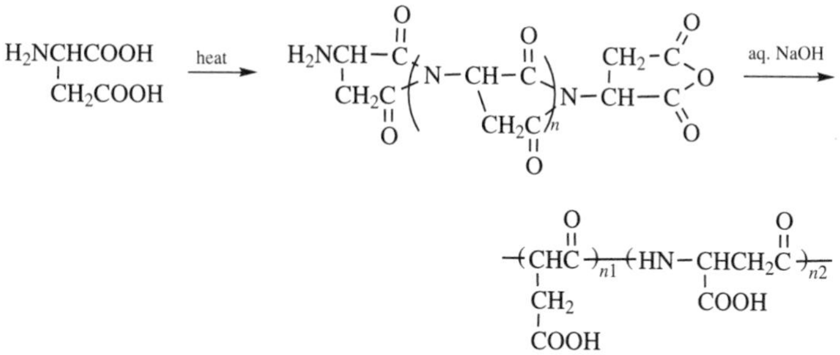

Ninhydrin-Color Reaction.   This reaction is commonly used for qualitative analysis of α-amino acids, peptides, and proteins.

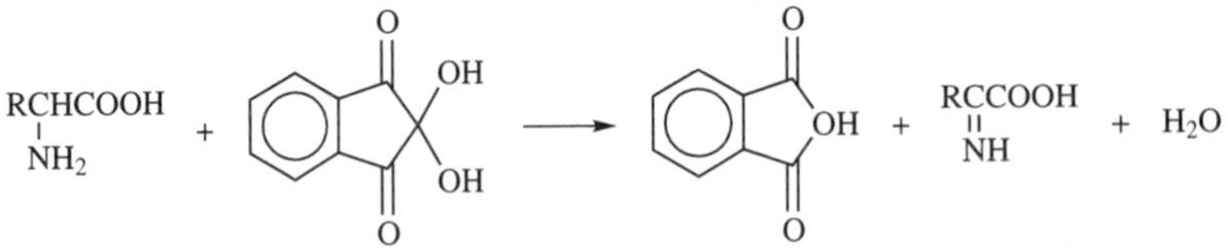

ninhydrin

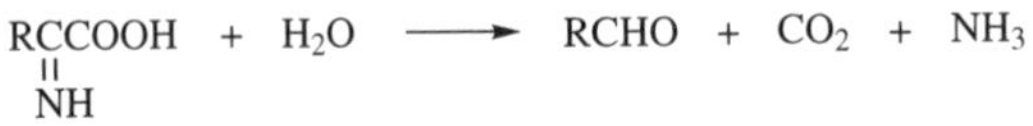

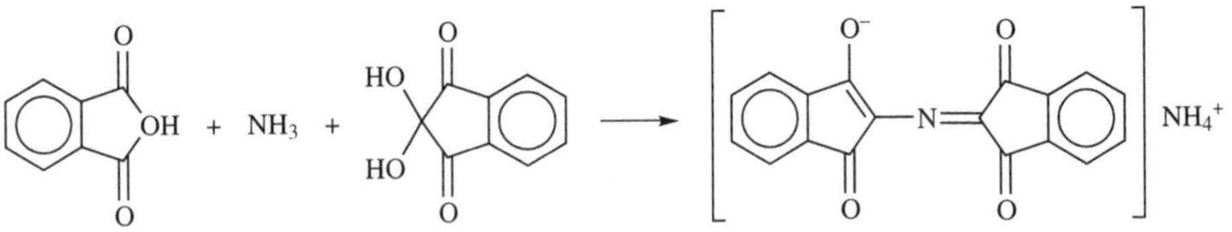

$$RCCOOH + H_2O \longrightarrow RCHO + CO_2 + NH_3$$

Ruhemann's Purple

3.4. Other Reactions.   *Salt Formation and Metal Chelation.*   Most α-amino acids form salts in alkaline and acidic aqueous solutions (49). For example, α-amino acids form inner complex salts with copper.

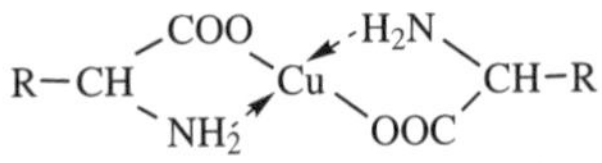

Benzenesulfonate compounds yield very insoluble salts which have been used for separation and identification of amino acids (50). Similarly, phosphotungstic acid forms insoluble salts with basic amino acids such as lysine, arginine, and cysteine.

Synthesis of Peptide.   There is continual progress in the improvement of instruments and reagents for peptide synthesis, especially "solid phase polymerization" (51) (see PROTEINS). This method is suitable for the synthesis of peptides with 20 ~ 30 amino acid units.

Induction of Asymmetry by Amino Acids.   No fewer than six types of reactions can be carried out with yields of 75–100% using amino acid catalysts, ie, catalytic hydrogenation, intramolecular aldol cyclizations, cyanhydrin synthesis, alkylation of carbonyl compounds, hydrosilylation, and epoxidations (52).

4. Synthesis of α-Amino Acids

Many methods for chemical synthesis of α-amino acids have been established. Because excellent reviews have been published (53), well-known reactions are introduced here only by their names and synthetic pathways.

4.1. Synthetic Pathways

Strecker Synthesis.

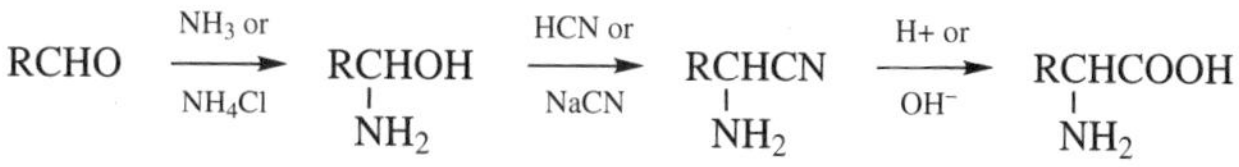

$$RCHO \xrightarrow[\text{NH}_4\text{Cl}]{\text{NH}_3 \text{ or}} \underset{\underset{NH_2}{|}}{RCHOH} \xrightarrow[\text{NaCN}]{\text{HCN or}} \underset{\underset{NH_2}{|}}{RCHCN} \xrightarrow[\text{OH}^-]{\text{H+ or}} \underset{\underset{NH_2}{|}}{RCHCOOH}$$

Bucherer Synthesis.

$$RCHO \xrightarrow[\text{NaCN}]{(NH_4)_2CO_3} \underset{\underset{NH-C}{|}}{RCH-C} \!\!\!\!\!\!\!\!\!\!\!\!\!\!\!\! \nearrow^{O} \!\!\!\! NH \xrightarrow{OH^-} \underset{\underset{NH_2}{|}}{RCHCOOH}$$

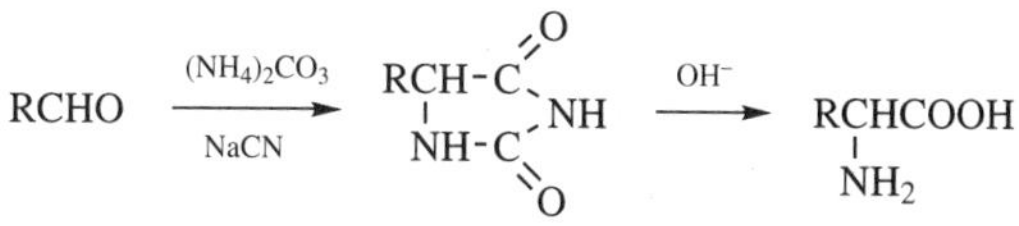

These two methods are popular for α-amino acid synthesis, and used in the industrial production of some amino acids since raw materials are readily available.

Amination of α-Halogeno Carboxylic Acids
Original Method

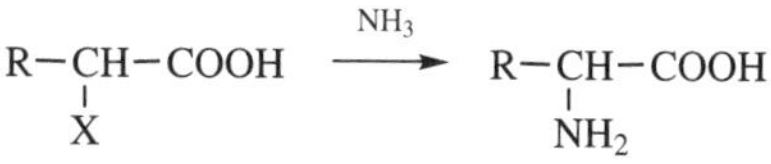

$$\underset{\underset{X}{|}}{R-CH-COOH} \xrightarrow{\text{NH}_3} \underset{\underset{NH_2}{|}}{R-CH-COOH}$$

Gabriel's Modification

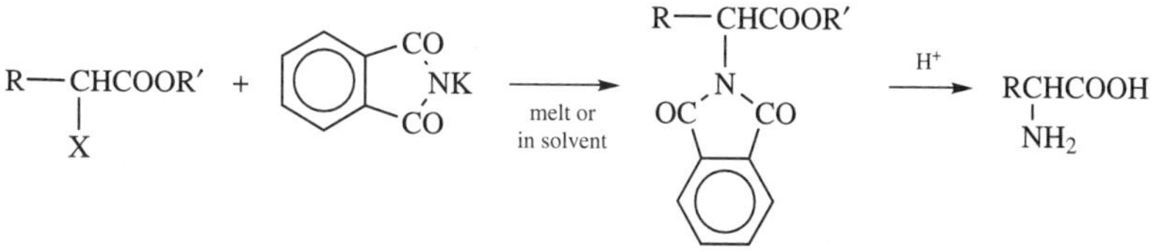

$$\underset{\underset{X}{|}}{R-CHCOOR'} + \text{(phthalimide)} \xrightarrow[\text{in solvent}]{\text{melt or}} \text{(phthaloyl intermediate)} \xrightarrow{\text{H}^+} \underset{\underset{NH_2}{|}}{RCHCOOH}$$

Alkylation of Active Methylene Compounds Erlenmeyer Synthesis and Others.
Hydantoin [461-72-3] (**3**), azlactone (**4**), diketopiperazine [106-57-0] (**5**), etc, are readily available, so that these methods are often utilized. In structure (**4**), R=CH$_3$ [24474-93-9] or R=C$_6$H$_5$ [1199-01-5].

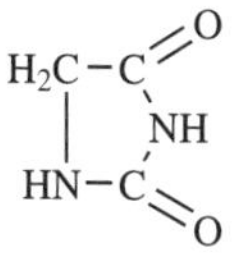

        (**3**) 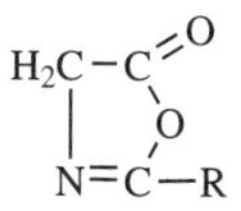   (**4**) 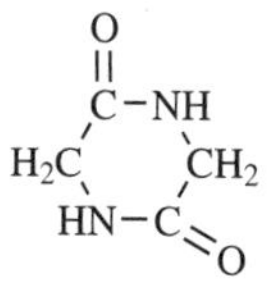   (**5**)

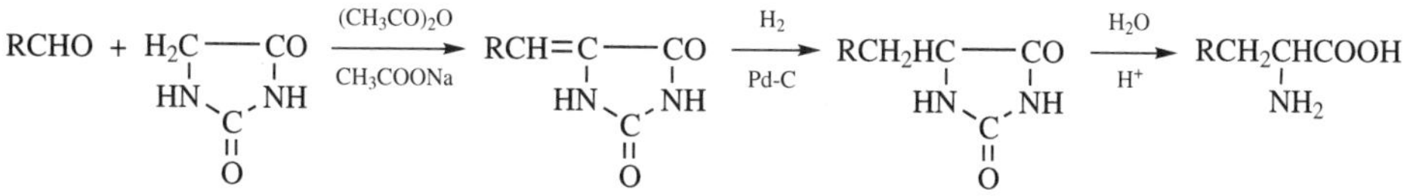

$$RCHO + H_2C\!\!-\!\!CO \xrightarrow[\text{CH}_3\text{COONa}]{(CH_3CO)_2O} RCH\!=\!C\!\!-\!\!CO \xrightarrow[\text{Pd-C}]{\text{H}_2} RCH_2HC\!\!-\!\!CO \xrightarrow[\text{H}^+]{\text{H}_2\text{O}} \underset{\underset{NH_2}{|}}{RCH_2CHCOOH}$$

Reaction with Alkyl Halide.   The active methylene group of an *N*-acyla-mino-malonic acid ester or *N*-acylamino cyanoacetic acid ester condenses readily with primary alkyl halides.

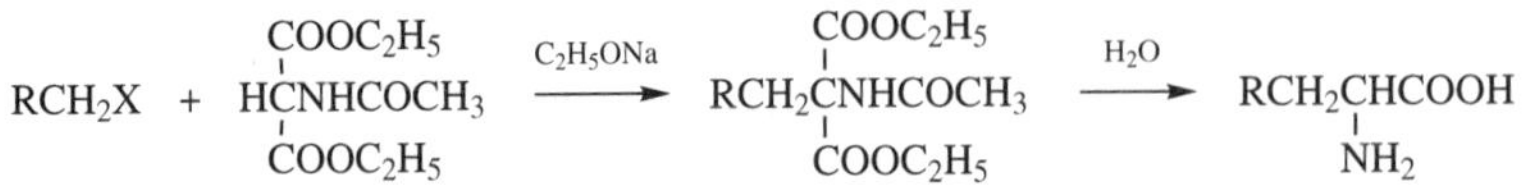

$$RCH_2X \;+\; \underset{\overset{|}{COOC_2H_5}}{\overset{COOC_2H_5}{HCNHCOCH_3}} \xrightarrow{\;C_2H_5ONa\;} \underset{\overset{|}{COOC_2H_5}}{\overset{COOC_2H_5}{RCH_2CNHCOCH_3}} \xrightarrow{\;H_2O\;} \underset{\overset{|}{NH_2}}{RCH_2CHCOOH}$$

Also, Michael addition reactions occur between *N*-acylaminomalonic acid esters and unsaturated compounds, ie, acrolein [107-02-8], acrylonitrile [107-13-1], acrylic acid esters, and amino acids result from hydrolysis of the addition products.

Reaction of Bisglycinatocopper(II).   Bisglycinatocopper(II) [13479-54-4] condenses with aliphatic aldehydes. Removal of copper from the condensate results in β-hydroxy-α-amino acid. This is a classical synthetic method of DL-threonine, but the formation of *allo*-isomer is unavoidable.

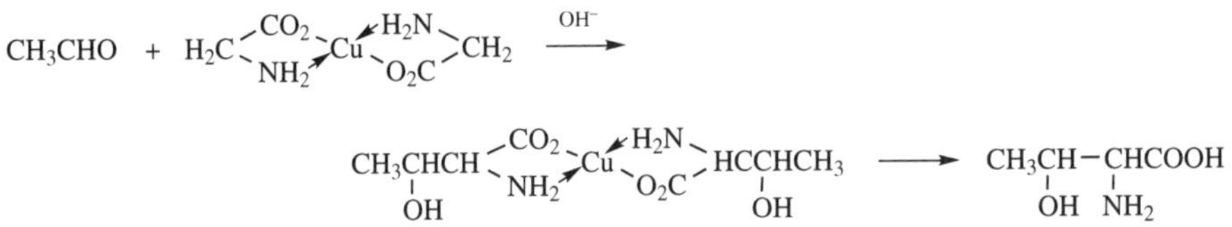

$$CH_3CHO \;+\; H_2C\!\!\underset{NH_2}{\overset{CO_2}{<}}\!\!Cu\!\!\underset{O_2C}{\overset{H_2N}{>}}\!\!CH_2 \xrightarrow{\;OH^-\;}$$

$$CH_3\underset{\overset{|}{OH}}{CHCH}\!\!\underset{NH_2}{\overset{CO_2}{<}}\!\!Cu\!\!\underset{O_2C}{\overset{H_2N}{>}}\!\!\underset{\overset{|}{OH}}{HCCHCH_3} \longrightarrow CH_3\underset{\overset{|}{OH}}{CH}\!-\!\underset{\overset{|}{NH_2}}{CHCOOH}$$

Amination of α-Keto Acids.   α-Keto acids are catalytically reduced

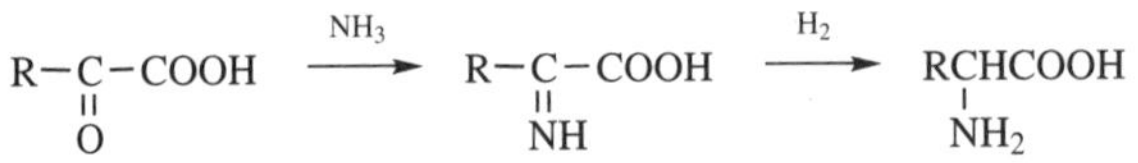

$$R\!-\!\underset{\overset{||}{O}}{C}\!-\!COOH \xrightarrow{\;NH_3\;} R\!-\!\underset{\overset{||}{NH}}{C}\!-\!COOH \xrightarrow{\;H_2\;} R\underset{\overset{|}{NH_2}}{CHCOOH}$$

in the presence of ammonia. α-Keto acids are readily prepared by hydrolysis of substituted hydantoins or double carbonylation of benzyl halide in the case of phenylpyruvic acid [156-06-9]. Enzymatic amination of α-keto acids has been developed by many research groups (54).

Reduction of α-Ketoxime

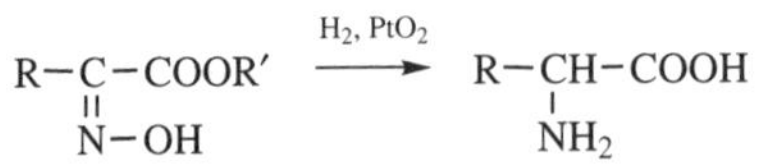

$$R\!-\!\underset{\overset{||}{N-OH}}{C}\!-\!COOR' \xrightarrow{\;H_2,\ PtO_2\;} R\!-\!\underset{\overset{|}{NH_2}}{CH}\!-\!COOH$$

Reduction of α-Nitro Carboxylic Acid

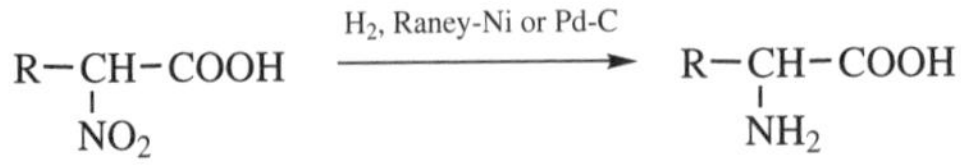

$$R\!-\!\underset{\overset{|}{NO_2}}{CH}\!-\!COOH \xrightarrow{\;H_2,\ Raney\text{-}Ni\ or\ Pd\text{-}C\;} R\!-\!\underset{\overset{|}{NH_2}}{CH}\!-\!COOH$$

Hofmann Degradation

$$R\underset{\overset{|}{CN}}{CH}\!-\!COOC_2H_5 \xrightarrow[H_2SO_4]{H_2O} R\underset{\overset{|}{CONH_2}}{CH}\!-\!COOC_2H_5 \xrightarrow[KOH]{KOBr} R\underset{\overset{|}{NCO}}{CH}\!-\!COOK \xrightarrow{\;H_2O\;} R\underset{\overset{|}{NH_2}}{CH}\!-\!COOH$$

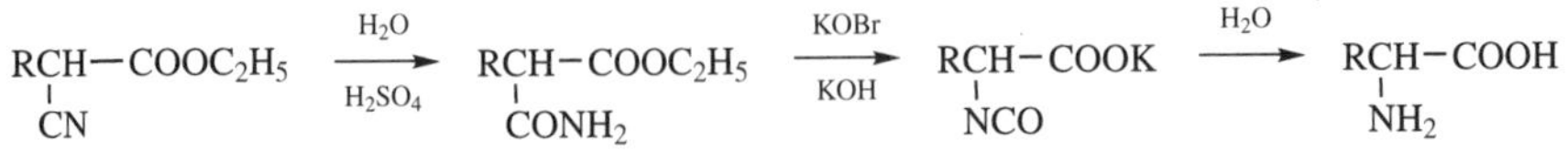

Schmidt Reaction

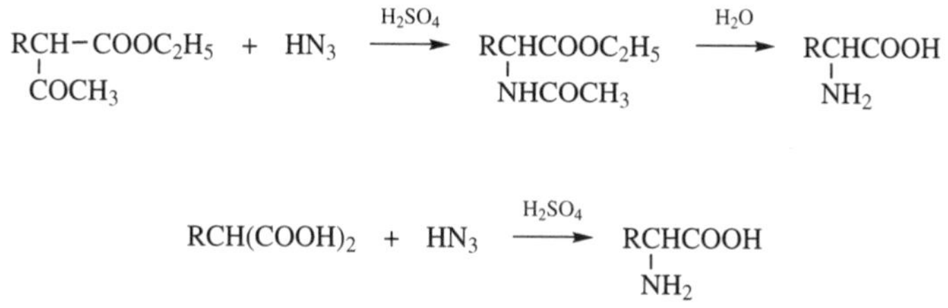

$$\underset{\overset{|}{COCH_3}}{RCH-COOC_2H_5} + HN_3 \xrightarrow{H_2SO_4} \underset{\overset{|}{NHCOCH_3}}{RCHCOOC_2H_5} \xrightarrow{H_2O} \underset{\overset{|}{NH_2}}{RCHCOOH}$$

$$RCH(COOH)_2 + HN_3 \xrightarrow{H_2SO_4} \underset{\overset{|}{NH_2}}{RCHCOOH}$$

Curtius Degradation

$$\underset{\overset{|}{COOC_2H_5}}{RCH-COOK} \xrightarrow{NH_2NH_2} \underset{\overset{|}{CONHNH_2}}{RCH-COOK} \xrightarrow{HNO_2} \underset{\overset{|}{CON_3}}{RCH-COOK} \xrightarrow[H_2O]{H^+} \underset{\overset{|}{NH_2}}{RCHCOOH}$$

$$\underset{\overset{|}{COOC_2H_5}}{RCHCN} \xrightarrow{NH_2NH_2} \underset{\overset{|}{CONHNH_2}}{RCHCN} \xrightarrow{HNO_2} \underset{\overset{|}{CON_3}}{RCHCN} \xrightarrow[H_2O]{H^+} \underset{\overset{|}{NH_2}}{RCHCOOH}$$

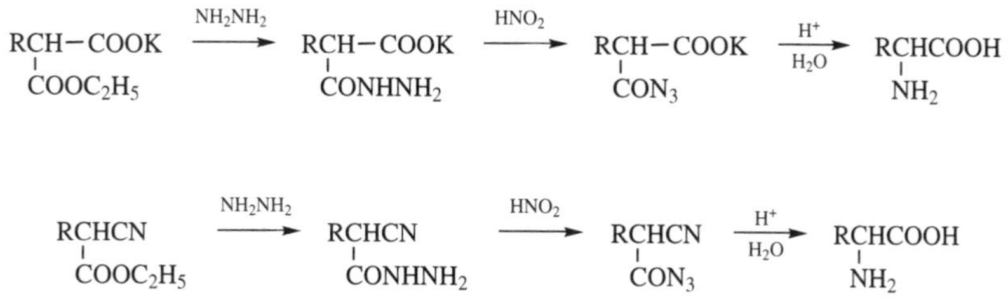

Amine Addition to Double Bond. Production of D,L-aspartic acid from maleic acid ester or fumaric acid ester is a typical example.

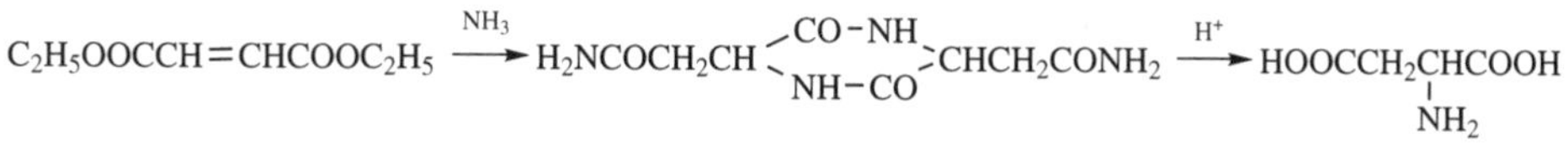

$$C_2H_5OOCCH=CHCOOC_2H_5 \xrightarrow{NH_3} H_2NCOCH_2CH\underset{NH-CO}{\overset{CO-NH}{<}}CHCH_2CONH_2 \xrightarrow{H^+} \underset{\overset{|}{NH_2}}{HOOCCH_2CHCOOH}$$

Carbonylation of Aldehyde. This method (55, 56) is noteworthy as an efficient one-step synthesis.*Wakamatsu Reaction.*

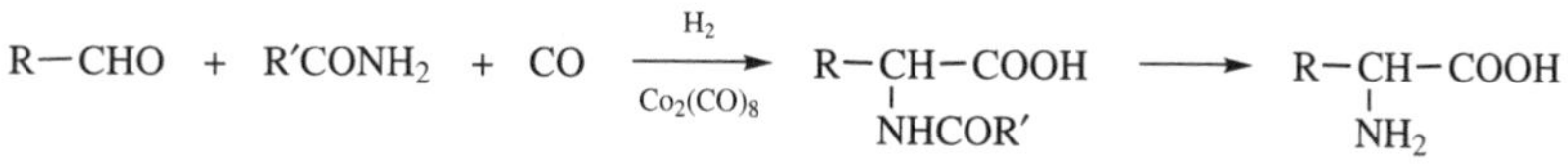

$$R-CHO + R'CONH_2 + CO \xrightarrow[Co_2(CO)_8]{H_2} \underset{\overset{|}{NHCOR'}}{R-CH-COOH} \longrightarrow \underset{\overset{|}{NH_2}}{R-CH-COOH}$$

Modified Method

$$C_6H_5CH\underset{O}{-\!\!-}CH_2 + CH_3CONH_2 + CO \xrightarrow[Co_2(CO)_8, Ti(O\text{-}iC_3H_7)_4]{H_2} \underset{\overset{|}{NHCOCH_3}}{C_6H_5CH_2CHCOOH}$$

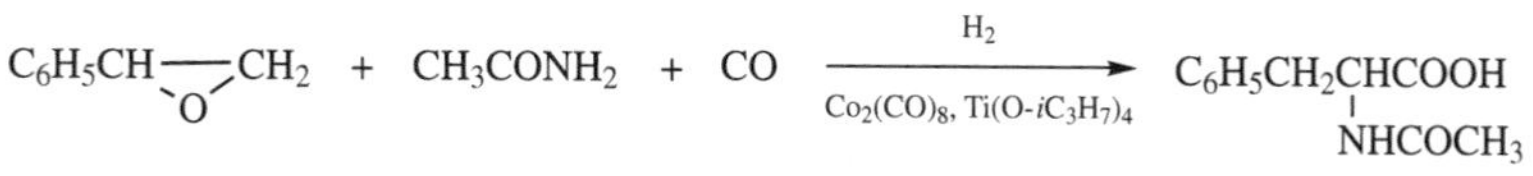

4.2. Optical Resolution. In many cases only the racemic mixtures of α-amino acids can be obtained through chemical synthesis. Therefore, optical resolution (57) is indispensable to get the optically active L- or D-forms in the production of expensive or uncommon amino acids. The optical resolution of amino acids can be done in two general ways: physical or chemical methods which apply the stereospecific properties of amino acids, and biological or enzymatic methods which are based on the characteristic behavior of amino acids in living cells in the presence of enzymes.

Crystallization Method.   Such methods as mechanical separation, preferential crystallization, and substitution crystallization procedures are included in this category. The preferential crystallization method is the most popular. The general procedure is to inoculate a saturated solution of the racemic mixture with a seed of the desired enantiomer. Resolutions by this method have been reported for histidine (58), glutamic acid (59), DOPA (60), threonine (61), *N*-acetyl phenylalanine (62), and others. In the case of glutamic acid, the method had been used for industrial manufacture (63).

Diastereoisomeric Salts.   The formation of salts of optically active bases with racemic acids or of optically active acids with racemic bases leads to diastereomeric mixtures which may be resolved by the differential solubility of the components of such mixtures (64), ie,

$$(+)B + (DL)A \longrightarrow (+)B \cdot (L)A + (+)B \cdot (D)A$$

or

$$(+)A + (DL)B \longrightarrow (+)A \cdot (L)B + (+)A \cdot (D)B$$

The salts in turn may be decomposed by a metathetical reaction involving a stronger base than (+)B or stronger acid than (+)A.

Typical examples of optically active materials for resolution are as follows:

Name	Configuration	CAS Registry Number
	Acidic	
camphorsulfonic acid	D	[3144-16-9]
camphoric acid	*R*	[124-83-4]
camphoric acid	*S*	[560-09-8]
tartaric acid	D	[147-71-7]
tartaric acid	L	[87-69-4]
dibenzoyltartaric acid	*R*	[2743-38-6]
dibenzoyltartaric acid	*S*	[17026-42-5]
malic acid	*R*	[636-61-3]
malic acid	*S*	[97-67-6]
mandelic acid	*R*	[611-71-2]
mandelic acid	*S*	[17199-29-0]
glutamic acid	L	[56-86-0]
	Basic	
brucine		[357-57-3]
cinchonidine	*R*	[485-71-2]
ephedrine	*R*	[299-42-3]
strychnine		[57-24-9]
morphine		[52-27-2]
α-methylbenzylamine	*R*	[3886-69-9]
α-methylbenzylamine	*S*	[2627-86-3]
1-(1-naphthyl)ethylamine	*R*	[3886-70-2]
1-(1-naphthyl)ethylamine	*S*	[10420-89-0]
1-phenyl-2-(*p*-toly)ethylamine	*R*	[30339-32-3]
1-phenyl-2-(*p*-toly)ethylamine	*S*	[30339-30-1]

This procedure is restricted mainly to aminodicarboxylic acids or diamino-carboxylic acids. In the case of neutral amino acids, the amino group or carboxyl group must be protected, eg, by N-acylation, esterification, or amidation. This protection of the racemic amino acid and deprotection of the separated enantio-mers add stages to the overall process. Furthermore, this procedure requires a stoichiometric quantity of the resolving agent, which is then difficult to recover efficiently. Practical examples of resolution by this method have been published (65,66).

Enzymatic Method. L-Amino acids can be produced by the enzymatic hydrolysis of chemically synthesized DL-amino acids or derivatives such as esters, hydantoins, carbamates, amides, and acylates (24). The enzyme which hydro-lyzes the L-isomer specifically has been found in microbial sources. The resulting L-amino acid is isolated through routine chemical or physical processes. The D-isomer which remains unchanged is racemized chemically or enzymatically and the process is recycled. Conversely, enzymes which act specifically on D-isomers have been found. Thus various D-amino acids have been produced (see Table 10).

In another procedure, D-amino acid oxidase (67) is useful to produce L-amino acids from DL-amino acids. α-Ketocarboxylic acids which are formed by the action of enzymes on D-amino acids, are aminated to form L-amino acids by coupling through the action of amino acid aminotransferases (68).

In these procedures, the choice of derivatives and enzyme is important. Sometimes it is possible to get a D-amino acid which remains in the microbial cul-ture supplemented with DL-amino acids (69).

Chromatographic Method. Progress in the development of chromato-graphic techniques (70), especially, in high performance liquid chromatography, or hplc, is remarkable (71). Today, chiral separations are mainly carried out by three hplc methods: chiral hplc columns, achiral hplc columns together with chiral mobile phases, and derivatization with optical reagents and separation on achiral columns. All three methods are useful but none provides universal application.

Chiral Hplc Columns. There are about 40 commercially available chiral columns which are suitable for analytical and preparative purposes (72). In spite of the large number of commercially available chiral stationary phases, it is difficult and time-consuming to obtain good chiral separation. In order to try a specific resolution meaningfully, a battery of chiral hplc columns is necessary and this is quite expensive.

Among various types of chiral stationary phases, the host-guest type of chiral crown ether is able to separate most amino acids completely (73).

Achiral Columns Together with Chiral Mobile Phases. Ligand-exchange chromatography for chiral separation has been introduced (74), and has been applied to the resolution of several α-amino acids. Prior derivatization is some-times necessary. Preparative resolutions are possible, but the method is sensitive to small variations in the mobile phase and sometimes gives poor reproducibility.

The principle of this method depends on the formation of a reversible dia-stereomeric complex between amino acid enantiomers and chiral addends, by coordination to metal, hydrogen bonding, or ion–ion mutual action, in the pre-sence of metal ion if necessary. L-Proline (75), L-phenylalanine (76), N-(p-toluene sulfonyl)-L-phenylalanine (77), L-histidine methyl ester (78), N-acetyl L-valine t-butyl amide (79), etc, are used as chiral addends.

   Derivatization with Optically Active Reagents and Separation on Achiral Columns. This method has been reviewed (80); a great number of homochiral derivatizing agents (HDA) are described together with many applications. An important group is the chloroformate HDAs. The reaction of chloroformate HDAs with racemic, amino-containing compounds yields carbamates, which are easily separated on conventional hplc columns, eg, Ref. (81).

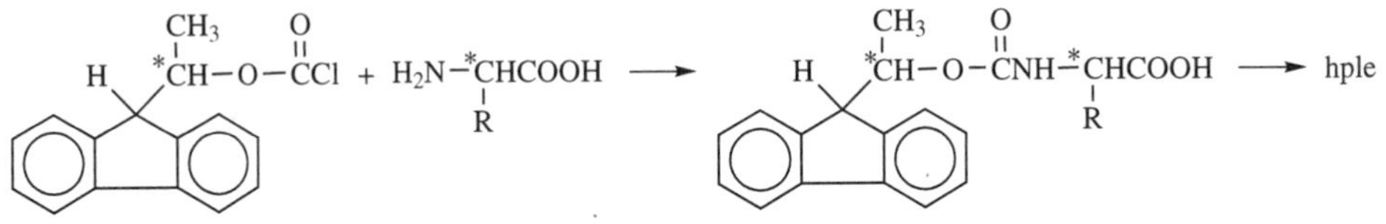

Gas chromatography (gc) is inferior to hplc in separating ability. With gc, it is better to use capillary columns and the application is then limited to analysis (82). Resolution by thin layer chromatography or tlc is similar to lc, and chiral stationary phases developed for lc can be used. However, tlc has not been studied as extensively as lc and gc. Chiral plates for analysis and preparation of micro quantities have been developed (83).

   A new technique referred to as the liquid membrane method has been developed (84). Immobilized liquid membranes are expected to become practical because of many advantages.

   Enzymatic hydrolysis of N-acylamino acids by amino acylase and amino acid esters by lipase or carboxy esterase (85) is one kind of kinetic resolution. Kinetic resolution is found in chemical synthesis such as by epoxidation of racemic allyl alcohol and asymmetric hydrogenation (86). New routes for amino acid manufacturing are anticipated.

   Asymmetric Synthesis. Asymmetric synthesis is a method for direct synthesis of optically active amino acids and finding efficient catalysts is a great target for researchers. Many excellent reviews have been published (87). Asymmetric syntheses are classified as either enantioselective or diastereoselective reactions. Asymmetric hydrogenation has been applied for practical manufacturing of L-DOPA and L-phenylalanine, but conventional methods have not been exceeded because of the short life of catalysts. An example of an enantioselective reaction, asymmetric hydrogenation of α-acetamidoacrylic acid derivatives, eg, Z-2-acetamidocinnamic acid [55065-02-6] (6), is shown below and in Table 4 (88).

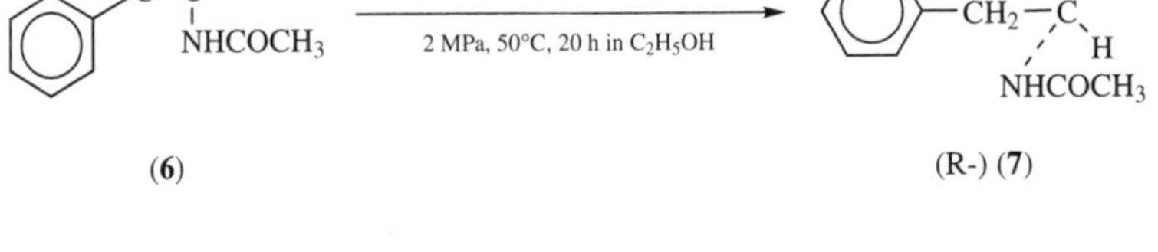

Table 4. **Asymmetric Hydrogenation of (Z)-2-Acetamidocinnamic Acid (6) to (R)-N-Acetylphenylalanine**[a,b]

Ligand	$\frac{[\text{Substrate}]}{[\text{Rh}]}$	Conversion, %	Enantiomeric excess, %
(2S, 4S)-BPPM	1000	100	78.0
	10000	11	79.7
(2S, 4S)-BCPM	10000	100	37.0
(2S,4S)-o-methoxy-BPPM	1000	100	98.0
	10000	64	98.9
(2S,4S)-m-methoxy-BPPM	1000	100	84.8
	10000	7	
(2S,4S)-p-methoxy-BPPM	1000	100	90.4
	10000	100	89.8

[a] Ref. 7.
[b] The R enantiomer is D-N-acetylphenylalanine [10172-89-1].

The ligands are phosphino derivatives of N-tert-butoxycarbonyl pyrrolidine, BPPM and BCPM:

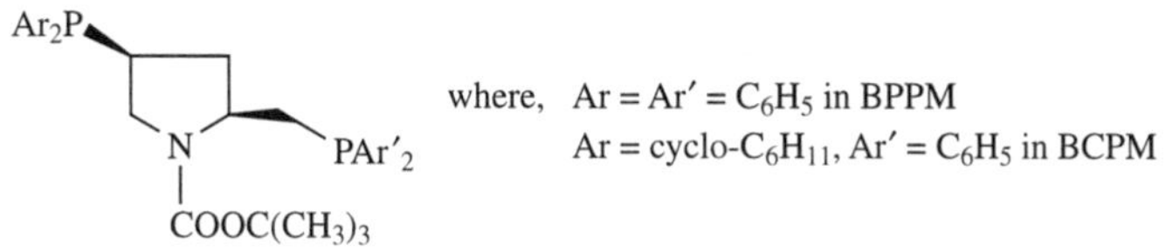

where, Ar = Ar′ = C_6H_5 in BPPM
Ar = cyclo-C_6H_{11}, Ar′ = C_6H_5 in BCPM

Recent developments in asymmetric synthesis include asymmetric amplifying effects and the phenomenon of a nonlinear effect in which the asymmetric reaction gives a product with very high enantiomeric excess (ee) by a chiral auxiliary of low ee (89). Such new techniques will hopefully be applied to the asymmetric synthesis of amino acids in the near future.

Alkylation of protected glycine derivatives is one method of α-amino acid synthesis (90). Asymmetric synthesis of a D-α-amino acid from a protected glycine derivative by using a phase-transfer catalyst derived from the cinchona alkaloids (8) has been reported (91).

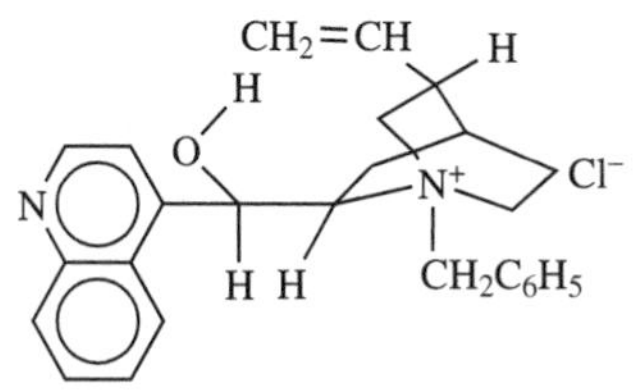

(8)

This catalyst (0.2 equivalents) is used in 50% aq NaOH in CH_2Cl_2 at 25°C for 15 h. The reaction may be represented

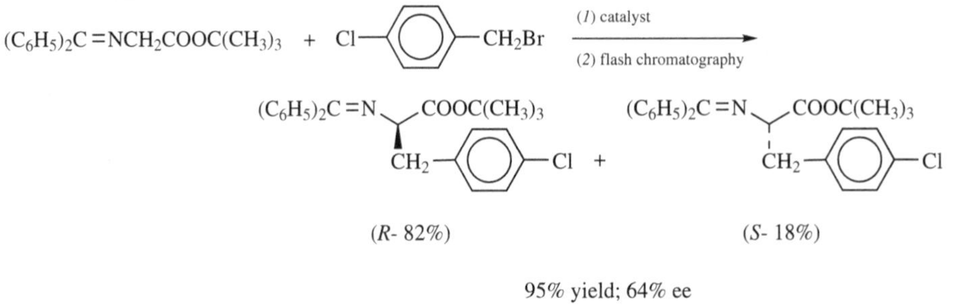

$$(C_6H_5)_2C=NCH_2COOC(CH_3)_3 \quad + \quad Cl-\!\!\!\left\langle\bigcirc\right\rangle\!\!\!-CH_2Br \quad \xrightarrow[\text{(2) flash chromatography}]{\text{(1) catalyst}}$$

(R- 82%)   (S- 18%)

95% yield; 64% ee

The initial product is recrystallized and filtered. The filtrate contains a 65% yield (99% ee) of the *R*-isomer. The crystals are racemic (32% yield; 8% ee).

5. Manufacture and Processing

Since the discovery of amino acids in animal and plant proteins in the nineteenth century, most amino acids have been produced by extraction from protein hydrolyzates. However, there are many problems in the efficient isolation of the desired amino acid in the pure form.

DL-Alanine is the first amino acid which was synthesized chemically (92). Glycine and DL-methionine have also been supplied by this method (20). However, amino acids formed by the chemical method are racemic, and it is necessary to resolve the mixture to get the L- or D-form amino acid which is usually demanded.

In the 1950s, a group of coryneform bacteria which accumulate a large amount of L-glutamic acid in the culture medium were isolated (21). The use of mutant derivatives of these bacteria offered a new fermentation process for the production of many other kinds of amino acids (22). The amino acids which are produced by this method are mostly of the L-form, and the desired amino acid is singly accumulated. Therefore, it is very easy to isolate it from the culture broth. Rapid development of fermentative production and enzymatic production have contributed to the lower costs of many protein amino acids and to their availability in many fields as economical raw materials.

5.1. Direct Fermentation Process. In this process, the microorganisms are cultured in the medium containing carbohydrates (eg, cane molasses, sucrose, glucose, starch hydrolyzate), acetic acid, alcohols (eg, ethanol, methanol) or hydrocarbons (eg, *n*-paraffin) as carbon sources, and nitrogen sources (eg, liquid ammonia, urea, ammonium salt), and minor nutrients [phosphate, sulfate, metal ions (K^+, Mg^{2+}, Fe^{2+}, Ca^{2+} etc.), and if necessary, other growth factors (eg, vitamins, amino acids)]. The amino acid-producing microorganisms metabolize the carbon source and the nitrogen source via the pathways shown in Figure 2 to overproduce and accumulate amino acids in the culture medium. Figure 3 exemplifies the time course of L-glutamic acid production with Corynebacterium glutamicum. A flow sheet of fermentative production of amino acids is shown in Figure 4. The amount and the kinds of amino acids accumulated depend on the trait of microbial strains used and the culture conditions (such as medium composition, pH, temperature, aeration, agitation) (22). At present, cane molasses

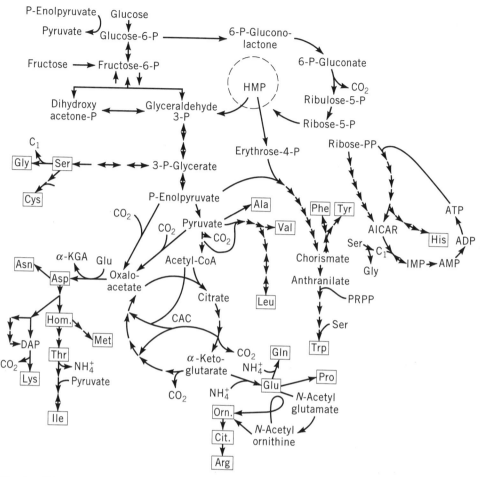

Fig. 2. Biosynthetic pathways for amino acids. HMP = hexose monophosphate pathway; CAC = citric acid cycle; P = phosphate; PP = pyrophosphate; AMP, ADP, and ATP = adenosine mono-, di-, and triphosphate; IMP = inosine 5′-monophosphate; AICAR = 5′-phosphoribosyl-5-amino-4-imidazolecarboxamide; DAP = diaminopimelic acid; PRPP = phosphoribosyl pyrophosphate; α-KGA = α-ketoglutaric acid; Orn = ornithine; Cit = citrulline; C_1 represents the one carbon unit lost to tetrahydrofolate as serine is converted to glycine.

and starch hydrolyzate have been used as carbon sources in the manufacture of amino acids. As can be seen in Table 5 which lists, the amino acid production from carbohydrate, the microorganisms which produce amino acids in high amounts (enough to use for manufacture), are very few: (1) So called "glutamic acid bacteria" (95) (eg, *Corynebaterium glutamicum*, *Brevibacterium flavum*, *Brevibacterium lactofermentum*); (2) Enteric bacteria (eg, *Serratia marcescens*, *E. coli*); and (3) *Bacillus* sp. (eg, *Bacillus subtilis*). Among these, the glutamic acid bacteria can produce some amino acids from acetic acid and ethanol in high amount and these have been used widely as the carbon source (96). Ethanol alone has not been used. The bacteria which can assimilate hydrocarbons (C_{11-14} kerosenes are favorable) have been isolated independently of the ones described

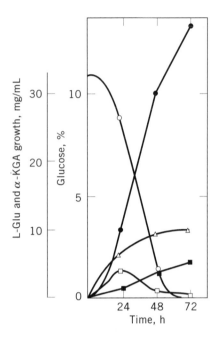

Fig. 3. L-Glutamic acid (and α-ketoglutaric acid, α-KGA) production by a wild type *Corynebacterium glutamicum* (93). The fermentation was carried out with a medium containing 10% glucose, 0.05% KH_2PO_4, 0.05% K_2HPO_4, 0.025% $MgSO_4 \cdot 7\ H_2O$, 0.001% $FeSO_4 \cdot 7\ H_2O$, 0.001% $MnSO_4 \cdot 4\ H_2O$, 0.5% urea, and 2.5 µg/L biotin under aeration and agitation. The pH of the medium was controlled by adding urea during fermentation. ●, L-glutamic acid; ○, glucose; □, lactic acid; △, growth (dry cell weight); ■, α-ketoglutaric acid.

above. As shown in Table 6, the mutant derivatives of *Arthrobacter, Corynebacterium, Brevibacterium, and Nocardia* species have been known to produce a large amount of many kinds of amino acids (111). However, these have not been used in actual production at present. The mutant derivatives of methanol assimilating bacteria belonging to *Pseudommonas, Arthrobacter, Methylomonas, Protaminobacter, Microcylus* (most of these are recently named *Methylobacillus glycogenes* (115)) have been isolated and their accumulation of L-glutamic acid (116), L-leucine, L-valine, L-isoleucine, and L-phenylalanine (115) have been reported. An L-lysine producing mutant of Bacillus sp., a thermophilic methylotroph, has been reported (117).

The pathway for amino acid biosynthesis is regulated at the key enzyme mainly by two regulation mechanisms—feedback inhibition (inhibition of enzyme activity usually by the end product of the pathway) and repression (repression of enzyme formation usually by the end product) (118). These prevent the overproduction of amino acids in the wild-type microbial cells. Amino acid production is attained by selecting favorable culture conditions, and/or changing the bacterial trait through mutational or other genetic treatment to overcome the feedback regulations and induce the overproduction and excretion of amino acids outside the cells (22). Production of L-glutamic acid (112), L-glutamine (119), and L-proline (120) by wild-type glutamic acid bacteria are typical examples of the

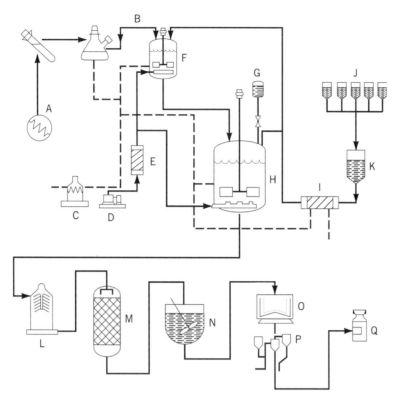

Fig. 4. Fermentative production of amino acids (94). A, pure culture; B, inoculation; C, boiler; D, air compressor; E, air filter; F, seed tank; G, ammonia water for pH control; H, fermenter; I, sterilizer; J, culture media; K, preparation tank; L, centrifugal separator; M, ion-exchange column; N, crystallizing tank; O, crystal separator; P, dryer; Q, amino acid production.

effect of controlling the culture condition. The limited addition of biotin [58-85-5], a growth factor, or the addition of penicillin (or other β-lactam antibiotics) to the culture induce the accumulation of a large amount (80 mg/mL or more) of L-glutamic acid in the culture medium. Other strains of glutamic acid bacteria accumulate L-proline or L-glutamine when NaCl is added to the medium in a high concentration (eg, 6%).

Many kinds of amino acids (eg, L-lysine, L-ornithine, L-phenylalanine, L-threonine, L-tyrosine, L-valine) are accumulated by auxotrophic mutant strains (which are altered to require some growth factors such as vitamins and amino acids) (Table 5, Primary mutation) (22). In these mutants, the formation of regulatory effector(s) on the amino acid biosynthesis is genetically blocked and the concentration of the effector(s) is kept low enough to release the regulation and induce the overproduction of the corresponding amino acid and its accumulation outside the cells (22).

The amino acids (eg, L-arginine, L-histidine) that are synthesized by a straight line biosynthetic pathway are overproduced by the regulatory mutants in which the feedback regulations are genetically released. These mutants are easily selected among the mutants that are resistant to the growth inhibition

Table 5. Fermentative Production of Amino Acids and Their Related Substances from Carbohydrates[a]

Microorganisms	Primary mutation[b]	Genetic markers[b] or culture conditions which enhance productivity	Yield, mg/mL (and or %)	References
L-Glutamic acid				
"glutamic acid bacteria" such as *Coryneb. glutamicum, Brevib. flavum, Brevib. lactofermentum, Microb. ammoniaphilum*	wild	penicillin is added, or biotin is limited	>100, 50%	21
Brevib. thiogenitalis	oleic acid⁻	production from acetic acid and Cu^{2+} is effective	66, 50%	
Coryneb. alkanolyticum	glycerol⁻		40%	
B. lactofermentum	temperature[s]		50%	
C. glutamicum	lysozyme[s]			
B. flavum	vitamin P[r]			97
(*N*-acetyl-L-glutamic acid) *C. glutamicum*	arg⁻		7.5	
(*N*-acetylglutamic-γ-semialdehyde) *C. glutamicum*	arg⁻		2.7	
(L-glutamic-γ-semialdehyde) *B. flavum*	pro⁻	sulfite is effective	13.2	98
L-Glutamine				
"glutamic acid bacteria"	wild	high concentration of NaCl is effective	>44	
Flavobacterium rigens	ser⁻	ammonium fumarate is added	25	
B. flavum	sulfaguanidine[r]		40	
(*N*-acetyl-L-glutamine) "glutamic acid bacteria"	wild		20%	
L-Proline				
B. flavum	ile⁻; 3,4-dehydro-Pro[r]		25.5	
Brevibacterium sp.	his⁻			
C. glutamicum	wild	high concentration of NH_4Cl is effective	40	
Kurthia cathenaforma	ser⁻	Asp is added	30	
Serratia marcescens	3,4-dehydro-Pro[r]	prodegradation⁻, thiazoline-4-carboxylate[r], azetidine-2-carboxylate[r] (transduction)	75	99
Coryneb. melassecola	tyr⁻, phe⁻			
L-Arginine				

Organism	Genotype	Selection / remarks	Yield	Ref.
B. flavum	2-thiazole-3-Ala^r	Guanine$^-$	35	
C. glutamicum	D-Ser^s	D-Arg^r, Arg·hydroxamate^r, 2-thiazole-3-Ala^r; contenious culture is effective	60	100
B. subtilis	Arg·hydroxamate^r	6-azauracil^r	28	
S. marcescens	Arg·degradation$^-$, argR$^-$, argA$^-$	succinate nonassimilating, 6-azauracil^r (transduction)	100	
L-Citrulline				
C. glutamicum, B. flavum	arg$^-$		16	
B. subtilis	arg$^-$		26	
L-Ornithine				
C. glutamicum, B. flavum, B. subtilis (N$^\delta$-acetyl-L-ornithine)	arg$^-$	Arg-analogue^r, 6-azauracil^r	36%	
Paracolobactrum coliforme	arg$^-$, uracil$^-$			
B. subtilis	Arg·hydroxamate^r			
Streptomyces virginae (N$^\alpha$-acetyl-L-ornithine)	lys$^-$		11	
C. glutamicum	arg$^-$		1.2	
L-Lysine				
C. glutamicum	hom$^-$/(Thr$^-$, Met$^-$)	leu$^-$, AEC^r, pyrimidine-analogue^r, Asp-analogue^r; contenious culture is effective	100	101
B. flavum	hom$^-$/(Thr$^-$, Met$^-$)	AEC^r, 2-fluoropyruvic acid^s, pyruvate kinase$^\pm$	40%	102
B. lactofermentum	AEC^r	leu$^-$, ala$^-$, 2-fluoropyruvic acid^s	50, 30%	
B. lactofermentum	AHV^r		16%	
S. marcescens	Lys decarboxylase$^-$, hydroxy-Lys^r, Lys-hydroxamate^r AEC^r		6.5	
Candida pelliculossa (N$^\epsilon$-acetyl-L-lysine)	AEC^r		3.2	
C. hydrocarboclastus (ε-Polylysine)		phe$^+$	23	
Streptomyces albus, S. noursei	wild			
α-ε-Diaminopimelic acid (DAP)				
E. coli, C. glutamicum, B. subtilis, Brevib. ammoniagenes (N-succinyl-DAP)	lys$^-$	AEC^r, hom$^-$/(Met$^-$, Thr$^-$); addition of Lys is effective	20.3	103
			25	

Table 5 (Continued)

Microorganisms	Primary mutation[b]	Genetic markers[b] or culture conditions which enhance productivity	Yield, mg/mL (and or %)	References
L-Homoserine				
A. aerogenes, S. marcescens	DAP^-		20	
C. glutamicum, E. coli, B. flavum (O-acetyl-L-homoserine)	thr^-		15	
Bacillus sp.	met^-		1	
L-Threonine				
E. coli	met^-, DAP^-	ile^-, AHV^r, Thr – degradation^-, rifampicin^r, Lys^r, Met^r, Hom.^r, Asp^r	76	104
E. coli	AHV^r	AEC^r, S-methyl-Cys · hydroxamate^r, ile^-, leu^-	55	
B. lactofermentum	AHV^r	AEC^r, S – methyl – Cys hydroxamate^r, ile^-, leu^-	25	
C. glutamicum	AHV^r	met^-	14	
S. marcescens	AHV^r	Thr – degradation^-, ile^-, AEC^r (transduction)	40	
Providencia rettgerii	ile^-, AHV^r	ethionine^r, Asp·hydroxamate^r, leu^-, Thr – degradation^-, Thia-Ile^r	82	105
B. flavum	AHV^r	dihydropicolinate synthase^-	16.7	
Proteus rettgerii	ile^-, AHV^r		13	
Candida guilliermondii	ile^-, met^-, trp^-		4	
L-Methionine				
C. glutamicum	thr^-, ethionine^r		2	
L-Aspartic acid				
B. flavum	diaminopurine^r		10	
B. flavum	glu^±			
D,L-Alanine				
C. gelatinosum, B. pentosaminoacidicum, Fusarium moniliforme etc	wild		40	
M. ammoniaphilum	Arg·hydroxamate	addition of lactic acid is effective	60	

(L-Alanine)				
Pseudomonas sp.	wild			
C. glutamicum	Ala racemase$^-$	addition of DL-Ala is effective	48	106
(D-Alanine)				
Coryneb. fasciens	wild			
B. lactofermentum	D-cyclo-Ser^r			
L-Isoleucine				
S. marcescens	Ile-hydroxamate^r	α-aminobutyric acid^r, lys C$^-$, thr A$^-$	25	
B. flavum	AHV^r	O-methyl-Thr^r, ethionine^r	34	
C. glutamicum	Thia-Ile^r	AHV^r, Aza-Leu^r, α-aminobutyric acid^r	10	
L-Leucine				
S. marcescens	α-aminobutyric acid^r	ile$^-$	15	
B. lactofermentum	2-thiazole-3-Ala^r	met$^-$, ile$^-$	30	
C. glutamicum	AEC^r		20	
L-Valine				
P. coliforme, E. coli, A. aerogenes, B. ammoniagenes	wild		7–12, 23%	
C. glutamicum	ile$^-$, leu$^-$		30	
S. marcescens	α–aminobutyric acid$^-$		8	
B. lactofermentum	2-thiazole-3-Ala^r		31	
C. glutamicum	AEC^r		20	
L-Tryptophan				
B. flavum	5-fluoro-Trp^r	*m-fluoro-Phe^r, phe$^-$, tyr$^-$, Aza-Ser^r*	10	
C. glutamicum	phe$^-$, tyr$^-$, 5-methyl-Trp^r	*Trp-hydroxamate^r, 4-methyl-Trp^r, ppc$^-$; gene technology is effective*	45	107
B. subtilis	5-fluoro-Trp^r	*indolemycin^r, Aza-Ser^r, 6-diazo-5-oxonorleucine^r*	21.5	
E. coli	5-fluoro-Trp^r	*gene technology is effective; addition of detergent is effective*	40	108
L-Tyrosine				
C. glutamicum	phe$^-$	*3-amino-Tyr^r, p-amino-Phe^r, p-fluoro-Phe^r, Tyr-hydroxamate^r*	17.6	
B. flavum	phe$^-$	*m-fluoro-Phe^r*		

Table 5   (Continued)

Microorganisms	Primary mutation[b]	Genetic markers[b] or culture conditions which enhance productivity	Yield, mg/mL (and or %)	References
L-Phenylalanine				
B. lactofermentum	tyr⁻	p-fluoro-Phe^r, 5-methyl-Trp^r, decoinine^r	25	
C. glutamicum	tyr⁻	p-fluoro-Phe^r	9.5	
L-Serine				
C. glycinophilum	sulfaguanidine^r		4.5	
L-Histidine				
C. glutamicum	2-thiazole-3-Ala^r; triazole-Ala^r; 2-fluoro-His^r	purine and pyrimidine analogues^r	15	
B. flavum	2-thiazole-3-Ala^r	sulfa drugs^r, AEC^r	10	
S. marcescens	2-methyl-His^r, Triazole-Ala^r	His degradation⁻ (transduction)	23	
B. subtilis	5-fluoro-Trp^r, tri-	dimethyltriazaindolidine^r, 2-thiazole-3-Ala^r	13.6	
Streptomyces coelicolor (L-Histidinol)	his⁻		3.5	
B. flavum	his⁻		8	
C. glutamicum	his⁻		10	

[a] Refs. (21, 22, 109).

[b] Abbreviations: AHV, α-amino-β-hydroxyvaleric acid; Hom, L-homoserine; AEC, (S-(2-aminoethyl)-L-cysteine; ppc, phosphoenolpyruvate carboxylase; the strain improvement largely depends on the transduction technology; ^s, sensitive; ^r, resistant; −, auxotroph or deficient; ±, leaky auxotroph; +, prototrophic revertant.

586

Table 6. **Amino Acid Production from Hydrocarbons**[a]

Amino acid produced	Microorganisms	Characteristics[b] of amino acid producers	Amount of accumulation, mg/mL
L-glutamic acid[c]	Arthrobacter paraffineus, Coryneb. hydrocarboclastus, Coryneb. alkanolyticum etc	wild	84
L-ornithine	C. alkanolyticum C. hydrocarboclastus, A. paraffineus	glycerol⁻ arg⁻/cit⁻	9
L-citrulline	Corynebacterium sp.	arg⁻	8
L-lysine (N-acetyl-L-lysine)[d]	Brevibacterium ketoglutamicum, Nocardia sp.	hom⁻	75
diaminopimelic acid	C. hydrocarboclastus	AEC[r], his⁻	41
L-serine	A. paraffineus	lys⁻	10
L-threonine	A. paraffineus	ile⁻	3
L-homoserine	A. paraffineus	ile⁻	27
DL-alanine	Corynebacterium sp.	thr⁻	12
L-valine	C. hydrocarboclastus	wild	4
L-isoleucine	A. paraffineus	ile⁻	5
	Microbacterium paraffinolyticum	val⁻, leu⁻	1.6
L-tyrosine[e]	A. paraffineus	phe⁻	18
L-phenylalanine[e]	A. paraffineus	tyr⁻	15

[a] Ref. (23, 111).
[b] Abbreviations: Hom, homoserine; AEC, S-(2-aminoethyl)-L-cysteine; r, resistant; −, auxotroph.
[c] Ref. 112.
[d] Ref. 113.
[e] Ref. 114.

of the amino acid analogue (whose chemical structure closely resembles the amino acid and falsely regulates the amino acid biosynthesis) (22). Many other amino acids (eg, L-isoleucine, L-leucine, L-lysine, L-proline, L-threonine, L-tryptophan, L-valine) are also overproduced by the analogue resistant mutants as well as the auxotrophic mutants (22).

Amino acid producing strains do not always produce amino acids efficiently enough to be useful for manufacture. Mutant strains are usually improved by combined additions of various genetic markers including other kinds of auxotrophy, analogue resistance, etc (Table 5). These mutations, which increase the amino acid productivity, not only include those which cause deregulation of biosynthesis, but also mutations which make the metabolic flow more efficient to produce the desired amino acid (22,102). The mutations which cause elimination of degradation enzymes and a permeability barrier for the amino acid are, if these are serious, important to improve the amino acid producers. L-Glutamic acid, L-glutamine, L-proline, L-arginine, L-ornithine, L-lysine, L-isoleucine, L-threonine, L-leucine, L-valine, L-tyrosine, L-phenylalanine, L-tryptophan, and L-histidine have been manufactured by fermentation processes.

Advanced biotechnologies such as cell fusion and gene technology are powerful tools offering improvements (121,122). By cell fusion technology, L-lysine and L-isoleucine production of glutamic acid bacteria has been improved

(123). The transduction method is very useful for strain breeding because it is very easy to introduce excellent genetic markers to the amino acid producers (124). The L-histidine, L-arginine, L-threonine, L-proline, and L-isoleucine producers of Serratia marcescens have been skillfully improved by this method using phage PS 20. The bred strains have been used in the manufacture of those amino acids.

Since 1982, the gene multiplication method with plasmid vectors has been applied to the breeding of amino acid producers. To the plasmid vector which can multiply in the amino acid producers, the DNA fragment carrying the genetic information of the other strains of the same or other microorganisms is joined enzymatically, and the resultant recombinant DNA is transferred to the amino acid producers to introduce the new markers due to the donor microorganisms (121,122,125). As exemplified in Table 7, great advances have been made in *C. glutamicum, B. lactofermentum, S. marcescens, B. subtils,* and *E. coli.*

Table 7. Breeding of Amino Acid Producers by Gene Technology[a]

Amino acid produced	Microorganisms	Gene donor	Cloned gene[b] or enzyme	Yield mg/mL	Reference
L-alanine	*E. coli*	*B. stearothermophilus*	Ala dehydrogenase		
D-alanine	*E. coli*	*Ochrobactrum anthropi*	D-aminopeptidase	200	128
L-aspartic acid	*E. coli*	*E. coli*	Asp A	[c]	129
	S. marcescens	*S. marcescens*	Asp A		
L-arginine	*C. glutamicum*	*E. coli*	Arg E, C, B, H		
	E. coli	*E. coli*	Arg A		
L-glutamic acid	*C. melassecola*	*C. melassecola*	Glu A, citrate dehydrogenase, ppc, aconitate dehydratase		
	C. glutamicum	*C. glutamicum*	Glu A		
L-histidine	*C. glutamicum*	*C. glutamicum*	His G, D, C, B	15	
	B. subtilis	*B. subtilis*	His D	8.8	
	S. marcescens	*S. marcescens*	His G, D, B	43	
L-isoleucine	*C. glutamicum*	*C. glutamicum*	Hom dehydrogenase	11	
	S. marcescens	*S. marcescens*	ilv A		
	B. flavum	*E. coli*	ilv A	21	
	C. glutamicum	*B. lactofermentum*	ilv B		
L-lysine	*E. coli*	*B. lactofermentum*	lys A, asd-1		
	E. coli	*Achromob. obae*	α-Amino-ε-caprolactum racemase		
	B. subtilis	*B. subtilis*	Lys A		
	C. glutamicum	*C. glutamicum*	Lys A, dap A,B,D,Y		
	B. flavum	*B. flavum*	ppc		
	C. glutamicum	*E. coli*	Asp A		

Table 7   (*Continued*)

Amino acid produced	Microorgan- isms	Gene donor	Cloned gene[b] or enzyme	Yield mg/mL	Refer- ence
L-phenylala- nine	*C. glutami- cum*	*C. glutamicum*	aro F, chorismate mutase, PRDH	28	
	C. glutami- cum	*E. coli*	aro G, Phe A		
	B. lactofer- mentum	*B. lactofermentum*	aro F, E, L, PRDH	21	
	E. coli	*E. coli*	aro F, Phe A	28.5	
	E. coli	*E. coli*	transaminase		
	E. coli	*C. freuindii*	transaminase		
	E. coli	*Paracoccus denitri- ficans*	transaminase		
	E. coli	*B. stearothermo- philus, Sporosar- cina ureae, B.sphaeroides*	Phe dehydrogenase		
	E. coli	*Rhodotorula rubra*	Phe ammonialyase		130
L-proline	*S. marces- cens*	*S. marcescens*	Pro A, B	75	99
L-serine	*E. coli*	*E. coli*	Gly A		
	E. coli	*E. coli*	Ser A, B, C	1.2	
	C. glutami- cum	*C. glutamicum*	Gly A		131
L-threonine	*E. coli*	*E. coli*	Thr A, B, C	55	132
	C. glutami- cum	*E. coli*	asp kinase, hom dehy- drogenase,hom kinase, Thr C	21	
	C. glutami- cum	*C. glutamicum*	hom dehydrogenase, hom kinase,Thr C	51	133
	B. lactofer- mentum	*B. lactofermentum*	ppc, hom dehydrogen- ase, homkinase	33	134
	B. flavum	*E. coli*	Thr B, C	27	
	S. marces- cens	*E. coli*	ppc	60	
L-tryptophan	*B. subtilis*	*B. subtilis*	Trp B, C, F		
	E. coli	*E. coli*	Trp A, B, E, Ser B		
	E. coli	*E. coli*	Trp A, E, R, tna A	40	108
	B. lactofer- mentum	*B. lactofermentum*	ant-PR transferase, aro B, L, E;Trp A, B, C, D, E, G	7.5	135
	C. glutami- cum	*C. glutamicum*	Trp E, aro F, chorismate mutase,PRDH	45	107
	E. coli	*Enterob. aerogenes*	Tna A		
	E. coli	*Alcalig. faecalis*	Tna A		
L-tyrosine	*C. glutami- cum*	*E. coli*	Aro F	9	
	B. lactofer- mentum	*B. lactofermentum*	Aro A		

[a] Ref. (121, 122, 126).

[b] Gene symbols are according to those of *E. coli.* (127). Abbreviations: Hom, Homoserine; Ant, Anthra- nilic acid; PR, Phosphoribosyl; ppc, Phosphoenolpyruvate carboxylase; PRDH, prephenate dehydrogenase.

[c] 80-fold activity.

Table 8. **Amino Acid Production by Semifermentation Process**[a]

Amino acid produced	Precursor added to the medium	Amount, mg/mL	Reference
D-alanine	DL-alanine	48.8	106
L-histidine	L-histidinol	4	
L-isoleucine	D-threonine	15	
	DL-α-aminobutyric acid	15.7	138
	DL-α-hydroxybutyric acid		
	DL-α-bromobutyric acid		
L-homoisoleucine	L-isoleucine	0.5	
L-methionine	L-hydroxy-4-methylthiobutyric acid	10.9	
	DL-5-(2-methylthioethyl)hydantoin	34	139
L-norleucine	L-norvaline	3	
L-norvaline	L-α-aminobutyric acid	5.5	
	D-threonine		
L-phenylalanine	acetoamidocinnamic acid	75.9	140
L-proline	L-glutamic acid	108.3	141
L-serine	glycine	16	142
	glycine + methanol	54.5	143
L-threonine	L-homoserine	16	
L-tryptophan	anthranilic acid	40	108
	indole	16.7	144

[a] Refs. 23,109.

5.2. Semifermentation Process. In this process, the metabolic intermediate in the amino acid biosynthesis or the precursor thereof is added to the medium, which contains carbon and nitrogen sources, and other nutrients required for growth and production, and the metabolite is converted to the amino acid during fermentation. Some part of the carbon skeleton, the amino donor, and the energy required to complete the amino acid formation are supplied *de novo*. L-Serine production from glycine and methanol (by methylotrophic bacteria, *Hyphomicrobium* sp. and *Pseudomonas* sp.), L-tryptophan production from anthranillic acid [118-92-3] (or indole [120-72-9]) (by *E. coli* and *B. subtilis*) and L-isoleucine production from DL-α-aminobutyric acid and ethanol (by *Brevibacterium* sp.) have been done commercially by this process (22) (Table 8).

5.3. Enzymatic Process. Chemically synthesized substrates can be converted to the corresponding amino acids by the catalytic action of an enzyme or the microbial cells as an enzyme source. L-Alanine production from L-aspartic acid, L-aspartic acid production from fumaric acid, L-cysteine production from DL-2-aminothiazoline-4-carboxylic acid, D-phenylglycine (and D-*p*-hydroxyphenylglycine) production from DL-phenylhydantoin (and DL-*p*-hydroxyphenylhydantoin), and L-tryptophan production from indole and DL-serine have been in operation as commercial processes. Some of the other processes shown in Table 9 are at a technical level high enough to be useful for commercial production (24). Representative chemical reactions used in the enzymatic process are shown in Figure 5.

Nonenzymatic fluorinated amino acids (136) have been developed by enzymatic and chemical methods as bioactive compounds since the antiviral effect of fluorinated alanine was found (137).

Table 9. Enzymatic Production of Amino Acids[a]

Amino acid produced	Substrate	Enzyme	Enzyme source	Reference
		L-Form amino acids		
L-amino acids	DL-acetylamino acids	L-aminoacylase	*Asp. oryzae*	25
	DL-amino acid carbamates		*Pseud. fluorescens*	145
L-alanine	L-aspartic acid	Aspartic-β-decarboxylase	*Pseud. dacunhae* etc.[b]	146
	pyruvic acid + NH_4^+ +NADH	Ala dehydrogenase		
L-aspartic acid	fumaric acid + NH_4^+	aspartase	*E. coli*[b]	
L-citrulline	L-arginine	Arg deiminase		
L-cysteine (derivatives)	β-chloro-DL-alanine + Na_2S (thiols)	Cys desulfhydrase	*Enterobacter cloacae*	147
	L-serine (derivatives) + H_2S	Trp synthase	*E. coli*[b]	
L-cysteine	DL-2-aminothiazoline-4-carboxylic acid	hydrolase + racemase	*Pseud. thiazolinophilum*[b]	
	acetylserine	sulfhydrylase		
L-cystathionine(deriva-tives)	L-acetylserine + H_2S	cystathionine-γ-synthase	*Streptomyces phaeochro-mogenes*	
	L-homoserine + L-cysteine (derivatives)	cystathionine-α-lyase	*Ervi carotovora*	168
L-homoserine (deriva-tives)	L-homoserine + thiol(s)	L-methionine-γ-lyase	*Clostridium, Pseudomonas*	168
L-leucine	α-ketoisocaproic acid + NH_4^+ + NADH	Leu dehydrogenase	*Bacillus sphaericus*[b]	148,149
	DL-leucine amide		*Coryneb. metharica*	150
L-3-methylvaline	α-aminoisocapronitril		*Nocardia* sp.	151
	3,3-dimethyl-2-oxobutyric acid + Asp	transaminase		
L-lysine	DL-α-aminocaprolactum	hydrolase+ Racemase	*Cryptococcus leurentii* *Achromobacter obae*[b]	
L-methionine	α,ε-diaminopimelic acid	DAP decarboxylase		
	DL-methionine + Asp	D-amino acid oxidase + transaminase	*Trigonopsis variabilis*	149
	α-keto-γ-methylthiobutyric acid + NH_4 + formic acid	Phe dehydrogenase+formate dehydrogenase	*Sporosarcina ureae*	
L-ethionine	L-methionine+ Na_2S	methioninase		
L-phenylalanine	*trans*-cinnamic acid + NH_4^+	Phe ammonialyase	*Rhodotolura glutinis,* *Endomyces linderi* etc.[b]	152

591

Table 9 (Continued)

Amino acid produced	Substrate	Enzyme	Enzyme source	Reference
	phenylpyruvic acid + Asp(or NH4-fumarate)	transaminase	Citrobacter freuindii, E. coli[b]	149,153
	phenylpyruvic acid +- NH$_4^+$ + formic acid	Phe dehydrogenase +formate dehydrogenase	Brevibacterium sp., Rodccussp., Sporosarcina ureae Coryneb. metharica Lactobacillus casei	
	DL-phenylalanine amide DL-phenyllactic acid + NH$_4^+$	D- and L-hydroxyisocaproate dehydrogenase + Phe dehydrogenase dehydrogenase hydantoinase + hydrolase acylase + Phedehydrogenase transaminase		
L-2-amino-4-phenylbutyric acid	DL-5-phenylhydantoin acetoamidocinnamic acid 2-keto-4-phenylbutyric acid (KPBA) + Asp KPBA + formic acid +NH$_4^+$	Phe dehydrogenase + formate dehydrogenase	Flavob. sp. Brevib. sp.	
L-α-methylphenylalanine	DL-α-methylphenylalanineamide	transaminase etc Phe ammonialyase	Micobacterium sp.	154
fluoro-L-phenylalanine	fluorophenylpyruvic acid +Asp fluoro-trans-cinnamic acid			
L-serine	glycine + formaldehyde DL-2-oxo-oxazolidine-4-carboxylic acid D-glyceric acid + NH$_4^+$	serine hydroxymethyltransferase[b] hydrolase + racemase glycerate dehydrogenase +Ala dehydrogenase	E. coli Pseud. testeroni	155 156
selenium amino acids	amino acid(s) + selenol(s)	methionine-γ-lyase		
L-tryptophan	indole+ pyruvic acid+NH$_4^+$ indole + DL-serine indole +L-serine	tryptophanase tryptophanase+ Ser racemase Trp synthase	Proteus rettgerii E. coliPseud. putida[b] E. coli	157
	DL-5-indorylmethylhydantoin indole-3-pyruvic acid +NH$_4^+$ + NADH	hydantoinase + hydrolase Phe dehydrogenase	Flavobacterium sp.[b]	158 149
5-hydroxy-L-tryptophan	5-hydroxyindole + pyruvic acid + NH$_4^+$	tryptophanase		

592

fluro-L-tryptophan	fluoroindole + pyruvic acid+NH$_4^+$	tryptophanase	*Erwinia herbicola*[b]	144
L-tyrosine	phenol + pyruvic acid +NH$_4^+$	β-tyrosinase		
3,4-dihydroxy-L-phenylalanine	cathecol + pyruvic acid +NH$_4^+$	β-tyrosinase		
L-valine	α-ketoisovaleric acid + NH$_4^+$ + formic acid	Phe dehydrogenase +formate dehydrogenase		160
		D-Form amino acids[c]		
D-amino acids	DL-acylamino acids	D-aminoacylase	*Arth. crystallopoietes*	
D-alanine	DL-alanine hydantoin	D-hydantoinase + D-N-carbamylamino acid amidohydrolas		
D-alanine amide	DL-alanine amide	D-amino acidaminopeptidase	*Arthrobacter sp.*	128
D-alanine (peptides)	DL-alanine (peptide) amide(s)		*Ochrobactrum anthropi; Rhodococcus erythropolis*[b]	
D-aspartic acid	DL-aspartic acid	asp-β-decarboxylase	*Pseudomonas putida*	
D-cysteine (derivatives)	3-chloro-DL-alanine + NaHS (derivatives)	3-chloro-D-Aladehydrochlorinaseord-cysteinedesulfhydrase		
D-glutamic acid	L-glutamic acid	Glu decarboxylase + Gluracemase	*E. coli* etc	161
D-methionine	DL-acetylmethionine	D-aminoacylase	*Lactobac. brevis*	162
D-phenylalanine	DL-phenylalanine amide	+ acylamino acidracemase + amidase	*Alcaligenes denitrificans-Streptomyces sp.*	
D-phenylglycine	DL-phenylglycine amide	amidase	*Ochrobactrum anthropi, Rhodococcus erythropolis*	
N-carbamyl-D-phenylglycine	DL-5-phenylglycine hydantoin	hydantoinase	*Pseud. putida*	
N-carbamyl-D-p-hydroxyphenylglycine	DL-p-hydroxyphenylglycinehydantoin	hydantoinase	*Pseud. putida*[b]	
p-hydroxy-D-phenylglycine	p-hydroxy-DL-phenylglycinehydantoin	hydantoinase	*Pseud. putida*	163
N-carbamyl-D-valine	DL-valine hydantoin	hydantoinase		

[a] Refs. 23,24.
[b] Chemical reactions are shown in Figure 6.
[c] Refs. 24,168,159.

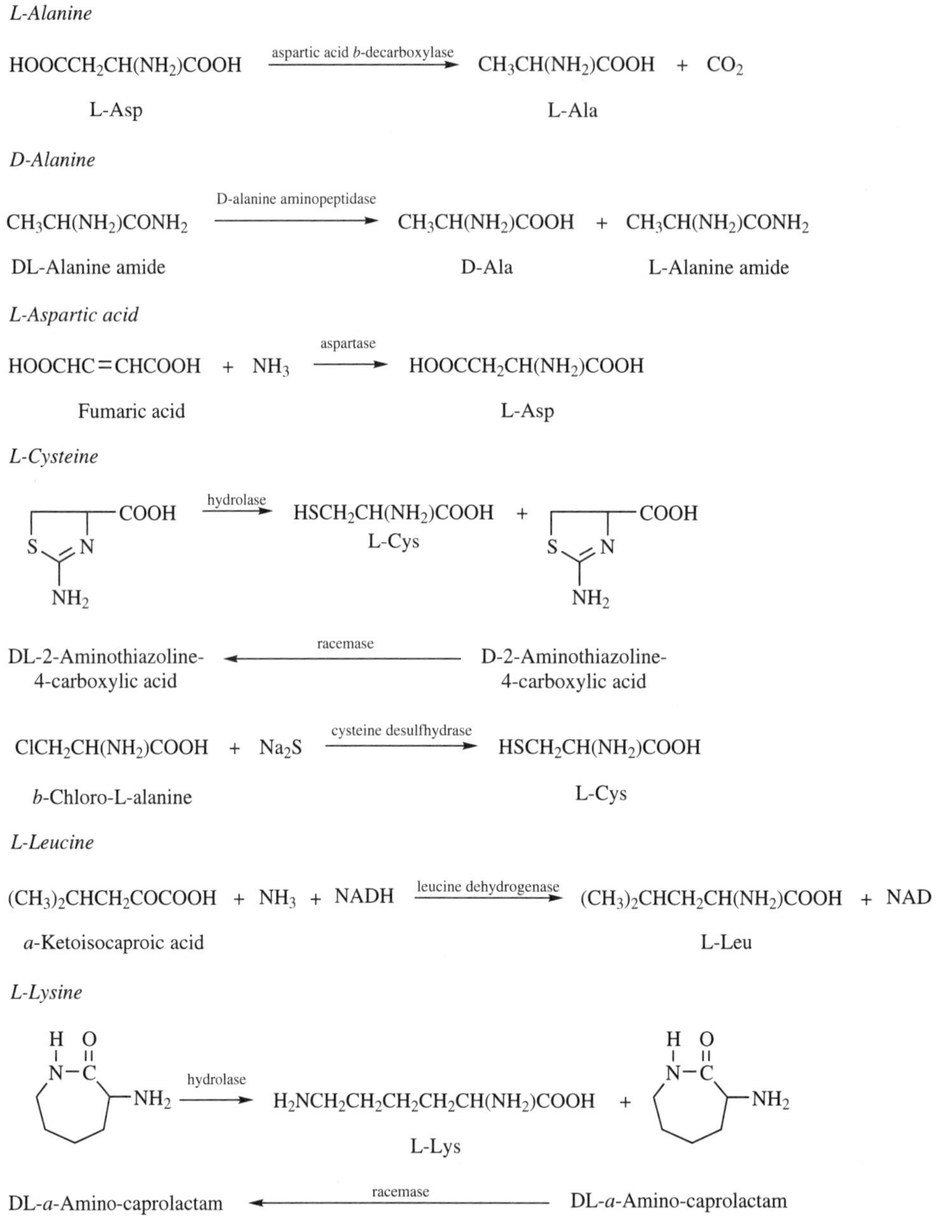

L-Alanine

$$HOOCCH_2CH(NH_2)COOH \xrightarrow{\text{aspartic acid } b\text{-decarboxylase}} CH_3CH(NH_2)COOH + CO_2$$

   L-Asp   L-Ala

D-Alanine

$$CH_3CH(NH_2)CONH_2 \xrightarrow{\text{D-alanine aminopeptidase}} CH_3CH(NH_2)COOH + CH_3CH(NH_2)CONH_2$$

DL-Alanine amide   D-Ala   L-Alanine amide

L-Aspartic acid

$$HOOCHC{=}CHCOOH + NH_3 \xrightarrow{\text{aspartase}} HOOCCH_2CH(NH_2)COOH$$

   Fumaric acid   L-Asp

L-Cysteine

DL-2-Aminothiazoline- $\xleftarrow{\text{racemase}}$ D-2-Aminothiazoline-
4-carboxylic acid   4-carboxylic acid

$$ClCH_2CH(NH_2)COOH + Na_2S \xrightarrow{\text{cysteine desulfhydrase}} HSCH_2CH(NH_2)COOH$$

b-Chloro-L-alanine   L-Cys

L-Leucine

$$(CH_3)_2CHCH_2COCOOH + NH_3 + NADH \xrightarrow{\text{leucine dehydrogenase}} (CH_3)_2CHCH_2CH(NH_2)COOH + NAD$$

a-Ketoisocaproic acid   L-Leu

L-Lysine

DL-*a*-Amino-caprolactam $\xleftarrow{\text{racemase}}$ DL-*a*-Amino-caprolactam

Fig. 5. Representative chemical reactions in the enzymatic production of amino acids.

L-Phenylalanine

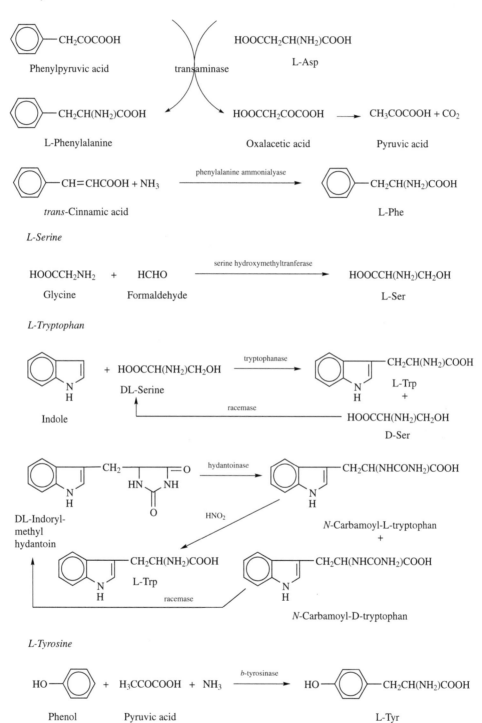

Fig. 5   (*Continued*)

5.4. Chemical Production.   Glycine, DL-methionine, and DL-alanine are produced by chemical synthesis. From 1964 to 1974, some glutamic acid was produced chemically (48). The synthetic amino acid with the largest production is DL-methionine from acrolein (see ACROLEIN AND DERIVATIVES). The industrial production method is shown in the following (164).

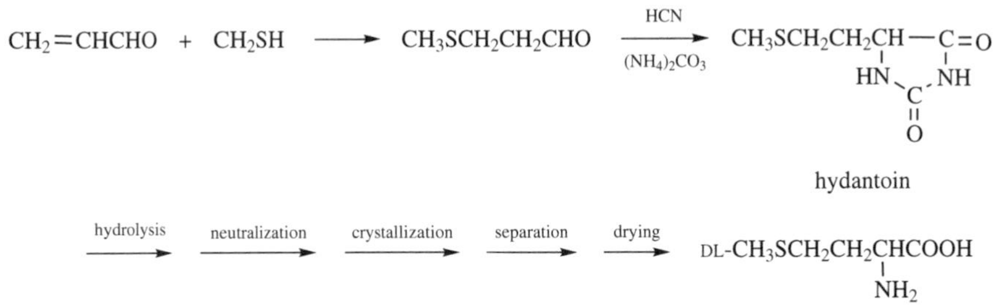

$$CH_2{=}CHCHO + CH_2SH \longrightarrow CH_3SCH_2CH_2CHO \xrightarrow[\text{(NH}_4)_2CO_3]{\text{HCN}} CH_3SCH_2CH_2CH{-}C{=}O$$

$$\underset{\substack{\parallel \\ O}}{HN\diagdown C \diagup NH}$$

hydantoin

$$\xrightarrow{\text{hydrolysis}} \xrightarrow{\text{neutralization}} \xrightarrow{\text{crystallization}} \xrightarrow{\text{separation}} \xrightarrow{\text{drying}} \text{DL-}CH_3SCH_2CH_2CHCOOH$$
$$\underset{NH_2}{|}$$

For glycine (165), two production methods have been employed; Strecker's process and amination of monochloroacetic acid.
Strecker's Process

$$CH_2O + HCN + NH_3 \longrightarrow H_2NCH_2CN \longrightarrow H_2NCH_2COOH$$

Monochloroacetic Acid Process

$$ClCH_2COOH + NH_3 \longrightarrow H_2NCH_2COOH$$

In some cases, Bucherer's process is employed also, but strict control of reaction conditions is needed because the reactivity of formaldehyde is different from other aldehydes. DL-Alanine (166) is produced by either Strecker's or Bucherer's process from acetaldehyde.

5.5. Production by Isolation.   Natural cysteine and cystine have been manufactured by hydrolysis and isolation from keratin protein, eg, hair and feathers. Today the principal manufacturing of cysteine depends on enzymatic production that was developed in the 1970s (167).

6. Economic Aspects

The United States amino acid market was estimated at 400×10^6 kg (885×10^6 lb) valued at $11.2 billion in 1999. Key products were methionine and lysine. Both are used as additives in animal feed. Demand should reach 0.5×10^9 kg (1.1×10^9 lb) in 2004, representing an increase in demand of 3.8%.

Table 10 lists the U. S. market demand of various amino acids by type (169). Table 11 lists the U. S. amino market demand by end use (169).

7. Analytical Methods

Methods have been developed for analysis or determination of free amino acids in blood, food, and feedstocks (170). In proteins, the first step is

Table 10. **U. S. Demand for Amino Acids by Type,** $\times 10^6$ **kg (10^6 lb)**[a]

Amino acids	1999	2004[b]	2009[b]
methionine	200(440)	231(510)	268(590)
lysine	133(293)	169(372)	203(448)
glutamic acid	49(107)	54(120)	59(130)
aspartic acid	7.7(17)	11(25)	19(42)
threonine	4.1(9)	6.3(14)	8.6(19)
phenylalanine	2.7(6)	3.2(7)	3.6(8)
other	5.9(13)	8.2(18)	11(24)
total	*401(885)*	*483(1065)*	*572(1260)*

[a] Ref. 169.
[b] Estimated.

hydrolysis, then separation if necessary, and finally, analysis of the amino acid mixture.

7.1. Protein Hydrolysis.   Acid hydrolysis of protein by 6 M HCl in a sealed tube is generally used (110°C, 24-h). During hydrolysis, slight decomposition takes place in serine (ca 10%) and threonine (ca 5%). Cystine and tryptophan in protein cannot be determined by this method because of complete decomposition.

For determination of tryptophan, 4 M methanesulfonic acid hydrolysis is employed (18). For cystine, the protein is reduced with 2-mercaptoethanol, the resultant cysteine residue is carboxymethylated with iodoacetic acid, and then the protein sample is hydrolyzed. Also, a one-pot method with mercaptoethanesulfonic acid has been developed for tryptophan and cystine (19).

$$R'\!-\!S\!-\!S\!-\!R'' \xrightarrow{\text{reduction}} R'\!-\!SH + R''\!-\!SH$$

$$R\!-\!SH + ICH_2COOH \xrightarrow{\text{carboxymethylation}} R\!-\!S\!-\!CH_2COOH$$

The automated amino acid analyzer depends on ion-exchange chromatography (171) and is now a routine tool for the analysis of amino acid mixtures (172). This most advanced machine can detect as little as 10 pmol in ninhydrin reaction analysis. One-half to two hours are required for each analysis. An analysis chart is shown in Figure 6.

Individual analyses for each amino acid have also been established (173), in particular, Edelhoch spectrometric analysis for tryptophan (174) and Ellman colormetric analysis (175) for cysteine are often used.

Table 11. **U. S Demand for Amino Acids by Market, 10^6**[a]

Market	1999	2004[b]	2009[b]
animal feed additives	1195	1545	1176
food and beverage additives	274	321	378
pharmaceuticals and neutracals	178	244	331
other	20	36	65

[a] Ref. 109.
[b] Estimated.

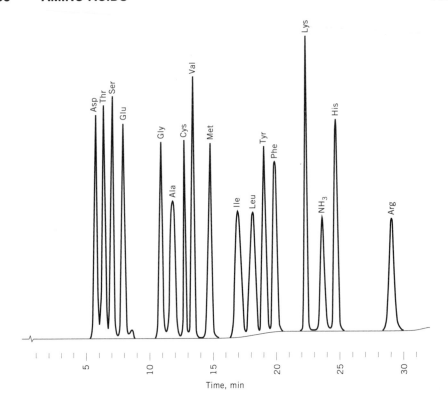

Name	RT	Height	Area	Ratio
Asp	5.81	32764	556900	1699.7
Thr	6.44	34036	592872	1741.8
Ser	7.10	35493	591477	1666.4
Glu	8.02	31380	568744	1812.4
Gly	10.92	28436	574689	2020.9
Ala	11.85	20527	529589	2579.9
Cys	12.81	28942	313874	1084.4
Val	13.45	38240	555963	1453.8
Met	14.77	28544	586978	2056.3
Ile	17.02	18485	538596	2913.6
Leu	18.17	17819	530384	2976.5
Tyr	19.08	27625	519627	1881.0
Phe	19.93	25061	538429	2148.4
Lys	22.33	44070	612254	1389.2
NH_3	23.69	16393	376220	2295.0
His	24.73	31333	586159	1870.7
Arg	29.13	16421	484608	2951.1

Fig. 6. Amino acid analysis by automated ion-exchange chromatography. Standard column, 4.6 mm ID × 60 mm; Ninhydrin developer. Computer print out indicates retention time (RT), height and area of peaks, and the ratio of the height of an amino acid in the sample to the height of a standard amino acid. The number of ng in a 2 nmol sample of each amino acid is also tabulated (not shown here).

7.2. Chromatographic Methods. *High Performance Liquid Chromatography (hplc).* Hplc is currently the fastest growing analytical method and is now available in many laboratories. DL-Analysis by hplc has already been described and hplc methods have been reviewed (176).

Gas Chromatography (gc). A principal advantage of gas chromatography has been the facility with which it can be combined with mass spectrometry for amino acid identification and confirmation of purity. The gc-mass spectrometry combination offers the advantage of obtaining structural information rather than the identification by retention time in hplc.

Successful analysis of amino acids with gas chromatography is dependent on the synthesis of derivatives that are stable, yet volatile (177). The first step is esterification. A variety of alcohols have been used for esterification, including methanol, *n*-propanol, 2-propanol, *n*-butanol, and isobutyl alcohol, as well as some optically pure alcohols, eg, (+)butan-2-ol, (+)octan-2-ol. The next step is *N*-acylation by the addition of acetic anhydride, trifluoroacetic anhydride (TFAA), pentafluoropropionic anhydride (PFPA) or heptafluorobutyric anhydride (HFBA), along with an appropriate solvent. Alkylsilylation which has the advantage of being a fast, one-step derivatization for all groups commonly encountered (NH, OH, SH, COOH) is a useful tool for mass spectrometry. Trimethyl-silylation also is well-suited to chromatographic studies.

The synthesis and the quantitative gas chromatographic analysis of stable, yet volatile, *N*-trifluoroacetyl-*n*-butyl esters of amino acids has been established (178). An extensive review of subsequent advances in gas chromatographic instrumentation has been provided (179).

Thin-Layer Chromatography (tlc). Tlc (180) is used widely for qualitative analysis and micro-quantity separation of amino acid mixtures. The amino acids detected are developed by ninhydrin coloring, except for proline and hydroxyproline. Isatin has been recommended for specific coloring of proline (181).

7.3. Colorimetric and Fluorimetric Analysis. The functional groups of amino acids exhibit little absorption of uv light from 210 to 340 nm where uv absorption spectrometry is most conveniently conducted. Thus color or fluorescence formation reactions are employed for amino acid detection (182).

The most widely applied colorimetric assay for amino acids relies upon ninhydrin-mediated color formation (183). Fluorescamine [38183-12-9] and *o*-phthalaldehyde [643-79-8] are popular as fluorescence reagents. The latter reagent, in conjunction with 2-mercaptoethanol, is most often used in post-column detection of amino acids separated by conventional automated amino acid analysis. More recently, determination by capillary zone electrophoresis has been developed and it is possible to determine attomole quantities of amino acids (184).

7.4. Spectrometric Analysis. Remarkable developments in mass spectrometry (ms) and nuclear magnetic resonance methods (nmr), eg, secondary ion mass spectrometry (sims), plasma desorption (pd), thermospray (tsp), two or three dimensional nmr, high resolution nmr of solids, give useful structure analysis information (185). Because nmr analysis of ^{13}C- or ^{15}N-labeled amino acids enables determination of amino acids without isolation from organic samples, and without destroying the sample, amino acid metabolism can be dynamically analyzed (186). Protein metabolism and biosynthesis of many important

metabolites have been studied by this method. Preparative methods for labeled compounds have been reviewed (168).

7.5. Enzymatic Determination and Microbial Assay.

In these methods, only the desired amino acid is detected in spite of the presence of other amino acids. No expensive tools are needed for these determinations. The required nutrients for microorganisms and practical operations for the microbial assay of amino acids have been reviewed (187,188).

Manometric determination of L-lysine, L-arginine, L-leucine, L-ornithine, L-tyrosine, L-histidine, L-glutamic acid, and L-aspartic acid has been reviewed (189). This method depends on the measurement of the carbon dioxide released by the L-amino acid decarboxylase which is specific to each amino acid.

A kit for the enzymatic determination of L-glutamic acid has been commercialized. Hydrogen peroxide, which is formed through the L-glutamic acid oxidase reaction, is determined by coupling with the peroxidase reaction which forms a blue color complex in the presence of 4-aminoantipyrine [83-07-8] and N-hydroxysulfopropylate. The oxidase reactions are applied to amino acid sensors, which have been commercialized recently for the determination of L-glutamic acid, L-lysine, or mixtures of L-amino acids. The amount of oxygen consumed by the reaction is electrically determined. It is possible to determine amino acids through the NAD-dependent dehydrogenase reaction which is specific to the corresponding amino acid (190). For the details of the enzymatic determination of amino acids, see reference 189.

8. Health and Safety Factors

8.1. Nutrition.

Protein amino acids, which are not synthesized by the body and should be supplied as nutrients to maintain life, are called essential amino acids (6). For humans, L-arginine, L-histidine, L-isoleucine, L-leucine, L-lysine, L-methionine, L-phenylalanine, L-valine, L-threonine, and L-tryptophan are essential amino acids. However, in adults, L-arginine and L-histidine are somewhat synthesized in cells. For histidine, there is evidence that it is dietetically essential for the maintenance of nitrogen balance (191). On the other hand, those amino acids which are synthesized in apparently adequate amounts are nonessential amino acids: L-alanine, L-asparagine, L-aspartic acid, L-cysteine, L-glutamic acid, L-glutamine, glycine, L-proline, L-serine, and L-tyrosine. Of these, L-tyrosine and L-cysteine are essential for children. Recent advances in nutritional studies of amino acids have led to development of amino acid transfusion (192).

The nutritional value of a protein can be improved by the addition of amino acids which are short in the protein (11). The amino acid score has been used to evaluate the nutritive value of food proteins. The reference pattern or scoring pattern of essential amino acids presented by the Food and Agriculture Organization of the World Health Organization (FAO/WHO) in 1973 (193) and by FAO/WHO/of the United Nations in 1985 (194) has been used to calculate the amino acid score. The proportion of each essential amino acid in a particular protein to that of the reference pattern is calculated and an essential amino acid that has a score of less than 100% is called the limiting amino acid of the protein. If there are two or more essential amino acids of this kind, they are called the first limiting amino acid, the second limiting amino acid, and so on in ascending order of their percentage. The amino acid scores and the limiting amino acids of many foods have been calculated (195). Almost all of the plant proteins have limiting amino acids. The nutritionally important cereal proteins are particularly deficient in L-lysine and are also low in L-threonine and L-tryptophan. The limiting amino acid in soybean meal is methionine. No animal proteins, except for those of shellfish, mollusks, and crustaceans, have limiting amino acids.

Before the reference pattern was established, it was usual to use the amino acid composition of egg protein as a standard. The ratio of the amount of limiting amino acid present in a protein to that in egg protein is the chemical score of that protein. For practical purposes, the chemical score and the amino acid score are quantitatively similar. Because these are based on the chemical analysis of protein, both ignore the biological availability of the essential amino acids. The biological value (BV, the ratio of nitrogen retained in the body to that absorbed), the Net Protein Utilization (NPU, the BV of the protein multiplied by the digestibility), and the Protein Efficiency Ratio (PER, the gain in weight per gram of dietary protein) are biological measures. For details of protein quality, see references 6 and 195.

Feeding standards, which have been instituted nationally, indicate the amount of the essential amino acids (together with other nutrients) for the rational breeding of domestic animals. The feeding standards of the National Research Council (NRC) of the United States and Agricultural Research Council (ARC) of the United Kingdom are well known (the former indicates the minimal amount and the latter shows the recommended amount). Japanese Feeding Standards have been instituted (196).

Amino acids essential for young rats (197) and fishes (198) have been reviewed. Rats preferably eat a diet with sufficient amounts of essential amino acids rather than one that is deficient (199). Each essential amino acid, consumed in self-selection, has been reviewed (200). A protein diet with an excess of essential amino acids has been described as a poor protein diet from investigations that showed remarkable growth inhibition and occurrence of fatty liver disease in rats (201). This is called amino acid imbalance (202).

8.2. Biosynthesis of Protein. The dynamic equilibrium of body protein was confirmed by animal experiments using ^{15}N-labeled amino acids in 1939 (203). The human body is maintained by a continuous equilibrium between the biosynthesis of proteins and their degradative metabolism where the nitrogen lost as urea (about 85% of total excreted nitrogen) and other nitrogen compounds is about 12 g/d under ordinary conditions. The details of protein biosynthesis in living cells have been described (2,6) (see also PROTEINS).

8.3. Toxicity of α-Amino Acid. LD_{50} values of α-amino acids are listed in Table 12. L-Lysine and L-arginine are mutually antagonistic. The addition of an excess of one reduces the biological value of protein and only the addition of the other overcomes the effect. The other antagonism occurs between the branched-chain amino acids (L-isoleucine, L-leucine, and L-valine). Pellagra is caused by L-leucine inhibition of niacin formation from L-tryptophan. People living on low protein nutrition and with only a marginally adequate intake of

Table 12. **Toxicity^a of Amino Acids**

Amino acid	LD$_{50}$^{b}	
	Oral	Intraperitoneal
L-Arg·HCl	12 g	
L-Cys	5580 mg	1620 mg
L-Cys·HCl		1250 mg^c
L-(Cys)$_2$	25 g	
L-His	7930 mg	
L-Ileu		6822 mg
L-Leu		5379 mg
D-Leu		6429 mg
L-Lys·HCl	10 g	4019 mg
L-Met	36 g	4328 mg
L-Phe		5287 mg
D-Phe		5452 mg
DL-Thr		3098 mg
L-Trp		1634 mg
D-Trp		4289 mg
L-Val		5390 mg
D-Val		6093

^{a} Rat, unless otherwise noted.
^{b} Ref. 203.
^{c} Mouse.

L-tryptophan and niacin, eg, living on sorghum as a dietary staple, are at a risk of developing pellagra if they have an excess intake of L-leucine (6).

In the case of hyperphenylalaninaemia, which occurs in phenylketonuria because of a congenital absence of phenylalanine hydroxylase, the observed phenylalanine inhibition of protein synthesis may result from competition between L-phenylalanine and L-methionine for methionyl-tRNA. Patients suffering from maple syrup urine disease, an inborn lack of branched chain oxo acid decarboxylase, are mentally retarded unless the condition is treated early enough. It is possible that the high level of branched-chain amino acids inhibits uptake of L-tryptophan and L-tyrosine into the brain. Brain injury of mice within ten days after their birth was reported as a result of hypodermic injections of monosodium glutamate (MSG) (0.5–4 g/kg). However, the FDA concluded that MSG is a safe ingredient, because mice are born with underdeveloped brains regardless of MSG injections (205).

Furthermore, the Joint Expert Committee on Food Additives (206) (JECFA) of the WHO and FAO of the United Nations issued the evaluation of the safety, stating that on the basis of the available data, the total dietary intake of glutamates arising from their use at the levels necessary to achieve the desired technological effect and from their acceptable background in food does not, in the opinion of the committee, represent a hazard to health.

The nephrotoxic amino acid, lysinoalanine [18810-04-3], formed upon alkaline treatment of protein, was reported in 1964 (207). Its toxicity seems to be mitigated in protein in that it is not released by normal digestion (208). Naturally occurring new amino acids, which can be classified as proteinaceous or non-proteinaceous, can, as in the case of those from some legumes, show a remarkable toxicity (209). For the details of amino acid toxicity, see reference 6. Enzyme inhibition by amino acids and their derivatives have been reviewed (210).

8.4. Metabolism of Amino Acids. The amino acids are metabolized principally in the liver to a variety of physiologically important metabolites, eg, creatine (creatinine), purines, pyrimidines, hormones, lipids, amino sugars, urea, ammonia, carbon dioxide, and energy sources. The reactions of transamination, deamination, and decarboxylation are important for amino acid metabolism. Glutamate–oxalacetate transaminase (GOT) and glutamate–pyruvate transaminase (GPT) are the most important transaminases. NAD-dependent glutamic acid dehydrogenase, amino acid oxidases, and aspartic acid ammonia-lyase catalyze the deamination of amino acids. Amino acid decarboxylase is important to amine formation. The products of these enzyme reactions are metabolized finally to carbon dioxide and water via the pathways for sugar metabolism or fatty acid metabolism. The amino acids which are degraded via the former or the latter pathway are called glucogenic (or glycogenic) amino acids or ketogenic amino acids, respectively (13).

8.5. As Neurotransmitters. Several amino acids serve as specialized neurotransmitters in both vertebrate and invertebrate nervous systems. These amino acids can be classified as inhibitory transmitters, such as γ-aminobutyric acid [56-12-2] (GABA) and glycine, and excitatory amino acids, examples of which are L-glutamic acid and L-aspartic acid. A number of other amino acids and their related substances occur in the brain and have some physiological activity. These include taurine [107-35-7], serine, proline, pipecolic acid [535-75-1], N-acetyl-aspartic acid [997-55-7], α- and β-alanines, and L-cysteine sulfinic acid [2381-08-0]. For more details about neurotransmitter amino acids, see reference 211.

8.6. Modification of Amino Acid in Protein Molecules. Protein kinases, whose activities are regulated by secondary messengers, such as cyclic nucleotide and Ca^{2+}, modify physiologically important proteins by phosphorylating the hydroxy moiety of serine, threonine, and tyrosine in protein molecules. Consequently, various cellular functions, cell growth, and cell differentiation are seriously affected (212). Because the intracellular pool of secondary messengers is under the control of hormones, growth factors, and neurotransmitters, protein phosphorylation is also regulated by these signal compounds.

The participation of protein kinases in oncogenesis has been suggested (213). Thus some oncogenes (src, fps, yes, mos) are known to encode for protein kinases. Formation of some cross-links within molecules of proteins which are physiologically important and have a slow turnover rate (such as collagen) is believed to correlate seriously with aging of animal tissue. Nonenzymatic glycation of protein by the Maillard reaction possibly acts by such cross-link formation. L-Lysine and L-hydroxylysine moieties in a collagen molecule are often oxidized to their corresponding aldehydes by the action of lysyloxidase. Aldehydes react with the amino groups of L-lysine or L-hydroxylysine in the adjacent collagen molecules to form Schiff-base cross-links among the different collagen molecules. β-Aminopropionitrile [151-18-8] inhibits the oxidase reaction. Pyridinoline [63800-01-1] (**9**) and histidinoalanine [65428-77-5] (10) are also found as cross-linkers in collagen. The relationship between the aging of animal tissues and these cross-linking agents has been discussed (214).

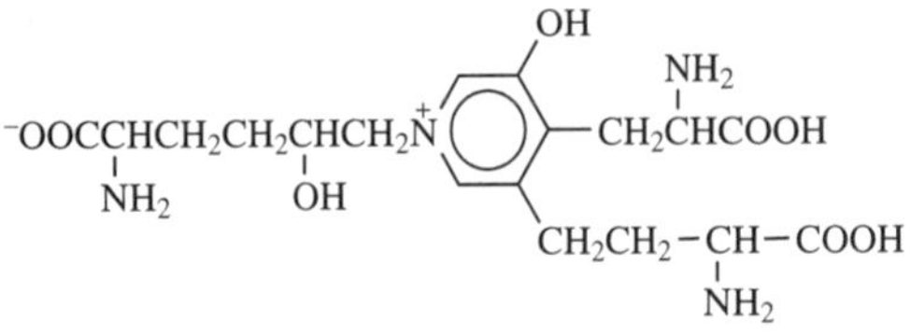

(9)

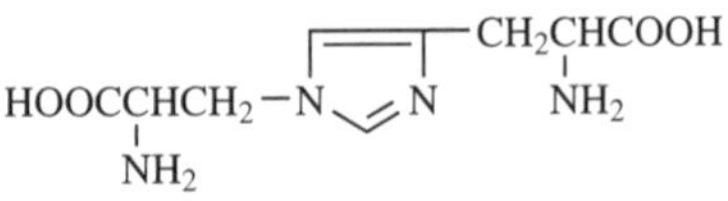

(10)

9. Uses

Amino acids are used in feeds (215), food (215), parenteral and enteral nutrition (192), medicine (216), cosmetics (217), and raw materials for the chemical industry (218).

9.1. In Feeds.   The agricultural products which are used as feedstuff for domestic animals are different, depending on the areas where they are used. These feedstuffs do not always meet the essential amino acid requirements for the economical growth of the animals and usually require DL-methionine and L-lysine supplements as the first and/or second limiting amino acid as shown in Table 13. The addition of these amino acids to the feeds saves the use of feed protein without affecting the growth response of animals. L-Tryptophan is the second limiting amino acid of maize and as the feed protein level becomes lower, its requirement increases. When the protein level decreases further, L-threonine becomes another limiting amino acid. In Western Europe where wheat and barley are the basis of feeds, L-threonine is the second limiting

Table 13. Limiting Amino Acids of Some Common Feedstuffs for Pig and Chicken[a].

Ingredient	Crude protein, %	Pig		Chicken	
		First	Second	First	Second
maize	8.9	Lys	Trp	Lys	Trp
sorghum	9.5	Lys	Thr	Lys	Arg
barley	11.1	Lys		Lys	Met
wheat	12.6	Lys	Thr	Lys	Thr
soybean meal	46.2	Met	Thr	Met	Thr
fish meal	64.3			Arg	
rapeseed meal	35.3	Met		Lys	Arg
peanut meal	47.4	Lys		Met	Lys
sunflower seed meal	31.7	Lys		Met	Thr
meat and bone meal	48.6	Lys		Trp	Met
cottenseed meal	64.3	Lys	Thr	Lys	Met
corn gluten meal	63.6	Lys	Trp	Lys	Trp

[a] Ref. 216

amino acid after L-lysine and both are usually added to the feedstuff. The issue most focused on at present is development of protected amino acids for ruminants (219). Protected methionine has already been commercialized. A method for supplying methionine to cows (220) and an aqueous lysine-containing animal feed supplement (221) have been described.

9.2. In Foods. Each amino acid has its characteristic taste of sweetness, sourness, saltiness, bitterness, or "umami" as shown in Table 14. Umami taste, which is typically represented by L-glutamic acid salt (and some 5′-nucleotide salts), makes food more palatable and is recognized as a basic taste, independent of the four other classical basic tastes of sweet, sour, salty, and bitter (222).

The existence of protein receptors in the tongues of mice and cows have been shown. Monosodium L-glutamate MSG [142-47-2] is utilized as a food flavor enhancer in various seasonings and processed foods. D-Glutamate is tasteless. L-Aspartic acid salt has a weaker taste of umami. Glycine and L-alanine are slightly sweet. The relationship between taste and amino acid structure has been discussed (223).

Aspartame (L-aspartyl-L-phenylalanine methyl ester [22839-47-0]) is about 200 times sweeter than sucrose. The Acceptable Daily Intake (ADI) has been

Table 14. **Taste Profiles of L-Amino Acids**[a]

L-Amino acid	Threshold-value, mg/dL	Sweet	Sour	Bitter	Salty	Umami
Gly	110	++				
Hyp	50	++		++		
β-Ala	60	++				+
Thr	260	++				
Pro	300	++		+++		
Ser	150	++				+
Lys·HCl	50	++		++		+
Gln	250	+				+
Phe	150			+++		
Trp	90			+++		
Arg	10			+++		
Arg·HCl	30	+		+++		
Ile	90			+++		
Val	150	+		+++		
Leu	380			+++		
Met	30			+++		+
His	20			++		
His·HCl	5		+++	+	+	
Asp	3		+++			+
Glu	5		+++			++
Asn	100		++	+		
MSG	30	+			+	+++
monosodium aspartate	100				++	++

[a] Ref. 215.
[b] Profiles of each basic taste intensity are expressed as follows: (+ + +) strongest, (++) stronger, and (+) detectable.

established by JECFA as 40 mg/kg/day. Structure-taste relationship of peptides has been reviewed (224). Demand for L-phenylalanine and L-aspartic acid as the raw materials for the synthesis of aspartame has been increasing. D-Alanine is one component of a sweetener "Alitame" (225). Derivatives of aspartame are also described as flavor modifiers (226) and sweeteners in chewing gum (227).

In traditional cooking of proteinaceous foods, the fundamental difference between Western and Oriental cultures is that the former cooks proteins with unseasoned fats and the latter cooks with many kinds of traditional seasonings that have tastes of amino acids. Western cultures have some traditional foods with amino acid taste such as cheese. Protein hydrolysates are popular as seasonings (228).

The enzymatic hydrolysates of milk casein and soy protein sometimes have a strong bitter taste. The bitter taste is frequently developed by pepsin [9001-75-6], chymotrypsin [9004-07-3], and some neutral proteases and accounted for by the existence of peptides that have a hydrophobic amino acid in the carboxylic terminal (229). The relation between bitter taste and amino acid constitution has been discussed (230).

Amino acids play a role in food processing in the development of a cooked flavor as the result of a chemical reaction called the nonenzymatic browning reaction (231).

Currently available proteins are all deficient to greater or lesser extent in one or more of the essential amino acids. The recently advanced plastein reaction (232) has made it possible to use protein itself as substrate and to attach amino acid esters to the protein with high efficiency. By this method, soy bean protein (which is deficient in methionine) has been improved to the extent of having covalently attached L-methionine at 11%.

9.3. In Parenteral and Enteral Nutrition. Amino acid transfusion has been widely used since early times to maintain basic nitrogen metabolism when proteinaceous food cannot be eaten. It was very difficult to prepare a pyrogen-free transfusion from protein hydrolysates. Since the advances in L-amino acid production, the crystalline L-amino acids have been used and the problem of pyrogen in transfusion has been solved. The formulation of amino acid transfusion has been extensively investigated, and a solution or mixture in which the ratio between essential and nonessential amino acid is 1:1, has been widespread clinically. Special amino acid mixtures (eg, branched chain amino acids-enriched solution) have been developed for the treatment of several diseases (192). An enteral nutritional composition containing methionine for clinical or dietary use has been described (233).

9.4. In Medicine. Many amino acids have been used or studied for pharmaceutical purposes. L-Glutamine has been used as a remedy for gastric and duodenal ulcers. L-DOPA [L-3-(3,4-dihydroxyphenyl)alanine] has been widely employed as an antiparkinsonism agent. L-α-MethylDOPA is an effective antihypertensive drug. L-Tryptophan and 5-hydroxy-L-tryptophan [4350-09-8] are effective as antidepressants. In animal experiments it was demonstrated that L-tryptophan induces sleep. Potassium aspartate [14007-45-5] is widely used for improving disturbances in electrolyte metabolism. Calcium aspartate [10389-09-0] is known as a calcium supplement. Glutamic acid hydrochloride is a gastric acidifier which acts to counterbalance a deficiency of hydrochloric acid in the gastric juice. L-Arginine and L-ornithine are used for ammonia detoxification. p-Hydroxy-D-phenylglycine, D-phenylglycine, D-cysteine (234), D-aspartic acid (235) are important as the side chains of β-lactam antibiotics (see ANTIBIOTICS, β-LACTAMS). D-Homophenylalanine (236) is a raw material for chemical synthesis of enalapril [75847-73-3], an inhibitor of angiotensin converting enzymes. D-Valine (237) is the raw material for the chemical synthesis of pyrethroid agricultural chemicals. For the details of the use of amino acids in medicine, see reference 138. Some of the recently reported medical applications include: glutamic acid decarboxylase for treating type I diabetes (238); compositions based on proline, glycine, and lysine for treating lesions and wounds (239); modulation of Bruton's tyrosine kinease and intermediates for the treatment of osteoporosis (240); and the regulation of T cell mediated immunity by tryptophan (241).

9.5. In Cosmetics.

Amino acids and their derivatives occur in skin protein, and they exhibit a controlling or buffering effect of pH variation in skin and a bactericidal effect (217). Serine is one component of skin care cream or lotion. N-Acylglutamic acid triethanolamine monosalt is used for shampoo. Glucose glutamate is a moisturizing compound for hair and skin (242). New histidine derivatives as free antiradical agents in cosmetics has been described (243).

Cysteine is used as a reductant for cold wave treatment in place of thioglycolic acid. N-Lauroylarginine ethyl ester [48076-74-0] is applied as the hydrochloride as a preservative. Urocanic acid [104-98-3] which is derived from histidine is used in skin cream as a uv absorber (244).

9.6. In Industrial Chemicals.

Recently, as some amino acids (eg, L-glutamic acid, L-lysine, glycine, DL-alanine, DL-methionine) have become less expensive chemical materials, they have been employed in various application fields. Poly(amino acid)s are attracting attention as biodegradable polymers in connection with environmental protection (245).

Surfactants. N-Acylglutamates, sodium N-lauroyl sarcosinate [137-16-6], and N-acyl-β-alanine Na salt are used in the cosmetic field as nontoxic surfactants (246). Some of them (eg, N-acylglutamic acid dibutylamide) are used as oil gellating agents to recover effluent oil in seas and rivers (247).

Liquid Crystals. Ferroelectric liquid crystals have been applied to LCD (liquid crystal display) because of their quick response (248). Ferroelectric liquid crystals have chiral components in their molecules, some of which are derived from amino acids (249). Concentrated solutions (10–30%) of α-helix poly(amino acid)s show a lyotropic cholesteric liquid crystalline phase, and poly(glutamic acid ester) films display a thermotropic phase (250). Their practical applications have not been determined.

Artificial Leather. Poly(γ-methyl glutamate) [29967-97-3] that has excellent weatherability, nonyellowing, high moisture permeability, and heat resistance, was developed as the original coating agent for artificial leather (85). To improve flexibility and stretch, a block copolymer with polyurethane was developed. Poly(L-leucine) [25248-98-0] is being tested as artificial skin or wound dressing (251).

Protected Amino Acids. Various types of protected amino acids for peptide synthesis are available commercially (252).

Isocyanate. Lysine has two amino groups in the molecule and diisocyanate is prepared by reaction with phosgene. Lysine triisocyanate [69878-18-8] (LTI) is developing on a commercial scale in Japan (253).

$$OCN-(CH_2)_4CHCOOCH_2CH_2-NCO$$
$$\underset{NCO}{|}$$

Hardeners and Vulcanizing Agents. For epoxy resins, acylhydrazide derivatives of amino acids are used (254).

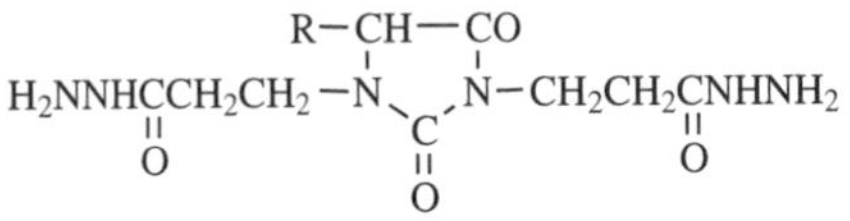

$$H_2NNHCCH_2CH_2-N\underset{\underset{O}{\overset{C}{||}}}{\overset{R-CH-CO}{\overset{|}{\underset{}{\smile}}}}N-CH_2CH_2CNHNH_2 \qquad H_2NNHCCH_2CH_2CHCNHNH_2$$
$$R = CH(CH_3)_2, CH_2CH_2SCH_3 \qquad \underset{\overset{||}{O}}{NHCC_6H_5}$$

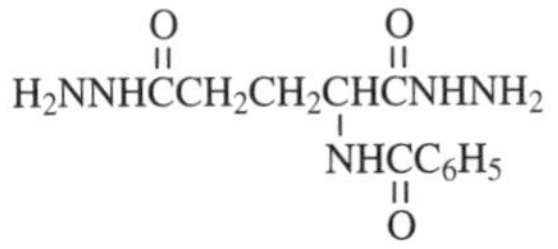

As vulcanizing agents, amino acids with or without sulfur are used for nipple rubber of babies' bottles and rubbers used in medical applications (255).

BIBLIOGRAPHY

"Amino Acids and Their Salts" in *ECT* 1st ed., Vol. 1, pp. 717–728, by M. S. Dunn, University of California; "Amino Acids" in *ECT* 1st ed., Suppl. 1, pp. 51–59, by H. T. Huang, Chas. Pfizer & Co., Inc.; "Amino Acids, Survey" in *ECT* 2nd ed., Vol. 2, pp. 156–197, by Maurice Vigneron, Société de Chimie Organique et Biologique A.E.C.; in *ECT* 3, Vol. 2, pp. 376–410, by A. Yamamoto, Kyowa Hakka Kogyo Co., Ltd; in *ECT* 4th ed., Vol. 2, pp. 504–571, by Kazumi Araki, University of East Asia and Toshitsugu Ozeki, Kyowa Hakko Kogyo Company; "Amino Acids, Survey" in *ECT* (online), posting date: December 4, 2000, by Kazumi Araki, University of East Asia and Toshitsugu Ozeki, Kyowa Hakko Kogyo Company.

CITED PUBLICATIONS

1. P. Hardy in G. C. Barrett, ed., *Chemistry and Biochemistry of the Amino Acids*, Chapman and Hall, London, 1985, Chapt. 2, pp. 6–24.
2. B. Alberts and co-eds., *Molecular Biology of the Cell*, 2nd ed., Garland Publishing, New York, 1989.
3. H. Braconnot, *Ann. Chim. Phys.* **13**, 113 (1820).
4. W. C. Rose, *J. Biol. Chem.* **112**, 275 (1935).
5. H. B. Vickery, *Adv. Protein Chem.* **26**, 82 (1972); *Ann. N. Y. Acad. Sci.* **41**, 87 (1941); *Chem. Rev.* **9**, 169 (1931).
6. D. A. Bender in Ref. 1, Chapt. 5, pp. 139–196.
7. S. Hunt in Ref. 1, Chapt. 4, pp. 55–138.
8. P. J. Lea, R. M. Wallsgrove, and B. J. Miflin in Ref. 1, Chapt. 6, pp. 197–226; E. A. Bell, *FEBS Lett.* **64**, 29 (1976).
9. N. Esaki and K. Soda, *Kagaku To Seibutsu* **20**, 425 (1982).
10. G. C. Barrett in Ref. 1, Chapt. 1, pp. 1–5.
11. N. S. Scrimshow and A. M. Altschul, eds., *Amino Acid Fortification of Protein Foods*, Massachusetts Institute of Technology Press, Cambridge, Mass., 1971.

12. R. Funabiki in A. Yoshida and co-eds., *Nutrition: Proteins and Amino Acids*, Japan Science Society Press, Tokyo, Japan, 1990, pp. 35–46.

13. D. A. Bender, *Amino Acid Metabolism*, 2nd ed., John Wiley & Sons, Inc., Chichester, 1985; P. K. Bondy and L. E. Rosenberg, *Metabolic Control and Disease*, 8th ed., W. B. Saunders, Philadelphia, Pa., 1980.

14. D. H. Speckman, W. H. Stein, and S. Moor, *Anal. Chem.* **30**, 1190 (1958).

15. A. P. F. Tuner, I. Karube, and G. S. Wilson, eds., *Biosensor: Fundamentals and Applications*, Oxford University Press, Oxford, 1987.

16. I. Karube in K. Aida and co-eds., *Biotechnology of Amino Acid Production*, Elsevier, Amsterdam, The Netherlands, 1966, pp. 81–89.

17. K. Yamaguchi, S. Ishino, K. Araki, and K. Shirahata, *Agric. Biol. Chem.* **50**, 2453 (1986).

18. R. J. Simpson, M. R. Neuberger, and T. Y. Liu, *J. Biol. Chem.* **251**, 1936 (1976).

19. H. Yamada and H. Tsugita, *Seikagaku* **62**, 857 (1990).

20. T. Matoba and co-workers, *Agric. Biol. Chem.* **46**, 465 (1982).

21. S. Kinoshita, S. Udaka, and M. Shimono, *J. Gen. Appl. Microbiol.* **3**, 193 (1957); S. Kinoshita in A. L. Demain and N. A. Solomon, eds., *Biology of Industrial Microorganisms*, Benjamin-Cummings Publishing Company, London, 1985, pp. 115–142; T. Asai, K. Aida, and K. Oishi, *Bull. Agric. Chem. Soc. Jpn.* **21**, 134 (1957).

22. K. Nakayama in G. Reed, ed., *Prescott & Donn's Industrial Microbiology*, 4th ed., AVI Publishing Company, Westport, Conn., 1982, 748–801; K. Araki in Ref. 12, 303–322; I. Shiio and S. Nakamori in J. O. Neway, ed., *Fermentation Process Development of Industrial Organisms*, Marcel Dekker, New York, 1989, pp. 133–168.

23. K. Aida and co-eds., *Biotechnology of Amino Acid Production*, Elsevier, Amsterdam, The Netherlands, 1986.

24. K. Soda, H. Tanaka, and N. Esaki in H. Dellweg, ed., *Biotechnology*, Vol. 3, Verlag Chemie, Weinheim, Germany, 1982, pp. 479–530; I. Chibata, T. Tosa, and T. Sato in J. F. Kennedy, ed., *Biotechnology*, Vol. 7a, Verlag Chemie, Weinheim, Germany, 1987, pp. 653–684; H. Yamada and T. Nagasawa in H. Yamada, ed., *Development of New Functions of Enzyme* (in Japanese), Kodansha Scientific, Tokyo, Japan, 1987, pp. 1–74; H. Yamada and S. Shimizu in J. Tramper, H. C. van der Plas, and P. Linko, eds., *Biocatalysts in Organic Syntheses*, Elsevier, Amsterdam, The Netherlands, 1986, p. 19.

25. T. Kaneko and co-eds., *Synthetic Production and Utilization of Amino Acids*, John Wiley & Sons, Inc., New York, 1974.

26. S. Yamaguchi, T. Yoshikawa, S. Ikeda, and T. Nimomiya, *J. Food Sci.* **38**, 846 (1971).

27. M. Tada, I. Shinoda, and H. Okai, *J. Agric. Food Chem.* **32**, 992 (1984); Y. Kawasaki and co-workers, *Agric. Biol. Chem.* **52**, 2679 (1988); M. Tamura and H. Okai, *J. Agric. Food Chem.* **38**, 1994 (1990).

28. D. Dorphin and co-workers, eds., *Glutathion: Chemical, Biochemical, and Medical Aspects*, Part B, John Wiley & Sons, Inc., New York, 1989.

29. G. V. Gurskaya, *The Molecular Structure of Amino Acid: Determination by X-Ray Diffraction Analysis*, Consultants Bureau, New York, 1968.

30. S. Hirokawa, *Acta Crystallog.* **8**, 637 (1955).

31. N. Hirayama and co-workers, *Bull. Chem. Soc. Jpn.* **53**, 30 (1980).

32. S. Budavari, ed., *The Merck Index*, 11th edition, Merck & Co., Inc., Rahway, N.J., 1989.

33. J. G. Grasselli, ed., *Atlas of Spectral Data and Physical Constants for Organic Compounds*, CRC Press, Cleveland, Ohio, 1973.

34. J. P. Greenstein and M. Wintz, *Chemistry of the Amino Acids*, Vol. 1, John Wiley & Sons, Inc., New York, 1961, pp. 46, 435, 523, 569.

35. *Ibid.*, p. 443.

36. *Data Sheets of Amino Acids*, Part 1 (1985), The Japan Essential Amino Acids Association, Tokyo, Japan, 1985; R. C. Weast, ed., *Handbook of Chemistry and Physics*, 67th edition, CRC Press, Inc., Boca Raton, Fla., 1986, p. C-702.

37. Ref. 34, pp. 486–491.

38. G. C. Barrett in Ref. 1, Chapt. 11, 354–375; T. Kaneko and T. Shiba in M. Kotake and co-eds., *Dai Yuki Kagaku* (Comprehensive Organic Chemistry in Series), Vol. 5, Asakura, Tokyo, Japan, 1959, 293–307.

39. M. Dymicky, *Org. Prep. Proceed. Int.* **12**, 207 (1980).

40. G. R. Waller and M. S. Feather in *The Maillard Reaction in Foods and Nutrition*, American Chemical Society, Washington, D.C. 1983; F. Hata and M. Oimomi, *Rinsho Kensa* **33**, 893 (1989).

41. G. S. Poindexter and A. I. Meyers, *Tetrahedron Lett.* , 3527 (1977).

42. M. Kubota, O. Nagase, and H. Yajima, *Chem. Pharm. Bull.* **29**, 1169 (1981).

43. E. R. Rothgery and L. F. Hohnstedt, *Inorg. Chem.* **10**, 181 (1971).

44. S. Kiyooka, F. Goto, and K. Suzuki, *Chem. Lett.* 1429 (1981).

45. F. M. F. Chen, K. Kuroda, and N. L. Benoiton, *Synthesis* 928 (1978).

46. Y. Fujimoto, ed., *Poly Amino Acid*, Kodansha, Tokyo, Japan, 1974; T. Endo, ed., *Amino Acid Polymers: Synthesis and Applications* (in Japanese), CMC, Tokyo, Japan, 1988.

47. J. Kohn and R. Langer, *J. Am. Chem. Soc.* **109**, 817 (1987).

48. Jpn. Kokai 48 51,995 (July 21, 1973), M. Teranishi and Y. Fujimoto (to Kyowa Hakko Kogyo); Jpn. Kokai 49 26,193 (Mar. 8, 1974); Ger. Offen. 2,253,190 (May 10, 1973).

49. Ref. 34, Chapt. 6, 569–682; R. W. Hay and K. B. Nolan, *Amino Acids Pept.* **20**, 297 (1989); O. Yamauchi and A. Kotani, *Nihon Kagaku Kaishi* (4), 369 (1988).

50. W. H. Stein and co-workers, *J. Biol. Chem.* **135**, 489 (1940); **139**, 481 (1941); and **143**, 121 (1942); W. H. Stein and co-workers, *Chem. Rev.* **30**, 423 (1942).

51. M. Bodansky, *Peptide Synthesis*, 2nd ed., John Wiley & Sons, Inc., New York, 1976; J. Meinhofer in Ref. 1, Chapt. 9, p. 297; G. R. Pettit, *Synthetic Peptides*, Vols. 1–4, Van Nostrand Reinhold, New York, 1980, Vols. 5, 6, Elsevier, New York, 1982; E. Shroeder and K. Luebke, *The Peptide*, Vol. 1, *Methods of Peptide Synthesis*, Academic Press, New York, 1965; N. Izumiya and co-workers, *Fundamentals and Experiments of Peptide Synthesis* (in Japanese), Maruzen, Tokyo, Japan, 1987; R. B. Merrifield, *J. Am. Chem. Soc.* **85**, 2149 (1963); G. Barany and R. B. Merrifield in E. Gross and J. Meinenhofer, eds., *The Peptides: Analysis, Synthesis, Biology*, Vol. 2, Academic Press, New York, 1980, pp. 1–284; G. R. Marshall, *Peptides: Chemistry and Biology*, Escom, Leiden, The Netherlands, 1988.

52. K. Dranz, A. Kleeman, and J. Martens, *Angew. Chem. Int. Ed. Engl.* **21**, 584 (1988); J. Martens, *Top. Curr. Chem.* **125**, 165 (1984); A. Mortreux and co-workers, *Bull. Soc. Chim. Fr.* (4), 631 (1987); A. M. Coppola and H. F. Schuster, *Asymmetric Synthesis, Construction of Chiral Molecules Using Amino Acids*, John Wiley & Sons, Inc., New York, 1987.

53. Ref. 34, Chapt. 8, pp. 697–714; G. C. Barrett in Ref. 1, Chapt. 8, pp. 246–296.

54. Jpn. Kokai 60 164,493 (Aug. 27, 1985), K. Araki and co-workers (to Kyowa Hakko Kogyo Company); Jpn. Kokai 61 15,697 (Jan. 23, 1986), K. Araki and H. Anazawa (to Kyowa Hakko Kogyo Company); Jpn. Kokai 61 141,893 (June 28, 1986); Jpn. Kokai 61 47,197 (Mar. 7, 1986) and Jpn. Kokai 61 56,088 (Mar. 20, 1986), M. Tsuji, T. Kitamura, and N. Yoshimura (to Kurare Company); Jpn. Kokai 61 202,695 (Sept. 8, 1966); Y. Muro, A. Nakayama, and T. Akashiba (to Showa Denko); Ger. Offen. 3,423,936 (Jan. 2, 1986), H. Voelskow and co-workers (to Hoechst); Eur. Pat. 132,999 (Feb. 13, 1985) and U. S. Pat. 4,600,692 (July 15, 1986); L. L. Wood and G. J. Calton (to Purification Engineering Company).

55. H. Wakamatsu, J. Uda, and N. Yamakami, *Chem. Commun.* 1540 (1971); K. Izawa, S. Nishi, and S. Asada, *J. Mol. Catal.* **41**, 935 (1987); M. Tanaka and T. Sakakura, *Petrotech.* **10**, 263 (1987).
56. I. Ojima and co-workers, *J. Organomet. Chem.* **279**, 203 (1985).
57. J. Jaques, A. Collet, and S. Willen, *Enantiomers, Racemate, and Resolutions*, John Wiley & Sons, Inc., New York, 1981; The Chemical Society of Japan, eds., *Kikan Kagaku Sosetsu* (No. 6, Resolution of Optical Isomers), Gakkai Shuppan Senta, Tokyo, Japan, 1989; G. C. Barrett in Ref. 1, Chapt. 10, pp. 338–353; S. Otsuka and T. Mukaiyama, *Progress of Asymmetric Synthesis and Optical Resolution* (in Japanese), Kagaku Dojin, Kyoto, Japan, 1982.
58. R. Duschinsky, *Chem. Ind.* **53**, 10 (1934).
59. T. Akashi, *Nihon Kagaku Zasshi* **83**, 417, 532 (1962); L. Velluz and G. Amiard, *Bull. Soc. Chim. Fr.* **20**, 903 (1953).
60. U.S. Pat. 3,405,159 (Oct. 8, 1968), K. H. Krieger, J. Lago, and J. A. Wantuck (to Merck & Co., Inc.).
61. F. Koegl, H. Erxleben, and G. J. van Veersen, *Z. Physiol. Chem.* **277**, 260 (1943).
62. T. Shiraiwa and co-workers, *Nihon Kagaku Kaishi*, 1189 (1983); H. Miyazaki, T. Shiraiwa, and H. Kurokawa, *Nihon Kagaku Kaishi*, 87 (1986).
63. H. Wakamatsu, *Food Eng.* **40**, 92 (1968).
64. P. Newman, *Optical Resolution Procedures for Chemical Compounds*, Vol. 1 (Amines and Related Compounds), Optical Resolution Information Center, New York, 1978; H. Nohira in Chemical Society of Japan, eds., *Kikan Kagaku Sosetsu* (No. 6, Resolution of Optical Isomers), Gakkai Shuppan Senta, Tokyo, Japan, 1989, Chapt. 4, pp. 45–54.
65. Ref. 34, Chapt. 9, pp. 722–727.
66. G. C. Barrett in Ref. 1, Chapt. 10, pp. 338–353.
67. P. Brodelius, K. Nilsson, and K. Mosback, *Appl. Biochem. Biotechnol.* **6**, 293 (1981).
68. K. Araki and H. Anazawa, *J. Synth. Org. Chem. Japan* (in Japanese) **46**, 160 (1988).
69. I. Chibata, K. Toi, and S. Yamada in Ref. 25, p. 17.
70. G. Blaschke, *Angew. Chem. Int. Ed. Engl.* **19**, 13 (1980); S. Hara and J. Cazes, eds., *J. Liq. Chromatogr.* **9** (2 and 3) (1986) (Special issues on optical resolution by liquid chromatography); M. Zief and L. J. Crane, eds., *Chromatogr. Sci.* **40** (1988) (all articles) (special issue on chromatographic chiral separations); A. M. Krstuloric, ed., *Chiral Separation by HPLC*, Ellis Horwood, Chichester, 1989; K. Mizuguchi and I. Okamoto, *Kagaku Keizai*, (2), 52 (1988).
71. D. Perret in Ref. 1, Chapt. 15, pp. 426–461.
72. R. Daeppen, H. Arm, and V. R. Meyer, *J. Chromatogr.* **373**, 1–20, 1986; Y. Okamoto in Chemical Society of Japan, eds., *Kikan Kagaku Sosetsu* (No. 6, Resolution of Optical Isomers), Gakkai Shuppan Senta, Tokyo, Japan, 1989, Chapt. 11, pp. 132–143; Chiral HPLC columns are available from Regis, Spelco, and AST in the United States; Merck, Nagel, and Serva in Germany; LKB in Sweden; and Daiseru, Sumitomo, and Toso in Japan.
73. T. Shimbo and co-workers, *J. Chromatogr.* **405**, 145 (1987); G. D. Y. Sogah and D. J. Cram, *J. Am. Chem. Soc.* **98**, 3038 (1976); D. J. Cram and co-workers, *J. Am. Chem. Soc.* **100**, 4555, 4569 (1978).
74. S. V. Rogozhin and V. A. Davankov, *J. Chem. Soc., Chem. Commun.* 490 (1971); V. A. Davankov and co-workers, *J. Chromatogr.* **82**, 359 (1973); N. Nimura in Chemical Society of Japan, eds., *Kikan Kagaku Sosetsu* (No. 6, Resolution of Optical Isomers), Gakkai Shuppan Senta, Tokyo, Japan, 1989, Chapt. 12, 144–153; P. E. Hare, *Chromatogr. Sci.* **40**, 165–177 (1988).
75. P. E. Hare and E. Gil-Ave, *Science* **204**(4398), 1226 (1979); E. Gil-Ave, A. Tishbee, and P. E. Hare, *J. Am. Chem. Soc.* **102**, 5115 (1980).

76. E. Oelrich, H. Preusch, and E. Wilheilm, *J. High Resolut. Chromatogr., Chromatogr. Commun.* **3**, 269 (1980); T. Arai, H. Koike, K. Hirata, and H. Oizumi, *J. Chromatogr.* **448**, 439 (1988).

77. N. Nimura and co-workers, *Anal. Chem.* **53**, 1380 (1981).

78. S. Lam, *J. Chromatogr. Sci.* **22**, 416 (1984).

79. A. Dobashi and S. Hara, *Anal. Chem.* **55**, 1805 (1983).

80. J. Gal, *LC-GC Magazine* **5**, 106 (1989).

81. S. Einarsson, B. Josefsson, P. Moeller, and D. Sanchez, *Anal. Chem.* **59**, 1191 (1987).

82. N. Oi in Chemical Society of Japan, eds., *Kagaku Sosetsu* (No. 6, Resolution of Optical Isomers), Gakkai Shuppan Senta, Tokyo, Japan, 1989, Chapt. 15, p. 176; M. H. Endel and P. E. Hare in Ref. 1, Chapt. 16, pp. 469–474.

83. M. Mack, H. E. Hauk, and H. Herbert, *J. Planar Chromatgr, Mod. TLC.* **1**, 304 (1988); M. Mack, H. E. Hauk, *Chromatographia* **26**, 197 (1988); K. Guenther, *J. Chromatogr.* **448**, 11 (1988); HPTLC plate "Chiral" is available from Merck & Company, Rahway, N.J.

84. D. S. Lingenfelter, R. C. Helgeson, and D. J. Cram, *J. Org. Chem.* **46**, 393 (1981); T. Yamaguchi and co-workers, *Chem. Lett.* 1549 (1985); T. Yamaguchi and co-workers, *Maku (Membrane)* **10**, 178 (1985); J. Rebek, Jr., and co-workers, *J. Am. Chem. Soc.* **109**, 2432 (1987).

85. T. Miyazawa and co-workers, *Chem. Lett.* 2219 (1989); N. Izumiyama and co-workers, *Bull. Chem. Soc. Jpn.* **61**, 575 (1988).

86. J. M. Brown, *Chem. Ind.* 612 (1988); *Angew. Chem. Int. Ed. Engl.* **26**, 190 (1987); J. M. Brown and I. Cutting, *J. Chem. Soc. Chem. Commun.* 578 (1985); J. M. Brown and A. P. James, *J. Chem. Soc. Chem. Commun.* 181 (1987); J. M. Brown, A. P. James, and L. M. Prior, *Tetrahedron Lett.* **28**, 2179 (1987); K. B. Sharpless and co-workers, *J. Am. Chem. Soc.* **103**, 6237 (1981); R. Noyori and co-workers, *J. Am. Chem. Soc.* **111**, 9134 (1989).

87. R. M. Williams, *Synthesis of Optically Active α-Amino Acids*, Pergamon Press, New York, 1989; H. Takahashi, T. Morimoto, and K. Achiwa, *J. Synth. Org. Chem. Jpn.* (in Japanese) **48**, 29 (1990); B. P. Vineyard, W. S. Knowles, and M. J. Sebacky, *J. Mol. Catal.* **19**, 159 (1983); D. Hoppe, *Nachr. Chem., Tech. Lab.* **30**, 782 and 852 (1982); *Chem. Abstr.* **97**, 216633j and **98**, 34910k; Y. Sugi, *J. Synth. Org. Chem. Jpn.* (in Japanese) **37**, 71 (1979); *Kagaku Kogyo* **32**, 169 (1981).

88. H. Takahashi and K. Achiwa, *Chem. Lett.* 305 (189); *Chem. Pharm Bull.* **36**, 3230 (1988).

89. H. B. Kagan and co-workers, *J. Am. Chem. Soc.* **108**, 2353 (1986); N. Oguni, Y. Matsuda, and T. Kaneko, *J. Am. Chem. Soc.* **110**, 7877 (1988); N. Oguni, T. Omi, Y. Yamamoto, and A. Nakamura, *Chem. Lett.* 841 (1983); N. Oguni, *Kagaku* **45**, 182 (1990); M. Kitamura, S. Suga, K. Kawai, and R. Noyori, *J. Am. Chem. Soc.* **108**, 6071 (1986); M. Kitamura, S. Okada, S. Suga, and R. Noyori, *J. Am. Chem. Soc.* **111**, 4028 (1989); R. Noyori and co-workers, *Pure Appl. Chem.* **60**, 1597 (1988).

90. S. Ikegami, T. Hayama, T. Katsuki, and M. Yamaguchi, *Tetrahedron Lett.* **27**, 3403 (1986); R. M. Williams, P. J. Sinclair, D. Zhai, and D. Chen, *J. Am. Chem. Soc.* **110**, 1547 (1988).

91. M. J. O'Donnell, W. D. Bennett, and S. Wu, *J. Am. Chem. Chem.* **111**, 2353 (1989).

92. A. Strecker, *Ann. Chem.* **75**, 27 (1850).

93. S. Kinoshita and K. Tanaka in K. Yamada and co-workers, *Microbial Production of Amino Acids*, John Wiley & Sons, Inc., New York, 1972.

94. S. Kinoshita, *Fermentation Industry*, 2nd. ed., (in Japanese), Dainihon Tosho Company, Tokyo, Japan, 1975, p. 83.

95. S. Kinoshita in A. L. Demain and N. A. Solomon, eds., *Biology of Industrial Microorganisms*, Benjamin-Cummings Publishing Company, London, 1985, pp. 115–142.

96. Y. Minoda in Ref. 23, pp. 51–66.
97. E. Nakazawa, E. Akutsu, and H. Kamiya, *Proc. Ann. Mtg. of Japanese Society of Fermentation Technology*, 1981, p. 46.
98. H. Morioka and co-workers, *Agric. Biol. Chem.* **53**, 911 (1989).
99. S. Komatsubara and co-workers, *J. Biotechnol.* **3**, 59 (1985).
100. T. Azuma and co-workers, *Proc. Ann. Mtg. of Japanese Society of Fermentation Technology*, 1987, p. 104.
101. T. Hirao and co-workers in Ref. 100.
102. S. Sugimoto and I. Shiio, *Agric. Biol. Chem.* **53**, 2081 (1989).
103. Jpn. Kokai 1 187,090 (July 26, 1989), M. Fujii and Y. Morita (to Chisso Company).
104. M. Furukawa and co-workers, *Appl. Microbiol. Biotechnol.* **29**, 253, 550 (1988).
105. M. Yamada, H. Tsutsui, and K. Shimoto, *Proc. Ann. Mtg. Japan Society of Bioscience Biotechnology and Agrochemistry*, 1990, p. 680; Jpn. Kokai 1 148,194 (June 9, 1989), M. Shirai and co-workers (to Toray Industrial Company).
106. Jpn Kokai 1 187,091 (July 26, 1989), Jpn Kokai 1 309,691 (Dec. 14, 1989), M. Takeuchi and T. Yonehara (to Toray Industrial Company); T. Yonehara, M. Takeuchi, and H. Tsutsui in Ref. 151.
107. K. Kino, M. Ikeda, and R. Katsumata in Ref. 151, p. 681.
108. S. Aiba and co-workers, *Proc. Ann. Mtg. of Japanese Society of Fermentation Technology*, 1987, p. 100.
109. K. Yamada and co-workers, *Microbial Production of Amino Acids*, John Wiley & Sons, Inc., New York, 1972.
110. R. Konno and Y. Yasumura, *Lab. Anim. Sci.* **38**, 292–295 (1988).
111. T. Tsunoda and S. Okumura in Japanese Society of Petroleum Fermentation, eds., *Sekiyu Hakko*, Saiwai Shobo, Tokyo, Japan, 1970, 123–219; K. Tanaka, *Hakko Kyokaishi* **26**, 227 (1968).
112. M. Kikuchi and Y. Nakao in K. Aida and co-eds., *Biotechnology of Amino Acid Production*, Kodansha, Ltd., Tokyo, Japan, 1986, 101–116.
113. F. Tomita, T. Suzuki, and A. Furuya, *Amino Acid, Nucleic Acid* **32**, 1 (1975).
114. K. Yamaguchi and co-workers, *Agric. Biol. Chem.* **37**, 2189 (1973).
115. S. Uragami, *Bioscience and Industry* **48**, 148 (1990).
116. K. Nakayama, *Genetics of Industrial Microorganisms*, Vol. 1 (Bacteria), Elsevier, Amsterdam, The Netherlands, 1973, p. 219.
117. F. J. Schendel and co-workers, *Appl. Environ. Microbiol.* **56**, 963 (1990).
118. K. M. Herrman and R. L. Somerville, eds., *Amino Acids: Biosynthesis and Genetic Regulation*, Addison-Wesley Publishing Company, Reading, Mass., 1983.
119. T. Tachiki and T. Tochikura in Ref. 23, pp. 121–130.
120. Y. Yoshinaga in Ref. 23, pp. 117–120.
121. I. Shiio and S. Nakamori in J. O. Neway, ed., *Fermentation Process Development of Industrial Organisms*, Marcel Dekker, New York, 1989, pp. 133–168.
122. T. Beppu in Ref. 23, pp. 24–35.
123. O. Tosaka and K. Takinami in Ref. 23, pp. 152–172; M. Karasawa and co-workers, *Agric. Biol. Chem.* **50**, 339 (1986); S. Furukawa and co-workers, *Appl. Microbiol. Biotechnol.* **29**, 248 (1988).
124. M. Kisumi in Ref. 23, pp. 14–23.
125. R. Katsumata and co-workers in O. M. Neijssel and co-workers, *Proc. 4th European Congress on Biotechnology*, Vol. 4, Elsevier, Amsterdam, The Netherlands, 1987, pp. 767–776.
126. R. Biegelis in H. J. Rehm and G. Reed, eds., *Biotechnology*, Vol. 7b, Verlag-Chemie, Weinheim, Germany, 1989, pp. 229–259.
127. B. T. Bachman, K. Brooks Low, and A. L. Taylor, *Bacteriol. Rev.* **40**, 116–167 (1976).

128. Y. Asano and co-workers, *Proc. Ann. Mtg. of Japanese Society of Fermentation Technology*, 1989, p. 25.
129. N. Nishimura and co-workers, *J. Ferment. Technol. Bioeng.* **67**, 107 (1987).
130. D. Filpula and co-workers, *Nucleic Acid Res.* **16**, 11381 (1988); Jpn. Kokai 63 291,583 (Nov. 29, 1988) and Jpn. Kokai 63 196,292 (Aug. 15, 1988), N. Fukuhara and co-workers (to Mitsui Toatsu Chemicals, Inc.).
131. H. Yokoi and co-workers in Ref. 105, p. 144.
132. E. Shimizu and co-workers, *Proc. Ann. Mtg. of Agric. Chem. Soc. Japan* 1983, p. 353.
133. R. Katsumata and co-workers, *Hakko to Kogyo* **45**, 371 (1987).
134. Y. Morinaga and co-workers, *Agric. Biol. Chem.* **51**, 93 (1987).
135. K. Matsui and co-workers, *Agric. Biol. Chem.* **52**, 1863 (1988).
136. K. Uchida and H. Tanaka, *J. Synth. Org. Chem. (Japan)* **46**, 977 (1988); K. Uchida and Y. Matsumura in N. Ishikawa, ed., *Synthesis and Function of Flurocompounds* (in Japanese), CMC, Tokyo, Japan, 1988, pp. 251–287.
137. J. Kollitsch, S. Marburg, and L. M. Perkins, *J. Org. Chem.* **41**, 3107 (1976).
138. M. Terasawa and co-workers, *Process Biochem.* **4**, 60 (1989).
139. T. Ishikawa and co-workers in Ref. 105, p. 134.
140. K. Nakamichi and co-workers, *Appl. Microbiol. Biotechnol.* **19**, 100 (1984); *Appl. Biochem. Biotechnol.* **11**, 367 (1985).
141. T. Nakanishi and co-workers, *J. Ferm. Technol.* **65**, 139 (1987).
142. K. Kubota and K. Yokozeki, *J. Ferm. Bioeng.* **67**, 387 (1989).
143. P. Sirirote and co-workers, *J. Ferm. Technol.* **64**, 389 (1989); M. Watanabe and co-workers, *J. Ferm. Technol.* **65**, 617 (1987); H. Yamada and co-workers, *Agric. Biol. Chem.* **59**, 17 (1986).
144. S. Oita, A. Yokita, and S. Takao, *J. Ferm. Bioeng.* **69**, 256 (1990).
145. C. Sambale and M. R. Kula, *J. Bacteriol.* **7**, 49 (1988).
146. N. Nishimura and co-workers, *J. Bacteriol.* **7**, 11 (1988).
147. K. Ishiwata and co-workers, *J. Ferm. Bioeng.* **67**, 169 (1989); K. Ishido and co-workers, *J. Ferm. Bioeng.* **68**, 84 (1989).
148. R. Wichman and co-workers, *Biotechnol. Bioeng.* **23**, 2789 (1981).
149. Y. Asano and A. Nagazawa, *Agric. Biol. Chem.* **51**, 2035 2621 (1987).
150. A. Miura and co-workers, *Agric. Biol. Chem.* **51**, 236 (1987).
151. Eur. Pat. 248,357 (Dec. 9, 1987) and Jpn. Kokai 62 296,886 (Dec. 24, 1987); J. Then and co-workers (to Hoechst A. G.).
152. N. Onisshi and co-workers, *Agric. Biol. Chem.* **51**, 291 (1987).
153. W. Hummer and co-workers, *Appl. Microbiol. Biotechnol.* **27**, 283 (1987).
154. Jpn. Kokai 63 123,398 (May 27, 1988), T. Uragami and co-workers (to Mitsubishi Gas Chemical Company).
155. H. Y. Hsaiao and Y. Wei, *Biotechnol. Bioeng.* **28**, 1510 (1986).
156. S. Furuyoshi and co-workers, *Agric. Biol. Chem.* **53**, 3075 (1989).
157. K. Soda, *Sei Kagaku* **46**, 203 (1974); N. Makiguchi and co-workers, *Proc. Ann. Mtg. of Japan Society Bioscience, Biotechnology, Agrochemistry*, 1985, p. 344.
158. Y. Nishida and co-workers, *Enz. Microbiol. Biotechnol.* **9**, 721 (1987).
159. Y. Asano, *Bioscience and Industry* (in Japanese) **48**, 131 (1990).
160. M. Sugie and H. Suzuki, *Agric. Biol. Chem.* **44**, 1089 (1980); K. Sakai and co-workers, *Appl. Environ. Microbiol.* **54**, 2767 (1988).
161. M. Yazaki and co-workers in Ref. 128, p. 67.
162. M. Moriguchi and K. Ideta, *Appli. Environ. Microbiol.* **54**, 2767 (1988); Jpn. Kokai 1 137,973 (May 30, 1989) and Eur. Pat. 304,021 (Feb. 22, 1989), T. Takahashi and K. Hatano (to Takeda Chemical Industry).
163. K. Yokozeki and co-workers, *Agric. Biol. Chem.* **51**, 715 (1987).

164. T. Yamagishi, *Japan Chem. Ind. Assoc. Monthly* (3), 163 (1967); R. Niklasson, *Kem. Tidskr.* **95**, 33 (1983); *Chem. Abstr.* **100**, 105476w (1984).

165. Fr. Pat. 1,237,327 (Nov. 26, 1960), H. M. Guinot (to Société de Produits Chimiques Industriels); U.S. Pat. 3,190,914 (June 22, 1965), R. E. Williams (to Rexall Drug and Chemical Company); Brit. Pat. 908,735 (Oct. 24, 1962), R. Imagawa, J. Kato, and I. Noda (to Ajinomoto Company); U.S. Pat. 3,167,582 (Jan. 26 1965), K. W. Saunders, W. H. Montgomery, and J. C. French (to American Cyanamid Company).

166. U.S. Pat. 2,557,920 (June 19, 1951); U.S. Pat. 2,642,459 (June 16, 1953); U.S. Pat. 2,700,054 (Jan. 18, 1955), H. C. White (to The Dow Chemical Company).

167. K. Sano and co-workers, *Appl. Environ. Microbiol.* **34**, 806 (1977); *Agric. Biol. Chem.* **42**, 2315 (1978); Jpn. Kokai 51 67,790 (June 11, 1976), K. Sano, K. Matsuda, and K. Mitsumoto (to Ajinomoto Company); Jpn. Kokai 53 56,388 (May 22, 1978), K. Sano, N. Yasuda, and K. Mitsumoto (to Ajinomoto Company); Jpn. Kokai 52 72,883 (June 17, 1977), Y. Anzai and co-workers (to Ajinomoto Company); Jpn. Kokai 53 112,811 (Oct. 2, 1978), E. Tsuchiya, T. Kitahara, and S. Asai (to Ajinomoto Company).

168. K. Soda and K. Yonaha in J. F. Kennedy, ed., *Biotechnology*, Vol. 7a, Verlag-Chemie, Weinheim, Germany, 1987, 605–652; J. Cox and co-workers, *Trend in Biotechnol.* **6**, 279 (1988); T. Oshima and co-workers, *Biotechnol. Bioeng.* **27**, 1616 (1985).

169. *Amino Acids to 2004*, Freedonia Group, 2001.

170. S. Blackman, *Amino Acid Determination Methods and Techniques*, Marcel Dekker, New York, 1968; A. Niederweiser and G. Pataki, eds., *New Techniques in Amino Acids, Peptide and Protein Analysis*, Ann Arbor Science Publishers, Ann Arbor, Mich., 1971; The Chemical Society of Japan, eds., *New Experimental Chemistry Series*, Vol. 1 (Biochemistry 1) (in Japanese), Maruzen, Tokyo, Japan, 1978, pp. 141–160.

171. S. Moore and W. H. Stein, *J. Biol. Chem.* **211**, 893 (1954); D. H. Spackman, W. H. Stein, and S. Moore, *Anal. Chem.* **30**, 1190 (1958); P. B. Hamilton, *Anal. Chem.* **35**, 2055 (1963).

172. P. E. Hare, P. A. St. John, and M. H. Engel in Ref. 1, Chapt. 14, 415–425; K. Takahashi in *New Biochemical Experiments Series*, Vol. 1, *Protein 1, Separation, Purification, and Properties* (in Japanese), Tokyo Kagaku Dojin, Tokyo, Japan, 1990, pp. 439–446.

173. Chemical Society of Japan, eds., *Experimental Chemistry Series*, Vol. 23, *Biochemistry 1* (in Japanese), Maruzen, Tokyo, Japan, 1957, pp. 1–244.

174. H. Edelhoch, *Biochemistry* **6**, 1948 (1967).

175. G. L. Ellman, *Arch. Biochem. Biophys.* **82**, 70 (1957).

176. D. Perrett in Ref. 1, Chapt. 15, 426–459; L. R. Snyder and J. J. Kirkland, *Introduction to Modern Liquid Chromatography*, John Wiley & Sons, Inc., New York, 1974; Kanto Branch of the Japan Society of Analytical Chemistry, eds., *High Performance Liquid Chromatography Handbook*, Maruzen, Tokyo, Japan, 1985.

177. M. H. Engel and P. E. Hare in Ref. 1, Chapt. 16, pp. 462–479.

178. C. W. Gehrke, K. C. Kuo, and R. W. Zumwalt, *J. Chromatogr.* **57**, 209 (1971); C. W. Gehrke and M. Takeda, *J. Chromatogr.* **76**, 63 (1973).

179. S. P. Cram, T. H. Rishy, L. R. Field, and W. L. Yu, *Anal. Chem.* **52**, 342R (1980).

180. E. Stahl, ed., *Thin-Layer Chromatography: A Laboratory Handbook*, Springer-Verlag, New York, 1965.

181. R. J. Elliott and D. L. Gardner, *Anal. Biochem.* **70**, 268 (1976).

182. G. A. Rosenthal in Ref. 1, Chapt. 20, pp. 573–590.

183. D. J. McCaldin, *Chem. Rev.* **60**, 39 (1960); P. J. Lamothe and P. G. McCornick, *Anal. Chem.* **45**, 1906 (1973).

184. *Chem. Eng. News*, 58 (Mar. 19, 1990); S. Stinson, *Chem. Eng. News* **35** (Nov. 5, 1990); W. G. Kuhn and E. S. Yeung, *Anal. Chem.* **60**, 1832 (1988); M. Yu and N. J. Dovichi, *Anal. Chem.* **61**, 37 (1989).

185. R. A. W. Johnson and M. E. Ross in Ref. 1, Chapt. 17, 480–524; G. C. Barrett and J. S. Davies in Ref. 1, Chapt. 18, pp. 525–544.

186. N. E. McKenzie and P. R. Gooley, *Med. Res. Rev.* **8**, 57 (1988); P. J. Hore, *J. Mag. Reson.* **55**, 283 (1983).

187. G. D. Shockman in F. Kavanaugh, ed., *Analytical Microbiology*, Vol. 1, Academic Press, New York, 1963, p. 568; A. E. Bolinder in Vol. 2, 1972, p. 480.

188. H. Itoh, T. Morimoto, and I. Chibata, *Anal. Biochem.* **60**, 573 (1974).

189. E. G. Gale in H. U. Bermeyer, ed., *Methods of Enzymatic Analysis*, Academic Press, New York, 1963, pp. 373–377; T. Tsunoda in K. Yamada and co-eds., *The Microbial Production of Amino Acids*, John Wiley & Sons, Inc., New York, 1972.

190. H. Misono and co-workers in M. Yamada, H. Tsutsui and K. Shimoto, *Proceed. Ann. Mtg. of Japan Society of Bioscience Biotechnology Agrochemistry*, 1990, p. 364.

191. J. D. Kopple and M. E. Swendseid, *J. Clin. Invest.* **55**, 881 (1975).

192. T. Muto and K. Yoshikawa in Ref. 12, pp. 259–272; A. Ichikawa and co-workers, *Japanese J. Parent Ent. Nutr.* **11**, 795 (1989).

193. Joint FAO/WHO Ad Hoc Expert Committee, *Report of FAO Nutrition Meeting*, Geneva, 1973, Series No. 52; *WHO Technical Report*, Series No. 522.

194. Joint FAO/WHO Expert Consultation, *WHO Technical Report*, Rome, 1985, Series No. 724.

195. M. Yamaguchi in Ref. 12, pp. 187–194.

196. A. Ariyoshi and I. Tasaki in Ref. 12, pp. 77–84.

197. A. Yoshida in Ref. 12, pp. 97–106.

198. T. Nose and T. Murai in Ref. 12, pp. 85–95.

199. Y. Yamamoto, M. Suzuki, and K. Muramatsu, *Agric. Biol. Chem.* **49**, 2859 (1985).

200. K. Muramatsu and Y. Yamamoto in Ref. 12, pp. 133–142.

201. C. A. Elvehjem, *Science* **101**, 283 (1945); *Fed. Proc.* **15**, 965 (1985).

202. A. E. Harper in H. N. Munro and J. B. Allison, eds., *Mammalian Protein Metabolism*, Vol. 2, Academic Press, New York, 1964, p. 87; J. C. Sanahuja in Ref. 11, p. 179.

203. G. L. Foster, R. Schoenheimer, and D. Rittenberg, *J. Biol. Chem.* **127**, 319 (1939); R. Schoenheimer, S. Ratner, and D. Rittenberg, *J. Biol. Chem.* **127**, 333 (1939) and **130**, 703 (1939); D. Rittenberg, D. Schoenheimer, and A. S. Keston, *J. Biol. Chem.* **128**, 603 (1939).

204. R. J. Lewis, Sr., *Sax's Dangerous Properties of Industrial Materials*, 10th ed., Wiley, New York, 2000.

205. *NAS/NRC Report*, Food and Drug Administration, Washington, D.C., Oct. 7, 1970.

206. Joint Expert Committee on Food Additives, World Health Organization, 1988.

207. Z. Bohak, *J. Biol. Chem.* **239**, 2878 (1964).

208. *Nutr. Rev.* **34**, 120 (1976).

209. J. F. Thompson, C. J. Morris, and I. K. Smith, *Ann. Rev. Biochem.* **38**, 137 (1969).

210. M. J. Jung in Ref. 1, Chapt. 7, pp. 227–245.

211. P. L. McGeer and E. G. McGeer in G. J. Siegel and co-eds., *Basic Neurochemistry*, 4th ed., Raven Press, New York, 1989, Chapt. 15, pp. 311–332.

212. T. Hunter, *Cell* **50**, 823 (1987).

213. D. A. Persons and co-workers, *Cell* **52**, 447 (1988).

214. D. Fujimoto in A. M. Kligman and Y. Takase, eds., *Cutaneous Aging*, University of Tokyo Press, Tokyo, Japan, 1988, pp. 263–274.

215. T. Kobayashi, Y. Doi, and T. Takami in Ref. 12, pp. 285–299.

216. I. Chibata and K. Kawashima in Ref. 12, pp. 273–284.

217. G. A. Nowak, *Parfuemerie und Kosmetik* **65**, 73 (1984); Cosmetics Special Report, *Chem. Week* **20** (Nov. 15, 1989); M. Suzuki, *Mol* (in Japanese) (12), 70 (1989).

218. T. Tsunoda in Bioindustry Development Center (BIDEC), eds., *From Fermentation to New Biotechnology* (in Japanese), Japan Science Society Press, Tokyo, Japan, 1989.

219. T. Klpfenstein and R. Stock, *Feedstuff*, **26** (Feb. 11, 1980).

220. U.S. Pat. Appl. 20020103258 (Aug. 1, 2002), J. C. Robert.

221. U.S. Pat. Appl. 20020106421 (Aug. 8, 2002), M. Binder and K. E. Uffmann.

222. Y. Kawamura and M. R. Kara, *Umami: A Basic Taste*, Marcel Dekker, New York, 1987.

223. R. S. Schallenberger, T. E. Acree, and C. Y. Lee, *Nature (London)* **221**, 555 (1969); Y. K. Lee and T. Kaneko, *Bull. Chem. Soc. Japan* **46**, 3494 (1973).

224. G. A. Crosby, *CRC Crit. Rev. Food Sci. Nutr.* **7**, 297 (1976); M. R. Coloninger and R. E. Baldwin, *Science* **170**, 80 (1982); H. Okai and I. Miyake, *Kagaku to Seibutsu* (in Japanese) **20**, 709 (1982); Y. Ariyoshi, Y. Hasellawa, and M. Ota, *Agric. Biol. Chem.* **54**, 1623 (1990).

225. Eur. Pat. 34,876 (Jan. 8, 1981) and U. S. Pats. 4,454,328 (June 12, 1980), 4,399,163 (Aug. 6, 1983) and 4,797,298 (June 10, 1989), T. Brennan and M. Hendrick (to Pfizer Corporation).

226. U.S. Pat. 2003008046 (Jan. 9, 2003), P. A. Gerlat and co-workers.

227. U.S. Pat. Appl. 20020164397 (Nov. 7, 2002), S. V. Ponakala (to the Nutrasweet Co.).

228. K. Prendergast, *Food Trade Rev.* **44**, 14 (1974); F. K. Imrie, *Chem. Ind. (London)* 584 (1976).

229. T. Matoba and T. Hata, *Agric. Biol. Chem.* **36**, 1423 (1972); H. D. Belitz and H. Weiser, *Z. Lebensm. Unters. Forsch.* **160**, 251 (1976).

230. K. Otagari and co-workers in T. Shiori, ed., *Peptide Chemistry 1981*, Protein Research Foundation, Osaka, Japan, 1982, p. 75; H. Okai and I. Miyake, *Kagaku to Seibutsu* (in Japanese) 20, 709 (1982).

231. J. E. Hodge, F. D. Mills, and B. E. Fisher, *Cereal Sci. Today* **17**, 34 (1972); T. M. Reynolds, *Food Technol. Aust.* 610 (1970).

232. M. Fujimaki and S. Arai in Ref. 12, pp. 221–230.

233. U.S. Pat. Appl. 20020142025 (Oct. 3, 2002), R. J. J. Hageman.

234. T. Nagasawa and co-workers, *Appl. Biochem. Biotechnol.* **13**, 147 (1988).

235. R. Yoshioka and co-workers, *Chem. Pharm. Bull.* **37**, 883 (1989); E. Diaconu and co-workers, *Rev. Med. Chir.* **92**, 359 (1988).

236. Jpn. Kokai 64 45,354 (Feb. 17, 1989), H. Kondo and co-workers (to Sagami Chemical Research Center).

237. S. Takahashi, *Hakkokogaku Kaishi* **61**, 139 (1983).

238. U.S. Pat. Appl. 20020131963 (Sept. 19, 2002), S. Baekkeskou and co-workers.

239. U.S. Pat. Appl. 20020013359 (Jan. 31, 2002), F. S. Dioguardi.

240. U.S. Pat. Appl. 20030040461 (Feb. 27, 2003), C. P. McAtee (to Bristol-Myers Squibb).

241. U.S. Pat. Appl. 20020155104 (Oct. 24, 2002), D. Munn and co-workers (to A. Mellor Medical College of Georgia Research Institute, Inc.).

242. U.S. Pat. 3,231,472 (Jan. 25, 1966), O. K. Jacobi and E. Engelmann (to Kolmar Laboratories).

243. U.S. Pat. Appl. 20020165165 (Nov. 7, 2002), M. Philippe (to l'Oreal).

244. Jpn. Kokai 58 164,504 (Sept. 29, 1983), S. Ishijima and M. Honma (to Ajinomoto Company).

245. T. Endo and H. Kubota, *Abstracts of the International Symposium on Biodegradable Polymers*, Oct. 1990, Tokyo, Japan, 114–119; Y. Saotome, T. Miyazawa, and T. Endo, *Chem. Lett.* 21 (1991).

246. M. Takehara and co-workers, *J. Am. Oil Chem. Soc.* **50**, 227 (1973); M. Takehara, *Yukagaku* (J. Japan Oil Chemists' Society) **34**, 964 (1985); S. Miyagishi, T. Asakawa,

and M. Nishida, *J. Colloid interface Sci.* **131**, 68 (1989); S. Osanai, Y. Yoshida, K. Fukushima, and S. Yosikawa, *Yukaguku* **38**, 633 (1989).

247. K. Tahara, H. Suzuki, and M. Honma, *Hyomen* (Surface) **19**, 688 (1981); M. Honma, *Gendai-Kagaku* (8), 58 (1987).
248. P. G. Genus, *The Physics of Liquid Crystals*, Clarendon Press, Oxford, England, 1974; G. J. Sprokel, ed., *The Physics and Chemistry of Liquid Crystal Devices*, Plenum Press, New York, 1980.
249. T. Sakurai in Ref. 46, pp. 143–154.
250. J. Watanabe and T. Nagase, *Macromolecules* **21**, 171 (1988); *Polymer J.* **19**, 781 (1987); J. Watanabe in Ref. 46, pp. 120–142; K. Hanabusa and co-workers, *J. Poly. Sci., Chem. Ed.* 28 825 (1990).
251. Y. Kuroyanagi in Ref. 85, pp. 260–276.
252. Catalogues of Chemical Reagent Suppliers, (Aldrich, Sigma, Wako, Kanto).
253. Jpn. Kokai 53 135,931 (Nov. 28, 1978), K. Ueda, R. Takigawa, and M. Saito (to Toray Industrial Company); semi-commercial sample is available from Kyowa Hakko Kogyo Company.
254. T. Tsunoda in Association of Amino Acid and Nucleic Acid, eds., *The 30th Anniversary Symposium on Amino Acid and Nucleic Acid*, Bioindustry Development Center, Tokyo, Japan, 1988, 49–62.
255. *Vulcanization of Rubber and Resin by Amino Acid*, Technical Bulletin, Sanyo Trading Company.

Kazumi Araki
University of East Asia

Toshitsugu Ozeki
Kyowa Hakko Kogyo Company

AMINO RESINS AND PLASTICS

1. Introduction

Amino resins are thermosetting polymers made by combining analdehyde with a compound containing anamino (—NH$_2$) group. Urea–formaldehyde (U/F) accounts for >80% of amino resins; melamine–formaldehyde accounts for most of the rest. Other aldehydes and other amino compounds are used to a very minor extent. The first commercially important amino resin appeared ~1930, or some 20 years after the introduction of phenol–formaldehyde resins and plastics (see Phenolic resins).

The principal attractions of amino resins and plastics are water solubility before curing, which allows easy application to and with many other materials, colorlessness, which allows unlimited colorability with dyes and pigments, excellent solvent resistance in the cured state, outstanding hardness and abrasion resistance, and good heat resistance. Limitations of these materials include release of formaldehyde during cure and, in some cases, such as in foamed insulation, after cure, and poor outdoor weatherability for urea moldings. Repeated cycling of wet and dry conditions causes surface cracks. Melamine moldings have relatively good outdoor weatherability.

Amino resins are manufactured throughout the industrialized world to provide a wide variety of useful products. Adhesives (qv), representing the largest single market, are used to make plywood, chipboard, and sawdust board. Other types are used to make laminated wood beams, parquet flooring, and for furniture assembly (see WOOD-BASED COMPOSITES AND LAMINATES).

Some amino resins are used as additives to modify the properties of other materials. For example, a small amount of amino resin added to textile fabric imparts the familiar wash-and-wear qualities to shirts and dresses. Automobile tires are strengthened by amino resins that improve the adhesion of rubber to tire cord (qv). A racing sailboat may have a better chance to win because the sails of Dacron polyester have been treated with an amino resin (1). Amino resins can improve the strength of paper even when it is wet. Molding compounds based on amino resins are used for parts of electrical devices, bottle and jar caps, molded plastic dinnerware, and buttons.

Amino resins are also often used for the cure of other resins such as alkyds and reactive acrylic polymers. These polymer systems may contain 5–50% of the amino resin and are commonly used in the flexible backings found on carpets and draperies, as well as in protective surface coatings, particularly the durable baked enamels of appliances, automobiles, etc.

The term amino resin is usually applied to the broad class of materials regardless of application, whereas the term aminoplast or sometimes amino plastic is more commonly applied to thermosetting molding compounds based on amino resins. Amino plastics and resins have been in use since the 1920s. Compared to other segments of the plastics industry, they are mature products, and their growth rate is only about one-half of that of the plastics industry as a whole. They account for ~3% of the U.S. plastics and resins production.

1.1. History. The basic chemistry of amino resins was established as early as 1908 (2), but the first commercial product, a molding compound, was patented in England (3) only in 1925. It was based on a resin made from an equimolar mixture of urea and thiourea and reinforced with purified cellulose fiber and was trademarked Beetle (indicating it could "beat all" others). Patent rights were acquired by the American Cyanamid Company along with the Beetle trademark. By 1930 a similar molding compound was being marketed in the United States. The new product was hard and not easily stained and was available in light, translucent colors; furthermore, it had no objectionable phenolic odor. The use of thiourea improved gloss and water resistance, but stained the steel molds. As amino resin technology progressed the amount of thiourea in the formulation could be reduced and finally eliminated altogether.

In the early 1920s, experimentation with urea–formaldehyde resins [9011-05-6] in Germany (4) and Austria (5,6) led to the discovery that these resins might be cast into beautiful clear transparent sheets, and it was proposed that this new synthetic material might serve as an organic glass (5,6). In fact, an experimental product called Pollopas was introduced, but lack of sufficient water resistance prevented commercialization. Melamine–formaldehyde resin [9003-08-1] does have better water resistance but the market for synthetic glass was taken over by new thermoplastic materials such as polystyrene and poly(methyl methacrylate) (see METHACRYLIC POLYMERS; STYRENE PLASTICS).

Melamine resins were introduced about ten years after the Beetle molding compound. They were very similar to those based on urea but had superior qualities. In Germany, Henkel was issued a patent for a melamine resin in 1936 (7). Melamine resins rapidly supplanted urea resins and were soon used in molding, laminating, and bonding formulations, as well as for textile and paper treatments. The remarkable stability of the symmetrical triazine ring made these products resistant to chemical change once the resin had been cured to the insoluble, cross-linked state.

Prior to the rapid expansion of thermoplastics following World War II, amino plastics served a broad range of applications in molding, laminating, and bonding. As the newer and more versatile thermoplastic materials moved into these markets, aminos became more and more restricted to applications demanding some specific property best offered by the thermosetting amino resins. Current sales patterns are very specific. Urea molding powders find application in moldings for electrical devices and in closures for jars and bottles. Melamine molding compound is used principally for molded plastic dinnerware. Urea resins have retained their use in electrical wiring devices because of good electrical properties, good heat resistance, and an availability of colors not obtainable with phenolics. Urea–formaldehyde resins are useful as closures because of their excellent resistance to oils, fats, and waxes often found in cosmetics, and their availability in a broad range of colors. Melamine plastic is used for molded dinnerware primarily because of outstanding hardness, water resistance, and stain resistance. Melamine–formaldehyde is the hardest commercial plastic material.

Aminoplasts and other thermosetting plastics are molded by an automatic injection molding process similar to that used for thermoplastics, but with an important difference (8). Instead of being plasticized in a hot cylinder and then injected into a much cooler mold cavity, the thermosets are plasticized in a warm cylinder and then injected into a hot mold cavity where the chemical reaction of cure sets the resin to the solid state. The process is best applied to relatively small moldings. Melamine plastic dinnerware is still molded by standard compression-molding techniques. The great advantage of injection molding is that it reduces costs by eliminating manual labor, thereby placing the amino resins in a better position to compete with thermoplastics (see PLASTICS PROCESSING).

The future for amino resins and plastics seems secure because they can provide qualities that are not easily obtained in other ways. New developments will probably be in the areas of more highly specialized materials for treating textiles, paper, etc, and for use with other resins in the formulation of surface coatings, where a small amount of an amino resin can significantly increase the value of a more basic material. Additionally, since amino resins contain a large proportion of nitrogen, a widely abundant element, they may be in a better position to compete with other plastics as raw materials based on carbon compounds become more costly.

2. Raw Materials

Most amino resins are based on the reaction offormaldehyde [50-00-0] with urea [57-13-6] or melamine [108-78-1]

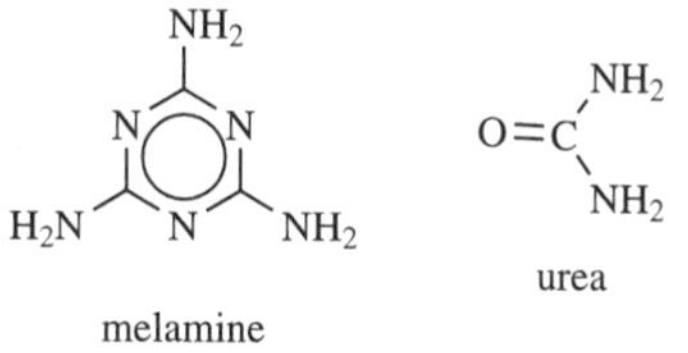

melamine

urea

Although formaldehyde will combine with many other amines, amides, and aminotriazines to form useful products, only a few are used and are of minor importance compared to products based on urea and melamine. Benzoguanamine [91-76-9], eg, is used in amino resins for coatings because it provides excellent resistance to laundry detergent, a definite advantage in coatings for automatic washing machines. Dihydroxyethylene urea [3720-97-6] is used for making amino resins that provide wash-and-wear properties in clothing. Glycoluril [496-46-8] resins provide coatings with high film flexibility.

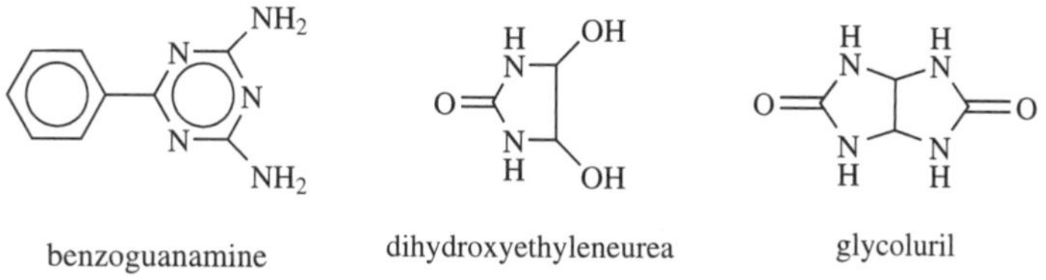

benzoguanamine dihydroxyethyleneurea glycoluril

Aniline–formaldehyde resins were once quite important because of their excellent electrical properties, but their markets have been taken over by newer thermoplastic materials. Nevertheless, some aniline resins are still used as modifiers for other resins. Acrylamide (qv) occupies a unique position in the amino resins field since it not only contains a formaldehyde reactive site, but also a polymerizable double bond. Thus it forms a bridge between the formaldehyde condensation polymers and the versatile vinyl polymers and copolymers.

In the sense that formaldehyde can supply a methylene link between two molecules, it is difunctional. Each amino group has two replaceable hydrogens that can react with formaldehyde; hence, it also is difunctional. Since the amino compounds commonly used for making amino resins, urea, and melamine contain two and three amino groups, they are polyfunctional and react with formaldehyde to form three-dimensional, cross-linked polymer structures. Compounds with a single amino group such as aniline or toluenesulfonamide can usually react with formaldehyde to form only linear polymer chains. However, in the presence of an acid catalyst at higher temperatures, the aromatic ring of aniline may react with formaldehyde to produce a cross-linked polymer.

2.1. Urea. Urea (carbamide) CH_4N_2O, is the most important building block for amino resins because urea–formaldehyde is the largest selling amino resin, and urea is the raw material for melamine, the amino compound used in the next largest selling type of amino resin. Urea is also used to make a variety of other amino compounds, such as ethyleneurea, and other cyclic derivatives used for amino resins for treating textiles. They are discussed later.

Urea is soluble in water, and the crystalline solid is somewhat hygroscopic, tending to cake when exposed to a humid atmosphere. For this reason, urea is frequently pelletized or prilled (formed into little beads) to avoid caking and making it easy to handle.

Only ~10% of the total urea production is used for amino resins, which thus appear to have a secure source of low cost raw material. Urea is made by the reaction of carbon dioxide and ammonia at high temperature and pressure to yield a mixture of urea and ammonium carbamate; the latter is recycled.

$$CO_2 + 2\,NH_3 \longrightarrow NH_2CONH_2 + H_2O \rightleftharpoons H_2NCOONH_4$$

2.2. Melamine. Melamine   (cyanurotriamide,2,4,6-triamino-*s*-triazine) $C_3H_6N_6$, is a white crystalline solid, melting at approximately 350°C with vaporization, only slightly soluble in water. The commercial product, recrystallized grade, is at least 99% pure. Melamine was synthesized early in the development of organic chemistry, but it remained of theoretical interest until it was found to be a useful constituent of amino resins. Melamine was first made commercially from dicyandiamide [461-58-5] (see Cyanamides), but is now made from urea, a much cheaper starting material (9–12) (see also Cyanuric and isocyanuric acids).

Urea is dehydrated to cyanamide which trimerizes to melamine in an atmosphere of ammonia to suppress the formation of deamination products. The ammonium carbamate [1111-78-0] also formed is recycled and converted to urea. For this reason the manufacture of melamine is usually integrated with much larger facilities making ammonia and urea.

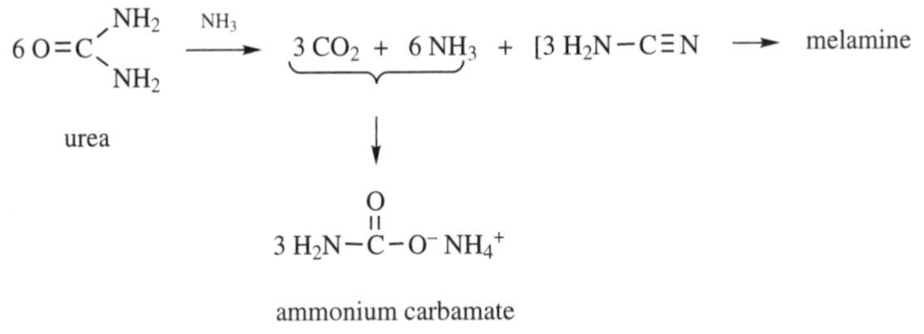

Since melamine resins are derived from urea, they are more costly and are therefore restricted to applications requiring superior performance. Essentially all of the melamine produced is used for making amino resins and plastics.

2.3. Formaldehyde. Pure formaldehyde, CH_2O, is a colorless, pungent smelling reactive gas (see Formaldehyde). The commercial product is handled either as solid polymer, paraformaldehyde (13), or in aqueous or alcoholic solutions. Marketed under the trade name Formcel, solutions in methanol, *n*-butanol, and isobutyl alcohol, made by Hoechst-Celanese, are widely used for making alcohol-modified urea and melamine resins for surface coatings and treating textiles.

Aqueous formaldehyde, known as formalin, is usually 37 wt % formaldehyde, though more concentrated solutions are available. Formalin is the general-purpose formaldehyde of commerce supplied unstabilized or methanol stabilized. The latter may be stored at room temperature without precipitation of solid formaldehyde polymers because it contains 5–10% methanol. The uninhibited type must be maintained at a temperature of at least 32°C to prevent the separation of solid formaldehyde polymers. Large quantities are often supplied in more concentrated solutions. Formalin at 44, 50, or even 56% may be used to reduce shipping costs and improve manufacturing efficiency. Heated storage tanks must be used. For example, formalin containing 50% formaldehyde must be kept at a temperature of 55°C to avoid precipitation. Formaldehyde solutions stabilized with urea (U) are used (14), and various other stabilizers have been proposed (15,16). With urea-stabilized formaldehyde (F) the user need only adjust the U/F ratio by adding more urea to produce a urea resin solution ready for use.

Paraformaldehyde [30525-89-4] is a mixture of polyoxymethylene glycols, $HO(CH_2O)_nH$, where n is from 8 to as much as 100. It is commercially available as a powder (95%) and as flake (91%). The remainder is a mixture of water and methanol. Paraformaldehyde is an unstable polymer that easily regenerates formaldehyde in solution. Under alkaline conditions, the chains depolymerize from the ends, whereas in acid solution the chains are randomly cleaved (17). Paraformaldehyde is often used when the presence of a large amount of water should be avoided as in the preparation of alkylated amino resins for coatings. Formaldehyde may also exist in the form of the cyclic trimer trioxane [110-88-3]. This compound is fairly stable and does not easily release formaldehyde, hence it is not used as a source of formaldehyde for making amino resins.

Approximately 25% of the formaldehyde produced in the United States is used in the manufacture of amino resins and plastics.

2.4. Other Materials. Benzoguanamine and acetoguanamine may be used in place of melamine to achieve greater solubility in organic solvents and greater chemical resistance. Aniline and toluenesulfonamide react with formaldehyde to form thermoplastic resins. They are not used alone, but rather as plasticizers (qv) for other resins including melamine and urea–formaldehyde. The plasticizer may be made separately or formed *in situ* during preparation of the primary resin.

Acrylamide [79-06-1] is an interesting monomer for use with amino resins; the vinyl group is active in free-radical catalyzed addition polymerizations, whereas the $-NH_2$ group is active in condensations with formaldehyde. Many patents describe methods of making cross-linked polymers with acrylamide by taking advantage of both vinyl polymerization and condensation with formaldehyde. For example, acrylamide reacts readily with formaldehyde to form *N*-methylolacrylamide [924-42-5], which gives the corresponding isobutyl ether with isobutyl alcohol.

$$CH_2{=}CH\overset{\overset{\displaystyle O}{\|}}{C}NH_2 \ + \ HCHO \ \longrightarrow \ CH_2{=}CH\overset{\overset{\displaystyle O}{\|}}{C}NHCH_2OH \ \xrightarrow{\ HOCH_2CH(CH_3)_2\ }$$

$$CH_2{=}CH\overset{\overset{\displaystyle O}{\|}}{C}NHCH_2OCH_2CH(CH_3)_2 \ + \ H_2O$$

This compound is soluble in most organic solvents and may be easily copolymerized with other vinyl monomers to introduce reactive side groups on the polymer chain (18). Such reactive polymer chains may then be used to modify other polymers including other amino resins. It may be desirable to produce the cross-links first. Thus, N-methylolacrylamide can react with more acrylamide to produce methylenebisacrylamide, a tetrafunctional vinyl monomer.

$$CH_2=CHCNHCH_2OH \ + \ CH_2=CHCNH_2 \ \longrightarrow \ CH_2=CHCNHCH_2NHCCH=CH_2 \ + \ H_2O$$

3. Chemistry of Resin Formation

The first step in the formation of resins and plastics from formaldehyde and amino compounds is the addition of formaldehyde to introduce the hydroxymethyl group, known as methylolation or hydroxymethylation.

$$R-NH_2 + HCHO \longrightarrow R-NH-CH_2OH$$

The second step is a condensation reaction that involves the linking together of monomer units with the liberation of water to form a dimer, a polymer chain, or a vast network. This reaction is usually referred to as methylene bridge formation, polymerization, resinification, or simply cure, and is illustrated in the following equation:

$$RNH-CH_2OH + H_2NR \longrightarrow RNH-CH_2-NHR + H_2O$$

Success in making and using amino resins largely depends on the precise control of these two chemical reactions. Consequently, these reactions have been much studied (19–30).

The first reaction, the addition of formaldehyde to the amino compound, is catalyzed by either acids or bases. Hence, it takes place over the entire pH range. The second reaction joins the amino units with methylene links and is catalyzed only by acids. The rates of these reactions have been studied over a broad range of pH (28). The results are presented in Figure 1.

The same study also examined some of the subsequent reactions involved in the formation of more complex U/F condensation products. Rate constants for these reactions at 35°C are shown in Table 1.

The methylol compounds produced by these reactions are relatively stable under neutral or alkaline conditions, but undergo condensation, forming polymeric products under acid conditions. Consequently, the first step in making an amino plastic is usually carried out under alkaline conditions. The amino compound and formaldehyde are combined and form a stable resin intermediate that may be used as an adhesive or combined with filler to make a molding compound. The second step is the addition of an acidic substance to catalyze the curing reaction, often with the application of heat to cure the amino resin to the solid cross-linked state. In this reaction, the methylol group is probably protonated and a molecule of water lost, giving the intermediate carbonium–imonium ion, which then reacts with an amino group to form a

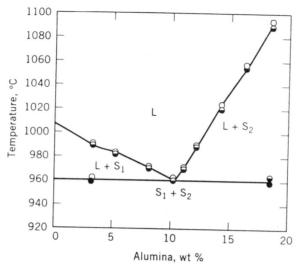

Fig. 1. Influence of pH on A, the addition reaction of urea and formaldehyde (1:1); and B, the condensation of methylolurea with the amino hydrogen of a neighboring urea molecule. Temperature = 35°C; 0.1 M aq.

methylene link.

$$RNHCH_2OH + H^+ \rightleftharpoons RNHCH_2\overset{+}{O}H_2$$

$$RNHCH_2\overset{+}{O}H_2 \rightleftharpoons RNH\overset{+}{C}H_2 + H_2O$$

$$RNH\overset{+}{C}H_2 + H_2NR' \rightleftharpoons RNHCH_2H_2\overset{+}{N}R'$$

$$RNHCH_2H_2\overset{+}{N}R' \rightleftharpoons RNHCH_2HNR' + H^+$$

In addition to the two main reactions, ie, methylolation and condensation, there are a number of other reactions important for the manufacture and uses of amino resins. For example, two methylol groups may combine to produce a dimethylene ether linkage and liberate a molecule of water:

$$2\,RNHCH_2OH \rightleftharpoons RNHCH_2{-}O{-}CH_2NHR + H_2O$$

Table 1. **Urea–Formaldehyde Reaction Rate Constants**

Reaction at 35°C and pH 4.0	k, L/(mol·s)
$U + F \rightarrow U/F$	4.4×10^{-4}
$U/F + U \rightarrow U{-}CH_2U$	3.3×10^{-4}
$U/F + U/F \rightarrow U{-}CH_2{-}U/F$	0.85×10^{-4}
$U/F_2 + U/F \rightarrow FU{-}CH_2{-}U/F$	0.5×10^{-4}
$U/F_2 + U/F_2 \rightarrow FU{-}CH_2{-}U/F_2$	$<3 \times 10^{-6}$

The dimethylene ether so formed is less stable than the diamino–methylene bridge and may rearrange to form a methylene link and liberate a molecule of formaldehyde.

$$RNH-CH_2O-CH_2NHR \longrightarrow RNH-CH_2-NHR + HCHO$$

The simple methylol compounds and the low molecular weight polymers obtained from urea and melamine are soluble in water and quite suitable for the manufacture of adhesives, molding compounds, and some kinds of textile treating resins. However, amino resins for coating applications require compatibility with the film-forming alkyd resins (qv) or copolymer resins with which they must react. Furthermore, even where compatible, the free methylol compounds are often too reactive and unstable for use in a coating-resin formulation that may have to be stored for some time before use. Reaction of the free methylol groups with an alcohol to convert them to alkoxy methyl groups solves both problems.

The replacement of the hydrogen of the methylol compound with an alkyl group renders the compound much more soluble in organic solvents and more stable. This reaction is also catalyzed by acids and usually carried out in the presence of considerable excess alcohol to suppress the competing self-condensation reaction. After neutralization of the acid catalyst, the excess alcohol may be stripped or left as a solvent for the amino resin.

The mechanism of the alkylation reaction is similar to curing. The methylol group becomes protonated and dissociates to form a carbonium ion intermediate that may react with alcohol to produce an alkoxymethyl group or with water to revert to the starting material. The amount of water in the reaction mixture should be kept to a minimum since the relative amounts of alcohol and water determine the final equilibrium.

Another way of achieving the desired compatibility with organic solvents is to employ an amino compound having an organic solubilizing group in the molecule, such as benzoguanamine. With one of the $-NH_2$ groups of melamine replaced with a phenyl group, benzoguanamine-formaldehyde resins [26160-89-4] have some degree of oil solubility even without additives. Nevertheless, benzoguanamine-formaldehyde resins are generally modified with alcohols to provide a still greater range of compatibility with solvent-based surface coatings. Benzoguanamine resins provide a high degree of detergent resistance together with good ductility and excellent adhesion to metal.

Displacement of a volatile with a nonvolatile alcohol is an important reaction for curing paint films with amino cross-linkers and amino resins on textile fabrics or paper. The following example is of a methoxymethyl group on an amino resin reacting with a hydroxyl group of a polymer chain.

$$RNHCH_2OCH_3 + HOCH_2 \text{—} \longrightarrow RNHCH_2OCH_2 \text{—} + CH_3OH$$

A troublesome side reaction encountered in the manufacture and use of amino resins is the conversion of formaldehyde to formic acid. Often the reaction mixture of amino compound and formaldehyde must be heated under alkaline conditions. This favors a Cannizzaro reaction in which two molecules of formaldehyde interact to yield one molecule of methanol and one of formic acid.

$$2\ HCHO + H_2O \longrightarrow CH_3OH + HCOOH$$

Unless this reaction is controlled, the solution may become sufficiently acidic to catalyze the condensation reaction causing abnormally high viscosity or premature gelation of the resin solution.

4. Manufacture

Precise control of the course, speed, and extent of the reaction is essential for successful manufacture. Important factors are mole ratio of reactants; catalyst (pH of reaction mixture); and reaction time and temperature. Amino resins are usually made by a batch process. The formaldehyde and other reactants are charged to a kettle, the pH adjusted, and the charge heated. Often the pH of the formaldehyde is adjusted before adding the other reactants. Aqueous formaldehyde is most convenient to handle and lowest in cost.

In general, conditions for the first part of the reaction are selected to favor the formation of methylol compounds. After addition of the reactants, the conditions may be adjusted to control the polymerization. The reaction may be stopped to give a stable syrup. This syrup could be an adhesive or laminating resin and might be blended with filler to make a molding compound (see also LAMINATES; REINFORCED PLASTICS). It might also be an intermediate for the manufacture of a more complicated product, such as an alkylated amino resin, for use with other polymers in coatings.

The flow sheet (Fig. 2) illustrates the manufacture of amino resin syrups, cellulose-filled molding compounds, and spray-dried resins.

In the manufacture of amino resins, every effort is made to recover and recycle the raw materials. However, there may be some loss of formaldehyde, methanol, or other solvent as tanks and reactors are vented. Some formaldehyde, solvents, and alcohols are also evolved in the curing of paint films and the curing of adhesives and resins applied to textiles and paper. The amounts of material evolved in curing the resins may be small so that it may be difficult to justify the installation of complex recovery equipment. However, in the development of new resins for coatings and for treating textiles and paper, emphasis is being placed on those compositions that evolve a minimum of by-products on curing.

5. Uses

5.1. Adhesive Resins. From antiquity, glues had been made almost entirely from materials of animal or vegetable origin, and were sensitive to moisture, oxidation, and bacterial or fungus attack. Because of these deficiencies, production of durable plywood, eg, was not possible. The modern plywood industry actually owes its growth to the availability of relatively low cost urea

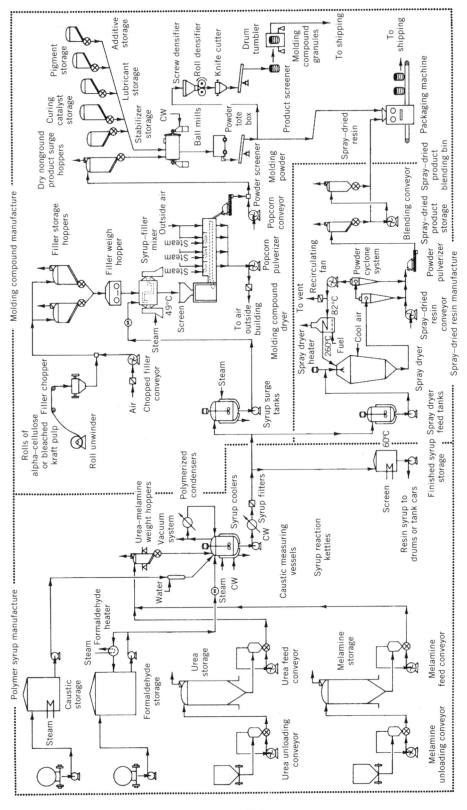

Fig. 2. Urea–formaldehyde and melamine–formaldehyde resin manufacture. CW = cold water. Courtesy of Stanford Research Institute.

adhesives. Plywood and chipboard or wood chip glues are often made at the plywood and chip board mill.

Urea and melamine adhesives represent products of very mature and overaged technologies. Essentially, they are simple reaction products of urea or melamine with formaldehyde; they may be liquids or powders. Liquids are converted to dry powders by "spray drying". Melamine–urea combinations generally are spray-dried powders of coreacted liquid melamine and urea-formaldehyde resins.

These adhesives are essentially for gluing wood. Urea–formaldehyde adhesives are used in the manufacture of plywood, in the fortification of starch adhesives for manufacture of paper bags and corrugated box boards, the production of "floral" and insulating foams, high quality sandpaper, and parquet flooring. Large volumes of urea resins are also sold in the United States for particle board bonding and other uses, at relatively low prices.

Melamine or melamine–ureas are used in the manufacture of truck and railroad flooring, laminated lumber, beams, exterior doors, marine plywood, toilet seats, and school furniture. The bonds in these products meet a variety of commercial, military, and federal specifications for exterior waterproof adhesives.

Ureas modified with furfuryl alcohol [98-00-0] find excellent acceptance for gluing or assembly of furniture, sporting goods, eg, tennis rackets, fishing rods, and water skiis, and for bonding decorative laminates to substrates such as particle board. Such ureas are called gap-filling adhesives. The modifier makes the resin less rigid, less highly cross-linked, and better able to relieve stresses imposed by shrinkage due to cure and loss of water in glue lines of nonuniform thickness.

In gluing, the adhesive must not saturate veneers or wood chips, but must remain in the glue line on the surface of the chips or between the plies. The adhesives are generally of high viscosity so that they remain in the glue line. Thickeners and extenders, such as powdered pecan shells and wheat flour, are often used.

Urea resin adhesives, by the use of the proper hardener, may be set either by heat or at room temperature. For room temperature curing, the hardener may be ammonium chloride, together with basic materials like calcium phosphate to neutralize excess acid that might damage the wood. Cold set or room temperature set adhesives are those that set satisfactorily at 20–30°C, whereas a hot set adhesive generally means one that is set >99°C. Those that cure well between these temperatures are often referred to as intermediate set. Cold set types naturally set much faster at intermediate temperatures. Cold pressing of plywood is usually done in a hydraulic press. The press may then be opened, the clamped assembly removed, and allowed to stay under the pressure of the clamps for at least 4 h to set the glue.

Very rapid cures may be achieved by applying heat in the form of microwave radiation (see MICROWAVE TECHNOLOGY). The moist amino resin adhesive absorbs the high frequency radiation more readily than dry wood, thereby concentrating the heat in the glue line where it is needed. Hot pressing may be conducted in a hydraulic press comprising a large number of steam heated platens usually 5 cm thick. Pressures usually range from 1 to 2 MPa (150–300 psi). Hot press temperatures for urea and melamine–urea are usually 115–132°C.

In plywood production with ureas, the spread veneers should be pressed as soon as possible. The time between spreading and pressing, usually called assembly time, should never exceed 1 h. With some formulations, the permissible assembly time may be no >15 min. Melamine formulations and uncatalyzed melamine–urea combinations, however, can be spread and stored for as much as 1 week before use.

Ureas are not satisfactory for prolonged water immersion or for continuous exposure to warm and excessively humid conditions, although they are fairly resistant to normal humidity. Somewhat more durable bonds are obtained by heat setting. Ureas can be extended with wheat or rye flour, using as much as 150 parts flour to 100 parts dry resin. The extended glue still retains a fair degree of moisture resistance. Melamine resins have excellent water resistance, but cannot be cured at room temperature. Durable laminated wood beams used in building construction usually employ microwave technology for heat curing.

Continuous production of urea–formaldehyde resins has been described in many patents. In a typical example, urea and formaldehyde are combined and the solution pumped through a multistage unit. Temperature and pH are controlled at each stage to achieve the appropriate degree of polymerization. The product is then concentrated in a continuous evaporator to ~60–65% solids (31).

5.2. Laminating Resins.

Phenolic and melamine resins are both used in the manufacture of decorative laminated plastic sheets for counters and table tops (see LAMINATES; REINFORCED PLASTICS). The phenolic is functional, being used in the backing or support sheets, whereas the melamine resin performs both decorative and functional roles in the print sheet and the protective overlay. Hardness, transparency, stain resistance, and freedom from discoloration are essential, in addition to a long-lasting working surface. Transparency is achieved because the refractive index of cured melamine-formaldehyde resin approaches that of the cellulose fibers and thus there is little scattering of light. Low cost and good mechanical properties are provided by the phenolic backing layers. In this instance, the combination of phenolic and amino resins achieves an objective that neither would normally be capable of performing alone. Developments in modified melamine resins have contributed to commercialization of premium priced through-color decorative laminates in which the dark color phenolic backing layers are replaced by color layers matching the full range of surface colors.

Phenolic resins are generally used in alcoholic solution, whereas melamine resins are best handled in water or water–alcohol mixtures. The paper or cloth web is passed through a dip tank containing resin solution, adjusted for pick-up on squeeze rolls, and then passed through a heated drying oven. Once dried, the treated paper or cloth is fairly stable and, stored in a cool place, it may be kept for several weeks or months before pressing into laminated plastic sheets.

A melamine laminating resin used to saturate the print and overlay papers of a typical decorative laminate might contain 2 mol of formaldehyde for each mole of melamine. In order to inhibit crystallization of methylol melamines, the reaction is continued until about one-fourth of the reaction product has been converted to low molecular weight polymer. A simple determination of free formaldehyde may be used to follow the first stage of the reaction, and the build-up of polymer in the reaction mixture may be followed by cloud-point dilution or viscosity tests.

A particularly interesting and useful test is run at high dilution. One or two drops of resin are added to a test tube half full of water. A cloudy streak as the drop sinks through the water indicates that the resin has advanced to the point where the highest molecular weight fraction of polymer is no longer soluble in water at that particular temperature. At this high dilution, the proportion of water to resin is not critical, hence the only measurement needed is the temperature of the water. The temperature at or below which the drops give a white streak is known as the hydrophobe temperature. This test is particularly useful with melamine resins.

Laminates are pressed in steam-heated, multiple-opening presses. Each opening may contain a book of as many as 10 laminates pressed against polished steel plates. Curing conditions are 20–30 min at ~150°C under a pressure of ~6900 kPa (1000 psi).

5.3. Molding Compounds. Molding was the first big application for amino resins, although molding compounds are more complex than either laminating resins or adhesives. A simple amino resin molding compound might be made by combining melamine with 37% formalin in the ratio of 2 mol of formaldehyde for each mole of melamine at neutral or slightly alkaline pH and a temperature of 60°C. The reaction should be continued until some polymeric product has been formed to inhibit crystallization of dimethylolmelamine upon cooling. When the proper reaction stage has been reached, the resin syrup is pumped to a dough mixer where it is combined with alpha-cellulose pulp, approximately one part of cellulose to each three parts of resin solids. The wet, spongy mass formed in the dough mixer is then spread on trays where it is combined with alpha-cellulose in a humidity-controlled oven to produce a hard, brittle popcorn-like intermediate. This material may be coarsely ground and sent to storage. To make the molding material, the cellulose–melamine resin intermediate is combined in a ball mill with a suitable catalyst, stabilizer, colorants, and mold lubricants. The materials must be ground for several hours to achieve the uniform fine dispersion needed to get the desired decorative appearance in the molded article. The molding compound may be used as a powder or it may be compacted under heat and pressure to a granular product that is easier to handle (32). A urea molding compound might be made in much the same way using a resin made with 1.3–1.5 mol of formaldehyde per 1.0 mol of urea.

Amino molding compounds can be compression, injection, or transfer molded. Urea molding compound has found wide use and acceptance in the electrical surface wiring device industry. Typical applications are circuit breakers, switches, wall plates, and duplex outlets. Urea is also used in closures, stove hardware, buttons, and small housings. Melamine molding compound is used primarily in dinnerware applications for both domestic and institutional use. It is also used in electrical wiring devices, ashtrays, buttons, and housings.

The emergence of a new amino application is rare at this point in its relatively long life, but one such has appeared and is growing rapidly. Because of the relative hardness of both urea and melamine moldings, a unique use has been developed for small, granular sized particles of cut up molded articles. It is the employment of a pressurized stream of plastic particles to remove paint without damaging the surface beneath, and can be compared to a sandblasting operation. This procedure is gaining wide acceptance by both commercial airlines and the military for the refinishing of painted surfaces. It does not harm the substrate and eliminates the use of chemicals formerly used in stripping paint.

To speed up the molding process, the required amount of molding powder or granules is often pressed into a block and prewarmed before placing it in the mold. Rapid and uniform heating is accomplished in a high frequency preheater; essentially an industrial microwave oven. The prewarmed block is then transferred to the hot mold, pressed into shape, and cured.

Production of decorated melamine plastic dinner plates makes use of molding and laminating techniques. The pattern is printed on the same type of paper used for the protective overlay of decorative laminates, treated with melamine resin and dried, and then cut into disks of the appropriate size.

To make a decorated plate, the mold is opened shortly after the main charge of molding compound has been pressed into shape, the decorative foil is laid in the mold on top of the partially cured plate, printed side down, and the mold closed again to complete the curing process. The melamine-treated foil is thus fused to the molded plate and, as with the decorative laminate, the overlay becomes transparent so that the printed design shows through yet is protected by the film of cured resin.

The excellent electrical properties, hardness, heat resistance, and strength of melamine resins makes them useful for a variety of industrial applications. Some representative properties of amino resin molding compounds, including the industrial-grade melamines, are listed in Table 2.

In 1989 quantity costs, which reflect the lowest cost, of urea molding compounds were ~$1.41/kg ($0.035/in.^{3}) for black and brown colors, $1.58/kg ($0.039/in.^{3}) for white and ivory; special colors are somewhat higher in price. The approximate cost of cellulose-filled melamine molding compound is $1.74/kg ($0.043/in.^{3}). Glass fiber-filled melamine sells for $7.70/kg ($0.22/in.^{3}).

Amino molding compounds are produced worldwide. The following list represents the major suppliers.

Company	Location	Product
American Cyanamid	USA	urea
		melamine
BIP Chemical	England	urea
		melamine
Budd Chemical	USA	urea
Carmel Chemical	Israel	urea
		melamine
Fiberite	USA	melamine
Perstrop	Sweden	urea
	USA	melamine
Plastics Mfg. Co.	USA	urea
		melamine
Plastics Engineering Co.	USA	melamine

Table 2. **Typical Properties of Filled Amino Resin Molding Compounds**

ASTM or UL test	Property	Urea	Melamine		
		Alpha-cellulose	Alpha-cellulose	Macerated fabric	Glass fiber
		Physical			
D792	specific gravity	1.47–1.52	1.47–1.52	1.5	1.8–2.0
D570	water absorption, 24 h, 3.2 mm thick, %	0.48	0.1–0.6	0.3–0.6	0.09–0.21
		Mechanical			
D638	tensile strength, MPa[a]	38–48	48–90	55–69	35–70
D638	elongation, %	0.5–1.0	0.6–0.9	0.6–0.8	
D638	tensile modulus, GPa[b]	9–9.7	9.3	9.7–11	16.5
D785	hardness, Rockwell M	110–120	120	120	115
D790	flexural strength, MPa[a]	70–124	83–104	83–104	90–165
D7900	flexural modulus, GPa[b]	9.7–10.3	7.6	9.7	16.5
D256	impact strength, J/m[c] of notch	14–18	13–19	32–53	32–1000
		Thermal			
C177	thermal conductivity, 10^{-4} W/(m·K)	42.3	29.3–42.3	44.3	48.1
D696	coefficient of thermal expansion, 10^{-5} cm/(cm·°C)	2.2–3.6	2.0–5.7	2.5–2.8	1.5–1.7
D648	deflection temperature at 1.8 MPa[a], °C	130	182	154	204
UL 94	flammability class	VO[d]	VO[d]		VO
	continuous no-load service temperature, °C	77[e]	99[e]	121	149–204

633

Electrical

D149	dielectric strength, V/0.00254 cm short time, 3.2 mm thick	330–370	270–300	250–350	170–300
	step by step	220–250	240–270	200–300	170–240
D150	dielectric constant, 22.8°C				
	at 60 Hz	7.7–7.9	8.4–9.4	7.6–12.6	9.7–11.1
	at 10^3 Hz		7.8–9.2	7.1–7.8	
D150	dissipation factor, 22.8°C				
	at 60 Hz	0.034–0.043	0.030–0.083	0.07–0.34	0.14–0.23
	at 10^3 Hz		0.015–0.036	0.03–0.05	
D257	volume resistivity, 22.8°C, 50% rh, $\Omega \cdot$cm	$0.5 - 5.0 \times 10^{11}$	$0.8 - 2.0 \times 10^{12}$	$1.0 - 3.0 \times 10^{11}$	$0.9 - 2.0 \times 10^{11}$
D495	arc resistance, s	80–100	125–136	122–128	180–186

[a] To convert MPa to psi, multiply by 145.
[b] To convert GPa to psi, multiply by 145,000.
[c] To convert J/m to ftlbf/in., divide by 53.38.
[d] Applies to specimens thicker than 1.6 mm.
[e] Based on no color change.

5.4. Coatings. Cured amino resins are far too brittle to be used alone as surface coatings for metal or wood substrates, but in combination with other film formers (alkyds, polyesters, acrylics, epoxies) a wide range of acceptable performance properties can be achieved. These combination binder coating formulations cure rapidly at slightly elevated temperatures, making them well suited for industrial baking applications. The amino resin content in the formulation is typically in the range of 10–50% of the total binder solids.

A wide selection of amino resin compositions is commercially available. They are all alkylated to some extent in order to provide compatibility with the other film formers, and formulation stability. They vary not only in the type of amine (melamine, urea, benzoguanamine, and glycoluril) used, but also in the concentration of combined formaldehyde, and the type and concentration of alkylation alcohol (*n*-butanol, isobutyl alcohol, methanol).

On curing, amino resins not only react with the nucleophilic sites (hydroxyl, carboxyl, amide) on the other film formers in the formulation, but also self-condense to some extent. Highly alkylated amino resins have less tendency to self-condense (33,34). Therefore they are effective cross-linking agents, but may require the addition of a strong acid catalyst to obtain acceptable cure even at bake temperatures of 120–177°C.

Amino resins based on urea have advantages in low temperature cure response and low cost. However, they are not as stable to ultraviolet (uv) radiation as melamine resins, and have poorer heat resistance; therefore, they have been successful primarily in interior wood finishes. Melamine resins, on the other hand, are uv stable, have excellent heat resistance, film hardness, and chemical resistance. They therefore dominate amino resin usage in (OEM) automotive coatings, general metals finishes, container coatings (both interior and exterior), and prefinished metal applications. Glycoluril resins have also found use in prefinished metal, primarily because of their high film flexibility properties. Unalkylated glycoluril resins are unique in that they are stable under slightly acidic conditions and have therefore found use in low temperature cure waterborne finishes. Benzoguanamine resins have historically been successful in appliance finishes because of their superior chemical resistance and specifically their detergent resistance. However, they have both poor uv resistance and economics, which have limited their use in other application areas.

When first introduced to the coatings industry, amino resin compositions were partially butylated and relatively polymeric in nature, with degrees of polymerization of 4–6. However, the dominant amino resin in today's industrial coating is based on a highly methylated, highly monomeric (degrees of polymerization of 1.4–2.6) melamine cross-linking agent. Variations of extent of methylolation and methylation exist along with a number of co-ethers where the melamine molecule is both methylated and either *n*-butylated or isobutylated. This type of composition dominates because it best addresses the pollution (low volatile organic compounds) and performance requirements of today's industrial finishes (see COATINGS, INDUSTRIAL).

Methylation provides fast cure response, improved exterior exposure, high weight retention on curing, and suitability for both solvent and waterborne systems. Waterborne systems, in most instances, provide lower pollution than solvent-based formulations. High monomer content reduces the viscosity of the amino resin, again lowering pollution particularly when used in solvent-based systems, and also improves film flexibility, and recoat adhesion. When some butylation is included as part of the alkylation, the viscosity of the amino resin is lowered, thereby lowering pollution (of the formulated coating), improving recoat adhesion, and improving wetting and flow characteristics. An amino resin is usually selected based on specific performance properties required or performance to be emphasized.

Stability in storage is an important property for coating systems containing amino resins. If the amino resin undergoes self-condensation or reacts at room temperature with the alkyd or other film-forming polymer, the system may become too viscous or thicken to a gel that can no longer be used for coating. Alkyds usually contain sufficient free carboxyl groups to catalyze the curing reaction when the coating is baked, but this may also cause the paint to thicken in storage. Partial neutralization of the acid groups with an amine can greatly improve storage stability yet allow the film to cure when baked, since much of the amine is vaporized with the solvent during the baking process. 2-Amino-2-methyl-1-propanol [124-68-5], triethylamine [121-44-8], and dimethylaminoethanol [108-01-1] are commonly used as stabilizers. Alcohols as solvent also improve storage stability. Catalyst addition just before the coating is to be applied permits rapid curing and avoids the problem of storage stability. A strong acid soluble in organic solvents such as p-toluenesulfonic acid is very effective and may be partially neutralized with an amine to avoid premature reaction.

A butylated urea–formaldehyde resin for use in the formulation of fast-curing baking enamels might be made beginning with the charge: urea (1.0 mol), paraformaldehyde (2.12 mol), and butanol (1.50 mol). Triethanolamine is added to make the solution alkaline ($\sim$1% of the weight of the urea) and the mixture is refluxed until the paraformaldehyde is dissolved. Phthalic anhydride is added to give a pH of 4.0 and the water is removed by azeotropic distillation until the batch temperature reaches 117°C. Cooling and dilution with solvent is done until the desired solids content is reached (35).

A highly methylated melamine–formaldehyde resin for cross-linking with little or no self-condensation might be made as follows (36). A solution of formaldehyde in methanol is charged to a reaction kettle and adjusted to a pH of 9.0–9.5 using sodium hydroxide. Melamine is then added to give a ratio of 1 mol of melamine for each 6.5 mol of formaldehyde and the mixture refluxed for ½ h. The reaction is then cooled to 35°C and more methanol added to bring the ratio of methanol per mole of melamine up to 11. With the batch temperature at 35°C, enough sulfuric acid is added to reduce the pH to 1.0. After holding the reaction mixture at this temperature and pH for 1 h, the batch is neutralized with 50% sodium hydroxide and the excess methanol stripped to give a product containing 60% solids which is then clarified by filtration. A highly methylated resin, such as this, may be used in water-based (37) or solvent-type coatings. It might also be used to provide crease resistance to cotton fabric.

The principal problems facing amino resins in the industrial coatings of the 1990s are formaldehyde emission and low temperature cure performance. Significant progress has been made in reducing the residual free formaldehyde in the amino resin, but formaldehyde generation on baking must still be addressed. Concerning low temperature. cure performance, emphasis is being placed on catalyst selection. Amino resins cure at bake temperatures as low as 71–82°C, but at these bake temperatures they require high concentrations of acid catalyst, which negatively affect hydrolysis resistance or water sensitivity of the cured film. The development of improved catalysts is the most promising solution to low temperature cure performance enhancement.

5.5. Textile Finishes. Most amino resins used commercially for finishing textile fabrics are methylolated derivatives of urea or melamine. Although these products are usually monomeric, they may contain some polymer by-product.

Amino resins react with cellulosic fibers and change their physical properties. They do not react with synthetic fibers, such as nylon, polyester, or acrylics, but may self-condense on the surface. This results in a change in the stiffness or resiliency of the fiber. Partially polymerized amino resins of such molecular size that prevents them from penetrating the amorphous portion of cellulose also tend to increase the stiffness or resiliency of cellulose fibers.

Monomeric amino resins react predominantly with the primary hydroxyls of the cellulose, thereby replacing weak hydrogen bonds with strong covalent bonds which leads to an increase in fiber elasticity. When an untreated cotton fiber is stretched or deformed by bending, as in forming a crease or wrinkle, the relatively weak hydrogen bonds are broken and then reform to hold the fiber in its new position. The covalent bonds that are formed when adjacent cellulose chains are cross-linked with an amino resin are five to six times stronger than the hydrogen bonds. Covalent bonds are not broken when the fiber is stretched or otherwise deformed. Consequently, the fiber tends to return to its original condition when the strain is removed. This increased elasticity is manifested in two important ways: (*1*) When a cotton fabric is cross-linked while it is held flat, the fabric tends to return to its flat condition after it has been wrinkled during use or during laundering. Garments made from this type of fabric are known as wash-and-wear, minimum care, or no-iron. (*2*) A pair of pants that is pressed to form a crease and then cross-linked tends to maintain the crease through wearing and laundering. This type of garment is called durable-press or permanent press (see TEXTILES).

This increased elasticity is always accompanied by a decrease in strength of the cellulose fiber that occurs even though weak hydrogen bonds are replaced by stronger covalent bonds. The loss of strength is not caused by hydrolytic damage to the cellulose. If the cross-linking agent is removed by acid hydrolysis, eg, the fiber will regain most, if not all, of its original strength. The loss in strength is believed to be due to intramolecular reaction of the amino resin along the cellulose chain to displace a larger number of hydrogen bonds, resulting in a net loss in strength. The intramolecular and intermolecular reactions (cross-linking) both occur at the same time.

Although there are many different amino resins used for textile finishing, all of them impart about the same degree of increase in elasticity when applied on an equal molar basis. Elasticity can be measured by determining the recovery from wrinkling. Although all these products impart about the same degree of improvement in elasticity, they also may impart many other desirable or undesirable properties to the fabric. The development of amino resins for textile finishing has been aimed toward maximizing the desirable properties and minimizing the undesirable ones. Most of the resins and reactants used in today's textile market are based on urea as a starting material. However, the chemistry differs considerably from that employed in early textile-finishing operations.

The first amino resins used commercially on textiles were the so-called urea–formaldehyde resins, dimethylolurea [140-95-4], or its mixtures with monomethylolurea [1000-82-4].

$$
\underset{\text{HOCH}_2\text{NH}-\overset{\displaystyle\text{O}}{\overset{\|}{\text{C}}}\text{NH}-\text{CH}_2\text{OH}}{}
$$

$$
\underset{\text{H}_2\text{N}-\overset{\displaystyle\text{O}}{\overset{\|}{\text{C}}}-\text{NHCH}_2\text{OH}}{}
$$

Their performance falls short of most present finishes, particularly in durability, resistance to chlorine-containing bleaches, and formaldehyde release, and they are not used much today. Both urea and formaldehyde are relatively inexpensive, and manufacture is simple; ie, 1–2 mol of formaldehyde as an aqueous solution reacts with 1 mol of urea under mildly alkaline conditions at slightly elevated temperatures.

Since the methylolurea monomers have limited water solubility ($\sim$30%), they were usually marketed in dispersed form as soft pastes containing 55–65% active ingredient in order to decrease container and shipping costs. By increasing the temperature and using slightly acid conditions, dimethyloureas can be made as a series of short polymers that have infinite water solubility and can be marketed at concentrations as high as 85%. However, because these result in increased fabric stiffness, they cannot be used interchangeably with the monomeric materials. Both forms polymerize readily in storage and, unless kept under refrigeration, become water insoluble within a few weeks at ambient temperatures.

To overcome stability and water solubility problems, methylolurea resins are frequently alkylated to block the reactive hydroxyl groups. For reasons of economy, the alkylating agent is usually methanol. In this process, 2 mol of aqueous formaldehyde reacts with 1 mol of urea under alkaline conditions to form dimethylolurea. Excess methanol is then added, and the reaction is continued under acid conditions to form methoxymethylurea. Both methylol groups can be methylated by maintaining low concentrations of water and using a large excess of methanol; however, methylation of only one of the methylol groups is sufficient to provide adequate shelf life and water solubility. Upon completion of the methylation reaction, the resin is adjusted to pH 7–10, and excess methanol and water are removed by distillation under reduced pressure to provide syrups of 50–80% active ingredients.

Like methylolureas, cyclic ureas are based on reactions between urea and formaldehyde; however, the amino resin is cyclic rather than linear. Many cyclic urea resins have been used in textile-finishing processes, particularly to achieve wrinkle resistance and shrinkage control, but the ones described below are the most commercially important. They are all in use today to greater or lesser extents, depending on specific end requirements (see also TEXTILES, FINISHING).

Ethyleneurea Resins. One of the most widely used resins during the 1950s and 1960s was based on dimethylolethyleneurea [136-84-5] (1,3-bis(hydroxymethyl)-2-imidazolidinone) commonly known as ethyleneurea resin. This resin [28906-87-8] is most conveniently prepared from urea, ethylenediamine, and formaldehyde. 2-Imidazolidinone [120-93-4] (ethyleneurea) is first prepared by the reaction of excess ethylenediamine [107-15-3] with urea (38) in an aqueous medium at ~116°C.

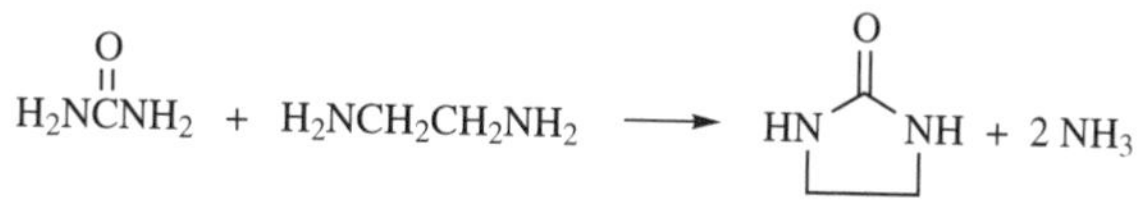

$$H_2NCNH_2 \ + \ H_2NCH_2CH_2NH_2 \ \longrightarrow \ \underset{\text{2-imidazolidinone}}{HN \diagup \overset{O}{\diagdown} NH} + 2\,NH_3$$

A fractionating column is required for the removal of ammonia and recycle of ethylenediamine. The molten product (mp 133°C) is then run into ice water to give a solution that is methylolated with 37% aqueous formaldehyde.

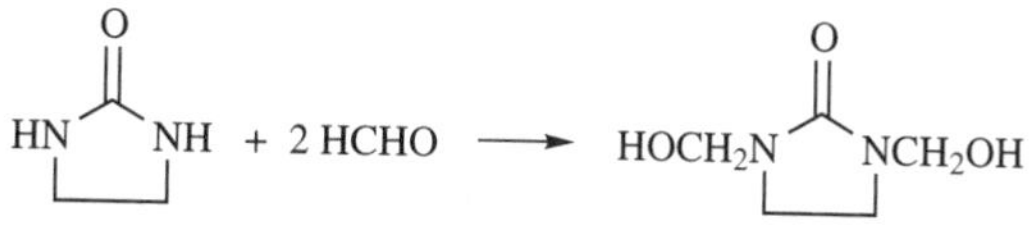

$$\underset{\text{ethyleneurea}}{HN \diagup \overset{O}{\diagdown} NH} + 2\,HCHO \ \longrightarrow \ \underset{\text{dimethyloethyleneurea}}{HOCH_2N \diagup \overset{O}{\diagdown} NCH_2OH}$$

The resin, generally a 50% solution in water, has excellent shelf life and is stable to hydrolysis and polymerization.

Propylene Urea Resins. In similar fashion to ethyleneurea, dimethylolpropyleneurea [3270-74-4] [1,3-bis(hydroxymethyl)tetrahydro-2-(1H)-pyrimidinone] is the basis of propyleneurea–formaldehyde resin [65405-39-2]. Its preparation is from urea, 1,3-diaminopropane [109-76-2], and formaldehyde.

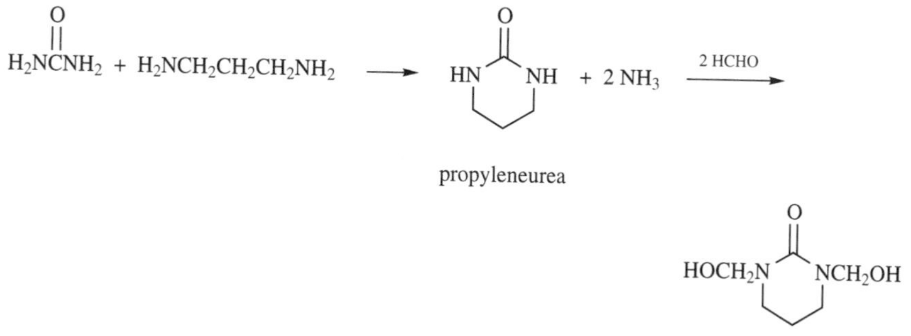

$$H_2NCNH_2 \ + \ H_2NCH_2CH_2CH_2NH_2 \ \longrightarrow \ \underset{\text{propyleneurea}}{HN \diagup \overset{O}{\diagdown} NH} + 2\,NH_3 \ \xrightarrow{2\ HCHO}$$

$$\underset{\text{dimethylolpropyleneurea}}{HOCH_2N \diagup \overset{O}{\diagdown} NCH_2OH}$$

This resin was temporarily accepted, primarily because of its improved resistance to acid washes. However, the relatively high cost of the diamine precluded widespread commercial acceptance.

Triazone.   Triazone is the common name for the class of compounds corresponding to the dimethylol derivatives of tetrahydro-5-alkyl-*s*-triazone. They can be made readily and cheaply from urea, formaldehyde, and a primary aliphatic amine. A wide variety of amines may be used to form the six-membered ring (39); however, for reasons of cost and odor, hydroxyethylamine (monoethanolamine) is used preferentially (see ALKANOLAMINES). Since the presence of straight-chain methylolureas causes no deleterious effects to the fabric finish, the triazones typically are prepared with less than the stoichiometric quantity of the amine. This results not only in a less costly resin but also in improved performance (40).

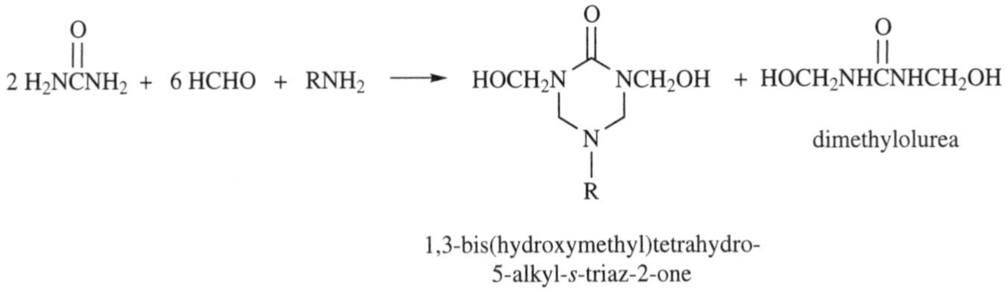

$$2\ H_2NCNH_2 + 6\ HCHO + RNH_2 \longrightarrow HOCH_2N \overset{O}{\overbrace{\qquad}} NCH_2OH + HOCH_2NHCNHCH_2OH$$

1,3-bis(hydroxymethyl)tetrahydro-
5-alkyl-*s*-triaz-2-one

dimethylolurea

The resin is simply prepared by heating the components together. Usually the urea and formaldehyde are first charged to the kettle and heated under alkaline conditions to give a mixture of polymethylolureas, followed by the slow addition of the amine with continued heating to form the cyclic compound. The order of addition can be varied as can the molar ratios to yield a range of chain-ring compound ratios. The commercial resin is usually sold as a 50% solids solution in water.

Uron Resins.   In the textile industry, the term uron resin usually refers to the mixture of a minor amount of melamine resin and so-called uron, which in turn is predominantly *N,N'*-bis(methoxymethyl)uron [7388-44-5] plus 15–25% methylated urea–formaldehyde resins, a by-product. *N,N'*-bis(methoxymethyl)-uron was first isolated and described in 1936 (41), but was commercialized only in 1960. It is manufactured (42) by the reaction of 4 mol of formaldehyde with 1 mol of urea at 60°C under highly alkaline conditions to form tetramethylolurea [2787-01-1]. After concentration under reduced pressure to remove water, excess methanol is charged and the reaction is continued under acidic conditions at ambient temperatures to close the ring and methylate the hydroxymethyl groups. After filtration to remove the precipitated salts, the methanolic solution is concentrated to recover excess methanol. The product (75–85% pure) is then mixed with a methylated melamine–formaldehyde resin to reduce fabric strength losses in the presence of chlorine, and diluted with water to 50–75% solids. Uron resins do not find significant use today due to the greater amounts of formaldehyde released from fabric treated with these resins.

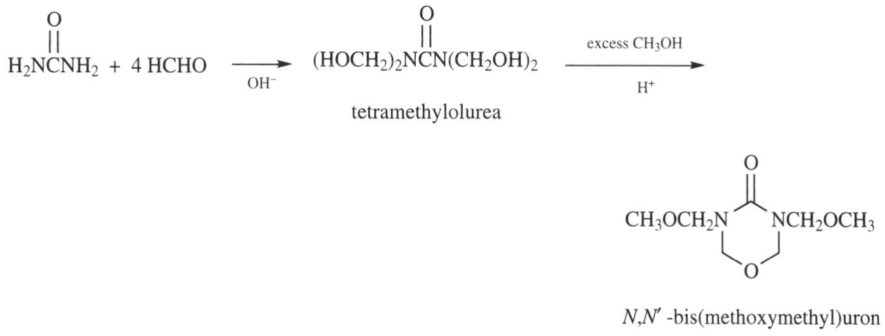

$$\text{H}_2\text{NCNH}_2 + 4\,\text{HCHO} \xrightarrow[\text{OH}^-]{} (\text{HOCH}_2)_2\text{NCN}(\text{CH}_2\text{OH})_2 \xrightarrow[\text{H}^+]{\text{excess CH}_3\text{OH}}$$

tetramethylolurea

$$\text{CH}_3\text{OCH}_2\text{N} \underset{\text{O}}{\overset{\text{O}}{\diamond}} \text{NCH}_2\text{OCH}_3$$

N,N'-bis(methoxymethyl)uron

Glyoxal Resins. Since the late 1960s, glyoxal resins have dominated the textile-finish market for use as wrinkle-recovery, wash-and-wear, and durable-press agents. These resins are based on 1,3-bis(hydroxymethyl)-4,5-dihydroxy-2-imidazolidinone, commonly called dimethyloldihydroxyethyleneurea [1854-26-8] (DMDHEU). Several methods of preparation are described in the literature (43). On a commercial scale, DMDHEU can be prepared inexpensively at high purity by a one-kettle process (44): 1 mol of urea, 1 mol of glyoxal [107-22-2] as 40% solution, and 2 mol of formaldehyde in aqueous solution are charged to the reaction vessel. The pH is adjusted to 7.5–9.5 and the mixture heated at 60–70°C. The reaction is nearly stoichiometric; excess reagent is not necessary.

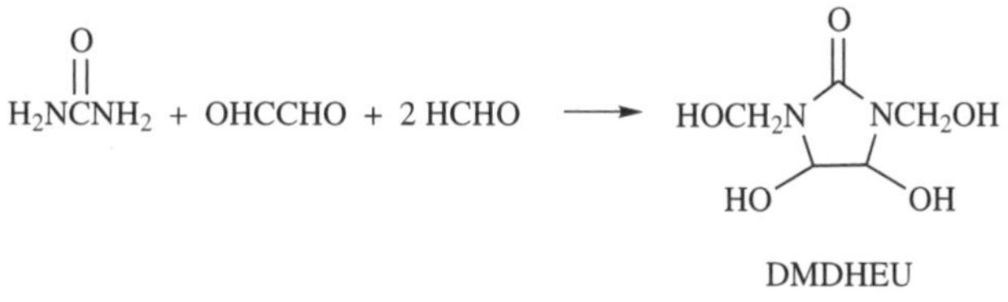

$$\text{H}_2\text{NCNH}_2 + \text{OHCCHO} + 2\,\text{HCHO} \longrightarrow \text{HOCH}_2\text{N} \overset{\text{O}}{\diamond} \text{NCH}_2\text{OH}$$

DMDHEU

Glyoxal resins are generally sold at 45% solids solutions in water. Resin usage for crease-resistant fabrics had increased to well over 60×10^6 kg by 1974 and over one-half of this was DMDHEU for durable-press garments. In the early 1980s, glyoxal resins modified with diethylene glycol [111-46-6] became prominent in the marketplace. These products are either simple mixtures of diethylene glycol and DMDHEU in water solution or the reaction product of diethylene glycol and DMDHEU. Rarely, ethylene glycol has been used in place of diethylene glycol. The diethylene glycol modified DMDHEU products have the advantage of releasing significantly less formaldehyde from the finished fabric after resin curing than fabric treated with DMDHEU. On the other hand, durable-press performance and shrinkage control are somewhat less with the glycol modified resins.

A less important glyoxal resin is (tetramethylolacetylenediurea) tetra-menthylolglycoluril [5395-50-6] produced by the reaction of 1 mol of glyoxal with 2 mol of urea, and 4 mol of formaldehyde.

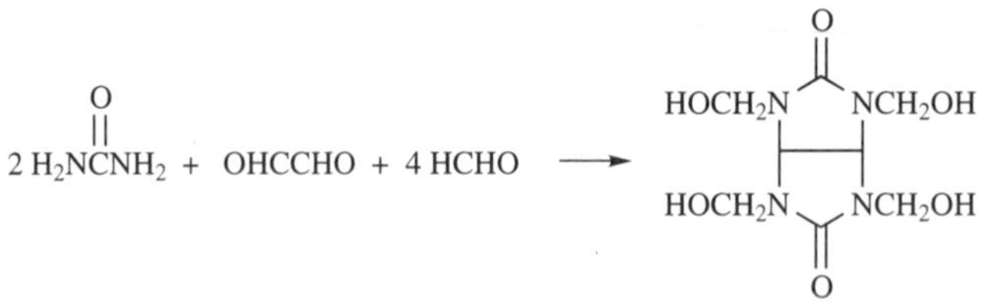

$$2\,H_2NCNH_2 \ + \ OHCCHO \ + \ 4\,HCHO \ \longrightarrow$$

(structure of tetramethylolglycoluril)

tetramethylolglycoluril

This resin was most popular in Europe, partly because of its lower requirements of glyoxal. However, because of increased availability and lower glyoxal costs plus certain application weaknesses, it has been generally replaced by DMDHEU.

Melamine–Formaldehyde Resins. The most versatile textile-finishing resins are the melamine–formaldehyde resins. They provide wash-and-wear properties to cellulosic fabrics, and enhance the wash durability of flame-retardant finishes. Butylated melamine–formaldehyde resins of the type used in surface coatings may be used in textile printing-ink formulations. A typical textile melamine resin is the dimethyl ether of trimethylolmelamine [1852-22-8] that can be prepared as follows:

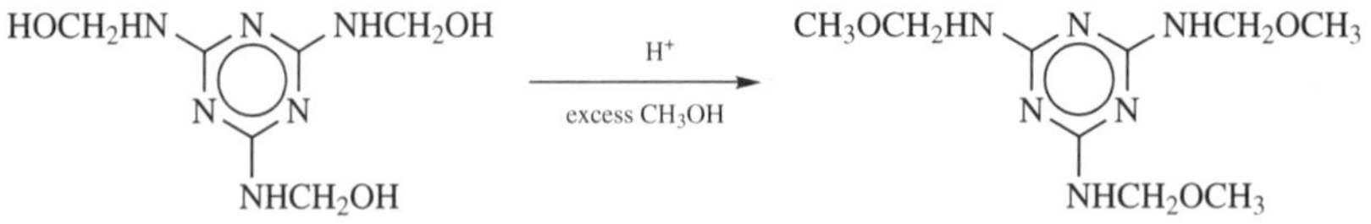

(reaction scheme: trimethylolmelamine + excess CH_3OH, H^+ → dimethyl ether of trimethylolmelamine)

Under alkaline conditions, 3 mol of formaldehyde react with 1 mol of melamine at elevated temperatures. Since water interferes with the methylation, methylolation is carried out in methanol with paraformaldehyde and by simply adjusting the pH to ∼4 with continued heating. After alkylation is complete the pH is adjusted to 8–10 and excess methanol is distilled under reduced pressure. The resulting syrup contains ∼80% solids.

Miscellaneous Resins. Much less important than the melamine–formaldehyde and urea–formaldehyde resins are the methylol carbamates. They are urea derivatives since they are made from urea and an alcohol (R can vary from methyl to a monoalkyl ether of ethylene glycol).

$$ROH \ + \ H_2NCNH_2 \ \longrightarrow \ ROCNH_2 \ + \ NH_3$$

Temperatures >of 140°C are required to complete the reaction and pressurized equipment is used for alcohols boiling below this temperature; provision must be made for venting ammonia without loss of alcohol. The reaction is straightforward and, in the case of the monomethyl ether of ethylene glycol [109-86-4], can be carried out at atmospheric pressure using stoichiometric quantities of urea and alcohol (45). Methylolation with aqueous formaldehyde is carried out at 70–90°C under alkaline conditions. The excess formaldehyde needed for complete dimethylolation remains in the resin and prevents more extensive usage because of formaldehyde odor problems in the mill.

Other amino resins used in the textile industry for rather specific properties have included the methylol derivatives of acrylamide (46), hydantoin [461-79-3] (47), and dicyandiamide (48).

Textiles are finished with amino resins in four steps. The fabric is (1) passed through a solution containing the chemicals, (2) through squeeze rolls (padding) to remove excess solution, (3) dried, and (4) heated (cured) to bond the chemicals with the cellulose or to polymerize them on the fabric surface.

The solution (pad bath) contains one or more of the amino resins described above, a catalyst, and other additives such as a softener, a stiffening agent, or a water repellant. The catalyst may be an ammonium or metal salt, eg, magnesium chloride or zinc nitrate. Synthetic fabrics, such as nylon or polyester, are treated with amino resins to obtain a stiff finish. Cotton (qv) or rayon fabrics or blends with synthetic fibers are treated with amino resins to obtain shrinkage control and a durable-press finish.

Normally, fabrics are treated in the sequence outlined above. The temperature of the drying unit is 100–110°C and the temperature of the curing unit can vary between 120 and 200°C but usually ranges from 150 to 180°C. The higher temperatures are employed to polymerize the resins on synthetic fabrics and at the same time to heat-set the fibers. Temperatures up to 180°C are used to allow the amino resins to react with cellulosic fibers alone or blended with synthetic fibers. The fabric is held flat but with minimum tension during drying and curing and always tends to become flat when creased or wrinkled during use or laundering. The resin-treated cellulose absorbs less water and swells less than untreated cellulose. This reduced swelling along with little or no tension induced during drying minimizes shrinkage during laundering.

The steps followed in the precure are repeated in the postcure process, except that after the drying step the goods are shipped to a garment manufacturer who makes garments, presses them into the desired shape with creases or pleats, and then cures the amino resin on the completed garment. It is important that the amino resins used in the postcure process should (1) not react with the fabric before it has been fashioned into a garment, and (2) release a minimum amount of formaldehyde into the atmosphere, especially while the goods are in storage or during the cutting and sewing operations. These requirements are met, at present, with the diethylene glycol modified DMDHEU resin.

Tire Cord.   Melamine resins are also used to improve the adhesion of rubber to reinforcing cord in tires. Textile cord is normally coated with a latex dip solution composed of a vinylpyridine–styrene–butadiene latex rubber containing resorcinol–formaldehyde resin. The dip coat is cured prior to use. The dip coat improves the adhesion of the textile cord to rubber. Further improvement in adhesion is provided by adding resorcinol and hexa(methoxymethyl) melamine (HMMM) [3089-11-0] to the rubber compound that is in contact with the textile cord. The HMMM resin and resorcinol cross-link during rubber vulcanization and cure to form an interpenetrating polymer within the rubber matrix that strengthens or reinforces the rubber and increases adhesion to the textile cord. Brass-coated steel cord is also widely used in tires for reinforcement. Steel belts and bead wire are common applications. Again, HMMM resins and resorcinol [108-46-3] are used in the rubber compound which is in contact with the steel cord to reinforce the rubber and increase the adhesion of the rubber to the steel cord. This use of melamine resins is described in the patent literature (49).

5.6. Amino Resins in the Paper Industry. Paper (qv) is a material of tremendous versatility and utility, prepared from a renewable resource. It may be made soft or stiff, dense or porous, absorbent or water repellent, textured or smooth. Some of the versatility originates with the fibers, which may vary from short and supple to long and stiff, but the contribution of chemicals should not be underestimated (see PAPERMAKING MATERIALS AND ADDITIVES).

Amino resins are used by the paper industry in large volume for a variety of applications. The resins are divided into two classes according to the mode of application. Resins added to the fiber slurry before the sheet is formed are called wet-end additives and are used to improve wet and dry strength and stiffness. Resins applied to the surface of formed paper or board, almost invariably together with other additives, are used to improve the water resistance of coatings, the sag resistance in ceiling tiles, and the scuff resistance in cartons and labels.

The requirements for the two types of resins are very different. Wet-end additives are used in dilute fiber slurries in small amounts. After the sheet is formed, most of the water is drained away and some of the remaining water is pressed out of the sheet before it is dried. The amino resin must be retained (absorbed) on the surface of the cellulose fibers so that it will not be washed away. On a typical paper machine, fiber concentration in the headbox would be ~1%. If the amount of wet-strength resin used is 1% of the weight of the fiber, the concentration of resin in the headbox would be only 0.01%. If no mechanism for attaching the resin to the fiber is provided, only a trace of the resin added to the slurry would be retained in the finished sheet. Good retention is achieved with the amino resin by making the resin cationic. Since the cellulose surface is anionic because of the carboxylic acid groups present, the cationic charge on the resin makes it substantive with the fiber leading to good retention of the resin when applied in the wet-end. Resins for application to the surface of preformed paper are not required to be substantive to cellulose and they may be formulated for adhesion, cure rate, viscosity, compatibility with other materials, etc, without concern for retention.

The integrity of a paper sheet is dependent on the hydrogen bonds that form between the fine structures of cellulose fibers during the pressing and drying operations (see CELLULOSE). The bonds between hydroxyl groups of neighboring fibers are very strong when the paper is dry but are severely weakened as soon as the paper becomes wet. Bonding between the hydroxyls of cellulose and water is as energetic as bonding between two cellulose hydroxyl groups. Consequently, ordinary paper loses most of its strength when it is wet or exposed to very high humidity. The sheet loses its stiffness and bursting, tensile, and tearing strength.

Many materials have been used over the years in an effort to correct this weakness in paper. If water can be prevented from reaching the sites of the bonding by sizing or coating the sheet, then a measure of wet strength may be attained. Water molecules are so small and cellulose so hydrophilic that this solution usually affords only temporary protection. Formaldehyde, glyoxal, poly-ethylenimine, and, more recently, derivatized starch (50) and derivatized catio-nic polyacrylamide resins (51) have been used to provide temporary wet strength. The first two materials must be applied to the formed paper but the other materials are substantive to the fiber and may be used as wet-end addi-tives. Carboxymethyl–cellulose– calcium chloride and locust bean gum–borax are examples of two-component systems applied separately to paper that were used to a limited extent before the advent of the amino resins. Today three major types of wet-strength resins are used in papermaking: polyamide–polya-mine resins cross-linked with epichlorohydrin (52) are used in neutral to alka-line papers; cationic polyacrylamide resins cross-linked with glyoxal are used for acid to neutral papers; and melamine–formaldehyde resins are used for acid papers.

During the thirty year period following the introduction of synthetic wet-strength additives to papermaking in 1942, most paper was made at acid pH. Low molecular weight (or even monomeric) trimethylolmelamine [1017-56-7], when dissolved in the proper amount of dilute acid and aged, polymerizes to a colloidal polymer that is retained well by almost all types of papermaking fiber, and produces high wet strength under the mild curing conditions easily attained on a paper machine (53, 54). This resin, introduced by American Cyana-mid in 1942, is still extensively used when rapid cure, high wet strength, and good dry strength are important in acid paper. Some processing improvements have been made, including a report (55) describing the formation of a stable mel-amine resin acid colloid using formic and phosphoric acids. The chemistry of this reaction is quite interesting.

Melamine–formaldehyde acts as an amine when dissolved in dilute acid, usually hydrochloric acid (HCl). During polymerization, between 20 and 80 monomeric units combine to form a polymer of colloidal dimensions (6–30 nm) with the elimination of water and HCl (56, 57). The development of cationicity is associated with the loss of HCl, since a unit of charge on the polymer is gen-erated for every mole of acid lost, and the pH decreases steadily during the poly-merization. In a typical formulation at 12% solids at room temperature, polymerization is complete in ~3 h. The initially colorless solution develops a light blue haze and shows a strong Tyndall effect.

Such a colloidal sol is highly substantive to all papermaking fibers, kraft, sulfite, groundwood, and soda. For its successful use in paper mills, the pH must be kept low, both to prevent precipitation of the resin in an unusable form and to promote curing of the resin; and the concentration of sulfates in the white water on the paper machine must not be allowed to exceed 100 ppm, again because the resin is precipitated in an inactive form by high concentrations of sulfates. High sulfate concentrations may build up in mills using large amounts of alum for setting size or sulfuric acid for controlling pH.

The problem of sulfate sensitivity was solved by adding formaldehyde to the aged colloid that improved wet-strength efficiency and reduced sensitivity to sul-fates (58). Later, equivalent results were obtained by adding the extra formalde-hyde before the colloid was aged. The additional formaldehyde acts like an acid during the aging process and, unless compensated for by a reduction in the amount of acid charged, lowers the pH to a point where polymerization to the

Table 3. **Formulations for Regular and HE Colloid Resins**

	Regular, MF_3	HE MF_8
water, $20° \pm 10°C$, kg	412.0	330.8
HCl, 20° Bé, kg (1.16 g/mL), kg	17.7	14.1
formaldehyde, 37%, kg		84.8
trimethylolmelamine, kg	45.4	45.4
Total	*475.1*	*475.1*

colloids is inhibited. The high efficiency (HE) resins have been used in mills with sulfate concentrations so high that use of regular trimethylolmelamine (MF_3) colloids would be uneconomical. Sulfate tolerance is a function of the amount of extra formaldehyde present. For best cost-performance, a family of HE colloids is necessary with composition varying from MF_4, for moderate sulfate concentrations, to MF_9, for very high sulfates.

Formulations for regular and HE colloids are shown in Table 3 (59). The materials are added in the order listed to a 454 L (120 gal) tank provided with good agitation and ventilation. Formaldehyde fumes are evolved even from the regular colloid. The colloids develop only after aging and freshly prepared solutions are ineffective for producing wet strength. Stability of the colloids depends on temperature and concentration. Colloids at 10–12% are stable at room temperature for at least 1 week; stability may be extended by dilution after the colloids have aged properly.

Both regular and HE colloids increase the wet strength of paper primarily by increasing adhesion between fibers; the strength of the individual fiber itself is unaffected (60). The resin appears to improve the adhesion between the fibers, whether they are wet or dry, by forming bonds that are unaffected by water. The excess formaldehyde in the HE colloid appears to function by increasing the amount of formaldehyde bound in the colloid (59). The regular colloid, starting with ~3 mol of formaldehyde per 1 mol of melamine, has ~2 mol bound in the colloid and 1 mol free. By mass action, the additional formaldehyde increases the amount of bound formaldehyde in the colloid. When an HE colloid is dialyzed or stored at very low concentrations (0.05%), it loses the extra bound formaldehyde and behaves as a regular colloid.

The first urea–formaldehyde resins used to any extent as wet-strength agents were anionic polymers made by the reaction of a urea resin with sodium bisulfite (61). Attempts to use nonionic urea–formaldehyde polymers were unsuccessful; the neutral charge on the polymer made it unsubstantive to fiber resulting in lack of retention. The sulfomethyl group introduced by reaction with $NaHSO_3$ gave the polymers strong anionicity but substantivity was largely restricted to unbleached kraft pulp. Lignin residues probably provided sites for absorption of the polymer. The use of alum as a mordant was essential, since both the resin and the fiber were anionic. The reaction of bisulfite with the urea–formaldehyde polymer may be represented as

$$R_2NCNHCH_2OH + NaHSO_3 \longrightarrow R_2NCNHCH_2OSO_2^-Na^+$$

In 1945, cationic urea resins were introduced and quickly supplanted the anionic resins, since they could be used with any type of pulp (62). Although they have now become commodities, their use in the industry has been steadily declining as the shift toward neutral and alkaline papermaking continues. They are commonly made by the reaction of urea and formaldehyde with one or more polyethylene–polyamines. The structure of these resins is very complicated and has not been determined. Ammonia is evolved during the reaction, probably according to the following:

$$R_2NCNH_2 + H_2N(CH_2CH_2NH)_xH \longrightarrow R_2NCNH(CH_2CH_2NH)_xH + NH_3$$

Formaldehyde may react with the active hydrogens on both the urea and amine groups, and therefore the polymer is probably highly branched. The amount of formaldehyde (2–4 mol per 1 mol urea), the amount and kind of polyamine (10–15%), and resin concentration are variable and hundreds of patents have been issued throughout the world. Generally, the urea, formaldehyde, polyamine, and water react at 80–100°C. The reaction may be carried out in two steps with an initial methylolation at alkaline pH, followed by condensation to the desired degree at acidic pH, or the entire reaction may be carried out under acidic conditions (63). The product is generally a syrup with 25–35% solids and is stable for up to 3 months.

The cationic urea resins are added to paper pulp preferably after all major refining operations have taken place. The pH on the paper machine must be acidic for reasonable rates of cure of the resin. Urea resins do not cure as rapidly as melamine–formaldehyde resins and the wet strength produced is not as resistant to hydrolysis. Furthermore, the resins are not retained as well as the melamine resins. On a resin-retained basis, however, their efficiency is as good. The lower retention of the urea–formaldehyde resins is due to their polydisperse molecular weight distribution. High molecular weight species are strongly absorbed on the fibers and are large enough to bridge two fibers. Low molecular weight species are not retained as well because of fewer charge sites. Attempts to improve the performance of urea–formaldehyde resins by fractionating the syrups by salt or solvent precipitation, or selective freezing or dialysis have been technically successful but economically impractical. The process for production of resins is sufficiently simple so that some paper mills have set up their own production units. With captive production, resins with higher molecular weights and lower stability may be tolerated.

The recovery of fiber from broke (off-specification paper or trim produced in the paper mill) is complicated by high levels of urea–formaldehyde and melamine–formaldehyde wet-strength resin. The urea resins present a lesser problem than the melamine resins because they cure slower and are not as resistant to hydrolysis. Broke from either resin treatment may be reclaimed by hot acidic repulping. Even the melamine resin is hydrolyzed rapidly under acidic conditions at high temperature. The cellulose is far more resistant and is not harmed if the acid is neutralized as soon as repulping is complete.

The TAPPI monograph (64) is an excellent source of additional information on technical and economic aspects of wet strength. An informative overview of the chemistry and mechanisms involved in wet strength chemistry can be found in 65.

Wet-strength applications account for the majority of amino resin sales to the paper industry but substantial volumes are sold for coating applications. The largest use is to improve the resistance of starch-clay coatings to dampness. In offset printing, which is becoming ever more important in the graphic arts, the printing paper is exposed to both ink and water. If the coating lifts from the paper and transfers to the plate, it causes smears and forces a shutdown for cleaning. A wide variety of materials have been added to the coatings to improve wet-rub resistance, including casein, soya protein, poly(vinyl acetate), styrene–butadiene latices, glyoxal–urea resins, and amino resins. Paper coatings are applied at as high a solids content as possible to ease the problem of drying. Retention is not a problem since the resin is applied to a preformed sheet. The important characteristics for coating resins are high solids at low viscosity, high cure rates, and high wet-rub efficiency. Urea and melamine resins or mixtures are sold as high solids syrups or dry powders. They are used with starch-pigment coatings with acidic catalysts, or with starch–pigment–casein (or protein) coatings usually without catalysts. The syrups are frequently methylated for solubility and stability at high solids. All of the resins are of intrinsically low molecular weight to reduce viscosity for ease of handling (see COATINGS, INDUSTRIAL).

Closely allied to resins for treating paper are the resins used to treat regenerated cellulose film (cellophane) that does not have good water resistance unless it is coated with nitrocellulose or poly(vinylidene chloride). Adhesion of the waterproofing coating to the cellophane film is achieved by first treating the cellophane with an amino resin. The cellophane film is passed through a dip tank containing ~1% of a melamine–formaldehyde acid colloid type of resin. Some glycerol may also be present in the resin solution to act as a plasticizer. Resins for this purpose are referred to as anchoring agents.

5.7. Other Uses. Water-soluble melamine–formaldehyde resins are used in the tanning of leather in combination with the usual tanning agents (see LEATHER). By first treating the hides with a melamine–formaldehyde resin, the leather is made more receptive to other tanning agents and the finished product has a lighter color. The amino resin is often referred to as a plumping agent because it makes the finished leather firmer and fuller.

Urea–formaldehyde resins are also used in the manufacture of foams. The resin solution containing an acid catalyst and a surface-active agent is foamed with air and cured. The open-cell type of foam absorbs water readily and is soft enough so that the stems of flowers can be easily processed into it. These features make the urea resin foam ideal for supporting floral displays. Urea–formaldehyde resin may also be foamed in place. A special nozzle brings the resin, catalyst, and foaming agent together. Air pressure is used to deposit the foam where it is desired, eg, within the outside walls of older houses to provide insulation. This application might be expected to grow as energy costs increase, if undesirable odors can be controlled.

Urea–formaldehyde resins are also used as the binder for the sand cores used in the molds for casting hollow metal shapes. The amino resin is mixed with moist sand and formed into the desired shape of the core. After drying and curing, the core is assembled into the mold and the molten metal poured in. Although the cured amino resin is strong enough to hold the core together while the hot metal is solidifying, it decomposes on longer heating. Later, the loose sand may be poured out of the hollow casting and recovered.

6. Regulatory Concerns

Both urea– and melamine–formaldehyde resins are of low toxicity. In the uncured state, the amino resin contains some free formaldehyde that could be objectionable. However, uncured resins have a very unpleasant taste that would discourage ingestion of more than trace amounts. The molded plastic, or the cured resin on textiles or paper may be considered nontoxic. Combustion or thermal decomposition of the cured resins can evolve toxic gases, such as formaldehyde, hydrogen cyanide, and oxides of nitrogen.

Melamine–formaldehyde resins may be used in paper that contacts aqueous and fatty foods according to 21 CFR 121.181.30. However, because a lower PEL has been established by OSHA, some mills are looking for alternatives. Approaches toward achieving lower formaldehyde levels in the resins have been reported (66, 67); the efficacy of these systems needs to be established. Although alternative resins are available, significant changes in the papermaking operation would be required in order for them to be used effectively.

7. Economic Aspects

Japan produces more amino resin than any other country; the United States is next, with the Union of Soviet Socialist Republics, France, the United Kingdom, and Germany following.

Many large chemical companies produce amino resins and the raw materials needed, ie, formaldehyde, urea, and melamine. Some companies may buy raw materials to produce amino resins for use in their own products, such as plywood, chipboard, paper, textiles, or paints, and may also find it profitable to market these resins to smaller companies. The technology is highly developed and sales must be supported by adequate technical service to select the correct resin and see that it is applied under the best conditions.

During the past 10 years there has been considerable change in the suppliers of amino resins as a result of acquisitions, spin-offs and withdrawal of some of the smaller companies from the business. The following is a representative list of those currently in the business:

Badische Aniline and Soda-Fabrik (BASF), Ludwigshafen, Germany; Berger International Chemicals, Newcastle upon Tyne, England; Borden Chemical Div., Columbus, OH; Casella Farbwerk Mainkur A.G., Frankfurt-Fechenheim, Germany; Cuyahoga Plastics, Cleveland, OH; Cytec Industries, West Patterson, NJ; DSM Coating Resins, Zwolle, Holland; Dainippon Ink, Ltd., Tokyo, Japan; Dynamit-Nobel, A.G., Troisdorf-Koln, Germany; Fiberite Corp, Winona, MN; Georgia-Pacific Corp., Atlanta, GA; Gulf Adhesives, Lansdale,

PA; Hitachi Chemical Co. Ltd, Tokyo, Japan; Matsushita Electric Works, Ltd., Osaka, Japan; McWhorter Technologies, Carpentersville, IL (Division of Eastman Chemical Co., Kingsport, TN); Melamine Chemicals, Donaldson, LA (Division of Borden Chemical, Columbus, OH); Mitsui-Toatsu Chemicals, Ltd, Tokyo, Japan; Mitsui-Cytec, Ltd., Tokyo, Japan; Montedison SpA, Milan, Italy; Pacific Resins, Tacoma, WA; Perstorp AB, Perstorp, Sweden; Perstorp Compounds, Inc., Florence, MA; Reichhold Chemicals, White Plains, NY (Division of Dainippon Ink, Tokyo, Japan); Solutia, Inc, St. Louis, MO; Sumitomo Ltd., Tokyo, Japan; Vianova Resins GmbH & Co. Wiesbaden, Germany (Division of Solutia, St. Louis, MO).

BIBLIOGRAPHY

"Amino Resins and Plastics" in *ECT* 1st ed., Vol. 1, pp. 741–771, by P. O. Powers, Battelle Memorial Institute; in "Amino Resins and Plastics" *ECT* 2nd ed., Vol. 2, pp. 225–258, by H. P. Wohnsiedler, American Cyanamid Company; in "Amino Resins and Plastics" *ECT* 3rd ed., Vol. 2, pp. 440–469, by I. H. Updegraff, S. T. Moore, W. F. Herbes, and P. B. Roth, American Cyanamid Company; *ECT* 4th ed., Vol. 2, pp. 604–637, by Laurence L. Williams, American Cyanamid Company; " Amino Resins and Plastics" in *ECT* (online), posting date: December 4, 2000, by Laurence L. Williams, American Cyanamid Company.

CITED PUBLICATIONS

1. R. Bainbridge, *Sail* 8(1), 142 (1977).
2. A. Einhorn and A. Hamburger, *Ber. Dtsch. Chem. Ges.* **41**, 24 (1908).
3. Brit. Pats. 248,477 (Dec. 5, 1924), 258,950 (July 1, 1925), 266,028 (Nov. 5, 1925), E. C. Rossiter (to British Cyanides Company, Ltd.).
4. Brit. Pats. 187,605 (Oct. 17, 1922), 202,651 (Aug. 17, 1923), 208,761 (Sept. 20, 1922), H. Goldschmidt and O. Neuss.
5. Brit. Pats. 171,096 (Nov. 1, 1921), 181,014 (May 20, 1922), 193,420 (Feb. 17, 1923), 201,906 (July 23, 1923), 206,512 (July 23, 1923), 213,567 (Mar. 31, 1923), 238,904 (Aug. 25, 1924), 270,840 (Oct. 1, 1924), 248,729 (Mar. 3, 1925), F. Pollak.
6. U.S. Pat. 1,460,606 (July 3, 1923), K. Ripper.
7. Ger. Pat. 647,303 (July 6, 1937), Brit. Pat. 455,008 (Oct. 12, 1936), W. Hentrich and R. Köhler (to Henkel and Co., GmbH).
8. R. Rager, *Mod. Plast.* **49**(4), 67 (1972).
9. E. Drechsel, *J. Prakt. Chem.* [2] **13**, 330 (1876).
10. U.S. Pat. 2,727,037 (Dec. 13, 1955), C. A. Hochwalt (to Monsanto Chemical Company).
11. Ger. Pat. 1,812,120 (June 11, 1970), D. Fromm, K. W. Leonhard, R. Mohr, M. Schwartzmann, and H. Woehrle (to Badische Anilin und Soda-Fabrik A.G.).
12. P. Ellwood, *Chem. Eng.* **77**(23), 101 (1970).
13. J. F. Walker, *Formaldehyde, American Chemical Society Monograph, No. 159,* 3rd ed., Reinhold Publishing Corp., New York, 1964.
14. *U.F. Concentrate-85,* Technical Bulletin, Allied Chemical Corp., New York, 1985.
15. U.S. Pat. 3,129,226 (Apr. 14, 1964), G. K. Cleek and A. Sadle (to Allied Chemical Corp.).
16. U.S. Pat. 3,458,464 (July 29, 1969), D. S. Shriver and E. J. Bara (to Allied Chemical Corp.).

17. Ref. 13, p. 151.
18. *N-(iso-butoxymethyl) acrylamide, Technical Bulletin PRC 126,* American Cyanamid Co., Wayne, N.J., Feb. 1976.
19. M. Gordon, A. Halliwell, and T. Wilson, *J. Appl. Polym. Sci.* **10**, 1153 (1966).
20. J. W. Aldersley, M. Gordon, A. Halliwell, and T. Wilson, *Polymer* **9**, 345 (1968).
21. I. H. Anderson, M. Cawley, and W. Steedman, *Br. Polym. J.* **1**, 24 (1969).
22. K. Sato, *Bull. Chem. Soc. Jpn.* **40**(4), 724 (1967) (in Eng.).
23. K. Sato and T. Naito, *Polym. J.* **5**, 144 (1973).
24. K. Sato and Y. Abe, *J. Polym. Sci. Polym. Chem. Ed.* **13**, 263 (1975).
25. V. A. Shenai and J. M. Manjeshwar, *J. Appl. Polym. Sci.* **18**, 1407 (1974).
26. A. Berge, S. Gudmundsen, and J. Ugelstad, *Eur. Polym. J.* **5**, 171 (1969).
27. A. Berge, B. Kvaeven, and J. Ugelstad, *Eur. Polym. J.* **6**, 981 (1970).
28. J. I. DeJong and J. DeJonge, *Rec. Trav. Chim.* **71**, 643, 661, 890 (1952); **72**, 88, 139, 202, 207, 213, 1027 (1953).
29. R. Steele, *J. Appl. Polym. Sci.* **4**, 45 (1960).
30. G. A. Crowe and C. C. Lynch, *J. Am. Chem. Soc.* **70**, 3795 (1948); **71**, 3731 (1949); **72**, 3622 (1950).
31. Brit. Pat. 829,953 (Mar. 9, 1960), E. Elbel.
32. U.S. Pats. 3,007,885 (Nov. 7, 1961), 3,114,930 (Dec. 24, 1963), W. N. Oldham, N. A. Granito, and B. Kerfoot (to American Cyanamid Co.).
33. U.S. Pat. 3,661,819 (May 9, 1972), J. N. Koral and M. Petschel, Jr. (to American Cyanamid Co.).
34. U.S. Pat. 3,803,095 (Apr. 9, 1974), L. J. Calbo and J. N. Koral (to American Cyanamid Co.).
35. W. Lindlaw, *The Preparation of Butylated Urea–Formaldehyde and Butylated Melamine Formaldehyde Resins Using Celanese Formcel and Celanese Paraformaldehyde,* Technical Bulletin, Celanese Chemical Co., New York, Table XIIA.
36. *Technical Bulletin S-23-8,* 1967, *Supplement to Technical Bulletin S-23-8,* 1968, Celanese Chemical Co., Example VIII.
37. W. J. Blank and W. L. Hensley, *J. Paint Technol.* **46**, 46 (1974).
38. U.S. Pat. 2,517,750 (Aug. 8, 1950), A. L. Wilson (to Union Carbide and Carbon Corp.).
39. U.S. Pat. 2,304,624 (Dec. 8, 1942), W. J. Burke (to E. I. du Pont de Nemours & Co., Inc.).
40. U.S. Pat. 3,324,062 (June 6, 1967), G. S. Y. Poon (to Dan River Mills Inc.).
41. H. Kadowaki, *Bull. Chem. Soc. Jpn.* **11**, 248 (1936).
42. U.S. Pat. 3,089,859 (May 14, 1963), T. Oshima (to Sumitomo Chemical Company, Ltd.).
43. U.S. Pats. 2,731,472 (Jan. 17, 1956), 2,764,573 (Sept. 25, 1956), Bruno V. Reibnitz and co-workers (to Badische Anilin-und Soda-Fabrik); U.S. Pat. 2,876,062 (Mar. 3, 1959), Erich Torke (to Phrix-Werke A.G.).
44. U.S. Pat. 3,487,088 (Dec. 30, 1969), K. H. Remley (to American Cyanamid Co.).
45. U.S. Pat. 3,524,876 (Aug. 18, 1970), J. E. Gregson (to Dan River Mills, Inc.).
46. U.S. Pat. 3,658,458 (Apr. 25, 1972), D. J. Gale (to Deering Milliken Research Corp.).
47. U.S. Pats. 2,602,017; 2,602,018 (July 1, 1952), L. Beer.
48. C. Hasegawa, *J. Soc. Chem. Ind. Jpn.* **45**, 416 (1942).
49. U.S. Pat. 3,212,955 (Oct. 19, 1965), S. Kaizerman (to American Cyanamid Co.).
50. U.S. Pat. 4,741,804 (May 3, 1988), D. B. Solarek and co-workers (to National Starch and Chemical Corp.).
51. U.S. Pat. 4,605,702 (Aug. 12, 1986), G. J. Guerro, R. J. Proverb, and R. F. Tarvin (to American Cyanamid Co.).
52. U.S. Pats. 2,926,116; 2,926,154 (Feb. 23, 1960); G. L. Keim (to Hercules Powder Co.).
53. U.S. Pat. 2,345,543 (Mar. 28, 1944), H. P. Wohnsiedler and W. M. Thomas (to American Cyanamid Co.).

54. C. G. Landes and C. S. Maxwell, *Pap. Trade J.* **121**(6), 37 (1945).
55. Ger. Pat. 2,332,046 (Jan. 23, 1975), W. Guender and G. Reuss (to Badische Anilin-und Soda-Fabrik A.G.).
56. J. K. Dixon, G. L. M. Christopher, and D. J. Salley, *Pap. Trade J.* **127**(20), 49 (1948).
57. Unpublished data, American Cyanamid Co.
58. U.S. Pat. 2,559,220 (July 3, 1951), C. S. Maxwell and C. G. Landes (to American Cyanamid Co.).
59. C. S. Maxwell and R. R. House, *TAPPI* **44**(5), 370 (1961).
60. D. J. Salley and A. F. Blockman, *Pap. Trade J.* **121**(6), 41 (1945).
61. U.S. Pat. 2,407,599 (Sept. 10, 1946), R. W. Auten and J. L. Rainey (to Resinous Products and Chemical Co.).
62. U.S. Pat. 2,742,450 (Apr. 17, 1956), R. S. Yost and R. W. Auten (to Rohm and Haas Co.).
63. U.S. Pat. 2,683,134 (July 6, 1954), J. B. Davidson and E. J. Romatowski (to Allied Chemical and Dye Corp.).
64. J. P. Weidner, ed., *Wet Strength in Paper and Paper Board, Monograph Series, No. 29,* Technical Association of Pulp and Paper Industry, New York, 1965.
65. K. W. Britt in J. P. Casey, ed., *Pulp and Paper,* Vol. III, John Wiley & Sons, Inc., New York, 1981, Chapt. 18.
66. W. Kamutzki, *Ind. Carta* **26**(6), 297 (1988).
67. W. Kamutzki, *Kunstharz-Nachr.* **24**, 9 (1987).

LAURENCE L. WILLIAMS
American Cyanamid Company

AMINOPHENOLS

1. Introduction

Aminophenols and their derivatives are of commercial importance, both in their own right and as intermediates in the photographic, pharmaceutical, and chemical dye industries. They are amphoteric and can behave either as weak acids or weak bases, but the basic character usually predominates. 3-Aminophenol (1) is fairly stable in air unlike 2-aminophenol (2) and 4-aminophenol (3) that easily undergo oxidation to colored products. The former are generally converted to their acid salts, whereas 4-aminophenol is usually formulated with low concentrations of antioxidants which act as inhibitors against undesired oxidation.

(1)   (2)   (3)

2. Physical Properties

The simple aminophenols exist in three isomeric forms depending on the relative positions of the amino and hydroxyl groups around the benzene ring. At room temperature they are solid crystalline compounds. In the past, the commercial-grade materials were usually impure and colored because of contamination with oxidation products, but now virtually colorless, high purity commercial grades are available. The partitioning of aminophenols between aqueous and organic solvent systems has been studied; 2-aminophenol behaves anomalously because of intramolecular hydrogen bonding (1,2). The solubilities of these compounds in common solvents of differing polarities (dielectric constants) are given in Table 1 and their spectral characteristics in Table 2. In acidic solution all isomers exhibit fluorescence. 4-Aminophenol shows two bands; one at 300 nm common to all the isomers, and the second at 370 nm attributed to the existence of an additional aqueous ionic species. Fluorescence also exists in neutral solution, but is abolished at high pH values (3–13).

2.1. 2-Aminophenol. This compound forms white orthorhombic bipyramidal needles when crystallized from water or benzene, which readily become yellow-brown on exposure to air and light. The crystals have eight molecules to the elementary cell and a density of 1.328 g/cm^3 (1.29 also quoted) (14–16). The molecules are hydrogen bonded from OH to N to form chains parallel to the b-axis; these chains are linked together to form sheets by NH to O hydrogen bonds essentially parallel to the c-axis. There are large cavities between the sheets permitting the intercalation of small foreign molecules (17) (see Tables 3–5).

2.2. 3-Aminophenol. This is the most stable of the isomers under atmospheric conditions. It forms white prisms when crystallized from water or toluene. The orthorhombic crystals have a tetramolecular unit and a density of 1.195 g/cm^3 (1.206 and 1.269 also quoted) (15,16) (see Tables 3–5).

Table 1. Solubility^a of Aminophenols in Common Solvents Arranged in Order of Increasing Polarity (Dielectric Constant)^b

Solvent	2-Aminophenol	3-Aminophenol	4-Aminophenol
benzene	1	1	0
toluene	1	1	1
acetonitrile	3	3	2
diethyl ether	2	3	1
chloroform	1	1	0
ethyl acetate	3	3	2
acetone	3	3	2
ethanol	2	3	1
dimethyl sulfoxide	3	3	3
water			
hot	2	3	2
cold	1	2	1

^{a} 0, insoluble; 1, slightly soluble; 2, soluble; 3, very soluble.
^{b} Eluotropic series.

Table 2. Spectral Characteristics of the Aminophenol Isomers

	2-Aminophenol	3-Aminophenol	4-Aminophenol
uv[a] nm absorption	233,285 (methanol) 229,281 (water) 235,288 (cyclohexane)	287 (methanol) 270 (0.1 M HCl) 234,284 (cyclohexane)	234,301 (methanol) 229,294 (water) 235,304 (cyclohexane)
emission	λ_{ex} λ_{em}	λ_{ex} λ_{em}	λ_{ex} λ_{em}
fluorescence	291 336 (ethanol) 286 338–344 (water) 283 330 (cyclohexane)	287 333 (ethanol) 286 331–334 (water) 290 320 (cyclohexane)	302 364.5 (ethanol) 301 367–374 (water) 270 330 (cyclohexane)
phosphorescence	291 440 (ethanol)	287 425 (ethanol)	302 470 (ethanol)
ir[b] cm^{-1}	3380,3300,1600,1510,1470,1270, 900,740	3370,3310,1600,1470,1390,1260, 1180,910	3050–2580,1500,1470,1240,970,830, 750
ms[c]	109 (100) 80 (39) 53 (12) 28 (11)	109 (100) 80 (23) 81 (10) 53 (6)	109 (100) 80 (31) 107 (23) 53 (20)
nmr[d] ppm	6.9–7.5,8.6 (TFA)	4.7,6.0,6.1,6.8,8.8 (DMSO)	7.1,7.4,8.7 (TFA)

[a] ultraviolet (uv) spectra: λ_{ex}, excitatory wavelength; λ_{em}, emission wavelength (3, 4, 7–9).

[b] Only infrared (ir) absorption bands reported as very strong are included (accuracy ± 10 cm^{-1}) (5, 6).

[c] Values quoted are ion (m/z) followed by relative abundance in parentheses. After m/z 109 and 80, the other ions may vary in abundance order dependent upon conditions employed (10).

[d] Proton chemical shift spectra over the range of 0–15 ppm (±0.1 ppm): TFA, trifluoroacetic acid; DMSO, dimethyl sulfoxide. When complex spectra caused by second-order effects or overlapping resonances were encountered, the range was recorded (11,12). Nuclear magnetic resonance = nmr.

Table 3. **General Properties of Aminophenols**

Property	2-Aminophenol	3-Aminophenol	4-Aminophenol
alternative names	2-hydroxyaniline 2-amino-1-hydroxy-benzene	3-hydroxyaniline 3-amino-1-hydroxy-benzene	4-hydroxyaniline 4-hydroxy-1-aminobenzene
C.I. designation	76,520		
CAS Registry Number	[95-55-6]	[591-27-5]	[123-30-8]
molecular formula	C_6H_7NO	C_6H_7NO	C_6H_7NO
molecular weight	109.13	109.13	109.13
melting point, °C	174	122–123	189–190[a]
boiling point, °C			
0.04 kPa			130[a], 110[b]
0.4 kPa			150
1.07 kPa			167
1.47 kPa	153[b,c]	164[c]	174
101.3 kPa			284
ΔH_f, kJ/mol[d]	-191.0 ± 0.9	-194.1 ± 1.0	-190.6 ± 0.9[e]

[a] Decomposes.
[b] Sublimes. To convert kPa to mm Hg, multiply by 7.5.
[c] Ref. 18.
[d] In the crystalline state (19). To convert kJ to kcal, divide by 4.184.
[e] -179.1 is also quoted (20).

2.3. 4-Aminophenol.

This compound forms white plates when crystallized from water. The base is difficult to maintain in the free state and deteriorates rapidly under the influence of air to pink–purple oxidation products. The crystals exist in two forms. The α-form (from alcohol, water, or ethyl acetate) is the more stable and has an orthorhombic pyramidal structure containing four molecules per unit cell. It has a density of 1.290 g/cm³ (1.305 also quoted). The less stable β-form (from acetone) exists as acicular crystals that turn into the α-form on standing: they are orthorhombic bipyramidal or pyramidal and have a hexamolecular unit (15,16,24) (see Tables 3–5).

Table 4. **Acid Dissociation Constants**[a] **of Aminophenols**[b]

Compound	pK_1	pK_2	Temperature,°C
2-aminophenol	4.72		21
	4.66[c]		25
		9.66	15
		9.71	22
3-aminophenol	4.17		21
	4.31c		25
		9.87	22
4-aminophenol	5.5		21
	4.86	10.60	30
	5.48c		25
		10.30	22

[a] In water unless otherwise noted.
[b] Refs. 21–23.
[c] 1 vol% ethyl alcohol in water.

Table 5. **Salts of the Aminophenols**

Salt	2-Aminophenol	3-Aminophenol	4-Aminophenol
hydrochloride			
CAS Registry Number	[51-19-4]	[51-81-0]	[51-78-5]
melting point, °C	207	224	306[a]
crystal form	needles	prisms	prisms
hydroiodide			
CAS Registry Number			[33576-76-0]
melting point, °C		209	
crystal form		prisms	
oxalate			
melting point, °C	167.5[a]	275	183
acetate			
CAS Registry Number	[97777-54-3]	[97777-55-4]	[13871-68-6]
melting point, °C	150[b]		183
chloroacetate			
melting point, °C			148
crystal form			needles
trichloroacetate			
CAS Registry Number	[97777-57-6]	[97777-56-5]	
melting point, °C			166
crystal form			needles
sulfate			
CAS Registry Number		[66671-80-5]	[54646-39-8]
melting point, °C		152	
crystal form		plates or needles	
hydrosulfate			
CAS Registry Number	[40712-56-9]		[15658-52-3]
melting point, °C			272
crystal form			needles

[a] Decomposes.
[b] The formate salt melts at 120°C.

3. Chemical Properties

The chemical properties and reactions of the aminophenols and their derivatives are to be found in detail in many standard chemical texts (25). The acidity of the hydroxyl function is depressed by the presence of an amino group on the benzene ring; this phenomenon is most pronounced with 4-aminophenol. The amino group behaves as a weak base, giving salts with both mineral and organic acids. The aminophenols are true ampholytes, with no zwitterion structure; hence they exist either as neutral molecules (4), or as ammonium cations (5), or phenolate ions (6), depending on the pH value of the solution.

The existence of half-salt complex cations B^+_2, formed by the association of an ammonium cation B^+, with a neutral molecule, B, has also been postulated. This association phenomenon is most apparent with 4-aminophenol, but is also displayed by the other isomers (23).

The aminophenols are chemically reactive, undergoing reactions involving both the aromatic amino group and the phenolic hydroxyl moiety, as well as substitution on the benzene ring. Oxidation leads to the formation of highly colored polymeric quinoid structures. 2-Aminophenol undergoes a variety of cyclization reactions.

3.1. Alkylation. All the possible mono-, di-, and trimethylated aminophenols are known. *N*-Monoalkylation occurs when the aminophenol is heated with the appropriate alkyl halide or with an alcohol and Raney nickel; equal or even better results can be achieved using aldehydes or ketones in place of the alcohol. Specific alkylation of the hydroxyl group to form methoxyanilines (anisidines) or ethoxyanilines (phenetidines) is difficult because of the reactivity of the amino group; mixed alkylated products usually are obtained. 3-Methoxyanilines may be prepared by methylation of 3-aminophenol under alkaline conditions, but it is more usual to protect the amino group and to methylate 3-acetylaminophenol, followed by hydrolysis. The other anisidines and phenetidines are prepared indirectly by reduction of the nitro analogue.

3.2. Acylation. Reaction conditions employed to acylate an aminophenol (using acetic anhydride in alkali or pyridine, acetyl chloride and pyridine in toluene, or ketene in ethanol) usually lead to involvement of the amino function. If an excess of reagent is used, however, especially with 2-aminophenol, O,N-diacylated products are formed. Aminophenol carboxylates (O-acylated aminophenols) normally are prepared by the reduction of the corresponding nitrophenyl carboxylates, which is of particular importance with the 4-aminophenol derivatives. A migration of the acyl group from the O to the N position is known to occur for some 2- and 4-aminophenol acylated products. Whereas ethyl 4-aminophenyl carbonate is relatively stable in dilute acid, the 2-derivative has been shown to rearrange slowly to give ethyl 2-hydroxyphenyl carbamate [35580-89-3] (26).

3.3. Diazonium Salt Formation. The aromatic amino group of aminophenols can be converted to the diazonium salt using sodium nitrite in aqueous acid, although difficulties may be encountered when the aminophenol is of low solubility or easily oxidized. Crystalline diazonium salts have been isolated using the hydrochloride or sulfate of the appropriate aminophenol under anhydrous conditions. Such diazo derivatives find extensive use in the dye industry (27,28).

3.4. Cyclization Reactions. 2-Aminophenol is particularly susceptible to cyclization and condensation reactions because of the close proximity of the amino and hydroxyl groups attached to the benzene ring. A nonspecific oxidative environment (ferric chloride, light, enzymes, autooxidation on silica thin-layer plates) gives 2-aminophenoxazin-3-one [1916-59-2] (7), further oxidation (ferric cyanide, heating in ethanolic potassium hydroxide) gives a pentacyclic structure, triphenoxdioxazine (benzoxazinophenoxazine) [258-72-0] (8). 2-Aminophenol and its derivatives are useful starting materials for the synthesis of phenoxazines, phenoxazones, benzoxazoles, and thiobenzoxazoles. Most of these condensation reactions involve heating at 200–300°C with a suitable catalyst (25).

(7)   (8)

3.5. Condensation Reactions.

Condensation of substituted benzaldehydes with 2-aminophenol in the presence of a catalyst (aluminum, iron, zinc or phosphorus chlorides) yields a Schiff base, with the elimination of water, in 52–88% yields (29). In general, substituted diphenylamines or diphenyl ethers are obtained from aminophenols and suitable reactants by elimination of ammonia or hydrogen chloride.

3.6. Reactions of the Benzene Ring.

Both the amino and hydroxyl groups attached to the benzene nucleus are electron-donating because of resonance effects, which predominate over electron-withdrawing inductive effects. Many substituted derivatives are known. The controlled interaction of aminophenols with chlorine or bromine in glacial acetic acid can give a variety of mono-, di-, tri-, or tetrahalogenated products. The use of concentrated sulfuric acid or oleum, with or without heat, gives aromatic sulfonic acids. The sulfonic acid group enters the 2- or 4-position relative to the hydroxyl group. Further treatment with oleum leads to the formation of disulfonated compounds. The carboxylation of 3-aminophenol leads to the formation of 4-aminosalicyclic acid.

4. Manufacture and Processing

Aminophenols are either made by reduction of nitrophenols or by substitution. Reduction is accomplished with iron or hydrogen in the presence of a catalyst. Catalytic reduction is the method of choice for the production of 2- and 4-aminophenol (see AMINES BY REDUCTION). Electrolytic reduction is also under industrial consideration and substitution reactions provide the major source of 3-aminophenol.

4.1. Reduction.

Iron Reduction. The reduction of nitrophenols with iron filings or turnings takes place in weakly acidic solution or suspension (30). The aminophenol formed is converted to the water soluble sodium aminophenolate by adding sodium hydroxide before the iron–iron oxide sludge is separated from the reaction mixture (31). Adjustment of the solution pH leads to the precipitation of aminophenols, a procedure performed in the absence of air because the salts are very susceptible to oxidation in aqueous solution.

Insoluble red lakes are formed as by-products that decrease yields when 2-nitrophenol [88-75-5] is reduced with iron. Consequently, the iron reduction of this nitro compound to 2-aminophenol is of minor industrial importance today.

Catalytic Reduction. Catalytic reduction usually takes place in solution, emulsion, or suspension in autoclaves or pressurized vessels; after the catalyst is added, the vessel is pressurized with hydrogen (32,33). Water and methanol are the preferred solvents. In water the addition of alkali hydroxide (34), alkali carbonate (35), or acid (36) has been recommended.

The chemical production of aminophenols via the reduction of nitrobenzene occurs in two stages. Nitrobenzene [98-95-3] is first selectively reduced with hydrogen in the presence of Raney copper to phenylhydroxylamine in an organic solvent such as 2-propanol (37). With the addition of dilute sulfuric acid, nucleophilic attack by water on the aromatic ring of N-phenylhydroxylamine [100-65-2] takes place to form 2- and 4-aminophenol. The by-product, 4,4′-diaminodiphenyl ether [13174-32-8], presumably arises in a similar manner from attack on the ring by a molecule of 4-aminophenol (38,39). Aniline [62-53-3] is produced via further reduction (40,41).

In past years, metals in dilute sulfuric acid were used to produce the nascent hydrogen reductant (42). Today, the reducing agent is hydrogen in the presence of a catalyst. Nickel, preferably Raney nickel (34), chromium or molybdenum promoted nickel (43), or supported precious metals such as platinum or palladium (35,44) on activated carbon, or the oxides of these metals (36,45), are used as catalysts. Other catalysts have been suggested such as molybdenum and platinum sulfide (46,47), or a platinum–ruthenium mixture (48).

The addition of "wetting agents" increases the aminophenol yield. These agents must be water soluble and stable in the presence of sulfuric acid (49). Quaternary ammonium salts that contain at least one alkyl group with at least ten carbon atoms are suitable (49,50). Dimethylalkylamine oxide increases the hydrogenation rate and improves the selectivity of the reaction for 4-aminophenol (51). The reaction temperature does not usually exceed 100–110°C, and is performed either at atmospheric or at a higher pressure, preferably around 2 MPa (20 atm) (up to 6 MPa). Hydrogen is added during the reaction as it is consumed. The addition of an inert organic solvent, not miscible with water, further increases the yield of 4-aminophenol and product quality (34,35), and the presence of an organic divalent sulfur compound inhibits the formation of aniline (40,41,52).

In another process variant, only 88% of the nitrobenzene is reduced, and the reaction mixture then consists of two phases; the precious metal catalyst (palladium on activated carbon) remains in the unreacted nitrobenzene phase. Therefore, phase separation is sufficient as work-up, and the nitrobenzene phase can be recycled directly to the next batch. The aqueous sulfuric acid phase contains 4-aminophenol and by-product aniline. After neutralization, the aniline is stripped, and the aminophenol is obtained by crystallization after the aqueous phase is purified with activated carbon (53).

Electrolytic Reduction. Electrochemical reduction is finding commercial favor and causes less concern over pollution than metal–acid reduction systems (42,54–56). Electrolysis of nitrobenzene, phenylhydroxylamine, or azoxybenzene [495-48-7] in deoxygenated acid solutions, using graphite or copper–mercury cathodes at potential differences of −300 to −600 mV and temperatures of 60–90°C, produces 4-aminophenol in yields of 65–99% (57,58). Aniline is produced as a by-product (59). The use of 2-nitrophenol yields 2-aminophenol (60); the process does not appear satisfactory for the production of 3-aminophenol (61). The use of bismuth, tin, or titanium salts as additives (62), electrolyte agitation (63), flow-through cells with rotating copper-amalgamated electrodes, and square-wave pulsed current control are increasing the efficiency, product selectivity, and scale up of the process (64,65). Electrocatalytic oxidation of aniline to 4-aminophenol by electrochemically activated molecular oxygen via direct electron transfer from the cathode, in the presence of iron compounds, is also possible (66).

4.2. Substitution. Substitution of various groups by amino or hydroxyl functions is industrially unimportant for the production of 2- and 4-aminophenol, but this type of reaction is used for the synthesis of 2- and 4-aminophenol derivatives. However, 3-aminophenol cannot be obtained easily by reduction. It is made by the reaction of 3-aminobenzenesulfonic acid [121-47-1] with sodium hydroxide under fusion conditions (5–6 h; 240–245°C). The product is purified by vacuum distillation (25).

In an alternative industrial process, resorcinol [108-46-3] is autoclaved with ammonia for 2–6 h at 200–230°C under a pressurized nitrogen atmosphere, 2.2–3.5 MPa (22–35 atm). Diammonium phosphate, ammonium molybdate, ammonium sulfite, or arsenic pentoxide may be used as a catalyst to give yields of 60–94% with 85–90% selectivity for 3-aminophenol (67,68). A vapor-phase system operating at 320°C using a silicon dioxide catalyst impregnated with gallium sesquioxide gives a 26–31% conversion of resorcinol with a 96–99% selectivity for 3-aminophenol (69).

The direct conversion of aniline into aminophenols may be achieved by hydrogen peroxide hydroxylation in SbF_5–HF at −20 to −40°C. The reaction yields all possible aminophenols via the action of $H_3O^+_2$ on the anilinium ions; the major product is 3-aminophenol (64% yield) (70,71). This isomer may also be made by the hydrolysis of 3-aminoaniline [108-45-2] in dilute acid at 190°C (72). Another method of limited importance, but useful in the synthesis of derivatives, is the dehydrogenation of aminocyclohexenones (73).

4.3. Purification. Contaminants and by-products that are usually present in 2- and 4-aminophenol made by catalytic reduction can be reduced or even removed completely by a variety of procedures. These include treatment with 2-propanol (74), with aliphatic, cycloaliphatic, or aromatic ketones (75), with aromatic amines (76), with toluene or low mass alkyl acetates (77), or with phosphoric acid, hydroxyacetic acid, hydroxypropionic acid, or citric acid (78). In addition, purity may be enhanced by extraction with methylene chloride, chloroform (79), or nitrobenzene (80).

Another method employed is the treatment of aqueous solutions of aminophenols with activated carbon (81,82). During this procedure, sodium sulfite, sodium dithionite, or disodium ethylenediaminotetraacetate (82) is added to increase the quality and stability of the products and to chelate heavy-metal ions that would catalyze oxidation. Addition of sodium dithionite, hydrazine (82), or sodium hydrosulfite (83) also is recommended during precipitation or crystallization of aminophenols.

Generally, aminophenols of high purity may be obtained by sublimation at reduced pressure. 3-Aminophenol may be purified by vacuum distillation and a colorless product obtained by adding sulfur dioxide during the distillation (84) or by collecting the distillate under a blanket of unreactive liquid of lower density such as water (85). During shipment contact with metal surfaces should be avoided as they promote oxidation. Transport of the technical grade material usually occurs in heavy duty, plastic-lined paper sacks. The fine chemicals are usually shipped in smaller quantities in brown, air-tight glass bottles.

5. Economic Aspects

Production figures for the aminophenols are scarce, the compounds usually being classified along with many other aniline derivatives (86). Most production of the technical grade materials (95% purity) occurs on-site as they are chiefly used as intermediate reactants in continuous chemical syntheses. World production of the fine chemicals (99% purity) is probably no more than a few hundred metric tons yearly, at prices ranging between $40–70 per kg in 2000 with 4-aminophenol being the least expensive.

6. Analytical and Test Methods, Storage

Aminophenols have been detected in waste water by investigating uv absorptions at 220, 254, and 275 nm (87). These compounds can also be detected spectrophotometrically after derivatization at concentrations of 1 part per 100 million by reaction in acid solution with N-(1-naphthyl)ethylenediamine [551-09-7] (88) or 4-(dimethylamino)benzaldehyde [100-10-7] (89), and the Schiff base formed can be stabilized in chloroform by chelation to increase detection limits (90).

Reaction with 1,3-benzenediamine-periodate (91) or with a hypochlorite–alkaline phenol (Berthelot) reagent enables the detection of both 2- and 4-aminophenol, the latter reagent giving distinguishable blue and dark green products, respectively (92). 4-Aminophenol itself has been shown to react in alkaline solution with both the 2- and 3-aminophenol isomers, a reaction exploited for their detection (93).

More specifically, 2-aminophenol can be detected in solution using an iron(II) sulfate–hydrogen peroxide reagent (94) or dimerized in acidic solution to 2-hydroxyisophenoxazine-3-one, an intensively colored dye (95). 3-Aminophenol has been analyzed colorimetrically by oxidation in base and subsequent extraction of a violet quinoneimide dye (96). A colorimetric method using 3-cyano-N-methoxypyridinium perchlorate as reagent detects 4-aminophenol in the presence of N-acetyl-4-aminophenol (97). 4-Aminophenol has also been detected spectrophotometrically after conversion to indophenol [500-85-6] with alkaline phenol, a method quoted as detecting as little as 10^{-18}mol/L (98), and fluorimetrically after reaction with 3-amino-2(1H)-quinolinethione to give a yellow-green fluorescent product (99).

Filter paper impregnated with dicarbonyl(benz-2,1,3-thiadiazole)rhodium chloride gives characteristic colorations with the aminophenol isomers after fixation and can be used as an indicator paper (100).

The potentiometric microdetection of all aminophenol isomers can be done by titration in two-phase chloroform–water medium (101), or by reaction with iodates or periodates, and the back-titration of excess unreacted compound using a silver amalgam and SCE electrode combination (102). Microamounts of 2-aminophenol can be detected by potentiometric titration with cupric ions using a copper-ion-selective electrode; the 3- and 4-aminophenol isomers could not be detected by this method (103). Polarographic detection of 4-aminophenol is possible after conversion to the diazonium salt with sodium nitrite (104) and this isomer can also be analyzed by voltametry (105).

Chromatographic methods for the separation, identification, and quantification of aminophenols also have been described (106). Thin-layer chromatography (tlc) provides a rapid and convenient method of separating the isomers from many derivatives, and subsequent spraying with a variety of chromogenic reagents gives additional information (107,108). Impregnation of plates with nitrite (109) or the use of high performance plates and subsequent densitometry (110) provide quantification to the level of 0.1 µg.

Several gas–liquid chromatographic procedures, using electron-capture detectors after suitable derivatization of the aminophenol isomers, have been cited for the determination of impurities within products and their detection within environmental and wastewater samples (111,112). Modern high pressure liquid chromatographic (hplc) separation techniques employing fluorescence (113) and electrochemical (114) detectors in the 0.01-µg range have been described and should meet the needs of most analytical problems (115,116). The use of selected-ion mass spectrometry also greatly increases detection sensitivity (117,118).

Under atmospheric conditions, 3-aminophenol is the most stable of the three isomers. Both 2- and 4-aminophenol are unstable; they darken on exposure to air and light and should be stored in brown glass containers, preferably in an atmosphere of nitrogen. The use of activated iron oxide in a separate cellophane bag inside the storage container (119), or the addition of stannous chloride (120), or sodium bisulfite (121) inhibits the discoloration of aminophenols. The salts, especially the hydrochlorides, are more resistant to oxidation and should be used where possible.

7. Health and Safety Factors

In general, aminophenols are irritants. Their toxic hazard rating is slight to moderate and their acute oral toxicities in the rat (LD_{50}) are quoted as 1.3, 1.0, and 0.375 g/kg body weight for the 2-, 3-, and 4-isomer, respectively (122). Repeated contamination may cause general itching, skin sensitization, dermatitis, and allergic reactions (123). Immunogenic conjugates are spontaneously produced upon exposure to 2- and 4-aminophenol (124). Methemoglobin formation with subsequent cyanosis is another possible complication (125). Inhalation of aminophenols causes irritation of the mucosal membranes and may precipitate allergic bronchial asthma. Thermal decomposition will release toxic fumes of carbon monoxide and nitrogen oxides.

2-Aminophenol is neuroactive, inducing spike discharges when instilled into the cerebroventricle of the rat (126). 4-Aminophenol is a selective nephrotoxic agent and interrupts proximal tubular function (127,129). Disagreement exists concerning the nephrotoxity of the other isomers although they are not as potent as 4-aminophenol (130,131). Respiration, oxidative phosphorylation, and ATPase activity are inhibited in rat kidney mitochondria (132). The aminophenols and their derivatives are inhibitors of 5-lipoxygenase (133) and prostaglandin synthetase (134) and are being investigated as therapeutic and prophylactic agents for leukotriene or prostaglandin-induced allergic bronchial, tracheal, and lung disease (135).

Teratogenic effects have been noted with 2- and 4-aminophenol in the hamster, but 3-aminophenol was without effect in the hamster and rat (136,137). 2-Aminophenol causes DNA damage in the presence of copper(II) ions (138). 4-Aminophenol is known to inhibit DNA synthesis and alter DNA structure in human lymphoblasts (139,140) and is mutagenic in mouse micronuclei tests (141). The aminophenols have been shown to be genotoxic, as evidenced by the induction of sister chromatid exchanges (142,143), but they also exert a protective effect against DNA interaction with other noxious chemicals (144). After assessment of available data a recent report stated that the aminophenols were safe as cosmetic ingredients in their present uses and concentrations (145).

Obviously, care should be taken in handling these compounds with the wearing of chemical-resistant gloves and safety goggles; prolonged exposure should be avoided. Contaminated clothing should be removed immediately and the affected area washed thoroughly with running water for at least 10 minutes.

Since the aminophenols are oxidized easily, they tend to remove oxygen from solutions. Hence, if they are released from industrial waste waters into streams and rivers, they will deplete the capacity of these environments to sustain aquatic life. Concern has also been raised that chlorination of drinking water may enhance the toxicity of aminophenols present as pollutants (146); chlorinated aminophenols are known to be more toxic (147).

The addition of slaked lime (148) and the initiation of polymerization reactions with H_2O_2 and ferric or stannous salts (149) are techniques employed to remove aminophenols from waste waters. The adsorption of aminophenols onto metal ferrocyanides and activated carbons from bituminous coal has met with limited success but also may provide a possible means of removal (150,151).

An enzymatic method using horseradish peroxidase to cross link and precipitate the compounds as insoluble polymers has also been studied (152). Biological degradation of 3-aminophenol has been carried out in aeration tanks using adapted microflora (>4 month period) from activated sludge (153); and other microbial degradation techniques are under investigation (154–159) including the immobilization of microbial cells onto calcium-alginate (polymannuronate) beads (160).

8. Uses

The aminophenols are versatile intermediates and their principal use is as synthesis precursors; their products are represented among virtually every class of stain and dye.

Both 2- and 4-aminophenols are strong reducing agents and are employed as photographic developers under the trade names of Atomal and Ortol (2-aminophenol); Activol, Azol, Certinal, Citol, Paranol, Rodinol, Unal, and Ursol P (4-aminophenol); they may be used alone or in combination with hydroquinone. The oxalate salt of 4-aminophenol is marketed under the name of Kodelon. They also act as corrosion inhibitors in paints (161) and as anticorrosion-lubricating agents in two-cycle engine fuels (162).

As a result of the close proximity of the amino and hydroxyl groups on the benzene ring and their ease of condensation with suitable reagents, 2-aminophenol is a principal intermediate in the synthesis of such heterocyclic systems as oxyquinolines, phenoxazines, and benzoxazoles. The last-named compounds have been used as inflammation inhibitors (163), and other derivatives have potential as antiallergic agents (164). In addition, 2-aminophenol is specifically used for shading leather, fur, and hair from grays to yellowish brown. It has also found application in the determination and extraction of certain precious metals (165,166).

3-Aminophenol has been used as a stabilizer of chlorine-containing thermoplastics (167), although its principal use is as an intermediate in the production of 4-amino-2-hydroxybenzoic acid [65-49-6], a tuberculostat. This isomer is also employed as a hair colorant and as a coupler molecule in hair dyes (168,169).

Nitrogen-substituted 4-aminophenols have long been known as antipyretics and analgesics, and the production of these derivatives represents significant use of this compound. 4-Aminophenol is also used as a wood stain, imparting a roselike color to timber (170), and as a dyeing agent for fur and feathers.

9. Derivatives

The derivatives of the aminophenols have important uses both in the photographic and the pharmaceutical industries. They are also extensively employed as precursors and intermediates in the synthesis of more complicated molecules, especially those used in the staining and dye industry. All of the major classes of dyes have representatives that incorporate substituted aminophenols; these compounds produced commercially as dye intermediates have been reviewed

(171). Details of the more commonly encountered derivatives of the aminophenols can be found in standard organic chemistry texts (25,172). A few examples, which have specific uses or are manufactured in large quantities, are discussed in detail in the following (see Table 6).

9.1. Derivatives of 2-Aminophenol. *2-Amino-4-nitrophenol.* This derivative, 2-hydroxy-5-nitroaniline (9), forms orange prisms from water. These prisms are hydrated with one water of crystallization, mp 80–90°C, and can be dehydrated over sulfuric acid to the anhydrous form, mp 143–145°C. The compound is soluble in ethanol, diethyl ether, acetic acid, and warm benzene and slightly soluble in water.

2-Amino-4-nitrophenol is produced commercially by the partial reduction of 2,4-dinitrophenol. This reduction may be achieved electrolytically using vanadium (173) or chemically with polysulfide, sodium hydrosulfide, or hydrazine and copper (174). Alternatively, 2-acetamidophenol or 2-methylbenzoxazole may be nitrated in sulfuric acid to yield a mixture of 4- and 5-nitro derivatives that are then separated and hydrolyzed with sodium hydroxide (175).

The major use of this compound is in the production of mordant and acid dyes. 2-Amino-4-nitrophenol also has found limited use as an antioxidant and light stabilizer in butyl rubbers and as a catalyst in the manufacture of hexadiene. The compound has been shown to be a skin irritant and continuous exposure should be avoided. Toxicological studies indicate that it is nonaccumulative (176) but suggest that it may be carcinogenic (177).

(9)　　　　　　　　(10)

2-Amino-4,6-dinitrophenol. This derivative (10), also known as picramic acid, forms dark red needles from ethanol and prisms from chloroform. The compound flashes at 210°C in contact with an open flame, ignites rapidly and burns relatively fast. 2-Amino-4,6-dinitrophenol is soluble in glacial acetic acid, water, ethanol, benzene, and aniline and is sparingly soluble in diethyl ether and chloroform.

The compound can be prepared from 2,4,6-trinitrophenol (picric acid [88-89-1]) by reduction with sodium hydrosulfide (178), with ammonia-hydrogen sulfide followed by acetic acid neutralization of the ammonium salt (179), with ethanolic hydrazine and copper (180), or electrolytically with vanadium sulfate in alcoholic sulfuric acid (173). Heating 4,6-dinitro-2-benzamidophenol in concentrated HCl at 140°C also yields picramic acid (181).

2-Amino-4, 6-dinitrophenol is an important intermediate in the manufacture of colorants, especially mordant dyes. It has also been used as an indicator dye in titrations (yellow with acid, red with alkali) and as a reagent for albumin determination. The compound induces sister chromatid exchange and micronuclei formation suggesting potential health hazards (182,183).

Table 6. **Derivatives of Aminophenols**

Common name	Structure number	CAS Registry Number	Molecular formula	Molecular weight	Melting point, °C	Boiling point[a], °C
Derivatives of 2-aminophenol						
2-amino-4-nitrophenol	(9)	[99-57-0]	$C_6H_6N_2O_3$	154.13	80–90	
2-amino-4,6-dinitrophenol	(10)	[96-91-3]	$C_6H_5N_3O_5$	199.13	169–170	
2-amino-4,6-dichlorophenol	(11)	[527-62-8]	$C_6H_5Cl_2NO$	168.15	95–96	
2,4-diaminophenol	(12)	[95-86-3]	$C_6H_8N_2O$	124.14	78–80[b]	
acetarsone	(13)	[97-44-9]	$C_8H_{10}AsNO_5$	275.08	240–250[b]	
Derivatives of 3-aminophenol						
3-(N,N-dimethylamino)phenol	(14)	[99-07-0]	$C_8H_{11}NO$	137.18	87	265–268 206 (13.3) 194 (6.7) 153 (0.7)
3-(N-methylamino)phenol	(15)	[14703-69-6]	C_7H_9NO	123.15		170 (1.6)
3-(N,N-diethylamino)phenol	(16)	[91-68-9]	$C_{10}H_{15}NO$	165.23	78	276–280 209–211 (1.6)
3-(N-phenylamino)phenol	(17)	[101-18-8]	$C_{12}H_{11}NO$	185.22	81.5–82	340
4-amino-2-hydroxybenzoic acid	(18)	[65-49-6]	$C_7H_7NO_3$	153.13	150–151	
Derivatives of 4-aminophenol						
4-(N-methylamino)phenol	(19)	[150-75-4]	C_7H_9NO	123.15	87	168–169 (2)
4-(N,N-dimethylamino)phenol	(20)	[619-60-3]	$C_8H_{11}NO$	137.18	75–76	101–103 (0.067)
4-hydroxyacetanilide	(21)	[103-90-2]	$C_8H_9NO_2$	151.15	169–171	
4-ethoxyacetanilide	(22)	[62-44-2]	$C_{10}H_{13}NO_2$	179.21	134–135	
N-(4-hydroxyphenyl)glycine	(23)	[122-87-2]	$C_8H_9NO_3$	167.16	245–247[b]	

[a] At 101.3 kPa = 760 mm Hg unless otherwise noted in parentheses. Values in parentheses are in the kPa. To convert kPa to mm Hg, multiply by 7.5.
[b] Decomposes.

2-Amino-4,6-dichlorophenol. This compound (11) forms long white needles from carbon disulfide, and aggregate spheres from benzene. It sublimes at 70–80°C (8 Pa = 0.06 mm Hg) and decomposes >109°C. It is freely soluble in benzene and carbon disulfide, and is sparingly soluble in petroleum ether, water, and ethanol. The free base is unstable and the hydrochloride salt (mp 280–285°C, dec) is employed commercially.

Industrial production is by reduction of the corresponding nitrophenol with iron or hydrazine (184,185). 2-Amino-4,6-dichlorophenol finds important use as an azo-dye intermediate (see Azo DYES).

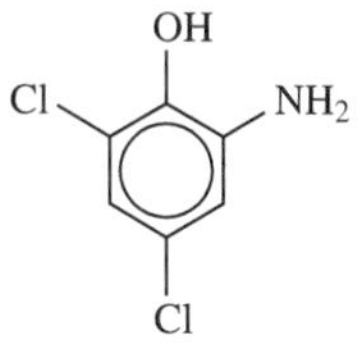

(11) 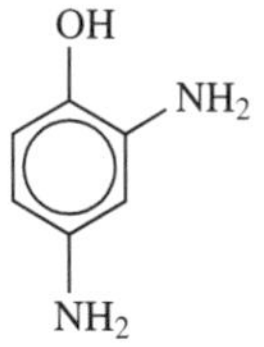   (12)

2,4-Diaminophenol. 4-Hydroxy-*m*-phenylenediamine [95-86-3] (12) forms leaflets that darken on exposure to air. It is soluble in acid, alkali, ethanol, and acetone and is sparingly soluble in chloroform, diethyl ether, and ligroin. 2,4-Diaminophenol usually is sold as the sulfate [74283-34-4] (Diamol) or dihydrochloride salt [137-09-7] (Acrol, Amidol). 2,4-Diaminophenol can be prepared from 2,4-dinitrophenol by catalytic hydrogenation or, less conveniently, by metal reduction in acid solution (Béchamp method) (186,187). Alternatively, electrolytic reduction and subsequent hydroxylation of 1,3-dinitrobenzene or 3-nitroaniline in sulfuric acid can be undertaken (188,189).

The dihydrochloride salt is used as a photographic developer. It also is employed as an intermediate in the manufacture of fur dyes, in hair dyeing, as a reagent in testing for ammonia and formaldehyde, and as an oxygen scavenger in water to prevent boiler corrosion (190).

Acetarsone. Acetarsone (3-acetamido-4-hydroxyphenyl arsonic acid) (13), also known as acetarsol, stovarsol, and Ehrlich 594, forms white prisms from water.

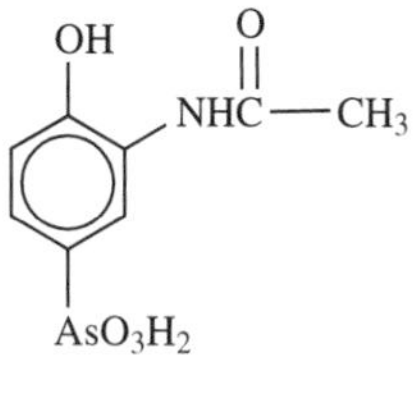

(13)

This compound is odorless with a faintly acidic taste; it is practically insoluble in water, ethanol and dilute acids but freely soluble in dilute aqueous alkali with dissociation constants, pK_a, 3.73, 7.9, 9.3. The compound is prepared by sodium hydrosulfite reduction of 3-nitro-4-hydroxyphenylarsonic acid [121-19-7] and

then acetylation in aqueous suspension with acetic anhydride at 50–55°C for 2 h
(191,192).

Salts of acetarsone are used in the treatment of intestinal amoebiasis, vagi-
nal trichomoniasis, and necrotizing ulcerative gingivitis (Vincent's angina). The
diethylamine salt (acetylarsan [534-33-8]) has antisyphilitic properties. Owing
to toxicity problems, safer drugs have been developed. Oral LD_{50} in rabbits is
150 mg/kg.

9.2. Derivatives of 3-Aminophenol.   *3-(N,N-Dimethylamino)phenol.*
3-Hydroxy-*N,N*-dimethylaniline (14) forms white needles and is soluble in alkali,
mineral acid, ethanol, diethyl ether, acetone, and benzene and practically insoluble in water.

It can be prepared by heating resorcinol with an aqueous solution of
dimethylamine and its hydrochloride at 200°C under pressure for 12 h (193).
The treatment of dimethylaniline with oleum at 55–60°C, followed by fusion
with sodium hydroxide at 270–300°C, also gives 3-(*N,N*-dimethylamino)phenol
(194). In addition, 3-aminophenol may be methylated with dimethyl sulfate
under neutral conditions, or its hydrochloride salt heated with methanol at
170°C under pressure for 8 h to give the desired product (195). The compound
is used primarily as an intermediate in the production of basic (Red 3 and 11)
and mordant (Red 77) dyes.

(14)   (15)   (16)   (17)

3-(N-Methylamino)phenol.   This derivative (15) is easily soluble in ethyl
acetate, ethanol, diethyl ether, and benzene. It is also soluble in hot water, but
only sparingly soluble in cold water. Industrial synthesis is by heating
3-(*N*-methylamino)benzenesulfonic acid with sodium hydroxide at 200–220°C
(196) or by the reaction of resorcinol with methylamine in the presence of
aqueous phosphoric acid at 200°C (197).

3-(N,N-Diethylamino)phenol.   This derivative (16) forms rhombic bipyramidal crystals. Industrial synthesis is analogous to the previously described
synthesis of 3-(*N,N*-dimethylamino)phenol from resorcinol and diethylamine,
by reaction of 3-(*N,N*-diethylamino)benzenesulfonic acid with sodium hydroxide,
or by alkylation of 3-aminophenol hydrochloride with ethanol.

3-(N-Phenylamino)phenol.   This phenol (17) is slightly soluble in ethanol,
diethyl ether, acetone, benzene, and water. The compound is made by heating
resorcinol and aniline at 200°C in the presence of aqueous phosphoric acid or
calcium chloride. In another process, 3-aminophenol is heated with aniline
hydrochloride at 210–215°C (198).

4-Amino-2-hydroxybenzoic acid.   This derivative (18) more commonly
known as 4-aminosalicylic acid, forms white crystals from ethanol, melts with
effervescence and darkens on exposure to light and air. A reddish-brown crystalline powder is obtained on recrystallization from ethanol–diethyl ether. The

compound is soluble in dilute solutions of nitric acid and sodium hydroxide, ethanol, and acetone; slightly soluble in water and diethyl ether; and virtually insoluble in benzene, chloroform or carbon tetrachloride. It is unstable in aqueous solution and decarboxylates to form 3-aminophenol. Because of the instability of the free acid, it is usually prepared as the hydrochloride salt, mp 224°C (dec), dissociation constant pK_a 3.25.

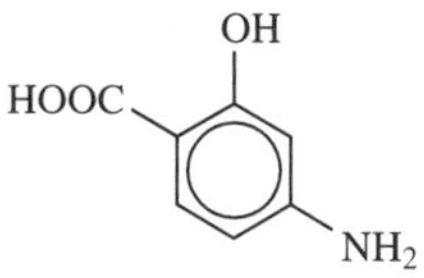

(18)

4-Amino-2-hydroxybenzoic acid is manufactured by carboxylation of 3-aminophenol under pressure with ammonium carbonate at 110°C (199) or with potassium bicarbonate and carbon dioxide at 85–90°C (200) with subsequent acidification.

The major use of this compound is as a bacteriostatic agent against tubercle bacilii. This compound also is used as an adjunct to streptomycin and isoniazid. The free acid and its sodium, potassium, and calcium salts are marketed under many trade names (eg, Aminox, Apacil, Deapasil, Paramycin, Parasalicil, Pasnodia, Rezipas, Sanipirol-4, Tubersan). Up to 10–15 g of the sodium salt may be administered each day, although prolonged use may give rise to toxic symptoms. Oral LD_{50} of the free acid in mice is 4 g/kg.

9.3. Derivatives of 4-Aminophenol. *4-(N-Methylamino)phenol.* This derivative, also named 4-hydroxy-*N*-methylaniline (19), forms needles from benzene that are slightly soluble in ethanol and insoluble in diethyl ether. Industrial synthesis involves decarboxylation of *N*-(4-hydroxyphenyl)glycine [122-87-2] at elevated temperature in such solvents as chlorobenzene–cyclohexanone (201,202). It also can be prepared by the methylation of 4-aminophenol, or from methylamine [74-89-5] by heating with 4-chlorophenol [106-48-9] and copper sulfate at 135°C in aqueous solution, or with hydroquinone [123-31-9] at 200–250°C in alcoholic solution (203).

Its chief use is as a component in photographic developers. Because the free compound is unstable in air and light, it is usually marketed as the sulfate salt [55-55-0], Metol, mp 260°C (dec). It also finds application as an intermediate for fur and hair dyes and, under certain circumstances, as a corrosion inhibitor for steel. Prolonged exposure to 4-(*N*-methylamino)phenol has been associated with the development of dermatitis and allergies.

4-(N,N-Dimethylamino)phenol. 4-Hydroxy-*N*,*N*-dimethylaniline (20) forms large rhombic crystals from diethyl ether–hexane or diethyl ether–ligroin. It forms a salt with sulfuric acid, mp 208–210°C (204).

Methylation of 4-aminophenol with a methyl halide under pressure produces 4-(*N*,*N*-dimethylamino)phenol. The competing product, 4-hydroxyphenyltrimethyl ammonium halide (or the corresponding base), also yields 4-(*N*,*N*-dimethylamino)phenol on distillation. Alternatively, it can be synthesized by dealkylation of 4-methoxy-*N*,*N*-dimethylaniline [701-56-4] with hydroiodic acid at reflux temperature for 10 h (205) or by the photodecomposition of 4-dimethy-laminobenzene diazonium tetrafluoroborate [24564-52-1] (204).

The compound is an intermediate in several synthetic reactions and recently has found extensive use in experimental toxicity studies in animals. It has been shown to cause methemoglobinemia; its metabolism in humans has been discussed (206,207).

(19)   (20)   (21)

4-Hydroxyacetanilide. This derivative (21), also known as 4-acetamido-phenol, acetaminophen, or paracetamol, forms large white monoclinic prisms from water. This compound is odorless and has a bitter taste. 4-Hydroxyacetanilide is insoluble in petroleum ether, pentane, and benzene; slightly soluble in diethyl ether and cold water; and soluble in hot water, alcohols, dimethylformamide, 1,2-dichloroethane, acetone, and ethyl acetate. The dissociation constant, pK_a, is 9.5 (25°C).

Production is by the acetylation of 4-aminophenol, which can be achieved with acetic acid and acetic anhydride at 80°C (208), with acetic acid anhydride in pyridine at 100°C (209), with acetyl chloride and pyridine in toluene at 60°C (210), or by the action of ketene in alcoholic suspension. 4-Hydroxyacetanilide also may be synthesized directly from 4-nitrophenol. The available reduction—acetylation systems include tin with acetic acid, hydrogenation over Pd—C in acetic anhydride, and hydrogenation over platinum in acetic acid (211,212). Other routes include rearrangement of 4-hydroxyacetophenone hydrazone with sodium nitrite in sulfuric acid and the electrolytic hydroxylation of acetanilide [103-84-4] (213).

4-Hydroxyacetanilide is used as an intermediate in the manufacture of azo dyes (qv) and photographic chemicals. The compound possesses antipyretic and analgesic properties and is used widely in this context. Typical formulations containing 4-hydroxyacetanilide include Acetalgin, Cetadol, Dirox, Febrilix, Hedex, Panadol, Tylenol, and Valadol. The oral LD_{50} in rats is 3.7 g/kg.

4-Ethoxyacetanilide. This compound (22), also known as phenacetin, is a white crystalline powder. The compound is odorless and has a slightly bitter taste. It is sparingly soluble in cold water and more soluble in hot water, ethanol, diethyl ether, and chloroform. At relative humidities between 15 and 90% the equilibrium moisture content is about 2% (25°C).

The main production route to 4-ethoxyacetanilide is by catalytic reduction of 4-nitrophenetole [100-29-8] with hydrogen and subsequent acetylation using acetic anhydride. The compound also can be synthesized by ethylating 4-nitro-phenol with ethyl sulfate in alkali, reducing the nitro group to an amino group with iron in acid and then acetylating by boiling with glacial acetic acid (214).

Alternatively, 4-aminophenol may be ethylated with ethyl iodide in alcoholic alkali and the resulting 4-phenetidine [156-43-4] then acetylated. The acetylation also may be carried out first, followed by the ethylation.

4-Ethoxyacetanilide possesses both antipyretic and analgesic properties, but it is of little value for the relief of severe pain. Its use for prolonged periods should be avoided because one of its minor metabolites (2-hydroxyphenetidine) is nephrotoxic and may be involved in the formation of methemoglobinemia. The oral LD_{50} in rats is 1.65 g/kg (215).

N-(4-Hydroxyphenyl)glycine. This derivative (23) forms aggregate spheres or shiny leaflets from water. It turns brown at 200°C, begins to melt at 220°C, and melts completely with decomposition at 245–247°C. The compound is soluble in alkali and mineral acid and sparingly soluble in water, glacial acetic acid, ethyl acetate, ethanol, diethyl ether, acetone, chloroform, and benzene.

N-(4-Hydroxyphenyl)glycine can be prepared from 4-aminophenol and chloracetic acid (216,217) or by alkaline hydrolysis of the corresponding nitrile with subsequent elimination of ammonia (218).

N-(4-Hydroxyphenyl)glycine is used as a photographic developer under the trade names of Glycine, Iconyl, and Monazol. It is also applied as a photoresist in the dye industry and serves as an intermediate in the production of 4-(*N*-methylamino)phenol (Metol) by liberation of CO_2. *N*-(4-Hydroxyphenyl)glycine is used in analytical chemistry for the determination of iron, phosphorus , and silicon, and as an acid indicator in bacteriology. Prolonged exposure to this compound may result in kidney damage (219).

BIBLIOGRAPHY

"Aminophenols" in *ECT* 1st ed., Vol. 1, pp. 737–741, by J. Werner, General Aniline & Film Corporation; in *ECT* 2nd ed., Vol. 2, pp. 213–225, by S. K. Morse, General Aniline & Film Corporation; in *ECT* 3rd ed., Vol. 2, pp. 422–440, by R. E. Faris, GAF Corporation; in *ECT* 4th ed., Vol. 2, pp. 580–604, by Stephen C. Mitchell, University of London and Rosemary Waring, University of Birmingham, England; "Aminophenols" in *ECT* (online), posting date: December 4, 2000, by Stephen C. Mitchell, University of London, Rosemary Waring, University of Birmingham, England.

CITED PUBLICATION

1. I. Y. Korenman, N. G. Sotnikova, G. S. Lineva, and L. E. Zadorkina, *Zh. Fiz. Khim.* **55**, 3081–3083 (1981).

2. I. Y. Korenman, N. G. Sotnikova, G. S. Lineva, and L. E. Zadorkina, *Zh. Prikl. Khim. (Leningrad)* **55**, 2637–2639 (1982).

3. D. V. S. Jain, F. S. Nandel, and P. Lata, *Indian J. Chem.* **21A**, 559–563 (1982).

4. G. Traverse, V. Vargas, and F. Parrini, *Bol. Soc. Chil. Quim.* **27**, 59–61 (1982).

5. G. Varsanyi, *Assignment for Vibrational Spectra of Seven Hundred Benzene Derivatives*, Akademiai Kiado, Budapest. Adam Hilger, London, 1974, pp. 136, 206, 253, 473, 476, 478.

6. C. J. Pouchert, ed., *The Aldrich Library of Infrared Spectra*, 3rd ed., Vol. 3, Aldrich Chemical Co., Milwaukee, Wis., 1981, pp. 718B, 720D, 725D.

7. W. F. Forbes and I. R. Leckie, *Can. J. Chem.* **36**, 1371–1380 (1958).

8. J. C. Dearden and W. F. Forbes, *Can. J. Chem.* **37**, 1294–1304 (1959).

9. *Sadtler Catalog of Ultraviolet Spectra, SAD No. 236, 1894, 3509.* The Sadtler Research Co., Philadelphia, Pa., 1976.

10. *Eight Peak Index of Mass Spectra*, 3rd ed., Vol. 1, Part 1. The Royal Society of Chemistry, The University, Nottingham, U.K., 1983, p. 67.

11. C. J. Pouchert and J. R. Campbell, eds., *The Aldrich Library of NMR Spectra*, Vol. 5, Aldrich Chemical Co., Milwaukee, Wis., 1974–1975, pp. 45C, 51B, 57A.

12. *Sadtler Catalog of NMR Spectra, SAD No. 717, 1176, 10220*, The Sadtler Research Co., Philadelphia, Pa., 1972.

13. J. G. Grasselli, ed., *CRC Atlas of Spectral Data and Physical Constants for Organic Compounds*, CRC Press, Cleveland, Ohio, 1973, p. B754.

14. S. Ashfaquzzaman and A. K. Pant, *Acta Crystallogr.* **B35**, 1394–1399 (1979).

15. W. A. Caspari, *Philos. Mag.* **4**, 1276–1285 (1927).

16. J. D. H. Donnay and H. M. Ondik, *Crystal Data; Determinative Tables*, 3rd ed., U.S. Department of Commerce, National Bureau of Standards & Joint Committee on Powder Diffraction Standards, Washington, D.C., 1972, pp. 0–23, 0–73, 0–97, 0–106.

17. J. D. Korp, I. Bernal, L. Aven, and J. L. Mills, *J. Cryst. Mol. Struct.* **11**, 117–124 (1981).

18. N. V. Sidgwick and R. K. Callow, *J. Chem. Soc. (Trans)* **125**, 522–527 (1924).

19. L. Nunez, L. Barrai, S. L. Gavilanes, and G. Pilcher, *J. Chem. Thermodyn.* **18**, 575–579 (1986).

20. M. S. Kharasch, *J. Res. Nat. Bur. Stand.* **2**, 359–379 (1929).

21. R. Kuhn and A. Wassermann, *Helv. Chim. Acta.* **11**, 1–30 (1928).

22. V. H. Veeley, *J. Chem. Soc.* **93**, 2122–2144 (1908).

23. G. Chuchani, J. A. Hernandez, and J. Zabicky, *Nature (London)* **207**, 1385–1386 (1965).

24. R. W. G. Wyckoff, in *Crystal Structures*, 2nd ed., Vol. 6, Part 1. Wiley-Interscience, New York, 1969, pp. 186–190.

25. A. R. Forester and J. L. Wardell, in *Rodd's* Chemistry of Carbon Compounds, S. Coffey, ed., 2nd ed., Vol. 3A, Elsevier Publishing Co., Amsterdam, The Netherlands, 1971, pp. 352–363, Chapt. 4.

26. J. H. Ransom, *Am. Chem. J.* **23**, 1–50 (1900).

27. K. H. Saunders, *The Aromatic Diazo Compounds and Their Technical Application*, Arnold, London, 1949, pp. 17–20.

28. H. Zollinger, *Azo and Diazo Chemistry*, Interscience, New York, 1961, pp. 51–53, 250–265.

29. M. Moazzam, Z. H. Chohan, and Q. Ali, *J. Pure Appl. Sci.* **6**, 29–31 (1987).

30. British Intelligence Objectives Subcommittees (BIOS), *Report 986*, London England, 1946, pp. 45–46, 412.

31. Czech Pat. 159,564 (Aug. 15, 1975), D. Kulda, J. Fuka, J. Ott, and Z. Misar, *Chem. Abstr.* **84**, 164381k (1976).

32. U.S. Pat. 3,535,382 (Oct. 20, 1970), B. B. Brown and F. A. E. Schilling (to CPC International, Inc.).

33. Eur. Pat. 41,837 (Dec. 16, 1981), W. R. Clingan, E. L. Derrenbacker, and T. J. Dunn (to Mallinckrodt, Inc.).

34. Ger. Pat. 1,244,196 (July 13, 1967), E. Tolksdorff (to Badische Anilin & Soda Fabrik, A.-G.).

35. Jpn. Pat. 75,135,042 (Oct. 25, 1975), K. Fujiwara and K. Tanaka (to Seiko Chemical Industry Co. Ltd.).

36. U.S. Pat. 3,079,435 (Feb. 26, 1963), M. Freifelder and R. M. Robinson (to Abbott Laboratory).

37. Fr. Pat. 1,392,098 (Mar. 12, 1965), L. Spiegler (to E.I. du Pont de Nemours & Co., Inc.).

38. I. Kukhtenko, *Zh. Org. Chim.* **7**, 330–333 (1971).

39. T. Sone, Y. Tohuda, T. Sakai, S. Shinkai, and O. Manabe, *J. Chem. Soc. Perkin Trans.* **II**, 298–302 1596–1598 (1981).

40. U.S. Pat. 4,571,437 (Feb. 18, 1986), D. C. Caskey and D. W. Chapman (to Mallinckrodt, Inc.).

41. Eur. Pat. 85511 (Aug. 10, 1983), D. C. Caskey and D. W. Chapman (to Mallinckrodt, Inc.).

42. U.S. Pat. 2,446,519 (Aug. 10, 1948), F. R. Bean (to Eastman Kodak Co.).

43. USSR Pat. 883017 (Nov. 23, 1981), I. I. Bat, G. A. Chistyakova, P. N. Ostrovskii, and V. V. Rebrova.

44. Fr. Pat. 1,354,430 (Mar. 6, 1964), J. Levy (to Universal Oil Products Co.).

45. U.S. Pat. 2,947,781 (Aug. 2, 1960), L. Spiegler (to E.I. du Pont de Nemours & Co., Inc.).

46. U.S. Pat. 3,953,509 (Apr. 27, 1976), N. P. Greco (to Koppers Co., Inc.).

47. U.S. Pat. 2,198,249 (Apr. 23, 1938), C. O. Henke and J. V. Vaughan (to E.I. du Pont de Nemours & Co., Inc.).

48. Brit. Pat. 1,181,969 (Feb. 18, 1970), N. R. W. Benwell and I. Buckland (to International Nickel, Ltd.).

49. U.S. Pat. 3,535,382 (Oct. 20, 1970), B. B. Brown and F. A. E. Schilling (to CPC International Inc.).

50. U. M. Dewal, R. D. Mhaskar, J. B. Joshi, and S. B. Sawant, *Indian Chem. J.* **15**, 29–32 (1980).

51. U.S. Pat. 4,307,249 (Dec. 22, 1981), E. L. Derrenbacker (to Mallinckrodt, Inc.).

52. Eur. Pat. 85890 (Aug. 17, 1983), D. C. Caskey and D. W. Chapman (to Mallinckrodt, Inc.).

53. Fr. Pat. 1,542,073 (Oct. 11, 1968), R. G. Benner.

54. M. Noel, P. N. Anantharama, and H. V. K. Udupa, *J. Appl. Electrochem.* **12**, 291–298 (1982).

55. E. Theodoridou and D. Jannakondakis, *Z. Naturforsch.* **36B**, 840–845 (1981).

56. K. S. Udupa, G. S. Subramanian, and H. V. K. Udupa, *Trans. Soc. Adv. Electrochem. Sci. Technol.* **7**, 49–50 (1972).

57. M. Levi, I. Pesheva, and M. Dolapchieva, *Elektrokhimiya*, **19**, 44–46 (1983).

58. H. V. K. Udupa, *Indian Chem. Eng.* **30**, 53–56 (1988).

59. C. Ravichandran, S. Chellammal, and P. N. Anantharaman, *J. Appl. Electrochem.* **19**, 465 (1989).

60. Indian Pat. 143,869 (Feb. 18, 1978), H. V. K. Udupa and P. N. Anantharaman (to Council of Scientific and Industrial Research, India).

61. E. Theodoridou and D. Jannakoudakis, *Z. Naturforsch.* **36B**, 840–845 (1981).

62. Indian Pat. 142,241 (June 18, 1977), H. V. K. Udupa, K. S. Udupa, and K. Jayaraman (to Council of Scientific and Industrial Research, India).

63. J. Marquez and D. Pletcher, *J. Appl. Electrochem.* **10**, 567–573 (1980).
64. Indian Pat. 161376 (Nov. 21, 1987), H. V. K. Udupa and N. N. Nagendra (to Council of Scientific and Industrial Research, India).
65. J. Marquez and D. Pletcher, *Acta Cient. Venez.* **37**, 391–393 (1986).
66. V. D. Sokolovskii, V. D. Belyaev, and G. P. Snytnikova, *React. Kinet. Catal. Lett.* **22**, 127–131 (1983).
67. Eur. Pat. 197,633 (Oct. 15, 1986), H. Harada, H. Maki, and S. Sasaki (to Sumitomo Chemical Co., Ltd.).
68. U.S. Pat. 4,675,444 (June 23, 1987), F. Matsunaga, E. Kato, T. Kimura, and Y. Isota (to Mitsui Petrochemical Industries Ltd.; Honshu Chemical Industry Co., Ltd.).
69. Jpn. Pat. 55/53246 and 55/53250 (Apr. 18, 1980) (to Mitsui Petrochemical Industries Ltd.).
70. J. C. Jacquesy, M. P. Jouannetaud, G. Morellet, and Y. Vidal, *Tetrahedron Lett.* **25**, 1479–1482 (1984).
71. J. C. Jacquesy, M. P. Jouannetaud, G. Morellet, and Y. Vidal, *Bull. Soc. Chim. Fr.* 625–629 (1986).
72. Fr. Pat. 1,354,430 (Mar. 6, 1964), J. Levy (to Universal Oil Products Co.).
73. U.S. Pat. 4,212,823 (July 15, 1980), W. H. Mueller (to Hoechst A-G).
74. Brit. Pat. 1,028,078 (May 4, 1966), N. Barton and C. Thomas (to Imperial Chemical Industries, Inc.).
75. Fr. Pat. 1,564,882 (Apr. 25, 1969), H. Daunis and M. Gominet (to Société des Usines Chimique Rhône-Poulenc).
76. Ger. Pat. 2,050,943 (Apr. 29, 1971), F. A. Baron and R. G. Benner (to Howard Hall & Co.).
77. Ger. Pat. 2,050,927 (Apr. 29, 1971), F. A. Baron and R. G. Benner (to Howard Hall & Co.).
78. Ger. Pat. 2,054,282 (June 9, 1971), F. A. Baron (to Howard Hall & Co.).
79. Ger. Pat. 2,103,548 (Aug. 5, 1971), K. C. Reid (to MacFarlan Smith, Ltd.).
80. U.S. Pat. 3,876,703 (Apr. 8, 1975), R. Harmetz, D. C. Ruopp, and B. B. Brown (to CPC International, Inc.).
81. Ger. Pat. 1,902,418 (Aug. 13, 1970), G. Lorenz, W. Schmidt, and M. Gallus (to Farbenfabriken Bayer A.-G.).
82. Brit. Pat. 1,038,005 (Aug. 3, 1966), J. D. Seddan (to Imperial Chemical Industries, Ltd.).
83. Pol. Pat. 46,829 (May 8, 1963), Z. Grzymalski and F. Mirek.
84. Jpn. Pat. 7,611,722 (Jan. 30, 1976), A. Yamamoto, A. Hirano, and S. Kawasaki (to Toaka Dyestuffs Manufacturing Co., Ltd.).
85. Ger. Pat. 1,104,970 (Apr. 20, 1961), K. Merkel and R. Scweizer (to Badische Anilin- & Soda-Fabrik A.-G.).
86. Y. A. Novikova, *Khim. Prom-st. Rubezhom*, **(2)** 46–61 (1980).
87. K. Urano, K. Kawamoto, and K. Hayashi, *Suishitsu Odaku Kenkyu* **4**, 43–50 (1981).
88. G. Norwitz and P. N. Keliher, *Anal. Chem.* **55**, 1229–1229 (1983).
89. Y. I. Korenman and N. G. Sotnikova, *Zh. Prikl. Khim. (Leningrad)* **56**, 2278–2280 (1983).
90. M. S. Mayadeo and R. K. Banavali, *Indian J. Chem.* **25A**, 789–790 (1986).
91. K. E. Rao and C. S. Sastry, *Mikrochim. Acta* **1**, 313–319 (1984).
92. T. T. Ngo and C. F. Yam, *Anal. Lett.* **17**, 1771–1782 (1984).
93. K. E. Rao, K. V. S. S. Murthy, and C. S. P. Sastry, *Indian Drugs* **20**, 58–61 (1982).
94. C. S. P. Sastry and K. V. S. S. Murthy, *Natl. Acad. Sci. Lett. (India)* **5**, 15–17 (1982).
95. N. H. Cnubben, B. Blaauboer, S. Juyn, J. Vervoot, and I. M. Rietjens, *Anal. Biochem.* **220**, 165–171 (1994).

96. A. Mazzeo-Farina, M. A. Ionio, and A. Laurenzi, *Ann. Chim. (Rome)* **71**, 103–109 (1981).
97. M. A. Korany, D. Heber, and J. Schnekenburger, *Talanta* **29**, 332–334 (1982).
98. K. Nagashima and S. Suzuki, *Bunseki Kagaku* **31**, 724–726 (1982).
99. T. Yoshida, H. Taniguchi, T. Yoshida, and S. Nakano, *Yakugaku Zasshi* **100**, 295–301 (1980).
100. USSR Pat. 1,354,075 (Nov. 23, 1987), Y. N. Kukushkin, N. K. Krylov, N. R. Popova, and V. A. Esaulova (to Leningrad Technological Institute).
101. R. A. Hux, S. Puon, and F. F. Cantwell, *Anal. Chem.* **52**, 2388–2392 (1980).
102. M. A. Zayed, H. Khalifa, and E. F. A. Nour, *Microchem. J.* **34**, 204–210 (1986).
103. W. Selig, *Microchem. J.* **28**, 126–131 (1983).
104. Czech Pat. 224,568 (Oct. 1, 1984), J. Lakomy and J. Kubias.
105. D. J. Miner, J. R. Rice, R. M. Riygin, and P. T. Kissinger, *Anal. Chem.* **53**, 2258–2263 (1981).
106. J. Gasparic, in I. M. Hais and K. Macek, eds., *Paper Chromatography: A Comprehensive Treatise*, Academic Press, New York, 1963, pp. 418–431.
107. L. Reio, *J. Chromatogr.* **1**, 338–373 (1958).
108. S. C. Mitchell and R. H. Waring, *J. Chromatogr.* **151**, 249–251 (1978).
109. R. S. Dhillon, J. Singh, V. K. Gautam, and B. R. Chhabra, *J. Chromatogr.* **435**, 256–257 (1988).
110. T. A. Kouimtzis, I. N. Papadoyannis, and M. C. Sofoniou, *Microchem. Acta* **26**, 51–54 (1981).
111. R. T. Coutts, E. E. Hargesheimer, F. M. Pasutto, and G. B. Baker, *J. Chromatogr. Sci.* **19**, 151–155 (1981).
112. Y. Osaki and T. Matsueda, *Bunseki Kagaku* **37**, 253–258 (1988).
113. B. Schultz, *J. Chromatogr.* **299**, 484–486 (1984).
114. M. Goto, Y. Koyanagi, and D. Ishii, *J. Chromatogr.* **208**, 261–268 (1981).
115. N. T. Bernabei, V. Ferioli, G. Gamberini, and R. Cameroni, *Farmaco Ed. Prat.* **36**, 249–255 (1981).
116. L. A. Stevenson, *IARC Sci. Publ.* **40**, 219–228 (1981).
117. N. Tanada, M. Kageura, K. Hara, Y. Hieda, M. Takamoto, and S. Kashimura, *Forensic Sci. Int.* **52**, 5–11 (1991).
118. N. Tanada, M. Kageura, K. Hara, Y. Hieda, M. Takamoto, and S. Kashimura, *Forensic Sci. Int.* **64**, 1–8 (1994).
119. Jpn. Pat. 80,818,43 (June 20, 1980) (to Mitsui Toatsu Chemicals, Inc.).
120. Jpn. Pat. 60/239447 (Nov. 28, 1985) (to Shiratori Pharmaceutical Co., Ltd.).
121. Jpn. Pat. 62/61956 (Mar. 18, 1987), H. Harada, H. Maki, and S. Sasaki (to Sumitomo Chemical Co., Ltd.).
122. K. L. Markaryan and E. A. Babayan, *Gig. Tr. Prof. Zabol.* **1**, 49–50 (1988).
123. H. Oshima, T. Tamaka, I. T. Oh, and M. Koga, *Contact Dermatitis* **45**, 359 (2001).
124. M. Cirstea, G. Suhaciu, M. Cirje, G. Petec, and A. Vacariu, *Rev. Roum. Morphol. Embryol. Physiol.* **17**, 91–96 (1980).
125. M. Akazawa, M. Takasaki, and A. Tomoda, *Tohoku J. Exp. Med.* **192**, 301–312 (2000).
126. Y. Nishijima, *Okayama Igakkai Zasshi* **104**, 471–482 (1992).
127. J. D. Tange, B. D. Ross, and J. G. G. Ledingham, *Clin. Sci. Mol. Med.* **53**, 485–492 (1977).
128. K. P. R. Gartland, F. W. Bonner, J. A. Timbrell, and J. K. Nicholson, *Arch. Toxicol.* **63**, 97–106 (1989).
129. H. Song and T. S. Chen, *J. Biochem. Mol. Toxicol.* **15**, 34–40 (2001).
130. J. F. Newton, C. H. Kuo, M. W. Gemborys, G. H. Mudge, and J. B. Hook, *Toxicol. Appl. Pharmacol.* **65**, 336–344 (1982).

131. I. Ito, *Tokyo Joshi Ika Daigaku Zasshi* **57**, 1655–1666 (1987).

132. C. A. Crowe and co-workers, *Xenobiotica* **7**, 345–356 (1977).

133. T. Miyamoto and T. Obata, *Int. Cong. Ser. Excerpta Med.* **623**, 78–80 (1983).

134. J. Baumann, F. Von Bruchhausen, and G. Wurm, *Pharmacology*, **27**, 267–280 (1983).

135. Jpn. Pat. 58/140016 (Aug. 19, 1983) to Ono Pharmaceuticals Co., Ltd.

136. J. V. Rutkowski and V. H. Ferm, *Toxicol. Appl. Pharmacol.* **63**, 264–269 (1982).

137. T. A. Re and co-workers, *Fundam. Appl. Toxicol.* **4**, 98–104 (1984).

138. Y. Ohkuma and S. Kawanishi, *Arch. Biochem. Biophys.* **389**, 49–56 (2001).

139. N. K. Hayward, M. F. Lavin, and P. W. Craswell, *Biochem. Pharmacol.* **31**, 1425–1429 (1982).

140. N. K. Hayward and M. F. Lavin, *Life Sci.* **36**, 2039–2046 (1985).

141. S. M. Sicardi, J. L. Martiarena, and M. T. Iglesias, *Acta Farm. Bonaerense* **6**, 71–75 (1987).

142. G. Kirchner and U. Bayer, *Hum. Toxicol.* **1**, 387–392 (1982).

143. J. A. Holme and co-workers, *Mutagenesis* **3**, 51–56 (1988).

144. I. Niculescu-Duvaz and co-workers, *Neoplasma* **35**, 539–548 (1988).

145. Cosmetic, Toiletry, and Fragrance Assoc. Inc., *J. Am. Coll. Toxicol.* **7**, 279–333 (1988).

146. Y. A. Rakhmanin and co-workers, *Gig. Sanit.* 4–7 (1985); *Chem. Abstr.* **103**, 17930w (1985).

147. E. V. Shtannikov and I. N. Lutsevich, *Gig. Sanit.* 19–22 (1983); *Chem. Abstr.* **99**, 189185x (1983).

148. S. B. Deshmukh, T. Swaminathan, P. V. R. Subrahmanyam, and B. B. Sundaresan, *J. Hazard. Mater.* **9**, 171–179 (1984).

149. Jpn. Pat. 62/193694 (Aug. 25, 1987), T. Nogami (to Goni Chemical Industry Co., Ltd.).

150. B. B. Tewara and A. Kamaludin, *J. Colloid Interface Sci.* **193**, 167–171 (1997).

151. C. Moreno-Castilla, J. Rivera-Ultrilla, M. V. Lopez-Ramon, and F. Carrasco-Marin, *Carbon* **33**, 845–851 (1995).

152. A. M. Klibanov, B. N. Alberti, E. D. Morris, and L. M. Felshin, *J. Appl. Biochem.* **2**, 414–421 (1980).

153. L. N. Zayidullina, E. A. Korneeva, and A. S. Lukyanova, *Vodosnabzh. Sanit. Tekh.* **8**, 12–13 (1981).

154. S. D. Deshpande, T. Chakrabarti, P. V. R. Subrahmanyam, and B. B. Sundaresan, *Water Res.* **19**, 293–298 (1985).

155. C. M. Aelion, C. M. Swindoll, and F. K. Pfaender, *Appl. Environ. Microbiol.* **53**, 2212–2217 (1987).

156. U. Lechner and G. Straube, *J. Basic Microbiol.* **28**, 629–637 (1988).

157. U. Lendenmann and J. C. Spain, *J. Bacteriol.* **178**, 6227–6232 (1996).

158. S. Takenaka, S. Murakami, R. Shinke, K. Hatakeyama, H. Yukawa, and K. Aoki, *J. Biol. Chem.* **272**, 14727–14732 (1997).

159. S. Takenaka, S. Murakami, R. Shinke, and K. Aoki, *Arch. Microbiol.* **170**, 132–137 (1998).

160. J. Beunink and H. J. Rehm, *Appl. Microbiol. Biotech.* **31**, 108–115 (1990).

161. J. D. Talati, G. A. Patel, and B. P. Patel, *Br. Corros. J.* **15**, 85–88 (1980).

162. U.S. Pat. 4,425,138 (Jan. 10, 1984), K. E. Davis (to Lubrizol Corp.).

163. M. Terashima, M. Ishi, and Y. Kanaoka, *Synthesis* **6**, 484–485 (1982).

164. Jpn. Pat. 61-126067 (June 13, 1986), H. Wakatsuka, T. Miyamoto, and Y. Arai (to Ono Pharmaceutical Co., Ltd.).

165. C. S. P. Sastry and K. V. S. S. Murty, *J. Indian Chem. Soc.* **60**, 72–75 (1983).

166. T. D. Alizade, G. A. Gamidzade, R. Y. Abdullaeva, and G. F. Askerov, *Nauchn. Tr. Azerb. Un-t. Ser. Khim. N.*, 18–22 (1979).

167. Eur. Pat. 48,222 (Mar. 24, 1982), H. O. Wirth, J. Buessing, and H. H. Friedrich (to CIBA-GEIGY AG).

168. U.S. Pat. 4,297,098 (Oct. 27, 1981), G. F. Dasher and T. J. Schamper (to Alberto-Culver Co.).

169. Ger. Pat. 3,421,694 (Dec. 13, 1984), A. Bugaut and A. Junino (to Oreal S.A.).

170. Jpn. Pat. 81,201,62 (May 12, 1981) (to Matsushita Electric Works Ltd.).

171. *Colour Index*, 3rd ed., The Society of Dyers and Colorists, Bradford, England, Vol. 4, 1971, pp. 4001–4863 and Vol. 6, 1975, pp. 6391–6410.

172. *Beilstein's* Handbuch der Organischen Chemie, Julius Springer, Berlin, 1918, Section 13, pp. 354–549 and Section 13 (2), pp. 164–308.

173. H. Hofer and F. Jacob, *Ber. Dtsch. Chem. Ges.* **41**, 3187–3199 (1908).

174. W. W. Hartmann and H. L. Siloway, *Org. Synth.* **25**, 5–7 (1945).

175. *P.B. Report 74,197*, U.S. Department Commission Office Technology Service, Washington, D.C., 1947, pp. 695–697.

176. M. P. Slyusar, *Nauchn. Tr. Ukr. Nauchno Issed. Inst. Gig. Tr. Profzabol.* **27**, 103 (1958).

177. F. Chen, S. Oikawa, Y. Hiraku, M. Murata, N. Yamashita, and S. Kawanishi, *Cancer Lett.* **126**, 67–74 (1998).

178. *P.B. Report 85,172*. U.S. Department Commission Office Technology Service, Washington, D.C., 1948 p. 242.

179. G. Egerer, *J. Biol. Chem.* **35**, 565–566 (1918).

180. S. Kubota, K. Nara, and S. Onishi, *Yakugaku Zasshi* **76**, 801 (1956).

181. H. Hubner, *Justus Liebigs Ann. Chem.* **210**, 328–396 (1881).

182. Z. C. Heng, J. Nath, X. Liu, and T. M. Ong, *Teratogen. Carcinogen. Mutagen.* **16**, 81 (1996).

183. Z. C. Heng, T. Ong, and J. Nath, *Mutat. Res.* **368**, 149–155 (1996).

184. J. Meyer, *Helv. Chim. Acta* **41**, 1890–1891 (1958).

185. L. Katz and M. S. Cohen, *J. Org. Chem.* **19**, 758–766 (1954).

186. L. Gattermann, *Ber. Dtsch. Chem. Ges.* **26**, 1844–1856 (1893).

187. E. A. Braude, R. P. Linstead, and K. R. H. Wooldridge, *J. Chem. Soc.* 3586–3595 (1954).

188. H. V. K. Udupa, G. Subramanian, and K. S. Udupa, *Res. Ind.* **5**, 309–310 (1960).

189. U.S. Pat. 2,525,515 (Oct. 10, 1950), F. R. Bean (to Eastman Kodak Co.).

190. U.S. Pat. 4,279,767 (June 21, 1981), J. A. Muccitelli (to Betz Laboratories, Inc.).

191. G. W. Raiziss and J. L. Gavron, *J. Am. Chem. Soc.* **43**, 582–585 (1921).

192. G. W. Raiziss and B. C. Fisher, *J. Am. Chem. Soc.* **48**, 1323–1327 (1926).

193. Ger. Pat. 121,683 (1901) (to Badische Anilin-und Soda-Fabrik).

194. Ger. Pat. 44,792 (1888) (to Gesellschaft fr Chemische Industrie).

195. M. L. Crossley, P. F. Dreisbach, C. M. Hoffmann, and R. P. Parker, *J. Am. Chem. Soc.* **74**, 573–578 (1952).

196. Ger. Pat. 48,151 (1889) (to Badische Anilin-und Soda-Fabrik).

197. U.S. Pat. 2,376,112 (May 15, 1945), F. R. Bean and T. S. Donovan (to Eastman Kodak Co.).

198. A. Calm, *Ber. Dtsch. Chem. Ges.* **16**, 2786–2814 (1883).

199. J. T. Sheenan, *J. Am. Chem. Soc.* **70**, 1665–1666 (1948).

200. H. Erlenmeyer, B. Prijs, E. Sorkin, and E. Suter, *Helv. Chim. Acta* **31**, 988–992 (1948).

201. U.S. Pat. 1,993,253 (Mar. 5, 1934), C. H. W. Whitaker (to Industrial Dyestuffs Co.).

202. U.S. Pat. 2,101,749 (Dec. 7, 1937), S. Norman (to Industrial Dyestuffs Co.).

203. R. N. Harger, *J. Am. Chem. Soc.* **41**, 270–276 (1919).

204. J. DeJonge and R. Dijsktra, *Rec. Trav. Chim. Pays-Bas* **68**, 426–429 (1949).

205. F. G. Bordwell and P. J. Boutan, *J. Am. Chem. Soc.* **78**, 87–91 (1956).

206. P. Jansco, L. Szinicz, and P. Eyer, *Arch. Toxicol.* **47**, 39–45 (1981).
207. J. E. Bright and T. C. Mars, *Arch. Toxicol.* **50**, 57–64 (1982).
208. H. Fierz-David and W. Kuster, *Helv. Chim. Acta* **22**, 82–112 (1939).
209. A. L. LeRosen and E. D. Smith, *J. Am. Chem. Soc.* **70**, 2705–2709 (1948).
210. V. R. Olsen and H. B. Feldman, *J. Am. Chem. Soc.* **59**, 2003–2005 (1937).
211. H. N. Morse, *Ber. Dtsch. Chem. Ges.* **11**, 232–233 (1878).
212. J. H. Burckhalter and co-workers, *J. Am. Chem. Soc.* **70**, 1363 (1948).
213. D. E. Pearson, K. Carter, and C. M. Greer, *J. Am. Chem. Soc.* **75**, 5905–5908 (1953).
214. U.S. Pat. 2,887,513 (May 19, 1959), C. M. Eaker and J. R. Campbell (to Monsanto Chemicals Co.).
215. P. K. Smith, *Acetophenetidin*, a monograph, Interscience, New York, 1958.
216. H. Vater, *J. Prakt. Chem.* **29(NF)**, 286–299 (1884).
217. R. Medola, H. S. Foster, and R. Brightman, *J. Chem. Soc.* **111**, 551–553 (1917).
218. L. Galatis, *Helv. Chim. Acta* **4**, 574–579 (1921).
219. B. N. Li, T. Kemeny, and J. Sos, *Acta Med. Acad. Sci. Hung.* **19**, suppl. 19, 111 (1963).

GENERAL REFERENCES

S. Coffey, ed., *Rodd's* Chemistry of Carbon Compounds, 2nd ed., Vol. 3A, Elsevier Publishing Co., Amsterdam, The Netherlands, 1971, pp. 352–363.

Colour Index, 3rd ed., Society of Dyers and Colorists, Bradford, England, Vol. 4, 1971, pp. 4001–4863 and Vol. 6, 1975, pp. 6391–6410.

Beilstein's Handbuch der Organischen Chemie, Julius Springer, Berlin, 1918, Section 13, pp. 354–549 and Section 13(2), pp. 164–308.

STEPHEN C. MITCHELL
PAUL CARMICHAEL
Imperial College London

ROSEMARY WARING
University of Birmingham

AMMONIA

1. Introduction

Ammonia [7664-41-7], NH_3, a colorless alkaline gas, is lighter than air and possesses a unique, penetrating odor. The preparation of ammonium salts dates back to the early Egyptians in the fourth century BC. Ammonia gas was first produced as a pure compound by Priestly in 1774.

In 1840, Justus von Liebig outlined the theoretical principles of plant nutrition and the role of fertilizers as essential plant nutrients. The second half of the century saw increasing use of fertilizers, however nitrogen sources were limited. Ammonium sulfate (a by product of coke ovens), Chilean saltpeter, Peruvian guano, crop rotations and other natural nitrogen sources were the main nitrogen fertilizers during this time.

The first nitric acid production facility based on the Ostwald process of oxidation of ammonia came on stream in 1906 at Gerthe, Germany. At the time, ammonia was still difficult to obtain, so the Ostwald process had to compete with other methods such as acidification of Chilean saltpeter and the electric arc process. During this same time period, the chemistry for production of synthetic ammonia by fixing atmospheric nitrogen was being worked out by Haber. Later in 1913, the first industrial ammonia process using the Haber-Bosch process was installed as a 30 metric ton/day plant at BASF's site in Oppau, Germany. A second ammonia plant went into operation in 1917 Leuna, which produced 350 metric ton/day by the end of World War I. The availability of low cost ammonia secured the long term viability of nitric acid production by oxidation of ammonia.

Both Haber and Bosch won the Nobel Prize for their work on ammonia. The development of the nitrogen fertilizer industry is a prime example of how people have technology to improve living standards. It is estimated that today as much as 40% of human's protein needs are derived indirectly from atmospheric nitrogen fixed by the Haber-Bosch process and its successors (1).

However, technology can also be used to the detriment of humankind. The ammonia and nitric acid technology enabled Germany to produce massive quantities of explosives during World War I even though the country was cut off from their traditional Chilean saltpeter nitrogen source. The Allies acquired the ammonia synthesis technology soon after World War I ended and thereafter began a swift expansion of world production. Today nearly 100×10^6 metric tons of synthetic ammonia and associated nitrogen chemicals are produced each year.

2. Occurrence

Ammonia occurs naturally throughout the universe. For example, ammonia is a minor component of the atmospheres of Jupiter and Saturn. On Earth, natural ammonia deposits contain salts rather than the base form. For example, ammonium chloride, [12125-02-9], NH_4Cl, deposits are found near volcanoes such as Vesuvius in Italy.

Nitrogen, in various forms, is an important element to life. It is a building block for amino and nucleic acids. Bound nitrogen cycles through the biosphere among plants, animals and micro-organisms. The ultimate source of all this nitrogen is the Earth's atmosphere. Atmospheric nitrogen can only be fixed to ammonia by a relatively few number of species of micro-organisms. Some are free living, such as the cyanobacteria; others are symbiotic with certain plants such as legumes.

Many important crops tend to take up nitrogen mainly in the nitrate form. This form of nitrogen is produced by soil organisms from various bound nitrogen sources, or is intentionally applied as a nitrate fertilizer. Unfortunately the nitrate ion is prone to groundwater leaching. Nitrate contamination of groundwater became a significant environmental issue in the 1980s. Improved farm management techniques have been the main tools used to deal with this issue.

Table 1. **Physical Properties of Anhydrous Ammonia**

Property	Value
molecular weight	17.03
boiling point, °C	−33.35
freezing point, °C	−77.7
critical temp, °C	133.0
critical pressure, kPa[a]	11,425
specific heat, J/(kg·K)[b]	
0°C	2097.2
100°C	2226.2
200°C	2105.6
heat of formation of gas, ΔH_f, kJ/mol[b]	
0 K	−39,222
298 K	−46,222
solubility in water, wt %	
0°C	42.8
20°C	33.1
40°C	23.4
60°C	14.1
specific gravity	
−40°C	0.690
0°C	0.639
40°C	0.580

[a] To convert kPa to psi, multiply by 0.145.
[b] To convert J to cal, divide by 4.184.

3. Physical Properties

Table 1 lists the important physical properties of ammonia; Table 2 gives the densities at 15°C; Figure 1 is a Mollier diagram giving additional thermodynamic data for ammonia. The flammable limits of ammonia in air are 16 to 25% by volume; in oxygen the range is 15 to 79%. Such mixtures can explode although ammonia–air mixtures are quite difficult to ignite. The ignition temperature is about 650°C.

Ammonia is readily absorbed in water to make ammonia liquor. Figure 2 summarizes the vapor–liquid equilibria of aqueous ammonia solutions and Figure 3 shows the solution vapor pressures. Additional thermodynamic properties may be found in the literature (2,3). Considerable heat is evolved during the solution of ammonia in water: approximately 2180 kJ (520 kcal) of heat is evolved upon the dissolution of 1 kg of ammonia gas.

Table 2. **Density of Aqueous Ammonia at 15°C**

Ammonia, wt %	Density, g/L
8	0.970
16	0.947
32	0.889
50	0.832
75	0.733
100	0.618

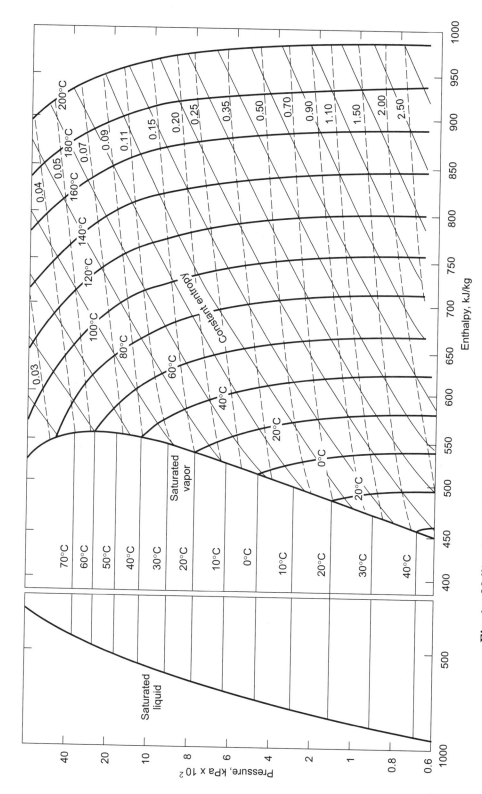

Fig. 1. Mollier diagram for ammonia. Numbers on dashed lines represent specific volume values in m³/kg. To convert from kPa to psi, multiply by 0.145. To convert kJ to kcal, divide by 4.184. Courtesy of Elliot Company.

681

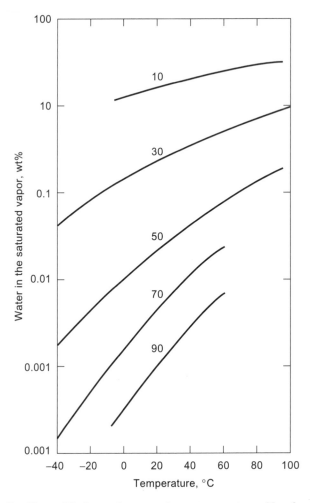

Fig. 2. Vapor–liquid equilibrium of ammonia–water system. Numbers represent the weight percent of ammonia in the liquid (2).

Ammonia is an excellent solvent for salts, and has an exceptional capacity to ionize electrolytes. The alkali metals and alkaline earth metals (except beryllium) are readily soluble in ammonia. Iodine, sulfur, and phosphorus dissolve in ammonia. In the presence of oxygen, copper is readily attacked by ammonia. Potassium, silver, and uranium are only slightly soluble. Both ammonium and beryllium chloride are very soluble, whereas most other metallic chlorides are slightly soluble or insoluble. Bromides are in general more soluble in ammonia than chlorides, and most of the iodides are more or less soluble. Oxides, fluorides, hydroxides, sulfates, sulfites, and carbonates are insoluble. Nitrates (eg, ammonium nitrate) and urea are soluble in both anhydrous and aqueous ammonia making the production of certain types of fertilizer nitrogen solutions possible. Many organic compounds such as amines (qv), nitro compounds, and aromatic sulfonic acids, also dissolve in liquid ammonia. Ammonia is superior to water

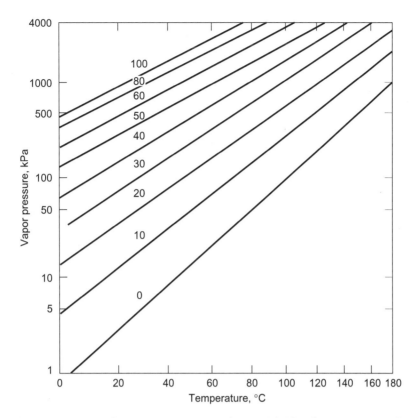

Fig. 3. Vapor pressure of aqueous ammonia solution (2). Numbers represent the weight percent of ammonia in the liquid. To convert kPa to psi, multiply by 0.145.

in solvating organic compounds such as benzene (qv), carbon tetrachloride, and hexane.

4. Chemical Properties

Ammonia is comparatively stable at ordinary temperatures, but decomposes into hydrogen and nitrogen at elevated temperatures. The rate of decomposition is greatly affected by the nature of the surfaces with which the gas comes into contact: glass is very inactive; porcelain and pumice have a distinct accelerating effect; and metals such as iron, nickel, osmium, zinc, and uranium have even more of an effect. At atmospheric pressure, decomposition begins at about 450–500°C, whereas in the presence of catalysts, it begins as low as 300°C and is nearly complete at 500–600°C. At 1000°C, however, a trace of ammonia remains. Ammonia decomposition, a source of high purity hydrogen and nitrogen for use in metals processing, can also be promoted electrically or photochemically.

Ammonia reacts readily with a large variety of substances (see AMMONIUM COMPOUNDS; AMINES BY REDUCTION; AMINES). Oxidation at a high temperature is one of the more important reactions, giving nitrogen and water. Gaseous ammonia is oxidized to water and nitrogen when heated to a relatively high temperature in the presence of oxides of the less positive metals, such as cupric oxide. Powerful oxidizing agents, eg, potassium permanganate, react similarly at ordinary temperatures.

$$2\,NH_3 + 2\,KMnO_4 \longrightarrow 2\,KOH + 2\,MnO_2 + 2\,H_2O + N_2 \tag{1}$$

The action of chlorine on ammonia can also be regarded as an oxidation reaction.

$$8\,NH_3 + 3\,Cl_2 \longrightarrow N_2 + 6\,NH_4Cl \tag{2}$$

A major step in the production of nitric acid [7697-37-2] (qv) is the catalytic oxidation of ammonia to nitric acid and water. Very short contact times on a platinum–rhodium catalyst at temperatures above 650°C are required.

$$4\,NH_3 + 5\,O_2 \longrightarrow 4\,NO + 6\,H_2O \tag{3}$$

$$2\,NO + O_2 \longrightarrow 2\,NO_2 \tag{4}$$

$$3\,NO_2 + H_2O \longrightarrow 2\,HNO_3 + NO \tag{5}$$

The neutralization of acids is of commercial importance. Three principal fertilizers, ammonium nitrate [6484-52-2], NH_4NO_3, ammonium sulfate [7782-20-2], $(NH_4)_2SO_4$, and ammonium phosphate [10361-65-6], $(NH_4)_3PO_4$, are made by reaction of the respective acids with ammonia.

The reaction between ammonia and water is reversible.

$$NH_3 + H_2O \rightleftharpoons NH_4^+ + OH^- \tag{6}$$

The aqueous solubility of ammonia decreases rapidly as temperature increases. The existence of undissociated ammonium hydroxide [1336-21-6], NH_4OH, in aqueous solution is doubtful although there are indications that ammonia exists in water in the form of the hydrates NH_3H_2O and $2\,NH_3 \cdot H_2O$. The ammonium ion, NH_4^+, behaves similarly to the alkali metal cations. Ammonia, a comparatively weak base, however, ionizes in water to a much lesser extent than sodium hydroxide. In a molar solution of aqueous ammonia, the concentration of the hydroxyl ion is about two-hundredths that of the hydroxyl ion concentration in a molar sodium hydroxide solution.

Aqueous ammonia also acts as a base precipitating metallic hydroxides from solutions of their salts, and in forming complex ions in the presence of excess ammonia. For example, using copper sulfate solution, cupric hydroxide, which is at first precipitated, redissolves in excess ammonia because of the formation of the complex tetramminecopper(II) ion.

$$CuSO_4 + 2\,NH_3 \cdot H_2O \longrightarrow Cu(OH)_2 + (NH_4)_2SO_4 \tag{7}$$

$$Cu(OH)_2 \longrightarrow Cu^{2+} + 2\,OH^- \tag{8}$$

$$4\,NH_3 + Cu^{2+} \rightleftharpoons \left[Cu(NH_3)_4\right]^{2+} \tag{9}$$

Potassium dissolves in liquid ammonia, but the conversion of a small amount of the metallic potassium to the metallic amide takes several days. By applying the same technique using sodium metal, sodium amide [7782-92-5], $NaNH_2$, a white solid, can be formed.

$$2\,Na + 2\,NH_3 \longrightarrow 2\,NaNH_2 + H_2 \tag{10}$$

Heating metallic lithium in a stream of gaseous ammonia gives lithium amide [7782-89-0] $LiNH_2$, which may also be prepared from liquid ammonia and lithium in the presence of platinum black. Amides of the alkali metals can be prepared by double-decomposition reactions in liquid ammonia. For example

$$NaI + KNH_2 \xrightarrow{\text{liq } NH_3} NaNH_2 + KI \tag{11}$$

Heating ammonia with a reactive metal, such as magnesium, gives the nitride.

$$3\,Mg + 2\,NH_3 \longrightarrow Mg_3N_2 + 3\,H_2 \tag{12}$$

Magnesium reacts slowly at lower temperatures to give the amide, as do all active metals; this reaction is catalyzed by transition metal ions. Aluminum nitride [24304-00-5], AlN, barium nitride [12047-79-9], Ba_3N_2, calcium nitride [12013-82-0], Ca_3N_2, strontium nitride [12033-82-8], Sr_3N_2, and titanium nitride [25583-20-4], TiN, may be formed by heating the corresponding amides.

Halogens react with ammonia. Chlorine orbromine liberate nitrogen from excess ammonia and give the corresponding ammonium salt. Substitution probably takes place first. The resulting trihalide combines loosely with another molecule of ammonia, to give $NCl_3 \cdot NH_3$ for example. These ammoniates are very unstable and decompose in the presence of excess ammonia to give the ammonium salt and nitrogen.

$$NCl_3 \cdot NH_3 + 3\,NH_3 \longrightarrow N_2 + 3\,NH_4Cl \tag{13}$$

The iodine compound is more stable and separates as so-called nitrogen triiodide monoammoniate [14014-86-9], $NI_3 \cdot NH_3$, an insoluble brownish-black solid, which decomposes when exposed to light in the presence of ammonia. In reactions of the halogens with the respective ammonium salts, however, the action is different. Chlorine replaces hydrogen and nitrogen chloride [10025-85-1], NCl_3, separates as oily, yellow droplets capable of spontaneous explosive decomposition.

$$NH_4Cl + 3\,Cl_2 \longrightarrow NCl_3 + 4\,HCl \tag{14}$$

The hydrogen of the ammonium salt is not replaced by bromine and iodine. These elements combine with the salt to form perhalides.

$$NH_4Br + Br_2 \longrightarrow NH_4Br_3 \tag{15}$$

A number of perhalides are known, and one of the most stable is ammonium tetrachloroiodide [19702-43-3], NH_4ICl_4. Ammonia reacts with chlorine in dilute solution to give chloramines, a reaction important in water purification (see CHLORAMINES AND BROMAMINES). Depending upon the pH of the water, either monochloramine [10599-90-3], NH_2Cl, ordichloramine [3400-09-7], $NHCl_2$, is formed. In the dilutions encountered in waterworks practice, monochloramine is nearly always found, except in the case of very acidic water (see BLEACHING AGENTS; WATER).

Ammonia reacts with phosphorus vapor at red heat to give nitrogen and phosphine.

$$2\,NH_3 + 2\,P \longrightarrow 2\,PH_3 + N_2 \tag{16}$$

Sulfur vapor and ammonia react to give ammonium sulfide and nitrogen; sulfur and liquid anhydrous ammonia react to produce nitrogen sulfide [28950-34-7], N_4S_4.

$$10\,S + 4\,NH_3 \longrightarrow 6\,H_2S + N_4S_4 \tag{17}$$

Ammonia and carbon at red heat giveammonium cyanide [12211-52-8], NH_4CN.

Ammonia forms a great variety of addition or coordination compounds (qv), also called ammoniates, in analogy with hydrates. Thus $CaCl_2{\cdot}6NH_3$ and $CuSO_4{\cdot}4NH_3$ are comparable to $CaCl_2{\cdot}6H_2O$ and $CuSO_4{\cdot}4H_2O$, respectively, and, when regarded as coordination compounds, are called ammines and written as complexes, eg, $[Cu(NH_3)_4]SO_4$. The solubility in water of such compounds is often quite different from the solubility of the parent salts. For example, silver chloride, AgCl, is almost insoluble in water, whereas $[Ag(NH_3)_2]Cl$ is readily soluble. Thus silver chloride dissolves in aqueous ammonia. Similar reactions take place with other water insoluble silver and copper salts. Many ammines can be obtained in a crystalline form, particularly those of cobalt, chromium, andplatinum.

Of major industrial importance is the reaction of ammonia and carbon dioxide giving ammonium carbamate [1111-78-0], $CH_6N_2O_2$.

$$2\,NH_3 + CO_2 \longrightarrow NH_2CO_2NH_4 \tag{18}$$

which then decomposes to urea (qv) and water

$$NH_2CO_2NH_4 \longrightarrow NH_2CONH_2 + H_2O \tag{19}$$

This is an example of an ammonolytic reaction in which a chemical bond is broken by the addition of ammonia. It is analogous to the hydrolysis reactions of water. An impressive number of inorganic and organic compounds undergo ammonolysis.

5. Source and Supplies

Ammonia is a world class commodity. It is manufactured in over 80 countries. Global ammonia production capacities are given in Table 3. China and the former USSR are currently the largest producers. Table 4 lists U.S. ammonia production capacities by company.

Table 3. **Estimated 1998 Global Ammonia Capacity By Region**[a]

Region	Capacity, 1000 t/yr	Share, %
United States	18,465	15
China	27,082	22
former USSR	20,927	17
South Asia	14,772	12
Western Europe	11,079	9
Near East	8,617	7
Central Europe	7,386	6
Indonesia and Japan	4,924	4
Mexico and Caribbean	3,693	3
other	6,155	5
Total	*123,100*	*100*

[a] Ref. 1.

Table 4. **Table 1999 U.S. Ammonia Producers and Capacities**[a]

Type	Company	Locations	Capacity, 1000 t/yr
Exisiting:	Agrium US	Borger, TX	430
	Air Products	Pace, FL	90
	Allied Signal	Hopewell, VA	535
	Avondale Ammonia	Waggaman, LA	400
	Borden	Geismar, LA	370
	CF Industries	Donaldson, LA	2,455
	Chevron	El Segundo, CA; Richmond, CA	35
	Coastal	Cheyenne, WY; Freeport, TX; St. Helens, OR	1,105
	Dakota Gasification	Beulah, ND	330
	DuPont	Beaumont, TX	430
	Farmland Industries	Beatrice, NE; Dodge City, KS; Enid, OK; Ft. Dodge IA; Lawrence, KS; Pollock, LA	2,230
	Green Valley	Creston, IA	50
	IMC	Faustina, LA	510
	Koch Nitrogen	Sterlington, LA	1,100
	LaRoche	Cherokee, AL	155
	Mississippi Chemical	Yazoo City, MS	645
	PCs	Augusta, GA; Geismar, LA; Lima, OH; Memphis, TN	2,500
	Royster-Clark	East Dubuque, IL	275
	J.R. Simplot	Pocatello, ID	110
	Terra	Blytheville, AR; Sergeant Bluff, IA; Verdigris, OK; Woodward, OK	2,060
	Triad Nitrogen	Donaldson, LA	925
	Ultramar Diamond Shamrock	Dumas, TX	135
	Unocal	Finely, WA; Kenai, AK	1,335
	Wil-Gro	Pryor, OK	255
		Total exisiting	*18,465*
Announced:	Farmland Industries	Coffeyville, KS	600
	Terra	Beaumont, TX	280
		Total announced	*880*

[a] Ref. 4

The current global situation is one of overcapacity and weak prices. The past few years have seen several U.S. plant closures and temporary shutdowns. Many of the plants in the United States were built in the 1960s and 1970s. Although U.S. natural gas prices are high compared to natural gas prices in remote regions, U.S. plants are often able to compete with imports on a cash basis since the original plant investment has been fully depreciated. Many U.S. plants have been modernized to upgrade technology and to expand production beyond their original nameplate capacity. However, low cost imports and soft demand for nitrogen fertilizers are expected to plague the U.S. ammonia industry.

In order to compete with existing facilities, new complexes must use modern technologies and have access to low cost feedstock. Usually new plants are built at world scale capacities to enjoy economies of scale. The trend of the past 10–20 years has been that new ammonia plants are built primarily in regions such as Trinidad, the Middle East, North Africa, etc, where large quantities of low cost natural gas are available. Countries in these areas often view a nitrogen fertilizer plant as a way to monetize remote natural gas which otherwise would have little value due to the lack of natural gas infrastructure and local markets.

6. Manufacture

The ammonia synthesis reaction is deceptively simple: nitrogen is combined with hydrogen in a 1:3 stoichiometric ratio to give ammonia with no by-products. The difficulty lies in how to obtain the hydrogen needed for the reaction. As shown in Figure 4, the hydrogen production method is the main source of distinction between the various ammonia production routes. In fact, the majority of the equipment in a typical ammonia plant is devoted to hydrogen production rather than ammonia synthesis. Most of the improvements in technology over the past 90 years have concerned the hydrogen production step.

The bulk of world ammonia production is based on steam reforming. Any hydrocarbon feed that can be completely vaporized can be used. Natural gas and naphtha are common. The bulk of existing and nearly all new steam reforming facilities use natural gas for feedstock. Economics and availability influence the decision on feedstock and production technology. The recent trend towards using ammonia and other synthesis gas products to monetize remote natural gas suggests that steam reforming of natural gas will maintain its dominant position in the foreseeable future.

The exact technology used within a facility depends upon the age of the plant, economics, and the overall fit of the ammonia facility within the entire nitrogen or syngas complex. A standard steam methane reforming plant with conventional gas purification is described in some detail. Other technology routes based on partial oxidation, water electrolysis plus predictions on future technologies are also briefly discussed.

6.1. Steam Reforming With Conventional Gas Purification. Today's prevailing plant design is the single-train plant in which all of the large equipment and machinery are single units. This gives lower capital costs but

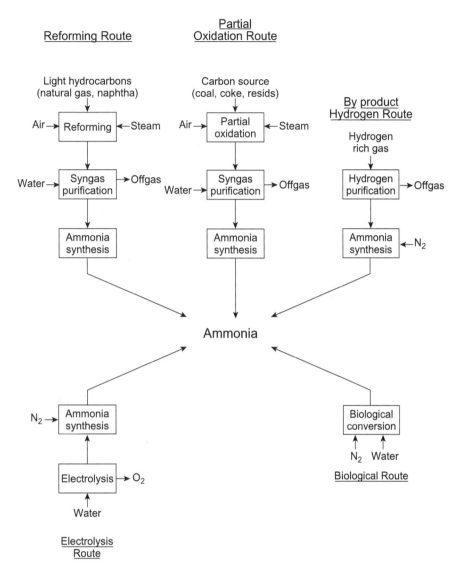

Fig. 4. Ammonia production routes.

reduces the ability to operate the plant at reduced rates during failures of major pieces of equipment. However, attention to equipment reliability issues and preventative maintenance has resulted in on-line factors greater than 95%. Reference 5 reports on the causes of unscheduled downtime for four large ammonia plants.

Figure 5 is a flow diagram for a steam reforming plant. The main steps are (1) feed preparation where the natural gas is compressed to reforming pressures and desulfurized to protect downstream catalysts, (2) reforming where the natural gas is mixed with steam and air to produce a crude syngas mixture, (3) Syngas clean-up where a purified stream of nitrogen and hydrogen is produced in the

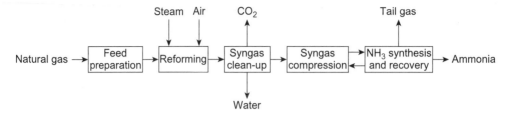

Fig. 5. Block flow diagram for ammonia production by steam reforming of methane.

correct stoichiometric ratio, (4) Syngas compression followed by (5) ammonia synthesis and recovery.

Figure 6 is a flow sheet for a Kellogg high capacity single-train ammonia plant using conventional technologies. Although these diagrams represent only one of many possible flow sheet variations, Figures 5 and 6 are used to illustrate a typical ammonia plant. Common variations are also discussed.

Feed Preparation. Steam reforming of methane is usually conducted at ca 3–4 MPa (30–40 atm). This pressure is a compromise leading to a minimum capital and/or operating cost for the overall process while satisfying mechanical constraints of the process equipment.

Usually downstream ammonia synthesis occurs at a much higher pressure than what is allowed by the mechanical limits of the steam reforming furnace, so the natural gas is compressed to reforming pressures. A second compression step occurs later in the flow sheet where the purified nitrogen and hydrogen mixture produced from the reforming operation is further compressed to ammonia synthesis pressure.

Conceptually, this two-step compression sequence is not very satisfactory since it leads to higher costs than a single compression step would require. However, current mechanical constraints and ammonia synthesis reaction kinetics force this compromise solution. Compression costs are roughly related to the number of moles processed. The advent of high pressure reforming in the 1960s lead to a significant decline in ammonia production cost since much of the compression was shifted upstream of the reformer where there are fewer moles of gas to process. Much work has been done to lower the ammonia synthesis pressure using higher activity catalysts. The long term goal is to eliminate completely the need for the intermediate syngas compressors and provide all of the process compression upstream of the reformer.

The feed gas compressor may or may not be needed depending upon plant location. The feed compressor is not needed if a high pressure gas transmission line supplies the plant. Much attention has been paid to all of the compressors in ammonia plants (natural gas feed, process air, syngas, and refrigeration) since they strongly influence plant reliability, capital and operating costs, and plant capacity.

Sweet natural gas typically contains 1–5 ppm of sulfur. About half of the sulfur is in the form of H_2S. This is just the residual sulfur left in the gas after sweetening at the gas plant. The other half of the sulfur is organic sulfur from odorants added to help locate pipeline leaks. The odorants may not be

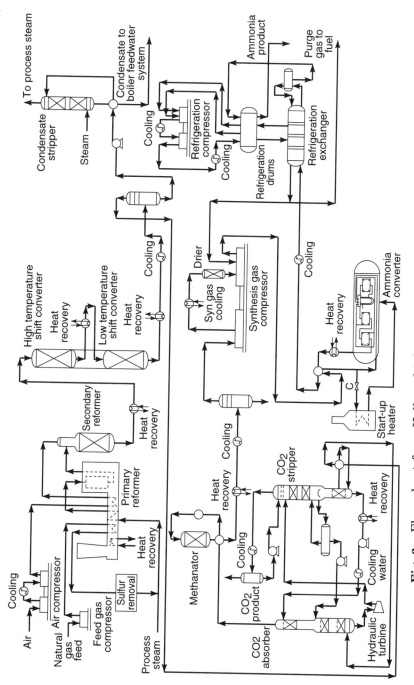

Fig. 6. Flow sheet for a Kellogg high capacity single-train ammonia plant.

present in the gas if the ammonia plant is located in a remote region close to the natural gas source.

The catalysts used in the downstream reactors are very sensitive to sulfur poisoning. Conventional steam reformers are designed for 2–3 years of operation between catalyst changeout. This translates into a total sulfur specification of less than 1 ppm. Even lower levels (ca 0.1 ppm) are required when low temperature shift catalysts are used. Most conventional plants meet these stringent requirements by pretreatment of the feed using a combination of hydrogenation to convert organic sulfur into H_2S followed by absorption of H_2S onto zinc oxide.

In the hydrogenation step, a small slipstream of product hydrogen is recycled back into the feed and the combined feed and recycle hydrogen is passed over a cobaltmolybdenum catalyst. For a mercaptan (or thiol), the reaction is

$$RCH_2SH + H_2 \longrightarrow H_2S + RCH_3 \tag{20}$$

Typically this reaction is done at 350–400°C . The H_2S is then absorbed on a bed of zinc oxide:

$$ZnO + H_2S \longrightarrow ZnS + H_2O \tag{21}$$

The zinc oxide beds are typically not regenerated but are replaced with new absorbent once exhausted.

Fixed bed desulfurization is impractical if the natural gas contains large amounts of sulfur. In this case, bulk sulfur removal and recovery in an acid gas absorption stripping system, followed by fixed bed residual clean up is usually employed.

Chlorides may be found in natural gas, particularly associated with offshore reservoirs. Modified alumina catalysts have been developed to irreversibly absorb these poisons from the feed gas. Some natural gases have also been found to contain mercury, which is a reformer catalyst poison when present in sufficiently large amounts. Activated carbon beds impregnated with sulfur have been found to be effective in removing this metal.

Reforming.   The steam reforming reactions for the alkanes present in natural gas are shown in equations 22–24, $\Delta H°$ values are at 25°C in kJ/kgmol; $\Delta S°$ values are at 25°C in kJ/K kgmol

$$CH_4 + 2\,H_2O \leftrightarrow CO + 3\,H_2 \quad \Delta H°,\,206,172;\,\Delta S°,\,214.6 \tag{22}$$

$$CH_3\,CH_3 + 2\,H_2O \leftrightarrow 2\,CO + 5\,H_2 \quad \Delta H°,\,347,338;\,\Delta S°,\,441.4 \tag{23}$$

$$CH_3\,CH_2\,CH_3 + 3\,H_2O \leftrightarrow 3\,CO + 7\,H_2 \quad \Delta H°,\,497,797;\,\Delta S°,\,671.3 \tag{24}$$

As indicated these are reversible equilibrium type reactions. The reforming reaction is highly endothermic, as shown by the large positive values of $\Delta H°$. Conventional primary steam reformers operate in the gas phase at high temperature (750–850°C) and medium pressures (3–4 MPa). Temperatures and pressures are dictated by chemistry, mechanical limitations and the design philosophy used for compression within the overall process train.

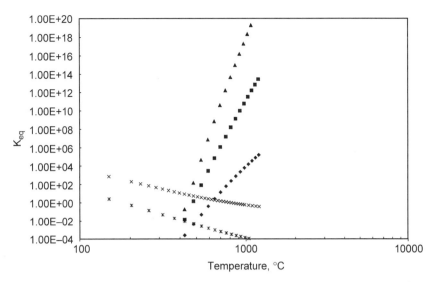

Fig. 7. Effects of temperature on equilibrium constants ◆ methane reforming; ■ ethane reforming; ▲ propane reforming; × shift; ✳ ammonia synthesis.

Temperature has a strong effect on the position of equilibrium for the reforming reactions. The equilibrium constant at any temperature can be approximated by:

$$\ln\left(K_{eq}\right) = \ln\left(\frac{[CO]^n[H_2]^{(2n+1)}}{[C_n\,H_{(2n+2)}][H_2O]^n}\right) = -\frac{\Delta H^\circ - T\Delta S^\circ}{RT}$$

where K_{eq} is the equilibrium constant, the terms in bracket represent concentrations, n is the number of carbon atoms in the hydrocarbon feed, and R is the gas constant. Figure 7 shows that higher temperatures lead to larger equilibrium constants. This means that higher temperatures favor hydrogen production. The figure also shows that at a given temperature, the equilibrium constant is larger as the molecular weight of the hydrocarbon feed increases. In fact, at conditions typically used for conventional primary steam reformers, about 95% of the methane is converted while nearly 100% conversions are obtained for the ethane and higher hydrocarbons in the feed. The values in Figure 7 deviate slightly from the above approximation since corrections were made to account for the effects of temperature on ΔH° and ΔS°.

The effect of pressure can be qualitatively examined using LeChatelier's principle. The left hand side of the reforming reactions contain fewer moles than the right, so a lower system pressure would favor hydrogen production. Operating pressures, however, are usually set by the compression philosophy rather than being based on the process chemistry. Compression costs can be significant and these are related to the number of moles processed. The hydrogen product contains more moles than the feed, so the compression philosophy is usually to operate the reforming section at as high a pressure as allowed by mechanical limitations of the primary reformer. This minimizes the cost associated with the downstream synthesis gas compressor and leads to lower overall costs.

The steam to carbon ratio (S/C ratio) of the reformer feed can also be manipulated. This ratio is defined as the number of moles of steam divided by the number of moles of carbon atoms in the feed. High S/C ratios favor hydrogen production but at the expense of larger equipment and higher utility costs. Low S/C ratios encourage carbon soot formation reactions. Conventional steam reformers are designed for S/C between 2 to 6, with 3 being a typical value.

Decades of R&D have gone into improving reforming catalysts. Commercial reforming catalysts typically contain nickel oxide (15–25 wt%) on a carrier such as α-alumina. Sometimes promoters such as uranium oxide or chromium oxide are also used. The nickel oxide must be reduced to metallic nickel prior to start-up since the catalyst is active only in this form. This requirement also explains why reforming catalyst are sensitive to sulfur poisoning since metallic nickel can react with sulfur compounds to form nickel sulfide. In addition to catalyzing the reforming reaction, commercial catalysts also catalyze the water-gas shift reaction described later, although the conditions required by the reforming reactions are such that the equilibrium for the shift reaction is to the left of its normal position in conventional shift reactors.

Conventional steam reformers use a fired heater to supply high temperature heat to drive the reforming reactions. The heat needed to drive the reforming reactions to a high degree of conversion must be supplied at a high temperature. Usually this is done by passing the feed gas through tubes in the radiant section of the furnace. The tubes are packed with catalyst to improve the reaction rate. The fuel to the reforming furnace is usually natural gas and process tail gases, although other fuel sources could be used. After leaving the radiant section of the furnace, the combustion gases still contain a large amount of available heat energy that can be recovered by producing high pressure steam. This high pressure steam can either be exported to other units in the facilities, used directly for process gas compression with steam driven turbines, or used for electricity production in a Rankine or similar cycle.

Preforming (not shown in Fig. 6) is another potential use for the heat available in the convection section of the furnace. In this scheme, the natural gas and steam mixture is preheated in the convection section of the reforming furnace and then past through an external adiabatic bed of reforming catalyst. The temperatures in preforming are not high enough to drive the reforming reactions to high conversion, thus the effluent from a preformer is usually further heated in the furnace convection section and then routed to the radiant section for primary reforming. The main advantage of preforming is that all of the ethane and higher hydrocarbons in the feed will be reformed prior to reaching the primary reformer. This results in more even operation of the reformer furnace in the face of feed composition changes. Preforming also allows for lower S/C ratios since coking is less likely to occur. Preforming uses a portion of the available heat in the convection section to drive the reforming reaction. Whether it is more cost effective to use the convection section heat for preforming or for steam/electricity production depends on the specific economic details for the plant.

The product gases from the primary reforming step contain about 5–8% unreacted methane. The gases are usually sent to a secondary reformer to complete the reforming reactions. A secondary reformer consists of a burner in a refractory lined vessel and another reforming catalyst bed. In primary reforming, the high temperature energy absorbed by the reaction is supplied indirectly by a fired heater that uses natural gas, fuel oil, or tail gases as the fuel. In secondary reforming, a portion of the feed is combusted in an oxygen starved environment. Burner exit temperatures are on the order of 1200°C. The hot gases are then passed over reforming catalysts and the temperature drops to ca 1000°C due to the endothermic heat of reaction. At this temperature, the methane content of the syngas has been reduced to ca 0.2–0.3%. The heat released during combustion is used to drive the reforming reactions. From a chemistry point of view, primary and secondary reforming are closely related with the main difference being the source of high temperature heat used to the drive the reforming reaction.

The combustion reactions for natural gas components are hown in equations 25–29. The $\Delta H°$ values are at 25°C in kJ/kgmol ard the $\Delta S°$ values at 25°C are in kJ/(K kgmol)

$$CH_4 + 3/2\,O_2 \;\leftrightarrow\; CO + 2\,H_2O \quad \Delta H°,\; -519,306\;\Delta S°,\; 81.3 \tag{25}$$

$$CH_3\,CH_3 + 5/2\,O_2 \;\leftrightarrow\; 2\,CO + 3\,H_2O \quad \Delta H°,\; -861,792\;\Delta S°,\; 219.3 \tag{26}$$

$$CH_3\,CH_2\,CH_3 + 7/2\,O_2 \;\leftrightarrow\; 3\,CO + 4\,H_2O \quad \Delta H°,\; -1,194,985\;\Delta S°,\; 360.4 \tag{27}$$

$$CO + 1/2\,O_2 \;\leftrightarrow\; CO_2 \quad \Delta H°,\; -282,995\;\Delta S°,\; -86.4 \tag{28}$$

$$H_2 + 1/2\,O_2 \;\leftrightarrow\; H_2O \quad \Delta H°,\; -241,826\;\Delta S°,\; -44.4 \tag{29}$$

Reactions 25–29 produce carbon monoxide, reaction 27 completes the oxidation to carbon dioxide, and reaction 29 is for the oxidation of hydrogen. The thermodynamic functions are evaluated for gaseous water. These reactions are irreversible and strongly exothermic as indicated by the large negative values for $\Delta H°$.

The oxygen used in the partial oxidation route can be supplied from compressed air, or from enriched oxygen steams obtained from membrane units, vacuum swing adsorption units or cryogenic air separation plants. The source of the oxygen used depends upon economics, process constrains, and specifications on hydrogen product purity. Frequently compressed air is used with the amount used being set by the nitrogen stoichiometry for ammonia synthesis.

It is possible to eliminate completely the primary reforming step and do all of the reforming in a special secondary reformer design called an autothermal reformer. The energy balance usually requires than either an oxygen enriched air source is needed or nitrogen in excess of that needed for ammonia synthesis has to be removed in the downstream syngas clean-up step.

Syngas Clean-up. The reformer effluent gas contains the desired hydrogen and nitrogen plus bulk impurities such as carbon monoxide, carbon dioxide, water, residual methane and trace impurities such as argon and helium. The bulk impurities have to be removed prior to ammonia synthesis. Oxygen containing compounds are serious catalyst poisons for the ammonia synthesis catalyst. A 10 ppm total is a typical maximum specification allowed for all oxygen containing compounds in the ammonia synthesis reactor feed. In addition, the economics for removal of trace inerts from the synthesis gas are sometimes justified since this reduces and potentially eliminates the need to purge the ammonia synthesis gas loop.

As shown in Figure 6, most conventional ammonia plants use the following combination of steps for syngas clean-up: high temperature and low temperature shift conversion to convert the bulk of the carbon monoxide into carbon dioxide, bulk water removal by cooling of the shifted gas followed by knockout of the water condensate, bulk carbon dioxide removal using an absorber–stripper system, methanation to convert residual carbon oxides back to essentially inert methane gas followed by desiccant drying to remove trace water.

The shift reactors are adiabatic packed bed reactors used to convert carbon monoxide into carbon dioxide and additional hydrogen. The stoichiometry and thermodynamics of the water-gas shift reaction (eq. 30) are ΔH° at 25°C, kJ/kgmol $= -41,169$; ΔS° at 25°C, kJ/(K kgmol) $= -42.0$.

$$CO + H_2O \leftrightarrow CO_2 + H_2 \tag{30}$$

Like the reforming reactions, the water-gas shift reaction is a reversible equilibrium type reaction. The reaction is exothermic, as indicated by the negative value of ΔH°.

Figure 7 shows the effect of temperature on the equilibrium constant. Lower temperatures shift the reaction to the right. Reforming catalyst will also catalyze the shift reaction, however the position of equilibrium favors the left hand components at the temperatures commonly encountered in reforming. Industrial shift catalysts do not catalyze reforming reactions. Pressure does not influence the thermodynamics of the shift reaction since the number of moles is the same on both sides of the equation. Pressure does, however, affect the kinetics of the shift reaction, with higher pressures leading to faster kinetic rates.

The shift reaction can occur without the presence of a catalyst. Depending upon pressure, temperatures in excess of 900°C will permit the shift reaction to proceed without a catalyst. The high temperatures encountered in partial oxidation and experimental plasma reactors are high enough that the uncatalyzed shift reaction has to be considered.

Decades of R&D have also gone into improving shift catalyst. In clean gas services, the high temperature shift (HT shift) catalysts are typically a chromium promoted iron oxide. The catalyst is supplied in a oxidized state and must be reduced *in-situ* prior to start-up. Like reforming catalysts, iron-based shift catalysts are poisoned by sulfur compounds. Dirty shift catalysts based on cobalt–molybdenum are used in raw gas processes containing elevated sulfur levels. These catalysts actually require sulfur in the feed to maintain their activity. High temperature shift reactions are carried out between 300–530°C. This range is set by the catalyst activity and the desired conversion. Typically, carbon monoxide content is ca 3 vol% after high temperature shift conversion. The reaction is commonly carried out in a packed-bed reactor. Since the reaction is exothermic, temperature rises on the order of 70°C are common. When large amounts of carbon monoxide are to be converted, the bed is broken-up into two or more vessels and intercoolers are used to limit the temperature rise and stay within the operating temperature range for the catalyst.

Low temperature shift (LT shift) catalysts are typically copper–zinc oxide catalysts supported on alumina. The catalyst is active in the temperature range of 200–270°C. The lower temperature results in a lowering of the carbon monoxide effluent to ca 0.2–0.3 vol%. LT shift catalysts are extremely sensitive to sulfur poisoning and feed specifications on sulfur content are lowered to <0.1 ppm sulfur when LT shift catalysts are used. Depending upon the catalyst, a small amount of methanol side product is also formed (6).

After shift, the syngas is cooled with a combination of process cross-exchange and ejection of excess heat to the atmosphere using either an air fin cooler, a cooling water cooled exchanger, or a combination of the two. Cooling the gas to near ambient temperature causes water to condense, which is easily removed from the system using a knockout drum. Steam reforming of methane is a net overall consumer of water. The condensate from the knockout drum is steam stripped to remove dissolved gases and then it is usually recycled back into the boiler feedwater system to reduce the total amount of make-up water required for the unit.

The effluent gases from the shift converters contain about 17–19 vol% (dry basis) carbon dioxide. The levels of carbon dioxide are reduced to the 100–2000 ppm range by bulk CO_2 removal. The mechanisms of CO_2 removal systems can be classified as reaction, combination reaction-physical, and physical absorption systems. The processes generally have an absorber–stripper configuration. Table 5 compares the main features of various CO_2 removal systems.

In many ammonia plants the carbon dioxide by product has value. It can be used as a feed for urea manufacture if a urea unit is part of the nitrogen complex. Carbon dioxide can also be used to adjust the C:H ratio in a methanol or other syngas plant if they are present as part of the industrial complex. Some ammonia plants further process the carbon dioxide to make it suitable for the carbonated beverage industry. Others sell byproduct CO_2 for use in enhanced oil recovery.

Methanation is the simplest method for lowering the residual carbon monoxide and carbon dioxide content down to the levels needed to prevent ammonia synthesis catalyst poisoning. Methanation is simply the reverse of the steam reforming and shift reactions (eq. 31 and eq. 32): $\Delta H°$ at 25°C, in kJ/kg mol; $\Delta S°$ at 25°C, in kJ/(K kgmol)

$$CO + 3\,H_2 \;\leftrightarrow\; CH_4 + H_2O \quad \Delta H°, -206,172; \; \Delta S°, -214.6 \tag{31}$$

$$CO_2 + H_2 \;\leftrightarrow\; CO + H_2O \quad \Delta H°, +41,169; \; \Delta S°, +42.0 \tag{32}$$

Normal operating temperatures and pressures are 250–350°C and 3 MPa. Most methanation catalysts contain nickel supported on alumina, kaolin, or calcium aluminate cement. Sulfur, arsenic and carryover liquids from the CO_2 removal systems are catalyst poisons.

Methanation of carbon monoxide is highly exothermic. Temperatures exceeding the safe limits of operation can quickly be reached in out-of-control

Table 5. **Acid Gas Treating Processes**[a]

Process	Solvent	Solution circulation	Acid gas content in treated gas, ppm
	Reaction systems		
MEA[b]	20% monoethanolamine	medium	<50
promoted MEA	25–35% monoethanolamine plus Amine Guard	medium	<50
DGA[c]	60% diglycolamine	medium	<100
MDEA[d]	40% methyldiethanolamine plus additives	medium	<50
Vetrocoke	K_2CO_3 plus As_2O_3–glycine	high	500–1000
Carsol	K_2CO_3 plus additives	high	500–1000
Catacarb	25–30% K_2CO_3 plus additives	high	500–1000
Benfield	25–30% K_2CO_3 plus diethanolamine and additives	high	500–1000
Flexsorb HP	K_2CO_3 amine promoted	high	500–1000
Lurgi	25–30% K_2CO_3 plus additives	high	500–1000
Alkazid	potassium salt of 2-(or 3-) methylaminopropionic acid	[e]	[e]
	Combination reaction-physical systems		
Sulfinol	sulfone and 1.1'-imminobis-2-propanol	medium	<100
TEA-MEA[f]	triethanolamine and monoethanolamine	high (TEA) low (MEA)	<50
MDEA–sulfinol–H_2O		medium	<50
	Physical absorption systems		
Purisol (NMP)	N-methyl-2-pyrrolidinone	medium	<50
Rectisol	methanol	medium	<10
Fluor Solvent	propylene carbonate	[g]	[g]
Selexol	dimethyl ether of polyethyleneglycol	[g]	[g]

[a] Ref. 7.
[b] MEA is monoethanolamine.
[c] DGA is diglycolamine.
[d] MDEA is methyldiethanolamine.
[e] Dependent on service.
[f] TEA–MEA is triethanolanine–monoethanolamine.
[g] Dependent on pressure.

situations. Control systems for methanation reactions are carefully designed to a prevent runaway reaction which could potentially damage the catalyst or the pressure vessel used for the reaction.

Methanation results in a slight loss of hydrogen. However, the low cost and simple operation make it the most common method for removal of carbon oxides. More sophisticated cryogenic systems have also been used to remove carbon oxides. These systems have the additional advantage of reducing inerts, resulting in better ammonia yields by either reducing or eliminating the purge from the ammonia synthesis loop. Cryogenic systems have also been used to eliminate surplus nitrogen. Adding the capability to remove surplus nitrogen allows larger amounts of air to be injected in the secondary or autothermal reformer step. This in turn, lowers or eliminates the cost of the primary reformer.

Trace water removal using molecular sieve desiccant dryers is the last purification step commonly used in commercial processes. Often, as shown in Figure 6, the dryers are located on an interstage stream of the syngas compressor. This location minimizes the size of the dryer. Final syngas water levels below 1 ppm are achieved.

The overall sequence of steps used for syngas clean-up is rather complicated. The front end of an ammonia plant is basically the same as a hydrogen plant with the exception of the secondary reforming step. Hydrogen is widely used in petroleum refining for upgrading fuels. Hydrogen production facilities installed in the refining industry after the mid-1980s have a much simplified gas clean-up method based on high temperature shift, cooling and knockout of bulk water followed hydrogen purification using pressure swing adsorption (PSA). The pressure swing adsorption unit produces two streams. One is an extremely high purity hydrogen product stream that is at a pressure only slightly lower than the feed pressure. The other stream, called PSA tail gas, is a mix of carbon oxides, water, methane and hydrogen. This tail gas stream is at a pressure only slightly higher than ambient. It is usually fed to the burners of the reforming furnace, where it provides up to 90% of the heat needed for the furnace. The remaining furnace duty is supplied by a trim of natural gas to improve controllability of the furnace. In any case, the pressure swing adsorption unit replaces many of the processing steps used in the syngas clean-up method described for conventional ammonia plants.

Figure 8 is a process flow sheet for an ammonia plant that uses pressure swing adsorption for syngas clean-up. Despite the fact that the PSA unit greatly simplifies the process flow sheet, PSA based units have not been widely used in recently constructed ammonia plants. One reason for this is that the carbon dioxide by-product of the conventional clean-up process has value for urea synthesis and other markets as discussed previously. The carbon dioxide by-product in the PSA route is at low pressure and is mixed with many other chemical species. If there is no market or internal use for CO_2 then the PSA-based clean-up route would be preferred. This would be the case for a stand alone ammonia plant. However, in general, ammonia units are part of a larger nitrogen fertilizer/syngas product complex and thus the CO_2 by-product has value. Strong proponents of the PSA-based syngas clean-up route argue that purified CO_2 can be recovered from the PSA tail gas, however this clearly adds complexity and reduces the overall benefits of the PSA-based method.

Synthesis Gas Compression.   Modern ammonia plants with capacities greater than 600 metric ton/day exclusively use centrifugal compressors for syngas compression. Usually more than one stage of compression is used with interstage cooling. The last compression stage also boosts the ammonia synthesis loop recycle gas to overcome pressure drop in the synthesis loop.

Typically, the syngas compressors are driven by steam turbines. The high pressure steam supplied to the turbine is generated by heat recovery in the reformer furnace convection section and elsewhere in the flow sheet. Integration of support utilities, mainly steam and electricity, into the ammonia production process is a significant part of the design and operation of ammonia facilities.

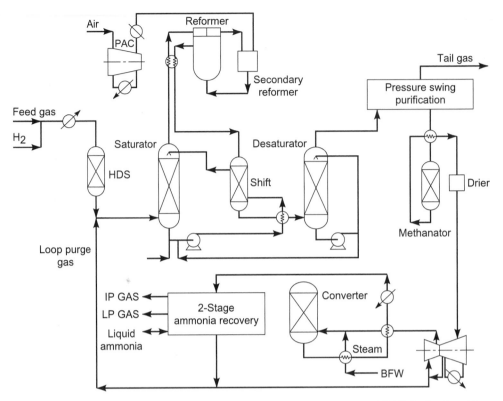

Fig. 8.   ICI-LCA process flow sheet; PAC, purified air compressor; HDS, hydrodesulfurization; IP, intermediate pressure; LP, liquefied petroleum; BFW, boiler feed water.

Ammonia Synthesis and Recovery.   The stoichiometry and thermodynamics of the ammonia synthesis reaction (eq. 33) are ΔH° at 25°C, kJ/kgmol $= -46,110$; ΔS° at 25°C, kJ/(K kgmol) $= -99.381$.

$$1/2\,N_2 + 3/2\,H_2 \;\leftrightarrow\; NH_3 \tag{33}$$

The thermodynamic values were taken from reference (8). The ammonia synthesis reaction is a reversible equilibrium type reaction. The reaction is exothermic, as indicated by the negative value of ΔH°.

Most commercial catalysts are just improved versions of the original iron catalyst used in the original Haber-Bosch process. Metallic iron, produced from magnetite, Fe_3O_4, is promoted with alkali in the form of potash and various metals such as aluminum, calcium, and magnesium. Recent commercial efforts have focused on fundamentally different catalysts. For example, Kellogg's Advanced Ammonia Process (KAAP) uses a promoted ruthenium catalyst with a high surface area graphite support. The catalyst has improved activity at high ammonia concentration with excellent low pressure and low temperature performance (9–11).

Ammonia synthesis typically occurs in the range of 350–550 °C. Figure 7 shows the effect of temperature on the equilibrium constant. Lower temperatures shift the reaction to the right. However, kinetics limit the lowest useful temperature to ca 350 °C. A quench converter is a common reactor configuration. A quench converter contains a series of adiabatic beds which are used to produce ammonia from syngas. Cold feed is injected between the beds to cool the reaction products, resulting in a saw-tooth temperature profile across the entire reactor. Typical per pass conversions are only ca 25–35%.

Pressure affects synthesis reactor performance. Following LeChatelier's principle, higher pressures improve equilibrium conversion. Higher pressures also improves kinetics. A wide range of pressures have been used in commercial processes, ranging from 8 to 40 MPa with most recent plants operating in the range of 15–25 MPa.

Pressure also affects the details of ammonia recovery. Ammonia is recovered by cooling the ammonia synthesis effluent to cause the ammonia product to condense into a liquid. The residual gas is then recycled back to the synthesis gas compressor where it compressed and then used to produce more ammonia product. Older plants using high ammonia synthesis loop pressures were able to use air fin coolers or cooling water to condense the product ammonia in the recovery step. However, today most plants need refrigeration systems to recover ammonia because of the lower synthesis pressures commonly used. Mechanical refrigeration using ammonia as the working fluid is most common. Absorption refrigeration systems using a water-ammonia mixture have also been used. These have the advantage of being able to use low temperature waste heat from the process (12,13).

The synthesis gas used for ammonia synthesis typically contains inerts such as methane from the methanation step, argon from the air used for secondary reforming and helium from the natural gas. These inerts have some solubility in the liquid ammonia product, but usually the solubility is too low to prevent build-up in the synthesis loop. In this case, the inerts are removed by taking a small gas purge from the ammonia synthesis loop. The purge rate can have an impact on the overall process economics. Often recovery of hydrogen from the purge stream using cryogenics, pressure swing adsorption or membrane processes is economically justified. The recovered hydrogen is recycled for additional ammonia production while the offgas containing the inerts are incinerated in the primary reforming furnace.

6.2. Other Production Routes. Partial oxidation of heavy feedstocks such as petroleum resids, petroleum coke, and coal is responsible for most of the non steam reformer based production of hydrogen used in ammonia synthesis. Partial oxidation routes are able to handle a wide variety of feedstocks. Solids handling equipment and the need to deal with soot by-products increase the complexity and capital costs for these routes. Ammonia synthesis is unique in syngas processes since air-blown rather than oxygen-blown units can be used for partial oxidation. Nitrogen associated with the oxygen in air is needed in the downstream ammonia synthesis step. Nitrogen is considered as an undesirable impurity in processes for most other synthesis gas products (eg, methanol). The requirement for large amounts of high purity oxygen has been the major hurdle preventing wider use of partial oxidation in syngas technologies. Recent advancements in air separation technologies based upon membranes and vacuum swing adsorption have resulted in renewed interest in industrial scale partial oxidation technologies for all synthesis gas products.

By-product hydrogen is generated in units that were designed for other purposes but happen to produce hydrogen as a by-product. Examples include many petroleum refining operations such as catalytic reforming, catalytic crackers, thermal crackers, and cokers. Other major by-product sources are: offgas from ethylene crackers, offgas from industrial electrolysis processes such as those used for caustic/chlorine manufacture, and tail gases from other synthesis gas processes such methanol production.

A very small portion of on-purpose hydrogen production comes from electrolysis. The electrolysis route is not currently used for ammonia production. For hydrogen, it is currently only competitive for small scale production and/or for high purity hydrogen applications such as hydrogen for use in semiconductor processing.

The electrolysis route fits well with the vision of using either hydrogen or ammonia as an energy carrier. In this view, renewable-based electricity generated is generated from hydroelectric, solar, geothermal, ocean thermal, or other sources. The electricity is used to split water, the hydrogen is stored and/or transported, and then electricity is regenerated from the hydrogen using a fuel cell. Temporary conversion of the hydrogen to ammonia followed by back-conversion to hydrogen is envisioned as one way to simplify storage and transport. In either case, the hydrogen or the ammonia is merely being used as an energy carrier rather than as an ultimate source of energy.

At the time of this writing, the hydrogen economy is still in the future. Production costs and lack of infrastructure remain as significant hurdles. Continued availability of low cost nonrenewable hydrocarbons and their associated existing and largely depreciated infrastructures will slow commercialization progress in this area.

Biological processes based on genetic engineering have the potential to eliminate the synthetic nitrogen fertilizer industry completely. The genes responsible for nitrogen fixation could potentially be isolated from symbiotic soil organisms and then transferred and expressed in non-leguminous crops. Alternatively, nitrogen fixation could be transferred and expressed in a wider range of natural soil bacteria. At this point in time, these are rather futuristic scenarios.

7. Storage and Shipment

Although much ammonia serves as feedstock for other processes, the largest single use in the United States is as a direct application fertilizer without further processing. This direct application consumption is mostly in the farm belt and ammonia produced in the Gulf Coast or other locations is shipped to terminal facilities and then distributed by retail outlets to the farmer.

Anhydrous ammonia is ordinarily stored in refrigerated tanks at the plant site and major distribution points. Atmospheric pressure and at temperature of ca -33°C are typical conditions. Very large quantities have to be stored since the fertilizer market is seasonal. Cylindrical tank of large (27,000–50,000 t) capacities, either of single-walled or doubled-walled construction, have been installed in primary market areas (14). Double integrity tanks have been designed with secondary containment for added safety (15).

Evaporating ammonia must be vented. The vented gas is reliquefied for recycle or absorbed in water to make aqua ammonia. Refrigerated storage tanks are insulated using great care to minimize heat loss and access of air and moisture to the insulation or metal surface. In double-wall tanks, the annular space is usually filled with perlite and the external surface of the outer tank is painted for corrosion protection.

Relatively small quantities of anhydrous ammonia are stored in spherical vessels at pressures above 275 kPa gauge (40 psig) and ambient temperature. Not more than 19 m³ (50,000 U.S. gal) of ammonia should be stored in an unrefrigerated tank (16). Storage tanks must not be filled to more than 56% of the water weight capacity of the container at 15 °C (17). There is a tendency to design foundations for the weight of the tank filled with water to hydrostatically test the tank completely. Tanks should be built of steel and designed for a minimum working pressure of 1724 kPa gauge (250 psig).

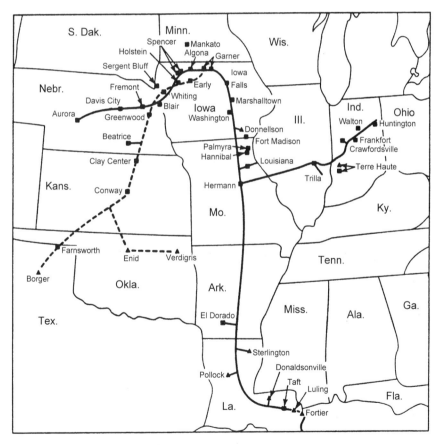

Fig. 9. Pipeline systems of transport for anhydrous ammonia within the United States where ▲ represnts an ammonia plant location; ■, storage terminals; (—), Gulf Central pipeline; and (—), MAPCO pipeline.

Ammonia is usually transported for long distances by barge, pipeline, rail, and for short distances, by truck. Factors that govern the type of carrier used in anhydrous ammonia transportation systems are distance, location of plant site in relation to consuming area, availability of transportation equipment, and relative cost of available carriers. Typical costs of pipeline, barge, and rail modes for long distance transport are $0.0153, $0.0161 and $0.0215 per ton per kilometer, respectively, for distances of about 1600 km (18). Short truck transportation costs are much higher. Costs are typically $0.0365/(t/km) for distances on the order of 160 km. Ammonia in the world trading markets costs on the order of $35/t to ship between the United States and Western Europe.

The primary ammonia pipelines in the United States are the Mid-America Pipeline Company (MAPCO) and Gulf Central (see Fig. 9) (see FERTILIZERS). The MAPCO pipeline (20) contains 1763 kilometers of 101 mm, 152 mm, and 203 mm pipe have a pumping capacity of 3885 metric tons per day and supporting terminal storage facilities. Peak delivery from the system is 4,216 metric tons per day. The Gulf Central system (21) contains 3220 km of 152 mm, 203 mm, and 254 mm pipe and has a pumping capacity of 2545 metric tons per day and supporting terminal storage facilities. The Tampa Bay Pipeline network services several ammonia plants along a 133 km route in Florida.

The other major ammonia pipeline is in the former Soviet Union. A 2400-km mainline extends from Togliatti on the Volga River to the Port of Odessa on the Black Sea. A 2200-km branch line extends from Gorlovka in the Ukraine to Panioutino. Mainline operating pressure is 8.15 MPa (1182 psi) and branch line operating pressure is 9.7 MPa (1406 psi) (22).

Export oriented ammonia producing countries utilize huge ocean going tankers that contain up to 50,000 t for distribution of ammonia. Co-shipment in refrigerated LNG tankers in usually done. Capacities for trucks and rail cars are normally 100–150 m^3. Barges can usually accommodate from 400–2000 t.

8. Economic Aspects

Examination of typical ammonia production economics explains the recent trend in new plant location. Table 6 provides a breakdown of the estimated manufacturing cost of ammonia in a 1000 metric ton per day reduced energy natural gas reforming plant built in early 1990 on the U.S. Gulf Coast. As can be seen from this analysis, the natural gas feedstock accounts for nearly 70% of cash operating costs. Natural gas feedstock and capital charges amount to over 93% of production costs.

Over the past decade inflation rates have been quite low. Natural gas prices have been more of a function of geography rather than time. A natural gas delivered price of $1.90/GJ ($2.00/MMBTU), on lower heating value (LHV) basis, was reported for February 1990 in the United States (24). In the energy-rich countries of the Near East, in the former USSR, Trinidad, and in North Africa, natural gas prices are on the order of $0.47/GJ ($0.5/MMBTU); however construction cost at these locations tends to be higher than on the U.S. Gulf Coast. Cost of transporting ammonia from these more remote locations to current agricultural centers also has to be taken into account; however, the rationale for siting of new nitrogen complexes near remote gas sources is clear.

Table 6. **Estimated Ammonia Manufacturing Cost**[a]

Factor	Unit rate/t	$US/units	Cost, $/t
Variable costs			
natural gas (feed-fuel), GJ[b]	28.45	1.90	54.06
boiler feed water make-up, m^3	1.1	0.25	0.28
cooling water circulation, m^3	210	0.01	2.10
(Δ temperature = 10°C)			
catalyst cost			1.7
Fixed costs[c]			
labor, personnel shift ($10/h)	6	10	1.44
supervision, 100% of labor			1.44
interest on working capital, %	10		1.30
indirect charges, %[d]	16		44.06
Total production cost			*106.38*
pretax return on investment (ROI), %	20		55.07
ammonia cost (FOB plant)			161.45
cash cost of production[e]			78.84

[a] Reduced energy ammonia process utilizing natural gas feed at a U.S. Gulf Coast location. Daily production of 1,000 t, yearly production of 345,000 t, and total investment of $99,500,000 (U.S. $95 million in fixed investment and $4.5 million in working capital).

[b] To convert GJ to Gcal, divide by 4.184.

[c] Fixed investment includes cooling tower, boiler feedwater treatment, raw water ammonia storage as minimum off-sites requirement.

[d] Indirect charges are 16% of fixed investment per ton per year: 10% depreciation; 1.5%, insurance; 3%, maintenance (labor, materials); 1.5%, general sales and marketing.

[e] Cash cost excludes 10% depreciation allowance and 20% pretax ROI.

Whereas manufacturing costs are strongly influenced by energy prices and capital investment, ammonia selling prices are determined by supply and demand. The profitability of an ammonia plant is determined by the margin between cost of production and ammonia price. In February 1990, the FOB price of ammonia on the U.S. Gulf Coast was $107 per short ton (24). Therefore a Gulf Coast producer with a reduced energy unit would have a $28 margin at the assumed natural gas price. However, most plants on the U.S. Gulf Coast were built in the late 1960s and early 1970s. Energy consumption for plants of this era is about 34.85 GJ/t (30 MMBtu/short ton) vs. 28.45 GJ/t (24.5 MMBtu/short ton) for the modern reduced energy designs, thus cash margins for older plants that did not modernize their technology were only $16/ton.

Most ammonia production is processed or used in the countries where it is produced. World trade of ammonia accounts for only 11% of world production; however, this statistic does not take into account ammonia present in derivatives such as urea. The major net exporters of ammonia in 1997 were Russia and the Ukraine accounting for 43% of world exports, Trinidad for 22% and the Middle East for 13%. The major net importers were the United States (43%) and Western Europe (26%) (1).

9. Grades and Specifications

Anhydrous and aqua ammonia are manufactured in various grades depending on use. Specifications are given in Table 7 and 8, respectively.

Table 7. **Anhydrous Ammonia Specifications**[a]

Material content	Commercial or fertilizer grade	Refrigeration grade	Metallurgical grade
ammonia, min wt %	99.5	99.98	99.99
water, max ppm by wt	5000	150	33
oil, max ppm by wt	5	3	2
noncondensable gases, max mL/g		0.2	10

[a] Ref. 25.

Table 8. **Aqua Ammonia Specifications**[a]

Grade	Ammonia, wt %
United States Pharmacopeia	
stronger ammonia water	28–30
ammonia test solution	9.5–10.5
normal (1 *N*) aqua ammonia	1.7
chemically pure	28
technical, Bé°[b]	
26	29.4
16	10.3
18	14
20	17.75

[a] Ref. 26, 27.
[b] Bé° = degree Baumé; specific gravity = 145/(145 − Bé°).

10. Analytical Methods

Anhydrous ammonia is normally analyzed for moisture, oil and residue. The ammonia is first evaporated from the sample and the residue tested (28). In most instances, the amount of oil and sediment in the samples are insignificant and the entire residue is assumed to be water. For more accurate moisture determinations, the ammonia can be dissociated into nitrogen and hydrogen and the dewpoint of the dissociated gas obtained. This procedure works well where the concentration of water is in the ppm range. Where the amount of water is in the range of a few hundredths of a percent, acetic acid and methanol can be added to the residue and a Karl Fischer titration performed to an electrometrically detected end point (29–32).

11. Environmental Concerns

Ammonia production by steam reforming of natural gas is a relatively clean operation and presents no unique environmental problems. NO_x, emissions from the flue gas of the primary reformer can be suitably treated by a combination of conventional control techniques including low NO_x burners, selective catalytic reduction and flue gas scrubbing.

Ammonia production using heavy feedstocks raises additional environmental issues. Particulate emissions from solids handling of the feeds must be controlled. The soot, ash and slag produced from these routes must also be disposed of in an environmentally acceptable manner. These feeds are also more likely to create liquid and gases by-products such as tars, phenols, sulfur, cyanides, etc, which must be handled properly.

12. Health and Safety Factors

The American Institute of Chemical Engineers sponsors a yearly symposium on plant safety for ammonia and related facilities. The published papers from these symposia are a rich source of information on the design and operations of safe facilities. Fire is the most frequent cause of safety incidents in ammonia production. In a recently published survey based on 89 ammonia plants operating in 25 countries, the average ammonia plant had one fire every 14.2 months over the three year survey period of 1994–1996 (33). The most common fuel sources were flanges leaking hydrogen rich gases and oil leaks associated with compressor and pump lubricating–sealing systems. The good news is that safety programs seem to be working since the percentage of plants reporting no fires has steadily increased over time. Survey results from 1994–1996 had 41% of plants reporting no fires compared to 30% for the period 1982–1985, 23% for the period 1977–1981 and 7% for the period 1973–1976.

The National Fire Protection Association (NFPA) ratings for anhydrous ammonia (34) are Health = 3, Fire = 1 and Reactivity = 0. Ammonia is a strong local irritant that also has a corrosive effect on the eyes and the membranes of the pulmonary system. Vapor concentrations of 10,000 ppm are mildly irritating to the skin, whereas 30,000 ppm may cause burns. The physiological effects from inhalation are described in Table 9. Prolonged, intentional exposure to high levels of ammonia is unlikely because its characteristic odor can be detect at levels as 1–5 ppm (35).

OSHA standards specify the threshold limit value (TLV) 8-h exposure to ammonia as 50 ppm (35 mg/m^3). However, the ACGIH recommends a TLV of 25 ppm (36). Respiratory protection should be provided for workers exposed to ammonia. Protective clothing such as rubber aprons, boots, gloves and goggles should be worn when handling ammonia.

Table 9. **Physiological Effects of Ammonia**[a]

Concentration, ppm	Effects
20–50	perceptible odor
40–100	eye and respiratory system irritation
400–700	severe eye and respiratory irritation; potential for permanent damage
1,700	convulsive coughing and bronchial spasms; half four exposure potentially fatal
5,000–10,000	death from suffocation

[a] Ref. 27.

Table 10. **1998 U.S. Ammonia Demand by Category**[a]

Category	Demand, 1000 t/yr	Share, %
fertilizers		
direct application	3,600	20
nitrogen derivatives	10,800	60
chemical Intermediates	3,420	19
Other	180	1
Total	*18,000*	*100*

[a] Ref. 4.

The flammable limits of ammonia in air are 16 to 25% by volume at atmospheric pressure; in oxygen the range is 15 to 79%. Ignition temperature is ca 650°C. Although ignition is difficult, it is not impossible as proved by an explosion resulting from a leak in an indoor ammonia refrigeration system (37). Ammonia does not produce unstable or hazardous decomposition products. However, contact with calcium, gold, mercury, silver or chlorates may result in explosive compounds.

13. Uses

Table 10 shows the break down of uses for ammonia in the United States. Nonfertilizer use accounts for about 20% of consumption. Globally, the chemical markets for ammonia are smaller, only accounting for about 15% of consumption. The vast majority of ammonia is used for nitrogen fertilizers. Three principal fertilizers, ammonium nitrate [6484-52-2], NH_4NO_3, ammonia sulfate [7782-20-2], $(NH_4)SO_4$, and ammonium phosphate [10361-65-6], $(NH_4)_3PO_4$, are made by reaction of the respective acids with ammonia. Table 11 gives a current global breakdown of nitrogen fertilizer products by type.

The catalytic oxidation of ammonia in the presence of methane is commercially used for the production of hydrogen cyanide.

Acrylonitrile, [107-13-1], $CH_2=CHC\equiv N$, is produced in commercial quantities almost exclusively by the ammoxidation of propylene.

Table 11. **World Nitrogen Fertilizer Consumption**[a]

Fertilizer	Amount, 1000 t/yr	Share, %
ammonium sulfate	2,610	3.2
urea	37,570	46.4
ammonium nitrate	7,100	8.8
calcium arnmoniun nitrate	3,580	4.4
direct ammonia application	4,180	5.2
nitrogen solutions	3,990	4.9
MAP/DAP & other NP compounds	6,220	7.7
NK/NPK compounds	6,510	8.0
others	9,190	11.4
Total	*80,950*	*100*

[a] Ref. 1.

Urea, formed from the reaction of ammonia with carbon dioxide, has enjoyed a long term steady rise in market share to point where today it is the most common nitrogen fertilizer. This increase in market share for urea has been at the expense of ammonium nitrate, ammonium sulfate and direct application of ammonia.

Caprolactam [105-60-2] is made from ammonia and cyclohexanone. Caprolactam is the monomer for nylon 6. By-product ammonium sulfate from this route is a major source of this fertilizer ingredient. Ammonia is also used in the production of pyridines, amines and amides.

BIBLIOGRAPHY

"Ammonia" in *ECT* 1st ed., Vol. 1, pp. 771–810, by R. M. Jones, Barrett Division, and R. L. Baker, Nitrogen Division, Allied Chemical & Dye Corporation; "By-Product Ammonia" *ECT* 1st ed. under "Ammonia," Vol. 1, pp. 778–780, by R. M. Jones, Barrett Division, and R. L. Baber, Nitrogen Division, Allied Chemical & Dye Corporation; "Ammonia" in *ECT* 2nd ed., Vol. 2, pp. 258–298, by G. L. Frear and R. L. Baber, Nitrogen Division, Allied Chemical Corporation; "Ammonia, By-Product" in *ECT* 2nd ed., Vol. 2, pp. 299–312, by Walter E. Carbone and Oscar F. Fissore, Wilputte Coke Oven Division, Allied Chemical Corporation; in *ECT* 3rd ed., Vol. 2, pp. 470–516, by J. R. Le Blanc, Jr., S. Madhavan, R. E. Porter, Pullman Kellogg Company. "Ammonia" in *ECT* 4th ed., Vol. 2, pp. 638–691, by T. A. Czuppon, S. A. Knez, and J. M. Rovner, M. W. Kellogg Company; "Ammonia" in *ECT* (online), posting date: December 4, 2000, by T. A. Czuppon, S. A. Knez, and J. M. Rovner, M. W. Kellogg Company.

CITED PUBLICATIONS

1. International Fertilizer Industry Association, Web site: www.fertilizer.org.
2. R. A. Macriss and co-workers, *Inst. Gas Technol., Chicago, Res. Bull.* **34** (Sept. 1964).
3. K. Jones, *The Chemistry of Nitrogen*, Pergamon Press, Oxford, UK, 1973, 199–227.
4. *Chemical Profiles - Ammonia*, Nov. 29, 1999, Schnell Publishing Company, Web site: www.chemexpo.com.
5. I. Dybkjaer, "Design and Operating Experience of Large Ammonia Plants", *Ammonia Plant Safety and Related Facilities*, American Institute of Chemical Engineers, Vol. **34**, 1994, pp. 199–209.
6. J. H. Cartensen, and S. Hammershoi, "Reducing Methanol Byproduct Formation Over The LTS Converter", *Ammonia Plant Safety and Related Facilities*, American Institute of Chemical Engineers, Vol. **39**, 1999, pp. 171–178.
7. O. J. Quartulli, *J. Hungarian Chem. Soc.* **30**(8), 404 (1975).
8. D. D. Wagman and co-workers, The NBS Tables of Chemical Thermodynamic Properties: Selected Values for Inorganic and C1 and C2 Organic Substances in SI Units, *J. Phys. Chem. Ref. Data* **11**(Suppl. 2) (1982).
9. U.S. Pat. 4,600,571 (1986), J. J. McCarroll, S. R. Tennison, and N. P. Wilkinson (to British Petroleum).
10. U.S. Pat. 4,568,530 (1986), G. S. Benner and co-workers (to M. W. Kellogg Corporation).
11. J. R. Le Blanc and P. J. Shires, *Kellogg's Advanced Ammonia Process*, Kellogg's Ammonia Club Meetings, San Francisco, Nov. 4–5, 1989.

12. M. J. P. Bogart, *Hydrocarbon Process* **57**(4), 145 (1978).
13. G. Holldorff, *Hydrocarbon Process* **58**(7), 149 (1979).
14. C. Clay Hale, *Plant Operations Prog.* **3**(3), 147–159 (July 1984).
15. A. Josefson, "Transportation, Storage, Handling of Ammonia", *Ammonia Plant Safety and Related Facilities* American Institute of Chemical Engineers, **27**, 1986, pp. 155–159.
16. American Standards Safety Requirements for the Storage and Handling of Anhydrous Ammonia, ASA K61.1- 1960, (Oct. 28, 1960), Boston, Mass., 1986, p. 16.
17. *Pamphlet G-2*, Compressed Gas Association, American National Standards Institute, New York, 1962.
18. Farmland Industries, Inc., private communication, March 7, 1990.
19. R. Vandoni and M. Laudy, *J. Phys. Chim.* **49**, 99–102 (1952).
20. MAPCO Corp., Tulsa, Okla., private communication, Jan. 1990.
21. C. Clay Hale and W. H. Lichtenberg, "U.S. Anhydrous Ammonia Distribution System in Transition", *Ammonia Plant Safety and Related Facilities*, American Institute of Chemical Engineers, Vol. **28**, 1988, pp. 76–90.
22. N. Hetland, *Pipeline Gas J.* **250**, 38 (Oct. 1978).
23. *DOT Hazardous Materials Code of Federal Regulations*, Title 49, Washington, DC.
24. *Green Markets*, McGraw-Hill Publications, New York, Feb. 26, 1990.
25. "Nitrogen Products" in *Chemical Economics Handbook*, SRI International, Menlo Park, CA., June 1988.
26. *U.S. Pharmacopeia XXII and National Formulary XVII*, U.S. Pharmacopeial Conventions, Inc., Washington, DC, 1990.
27. *Matheson Gas Data Book*, 5th ed., Matheson Gas Products, Rutherford, N.J., 1971.
28. *Joint Army-Navy Specification JAN-A-182*, U.S. Government Printing Office, Washington, D.C., April 30, 1945.
29. E. D. Peters and J. L. Jangnickel, *Anal. Chem.* **27**, 450 (1955).
30. L. F. Fieser, *Experiments in Organic Chemistry*, 3rd ed., D. C. Heath and Co., 1955, p. 289.
31. *Official Methods of Analysis*, 9th ed., Association of Official Agricultural Chemists, Washington, D.C., 1960.
32. J. Mitchell and D. M. Smith, *Aquametry*, Interscience Publishers, New York, 1948.
33. G. P. Williams, "Safety Performance in Ammonia Plants: Survey VI", *Ammonia Plant Safety and Related Facilities*, American Institute of Chemical Engineers, Vol. **39**, 1999, pp. 1–7.
34. *ANSI/NFPA 49 Hazardous Chemicals Data*, 1991 edition, National Fire Protection Association, Quincy, MA, Feb. 8, 1991.
35. *Occupational Health Guideline for Ammonia*, U.S. Dept. of Health and Human Services, Washington, D.C., 1978.
36. Documentation of Threshold Limit Values for Substances in Workroom Air, *American Conference of Governmental Industrial Hygienists*, Cincinnati, Ohio, 1974.
37. M. H. McRae, "Anhydrous Ammonia Explosion in an Ice Cream Plant", *Ammonia Plant Safety and Related Facilities*, American Institute of Chemical Engineers, Vol. **27**, 1986, pp. 1–4.

TIM EGGEMAN
Neoterics International

AMMONIUM COMPOUNDS

1. Introduction

There are a considerable number of stable crystalline salts of the ammonium ion [14798-03-9], NH_4^+. Several are of commercial importance because of large scale consumption in fertilizer and industrial markets. The ammonium ion is about the same size as the potassium and rubidium ions, so these salts are often iso-morphous and have similar solubility in water. Compounds in which the ammonium ion is combined with a large, uninegative anion are usually the most stable. Ammonium salts containing a small, highly charged anion generally dissociate easily into ammonia (qv) and the free acid (1). At about 300°C most simple ammonium salts volatilize with dissociation, for example

$$NH_4Cl(s) \longrightarrow NH_3(g) + HCl(g)$$

Exceptions are salts of oxidizing anions, which decompose with oxidation of the ammonium ion to nitrous oxide [10024-97-2], N_2O, or nitrogen, N_2.

A number of simple, standard methods have been developed for the analysis of ammonium compounds, several of which have been adapted to automated or instrumental methods. Ammonium content is most easily determined by adding excess sodium hydroxide to a solution of the salt. Liberated ammonia is then distilled into standard sulfuric acid and the excess acid titrated. Other methods include colorimetry (2) and the use of a specific ion electrode (3).

Covered in detail in this article are the quaternary ammonium compounds. Covered elsewhere are: ammonium borates [12007-89-5], $NH_4B_5O_8$, (see BORON COMPOUNDS); ammonium chromate [7788-98-9], $NH_4Cr_2O_4$, and ammonium dichromate [7789-09-5], $NH_4Cr_2O_7$, (see CHROMIUM COMPOUNDS); ammonium cyanide [12211-52-8], NH_4CN, (see CYANIDES); ammonium glutamate [20806-32-0], $NH_4C_5H_8NO_4$, (see AMINO ACIDS); ammonium hydroxide [1336-21-6], NH_4OH, (see AMMONIA); ammonium molybdate [13106-76-8], $(NH_4)_2 MoO_4$, (see MOLYBDE-NUM COMPOUNDS); ammonium oxalate [1113-38-8], $(NH_4)_2C_2O_4$, (see OXALIC ACID); ammonium phosphate dibasic [7783-28-0], $(NH_4)_2HPO_4$, and ammonium phosphate monobasic [7722-76-1], $(NH_4)H_2PO_4$, (see PHOSPHORIC ACID AND THE PHOS-PHATES); ammonium sulfamate [7773-06-0], $NH_4SO_3NH_2$, (see SULFAMIC ACID AND SULFAMATES); and ammonium thiosulfate [7783-18-8], $(NH_4)_2S_2O_3$ (see THIOSUL-FATES).

2. Ammonium Acetates

Both normal or neutral ammonium acetate [631-61-8], $NH_4C_2H_3O_2$, and the acid salt are known. The normal salt results from exact neutralization of acetic acid using ammonia; the acid salt is composed of the neutral salt and acetic acid.

The normal salt, CH_3COONH_4, is a white, deliquescent, crystalline solid, formula wt 77.08, having a specific gravity of 1.073. It is quite soluble in water or ethanol: 148 g dissolve in 100 g of water at 4°C. The salt's solubility in water increases only slightly as temperature increases up to about 25°C. The specific gravity of aqueous neutral ammonium acetate ranges from 1.022 to 1.092 as solution concentration increases from 10 to 50 wt % (4). The normal salt melts at 114°C, but decomposes before reaching its boiling point.

Ammonium acetate solutions formed by neutralizing acetic acid using ammonium hydroxide are essentially neutral. Thus, these solutions are suitable for standardization of electrodes, and for use as titration standards. Solutions must be used while fresh, however, as they become acidic on standing.

Isolation of dry, normal ammonium acetate, prepared by neutralizing acetic acid with anhydrous ammonia or ammonium carbonate, is difficult because of ammonia loss during evaporation of water. Consequently, commercial grades of ammonium acetate are often mixtures of the neutral and acid salts, or are supplied as ammonium acetate solution [8013-61-4].

The acidic double salt of ammonium acetate and acetic acid [25007-86-7], $CH_3COONH_4 \cdot CH_3COOH$, is made by dissolving the neutral salt in hot acetic acid or by distilling the neutral salt. During distillation the acid salt is formed as a heavy oil that solidifies on cooling. It crystallizes as long, deliquescent needles that melt at 66°C. Acid ammonium acetate is readily soluble in both water and alcohol.

Ammonium acetate has limited commercial uses. It serves as an analytical reagent, and in the production of foam rubber and vinyl plastics; it is also used as a diaphoretic and diuretic in pharmaceutical applications. The salt has some importance as a mordant in textile dyeing. In a hot dye bath, gradual volatilization of ammonia from the ammonium acetate causes the dye solution to become progressively more acidic. This increase in acidity enhances the color and permanence of the dyeing process. Ammonium acetate is a poison by intravenous route. When heated to decomposition, it emits toxic fumes (5).

3. Ammonium Carbonates

The earliest mention of an ammonium carbonate, salt of hartshorn, appears in English manuscripts of the 14th century. As the name implies, the material was obtained by dry distillation of animal waste such as horn, leather, and hooves. Although many salts have been described in the literature for the ternary $NH_3–CO_2–H_2O$ system, most, except for ammonium bicarbonate [1066-33-7], NH_4HCO_3, ammonium carbonate [506-87-6], $(NH_4)_2CO_3$, and ammonium carbamate [1111-78-0], $NH_4CO_2NH_2$, are mixtures (6,7).

3.1. Ammonium Bicarbonate.   Ammonium bicarbonate, also known as ammonium hydrogen carbonate or ammonium acid carbonate, is easily formed. However, it decomposes below its melting point, dissociating into ammonia, carbon dioxide, and water. If this process is carefully controlled, these compounds condense to reform ammonium bicarbonate. The vapor pressures of dry ammonium bicarbonate are shown below (8). (To convert kPa to mm Hg, multiply by 7.5.)

Temperature, °C	Vapor Pressure, kPa	Temperature, °C	Vapor Pressure, kPa
25.4	7.85	50.0	52.65
34.2	16.26	55.8	82.11
40.7	26.79	59.3	108.64
45.0	37.06		

Ammonium bicarbonate, sp gr 1.586, formula wt 79.06, is the only compound in the NH_3–CO_2–H_2O system that dissolves in water without decomposition. Solubility in 100 g of H_2O ranges from 11.9 g at 0°C to 59.2 g/100 g of H_2O at 60°C (9). The heat of formation from gaseous ammonia and carbon dioxide and liquid water is 126.5 kJ/mol (30.2 kcal/mol). Ammonium bicarbonate is manufactured by passing carbon dioxide gas countercurrently through a descending stream of aqua ammonia. The reaction is normally carried out in a packed tower or absorption column. Because the reaction is exothermic, cooling the lower portion of the tower is advisable. Concentration of the solution is monitored by determining specific gravity. When the solution is sufficiently saturated, crystallization of ammonium bicarbonate occurs. The crystals are separated from the mother liquor by filtration or centrifugation and washing, and are dried using 50°C air.

Ammonium bicarbonate is produced as both food and standard grade and the available products are normally very pure. Although purification is possible by sublimation at low temperatures, it is more economical to prepare the desired product directly by using ammonia and carbon dioxide of high purity.

When heated, ammonium bicarbonate decomposes to ammonia, water, and carbon dioxide, leaving no solid residue. This property explains its use as a leavening agent for certain baked goods (see BAKERY PROCESSES AND LEAVENING AGENTS). At about 150°C, 1 kg yields 1.3 m^3 (46.2 ft^3) of STP gas. Ammonium bicarbonate is also used in pharmaceuticals (qv), in production of ammonium salts, and in formulation of fire-extinguishing agents (qv). It is effective in scale removal, for example, in heat-exchanger tubes by dissolving thin layers of calcium sulfate.

$$CaSO_4 + 2\,NH_4HCO_3 \longrightarrow CaCO_3 + (NH_4)_2SO_4 + CO_2 + H_2O$$

For thicker layers of scale, alternate treatments using dilute hydrochloric acid appear desirable (10). Its DOT number is NA9081. Ammonium bicarbonate is poisonous by intravenous route. When heated to accomposition, it emits toxic fumes (5).

3.2. Ammonium Carbonate.   Normal ammonium carbonate, mp 43°C, formula wt 96.09, is a crystalline solid. The commercial product may be produced by passing carbon dioxide into an absorption column containing aqueous ammonia solution and causing distillation. Vapors containing ammonia, carbon dioxide, and water condense to give a solid mass of crystals. Ammonium carbonate is the principal ingredient of smelling salts because of its characteristic strong ammonia odor. It is also used for other medicinal purposes and as a leavening agent.

Ammonium carbonate is poisonous by subcutaneous and intravenous routes. When heated to decomposition, it emits toxic fumes (5).

4. Ammonium Citrate

Diammonium citrate [3012-65-5], $(NH_4)_2C_6H_6O_7$, mol wt 226.19, is soluble in an equal weight of water, but is only slightly soluble in ethanol. The pH of a 0.1 M solution is 4.3. It is made by neutralization of citric acid with ammonia; the crystalline or granular product is used as a chemical reagent and pharmaceutically as a diuretic.

5. Ammonium Halides

Ammonium chloride [12125-02-9], NH_4Cl, ammonium bromide [12124-97-9], NH_4Br, ammonium fluoride [12125-01-8] and ammonium iodide [12027-06-4], NH_4I, are crystalline, ionic compounds of formula wts 53.49, 97.94, and 144.94, respectively. Their densities d_4^{20} systematically follow the increase in formula weight: 1.53, 2.40, and 2.52. All three exist in two crystal modifications (12): the chloride, bromide, and iodide have the CsCl structure below temperatures of 184.5, 137.8, and $-17.6°C$, respectively; each reversibly transforms to the NaCl structure at higher temperatures.

Ammonium fluoride is a white, deliquescent, crystalline salt. It tends to lose ammonia gas to revert to the more stable ammonium bifluoride. Its solubility in water is 45.3 g/100 g of H_2O at 25°C and its heat of formation is -466.9 kJ/mol (-116 kcal/mol). Ammonium fluoride is available principally as a laboratory reagent. If it is needed in large quantities, one mole of aqueous ammonia can be mixed with one mole of the more readily available ammonium bifluoride (11).

The solubility of the ammonium halides in water also increases with increasing formula weight. For ammonium chloride, the integral heat of solution to saturation is 15.7 kJ/mol (3.75 kcal/mol); at saturation, the differential heat of solution is 15.2 kJ/mol (3.63 kcal/mol). The solubility of three salts is given in Table 1 (8).

All ammonium halides exhibit high vapor pressures at elevated temperatures, and thus, sublime readily. The vapor formed on sublimation consists not of discrete ammonium halide molecules, but is composed primarily of equal volumes of ammonia and hydrogen halide. The vapor densities are essentially half that expected for the vaporous ammonium halides. Vapor pressures at various temperatures are given in Table 2 (13). Latent heats of sublimation, assuming complete dissociation of vapors and including heats of dissociation are 165.7, 184.1, and 176.6 kJ/mol (39.6, 44.0, and 42.2 kcal/mol), for NH_4Cl, NH_4Br, and NH_4I, respectively.

The heat of formation of ammonium chloride from the elements is 317 kJ/mol (75.8 kcal/mol); it is 175 kJ/mol (41.9 kcal/mol) from gaseous ammonia and gaseous hydrogen chloride. The heat of formation of ammonium bromide from the elements, bromine in the liquid form, is 273 kJ/mol (65.3 kcal/mol); for ammonium iodide, the corresponding heat of formation is 206 kJ/mol (49.3 kcal/mol). Iodine is in the solid state.

Aqueous solutions of ammonium halides, like the other ammonium salts of strong acids, are acidic; on storage and exposure these solutions tend to become more acidic through ammonia loss. They also have a pronounced tendency to

Table 1. **Solubilities of Ammonium Halides**

Temperature, °C	Solubility, g/100 g water			Temperature, °C	Solubility, g/100 g water		
	NH_4Cl	NH_4Br	NH_4I		NH_4Cl	NH_4Br	NH_4I
0	29.4	60.6	154.2	60	55.3	107.8	208.9
20	37.2	75.5	172.3	80	65.6	126.0	228.8
40	45.8	91.1	190.5	100	77.3	145.6	250.3

Table 2. **Vapor Pressures of Ammonium Halides**

Temperature, °C	Vapor pressure, kPa[a]		
	NH_4Cl	NH_4Br	NH_4I
250	6.5		
280	17.9		
300	33.5	7.3	
320	60.9	13.3	
338	101.1		
360		41.2	31.3
380		73.4	54.1
395		101.1	
400		115.4	89.8
405			101.1

[a] To convert kPa to mm Hg, multiply by 7.5.

attack ferrous and other metal surfaces, especially those of copper and copper alloys.

5.1. Ammonium Chloride. *Manufacture.* The history of ammonium chloride manufacture is linked to the birth of the soda and synthetic ammonia industries. Consequently this halide has always been a by-product in great supply. Production by direct reaction of ammonia and hydrochloric acid is simple but usually economically unattractive; a process based on metathesis or double decomposition is generally preferred.

Several commercial grades are available: fine crystals of 99 to 100% purity, large crystals, pressed lumps, rods, and granular material.

Double-Decomposition Methods. Double-decomposition processes all involve the reaction of sodium chloride, the cheapest chlorine source, with an ammonium salt. The latter may be supplied directly, or generated *in situ* by the reaction of ammonia and a supplementary ingredient. Ammonium chloride and a sodium salt are formed. The sodium salt is typically less soluble and is separated at higher temperatures; ammonium chloride is recovered from the filtrate by cooling.

Ammonia-Soda Process. Ammonium chloride is made as a by-product of the classic Solvay process, used to manufacture sodium carbonate (qv) (14,15). The method involves reaction of ammonia, carbon dioxide, and sodium chloride in water

$$NH_3 + CO_2 + NaCl + H_2O \longrightarrow NaHCO_3 + NH_4Cl$$

Sodium bicarbonate precipitates from solution and is recovered by filtration. Ammonium chloride is then crystallized from the filtrate, separated, washed, and dried. The exact proportion of ammonium chloride recovered depends on the relative demands for sodium carbonate and ammonium chloride. If economic conditions require, part of the ammonia can be recovered and returned to the brine-ammoniation step by distillation of the ammonium chloride solution in the presence of lime. The spent calcium chloride liquor, a final product in manufacture of sodium carbonate by the ammonia–soda process, can also be used to obtain ammonium chloride. This liquor is treated with ammonia and carbon dioxide

$$CaCl_2 + 2\,NH_3 + CO_2 + H_2O \longrightarrow CaCO_3 + 2\,NH_4Cl$$

Calcium carbonate is removed by filtration leaving an ammonium chloride solution.

Ammonium Sulfate–Sodium Chloride Process. Ammonium sulfate, a readily available by-product, has been much used to make ammonium chloride by a double decomposition reaction with sodium chloride.

$$(NH_4)_2SO_4 + 2\,NaCl \longrightarrow Na_2SO_4 + 2\,NH_4Cl$$

The ammonium sulfate and sodium chloride are simultaneously dissolved, preferably in a heel of ammonium chloride solution. The sodium chloride is typically in excess of about 5%. The pasty mixture is kept hot and agitated vigorously. When the mixture is separated by vacuum filtration, the filter and all connections are heated to avoid crust formation. The crystalline sodium sulfate is washed to remove essentially all of the ammonium chloride and the washings recycled to the process. The ammonium chloride filtrate is transferred to acid resistant crystallizing pans, concentrated, and cooled to effect crystallization. The crystalline NH_4Cl is washed with water to remove sulfate and dried to yield a product of high purity. No attempt is made to recover ammonium chloride remaining in solution. The mother liquor remaining after crystallization is reused as a heel.

Ammonium Sulfite–Sodium Chloride Process. Ammonium chloride has been produced by the reaction of ammonium sulfite [10196-04-0], NH_4SO_3, and sodium chloride in a large Canadian plant (16). Ammonium sulfite is never actually isolated, rather ammonia and sulfur dioxide react in water with sodium chloride.

$$2\,NaCl + SO_2 + 2\,NH_3 + H_2O \longrightarrow Na_2SO_3 + 2\,NH_4Cl$$

This process is only practical when the raw materials are readily available and of high purity.

The sodium sulfite precipitates first and is removed by centrifugation, washed with water, and dried. The mother liquor containing ammonium chloride is sent to crystallizing tanks and the salt thus formed is washed and dried, giving a product said to analyze well over 99%.

Direct Neutralization. Because of the availability of by-product ammonium salts, the double decomposition routes are usually more favorable economically for ammonium chloride manufacture. However, where surplus hydrogen chloride is available, the direct neutralization process has been used (17)

$$NH_3(g) + HCl(g) \longrightarrow NH_4Cl \qquad \Delta H = -175.7\ \text{kJ/mol}\ (-42\ \text{kcal/mol})$$

The reaction is very exothermic and the heat generated is used to evaporate a large part of the water present when aqueous hydrochloric acid is used. Batch or continuous crystallization is then employed to recover the ammonium chloride.

A Brazilian company had reported producing ammonium chloride from hydrogen chloride gas (18). Hydrogen chloride is mixed with air and introduced into a saturated ammonium chloride suspension at 80°C. Excess ammonia is added to a conical section of the saturator to maintain a pH of 8. The ammonium chloride is recovered from the suspension by thickening in a hydroclone, followed by centrifugation and drying. Mother liquor and the water used to scrub waste gases, are returned to the saturator.

Uses. Ammonium chloride is used as a nitrogen source for fertilization of rice, wheat, and other crops in Japan, China, India, and Southeast Asia. Japan is a large producer, much of which is as by-product.

Ammonium chloride has a number of industrial uses, most importantly in the manufacture of dry-cell batteries, where it serves as an electrolyte. It is also used to make quarrying explosives, as a hardener for formaldehyde-based adhesives, as a flame suppressant, and in etching solutions in the manufacture of printed circuit boards. Other applications include use as a component of fluxes in zinc and tin plating, and for electrolytic refining of zinc.

Technical grade ammonium chloride is widely available as the crystalline salt; technical rods and granules are also available. Its DOT number is NA9085.

5.2. Ammonium Bifluoride. *Properties.* Ammonium bifluoride, NH_4HF_2, is a colorless, orthorhombic crystal (19). The compound is odorless; however, less than 1% excess HF can cause an acid odor. The salt has no tendency to form hydrates yet is hygroscopic if the ambient humidity is over 50%. A number of chemical and physical properties are listed in Table 3.

Ammonium bifluoride dissolves in aqueous solutions to yield the acidic bifluoride ion; the pH of a 5% solution is 3.5. In most cases, NH_4HF_2 solutions react readily with surface oxide coatings on metals; thus NH_4HF_2 is used in

Table 3. **Properties of Ammonium Bifluoride, NH_4HF_2**

Property	Value	Reference
melting point, °C	126.1	20
boiling point, °C	239.5	20
index of refraction, n_D	1.390	20
solubility at 25°C, wt %		
water	41.5	20
90% ethanol	1.73	20
specific gravity	1.50	20
standard heat of formation, kJ/mol[a]	−798.3	21
heat of fusion, kJ/mol[a]	19.1	21
heat of vaporization, kJ/mol[a]	65.3	21
heat of solution, kJ/mol[a]	20.3	21
heat of dissociation,[b] kJ/mol[a]	141.4	21
heat capacity, C_p, J/(mol·K)[a] at 25°C	106.7	21
vapor pressure,[c] $\log P_{Pa} = a - bT^{-1}$		
153–207°C	$a = 11.72, b = 3370$	22
207–245°C	$a = 9.38, b = 2245$	22

[a]To convert kJ to kcal, divide by 4.184.
[b]$NH_4HF_2 \longrightarrow NH_3 + 2\,HF$.
[c]To convert Pa to mm Hg, multiply by 7.5×10^{-3}.

pickling solutions (see METAL SURFACE TREATMENTS). Many plastics, such as poly-ethylene, polypropylene, unplasticized PVC, and carbon brick, are resistant to attack by ammonium bifluoride.

Manufacture.   Anhydrous ammonium bifluoride containing 0.1% H_2O and 93% NH_4HF_2 can be made by dehydrating ammonium fluoride solutions and by thermally decomposing the dry crystals (23). Commercial ammonium bifluoride, which usually contains 1% NH_4F, is made by gas-phase reaction of one mole of anhydrous ammonia and two moles of anhydrous hydrogen fluoride (24); the melt that forms is flaked on a cooled drum.

Production of bifluoride from fluoride by-products from the phosphate industry (25) has had little if any commercial significance.

Health and Safety Factors.   Ammonium bifluoride, like all soluble fluor-ides, is toxic if taken internally. When heated to decomposition, it emits very toxic fumes (5). Hydrofluoric acid burns may occur if the material comes in con-tact with moist skin. Ammonium bifluoride solutions should be thoroughly washed from the skin with mildly alkaline soap as soon as possible; however, if contact has been prolonged, the affected areas should be soaked with 0.13% solution of Zephiran chloride, or 0.2% Hyamine 1622 or calcium gluconate, the treatment recommended for hydrofluoric acid burns. If any of these solutions come in contact with the eyes, they should be washed with water for at least 10 min and a physician should be consulted.

Uses.   Ammonium bifluoride solubilizes silica and silicates by forming ammonium fluorosilicate [16919-19-0], $(NH_4)_2SiF_6$. Inhibited 15% hydrochloric acid containing about 2% ammonium bifluoride has been used to acidize oil wells in siliceous rocks to regenerate oil flow (26) (see PETROLEUM). Ammonium fluoride solution is made on-site near the well bore from ammonium bifluoride and ammonia mixed with methyl formate to prevent rapid consumption of most of the HF (27). The use of ammonium bifluoride is important in locations where dissolved silicates foul boiler tubes with scale that cannot be removed using usual cleaning aids (28). Ammonium bifluoride is also used as an etching agent for silicon wafers.

Rapid frosting of glass is accomplished in a concentrated solution of ammo-nium bifluoride and hydrofluoric acid with nucleating agents that assure uni-form frosts (29). A single dip in an aqueous solution of NH_4HF_2, HF, and sorbitol at $<20°C$ for less than 60 s produces the low specular-reflecting finish on television face plates and on (qv) for picture framing (30). Treating glass, eg, often badly weathered window panes, with 2–5% solutions of ammonium bifluoride results in a polishing effect. Glass ampuls for parenteral solutions (31) and optical lenses (32) are best cleaned of adhering particulate matter in dilute ammonium bifluoride solutions.

Ammonium bifluoride is used as a sour or neutralizer for alkalies in com-mercial laundries and textile plants. Treatment also removes iron stain by form-ing colorless ammonium iron fluorides that are readily rinsed from the fabric (33).

Ammonium fluorides react with many metal oxides or carbonates at ele-vated temperatures to form double fluorides; eg,

$$2\,NH_4HF_2 + BeO \xrightarrow{-H_2O} (NH_4)_2BeF_4 \xrightarrow{\Delta} BeF_2 + 2\,NH_3 + 2\,HF$$

The double fluorides decompose at even higher temperatures to form the metal fluoride and volatile NH_3 and HF. This reaction produces pure salts less likely to be contaminated with oxyfluorides. Beryllium fluoride [7787-49-7], from which beryllium metal is made, is produced this way (34) (see BERYLLIUM AND BERYLLIUM ALLOYS). In pickling of stainless steel and titanium, NH_4HF_2 is used with high concentrations of nitric acid to avoid hydrogen embrittlement. Ammonium bifluoride is used in acid dips for steel (qv) prior to phosphating and galvanizing, and for activation of metals before nickel plating (35,36). Ammonium bifluoride also is used in aluminum anodizing formulations. Ammonium bifluoride is used in treatments to provide corrosion resistance on magnesium and its alloys (37). Such treatment provides an excellent base for painting and good abrasion resistance, heat resistance, and protection from atmospheric corrosion. A minor use for ammonium bifluoride is in the preservation of wood (qv) (38).

5.3. Ammonium Bromide and Iodide. *Manufacture.* Ammonium bromide and Ammonium iodide are manufactured either by the reaction of ammonia with the corresponding hydrohalic acid or, more economically, by the reaction of ammonia with elemental bromine or iodine. In the latter reaction, an excess of ammonia must be used.

$$8\,NH_3 + 3\,Br_2 \longrightarrow 6\,NH_4Br + N_2$$

For ammonium bromide, another method involving reaction of an aqueous bromine solution and iron filings has been used. The solution of ferrous and ferric bromide thus formed then reacts with ammonia to precipitate hydrated oxides of iron. Ammonium bromide can be recovered by crystallization from the concentrated liquor.

Uses. Ammonium bromide is available as a dry technical grade or as 38 to 45% solutions. It is used to manufacture chemical intermediates, and in photographic chemicals; it also has some flame retardant applications.

Ammonium iodide has limited use in photographic and pharmaceutical preparations.

6. Ammonium Nitrate

Ammonium nitrate [6484-52-2], NH_4NO_3, formula wt 80.04, is the most commercially important ammonium compound both in terms of production volume and usage. It is the principal component of most industrial explosives and nonmilitary blasting compositions; however, it is used primarily as a nitrogen fertilizer. Ammonium nitrate does not occur in nature because it is very soluble. It was first described in 1659 by the German scientist Glauber, who prepared it by reaction of ammonium carbonate and nitric acid. He called it nitrium flammans because its yellow flame (from traces of sodium) was different from that of potassium nitrate.

Ammonium nitrate fertilizer incorporates nitrogen in both of the forms taken up by crops: ammonia and nitrate ion. Fertilizers (qv) containing only ammoniacal nitrogen are often less effective, as many important crops tend to take up nitrogen mainly in the nitrate form and the ammonium ions must be transformed into nitrate by soil organisms before the nitrogen is readily available. This transformation is slow in cool, temperate zone soils. Thus, ammonium nitrate is a preferred source of fertilizer nitrogen in some countries.

One general disadvantage of nitrogen fertilizers, and ammonium nitrate in particular, is that the nitrate ion is more prone to leach through the soil profile and enter the groundwater. The presence of nitrate in groundwater became an important environmental issue in the 1980s (see GROUNDWATER MONITORING). Efforts by farmers to address this problem have focused on improved management techniques, such as multiple or delayed applications, accurate and efficient placement, and the use of cover crops to take up residual nitrogen and reduce erosion. Controlling the dissolution and release of ammonium nitrate into the soil by use of rosin-coated fertilizer (39) has also been suggested.

6.1. Physical and Chemical Properties. Ammonium nitrate is a white, crystalline salt, $d_4^{20} = 1.725$, that is highly soluble in water, as shown in Table 4 (8). Although it is very hygroscopic, it does not form hydrates. This hygroscopic nature complicates its usage in explosives, and until about 1940, was a serious impediment to its extensive use in fertilizers. The solid salt picks up water from air when the vapor pressure of water exceeds the vapor pressure of a saturated aqueous ammonium nitrate solution (see Table 5).

The boiling point of ammonium nitrate–water solutions, given in Table 6, indicates the temperatures required for removing water (40).

Solid ammonium nitrate occurs in five different crystalline forms (41) (Table 7) detectable by time–temperature cooling curves. Because all phase

Table 4. Solubility of Ammonium Nitrate

Temperature, °C	Solubility of NH_4NO_3, g/100 g		Temperature, °C	Solubility of NH_4NO_3, g/100 g	
	Water	Soln		Water	Soln
0	118	54.2	60	410	80.4
10	150	60.0	70	499	83.3
20	187	65.2	80	576	85.2
30	232	69.9	90	740	88.1
40	297	74.8	100	843	89.4
50	346	77.6			

Table 5. Vapor Pressure of Ammonium Nitrate Solutions

Temperature, °C	Vapor pressure, kPa[a]	
	Water	Saturated NH_4NO_3 soln
10	1.2	0.85
20	2.3	1.5
30	4.2	2.5
40	7.4	3.9

[a]To convert kPa to mm Hg, multiply by 7.5.

Table 6. **Boiling Point of Ammonium Nitrate Solutions**

NH_4NO_3, wt %	Bp, °C	NH_4NO_3, wt %	Bp, °C	NH_4NO_3, wt %	Bp, °C
10	101	60	113.5	94	165
20	102.5	70	119.5	95	170
30	104	80	128.5	96	182
40	107.5	85	136	98	203
50	109.5	90	157	99	222

changes involve either shrinkage or expansion of the crystals, there can be a considerable effect on the physical condition of the solid material. This is particularly true of the 32.3°C transition point which is so close to normal storage temperature during hot weather.

The specific heat of solid β-phase ammonium nitrate is 1.70 J/g (0.406 cal/g) between 0 and 31°C; the specific heats of aqueous NH_4NO_3 solutions are shown in Table 8 (9,42). The coefficient of expansion is 0.000920 at 0°C, 0.000982 at 20°C, and 0.001113 at 100°C; the heat of formation from the elements is 364 kJ/mol (87.1 kcal/mol).

Ammonium nitrate has a negative heat of solution in water, and can therefore be used to prepare freezing mixtures. Dissolution of ammonium nitrate in anhydrous ammonia, however, is accompanied by heat evolution. In dilute solution the heat of neutralization of nitric acid using ammonia is 51.8 kJ/mol (12.4 kcal/mol).

Decomposition and Detonation Hazard. Ammonium nitrate is considered a very stable salt, even though ammonium salts of strong acids generally lose ammonia and become slightly acidic on storage. For ammonium nitrate, endothermic dissociation from lowering pH occurs above 169°C.

$$NH_4NO_3 \longrightarrow HNO_3 + NH_3 \qquad \Delta H = 175 \text{ kJ/mol (41.8 kcal/mol)}$$

Table 7. **Crystalline Forms of Ammonium Nitrate**

Designation	Temperature range, °C	Crystal system
α	<−18	tetragonal
β	−18 − 32.1	rhombic
γ	32.1−84.2	rhombic
δ	84.2−125.2	tetragonal
ε	125.2−169.6	cubic

Table 8. **Specific Heats of Aqueous Solutions of Ammonium Nitrate at 100°C**

NH_4NO_3, wt %	Specific heat, J/g[a]	NH_4NO_3, wt %	Specific heat, J/g
10	3.9	70	2.4
30	3.4	90	1.9
50	2.9		

[a]To convert J to cal, divide by 4.184.

When the salt is heated to temperatures from 200 to 230°C, exothermic decomposition occurs (4,43). The reaction is rapid, but it can be controlled, and it is the basis for the commercial preparation of nitrous oxide [10024-97-2].

$$NH_4NO_3 \longrightarrow N_2O + 2\,H_2O \qquad \Delta H = -37 \text{ kJ/mol } (-8.8 \text{ kcal/mol})$$

Above 230°C, exothermic elimination of N_2 and NO_2 begin.

$$4\,NH_4NO_3 \longrightarrow 3\,N_2 + 2\,NO_2 + 8\,H_2O \qquad \Delta H = -102 \text{ kJ/mol } (-24.4 \text{ kcal/mol})$$

The final violent exothermic reaction occurs with great rapidity when ammonium nitrate detonates.

$$2\,NH_4NO_3 \longrightarrow 2\,N_2 + 4\,H_2O + O_2 \qquad \Delta H = -118.5 \text{ kJ/mol } (-28.3 \text{ kcal/mol})$$

Ammonium nitrate is normally classified as an oxidizing agent. The pure salt is not classed as an explosive because it is difficult to detonate. Spark, flame, or friction do not cause detonation, and ammonium nitrate is relatively insensitive to shock. However, a variety of substances, such as chloride and oil, are known to sensitize the material, so manufacturers strive to eliminate such substances from their processes.

When used in blasting, ammonium nitrate is mixed with fuel oil and sometimes sensitizers such as powdered aluminum. Lower density ammonium nitrate is preferred for explosive formulation, because it absorbs the oil more effectively. When detonated,these mixtures have an explosive power of 40 to 50% that of TNT (see EXPLOSIVES).

6.2. Manufacture. Historically, ammonium nitrate was manufactured by a double decomposition method using sodium nitrate and either ammonium sulfate or ammonium chloride. Modern commercial processes, however, rely almost exclusively on the neutralization of nitric acid (qv), produced from ammonia through catalyzed oxidation, with ammonia. Manufacturers commonly use onsite ammonia although some ammonium nitrate is made from purchased ammonia. Solid product used as fertilizer has been the predominant form produced. However, sale of ammonium nitrate as a component in urea–ammonium nitrate liquid fertilizer has grown to where about half the ammonium nitrate produced is actually marketed as a solution.

Three steps are essential to ammonium nitrate manufacture: neutralization of nitric acid with ammonia to produce a concentrated solution; evaporation to give a melt; and processing by prilling or granulation to give the commercial solid product.

Neutralization. The reaction of ammonia and nitric acid is highly exothermic and the heat released evaporates water, most commonly concentrating the reaction mixture to 83–87% ammonium nitrate. Both reactants are also volatile at the resulting temperatures, thus close control of reactor conditions is necessary to prevent loss of material. Strict temperature regulation in the entire reactor, normally achieved by regulating the addition of feeds and by removal of heat, is particularly important. The heat removed can be recovered for acid preheating, ammonia evaporation, or evaporation of additional water. To avoid localized overheating, reactors are also designed for excellent mixing and utilize automatic pH control.

Neutralizers can be of three designs, depending on the temperature in the reactor zone. They may operate under, exactly at, or above the atmospheric boiling point of the contained ammonium nitrate solution.

Vacuum flash processes, which operate under the atmospheric boiling point of the solution, include the Uhde–I.G. Farbenindustrie process and the closely related Kestner process (44). In these, ammonia, nitric acid, and recirculated ammonium nitrate solution are fed into the neutralizer. Hot solution overflows to an intermediate tank and then to a flash evaporator kept at 18–20 kPa (0.18–0.2 atm) absolute pressure. Partial evaporation of water at this point cools and concentrates the solution, part of which is routed to evaporation. The rest is circulated to the neutralizer.

The ICI process is an example of neutralization at atmospheric pressure. Nitric acid feed is preheated by part of the vapors produced in the neutralizer and is then split into two streams. Recycled, undersized product is dissolved in one stream, conditioning material in the other. The recombined streams are added to a two stage neutralizer along with ammonia and recirculated solution to give 87 to 89% ammonium nitrate feed for evaporation. The C&I–Girdler-Cominco process is similar in principle; the Pintsch-Bamag (45) process uses a two-stage neutralizer without recirculation.

The Fauser process (46), which operates above atmospheric pressure, was an early attempt to fully utilize the heat of neutralization. The neutralizing zone of the enclosed reactor operates at 500–600 kPa (5 to 6 atm). Reactants enter at the bottom of this chamber and hot ammonium nitrate streams upward, where it is discharged continuously into an outer vessel operated at atmospheric pressure. The arriving solution loses part of its water; subsequently most is recirculated through the outer vessel to the lower neutralizing space, while part is removed for further processing. Ammonia and nitric acid feed streams are preheated by partial utilization of the steam from the outer vessel.

The Stamicarbon (44) and Kaltenbach high concentration processes are designed to use the evaporated water vapor produced by pressure neutralization to heat the evaporator used for concentration. The Kaltenbach neutralizer operates at 350 kPa (3.5 bar) and 175°C, and produces steam used to concentrate the solution to 95% in a vacuum evaporator. A recent variation uses a final atmospheric evaporator to produce a 99.7% melt (44).

Other processes, including the Société Belge de l'Azote and Union Chimique Belge (44) are designed to achieve even higher heat recovery which is inherently possible through use of pressure neutralization.

Concentration. Evaporation procedures depend on the concentration of the solution produced during neutralization and the water content required for the subsequent production of solid product. Neutralizer solutions can contain as little as 2% and as much as 25% water; feeds to drum granulators can contain 5% water, prill towers 0.3 to 0.5% water.

Since about 1965, efficient vacuum evaporators have been used in most plants. Second stage evaporators, where the ammonium nitrate is concentrated to more than 99%, are designed to retain only a small volume of melt, have short residence times, and are protected from overheating and contamination by sensitizers. Falling film units are especially suited for this application.

Solid Finished Product.   The final step in ammonium nitrate fertilizer manufacture is the production of a uniformly sized, abrasion and crush-resistant, and free-flowing solid possessing good storage properties. Besides being hygroscopic, ammonium nitrate is subject to degradation, or "sugaring", which occurs during storage as temperatures fluctuate across the 32.1°C crystal phase change. Most manufacturers seek to overcome this latter problem by adding a stabilizing agent to the melt.

Graining, flaking, and spraying have all been used to make solid ammonium nitrate particles. Most plants have adopted various prilling or granulation processes. Crystallized ammonium nitrate has been produced occasionally in small quantities for use in specialty explosives. The Tennessee Valley Authority developed and operated a vacuum crystallization process (47), but the comparatively small crystals were not well received as a fertilizer. The process was subsequently modified to pan granulation (48).

Prilling Process.   Prilling is the formation of a rounded, granular solid by allowing molten droplets to fall through a fluid cooling medium. Prilling of ammonium nitrate involves spraying the concentrated (96% or 99 + %) solution into the top of a large tower. The descending droplets are cooled by an upward flow of air, solidifying into spherical prills that are collected at the bottom. The process yields particles that vary in size depending on the residual moisture of solution, air temperature, and flow rate.

The 96–97% ammonium nitrate solutions are sprayed into towers 33 to 60 m high to produce low density 770 kg/m^3 (48 lbs/ft^3) prills favored for use in ammonium nitrate–fuel oil blasting agents. A drying step is required after prilling. This porous product promotes propagation of detonation and allows for a higher fuel oil loading.

The 99.7–99.8% solutions are sprayed into towers only 20 to 30 m high (49) to give high density 860 kg/m^3 (54 lbs/ft^3) prills preferred by the fertilizer industry. These prills, which require cooling but no drying, are sometimes coated using 2.5–3% of activated clay or diatomaceous earth, although some producers add chemical additives to the melt, giving a product needing no surface conditioner.

Granulation Processes.   In the early 1970s, production of large-particle ammonium nitrate using the spheroidizer granulation process was adopted by several manufacturers. This process, developed by Cominco, Ltd. (Canada) and the C&I–Girdler Corporation, is an adaptation of one used for production of other fertilizers (50). Ammonium nitrate is layered in onion-skin fashion on small seed particles by spraying a 99 + % solution on a dense cascading curtain of granules. This process is carried out in a 3.5–4.5 m diameter by 14–18 m rotating drum having specially designed flights.

The well-rounded particles produced are screened and cooled, giving granules having a moisture content of about 0.1%; they do not require a conditioner. These granules also have a higher crushing strength than prills, and are less subject to breakdown in storage and handling. The process is used by several ammonium nitrate producers in the United States; most calcium ammonium nitrate [39368-85-9] and ammonium nitrate sulfate

[12436-94-1] producers outside of the United States also use this granulation technique.

A fluid bed granulation process developed by Nederlandse Stikstof Maatschappij (NSM) is also available (51). The granules grow from single seed particles. These are fed to a baffled, rectangular vessel, where they are fluidized by a flow of preheated air. A 97% ammonium nitrate solution is sprayed upward into this bed of particles, continuously coating them. Efficient air contact during the 15 min residence time allows evaporation of water as the particles grow in size. The largest granules finally settle and flow from the bottom of the vessel. Pollution control is effected by a wet scrubber and presents few problems.

Blungers and classic drum granulator equipment have also been adapted to the production of granular ammonium nitrate. In the Norsk Hydro Fertilizer, Ltd. process a 92–95% solution is sprayed onto a bed of recycled fines inside a drum granulator (52). Granulator temperatures are held between 85 and 100°C; a small amount of ammonia is also fed to the rolling bed. Material discharged from the granulator is dried either by indirectly heated air or by a gas-fired burner.

Granulation processes offer a number of important advantages. The most significant are decreased pollution problems and the ability to produce granules of almost any reasonable size allowing close size matching with granular ammonium phosphates and potassium chloride in the preparation of NPK fertilizers (48).

6.3. Quality of Product. Ammonium nitrate, commonly made from pure synthetic raw materials, is itself of high purity. If the product is intended for use in explosives, it should be at least 99% ammonium nitrate and contain no more than 0.15% water. It should contain only small amounts of water-insoluble and ether-soluble material, sulfates and chlorides, and should not contain nitrites. The solid product ought to be free from alkalinity, but be only slightly acidic.

If the product is to be used in the manufacture of nitrous oxide, an anesthetic gas, a purity of not less than 99.5% is required. The salt must be almost completely free of contaminating organic matter, iron, sulfate, and chloride.

6.4. Health and Safety Factors. Ammonium nitrate can be considered a safe material if treated and handled properly. Potential hazards include those associated with fire, decomposition accompanied by generation of toxic fumes, and explosion. It is also an allergien (5).

Although ammonium nitrate does not itself burn, it is a strong oxidizer capable of supporting the combustion of numerous substances when heated. It can support and intensify a fire even when air is excluded. Fires involving ammonium nitrate also present a toxic hazard from the release of nitrogen oxides, even though the solid itself is generally considered not to be toxic.

Pure ammonium nitrate is a relatively insensitive explosive material, requiring high initiation energy, but when detonated it can have about 70% of the disruptive strength of nitroglycerine. In 1947, after vigorous fires, two explosions occurred on freighters loaded with heavy paper containers of fertilizer grade ammonium nitrate. These explosions resulted in the proposal of precautions for the handling and transportation of ammonium nitrate (53). Thus, ammonium nitrate must be considered a high explosive under the following three conditions: bolstering by a high velocity explosive, confinement at elevated temperatures, and presence of oxidizable materials.

6.5. Economic Aspects and Uses.   Before World War II most ammonium nitrate was used as an ingredient in high explosives. Subsequently its use as a fertilizer grew rapidly, absorbing about 90% of production in 1975. Ammonium nitrate lost its position as fertilizer to urea. Consumption peaked at 18×10^6 t in 1988 and declined by 12% to 16×10^6 t in 1998. Consumption declined because of changes in agricultural subsidy policies. In 1998, Western Europe, the United States, and the former USSR accounted for 75% of world production in 1998. Marginal growth at a rate of 0.3% is expected. Some of the growth is expected in the former USSR (54).

The DOT label required is OXY; the United Nations number is UN 1842.

Most ammonium nitrate manufactured for the explosives market is used in blasting agents prepared by adding a fuel component, such as diesel oil, to the prilled product. Ammonium nitrate can be hardened for use in explosives (55). Much is consumed in coal mining; the remaining explosive markets are metal mining, nonmetal mining, quarrying, and highway construction.

A small but important use of ammonium nitrate is in the production of nitrous oxide. The gas is generated by controlled heating of ammonium nitrate above 200°C. Nitrous oxide is used primarily as an anesthetic and as an aerosol propellant for food products (see ANESTHETICS; AEROSOLS).

7. Ammonium Nitrate Limestone

Many plants outside of North America prill or granulate a mixture of ammonium nitrate and calcium carbonate. Production of this mixture, often called calcium ammonium nitrate, essentially removes any explosion hazard. In many cases calcium nitrate recovered from acidulation of phosphate rock (see PHOSPHORIC ACID AND THE PHOSPHATES) is reacted with ammonia and carbon dioxide to give a calcium carbonate–ammonium nitrate mixture containing 21 to 26% nitrogen (45).

8. Ammonium Nitrite

Ammonium nitrite [13446-48-5], NH_4NO_2, a compound of questionable stability, can be prepared by reaction of barium nitrite and aqueous ammonium sulfate. After removal of the precipitated barium sulfate by filtration, the ammonium nitrite can be recovered from solution. The salt is a powerful oxidizer and there is a severe explosion hazard when shocked or exposed to heat (60–70°C). When heated to decomposition, it emits toxic fumes (5).

9. Ammonium Sulfate

Ammonium sulfate [7783-20-2], $(NH_4)_2SO_4$, is a white, soluble, crystalline salt having a formula wt of 132.14. The crystals have a rhombic structure; d_4^{20} is 1.769. An important factor in the crystallization of ammonium sulfate is the sensitivity of its crystal habit and size to the presence of other components in the crystallizing solution. If heated in a closed system ammonium sulfate melts at $513 \pm 2°C$ (14); if heated in an open system, the salt begins to decompose at $100°C$, giving ammonia and ammonium bisulfate [7803-63-6], NH_4HSO_4, which melts at $146.9°C$. Above $300°C$, decomposition becomes more extensive giving sulfur dioxide, sulfur trioxide, water, and nitrogen, in addition to ammonia.

The solubility of ammonium sulfate in 100 g of water is 70.6 g at $0°C$ and 103.8 g at $100°C$. It is insoluble in ethanol and acetone, does not form hydrates, and deliquesces at only about 80% relative humidity. The integral heat of solution of ammonium sulfate to saturation in water is 6.57 kJ/mol (1.57 kcal/mol) at $30°C$; at saturation at the same temperature, the differential heat of solution is 6.07 kJ/mol (1.45 kcal/mol).

9.1. Manufacture. Ammonium sulfate is produced from the direct neutralization of sulfuric acid with ammonia; the heat of reaction is sufficient to evaporate all water if the concentration of the acid is 70% or higher.

$$2\,NH_3(g) + H_2SO_4(aq) \longrightarrow (NH_4)_2SO_4(s) \qquad \Delta H = -274 \text{ kJ/mol } (-65.5 \text{ kcal/mol})$$

Ammonium sulfate is also recovered as a by-product in large amounts during the coking of coal, nickel refining, and organic monomer synthesis, particularly during production of caprolactam (qv). About four metric tons of ammonium sulfate are produced per ton of caprolactam which is an intermediate in the production of nylon.

Some companies have used the Merseburg process to manufacture ammonium sulfate from gypsum, but the process is only economically attractive where sulfur is unavailable or very expensive (56), and is thus not used in the United States. Ammonium carbonate, formed by the reaction of ammonia and carbon dioxide in an aqueous medium, reacts with suspended, finely ground gypsum. Insoluble calcium carbonate and an ammonium sulfate solution are formed.

$$(NH_4)_2CO_3 + CaSO_4 \cdot 2H_2O \longrightarrow (NH_4)_2SO_4 + CaCO_3 + 2\,H_2O$$

After removal of the calcium carbonate, the sulfate is recovered by evaporation and crystallization.

Some sulfuric acid producers use ammoniacal solutions to scrub tail gases from stacks in order to conform to federal and state regulations on sulfur dioxide emissions; ammonium sulfate is recovered as a product. Several similar scrubbing processes for removal of sulfur dioxide from electric power plant stack gas have been studied (57). However, such scrubbing operations could generate huge tonnages of ammonium sulfate, flooding local markets because long-distance shipment of the product is generally uneconomical.

9.2. Economic Aspects and Uses. Almost all ammonium sulfate is used as a fertilizer; for this purpose it is valued both for its nitrogen content and for its readily available sulfur content. In North America ammonium sulfate is largely recovered from caprolactam production.

Ammonium sulfate is a good fertilizer for rice, citrus, and vines, and can be especially useful for some sulfur-deficient or high pH soils. Nonfertilizer uses include food processing, fire control, tanning, and cattle feed.

In general, industrialized nations account for world production. World production in 1999 was 3.5×10^6 t. In 1999, Western Europe, the United States, and the former USSR accounted for 44% of world capacity and 49% world production. Southeast Asia was the largest consumer in 1999 and accounted for 19% of consumption (58).

9.3. Health and Safety Factors. Ammonium sulfate is moderately toxic by several routes. Effects of human ingestion are hypermotility, diarrhea, nausea or vomiting. Reactions with sodium hypochlorite gives the unstable explosive nitrogen trichloride. When heated to decomposition, it emits very toxic fumes (5).

10. Ammonium Sulfides

Ammonia combines with hydrogen sulfide, sulfur, or both, to form various ammonium sulfides and polysulfides. Generally these materials are somewhat unstable, tending to change in composition on standing. Ammonium sulfides are used by the textile industry.

10.1. Ammonium Sulfide. Ammonium sulfide [12135-76-1], $(NH_4)_2S$, can be produced by the reaction of hydrogen sulfide with excess ammonia

$$2\,NH_3 + H_2S \longrightarrow (NH_4)_2S$$

Solid ammonium sulfide decomposes to ammonia and ammonium hydrosulfide at about $-18°C$; consequently it is normally marketed as a 40–44% aqueous solution.

10.2. Ammonium Hydrosulfide. The reaction of equimolar amounts of ammonia and hydrogen sulfide results in the formation of ammonium hydrosulfide [12124-99-1], NH_4HS, which is also produced by the loss of ammonia from ammonium sulfide. The hydrosulfide is very soluble in water, liquid ammonia, liquid hydrogen sulfide, and alcohol. Solid ammonium hydrosulfide has a high vapor pressure, 99.7 kPa (748 mm Hg) at 32.1°C, and sublimes easily at ordinary temperatures. Vapors from the hydrosulfide, composed of ammonia and hydrogen sulfide, are very toxic.

11. Quaternary Ammonium Compounds

There are a vast number of quaternary ammonium compounds or quaternaries (59,60). Many are naturally occurring and have been found to be crucial in biochemical reactions necessary for sustaining life. A wide range of quaternaries are also produced synthetically and are commercially available. Over 344,000 metric tons of quaternary ammonium compounds are produced annually in the United States. The economic value is estimated at $810 MM (61). These have many diverse applications. Most are eventually formulated and make their way to the marketplace to be sold in consumer products. Applications range from cosmetics (qv) to hair preparations (qv) to clothes softeners, sanitizers for eating utensils, and asphalt emulsions.

Most quaternary ammonium compounds have the general formula $R_4N^+ \, X^-$ and are a type of cationic organic nitrogen compound. The nitrogen atom, covalently bonded to four organic groups, bears a positive charge that is balanced by a negative counterion. Heterocyclics, in which the nitrogen is bonded to two carbon atoms by single bonds and to one carbon by a double bond, are also considered quaternary ammonium compounds. The R group may either be equivalent or correspond to two to four distinctly different moieties. These groups may be any type of hydrocarbon: saturated, unsaturated, aromatic, aliphatic, branched chain, or normal chain. They may also contain additional functionality and heteroatoms. Examples include methylpyridinium iodide [930-73-4] (59), benzyldimethyloctadecylammonium chloride [122-19-0] (61), and di(hydrogenated tallow)alkyldimethylammonium chloride [61789-80-8] (62), where $R=C_{14}-C_{18}$.

11.1. Nomenclature.
Quaternary ammonium compounds are usually named as the substituted ammonium salt. The anion is listed last (62). Substituent names can be either common (stearyl) or IUPAC (octadecyl). If the long chain in the compound is from a natural mixture, the chain is named after that mixture, eg, tallowalkyl. Prefixes such as di- and tri- are used if an alkyl group is repeated. Complex compounds usually have the substituents listed in alphabetical order. Some common quaternary ammonium compounds and their applications in patent literature are listed in Table 9.

11.2. Physical Properties.
Most quaternary compounds are solid materials that have indefinite melting points and decompose on heating. Physical properties are determined by the chemical structure of the quaternary ammonium compound as well as any additives such as solvents. The simplest quaternary ammonium compound, tetramethylammonium chloride [75-57-0], is very soluble in water (211) insoluble in nonpolar solvents. As the molecular weight of the quaternary compound increases, solubility in polar solvents decreases and solubility in nonpolar solvents increases (212–214). For example, trimethyloctadecylammonium chloride [112-03-8] is soluble in water up to 27%, whereas dimethyldioctadecylammonium chloride [107-64-2] has virtually no solubility in water. Appropriately formulated, however, this latter compound can be dispersed in water at relatively high (~15%) levels.

The ability to form aqueous dispersions is a property that gives many quaternary compounds useful applications. Placement of polar groups, eg, hydroxy or ethyl ether, in the quaternary structure can increase solubility in polar solvents.

Higher order aliphatic quaternary compounds, where one of the alkyl groups contains ~10 carbon atoms, exhibit surface-active properties (215,216). These compounds compose a subclass of a more general class of compounds known as cationic surfactants (qv). These have physical properties such as substantivity and aggregation in polar media (217) that give rise to many practical

Table 9. **Selected Quaternary Ammonium Compounds and Their Applications**

Quaternary	CAS Registry Number	Industry	Application and function	Comments	References
many compounds claimed, as an example: polypropoxylated (6) choline chloride		agricultural	surfactant	key component in glyphosphate composition	63–66
tetrabutylammonium hydroxide	[2052-49-5]	chemical	phase-transfer catalyst	used as a catalyst for the production of gem-dichloro compounds	67
alkyltrimethyl or dialkydimethyl type quaternaries, as an example dicocoalkyldimethylammonium chloride	[61789-77-3]	chemical	emulsifier	emulsifier for silanes useful as masonry water repellents	68
many compounds claimed, as an example: dicocoalkyldimethyl-ammonium chloride	[61789-77-3]	chemical	complex agent	complexed with anionic dyes to produce a formulation free of inorganic salts	69,70
many perfluoroalkyl quaternary ammonium compounds claimed, as an example: di(4,4,5,5,6,6,7,7,7-nonafluoro-heptyl)dimethylammonium chloride		chemical	phase-transfer catalyst	a family of novel quaternaries useful as phase-transfer catalysts especially in basic media	71
many diquaternary ammonium compounds claimed, as an example: 1,3-bis(dipentylethyl-ammonium)propane dibromide		chemical	catalysts	catalyst for the interfacial polymerization polycarbonate preparation	72
many compounds claimed, as an example: tridecylmethylammonium chloride	[5137-56-4]	chemical	component in catalyst	key component in a catalyst composition containing a zirconium compound	73
di(hydrogenated tallowalkyl)dimethyl ammonium chloride and others	[61789-80-8]	chemical	nonvolatile compositions	compositions containing vegetable oils as diluents useful in organoclays	74

Compound	CAS Number	Function	Industry	Application	Ref.
many compounds claimed, as an example: diallyldimethylammonium chloride	[7398-69-8]	antimicrobial	chemical	antibacterials for polymer latexes and resins	75–77
benzyltrimethylammonium chloride and others	[56-93-9]	surfactant	defense	component in a formulation to neutralize chemical and biological warfare agents	78,79
many compounds claimed, as an example: benzyldimethyl[2-(3,5-di-*tert*-butyl 4-hyroxybenzoyloxy)-ethyl]ammonium *m*-nitrobenzenesulfonate		charge control agent	electronics	component in toner compositions	80–84
benzylacetyldimethylammonium chloride		dispersant	electronics	component in electrostatic liquid developer formulation useful in color copying	85
mixtures of quaternary ammonium hydroxides and halides		dissolution agent	electronics	components in positive photoresist formulation to improve the dissolution selectivity between exposed and unexposed portion of photoresist	86
polyquaternary ammonium compounds		binder	electronics	key component in information storage layer of electronic recording medium	87
many compounds claimed, as example: methyltrioctylammonium chloride and tridecylmethylammonium chloride	[5137-55-3] and [5137-56-4]	surfactant	electronics	component in blocking layer of an electrographic photosensitive material	88
mixtures of quaternary ammonium hydroxides and carbonates, as an example: tetramethyl ammonium hydroxide and tetramethylammonium hydrogen carbonate	[75-59-2]	surfactant/ buffering agent	electronics	components in a developing solution for producing printed circuit boards	89

Table 9 (*Continued*)

Quaternary	CAS Registry Number	Industry	Application and function	Comments	References
As an example: tetrabutylammonium tetrafluoroborate	[429-42-5]	electronics	charge control agent	additive for the preparation of phase change inks with increased specific conductance	90
many compounds claimed, as an example: tetradecyltrimethylammonium bromide	[1119-97-7]	electronics	biocides	biocides for color reversal photographic film and photographic reversal bath	91–93
tetramethylammonium hydroxide	[75-59-2]	electronics	surface active agent	in formulation for polishing semiconductor wafers	94
quaternary ammonium hydroxide		electronics	surfactant	key component in removing agent formulation for producing semiconductor integrated circuits	95
quaternary ammonium hydroxide		electronics	surfactant	components in cleaning formulations for semiconductor devices	96,97
quaternary ammonium hydroxide		electronics	surfactant	component in cleaning composition for removing plasma etching residues	98
tetraalkylammonium halide		food	antimicrobial	compositions for removal and prevention of microbial contamination	99,100
many compounds claimed		household	antimicrobial	disenfecting component in cleaning composition	101–109
hexadecylpyridinium chloride	[123-03-5]	household	antimicrobial	antimicrobial is dispersed throughout plastic toothbrush	110

732

compound	CAS	sector	application	description	ref
dimethyldialkylquaternary ammonium		household	antimicrobial	claimed synergy when used in combination with water soluble anionic surfactant	111
pentamethyltallowalkyl-1,3-propanediammonium chloride	[68607-29-4]	household	antimicrobial	used in a composition for deodorizing footware	112
hexadecyltrimethylammonium chloride	[112-02-7]	household	surfactant	component in vicoselastic thickening system for opening drains	113
ethoxylated quaternary ammonium compounds		household	surfactant	used in a formulation to remove road-film	114,115
trialkylammoniumacetylpyrrolidone chloride		household	bleach activator	in a detergent or cleaning formulation	116
ester based quaternary ammonium compounds		household	fabric Softening	for the preparation of rinse cycle fabric softening formulations	117–136
amidoamine and branched quaternary ammonium compounds		household	fabric softening	for the preparation of rinse cycle fabric softening formulations	137,138
ester based quaternary ammonium compounds		household	antistatic	for use in dryer-activated fabric conditioning and antistatic compositions	139–144
alkoxylated quaternary ammonium compounds		household	detergents	for improved performance in detergent formulations such as optimum grease and soil removal, enhancement of bleach efficacy, and better cold temperature performance	145–153
many compounds claimed, as an example: hexadecyltrimethylammonium bromide	[57-09-0]	household	antimicrobial	for use in liquid laundry detergent composition	154,155
ester based quaternary ammonium compounds		household	fabric softening	for use in liquid detergent formulations that soften fabric	155–158

Table 9 (*Continued*)

Quaternary	CAS Registry Number	Industry	Application and function	Comments	References
2-hydroxyethyltrialkyl ammonium halide		mining	complexing agent for gold anions	quat is adsorbed onto porous polymer resin and forms ion pairs with gold anions	159
quaternary ammonium salts and mixtures		mining	froth flotation	recovery of petalite free of feldspar	160
many compounds claimed, as an example: didecyldimethylammonium chloride	[7173-51-5]	other	antimicrobial	active component in antimicrobial formulations	161–164
pyridinium halides such as decyl-pyridinium bromide		other	anticorrosion	functions as corrosion inhibitor to protect metal surfaces from acid	165,166
polyquaternary ammonium		other	superabsorbent	cationic copolymer with improved water absorbing properties	167
unsaturated quaternary ammonium compounds		other	surfactant	component in a photo-curable antifogging composition for glass	168
polyquaternary ammonium		other	pesticide	to control the infestation of snails in aqueous systems	169
2-ethylhexylhydrogenated-*t*-allowalkyl-dimethyl ammonium methosulfate	[308074-31-9]	organoclay	modify smectite-type clay	organoclay product containing branched chain quaternary ammonium compounds	170
alkyl quaternary ammonium salt		organoclay	modify mineral clay	organoclay composition comprising a mineral clay treated with an alkyl quaternary ammonium salt	171
ester based quaternary ammonium compounds		organoclay	modify smectite-type clay	organoclays useful for non-aqueous systems like paints, inks and coatings	172

734

Area	Application	Description	CAS number	Compound	Ref.
organoclay	modify smectite-type clay	organoclays useful for nanocomposites and rheological additives	[61789-80-8]	preferred is dimethyldi(hydrogenated tallowalkyl)ammonium chloride	173
paper	active in paper softening composition	key component to prepare tissue with a soothing feeling	[112-03-8]	many compounds claimed, as an example: octadecyltrimethyl-ammonium chloride	174
paper	component in tissue paper web	acts as an antimigration material for emollient lotion	[61789-81-9]	many compounds claimed, as an example: dimethyldi(hydrogenated tallowalkylalkyl)ammonium methosulfate	175
paper	softening	for the manufacture of soft absorbent paper products such as paper towels, facial tissue and toilet tissue		ester based quaternary ammonium compounds	176–180
personal care	thickeners and dispersants	quaternary ammonium copolymers useful in cosmetic compositions		polyquaternary ammonium	181,182
personal care	conditioning shampoo	conditioning component	[57905-74-3]	many compounds claimed, as an example: N-methyl-N,N-bis(2-($C_{16/18}$-acyloxy)ethyl)-N-hydroxyethylammonium methosulfate	183–188
personal care	hair conditioner	active ingredient in formulation		as an example: 1,2-ditallowalkyloxy-3-trimethylammonium-propane chloride	189,190
personal care	delivery agent	used in a formulation to deliver the active component to the hair or skin		unsaturated quaternary ammonium compounds	191,192
personal care	antimicrobial	used as a disinfectant for ophthalmic compositions		polyquaternary ammonium	193
petroleum	drilling fluids	drilling fluid compositions having special rheological properties	[61789-80-8]	as an example: dimethyldi(hydrogenated tallowalkyl)ammonium chloride	194,195
petroleum	key component of a process	method for scavenging mercaptans in hydrocarbon fluid		quaternary ammonium hydroxides	196,197

Table 9 (Continued)

Quaternary	CAS Registry Number	Industry	Application and function	Comments	References
dicocoalkyldimethylammonium chloride	[61789-77-3]	petroleum	oil spill recovery	oil spill rediediation agent containing organoclay and waste paper	198
eg, tetrabutylammonium bromide	[1643-19-2]	petroleum	phase-transfer catalyst	method of removing contaminants from petroleum distillates and contaminants from used oils	199–201
ester based quaternary ammonium compounds		petroleum	surfactant	used in a composition for enhanced recovery of crude oils	202
chloromethyl(8-hexadecenyl)dimethylammonium chloride		pharmaceutical	antitumorous	active component in composition	203
cationic lipids		pharmaceutical	receptor	improved cell targeting ability for the delivery of molecules into cells	204
polyquaternary ammonium		pharmaceutical		used to lower cholesterol levels	205
hydrogenatedtallowalkyltrimethyl ammonium chloride	[61788-78-1]	remediation	component in an organoclay	organoclay adsorbs dissolved heavy metals including lead and radiactive contaminants from aqueous solutions	206
many compounds claimed, as an example: benzyltrimethylammonium chloride	[56-93-9]	remediation	component in an organoclay	removal of aromatic petroleum-based contaminants from water	207
trimethylbetahydroxyethyl ammonium hydroxide and others	[123-41-1]	rubber	accelerator	for vulcanization of rubber with non-toxic material	208
dialkyldimethyl quaternary ammonium compounds		wood	biocides	biocidal component in a formulation to waterproof and preserve wood	209,210

applications. The aqueous phase behavior (218) and adsorption at interfaces (219) of cationic surfactants has been discussed. In some cases the ammonium compounds are referred to as inverse soaps because the charge on the organic portion of the molecule is cationic rather than anionic.

11.3. Chemical Properties. Reactions of quaternaries can be categorized into three types (220): Hofmann eliminations, displacements, and rearrangements. Thermal decomposition of a quaternary ammonium hydroxide to an alkene, tertiary amine, and water is known as the Hofmann elimination (eq. 1a) (221). This reaction has not been used extensively to prepare olefins. Some cyclic olefins, however, are best prepared this way (222). Exhaustive methylation, followed by elimination, is known as the Hofmann degradation and is important in the structural determination of unknown amines, especially for alkaloids (qv) (223).

$$CH_3CH_2CH_2\overset{+}{N}(CH_3)_3 \;+\; {}^-OH \quad\xrightarrow[\text{displacement}]{\text{elimination}}$$

$$CH_3CH_2{=}CH_2 \;+\; N(CH_3)_3 \;+\; H_2O \qquad (1a)$$

$$CH_3CH_2CH_2N(CH_3)_2 \;+\; CH_3OH \qquad (1b)$$

Displacement of a tertiary amine from a quaternary (eq. 1b) involves the attack of a nucleophile on the α-carbon of a quaternary and usually competes with the Hofmann elimination (224). The counterion greatly influences the course of this reaction. For example, the reaction of propyltrimethylammonium ion with hydroxide ion yields 19% methanol and 81% propylene, whereas the reaction with phenoxide ion yields 65% methoxybenzene and 15% propylene (225).

The Stevens rearrangement (eq. 2) is a base-promoted 1,2-migration of an alkyl group from a quaternary nitrogen to carbon (226,227). The Sommelet-Hauser rearrangement (eq. 3) is a base-promoted 1,2-migration of a benzyl group to the ortho position of that benzyl group (228,229).

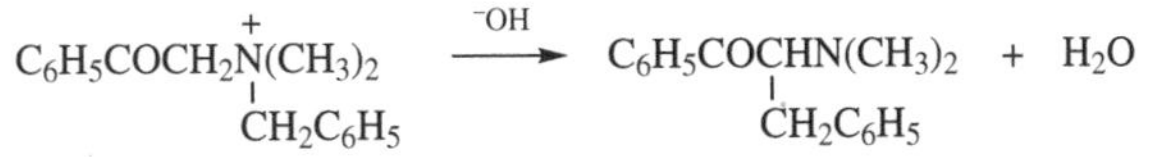

$$C_6H_5COCH_2\overset{+}{\underset{\underset{CH_2C_6H_5}{|}}{N}}(CH_3)_2 \xrightarrow{\;{}^-OH\;} C_6H_5COCH\underset{\underset{CH_2C_6H_5}{|}}{N}(CH_3)_2 \;+\; H_2O \qquad (2)$$

$$C_6H_5CH_2\overset{+}{\underset{\underset{CH_2}{|}}{N}}(CH_3)_2 \xrightarrow[\text{NH}_3\,(l)]{\text{NaNH}_2} \qquad (3)$$

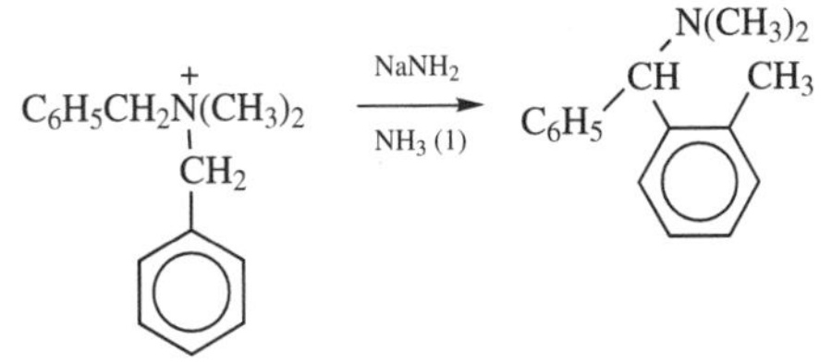

11.4. Naturally Occurring Quaternaries. Naturally occurring quaternary ammonium compounds have been reviewed (230). Many types of aliphatic, heterocyclic, and aromatic derived quaternary ammonium compounds are produced both in plants and invertebrates. Examples include thiamine (vitamin B₁) (231) (see VITAMINS); choline (qv) [62-49-7] (4); and acetylcholine (64). These have numerous biochemical functions. Several quaternaries are precursors for active metabolites.

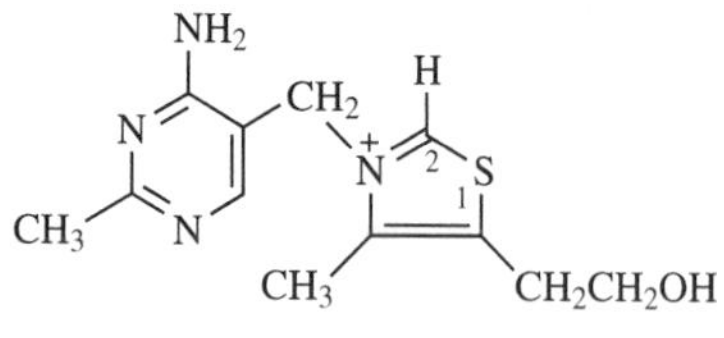

(4) 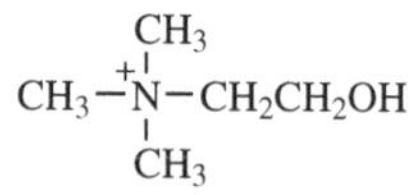   (5) 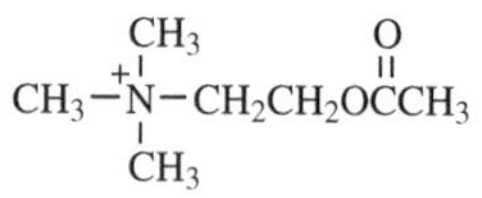   (6)

Thiamine (231) functions as a coenzyme in several enzymatic reactions in which an aldehyde group is transferred from a donor to a receptor molecule. The thiazole ring is the focus of this chemistry. Thiamine also serves as a coenzyme in the pyruvate dehydrogenase and α-ketoglutarate dehydrogenase reactions. These take place in the main pathway of oxidation of carbohydrates (qv) in cells.

Choline functions in fat metabolism and transmethylation reactions. Acetylcholine functions as a neurotransmitter in certain portions of the nervous system. Acetycholine is released by a stimulated nerve cell into the synapse and binds to the receptor site on the next nerve cell, causing propagation of the nerve impulse.

Biochemically, most quaternary ammonium compounds function as receptor-specific mediators. Because of their hydrophilic nature, small molecule quaternaries cannot penetrate the alkyl region of bilayer membranes and must activate receptors located at the cell surface. Quaternary ammonium compounds also function biochemically as messengers, which are generated at the inner surface of a plasma membrane or in a cytoplasm in response to a signal. They may also be transferred through the membrane by an active transport system.

General types of physiological functions attributed to quaternary ammonium compounds are curare action, muscarinic–nicotinic action, and ganglia blocking action. The active substance of curare is a quaternary that can produce muscular paralysis without affecting the central nervous system or the heart. Muscarinic action is the stimulation of smooth-muscle tissue. Nicotinic action is primary transient stimulation and secondary persistent depression of sympathetic and parasympathetic ganglia.

11.5. Synthesis and Manufacture. A wide variety of methods are available for the preparation of quaternary ammonium compounds (232–234). Significantly fewer can be used on a commercial scale. A summary of the most commonly used commercial methods is given herein.

Quaternary ammonium compounds are usually prepared by reaction of a tertiary amine and an alkylating agent (eq. 4). The most widely used alkylating

Table 10. Typical Alkylating Agents for the Preparation of Quaternaries

Alkylating Agent	Chemical Formula	CAS Registry Number	Final Quaternary
methyl chloride	CH_3Cl	[74-87-3]	$R_3N^+CH_3$ Cl^-
dimethyl sulfate	$(CH_3)_2SO_4$	[77-78-1]	$R_3N^+CH_3$ CH_3SO_4
diethyl sulfate	$(CH_3CH_2)_2SO_4$	[64-67-5]	$R_3N^+CH_3$ $CH_3CH_2SO_4$
benzyl chloride	$C_6H_5CH_2Cl$	[100-44-7]	$R_3N^+CH_2C_6H_5$ Cl^-

agents are listed in Table 10. Some of these alkylating reagents pose significant health concerns and require special handling techniques. Alkylation reactions are usually run at moderate (60–100°C) temperatures. When methyl chloride is used, the reactions are often performed under moderate [415–790 kPa (60–115 psi)] pressures.

$$R-\underset{\underset{}{\overset{\overset{R'}{|}}{N}}}{}-R'' \;+\; R'''X \;\longrightarrow\; R-\underset{\underset{R'''}{|}}{\overset{\overset{R'}{+|}}{N}}-R''\;\; X^- \tag{4}$$

Equation 4 can be classified as S_N2, ie, substitution nucleophilic bimolecular (235). The rate of the reaction is influenced by several parameters: basicity of the amine, steric effects, reactivity of the alkylating agent, and solvent polarity. The reaction is often carried out in a polar solvent, eg, isopropyl alcohol, which may increase the rate of reaction and make handling of the product easier.

Primary and secondary amines are usually converted to tertiary amines using formaldehyde and hydrogen in the presence of a catalyst (eqs. 5 and 6). This process, known as reductive alkylation (236), and is attractive commercially. The desired amines are produced in high yields and without significant by-product formation. Quaternization by reaction of an appropriate alkylating reagent then follows.

$$RNH_2 + 2\,CH_2O \xrightarrow[\text{H}_2/\text{catalyst}]{} RN(CH_3)_2 \tag{5}$$

$$R_2NH + CH_2O \xrightarrow[\text{H}_2/\text{catalyst}]{} R_2NCH_3 \tag{6}$$

Dialkyldimethyl and alkyltrimethyl quaternaries can be prepared directly from secondary and primary amines as shown in eqs. 7 and 8, respectively. This process, known as exhaustive alkylation, is usually not the method of choice on a commercial scale. This technique requires the continuous addition of basic material over the course of the reaction to prevent the formation of amine salts (237–238). Furthermore, products such as inorganic salt and water must be removed from the quaternary. The salt represents a significant disposal problem.

$$R_2NH + 2\,CH_3Cl + NaOH \rightarrow R_2N^+(CH_3)_2 + NaCl + H_2O \tag{7}$$

$$RNH_2 + 3\,CH_3Cl + 2\,NaOH \rightarrow RN^+(CH_3)_3 + 2\,NaCl + 2\,H_2O \tag{8}$$

Synthesis and Manufacture of Amines. The chemical and business segments of amines (qv) and quaternaries are so closely linked that it is difficult to consider these separately. The majority of commercially produced amines originate from three amine raw materials: natural fats and oils, α-olefins, and fatty alcohols. Most large commercial manufacturers of quaternary ammonium compounds are fully back-integrated to at least one of these three sources of amines. The amines are then used to produce a wide array of commercially available quaternary ammonium compounds. Some individual quaternary ammonium compounds can be produced by more than one synthetic route.

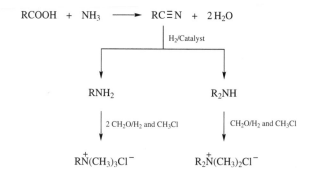

Fig. 1.   Quaternaries from fatty nitriles where R is a fatty alkyl group.

Nitrile Intermediates.   Most quaternary ammonium compounds are produced from fatty nitriles (qv), which are in turn made from a natural fat or oil-derived fatty acid and ammonia (qv) (Fig. 1) (see FATS AND FATTY OILS) (239). The nitriles are then reduced to the amines. A variety of reducing agents may be used (240). Catalytic hydrogenation over a metal catalyst is the method most often used on a commercial scale (241). Formation of secondary and tertiary amine side-products can be hindered by the addition of acetic anhydride (242) or excess ammonia (243). In some cases secondary amines are the desired products.

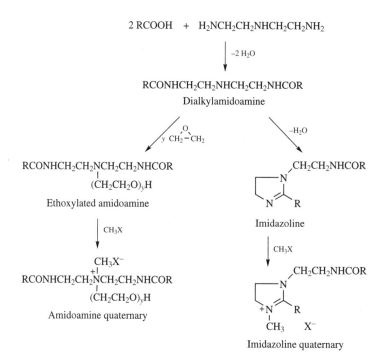

Fig. 2.   Quaternaries from amidoamines and imidazolines where R is a fatty alkyl group.

Fats, Oils, or Fatty Acids. The primary products produced directly from fats, oils, or fatty acids without a nitrile intermediate are the quaternized amidoamines, imidazolines, and ethoxylated derivatives (Fig. 2). Reaction of fatty acids or tallow with various polyamines produces the intermediate dialkylamidoamine. By controlling reaction conditions, dehydration can be continued until the imidazoline is produced. Quaternaries are produced from both amidoamines and imidazolines by reaction with methyl chloride or dimethyl sulfate. The amidoamines can also react with ethylene oxide (qv) to produce ethoxylated amidoamines that are then quaternized.

These compounds and their derivatives can be manufactured using relatively simple equipment compared to that required for the fatty nitrile derivatives. Cyclization of amidoamines to imidazolines requires higher reaction temperatures and reduced pressures. Prices of imidazolines are therefore high.

Olefins and Fatty Alcohols. Alkylbenzyldimethylammonium (ABDM) quaternaries are usually prepared from α-olefin or fatty alcohol precursors. Manufacturers that start from the fatty alcohol usually prefer to prepare the intermediate alkyldimethylamine directly by using dimethylamine and a catalyst rather than from fatty alkyl chloride. Small volumes of dialkyldimethyl and alkyltrimethyl quaternaries in the C_8–C_{10} range are also manufactured from these precursors (Fig. 3).

Quaternized Esteramines. Esterquaternary ammonium compounds or esterquats can be formulated into products that have good shelf stability (244). Many examples of this type of molecule have been developed (see Fig. 4).

Quaternized esteramines are usually derived from fat or fatty acid that reacts with an alcoholamine to give an intermediate esteramine. The esteramines are then quaternized. A typical reaction scheme for the preparation of a diester quaternary is shown in equation 9 (246), where R is a fatty alkyl group. Reaction occurs at 75–115°C in the presence of sodium methoxide catalyst. Free fatty acids (254) and glycerides (255) can be used in place of the fatty acid methylester.

$$2\,RCOOCH_3 \; + \; N(CH_2CH_2OH)_3 \;\xrightarrow{-2\,CH_3OH}\; \underset{CH_2CH_2OH}{N(CH_2CH_2OOCR)_2} \;\xrightarrow{(CH_3)_2SO_4}\; (8, X - CH_3SO_4) \qquad (9)$$

11.6. Economic Aspects.

A summary of list prices and suppliers for selected quaternaries is given in Table 11. Other commercial products include

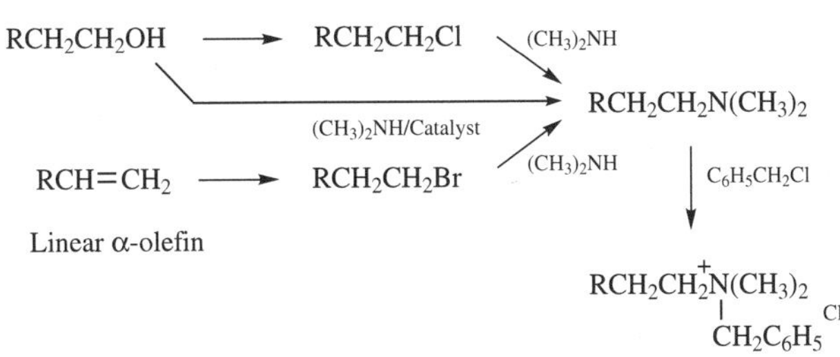

Fig. 3. Quaternaries from α-olefins of fatty alcohols where R is a fatty alkyl group. The product is alkylbenzyldimethyl quaternary.

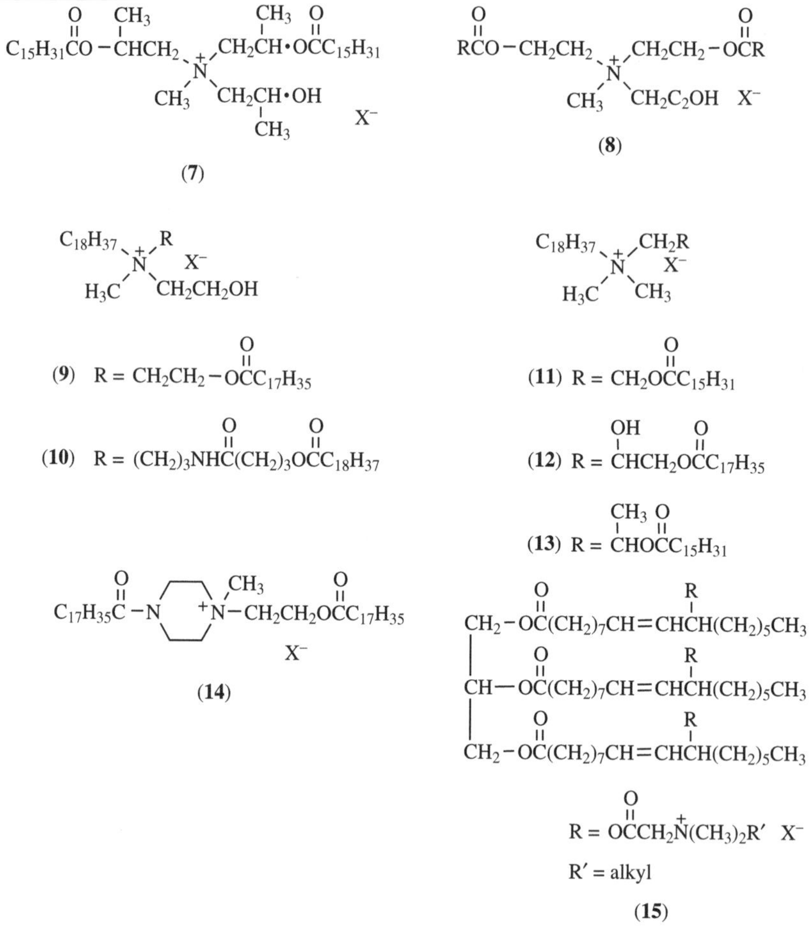

$$C_{15}H_{31}\overset{O}{\overset{\|}{C}}O-\overset{CH_3}{\overset{|}{C}}HCH_2\overset{+}{\underset{N}{N}}CH_2CH\cdot O\overset{O}{\overset{\|}{C}}C_{15}H_{31}$$

(structures around (7), (8))

(7)

(8)

$$\begin{array}{c} C_{18}H_{37}\,\diagdown\,\overset{+}{N}\diagup R \\ H_3C\diagup\quad\diagdown CH_2CH_2OH \end{array}\quad X^-$$

(9) R = $CH_2CH_2-\overset{O}{\overset{\|}{O}}\overset{}{C}C_{17}H_{35}$

(10) R = $(CH_2)_3NH\overset{O}{\overset{\|}{C}}(CH_2)_3O\overset{O}{\overset{\|}{C}}C_{18}H_{37}$

$$\begin{array}{c} C_{18}H_{37}\,\diagdown\,\overset{+}{N}\diagup CH_2R \\ H_3C\diagup\quad\diagdown CH_3 \end{array}\quad X^-$$

(11) R = $CH_2O\overset{O}{\overset{\|}{C}}C_{15}H_{31}$

(12) R = $\overset{OH}{\overset{|}{C}}HCH_2O\overset{O}{\overset{\|}{C}}C_{17}H_{35}$

(13) R = $\overset{CH_3}{\overset{|}{C}}HO\overset{O}{\overset{\|}{C}}C_{15}H_{31}$

$$C_{17}H_{35}\overset{O}{\overset{\|}{C}}-N\diagup\bigcirc\diagdown\overset{CH_3}{\overset{+}{N}}-CH_2CH_2O\overset{O}{\overset{\|}{C}}C_{17}H_{35}$$

X⁻

(14)

$CH_2-O\overset{O}{\overset{\|}{C}}(CH_2)_7CH=CHCH(CH_2)_5CH_3$ R
$CH-O\overset{O}{\overset{\|}{C}}(CH_2)_7CH=CHCH(CH_2)_5CH_3$ R
$CH_2-O\overset{O}{\overset{\|}{C}}(CH_2)_7CH=CHCH(CH_2)_5CH_3$ R

R = $O\overset{O}{\overset{\|}{C}}CH_2\overset{+}{N}(CH_3)_2R'$ X⁻

R' = alkyl

(15)

Fig. 4. Quaternary esteramines: (**7**) (245); (**8**) (244,246); (**9**) (247,248); (**10**) (249); (**11**) (250); (**12**) (251); (**13**) (196); (**14**) (252); and (**15**) (253).

diallyldimethylammonium chloride [7398-69-8], produced by Ciba Specialty Chemicals; di(hydrogenated tallow)alkyldimethylammonium methosulfate [61789-81-9], produced by Akzo Nobel; and tetrabutylammonium bromide [1643-19-2], produced by Cognis Corp., RSA Corp and Sachem Inc. The leading producers of phase-transfer quaternaries are Eastman Kodak Company, Hexcel Corporation, RSA, Chemical Dynamics Corporation, Lindan Chemicals, Henkel Corporation, and Akzo Nobel. From 2,300 to 11,300 metric tons of quaternaries were used as phase-transfer catalysts during 1991 (257). The principal producers of perfluorinated quaternaries are Ciba, 3M Specialty Chemicals, and E. I. du Pont de Nemours & Company, Inc.

The leading U.S. manufacturers in terms of volumes of fatty nitrogen derivatives are Akzo Nobel and Goldschmidt-Degussa. The combined annual production of these two companies accounts for ~80% of the fatty nitrogen products manufactured in the United States. The remaining production is divided

Table 11. **1998 Prices and Suppliers of Selected Quaternaries and Tertiary Amine Precursors**[a]

Material	CAS Registry Number	Percent active as supplied	$/kg	Manu-facturer[b]
methyl bis(tallowamido ethyl)-2-tallow imidazolinium methosulfate	[68122-86-1]	75	1.76	A, Cr, G, S
methyl bis(tallowalkylami-doethyl)-2-hydroxyethylam-monim methosulfate	[68410-69-5]	90	1.76	Cr, L, S
di(hydrogenated tallow)-alkyldimethyl ammonium chloride	[61789-80-8]	80	2.29	A, G, R
tallowalkyl diamine	[61791-55-7]	100	2.67	A, C, T
(hydrogenated tallow)alkyl amine	[61788-45-2]	100	2.45	A, C, G
(hydrogenated tallow)alkyl amine, distilled	[61788-45-2]	100	2.98	A
methyldi(hydrogenated tallow)alkyl amine	[61788-63-4]	100	2.27	A, C, G
cocoalkylamine	[61788-46-3]	100	3.48	A, C, Cl, G
alkyl(C12-C16)dimethyl amine	[61788-93-0]	100	2.65	A, C, L
alkyl(C12-C16)benzyldimethyl ammonium chloride	[61789-71-7]	50	2.98	A, Al, L, M, R, S

[a]Ref. 256.
[b]Manufacturer: A = Akzo Nobel Chemicals Inc; Al = Alfachem; C = Atofina North America; Cl = Clar-Clariant Corp; Cr = Croda Inc; G = Goldschmidt-Degussa Chemical Group; L = Lonza Inc; M = Mason Chemical Co; R = Rhodia; T = Tomah Products Inc.

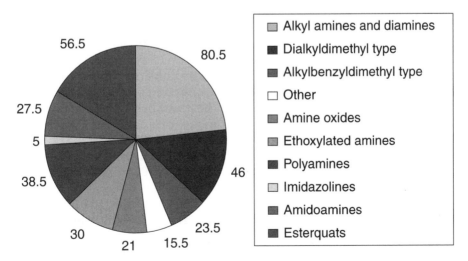

Fig. 5. North American consumption of quaternaries by product type (1997 total is 344 M metric tons).

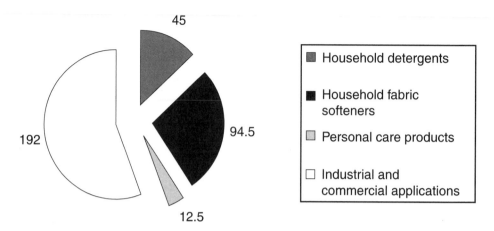

Fig. 6. North American consumption by market (1997 total is 344 M metric tons).

among smaller-volume producers. Nearly all fatty nitrile plants require considerable technological expertise and capital investment to operate.

Figure 5 shows the consumption of quaternaries in North America during 1997 as a function of product type. Alkyl amines and diamines accounted for 80,000 metric tons during 1997. The next largest volume type of quaternary ammonium compound was esterquats accounting for over 56,000 metric tons. The single largest market for quaternary ammonium compounds is as fabric softeners. Both household and industrial type formulations are produced, the household market is larger. In 1997 the household market accounted for over 94,000 metric tons of quaternaries in the United States (258). Consumption of these products is increasing at an annual rate of ∼2–3%. The hair care market consumed over 12,000 metric tons of quaternary ammonium compounds in 1997 (259). The annual consumption for organoclays is estimated at 18,000 metric tons (260).

Figure 6 shows the consumption of quaternaries in North America during 1997 as a function of market segment. Industrial and commercial applications is made up of several market segments including I&I cleaning, agricultural emulsions, textiles, oilfield, mining, and asphalt.

11.7. Analytical Methods. There are no universally accepted wet analytical methods for the characterization of quaternary ammonium compounds. The American Oil Chemists' Society (AOCS) has established, however, a number of applicable tests (261). These include sampling, color, moisture, amine value, ash, iodine value, average molecular weight, pH, and flash point.

Numerous "wet chemical" methods have been developed for the determination of the activity of quaternary ammonium samples (262–264). First is the partition titration in immiscible solvent systems, usually chloroform and water, using an anionic surfactant such as sodium lauryl sulfate [151-21-3] and an anionic dye indicator (265). This process, essentially a microtitration, is often referred to as the Epton titration (266,267). The end point requires considerable practice to detect. Second is the direct titration of the long-chain cation with sodium tetraphenylboron [143-66-8] using an anionic indicator (268). This macro method is convenient and relatively simple to perform. Third is the titration of the halide anion (Volhardt type) or perchloric acid titration in acetic anhydride (269). The fourth category includes colorimetric methods using anionic dyes or indicators and partition solvent systems. These partition/colorimetric analytical methods, which have found widespread use in environmental analysis, have long been used for determining small amounts of quaternary ammonium compounds (270). Although they are not specific, these methods do indicate the presence of long-chain cationics that have at least one chain composed of eight or more carbon atoms. The use of electrodes to determine the end point of the activity titration is becoming more popular in the industry (271). The change is prompted by several factors. Among them are a trend toward minimizing the use of chlorinated solvents required for many partition-based methods and the increased use of the esterquat molecule, which does not give a sharp visual endpoint. Both platinum sheet and platinum bulb electrodes are used. Ion specific electrodes offer a high degree of specificity but have the drawback of being sensitive to potential interferences (272). Where relatively pure compounds are used, ion specific electrodes have begun to gain acceptance (273,274).

The chain length composition of quaternaries can be determined by gas-chromatography (qv) (275). Because of low volatility, quaternaries cannot be chromatographed directly, but only as their breakdown products. Quaternary ammonium salts can be analyzed by injecting them directly dimethylsilicone based capillary columns. In most cases, long-chain monoalkyl quaternaries break down to form two sets of peaks: short-chain alkyl halides and the long-chain tertiary amines. Long-chain dialkyl quaternaries elute on the basis of total carbon atoms present in the resulting methyldialkyl tertiary amine.

Mass spectral analysis of quaternary ammonium compounds can be achieved by fast-atom bombardment (fab) ms (276,277). This technique relies on bombarding a solution of the molecule, usually in glycerol [56-81-5] or *m*-nitrobenzyl alcohol [619-25-0], with argon and detecting the parent cation plus a proton (MH^+). A more recent technique has been reported (278), in which information on the structure of the quaternary compounds is obtained indirectly through cluster-ion formation detected via liquid secondary ion mass spectrometry (lsims) experiments.

Liquid chromatography has been widely applied for analysis of quaternaries. Modified reverse-phase columns can provide chain length information, whereas normal-phase chromatography results in groupings of alkyl distributions. Quaternary ammonium compounds can be separated into their mono-, di-, and trialkyl components on a normal-phase silica or alumina column using a conductivity detector (279). Solvent systems generally include tetrahydrofuran, methanol, and acetic acid. Because the conductivity detector is only sensitive to ionic species, solvents and nonionic components of the sample are not seen by the detector. Alternative columns are amino, modified silica, or cation exchange. Evaporative laser light scattering detectors (elsd) have also been utilized for the nonvolatile, nonultraviolet-absorbing quaternaries (286).

Nuclear magnetic resonance (nmr) spectroscopy is useful for determining quaternary structure. The ^{15}N nmr can distinguish between quaternary ammonium compounds and amines, whether primary, secondary, or tertiary, as well as provide information about the molecular structure around the nitrogen atom.

The ^{13}C nmr can distinguish among oleic, tallow, and hydrogenated tallow sources (281). In addition to information on chemical structure, nmr is useful for understanding the physical chemistry of cationic surfactants in solution. For example, pulsed field gradient self-diffusion nmr experiments have shown that water permeates readily through vesicle membranes in dispersions of dialkyl quaternary ammonium compounds (282).

11.8. Health and Safety Factors.   Acute oral toxicity data (albino rats) show most structures to have an LD_{50} in the range of 100–5000 mg kg (283). Many quaternaries are considered to be moderately to severely irritating to the skin and eyes.

Some quaternary ammonium compounds are potent germicides (212, 284,285), toxic in small (mg L range) quantities to a wide range of microorganisms. Bactericidal, algicidal, and fungicidal properties are exhibited. Ten-minute-contact kills of bacteria are typically produced by quaternaries in concentration ranges of 50–333 mg L (286). Toxicity of quaternary ammonium salts to aquatic organisms is measured in standard laboratory toxicity tests with aqueous medium containing the test chemical (287).

Acute toxicity at low (1 mg L) levels has been reported in invertebrates, snails, and fish (288,289). In plant systems, growth inhibition of green algae and great duckweed occurs at 3–5 mg L (290).

Over the last decade, considerable advances have been made in understanding the metabolic pathway of quaternary ammonium compounds. The general metabolic pathway of several quaternary ammonium materials is given in Figure 7. All research to date shows the main pathway is cleavage of the calkyl-N bonds to produce smaller molecular weight amines and alkanals. The alkanals are degraded by β-oxidation. The amines are readily degraded by microorganisms. The results obtained with ester quats are consistent (64–66).

Numerous internationally recognized standardized methods are available for assessing the biodegradation of chemicals under aerobic conditions. There are three testing method levels: ready biodegradeability, inherent biodegradeability, and simulation of biological treatment systems.

Ready biodegradeability tests are used primarily for regulatory purposes. These include the Closed Bottle test, MITI I test, and the Sturm test. All these tests are usually performed within a 28-day period. A biodegradation percentage of >60% is interpreted as complete. The biodegradation of surfactants in these tests has been reviewed extensively (67,68). The biodegradation classification of some selected quaternaries in given in Table 12.

There are three standard test methods to assess inherent biodegradeability. These are the MITI II test, Zahn-Wellens and SCAS test. Sludge (CAS) and Semi-Continuous Activated Sludge (SCAS) tests are used to simulate behavior of materials in biological treatment systems. The CAS test is believed to give realistic results similar to full-scale treatment. Results obtained in the CAS and SCAS tests with cationic surfactants are summarized in Table 13. These studies show extensive removal.

A dynamic equilibrium exists between quaternary ammonium species in the aqueous phase and those existing as a soild after absorption. Thus, only the fraction in the aqueous phase is "available" at any given time to be biodegraded. Bioavailability is defined as that fraction of material that is readily

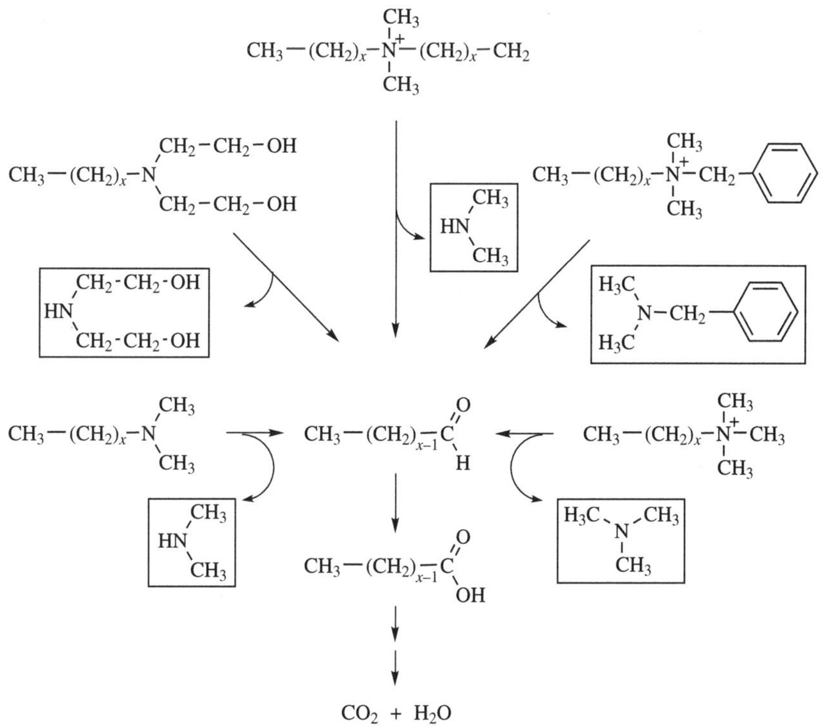

$$CH_3-(CH_2)_x-\overset{\overset{\displaystyle CH_3}{|}}{\underset{\underset{\displaystyle CH_3}{|}}{N^+}}-(CH_2)_x-CH_2$$

$$CH_3-(CH_2)_x-N\overset{\displaystyle CH_2-CH_2-OH}{\underset{\displaystyle CH_2-CH_2-OH}{<}}$$

$$HN\overset{\displaystyle CH_2\cdot CH_2\cdot OH}{\underset{\displaystyle CH_2\cdot CH_2\cdot OH}{<}}$$

$$HN\overset{\displaystyle CH_3}{\underset{\displaystyle CH_3}{<}}$$

$$CH_3-(CH_2)_x-\overset{\overset{\displaystyle CH_3}{|}}{\underset{\underset{\displaystyle CH_3}{|}}{N^+}}-CH_2-\langle\text{phenyl}\rangle$$

$$\overset{\displaystyle H_3C}{\underset{\displaystyle H_3C}{>}}N-CH_2-\langle\text{phenyl}\rangle$$

$$CH_3-(CH_2)_x-N\overset{\displaystyle CH_3}{\underset{\displaystyle CH_3}{<}} \longrightarrow CH_3-(CH_2)_{x-1}-\overset{\overset{\displaystyle O}{\|}}{C}-H \longleftarrow CH_3-(CH_2)_x-\overset{\overset{\displaystyle CH_3}{|}}{\underset{\underset{\displaystyle CH_3}{|}}{N^+}}-CH_3$$

$$HN\overset{\displaystyle CH_3}{\underset{\displaystyle CH_3}{<}}$$

$$CH_3-(CH_2)_{x-1}-\overset{\overset{\displaystyle O}{\|}}{C}-OH$$

$$\overset{\displaystyle H_3C}{\underset{\displaystyle H_3C}{>}}N-CH_3$$

$$CO_2 + H_2O$$

Fig. 7. General metabolic pathway of quaternary ammonium compounds.

accessible to microbial degradation. Recently a method was developed to test the hypothesis that water-insoluble dialkyldimethylammonium salts are available to microorganisms when desorbed. The method consists of a flow-through system comprising a storage vessel with aerated river water, a pump, a column with a quaternary ammonium salt adsorbed on silica gel, and a collecting vessel. Aerated river water was pumped through the column packed with particles of silica gel with adsorbed quaternary compounds. Biodegradation of dicocodimethylammonium chloride was evident from the marked decline in the quaternary ammonium salt concentration in the effluent after a few days (71). The concept of bioavability helps to explain earlier observations for dicocodimethylammonium chloride obtained during prolonged Closed Bottle testing (70).

Most uses of quaternary ammonium compounds can be expected to lead to these compounds' eventual release into wastewater treatment systems except for those used in drilling muds. Useful properties of the quaternaries as germicides can make these compounds potentially toxic to sewer treatment systems. It appears, however, that quaternary ammonium compounds are rapidly degraded in the environment and strongly sorbed by a wide variety of materials. Quaternaries appear to bind anionic compounds and thus are effectively removed from wastewater by producing stable, lower toxicity compounds (291). Under normal circumstances these compounds are unlikely to pose a significant risk to microorganisms in wastewater treatment systems (292). Microbial populations acclimate readily to low levels of quaternary compounds and biodegrade them.

Table 12. **Biodegradation Data for Selected Quaternaries**

Cationic compound	CAS Registry Number	Classification	Percent and degraded time	Reference
octyltrimethylammonium chloride		readily	> 70% in 10 days	100
octadecyltrimethylammonium chloride	[112-03-8]	readily	60%[a] in 10 days	AN unpubl
didecyldimethylammonium chloride	[7173-51-5]	readily	> 70% in 28 days	AN unpubl
didodecyldimethylammonium chloride	[3401-14-9]	readily	> 60%[a]	AN unpubl
polyoxyethylene(15)-cocoalkymethylammonium chloride	[61791-10-4]	inherently	30% in 28 days	62
polyoxyethylene(15)-cocoalkyl amine	[61791-14-8]	inherently		79
bis(2-hydroxyethyl)octadecyl amine	[10213-78-2]	readily	> 70% in 28 days	78,79
alkylbenzyldimethylammonium salts		readily	> 60% in 10 days	100,101
benzyldecyldimethylammonium chloride	[965-32-2]	readily	> 60% in 6 days	100,101
primary fatty amines C8–C18		readily	> 60% in 12 days	70
alkyldimethylamines C12–C18		readily	> 60%	70,76
di-(tallowalkyl)ester of di-2-hydroxyethyldimethylammonium chloride		readily	80% in 28 days	80
N-methyl-N,N-bis(2-$C_{16/18}$-acyloxy)ethyl)-N-(2-hydroxyethyl ammonium methosulfate)	[157905-74-3]	readily		102
di-(hydrogenated tallowalkyl)ester of 2,3-dihydroxypropyltrimethylammonium chloride		readily	85%	22

[a]Test conducted with humic acid present.

Environmental toxicity and stability of these compounds have been described (292,293).

Newer classes of quaternaries, eg, esters (294) and betaine esters (295), have been developed. These materials are more readily biodegraded. The mechanisms of antimicrobial activity and hydrolysis of these compounds have

Table 13. Removal/Biodegradation of Cationic Surfactants in Continuous Flow Activated Sludge (CAS) and Semicontinuous Activated Sludge (SCAS) Treatment Systems

Test compound	CAS Registry Number	Test	Removal (%)	Reference
hexadecyltrimethylammonium bromide	[57-09-0]	CAS	91–98	109
hexadecyltrimethylammonium bromide	[57-09-0]	CAS	98–99	110
hexadecyltrimethylammonium bromide	[57-09-0]	CAS	100	111
octadecyltrimethylammonium chloride	[112-03-8]	SCAS	98	107
didecyldimethylammonium chloride	[7173-51-5]	CAS	95	109
dioctadecyldimethylammonium chloride	[107-64-2]	CAS	95	109
dioctadecyldimethylammonium chloride	[107-64-2]	CAS	91–93	112
benzyldodecyldimethylammonium chloride	[7281-04-1]	CAS	96	109
tetradecylbenzyldimethylammonium chloride	[139-08-2]	CAS	>70	113
cocobenzyldimethylammonium chloride	[61789-71-7]	CAS	94	114
di-(tallow fatty acid) ester of di-2-hydroxyethyldimethyammonium chloride		CAS	>99	80
N-methyl-N,N-bis(2-(C$_{16/18}$-acyloxy)ethyl)-N-hydroxyethylammonium methosulfate	[157905-74-3]	CAS	>90	102
ditallow ester of 2,3-dihydroxypropane-trimethylammonium chloride		CAS	>99	81

been studied (295). Applications as surface disinfectants, antimicrobials, and *in vitro* microbiocidals have also been reported. Examples of ester-type quaternaries are shown in Figure 4.

11.9. Uses. Uses of quaternary ammonium compounds range from surfactants to germicides and encompass a number of diverse industries (see Table 9).

Fabric Softening. The use of quaternary surfactants as fabric softeners and static control agents can be broken down into three main household product types: rinse cycle softeners; tumble dryer sheets; and detergents containing softeners, also known as softergents. Rinse cycle softeners are aqueous dispersions of quaternary ammonium compounds designed to be added to the wash during the last rinse cycle (191,195,197,203). Original products contained from 3–8% quaternary ammonium compound, typically di(hydrogenated tallow)alkyldimethylammonium chloride [61789-80-8] (DHTDMAC). During the 1980s and early 1990s, rinse cycle softeners went through significant changes as the active concentration of the dispersions was increased to reduce packaging and solid waste. Refills were introduced that had cationic activity of 16–27%. These products are meant to be added to a regular-strength softener bottle [typically 64 oz (0.9 kg)] and diluted. As of 1995, the latest innovation was the ultrafabric softener. This product, also formulated to contain from 16–27% cationic activity, is designed to be used without dilution. The actual dosage to the washer was decreased to approximately one fluid ounce (29 mL) from the older standard of 3–4-fluid oz (87–116 mL). Although DHTDMAC is a widely employed softener, the use of imidazoline and amidoamine quaternaries has increased because these latter are easier to formulate into high active systems. The combination of DHTDMAC, imidazoline, and amidoamine is used to maximize softening performance and facilitate handling and formulation.

In 1991, the European fabric softener market took a sharp turn. Producers in Germany, the Netherlands, and later in Austria and Switzerland voluntarily gave up the use of DHTDMAC (290) because of pressure from local environmental authorities, who gave an environmentally hazardous classification to DHTDMAC. A number of esterquats were developed as candidates to become successors to DHTDMAC (see Fig. 8). The ester group facilitates biodegradation. The esterquat is now the molecule of choice in North America as well, accounting for well over the half the volume of cationic surfactants used in this market segment. Figure 8 shows consumption of fabric softening actives by type for 1997.

Tumble dryer sheets contain a quaternary ammonium compound formulation applied to a nonwoven sheet typically made of polyester or rayon (190,192) (see NONWOVEN FABRICS). These sheets are added with wet clothes to the tumble dryer. Although these products afford some softening to the clothes, their greatest strength is in preventing static charge buildup on clothes during the drying cycle and during wear. A nonionic surfactant, such as an ethoxylated alcohol or fatty acid, is typically used in combination with the quaternary ammonium compound. The nonionics are known as release agents (qv) or distribution agents. More efficient transfer of the quaternary from the substrate to the drying fabric can be obtained.

Detergents containing softeners are also produced (194) (see DETERGENCY). These softergents are made from complex formulas in order to accomplish both detergency and softening during the wash cycle. These formulations typically

Fig. 8.   Fabric softening consumption of cationics by type (1997 total is 94.5 M metric tons).

contain quaternary ammonium compounds mixed with other materials such as clays (qv) for softening, in conjunction with the typical nonionic and anionic cleaning surfactant. The consumer benefits of quaternaries are fabric softening, antistatic properties, ease of ironing, and reduction in energy required for drying.

Hair Care.   Quaternary ammonium compounds are the active ingredients in hair conditioners (88–97,297). Quaternaries are highly substantive to human hair because the hair fiber has anionic binding sites at normal pH ranges. Surface analysis by X-ray photoelectron spectroscopy (XPS) has shown specific 1:1 (ionic) interaction between cationic alkyl quaternary surfactant molecules and the anionic sulfonate groups present on the hair surface (298). The use of quaternaries as hair conditioners can be broken down into creme rinses and shampoo conditioners.

Creme rinses are applied to the hair after washing. Frequently used quaternaries in creme rinses are dodecyltrimethylammonium chloride [112-00-5], dimethyloctadecyl(pentaethoxy)ammonium chloride, benzyldimethyloctadecyl-lammonium chloride [122-19-0], and dimethyldioctadecylammonium chloride [107-64-2] (89–92,94).

Conditioning shampoos are formulations that contain anionic surfactants for cleaning hair and cationic surfactants for conditioning (97,259,297). The quaternary ammonium compounds most often used are either trihexadecylmethyl-lammonium chloride [71060-72-5], ethoxylated quaternaries, or one of the polymeric quaternaries. The polymeric quaternaries have either a natural or a synthetic backbone and numerous quaternary side functions. The polymer may offer an advantage by showing a high degree of affinity to the human hair surface and providing better compatibility with the other ingredients of conditioner shampoos (299).

Regardless of how the conditioner is applied or what the structure of the quaternary is, benefits provided to conditioned hair include the reduction of combing forces, increased luster, and improved antistatic properties.

Germicides.   The third largest market for quaternaries is sanitation (300). Generally, quaternaries offer several advantages over other classes of sanitizing chemicals, such as phenols, organohalides, and organomercurials, in that quaternaries are less irritating, low in odor, and have relatively long activity (301). The first use of quaternaries in the food industry occurred in the dairy industry for the sanitization of processing equipment. Quaternaries find use as disinfectants and sanitizers in hospitals, building maintenance, and food processing (qv) (302–310); in secondary oil recovery for drilling fluids (152–156); and in cooling water applications (see DISINFECTANTS AND ANTISEPTICS; PETROLEUM). Quaternaries have also received extensive attention for use as a general medicinal antiseptics (161) and in the pharmaceutical area as skin disinfectants and surgical antiseptics (162). In addition, quaternaries have been used in the treatment of eczema and other dermatological disorders as well as in contraceptive formulations (see CONTRACEPTIVES) (163) and ocular solutions for contact lenses (qv) (164).

Alkylbenzyldimethyl quaternaries (ABDM) are used as disinfectants (107) and preservatives. The most effective alkyl chain length for these compounds is between 10 and 18 carbon atoms. Alkyltrimethyl types, alkyldimethylbenzyl types, and didodecyldimethylammonium chloride [3401-74-9] exhibit excellent germicidal activity (302–310). Dialkyldimethyl types are effective against anaerobic bacteria such as those found in oil wells (152–156). One of the most effective and widely used biocides is didecyldimethylammonium chloride [7173-57-5].

Organoclays. Another large market for quaternary ammonium salts is the manufacture of organoclays, ie, organomodified clays (260). Clay particles are silicate minerals that have charged surfaces and that attract cations or anions electrostatically. Organoclays are produced by ion-exchange (qv) reaction between the quaternary ammonium salt and the surface of the clay particles (137,138,174,311). The quaternary ammonium salt displaces the adsorbed cations, usually sodium or potassium, producing an organomodified clay. The new modified clay exhibits different behavior from that of the initial clay. Most importantly, it is preferentially wet by organic liquids and not by water.

The main use of these clays is to control, or adjust, viscosity in nonaqueous systems. Organoclays can be dispersed in nonaqueous fluids to modify the viscosity of the fluid so that the fluid exhibits non-Newtonian thixotropic behavior. Important segments of this area are drilling fluids, greases (137,138), lubricants, and oil-based paints. Quaternaries used to produce organoclays are dimethyldi (hydrogenated tallow)alkylammonium chloride [61789-80-8], dimethyl(hydrogenated tallow)alkylbenzylammonium chloride [61789-72-8], and methyldi(hydrogenated tallow)alkylbenzylammonium chloride [68391-01-5].

Miscellaneous Uses. *Phase-Transfer Catalysts.* Many quaternaries have been used as phase-transfer catalysts (PTC). A PTC increases the rate of reaction between reactants in different solvent phases. Usually, water is one phase and a water-immiscible organic solvent is the other. An extensive amount has been published on the subject of PTC (257). Both the industrial applications in commercial manufacturing processes (312) and their synthesis (313) have been reviewed. Common quaternaries employed as phase-transfer agents include benzyltriethylammonium chloride [56-37-1], tetrabutylammonium bromide [1643-19-2], tributylmethylammonium chloride [56375-79-2], and hexadecylpyridinium chloride [123-03-5].

Polyamine-Based Quaternaries. Another important class of quaternaries are the polyamine based or polyquats. Generally, polyamine-based quaternaries have been used in the same applications as their monomeric counterparts (314). Discussions, including the use of polymeric quaternaries in laundry formulations (315) and in the petroleum industry as damage control agents (316), have been published.

Perfluorinated Quaternaries. Perfluorinated quaternaries are another important, but smaller, class of quaternary ammonium compounds. In general, these are similar to their hydrocarbon counterparts but have at least one of the hydrocarbon chains replaced with a perfluoroalkyl group. These compounds are generally much more expensive than hydrocarbon-based quaternaries, so they must offer a significant performance advantage if they are to be used. Production volumes of perfluorinated quaternary ammonium compounds are significantly smaller than those of other classes. Many of these quaternaries have proprietary chemical structures. They are used in water-based coating applications to promote leveling, spreading, wetting, and flow control.

BIBLIOGRAPHY

"Ammonium Compounds" in *ECT* 1st ed., Vol. 1, pp. 810–826, by W. C. Holmes; in *ECT* 2nd ed., Vol. 2, pp. 313–331, by W. C. Holmes, Consultant; in *ECT* 3rd ed., Vol. 2, pp. 516–536, by R. D. Young, Tennessee Valley Authority; in *ECT* 4th ed., Vol. 2, pp. 692–708, by Charles W. Weston, Freeport Research ard Engineering Company; "Ammonium" under "Fluorine Compounds, Inorganic," in *ECT* 1st ed., Vol. 6, p. 676, by G. C. Whitaker, The Harshaw Chemical Co.; "Ammonium Fluoride" under "Fluorine Compounds, Inorganic," in *ECT* 2nd ed., Vol. 9, pp. 548–549, by G. C. Whitaker, The Harshaw Chemical Co.; "Ammonium" under "Fluorine Compounds, Inorganic," in *ECT* 3rd ed., Vol. 10, pp. 675–678, by H. S. Halbedel and T. E. Nappier, The Harshaw Chemical Co.; in *ECT* 4th ed., Vol. 11, pp. 287–290, by John R. Papcun, Atotech; "Quaternary Ammonium Compounds" in *ECT* 2nd ed., Vol. 19, pp. 859–865, by R. A. Reck, Armour Industrial Chemical Co.; in *ECT* 3rd ed., Vol. 19, pp. 521–531, by R. A. Reck, Armak Co.; in *ECT* 4th ed., Vol. 20, pp. 739–767, by Maurice Dery, Akzo Nobel Chemicals, Inc.; "Ammonium Compounds" in *ECT* (online), posting date: December 4, 2000, by Charles W. Weston, Freeport Research and Engineering Company.

CITED PUBLICATIONS

1. W. L. Jolly, *The Inorganic Chemistry of Nitrogen*, W. A. Benjamin, Inc., 1964, pp. 40–42.
2. C. W. Gehrke, L. L. Wall, J. S. Killingley, and K. Inada, *J. Assoc. Off. Anal. Chem.* **53**, 124 (1971).
3. F. J. Johnson and D. L. Miller, *J. Assoc. Off. Anal. Chem.* **57**, 8 (1974).
4. G. Feick and R. M. Hainer, *J. Am. Chem. Soc.* **76**, 5860 (1954).
5. R. J. Lewis, Sr., *Dangerous Properties of Industrial Materials*, Vol. 2, John Wiley & Sons, Inc., New York, 2000.
6. E. Terres and H. Weiser, *Z. Elektrochem.* **27**, 177–244 (1921).
7. E. Janecke, *Z. Elektrochem.* **35**, 723–727 (1929).
8. *Gmelins Handbuch der anorganischen Chemie*, 8th ed., Deutsche, Vol. 8, Chemische Gesellschaft, Verlag Chemie, Berlin, Germany, 1936, pp. 1–602.
9. J. W. Mellor, *Comprehensive Treatise on Inorganic and Theoretical Chemistry*, Vol. 2, Longmans Green and Co., New York, 1927, pp. 837 and 843.
10. W. C. Eichelberger, *Ind. Eng. Chem.* **48**(3), 102A (1956).
11. Fr. Pat. 1,546,234 (Nov. 15, 1968), (to Farbenfabriken Bayer A.-G.).
12. A. F. Wells, *Structural Inorganic Chemistry*, 5th ed., Clarendon Press, Oxford, UK, 1984, pp. 362–363.
13. A. Seidell, ed., *Solubilities of Inorganic and Organic Compounds*, Vol. 1, Van Nostrand Co., Inc., New York, 1958.
14. T. P. Hou, "Alkali and Chlorine Production," in R. Furnas, ed., *Manual of Industrial Chemistry*, 6th ed., Vol. 1, Van Nostrand Co., Inc., New York, 1942, pp. 402–459.
15. T. Miyata, *Chem. Ind. (London)* (4) 142–145 (Feb. 21, 1983).
16. E. J. R. Cook, *Can. Chem. Process Ind.* **29**, 221 (1945).
17. V. Sauchelli, ed., *Fertilizer Nitrogen, Its Chemistry and Technology*, Reinhold Publishing Corp., New York, 1964, pp. 237–241.
18. A. W. Bamforth and S. R. S. Sastry, *Chem. Proc. Eng. (Bombay)* **53**(2), 72–74 (1972).
19. O. Hassel and H. Luzanski, *Z. Kristallogr.* **83**, 448 (1932).
20. R. C. Weast, ed., *Handbook of Chemistry and Physics*, 59th ed., The Chemical Rubber Co., Cleveland, Ohio, 1978.

21. H. Schutza, M. Eucken, and W. Namesh, *Z. An. All. Chem.* **292**, 293 (1957).

22. L. N. Lazarev and B. V. Andronov, *J. Appl. Chem. USSR* **46**, 2087 (1973). H. C. Hodge and F. A. Smith, in J. H. Simon, ed., *Fluorine Chemistry*, Vol. 4, Academic Press, Inc., New York, 1965, p. 192.

23. U.S. Pat. 3,310,369 (Mar. 21, 1967), J. A. Peterson (to Hooker Chemical Corp.).

24. U.S. Pat. 2,156,273 (Apr. 28, 1939), A. R. Bozarth (to Harshaw Chemical Co.).

25. U.S. Pat. 3,501,268 (Mar. 17, 1970), R. J. Laran, A. P. Giraitix, and P. Kobetz (to Ethyl Corp.).

26. H. K. van Poolen, *Oil Gas J.* **65**, 93 (Sept. 11, 1967).

27. U.S. Pat. 3,953,340 (Apr. 27, 1976), C. C. Templeton, E. H. Street, Jr., and E. A. Richardson (to Shell Oil Co.).

28. W. S. Midkiff and H. P. Foyt, *Mater. Perform.* **17**(2), 17 (1978).

29. *Glass Frosting and Polishing Technical Service Bulletin 667*, Harshaw Chemical Co., Solon, Ohio.

30. U.S. Pat. 3,373,130 (Mar. 19, 1968), E. E. Junge and J. Chabal (to PPG Industries).

31. A. L. Hinson, *Bull. Parenter. Drug. Assoc.* **25**, 266 (1971).

32. R. L. Parkes and M. R. Browne, *Appl. Opt.* **17**, 1845 (1978).

33. *Control of Souring Operations, Special Report #7*, American Institute of Laundering, Joliet, Ill.

34. Brit. Pat. 833,808 (Apr. 27, 1960), A. R. S. Gough and E. W. Bennet (to the United Kingdom Atomic Energy Commission).

35. U.S. Pat. 3,767,582 (Oct. 23, 1973), G. A. Miller (to Texas Instruments, Inc.).

36. U.S. Pat. 3,296,141 (Jan. 3, 1967), W. A. Lieb and E. Billow (to R. O. Hull Co.).

37. L. F. Spencer, *Met. Finish.* **68**(10), 52 (1970); H. K. DeLong, *Met. Prog.* **97**, 105 (June 1970); *Met Prog.* **98**, 43 (Mar. 1971); W. F. Higgins, *Light Met. Age* **17**(12), 8 (1959); A. E. Yaniv and H. Schick, *Plating* **55**, 1295 (1968).

38. E. Panck, *Am. Wood Preservers Assoc.* **59**, 189 (1963).

39. S. Jimenez, M. C. Cartagena, and A. Vallejo, *Agrochimica*, **32**(4), 245–252 (1988).

40. K. Drews, "Die technischen Ammoniumsalze," in *Sammlung Chemischer and Chemischtechnischer Vortrage*, Pummerer-Erlangen, Enke, Stuttgart, Germany, 1938, 200 pp.

41. S. B. Hendricks, E. Posnjak, and F. C. Kracek, *J. Am. Chem. Soc.* **54**, 2766–2786 (1932).

42. D. F. Othmer and G. J. Frohlich, *AIChE J.* **6**, 210–214 (1960).

43. G. Hansen and W. Berthold, *Chem. Zeitung* **96**, 449–455 (1972).

44. G. D. Honti, "Production of Ammonium Nitrate," in C. Keleti, ed., *Nitric Acid and Fertilizer Nitrates*, Marcel Dekker, Inc., New York, 1985, pp. 208–222.

45. "Pintsch-Bamag's Continuous Process for Production of Calcium Ammonium Nitrate," *Nitrogen* (28), 28 (1964).

46. J. J. Dorsey, Jr., *Ind. Eng. Chem.* **47**, 11 (1955).

47. W. C. Saeman, I. W. McCamy, and E. C. Houston, *Ind. Eng. Chem.* **44**, 1912 (1952).

48. R. D. Young and I. W. McCamy, *Can. J. Chem. Eng.* **45**, 50 (1967).

49. R. W. R. Carter and A. J. Roberts, "The Production of Ammonium Nitrate Including Handling and Safety," *The Fertilizer Society of London Proceedings*, No. 110, Oct. 23, 1969.

50. R. M. Reed and J. C. Reynolds, *Chem. Eng. Prog.* **69**, 62 (1973).

51. J. P. Bruynseels, *Hydrocarbon Process* **60**, 203 (1981).

52. T. Heggeboe, "Granular Ammonium Nitrate," in C. Keleti, ed., *Nitric Acid and Fertilizer Nitrates*, Marcel Dekker, Inc., New York, 1985, pp. 251–259.

53. National Board of Fire Underwriters, *Chem. Eng. News* **25**, 1594 (1947).

54. D. H. Lauriente, "Ammonium Nitrate," *Chemical Economics Handbook*, SRI, Menlo Park, CA, Oct. 2000.

55. U.S. Pat. Appl. 20010000200 (April 12, 2001), A. K. Chattopadhyay.

56. "Conversion of Gypsum or Anhydrite to Ammonium Sulfate," *Nitrogen* **46**, 21 (1967).

57. L. B. Hein, A. B. Phillips, and R. D. Young, "Recovery of Sulfur Dioxide from Coal Combustion Stack Gases," in F. S. Mallette, ed., *Problems and Control of Air Pollution*, Reinhold Publishing Corporation, New York, 1955, pp. 155–169.

58. D. H. Lauriente and Y. Sakuma, "Ammonium Sulfate," *Chemical Economics Handbook*, SRI, Menlo Park, CA, Dec. 2000.

59. E. Jungermann, ed., *Cationic Surfactants*, Marcel Dekker, Inc., New York, 1969, chapts. 2–5.

60. D. Rubingh and P. Holland, eds., *Cationic Surfactants*, Marcel Dekker, Inc., New York, 1991.

61. *Specialty Chemicals*, SRI International, Menlo Park, Calif., 1998, p. 129.

62. J. Fletcher, O. Dermer, and R. Fox, *Nomenclature of Organic Compounds*, ACS, Washington, D.C., 1974, pp. 189–195.

63. Ger. Pat. 2,506,834 (Aug. 21, 1975), R. Ford and T. Tadros (to Imperial Chemical Industries Ltd.).

64. U.S. Pat. 5,078,781 (Jan. 7, 1992), C. Finch (to Imperial Chemical Industries Ltd.).

65. Ger. Pat. 2,304,204 (Aug. 2, 1973), N. Drewe, R. Parker, and T. Tadros (to Imperial Chemical Industries Ltd.).

66. Ger. Pat. 2,352,334 (May 9, 1974), F. Hauxwell, H. Murton, and R. Brain (to Imperial Chemical Industries Ltd.).

67. U.S. Pat. 2,844,466 (July 22, 1958), A. Rogers and co-workers (to Armour and Co.).

68. U.S. Pat. 3,506,433 (Apr. 14, 1970), W. Abramitis and R. A. Reck (to Armour and Co.).

69. U.S. Pat. 4,765,823 (Aug. 23, 1988), K. Lurssen (to Bayer AG).

70. U.S. Pat. 3,698,951 (Oct. 17, 1972), M. Bennett (to Tate and Lyle Ltd.).

71. U.S. Pat. 5,017,612 (May 21, 1987), J. Nayfa (to J. Nayfa).

72. U.S. Pat. 2,740,744 (Apr. 3, 1956), W. Aoramitis and co-workers (to Armour and Co.).

73. U.S. Pat. 3,725,014 (Apr. 3, 1973), R. Poncha and co-workers (to Allied Chemical Corp.).

74. U.S. Pat. 5,210,250 (Aug. 27, 1992), I. Watanuki and co-workers (to Shin-Etsu Chemical Co.).

75. U.S. Pat. 4,398,958 (Aug. 16, 1983), R. Hodson and co-workers (to Cempol Sales Ltd.).

76. U.S. Pat. 4,636,373. (Aug. 30, 1987), M. Rubin (to Mobil Oil Corp.).

77. U.S. Pat. 4,640,829 (Aug. 30, 1987), M. Rubin (to Mobil Oil Corp.).

78. U.S. Pat. 4,642,226 (Mar. 29, 1985), R. Calvert and co-workers (to Mobil Oil Corp.).

79. U.S. Pat. 5,191,085 (Mar. 2, 1993), H. Jakob and co-workers (to Degussa AG).

80. U.S. Pat. 4,661,547 (Apr. 28, 1987), M. Harada, K. Tsukanoto, and E. Tomohiro (to Nippon Rubber Co., Ltd.).

81. U.S. Pat. 4,064,067 (Dec. 20, 1977), A. Lore (to E. I. du Pont de Nemours & Co., Inc.).

82. U.S. Pat. 4,771,088 (Jan. 15, 1987), A. Pekarkik (to Glidden Co.).

83. U.S. Pat. 4,840,980 (Feb. 29, 1988), A. Pekarkik (to Glidden Co.).

84. U.S. Pat. 4,208,485 (June 17, 1980), R. Nahta (to GAF Corp.).

85. U.S. Pat. 4,230,746 (Oct. 28, 1980), R. Nahta (to GAF Corp.).

86. U.S. Pat. 4,198,316 (Apr. 15, 1980), R. Nahta (to GAF Corp.).

87. U.S. Pat. 4,226,624 (Oct. 7, 1980), J. Ohr (to U.S. Navy).

88. U.S. Pat. 4,677,158 (Nov. 12, 1985), S. Tso and co-workers (United Catalysts Inc.).

89. U.S. Pat. 5,019,376 (May 28, 1991), H. Vick (to S.C. Johnson & Son, Inc.).

90. U.S. Pat. 4,818,523 (Apr. 4, 1989), J. Clarke and co-workers (to Colgate-Palmolive Co.).

91. U.S. Pat. 4,886,660 (Dec. 12, 1989), A. Patel and co-workers (to Colgate-Palmolive Co.).

92. U.S. Pat. 4,144,326 (Mar. 13, 1979), O. Luedicke and F. Gichia (to American Cyanamide Co.).
93. U.S. Pat. 5,034,219 (July 23, 1991), V. Deshpande and J. Walts (to Sterling Drug Inc.).
94. U.S. Pat. 4,719,104 (Jan. 12, 1988), C. Patel (to Helene Curtis, Inc.).
95. U.S. Pat. 3,980,091 (Sept. 14, 1976), G. Dasher and co-workers (to Alberto Culver Co.).
96. U.S. Pat. 4,919,846 (Apr. 24, 1990), Y. Nakama and co-workers (to Shiseido Co., Ltd.).
97. U.S. Pat. 3,577,528 (May 4, 1971), E. McDonough (to Zotos International).
98. U.S. Pat. 5,004,737 (Apr. 2, 1991), Y. Kim and B. Ha (to Pacific Chemical Co. Ltd.).
99. U.S. Pat. 5,015,469 (May 14, 1991), T. Yoneyama and co-workers (to Shiseido Co., Ltd.).
100. U.S. Pat. 4,069,347 (Jan. 17, 1978), J. McCarthy and co-workers (to Emery Industries, Inc.).
101. U.S. Pat. 5,035,826 (Sept. 22, 1991), P. Dorbut, M. Mondin, and G. Broze (to Colgate-Palmolive Co.).
102. U.S. Pat. 4,284,435 (Aug. 18, 1981), D. Fox (to S. C. Johnson & Son, Inc.).
103. U.S. Pat. 3,969,281 (July 13, 1976), T. Sharp (to T. Sharp.).
104. U.S. Pat. 4,011,097 (Mar. 8, 1977), T. Sharp (to T. Sharp.).
105. U.S. Pat. 4,541,945 (Sept. 17, 1985), J. Anderson and S. Seiglle (to Amchem Products).
106. U.S. Pat. 4,636,330 (Jan. 13, 1987), J. Melville (to Lever Brothers Co.).
107. U.S. Pat. 4,576,729 (Mar. 18, 1986), L. Paszek and B. Gebbia (to Sterling Drug, Inc.).
108. U.S. Pat. 4,126,586 (Nov. 21, 1978), M. Curtis, R. Davies, and J. Galvin (to Lever Brothers Co.).
109. U.S. Pat. 4,814,108 (Mar. 21, 1989), J. Geke and H. Rutzen (to Henkel Kommanditgesellschaft auf Aktien).
110. U.S. Pat. 4,678,605 (July 7, 1987), J. Geke and S. Seiglle (to Henkel Kommanditgesellschaft auf Aktien).
111. U.S. Pat. 5,127,991 (July 7, 1992), S. Lai and C. Smith (to AT&T Bell Laboratories).
112. U.S. Pat. 5,185,235 (Mar. 30, 1990), H. Sato, K. Tazawa, and T. Aoyama (to Tokyo Ohka Kogyo Co., Ltd.).
113. U.S. Pat. 4,892,649 (Jan. 9, 1990), J. Mehaffey and T. Newman (to Akzo America Inc.).
114. U.S. Pat. 4,995,965 (Feb. 26, 1991), J. Mehaffey and T. Newman (to Akzo American Inc.).
115. U.S. Pat. 3,976,565 (Aug. 24, 1976), V. Petrovich (to V. Petrovich).
116. U.S. Pat. 4,098,686 (July 4, 1978), V. Petrovich (to V. Petrovich).
117. U.S. Pat. 4,225,428 (Sept. 30, 1980), V. Petrovich (to V. Petrovich).
118. U.S. Pat. 3,979,207 (Sept. 7, 1976), J. MacGregor (to Mattey Rustenburg Refiners).
119. U.S. Pat. 4,306,081 (Dec. 15, 1981), G. Rich (to Albee Laboratories, Inc.).
120. U.S. Pat. 4,289,530 (Sept. 15, 1981), G. Rich (to Albee Laboratories, Inc.).
121. U.S. Pat. 4,351,699 (Sept. 28, 1982), T. Osborn (to Procter & Gamble Co.).
122. U.S. Pat. 4,441,962 (Apr. 10, 1984), T. Osborn (to Procter & Gamble Co.).
123. U.S. Pat. 5,240,562 (Aug. 31, 1993), D. Phan and P. Trokhan (to Procter & Gamble Co.).
124. U.S. Pat. 3,916,058 (Oct. 28, 1975), P. Vossos (to Nalco Chemical Co.).
125. U.S. Pat. 4,119,486 (Oct. 10, 1978), R. Eckert (to Westvaco Corp.).
126. U.S. Pat. 5,013,404 (May 7, 1991), S. Christiansen, T. Littleton, and R. Patton (to Dow Chemical Co.).

127. U.S. Pat. 5,145,558 (Sept. 8, 1992), S. Christiansen, T. Littleton, and R. Patton (to Dow Chemical Co.).

128. U.S. Pat. 4,134,786 (Jan. 16, 1979), J. Fletcher and M. Humphrey (to J. Fletcher and M. Humphrey).

129. U.S. Pat. 4,343,746 (Aug. 10, 1982), J. Anglin, Y. Ryu, and G. Singerman (to Gulf Research & Development Co.).

130. U.S. Pat. 4,400,282 (Aug. 23, 1983), J. Anglin, Y. Ryu, and G. Singerman (to Gulf Research & Development Co.)

131. U.S. Pat. 4,343,747 (Aug. 10, 1982), J. Anglin, Y. Ryu, and G. Singerman (to Gulf Research & Development Co.).

132. U.S. Pat. 4,364,742 (Dec. 21, 1982), K. Knitter and J. Villa (to Diamond Shamrock Corp.).

133. U.S. Pat. 4,364,741 (Dec. 21, 1982), J. Villa (to Diamond Shamrock Corp.).

134. U.S. Pat. 4,398,918 (Aug. 16, 1983), T. Newman (to Akzona Inc.).

135. U.S. Pat. 4,478,602 (Oct. 23, 1984), E. Kelley, W. Herzberg, and J. Sinka (to Diamond Shamrock Chemicals Co.).

136. U.S. Pat. 4,575,381 (Mar. 11, 1986), R. Corbeels and S. Vasconcellos (to Texaco Inc.).

137. U.S. Pat. 4,317,737 (Mar. 2, 1982), A. Oswald, G. Hatting, and H. Barnum (to Exxon Research and Engineering Co.).

138. U.S. Pat. 4,365,030 (Dec. 21, 1982), A. Oswald and H. Barnum (to Exxon Research & Engineering Co.).

139. U.S. Pat. 4,828,724 (May 9, 1989), C. Davidson (to Shell Oil Co.).

140. U.S. Pat. 3,974,220 (Aug. 10, 1976), L. Heiss and M. Hille (Hoechst AG).

141. U.S. Pat. 4,206,079 (June 3, 1980), R. Frame (to UOP Inc.).

142. U.S. Pat. 4,260,479 (Apr. 7, 1981), R. Frame (to UOP Inc.).

143. U.S. Pat. 4,290,913 (Sept. 22, 1981), R. Frame (to UOP Inc.).

144. U.S. Pat. 4,295,993 (Oct. 20, 1981), D. Carlson (to UOP Inc.).

145. U.S. Pat. 4,354,926 (Oct. 19, 1982), D. Carlson (to UOP Inc.).

146. U.S. Pat. 4,337,141 (June 19, 1982), R. Frame (to UOP Inc.).

147. U.S. Pat. 4,124,493 (Nov. 7, 1978), R. Frame (to UOP Inc.).

148. U.S. Pat. 4,474,622 (Oct. 2, 1984), M. A. Forster (Establisments Somalor-Ferrari Somafer SA).

149. U.S. Pat. 4,376,040 (Mar. 8, 1993), G. Sader (to G. Sader).

150. U.S. Pat. 5,200,062 (Apr. 6, 1993), M.-A. Poirier and J. Gilbert (to Exxon Research and Engineering Co.).

151. U.S. Pat. 4,514,286 (Apr. 30, 1985), S. Wang, G. Roof, and B. Porlier (to Nalco Chemical Co.).

152. U.S. Pat. 4,427,435 (Jan. 24, 1984), J. Lorenz and R. Grade (Ciba-Geigy Corp.).

153. U.S. Pat. 4,470,918 (Apr. 19, 1984), B. Mosier (to Global Marine, Inc.).

154. U.S. Pat. 4,560,761 (Dec. 24, 1985), E. Fields and M. Winzenburg (to Standard Oil Co.).

155. U.S. Pat. 6,492,322 (Feb. 10, 2002), H. A. Cooper and co-workers (to Procter & Gamble Co.).

156. U.S. Pat. 4,526,986 (July 2, 1985), E. Fields and M. Winzenburg (to Standard Oil Co.).

157. U.S. Pat. 4,857,525 (Aug. 15, 1989), M. Philippe and co-workers (to L'Oreal).

158. U.S. Pat. 5,001,156 (Mar. 19, 1991), M. Philippe and co-workers (to L'Oreal).

159. U.S. Pat. 4,321,277 (Mar. 23, 1982), V. Saurino (to Research Lab Products, Inc.).

160. U.S. Pat. 5,165,918 (Jan. 5, 1990), B. Heyl, L. Winterton, and F. P. Tsao (Ciba-Geigy Corp.).

161. U.S. Pat. 5,244,666 (Sept. 14, 1993), J. Murley (to Consolidated Chemical, Inc.).

162. U.S. Pat. 4,941,989 (July 17, 1990), D. Kramer and P. Snow (to Ridgely Products Co., Inc.).

163. U.S. Pat. 5,132,050 (July 21, 1992), R. Baker and J. Thompson (to Lexmark International, Inc.).

164. U.S. Pat. 4,965,168 (Oct. 23,1990), H. Yoshida, T. Kamada, and O. Kainuma (to Nikken Chemical Laboratory Co., Ltd.).

165. U.S. Pat. 5,035,993 (July 30, 1991), S. Hirano, A. Murai, and S. Suzuki (Fuji Photo Film Co., Ltd.).

166. U.S. Pat. 4,828,973 (May 9, 1989), S. Hirano, A. Mural, and S. Suzuki (Fuji Photo Film Co., Ltd.).

167. U.S. Pat. 4,471,044 (Sept. 11, 1984), R. Parton, W. Gaugh, and K. Wiegers (to Eastman Kodak Co.).

168. U.S. Pat. 4,628,068 (Dec. 9, 1986), H. Kesling and J. Harris (to Atlantic Richfield Co.).

169. U.S. Pat. 4,603,149 (July 29, 1986), H. Kesling and J. Harris (to Atlantic Richfield Co.).

170. U.S. Pat. 4,599,366 (July 8, 1986), H. Kesling and J. Harris (to Atlantic Richfield Co.).

171. U.S. Pat. 4,622,345 (Nov. 11, 1986), H. Kesling and J. Harris (to Atlantic Richfield Co.).

172. U.S. Pat. 5,110,835 (May 5, 1992), M. Walter, K.-H. Wassmer, and M. Lorenz (to BASF AG).

173. U.S. Pat. 4,410,462 (Oct. 18, 1983), W. Kroenke (to B. F. Goodrich Co.).

174. U.S. Pat. 3,974,125 (Aug. 10, 1976), A. Oswald and H. Barnum (to Exxon Research and Engineering Co.).

175. U.S. Pat. 3,981,679 (Sep. 21, 1976), N. Christie and J. Karnilaw (to Diamond Shamrock Corp.).

176. U.S. Pat. 4,441,884 (Apr. 10, 1984), H.-P. Baumann and U. Mosimann (to Sandoz Ltd.).

177. U.S. Pat. 4,104,175 (Aug. 1, 1978), M. Martinsson and K. Hellsten (to Modokemi Aktiebolag).

178. U.S. Pat. 3,972,855 (Aug. 3, 1976), M. Martinsson and K. Hellsten (to Modokemi Aktiebolag).

179. U.S. Pat. 4,104,443 (Aug. 1, 1978), B. Latha, C. Stevens and B. Dennis (to J. P. Stevens & Co., Inc.).

180. U.S. Pat. 4,406,809 (Sept. 27, 1983), K. Hasenclever (to Chemische Fabrik Kreussler & Co., GmbH).

181. U.S. Pat. 6,525,034 (Feb. 25, 2003), D. M. Dalyrmple, M. Manning and F. Mert (to Goldschidt Chemical).

182. U.S. Pat. 4,416,787 (Nov. 22, 1983), R. Marshall, W. Archie, and K. Dardoufas (to Allied Corp.).

183. U.S. Pat. 3,756,835 (Sept 4, 1973), R. Beffy and H. Nemeth (to Akzona Inc.).

184. U.S. Pat. 3,497,365 (Feb. 24,1970), J. Roselle and W. Wagner (to Armour Industrial Chemicals, Co.).

185. U.S. Pat. 3,551,168 (Dec. 29, 1970), J. Roselle and W. Wagner (to Armour Industrial Chemicals, Co.).

186. U.S. Pat. 5,167,827 (Jan. 8, 1992), B. Glatz (to Hewlett-Packard Co.).

187. U.S. Pat. 3,625,891 (Dec. 7, 1991), M. Walden, W. Springs, and A. Mariahazey-Westchester (to Armour Industrial Chemical Co.).

188. U.S. Pat. 3,505,221 (Apr. 7, 1970), M. Walden, Northbrook, and A. Mariahazey-Westchester (to Armour Industrial Chemical Co.).

189. U.S. Pat. 3,573,091 (Mar. 30, 1971), M. Walden, Northbrook, and A. Mariahazey-Westchester (to Armour and Co.).

190. U.S. Pat. 4,096,071 (June 20, 1978), A. Murphy (to Procter & Gamble Co.).
191. U.S. Pat. 4,203,852 (May 20, 1980), J. Johnson and W. Chirash (to Colgate-Palmolive Co.).
192. U.S. Pat. 4,255,484 (Mar. 10, 1981), F. Stevens (to A. E. Staley Manufacturing Co.).
193. U.S. Pat. 4,772,404 (Sept. 20, 1988), D. Fox, M. Sullivan, and A. Cuomo (to Lever Brothers Co.).
194. U.S. Pat. 4,806,260 (Feb. 21, 1989), G. Broze and D. Bastin (to Colgate-Palmolive Co.).
195. U.S. Pat. 4,844,822 (July 6, 1987), P. Fox and B. Felthouse (to Dial Corp.).
196. U.S. Pat. 4,808,321 (Feb. 28, 1989), D. Walley (to Procter and Gamble Co.).
197. U.S. Pat. 4,895,667 (Jan. 23, 1990), P. Fox and B. Felthouse (to Dial Corp.).
198. U.S. Pat. 4,970,008 (Nov. 13, 1990), T. Kandathil (to T. Kandathil).
199. U.S. Pat. 4,948,520 (Aug. 14, 1990), H. Sasaki (to Lion Corp.).
200. U.S. Pat. 4,986,922 (Jan. 22, 1991), S. Snow and L. Madore (to Dow Corning Corp.).
201. U.S. Pat. 5,132,425 (July 21, 1992), K. Sotoya, U. Nishimoto, and H. Abe (to Kao Corp.).
202. U.S. Pat. 5,151,223 (Sept. 29, 1992), H. Maaser (to Colgate-Palmolive Co.).
203. U.S. Pat. 5,259,964 (Nov. 9, 1993), N. Chavez and I. Oliveros (to Colgate-Palmolive Co.).
204. U.S. Pat. 5,221,794 (June 22, 1993), J. Ackerman, M. Miller, and D. Whittlinger (to Sherex Chemical Co., Inc.).
205. U.S. Pat. 4,256,824 (Mar. 17, 1981), C. Lu (to Xerox Corp.).
206. U.S. Pat. 4,323,634 (Apr. 6, 1982), T. Jadwin (to Eastman Kodak Co.).
207. U.S. Pat. 4,604,338 (Aug. 5, 1986), R. Gruber, R. Yourd, and R. Koch (to Xerox Corp.).
208. U.S. Pat. 4,560,635 (Dec. 24, 1985), T. Hoffend and A. Barbetta (to Xerox Corp.).
209. U.S. Pat. 3,985,663 (Oct. 12, 1976), C. Lu and R. Parent (to Xerox Corp.).
210. U.S. Pat. 4,059,444 (Nov. 22, 1977), C. Lu and R. Parent (to Xerox Corp.).
211. *CRC Handbook of Chemistry and Physics*, 61 st ed., CRC Press, Inc., Boca Raton, Fla., 1981, p. C-586.
212. R. Shelton and co-workers, *J. Am. Chem. Soc.* **66**, 753 (1946).
213. R. Reck, H. Harwood, and A. Ralston, *J. Org. Chem.* **12**, 517 (1947).
214. A. Ralston and co-workers, *J. Org. Chem.* **13**, 186 (1948).
215. Ref. 59, Chapts. 7 and 8.
216. Ref. 61, Chapts. 1–7.
217. Ref. 60, Chapts. 9, 10, and 11.
218. Ref. 60, Chap. 1.
219. Ref. 60, Chap. 3.
220. E. White and D. Woodcock, in Patai, ed., *The Chemistry of the Amino Group*, Wiley-Interscience, New York, 1968, pp. 409–416.
221. A. Hofmann, *Ann. Chem.* **78**, 253 (1851); H. Hofman, *Ann. Chem.* **79**, 11 (1851).
222. J. March, *Advanced Organic Chemistry*, 3rd ed., John Wiley & Sons, Inc., New York, 1985, pp. 906–908.
223. A. Cope and F. Trumbull, *Org. Reactions* **11**, 317 (1960).
224. J. Baumgarten, *J. Chem. Ed.* **45**, 122 (1968).
225. W. Hawart and C. Ingold, *J. Chem. Soc.*, 997 (1927).
226. T. Stevens and co-workers, *J. Chem. Soc.*, 3193 (1928); G. Wittig, R. Mangold, and G. Felletschin, *Ann. Chem.* **560**, 117 (1948).
227. T. Stevens, *J. Chem. Soc.* (2), 2107 (1930); R. Johnson and T. Stevens, *J. Chem. Soc.* (4), 4487 (1955).
228. M. Sommelet, *Compt. Rend.* **205**, 56 (1937); S. Kantor and C. Hauser, *J. Am. Chem. Soc.* **73**, 4122 (1951); C. Hauser and D. Van Eenam, *J. Am. Chem. Soc.* **79**, 5513 (1957); W. Beard and C. Hauser, *J. Am. Chem. Soc.* **25**, 334 (1961).

229. H. Zimmerman, in de Mayo, ed., *Molecular Rearrangements*, Vol.1, John Wiley & Sons, Inc., New York, 1963, p. 387.
230. U. Anthoni and co-workers, *Comp. Biochem. Physiol.* **99B**, 1–18 (1991).
231. U.S. Pat. 5,008,410 (Apr. 16, 1991), K. Tashiro and K. Tanaka (to Sumitomo Chemical Co., Ltd.).
232. B. Challis and A. Butler, in Ref. 220, pp. 290–300.
233. M. Gibson, in Ref. 220, pp. 44–55.
234. Ref. 222, p. 364.
235. C. Ingold, *Structure and Mechanism in Organic Chemistry*, Bell, London, 1953, Chapt. 7; A. Streitwieser, *Chem. Rev.* **56**, 571 (1956).
236. W. Emerson, *Organic Reactions*, Vol. IV, John Wiley & Sons, Inc., New York, 1948, Chapt. 3.
237. H. Sommer and L. Jackson, *J. Org. Chem.* **35**, 1558 (1970).
238. H. Sommer, H. Lipp, and L. Jackson, *J. Org. Chem.* **36**, 824 (1971).
239. S. Billenstein and G. Blaschke, *J. Am. Oil Chem. Soc.* **61**, 353 (1984).
240. R. Schroter, in Houben-Weyl, ed., *Methoden der Organischen Chemie*, Vol. XIII, 4th ed., George Thieme Verlag, Stuttgart, Germany, 1957, Chapt. 4; M. Rabinovitz, in Z. V. C. Rappoport, *The Chemistry of the Cyano Group*, Wiley-Interscience, New York, 1970, pp 307–340; C. De Bellefon and P. Fouilloux, *Catal. Rev.-Sci. Eng.* **36**, 459 (1994).
241. P. Rylander, *Catalytic Hydrogenation Over Platinum Metals*, Academic Press, Inc., New York, 1967, pp. 203–226.
242. F. Gould, G. Johnson, and A. Ferris, *J. Org. Chem.* **25**, 1658 (1960).
243. M. Freifelder, *J. Am. Chem. Soc.* **82**, 2386 (1960).
244. U.S. Pat. 4,767,547 (Aug. 30, 1988), T. Straathof and A. Konig (to Procter & Gamble Co.).
245. U.S. Pat. 4,787,491 (Dec. 6, 1988), N. Chang and D. Walley (to Procter & Gamble Co.).
246. U.S. Pat. 3,915,867 (Oct. 28, 1975), H. Kang and R. Peterson, and E. Knaggs (to Stepan Chemical Co.).
247. U.S. Pat. 4,339,391 (July 13, 1982), E. Hoffmann and co-workers (to Hoechst AG).
248. U.S. Pat. 4,963,274 (Oct. 16, 1990), W. Ruback and co-workers (to Hüls AG).
249. R. Lagerman and co-workers, *J. Am. Oil Chem. Soc.* **71**, 97 (1994).
250. U.S. Pat. 5,066,414 (Nov. 19, 1991), N. Chang (to Procter & Gamble Co.).
251. U.S. Pat. 4,840,738 (June 20, 1981), F. Hardy and D. Walley (to Procter & Gamble Co.).
252. U.S. Pat. 5,128,473 (July 7, 1992), F. Friedli and M. Watts (to Sherex Chemical Co.).
253. U.S. Pat. 4,857,310 (Aug. 15, 1989), A. Baydar (to Gillette Co.).
254. U.S. Pat. 4,830,771 (May 16, 1989), W. Ruback and J. Schut (to Hüls AG).
255. Can. Pat. 1,312,619 (Jan. 12, 1993), M. Hofinger and co-workers (to Hoechst AG).
256. *Synthetic Organic Chemicals*, U.S. International Trade Commission, Washington, D.C., 1994.
257. C. Starks, *Ind. Appl. Surfactants II* **77**, 165 (1990); C. Starks, ed., *Phase-Transfer Catalysis: New Chemistry, Catalysts and Applications*, American Chemical Society, Washington, D.C., 1987; E. Dehmlov, *Phase-Transfer Catalysis*, Verlag Chemie, Deerfield Beach, Fla., 1983; M. Halpern, *Phase-Transfer Catalysis in Ullman's Encyclopedia of Industrial Chemistry*, Vol. A19, VCH V6, New York, 1991; M. Halpern, *Phase-Transfer Catalysis Commun.* 1, 1 (1995). *Specialty Surfactants Worldwide* in *Specialty Chemicals*, SRI International, Menlo Park, Calif., 1989, pp. 81–94.
258. Ref. 61, p. 657.5000yz; W. Evans, *Chem. Ind.* **27**, 893 (1969); R. Puchta, *J. Am. Oil Chem. Soc.* **61**, 367 (1984).

259. *Cosmetic Chemicals, in Specialty Chemicals,* SRI International, Menlo Park, Calif., 1992, p.73.

260. Technical data, Akzo Nobel Chemicals Inc., Dobbs Ferry, N.Y., Apr. 1995; for a review, see W. Mardis, *J. Am. Oil Chem. Soc.* **61**, 382 (1984).

261. *Official Methods and Recommended Practices of American Oil Chemists' Society,* 4th ed., American Oil Chemists Society, Champaign, Ill., 1990, pp. S 4c-64 and Section T; L. Metcalfe, *J. Am. Oil Chem. Soc.* **61**, 363 (1984).

262. Ref. 59, Chapt. 13.

263. M. Rosen and H. Goldsmith, *Systematic Analysis of Surface Active Agents,* Interscience, New York, 1960.

264. J. Morelli and G. Szajer, *J. Surf. Det.* **4**, 75 (2001).

265. D. Herring, *Lab. Practice* **II**, 113 (1962).

266. S. Epton, *Nature (London)* **160**, 795 (1947).

267. S. Epton, *Trans. Faraday Soc.* **44**, 226 (1948).

268. Ref. 263, p. 455.

269. Ref. 263, p. 445.

270. For a review on the determination of surfactants in the environment, see G. Kloster, in M. Schuwuger, ed., *Detergents in the Environment,* Marcel Dekker, Inc., New York, 1997.

271. Ref. 264, pp. 76–77. (morelli article)

272. W. Freesenius', *Z. Anal. Chem.* **312**, 419 (1982).

273. W. Straw, *Anal. Proc.* **22**, 142 (1985).

274. N. Buschmann and R. Schulz, *Tenside Surf. Det.* **29**, 128 (1992).

275. Ref. 59, p. 430.

276. M. Bambagiottialberti and co-workers, *Rapid Commun. Mass Spectrom.* **8**, 439 (1994).

277. A. Tyler, L. Romo, and R. Cody, *Int. J. Mass. Spectrom. Ion Process.* **122**, 25 (1992).

278. D. Fisher and co-workers, *Rapid Commun. Mass Spectrom.* **8**, 65 (1994).

279. V. Wee and J. Kennedy, *Anal. Chem.* **54**, 1631 (1982).

280. G. Szajer, in Perkins, ed., *Analyses of Fats, Oils and Lipoproteins*, American Oil Chemists Society, Champaign, Ill., 1991, Chapt. 18; T. Schmitt, *ibid.*, Chapt. 19.

281. F. Mozayeni, in J. Cross and E. Singer, eds., *Cationic Surfactants*, Marcel Dekker, Inc., New York, 1994, Chapt. 11.

282. J. Bender, Ph.D. Thesis, *Physical Chemistry of Softeners Water Transport Properties Studied by Self-diffusion NMR*, Department of Applied Surface Chemistry, Chalmers University of Technology, Gotenborg, Sweden, 2001.

283. Technical data, *Toxicity,* Akzo Nobel Chemicals Inc., Dobbs Ferry, N.Y., Apr. 1995.

284. R. Shelton and co-workers, *J. Am. Chem. Soc.* **66**, 755 (1946).

285. Ref. 284, p. 757.

286. W. Sexton, *Chemical Constitution and Biological Activity*, Van Nostrand, Princeton, N.J., 1963.

287. M. Lewis, *Ecotox Env. Safety* **20**, 123–140 (1990).

288. A. Vallejo-Freire, *Science* **114**, 470 (1954).

289. K. Biesinger and co-workers, *J. Water Pollut. Control Fed.* **48**, 183 (1976).

290. J. Walker and S. Evans, *Marine Pollut. Bull.* **9**, 136 (1978).

291. U.S. Pat. 4,204,954 (May 27, 1980), J. Jacob (to Chemed Corp.) for a patented procedure to remove quaternaries from wastewater.

292. R. Boethling, *Water Res.* **18**, 1061 (1984); L. Huber, *J. Am. Oil Chem. Soc.* **61**, 377 (1984).

293. J. Cooper, *Ecotoxicity Environ. Safety* **16**, 65 (1988).

294. W. Ruback, *Chem. Today*, 15 (May/June 1994).

295. L. Edebo and co-workers, *Ind. Appl. Surfactants* **III**, 184 (1992).

296. H. Berenbold, *Inform* **5**, 82 (1994).
297. E. Spiess, *Perfumerie Kusmetik* **72**, 370 (1991); M. Jurczyk, D. Berger, and D. Damaso, *Cosmetics Toiletries* **106**, 63 (1991).
298. B. Beard and J. Hare, *J. Surf. Det* **5**, 145 (2002).
299. C. Reich and C. Robbins, *J. Soc. Cosmet. Chem.* **44**, 263 (1993).
300. P. D'Arcy and E. Taylor, *J. Pharm. Pharmacol.* **14**, 193 (1962); P. Schaeufele, *J. Am. Oil Chem. Soc.* **61**, 387 (1984).
301. Ref. 59, Chapt. 14.
302. U.S. Pat. 3,970,755 (July 20, 1976), E. Gazzard and M. Singer (to Imperial Chemical Industries Ltd.).
303. U.S. Pat. 4,272,395 (June 9, 1981), R. Wright (to Lever Brothers Co.).
304. U.S. Pat. 4,444,790 (Apr. 24, 1984), H. Green, A. Petrocci, and Z. Dudzinski (to Millmaster Onyx Group, Inc.).
305. U.S. Pat. 4,868,217 (Sept. 19, 1989), S. Araki and co-workers (to Eisai Co., Ltd. Kao Corp.).
306. U.S. Pat. 4,800,235 (Jan. 24, 1989), T. La Marre and C. Martin (to Nalco Chemical Co.).
307. U.S. Pat. 4,847,089 (July 11, 1989), D. Kramer and P. Snow (to D. N. Kramer).
308. U.S. Pat. 4,941,989 (July 17, 1990), D. Kramer and P. Snow (to Ridgely Products Co., Inc.).
309. U.S. Pat. 4,983,635 (Jan. 8, 1991), H. Martin (to M. Howard).
310. U.S. Pat. 5,049,383 (Sept. 17, 1991), H.-U. Huth, H. Braun and F. Konig (to Hoechst AG).
311. U.S. Pat. 5,286,109 (Dec. 7, 1993), S. Boyd (to S. Boyd).
312. H. Freedman, *Pure Appl. Chem.* **58**, 857 (1986).
313. J. D'Souza and N. Sridhar, *J. Sci. Ind. Res.* **42**, 564 (1983).
314. J. Salamone and W. Rice, in J. I. Kroschwitz, ed., *Encyclopedia of Polymer Science and Engineering,* 2nd ed., John Wiley & Sons, Inc., New York, 1988.
315. R. McConnell, *Soap, Cosmetics, Chem. Special.*, 37 (Apr. 1989).
316. J. K. Borchardt, *Proceedings of the Symposium on Advances in Oil Field Chemistry,* Toronto, Canada, June, 5–11, 1988.

CHARLES W. WESTON
Freeport Research and Engineering Company
JOHN R. PAPCUN
Atotech
MAURICE DERY
Akzo Nobel Chemicals, Inc.

AMYL ALCOHOLS

1. Introduction

Amyl alcohol describes any saturated aliphatic alcohol containing five carbon atoms. This class consists of three pentanols, four substituted butanols, and a disubstituted propanol, ie, eight structural isomers $C_5H_{12}O$: four primary, three secondary, and one tertiary alcohol. In addition, 2-pentanol, 2-methyl- 1-butanol, and 3-methyl-2-butanol have chiral centers and hence two enantiomeric forms.

The odd-carbon structure and the extent of branching provide amyl alcohols with unique physical and solubility properties and often offer ideal properties for solvent, surfactant, extraction, gasoline additive, and fragrance applications. Amyl alcohols have been produced by various commercial processes in past years. Today the most important industrial process is low pressure rhodium-catalyzed hydroformylation (oxo process) of butenes.

Mixtures of isomeric amyl alcohols (1-pentanol and 2-methyl-1-butanol) are often preferred because the different degree of branching imparts a more desirable combination of properties; they are also less expensive to produce commercially.

2. Physical Properties

With the exception of neopentyl alcohol (mp 53°C), the amyl alcohols are clear, colorless liquids under atmospheric conditions, with characteristic, slightly pungent and penetrating odors. They have relatively higher boiling points than ketonic or hydrocarbon counterparts and are considered intermediate boiling solvents for coating systems (Table 1) (1–16).

Commercial primary amyl alcohol is a mixture of 1-pentanol and 2-methyl-1-butanol, in a ratio of ca 65 to 35. Typical physical properties of this amyl alcohol mixture are listed in Table 2 (17).

Like the lower alcohols, amyl alcohols are completely miscible with numerous organic solvents and are excellent solvents for nitrocellulose, resin lacquers, higher esters, and various natural and synthetic gums and resins. However, in contrast to the lower alcohols, they are only slightly soluble in water. Only 2-methyl-2-butanol exhibits significant water solubility. As associated liquids, amyl alcohols form azeotropes with water and/or a variety of organic compounds (Table 3).

Figures 1 and 2 show the relationship of viscosity and surface tension with temperature for various amyl alcohols. Curves for lower and higher alcohol homologues are shown for comparison.

The physical characteristics of *tert*-amyl alcohol diverge from the standard trends for the other alcohols; it has a lower boiling point, higher melting point, higher vapor pressure, and low surface tension. Most notably, organic molecules are highly soluble in *tert*-amyl alcohol.

3. Chemical Properties

The amyl alcohols undergo the typical reactions of alcohols which are characterized by cleavage at either the oxygen–hydrogen or carbon–oxygen bonds.

3.1. Dehydration. Dehydration of amyl alcohols is important for the preparation of specialty olefins and where it may produce unwanted by-products under acidic reaction conditions. Olefin formation is especially facile with

Table 1. The Amyl Alcohols and Some of Their Physical Properties

Properties	1-Pentanol	2-Pentanol	3-Pentanol	2-Methyl-1-butanol	3-Methyl-1-butanol	2-Methyl-2-butanol	3-Methyl-2-butanol	2,2-Dimethyl-1-propanol
CAS Registry Number	[71-41-0]	[6032-29-7]	[584-02-1]	[137-32-6]	[123-51-3]	[75-85-4]	[598-75-4]	[75-84-3]
common name	n-amyl alcohol	sec-amyl alcohol			isoamyl alcohol	tert-amyl alcohol		neopentyl alcohol
critical temperature, °C	315.35	287.25	286.45	291.85	306.3	272.0	300.85	276.85
critical pressure, kPa[a]	3868.	3710.	3880.	3880.	3880.	3880.	3960.	3880.
critical specific volume, mL/mol	326.5	328.9	325.3	327	327	327	327	327
critical compressibility	0.25810	0.26188	0.27128	0.27009	0.26335	0.27992	0.27133	0.27745
boiling point at pressure, °C								
101.3 kPa[a]	137.8	119.3	115.3	128.7	130.5	102.0	111.5	113.1
40 kPa	111.5	93.8	90.9	103.5	105.6	78.3	87.2	89.0
1.33 kPa	44.6	32.0	27.7	40.2	43.0	21.0	26.0	25.0
vapor pressure[b] kPa[a]	0.218	0.547	0.761	0.274	0.200	1.215	0.810	0.929
melting point, °C	−77.6	−73.2	−69.0	<−70	−117.2	−8.8 −88	forms glass	54.0
heat of vaporization at normal boiling point, kJ/mol[c]	44.83	43.41	42.33	44.75	43.84	40.11	41.10	41.35
ideal gas heat of formation[d], kJ/mol[c]	−298.74	−313.80	−316.73	−302.08	−302.08	−329.70	−314.22	−319.07

liquid density[b], kg/m³	815.1	809.4	820.3	819.1	810.4	809.6	818.4	851.5[e]
liquid viscosity[b], mPa·s(=cP)	4.06	4.29	6.67	5.11	4.37	4.38	3.51d	2.5[e]
surface tension[b] mN/m(=dyn/cm)	25.5	24.2	24.6	25.1[d]	24.12	22.7	23.0d	14.87[e]
refractive index[d]	1.4080	1.4044	1.4079	1.4086	1.4052	1.4024	1.4075	1.3915
solubility parameter[d] (MJ/ m³)$^{0.5}$[f]	22.576	21.670	21.150	22.274	22.322	20.758	21.607	19.265[e]
solubility in water[b], wt %	1.88	4.84	5.61	3.18	2.69	12.15	6.07	3.74
solubility of water in[b], wt %	9.33	11.68	8.19	8.95	9.45	24.26	11.88	8.23

[a] To convert kPa to mm Hg, multiply by 7.5.
[b] At 20°C unless otherwise noted.
[c] To convert kJ/mol to cal/mol, multiply by 239.
[d] At 25°C.
[e] At the melting point.
[f] To convert (MJ/m³)$^{0.5}$ to (cal)$^{0.5}$, divide by 2.045.

Table 2. **Physical Properties of Primary Amyl Alcohol, Mixed isomers**[a]

Property	Value[b]
molecular weight	88.15
boiling point at 101.13 kPa[c], °C	133.2
freezing point, °C	−90[d]
specific gravity 20/20 °C	0.8155
absolute viscosity at 20°C, mPa·s(=cP)	4.3
vapor pressure at 20°C, kPa[c]	0.27
flash point (closed cup), °C	45
solubility at 20°C, by wt %	
in water	1.7
water in	9.2

[a] Ref. 17.
[b] 65/35 blend, ie, a mixture of 1-pentanol and 2-methyl-1-butanol, 65/35 wt %, respectively.
[c] To convert kPa to mm Hg, multiply by 7.5.
[d] Sets to glass below this temperature.

secondary or tertiary amyl alcohols under acidic conditions. The reverse reaction, hydration of olefins, is commonly used for the preparation of alcohols.

An example of a specialty olefin from an amyl alcohol is Phillips Petroleum's process for 3-methyl-1-butene (used in the synthesis of pyrethroids) from the catalytic dehydration of 3-methyl-1-butanol (21,22). The process affords 94% product selectivity and 94% alcohol conversion at 310°C and 276 kPa (40 psig).

Dehydration of 1-pentanol or 2-pentanol to the corresponding olefins has been accomplished, in high purity and yields, by vapor-phase heterogeneous catalyzed processes using a variety of catalysts including neutral gamma–Al$_2$O$_3$ catalyst doped with an alkali metal (23), zinc aluminate (24,25), lithiated clays (26), Ca$_3$(PO$_4$)$_2$ (27) and montmorillonite clays (28). Dehydration of 2-methyl-1-butanol occurs over zinc aluminate catalyst at 270–370°C to give the expected 2-methyl-1-butene in high selectivites (24). The Al$_2$O$_3$ catalyzed process can be optimized to give di-n-pentyl ether as the exclusive product (23). Dehydration of 1-pentanol over an alkali metal promoted Al$_2$O$_3$ catalyst at 300–350°C

Table 3. **Azeotropic Mixtures of Amyl Alcohols with Water**[a]

	Boiling point of azeotrope, °C[b]	Alcohol in azeotrope, wt %	Wt % alcohol	
			Upper layer	Lower layer
1-pentanol	95.93	43.65	84.29	2.54
2-pentanol	91.7	63.5		
3-pentanol	91.5	65	90.1	5.5
2-methyl-1-butanol	93.8	58.5		
3-methyl-1-butanol	94.82	50.67	84.15	2.96
2-methyl-2-butanol	87.35	72.5		
3-methyl-2-butanol	91.0	67.0		

[a] Refs. (18–20).
[b] At 101.13 kPa = 1 atm.

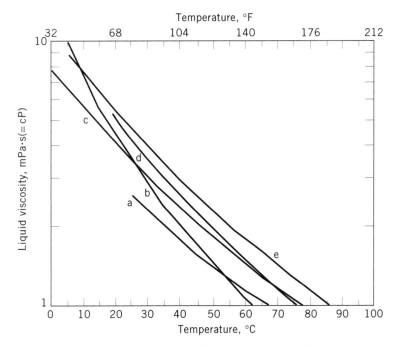

Fig. 1. Viscosities of amyl alcohols compared with 1-butanol and 1-hexanol (15). a, 1-butanol; b, 2-methyl-2-butanol; c, 1-pentanol; d, 2-methyl-1-butanol; e, 1-hexanol.

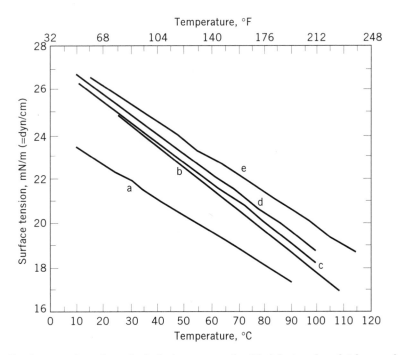

Fig. 2. Surface tension of amyl alcohols compared with 1-butanol and 1-hexanol (15). a, 2-methyl-2-butanol; b, 2-methyl-1-butanol; c, 1-butanol; d, 1-pentanol; e, 1-hexanol.

provides 1-pentene at selectivities of 92% (29,30). Purification produces polymerization grade (99.9% purity) 1-pentene. A flow chart has been shown for a pilot-plant process (29).

Koch Reaction.   C-6-neoacids are readily available from amyl alcohols by the Koch reaction. Greater than 95% 2,2-dimethylbutyric acid [595-37-9] was obtained from 2-methyl-1-butene at 304 kPa (3 atm) CO and 35°C for 1 h with cupric oxide and sulfuric acid catalyst (31). Likewise, 2,2-dimethylbutyric acid can be obtained in high yield (75–80%) from 1- or 2-pentanol or neopentyl alcohol from the Koch-Haaf reaction (32,33). *tert*-Amyl alcohol gives a mixture of trimethylacetic acid [75-98-9] (pivalic acid), 2,2-dimethylbutyric acid, C-7 acids, and C-11 acids under similar Koch-Haaf conditions (33).

3.2. Esterification.   Extensive commercial use is made of primary amyl acetate, a mixture of 1-pentyl acetate [28-63-7] and 2-methylbutyl acetate [53496-15-4]. Esterifications with acetic acid are generally conducted in the liquid phase in the presence of a strong acid catalyst such as sulfuric acid (34). Increased reaction rates are reported when esterifications are carried out in the presence of heteropoly acids supported on macroreticular cation-exchange resins (35) and zeolite (36) catalysts in a heterogeneous process. Judging from the many patents issued in recent years, there appears to be considerable effort underway to find an appropriate solid catalyst for a reactive distillation esterification process to avoid the product removal difficulties of the conventional process.

Reaction of 1-pentanol with propionic acid provides 1-pentyl propionate [624-54-4], a new coatings solvent for automotive refinish and OEM paints, appliances, and for higher-solids systems (37). The esterification of 1-pentanol with formic acid to 1-pentyl formate [638-49-3] is conducted by concomitant removal of by-product water by azeotropic distillation with diethyl ether (38).

Alkali metal xanthates are prepared in high yield from reaction of amyl alcohols with alkali metal hydroxide and carbon disulfide (39–42). The xanthates are useful as collectors in the flotation of minerals and have minor uses in vulcanization of rubber and as herbicides (39,41).

Zinc dithiophosphates, which serve as antioxidants (qv) and antiwear agents in lubricants, are prepared by reaction of amyl alcohol and phosphorus pentasulfide followed by treatment with zinc sulfate (43).

Esters of *tert*-amyl alcohol can be obtained by acylation of 2-methyl-2-butene in the presence of trifluoromethanesulfonic acid (44). The esters produced, in high yields, from reaction of amyl alcohols with carboxylic anhydrides, are used as intermediates for preparation of pyrylium salts (45, 46) and alkaloids (47). Triazoles prepared by acylation of 3-methyl-1-butanol are useful as herbicides (48).

3.3. Oxidation.   Oxidation of the *n*-amyl alcohols produces aldehydes, which after continued oxidation can yield acids. This route to aldehydes has little merit. However, oxidative esterifications with alkali metal hypohalites (eg, calcium chlorite, $Ca(OCl)_2$) (49), bromates (eg, sodium bromate, $NaBrO_3$) (50) and halites (eg, sodium bromite, $NaBrO_2$) (51–53)) have commercial potential. For example, a 90% yield of pentyl pentanoate [2173-56-0] was obtained from treatment of 1-pentanol with bromous acid or its salt in a solvent at pH 7–12 (53); sodium hypochlorite gave 87% yield of 3-methylbutyl isovalerate [659-70-1] from 3-methyl-1-butanol (49). Reaction is believed to involve dehydrogenation of an intermediate hemiacetal which, in turn, is formed from reaction of the first formed aldehyde with starting alcohol. Oxidation of secondary amyl alcohols afford high yields of ketones, eg, 3-pentanol gave 97% yield of diethyl ketone by oxidation with calcium hypochlorite (49).

Other examples of oxidation of amyl alcohols using hydrogen peroxide (54), $RuO_2/NaClO$ (55), $BaMnO_4$ (56), and chromic acid (57, 58) have been described for laboratory synthesis, but have not been utilized commercially.

3.4. Amination. Amyl alcohols can react with ammonia or alkylamines to form primary, secondary, or tertiary-substituted amines. For example, 3-methyl-butylamine [107-85-7] is produced by reductive ammonolysis of 3-methyl-1-butanol over a Ni catalyst at 150°C (59). Some diisoamyl- and triisoamylamines are also formed in this reaction. Good selectivities (88%) of neopentylamine [5813-64-9] are similarly produced by reductive ammonolysis of neopentyl alcohol (60).

Reaction of *tert*-amyl alcohol with urea in the presence of sulfuric acid gives a monoalkylated urea (61,62). Monoalkyl ureas are used to prepare uracil derivatives which are useful as herbicides, fungicides, and plant growth regulators (61).

3.5. Etherification. Ethers of amyl alcohols have been prepared by reaction with benzhydrol (63), activated aromatic halides (64), dehydration-addition reactions (65), addition to olefins (66–71), alkoxylation with olefin oxides (72,73) and displacement reactions involving their alkali metal salts (74–76).

The product benzhydrol pentyl ether [42100-71-0] from etherification of benzhydrol with 1-pentanol is utilized as low odor carrier oils for pressure-sensitive carbonless papers (63) and alkoxylated products are good surfactants (72). The *tert*-amyl functionality imparts pleasant aromas to compounds, eg, amyl substituted bicyclic polymethylated naphthalenones and indanones give fruity amberlike or woody aromas (69), and in acylhexamethylindans they provide a musklike odor (71), useful in perfumes. Ethers of neopentyl alcohol and 4-chlorobenzenethiol (4-thiophenyl neopentyl ether) are intermediates in the synthesis of thioether herbicide and plant growth regulators (64). Alkali *tert*-amyl alcoholates are useful in the preparation of colorfast pigments such as diphenylpyrrolopyroledione (75,76).

3.6. Condensation. The neopentyl trimethylolpropane carbonate [65332-76-5] formed from condensation of the trischloroformate of trimethylolpropane and neopentyl alcohol, is a clear yellow oil, useful as lubricant (77).

Methyl *t*-butyl ketone [1634-04-4] (pinacolone) has been prepared in 74% yield by reaction of *tert*-amyl alcohol with formaldehyde in the presence of strong acid catalyst (78,79).

The Guerbet reaction can be used to obtain higher alcohols; 2-propyl-1-heptanol [10042-59-8] from 1-pentanol condensation and 6-methyl-4-nonanol from 2-pentanol (80–83). Condensations with alkali phenolates as the base, instead of copper catalyst, produce lower amounts of carboxylic acids and require lower reaction temperatures (82,83). The crossed Guerbet reaction of 1-pentanol with methanol in the presence of sodium methoxide catalyst afforded 2-heptanol in selectivities of about 75% (84).

Friedel-Crafts alkylation of benzene with isomeric amyl alcohols proceeds with some rearrangement. For example, both 2- and 3-pentanol gave the identical product mixture (60% 2-phenylpentane, 31% 3-phenylpentane, and 9% *tert*-pentylbenzene) from reaction with benzene in the presence of BF$_3$ catalyst (85).

Reduction of 1-pentanol to pentane reportedly occurs during condensation of samarium with excess alcohol (86).

3.7. Other Reactions. Primary amyl alcohols can be halogenated to the corresponding chlorides by reaction with hydrogen chloride in hexamethylphosphoramide (87). Neopentyl chloride [753-89-9] is formed without contamination by rearrangement products. A convenient method for preparing *tert*-amyl bromide and iodide involves reaction of *tert*-amyl alcohol with hydrobromic or hydroiodic acid in the presence of Li or Ca halide (88). The metal halides increase the yields (85–95%) and product purity.

A method for the preparation of dipentyl carbonate from reaction of 1-pentanol with CO$_2$ in the presence of diethyl azodicarboxylate in high yields (81%) has been reported (89).

Neopentyl alcohol is useful for preparation of masked polyol silicate esters, capable of releasing the polyol under moisture conditions, in moisture-curable one-component liquid polyurethane compositions (90).

4. Manufacture

Three significant, commercial processes for the production of amyl alcohols include separation from fusel oils, chlorination of C-5 alkanes with subsequent hydrolysis to produce a mixture of seven of the eight isomers (91), and a low pressure oxo process, or hydroformylation, of C-4 olefins followed by hydrogenation of the resultant C-5 aldehydes.

The oxo process is the principal one in practice today; only minor quantities, mainly in Europe, are obtained from separation from fusel oil. *tert*-Amyl alcohol is produced on a commercial scale in lower volume by hydration of amylenes (92).

4.1. Fusel Oils. The original source of amyl alcohols was from fusel oil which is a by-product of the ethyl alcohol fermentation industry. Refined amyl alcohol from this source, after chemical treatment and distillation, contains about 85% 3-methyl-1-butanol and about 15% 2-methyl-1-butanol, both primary amyl alcohols. Only minor quantities of amyl alcohol are supplied from this source today. A German patent discloses a distillative separation process for recovering 3-methyl-1-butanol from fusel oil (93).

4.2. Chlorination-Hydrolysis Process. The first synthetic production of amyl alcohols was begun during the 1920s by the so-called chlorination-hydrolysis process in which a mixture of amyl chlorides is first produced by the continuous vapor-phase chlorination of a mixture of pentane and isopentane in the absence of light and catalysts. Hydrolysis of the chlorides with aqueous caustic at high temperatures produces a mixture of seven of the eight amyl alcohol isomers; formation of neopentyl alcohol is negligible. In contrast to the fusel oil and the oxo processes, this source provides significant quantities of the three secondary amyl alcohols, especially 2-pentanol. An excellent review with references and a process schematic has been published (91).

4.3. Oxo Process. Because of catalytic advances made since the 1970s in "oxo" chemistry and the comparatively high cost and waste disposal problems associated with the chlorination-hydrolysis process, the oxo process (qv) is now the principal source of amyl alcohols. In the low pressure hydroformylation (oxo) process for the production of amyl alcohols, 1-butene, 2-butene, and 2-methyl-propylene (isobutylene) react with a mixture of carbon monoxide and hydrogen in the presence of a suitable metal catalyst (rhodium) to form an isomeric mixture of aldehydes with one more carbon atom than the olefin. Once made, the 1-pentaldehyde [110-62-3] (*n*-valeraldehyde), 2-methylbutyraldehyde [96-17-3] and 3-methylbutyraldehyde [590-86-3] (isovaleraldehyde) are hydrogenated to the corresponding amyl alcohols.

$$
\begin{array}{c}
CH_3CH_2CH{=}CH_2 \\
+ \\
CH_3CH{=}CHCH_3 \; + CO + H_2 \;\; \xrightarrow[\text{catalyst}]{\text{Rh/P}} \;\; CH_3CH_2CH_2CH_2CHO \\
+ \\
(CH_3)_2C{=}CH_2
\end{array}
\qquad
\begin{array}{c}
CH_3CH_2CH_2CH_2CHO \\
+ \\
CH_3CH_2\underset{\underset{CH_3}{|}}{CH}CHO \\
+ \\
CH_3\underset{\underset{CH_3}{|}}{CH}CH_2CHO
\end{array}
\;\; \xrightarrow{H^+} \;\;
\begin{array}{c}
\text{1-pentanol} \\
+ \\
\text{2-methyl-1-butanol} \\
+ \\
\text{3-methyl-1-butanol}
\end{array}
$$

Prior to 1975, reaction of mixed butenes with syn gas required high temperatures (160–180°C) and high pressures 20–40 MPa (3000–6000 psi), in the presence of a cobalt catalyst system, to produce *n*-valeraldehyde and 2-methylbutyraldehyde. Even after commercialization of the low pressure oxo process in 1975, a practical process was not available for amyl alcohols because of low hydroformylation rates of internal bonds of isomeric butenes (91,94). More recent developments in catalysts have made low pressure oxo process technology commercially viable for production of low cost *n*-valeraldehyde, 2-methylbutyraldehyde, and isovaleraldehyde, and the corresponding alcohols in pure form.

Conventional triorganophosphite ligands, such as triphenylphosphite, form highly active hydroformylation catalysts (95–99); however, they suffer from poor durability because of decomposition. Diorganophosphite-modified rhodium catalysts (94,100,101), have overcome this stability deficiency and provide a low pressure, rhodium catalyzed process for the hydroformylation of low reactivity olefins, thus making lower cost amyl alcohols from butenes readily accessible. The diorganophosphite-modified rhodium catalysts increase hydroformylation rates by more than 100 times and provide selectivities not available with standard phosphine catalysts. For example, hydroformylation of 2-butene with 1,1'-biphenyl-2,2'-diyl 2,6-di-*tert*-butyl-4-methylphenylphosphite ligand (P/Rh ratio of 10:1) achieves a rate of 12.85 mol/(L·h) at 130°C, 0.2 MPa (30 psi) H_2, 0.2 MPa CO, producing a linear:branched C-5 aldehyde molar ratio of 1.13 (102). Hydroformylation of 2-butene provides a higher linear:branched aldehyde ratio (2.30) at 14.50 mol/(L·h). Isobutene produces only 3-methylbutyraldehyde at 1.65 mol/(L·h). Thus, new catalyst technologies have made it possible to increase the ratio of 2-methyl-1-butanol to 1-pentanol from 1- or 2-butene feedstock and provide commercially acceptable routes to 3-methyl-1-butanol at high rates and selectivities (103).

4.4. Hydration of Olefins. Several patents disclose the production of amyl alcohols by hydration of C-5 olefins (92,104–111). The Dow Chemical

Company was the first to come on stream with such a process to produce *tert*-amyl alcohol. In the Dow process (92) *tert*-amyl alcohol is prepared by hydrating 2-methyl butenes in acetone over a cation-exchange resin catalyst distributed through all three reactors of a 3-stage reactor system. A single phase solution containing the reactants and solvent are fed to the first reactor, and the first reactor effluent and additional butene feed is then fed sequentially to the second and third reactors. The effluent from the third reactor is treated with an anion-exchange resin to remove acidic impurities and distilled to recover acetone solvent and remove part of the unreacted hydrocarbon overhead. The bottoms are distilled to recover acetone, unreacted butenes (which are recycled), and product alcohol. Solvent requirements are lowered by introducing additional butene before the second reactor.

4.5. Reduction of Acids.   Patents claim catalysts for the hydrogenation of neoacids in the vapor-phase to the neoalcohols in good yields. For example, neopentyl alcohol has been prepared by passing pivalic acid (obtained by the Koch reaction of isobutylene) over a $CuO/ZnO/Al_2O_3$ catalyst at 250°C and 6.9 MPa (70.3 kg/cm^2) in 100% selectivity (112). The neoalcohol was also produced in selectivities of 99% by employing zirconium hydroxide catalysts (113, 114). The rates of the latter process, however, are reportedly low at liquid hourly space velocity (LHSV) of <1 kg/catalyst–h. A catalyst from Re_2O_7 and OsO_4 gave 94% neoalcohol at 100% pivalic acid conversion at 10 MPa (100 atm) hydrogen and 90°C (115). High yields are also obtained by the vapor-phase reduction of pivalic acid [75-98-9] with 2-propanol in the presence of zirconium oxide catalyst (116,117). Reduction with lithium aluminum hydride (118,119) also gives high yields but is less practical for commercial scale production. Production of neopentyl alcohol appears to be commercially viable, but this alcohol seems to be lacking significant large-scale applications.

4.6. Other Methods.   As part of a considerable effort during the 1980s to make chemicals, such as acetic acid, ethanol, and ethylene glycol, from C-1 chemistry, some approaches to higher alcohols by either homologation of inexpensive alcohols (such as methanol) with syn gas or from syn gas alone were made. Essentially, this was a search for appropriate catalysts to provide desired conversions and selectivities to selected alcohols. For example, novel $Ru–Mo–Na_2O$ catalysts supported on Al_2O_3 (120) and silica-supported molybdenum catalysts, promoted by KCl, (121) have high activity and good selectivities (up to about 70%) for a mixture of C-1–C-5 primary alcohols from syn gas. High temperatures (250–300°C) and pressures (10 – 50 MPa = 1450 – 7250 psi) are required. A variety of other complex metal catalyst systems have been described such as Cu–Cr–Co–Rh (122), Cu–Co–Zn–Al_2O_3–K (123), Fe–Cr_2O_3 (124), Cu–ZnO (125), Cu–U–Al–K (126), Mo–W–Rh–K (127), and the like. Low yields to 1-pentanol are generally obtained. Efforts to homologize methanol to higher alcohols have been much less vigorous. Yields of C-5 alcohols were not much higher than from syn gas alone, but more conventional catalyst compounds of Co, Rh, Ru, and La were employed (128–131). There are several reviews on the subject of C-1 chemistry (132,133).

Recent patents involve the separation of 2-methyl-1-butanol and 3-methyl-1-butanol from 1-pentanol by either extractive distillation or azeotropic. References 134 and 135 are examples.

5. Shipping and Storage

Amyl alcohols are best stored or shipped in either aluminum, lined steel, or stainless steel tanks. Baked phenolic is a suitable lining for steel tanks. Plain steel tanks can also be used for storage or shipping. However, storage of aqueous solutions can cause rusting. Also, the alcohols are sufficiently hygroscopic so that moisture pick-up can cause rusting of plain steel storage tanks. Storage and transfer under dry nitrogen is recommended. Storage and handling facilities should be in compliance with the OSHA "Flammable and Combustible Liquids" regulations (17). Piping and pumps can be made from the same metals as used for storage tanks. The freezing points of amyl alcohols are low and they remain fluid at cold outside temperatures, thus allowing storage facilities above or below ground.

6. Economic Aspects

All eight amyl alcohol isomers are available from fine chemical supply firms in the United States. Five of them, 1-pentanol, 2-pentanol, 2-methyl-1-butanol, 3-methyl-1-butanol, and 2-methyl-2-butanol (*tert*-amyl alcohols) are available in bulk in the United States; in Europe all but neopentyl alcohol are produced. In 2001, 8.4×10^6 t of oxo chemicals were produced. Oxo chemicals are expected to grow at a rate of 1.4% per year through 2007 (136).

7. Specifications

Typical specifications for commercially available amyl alcohols (137) are given in Table 4. The impurities typically present are other monomeric alcohols, dimeric alcohols, acetals, and several miscellaneous substances. The alcohols are substantially free of suspended matter.

8. Health and Safety Factors

The main effects of prolonged exposure to amyl alcohols are irritation to mucous membranes and upper respiratory tract, significant depression of the central nervous system, and narcotic effects from vapor inhalation or oral absorption. All the alcohols are harmful if inhaled or swallowed, appreciably irritating to the eyes and somewhat irritating to uncovered skin on repeated exposure (138, 139). Prolonged exposure causes nausea, coughing, diarrhea, vertigo, drowsiness, headache, and vomiting. Table 5 shows toxicity effects of amyl alcohols from animal studies (139). The toxicity of 3-methyl-2-butanol and 2,2-dimethyl-1-propanol has not been thoroughly investigated.

The C-5 alcohols are more toxic and narcotic than the lower homologues. Toxicity to rats from amyl alcohols decreases in the order: tertiary, secondary, primary. Toxicity of 3-methyl-1-butanol appears to have been studied the

Table 4. **Specifications of Commercial Amyl Alcohols**[a]

	1-Pentanol	2-Methyl-1-butanol	Primary amyl alcohol mixed isomers
purity, wt % min			
1-pentanol	99.0, min	1, max	50–70
2-methyl-1-butanol		98.0, min	
amyl alcohols			98.0 min
3-methyl-1-butanol			0.01, max
2- and 3-methyl-1-butanols	0.5, max		
aldehydes, wt % max[b]		0.20	0.20
total carbonyl, wt %, max[b]	0.05		
unsaturation, wt %, max[c]		0.5	
distillation, °C[d]			
initial, min	135.0	126.0	127.5
dry point, max	139.0	139.0	139.0
acidity, wt %	0.005[e]	0.01[f]	0.01[f]
water, wt %, max	0.2	0.3	0.20
specific gravity (20/20°C)			
min	0.814	0.815	0.812
max	0.818	0.822	0.819
color, Pt-Co	15	15	15

[a] Ref. 137.
[b] Calculated as C-5 aldehyde.
[c] As pentenal.
[d] At 101.3 kPa at 1 atm.
[e] Calculated as valeric acid.
[f] Calculated as acetic acid.

Table 5. **Toxicity of Amyl Alcohols**

	LD_{50} rats[a], mg/kg	LD_{50} rabbits[b], g/kg	LCL_o rats[c], ppm/4 h	Eye injury, rabbits
1-pentanol	2200[d]	3600[d]	14000	20 mg/24 h moderate
2-pentanol	1470[e]			20 mg/24 h moderate
3-pentanol	1870[d]	2520[d]		3 mg open severe
2-methyl-1-butanol	4920[d]	3540[f]		
3-methyl-1-butanol	1300[e]	3212[d]	150[g]	20 mg/24 h moderate
2-methyl-2-butanol	1000			

[a] Ingestion.
[b] Skin penetration.
[c] Inhalation.
[d] No toxic effects noted.
[e] Toxic effects not yet reviewed.
[f] Primary irritation.
[g] Sensory organs, lungs, and respiration irritation.

Table 6. **Flammability Limits of Amyl Alcohols**[a]

Alcohol	Flash point, °C		Flammability limits in air, vol%		Auto ignition temperature, °C
	Open cup	Closed cup	Lower	Upper	
1-pentanol	58	33	1.2	10	300.0
2-pentanol	42	34	1.5	9.7	343.33
3-pentanol	39	41	1.2	9.0	435.0
2-methyl-1-butanol	50	50	1.4	9.0	385
3-methyl-1-butanol	56	43	1.2	9.0	350
2-methyl-2-butanol	24	19	1.2	9.0	437.22
3-methyl-2-butanol	35	39	1.5	9.9	436.85
2,2-dimethyl-1-propanol		37	1.5	9.1	

[a] Refs. (5, 142–150).

most. This alcohol caused a slight increase in cancerous tumors compared to controls in two studies (140,141). The tumors were located primarily in the stomach and liver.

Odor data for the various amyl alcohols is limited. The lowest perceptible limit for 1-pentanol and *tert*-amyl alcohol are 10 and 0.04 ppm, respectively (138). *tert*-Amyl alcohol has a threshold value of 2.3 ppm (and a 100% recognition level of 0.23 ppm); 3-methyl-1-butanol has an odor threshold of 7.0 ppm. The odor of 1-pentanol has been described as sweet and pleasant whereas that of 3-methyl-2-butanol is sour (138).

OSHA PEL TWA for 2-pentanol is 100 ppm (360 mg/m^3); STEL is 125 ppm (450 mg/m^3). NIOSH REL (sec-isoamyl alcohol) TWA is 100 ppm. STEL is 125 ppm (151).

There is no ACGIH standard for 1-pentanol (152).

All of the amyl alcohols are TSCA and EINECS (European Inventory of Existing Commercial Chemical Substances) registered.

The amyl alcohols are readily flammable substances; *tert*-amyl alcohol is the most flammable (closed cup flash point, 19°C). Their vapors can form explosive mixtures with air (Table 6) (5,142–150).

9. Uses

Solvents and coatings are the biggest market for C5 alcohols (136). 1-Pentanol and 2-methyl-1-butanol is used for zinc diamyldithiophosphate lubrication oil additives (136) as important corrosion inhibitors and antiwear additives. Amyl xanthate salts are useful as frothers in the flotation of metal ores because of their low water solubility and miscibility with phenolics and natural oils. Potassium amyl xanthate, a collector in flotation of copper, lead, and zinc ores, is no longer produced in the United States (136).

Another significant application for amyl alcohols is for production of amyl acetates.

As solvents, the amyl alcohols are intermediate between hydrocarbon and the more water-miscible lower alcohol and ketone solvents. For example, they are good solvents and diluents for lacquers, hydrolytic fluids, dispersing agents in textile printing inks, industrial cleaning compounds, natural oils such as linseed and castor, synthetic resins such as alkyds, phenolics, urea–formaldehyde maleics, and adipates, and naturally occurring gums, such as shellac, paraffin waxes, rosin, and manila. In solvent mixtures they dissolve cellulose acetate, nitrocellulose, and cellulosic ethers.

Primary amyl alcohol and its acetate ester are considered important medium-boiling nitrocellulose lacquer coating solvents. This mixture is a latent solvent and coupling agent for nitrocellulose lacquers. In blends, the solvent mixture modifies evaporation rate, lowers viscosities, improves blush resistance, and imparts improved flow and leveling which results in high gloss lacquer coatings.

The principal component of primary amyl alcohol, 1-pentanol, although itself a good solvent, is useful for the preparation of specific chemicals such as pharmaceuticals and other synthetics (154).

Growth applications for amyl alcohols appear to be shifting toward higher boiling esters as plasticizers, perfumes, fragrances, and production of fine chemicals.

tert-Amyl alcohol is employed in formulations for stabilizing 1,1,1-trichloroethane (a replacement for trichloroethylene) which is used for degreasing metals, especially aluminum, copper, zinc, and iron and their alloys (155–158). The *tert*-amyl alcohol in stabilizing formulations allowed only negligible reaction between the 1,1,1-trichloroethane and metal. *tert*-Amyl alcohol is also used for stabilizing 1,1,1-trichloroethane mixtures for rosin flux removal compositions, eg, ionic and nonionic fluxes from circuit boards (159,160) and in stabilizer compositions for 1,1,1-trichloroethane for dry cleaning applications where it is durable against repeated use, without causing corrosion of the dry cleaner metal components and conforms with environmental and health standards enacted by the Occupational Safety and Health Act (161–163). *tert*-Amyl alcohol also has solvent use in the preparation of epoxy-containing novolak resins with low chloride content (164) and in mixtures with surfactants in enhanced petroleum recovery by flooding (165,166) (see PETROLEUM, ENHANCED OIL RECOVERY).

Other applications of amyl alcohols include their use as flavor and fragrance chemicals. Amyl isovalerate and amyl salicylate consume amyl alcohols (136). Isoamyl salicylate is used to a large extent in soap and cosmetic fragrances because of its cost effectiveness (167). Isoamyl alcohol is used as the extracting solvent for purification of wet process phosphoric acid. Total upgrading of 28–54% wet process phosphoric acid is achieved using this alcohol (168,169). *t*-Amyl methyl ether (TAME) is a useful gasoline additive as an octane booster (170,171). Amyl cinnamic aldehyde is an important ester of amyl alcohol (172). 1-Pentanol is used as an alcohol cosurfactant in a variety of applications (173,174). Amyl alcohols are used for the preparation of a variety of herbicides, fungicides, and pesticides. Amyl alcohols are a superior medium, compared to either benzyl alcohols or dichloromethane, for preparation of magnesia suspensions for electrophoretic deposition (175). Isoamyl alcohol is an intermediate in the synthesis of pyrethroids (21,22).

BIBLIOGRAPHY

"Amyl Alcohols" in *ECT* 1st ed., Vol. 1, pp. 844–849, by C. K. Hunt, Sharples Chemicals, Inc., "Amyl Alcohols" in *ECT* 2nd ed., Vol. 2, pp. 374–379, by L. A. Gillette, Pennsalt Chemicals Corporation; in *ECT* 3rd ed., Vol. 2, pp. 570–573, by P. D. Sherman, Jr., Union Carbide Corporation; in *ECT* 4th ed., Vol. 2, pp. 709–728, by Anthony J. Papa, Union Carbide Chemicals and Plastics Company Inc.

CITED PUBLICATIONS

1. Thermodynamics Research Center, *Selected Values of Properties of Chemical Compounds*, Data Project (loose-leaf data sheets), extant, Texas A&M University, College Station, Tex., 1980.
2. R. Wilhoit and B. J. Zwolinski, *J. Phys. Chem. Ref. Data* **2** (Suppl. No. 1) (1973).
3. J. J. Jasper, *J. Phys. Chem. Ref. Data* **1**(4), 841 (1972).
4. A. P. Kudchadker, G. H. Alani, and B. J. Zwolinski, *Chem. Rev.* **68**(6), 659 (1968).
5. J. S. Riddick and W. B. Bunger, *Organic Solvents: Physical Properties and Methods of Purification*, 3rd ed., Wiley-Interscience, New York, 1970.
6. J. Timmermans, *Physico-Chemical Constants of Pure Organic Substances*, 2nd ed., 2 vols., Elsevier, New York, 1965.
7. A. L. Lydersen, *Estimation of Critical Properties of Organic Compounds*, Report #3, College Engineering Experiment Station, University of Wisconsin, Madison, Wis., Apr. 1955.
8. B. J. Zwolinski and R. Wilhoit, "Heats of Formation and Heats of Combustion" in D. E. Gray, ed., *American Institute of Physics Handbook*, 3rd ed., McGraw-Hill, New York, 1972, pp. 4, 316, 342.
9. R. F. Fedors, *AIChE J.* **25**, 202 (1979).
10. *CHETAH-The ASTM Chemical Thermodynamic and Energy Release Evaluation Program*, ASTM Data Series Publication DS 51, American Society for Testing Materials, Philadelphia, 1974, original, updated.
11. Yu. V. Efremov, *Russ. J. Phys. Chem.* **40**(6), 667 (1966).
12. G. A. Pope, *Dissertation Abstract International* **33**(04-B), 1587 (1972).
13. S. Sugden, *J. Chem. Soc. Trans. (London)* **125**, 32 (1924).
14. *Physical Property Data*, unpublished, Union Carbide Chemicals and Plastics Corp., Danbury, Conn.
15. G. Sorensen and J. M. W. Arit, *Liquid-Liquid Equilibrium Data Collection, Binary Systems*, DECHEMA Chemistry Data Series, Vol. 5, part 1, Schon & Wetzel GmbH, Frankfurt/Main, Germany, 1979.
16. *TRC Thermodynamic Tables-Non-Hydrocarbons*, (loose-leaf data sheets, extant, 1989, suppl. #62), Thermodynamic Research Center, Texas A&M University System, College Station, Tex., Dec. 31, 1989.
17. *UCAR Alcohols for Coatings Applications, Brochure F-48588*, Solvents and Coatings Materials Division, Union Carbide Chemicals and Plastics Corp., Danbury, Conn., Sept. 1984.
18. T.-H Cho and co-workers, *Kagaku Kogaku Ronbunshu* **10**, 181 (1984), tabulated in Dechema I, 1b Suppl. 2.
19. L. H. Horsley, ed., *Azeotropic Data III, Adv. in Chem. Ser. #116*, American Chemical Society, Washington, D.C., 1973.
20. *Tables of Azeotropic Data*, Union Carbide Chemicals and Plastics Corp., 1957, unpublished.

21. *Chem. Eng. Prog.* **81**, 52–55 (May 1985).
22. *Chem. Eng. News* **63**, 27 (Apr. 22, 1985).
23. W. T. Reichle, *Res. Discl.* **283**, 717 (1987).
24. U.S. Pat. 4,260,845 (Apr. 7, 1981), T. K. Shioyama (to Phillips Petroleum Co.).
25. Brit. Pat. 2,181,070 (Apr. 15, 1987), C. S. John (to Shell International Research Maatschappi j B. V.).
26. Eur. Pat. 284,397 (Sept. 17, 1988), M. P. Atkins, J. Williams, J. A. Ballantine, and J. H. Purnell (to British Petroleum Co.).
27. V. T. Chalkina, S. Ya. Sklyar, S. A. Mannanova, and M. F. Sharnina, *Dokl. Neftekhim. Sekts.—Bashk. Resp. Pravl. Vses. Khim. O-va. im D. I. Mendeleeva*, 65–68 (1976).
28. J. A. Ballantine and co-workers, *J. Mol. Catal.* **26**(1), 37 (1984).
29. A. V. Timofeev and L. Ya. Romanchenko, *Khim. Prom-st (Moscow)*, (6), 333 (1985).
30. T. G. Min'ko and A. V. Timofeev, *Khim. Prom-st (Moscow)* (10), 595 (1986).
31. Jpn. Kokai 49,061,113 (June 13, 1974), Y. Soma and H. Sano (to Agency of Industrial Sciences and Technology).
32. H. Langhals, I. Mergelsberg, and C. Ruechardt, *Tetrahedron Lett.* **22**(25), 2365 (1981).
33. W. Haaf, *Org. Synth.* **5**, 739 (1973).
34. Eur. Pat. 158,499 A2 (Oct. 16, 1985), J. Russell and A. J. Stevenson (to BP Chemicals Ltd.).
35. T. Baba and Y. Ono, *Appl. Catal.* **22**(2), 321 (1986).
36. Z. Chen, *Kexue Tongbao* **30**(5), 616 (1985).
37. *UCAR n-Pentyl Propionate*, Brochure F-60454A, Union Carbide Chemicals and Plastics Corp., Danbury, Conn., May 1987.
38. Ger. Pat. 2,917,087 (Nov. 6, 1980), W. Werner (to Federal Republic of Germany).
39. U.S. Pat. 3,864,374 (Feb. 4, 1975), C-J. Chia, J. E. Currah, and G. R. Lusby (to Canadian Industries Ltd.).
40. J. Sejbl, *Ruby* **31**(5), 163 (1985).
41. Ger. Pat. 2,149,726 (Apr. 6, 1972), G. J. Novak and A. J. Robertson (to American Cynamid Co.).
42. South Afr. Pat. 8,004,227 (Aug. 26, 1981), D. J. Gannon and D. T. F. Fung (to CIL Inc., Canada).
43. A. D. Braazier and J. S. Elliott, *J. Inst. Pet.* **53**(518), 63 (1967).
44. C. Roussel, H. G. Rajoharison, L. Bizzari, and L. Shaimi, *J. Org. Chem.* **53**(3), 683 (1988).
45. Eur. Pat. 96,641 A2 (Dec. 21, 1983), H. G. Rajoharison, C. Roussel, and M. Ader (to Eastman Kodak Co.).
46. H. G. Rajoharison, C. Roussel, and J. Metzger, *J. Chem. Res. Synop.* (7), 186 (1981).
47. R. A. Aitken, J. Gopal, and J. A. Hirst, *J. Chem. Soc. Chem. Commun.* (10), 632 (1988).
48. Jpn. Kokai Tokkyo Koho 63 230,676 (Sept. 27, 1988), T. Shida and co-workers (to Kureha Chemical Industry Co., Ltd.).
49. S. O. Nwaukwa and P. M. Keehn, *Tetrahedron Lett.* **23**(1), 35 (1982).
50. S. Kajigaeshi and co-workers, *Bull. Chem. Soc. Jpn.* **59**(3), 747 (1986).
51. T. Kageyama, Y. Ueno, and M. Okawara, *Synthesis* (10), 815 (1983).
52. T. Kageyama and co-workers, *Chem. Lett.*, **7**, 1097 (1983).
53. Jpn. Kokai Tokkyo Koho 60 123,432 (July 2, 1985) (to Nippon Silica Industries Co., Ltd.).
54. C. M. Ashraf, I. Ahmad, and F. K. N. Lugemwa, *Arab Gulf J. Sci. Res.* **2**(1), 39 (1984).
55. Jpn. Kokai Tokkyo Koho 57 048,931 (Mar. 20, 1982) (to Nippon Zeon Co., Ltd.).
56. H. Firouzabadi and Z. Mostafavipoor, *Bull. Chem. Soc. Jpn.* **56**(3), 914 (1983).
57. J. Lou, *Chem. Ind. (London)* **10**, 312 (1989).

58. S. J. Flatt, G. W. J. Fleet, and B. J. Taylor, *Synthesis* (10), 815 (1979).

59. Jpn. Kokai Tokkyo Koho 63 275,547 (Nov. 14, 1988), L. Fukuya and H. Fukuda (to Daicel Chemicals Industries, Ltd.).

60. Eur. Pat. 22,532 (Jan. 21, 1981), F. Werner and co-workers (to Bayer A.-G.).

61. Hung. Pat. 42,456 (July 28, 1987), J. Kreidl and co-workers (to Richter, Gedeon, Vegyeszeti Gyar Rt.).

62. U.S. Pat. 3,673,249 (June 27, 1972), F. M. Furman (to American Cynamid Co.).

63. D. R. Nutter, *Res. Discl.* **195**, 270 (1980).

64. Ger. Pat. 3,643,851 (June 30, 1988), K. Sasse and co-workers (to Bayer A.-G.).

65. A. M. Habib, A. A. Saafan, A. K. Abou-Seif, and M. A. Salem, *Colloids Surf.* **29**(4), 337 (1988).

66. K. G. Sharonov, A. M. Rozhnov, V. I. Barkov, and R. I. Cherkasova, *Zh. Prikl. Khim. (Leningrad)* **60**(2), 359 (1987).

67. A. M. Habib, M. F. Abd-El-Megeed, A. Saafan, and R. M. Issa, *J. Inclusion Phenom.* **4**(2), 185 (1986).

68. V. Macho, M. Kavala, M. Polievka, M. Okresa, and W. Piecka, *Ropa Uhlie* **24**(7), 397 (1982).

69. Ger. Pat. 2,330,648 (Jan. 10, 1974), J. B. Hall, M. G. J. Beets, L. K. Lala, and W. I. Taylor (to International Flavors and Fragrances, Inc.).

70. U.S. Pat. 3,927,083 (Dec. 16, 1975), J. B. Hall, L. K. Lala, M. G. J. Beets, and W. I. Taylor (to International Flavors and Fragrances, Inc.).

71. Ger. Pat. 2,427,465 (Dec. 19, 1974), H. Miki and H. Hasui (to Mitsui Petrochemical Industries, Ltd.).

72. Rom. Pat. 83,751 (Apr. 30, 1984), A. Vizitiu and R. Manea (to Intreprinderea de Detergenti).

73. J. Chelebicki, *Rocz. Chem.* **49**(1), 207 (1975).

74. Jpn. Kokai Tokkyo Koho 61 010,525 (Jan. 18, 1986), T. Asano, M. Umemoto, and N. Sato (to Mitsui Toatsu Chemicals, Inc.).

75. Eur. Pat. 192,608 A2 (Aug. 27, 1986), W. Surber (to CIBA-GEIGY A.-G.).

76. Eur. Pat. 190,999 A2 (Aug. 13, 1986), F. Baebier (to CIBA-GEIGY A.-G.).

77. Fr. Pat. 2,321,477 (Mar. 18, 1977), D. Boutte and J. P. Senet (to Société Nationale des Poudres et Explosifs).

78. Jpn. Kokai Tokkyo Koho 54 066,613 (May 29, 1979), H. Tanaka, H. Tsuchiya, and S. Kyo (to Kuraray Co., Ltd.).

79. Jpn. Kokai Tokkyo Koho 54 148,711 (Nov. 21, 1979), S. Kyo, H. Tsuchiya, and H. Tanaka (to Kuraray Co., Ltd.).

80. P. L. Burk, R. L. Pruett, and K. S. Campo, *J. Mol. Catal.* **33**(1), 15 (1985).

81. P. L. Burk, R. L. Pruett, and K. S. Campo, *J. Mol. Catal.* **33**(1), 1 (1985).

82. Jpn. Kokai 52,077,002 (June 29, 1977), K. Ota and T. Kito (to Showa).

83. T. Kito, K. Ota, M. Takata, and M. Ariyoshi, *Yuki Gosei Kagaku Kyokai Shi* **36**(4), 331 (1978).

84. J. Sabadie and H. Descotes, *Bull. Soc. Chim. Fr.* **9–10**(2), 253 (1983).

85. A. A. Khalaf and R. M. Roberts, *Rev. Roum. Chim.* **30**(6), 507 (1985).

86. G. B. Sergeev, V. V. Zagorskii, and M. V. Grishechkina, *Metalloorg. Khim.* **1**(5), 1187 (1988).

87. R. Fuchs and L. L. Cole, *Can. J. Chem.* **53**(23), 3620 (1975).

88. H. Masada and Y. Murotani, *Bull. Chem. Soc. Jpn.* **53**(4), 1181 (1980).

89. W. A. Hoffman, III, *J. Org. Chem.* **47**, 5209 (1982).

90. Jpn. Kokai Tokkyo Koho 63 191,820 (Aug. 9, 1988), E. Morita and Y. Komizo (to Auto Chemical Industry Co., Ltd.).

91. Kirkpatrick Chemical Engineering Achievement Award, *Chem. Eng.* **84**, 110 (Dec. 5, 1977).

92. U.S. Pat. 4,182,920 (Jan. 8, 1980), J. H. Giles, J. H. Stultz, and S. W. Jones (to The Dow Chemical Company).
93. Ger. Pat. 255,526 (Apr. 6, 1988), M. Steinbrecker and co-workers (to VEB Leuna-Werke "Walter Ulbricht").
94. *Chem. Eng. News*, 27 (Oct. 10, 1988).
95. B. L. Booth, M. J. Else, R. Fields, and R. N. Hazeldine, *J. Organomet. Chem.* **27**, 119 (1971).
96. Y. Matsui, H. Taniguchi, K. Terada, and M. Irinchijima, *Bull. Jpn. Pet. Inst.* **19**, 68 (1977).
97. U.S. Pat. 3,527,809 (1970), R. L. Pruett and J. A. Smith (to Union Carbide Corp.).
98. P. W. N. M. Van Leeuwen and C. F. Roobeek, *J. Organomet. Chem.* **258**, 343 (1983).
99. U.S. Pat. 4,567,306 (1986), A. J. Dennis, G. E. Harrison, and J. P. Wyber (to Davy McKee (London) Ltd.).
100. U.S. Pat. 4,599,206 (1986), E. Billig and co-workers (to Union Carbide Corp.).
101. U.S. Pat. 4,717,775 (1988), E. Billig and co-workers (to Union Carbide Corp.).
102. U.S. Pat. 4,789,753 (1988), E. Billig and co-workers (to Union Carbide Corp.).
103. U.S. Pat. 4,769,498 (Sept. 6, 1988), E. Billig, A. G. Abatjoglou, and D. R. Bryant (to Union Carbide Corp.).
104. Z. Prokop and K. Setinek, *Collect. Czech. Chem. Commun.* **52**(5), 1272–1279 (1987).
105. Eur. Pat. 127,486 (Dec. 5, 1984), Y. Okumura, S. Kamiyama, H. Furukawa, and K. Kaneko (to Toa Nenryo Kogyo K. K.).
106. U.S. Pat. 3,285,977 (Nov. 15, 1966), A. M. Henke, R. C. Odioso, and B. K. Schmid (to Gulf Research and Development Co.).
107. Ger. Pat. 3,801,275 (July 27, 1989), R. Malessa and B. Schleppinghoff (to EC Erdoel-chemic GmbH).
108. Ger. Pat. 3,801,273 (July 27, 1989), R. Malessa, C. Gabel, H. V. Scheef, and M. Lux (to EC Erdoelchemic GmbH).
109. Eur. Pat. 325,144 (July 26, 1989), R. Malessa and B. Schleppinghoff (to EC Erdoel-chemic GmbH).
110. Jpn. Kokai 47,039,013 (Dec. 6, 1972), A. Nambu (to Japan Oil Co.).
111. U.S. Pat. 3,285,977 (Nov. 15, 1966), A. M. Henke, R. C. Odioso, and B. K. Schmid (to Gulf Research and Development Co.).
112. Eur. Pat. 180,210 (May 7, 1986), S. A. Butler and J. Stoll (to Air Products and Chemicals, Inc.).
113. Eur. Pat. 285,786 (Oct. 12, 1988), H. Matsushita, M. Shibagaki, and K. Takahashi (to Japan Tobacco Co.).
114. Jpn. Kokai Tokkyo Koho 62 108,832 (May 20, 1987), T. Maki, M. Nakajima, T. Yokoyama, and T. Setoyama (to Mitsubishi Chemical Industries Co., Ltd.).
115. Jpn. Kokai Tokkyo Koho 62 210,056 (Sept. 16, 1987), Y. Kajiwara, I. Y. Inamoto, and T. Yoshiaki (to Kao Corp.).
116. K. Takahashi, M. Shibagaki, H. Kuno, and H. Matsushita, *Chem. Lett.* (7), 1141 (1989).
117. M. Shibagaki, K. Takahashi, and H. Matsushita, *Bull. Chem. Soc. Jpn.* **61**(9), 3283 (1988).
118. A. F. D'Adams and R. H. Kienle, *J. Amer. Chem. Soc.* **77**, 4408 (1955).
119. N. Kornblum and D. C. Iffland, *J. Amer. Chem. Soc.* **77**, 6653 (1955).
120. T. Tatsumi, A. Muramatsu, and H. Tominaga, *Chem. Lett.* **5**, 685 (1984).
121. T. Tatsumi, A. Muramatsu, T. Fukunaga, and H. Tominaga, *Chem. Lett.* **6**, 919 (1986).
122. U.S. Pat. 4,537,909 (Aug. 27, 1985), F. N. Lim and F. Pennella (to Phillips Petroleum Co.).
123. W. X. Pan, R. Cao, and G. L. Griffin, *J. Catal.* **114**(2), 447 (1988).

124. M. Van der Riet, R. G. Copperthwaite, S. F. Demarger, and G. J. Hutchings, *J. Chem. Soc. Chem. Commun.* (10), 687 (1988).

125. D. J. Elliott, *J. Catal.* **111**(2), 445 (1988).

126. U.S. Pat. 4,677,091 (June 30, 1987), T. J. Mazanec and J. G. Frye, Jr., (to Standard Oil Co.).

127. U.S. Pat. 4,670,473 (June 2, 1987), R. H. Walker, D. A. Palmer, D. M. Salvatore, and E. J. Bernier (to Amoco Corp.).

128. Ger. Pat. 3,641,774 (June 30, 1988), W. Hilsebein, E. Supp, W. Friedrichl, and P. Koenig (to Metallgesellschaft A.-G.).

129. Jpn. Kokai Tokkyo Koho 63 190,837 (Aug. 8, 1988), Y. Isogai, A. Uda, K. Tanaka, and M. Hosokawa (to Agency of Industrial Sciences and Technology).

130. U.S. Pat. 4,533,775 (Aug. 6, 1985), J. R. Fox, F. A. Pesa, and B. S. Curatolo (to Standard Oil Co.).

131. U.S. Pat. 4,540,836 (Sept. 10, 1985), D. M. Fenton (to Union Oil Company of California).

132. T. Tatsumi, A. Muramatsu, K. Yokota, and H. Tominaga in D. M. Bibby, C. D. Chang, and R. F. Howe, eds., *Methane Conversion*, Elsevier Science Publishers, Amsterdam, The Netherlands, 1988.

133. Research Association for C1 Chemistry, eds., *Progress in C1 Chemistry in Japan*, Elsevier Science Publishers, Amsterdam, The Netherlands, 1989.

134. U.S. Pat. 6,024,841 (Feb. 15, 2000), L. Berg.

135. U.S. Pat. 5,779,862 (July 14, 1998), L. Berg.

136. *Chemical Economics Handbook*, SRI International, Menlo Park, Calif., Nov. 2002.

137. *Product Specifications* (1-6A5-1f, 1-6B5-1.1b, and 1-6E5-1h) Union Carbide Chemicals and Plastics Corp., Danbury, Conn., Jan. 20, 1989.

138. V. K. Rowe, S. B. McCollister, G. D. Clayton, and F. E. Clayton, eds., *Patty's Industrial Hygiene and Toxicology*, 3rd ed., Vol. 2C, Wiley-Interscience, New York, 1982, 2471–4588.

139. D. V. Sweet, ed., *Registry of Toxic Effects of Chemical Substances*, Publication No. 87-114, Vol. 4, National Institute for Occupational Safety and Health, DHHS (NIOSH), Washington, D.C., Apr. 1987.

140. *Amyl Alcohols or Pentanols (C5H12O). Toxicology Card No. 206*, Cahiers de Notes Documentaires, No. 118, 1st quarter, National Institute for Research and Safety, Paris, 1985, 143–146.

141. W. E. T. Gibel and co-workers, *Z. Exp. Chir.* **7**(4), 235 (1974).

142. J. A. Riddick, W. B. Bunger, and T. K. Sakano, *Techniques of Chemistry: Organic Solvents*, Vol. 2, 4th ed., John Wiley & Sons, Inc., New York, 1986, p. 887.

143. G. H. Tyron, ed., *Fire Protection Handbook*, 12th ed., National Fire Protection Association, Boston, Mass., 1962.

144. *Fire Protection Guide on Hazardous Materials*, 7th ed., National Fire Protection Association, Boston, Mass., 1978.

145. N. V. Steere, ed., *Handbook of Laboratory Safety*, 2nd ed., CRC Press, Boca Raton, Fla., 1982.

146. N. I. Sax, *Dangerous Properties of Industrial Materials*, 6th ed., Van Nostrand Reinhold Co., Inc., New York, 1984.

147. Yu. N. Shebeko, A. V. Ivanov, and T. M. Dmitrieva, *Sov. Chem. Ind.* **15**(3), 311 (1983).

148. D. P. Danner and T. E. Daubert, *Manual for Prediction Chemical Process Design Data*, AIChE, New York, (extant 1987).

149. N. I. Sax, *Dangerous Properties of Industrial Materials*, 5th ed., Van Nostrand Reinhold Co., Inc., New York, 1979.

150. G. Weiss, ed., *Hazardous Chemical Data Book*, Noyes Data Corp., Park Ridge, N.J., 1980.

151. R. J. Lewis, Sr., *Sax's Dangerous Properties of Industrial Materials*, 10th ed., Vol. 3, John Wiley & Sons, Inc., New York, 2000.

152. C. Bevan in E. Bingham, B. Cohrsson, C. N. Powell, eds., *Patty's Toxicology*, 5th ed., Vol. 6, John Wiley & Sons, Inc., New York, 2001, p. 429.

153. H. Bieber, *Encycl. Chem. Process Des.* **3**, 278 (1977).

154. "Monohydric Alcohols," in I. Mellan, ed., *Source Book of Industrial Solvents*, Vol. 3, Reinhold Publishing Corp., New York, 1959.

155. U.S. Pat. 4,327,232 (Apr. 27, 1982), A. Pryor and N. Ishibe (to The Dow Chemical Company).

156. U.S. Pat. 4,324,928 (Apr. 13, 1982), A. Pryor and N. Ishibe (to The Dow Chemical Company).

157. U.S. Pat. 4,309,301 (Jan. 5, 1982), N. Ishibe and T. G. Metcalf (to The Dow Chemical Company).

158. U.S. Pat. 4,115,461 (Sept. 19, 1978), D. R. Spencer and W. L. Wesley (to The Dow Chemical Company).

159. U.S. Pat. 4,524,011 (June 18, 1985), E. L. Tasset, S. M. Dallessand, and W. F. Richey (to The Dow Chemical Company).

160. Eur. Pat. 108,422 (May 16, 1984), E. L. Tasset, W. F. Richey, and S. M. Dallessandro (to The Dow Chemical Company).

161. U.S. Pat. 3,974,230 (Aug. 10, 1976), W. L. Archer and D. R. Spencer (to The Dow Chemical Company).

162. Jpn. Kokai Tokkyo Koho 54 157,108 (Dec. 11, 1979), T. Mizushiro, T. Kaneko, and Y. Sugawara (to Asahi-Dow, Ltd.).

163. Jpn. Kokai Tokkyo Koho 54 157,107 (Dec. 11, 1979), T. Mizushiro, Y. Sugawara, and T. Kaneko (to Asahi-Dow, Ltd.).

164. U.S. Pat. 4,778,863 (Oct. 18, 1988), C. S. Wang and Z. K. Liao (to The Dow Chemical Company).

165. Ger. Pat. 3,720,330 (Dec. 29, 1988), D. Balzer and H. Lueders (to Huels A.-G.).

166. C. J. Glover, M. C. Puerto, J. M. Maerker, and E. L. Sandvik, *Soc. Pet. Eng. J.* **19**(3), 183 (1979).

167. *Perfumer and Flavorist*, 60, 61 (Nov. 1981).

168. *Eur. Chem. News*, 221 (Aug. 2, 1974).

169. *Phosphorus Potassium*, 10 (Sept.–Oct. 1987).

170. *Chem. Eng. News*, 35, 36 (June 25, 1979).

171. *Hydrocarbon Process.*, 51–521 (Dec. 1985).

172. *Chem. Mark. Rep.*, 65 (Oct. 20, 1975).

173. *Chem. Eng. (N.Y.)* **90**(7), 19 (Apr. 4, 1983).

174. *Chem. Week*, 60 (Nov. 1, 1978).

175. D. U. K. Rao and E. C. Subbarao, *Am. Cer. Soc. Bull.*, 467–469 (Apr. 1979).

ANTHONY J. PAPA
Union Carbide Chemicals and Plastics Company Inc.

ANILINE AND ITS DERIVATIVES

1. Introduction

Aniline (benzenamine) [62-53-3] is the simplest of the primary aromatic amines. It was first produced in 1826 by dry distillation of indigo. In 1840 the same oily liquid was obtained by heating indigo with potash, and it was given the name aniline. The structure of aniline was established in 1843 with the demonstration that it could be obtained by reduction of nitrobenzene.

Aromatic amines can be produced by reduction of the corresponding nitro compound, the ammonolysis of an aromatic halide or phenol, and by direct amination of the aromatic ring. At present, the catalytic reduction of nitrobenzene is the predominant process for manufacture of aniline. To a smaller extent aniline is also produced by ammonolysis of phenol.

Important analogs of aniline include the toluidines, xylidines, anisidines, phenetidines, and its chloro-, nitro-, N-acetyl, N-alkyl, N-aryl, N-acyl, and sulfonic acid derivatives.

2. Physical Properties

Pure, freshly distilled aniline is a colorless, oily liquid that darkens on exposure to light and air. It has a characteristic sweet, aminelike aromatic odor. Aniline is miscible with acetone, ethanol, diethyl ether, and benzene, and is soluble in most organic solvents. Its solubility characteristics in water are as follows:

Temperature, °C	Parts aniline per 100 parts water	Parts water per 100 parts aniline
25	3.5	5.0
90	6.4	9.9

The physical properties of aniline are given in Table 1 and vapor pressure data in Table 2.

3. Chemical Properties

Aromatic amines are usually weaker bases than aliphatic amines as illustrated by the difference in pK_a of the conjugate acids of aniline (1), pK_a = 4.63 and cyclohexylamine (2), pK_a = 10.66.

(1)   (2)

Table 1. **Physical Properties of Aniline**

Property	Value
molecular formula	C_6H_7N
molecular weight	93.129
boiling point, °C	
101.3 kPa[a]	184.4
4.4 kPa[a]	92
1.2 kPa[a]	71
freezing point, °C	−6.03
density, liquid, g/mL	
20/4°C	1.02173
20/20°C	1.022
density, vapor (at bp, air = 1)	3.30
refractive index, n^{20}_D	1.5863
viscosity, mPa·s(=cP)	
20°C	4.35
60°C	1.62
enthalpy of dissociation, kJ/mo[b]	21.7
heat of combustion, kJ/mol[b]	3394
ionization potential, eV	7.70
dielectric constant, at 25°C	6.89
dipole moment at 25°C (calcd), C·m[c]	5.20×10^{-30}
specific heat at 25°C, J/(g·K)[b]	2.06
heat of vaporization, J/g[b]	478.5
flash point, °C	
closed cup	70
open cup	75.5
ignition temperature, °C	615
lower flammable limit, vol %	1.3

[a] To convert kPa to mm Hg, multiply by 7.5.
[b] To convert J to cal, divide by 4.184.
[c] To convert C·m to debye, multiply by 3×10^{29}.

This is due to a resonance effect. Aniline is stabilized by sharing its nitrogen lone-pair electrons with the aromatic ring. In the anilinium ion, the resonance stabilization is disrupted by the proton bound to the lone pair.

Aromatic amines form addition compounds and complexes with many inorganic substances, such as zinc chloride, copper chloride, uranium tetrachloride, or boron trifluoride. Various metals react with the amino group to form metal anilides; and hydrochloric, sulfuric, or phosphoric acid salts of aniline are important intermediates in the dye industry.

Table 2. **Vapor Pressure of Aniline**

Temperature, °C	Vapor pressure, kPa[a]	Temperature, °C	Vapor pressure, kPa[a]
175	80	139	28
162	53	119	13
151	40	102	7

[a] To convert kPa to mm Hg, multiply by 7.50.

3.1. N-Alkylation. A number of methods are available for preparation of N-alkyl and N,N-dialkyl derivatives of aromatic amines. Passing a mixture of aniline and methanol over a copper–zinc oxide catalyst at 250°C and 101 kPa (1 atm) reportedly gives N-methylaniline [100-61-8] in 96% yield (1). Heating aniline with methanol under pressure or with excess methanol produces N,N-dimethylaniline [121-69-7] (2,3).

In the presence of sulfuric acid, aniline reacts with methanol to form N-methyl- and N,N-dimethylaniline. This is a two-step process

$$C_6H_5NH_2 + CH_3OH \longrightarrow C_6H_5NHCH_3 + H_2O$$

$$C_6H_5NHCH_3 + CH_3OH \longrightarrow C_6H_5NH(CH_3)_2 + H_2O$$

and a study of its kinetics (4) shows that reaction rate is proportional to the concentration of aniline in the first step and the concentration of N-methylaniline in the second step. With 50% excess methanol, the reaction equilibrium reaches 99% N,N-dimethylaniline at 200°C. Reaction is clean with little by-product formation up to 230°C. At higher temperatures, ring alkylation and formation of formaldehyde, from oxidation of methanol, are observed.

Other catalysts that can be used are boron trifluoride (5), copper–chromium oxides (6), phosphoric acid (7), and silica–alumina (8). Under similar conditions, ethanol yields N-ethylaniline [103-69-5] and N,N-diethylaniline [91-66-7] (9, 10).

N-Alkylation can also be carried out with the appropriate alkyl halide or alkyl sulfate. Reaction of aniline with ethylene, in the presence of metallic sodium supported on an inert carrier such as carbon or alumina, at high temperature and pressure yields N-ethyl- or N,N-diethylaniline (11). At pressures below 10 MPa (100 atm), the monosubstituted product predominates.

Mixtures of N-alkylanilines can usually be separated by fractional distillation. Mixtures of the methyl or ethyl derivatives have also reportedly been separated by converting the N-ethyl or the N-methyl derivative to the nonvolatile salt with p-toluenesulfonic acid (12) or phthalic anhydride (13), followed by distillation.

Catalytic alkylation of aniline with diethyl ether, in the presence of mixed metal oxide catalysts, preferably titanium dioxide in combination with molybdenum oxide and/or ferric oxide, gives 63% N-alkylation and 12% ring alkylation (14).

Diphenylamine [122-39-4] is produced by heating aniline with aniline hydrochloride at 290°C and 2 MPa (21 atm) in an autoclave (15).

$$C_6H_5NH_2 + C_6H_5NH_2HCl \longrightarrow C_6H_5NHC_6H_5 + NH_3 + HCl$$

The by-product ammonia is vented from the reactor during the course of the reaction. Unconverted aniline is distilled off at the end of the reaction and the diphenylamine is washed with aqueous hydrochloric acid to remove trace amounts of aniline. The product is then washed with water and purified in a refining still.

The use of ammonium fluoroborate (NH_4BF_4) as a catalyst for this reaction, is claimed to be advantageous, since the catalyst can be recycled and is noncorrosive to ferrous metals (16).

Diphenylamine can also be produced by passing the vapors of aniline over a catalyst such as alumina, or alumina impregnated with ammonium fluoride (17). The reaction is carried out at 480°C and about 700 kPa (7 atm). Conversion per pass, expressed as parts diphenylamine per 100 parts of reactor effluent, is low (18–22%), and the unconverted aniline must be recycled. Other catalysts disclosed for the vapor-phase process are alumina modified with boron trifluoride (18), and alumina activated with boric acid or boric anhydride (19).

3.2. Ring Alkylation. The aromatic ring undergoes alkylation under certain conditions. For example, 2-ethylaniline [103-69-5], 2,6-diethylaniline [579-66-8], or a mixture of the two are obtained in high yield when aniline is heated with ethylene in the presence of aluminum–anilide catalyst (formed by heating aluminum and aniline) at 330°C and 4–5 MPa (40–50 atm) (20). *N*-Ethylaniline is alkylated in a similar manner, but at 205°C yields only *N*-ethyl-2-ethylaniline [578-54-1]. Other olefins can also be used to form ring-alkylated products, but the reaction rate decreases as the molecular weight of the olefin increases. A number of patents, claiming a variety of catalysts, have been issued in this area (21–24).

N-Alkylaniline and *N,N*-dialkylaniline hydrochlorides can be rearranged to *C*-alkyl anilines by heating the salts to 200–300°C. In this reaction, known as the Hofmann-Martius rearrangement, the alkyl group preferentially migrates to the para position. If this position is occupied, the ortho position is alkylated.

3.3. Acylation. Aromatic amines react with acids, acid chlorides, anhydrides, and esters to form amides. In general, acid chlorides give the best yield of the pure product. The reaction with acetic, propionic, butanoic, or benzoic acid can be catalyzed with phosphorus oxychloride or trichloride.

N-Phenylsuccinimide [83-25-0] (succanil) is obtained in essentially quantitative yield by heating equivalent amounts of succinic acid and aniline at 140–150°C (25). The reaction of a primary aromatic amine with phosgene leads to formation of an arylcarbamoyl chloride, that when heated loses hydrogen chloride to form an isocyanate. Commercially important isocyanates are obtained from aromatic primary diamines.

Conversion of aniline to acetanilide [103-84-4], by reaction with acetic anhydride, is a convenient method for protecting the amino group. The acetyl group can later be removed by acid or base hydrolysis.

3.4. Condensation. Depending on the reaction conditions, a variety of condensation products are obtained from the reaction of aromatic amines with aldehydes, ketones, acetals, and orthoformates.

Primary aromatic amines react with aldehydes to form Schiff bases. Schiff bases formed from the reaction of lower aliphatic aldehydes, such as formaldehyde and acetaldehyde, with primary aromatic amines are often unstable and polymerize readily. Aniline reacts with formaldehyde in aqueous acid solutions to yield mixtures of a crystalline trimer of the Schiff base, methylenedianilines, and polymers. Reaction of aniline hydrochloride and formaldehyde also yields polymeric products; and under certain conditions, the predominant product is 4,4'-methylenedianiline [101-77-9] (26), an important intermediate for 4,4'-methylenebis(phenylisocyanate) [101-68-8], or MDI.

3.5. Cyclization. Aniline, nitrobenzene, and glycerol react under acid catalysis (Skraup synthesis) to form quinoline [91-22-5] (27).

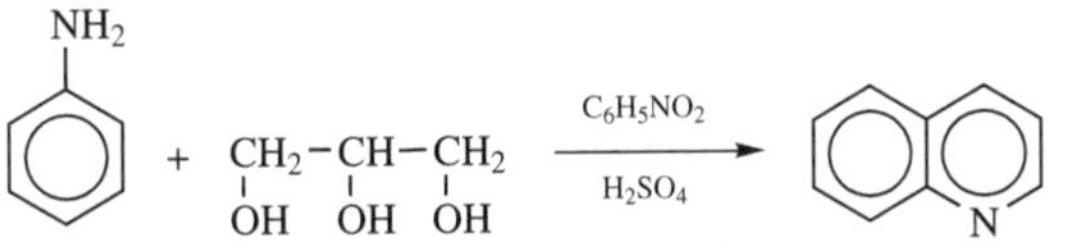

By substituting paraldehyde for glycerol, 2-methylquinoline [27601-00-9] may be synthesized. The Skraup synthesis is regarded as an example of the broader Doebner-von Miller synthesis. In the case of the Skraup synthesis, the glycerol undergoes an acid-catalyzed dehydration to provide a small concentration of acrolein that is the reactive species. If acrolein itself is used as a reactant, it would polymerize. Crotonaldehyde is the reactive intermediate in the Doebner-von Miller synthesis (28).

Many substituted quinolines are intermediates for antimalarials. The 2,4-di-substituted quinolines are produced from aniline and 1,3-diketones by the Combes quinoline synthesis (28). The reaction of aniline with nitrobenzene in the presence of dry sodium hydroxide at 140°C leads to formation of phenazine [92-82-0] and by-products (Wohl-Aue synthesis) (29).

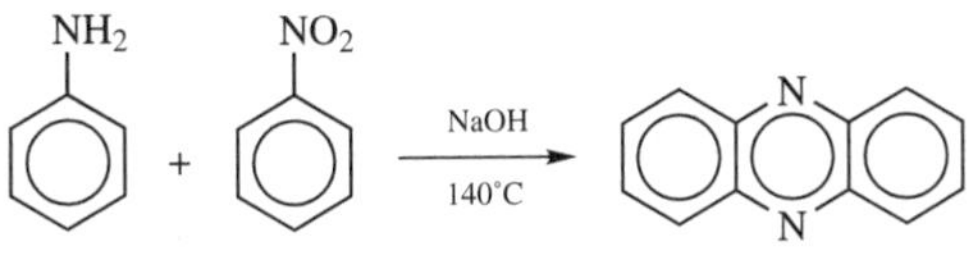

Aromatic amines react with 1-haloketones or 1-hydroxyketones to yield substituted indoles. This reaction is known as the Bischler indole synthesis (30).

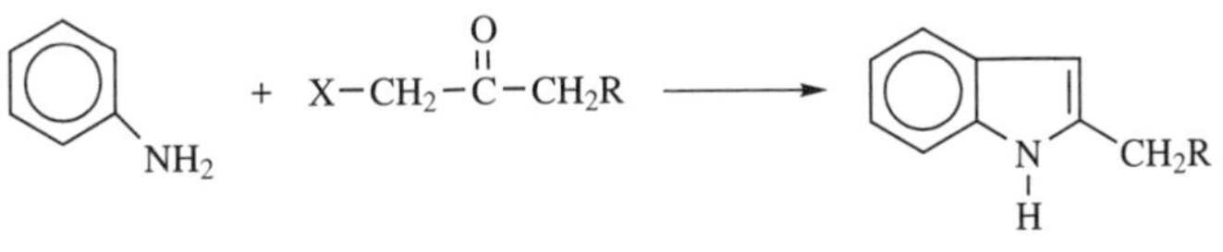

3.6. Reaction with Nitrous Acid. Primary, secondary, and tertiary aromatic amines react with nitrous acid to form a variety of products. Primary aromatic amines form diazonium salts.

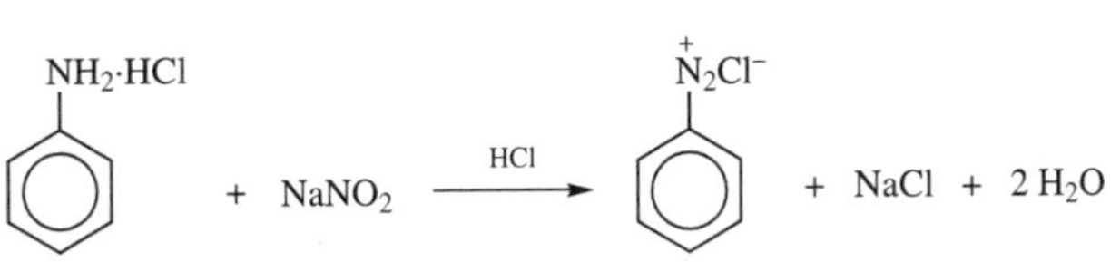

Secondary aromatic amines form N-nitrosamines and tertiary aromatic amines undergo ring nitrosation to yield C-nitroso products.

Aromatic diazonium salts are usually not isolated and react further to yield the desired product. Their reactions can be grouped in two categories: (*1*) those in which nitrogen is lost and the diazonium group is replaced by hydroxyl, hydrogen, nitrile, halogen, or aryl groups; and (*2*) those in which the diazonium ion can act as an electrophile in aromatic substitution reactions, that is, coupling reactions, with an activated aromatic compound.

Coupling of the diazonium salts with phenols and amines forms the basis for manufacture of a number of commercial dyes.

3.7. Oxidation.   Aromatic amines can undergo a variety of oxidation reactions, depending on the oxidizing agent and the reaction conditions. For example, oxidation of aniline can lead to formation of phenylhydroxylamine, nitrosobenzene, nitrobenzene, azobenzene, azoxybenzene or p-benzoquinone. Oxidation was of great importance in the early stages of the development of aniline and the manufacture of synthetic dyes, such as aniline black and Perkin's mauve.

Nitroso compounds are formed selectively via the oxidation of a primary aromatic amine with Caro's acid [7722-86-3] (H_2SO_5) or Oxone (Du Pont trademark) monopersulfate compound ($2KHSO_5 \cdot KHSO_4 \cdot K_2SO_4$); aniline black [13007-86-8] is obtained if the oxidation is carried out with salts of persulfuric acid (31). Oxidation of aromatic amines to nitro compounds can be carried out with peroxytrifluoroacetic acid (32). Hydrogen peroxide with acetonitrile converts aniline in a methanol solution to azoxybenzene [495-48-7] (33), perborate in glacial acetic acid yields azobenzene [103-33-3] (34).

$$C_6H_5NH_2 \; + \; H_2O_2 \quad \xrightarrow{\text{CH}_3\text{CN}} \quad \underset{\displaystyle \downarrow \atop \displaystyle O}{C_6H_5-N=N-C_6H_5}$$

Oxidation of aniline with a mixture of manganese dioxide and sulfuric acid has been used commercially for production of p-benzoquinone [106-51-4] (35).

3.8. Halogenation.   The presence of the amino group activates the ortho and para positions of the aromatic ring and, as a result, aniline reacts readily with bromine or chlorine. Under mild conditions, bromination yields 2,4,6-tribromoaniline [147-82-0].

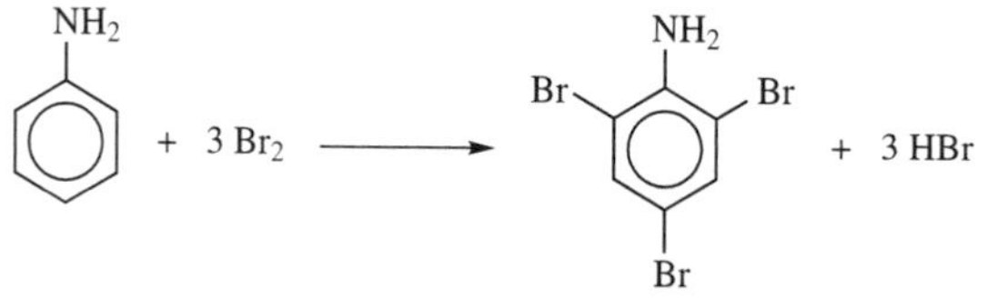

Controlled halogenation can be achieved by halogenation of the N-acetyl derivative of the aromatic amine, followed by hydrolysis of the acetyl group.

Chlorine in the presence of hydrogen chloride in an anhydrous organic solvent yields 2,4,6-trichloroaniline [634-93-5] (36,37). A mixture of aniline vapor and chlorine, diluted with an inert gas, over activated carbon at 400°C yields o-chloroaniline [95-51-2] (38). Aniline when treated with chlorine gas, in an aqueous mixture of sulfuric acid and acetic acid, at 105–115°C gives an 85–95% yield of p-chloranil [118-75-2] (39).

3.9. Sulfonation.   Aniline reacts with sulfuric acid at high temperatures to form p-aminobenzenesulfonic acid (sulfanilic acid [121-57-3]). The initial product, aniline sulfate, rearranges to the ring-substituted sulfonic acid (40). If the para position is blocked, the o-aminobenzenesulfonic acid derivative is isolated. Aminosulfonic acids of high purity have been prepared by sulfonating a mixture of the aromatic amine and sulfolane with sulfuric acid at 180–190°C (41).

In the reaction of anilinium sulfate [542-16-5] with fuming sulfuric acid, the major products are m- and p-aminobenzenesulfonic acid with less than 2% of the ortho isomer. With excess concentrated sulfuric acid (96.8–99.9%) at 60–100°C, the sulfate salt gives mainly the ortho and para isomers of the sulfonic acid (42).

When p-aminobenzenesulfonic acid is heated with sulfuric acid and phosphorus pentoxide, 2-amino-1,3,5-benzenetrisulfonic acid [64775-08-4] is the product (43).

3.10. Nitration. Direct nitration of aromatic amines with nitric acid is not a satisfactory method, because the amino group is susceptible to oxidation. The amino group can be protected by acetylation, and the acetylamino derivative is then used in the nitration step. Nitration of acetanilide in sulfuric acid yields the 4-nitro compound that is hydrolyzed to p-nitroaniline [100-01-6].

Nitration of aromatic amines with urea nitrate in sulfuric acid is reported to yield the p-nitro derivative exclusively (44). When the para position is blocked, the meta product is obtained in excellent yield.

3.11. Reduction. Hydrogenation of aromatic amines leads to formation of cycloalkylamines, dicycloalkylamines, or both, depending on the reaction conditions and the type of catalyst used. Hydrogenation of aniline in the liquid phase at 25 MPa (250 atm) over a cobalt–alumina catalyst at 140°C yields cyclohexylamine [108-91-8] in 80% yield (45). Dicyclohexylamine [101-83-7] is produced when aniline is hydrogenated in the vapor phase over a nickel-on-pumice catalyst (46). When aniline is hydrogenated at 160–200°C in the presence of a ruthenium–palladium catalyst, supported on γ-alumina impregnated with sodium hydroxide, the product mixture is 19.3% cyclohexylamine and 80.3% dicyclohexylamine (47). Hydrogenation with a similar ruthenium–palladium catalyst supported on γ-alumina treated with manganese and chromium salts gives 91.1% cyclohexylamine and 8.8% dicyclohexylamine (48).

Addition of ammonia to the hydrogenator, 1.5 to 5 parts per 100 parts aniline, has been reported to selectively yield cyclohexylamine (49). The reduction is carried out at 160 to 180°C and 2 to 5 MPa (20 to 50 atm) with a ruthenium-on-carbon catalyst.

Cyclohexylamine is also manufactured on a commercial scale by hydrogenation of nitrobenzene without intermediate separation of aniline.

4. Manufacturing and Processing

The predominant process for manufacture of aniline is the catalytic reduction of nitrobenzene with hydrogen. The reduction is carried out in the vapor phase (50–55) or liquid phase (56–60). A fixed-bed reactor is commonly used for the vapor-phase process and the reactor is operated under pressure. A number of catalysts have been cited and include copper, copper on silica, copper oxide, sulfides of nickel, molybdenum, tungsten, and palladium–vanadium on alumina or lithium–aluminum spinels. Catalysts cited for the liquid-phase processes include nickel, copper or cobalt supported on a suitable inert carrier, and palladium or platinum or their mixtures supported on carbon.

As an example of the vapor-phase process, nitrobenzene vapor and hydrogen are mixed and passed through a fluidized bed of copper-on-silica catalyst at 280–290°C and 500 kPa (5 atm). The reaction gas is filtered, cooled to condense aniline and water, and the excess hydrogen is recycled. The yield reported is 99.5% (42). The catalyst bed is regenerated by passing hot air through the bed to burn off the carbonaceous deposits. The aniline and water condensate are separated, and the latter is sent to the wastewater column. The organic phase is dried and the product is distilled.

Du Pont uses a liquid-phase hydrogenation process that employs a palladium–platinum-on-carbon catalyst. The process uses a plug-flow reactor that achieves essentially quantitative yields, and the product exiting the reactor is virtually free of nitrobenzene.

The old Bechamp batch process for reduction of nitrobenzene (iron–hydrochloric acid) is obsolete; however, Mobay Chemical Corporation is operating a plant using this process for production of pigment grade iron oxide as well as aniline.

The ammonolysis of phenol (61–65) is a commercial process in Japan. Sunoco (formerly Aristech Chemical Corporation) currently operates a plant at Haverhill, Ohio to convert phenol to aniline. The plant's design is based on Halcon's process (66). In this process, phenol is vaporized, mixed with fresh and recycled ammonia, and fed to a reactor that contains a proprietary Lewis acid catalyst. The gas leaving the reactor is fed to a distillation column to recover ammonia overhead for recycle. Aniline, water, phenol, and a small quantity of by-product diphenylamines are recovered from the bottom of the column and sent to the drying column, where water is removed.

A key feature of the Halcon process is the use of low pressure distillation (less than 80 kPa = 12 psi) to break the phenol–aniline azeotrope and allow economical separation of aniline from phenol (67).

5. Economic Aspects

Table 3 lists the U.S. producers of aniline and their capacities.

Other suppliers include BASF, Ludwigshafen, Germany and Narmada Chematur Petrochemicals, Gujrat, India.

Table 3. **U.S. Producers of Aniline and their Capacities**[a]

Producer	Capacity $\times 10^6$ kg ($\times 10^6$ lb)
BASF, Geismar, La.	206 (455)
Bayer, New Martinsville, W.Va.	18 (40)
DuPont, Beaumont, Tex.	127 (280)
First Chemical, Baytown, Tex.	113 (250)
First Chemical, Pascagoula, Miss.	154 (340)
Rubicon, Geismar, La.	395 (870)
Sunoco, Ironton, Ohio	68 (150)
Total	*1081 (2385)*

[a] Ref. 68.

Table 4. **Specifications for Commercial-Grade Aniline**

| Property | Value | | |
	Minimum	Maximum	Typical analysis
aniline, %	99.9		99.95
nitrobenzene, ppm		2.0	0.1
water, %		0.05	0.03
color[a], APHA		100	30

[a] When freshly distilled, aniline is a colorless oily liquid which darkens on exposure to light or air.

Demand in 2000 was 823×10^6 kg (1815×10^6 lb). Projected demand for 2004 is 959×10^6 kg (2115×10^6 lb). Demand equals production plus imports less exports. Although somewhat depressed by comparison to its performance in recent years, aniline should continue to produce better than GDP growth.

Prices have remained stable over the period 1995–2000. List price was $0.20–0.23/kg ($0.45–0.50/lb) tanks, fob, Current price $0.37–0.39/lb tanks, fob.

6. Specifications

A number of chemical methods can also be used for identification of aniline (69). Table 4 gives the specifications for a typical commercial-grade aniline.

7. Analytical Methods

7.1. Infrared. The infrared spectra of primary amines show a characteristic absorption doublet in the region of $3500 - 3300$ cm^{-1} because of the N–H stretching vibrations. Primary amines also show an N–H scissoring absorption in the range of $1650 - 1590$ cm^{-1}. Secondary amines have a single N–H stretching absorption band in the $3500 - 3300$ cm^{-1} range and the bands associated with the bending vibrations are weak or absent. Tertiary amines, as expected, show no absorption in the N–H region and thus are difficult to identify by infrared spectroscopy.

7.2. Ultraviolet. Benzene has a series of relatively low intensity absorption bands in the region of 230 to 270 nm. When there is a substituent on the ring with nonbonding electrons, such as an amino group, there is a pronounced increase in the intensity of these bands and a shift to longer wavelength. Aniline shows an absorption band at 230 nm ($\varepsilon = 8600$) and a secondary band at 280 nm ($\varepsilon = 1430$). Protonation of the amino groups reduces these effects and the spectrum resembles that of the unsubstituted benzene.

7.3. Nuclear Magnetic Resonance. The nmr spectrum of aromatic amines shows resonance attributable to the N–H protons and the protons of any *N*-alkyl substituents that are present. The N–H protons usually absorb in the δ 3.6–4.7 range. The position of the resonance peak varies with the concentration of the amine and the nature of the solvent employed. In aromatic amines, the resonance associated with N–CH protons occurs near δ 3.0, somewhat further downfield than those in the aliphatic amines.

7.4. Gas Chromatography. Aniline and many of its derivatives are volatile and can be analyzed by gas–liquid chromatography. The method offers a rapid and accurate procedure for determination of aniline in mixtures and is the method of choice for quality control used by producers of aniline.

8. Storage and Handling

The flash point of aniline (70°C) is well above its normal storage temperature; but, aniline should be stored and used in areas with minimum fire hazard (70). Air should not be allowed to enter equipment containing aniline liquid or vapor at temperatures equal to or above its flash point.

Strong oxidizing agents, such as nitric acid, perchloric acid, or ozone may cause aniline to oxidize spontaneously. Hexachloromelamine [2428-04-8] and trichloromelamine [12379-38-3] react violently with aniline, and in confined conditions the mixtures will explore or catch fire.

Aniline is slightly corrosive to some metals. It attacks copper, brass, and other copper alloys, and use of these metals should be avoided in equipment that is used to handle aniline. For applications in which color retention is critical, the use of 400-series stainless steels is recommended.

Aniline is shipped in tank truck and tank car quantities and is classified by the U.S. Department of Transportation (DOT) as a Class B poison (UN 1547), and must carry a poison label.

Wastes contaminated with aniline may be listed as RCRA Hazardous Waste, and if disposal is necessary, the waste disposal methods used must comply with U.S. federal, state, and local water pollution regulations. The aniline content of wastes containing high concentrations of aniline can be recovered by conventional distillation. Biological disposal of dilute aqueous aniline waste streams is feasible if the bacteria are acclimated to aniline. Aniline has a 5-day BOD of 1.89 g of oxygen per gram of aniline.

Aniline can be safely incinerated in properly designed facilities. It should be mixed with other combustibles such as No. 2 fuel oil to ensure that sufficient heating values are available for complete combustion of aniline to carbon dioxide, water, and various oxides of nitrogen. Abatement of nitrogen oxides may be required to comply with air pollution standards of the region.

9. Health and Safety Factors

Aniline is highly toxic and may be fatal if swallowed, inhaled, or absorbed through the skin. Aniline vapor is mildly irritating to the eye, and in liquid form it can be a severe eye irritant and cause corneal damage. The first sign of aniline poisoning is cyanosis, a bluish tinge to the lips and tongue, caused by conversion of the blood hemoglobin to methemoglobin. As methemoglobin concentration of the blood rises above a certain level, death may result from anoxia.

The U.S. Department of Labor (OSHA) has ruled that an employee's exposure to aniline in an 8-h work shift of a 40-h work week shall not exceed an 8-h time-weighted average (TWA) of 5 ppm vapor in air, 2 ppm skin. The American Conference of Governmental Industrial Hygienists (ACGIH) recommends a threshold limit value (TLV) of 2 ppm aniline vapor in air, TWA for an 8-h work day (71).

Based on tests with laboratory animals, aniline may cause cancer. The National Cancer Institute (NCI) and the Chemical Industry Institute of Toxicology (CIIT) conducted lifetime rodent feeding studies, and both studies found tumors of the spleen at high dosage (100–300 mg/kg per day of aniline chloride). CIIT found no tumors at the 10–30 mg/kg per day feeding rates. The latter value is equivalent to a human 8-h inhalation level of 17–50 ppm aniline vapor. In a short term (10-d) inhalation toxicity test by Du Pont, a no-effect level of 17 ppm aniline vapor was found for rats. At high levels (47–87 ppm), there were blood-related effects which were largely reversible within a 13-d recovery period (70).

In view of the above, aniline should be handled in areas with adequate ventilation and skin exposure should be avoided by wearing the proper safety equipment. Recommended personal protective equipment includes hard hat with brim, chemical safety goggles, full length face shield, rubber gauntlet gloves, rubber apron, and rubber safety shoes or rubber boots worn over leather shoes.

10. Uses

The major uses of aniline are in the manufacture of polymers, rubber, agricultural chemicals, dyes and pigments, pharmaceuticals, and photographic chemicals. Production of MDI (4,4-methylene diphenyl diisoyanate) accounts for 85% of aniline use. Other uses: rubber processing chemicals, 9%; herbicides, 2%; dyes and pigments, 2%; speciality fibers, 1%; miscellaneous including explosives, epoxy curing agents, and pharmaceuticals, 1%.

New uses for aniline described in recent patents include; aniline disulfide derivatives for treating allergic diseases (72), aniline compound in a hair dye composition and method of dyeing hair (73), and fluorine-containing aniline compounds as a starting material for insecticides (74).

The major consuming use MDI is tied to depressed economic conditions, but MDI growth continues to expand as new uses of polyurethanes are promoted outside traditional construction and refrigeration areas.

11. Methylenedianiline

Commercial production of 4,4'-methylenedianiline [101-77-9] (4,4'-MDA) is carried out by the acid catzlyzed reaction of formaldehyde with aniline. All processes produce polymeric MDA (PMDA), which consists of mixtures of isomers and oligomers of MDA. The amounts of MDA and oligomers are varied to produce products that have different applications. The isomeric distribution and the amount of MDA in the PMDA can be varied within wide ranges, depending on the needs of the consumer. More than 99% of the manufactured PMDA products are used in reactions with phosgene to produce the corresponding isocyanates for use in polyurethanes. The resultant polymeric isocyanates (PMDI) are either sold commercially or are purified to isolate 4,4'-methylenediphenyldiisocyanate (MDI) [101-68-8]. Only 15–20% of the total PMDI manufactured in the United States is consumed in the monomeric form. All of the MDA and PMDA produced is consumed in industries that are "destructive" of MDA's chemical identity. Thus MDA loses its unique chemical identity and is not encountered by household consumers.

The term MDA is sometimes used for pure 4,4'-MDA as well as the oligomeric mixture PMDA. Similar inconsistencies are encountered for the isocyanate derivatives (MDI and PMDI). Synonyms for 4,4'-methylenedianiline include MDA, 4,4'-MDA, methylenedianiline, 4,4'-methylenebisaniline, dianilinomethane, 4-(4'-aminobenzyl)aniline, 4,4'-diaminodiphenyl-methane, 4,4'-methylene-bis(benzenamine), bis(p-aminophenyl)methane, DADPM, DAPM, and DDM. The p,p'-and 4,4'-designations are used interchangeably. Synonyms for the oligomeric MDA mixtures include polyaminopolyphenylamine, polymethylenepolyphenylamine, and PMDA. In this section MDA will stand for 4,4'-MDA and/or its isomers, and PMDA is a mixture of MDA and MDA oligomers.

11.1. Physical Properties. The physical and chemical properties of 4,4'-MDA and a typical PMDA are listed in Table 5.

Table 5. **Physical and Chemical Properties of MDA and PMDA**

Property	4,4'-MDA	PMDA[a]
CAS Registry Number	[101-77-9]	[25214-70-4]
molecular formula	$C_{15}H_{14}N_2$	
RTECS accession number[b]	BY5425000	
molecular weight	198.3	
active hydrogen equivalent weight	49.6	51
melting point, °C	93	60–80
heat of vaporization, kJ/mol[c]	95.4	
specific heat, J/(g · °C)	2.1	2.1
heat of fusion, kJ/mol[c]	19.6	~19.6
flash point, °C	227	238
boiling point, °C	238 at 1.33 kPa[d]	398 at 101.3 kPa[d]
vapor pressure, Pa[d]	2.7×10^{-5} at 25°C	1.3 at 100°C
density, g/mL	1.070 at 103°C	1.07 at 70°C
viscosity, mPa · s (= cP)	8.3 at 100°C	80 at 70°C
approximate solubility, g/100 mL solvent at 25°C		
acetone	273.0	
benzene	9.0	
carbon tetrachloride	0.7	
ethyl ether	9.5	
methanol	143.0	
water	0.1	

[a] For PMDA containing approximately 70% MDA.
[b] Registry of Toxic Effcts of Chemical Substances.
[c] To conert kJ to kcal, divide by 4.184.
[d] To convert kPa to mm Hg, multiply by 7.5.

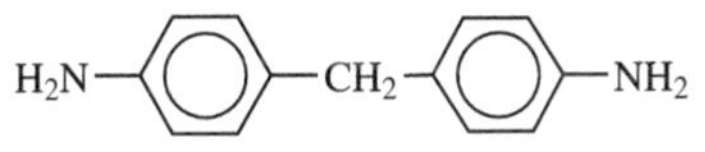

H₂N—⟨◯⟩—CH₂—⟨◯⟩—NH₂

4,4′-methylenedianiline (MDA)

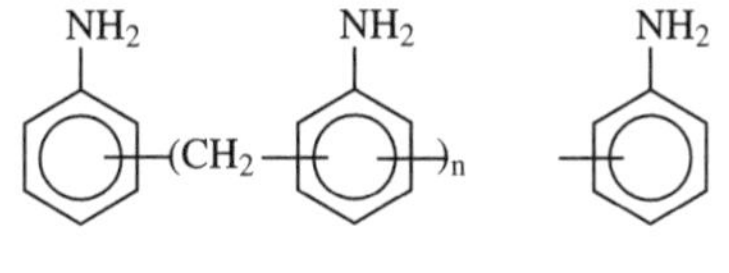

polymeric MDA (PMDA)

Purified 4,4′-MDA is a light tan to white crystalline solid with a faint aminelike odor. It slowly oxidizes in air with a darkening in color. PMDA mixtures are yellow to brown supercooled liquids or waxy solids.

11.2. Chemical Properties. MDA reacts similarly to other aromatic amines under the proper conditions. For example, nitration, bromination, acetylation, and diazotization (75–77) all give the expected products. Much of the chemistry carried out on MDA takes advantage of the difunctionality of the molecule in reacting with multifunctional substrates to produce low and high molecular weight polymers.

The most important commercial process is the reaction of MDA with an excess of phosgene to form the corresponding isocyanate, 4,4′-methylene-diphenyl-diisocyanate, MDI, $C_{15}H_{10}N_2O_2$. The reaction proceeds through the formation of a primary carbamyl chloride that is decomposed with heating and the removal of HCl.

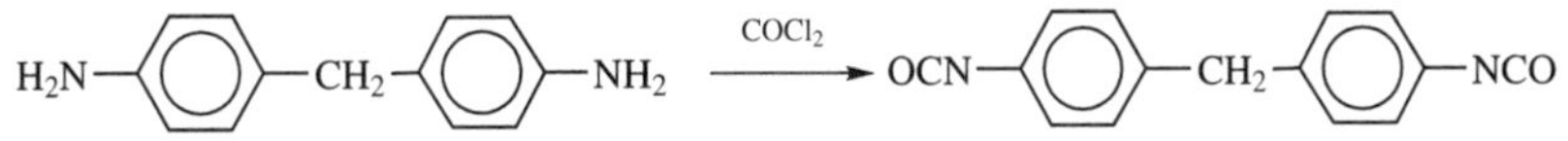

$$H_2N-\langle\bigcirc\rangle-CH_2-\langle\bigcirc\rangle-NH_2 \xrightarrow{COCl_2} OCN-\langle\bigcirc\rangle-CH_2-\langle\bigcirc\rangle-NCO$$

MDA reacts with acid anhydrides to form amides. In the reaction with maleic anhydride both of the amino hydrogens are replaced to form the imide, N,N'-(methylenedi-p-phenylene) dimaleimide [1367-54-5] $C_{21}H_{14}N_2O_4$.

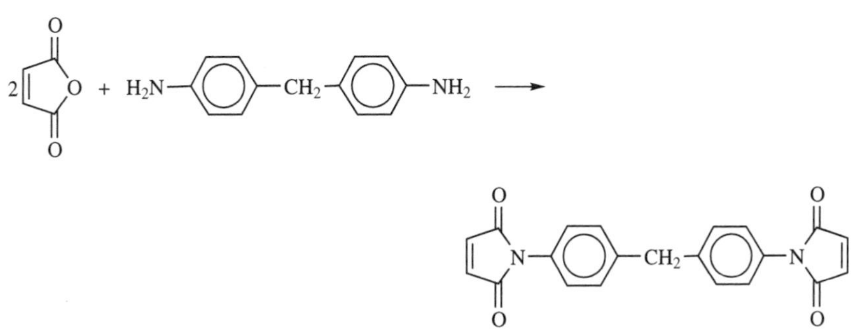

The bismaleimide can then be polymerized by reaction with additional amine to form polyaminobismaleimide or by radiation-induced homopolymerization to form polybismaleimide (78).

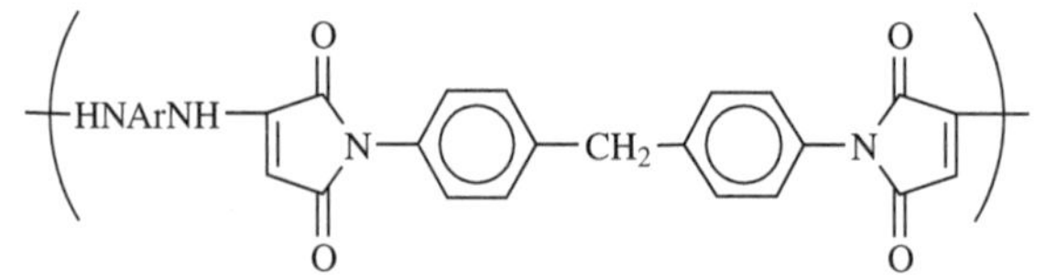

polyaminobismaleimide

The reaction of diphenic anhydride with excess MDA proceeds through an imide ring opening to produce a linear polymer, or it can react with PMDA to form a cross-linked polymer (79).

All of the amine hydrogens are replaced when MDA or PMDA reacts with epoxides to form amine based polyols. These polyols can be used in reactions with isocyanates to form urethanes or with additional epoxide to form cross-linked thermoset resins.

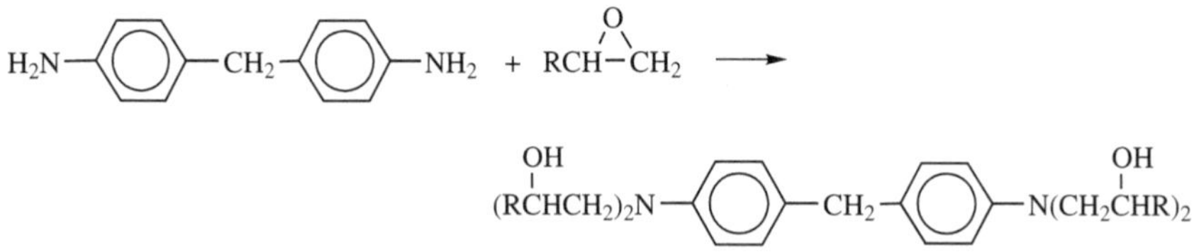

High temperature and high pressure reactions of MDA with hydrogen in the presence of noble metal catalysts convert 4,4'-MDA into bis(4-aminocyclohexyl)-methane (H_{12}MDA) [1761-71-3] ($C_{13}H_{26}N_2$). The products are a mixture of cis and trans isomers that can be controlled to some extent by the proper choice of catalyst and reaction conditions (80–86).

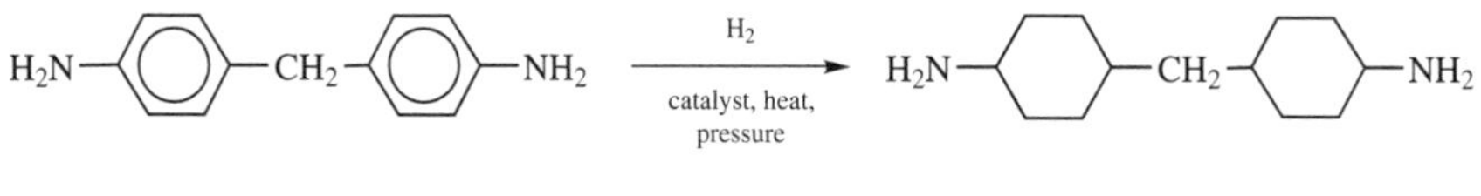

11.3. Manufacture and Processing.   MDA and oligomers (PMDA) are produced by the acid catalyzed condensation of aniline [62-53-3] (C_6H_7N) with formaldehyde [50-00-0] (CH_2O). The reaction does not lead to a single product, but to a mixture of 4,4'-, 2,4'-, and 2,2'-isomers and oligomeric MDAs. The amounts of MDA isomers and oligomers formed depend on the ratios of aniline, formaldehyde, and acid used, as well as the reaction temperature and time. Figure 1 shows a simpplified pathway to the formation of 4,4'-MDA. Similar routes can be drawn for the formation of 2,4'-MDA, 2,2'-MDA, and higher oligomers. The initial reaction of aniline with formaldehyde produces N-methylolaniline [61224-32-6] (C_7H_9NO). This product loses water rapidly to form the Schiff base (intermediate A) (86,87). The Schiff base reacts with aniline to form two types of aminals, linear (LA) and cyclic (HHT). The relative amounts of the aminals formed depend on the ratio of aniline to formaldehyde. For example, when aniline: formaldehyde molar ratios of >2:1 are employed, the predominant

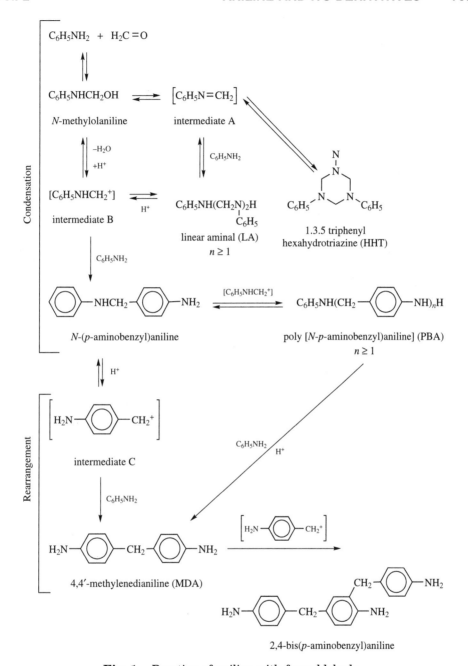

Fig. 1. Reaction of aniline with formaldehyde.

product is the linear aminal with $n = 1$. If the reaction is carried out in base or in neutral solution, the reactions stop at this stage. A process for preparing methylenedianiline by reacting aniline and formaldehyde with or without are and catalyst has been reported (88).

MDA does not form in the absence of acid. The aminals and the N-methylolaniline decompose rapidly in acid solution to form an anilinium ion (intermediate B). This reactive intermediate combines with aniline to form $N(p$-aminobenzyl)aniline [17272-83-2] ($C_{13}H_{14}N_2$) which reacts with intermediate B in the presence of acid to form oligomeric benzylamines (PBA) that exist in equilibrium with the monomer. The formation of this equilibrium mixture completes the condensation phase of the synthesis. Almost all of the side reactions take place during the condensation phase of the reaction. The typical side products formed are the N-methyl and quinazoline derivatives of aniline and MDA (86,89). Commercial processes have been successful in minimizing these side reactions.

The final step (rearrangement stage of the reaction) is decomposition to form the p-aminobenzyl carbonium ion (intermediate C) and alkylation of aniline to form 4,4'-MDA. In this step all of the secondary amine intermediates are converted to primary amine final products. Direct alkylation of PBA with aniline has also been hypothesized to form MDA without the formation of intermediate C (90). The formation of MDA is not reversible under normal reaction conditions. Almost all of the other reactions depicted are reversible to some extent. From a commercial standpoint the most important reactions taking place are the formation of oligomers, ie, the reaction of MDA with intermediate C or PBA. It is these reactions, which cannot be suppressed, that are responsible for the current development of MDA technology. MDA is formed slowly during the reaction and therefore is susceptible to further alkylation to form 2,4-bis(p-aminobenzyl)aniline [25834-80-4] ($C_{20}H_{21}N_3$). Further alkylations produce an oligomeric mixture.

The reaction of aniline with formaldehyde can be carried out in a single reactor (Fig. 2). However, most commercial processes probably use multiple reactors, which provide greater control of the MDA isomer distribution and oligomeric content of the final product (90–93). Use of hydrochloric acid and high reaction temperatures necessitates the use of corrosion resistant metallurgy. Normally the acid is first mixed with excess aniline, which causes an exotherm. Formaldehyde is then added, with efficient agitation and at low temperatures ($<50°C$), to the anilinaniline hydrochloride solution. The reaction is usually staged to control the condensation and rearrangement steps. The final reaction temperatures are normally 80–120°C. After completion of reaction, the acidic PMDA is treated with aqueous sodium hydroxide to neutralize the excess acid. A large amount of salt is formed during this step; thus the plants must be located near an outlet capable of handling the generated salt water (normally a seacoast).

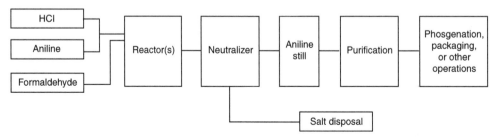

Fig. 2.   Methylenedianiline process.

Processes that recycle the acid and eliminate the salt disposal problem have been patented (94,95). The organic layer is then washed with water and stripped to remove unreacted aniline and water. The unreacted aniline is recycled back to the beginning of the reaction. The product may be purified to isolate pure 4,4'-MDA, packaged for shipment, or treated with phosgene to produce the corresponding isocyanate. The 4,4'-MDA is normally sold in flaked or granular form in lined steel drums. Depending on the MDA content, PMDA is sold as a waxy solid or a yellow to brown viscous supercooled liquid in steel drums.

Process parameters can be varied to change the MDA isomer distribution and oligomeric content of PMDA products. Generally, aniline to formaldehyde molar ratios of 2 to 5 are used. To increase the MDA content, higher ratios of aniline to formaldehyde are employed. Increasing the acid to aniline ratio also increases the 4,4'-MDA content of the diamine fraction. Historically, the polyurethane industry consumes as much of the 4,4'-MDI isomer as possible. There has been an increasing demand for higher 2,4'-MDI and 2,4'-PMDI products to be used as replacements for toluenediisocyanate (TDI). Low acid and high aniline to formaldehyde ratios increase the 2,4'-MDA content of the diamine fraction. At the lower aniline to formaldehyde ratios the tendency is to form higher oligomers. The 2,4'-MDA is more reactive toward alkylation than the 4,4'-MDA isomer. Commercial processes do not employ molar excesses of acid to aniline. Acid catalysis is necessary to form MDA, and HCl is the acid of choice. Silica (96), clay (97–99), ethanesulfonic acid (100), and tungsten (101,102) are among the catalysts that have been patented. All of these processes possess some commercial drawbacks, eg, high expense, severe reaction conditions, and/or poor yields of 4,4'-MDA. High reaction temperatures result in a decrease in the MDA content of the product. Reaction time and water content also affect the oligomeric distribution of the product; both should be minimized for maximum MDA formation (103). Water is introduced into the reaction from the hydrochloric acid used to make the aniline hydrochloride solution, from the formaldehyde, which is normally sold in water solution (formalin), and as a reaction product in the condensation phase of the reaction.

Commercially, the PMDA mixtures are normally treated with phosgene to produce the corresponding isocyanates. These isocyanate mixtures, commonly called polymeric MDI (PMDI), are sold directly and have varied chemical compositions. The 4,4'-MDI can be separated from the PMDI products by distillation or crystallization (103,104). The amount of 4,4'-MDI that is removed depends on marketing conditions. The residues are also viable commercial products.

Commercially, a small amount of the 4,4'-MDA is isolated by distillation from PMDA. Depending on the process employed, the removal of MDA can be partial (as is done with the isocyanates) or total. Partial removal of MDA gives some processing latitude but yields of 4,4'-MDA are reduced. Distillation residues from PMDA manufacture that contain less than 1% MDA pose a disposal problem. Processes for the regeneration of MDA by heating these residues in the presence of aniline and an acid catalyst have been patented (105–107). Waste disposal of PMDA is expensive and reclamation processes could become commercially viable. The versatility of the isocyanate process, however, can be used to avoid the formation of low MDA content distillation residues.

Table 6. **U.S. Producers of MDI (Methyl Diphenyl Diisocyanate) and Their Capacities**[a]

Producer	Capacity $\times 10^6$ kg ($\times 10^6$ lb)
BASF, Geismar, LA	261 (575)
Bayer, Baytown, TX	229 (505)
Bayer, New Martinsville, WV	73 (160)
Dow Chemical, La Porte, TX	181 (400)
Rubicon, Geismar, LA	381 (840)
Total	*1125 (2480)*

[a] Ref. 108

11.4. Economic Aspects.   Table 6 lists the U.S. producers of MDI and their capacities. Quantities are per year of polymeric and pure MDI. A mixture of MDI and its dimer and trimer are produced. Pure MDI is distilled from the mixture and is used for reaction injection molding, thermoplastic elastomers, and adhesives. The polymeric form is used to make rigid and semi-rigid polyurethane foams. The market split is 80% polymeric and 20% pure MDI (108).

Demand has been depressed compared to recent years, but MDI is expected to grow. The major consuming area, rigid foams, is expected to expand, however, as new uses outside the traditional construction and refrigeration sectors are found.

Demand in 2000 was 626×10^6 kg (1382×10^6 lb), projected for 2004 is 773×10^6 kg (1705×10^6 lb). Demand is production plus imports minus exports. Expected growth is 4.1% through 2004 (108).

Prices (1995–2000) \$0.60–0.64/kg (\$1.34–1.41/lb) list, polymeric, MDI bulk cl frt alld. Current price is \$0.64/kg (\$1.41/lb), market prices are below list at \$0.25–0.34/kg (\$0.55–0.75/lb) (108).

11.5. Specifications and Standards.   The 4,4′-MDA is sold commercially with a diamine assay of 98–99%. The major impurity the 2,4′-MDA isomer, which can be present in amounts up to 3%. PMDA products are normally defined by hydrogen equivalent weight and viscosity. Typical products exhibit a 50 hydrogen equivalent weight and a viscosity of $80 - 140$ mPa $\cdot$ s($=$ cP) at 70°C. PMDA products normally contain, in addition to the isomers and oligomers of MDA, small amounts of aniline, water, chlorides, and various alkylated amines. All MDA products should be stored in sealed containers in a cool dry area.

11.6. Analytical Methods.   The characterizations of MDA and PMDA are similar to those normally used for aromatic amines. In the manufacture of PMDA, the MDA isomer distribution and the formation of side products is determined primarily by gas chromatography (109,110). The amine content is determined by acid titration or diazotization. PMDA oligomeric distributions are determined by hplc or gpc techniques (111) and are estimated by viscosity measurements. Liquid chromatographic and spectrophotometric monitoring methods have been developed to determine ppm to low ppb quantities in workplace environments (111,112) urine (113), reacted polyurethanes (114), and epoxy systems (115). All of the environmental methods of analysis employ a collection technique that is sensitive to the presence of aerosols and are capable of quantitating MDA down to 10 ppb (v/v). OSHA recommends using Method #57 for monitoring purposes (116).

11.7. Health and Safety Factors.　All of the toxicity data on MDA have been collected using either 4,4'-MDA or the corresponding hydrochloride salt. The information discussed in this section can also be used for commercial products containing MDA or PMDA. Because MDA is a potentially hazardous chemical, worker exposure shduld be kept to a minimum. For complete health and safety information on MDA consult references 120 and (117–119).

The recommended threshold limit value (TLV) of MDA is 100 ppb for an 8-h time-weighted average (TWA) skin. MDA is a suspected human carcinogen (116,118,119). The oral LD_{50} = 830 mg/kg. In May of 1989 OSHA proposed a new standard for regulating MDA. The proposal cites a permissible exposure limit (PEL) for occupational exposure of MDA to an 8-h TWA of 10 ppb and a STEL (short-term exposure limit) of 100 ppb for a 15-min TWA (121,122). The standard does not apply if initial monitoring shows less than the action level of 5 ppb for an 8-h TWA (airborne) and if no dermal exposure is likely. The employer is required to implement engineering and work practices (eg, respirators) to maintain employee exposure levels at less than or equal to the permissible exposure limit. The proposal also includes having a changing room and showers for changing all contaminated clothing after a work shift; employer laundering of clothing; removal of clothing prior to eating, drinking, smoking, etc; and washing hands and face prior to eating. If food and beverages are consumed in the work area, the employer is expected to provide a positive pressure eating area. All surfaces must be maintained as free as possible from visible accumulations of MDA. The employer is responsible for conducting medical surveillance and record keeping, as well as for conducting periodic monitoring of the area according to existing exposure level guidelines.

The major exposure route in workers who experience MDA poisoning is by skin contact. If there is a likelihood of skin or eye contact, proper protective clothing (including gloves, head coverings, impervious shoes, aprons, coveralls, or other full body clothing) must be worn. Skin absorption is increased if the MDA is dissolved in an organic solvent. Face shields and/or goggles should be worn where appropriate. Care must be taken to minimize contamination of individuals and their personal environment from exposure to MDA. The low vapor pressure of MDA poses a minimum risk of MDA inhalation. However, grinding and drumming of MDA produces dust and vapors, which can cause inhalation problems. In addition, many operations require heating MDA to keep the material molten. It is therefore recommended that respirators be worn while handling MDA to keep exposures to a minimum.

There are no reports of cancer in humans as a result of MDA exposure. Acute exposures of MDA have caused epigastric pain, fever, jaundice, and other symptoms consistent with hepatitis. These effects, however, appear to be reversible. No liver damage has yet been observed from chronic exposure to MDA. Skin sensitivity and staining occurs in some individuals. In animals, chronic administration of MDA produces hepatic cirrhosis, liver lesions, and enlargement of the spleen, liver, and kidneys. MDA has also proved to be carcinogenic in rats and mice, causing liver and thyroid tumors. No cancers have been formed in dogs. Structurally similar compounds have been shown to cause cancer in laboratory animals (119). If appreciable amounts of MDA are swallowed, vomiting should be induced. If eye exposure does occur, the eyes should be flushed with water for 30 min. Skin exposures should be washed with soap and water. Medical attention should be obtained promptly whenever an exposure occurs. There are always dangers involved with the misuse of any chemical. However, with attention to proper work practices and policies, this valuable intermediate can be utilized safely.

11.8. Uses.   More than 99% of all the PMDA produced is used directly in the manufacture of the corresponding isocyanates (MDI and PMDI). Two types of isocyanates are sold, monomeric (MDI) and polymeric (PMDI). The PMDI products are available commercially in varying viscosities from 50 to 2000 mPa · s (= cP), containing from 25 to 60% monomeric MDI. The major use for PMDI products (80%) is in the manufacture of rigid foam, which is primarily used in housing and refrigeration insulation applications. (construction, 50%; refrigeration, 12%; packaging 18%; tank and pipe insulation, 3%; other uses include transportation marine flotation and furniture, 7%) (108). Other uses for PMDI products include reaction-injection molding applications, 13%; cast elastomers, 2%; miscellaneous uses are in thermoplastic resins and foundry core binders, 5% (108). MDI contains approximately 98% of the 4,4'-isomer and is used to manufacture thermoplastic resins (eg, films, gaskets, tubing), spandex fibers, and coatings. Because MDI is a solid at room temperature, a significant portion of the commercially produced pure MDI is converted to liquid products by either modifying the MDI with carbodiimide linkages or reaction with polyols to produce isocyanate terminated prepolymers. These products are used in automotive reaction injection molding (RIM), coatings, and recreation and military applications. MDI and PMDI products containing more than 10% of the 2,4'-MDI isomer are becoming available for use in nonrigid foam applications. These products are able to compete with toluenediisocyanate (TDI) in physical properties of the final products without the severe handling limitations normally needed for TDI. Nonisocyanate uses for MDA or PMDA include epoxy resin curing agents, wire coating applications, plastic fibers, polyurethane coreactants, an intermediate for pigments and dyes, intermediates for the preparation of polyamide–imide resins, reinforced composite materials, and military applications. Perhydrogenated 4,4'-MDA (H_{12}MDA or PACM) is used in light-stable elastomers and coatings. The H_{12}MDA is converted to the corresponding isocyanate, bis(4-isocyanatocyclohexyl) methane [5124-30-1] ($C_{15}H_{22}N_2O_2$) or H_{12}MDI, and is used in automotive safety glass and biomedical applications.

12. Other Derivatives

Most derivatives of aniline are not obtained from aniline itself, but are prepared by hydrogenation of their nitroaromatic precursors. The exceptions, for example, N-alkylanilines, N-arylanilines, sulfonated anilines, or the N-acyl derivatives, can be prepared from aniline and have been discussed. Nitroanilines are usually prepared by ammonolysis of the corresponding chloronitrobenzene. Special isolation methods may be required for some derivatives if the boiling points are close and separation by distillation is not feasible. Table 7 lists some of the derivatives of aniline that are produced commercially.

Table 7. Aniline Derivatives

Class of compound and common name	Molecular formula	CAS Registry Number	Condensed structural formula	Appearance	Melting point, °C	Boiling point, °C	Commercial derivatives and uses
salts							
aniline hydrochloride	$C_6H_7N \cdot ClH$	[142-04-1]	$C_6H_5NH_2 \cdot HCl$	white solid	198	245	Aniline black
aniline sulfate	$C_6H_7N \cdot \tfrac{1}{2}H_2O_4S$	[542-16-5]	$(C_6H_5NH_2)_2 \cdot H_2SO_4$	white crystals			sulfanilic acid
N-alkyl, N-aryl							
N-methylaniline	C_7H_9N	[100-61-8]	$C_6H_5NHCH_3$	yellow liquid	−57	194.6	vanillin; Michler's ketone; alkylating agents; dyes
N,N-dimethylaniline	$C_8H_{11}N$	[121-69-7]	$C_6H_5N(CH_3)_2$	yellow liquid (darkens in air)	2	193–194	
N-ethylaniline	$C_8H_{11}N$	[103-69-5]	$C_6H_5NHC_2H_5$	colorless liquid (darkens in air)	−63.5	204.7	explosive stabilizer; dyes
N,N-diethylaniline	$C_{10}H_{15}N$	[91-66-7]	$C_6H_5N(C_2H_5)_2$	pale yellow liquid	−38.8	215–216	alkylating agent
N-benzyl-N-ethylaniline	$C_{15}H_{17}N$	[92-59-1]					
$C_6H_5N(C_2H_5)CH_2C_6H_5$	light yellow oil	314		triphenylmethane dyes			
diphenylamine	$C_{12}H_{11}N$	[122-39-4]	$C_6H_5NHC_6H_5$	white crystals, floral odor	54–55	302	rubber antioxidants; phenothiazine
C-alkyl							
o-toluidine	C_7H_9N	[95-53-4]	$H_3CC_6H_4NH_2$	yellow liquid (darkens in air)		200–202	triphenylmethane dyes; safranine colors
m-toluidine		[108-44-1]		colorless liquid	−30.4	203–204	dyes

Table 7 (Continued)

Class of compound and common name	Molecular formula	CAS Registry Number	Condensed structural formula	Appearance	Melting point, °C	Boiling point, °C	Commercial derivatives and uses
p-toluidine		[106-49-0]		white crystals	44–45	200–201	Basic Red 9; Acid Green 25
2,3-xylidine	$C_8H_{11}N$	[87-59-2]	$(H_3C)_2C_6H_3NH_2$	liquid		221–222	Solvent Orange 7; Direct Violet 14
2,4-xylidine		[95-68-1]		liquid	16	214	
2,5-xylidine		[97-78-3]		oily liquid	15.5	213.5	*p*-xyloquinone; Red 26; Direct Violet 7
2,6-xylidine		[87-62-7]		colorless liquid	11–12	216–217	formerly in dyes synthetic riboflavin
3,4-xylidine		[95-64-7]		solid	51	226	
3,5-xylidine		[108-69-0]		oil	9.8	220–221	azo dyes
C-alkoxy							
o-anisidine	C_7H_9NO	[90-04-0]	$H_3COC_6H_4NH_2$	yellow liquid (darkens in air)	5–6	225	guaiacol synthesis; Direct Red 24; Solvent Red 1
m-anisidine		[536-90-3]		oily liquid	−1 – 1	251	
p-anisidine		[104-94-9]		white solid	57	243	dyes
o-phenetidine	$C_8H_{11}NO$	[94-70-2]	$H_5C_2OC_6H_4NH_2$	oily liquid		231–233	dyes
p-phenetidine		[156-43-4]		liquid (darkens in air)	3–4	254–255	phenacetin; phenocoll; rubber antioxidant; dyes
p-cresidine	$C_8H_{11}NO$	[120-71-8]	$H_3CO(CH_3)C_6H_3NH_2$	white crystals	52–54	235	FD&C Red 40
N-acyl							
formanilide	C_7H_7NO	[103-70-8]	$HCONHC_6H_5$	white crystals	50	271	analgesic and antipyretic
acetanilide	C_8H_9NO	[103-84-4]	$CH_3CONHC_6H_5$	colorless crystals	114.3	304	intermediate for sulfa drugs; hydrogen peroxide stabilizer; azo dyes

					mp, °C	bp, °C	
acetoacetanilide	$C_{10}H_{11}NO_2$	[102-01-2]	$CH_3COCH_2CONHC_6H_5$	white crystals	86		intermediate for pyrazolones and pyrimidines; Hansa yellows; benzidine yellow pigments
chloroanilines							
2-chloroaniline	C_6H_6ClN	[95-51-2]	$ClC_6H_4NH_2$	colorless liquid	−14	208–210	dyes
3-chloroaniline		[108-42-9]		colorless liquid	−10	230–231	dyes
4-chloroaniline		[106-47-8]		colorless liquid	72.5	232	azoic dye coupling Component 10 and 15
2,5-dichloroaniline	$C_6H_5Cl_2N$	[95-82-9]	$Cl_2C_6H_3NH_2$	needle crystals	51	251	dyes
3,4-dichloroaniline		[95-76-1]		white crystals	71.5	272	herbicides; dyes
sulfonated anilines							
orthanilic acid	$C_6H_7NO_3S$	[88-21-1]	$H_2NC_6H_4SO_3H$	colorless crystals		>320 dec	dyes
metanilic acid		[121-47-1]		white crystals	dec		Acid Yellow 36, Direct Yellow 44 dyes
sulfanilic acid		[121-57-3]		white crystals	288		Acid Orange 1; Food Yellow 3 dyes
nitroanilines							
2-nitroaniline	$C_6H_6N_2O_2$	[88-74-4]	$O_2NC_6H_4NH_2$	golden crystals	71–72	284	Vat Orange 7 and Red 14 dyes
3-nitroaniline		[99-09-2]		yellow crystals	114	305–307 dec	synthetic intermediate; dyes
4-nitroaniline		[100-01-6]		pale yellow crystals	148–149	332	intermediate; dyes; intermediate for 1,4-phenylenediamine
2,4-dinitroaniline	$C_6H_5N_3O_4$	[97-02-9]	$(O_2N)_2C_6H_3NH_2$	yellow crystals	187–188		Pigment Orange 5; dyes
2,4,6-trinitroaniline	$C_6H_4N_4O_6$	[489-98-5]	$(O_2N)_3C_6H_2NH_2$	yellow solid	192–195	explodes	explosives; detonators

BIBLIOGRAPHY

"Aniline and Its Derivatives" in *ECT* 1st ed., Vol. 1, pp. 914–929, by A. W. Dawes and Jesse Werner, General Aniline & Film Corp; in *ECT* 2nd ed., Vol. 2, pp. 411–427, by C. S. Kouris and J. Northcott, Allied Chemical Corp.; in *ECT* 3rd ed., Vol. 2, pp. 309–321, by J. Northcott, Allied Chemical Corp., in *ECT* 4th ed., Vol. 2, pp. 426–442, by Bijan Amini, E. I. du Pont de Nemours & Co., Inc.; "Amines, Aromatic (Methylenedianiline)" in *ECT* 3rd ed., Vol. 2, pp. 338–348, by William M. Moore, Dow Chemical U.S.A.; in *ECT* 4th ed., Vol. 2, pp. 461–473, by Steven Lowenkron, The Dow Chemical Company; "Amines, Aromatic, Aniline and Its Derivatives" in *ECT* (online), posting date: December 4, 2000, by Bijan Amini, E. I. du Pont de Nemours and Co., Inc.

CITED PUBLICATIONS

1. U.S. Pat. 2,580,284 (Dec. 25, 1951), T. J. Deahl, F. H. Stross, and M. D. Taylor (to Shell Development Co.).
2. U.S. Pat. 2,991,311 (July 4, 1961), M. Thoma (to Biller, Michaelis & Co.).
3. U.S. Pat. 4,062,893 (Dec. 13, 1977), T. Kyuma and M. Nakazawa (to New Japan Chemical Co., Ltd.).
4. R. N. Shreve, G. N. Vriens, and D. A. Vogel, *Ind. Eng. Chem.* **42**, 791 (1950).
5. U.S. Pat. 2,391,139 (Dec. 18, 1945), J. B. Dickey and J. G. McNally (to Eastman Kodak).
6. U.S. Pat. 3,819,709 (June 25, 1974), K. Murai, G. Akazome, T. Kyuma, and M. Nakazawa (to New Japan Chemical Co., Ltd.).
7. U.S. Pat. 3,969,411 (July 13, 1976), J. Schneider (to Bayer AG).
8. Jpn. Pat. 7,481,331 (Aug. 6, 1974), K. Murai, T. Kyuma, and M. Nakazawa (to New Japan Chemical Co., Ltd.).
9. U.S. Pat. 3,957,874 (May 8, 1976), T. Dockner and H. Krug (to BASF AG).
10. A. K. Bhattacharaya and K. D. Naudi, *Ind. Eng. Chem. Prod. Res. Dev.* **14**(3), 162 (1975).
11. U.S. Pat. 2,984,687 (May 16, 1961), D. L. Esmay and P. Fotis (to Standard Oil Co.).
12. P. H. Groggins, *Aniline and Derivatives*, Van Nostrand Rheinhold Co., New York, 1924, p. 177.
13. U.S. Pat. 1,908,951 (May 16, 1933), E. C. Britton and R. D. Holmes (to The Dow Chemical Company).
14. U.S. Pat. 4,721,810 (Jan. 26, 1988), D. C. Hargis (to Ethyl Corporation).
15. Ref. 12, p. 164.
16. U.S. Pat. 3,071,619 (Jan. 1, 1963), H. J. Kehe, R. T. Johnson, W. J. Driscoll, and R. R. Bloor (to B. F. Goodrich Co.).
17. U.S. Pat. 2,968,676 (Jan. 17, 1961), A. G. Potter and R. G. Weyker (to American Cyanamid Co.).
18. U.S. Pat. 4,814,504 (Mar. 21, 1989), R. E. Malz, Jr. (to Uniroyal Chemical Co., Inc.).
19. Ger. Pat. 1,187,628 (Feb. 25, 1965), A. Schultz (to BASF AG).
20. G. G. Ecke, J. P. Napolitano, and A. J. Kolka, *J. Org. Chem.* **21**, 711 (1956).
21. U.S. Pat. 3,923,892 (Dec. 2, 1975), O. E. H. Klopfer (to Ethyl Corporation).
22. U.S. Pat. 4,128,582 (Dec. 5, 1978), L. J. Governale and J. C. Wollensak (to Ethyl Corporation).
23. U.S. Pat. 3,275,690 (Sept. 27, 1966), R. Stroh, H. Haberland, and W. Hahn (to Bayer AG).
24. Eur. Pat. Appl. 286029 A1 (Oct. 12, 1988), R. Pierantozzi (to Air Products and Chemicals, Inc.).

25. N. Menschutkin, *Justus Liebigs Ann. Chem.* **162**, 165 (1872).

26. Fr. Pat. 2,057,373 (June 25, 1971) (to E. I. du Pont de Nemours & Co., Inc.).

27. G. Jones, *Quinolines*, Part 1, John Wiley & Sons, Inc., New York, 1977, p. 100.

28. Ref. 27, p. 101.

29. G. A. Swan and D. G. I. Felton, *Phenazines*, Interscience Publishers, Inc., New York, 1957, p. 7.

30. W. A. Remers and R. K. Brown in W. J. Houlihan, ed., *Indoles*, Part 1, John Wiley & Sons, Inc., New York, 1972, p. 317.

31. R. J. Kennedy and A. M. Stock, *J. Org. Chem.* **25**, 1901 (1960).

32. W. D. Emmons and A. F. Ferris, *J. Am. Chem. Soc.* **75**, 4623 (1953).

33. G. B. Payne, P. H. Deming, and P. H. Williams, *J. Org. Chem.* **26**, 659 (1961).

34. S. M. Mehta and M. V. Vakilwala, *J. Am. Chem. Soc.* **74**, 563 (1952).

35. U.S. Pat. 2,343,768 (Mar. 7, 1944), C. F. Gibbs (to B. F. Goodrich Co.).

36. U.S. Pat. 2,675,409 (Apr. 13, 1954), H. D. Orloff and J. F. Napolitano (to Ethyl Corporation).

37. U.S. Pat. 4,447,647 (May 8, 1984), F. Werner, K. Mannes, and V. Trescher (to Bayer AG).

38. H. Brintzinger and H. Orth, *Monatsch Chem.* **85**, 1015 (1954).

39. U.S. Pat. 2,422,089 (June 10, 1947), H. H. Fletcher (to U.S. Rubber Co.).

40. U.S. Pat. 4,681,710 (July 21, 1987), R. Miller, J. L. Bollini, and R. Schneider (to CIBA-GEIGY Corporation).

41. Eur. Pat. Appl. 289,689 A1 (Nov. 9, 1988), P. Rossi and M. Quintini (to Appital Porduzioni Industriali S.P.A.).

42. P. K. Maarsen and H. Cerfontain, *J. Chem. Soc. Perkin Trans.* **2**(8), 1008 (1977).

43. S. C. J. Olivier, *Rec. Trav. Chim.* **39**, 194 (1920).

44. T. P. Sura, M. M. Ramana, and N. A. Kudav, *Synth. Commun.* **18**, 2161 (1988).

45. Brit. Pat. 630,859 (Oct. 24, 1949), J. G. M. Bremner and F. Starkey (to Imperial Chemical Industries, Ltd.).

46. Ger. Pat. 805,518 (May 21, 1951), O. Stichnoth (to BASF AG).

47. Eur. Pat. Appl. 324,984 (July 26, 1989), O. Immel and H. H. Schwarz (to Bayer AG).

48. Ger. Pat. 3,801,755 (July 27, 1989), O. Immel, H. H. Schwarz, and R. Thiel (to Bayer AG).

49. U.S. Pat. 4,384,142 (May 17, 1983), H. L. Merten and G. R. Wilder (to Monsanto Co.).

50. U.S. Pat. 2,891,094 (June 16, 1959), O. C. Karkalitis, Jr., C. M. Vanderwaart, and F. Houghton (to American Cyanamid Co.).

51. Ger. Pat. 2,320,658 (Oct. 25, 1973), M. Sada, T. Zinno, Y. Otsuka, and T. Ohrui (to Sumitomo Chemical Co., Ltd.).

52. Ger. Pat. 1,133,394 (July 19, 1962), H. J. Pistor, H. Sperber, G. Poehler, and A. Wegerich (to BASF AG).

53. U.S. Pat. 3,538,018 (Nov. 3, 1970), K. Pilch and H. Sperber (to BASF AG).

54. U.S. Pat. 3,882,048 (May 6, 1975), H. Thelen and K. Halcour (to Bayer AG).

55. U.S. Pat. 4,265,834 (May 5, 1981), U. Birkenstock, B. Lachmann, J. Metten, and H. Schmidt (to Bayer AG).

56. U.S. Pat. 2,292,879 (Aug. 11, 1943), M. A. Kise (to Solvay Process Co.).

57. U.S. Pat. 2,823,235 (Feb. 11, 1958), D. P. Graham and L. Spiegler (to E. I. du Pont de Nemours & Co., Inc.).

58. U.S. Pat. 3,270,057 (Aug. 30, 1966), E. V. Cooke and H. J. Thurlow (to Imperial Chemical Industries, Ltd.).

59. U.S. Pat. 4,777,295 (Oct. 11, 1988), M. V. Twigg (to Imperial Chemical Industries).

60. Jpn. Pat. 89,036,459 (July 31, 1989) (to Mitsui Toatsu Chemical Co., Inc.).

61. U.S. Pat. 3,860,650 (Jan. 14, 1975), M. Becker and S. Khoobiar (to Halcon International, Inc.).

62. U.S. Pat. 3,965,182 (June 22, 1976), C. J. Worrel (to Ethyl Corporation).
63. Jpn. Pat. 7429,176 (Aug. 1, 1974), Y. Shinohara and A. Niiyama (to Mitsui Petro-chemicals Industries Ltd.).
64. Jpn. Kokai 7367,229 (Sept. 13, 1973), T. Kiyoura and T. Takahashi (to Mitsui Toatsu Chemicals, Inc.).
65. Intl. Pat. Appl. 88 03,920 (June 2, 1988), M. Yasuhara, Y. Tatsuki, M. Nakamura, and F. Matsunaga (to Mitsui Petrochemicals Industries, Ltd.).
66. I. McKechnie, F. Bayer, and J. Drennan, *Chem. Eng. New York* **87**(26), 26 (1988).
67. U.S. Pat. 3,682,782 (Aug. 8, 1972), C. Y. Choo (to Halcon International Inc.).
68. "Aniline, Chem Profile," *Chemical Week*, Jan. 21, 2002.
69. J. W. McDowell and J. Northcott, in C. L. Hilton, ed., *Encyclopedia of Industrial Chemical Analysis*, Vol. 5, Wiley-Interscience, New York, 1967, 421–459.
70. *Du Pont Aniline Properties, Uses, Storage, and Handling* Bulletin, E. I. du Pont de Nemours & Co., Inc., 1983.
71. R. J. Lewis, Sr., ed., *Sax's Dangerous Properties of Industrial Materials*, Vol. 2, 10th ed., John Wiley & Sons, Inc., New York, 2000, p. 262.
72. U.S. Pat. Appl. 20020013311 (Jan. 31, 2002); 20020115658 (Aug. 22, 2002), M. R. Helberg, Z. Feng, and S. T. Miller.
73. U.S. Pat. Appl. 20020197223 (Dec. 26, 2002), K. Imura (to Kanagawa Ken, Japan).
74. U.S. Pat. Appl. 20010041814 (Nov. 15, 2001), M. Tohnishi and co-workers.
75. H. King, *J. Chem. Soc.* **117**, 988 (1920).
76. G. D. Parkes and R. H. H. Morley, *J. Chem. Soc.* **139**, 315 (1936).
77. C. L. Butler, Jr, and R. Adams, *J. Am. Chem. Soc.* **47**, 2610 (1925).
78. M. A. J. Mallet, *Mod. Plast.* **50**, 78 (1973).
79. M. E. B. Jones and M. S. Chisholm, *Polymer* **29**, 1699 (1988).
80. U.S. Pat. 2,606,927 (Aug. 12, 1952), A. E. Barkdoll and G. M. Whitman (to E. I. du Pont de Nemours & Co., Inc.).
81. U.S. Pat. 2,606,928 (Aug. 12 1952), A. E. Barkdoll, C. D. Bell, and E. R. Graef (to E. I. du Pont de Nemours & Co., Inc.).
82. U.S. Pat. 2,606,924 (Aug. 12, 1952), G. M. Whitman (to E. I. du Pont de Nemours & Co., Inc.).
83. U.S. Pat. 2,494,563 (Jan. 17, 1950). W. Kirk, Jr., C. Hundred, R. S. Schreiber, and G. M. Whitman (to E. I. du Pont de Nemours & Co., Inc.).
84. A. E. Barkdoll, H. W. Gray, and W. Kirk, Jr., *J. Am. Chem. Soc.* **73**, 741 (1951).
85. Brit. Pat. 619,706 (Mar. 14, 1949), W. Kirk, Jr., R. S. Schreiber, and G. M. Whitman (to E. I. du Pont de Nemours & Co., Inc.).
86. Ger. Pat. 842,200 (June 23, 1952), O. Stichnoth and L. Wolf (to BASF).
87. E. C. Wagner, *J. Org. Chem.* **19**, 1862 (1954).
88. U.S. Pat. Appl. 20020132953 (Sept. 19, 2002) E. Strafer and co-workers.
89. M. M. Sprung, *Chem. Rev.* **26**, 297 (1940). C. Ringel, H. Böhm, H. Kroschwitz, and D. Scheller, *SYSpur Reporter* 1981, **19**, 143 (1981).
90. H. J. Twitchett, *Chem. Soc. Rev.* **3**, 209 (1974).
91. U.S. Pat. 3,954,867 (May 4, 1976), B. D. Funk, Jr., J. Mongiello, and W. S. Rabourn (to the Upjohn Co.).
92. U.S. Pat. 3,931,320 (Jan. 6, 1976), W. Eifler, R. Raue, E. H. Rohe, and J. Finkel (to Bayer AG).
93. Brit. Pat. 1,378,423 (Dec. 27, 1974), R. S. Mason, P. D. Rough, and H. W. Twitchett (to Imperial Chemical Industries, Ltd.).
94. U.S. Pat. 3,476,806 (Nov. 4, 1969), H. O. Wolf (to E. I. du Pont de Nemours and Co., Inc.).
95. U.S. Pat. 4,061,678 (Dec. 6, 1977), H. Knöfel and G. Ellendt (to Bayer AG).
96. Ger. Pat. 2,238,920 (Mar. 7, 1974), H. Knöfel (to Bayer AG).

97. Belg. Pat. 677,830 (Dec. 1, 1965), R. B. Lund and J. Vitrone (to Allied Chemical).
98. U.S. Pat. 4,071,558 (Jan. 31, 1978), F. E. Bentley (to Texaco Development Corp.).
99. U.S. Pat. 4,039,580 (Aug. 2, 1977), F. F. Frulla, A. A. R. Sayigh, H. Ulich, and P. J. Whitman (to the Upjohn Co.).
100. U.S. Pat. 4,039,581 (Aug. 2, 1977), F. F. Frulla, A. A. R. Sayigh, H. Ulrich, and P. J. Whitman (to the Upjohn Co.).
101. U.S. Patent 4,212,821 (July 15, 1980), E. T. Marquis and W. H. Brader, Jr. (to Texaco Development Corp.).
102. U.S. Pat. 4,286,107 (Aug. 25, 1981), E. T. Marquis and L. W. Watts, Jr. (to Texaco Inc.). U.S. Pat. 4,284,816 (Aug. 18, 1981), E. T. Marquis and L. W. Watts, Jr. (to Texaco Inc.).
103. U.S. Pat. 3,277,173 (Oct. 4, 1966), E. L. Powers and I. B. Van Horn (to Mobay Chemical Company).
104. U.S. Pat. 3,892,634 (July 1, 1975), J. D. Hajek and H. R. Steele (to the Upjohn Co.).
105. Brit. Pat. 1,423,993 (Feb. 4, 1976), D. Adsley and R. C. Smith (to Imperial Chemical Industries, Ltd.).
106. Brit. Pat. 1,127,347 (Sept. 18, 1968), J. A. Hall and B. L. Yates (to Imperial Chemical Industries Ltd.).
107. U.S. Pat. 4,089,901 (May 16, 1978), P. Ziemek, R. Raue, and H. J. Buysch (to Bayer AG).
108. MDI (Methyl diphenyl diisocyanate), Chemical Profile, *Chemical Market Reporter*, Dec. 24 2001.
109. *An Estimate of Industry Regulatory Compliance Costs and Economic Impact for Meeting an OSHA Workplace Exposure Standard for 4,4'*-Methylenedianiline (MDA), Heiden Associates, Inc., Washington, D.C., Dec. 12, 1986.
110. I. P. Krasnova, E. N. Boitsov, Y. N. Golovistikov, and B. M. Tsigin, *Zsvod. Lab.* **43**, 1324 (1977).
111. P. Falke, R. Tenner, and H. Knopp, *J. Prakt. Chem.* **328**, 142 (1986).
112. C. C. Anderson and E. C. Gunderson, *Methods Validation Study of High Performance Liquid Chromatographic Technique for Determining the MPDA and MDA Content of Air Samples*, SRI International, Menlo Park, Calif., 1986.
113. J. Cocker, L. C. Brown, H. K. Wilson, and K. Rollins, *J. Anal. Toxicol.* **12**(1), 9 (1988).
114. D. A. Ernes and D. T. Hanshumaker, *Anal Chem.* **55**, 408 (1983).
115. R. E. Smith and W. R. Davis, *Anal. Chem.* **56**, 2345 (1984).
116. C. J. Elskamp, *OSHA Analytical Method #57—4,4'*-Methylenedianiline (MDA), OSHA Analytical Laboratory, Salt Lake City, Utah, Jan. 1986.
117. *Federal Register* **54**, 20672–20744 (May 12, 1989).
118. *Documentation of the Threshold Limit Values and Biological Exposure Indices*, 5th ed., American Conference of Governmental Industrial Hygienists, Cincinnati, Ohio, 1989.
119. N. M. Bernholc, S. C. Morris, III, and J. E. Brower, *Biological Effects Summary Report 4,4'*-Methylenedianiline (BNL-51903), Brookhaven National Laboratory, Upton, N.Y., 1985.
120. *NIOSH Current Intelligence Bulletin—4,4'*-Methylenedianiline (MDA), 47 (July 25 1986).
121. *Threshold Limit Values and Biological Exposure Indices for 1989–1990*, American Conference of Governmental Industrial Hygienists, Cincinnati, Ohio, 1989.
122. R. J. Lewis, Sr., *Sax's Dangerous Properties of Industrial Materials*, 10th ed., Vol. 3, John Wiley & Sons, Inc., New York, 2000, p. 2438.

BIJAN AMINI
E. I. du Pont de Nemours & Co., Inc.

ANTIAGING AGENTS

1. General Considerations

Aging in humans is associated with a decline in physical vigor and function, with progressive deterioration in most major organ systems including central nervous system and immune system functions. The current view is that molecular damage and disorder that occur with age in macromolecules are largely responsible for the age-related changes observed at the organism level. Such damage to macromolecules may be caused by free radicals and by the formation of advanced glycation end-products, to cite two examples. It is still unclear, however, whether AGEs, which accumulate to high levels in many age-related chronic diseases, are the cause or the consequences of the diseases. Some inhibitors of glycation and free radical formation are showing promise against chronic conditions, particularly diabetes and its related complications. The most impressive results in extending the life span of animals and preventing the incidence of age-related conditions are observed with the regimen of reduced caloric intake or caloric restriction. Since this regimen has not yet been verified to work in humans and since it is a very difficult diet to maintain, the search for pharmacological mimics of calorie restriction is the focus of much effort in drug discovery. Aging is associated with an increase in oxidative damage to cellular macromolecules probably arising from the electron transport system as part of the day-to-day metabolic process. Agents that can lower oxidative stress and damage are beginning to show potential as calorie restriction mimics. Several hormones have been shown to improve certain changes associated with human aging, such as body mass composition. Agents that have been studied in specific organ systems as they relate to prevention and treatment of age-related changes are discussed for each system. For skin aging, basic cell functions are reduced with advancing age. Therefore, an effective antiaging agent should provide acceleration of mitochondrial activity, enhanced cell proliferation, and increased matrix component synthesis in dermal fibroblasts. Several agents are discussed which show activity in one or more of these areas. Much progress has been made in understanding the molecular basis of hair loss. Two drugs are now approved and marketed in the US for prevention and treatment of hair loss. The central nervous system suffers from a gradual decline in cognition, behavior, and function with advancing age. Alzheimer's disease (AD) is the most common form of dementia in the elderly. Currently, available treatments for AD only diminish certain symptoms but cannot halt the dementing process. New therapies currently being developed for AD include agents that target amyloid β peptide and downstream pathological changes as well as agents that increase the activity of the cholinergic transmitter system. Newer inhibitors of cholinesterase are more selective and show fewer side effects than the first generation series of inhibitors.

Antiinflammatory and immunotherapeutic approaches to the treatment and prevention of AD are also being investigated. Age is a major risk factor for the development of chronic diseases of the cardiovascular system and such disease are a direct result of atherosclerosis. Since atherosclerosis has both an autoimmune and an inflammatory component, new approaches for the treatment and prevention of heart disease are beginning to focus on these areas as well as the traditional risk factors such as dyslipidemia. The human immune system gradually declines with age and the changes that occur result in increased susceptibility to infectious disease, cancer and autoimmune disorders in the elderly. It is believed that pharmacologic intervention to restore immune function in the elderly will provide widespread benefits in helping to maintain health in advanced age. Age-related conditions that involve the musculoskeletal system include osteroarthritis and rheumatoid arthritis as well as osteoporosis. Several biologicals are now in use for treatment of rheumatoid arthritis. Future treatments will likely involve the inhibition of proinflammatory cytokines by small molecules. Finally, age is a major risk factor for developing cancer in humans. Many agents are being investigated for their ability to inhibit the formation and progression of various types of cancer, including breast, prostate, and colorectal cancers. It is important to distinguish between the intrinsic process of aging and the diseases and conditions associated with aging. Important also is the distinction between aging, longevity, and maximum life span. Theoretically, if one slows the body's rate of aging, one can increase the maximum life span. Last year however, researchers on aging noted in a position statement that no treatment on the market today has been proven to slow human aging (1). Medical interventions for age-related diseases do result in an increase in life expectancy, but none have been proven to modify the underlying processes of aging. This article discusses agents relative to the prevention and treatment of age-related changes and diseases.

The current view of scientists is that random damage that occurs within cells and among extracellular molecules is responsible for many of the age-related changes that are observed in organisms. Molecular disorder occurs and accumulates within cells and their products because this occurrence outpaces the cell's repair mechanisms.

1.1. Advanced Glycation End-Products. There is considerable evidence for molecular disorder and damage to macromolecules such as proteins and cell membrane lipids being caused by so-called advanced glycation end-products (AGEs) and by free radicals. The AGEs are complex components formed nonenzymatically during the Maillard reaction involving monosaccharides and the amino groups of proteins, which alter protein structure and functions. The AGEs affect the biochemistry and physical properties of proteins and the extracellular matrix. The AGEs have been implicated in the pathogenesis of diabetic complications, atherosclerosis, arthritis, Alzheimer's disease and in the process of normal aging. AGEs accumulate to high levels in these age-related chronic diseases (2). However, it is not clear whether AGEs are the cause or the consequences of the age-related complications (3). A large number of compounds have been reported as inhibitors of glycation and AGE-protein crosslink formation (4). Among those most studied is aminoguanidine [79-17-4], also known as pimagedine as the HCl salt.

Aminoguanidine has shown promise at slowing or preventing the cross-linking of collagen molecules to each other in vascular wall and myocardium tissue in aging and diabetic animals (5). The acetamidoethyl derivative of amino-guanidine, ALT-946 [192511-71-0], is about fivefold more potent than aminoguanidine as an inhibitor of cross-linking of a glycated protein to rat collagen (6).

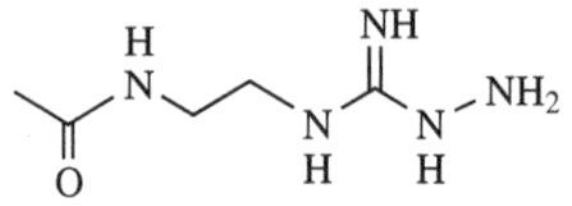

ALT-946

In addition, aminoguanidine inhibits the development of retinopathy in diabetic animals by a mechanism that probably involves, at least in part, the inhibition of retinal nitric oxide production (7,8). Other effective compounds studied for inhibition of AGEs include benfotiamine [22457-89-2], the thiazolidine OPB-9195 [163107-50-4], and pyridoxamine [85-87-0].

benfotiamine

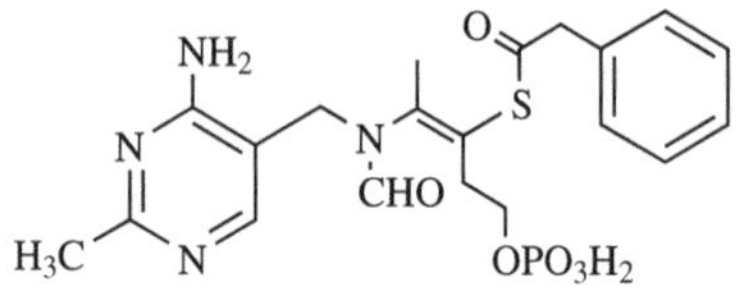

OPB-9195 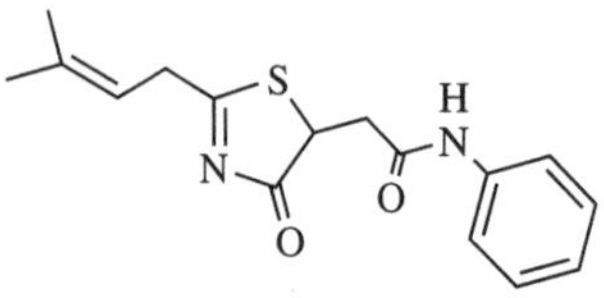   pyridoxamine

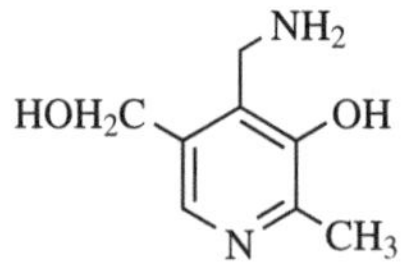

Benfotiamine has been shown to block three of the major biochemical pathways implicated in the pathogenesis of hyperglycemia induced vascular damage (the hexoseamine pathway, the AGE formation pathway and the diacylglycerol–protein kinase C pathway) and to prevent experimental diabetic retinopathy in rats (9). The compound OPB-9195 has been reported to reduce blood pressure and oxidative damage in the genetic hypertensive rat (10) and to be beneficial for the reduction of serum AGE and prevention of diabetic neuropathy (11). Pyridoxamine likewise is able to inhibit the development of retinopathy in experimental diabetes (12) and inhibits the formation of both AGEs and advanced lipoxidation end–products (ALEs) (13). Data suggests that the AGE/ALE inhibitory activity and therapeutic effect of pyridoxamine depend, at least in part, on its ability to trap reactive carbonyl intermediates that are formed during autooxidation of carbohydrates and peroxidation of lipids that lead to AGE/ALE formation, thereby inhibiting chemical modification of tissue proteins (14).

1.2. Caloric Restriction. Scientists first recognized >60 years ago the value of a low-calorie yet nutritionally balanced diet not only in prolonging the mean and maximum life span in a variety of animal species compared to free-feeding animals, but also in reducing the incidence of age-related conditions (15). This regimen of reduced caloric intake has produced the most impressive results when the regimen was begun early in the life of the animal and involved a reduction of ~30–50% in total calories. Although caloric restriction might extend the longevity of humans because it does so in many other animal species, there is no study in humans that has proven that it will work, although there are long-term nonhuman primate studies currently being conducted (16–18). Indeed, a calorically restricted diet, to be effective, must approach levels that most people would find intolerable. For this reason, the search for pharmacological mimetics of caloric restriction is the subject of many investigations. In this regard, there is considerable current interest in the search for the biological mechanisms underlying the observed retardation of aging and diseases by caloric restriction since identification of such should help in the discovery of suitable mimetics. Two areas receiving considerable attention are oxidative stress and regulation of glucose metabolism (19). These mechanisms are not mutually exclusive. For example, AGEs may be formed by oxidative and nonoxidative reactions (2). Caloric restriction, resulting in lower levels of blood glucose and insulin over long periods of time, may be the best prevention of AGE formation since hyperglycemia is the major cause of AGEs. Agents that can effectively lower the levels of circulating glucose and insulin or increase insulin sensitivity are under investigation. One of the most extensively studied agents as a caloric restriction mimetic is 2-deoxy-D-glucose [154-17-6] (2DG) (15).

2DG works by interfering with the way cells process glucose. Cells use glucose from the diet to generate ATP. By limiting food intake, caloric restriction minimizes the amount of glucose entering cells and decreases ATP generation. When 2DG is administered to animals that eat normally, glucose reaches cells in abundance, but the agent prevents most of the glucose from being processed and thus reduces ATP synthesis. Studies in animals have shown that although 2DG can mimic many of the effects seen in caloric restriction, the range between biologically effective dose and toxicity for 2DG is very narrow, which precludes its use in humans (20). Two additional agents that have shown some promise as potential mimetics are iodoacetate [64-69-7], an inhibitor of glyceraldehydes-3-phosphate dehydrogenase, and phenformin [114-86-3], a down-regulator of N-methyl-D-aspartate receptor expression (21). However, studies are still preliminary for these agents.

1.3. Reactive Oxygen Species. Recent work supports the notion that free radicals such as reactive oxygen species (ROS) play a central role in both the formation of AGEs and in AGE-induced pathological alterations in gene expression (22). In general, the free-radical theory of aging postulates that free radical reactions are responsible for the progressive accumulation of changes with time resulting in the ever-increasing likelihood of disease and death that accompanies advancing age (23). At the same time, free radicals play an important role in normal physiological processes (eg, the immune response and cell communication). At present, there is relatively little evidence from human studies that supplements containing antioxidants lead to a reduction in either the risk of age-related conditions or the rate of aging (1).

The role of oxidative stress of mitochondrial origin in aging is being studied extensively. Many studies have shown that aging is associated with an increase in oxidative damage to cellular macromolecules probably arising from the electron transport system as a normal consequence of energy metabolism (24). Likewise, there is much interest in the possibility that caloric restriction may act to retard aging by lowering oxidative stress–damage (25). It is possible, as preliminary data suggests, that α-phenyl-*tert*-butyl nitrone [3376-24-7] and other nitrone-based free-radical trapping agents (also known as spin-traps), as well as other antioxidants, will act as caloric restriction mimetics to prolong both average and maximum life span (19).

1.4. Telomeres.   Telomeres, the repeated hexameric sequence of nucleotides TTAGGG at the ends of chromosomes, shortens in many normal human cells with increased cell divisions. Thus, normal cells undergo a finite number of cell divisions and ultimately enter a nondividing state called replicative senescence. Solid scientific evidence has shown that telomere length plays a role in determining cellular life span in several normal human cell types (26,27). It was found that telomeres are synthesized de novo by the terminal transferase enzyme telomerase (28) and this is the only known reverse transcriptase that is necessary for normal cell activity (29). Telomerase has been found to occur in extracts of most immortal cell lines and ~ 90% of all human tumors studied, unlike normal cultured cell strains (30). Moreover, the level of telomerase activity found in normal cell populations is, per cell, significantly less than that found in cancer cells (30). In 1998 it was reported that normal, human cell strains could be immortalized with apparent retention of their normal properties by transfecting them with vectors encoding the human telomerase catalytic subunit, providing direct evidence for the role of telomerase shortening in cell senescence and telomerase expression in cell immortality (31). However, increasing the number of times a cell can divide may predispose cells to tumor formation (32,33).

1.5. Hormones.   A number of hormones, most notably human growth hormone, have been shown to improve some of the physiological changes associated with human aging. A study in older men showed that administration of growth hormone resulted in an increase in lean body mass and a decrease in adipose mass, but there was no assessment of muscle strength or quality of life (34). A more recent study, however, showed that administration of physiological doses of human growth hormone administration to healthy older men for 6 months also resulted in the same observations of body mass composition improvement, but functional ability did not improve and side effects occurred frequently (35). The steroid hormone dehydro-3-epiandrosterone [53-43-0] (DHEA) is normally synthesized in large quantities by the adrenal gland but serum concentrations decline with advancing age (36).

DHEA supplementation in elderly people has been advertised as an anti-aging medication. The DHEA might be useful for improving psychological well being in the elderly, reducing disease activity in people with mild to moderate systemic lupus erythematosus, improving mood in the clinically depressed and improving various parameters in women with adrenal insufficiency (36). However, subjects with a physiological, age-related decline in DHEA secretion show little benefit from DHEA administration (37). Although many other claims have been made for DHEA in diverse conditions, such as aging, dementia, and AIDS, no well-designed clinical trials have clearly substantiated the utility and safety of long-term DHEA supplementation and there is currently not enough evidence to recommend it in advanced age.

2. Organ Systems

In connection with the general considerations discussed above regarding age-related changes and diseases, there are also organ-specific considerations of note. Following are discussions of agents that have been studied in specific organ systems as they relate to prevention and treatment of age-related changes and diseases.

2.1. Skin and Hair. Skin aging, which includes photoaging and intrinsic aging, causes the formation of wrinkles and sagging. Dermal matrix components change qualitatively and quantitatively over time in aging skin (38). In addition, basic cell functions such as proliferation, mitochondrial respiration, and production of matrix components in dermal fibroblasts are reduced with aging. Thus, an effective antiaging agent for skin should provide acceleration of mitochondrial activity, enhanced cell proliferation, increased matrix component synthesis and improvement of collagen bundle fiber (38). Many cosmetic formulations have been marketed, with claims of preventing or reducing wrinkles and lines, containing a variety of agents. For example, N-amidino-L-proline [35404-57-0] is claimed to improve skin elasticity (39), phytosterol [83-46-5] for enhancing collagen I in human fibroblast cells (40), N,N-dimethyldodecyl amine oxide [1643-20-5] containing composition for the activation of corneum protease activity resulting in higher turnover and repair rates of the stratum corneum (41), and azelaic acid [123-99-9] for inhibition of collagenase activity (42). The effects of glycolic acid [79-14-1] on metabolic activity of human fibroblasts was also studied and results showed that cell growth was stimulated, dermal thickness was enhanced, and wrinkle-depth was repaired (43).

N-amidino-L-proline

Phytosterol

N,N-dimethyldodecyl amine oxide

azeleic acid

glycolic acid

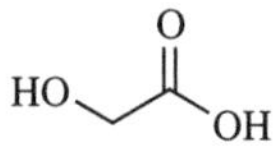

Skin wrinkles can also be the result of repetitive muscle activity. Clostridium botulinum toxin (Botox) type A [93384-43-1] has been widely used aesthetically for the past 15 years for facial skin rejuvenation (44). Botox works by clinically paralyzing the facial muscles underlying the lines and wrinkles on the surface with restoration of muscle activity usually commencing between 3 and 4 months after injection (45). A review of the correct use and complications of Botox has appeared (46).

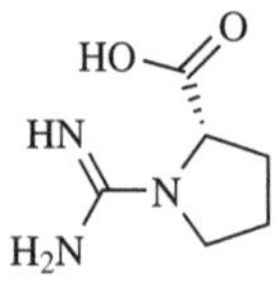

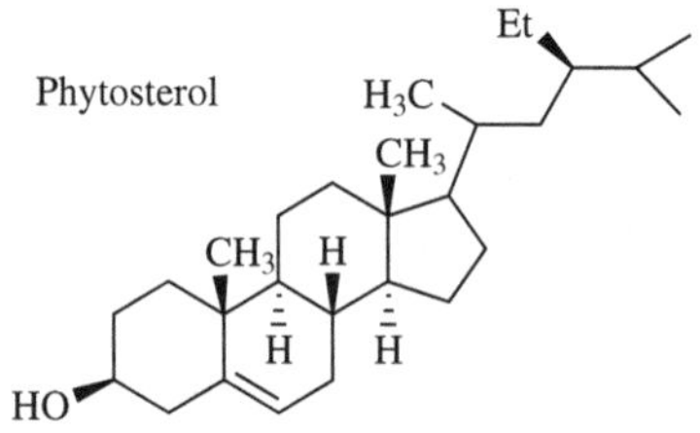

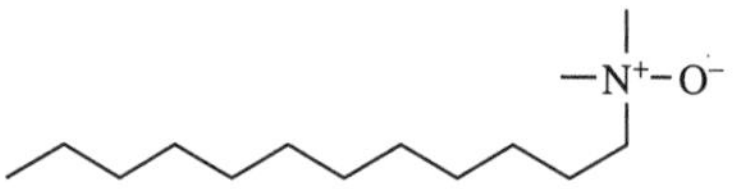

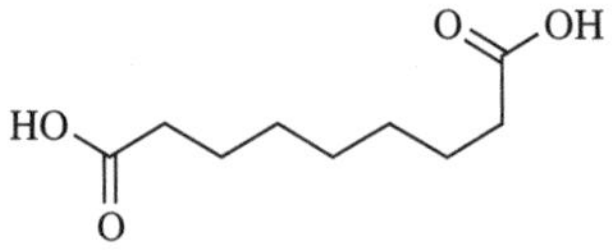

Hair loss (androgenetic alopecia) occurs in men and women, and is characterized by the loss of hair from the scalp in a defined pattern (47). In men it is often referred to as male-pattern baldness and affects up to 80% of men by age 80 (48). The involvement of androgens in androgenetic alopecia has been established for some time, and is well accepted. Eunuchs, who lack androgens, do not bald (49). Individuals who lack a functional androgen receptor are androgen insensitive and develop as females; again, these individuals do not bald (50). Likewise, no baldness is seen in individuals who lack 5α-reductase, the enzyme that converts testosterone to the potent androgen dihydrotestosterone (DHT) (51). The exact mechanism(s) through which androgens act to cause baldness remain unclear; however, given that the complex formed between the androgen receptor (AR) and androgen acts as a transcription factor, it is likely that genes controlling hair follicle cycling are regulated by androgen (47). Without treatment, androgenetic alopecia is a progressive condition. Apart from various camouflage and surgical options, currently only two pharmaceutical agents are approved for the treatment of androgenetic alopecia in males: topical minoxidil [38304-91-5] and oral finasteride [98319-26-7].

minoxidil   finasteride

Minoxidil is a vasodilator originally used to treat high blood pressure but a topical formulation was developed when patients treated with the drug showed increased hair growth (52). Finasteride is a synthetic azo-steroid and is a highly selective and potent 5α-reductase type-2 inhibitor (53), thus lowering DHT levels in scalp and serum by >60% at a daily dose of only 1 mg (54). Future potential therapies may include the use of androgen-receptor blockers, but only in scalp follicles due to the potential risks of gynaecomastia, feminisation and impotence if administered systemically (47).

2.2. Central Nervous System. As the world population is aging, there are increasing incidences of dementia reported worldwide, with Alzheimer's disease (AD) being the most common form of dementia in the elderly. It is characterized by a gradual decline in three domains: cognition, behavior, and function (55). Ideally, an effective treatment would target all three types of impairment. However, available treatments for AD diminish only certain symptoms and cannot halt the dementing process. As scientists uncover the pathogenic mechanisms of AD, additional treatments will likely emerge. The AGE accumulation (discussed above) in the CNS, eg, may be related to the aging process and the degenerating process of AD neurons (56). New therapies currently being developed include therapeutic agents that target amyloid β peptide and downstream pathological changes (57). Inhibition of the formation of amyloid peptide from its precursor protein is an attractive target for blocking the cascade process leading to the development of neurodegenerative disease. Bafilomycin A [116764-51-3], eg, and its analogues are of interest, since they very effectively and selectively block the formation of amyloid peptide by an indirect inhibition of β-secretase activity (58). Other strategies in the search for new therapeutic approaches include: agents that compensate for the lowered activity of the cholinergic system, agents that protect nerve cells from the toxic metabolites formed in neurodegenerative processes, agents that affect the process of formation of neurofibrillary tangles, and antiinflammatory agents that prevent the negative response of nerve cells to the pathological process (59). Currently, there are only a few therapeutic agents on the market for the treatment of AD. The main pharmacological effect of most of the agents is to improve the cognitive functions decreased in AD due to hypofunction of the cholinergic transmitter system (59). The cholinesterase inhibitors are the first and most developed group of drugs for AD treatment whose main mechanism of action is thought to be increased activation of cholinergic receptors (which is decreased in AD pathology) due to an increase of both concentration and duration of action of the neurotransmitter acetylcholine. First generation cholinesterase inhibitors include physostigmine [57-47-6], tacrine [321-64-2], and amiridine [90043-86-0], which show low selectivity for this enzyme, also inhibiting another enzyme in this group, butyrylcholinesterase (60). These compounds have significant side effects, such as sedation and hepatotoxicity, which limit their use (61).

Newer inhibitors are more selective and include Aricept [120011-70-3] (62), Galanthamine [357-70-0] (63), and Eptastigmine [101246-68-8] (64). One drug, Rivastigmine [123441-03-2], successfully targets acetylcholinesterase in the brain as opposed to peripheral forms (65). These second generation inhibitors show considerably fewer side effects than their first generation predecessors.

The close correlation between AD and the accumulation of reactive oxygen species (ROS) in CNS cells has been known for a long time. The role of ROS in AD has been thoroughly reviewed (66). The oxidative stress theory as it relates to AD involves homeostasis of intracellular calcium as well as the aggregation of protein, in this case, of amyloid peptide fibril (67). While nerve cells have an endogenous system that protects them from excessive ROS levels, an alternative approach to block the effects of ROS is the application of "external" antioxidants and/or the stimulation of the intracellular (endogenous) antioxidant systems.

Examples of such external antioxidants are the herbal triterpene Celastrol [34157-83-0] (68) and the vitamin E analog Raxofelast [128232-14-4] (69).

celastrol

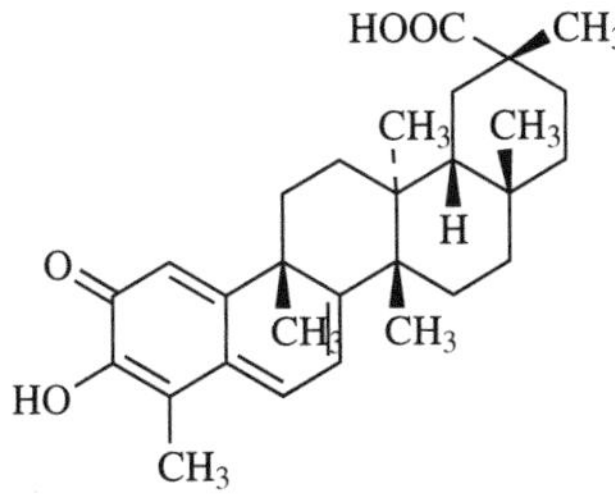

raxofelast

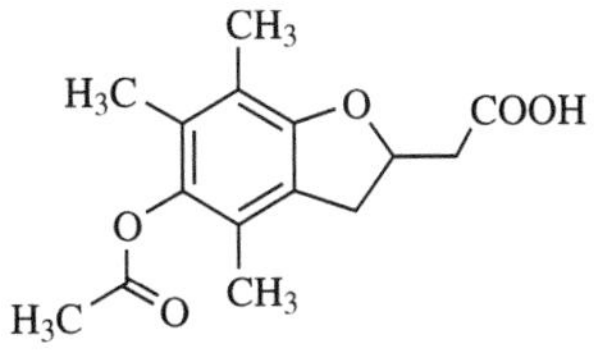

One of the promising approaches in the development of preventive therapies of AD is a design of agents based on derivatives and analogs of melatonin [73-31-4]. Melatonin is an endogenous hormone that has been shown to be effective against oxidative stress in the CNS (70). It has been shown in cell culture experiments and in animals that melatonin has a complex neuroprotective effect that includes both a specific antiamyloid component and a nonspecific geranto-protective effect because of its strong radical-blocking properties. These protective functions together with the fact that melatonin levels decrease with age and in AD, suggest that this endogenous bioregulator has an important preventive function against the pathogenesis of AD. Melatonin is registered in the United States and most European countries as a food additive or supplement. A precursor of melatonin, N-acetylserotonin [1210-83-9], is a more effective agent than melatonin due to its superior properties of radical-scavenging ability (71), inhibition of lipid peroxidation (72), and antiamyloid activity (73). Dimebon [3613-73-8], a structurally rigid analogue of melatonin, is currently under development as a new therapeutic agent for the treatment of AD (74).

melatonin

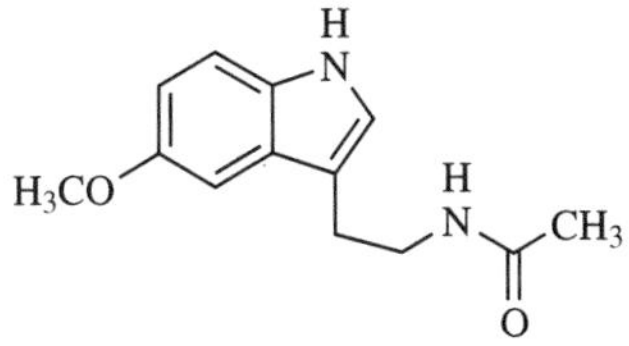

N-acetylserotonin

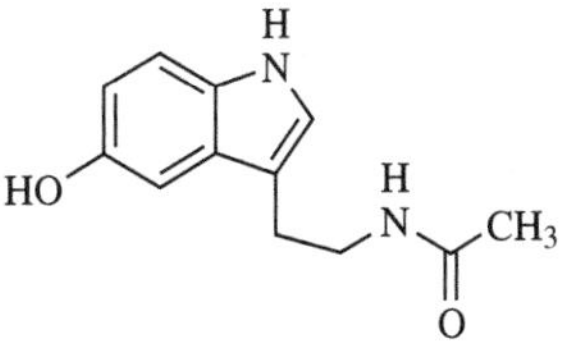

dimebon

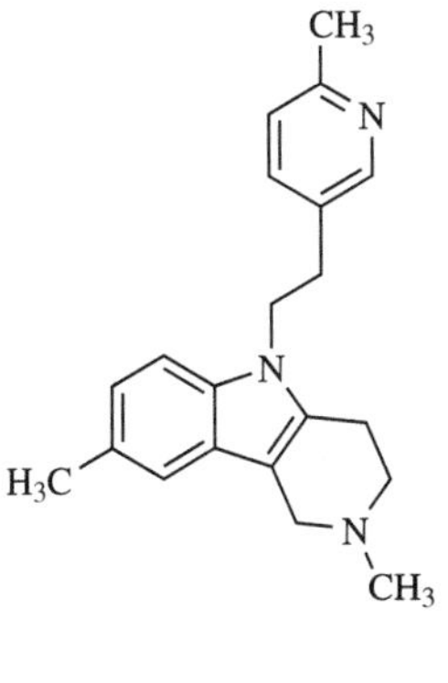

Antiinflammatory agents are predicted to be of use in AD therapy because neurodegenerative changes in the AD brain are accompanied by inflammatory reactions of the CNS. Nonsteroidal antiinflammatory agents (NSAIDS) have been shown to decrease the risk of developing AD in epidemiological studies (75). Ibuprofen [15687-27-1] was the first in the series of NSAIDS to be suggested for AD therapy (76) and the activity of these types of compounds is thought to be due primarily to the nonspecific inhibition of the cyclooxygenases (COXs). Other promising NSAIDS being studied in AD are Naproxen [22204-53-1] and Rofecoxib (Vioxx) [162011-90-7] (77).

Finally, since amyloid-β peptide plays a key role in the pathogenesis of AD, the possibility of using immunotherapeutic approaches against this peptide have been investigated. Immunization with amyloid peptide or with antibodies to the peptide cleared or prevented amyloid peptide-containing plaque deposits in the brains of transgenic AD mouse models (78). Recently, a clinical trial of a vaccine consisting of a 42-amino acid form of amyloid peptide (called AN-1792 [401586-29-6]) in 360 patients with mild to moderate AD was halted after 15 patients developed meningoencephalitis (79). A neuropathological examination of the brain of a patient that died in the trial showed intriguing evidence of an effective immune response against amyloid peptide with virtually total clearance of amyloid peptide deposits from much of the cerebral cortex (80). These data suggest a powerful effect of the vaccination and provide the strongest evidence to date that an induced immune response can affect amyloid peptide pathology in human AD. However, these data do not prove the effectiveness of the vaccine against AD. It is still unknown whether symptoms improve after clearance of amyloid peptide and cognitive testing data acquired during the trial is still unavailable. There are many aspects of AD vaccination immunotherapy suggesting that this field is not yet prepared to move forward with this therapeutic approach in humans. It has been proposed that, although inflammation may be a component of AD vaccination therapy, it is short-lived phenomena and potentially integral to the eventual benefit of vaccination treatment. Although the experimental and Phase I clinical vaccination immunotherapy studies of AN-1792 were successful, there is still a need for a greater understanding of the inflammatory consequences of autoimmunization in both mice and eventually in humans (81).

2.3. Cardiovascular System. There is a large body of evidence that biological aging is related to a series of long-term catabolic processes resulting in decreased function and structural integrity of several physiological systems, among which is the cardiovascular system (82). Some of the cardiovascular deficits that accompany aging in health can be retarded by physical conditioning. Likewise, growth hormone (discussed above) has been found to exert potent effects on cardiovascular function in young animals and reverses many of the deficits in cardiovascular function in aged animals and humans (82). There is also the questionable role of AGEs in cardiovascular disease and aging as discussed above. The formation of AGEs on vascular wall and myocardial collagen causes cross linking of collagen molecules leading to loss of collagen elasticity, and subsequently to a reduction in arterial and myocardial compliance. Aminoguanidine, an inhibitor of AGE formation discussed above, is effective in slowing or preventing arterial stiffening and myocardial diastolic dysfunction in aging and diabetic animals. In aged and diabetic animals, agents that can chemically break pre-existing crosslinking of collagen molecules are capable of reverting indexes of vascular and myocardial compliance to levels seen in younger or non-diabetic animals (5). These studies suggest that collagen crosslinking is a major mechanism that governs aging and diabetes-associated loss of vascular and cardiac compliance. The development of AGEs cross-link breakers may have an important role for future therapy of isolated systolic hypertension and diastolic heart failure in these conditions. Aging of the population will undoubtedly result in a concomitant increase in the incidence of the most common chronic cardiovascular diseases, including coronary artery disease, heart failure, myocardial infarction, and stroke (83). These diseases are direct consequences of atherosclerosis (AS), a multifactorial process that is described as both an autoimmune and inflammatory condition (84). The immune system plays a major role in the development and progression of AS involving macrophages and activated lymphocytes. Once AS is regarded as an autoimmune and inflammatory condition, aims for its prevention and treatment should be focused not only on control of traditional risk factors for cardiovascular disease, but on immune modulation of the process as well (84). In this regard, new guidelines from the Adult Treatment Panel III (ATP III) of the National Cholesterol Education Program recommend blood lipid management beyond low density lipoprotein (LDL) lowering, including aggressive treatment of elevated triglycerides since recent studies show that elevated triglycerides significantly increase cardiovascular disease risk (85). High density cholesterol (HDL), on the other hand, appears to play a protective role against development of AS by several mechanisms, including "reverse cholesterol transport", inhibition of oxidation or aggregation of LDL, and modulation of inflammatory responses to favor vasoprotection (86,87). Thus, raising HDL while lowering LDL levels would be beneficial for prevention and treatment of AS. The beneficial effects of 3-hydroxy-3-methylglutaryl CoA (HMG-CoA) reductase inhibitors (statins) in the treatment and prevention of cardiovascular disease have generally been attributed to their ability to lower cholesterol biosynthesis. The three most studied and widely used statins include atorvastatin [134523-03-8], simvastatin [79902-63-9], and pravastatin [81093-37-0].

atorvastatin

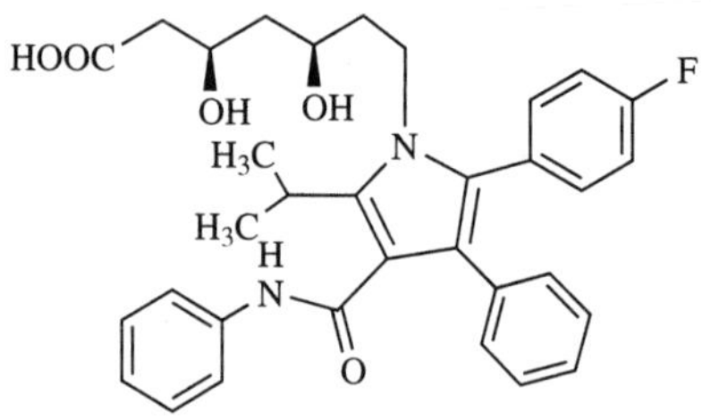

simvastatin

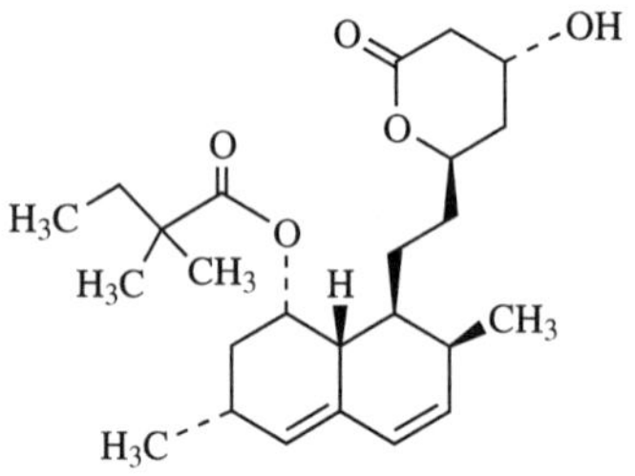

pravastatin

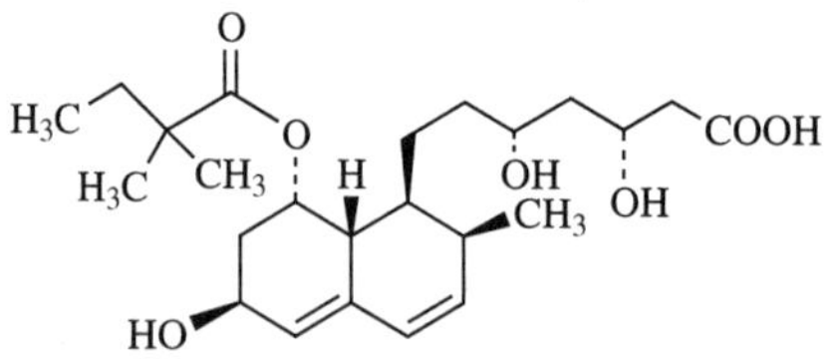

Besides their cholesterol-lowering properties, these statins show similar beneficial effects that include modification of thrombus formation and degradation, antiinflammatory response, plaque stabilization, and improved endothelial function (88). The statins have proven to be antiinflammatory by virtue of their effects on leukocyte adhesion and migration to sites of inflammation (89). Moreover, statins have been shown to lower C-reactive protein, a marker of systemic inflammation (90). This observation is significant because a growing body of evidence indicates that inflammation plays a substantial role in plaque progression and rupture and C-reactice protein appears to be a better biomarker than cholesterol levels for predicting cardiovascular risk in healthy persons as well as in persons with established cardiovascular disease (90). Another biomarker that has been shown in epidemiological studies to be an independent risk factor for AS, thrombosis, and hypertension is elevated homocysteine (91). Statins do not influence homocysteine plasma levels (92) but supplementation with B vitamins, and especially folic acid [75708-92-8], has been shown to be beneficial for treating hyperhomocysteinemia (93).

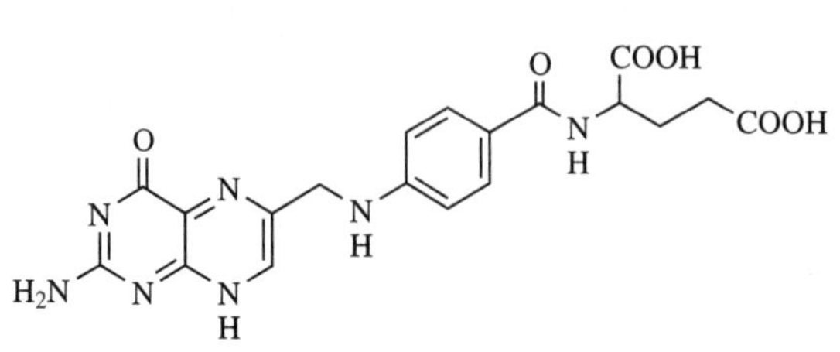

Finally, diet and exercise are well known to prevent or decrease the risk of cardiovascular diseases of all kinds. In addition to preventing obesity, which is a major risk factor for cardiovascular disease, physical activity slows down the age-associated loss of cardiopulmonary efficiency and has been shown to help prevent illness in old age (94). A diet that includes the "good" fats is also beneficial. Epidemiological studies over the past 40 years suggest that omega-3 fatty acids derived from fish and fish oil decrease the risk of coronary heart disease, hypertension and stroke, and their complications (95). Moreover, current evidence suggests that individuals with existing coronary artery disease may reduce their risk of sudden cardiac death by increasing their intake of long-chain omega-3 fatty acids by ~1 g/day (96).

2.4. Immune System. The functional capacity of the immune system gradually declines with age (97). The T lymphocytes are more severely affected than the B lymphocytes or antigen-presenting cells. The age-related alterations that occur in the immune system may be referred to as immunosenescence and involve both the innate and the adaptive immune responses. These alterations account for the increased susceptibility to certain microbial infections, autoimmune diseases, or malignancies in the elderly and contribute to increased morbidity and mortality with age. Hence, the restoration of immunological function is expected to have a beneficial effect in reducing pathology and maintaining a healthy condition in advanced age. A few intervention strategies are discussed here briefly. In animal studies, caloric restriction (discussed above) has been shown to be the most powerful modulator of the aging process (98) including its action on the immune system. Animal and human studies show that DHEA (discussed above) treatment results in stimulation of immune responses probably by induction of interleukin 2 (IL-2) production by T cells (99). Certain substituted guanosines, such as 7-thia-8-oxoguanosine [122970-40-5], 7-deazaguanosine [62160-23-0], and loxoribine [121288-39-9], activate the innate immune system through Toll-like receptor 7 in mouse and human cells (100).

R = D-ribofuranose

A synthetic thymic dipeptide, pidotimod [121808-62-6], stimulated the production of IL-2 in peripheral blood lymphocytes and splenocytes from old but not young rats (101). This has implications for restoration of immune function in the elderly.

pidotimod

2.5. Musculoskeletal System.   Several age-related conditions affecting the bones and joints are observed in both men and women. These include, among others, arthritis, both osteoarthritis (OA) and rheumatoid arthritis (RA), and bone loss including osteoporosis and fractures. A substantial part of the age-related decline in functional capabilities of the musculoskeletal system is not due to aging per se but to decreased and insufficient physical activity (102). Therefore, musculoskeletal disease prevention in the elderly must include physical exercise. Arthritis affects a large segment of the older population and is the leading cause of disability in the general population. Approximately 3% of the U.S. population suffers from RA (103) and RA is a systemic disease characterized by a chronic inflammatory reaction in the joint synovium, degeneration of cartilage, and erosion of adjacent joint bone. Many proinflammatory cytokines are expressed in diseased joints. One of the most effective agents for treatment of rheumatoid arthritis is etanercept (Enbrel [185243-69-0]) which is a biological disease-modifying antirheumatic drug (DMARD) that works by blocking the proinflammatory cytokine tumor necrosis factor-alpha (TNF-α) (104). Etanercept is a soluble TNF receptor fusion protein and is more effective than methotrexate [59-05-2], another DMARD and anticancer agent, at reducing the number of new erosions and joint-space narrowing in patients with active early rheumatoid arthritis (105).

methotrexate

Another biological that blocks TNF-α is infliximab (Remicade [170277-31-3]). Infliximab is a monoclonal antibody to TNF and is used effectively in Crohn's disease as well as RA (106). Combinations of methotrexate and these anti-TNF biologicals appear to be particularly effective in reducing the signs and symptoms of RA and, most importantly, in protecting joints against progressive structural damage (107). There is some evidence that omega-3 fatty acids in the diet can lessen the severity of rheumatoid arthritis as well as reduce bone loss in postmenopausal women (108). Whether or not a diet that includes omega-3 fatty acids is useful in prevention of RA or bone loss is not yet known. Glucosamine [3416-24-8], or its sulfate [29031-19-4], has been widely used to treat osteoarthritis in humans and is thought to work by suppressing neutrophil function and activation (109).

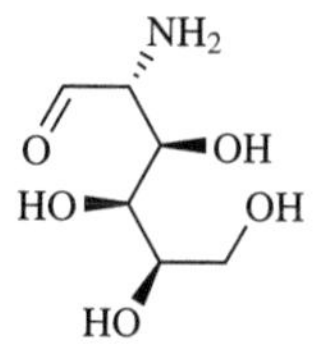

In addition, glucosamine also has been shown to inhibit inducible nitric oxide production via inhibition of inducible nitric oxide synthase expression (110) and thereby shows anti-inflammatory activity.

3. Cancer Prevention

Cancer is primarily a disease of aging even though there are many examples of cancer afflicting young individuals. For example, age is clearly the single most important risk factor for development of prostate cancer in men and breast cancer in women (111). Management options for women at high risk for breast cancer include close surveillance, chemoprevention, and prophylactic mastectomy. The optimal method remains to be determined. Chemoprevention refers to the use of specific natural or synthetic chemical agents to reverse, suppress, or prevent the progression to invasive cancer (112). The ideal chemopreventive agent is safe and nontoxic over the long term. Prevention of breast cancer is still under clinical investigation with only one drug, tamoxifen [10540-29-1], showing benefit in high risk patients (113). Tamoxifen is a nonsteroidal antiestrogen originally developed in 1996 as a contraceptive. Raloxifene [84449-90-1] is another nonsteroidal antiestrogen that is also being studied for potential chemoprevention of breast cancer.

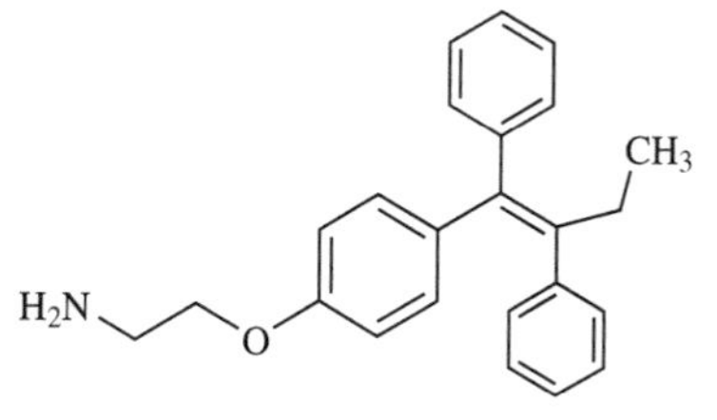

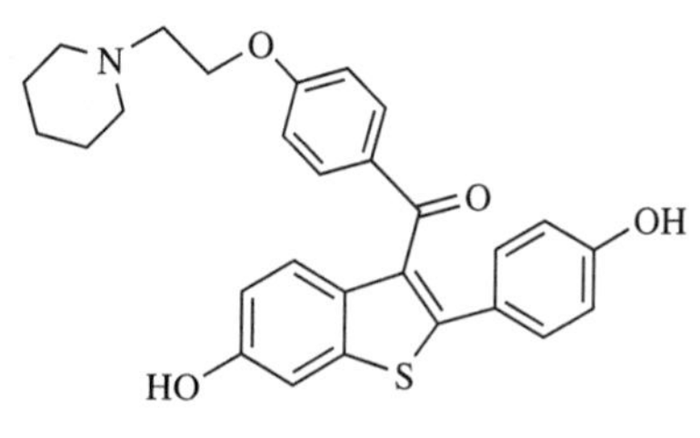

Both tamoxifen and raloxifene maintain bone density and raloxifene is now used to prevent osteoporosis and is also being tested as a preventive for coronary heart disease (114). Prostate cancer is the second leading cause of cancer death in the United States, largely because of the limitations of our current therapeutic options, especially once the cancer has metastasized (115). It is well established that the prostate is hormonally influenced. There is evidence suggesting that androgenic influences over a period of time encourages the process of prostate carcinogenesis (116). Moreover, early prostate tumors are often androgen dependent but androgen insensitive tumors inevitably develop that then have a very poor prognosis. This fact underscores the need for prevention strategies such as chemoprevention. Antiandrogens are among the promising chemopreventive agents for prostate cancer since prostate epithelium is androgen dependent. Studies of prostate biology support the concept that dihydrotestosterone (DHT) is the principal androgen responsible for normal and hyperplastic growth of the prostate gland, and therefore, inhibitors of DHT formation should be useful as chemopreventive agents for prostate cancer (117). As discussed above, 5α-reductase inhibitors, such as finasteride, are useful for inhibiting the formation of DHT. A large, randomized clinical chemoprevention trial supported by the National Cancer Institute is being conducted to test the efficacy of finasteride for prevention of prostate cancer incidence (118). Other trials that are ongoing include micronutients and phytochemicals such as lycopene [502-65-8], soy isoflavones, and vitamin E and selenium [7782-49-2], alone or in combination (118).

Lycopene is a carotenoid derived largely from tomato-based products. Recent epidemiological studies have suggested a potential benefit of this natural product against the risk of prostate cancer, especially the more advanced and aggressive form (119). In some studies, risk appeared to be lowered by 30–40% with high tomato or lycopene consumption. However, results are difficult to interpret until randomized controlled dietary intervention studies can be conducted (120).

An impressive body of epidemiological data suggests an inverse relationship between colorectal cancer risk and regular use of nonsteroidal antiinflammatory drugs (NSAID), including aspirin [50-78-2] (121). Clinical trials with NSAIDs have demonstrated that NSAID treatment caused regression of preexisting colon adenomas in patients with familial adenomatous polyposis. In addition, several phytochemicals with antiinflammatory activity and NSAIDs act to retard, block or reverse colon carcinogenesis (121). Thus, these agents exert tumor-suppressive activity on premalignant lesions (polyps) in humans and on established experimental tumors in mice. Trials using the NSAID sulindac [38194-50-2] reduced the number of polyps in patients with familial adenomatous polyposis (122). Some of the tumor-suppressive effects of NSAIDs depend on the inhibition of cyclooxygenase-2 (COX-2), a key enzyme in the synthesis of prostaglandins and thromboxane, which is highly expressed in inflammation and cancer (123). The selective COX-2 inhibitor celecoxib [169590-42-5] has been approved by the FDA for adjuvant treatment of familial adenomatous polyposis, and a large number of prevention and treatment trials of colorectal, prostate and breast cancer have been started (122). Green tea and (−)-epigallocatechin [989-51-5] are now acknowledged cancer preventives in Japan and there is new evidence that green tea and sulindac, in combination, have synergistic cancer preventive effects (124).

aspirin

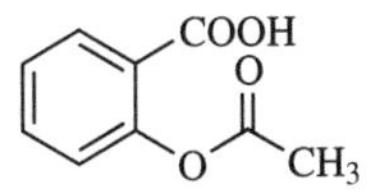

sulindac

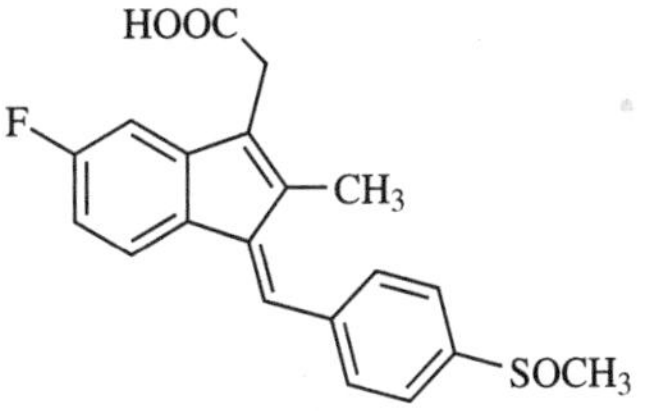

celecoxib

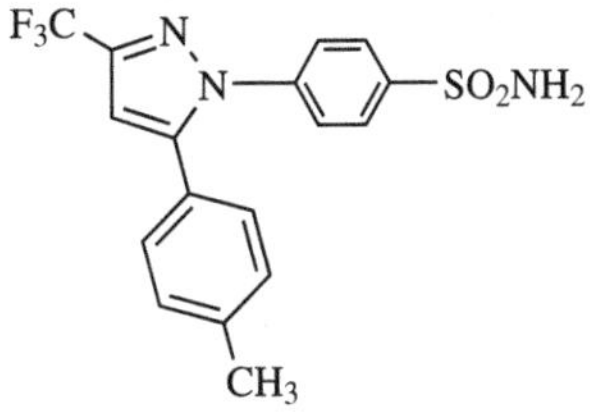

(−)-epigallocatechin

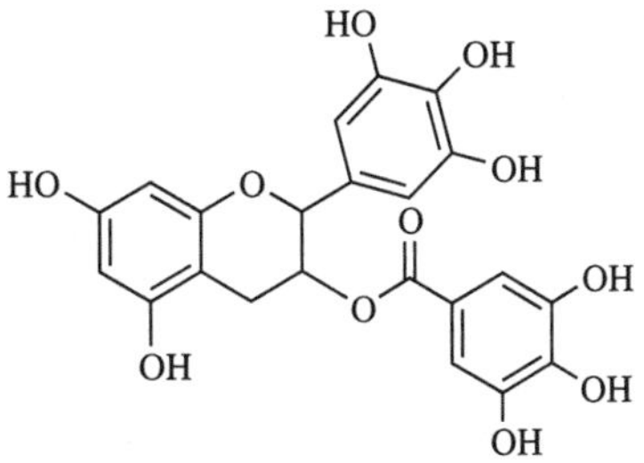

4. Applications

While there is a large number of products currently being sold by antiaging entrepreneurs who claim that it is now possible to slow, stop, or even reverse human aging, they have no scientifically demonstrated efficacy (1). Most biogerantologists believe that our rapidly expanding scientific knowledge holds the promise that methods may eventually be discovered to slow the rate of aging. If successful, these interventions are likely to postpone age-related diseases and disorders and extend the period of healthy life. Because aging is the greatest risk factor for the leading causes of death and other age-related pathologies, successful efforts to slow the rate of aging would certainly have dramatic health benefits for the population, by far exceeding the anticipated changes in the health and length of life that would result from complete elimination of heart disease, cancer, stroke, and other age-associated diseases and disorders (1).

BIBLIOGRAPHY

"Antiaging Agents" in *ECT* 4th ed., Vol. 2, pp. 815–829, by W. B. Jolley, ICN Pharmaceuticals, Inc., and H. B. Cottam, University of California, San Diego; "Antiaging Agents" in *ECT* (online), posting date: December 4, 2000, by W. B. Jolley, ICN Pharmaceuticals, Inc., and H. B. Cottam, University of California, San Diego.

CITED PUBLICATIONS

1. S. J. Olshansky, L. Hayflick, and A. Carnes Bruce, *J. Gerontol. Ser. A, Biol. Sci. Med. Sci.* **57**, B292 (2002).
2. J. W. Baynes, *Exp. Gerontol.* **36**, 1527 (2001).
3. R. Singh, A. Barden, T. Mori and L. Beilin, *Diabetologia* **44**, 129 (2001).
4. S. Rahbar, and J. L. Figarola, *Curr. Med. Chem.: Immunol., Endocrine Metabolic Agents* **2**, 135 (2002).
5. D. Aronson, *J. Hypertension* **21**, 3 (2003).
6. P. C. Ulrich and D. R. Wagle, (1997) in PCT Int. Appl. (Alteon Inc., USA). Wo, pp. 40.
7. T. S. Kern and R. L. Engerman, *Diabetes* **50**, 1636 (2001).
8. Y. Du, M. A. Smith, C. M. Miller, and T. S. Kern, *J. Neurochem.* **80**, 771 (2002).
9. H.-P. Hammes, X. Du, D. Edelstein, T. Taguchi, T. Matsumura, Q. Ju, J. Lin, A. Bierhaus, P. Nawroth, D. Hannak, M. Neumaier, R. Bergfeld, I. Giardino, and M. Brownlee, *Nature Med. (New York, NY, United States)* **9**, 294 (2003).
10. K.-I. Mizutani, K. Ikeda, K. Tsuda, and Y. Yamori, *J. Hypertension* **20**, 1607 (2002).
11. R. Wada, Y. Nishizawa, N. Yagihashi, M. Takeuchi, Y. Ishikawa, K. Yasumura, M. Nakano, and S. Yagihashi, *Eur. J. Clin. Inv.* **31**, 513 (2001).
12. A. Stitt, T. A. Gardiner, N. L. Anderson, P. Canning, N. Frizzell, N. Duffy, C. Boyle, A. S. Januszewski, M. Chachich, J. W. Baynes, and S. R. Thorpe, *Diabetes* **51**, 2826 (2002).
13. J. M. Onorato, A. J. Jenkins, S. R. Thorpe, and J. W. Baynes, *J. Biol. Chem.* **275**, 21177 (2000).
14. P. A. Voziyan, T. O. Metz, J. W. Baynes, and B. G. Hudson, *J. Biol. Chem.* **277**, 3397 (2002).
15. M. A. Lane, D. K. Ingram, and G. S. Roth, *Sci. Am.* **287**, 36 (2002).
16. D. K. Ingram, R. G. Cutler, R. Weindruch, D. M. Renquist, J. J. Knapka, M. April, C. T. Belcher, M. A. Clark, C. D. Hatcherson, and B. M. Marriott, *J. Gerontol.* **45**, B148 (1990).
17. J. Wanagat and R. Weindruch, *Sci. Geriatrics (2nd Ed.)* **1**, 153 (2001).
18. J. J. Ramsey, R. J. Colman, N. C. Binkley, J. D. Christensen, T. A. Gresl, J. W. Kemnitz, and R. Weindruch, *Exper. Gerontol.* **35**, 1131 (2000).

19. R. Weindruch, K. P. Keenan, J. M. Carney, G. Fernandes, R. J. Feuers, R. A. Floyd, J. B. Halter, J. J. Ramsey, A. Richardson, G. S. Roth, and S. R. Spindler, *J. Gerontol., Ser. A: Biol. Sci. Med. Sci.* **56A**, 20 (2001).

20. A. Lane Mark, J. Mattison, K. Ingram Donald, and S. Roth George, *Microsc. Res. Tech.* **59**, 335 (2002).

21. M. P. Mattson, W. Duan, and Z. Guo, *J. Neurochem.* **84**, 417 (2003).

22. M. Brownlee, *Metabolism, Clin. Exper.* **49**, 9 (2000).

23. D. Harman, *Intern. Cong. Ser.* **782**, 3 (1988).

24. K. B. Beckman and B. N. Ames, *Physiol. Rev.* **78**, 547 (1998).

25. R. S. Sohal and R. Weindruch, *Science* **273**, 59 (1996).

26. C. B. Harley, A. B. Futcher, and C. W. Greider, *Nature (London)* **345**, 458 (1990).

27. R. C. Allsopp, H. Vaziri, C. Patterson, S. Goldstein, E. V. Younglai, A. B. Futcher, C. W. Greider, and C. B. Harley, *Proce. Natl. Acad. Sci. U. S. Am.* **89**, 10114 (1992).

28. G. B. Morin, *CELL* **59**, 521 (1989).

29. L. Hayflick, *Br. J. Cancer* **83**, 841 (2000).

30. C. P. Chiu and C. B. Harley, *Proc. Soc. Biol. Med.* **214**, 99 (1997).

31. A. G. Bodnar, M. Ouellette, M. Frolkis, S. E. Holt, C. P. Chiu, G. B. Morin, C. B. Harley, J. W. Shay, S. Lichtsteiner, and W. E. Wright, *Science* **279**, 349 (1998).

32. T. de Lange and T. Jacks, *Cell* **98**, 273 (1999).

33. J. Wang, G. J. Hannon and D. H. Beach, *Nature (London)* **405**, 755 (2000).

34. D. Rudman, A. G. Feller, H. S. Nagraj, G. A. Gergans, P. Y. Lalitha, A. F. Goldberg, R. A. Schlenker, L. Cohn, I. W. Rudman, and D. E. Mattson, *New England J. Med.* **323**, 1 (1990).

35. M. A. Papadakis, D. Grady, D. Black, M. J. Tierney, G. A. Gooding, M. Schambelan, and C. Grunfeld, *Ann. Inter. Med.* **124**, 708 (1996).

36. M. D. Johnson, R. A. Bebb and S. M. Sirrs, *Ageing Res. Rev.* **1**, 29 (2002).

37. B. Allolio and W. Arlt, *Trends Endocrinol. Metabolism* **13**, 288 (2002).

38. Y. Okano, *Nippon Koshohin Kagakkaishi* **23**, 340 (1999).

39. T. Fujimura, K. Tsukahara, and H. Kawada, *Fragrance J.* **30**, 38 (2002).

40. N. Sonehara and T. Nishiyama (2001), in *Jpn. Kokai Tokkyo Koho* (Shiseido Co., Ltd., Japan). Jp, pp. 11.

41. J. R. Schiltz (2001), in *PCT Int. Appl.* (Mary Kay, Inc., USA). Wo, pp. 34.

42. S. Sakaki and H. Masaki (1997), in *Jpn. Kokai Tokkyo Koho* (Noevir K. K., Japan). Jp, pp. 5.

43. S. Sakaki, T. Arashima, A. Iwamoto, Y. Okano, and H. Masaki, *J. SCCJ* **30**, 71 (1996).

44. N. Lowe, *Round Table Series—Royal Society of Medicine Press* **74**, 23 (2002).

45. J. C. Cather, J. C. Cather, and A. Menter, *Dermatologic Clinics* **20**, 749 (2002).

46. H. A. Khawaja and E. Hernandez-Perez, *Inter. J. Dermatol.* **40**, 311 (2001).

47. J. A. Ellis, R. Sinclair and S. B. Harrap (2002), *Expert Reviews in Molecular Medicine [online computer file]*, No pp. given.

48. J. A. Ellis, M. Stebbing, and S. B. Harrap, *J. Investigative Dermatol.* **116**, 452 (2001).

49. J. B. Hamilton, *Am. J. Anatomy* **71**, 451 (1942).

50. J. E. Griffin and J. D. Wilson (1989), in *The Metabolic Basis of Inherited Diseases.*, ed. C. R. Scriver, et al., eds (McGraw Hill, New York), pp. 1919–1944.

51. J. Imperato-McGinley, L. Guerrero, T. Gautier, and R. E. Peterson, *Science* **186**, 1213 (1974).

52. B. L. Devine, R. Fife, and P. M. Trust, *Br. Med. J.* **2**, 667 (1977).

53. E. Olsen, *Aust. J. Dermatol.* **38**, A316 (1997).

54. L. Drake, M. Hordinsky, V. Fiedler, J. Swinehart, W. P. Unger, P. C. Cotterill, D. M. Thiboutot, N. Lowe, C. Jacobson, D. Whiting, S. Stieglitz, S. J. Kraus, E. I. Griffin, D. Weiss, P. Carrington, C. Gencheff, G. W. Cole, D. M. Pariser, E. S. Epstein, W.

Tanaka, A. Dallob, K. Vandormael, L. Geissler, and J. Waldstreicher, *J. Am. Acad. Dermatol.* **41**, 550 (1999).

55. G. W. Small, *Am. J. Med.* **104**, 32S (1998); discussion 39S–42S.

56. A. Takeda, T. Yasuda, T. Miyata, K. Mizuno, M. Li, S. Yoneyama, K. Horie, K. Maeda, and G. Sobue, *Neurosci. Lett.* **221**, 17 (1996).

57. T. E. Golde, *J. Clin. Investig.* **111**, 11 (2003).

58. M. D. Kane, R. D. Schwarz, L. St. Pierre, M. D. Watson, M. R. Emmerling, P. A. Boxer, and G. K. Walker, *J. Neurochem.* **72**, 1939 (1999).

59. S. O. Bachurin, *Med. Res. Rev.* **23**, 48 (2003).

60. R. Cocabelos, A. Nordberg, J. Caamano, A. Franco-Maside, L. Fernandez-Novoa, M. J. Gomez, X. A. Alvarez, M. Takeda, J. Prous, Jr., T. Nishimura, and B. Winblad, *Drugs Today* **30**, 259 (1994).

61. P. B. Watkins, H. J. Zimmerman, M. J. Knapp, S. I. Gracon, and K. W. Lewis, *JAMA* **271**, 992 (1994).

62. H. M. Bryson and P. Benfield, *Drugs Aging* **10**, 234 (1997); discussion 240–241.

63. L. J. Scott and K. L. Goa, *Drugs* **60**, 1095 (2000).

64. B. P. Imbimbo, P. Martelli, W. M. Troetel, F. Lucchelli, U. Lucca, and L. J. Thal, *Neurology* **52**, 700 (1999).

65. A. Enz, R. Amstutz, H. Boddeke, G. Gmelin, and J. Malanowski, *Progress Brain Res.* **98**, 431 (1993).

66. C. Behl, *Prog. Neurobiol.* **57**, 301 (1999).

67. B. J. Tabner, S. Turnbull, O. El-Agnaf, and D. Allsop, *Curr. Top. Med. Chem.* **1**, 507 (2001).

68. A. C. Allison, R. Cacabelos, V. R. Lombardi, X. A. Alvarez, and C. Vigo, *Prog. Neuropsychopharmacology Biol. Psychiatry* **25**, 1341 (2001).

69. F. N. Bolkenius, J. Verne-Mismer, J. Wagner, and J. M. Grisar, *Eur. J. Pharmacol.* **298**, 37 (1996).

70. R. Reiter, L. Tang, J. J. Garcia, and A. Munoz-Hoyos, *Life Sci.* **60**, 2255 (1997).

71. A. Wolfler, P. M. Abuja, K. Schauenstein, and P. M. Liebmann, *FEBS Lett.* **449**, 206 (1999).

72. B. Longoni, W. A. Pryor, and P. Marchiafava, *Biochem. Biophys. Res. Communi.* **233**, 778 (1997).

73. S. Bachurin, G. Oxenkrug, N. Lermontova, A. Afanasiev, B. Beznosko, G. Vankin, E. Shevtsova, T. Mukhina, and T. Serkova, *Ann. N. Y. Acad. Sci.* **890**, 155 (1999).

74. S. Bachurin, E. Bukatina, N. Lermontova, S. Tkachenko, A. Afanasiev, V. Grigoriev, I. Grigorieva, Y. Ivanov, S. Sablin, and N. Zefirov, *Ann. N. Y. Acad. Sci.* **939**, 425 (2001).

75. M. Hull, K. Lieb, and B. L. Fiebich, *Expert Opinion Investigational Drugs* **9**, 671 (2000).

76. G. P. Lim, F. Yang, T. Chu, P. Chen, W. Beech, B. Teter, T. Tran, O. Ubeda, K. H. Ashe, S. A. Frautschy, and G. M. Cole, *J. Neurosci.* **20**, 5709 (2000).

77. H. Blain, J. Y. Jouzeau, A. Blain, B. Terlain, P. Trechot, J. Touchon, P. Netter, and C. Jeandel, *Presse Medicale* **29**, 267 (2000).

78. J. McLaurin, R. Cecal, M. E. Kierstead, X. Tian, A. L. Phinney, M. Manea, J. E. French, M. H. L. Lambermon, A. A. Darabie, M. E. Brown, C. Janus, M. A. Chishti, P. Horne, D. Westaway, P. E. Fraser, H. T. J. Mount, M. Przybylski, and P. St George-Hyslop, *Nat. Med. (New York, NY, United States)* **8**, 1263 (2002).

79. E. Check, *Nature (London)* **415**, 462 (2002).

80. J. A. R. Nicoll, D. Wilkinson, C. Holmes, P. Steart, H. Markham, and R. O. Weller, *Nat. Med. (New York, NY, United States)* **9**, 448 (2003).

81. G. M. Pasinetti, L. Ho, and P. Pompl, *Neurobiol. Aging* **23**, 683 (2002).

82. A. S. Khan, D. C. Sane, T. Wannenburg, and W. E. Sonntag, *Cardiovascular Res.* **54**, 25 (2002).

83. O. Bonow Robert, A. Smaha Lynn, C. Smith Sidney, Jr., A. Mensah George, and C. Lenfant, *Circulation* **106**, 1602 (2002).

84. Y. Sherer and Y. Shoenfeld, *Autoimm. Rev.* **1**, 21 (2002).

85. O. Bonow Robert, *Circulation* **106**, 3140 (2002).

86. O. Stein and Y. Stein, *Atherosclerosis (Shannon, Ireland)* **144**, 285 (1999).

87. J. X. Rong and E. A. Fisher, *Ann. Med. (Helsinki)* **32**, 642 (2000).

88. C. Joukhadar, N. Klein, M. Prinz, C. Schrolnberger, T. Vukovich, M. Wolzt, L. Schmetterer, and G. T. Dorner, *Thrombosis Haemostasis* **85**, 47 (2001).

89. G. Weitz-Schmidt, *Trends Pharmacol. Sci.* **23**, 482 (2002).

90. M. Ridker Paul, *Clin. Cardiology* **26**, III39 (2003).

91. L. E. Spieker, T. F. Luscher, and G. Noll, *HeartDrug* **1**, 160 (2001).

92. G. Miltiadous, J. Papakostas, G. Chasiotis, K. Seferiadis, and M. Elisaf, *Atherosclerosis (Shannon, Ireland)* **166**, 199 (2003).

93. W. G. Haynes, *Cardiovascular Drugs Therapy* **16**, 391 (2003).

94. E. Lang and B. M. Lang, *Z. Gerontol.* **26**, 429 (1993).

95. A. P. Simopoulos, *Environ. Health Preventive Med.* **6**, 203 (2002).

96. W. S. Harris, Y. Park, and W. L. Isley, *Curr. Opin. Lipid.* **14**, 9 (2003).

97. B. Grubeck-Loebenstein and G. Wick, *Adv. Immunol.* **80**, 243 (2002).

98. M. Pahlavani, *Frontiers Biosc. [Electronic Publication]* **5**, D580 (2000).

99. O. Khorram, *Dehydroepiandrosterone (DHEA)*, 57 (2000).

100. J. Lee, T.-H. Chuang, V. Redecke, L. She, P. M. Pitha, D. A. Carson, E. Raz, and H. B. Cottam, *Proc. Nat. Acad. Sci. United States of Am.* **100**, 6646 (2003).

101. A. Chiarenza, M. P. Iurato, N. Barbera, L. Lempereur, G. Cantarella, U. Scapagnini, and R. Bernardini, *Pharmacol. Toxicol. (Copenhagen)* **74**, 262 (1994).

102. I. Vuori, *Res. Quarterly Exercise Sport* **66**, 276 (1995).

103. R. M. Fleischmann, S. W. Baumgartner, E. A. Tindall, A. L. Weaver, L. W. Moreland, M. H. Schiff, R. W. Martin, and G. T. Spencer-Green, *J. Rheumatol.* **30**, 691 (2003).

104. A. Alldred, *Expert Opinion Pharmacotherapy* **2**, 1137 (2001).

105. J. M. Bathon, R. W. Martin, R. M. Fleischmann, J. R. Tesser, M. H. Schiff, E. C. Keystone, M. C. Genovese, M. C. Wasko, L. W. Moreland, A. L. Weaver, J. Markenson, and B. K. Finck, *New England J. Med.* **343**, 1586 (2000).

106. E. Valle, M. Gross, and S. J. Bickston, *Expert Opinion Pharmacotherapy* **2**, 1015 (2001).

107. P. C. Taylor, *Mol. Biotechnol.* **19**, 153 (2001).

108. G. Fernandes (2003) *Abstracts of Papers, 225th ACS National Meeting, New Orleans, LA, United States, March 23-27, 2003*, AGFD-019.

109. I. Nagaoka, J. Hua, S. Suguro, and K. Sakamoto, *Ensho, Saisei* **22**, 461 (2002).

110. C. J. Meininger, K. A. Kelly, H. Li, T. E. Haynes, and G. Wu, *Biochem. Biophys. Res. Commun.* **279**, 234 (2000).

111. H. Sakorafas George, *Cancer Treatment Rev.* **29**, 79 (2003).

112. R. S. Sandler, *Important Adv. Oncol.* 123 (1996).

113. K. A. Carolin and H. A. Pass, *Crit. Rev. Oncol./Hematol.* **33**, 221 (2000).

114. V. C. Jordan, *J. Steroid Biochem. Mol. Biol.* **74**, 269 (2000).

115. J. M. Kaminski, J. B. Summers, M. B. Ward, M. R. Huber, and B. Minev, *Cancer Treatment Rev.* **29**, 199 (2003).

116. O. W. Brawley, *Urologic Oncol.* **21**, 67 (2003).

117. G. J. Gormley, *J. Cellular Biochem.* 113 (1992).

118. O. Kucuk, *Develop. Oncol.* **81**, 331 (2002).

119. E. Giovannucci, *Exp. Biol. Med. (Maywood, NJ, United States)* **227**, 852 (2002).

120. C. Miller Elizabeth, E. Giovannucci, W. Erdman John, Jr., R. Bahnson, J. Schwartz Steven, and K. Clinton Steven, *Uro. Clinics North Am.* **29**, 83 (2002).
121. B. S. Reddy and C. V. Rao, *J. Environ. Pathol., Toxicol. Oncol.* **21**, 155 (2002).
122. K.-E. Giercksky, *Best Practice Res., Clin. Gastroenterol.* **15**, 821 (2001).
123. O. Dormond and C. Ruegg, *Drug Resistance Updates* **4**, 314 (2001).
124. H. Fujiki, M. Suganuma, K. Imai, and K. Nakachi, *Cancer Lett. (Shannon, Ireland)* **188**, 9 (2002).

HOWARD B. COTTAM
University of California